The Frontiers Collection

The books in this collection are devoted to challenging and open problems at the forefront of modern science and scholarship, including related philosophical debates. In contrast to typical research monographs, however, they strive to present their topics in a manner accessible also to scientifically literate non-specialists wishing to gain insight into the deeper implications and fascinating questions involved. Taken as a whole, the series reflects the need for a fundamental and interdisciplinary approach to modern science and research. Furthermore, it is intended to encourage active academics in all fields to ponder over important and perhaps controversial issues beyond their own speciality. Extending from quantum physics and relativity to entropy, consciousness, language and complex systems—the Frontiers Collection will inspire readers to push back the frontiers of their own knowledge.

More information about this series at https://link.springer.com/bookseries/5342

Shyam Wuppuluri · Ian Stewart
Editors

From Electrons to Elephants and Elections

Exploring the Role of Content and Context

Editors
Shyam Wuppuluri
Einstein Forum
Potsdam, Germany

Ian Stewart
Mathematics Institute
University of Warwick
Coventry, UK

ISSN 1612-3018 ISSN 2197-6619 (electronic)
The Frontiers Collection
ISBN 978-3-030-92191-0 ISBN 978-3-030-92192-7 (eBook)
https://doi.org/10.1007/978-3-030-92192-7

This Springer imprint is published by the registered company Springer Nature Switzerland AG
The registered company address is: Gewerbestrasse 11, 6330 Cham, Switzerland

Foreword

Natural reality—and in consequences our knowledge of it—is organised hierarchically, in successive level of increasing aggregation, complexity: as per particles, atoms, molecules, condensed matter or cells, systems, organisms, communities or words, sentences, paragraphs, books and the like. Every level of such a systemic hierarchy has its characteristic regularities and laws descriptive of the phenomena encompassed at this level. Often, the phenomenal occurring at one level can be derived from and explained by those of its lower-level constituents. But not always. For everywhere, one encounters conditions and modes of operation that are *emergent* in being characteristic innovations not accounted for in terms of the modus operandi of lower-level phenomena. Thus, not all group behaviour can be accounted for by individual psychology, nor is all of social choice reducible to individual preferences. Systemic complexity stands in the way of constituent-geared 'reductive explanations'. (Even in mathematical logic, we encounter systems that are omega-inconsistent in that while each individual constituent possessed a critical factor the systemic generalist that this factor is all pervasive is false and the move from each to every inappropriate.)

Such reduction-resistant emergent higher-level phenomena have profound implications for our understanding of science. For the task of scientific understanding here becomes far more challenging. When our cognitive grasp of the lowest level of the hierarchy is sufficient, we need not worry about those super-ordinate levels. However interesting in itself, systemic integration and holistic syntheses are now dispensable. But since this is not in general the case, the demands of understanding grow exponentially with hierarchical assent, and the challenges of scientific understanding multiply accordingly.

The phenomenology of emergence accordingly transforms our understanding of science itself. The classical Greek model of geometry as a paradigm science where everything needful is achieved by the understanding of the basics becomes untenable as entirely different models of scientific understanding are called for.

The illuminating discussions of the present volume show how this situation arises recurrently through the range of scientific endeavour, constituting an illuminating Leitmotif across the entire range of inquiry into the ways of the natural and social realms.

Pittsburgh, USA
February 2022

Nicholas Rescher

Acknowledgements

We are but dwarfs mounted on the shoulders of giants, so that we can see more and further than they; yet not by virtue of the keenness of our eyesight, nor through the tallness of our stature, but because we are raised and borne aloft upon that giant mass.

—Bernard of Chartres

Assembling a volume with the length and breadth of this is no cakewalk. Let me begin by thanking everyone who has helped me do so. I shall do this in a chronological way. When I began working on this project for my Albert Einstein fellowship—with a preliminary structure and list of interested authors in my hand—I approached Prof. Ian Stewart to ask if he would be interested in joining this endeavour as a co-editor. Not only did he graciously agree to be a part of this project by commenting that the title I choose 'From Electrons, Elephants and Elections' has certain *Je ne sais quoi* to it, but given his association with diverse domains of knowledge, he could quickly understand the *content* and *context* of this project, reminiscing and indeed mentioning about the content/context relationship he has been reflecting upon since a very long time. His encouragement, support and kindness are acknowledged herewith. I would then like to acknowledge Prof. Nicholas Rescher for graciously providing me with a foreword for this volume. Dr. Angela Lahee has been a constant source of encouragement since the time I started to publish with Springer. Her valuable inputs, her constant support and her kindness are acknowledged herewith. I would also like to thank Springer editorial staff for their support during the typesetting.

Given the unprecedented times during which this volume was assembled, I reserve my heartfelt gratitude to all of the authors for their efforts and time. My sincere thanks to Prof. John Heil, Prof. Gianfranco Minati, Prof. Luciano Boi and Prof. Ehtibar Dzhafarov for being extremely supportive and encouraging. I would like to thank Prof. Jose Acacio de Barros, Prof. Decio Krause and Prof. Nana Last for their willingness to work on another volume of mine—it has been a pleasure collaborating with them. I am very grateful to Prof. Graham Priest for agreeing to contribute an article to the volume. His books are storehouses of practical wisdom to me and to many others, from time to time. I would also like to thank Prof. Tim Maudlin for his kindness and for his contribution. My deep gratitude to Prof. Denis Noble for his support, words of wisdom and warmth. I would also like to acknowledge the kindness of Prof. Leonardo Chiatti and am very grateful for his association and friendship. Much of this work has been done during my Albert Einstein fellowship and in the context of that I would like to thank Prof. Susan Neiman and other members of the board for having faith in me. I am particularly indebted to Susan for patiently replying and sharing her inspirational perspectives on various things in life. I thank the staff of Einstein Forum for their hospitality and warmth: which I can never forget! I am also thankful to Prof. David Shulman for encouraging me and for being an inspiring peace activist—we all look up to! Prof. Lorraine Daston and Prof. Wendy Doniger have been so generous towards me and I can't thank them enough. I am also very indebted to Dr. Krystyna Kauffmann for making my stay at Caputh unforgettable and I also thank the people of Caputh and around for treating me with love and compassion. Finally, words cannot express my gratitude towards Thích Nhất Hạnh and Tâm Liên Đài for inspiring me to witness the intricate interconnectedness of reality. During the moments of despair, I keep returning to Nhất Hạnh's books to find hope, solace and meaning—and they acted as my refuge all through. 'The wave lives the life of a wave, and at the same time, the life of water. When you breathe, you breathe for all of us' Nhất Hạnh writes. May we continue to breathe so, reassuring each other that all of our lives matter and spread joy and well-being everywhere we go.

Einsteinhaus, Caputh Shyam Wuppuluri

Introduction

The Context for this Volume

Electrons, elephants, elections... what connects them? Aside from all three of the words occurring within a span of ten pages in the *Oxford English Dictionary*? Their etymology sheds little light. 'Electron' comes from the ancient Greek *ἤλεκτρον*, meaning amber. 'Elephant' derives from the Latin *elephantus*, whose earlier origins are shrouded in mystery—possibly African. 'Election' comes from Latin *eligere*: to pick out. That makes more sense.

What brings them together in this collection of essays is a deep philosophical issue: the relation between fundamental physics and the world in which we live. Many of us have seen an elephant, perhaps ridden on one; many of us have taken part in an election, as either voter or candidate. Many have seen, or own, amber, but no one has seen an electron. Not even physicists. We infer the existence of electrons indirectly and deduce that deep down inside every elephant there are enormous numbers of electrons. Similarly, deep down inside every election are numerous voters, and again physics tells us that deep down inside every voter there are enormous numbers of electrons. An elephant is estimated to contain about 10^{31} electrons, mostly bound up in its atoms; a voter has a mere 10^{29}.

Today, none of this is particularly contentious. What remains debatable is the argument that since particles like electrons underpin all material objects—in particular elephants, electors, and through their actions, elections—the entire material world, in all its richness and variety, is a consequence of 17 types of 'fundamental' particles, interacting in vast numbers according to a small number of basic rules. We used to call such rules 'laws of nature', but we no longer use that term because it claims too much. Now what once were laws are often demoted to the status of models. Indeed, the most fundamental rules in today's particle physics constitute the drably named standard model.

The chain of reasoning that leads from lowly electrons to majestic pachyderms—let us leave elections aside for the moment to focus on a single topic—is long and convoluted. We can traverse it 'bottom-up': electrons and similar particles combine to make atoms, which make molecules such as amino acids and DNA, which make proteins and other biomolecules, which make chromosomes and mitochondria and nuclei, which make cells, which make organs such as hearts, lungs, skin and trunks, which make an elephant. Alternatively, we can view the same sequence 'top-down', passing from elephant to electron. Roughly speaking, most scientific discoveries are made top-down, and most scientific explanations are bottom-up. So, our title is about reasoning processes in science and their contribution to human knowledge and because much the same philosophical issues arise in any area of human intellectual enquiry, the contributions of the humanities, social sciences and arts.

Read in either direction, this type of causal sequence exemplifies the method of *reductionism*, in which the behaviour of a whole is inferred from that of its constituent parts, and how these parts interact. The scientific method has evolved to be heavily reductionist. Indeed, the sense in which 'fundamental' particles are fundamental reflects reductionist methodology. Immediate questions are: Does it have to be that way? What alternatives are there? How successful is reductionism anyway? Does it really construct a connected causal chain from 10^{31} electrons to one elephant? If you know enough about electrons, can you predict an elephant?

Such questions inspire other related ones and focus attention on a loosely connected set of abstract concepts: content, context, emergence, reductionism, holism. They draw attention to big issues of human existence: the nature of life, consciousness, free will, language, the bizarrely indeterminate world of the quantum, and the vast and enigmatic cosmos of which we experience only a minuscule part. How do we currently go about understanding these issues? Are there hidden gaps in our knowledge? How might we proceed instead? You will find a rich variety of analyses of all of these things within this volume. A few pages later, I will summarise a few key features, but first, let me indulge in some personal remarks, to help establish a context for this book.

About 30 years ago, I became aware that many of the questions that intrigued me could all be viewed as variations on a deeper theme: the relation between *content* and *context*. It was a time when mathematicians and scientists were coming to grips with the natural world in a radically new way: the behaviour of nonlinear dynamical systems. Very broadly, these are systems in which twice the input does not yield twice the output. However, the real world is often nonlinear. If one pill is good for you, two need not be twice as good for you; they might even kill you. If one bag of fertiliser increases a farmer's yield of crops by ten per cent, two bags do not necessarily increase it by twenty per cent and might even decrease it. An extra road can cause more traffic congestion, not less. A global increase in mean temperatures does not make everywhere on the planet slightly warmer, all the time: it causes massive heatwaves, cold snaps and huge floods. But evolutionary survival often depends on a rapid response rather than an accurate one, so default human thinking is often quick, dirty and relentlessly linear.

Moreover, before the twentieth century, most models in applied mathematics and mathematical physics were based on linear mathematics. The reasons are straightforward. Linear problems are, by and large, mathematically tractable. Nonlinear ones are not, at least, with the traditional methods of explicit formulas and calculations done by hand. So, science was full of linear approximations to reality: small swings of a pendulum, shallow water waves, slowly moving objects. Even elephants whose mass can be neglected. There were some successes in the nonlinear realm, of course; the poster child is probably Newton's law of gravitation and his explanation of Johannes Kepler's discovery that Mars's orbit is an ellipse. But such successes were rare compared to the enormous progress being made in the linear realm.

Around 1960, all that changed. Theoretical advances made in the late nineteenth and early twentieth centuries, such as Henri Poincaré's discovery of chaos in the three-body problem, suddenly became practical because of the rapid increase in speed and power of electronic computers. Models that could not be solved by traditional methods yielded easily to new technology and brute-force calculation. The counterintuitive nature of nonlinear systems promptly became unavoidable: it stared you in the face from your computer screen, even if at first it seemed largely incomprehensible. Simple equations led to behaviour of huge complexity—in a mathematical ideal, infinitely complex. Jargon like 'chaos' and 'complex adaptive systems' became widespread. Scientists came to realise that the world is far more complicated, far more interesting, but far more puzzling, than they had imagined. Attention turned in particular to complex systems and *emergent* phenomena, where an interconnected system behaves in ways that seem to transcend the behaviour of its components. For example, an elephant's manifest attributes are apparently unrelated to the quantum physics of electrons, even though most physicists would maintain that in principle those attributes are consequences of the rules of the quantum world. More precisely, an elephant cannot flout the quantum rule book—which is entirely reasonable, but not quite the same thing.

Much of today's science is reductionist. The metaphor of 'levels of description' is rampant, with elephants at the top and electrons at the bottom. Reductionist explanations are therefore said to be 'bottom-up': behaviour on one level is explained by reducing it to interacting components on a lower level. Chemistry is reduced to molecules, which are reduced to atoms, which are reduced to electrons, protons and neutrons, which are reduced to quarks. Ecosystems are explained in terms of populations of organisms, organisms are explained in terms of genetics, and genetics is explained in terms of DNA and proteins, which, in turn, are explained by molecules, atoms, electrons, etc. Causality runs from bottom to top: electrons determine how an elephant behaves, but elephants cannot influence the physics of electrons. So it is said.

In practice, such reductions are easier to accomplish than the converse: showing that the lower-level system really does explain the higher-level system from which it was reduced. For example, it is not known how to calculate the properties of atoms beyond hydrogen by a rigorous use of quantum equations. Not even helium. The behaviour of a molecule cannot be inferred with complete confidence from its atoms. How a protein folds cannot be calculated rigorously from its amino acid sequence. Deducing the phenotype of an organism from its genome seems hopeless. The dynamics of galaxies cannot be calculated, even numerically, from the forces exerted by gravity on its constituent stars, even if we simplify them to point masses. The chain from electron to elephant may consist of massive links, but their connections are weak.

The concept of emergence reflects this difficulty. In an extreme form, it denies any causal connection between electron and elephant, in which case it disputes the basis of reductionism. In its most common form, it accepts that in principle there is a causal connection, but one that is so complicated that in practice it would be impossible to describe it in detail. I prefer a less extreme definition: emergence is a causal network whose behaviour is deterministic but computationally intractable: even if in principle 10^{31} electrons imply an elephant, no one could possibly do the calculations using the equations of particle physics. Whatever the definition, there is often convincing evidence to *suggest*, with high plausibility, that the low-level description does imply what is observed on the higher level. However, such deductions are always partial, lack mathematical rigour and are based on approximations and assumptions whose validity is open to question. Ultimately, what ought to be rigorous logical deductions are replaced by leaps of faith. 'Handwaving', in mathematicians' parlance.

An example contributing the third component of our title is an *election*. In principle, each voter's decision is the result of a complex flow of electrons (and other stuff) in his or her brain. With enough prior knowledge, we merely have to solve Schrödinger's equation for a sufficiently large system to predict each voter's decision. Somehow the final result emerges from the flows of all those electrons. However, no one in their right mind would use that method to try to forecast who will become the next President. The calculations would be almost infinitely beyond the capabilities of even the fastest supercomputer. There may well be a clear chain of widely accepted scientific theories that leads from electrons to the winning candidate, but every link in that chain rests on unproved beliefs.

An alternative approach, taking pretty much the opposite approach, is *holistic* (or *top-down*) thinking. Consider an ecosystem or an organism *as a whole*, and seek patterns on the same scale. Model a galaxy not as 400 billion point masses (stars) but as an object in its own right with its own properties (spiral and elliptic). Model a hurricane not as a continuum of tiny fluid particles, but as a structured form of moist air and water vapour. Model an election in terms of probability distributions of voter preferences and mass psychology. Model an elephant as an intelligent autonomous agent interacting with its environment. Seek patterns on the same scale as the phenomenon of interest.

Reductionism is largely about content, whereas holistic models are more attuned to context. The articles in this volume examine the relationship between the two, along with issues that intertwine with them, such as emergence, from an enormous variety of points of view. The key point, to my mind at least, is that reductionism and holism, content and context, are not opposing philosophies. They work best in tandem. Why restrict yourself to one tool when two are available?

Blind Men and Elephants

Back to 1990 and my philosophical epiphany. My interest in the mathematics of chaos led to a meeting with the biologist Jack Cohen, which started a collaboration that continued until his death in 2018. Despite working in different fields, we found many common interests, ranging from science fiction fandom to the philosophy of science. Together, we examined issues such as chaos, complexity and emergence in several joint books, starting with *The Collapse of Chaos* and *Figments of Reality*. It all began at that first meeting when Jack asked me an intriguing question: If nonlinear systems are enormously complex and unpredictable, but also ubiquitous, how can living organisms possibly work? This innocent question led into much deeper waters than either of us expected, and before examining the articles contributing to this book, I want to set the scene by sharing one of Jack's early insights. It was inspired by a cartoon he had come across in a science fiction magazine. Not this cartoon: the next one. But we have to start with this one.

Blind men and the elephant, version 1

You probably know the story. It originated on the ancient Indian subcontinent, and one of the earliest versions occurs in the Buddhist text *Udana 6.4*. Several blind men, who have never before encountered an elephant, touch an unknown object and report their deductions. One declares it to be a rope, another a tree, the third maintains that it is a wall, the fourth a leaf, the fifth a spear and the sixth a snake.

'Aha!' says a wise man passing, 'I know what it is! It's an elephant'.

To quote Wikipedia: 'The moral of the parable is that humans have a tendency to claim absolute truth based on their limited, subjective experience as they ignore other people's limited, subjective experiences which may be equally true'. It is a neat piece of social relativism, but I think it misses the point. The lesson of the blind men and the elephant depends on how wise the wise man really was.

Let me contrast this interpretation with the cartoon in the science fiction magazine. This cartoon told much the same story, but with a twist. Four blind men each declare the object to be an elephant. Actually, one holds a snake, another is touching a tree, the third is touching a wall, and the fourth is holding a rope.

Blind men and the elephant, version 2

Both cartoons, so Jack told me, are about content and context. The content—the experiences of the blind men—is the same in both cases. But in the first cartoon, the context they are assuming does not include elephants, so they interpret each separate experience as best they can, in the context of what they do know. In the second cartoon, the context does include elephants, so each blind man interprets his experience as an elephant, even though it is not. If the wise man assumes the right context, his synthesis is correct; if not, it is not.

Science, Jack suggested, can be seen in both ways. Substitute 'electron' for 'elephant', and the first cartoon illustrates the early days of particle physics, when the concept of an electron had not yet crystallised. The experience for an ancient Greek was the ability of amber to attract small objects when rubbed against fur. (As I said at the start, the ancient Greek word for amber is *ἤλεκτρον*.) In the 1700s, Charles François du Fay found that while amber rubbed with wool attracts a charged gold leaf, glass rubbed with silk repels it. His interpretation was of two fluids: *vitreous fluid* emanating from glass rubbed with silk and *resinous fluid* from amber rubbed with wool. These fluids, which neutralise each other when combined, he called *electricity*. Benjamin Franklin's view, ten years later, was that electricity is a single fluid that can exist in two opposite states: positive and negative. The fluid 'carries' a charge and can have an excess of it or a deficit. George Stoney, doing experiments in electrolysis, decided that the electric 'fluid' is not a continuum: there is a 'single definite quantity' of electricity, the charge on a single ion. But he also thought these charges are permanently attached to atoms. In 1881, Hermann von Helmholtz interpreted the observations as evidence for 'behaviour like atoms of electricity'. Stoney named these elementary charges *electrolions*, but soon changed the name to *electrons*.

As in the first cartoon, the blind men had interpreted their observations very differently. One had perceived an attractive force, another two fluids, a third one fluid with two opposite properties, a fourth something more granular than a fluid, a fifth something *like* a lot of atoms, but different. At that point, a passing wise man (Sir Joseph John Thomson, building on the work of dozens of others) cut through the confusion and synthesised all of these disparate observations into a single *thing*: an exceedingly diminutive particle equipped with a negative electrical charge. Now the blind men were finally able to perceive their metaphorical elephant as an entity in its own right, an electron. It was the same name that Stoney had used, but yet again it was being interpreted differently.

That was not the end of the story. With the advent of quantum mechanics, light, hitherto considered to be a wave, sometimes behaved like a particle, the photon. Conversely an electron, thought to be a particle, could also behave like a wave. Matter on its smallest scales could be both wave *and* particle, though never both at once. Today's view is even subtler and technically different. The wise man has moved on, and his homely wisdom is now hedged with obscure caveats and shrouded in mathematical formulas.

The second cartoon invites us to question this entire train of reasoning. The resolution of the different perceptions centres upon their content. The unspoken assumption is that the same 'it' explains all of the observations, and that the very different contexts for the experiments—rub amber with wool, perform electrolysis—do not affect the thing itself, only how we observe 'it'. If this assumption is wrong, then we are equating several unknowns and creating a phantom. The philosophical question is whether 'the' electron actually exists. Is there truly a single object that has all of the different attributes experienced by all those scientists over the ages—or is the world more like the second cartoon, with different attributes being combined to offer the illusion of a single underlying cause?

Physicists like unity. Albert Einstein spent much of his later life trying to combine the two great physical theories, relativistic gravity and quantum physics, into a single unified field theory. He never succeeded, and neither has anyone else, yet. But the search continues with, if anything, even greater fervour: superstrings, quantum gravity, etc. On the other hand, the universe might be what Jack and I call the 'glass menagerie': each phenomenon is explicable by a theory of limited validity, but no underlying unification of those theories is possible. As Banesh Hoffmann wrote in *The Strange Story of the Quantum*, at one time scientists 'went around with woebegone faces sadly complaining that on Mondays, Wednesdays, and Fridays they must look on light as a wave; on Tuesdays, Thursdays, and Saturdays, as a particle. On Sundays they simply prayed'. In a glass menagerie world, that is the best we can ever do.

For electrons, the verdict of physics is clear and resounding: electrons are real objects—in some sense. Philosophers might dispute this conclusion and make a reasonable case for doing so, or question what that sense is, but the same concept explains so many diverse observations and experiments that from the point of view of physics, 'the' electron clearly exists. More precisely, the unified concept of an electron provides a consistent mathematical context in which all of the observations and experiments make sense. And that is what physicists mean by 'exist'. Although we cannot observe an electron directly, we can infer its existence in this sense using several independent lines of evidence. Just as we infer that the Sun's interior is extremely hot without ever having been there, even though for all we can tell it might be hollow, cold and surrounded by an alien force field that emits the appropriate radiation.

Less contentiously, we can ask similar questions about other basic concepts of modern science where the answers are less clear-cut. Fervently held theories where today's trumpeting elephants might turn into tomorrow's brick walls and ropes are not hard to find. Historical examples abound. Early experiments on how materials burn were interpreted as evidence for an unknown substance, *phlogiston*, emitted by a burning object. The phlogiston theory, proposed in 1667, was the conventional wisdom for just over a century. Then, it turned out that burnt matter weighs more than unburnt, provided all the products of burning are accounted for. Phlogiston was, in effect, negative oxygen. Indeed, for a time oxygen was named 'dephlogisticated air'.

More recently, cosmologists have spent years (and huge sums of cash) seeking direct evidence for 'dark matter', which they believe must exist in order to explain various gravitational anomalies in galaxies and other celestial objects. Even though dark matter must outweigh conventional 'baryonic' matter fivefold, no dark matter has yet been found. Some of the anomalies may have other explanations, some may even be modelling errors, but right now, cosmology's walls and ropes and snakes have been lumped together into an entity named 'dark matter'. This may well be a stroke of genius akin to the electron, or it may be cosmic phlogiston. Only time will tell.

Science has a habit of focusing on content rather than context. The successes of molecular biology in the second half of the twentieth century led to an overemphasis on genes as the sole explanation for the form and behaviour of organisms, only now being reversed. The relevant context for dark matter is almost universally assumed to be Albert Einstein's theory of general relativity, or, when appropriate, the older Newtonian theory of gravity, which is often entirely adequate for calculations. If you do not question this context, then the observations that led to the hypothesis of dark matter are puzzling. Within that context, the most plausible explanation is invisible matter formed by a hitherto unknown particle or particles, so everyone looks for this amazing new kind of matter. Few question the context, and those that do are largely ignored, even when they suggest plausible alternatives or point to difficulties in the dark matter theory.

To close this part of the discussion, let me describe a third cartoon. It would be hard to draw, because it involves 10^{31} blind men observing an elephant. Every single one of them says 'it's an electron!' (We could add similar numbers who perceive a proton or a neutron to obtain a more complete description. Or reduce it all to quarks and the like.) They are arguably *correct*, but that leaves open the key question: What does this tell us about the elephant?

The Content of this Volume

The content/context relationship is not just important in physical and biological sciences. It is if anything even more important in social science, the humanities and the arts. And this, finally, brings me to the main point of this introduction. I have examined the context for this volume; now I want to summarise its content.

The contributors include philosophers, scientists of all kinds, mathematicians, logicians, specialists in the humanities and others from a wide variety of disciplines. Their essays cut across traditional subject boundaries, mix speculation with traditional wisdom, question things long ago accepted as fact by most practitioners and interpret old evidence in new ways. The viewpoints represented are equally diverse. Any order imposed on such an eclectic collection must be artificial, but I have bitten the bullet and classified them into twelve overlapping categories. Many essays address several of these in combination, in which case I will locate them in the most prominent or most convenient category. The categories are: reductionism, holism, context, emergence, causality, probability, physics/mathematics, quantum, computation, biology/medicine, sociology and art. Important strands that cut across this classification include philosophy and creativity.

Reductionism. As already remarked, until recently most science has been reductionist in spirit, almost by default. Looking inside things to understand how they function has led to remarkable successes. Nevertheless, most authors in this volume are critical of reductionism as an overarching philosophy.

Mario de Caro makes a valuable distinction between reductionism as a guide to methodology and reductionism as a sweeping claim that reductions are always possible in principle. His example is the reduction of chemistry to physics, probably the case that ought to be most clear-cut; he argues that it is anything but. He also traces this extreme form of reductionism to the monist belief that everything in the universe ought to be explicable in the same way. Gianfranco Minati asks whether the standard approach to reductionism in terms of well-defined 'levels of description' is tenable, and suggests that incompleteness is one reason for rejecting this metaphor. In contrast, Terry Horgan makes a case in favour of reductionism, arguing that micro-physical phenomena are primary, so that the history of the cosmos must be an inevitable consequence of its microscopic history.

Holism. This is a standard alternative to reductionism, and like everything in this volume, the term is open to many interpretations. Urging people to focus on 'the whole' is easy, and it is difficult to argue against that advice, but deciding what 'the whole' actually is, and figuring out how to embrace it and understand it, is trickier.

John Heil points to the gap between the aspirations of reductionism and its achievements, considers holistic alternatives and suggests that although holism need not involve emergence, emergence is a key reason for rejecting strong reductionism. In the opposite direction, Sven Ove Hansson criticises misuses of the term 'holism' to justify pseudo-scientific claims, such as rejecting vaccinations to improve a child's prospects after reincarnation. He traces such attitudes to the belief that 'the whole' can be identified and forms a closed system, which need never be modified.

Context. It is becoming increasingly apparent that many real-world systems cannot be understood purely in terms of their content. This challenges standard reductionism and emphasises the role of context.

An essay by the late Jack Cohen, reproduced in edited form from his 1990 Novacon Special *Are You Content in Your Context,* explains his belief that context is at least as important as content. He contrasts information with meaning, taking genetics as an example, and explores the role of context in science fiction. I have added editorial notes for readers who are not 'fen' (SF fans in their jargon). Erich Rast relates systems of mathematical logic to context dependence, especially when investigating the relation between interdependent theories and meaning. Otávio Bueno also considers a mathematical topic: the extent to which theorems depend on their context, a relationship that is often denied by mathematicians. Hildegard Meyer-Ortmanns examines emergent phenomena in different contexts that have a unified underlying mathematical explanation, provided by modern theories of nonlinear dynamics. Mathematics provides the content; the interpretation of its variables adapts the results to different contexts. Robert Bishop explores a similar theme, emphasising the importance of context for all natural phenomena and showing how contextual emergence can capture the interplay between constituents and context.

Emergence. This topic is central to many of the issues discussed in this volume and appears in many essays discussed under other headings. What is emergence to one thinker is nothing of the kind to another, but all uses refer to systems whose behaviour appears to transcend that of its components. The differences arise from the meaning assigned to 'transcend' and the extent to which appearances are valid.

James Miller considers emergence in language. Here, some properties, such as truth, make sense only for high-level structures such as sentences—a single word cannot be categorised as true or false. He asks to what extent such ontologically emergent properties are inherent in lower-level forms. Arturo Carsetti also deals with linguistics, focusing on the origin of meaning, human cognition and creativity. Timothy O'Connor points out that although the successes of reductionist science demonstrate that everything we observe arises through elementary physical processes, it is wrong to deduce that the behaviour of more complex objects is merely a coarse-grained version of some of those basic processes. When new configurations of basic elements arise, they can create genuinely new processes and powers. Carl Gillett surveys historical disputes between proponents of reductionism and emergentism in science, with explicit examples such as protein form and function. He argues that the principle of parsimony does not justify reductionism, and offers a way to resolve such disputes. Alexander Carruth asks when apparently emergent phenomena can be meaningfully related to the behaviour of components' parts when the behaviour concerned is micro-latent: comes into play only in specific complex circumstances. If so, the phenomenon is often considered not to be truly emergent. He proposes a compromise position. Michael Silberstein investigates the role of contextual emergence in network neuroscience, cognition and psychology, emphasising the fundamental importance of multiscale contextual constraints. Michael Tye explores the role of emergence in consciousness by asking whether, in principle, we could construct an elephant by acquiring a supply of the relevant subatomic particle and assembling them to make an exact copy of an elephant. He suggests that the 'hard problem' of consciousness could be an obstacle, and that consciousness is not an emergent phenomenon but a basic feature of the universe. Conscious electrons give rise to a conscious elephant.

Causality. Philosophers have debated the meaning and nature of causality since at least the time of the ancient Greeks. Proximate causes (go out in the rain and you will get wet) differ from ultimate ones (water is formed from hydrogen and oxygen; hydrogen goes back to the Big Bang; oxygen was created by nuclear reactions in stars, etc.) Reductionism seeks ultimate causes in the micro-structure of the universe, but its grip on the causal chain is weaker than it likes to pretend. Context is as much part of causality as content. Relativity and quantum theory have cast their own peculiar doubts on naive views of causality.

Graham Priest surveys historical views of causation in Buddhist philosophy, paying special attention to questions of reductionism and holism. George Ellis and Jonathan Kopel tackle causality from a different viewpoint: how the interplay of upward emergence and downward causation combine to create complexity through adaptive modules. Purely bottom-up explanations of emergence, from lower levels to higher ones, assume that the underlying physics is complete, but this never happens. Causal closure also involves top-down effects, from higher levels to lower. COVID-19 provides an example. Tim Maudlin examines 'top-down' causality, a concept that depends on what is top, what is bottom and what is causality. Reference to Aristotle sheds some light on these issues.

Probability. Our most powerful mathematical tool for resolving and quantifying uncertainty is probability theory. The techniques are well established, but their interpretation is not—witness the ongoing debate between Frequentists (probability is a long-term proportion) and Bayesians (probability is a degree of belief). As is often the case, we know how to compute probabilities; we just do not know what they *are*. Three essays discuss how context affects our view of probabilities and suggests new techniques and concepts.

Ehtibar Dzhafarov reviews work on the logic of contextuality, applying it to basic philosophical issues in probability, such as the nature of random variables, and stochasticity versus determinism. He also describes new ideas applying contextual logic to Bayesian probability, visual illusions and logical paradoxes. Sergio Chibbaro, Lamberto Rondoni and Angelo Vulpiani study the relation between the conceptual notions of probability and the interpretation of experimental results, in the particular context of statistical mechanics and Hamiltonian systems. This sheds light on the longstanding problem of irreversible macroscopic change—such as the increase in entropy—occurring in microscopically reversible systems. Andrei Khrennikov reviews applications of quantum probability outside physics. Being context-dependent, this conception of probability can potentially shed light on other context-dependent systems, such as human beings. Such applications involve inherent ambiguities and uncertainties, but they also motivate a calculus of contextual probabilities.

Physics and Mathematics. Several authors locate their discussions in specific areas of mathematics and science, lending a more concrete aspect to the issues raised. In this category are essays featuring metals, electrons, quantum gravity, particle physics and chemical molecules. In each case, the crucial issue is the connection between micro- and macro-structure—including whether it aids understanding.

Tom Lancaster focuses on condensed matter physics: to be precise, the physics of metals. Metals have made huge impacts on human civilisation, but only recently have a deep understanding of metals from the point of view of quantum physics appeared. The electron ceases to be viewed as a simple particle; instead, it is a quasiparticle, 'dressed up' in interactions with the rest of the system—a context for the electron. This example provides lessons for other investigations of the universe. Karen Crowther examines recent developments in quantum gravity, suggesting that space-time in Einstein's sense is not fundamental. Instead, it emerges from relations between entities that are not themselves spatiotemporal. Attempts to construct such theories use both top-down and bottom-up strategies. However, the tidy metaphor of 'levels' that is typical does not necessarily make sense in this area. Leonardo Chiatti's chosen area is particle physics, normally viewed as a central example of reductionism. He points out that the 'real' nature of fundamental particles is not understood; indeed, the usual approach to quantum models is not to ask such questions. These issues allow an exploration of the relation between local and global structures and processes.

Quantum. Several authors mentioned above consider questions in quantum physics, but the articles in the next group are more explicitly focused on the deep scientific and philosophical issues posed by the radical nature of quantum mechanics. Among these are apparently irreducible uncertainty, the meaning, if any, of the wave function, and the problem of quantum measurement: How does the crisp, simple mathematical definition of an observation relate to the complex macroscopic apparatus used in practice?

Ignacio Licata discusses limitations to science, exemplified by the process by which an observable is measured in an experiment—the source of much debate in quantum theory. This process is affected by the choice of model as well as the actual experiment, because the model provides a context for interpreting the observations. The structure of science involves complex networks of inference, all with systemic uncertainty. Arkady Plotnitsky provides a historical survey of how quantum theory changed scientists' views of space, time and matter. Quantum uncertainty imposes limits on the applicability of reductionist views, because there is no deterministic link from bottom to top. On the other hand, quantum theory does not lend itself to a holistic approach either. So, it represents a distinct way of thinking in both physics and philosophy. Michel Planat tackles the problem of quantum observation in a novel, mathematically technical, way. He uses the 'language' of 'words' (sequence symbols) in a free group. Commutators in the group correspond to the commutators of operators that are basic to the quantum formalism. Words correspond to paths in a suitable 4-dimensional manifold, a kind of exotic space-time. Measurements correspond to related but distinct manifolds. The outcome is a kind of quantum logic, 'exotic non-contextuality'. Luciano Boi reviews current ideas in topological quantum field theories and string theory, from the point of view of geometry, topology and invariants. He discusses relations to recent breakthroughs in knot theory, which suggest connections between knot and link invariants and physical observables—a modern slant on an old suggestion of Peter Guthrie Tait, long discredited in its original 'vortex atom' form. He also considers the fuzziness of space-time. José Acacio de Barros, Federico Holik and Decio Krause discuss how context and content cause problems in quantum mechanics, because quantum properties depend on context. They explore the strange implications of this issue for elementary particles, which make them unsuitable for study using standard mathematics and thus require new mathematical ideas.

Computation. It might seem strange that something as concrete as computation should lead to deep philosophical issues, but the history of the subject argues that this is inevitable when suitable questions are asked. The nature of computability is one, and it led Alan Turing to his discovery that the halting problem is algorithmically undecidable. His interest in artificial intelligence is now flowering, with methods such as deep learning, and algorithms are beginning to affect daily life in myriad ways. Applications of computation are now posing serious social and philosophical questions. Not to mention the possibility of quantum computers...

Samson Abramsky offers a perspective on paradoxes and their resolution, the concept of partiality. He illustrates this in two settings: classical computation and the Church–Turing thesis that all sufficiently flexible computational systems are equivalent, and quantum computation, where partiality is a consequence of the mathematical formalism. Ilkka Niiniluoto examines another computational issue of great current interest, deep learning and artificial intelligence. He considers the training processes of AI from a philosophical perspective and reconsiders Hubert Dreyfus's criticism of AI. Thomas Filk analyses a related mathematical structure, the neural network, an archetypal complex adaptive system. A neural network can be viewed as interpolating between content and context, depending on its size. Recurrent networks behave like non-classical mechanics, with memory effects, and can even be trained to violate Bell inequalities of quantum mechanics, normally considered as ruling out deterministic explanations of quantum uncertainty. Genetic algorithms resemble Darwinian evolution.

Biology and Medicine. Nowhere is the challenge facing reductionist explanations more apparent than in the life sciences, where even the basic question 'what is life?' opens up Pandora's box ('Pandora's warehouse' might be a more appropriate metaphor). Despite decades of major advances in molecular biology, even a single cell has, if anything, become more enigmatic than ever. Let alone an organism or an ecosystem. This is not to denigrate the astonishing advances opened up by the discovery of the double helix, but it shows how much further we have yet to travel.

John Bickle explores a problem in which traditional holistic psychological ideas interface with the neuroscientific field of molecular and cellular cognition. That is, how high-level brain functions emerge from neurons and biochemistry. He observes that in this case, the links between the different 'levels' leave much to be desired. Raymond and Sir Denis Noble provide a layman's introduction to the idea that no specific causal level can be assumed primary, contrary to the reductionist assumption that micro-levels are privileged in terms of explanation. In his view, agency and consciousness are real, not illusions, because they result from functional constraints at higher levels. Daniel Dennett offers some speculative but insightful ideas about the machinery in the brain that is responsible for intelligent action, suggesting that foresight and self-monitoring are crucial to intelligent behaviour. The rational deliberator mostly worries about problems that never happen, *because* they worry about them and take action to prevent them. He explores implications of these ideas for the design of intelligent robots. Luciano Boi examines one of the most perplexing examples of just this issue: how an organism's genetics and molecular structure relate to its growth and evolution. He proposes that morphology on a macro-level has its own structures and patterns, which, although they may ultimately derive from genes and molecules, can be understood and used without reference to the micro-level. In effect, the proposal is to replace genetic 'information' by nonlinear dynamics. Marta Bertolaso and Héctor Velázquez observe that the tension between reductionist and holistic methods is becoming increasingly apparent in the life sciences, usually viewed as a dependency between content and context. Complexity is ubiquitous, and investigating the nature and origins of life inevitably leads to this tension. He suggests that the focus should shift from universal reductionist principles, such as basing everything on molecules, to 'a philosophy of the particular', and offers examples such as cancer. Marco Buzzoni, Luigi Tesio and Michael Stuart take the discussion further into medical territory with the need for doctors to combine knowledge and methods from both macro-level (the patient) and micro-level (drugs and genetics). The basic questions are 'what is disease?' and 'how should we treat it?' They suggest that the experimental paradigm can profitably be applied to holistic 'complementary' medicine, a proposal rejected or ignored by most of its practitioners. Manuel Rebuschi concentrates on a single example: schizophrenia. Here, it is necessary to interpret conversations with the patient, which involve context. Analysis of fictional conversations shed light on both the disease and how to manage it.

Sociology. 'Know then thyself, presume not God to scan; the proper study of mankind is man'. So wrote the poet Alexander Pope in *An Essay on Man: Epistle II* in 1733-4. Generations of scientists have studied anything but humanity; others have studied it, but not on its own level. Social scientists fill the gap and correctly remind us that the word 'Man' here is sexist. (My excuse is that this is what Pope wrote. Blame him.) Sociology offers penetrating insights into the human condition, but is bedevilled by huge and unavoidable obstacles, such as the difficulty of performing controlled experiments. It therefore faces issues such as context and emergence head-on.

Friedel Weinert examines the debate between two schools of thought on the level at which social science explanations should be stated. Is the key factor the behaviour of an individual in a social unit, or are social factors that affect that individual the important ones? A compromise position is to select the level according to the nature of the problem; however, in some cases the macro-level is unavoidable. Diederik Aerts and Massimiliano Sassoli de Bianchi investigate connections between basic science, such as thermodynamics and evolution, and aspects of matter, life and human culture. Following Erwin Schrödinger's pioneering footsteps in *What is Life?* they discuss how the second law of thermodynamics relates to living creatures and culture. Annika Döring and José García also discuss culture: specifically, the pitfalls of unrecognised cultural differences. They contrast Okakura Kakuzo's *The Book of Tea* with Martin Heidegger's *Being and Time*. The two works have quite of a lot of content in common, leading some to suspect that Heidegger drew on Kazuko's work. Here, it is pointed out that this resemblance is superficial: the two authors use similar words but with radically different meanings; they have similar contents, but different contexts. Understanding differences between Eastern and Western culture can correct this misperception.

Art. Historically, art and science were intimately related for much of human history. The early cave artists had to make their own pigments and oil lamps, not just sketch the outlines of mammoths. During the Enlightenment, scientists and artists routinely met together, and each influenced the other. In the twentieth century, the two areas of activity started to decouple in various ways: schools often had a science stream and an arts stream, for instance. Most artists and scientists still accept this separation, but a growing band is once more bridging the gap.

Nana Last tells us about the huge changes in architectural thinking inspired by the shift from postmodernist architecture-as-text to the digital age's architecture-as-algorithm. This has transformed the design process by making it possible to transfer information from a late detail to the overall conception, instead of starting with an overall plan and adding details later. John Barrow's essay was written shortly before his untimely death in 2020 and reflects his lifelong interest in connections between science and art. Here, he observes that both activities involve complexity, and considers one of physics' favourite complex systems: the sandpile. This is an example of robust simplicity on the level of the whole pile, resulting from huge complexity on the level of its individual sand grains. He suggests that the appeal of musical performances rests in a similar overall robustness, made all the more interesting by differences in fine structure.

Simplex, Complex, Multiplex

Each of us will have our own opinions about which authors and which assertions we agree with, and logic indicates that we cannot securely agree with them all. That is why this volume exists. It is why we all need to read every essay and try to understand the line of reasoning.

To wrap everything up in one tidy package, I will take inspiration from the other element of my collaboration with Jack Cohen: science fiction. One of the great science fiction authors, most active in the 1960s and 1970s, is Samuel R. Delany. His books are not the clichéed shoot-em-up space battles that those unfamiliar with the genre believe to constitute the whole of 'sci-fi'. Actually, very little of it fits that description. Like most of the best SF, Delany's novels are thoughtful and insightful explorations of the human condition, albeit in imaginative contexts. The novel that bears most keenly on our topic is *Empire Star*, published in 1966, and I recommend getting hold of a copy—especially if you have never read any SF—because it has a lot to say about the topic of this volume. It is short, clever, imaginative and superficially circular in structure, just as James Joyce's *Finnegans Wake* starts by completing an unfinished sentence with which it ends. However, Delany's structure is less simplex. In fact, his novel explicitly declares itself to be multiplex. And thereby hangs a tale...

The story centres on a boy called Jo who lives on a backwoods planet called Rhys. He encounters Jewel, who is a crystallised Tritovian and the sole survivor of a crashed organiform cruiser that had been en route to Empire Star, the administrative centre of the galaxy. One of the crew lives just long enough to tell Jo to take a message to Empire Star, but not long enough to tell him what the message is. Much later he finds out: 'Someone has come to free the Lll'. These immensely intelligent creatures are kept in slavery because of their ability to rebuild entire civilisations, and as payback they make everyone who owns them irredeemably sad. Jo also finds out that he should not deliver this message until someone *has* come to free the Lll, and this is not clear.

Anyway, that is the frame—the context. The specific element of content that I want to discuss, one small detail within Delany's widescreen baroque parable, is a running theme about the mental processes of sentient beings. Namely, they come in three main types: simplex, complex and multiplex. One test is to ask someone what is the most important thing in the world. If they are simplex, they will answer. Another test is whether they ask questions. Simplex minds seldom do. Thus, the inhabitants of the Geodetic Survey Station, in the midst of a vast project to catalogue all of the knowledge in the universe alphabetically, are enormously intelligent but hopelessly simplex. As proof, their project means everything to them and they never ask themselves whether it is sensible, feasible or worthwhile. A complex intellect appreciates that different people and cultures can have different opinions about the same thing, without one being wrong and the other right, and that new knowledge can appear at any time. Multiplex minds contemplate things simultaneously from many points of view. When faced with a mass of apparently contradictory or confusing information, the multiplex personality orders their perceptions until they figure out the correct question to ask.

Reductionism and holism *alone* are both simplex, as are all *isms*. To assert the superiority of one way of thinking, to the exclusion of all others, is irreducibly simplex, however cleverly the claim is expressed. As Delany's character Lump—the linguistic ubiquitous multiplex—says, 'intelligence and plexity do not necessarily go together'. Most of the essays in this book are complex, and the book itself is most definitely so, since some authors flatly contradict others. Indeed, the entire collection, approached in an appropriate frame of mind, is undoubtedly multiplex. Merely order your perceptions multiplexually, and you will understand how everything fits together.

Ian Stewart

References

Delany, S. R. (1966). *Empire star*. New York: Ace Books.
Heidegger, M. (1996). *Being and time* (English translation). New York: State University of New York Press.
Hoffmann, B. (1959). *The strange story of the quantum*. Harmondsworth: Penguin.
Kakuzo, O. (1906). *The book of tea* (English translation). New York: Duffield.
Schrödinger, E. (1944). *What is life?*. Cambridge: Cambridge University Press.
Stewart, I. & Cohen, J. (1994). *The collapse of chaos*. New York: Viking.
Stewart, I. & Cohen, J. (1997). *Figments of reality*. Cambridge: Cambridge University Press.

Contents

Humanities and Social Sciences

Setting the Context

Are You Content in Your Context?

Jack Cohen

Cover art by dave mooring for cohen (1990).

Science fiction readers enjoy performing an intellectual act, a literary contortion, that isn't understood, *can't* be understood, by nearly everyone they know. That's not news. People ask their authors 'Where do you get those crazy ideas?'. People ask

J. Cohen (Deceased) (✉)
Coventry, England

S. Wuppuluri and I. Stewart (eds.), *From Electrons to Elephants and Elections*, The Frontiers Collection, https://doi.org/10.1007/978-3-030-92192-7_1

you 'What do you get out of those crazy, impossible stories?', don't they? I've got a new slant on this, and I want to take you round the houses a bit to explain it.

Firstly, let's have some thoughts about 'messages' and their 'meanings'. There are two rather different ways of thinking about this, which many of us, and our authors, mix up. The kind of usage that worries about 'information content', about 'bit-strings' and other computer jargon, 'noise', 'bandwidth', and the rest of the radio-transmitter buzzwords, is very different from the everyday, or the literary, uses of 'meaning'. Very often, the obvious technological uses of the first kind lead us to think that the message *is* the object, or at least the important essence of the object. Tom Easton's near-future biotechnology stories[1] talk of the genes as if they *are* the characters of the animals. He gets a gas-bag shape from a jellyfish genome, pockets from a kangaroo, and so on, and sticks them together as if he's playing with Lego™. Many of our stories talk of the 'plans for' a new super-weapon as if they are the weapon itself, with characters and hit-men striving to get the microfilm or the tape. But it's important to realise that what's 'on' the tape is really just a sequence of numbers, expressed as differences in optical density or magnetism. This sequence of numbers actually doesn't have any unique meaning: if you don't know whether to put the tape in a video-player or a computer or a Walkman[2], you don't know if the sequence is a film of octopuses mating, the digits of $\surd 2$ from the 10^8th onwards, or Beethoven's Ninth Symphony.

Indeed, with a little imagination you could design players that would produce each of these from special bits of tape that had a recording *made from one of the others*. They would, perhaps, be like an old-fashioned jukebox, where an arbitrary number resulted in the sound (more rather than less...) of a record; or think of those computer opening-screens where you have to put in numbers at the keyboard to make the computer load the program you want to use. (There is no reason in principle why it shouldn't be a *long* sequence of numbers.) This should make the point that the technical use of a 'message' (really 'data-string') doesn't tie meaning to it any less fuzzily than ordinary usage.

Here are some situations in the ordinary world that demonstrate the same thing.[3] You get a letter, which says 'If I don't ring you on Monday night, that means Aunt Minnie will be coming on the 6.18 from London, and you should pick her up and take her to 6, Orchard Crescent... etc. etc.' Or someone says 'You will find that information on pages 1130–67 of the *Encyclopaedia Britannica.*' Or someone says 'Take the train from Canton, change at Hang Chou ...'. In all these cases, there is much more information passage in the background than in the message, more in the Context than in the Content (ah, he's getting there now...). In the first example, *one* bit of information (phone or not) leads you to all the things the letter tells you to do for Aunt Minnie; in the second example a simple page reference takes you to all the information on those pages, unambiguously; but the third example, which looks so simple, is the most informative for my purposes. It assumes a whole lot of context: that Canton is in China, how to 'change trains' (!), what instructions are like in the English language, the language itself.

Of course, all my examples assume English, and trains, and libraries, books and numbered pages, without which each message would be as meaningless as our piece of tape with a series of numbers on it. Every message *assumes* a context, and 'meaning' can be got across with a simple message, a trigger, together with much (immediate) context, as with Aunt Minnie; or it may be spelled out more explicitly in the message itself, as with an encyclopaedia—but even the encyclopaedia needs much more context than content. Think about it.

What has this to do with science fiction/fantasy/mundane[4] literature? Let's think of the 'world of the detective story'. In that world, a country house with upper-class guests is *always* the scene of a murder. There are several other contexts, like trains or police stations or seashores, where the 'detective story' context gives the reader clues. Again, the author's message can be played out against an assumed, common, contextual background known to both reader and author (but rarely to the characters—hence some of the humour in *Moonlighting*[5]). The 'message', the story that is given to us in this setting, can be very tortuous; for example the detective, or the first person narrator, could be the murderer. But the context is a 'given', just as it is in Mills and Boon (railway bookstore romance) stories, (most) fairy stories, kitchen-sink dramas, and nearly all mundane literature. The usual terrestrial background, death and taxes, clothes and kids, money and malingering, are assumed to be common to reader and author. Westerns are a bit different, in that a common historical setting has been 'improved' as a backcloth, but it's still carefully bedded in a 'real' context.

All these mundane stories retain the standard *context*, and different stories change only the message, the *content*. In science fiction and fantasy stories, though, it's the context that is changed, while the storyline, the message, often has just a simple mythic structure (how many Cinderellas, how many Frog Princes, have you read?) This demands much more of the reader, of course: the most difficult action it demands is to be *prepared* to change your context, in almost any direction, and then to find interest in the context as much as in the content of the story.

Another digression here, as to our old science fiction versus fantasy argument[6]: my position in this is (as you might have expected) simple masquerading as subtle. Fantasy stories seem to me to have a 'closed' universe: a simple, describable context in which everything can be answered by the author, in principle. They are like fairy stories, in that only the actions and beings of the participants are germane to the story. It doesn't matter what g is, or how many days there are in a year, or whether ice floats on water, except insofar as it affects decisions by the characters. 'Spells', 'bargains', 'oaths', 'powers', are all interpersonal *constraints* on the systems—characters can do *less* because of them.

Science fiction, in contrast, has 'open' contexts: the real world *plus* a lot more thinking about it—Ahead in time, or Away in space, or If something else had happened. We science fiction addicts, readers and authors, assume the common context (with the scientific knowledge left in, not taken out as it is in mundane stories...) and then *you* are required to take it a fuller, more extended context from your own knowledge of contexts. That's why anyone can read Tolkien without having read *anything* else; but to read and enjoy science fiction, you need to have read a lot... including science fiction, indeed! You need to know a lot more about the universe as it is; it's why you get more out of *Alice Through the Looking-Glass*, or indeed *Gödel, Escher, Bach* (Hofstadter, 1979), each further time you read them. But your closed-minded friends say they don't see what you see in them. And the same goes for Brin, Simak, or Heinlein.[7]

At a recent Con[8] this closed/open difference was made absolutely explicit for me when a group exercise in First Contact, a kind of role-playing game, failed. It failed because we asked 'real' questions: 'What was the actual message?' (so we could do our own cryptography on it—it doesn't have 'only one interpretation', as we were told); 'Was the carrier polarized?'; 'What was the frequency it was sent at?'; 'FM or AM or what?', and so on.[9] The organisers couldn't answer these, *as the real universe would have*, because they hadn't thought of them, or their corollaries. In role-playing games, which are the purest of fantasy in my sense, you need *only* what you're told. In contrast, in the real world—and when you read science fiction—you need to know everything; in fact, in science fiction stories, as in life itself, you *don't* know nearly all of it. What we enjoy is that the real world always answers when you ask the questions…

So a science fiction story demands a lot. It demands a disciplined exercise of an expanded imagination—just as a good teacher does. (Fantasy, perhaps, exercises a contracted, a restrained imagination.) It fails if it just sets an undemanding context ('Space' has now become common, no-effort intellectual property) and sets undemanding stories in that context. The 'space soap' *Jupiter Moon*[10] is mostly that, with a couple of touches that could not have happened in a small-time college in Oklahoma[11]; so are most of the *Star Trek* episodes of the first series (but a few are truly context-expanding, and perhaps needed the others to set the stage). Annie McCaffrey's Pern stories[12] *read* like fantasy, but are actually mind-expanding and can be set in a real, puzzling world, as we tried to do with *Dragonsdawn* (McCaffrey, 1988) and its sequels.[13] Harrison's *West of Eden* series[14], or Niven's *Heorot*[15] (to take examples at random!)[16], demand much more of the reader than to follow the story—the reader must invent, organise, criticise the contexts. That is why SF[17] will always be a minority pursuit: only a minority can stretch that far out of the ordinary. Most scientists don't have this ability, of course, just as most playwrights don't, or most people, for that matter. Those few scientists who read SF are notable for their ideas, their interesting lectures, their success at changing people's minds. The others *do* science, like people *do* knitting, in the mundane context that isn't enough for you lot.

Here's a very fannish example deriving from a conversation with Al Johnston and Bernie.[18] The *Rocky Horror* film[19] is, to us, much more than a simple cinematic offering. Any of you who've been along to a showing at a Con know that there is a complex ritual involving water-pistols, rice, and a variety of communal and individual fannish acts, guaranteed to bring despair to the heart and pocket-book of the most easygoing cinema proprietor. But to the mundane public, it's a piece of art to *receive*, to enjoy, to wonder at, to understand, to recoil from, to hate, to ban.[20] We use it differently: we create a ritual context for it, in which its oddity is contained within our own even odder rituals. This, as Al saw straight away (straight away when I had explained my context/content hang-ups) is our SF determination to give our expanded context to *anything* the Universe throws at us.[21].

Our authors challenge us with changes of context all the time. Perhaps, they suggest, the Wandering Jew[22] idea is based in reality, so we get Lazarus Long[23] or *Boat of a Million Years.*[24] Sometimes the context-addition is just a grace-note: an apprentice torturer,[25] or a dragon more ethical than his George.[26] As soon as I saw the context-change idea, I *understood* why I like SF and opera; I hope the idea gives you pleasure, too. So what is my message to you? *You* are context people, not message people; see yourselves as revolutionaries, not make-do-and-mend-aries. If you are secret-aries, show how you can do it *differently*, openly; if you are scientists, think theory—context, paradigm—as well as practice; if you are engineers, engineer the environment as well as the artefact. Most importantly, if you're a teacher and/or parent, a wholesale or retail purveyor of world-views and attitudes, make them other-world-views and beatitudes.

1 Editorial Notes

Jack Cohen, who died in 2018, originally wrote this essay as a Novacon Special, published in a limited edition of 450 by The Birmingham Science Fiction Group in 1990. In that year, Jack was Guest of Honour at the Group's annual Convention, Novacon 20. All GoH's are invited to contribute a special story or essay, printed as a chapbook (small booklet). Although aimed at science fiction fans, with their own jargon and in-jokes, the essay contains some important thoughts about content and context; in particular as regards the distinction between information and meaning. We have therefore reproduced the essay, with the permission of his daughter Rebecca Cohen and the Birmingham Science Fiction Group, to make it accessible to a wider audience. We have retained the original style and period flavour, but the text has been edited slightly to remove material not pertinent to the present volume. For the benefit of mundanes (see Note 4 below) and those who are too young to know what a Walkman is, we have added the following explanatory notes.

Notes

1. Thomas A. Easton wrote a series of magazine stories that led to the novels *Sparrowhawk* (1990), *Greenhouse* (1991), *Woodsman* (1992), *Tower of the Gods* (1993) and *Seeds of Destiny* (1994). They centre on a biotech revolution in which genetically modified animals, 'genimals', displace machines.
2. The Sony Walkman, first produced in 1979, was a personal music player using cassette tapes: an early forerunner of the iPod.
3. A more extensive discussion of these ideas can be found in Stewart and Cohen (Stewart & Cohen, 1994, 1997).
4. 'Mundane' is the term used in science fiction fandom for anything that is not science fiction, or not clued up about it.
5. *Moonlighting* is a quirky American TV series of the late 1980s, about a private detective agency, starring Sybill Shepherd, Bruce Willis, and Allyce Beasley.

6. Initially the main modern novels in the fantasy genre were J.R.R. Tolkien's *The Hobbit* and *Lord of the Rings*. After a lengthy gap, the genre began to take off, starting with the (almost unreadable) books of Stephen Donaldson in the series *The Chronicles of Thomas Covenant*, namely *Lord Foul's Bane, The Illearth War*, and *The Power that Preserves*. As fantasy blockbusters began to displace SF books in bookshops, SF fans initially went to great lengths to distinguish the two genres and proclaim the supremacy of SF over fantasy. In 1990, the controversy was very much alive. Today there has been so much genre-bending that the distinction has become extremely fuzzy, both genres occupy the same shelves in shops, and many authors have worked in both—notably George R.R. Martin with his *Game of Thrones* book and TV series.
7. David Brin, Clifford Simak, and Robert A. Heinlein: prominent science fiction authors.
8. 'Con' is fannish for 'convention'.
9. The technical terms are references to radio transmission.
10. *Jupiter Moon* is an SF 'soap opera' TV series broadcast in the 1990s, taking place in the spaceship *Ilea* orbiting the Jovian satellite Callisto.
11. Much of the action in *Jupiter Moon* takes place in a university on the *Ilea*, and centres on the daily lives of its students.
12. 'Annie' McCaffrey was Jack's name for SF author Anne McCaffrey, a friend of his. Her spectacularly successful series of books, set on the planet Pern, featured a society whose telepathic dragons fought the menace of Thread, a dangerous plant that fell from the sky when the Red Star made its regular appearances.
13. Among many other activities, Jack was a consultant to SF authors, in particular to Anne McCaffrey, who asked him to come up with scientifically plausible explanations of the main features of the Pern setting, starting with *Dragonsdawn*. This book, ninth in the series in terms of publication date, is a prequel that describes the initial colonisation of the planet Pern. The task was tricky because the earlier books had been based on McCaffrey's sense of narrative, rather than any consistent exercise in world-building.
14. *West of Eden* (1984) is the first of a trilogy written by Harry Harrison, set in a parallel universe in which the dinosaurs were not wiped out in the Cretaceous-Palaeogene extinction event, and evolved into the intelligent reptilian Yilané. The sequels are *Winter in Eden* (1986) and *Return to Eden* (1988).
15. The *Heorot* series, written by Larry Niven, Jerry Pournelle, and Steven Barnes, is notable for its focus on ecological issues. *The Legacy of Heorot* (1987) begins with the colonisation of Avalon, an island off the coast of a continent. The colonists discover the existence of grendels, fierce predators able to move with astonishing speed. A sequel *Beowulf's Children* (1995) takes the characters to the mainland. Shortly before his death in 2018, Jack acted as a consultant for the third volume *Starborn and Godsons* (2020).
16. By a strange coincidence, these 'random examples' are also by SF authors who built on Jack's biological expertise when plotting them.

17. SF, S-F, and s-f are abbreviations of 'science fiction' acceptable to fans.[27] In the past, the now common 'sci-fi' was definitely not: fans pronounced it 'skiffy' and considered its use to be evidence of ignorance of SF, even though it was first proposed by the SF superfan Forrest J. Ackerman in 1954. However, 'sci-fi' has now become more acceptable, even to fans. Its use is standard with reference to the movie industry, and universal in the mundane media.
18. Al Johnston was a regular attendee at Novacon, the annual convention of the Birmingham Science Fiction Group. Bernie Evans was a prominent member of the Group, and a member of the organising committee for Novacons 18, 19, 21, 22, 23 (respectively 1988, 1989, 1991, 1992, 1993). Jack was Guest of Honour at Novacon 20 (1990).
19. *The Rocky Horror Picture Show* is a comedy musical parody of bad horror movies, released in 1975. The main character, Dr. Frank N. Furter, is an alien tranvestite.
20. *The Rocky Horror Picture Show* was dismissed by critics, but became iconic when audiences started to take part in 1976—talking back at the screen, dressing as the characters, even miming the action as the film was being shown. Today its transvestite theme raises fewer eyebrows than it did at the time.
21. Some typical examples: the biannual Discworld Conventions, based on the late Sir Terry Pratchett's bestselling humorous fantasy books, have featured a specially adapted performance of *The Rocky Horror Picture Show*, and a parody of Queen's *Bohemian Rhapsody*.
22. The Wandering Jew is a mythical immortal who taunted Jesus at the crucifixion, and was punished by being made to wander through the world until the Second Coming. The myth became widespread in 13th Century Europe.
23. Several books by Robert A. Heinlein feature the character Lazarus Long, created by a selective breeding programme aimed at longevity. He eventually reached the ripe old age of 2000 years, helped by a few rejuvenation treatments. The books are *Methuselah's Children*, *Time Enough for Love*, *The Number of the Beast*, *The Cat Who Walks Through Walls*, and *To Sail Beyond the Sunset*.
24. *The Boat of a Million Years* is a 1989 novel by Poul Anderson, featuring eleven immortals living at various times in history.
25. In Gene Wolfe's series *The Book of the New Sun* and *The Urth of the New Sun*, the character Severian, a journeyman in a Torturers' Guild, has been exiled for showing mercy to a 'client'.
26. *The Dragon and the George* (1976) is a humorous novel by Gordon R. Dickson, the first of a series of nine. It features a knight and his sidekick, a Minnesota history professor's assistant whose mind has been transferred into a dragon's body.
27. The canonical fannish plural of 'fan' is 'fen', by analogy with man/men.

References

Cohen, J. (1990). *Are you content in your context?*, Novacon Special, The Birmingham Science Fiction Group.
Hofstadter, D. (1979). *Gödel, Escher, bach—an eternal golden braid*. Basic Books.
McCaffrey, A. (1988). *Dragonsdawn*. Ballantine Books.
Stewart, I., & Cohen, J. (1994). *The collapse of chaos*. Viking.
Stewart, I., & Cohen, J. (1997). *Figments of reality*. Cambridge University Press.

The Incremental Chain of Being

John Heil

1 How Came We to This?

Is reality stratified? Many serious people, including scientists, philosophers, and ordinary citizens, think so. Talk of higher- and lower-level phenomena is ubiquitous, so much so that it has taken on the character of an ideology. The cosmos is regarded, not as a uniform patchwork, but as organised into a hierarchy of autonomous or simi-autonomous domains, each with its distinctive objects, properties, laws. Explanatory practices in the various sciences are tailored to these. The levels picture supplanted conceptions of 'flat' universe according to which, when all is said and done, everything is reducible to physics. A multi-tiered universe is an accepted consequence of arguments against reductionism.

I believe that it is worth asking, especially now, what brought us to this juncture. The question is of more than historical interest. Revisiting considerations that led us to take up a particular ideology, can yield a fresh appreciation of the nature and credibility of that ideology. Ideologies can owe their influence to their being deployed unreflectively. On reflection, something everyone knows can cease to seem inevitable.

In what follows, I look back on some of the ideas and arguments that led to the widespread acceptance of the idea that we inhabit a hierarchical cosmos. The discussion is not meant to be exhaustive, only suggestive. I myself am convinced that the original arguments for levels were defective, and that they have not improved with age. Succinctly put: we were duped into drawing robust metaphysical conclusions from patently linguistic premises. My hope is that I can say enough to persuade you that the hierarchical, levels picture is, if not flatly wrong, at the very least optional.

J. Heil (✉)
Washington University, CB 1073, One Brookings Drive, St. Louis 63130-4899, US
e-mail: jh@wustl.edu

S. Wuppuluri and I. Stewart (eds.), *From Electrons to Elephants and Elections*,
The Frontiers Collection, https://doi.org/10.1007/978-3-030-92192-7_2

2 Broad Brush History

Impressed by advances in physics, many philosophers in the first half of the twentieth century embraced some form of reductionism. The sciences were regarded as unified by virtue of being reducible to physics: truths belonging to a higher-level science, such as biology, could in principle, if not in practice, be extracted from truths of physics. The truths in question included laws and explanations couched in terms of these laws. If there are biological laws these are, again, in principle, derivable from laws governing the fundamental particles. Owing to the complexity of organisms, the derivations required might be practically unattainable, but scientific practice is replete with compromises.

The reductionists' point was not to encourage scientists to undertake the reductions. All that was required was their in-principle attainability. As a practical matter, we treat the various sciences as autonomous, all the while recognising that, in God's mind, it all boils down to physics.

The de facto autonomy of the several sciences was bolstered by their being successful in their own terms, but the autonomy was taken to be de facto only. Complex things are made up of simpler things that obey their own laws. Laws applicable to complex things must in some way issue from the fundamental laws.

Reductionist programs in the sciences and in philosophy came under fire from many quarters. Behaviourists, for instance, had long dreamed of analysing talk of states of mind into to talk of behaviour. Talk of dodgy unobservable mental phenomena could be replaced with more respectable talk of behaviour and dispositions to behave (Wittgenstein 1953; Ryle, 1949).

It became clear, however that barriers to such analyses were not simply matters of detail or practicality. Proposed analyses had a disquietingly open-ended character. This suggested that the envisaged reductions were, not simply hard to come by, but wrong-headed.

Not all reductionist programmes appealed to analytical manoeuvres, however. Philosophers defending materialism, for instance, did not argue that mental terms could be analysed in a physical vocabulary, but that mental states are as a matter of fact *nothing but* brain states. Mind–brain identity theorists, including U.T. Place and J. J. C. Smart, regarded the identification of mental states with physical states as an empirical hypothesis, one subject to norms by which any other empirical hypothesis is evaluated (Place, 1956; Smart, 1959). Start with the idea that we have empirical evidence of correlations among states of mind and physical, brain states. How might these be explained?

Dualism affords one kind of answer: mental states, while not themselves identifiable with brain states, nevertheless enjoy an especially intimate relation to brain states. Traditional dualism came in a various flavours. Interactionists held that minds and brains causally interacted, epiphenomenalist regarded states of mind as by-products of complex physical mechanisms that, like the shadow cast by a speeding locomotive, played no part in their operation. Neuroscientists spoke breezily of neural mechanisms as the 'substrate' of consciousness.

Smart and Place argued that a much simpler explanation of the empirical data was available: purported correlations between states of mind and brain states were artefacts of the way minds and brains are studied. Imagine that a neuroscientist observes goings-on in your brain while you report your states of mind. What is correlated are not mental states and brain states, but *reports* of mental states and observations of brain states. Nothing stands in the way of identifying what is reported with what is observed. Under the circumstances, the hypothesis that mental states are (identical with) states of the brain should be accepted on the grounds that it is more parsimonious than dualism.

Proponents of various forms of dualism responded to the mind–brain identity theory in various ways, but, at the time, the most serious challenge came from functionalism. Functionalists argued that the idea that mental states are brain states involves a kind of category mistake. The identification fails, not because mental states are nonphysical, as proponents of dualism would have it, but because they are *functional* states (Fodor, 1968, 1981; Putnam, 1967).

Functionalism was a response not only to the kind of analytical reduction associated with behaviourism, but also to the nonanalytical species of reduction offered by the mind–brain identity theory. States of mind, although likely dependent on brain states, could not be identified with brain states, not even manageable disjunctions of such states. This is where it all becomes murky.

A functional state is a state, the identity of which, is bound up with its *causal role.* Your being in pain, for instance, is for you to be in a state with the right input–output profile. A pain state, for instance, might be a state brought about by tissue damage, excessive heat or pressure and subsequently produces aversive behaviour. The state you are in when you are in pain need not be physically anything like the state some other creature is in when that creature is in pain. A silicon-based creature might undergo pains provided only that it could be in a state that played the pain role.

Types of mental state, then, like functional states generally, are not identifiable with types of physical state. Many different kinds of physical state could satisfy the job description of any given mental state. Indeed, if there are nonphysical beings they could undergo pains provided they were organised in the right way. Functionalism is compatible with, but does not entail, materialism.

The problem with the identity theory is not simply that there is no one–one, or even one–many relation between states of mind and brain states, but rather the kinds of state capable of playing the role definitive of any mental state are in principle open-ended. Reduction is not on the cards.

3 Antireductionism and Levels

These kinds of antireductionist argument were expanded to include, not simply psychology, but to include, as well, the special sciences, the social sciences, along with many everyday human institutions (Fodor, 1997). Higher-level sciences could not be reduced to sciences at lower levels. Talk of higher- and lower-levels here is vaguely mereological. Objects at home in a higher-level science are by and large made up of parts that make up the subject matter of lower-level sciences, and the higher-level objects are themselves parts of objects making up the subject matter of still higher-level sciences.

Proponents of levels would agree that this falls well short of establishing a metaphysics of levels. At best it portends a division of labour in which distinct sciences are occupied with larger or smaller entities. More is required to move from this relatively uncontroversial position to a hierarchy of levels of being.

The impetus for the hierarchical scheme came, in large measure, from early attempts to address the metaphysics of functionalism. Recall that a functional state is a state that plays a particular causal role. Because many different kinds of physical state could play the same role, there was no prospect of identifying types of functional state with types of physical state, hence the slogan, 'no type–type identity'.

Philosophers advancing these arguments used 'type' and 'property' interchangeably, concluding that, if mental properties, or biological properties, or any other properties are functional properties, they are not candidates for identification with physical properties.

For many readers this is familiar territory. Familiarity has a philosophical downside, however. Too often, we feel comfortable with a familiar doctrine, not because we understand it and find it compelling, but because we can recite it by heart. When I reflect on my own philosophical development, I realise that, when it came to arguments for hierarchies and levels of entity, I could talk the talk, but I really did not understand what I was talking about. I chalked my lack of understanding up to a failure on my part to see what everyone else saw. This, together with the fact that no one was likely to call my bluff enabled me to repress a latent uneasiness and move on. Eventually I came to appreciate the Socratic point that, without uneasiness, philosophy dies.

Apologies for waxing autobiographical, but I am leading up to the question, what *is* a functional state? Being in a state is a matter of having a particular property, so to be in a functional state—to be in pain, for instance—is to have (or 'instantiate' or 'exemplify') a functional property. When you are in pain, when the pain property is on the scene, you are in a physical state that plays the pain role. You are in pain, you have the property of being in pain, by virtue of being in that state. Another creature might be in pain by virtue of being in a very different kind of physical state.

Here, distinct physical states are associated with a single property: the pain property. You and that other creature are each in pain. Are your respective pain states identifiable with (reducible to) your respective physical states? How could they be? You share the property of being in pain, but the physical states responsible for your both being in pain are different. Thus, the property of being in pain cannot be identified with the property of being in a particular kind of physical state.

This line of reasoning spawned the idea that physical states, while not identifiable with functional states, nevertheless 'implement' or 'realise' functional states. You and the other creature's respective physical states realise your respective pain states: one mental state, many physical realisers. Psychological states are *multiply realisable* higher-level states.

You can see how this works in the case of computing machines and their programs. Machines with very different kinds of physical architecture can perform the same computation: solving a particular differential equation, for instance. Their performing the computation requires transitions among the machines' hardware states. The machine's performing the computation is not identifiable with those hardware transitions, however. The very same computation could be implemented in different ways on machines with different physical architectures. Computational processes, then, must be higher-level occurrences implemented or realised by assorted lower-level processes.

Once you accept this point, a whole new way of seeing things opens before you. You can see that what goes for computational states and processes, goes for psychological states and processes generally. But why stop there? The same move extends smoothly to chemical, biological, and endless other higher-level states and processes. By virtue of being multiply realisable, higher-level items across the board cannot be reduced to—identified with—those at lower levels.

This moves us a step closer to the levels picture, but getting all the way there requires understanding how philosophers came to understand the relation higher-level items bear to their lower-level realisers. This brings us back to functionalism.

4 The Metaphysics of Functionalism

Early on, Ned Block distinguished two distinct species of functionalism (Block, 1980; see also Shoemaker, 1981). One of these, 'functional state specifier' functionalism, was associated with the work of David Armstrong and David Lewis (Armstrong, 1968; Lewis, 1966). For Armstrong and Lewis, functionalism includes both an analytical and an empirical component. Mental predicates, they thought, could be given functional analyses, but discovering what answers to these predicates is an empirical matter.

So, for Armstrong and Lewis, when you say that Gus is in pain, you are saying that Gus is in some state with the right input–output profile. To a first approximation, a state counts as a pain state when it is brought about by tissue damage, excessive heat or pressure and leads to aversive behaviour. So far so good. Armstrong and Lewis, unlike most other functionalists, held that functional states in general, and mental states in particular, were to be *identified with* whatever physical state played the right functional role: a mental state *is* its realiser.

Because states capable of playing a given role could differ across species, and even across individuals, the upshot is a kind of 'token identity' theory: every particular mental state is identified with some—presumably physical—realising state or other. If a state is an object's having a property (at a time), then this is not property or type identity. Mental types do not align with physical types, however mental terms—mental predicates—hold true of individuals by virtue of those individuals' being in particular physical states.

This is a natural enough way of understanding functionalism, so you might be surprised to learn that, when it came to functionalists' metaphysics, Armstrong and Lewis were a distinct minority. Mainstream functionalists explicitly rejected token identity, in part because it smacked of reductionism or 'eliminativism'. If *all there is* to being in a mental state is to be in some physical state, then the mental is nothing but the physical. But, they argued, reductionism is a nonstarter: there are plenty of reasons to doubt that the mental is reducible to the physical.

What is the alternative? Mainstream functionalists contended that Armstrong–Lewis style functionalism stemmed from a confusion. A functional state or property is not identifiable with its physical realiser. A functional property is a 'second order' property: the property of having a property that plays the right functional role.

A second-order property is not, as the label suggests, a property of a property, but the property of *having* a property. (Because referring to mental properties as second-order properties is potentially confusing, I prefer to speak of higher-*level* properties.) Block dubbed this brand of functionalism 'functional state identity functionalism', a somewhat confusing characterisation in the context.

If you found the foregoing hard to follow, that is unsurprising. Suffice it to say that the resulting picture issued in property levels. To have a mental property, the property of feeling frightened, for instance, would be to have the property of having some physical property—to be in a physical state—with the functional profile characteristic of fear. The state might be one caused by occurrences deemed frightening, for instance, that disposes its possessor to flee the occurrence. The realising state could vary across individuals and across species.

The question remains, given that mental properties are not identifiable with their realisers, how precisely are the two related? Many philosophers accepted Donald Davidson's (1970) contention that the mental 'supervened' on the physical and supposed supervenience to be a relation among property 'families' (Kim, 1978, 1979). Thus, if the family of mental properties supervenes on the family of physical properties, mental properties are distinct from, but dependent on, physical properties. This is what it is for a physical property to realise a mental property.

Supervenience thus construed was said to be *nonreflexive*—nothing supervenes on itself –*asymmetrical*—if the *A*s supervene on the *B*s, the *B*s do not, indeed could not, supervene on the *A*s—and transitive—if the *A*s supervene on the *B*s, and the *B*s supervene on the *C*s, the *A*s supervene on the *C*s.

This purely formal characterisation of supervenience encompasses many different kinds of relation (see Kim, 1990). These included causal relations (as when the *B*s causally suffice for the *A*s), constitution relations (when the *A*s are made up of the *B*s), and cases in which two events are caused by a third event (when the *A*s and *B*s are both effects of *C*s).

None of these familiar relations was what proponents of multiple realisability had in mind, however. The realising relation was something like the causal relation but synchronic: once a realiser is on the scene, so is whatever it realises. The relation is evidently *sui generus*. It is what you have when a higher-level property is synchronically dependent on, but distinct from some lower-level realising property.

Now, however, difficulties arise. Suppose your feeling of fear is dependent on, but distinct from, a physical state with the right causal profile. In that case it would be the realiser, the physical state, that is brought about and in turn brings about the relevant physical occurrences. When your fear leads you to flee, it is not your *fear* that is responsible for your fleeing, but its physical realiser. And so it is for functional states generally. Functional states are epiphenomenal, and, given that mental states are functional states, mental states are epiphenomenal (Jackson, 1996; Kim, 1993; McLaughlin, 1989).

You might regard epiphenomenalism as unobjectionable when it comes to conscious states of mind, but the problem generalises in ways most philosophers and nonphilosophers would find unacceptable. Higher-level sciences are concerned with higher-level states and properties. These figure centrally in laws and causal explanations offered by the sciences. Are all these epiphenomenal?

Physics aside, most of what is of interest in the various sciences resides at higher levels. Trees cast shadows, shed leaves, provide refuge for organisms of all sorts. The real causal work is in the hands of items at lower levels, however, ultimately in the hands of the quarks and leptons. Antireductionist arguments aimed at preserving the autonomy of the several sciences and gave birth to a hierarchy of levels culminated, ironically, in physics' being put back in the driver's seat.

Much has been written about the problem of causal efficacy of higher-level phenomena, but I will spare you the details (see, for instance, Jackson & Pettit, 1990). Suffice it to say that none of the many attempts to provide for the causal relevance of items at higher-levels has achieved anything close to universal acceptance. You might have thought that this would send the architects of the hierarchical picture back to the drawing board. As is so often the case in philosophy, however, rather than re-examining the foundations of the edifice, philosophers relegated the problems to the background.

The sense was that this is just a problem we must all learn to live with: what is a problem for everyone is a problem for no one. After all, the hierarchical, levels model is mandated by the success of the sciences in getting at the causal structure of the universe. Philosophers are in no position to challenge the standing of higher-level sciences. The sciences have paid their dues. If philosophers are unable to account for the causal efficacy of higher-level items, that is their problem, not one that should trouble the sciences.

5 Davidsonian Supervenience

This is yet another case in which apparently innocent philosophical presuppositions turn out to have unwelcome downstream consequences. In this instance, the assumption is that there are higher-level properties required by the sciences that depend on distinct lower-level properties. The hierarchical picture was largely the product of philosophers' tendency to move directly from claims about the non-reducibility of predicates and explanations, to claims about states and properties. If you look at influential arguments against reduction, you will find that they concern taxonomies. Biological or psychological taxonomies, for instance, cannot be derived from or recreated in a taxonomy belonging to physics. That seems right. But what follows? If talk of *B*s cannot be substituted for talk of *A*s, does it follow that the *A*'s and *B*s belong to distinct categories of entity? Do irreconcilable taxonomic differences call for distinct levels of being?

Why not suppose, instead, that there is just the one level describable in many different non-equivalent ways? If you distinguish predicates from properties, what antireductionist arguments show is that predicates belonging to taxonomies associated with the various sciences, cannot be replaced by predicates belonging to lower-level taxonomies without a loss of information. The decisive move is to assume that distinct families of predicate correspond distinct families of property, an especially unfortunate consequence of linguisticised metaphysics. 'The decisive movement in the conjuring trick has been made, and it was the very one that we thought quite innocent' (Wittgenstein 1953/1967: §3.08).

I can illustrate what I have in mind by returning to Davidson and the Davidsonian precept that the mental supervenes on the physical. Philosophers delving into the metaphysics of mental properties assumed that, in speaking of the mental supervening on the physical, Davidson must be referring to families of property (or, if he was not, this was due solely to an unbecoming ambivalence about properties). Were that so, then, given the irreducibility of the mental to the physical, mental properties must be distinct from physical properties.

Philosophers occasionally speak of the reducibility of properties as in 'mental properties are (or are not) reducible to physical properties'. Reduction is a relation holding among predicates (or laws, or explanations), however, not a relation among properties. (What would it be to reduce one property to *another* property?).

In arguing that the mental supervened on the physical, Davidson was contending that you could not have a mental difference without having a physical difference. This is a defeasible substantive thesis, not a product of conceptual analysis. If it is true that I am in pain and you are not, we must differ in some physical way.

Many of Davidson's readers followed Kim in interpreting this as a claim about a dependence relation among properties: mental properties, while distinct from physical properties, nevertheless depended on physical properties. If I have a mental property that you lack, we must differ in some nonmental way.

Numberless philosophers, this author included, devoted inordinate amounts of time and effort to the task of working out the metaphysics of supervenience. The upshot was a growing sentiment that supervenience was a *sui generis* relation, not further explicable. There had to be *some* kind of dependence relation between mental properties and physical properties, and supervenience was as good a name for that relation as any.

As a matter of fact, in invoking supervenience, Davidson was not advancing a metaphysical doctrine about families of property. Davidson rarely spoke of properties but, as a student of Quine's, when he did, he was not using 'property' in an ontologically serous sense. For Davidson, to say that a ball has the property of sphericality, is to say no more than 'is spherical' applies truly to the ball. A view of this kind is silent on the implications of the supervenience thesis, if any, for the metaphysics of properties.

Davidson's inspiration was R. M. Hare, who contended that there could be no moral difference without a nonmoral, natural difference (see Hare, 1952). Hare's focus was on moral judgements, not properties. Indeed, Hare was an anti-realist about moral properties. His idea was that, if you judge that one person is good and another not, there must be some nonmoral difference between the two. This is a defeasible substantive thesis about constraints on moral judgements, not a commitment to distinct families of property.

In similar fashion, Davidson took mental–physical supervenience to express a constraint on mental and physical judgements or descriptions. When you truly ascribe a mental state to me, that very state could be picked out using a non-mental, physical vocabulary. There are not two states, one lower- and the other higher-level. There is one state describable in conceptually orthogonal ways.

Supervenience amounts to the precept that whatever makes true the ascription of a state of mind, answers as well to a nonmental, physical description. This not the thesis that when you are in a position to ascribe a particular mental state to me, you are also in a position to offer a physical description of that state. All that follows is that there is some way of picking out the state in a nonmental vocabulary.

This is one aspect of Davidson's contention that something is mental or physical 'only as described' (1970: 215). The mental-physical distinction is not a metaphysical distinction between incommensurable families of property, it is a distinction between two incommensurable ways of describing the cosmos. For Davidson, the mental–physical distinction is not what the scholastics called a real distinction, but a distinction of reasoned reason, what today would be called a conceptual distinction.

Davidson's motive for embracing supervenience was straightforward. States of mind—beliefs, desires, intentions—are caused by physical occurrences and figure causally in the production of actions. This is a conceptual point. Ascriptions of states of mind are ascriptions of states that are caused by and cause physical events. It is true that you perceive a tree only if a tree figures causally in the production of your perceptual state. You act on reasons only if those reasons—in the form of beliefs, desires, and intentions—are causally responsible for your body's moving as it does when you act. The identity of a mental state is bound up with its causes and effects. Mental concepts are causally loaded. This might remind you of functionalism, but, notwithstanding important similarities, Davidson was not committed to the functionalist tenet that mental concepts could be given exhaustive functional analyses.

The causally loaded character of ascriptions of states of mind converges with the fact that we have every reason to think there is, whether we could give it or not, a physical explanation for your going into perceptual states, entertaining beliefs about trees, and for your bodily motions. Although there is no prospect of analysing or paraphrasing mental concepts in nonmental terms, this is no barrier to the idea that, whenever mental ascriptions are true, their truthmakers could be given nonmental, physical descriptions.

Not only is Davidson's supervenience not a kind of dependence relation among properties or property families, it is not even asymmetrical. The mental supervenes on the physical, and the physical supervenes on the mental. Anything answering to a mental description answers, as well, to some physical description or other, and anything that could be given a physical description could, with sufficient ingenuity, be described in a mental vocabulary. The symmetrical nature of supervenience reflects the fact that 'mental' and 'physical' designate different ways of characterising one and the same cosmos.

Owing to the incommensurability of the vocabularies to which the mental and physical terms belong, there is no one–one, or even one–many, mapping of mental terms onto physical terms. Davidson calls this *anomalous monism*: monism because there is just the one cosmos describable in different ways; anomalous because the modes of description are incommensurable.

6 Dénoument

Even if Davidson himself did not regard supervenience as a dependence relation among families of property, why think he is right? Why not think that, whatever Davidson himself believed, it is useful—perhaps even inevitable—that we take 'supervenience' as a name for a relation among property families. Davidson might eschew the metaphysics of properties, but this saddles the rest of us with the problem of understanding the relation between mental and physical properties, which are, after all, distinct species of property.

A better question might be, why imagine that the proponents of the hierarchical picture are right? Their case rests on arguments that start with plausible taxonomic premises which are then given a patently metaphysical spin. The conclusions are not supported by the arguments, and they provide no illumination concerning the relation between higher- and lower-level entities. In addition, they incur significant metaphysical costs, one of which, the problem of the causal relevance of the mental, surfaced earlier: how could higher-level properties, mental or otherwise, enter into the causal fray? Such properties would, instead, be undercut or pre-empted by their lower-level realisers.

This problem does not so much as arise for Davidson. If every state answering to a mental description answers to a nonmental, physical description, you would have one and the same state, differently characterised, making it true that you are afraid and bringing about your fleeing. To ask whether a state figured in a causal relation because it answered to a mental description or because it answered to a physical description is to court confusion.

The point extends beyond the psychological domain. Biology is not reducible to physics: biological taxonomies, laws, and explanations are orthogonal to the taxonomies, laws, and explanations deployed by physicists. Still, whatever could be given a biological description, answers as well to a description couched in the language of physics. 'This is an emu' is true of a particular creature because that creature has what it takes to be an emu. But that same creature could be described in a vocabulary mandated by physics. And so it is for other higher-level sciences. This is not reduction, there is no suggestion that you could recast talk of emus in terms of quarks and leptons.

In invoking Davidson, my aim has been to illustrate a general point: nothing is gained by interpreting taxonomic hierarchies as evidence for levels of being. You can honour antireductionist sentiments without embracing a hierarchical ontology. Doing so has the advantage of making sense of scientific practice without incurring the problems and mysteries accompanying a multi-tiered ontology.

Chemistry, biology, psychology, and economics do not concern distinct realities. What they afford are different, largely autonomous ways of addressing a single reality. This takes nothing away from the special sciences, the pronouncements of which are often enough true. To dispense with a tendentious philosophical reconstruction of relations among the sciences, is not to dispense with the sciences.

References

Armstrong, D. M. (1968). *A materialist theory of the mind*. Routledge and Kegan Paul.

Block, N. J. (1980). (1980) 'What is Functionalism.' In N. J. Block (Ed.), *Readings in philosophy of psychology* (Vol. 1, pp. 171–184). Harvard University Press.

Davidson, D. (1970). 'Mental Events'. In L.Foster, & J. Swanson (eds.) *Experience and theory*. University of Massachusetts Press: 79–101. Reprinted in *Essays on Actions and Events*. Clarendon Press (1980), 207–225.

Fodor, J. A. (1968). *Psychological explanation: An introduction to the philosophy of psychology*. Random House.

Fodor, J. A. (1981). The mind-body problem. *Scientific American, 244*, 114–123.

Fodor, J. A. (1997). 'Special sciences: Still autonomous after all these years'. *Philosophical Perspectives* 11: 149–63; reprinted in *In Critical condition: Polemical essays on cognitive science and the philosophy of mind*. MIT Press (1998), 9–24

Hare, R. M. (1952). *The language of morals*. Oxford University Press.

Jackson, F. C. (1996). Mental causation. *Mind, 105*, 377–441.

Jackson, F. C., & Pettit, P. (1990). In defense of folk psychology. *Philosophical Studies, 59*, 31–54.

Kim, J. (1978). Supervenience and Nomological Incommensurables. *American Philosophical Quarterly, 15*, 149–156.

Kim, J. (1979). Causality, identity, and supervenience. *Midwest Studies in Philosophy, 4*, 31–49.

Kim, J. (1990). 'Supervenience as a philosophical concept'. *Metaphilosophy* 12: 1–27. Reprinted in *Supervenience and mind: Selected philosophical essays*. Cambridge University Press (1993), 131–60.

Kim, J. (1993) 'The non-reductivist's troubles with mental causation'. In J. Heil, & A. R. Mele (eds.) *Mental causation*. Clarendon Press: 189–210. Reprinted in *Supervenience and Mind: Selected Philosophical Essays*. Cambridge University Press (1993): 336–57.

Lewis, D. K. (1966) 'An argument for the identity theory'. *Journal of Philosophy 63*, 17–25. Reprinted in *Philosophical Papers*, vol. 1. Oxford University Press (1983), 99–107.

McLaughlin, B. P. (1989). Type epiphenomenalism, type dualism, and the causal priority of the physical. *Philosophical Perspectives, 3*, 109–135.

Place, U. T. (1956). Is consciousness a brain process? *The British Journal of Psychology, 47*, 44–50.

Putnam, H. (1967) 'Psychological predicates'. In W. H. Capitan, & D. D. Merrill (eds.) *Art, mind, and religion*. University of Pittsburgh Press, 37–48. Reprinted as 'The nature of mental states' in

Mind, language, and reality: Philosophical papers, vol. 2. Cambridge University Press (1975), 429–40.
Ryle, G. (1949). *The concept of mind*. Hutchinson.
Shoemaker, S. (1981). 'Some varieties of functionalism'. *Philosophical Topics, 12*, 83–118. Reprinted in *Identity, cause, and mind: Philosophical essays*. Cambridge University Press (1984), 261–86.
Smart, J. J. C. (1959). Sensations and Brain Processes. *Philosophical Review, 68*, 141–156.
Wittgenstein, L. (1953/1968) *Philosophical investigations*. G. E. M. Anscombe, trans. Basil Blackwell.

Does Linguistics Need (Weak) Emergence?

J. T. M. Miller

There are many different sorts of linguistic objects: words, sentences, paragraphs, phonemes, morphemes, and many more. There are also linguistic properties. That is, there are properties that are primarily (perhaps even uniquely) instantiated by linguistic objects. Such properties include spellings, pronunciations, meanings, and various different grammatical properties.

Interestingly, some of these properties seem to only be attributable to 'higher-level' objects. For example, single words cannot have certain complex grammatical properties. I have, elsewhere, explored whether some of these linguistic properties instantiated by higher-level objects are ontologically emergent properties (Miller, 2017). In that paper, I explored whether at certain levels of complexity within linguistic derivations, new properties come into existence which contribute novel causal powers to the object that instantiates the property. If they exist, these ontologically emergent linguistic properties cannot be reduced to patterns of lower-level properties, or to additive properties that arise due to the nature of the lower-level properties.

This chapter extends the discussion begun in that earlier work. Though I explore the same phenomena—that of the truth-evaluability of sentences—I hope that this chapter is readable independently. My aim here is to re-evaluate the failure of reduction that I argued for previously and consider whether the observations might support positing an alternative conceptions of emergence, that of weak or epistemic emergence.

The chapter is structured as follows. In section one I outline the concept of emergence, and in particular the understanding of weak emergence that I will adopt in this chapter. In section two, I defend the idea that there are 'levels' in linguistics such that different linguistic objects (and the properties they instantiate) might be taken to be 'higher' or 'lower' level with respect to other linguistic objects. In section three, I rehearse some of the claims found in Miller (2017), most centrally the seemingly

J. T. M. Miller (✉)
Department of Philosophy, Durham University, 50 Old Elvet, Durham D1 3HN, UK
e-mail: james.miller@durham.ac.uk

S. Wuppuluri and I. Stewart (eds.), *From Electrons to Elephants and Elections*,
The Frontiers Collection, https://doi.org/10.1007/978-3-030-92192-7_3

failure of reductionism to account for the property of truth-evaluability instantiated by some (but not all) sentences. In section four, I consider whether weak emergence might allow us to solve the problems facing reductionism, thereby avoiding a commitment to truth-evaluability as being a strongly emergent property. I close in section five by reflecting briefly on what this means for linguistics more widely.

1 What is Weak Emergence?

Emergence is often understood with reference to a background commitment to reality being structured hierarchically in levels. This commitment to levels means supporters of emergence hold that reality is ordered and structured, with different entities at each of the levels. Different levels of reality then reflect the relative fundamentality of the entities within each level, with those entities at the 'bottom' level being the most fundamental.

Making use of this notion of levels, those that argue that there are emergent entities (typically, but not always properties)[1] hold that emergent entities are higher level entities that are something 'over and above' the lower-level entities that compose them. Emergent entities are not merely additive entities—not merely the sum of their parts. Rather, the idea is that emergent entities are 'both distinct from and novel with respect to the base phenomena from which they emerge, whilst nevertheless being dependent upon the base phenomena' (Carruth, 2019: 87).

Emergent properties are therefore possessed (or instantiated) by 'higher-level' complex systems or objects. That is, they are properties that are possessed by higher-level objects that are themselves composed by lower-level (or more fundamental) objects that possess (or instantiate) other lower-level properties. For example, one of the most discussed potential cases of emergence is that of consciousness. The idea is that consciousness is an emergent property that can be said to 'arise' out of more fundamental properties—such as the neurophysiological properties of the brain. Consciousness, if it is emergent, is therefore a novel property of the complex object that is the human brain (or mind), and is irreducible to lower-level properties of the brain (see O'Connor 2020).

Beyond this very broad characterisation, the precise details of putative cases of emergence have been understood in multiple different ways.[2] One major distinction relevant to this paper is between *ontological* (or strong) and *epistemic* (or weak) emergence. Again very broadly characterised, and not uncontroversially so, the difference between ontological and epistemic emergence lies in whether the emergent entities are (merely) part of our explanatory practices, or are a novel part

[1] For some exceptions to this, and further discussion of the plausibility of those arguments, see Hasker 2017, Nida-Rümelin 2007, O'Connor and Churchill 2010, and Rickabaugh 2018, and Zimmerman 2010.

[2] See Van Gulick 2001, and Gibb, Hendry, & Lancaster (2019) for overviews of the wide range of topics within the field.

of the ontology of world (in addition to the lower-level entities from which they emerge), though the precise details have been cashed out in many different ways. For example, Chalmers understands the distinction through the epistemic notion of deducibility such that:

> a high-level phenomenon is strongly emergent with respect to a low-level domain when the high-level phenomenon arises from the low-level domain, but truths concerning that phenomenon are not deducible even in principle from truths in the low-level domain. (Chalmers, 2006: 244)

And,

> 'a high-level phenomenon is weakly emergent with respect to a low-level domain when the high-level phenomenon arises from the low-level domain, but truths concerning that phenomenon are unexpected given the principles governing the low-level domain.' (Chalmers, 2006: 244)

Or, in the words of Bishop and Silberstein:

> ontologically emergent entities or properties are thought to be the result of irreducible bridge laws or causal powers to produce qualitatively new phenomena. Such strongly emergent phenomena are also said by some to possess novel "downward" causal powers that constrain the behavior of other phenomena at smaller spatial and temporal scales. (Bishop & Silberstein, 2019: 154).

And weakly emergent entities:

> fail to be predictable, derivable, explainable or characterizable in terms of the "more basic" entities and properties out of which the emergents arise. (Bishop & Silberstein, 2019: 154)[3]

What is clear about weak emergence understood in these ways is precisely how weak it is. Weak emergence is *epistemic* in nature, relating to our ability to predict, derive, or explain the phenomena we observe. It is about our epistemic capabilities, and not (necessarily) about the nature of the entities themselves. Indeed, this conception of emergence is so weak that failure of predictability is often (though not always) taken to be consistent with some form of ontological reductionism. As Wong puts it, 'unpredictability, however, is not supposed to introduce any new ontology into the world; it is merely epistemic' (2019: 180).

Another example of the consistency of weak emergence and reductionism comes from Bedau (1997) who argues that an epistemic (or weakly) emergent property (or state) is a macroscopic or higher-level property that could only be understood when we understand the nature or behaviours of the lower-level property if we have already modelled the higher-level property. Under this conception, there is a failure of *prediction* in epistemic emergence cases as knowledge of the lower-level states would not allow us to predict the higher-level (or macro) behaviour that we observe. However,

[3] Bishop and Silberstein do not accept these characterisations of emergence, favouring instead an alternative account of what it is for some entity to be emergent that they call 'ontological contextual emergence'. See Bishop 2005, 2010; Silberstein 2002, 2012, 2017. It would be an interesting task to consider that form of emergence to see what it would say about the sorts of cases that arise from linguistics, but I leave that to future work.

there is no irreducibility requirement. Higher-level properties *may* be reducible to lower-level ones, but to do this we must first study the higher-level properties. A study of the lower-level alone cannot lead us to understand fully the observed higher-level.

And, again, in the work of Butterfield who understands emergence in the following way:

> properties or behaviour of a system which are novel and robust relative to some appropriate comparison class. Here 'novel' means something like: 'not definable from the comparison class', and maybe 'showing features (maybe striking ones) absent from the comparison class'. And 'robust' means something like: 'the same for various choices of, or assumptions about, the comparison class'. Often these words are made more precise by the fact that the system is a composite. So the idea is that its properties and behaviour are novel and robust compared to those of its component systems (2011: 3)

Butterfield goes on to argue that this notion of emergence is consistent with an account of reduction as the "deduction of one theory from another", nor makes emergence a case of "mere supervenience" (2011: 3–4). The consistency of reduction and emergence comes from the higher-level showing properties that are novel with respect to the properties of the component systems, yet following our knowledge of the higher-level properties we can then learn to deduce the presence of the higher-level properties given certain arrangements or behaviours of the lower-level components system. Put an alternative way, the requirement is that we can only know about the higher-level properties through first recognising their existence at the higher level. Once we have recognised the property at the higher-level, we might be able to subsequently 'backwards engineer' the property and thus deduce the property from the lower level, thus satisfying Butterfield's conception of reduction. So understood, emergence asks very little of the world.

It is this weak conception of emergence that I have in mind what I ask my main question in this chapter: does linguistics need (weak) emergence? My aim is to argue that the answer to this question is yes. Thus, in this chapter, I will take a phenomenon to be weakly emergent, when such entities 'fail to be predictable, derivable, explainable or characterizable in terms of the "more basic" entities and properties out of which the emergents arise' (Bishop & Silberstein, 2019: 154), and will argue in the rest of this chapter that there is at least one case where linguistics should embrace this sort of emergence, before reflecting on what this might mean for linguistics more widely.

2 Levels in Linguistics

Amongst the various sorts of linguistic entities, there are some that are seemingly indisputably 'higher-level' entities. Sentences, for example, are complex entities, typically taken to be composed of words. Words are, in this context, 'lower-level' entities. They are simpler than sentences, and when put together following various rules we are able to bring into existence sentences as well as various other even more complex higher-level linguistic entities such as paragraphs, chapters, and, possibly, books.[4]

This is not to say that words are absolutely or universally to be classified as being lower-level. Pre-theoretically, words are the fundamental or atomic elements of our language system. That is, our pre-theoretic view is normally that the mental lexicon contains atomic units which are words. What this would mean is that our minds contain a store of words, available for us to draw down and place into syntactic arrangements to create sentences and other more complex linguistic entities.

This atomic conception of words, though, is widely rejected in the linguistics literature. Rather, relative to other entities, words are themselves taken to be higher-level, complex objects, with the composition of words specifically studied in the field of morphology. Indeed, it is not unheard in the literature to defend an even stronger view which denies the existence of words all together. Boeckx, for example, has called words the 'phlogiston of linguistics' (2008: 68): a pre-theoretical posit, that now that our scientific understanding has progressed, can be done away with. At the very least, even if not outright denying the existence of words, it is clear that words are not the sorts of things ordinary speakers have typically taken them to be—they are not the atomic elements of language. Though if words are not the atomic elements of the language faculty, then what are? That is, what are the smallest objects posited within linguistics?

Sprouse and Lau state that 'it is fair to say that there is some degree of consensus that the basic units are bundles of features' (2013). It is these 'bundles of features' that are the objects stored in the mental lexicon ready to be accessed when required as part of a linguistic derivation. For our purposes here, we can take the term 'feature' to be synonymous with the more common philosophical term of 'property' (and I will use the two interchangeably throughout). Thus, the consensus position is that the mental lexicon contains units that are bundles of properties, where those properties determine (at least) the phonological and conceptual information relating to those units and the ways that those units subsequently behave within a linguistic derivation.[5] For ease, I will follow Borer (2005) and call these units 'listemes'.

This is of course not to say that all agree about the precise nature of listemes. For example, whilst there is agreement that the primitive elements are listemes (as bundles of features/properties), there is disagreement about what features/properties listemes have. For example, Chomsky (1995) suggests that listemes need to possess 'formal' features—roughly speaking, syntactic information that identifies that particular listemes as being of a certain syntactic category. Borer denies this, at least for what she calls 'substantive' listemes which are category-less in her conception of the lexicon, only gaining a syntactic category as a result of the process of being merged within the syntactic processing that results in phrases and other more complex

[4] I say possibly not because I am raising some doubts that books exist, but rather that there could be doubts that books are linguistic entities. Certainly books do not seem to be *necessarily* linguistic entities as picture books are books that at least could contain no linguistic entities at all.

[5] See Miller (2021) for an extension of this idea to words more generally, arguing that words should also be characterised as being bundles of properties.

linguistic entities. For Borer, leaving aside various complexities, such listemes are sound-meaning pairs—a combination of some phonological information and some conceptual information only (Borer, 2005: 15).

These debates over the precise nature of listemes are not my focus here. The significance of this in this paper is that in order to understand emergence claims, we need to understand levels within the specific domain under discussion. Words are, within the domain of language, higher-level entities, and listemes 'lower-level'. Indeed, it is plausible that within the language faculty in humans, listemes, and the properties that they possess, are the lowest-level entities. That is, listemes that instantiate certain (potentially limited) properties, unlike words, at least have a claim to be the smallest elements of the language faculty. Or, putting this in the terminology common in debates about emergence, listemes are lower-level entities, with all other linguistic entities being relatively higher-level.

An important caveat is needed at this point, before returning to questions of emergence more explicitly. I have suggested that listemes are lower-level entities, with words, sentences, etc. as being relatively higher-level. This claim is strictly one about the ontology of the language faculty as it appears in humans, and not a wider claim about the ontology of the mind. The language faculty is undoubtedly just one part of a much bigger and more complex cognitive architecture. Minimally, the language faculty must have interfaces with broader conceptual systems and systems that control sensory-motor functions that ensure that any linguistic structures created within the language faculty can be suitably externalised. It is possible that the language faculty interfaces with far more distinct aspects of our cognitive systems too.

For my purposes here, which is solely a discussion about the possibility of emergence within the language faculty, I am interested only in the ontology of the language faculty. It may be that a full understanding of the entire cognitive architecture of humans will conclude that listemes and their properties that I have claimed are the smallest elements of the language faculty are in fact complex outputs of various other aspects of cognition. Indeed, I think this is likely in some cases, particular for listemes through which conceptual information enters into the language faculty.

However, as I am interested in properties of *linguistic entities*, and on the assumption that (at least in humans) only the language faculty produces or functions over linguistic entities, whether or not listemes are lower or higher-level entities relative to the wider cognitive architecture is not important. All claims of emergence require are levels, and the above suggests that at least within the language faculty, such levels exist, with listemes at the 'bottom' and other more complex linguistic entities at higher levels.

Returning to our main focus, what should we look for when seeing if there are possible cases of emergence in linguistics? Using our definition above of (weak) emergence, we are after cases where higher-level phenomena 'fail to be predictable, derivable, explainable or characterizable in terms of the "more basic" entities and properties out of which the emergents arise' (Bishop & Silberstein, 2019: 154).

Applying this to the ontology of language that we have sketched thus far, we are looking for cases where certain phenomena or properties of higher-level entities—words, sentences, etc.—fail to be predictable, derivable, explainable or characterizable in terms of the more basic entities—listemes. As already noted, there are various different properties of listemes, potentially including, depending on the wider conception of the mental lexicon, semantic features, syntactic features, and phonetic features. Possible candidates for emergence would therefore be any property of a higher-level that cannot be reduced to some such property of a listeme, or to the interaction of such properties of a listeme.

So understood, there might be various candidates for emergence in linguistics. Are lower-level properties sufficient to fully characterize the nature of words as they appear in ordinary language? Might it be the case that there are certain characteristics that *only* arise when certain properties of listemes interact, but are unpredictable from considering those features alone? Depending on the role within our theory of syntactic processes, it is distinctly plausible that certain properties arise due to the ways that listemes are affected by syntactic operations. The more properties we posit listemes as having, the more that we can predict the behaviour of higher-level entities, though (as I will return to below) positing additional properties purely to account for otherwise unpredictable characteristics of higher-level entities may be ad hoc.

If we are considering if there is emergence in linguistics we should not focus solely on features and listemes though. The discussion in this section was only intended to sketch a broad picture of the ontology of language, and illustrate that there are levels in our current linguistic theories. That is, that there are at least distinct levels where we find distinct objects—listemes, words, phrases, sentences, etc.—each of which is 'higher-level' with respect to the entity preceding it in that list. Emergence, if it exists in linguistics at all, may occur between any of these levels.

As it happens, the case that I want to focus the rest of this paper around is not found at between the levels of features and listemes, but concerns a property of certain phrases or sentences. But, that I focus on a property of phrases and sentences does not rule out that there might be other cases of possible emergence. The example I will outline here is the same property that I discuss in Miller (2017)—the property of truth-evaluability. I suspect that there are many other properties that are at least plausible candidates from being emergent between the different levels found in linguistics, but the discussion of other cases will have to be left for future work.

3 The Case for the Failure of Reduction

We might disagree about which entities can have the property of being true or false. A list of what entities, linguistic and non-linguistic, can be true or false will depend significantly on the results of further disputes, but certain linguistic entities indisputably have the property of being true or false.[6] Sentences, for example, are linguistic entities that can be true or false, as in (1):

(1) The apple is on the table.

There are of course cases where we do not know if a certain sentence is true or false. For example, I may not know whether the sentence 'There are exactly ten coins in my pocket' is true or not. Depending on our theory of truth, there might be various ways that we can work out if a given sentence is true or false. If we adopt a correspondence theory of truth, we might try to work out whether sentence corresponds with the world in a suitable way. Other theories of truth will provide other ways to determine whether a given sentence is true or false.

Some linguistic entities, though, cannot be true or false. Single words (with a potential exception to be discussed below) cannot be true or false. For example, (2) cannot, by itself be true or false. Nor can certain sub-sentential phrases, such as determiner phrases as in (3), or verb phrases as in (4), nor can morphemes, whether they are free as in (5), or bound as in (6).

(2) table
(3) a table
(4) kicked the ball
(5) town
(6) -ing

These examples illustrate the difference between truth, and truth-evaluability. I am interested in the latter here: the property that results in (1) being truth-evaluable, while (2)–(6) are not. That is, what makes it the case that we can ask whether (1) is true or false, irrespective to what the answer to that question is, while we cannot sensibly ask whether (2)–(6) are true or false. All of them are linguistic entities, thus we need to understand why only some of them have this property.

Interestingly, truth-evaluability is a property of certain linguistic entities independent of our view about how language connects with the world. What I mean by this is that supporters of all theories of truth should accept that (1) can be true or false, while (2)–(6) cannot be. This suggests strongly that truth-evaluability is a property that is internal to the object. By 'internal' I mean that it is a property of the linguistic entity, and not a result of some relation that the linguistic entity stands in to some further entities (including the world). Certain linguistic entities, such as (1), are truth-evaluable, independent of whether they are true or false, and those entities have this property independent of any further relations that we might posit as holding between

[6] For example, can pictures be true? Propositions are normally accepted as being able to be true or false, but propositions are not normally taken to be linguistic entities. These further issues about what other objects can have the property of being true or false will not be important here; I am only focusing instead on entities that are clearly linguistic, such as sentences.

linguistic entities and the world.[7] This hopefully suffices to get an initial grasp on the property I am considering.

What further can we say about the property? First, the examples in (1)–(6) also suggest that truth-evaluability is a higher-level property. This is because the only linguistic entity that has the property of is a sentence, and none of the lower-level entities can possess the property. This is important as it is a requirement for a possible case of emergence that the putative emergent property is higher-level. We have noted this above—if emergent properties exist, then they need to be higher-level properties that are dependent on, but distinct from lower-level properties. (1)–(6) show, minimally, that truth-evaluability is a higher-level property in that it is instantiated by (relatively) higher-level linguistic entities (i.e. sentences), and not lower-level entities (such as words, phrases, or morphemes).

Interestingly, being a sentence is only a necessary condition on being truth-evaluable, not a sufficient condition as shown by (7) and (8).

(7) What is on the table?
(8) Put the apple on the table!

Both (7) and (8) are well-formed grammatical sentences, unlike (2)–(6), and yet neither can be coherently thought of as true or false. This suggests that truth-evaluability requires a particular form of structural complexity (see Hinzen, 2009, 2013, 2014), and not merely a well-formed sentence.

Semantic content is also not relevant to truth-evaluability, as shown by certain nonsense sentences that are still truth-evaluable. For example, (9) is truth-evaluable. We may not know whether it is true or false, or we might think that the lack of meaning for the terms mean that we can never know whether it is true of false. But the requisite structural complexity is present for (9) to be truth-evaluable.

(9) All mimsy were the borogoves.

Summarising, the above examples show that truth-evaluability is a property only possessed by certain higher-level linguistic entities, such as sentences.[8] However, not just any sentence is truth-evaluable, and whether a sentence is truth-evaluable is independent of the coherence of the semantic content expressed by the sentence. We might think (9) is ultimately meaningless, but it still possesses the property of being truth-evaluable,[9] unlike (7) and (8), neither of which are truth-evaluable.

[7] There is a similar notion of 'truth-aptness' discussed by some philosophers. Truth-aptness has often been invoked to make claims about what linguistic structures aim to express facts about the world, for example in the context of non-cognitivist ethics (see Jackson, Oppy and Smith 1994). I am not interested in the relation between language and the world in this paper, only in the internal property of certain linguistic structures to be evaluated for truth, hence I will use the more neutral notion of truth-evaluability to avoid any confusion with other debates.

[8] Or matrix clauses more precisely.

[9] Note, the meaningless of (9) might lead some to think that it has no truth value if we hold that meaningless sentences necessarily have no truth value. However, there is a difference between an entity lacking a truth-value and being truth-evaluable. I contend that (9) possesses the latter property, irrespective of our views about whether it is true, false, or has a gappy truth-value.

Given that truth-evaluability is a higher-level property, is it a possible case of emergence? One prima facie way to argue that it is emergent would require showing the failure of reductionism with respect to truth-evaluability. To do that, there are two considerations that need to be made. First, does the property ever get instantiated at the lower-level? And, second, is the property plausibly structural? If the answer to either of these questions is yes, then there would seem to be no case for emergence (of any sort). In what remains of this section, I will argue that the answer to both questions is no.

First, is truth-evaluability instantiated by lower-level entities? From the above cases, we have seen already support for the claim that the property is not instantiated at the lower-level as single words, isolated phrases, and morphemes (bound or unbound) are incapable of instantiating the property. Putting this another way, (2)–(6) are evidence that no lower-level linguistic entity can be truth-evaluable. The property only can be instantiated by (relatively) higher-level entities, such as sentences.

Above, I mentioned in passing a possible counter-example to this. The possible counter-examples are certain single words that may on first glance appear to be truth-evaluable. For example, say that one person asks another what the weather outside is like, and the response given is 'Sunny'. In this case, we would appear to have a single word that is truth-evaluable as it is possible to evaluate the response ('Sunny') to determine whether it is true or false. Thus, we would appear to have single words that can be truth-evaluable, contra my claim that only higher-level linguistic entities can instantiate the property.

However, as I argued in previous work (Miller, 2017: 119), such cases are misleading. This is because the full linguistic analysis of such single word responses would involve positing anaphoric structure in the response inherited from the question. Thus, the linguistic analysis of the response 'Sunny' would hold that the response had the underlying structure present in (10), which, in English, does not need to be morphologically expressed.

(10) It is sunny outside.

(10), though, clearly is a sentence, and hence is a higher-level entity. That English (and other languages) does not require elements to be spoken does not rule out that such structures are still operating, supporting the claim that truth-evaluability can only be instantiated by higher-level linguistic entities.

What about the property being structural? By structural, I mean properties that arise due to the nature of lower-level entities, even when those lower-level entities do not possess those structural properties themselves. For example, my table is rectangular. However, at an atomic level, its parts likely do not instantiate the property of 'being rectangular'. Analogous cases come from the weights of composite objects. No particle instantiates the property of being 10 kg, but the object that the particles compose may instantiate that property. Such cases are widely accepted as not being cases of emergence. The reason is that such properties are 'additive' in nature (McLaughlin, 1992:89). While the higher-level entity does appear to have a novel property, that novelty can be explained by considering the combined effects of the lower-level properties—e.g. all of the weights of the particles combined that together weigh 10 kg.

Is truth-evaluability structural in this sense? Again, the answer is no.[10] The evidence for this comes from the fact that the same structure has different properties depending on whether it is a matrix clause, or whether it is embedded within some further structure. To see this, consider the examples, borrowed from Miller (2017: 123).

(11) Caesar destroyed Syracuse.
(12) Mary believes that [Caesar destroyed Syracuse].

(11) is clearly truth-evaluable, and so is (12). But, this case is important as it is one where we have the same structure appearing once as a matrix clause, and again embedded within a more complex structure. Language is an ordered and structured entity, and that ordering and structure is governed by various rules studied by linguistics. Given this, it is reasonable to suppose that the structure instantiated in (11) and the structure instantiated in the embedded clause of (12) is identical. But, unlike in (11), that same structure when it appears in (12), is not a truth-functional ingredient within that larger structure.

This suggests that truth-evaluability is not a structural property. If it were, then the same structure should be truth-evaluable whenever it exists, but in (11) and (12) we have a case where that is not true. This is of course not to deny that (12) *as a whole* is truth-evaluable. It absolutely is. It is only the claim that the structure that appears in (11) is identical to the structure that appears in the embedded clause in (12), and as it is truth-evaluable in one but not the other, that is at least initial evidence that truth-evaluability cannot be a structural property akin to shape or weight.

These arguments suggest that reductionism fails in the case of truth-evaluability. In Miller (2017) I argued that this gave us some reason to hold that truth-evaluability is a strongly emergent property. That is, I argued that this failure of reduction suggests that the property is a metaphysically novel and irreducible property instantiated at the higher-level. In the following section, I weaken this conclusion somewhat, and instead argue that the evidence is also compatible with the conclusion that truth-evaluability is weakly emergent, in the sense outline in section one.

4 Truth-Evaluability as Weak Emergent

I think that the above arguments are sufficient for us to reject any simplistic reductionist view. What I mean by this is that I think the above shows that it is unlikely that we will simply 'find' (or are justified in positing) a property at the lower-level that accounts for, or explains, the property of truth-evaluability as it appears at the higher-level. In Miller (2017) I argued that this failure of reduction was evidence

[10] The case against truth-evaluability as a structural property is more complex than I can summarise within the word limit here. For the full argument, see Miller 2017: 122–132.

for strong emergence in linguistics. In this section, I will instead sketch a view under which truth-evaluability is weakly emergent, contra the claims I made previously that the only alternative considering the failure of reductionism was positing strongly emergent properties.

In section one, weak emergence was characterised as being cases where entities 'fail to be predictable, derivable, explainable or characterizable in terms of the "more basic" entities and properties out of which the emergents arise' (Bishop & Silberstein, 2019: 154). Could truth-evaluability be weakly emergent? I think the answer is yes (or, at least, possibly yes). To see this, consider one further route of response for the reductionist: to argue that there is a property instantiated by listemes that accounts for truth-evaluability. That is, the reductionist could argue that in addition to semantic, formal, phonetic, and other sorts of features/properties that we might posit as being instantiated by listemes, there is also a 'truth-evaluability' property. This would then be a property instantiated at the lower-level, undermining any need to posit emergence.

It is worth pausing to distinguish this possible response from that of whether truth-evaluability is a structural property. In the structural case, the idea is that the property is one that comes into existence in line with certain structural complexity. Such structural properties would not therefore be a property of a listeme. They could only be a property of certain structures. This is distinct from the claim that there might be a property of truth-evaluability instantiated by listemes in that this claim instead posits that truth-evaluability is there all along. The idea being considered is that truth-evaluability is a property of certain listemes, but one that does not manifest its causal powers unless in the presence of other properties.

In Miller (2017), I argued that there was a problem with this reductionist response in that such a posit would appear to be ad hoc. There, I argued that the only reason that we would posit a lower-level property that accounts for truth-evaluability is because we want to avoid positing a strongly emergent property. After all, what other reason could there be? The above examples suggested that there is no independent evidence of truth-evaluability as a property of lower-level entities, so the only reason that we might insist that there really is some relevant lower-level property would be because we do not want to posit strongly emergent properties.

On reflection, I think this argument is perhaps a bit too quick. Certainly it is the case that given the evidence about single words not being truth-evaluable, any truth-evaluability property posited as being instantiated by listemes certainly cannot be exactly like other properties instantiated by listemes. Rather this property would need to be one that only bestows on the linguistic entity that contains it a property once there is some complexity present. What this means is that it would be a property such that it is only in the presence of some other feature or some complex set of features that its existence becomes apparent. The idea that there are certain properties of objects that only bestow causal powers on objects in the presence of other properties, either of the same object or distinct objects, is not new. There are many properties that only reveal themselves in the presence of other properties, or indeed in the absence of certain other properties.[11] There even seem to be other properties like this in language. Certain formal (or grammatical) properties that are (not indisputably) taken to be present in listemes for example. It is clear that many grammatical structures are only possible given the presence of multiple words, and complex relations between those words, so any formal properties of listemes can only become apparent once listemes interact or stand in certain relations to each other. It might be argued that the property that gives rise to truth-evaluability is similar.

Another way to consider this is to imagine that we had knowledge of *only* of the listemes within a language. That is, imagine that we only had knowledge of the lower-level entities, and had no way to investigate higher-level entities directly. Despite this limitation, in such a situation, we would know quite a lot about the language. For example, we would know (much of the) semantic information, as it is generally agreed that listemes contain semantic features/properties that introduce semantic information into the language faculty. If Chomsky is right, listemes also contain 'formal' features, such that the elements of the mental lexicon are pre-ordered into syntactic categories. Knowing about such properties would mean that we could derive solely from our knowledge of listemes some awareness of the syntactic properties of more complex entities. Knowing solely about the listemes of a language would therefore result in us knowing quite a lot about how language works and is structured. However, it is plausible that there are limits to that knowledge in that there might be properties of the higher-level linguistic entities that cannot be predicted or derived from knowledge of the lower-level entities. If there are properties of language like this, then these would be weakly emergent properties under the characterisation provided above.

Could truth-evaluability be a property like this? Prima facie, the answer seems to be yes. Truth-evaluability, as already noted, cannot be derived or predicted from the properties of lower-level objects, but can be easily recognised once we consider the higher-level objects directly. Truth-evaluability fails to be predictable, derivable, explainable or characterizable in terms of the 'more basic' entities and properties out of which the property arises. The listemes that are part of the mental lexicon are the 'more basic' entities that properties like truth-evaluability arise from, and yet complete knowledge of those elements alone would not result in knowledge about which linguistic entities were truth-evaluable. Truth-evaluability is a highly plausible candidate for being weakly emergent.

What does this mean for reductionism? As discussed in section one, weak emergence of the sort that I have used here is compatible with a certain form of ontological reduction. This is because weak emergence is significantly *epistemic*, and requires only that we can know about the higher-level properties only through first recognising their existence at the higher level. We have seen that this could be the case for truth-evaluability. We can only identify the property at the higher-level—i.e., that of sentences (or at least matrix clauses). If the reductionist wants to maintain

[11] For example, the property of a match to light on fire cannot be manifested if the match is underwater. The presence of the water, given the properties of the water, restrict the match's ability to light.

their opposition to positing truth-evaluability as strongly emergent, then accepting that the property is weakly emergent is a viable alternative. Certainly this would not be the *same* as simply reducing the property, but it would not carry with it the same sorts of ontological consequences that the reductionist objects to when we posit strongly emergent properties. Truth-evaluability would not be a wholly novel property, existing in addition to the properties posited at the lower-level. Rather, it would be a property that is grounded in the lower-level properties from which it arises, but cannot be predicted from an observation of those lower-level properties in isolation. If this suffices for ontological reduction as some have maintain (e.g. Butterfield, 2011), then the potential for an ontological reduction of truth-evaluability relies on accepting the weak emergence of truth-evaluability.

To summarise the main points of this section. In Miller (2017) I argued that the only alternative to reduction was strong emergence. Given the argued failure of reduction, I therefore concluded that truth-evaluability must be strongly emergent. This section has sketched an alternative. If we instead start with the idea that emergence should be understood in a weak way, we can avoid the conclusions I drew in that prior work. Under this approach, truth-evaluability is a weakly emergent property in that it is a property that fails to be predictable or derivable by considering the nature of the lower-level linguistic entities in isolation. The main benefit of this view is that it is (at least potentially) compatible with *ontological* reduction. Therefore, if I am right, those that want to uphold a form of ontological reduction about language can do so just so long as they also allow that in linguistics there are weakly emergent properties.

5 Weak Emergence and Linguistics

Emergence is a topic of great interest in many sciences, but there is little written about emergence specifically in the context of linguistics. Perhaps this is because weak emergence is often taken to be so prevalent. Still, even those things that seem to be obvious need to be argued for eventually, so here I have tried to sketch the view. I have suggested that simplistic reductionism does not work in the case of truth-evaluability, and that taking the property to be weakly emergent at least allows us to avoid strong emergence in this case.

Answering the further question in the title of this chapter—of how far linguistics 'needs' emergence—will depend on how many higher-level properties that are part of the subject matter of linguistic study can (or must) be analysed in a similar way. There are a lot of higher-level properties in language. Properties that can only be rightly attributed to words, phrases, or sentences. There may even properties that can only be attributed to paragraphs and chapters. To give just one further example of a higher-level property of language, consider the property of '(un)grammaticality'. This property is central to a lot of linguistic theorising. Many linguistic theories begin with a consideration of the (un)grammaticality of certain linguistic entities, primarily again higher-level entities such as sentences. One aim of linguistics is to explain that (un)grammaticality in a rigorous and systematic way.

It is likely that any analysis of the property of (un)grammaticality will turn out to be very complex, and it is certain that it is a multiply realisable property—there are multiple different structures that can instantiate the property of 'being grammatical'. But, investigating the underlying lower-level arrangements that give rise to (un)grammaticality relies on a prior recognition of the property at the higher-level. As we have seen, at least on some understandings, this is would be in line with taking the property of (un)grammaticality as being weakly emergent. If such a central property to linguistics is weakly emergent then there is a real sense in which positing weak emergence is required to engage in rigorous and systematic linguistic research. This lends itself to a (tentative conclusion) that weak emergence may be expected to be found in many more cases than just truth-evaluability.

This (tentative) conclusion might not be surprising. Much of the data of linguistics comes from the observation, analysis, and investigation of linguistic entities like words, phrases, and sentences, and, as we saw in section two, these are higher-level entities. This focus on *macro* linguistic entities arises because the central aim of linguistics is to understand the nature of language which is used (and understood) by humans at that higher-level—at the level of words, phrases, and sentences. It is widely agreed that speakers do not have access to their own listemes in anything like a conscious way. We experience language constantly in our lives, but the vast majority of that experience (including of our own language use) is experience of higher-level entities like words, phrases, and sentences. This paper has only directly argued that truth-evaluability is weakly emergent. The next question in this regard is 'how unique is truth-evaluability?'. Given the large number of higher-level linguistic properties and objects, it seems somewhat plausible that the answer to that question is 'not very'.

References

Bedau, M. A. (1997). Weak emergence. *Philosophical Perspectives, 11*, 375–399.

Bishop, R. C. (2005). Patching physics and chemistry together. *Philosophy of Science, 72*, 710–722.

Bishop, R. C. (2010). Whence chemistry? Reductionism and neoreductionism. *Studies in History and Philosophy of Modern Physics, 41*, 171–177.

Bishop, R., & Silberstein, M. (2019). Complexity and Feedback. In S. Gibb, R. Hendry, & T. Lancaster (Eds.), *The routledge handbook of emergence* (pp. 145–156). Routledge.

Boeckx, C. (2008). *Bare syntax*. Oxford University Press.

Borer, H. (2005). *In name only*. Oxford University Press.

Butterfield, J. (2011). Emergence, Reduction and Supervenience: A Varied Landscape. *Foundations of Physics, 41*, 920–959.

Carruth, A. (2019). Strong emergence and Alexander's dictum. In S. Gibb, R. Hendry, & T. Lancaster (Eds.), *The routledge handbook of emergence* (pp. 87–98). Routledge.

Chalmers, D. J. (2006). Strong and Weak Emergence. In P. Clayton & P. Davies (Eds.), *The re-emergence of emergence: The emergentist hypothesis from science to religion* (pp. 244–254). Oxford University Press.

Chomsky, N. (1995). *The minimalist program.* MIT Press.

Gibb, S., Hendry, R., & Lancaster, T. (Eds.). (2019). *The routledge handbook of emergence.* Routledge.

Hasker W. (2017). 'Emergent dualism'. In: Y. Nagasawa, B.Matheson (Eds.) *The palgrave handbook of the afterlife.* Palgrave Macmillan.

Hinzen, W. (2009). "Hierarchy, merge, and truth". In: M. Piattelli-Palmarini, J. Uriagereka, P. Salaburu (Eds.) *Of minds and language: A Dialogue with Noam Chomsky in the Basque Country,* Oxford University Press.

Hinzen, W. (2013). "Naturalism pluralism about truth". In: N. Pedersen, & C. Wright (Eds.), *Truth and pluralism,* OUP.

Hinzen, W. (2014). "Recursion and truth". In: T. Roeper, P. Speas (Eds.), *Recursion: Complexity in cognition,* Studies in Theoretical Psycholinguistics 43, Springer.

Jackson, F., Oppy, G., & Smith, M. (1994). Minimalism and truth aptness. *Mind, 103*(411), 287–302.

McLaughlin, B. (1992). "The rise and fall of British emergentism". In: A. Beckermann, H. Flohr, & J. Kim (Eds.), *Emergence or reduction? Essays on the prospects for non-reductive physicalism,* Walter de Gruyter.

Miller, J. T. M. (2017). Language and ontological emergence. *Philosophica, 91*(1), 105–143.

Miller, J. T. M. (2021). A bundle theory of words. *Synthese, 198*(6), 5731–5748. https://doi.org/10.1007/s11229-019-02430-3

Nida-Rümelin, M. (2007). Dualist Emergentism. In B. McLaughlin & J. Cohen (Eds.), *Contemporary debates in philosophy of mind* (pp. 269–286). Blackwell.

O'Connor, T., & Churchill, J. R. (2010). "Nonreductive physicalism or emergent dualism? The argument from mental causation", in *The waning of materialism: New essays,* R. C. Koons, & A. George Bealer (Eds.), Oxford University Press, 261–279.

O'Connor, T. (2020). 'Emergent properties', *The stanford encyclopedia of philosophy,* E.N. Zalta (Ed.).

Rickabaugh, B. (2018). 'Against emergent dualism', in *The blackwell companion to substance dualism.* J. Loose, A. Menuge, & J. P. Moreland (Eds.). Wiley Blackwell.

Silberstein, M. (2002). Reduction, emergence and explanation. In P. Machamer & M. Silberstein (Eds.), *The blackwell guide to the philosophy of science* (pp. 80–107). Blackwell.

Silberstein, M. (2012). Emergence and reduction in context: philosophy of science and/or analytic metaphysics. *Metascience, 21,* 627–642.

Silberstein, M. (2017). Contextual emergence. *Philosophica, 91,* 145–192.

Sprouse, J., & Lau, E. (2013). 'Syntax and the brain'. In M. den Dikken (Ed.), *The handbook of generative syntax.* Cambridge University Press.

Van Gulick, R. (2001). Reduction, emergence and other recent options on the mind/body problem: A philosophic overview. *Journal of Consciousness Studies, 8*(9–10), 1–34.

Wong, H. Y. (2019). Emergent dualism in the philosophy of mind. In S. Gibb, R. Hendry, & T. Lancaster (Eds.), *The routledge handbook of emergence* (pp. 179–186). Routledge.

Zimmerman, D. (2010). From property dualism to substance dualism. *Aristotelian Society Supplementary, 84*(1), 119–150.

Contextual Meaning and Theory Dependence

Erich H. Rast

1 Introduction

Natural language elicits many forms of context dependence. Many of them are overt. For example, indexicals depend on the deictic center I-here-now. However, there are more subtle forms of context dependence in natural language that are less regulated by meaning rules and more pragmatic. Roughly speaking, hearers arrive at an interpretation of what the speaker said based on what they believe the speaker assumes in the context of a conversation. I argue in this article that this interpretation process requires speakers to be able to track other speakers' theories, and that epistemic agents generally must have the ability to consider and compartmentalize theories without necessarily endorsing them.

In Sect. 2, a brief overview of select phenomena of linguistic context dependence is provided and it is argued that these are overall tractable by understanding interpretation in a context as an inference from often truth-conditionally incomplete to more specific semantic representations. However, there is a more profound and philosophically more challenging context dependence that can be described as a dependence of concepts and lexical meaning on theories. This is laid out in Sect. 3, in which several problems are discussed that result from the interdependence between lexical meaning and theories. I argue in Sect. 4 that these problems can be solved by rejecting global meaning holism in favor of local meaning holism and by acknowledging rational epistemic agents' ability to compartmentalize and keep track of theories.

2 Semantic Contextualism: a Brief Overview

This section provides a brief overview of linguistic context dependence. Much of the modeling of context dependence in the philosophy of language and epistemology is based on Kaplan (1989)'s *Logic of Demonstratives*, which lead to different versions of

E. H. Rast (✉)
IFILNOVA Institute of Philosophy, Universidade Nova de Lisboa, Av. de Berna 26C, 1069-061 Lisbon, Portugal
e-mail: erich@snafu.de

S. Wuppuluri and I. Stewart (eds.), *From Electrons to Elephants and Elections*,
The Frontiers Collection, https://doi.org/10.1007/978-3-030-92192-7_4

'two-dimensional' semantics (Chalmers, 2006) and corresponding forms of alethic contextualism and relativism. These approaches are the topic of the next section. Section 2.2 addresses their shortcomings and promotes the alternative view that many forms of linguistic context dependence are better described as an inference from potentially truth-functionally incomplete to more specific semantic contents instead of using parameterized modal logics.

2.1 Parameterized Contexts

Indexicals and demonstratives are typical overtly context-sensitive expressions. The reference of indexicals like *here*, *I*, and *now* depends on features of the utterance context. Their linguistic meaning partly mandates the resolution of this context dependence. For example, under normal circumstances, *I* refers to the speaker of the utterance, *now* to the time of the utterance, and *here* to the place of utterance. Such a rule 'picks out' the respective referent in a given context of utterance, thereby resolving the context dependence semantically. The result of this enrichment process is a proposition that is true or false in the given circumstances of evaluation. Absolute tenses are also often used indexically. To fully understand a use of the present tense a hearer may have to know the time of utterance, for instance.[1]

Understanding utterances with indexicals comes to a degree because the corresponding contextually-provided referents may be determined more or less precisely. In a sense, a hearer understands an utterance of *Yesterday, Bob had an accident* without knowing what day of the month or week it is; the accident happened the day before whatever day is now. However, this minimal understanding may turn out to be insufficient for a given communicative task. For example, when filling out an insurance form, merely knowing that something happened the day before the day on which the utterance took place might not suffice because a calendrical date is expected. A more precise understanding of the utterance could be paraphrased as *Bob had an accident on Friday, the 13th of November 2020*.[2]

Every indexical allows for such grades of understanding. Sometimes when interpreting a use of *I* it may suffice to know that someone spoke, whoever that may have been; in other cases, the hearer must spatiotemporally locate the speaker before they can rightfully be said to have understood the utterance as a whole. However, the reference to the deictic center acts as a hard constraint in any of those cases. The deictic center usually consists of the speaker, the time of utterance, the actual world, and the place of utterance. It can be shifted in some languages in indirect speech reports and for certain expressions like *local*, *around*, and medical uses of *right* and *left*.[3] The

[1] There are also non-indexical uses of absolute tenses, such as the use of the English present tense in a generic like *Cats are mammals*.

[2] Cf. Perry (2001) on incremental truth-conditions.

[3] See Schlenker (2000, 2003) on shifting first person pronouns, and, more generally, Fillmore (1997), Lyons (1977, p. 579), Levinson (1983, p. 64).

hearer, and sometimes even the speaker, can be wrong about this reference, and in that case, they fail to grasp the semantic content of the utterance. This dependence on a fact about the world that is independent of the speaker's intentions is characteristic of indexicals. Other context-dependent expressions need not involve a deictic center in this way.

In early approaches to indexicals such as Reichenbach (1947), Burks (1949), and Bar-Hillel (1954) a crucial question was whether these could be eliminated from a language that would serve as a foundation for all science. Bar Hillel argued that even though sentences containing indexicals can be substituted with sentences containing no indexicals, the reference to a conventionally fixed origin of a coordinate system cannot be eliminated. In tense logic, Prior (1957, 1967, 2003) famously argued that the logic of becoming and going expressed by operator-based tense logic could not be replaced without significant loss of expressivity by statements that quantify over points in time or time intervals directly and thereby lead to eternally true or false propositions. His arguments for this view were metaphysical and partly hinged on a specific interpretation of McTaggart's Paradox (McTaggart, 1908). In the Philosophy of Language the irreducibility of the basic indexicals *I*, *here*, and *now* was brought up by Castañeda (1967, 1989a, b) and Perry (1977, 1979, 1998a, b), and has been discussed in numerous follow-up publications. In this debate, the key question was whether thoughts, corresponding truth-functionally complete propositions, and broadly-conceived epistemic states that would ordinarily be expressed using indexicals like *now* and *I*, could be expressed by expressions only containing third person referential terms such as proper names and definite descriptions. There is a certain consensus in the literature that at least *now* and *I* are irreducible in cognition in terms of their expressive power for explaining behavior, which lead to various theories of de se belief attributions that take into account the 'essential indexicality' of these indexicals.[4]

A more recent debate started with Recanati (2004b) versus Cappelen and Lepore (2004). It addresses the more general question about linguistic context dependence's pervasiveness and what this means for literal meaning. Much of this discussion concerns the extent to which double-index modal logics can adequately represent linguistic context sensitivity. As part of the philosopher's toolbox, based on Kaplan (1989) and Lewis (1980), various modal logics and their interpretations are used in which contexts and circumstances of evaluation (CEs) are reified as parameters relative to which truth-in-a-model is determined. In Kaplan's two-layered account, for instance, the linguistic meaning (the character) of an expression is a function that in a context yields an intension (the semantic content), which is, in turn, a function that in given circumstances of evaluation yields an extension. Based on such modal logics with contexts and CEs, various contextualist and relativist positions have been developed and contrasted with Cappelen and Lepore's semantic invariantism on the one hand and Recanati's more radical contextualism on the other hand.

---

[4] See, for instance, Kaplan (1989), Lewis (1979b), Cresswell and von Stechow (1982), von Stechow (1984).

Going into the details of this complex debate would go beyond the scope of this contribution. Only a brief summary can be given. According to invariantism, simple clauses are not context-dependent except for the obvious and overt context dependence of indexicals. Cappelen and Lepore (2005) even go so far as to claim that a giraffe can be tall simpliciter. Others such as Recanati (2006) and MacFarlane (2007) found such an approach unsatisfactory. According to the radical contextualism of Recanati (2004a), the literal meaning on which such invariantist positions are based is an 'idle wheel'; instead, according to Recanati pragmatic modulation functions may change linguistic meaning on the fly during semantopragmatic construction of sentence-level content.

In contrast to this, indexicalists like Stanley (2004, 2005) model context-dependent expressions with open argument places bound either by semantic or by pragmatic processes. This use of open argument places makes their accounts slightly different from two-dimensional moderate contextualists who continue to use double-index modal logics to model a richer set of contextual variances than those elicited by overt indexicals. In the approaches based on modal logics with multiple parameters, these parameters are enriched with whatever additional ingredients are needed to get the semantics of context-sensitive expressions right that do not overtly depend on the deictic center. Usually, they are modeled as n-tupels containing all the needed ingredients.[5]

Broadly speaking, two-dimensional accounts come in three different varieties. According to classical contextualism, in a Kaplan-style two-layered modal logic the semantic content of the expression is fixed by some mechanism that takes into account features of the context parameter. If a context-sensitive expression is modeled in this way, then varying contexts will yield different semantic contents. This is the classical model of indexicals. In contrast to this, according to the nonindexical contextualism of MacFarlane (2009) the extension of an expression may depend on the context although the semantic content remains context-invariant. This means that the semantic content—i.e., the proposition expressed by the sentence in a context—is itself context-sensitive. Finally, according to full-fledged alethic relativism in a two-dimensional framework, the semantic content may yield different extensions not depending on the context but depending on non-traditional features of the circumstances of evaluation. Tense operators and modalities work in that way in traditional double-index modal logics because these operators implicitly quantify over time and possible worlds. In the debate between contextualists such as de Sa (2008, 2009) and relativists like MacFarlane (2008, 2012, 2014), relativists have argued that many more expressions may be truth-relative in this sense.

Within this discussion, some authors suggested that certain predicates of personal taste give rise to faultless disagreement between speakers that only a relativist semantics can adequately represent. In such a theory, the extension associated with

[5] Cresswell (1996) argues that modal logics with finitely many parameters are not expressive enough to deal with the indexical context dependence of arbitrarily long sentences. Instead, full quantification over reified contexts is needed. This argument has largely been ignored by the philosophy of language community.

a semantic content in given circumstances of evaluation is not just relative to times and possible worlds, but also relative to very nontraditional constituents of circumstances of evaluation parameters such as persons. For example, Lasersohn (2005) argues that the predicate *fun* is sensitive to an assessor (or, judge, in his parlance) in given circumstances of evaluation. Regardless of who is the speaker of an utterance, in this assessor-relativism an utterance of *Roller coasters are fun* may be true relative to one and false relative to another assessor. Consequently, two people may disagree about an utterance containing such an expression without one being at fault. They may both be right even when they seemingly contradict each other and one of them negates the other's statement. Relative to one assessor the semantic content of the proposition may be true and relative to another assessor the semantic content of its negation may be true. To do justice to this position, it is worth noting that each of the assessors may still be mistaken in such an approach. For example, an assessor might erroneously believe that roller coasters are fun (relative to her); actually riding a roller coaster would make her realize that she was wrong right from the start.

The differences between parameter-based traditional contextualism, nonindexical contextualism, and relativism primarily hinge on the role given to semantic content in theorizing. The idea behind relativist faultless disagreement is that two assessors who disagree faultlessly disagree *about* the same semantic content of an utterance. The relativist argues against the contextualist that two interlocutors would disagree about two different contents according to the contextualist two-dimensional semantics. If the assessor in one context is John and the assessor in another one is Mary, then under a contextualist semantics the content of Mary's beliefs would be the proposition that roller coasters are fun for Mary, and the content of John's beliefs would be the proposition that roller coasters are fun for John. According to the relativist, this cannot count as disagreement because the contents of their beliefs remain compatible with each other. This relativist standard objection to contextualism will play a role in the second part of this article and should be kept in mind.

However, if the peculiar notion of semantic content is not available because the model is not two-layered, if attitudes are modeled in another way, if incompatible contents are not taken as a necessary condition for disagreement, or if the disagreement is modeled on the basis of other content—such as content expressed by pragmatic presuppositions or any other pragmatically derived, non-literal speech act content—then the differences between parameter-based contextualism and relativism become less critical. Both theories have in common that they model contextual variance in a truth-conditional setting. If a context dependence is linguistically mandated like in the case of the truth-conditions expressed by the use of an indexical, then to some extent these parameterized approaches to context dependence model the contextual resolution process. For example, for *yesterday* the linguistically mandated reference rule is *the day before the day of the utterance*. It generally picks out the right referent and can be formalized in a double-index modal logic in which the date of the utterance is stored in the context parameter (provided that date calculations are available). Likewise, a relativist semantics for predicates like *being fun* and *tasty* states that utterances containing these expressions are true or false relative to an assessor and the suggested interpretation of the semantic apparatus is that respective assessors may differ from the speaker of the utterance.

2.2 *Semantic Underdetermination and Interpretation as Inference*

Although it is adequate for indexicals at a high level of abstraction, modeling other context-sensitive expressions *as if they were* indexicals can be misleading and inadequate. Many forms of linguistic context dependence are pragmatic, and sentence-level content is often semantically underdetermined. For example, there is no linguistic rule in the meaning of *ready* that determines what a person is ready for. The hearer must figure out what the speaker means by an utterance of (1) *He's ready*. Dubbed 'contextuals' by Rast (2014), such expressions require some additional interpretation; in the case of *ready*, there is a syntactically optional complement clause that is not optional from the perspective of sentence-level semantics. This is similar to cases such as *to buy* which also has syntactically optional argument places for a seller and a price, but from a semantic perspective requires these ingredients to differentiate it from other transfer verbs like *to obtain*, *to pay*, and *to borrow*.[6] Other expressions suggest a default interpretation, sometimes very strongly, but neither require it semantically nor syntactically. For instance, (2) *John had breakfast* can be meant to convey that John has had breakfast for the first time in his life, but by default it is taken to express the proposition that John had breakfast at the day of the utterance of (2). Indexicals are also often contextual in this sense in addition to their dependence on the deictic center. For instance, the place denoted by a use of *here* can only be determined on the basis of what has been said so far and assumptions about what the speaker wants to convey, as the place denoted by a use of *here* only needs to contain the deictic center as a mereological part and may otherwise be almost arbitrarily small or large. Depending on what has been said so far and the speaker's intentions, a use of *here* may be intended to convey locations such as *here in this box (where the speaker is crouching)*, *here in this room*, *here in this building*, *here in this city*, *here in this country*, *here on this continent*, *here on this planet*, and *here in this part of the Milky Way*.

Essentially three approaches have been proposed to deal with these forms of partly conventionalized, yet ultimately pragmatic context dependence. According to Bach (2005), utterances often express only propositional radicals by virtue of conventionalized meaning provided by a shared lexicon. What the speaker meant needs to be inferred from these truth-conditionally incomplete representations by a Gricean interpretation process. Rast (2014) suggests a variation of this approach that models the missing contextual factors as open argument places over which one may existentially quantify to obtain a minimal form of content. For example, the 'existential completion' of (1) is *John is ready for something*. Based on such

[6] See Jackendoff (1987, p. 381/2), Jackendoff (1990, pp. 189–194).

representations, abductive inference may yield more specific content such as *John is ready to call a cab*. This inference is derived from what has been said so far, from the topic of the conversation and question under discussion, from the interpreter's assumptions about what the speaker believes, and from general common-sense world knowledge. While the mechanisms laid out by Rast (2010, 2014) are very limited, the approach in general is based on the idea of considering interpretation as an inference to the best explanation (IBE). Relevance theory of Sperber and Wilson (1986, 2004) is a third, psychologically motivated approach to interpretation. It is based on bounded rationality. Hearers draw inferences about what the speaker wants to convey but this process competes with economy constraints. As long as logical and set-theoretic representations of semantic content are used, these three approaches can be linked up with the modeling of pragmatic context, common ground, and linguistic score-keeping at discourse level such as Stalnaker (1978), Lewis (1979a), Barwise and Perry (1983), Stokhof and Groenendijk (1991), Kamp and Reyle (1993), Asher and Lascarides (2003), and Ginzburg (2012).

The key to making any of these approaches fruitful is to represent semantic underdetermination of conventionalized meaning in a way that allows the interpreter to infer what the speaker meant based on existing beliefs about the speaker, what has been said so far, the common ground, general world knowledge, and knowledge about the particular communication situation. The approaches primarily differ in the extent to which they are motivated from empirical psychology. Relevance theory strives for empirical adequacy, whereas the Gricean model describes ideal communication situations and ideal interpretation. The IBE approach's degree of idealization lies in-between. It is based on broadly-conceived logical inference mechanisms from graded belief representations of common-sense ontologies and situational knowledge. All three approaches can be adopted for varying assumptions about the degree of conventionalization of meaning in a shared lexicon.

However, existential completions and Bach's propositional skeletons have to rely on mechanisms that allow for a finite number of existing argument slots to be 'filled in' by the interpretation process. Radical contextualists like Recanati (2004a) do not believe that such mechanisms suffice in general to adequately describe linguistic context dependence because they might not capture creative and poetic language use. Moderate contextualists in turn consider radical contextualism too general and unconstrained since pragmatic modulation functions can, in theory, turn any meaning into another meaning during semantic composition. The position of moderate contextualism is that the number of conventionalized contextual factors—those that are marked in a shared lexicon by semantic argument structures of words—may be large and require a decent amount of sophisticated semantic analysis, yet their number is ultimately finite. Likewise, it is stipulated that the number of rule-governed, broadly-conceived linguistically regulated pragmatic interpretation patterns is also finite.

Speakers and interpreters may occasionally allow contextual shortcuts whose understanding requires general intelligence instead of fixed, rule-based mechanisms. For example, a polite speaker of Japanese may leave out almost any part of speech. Understanding such an utterance and the meaning of not verbalizing part of the speech requires more than just linguistic skills and knowledge, and it is doubtful whether an inference to the best explanation mechanism could adequately explain such cases in sufficient detail in a rule-based manner. However, the existence of such phenomena does not speak against moderate contextualism in the same sense as not understanding someone's explanation of a mathematical problem does not speak against semantics. Understanding an utterance often requires intelligent reasoning that goes far beyond of a speaker's linguistic competence and what can reasonably be expected to be dissected by linguistic theorizing.

The problem of linguistic context dependence is thus principally solvable from the perspective of moderate contextualism. The challenges are in the detail, such as how to find an adequate semantic representation that allows for fruitful descriptions of the inferences that take place when a hearer interprets an utterance, systematic ways of cataloging a language's context-dependent expressions, and how to describe and model these inferences at a desired level of idealization. However, another potential source of context dependence is neither modeled by parameterized modal logics nor by the above mentioned inferential approaches: the possible dependence of meanings and concepts on background theories, opinions, and world views. This context dependence is the subject of the remainder of this article.

3 The Problems of Theory Dependence

Recently, there has been a renewed interest in philosophical aspects of discussions about word meaning. Plunkett and Sundell (2013, 2019) and Plunkett (2015) have argued that disputes are often implicitly about the meaning of words, the adequacy of using words in context, and the appropriateness of contextual norms. If some such disputes concern word meaning, then one may ask how speakers can understand each other if they presume different word meanings from the start. If, in turn, two speakers defend different theories about a particular topic and these theories characterize or define the word under dispute in different ways, then this leads to various *problems of theory dependence*. The topic has a long tradition in analytic philosophy. The role that theory dependence plays for lexical meaning is crucial for assessing Moore's thesis of 'good' as a primitive and the Paradox of Analysis (Moore, 1903), as well as for a later debate between Quine and Carnap about the internal/external distinction of theories and the notion of analyticity in studies by Carnap (1950) and Quine (1960, 1951).

3.1 The Problems

In what follows, the word *theory* shall be understood in the broadest possible sense as including all kinds of nonscientific beliefs, opinions, and world views in addition to scientific theories, approaches, models, and hypotheses. Given that broad understanding, the problems of theory dependence may be summarized as follows. Every theory either directly defines the meanings of words mentioned in it or indirectly characterizes the meanings of words used to formulate it by law-like statements in which those words are used *or* mentioned. Therefore, a definitional account of the meaning of a word central to a theory is directly or indirectly restricted by that theory. So in the context of two different theories, the meanings of words that are central to those theories are restricted in different ways, and, in the worst case, cannot even mean the same because those theories define or indirectly characterize their meanings in different ways.[7]

For example, if it follows from a physical theory that atoms can be split, then an adequate characterization of the meaning of the word *atom* cannot attribute the property of being indivisible to atoms. As another example, Arianism is the Christian belief that Jesus (God the Son) is not co-eternal with God the Father. Someone who believes this doctrine cannot at the same time believe in the trinity, that God the Father, God the Son, and the Holy Spirit are of the same essence. The Arian doctrine thus affects the possible theological characterizations of both *Jesus* and *God*. Historically, the conflict between Arians and Trinitarians led to persecution and violent clashes during the 4th Century AD, and ultimately the official church position was to declare Arianism a heresy. As a third example, consider competing theories of social institutions. According to Searle (1995, 2005), "...an institution is any system of constitutive rules of the form *X counts as Y in C*" (Searle, 2005, p. 10). In contrast to this, Guala (2016) argues that institutions are systems of regulative rules that lead to game-theoretic equilibria. According to Searle, this thesis is incompatible with his definition because, in his account, constitutive rules cannot be reduced to regulative rules. If Searle is right, then *institution* cannot mean the same in both theories.

If the meaning of a central word differs from theory to theory, or at least possible ways of understanding its meaning are restricted in different and sometimes mutually incompatible manners, then two follow-up problems occur. First, it is no longer clear how two competing theories can be about the same topic. For example, why would a theory according to which atoms are indivisible be about what we nowadays call atoms? Related to this, if two agents endorse two different theories A and B and talk about a term central to those theories, then it is no longer clear how they disagree. The problem is the same as in the relativist critique of contextualism. If two interlocutor's beliefs are such that a certain word has a different meaning, because they endorse different explicit definitions of it or their beliefs characterize its meaning in substantially different ways, then the semantic contents of their beliefs also differ. So why do they not just talk at cross purposes?

Semantic externalists may reply to these worries that only defending a theory or having opinions cannot directly influence public language meaning. According

---

[7] Since the discussion in what follows mostly concerns lexical meaning, *word* is used for the linguistic entities under consideration. These are usually nouns (general terms) in examples, but for brevity *word* is also used in a looser sense as a shortcut for linguistic expressions in general. This may include compound nouns, nouns with participial phrases, noninflected verb phrases, and phraseologisms, for instance.

to the most extreme form of externalism, there is no influence at all. The noun *atom* stands for atoms. Whatever theory of atoms we build and whatever beliefs we hold about atoms does not influence what atoms are. The problem with this view is that it conflates word extension with meaning and consequently does not explain meaning change at all. The meaning of *atom* could only change if atoms change, yet it seems that this meaning has changed over the past centuries. A more realistic form of externalism by Cappelen (2018) acknowledges that word meanings change over time, but not fast and not in a way that is under our control. Instead, meaning change is governed by hard to understand, long-term processes within a large speaker community, based on slowly changing patterns of use. These changes may be triggered by changing world views, theories and opinions of all kind that come to be believed by larger groups of speakers, but not merely by discussions between individual speakers.

This *lack of direct control thesis* is a valid point about public language meaning. However, it does not touch the problem's core. Surely, some sort of meanings are discussed in an explicitly metalinguistic dispute in which words are mentioned. If so, then at least some implicit metalinguistic disputes discussed by Plunkett and Sundell (2013) also have to concern word meanings. After all, any such implicit dispute could be turned into an explicit one at any time simply by mentioning the disputed word instead of (seemingly) using it. Maybe the meanings in such disputes are not meanings of public language expressions, and instead the underlying concepts or the meanings of words of idiolects and sociolects change. For instance, Ludlow (2014) argues with many examples that interlocutors adapt their 'microlanguages' to each other in conversations. So even if one does not buy into the theory dependence problem as a thesis about public language, the problem remains at the level of idiolects and concept systems that differ between speakers, whether or not these coincide with public language.

To illustrate this point, consider two early 19th Century physicists discussing and disagreeing about two wave theories of light that are both derived from Augustin Fresnel's theory of luminiferous aether but differ in various details. Neither the correctness of their theories nor the public language meaning of *aether* should have a substantial bearing on the meanings they associate with the word in the context of this discussion. It remains a problem to explain how they disagree about the same topic and why they are not just talking at cross purposes, if they indirectly characterize aether differently or even use different explicit definitions of *aether*. Likewise, consider two ancient fishers discussing whales. Both agree that whales are fish. One of them argues that they are the largest fish of the sea and being the largest fish of the sea is the whale's defining feature. The other one disagrees and claims to have seen larger fish; he thinks that being a fish with a blowhole is the defining feature of a whale. They have false beliefs about whales, some of which enter their putative definitions, and so their concept systems cannot represent public language meaning from the perspective of semantic externalism. Nevertheless, one might ask how their disagreement can be spelled out in terms of these flawed concepts, given that their conceptual systems differ with respect to the concepts they erroneously associate with the word *whale*.

3.2 *Definitional Meaning Does Not Imply an Epistemic Priority of Analyticity*

A popular reply to the problems of theory dependence is to reject any definitional account of word meaning. In further support of this position, one might first argue based on externalist arguments by Kripke (1972), Putnam (1975), and Burge (1979) that both public language meaning and thought contents are individuated externally by facts of the shared environment. As a classical example, *water* denotes H_2O because it is a natural kind term whose meaning is fixed indexically by virtue of the fact that water is mostly composed of H_2O. Correspondingly, if someone thinks about water, then the contents of that person's thoughts are also individuated externally. As Putnam's Twin Earth example is supposed to show, thinking about water is not the same as thinking about a colorless, odorless, transparent liquid, for instance.

As a bonus, it appears as if such a form of externalism also fared well with Quine's arguments against analyticity. I will argue below that this is not the case but let us consider the argument first. It seems to be very popular. In a definitional theory of concepts or lexical meaning according to which concepts or word meanings are characterized by the theories (in a broad sense) to which these are central, it seems that law-like statements that are taken to be definitory for a word or concept (whether in individual cognition or as a thesis about public language meaning) would make certain statements analytically true that are not. For example, if the property of being the smallest indivisible building blocks of nature with the characteristic properties of chemical elements takes part of a definition of *atom*, then it seems that *Atoms are indivisible* is analytically true. According to Quine (1951) such a notion of analyticity is ill-conceived and hinges on a notion of meaning, which, in turn, circularly presumes analyticity.

Although this sort of externalism may be appropriate for specific words of a public language in a truth-conditional setting, it comes with too many problems as a general theory of lexical meaning. First, in practice, word meanings are not always indirectly characterized but also sometimes defined explicitly. In that case, the meaning of the word under consideration clearly depends on an underlying theory, namely the one that simultaneously lends credibility and adequacy to the definition and uses it. Moreover, there is a gradual scale between the indirect characterization of word meaning and explicit definitions. Often a word is used in ways that amount to defining without making the definition explicit. One may define what *triangle* means more or less precisely, or one may understand it more intuitively based on examples. It is hard to say where the supposed externalist individuation starts and where it ends. Semantic externalists have mostly only provided convincing stories for everyday nouns for empirical objects such as *water*, *tiger*, and *pencil*, and their accounts remain mysterious for words like *democracy*, *dark matter*, *triangle*, and *institution*.

Second, as mentioned above, the rationality of metalinguistic disputes becomes questionable without a definitional approach to word meaning in idiolects (or, theories) and concepts. If these are externally individuated, then why and how could they be disputed? Notice that even though Plunkett and Sundell (2013) argue that metalinguistic disputes can be substantive, some of them are also not substantive. Suppose John and Mary argue about what counts as a chair, and after a while, they agree that stools with only one leg ought not be called *chairs*. If the meaning of *chair* was externally individuated, then this whole discussion would be irrational and pointless. However, although it may be pointless and not substantive, it is clearly rational and concerns the question of which minimal number of legs has an adequately definitory quality for *chair*.

We frequently dispute word meanings and propose various characterizing properties, which are derived from, and relative to, a supporting background theory. It is hard to see how this practice could be based on a systematic error. This does not mean that we should not embrace externalism, it means that we should embrace externalism *and* the theory dependence of word meaning and concepts. The underlying theories are hopefully about reality and not just about figments of our mind. Nevertheless, within each theory words may get their meaning relative to that theory by indirect characterization or explicit definition.[8] Indexicalist externalism makes sense for a limited number of natural kind terms because the underlying theories are particularly well-confirmed. It does not scale to theories about more contentious topics.

What about the analyticity objection, then? None of Quine's points against analyticity show in my opinion that there is something wrong with a definitional theory of word meaning and concepts. The lesson to learn from Quine (1951) is rather to be careful not to give epistemic priority to any allegedly analytic inference. For even if we appear to arrive at certain conclusions solely by word meaning, this is never the case. From the present point of view, there is no such thing as 'the' meaning of a word. Words get their meanings relative to the theories in which they are used. If such a theory is based on empirical evidence, then whatever we believe in having derived solely on the basis of word meaning hinges on the adequacy and merits of the theory and its supporting evidence. As a pragmatist naturalist, Quine believed that any theory is revisable and needs to be judged on its scientific merits (in proper scientific contexts). Even mathematics is revisable in that sense. From that perspective, seemingly analytic judgments are theory-relative and revisable like any other judgment. If, contrary to this, there was a non-theory dependent word meaning, then analytic judgment could have some epistemic priority. However, according to Quine, any such meaning would be a dubious stipulation and presume an equally dubious notion of analyticity. We cannot attribute any epistemic priority to inferences seemingly derived only on the basis of word meaning because according to the Quine/Duhem Thesis the underlying theories are confirmed or falsified holistically.

From all of this it follows that it is possible to consistently deny the usefulness of analyticity as an epistemic notion without giving up definitional word meaning

[8] By mentioning indirect characterizations and explicit definitions in this way, I do not want to presume that these are unique phenomena. There is not only a gradual transition between them, they are also umbrella terms for many different, yet related practices such as stipulating meaning postulates, operational definitions, definition as abbreviation, definition by example, definition by systems of axioms, providing prototypical information, specification, abstraction, and so forth.

and the thesis that theories characterize the meaning of words that play a central role in them. We may even continue to speak of analytic judgments (although Quine would not endorse this), as long as no special epistemic priority is given to them. For example, it is perfectly fine to contemplate whether *bachelor* means *unmarried man* or whether additional conditions need to be met, and it may even be true that under the first definition every bachelor is an unmarried man and vice versa. Talk like this is fine, as long as one keeps in mind that such considerations tell us nothing about the adequacy of that definition and its underlying theory, about the existence of bachelors and unmarried men, and about what other properties bachelors might have. The truth of the analytic statement hinges on the confirmation or falsification of the supporting theory.

If a complete rejection of explicit definitions and implicit characterizations of meanings is implausible for idiolects and concept systems, and if the theory dependence of word meaning and concepts remains compatible with Quine's arguments against analyticity, then the problems mentioned above cannot be ruled out that easily. When are two theories about the same topic or concern the same central words? How can two speakers advocating competing, mutually incompatible theories or world views be said to disagree and talk about the same things?

4 Tackling the Problems

There are several ways to tackle the problem of topic equality of theories. First, there are good reasons to assume we associate some minimal meanings with expressions that are not necessarily truth-functionally complete and represent 'everyday', common-sense word meanings. Rast (2017b, a) suggests the term *core meaning* for these and contrasts them with *noumenal meaning*, which represents what a word *really* means according to our current best understanding and theorizing. For instance, even speakers in the past who were not in a position to know that water consists of H_2O associated with it the core meaning of being a transparent, drinkable liquid. Likewise, we can recognize animals by the way they look under normal circumstances. The core meaning of *whale* is to look like a whale. So if two people disagree about the noumenal meaning of a word, for example, whether whales should be classified as fish or mammals, they may continue to talk about the same topic as long as they sufficiently agree about the associated truth-conditionally incomplete core meaning. Second, Rast (2020) lays out that words can also be associated with measurement operations. Competing theories are about the same topic if associated measurement operations (which may differ across agents and theories) roughly pick out the same extension.[9] Third, unless a noun is further qualified and distinguished from other uses, the same noun in two different theories A and B is supposed to stand for the same kind of entities in both A and B. Certain words, usually nouns

[9] Without emphasizing measurement operations, Cappelen (2018) also advocates such an extensional notion of topic equality.

plus qualifying adjectives, act as fixed points around which varying theories are constructed. Choosing the same words for such alleged fixed points tells speakers that two theories are supposed to be about the same kind of entities. This nominal topic equality is a fallible stipulation, but supporting theories are fallible, too, and in a sense also mere stipulations.

The most important mechanism is measurement because measuring roughly the same kind of entities warrants topic continuity; the others do not warrant but rather stipulate it. Taken together, these three mechanisms suffice to explain putative and real topic continuity. However, having an account of topic equality does not solve the problem of a potential drift of word meaning and concepts across speakers. How do we understand each other, if our background theories, opinions, and world views differ from each other and influence lexical meaning? I believe that the best answer to this question is twofold. First, as argued in the next section, it is only pressing when global meaning and concept holism is assumed. Instead, we should embrace local meaning and concept holism. Second, at least up to a certain degree we are able to, and *have* to be able to, track and entertain different opinions, world views, and theories without endorsing them. Hence, theory dependence is less of a problem for mutual understanding than one might think at first glance. This topic is addressed in Sect. 4.2.

4.1 The Case for Local Holism

Holism is best understood in opposition to atomism and the arguments that speak against it. A central thesis of semantic atomism is that the meanings of simple words are not generally composed of other words' meanings. A reasonable semantic atomism may acknowledge that there are more complex, morphologically derived words whose meanings are composed out of their parts' meanings. For instance, *consequential* may have a primitive meaning whereas *inconsequential* may have a meaning derived from the former. However, this must be limited to complex words. If the meanings of all words are decomposable into logical combinations of the meanings of other words, then the meaning of every word hinges on the meaning of those other words, which is a form of holism. So the semantic atomist has to assume primitive, non-decomposable meanings, or that simple words have no meaning at all and only serve as syntactic anchoring points in a computational theory of cognition, or—as the more common, externalist response—allow talk about 'meaning' only in a derived sense, for example by assuming that the extensions of simple words individuate their meanings. One form of semantic atomism can be found in works by Fodor (1975, 1987), while Fodor and Lepore (1992) thoroughly discuss arguments against holism without presuming Fodor's contested theory of cognition.

Atomism would provide an elegant solution to the problems of theory dependence if there were not such good counter-arguments against it, whether it concerns public language, idiolects, or concepts. First of all, if the meaning of a word is primitive, then how does it change? This is a generalized form of the earlier argument against indexicalist externalism. Word meanings change within discussions when they are defined explicitly. Consequently, they should also sometimes change when they are characterized implicitly. Likewise, concepts such as the concept of holy trinity can change over time even if they do not match the established current public language word meaning. If they can change over time because our conception of reality changes, then it seems equally reasonable to assume that they can also differ synchronously when different theories of reality are considered, defended, and supposed. There *are* metalinguistic discussions.

Atomists have a hard time explaining such negotiated concept and meaning changes because they do not allow for the logical decompositions of lexical meanings under dispute in metalinguistic discussions. Atomism is also questionable from the perspective of the inferences that can be drawn from word use. Suppose a fixed number of words has a primitive meaning that cannot be further dissected. Suppose α is such a word. This word α will have one set of consequences relative to theory A and may have another set of consequences relative to theory B. Shouldn't at least some such consequences count as an aspect of the word's meaning? It is hard to see how these different consequences could never be the result of different meanings. Another point against atomism is that some seemingly substantive theses can be turned into explicitly metalinguistic theses and vice versa, and the difference between them is only whether speakers quote linguistic material or not; some ways of talking are even in-between the two. Consider the following examples[10]:

(3) a. Every atom is indivisible.
b. Atoms are indivisible.
c. Being an atom entails being indivisible.
d. Being indivisible is a defining feature of atoms.
e. An essential aspect of the meaning of *atom* is that they are indivisible.
f. *atom* means *being a smallest indivisible building block of nature with the characteristic properties of a chemical element.*

Implicit to semantic atomism is the claim that examples like (3a)–(3d) do not concern the meaning of *atom*. Is this really plausible? Although only (3e) and (3f) explicitly mention words, in ordinary conversations the dependence on the natural language is often irrelevant, and all of the above statements characterize atoms in similar ways. In practice, we often define words without mentioning them at all. Even a simple use of a generic like in (3b) can have a 'metalinguistic flavor' in a context where a characterization, explanation, or definition of a word is expected. Neither is an explicitly metalinguistic definition like in (3f) arbitrary, nor does the use/mention distinction clearly indicate whether a word is defined or characterized, or whether a world-level claim is made. This does not mean that the choice between explicit definition and indirect characterization is unimportant or that every law-like statement in which a word is used has a definitory quality for that word. An

[10] Clearly, there are two uses of *atom* in the text. The examples are about physical theories, whereas *atomism* suggests a mereological use. Which one is meant is clear from the context. This is not another example of theory dependence but merely a case of ambiguity.

explicit definition may indicate particular methodological preferences, that it is only conventional or operational and later to be revised, or that a term is theoretical. Nevertheless, semantic atomism presumes a too large divide between the explicit definitions of complex words and the meaning of supposedly primitive words. There is no such gap in practice.

So if we reject semantic atomism, how can holism deal with the problem of theory dependence? To answer this question, holism has to be characterized in more detail. First, holism can apply to words in public language. In this view, the meaning of a public language word depends on and is partially constituted by the meanings of other public language words; if the meaning of a word α changes, then the meanings of words change that are partly constituted by the meaning of α. This is semantic holism as the counterpart to semantic atomism. An analogous thesis may be formulated for idiolects and sociolects, which we may call meaning holism in general. Finally, concept holism concerns individual agents' concept systems, where a concept is a meaning-like representation that is not necessarily associated with a word. For instance, a sculptor may have a concept for a particular shape, may be able to recognize it and use it while sculpting, without naming it and without there being a name for it in public language. Concept holism is the thesis that a concept c changes whenever other concepts change that partially constitute c.

Since some externalists deny that concepts exist, and it is also controversial whether public language meaning can change in the way relevant for the theory-dependence problem, I will focus in the following discussion on idiolectal meaning holism and for simplicity sometimes abbreviate it as *holism*. What can be said about this type of holism can also be said about the others. The focus shall also be on meaning change. Most of what can be said about meaning change can be transferred to the case when two agents disagree. The main difference between the two cases is that two agents may also differ in *other* beliefs that are peripheral or unimportant to the theories under consideration. This complicates matters, but not in a way relevant for what follows.

Consider the theory change scenario. An agent endorses a theory A but then for some reason starts to suspect that A is not the right theory and endorses theory B instead. Provided that A and B are not compatible with each other (they cannot simultaneously be true), the agent first has to retract A from his total belief base K and then integrate B into K. Although there are well-established formal theories for modeling these kind of processes such as AGM belief revision (Alchourrón et al., 1985) and KM update (Katsuno and Mendelzon, 1992), realistically speaking only some aspects of theory change can be modeled formally. The process is inherently creative and involves theory discovery of B. The retraction of A might not be minimal, it may be based on a shift in perspective and a massive re-evaluation of more beliefs in K than merely those required for A. Therefore, we cannot assume that those beliefs in K which are prima facie independent from A remain constant during such a revision. In any case, however, some statements involving words used in both K plus A and K revised by B will likely have different consequences before and after revision. If the effect can be isolated to only one word, then this word's inferential meaning has changed. Whether we are willing to say that its purported idiolectal meaning has also changed depends on whether the law-like statements responsible for the inferential meaning change count as attributing a definitory quality or are of a more accidental character.

For the current purposes, two versions of holism have to be distinguished. According to global holism, whenever the meaning of one word changes relative to a belief base, then the meaning of *all other* words changes, too. Analogously, in the two-agent case, every agent associates a slightly different meaning with each word in their idiolect, or they have slightly different concept systems. Why would this be the case? Generally, the idea behind this position is that words are only meaningful in larger units like sentences and discourse fragments, and that their meaning has been learned and is indirectly constituted by the network of law-like semantic relations and constraints between words. As Lepore and Fodor (1993) put it, "…meaning holism says that what the word 'dog' means in your mouth depends on the *totality* of your beliefs about dogs, including, therefore, your beliefs about whether Lincoln owned one. It seems to follow that you and I mean different things when we say 'dog'; hence that if you say 'dogs can fly' and I say 'dogs can't fly' we aren't disagreeing." (Lepore and Fodor, 1993, p. 638) Correspondingly, each concept in a concept system depends on other concepts in this view, and no two agents can learn and internalize exactly the same concept.

Arguments by Davidson (1967, 1973) are sometimes advanced in support of global holism. As a twist on Quine (1960), Davidson suggests to define truth-conditions for a language by Tarski-sentences of the form *'S' is true in language L iff. T*, where *T* specifies the truth-conditions for the sentence mentioned on the condition's left hand side.[11] In a radical interpretation situation, when a speaker of *L* would utter a sentence *S* and we have to figure out what this utterance means, we have to apply the Principle of Charity and assume that this speaker's beliefs are mostly true.[12] Based on this assumption, we can make sense of another person's rationality in a radical interpretation situation by attributing beliefs and desires to that person and associating them with our assumptions about what their utterances mean. However, the Principle of Charity can only get one so far. Since the beliefs of the interpreter and the interpreted person only roughly converge, understanding of the other person's language will only ever be a rough approximation in a radical interpretation situation. Moreover, since beliefs depend on each other just like the statements of a theory, the recovery of the other person's language in a radical interpretation situation seems to imply global holism even when the Principle of Charity is applied.

In contrast to global holism, local holism is the position that a meaning change of a word may trigger some finitely many meaning changes but that this does not imply that the whole idiolect changes. For example, suppose John calls any apple or pear an *apple*. He has a persistent misconception that pears were once similar to peaches but have long gone extinct. John later learns about pears and how to distinguish apples and pears by taste and shape like most speakers of English. The

[11] See Davidson (1973, p. 318).

[12] See Davidson (1967, pp. 312–313), cf. Davidson (1973, pp.323–324).

change affects *pear* and *apple* in John's idiolect, as well as the concepts of being a pear and being an apple. The incorrect pear concept is eliminated, and a more adequate one is internalized. In terms of theories, we may say that John learns better pear and apple theories. According to local holism, this change might affect related concepts and word meanings such as the meaning of *apple pie* (it's not the same as a pear pie), *juice* (pear juice exists), and *fruit* (pears are fruits, they are not extinct, and taste such-and-such). It will not affect every other word, though. For example, John's idiolectal meanings of *and*, *relation*, *democracy*, *dog*, and *greater than* are not affected. They are not just affected in a barely noticeable and neglectable way. They are not affected at all.

Theory dependence is a huge problem for the global holist. Since people have different beliefs about all kinds of topics, and every difference of beliefs leads to differences in idiolects and corresponding concept systems, even with a generous application of the Principle of Charity two interlocutors will likely talk at cross purposes and fail to fully understand each other. The farther the theories they endorse are apart from each other, the less they understand each other when discussing a topic common to those theories. So it seems at first glance. On a closer look, however, it turns out that the arguments for global holism are relatively weak. There are good reasons for rejecting global holism and accepting local holism instead.

Fodor and Lepore (1992) lay out in detail why many of the arguments *for* global holism based on Quine (1960) and Davidson (1967, 1973) are not conclusive. One of their points is that language learners and field linguists are never in a radical interpretation situation.[13] The environment is shared, the agents' cognition works in similar ways, and inadequate interpretations of utterances can be corrected over time. Speakers also share common features of their perceptions. For instance, a child learning the word *rabbit* from watching a living rabbit sees a rabbit and not rabbit slices like in the famous Gavagai example of Quine (1960). Radical interpretation scenarios are radically skeptic from an epistemic point of view, but successful language learners are not and cannot be radical skeptics. Other arguments by Fodor and Lepore (1992) also undermine the support that radical interpretation and the Quine/Duhem thesis seem to lend to global holism, but addressing them here would go beyond the scope of this contribution.

There is one positive logical argument against global holism that Fodor and Lepore do not endorse. When speakers adapt idiolectal meaning and related concepts to one another, only word meanings and concepts *central* to a given topic need to be revised. For example, children who learn what pears are and how they differ from apples only need to revise fruit- and nutrition-related concepts. There is no need or reason in such a case to revise unrelated concepts like being a tire or being a tiger. There are essentially two reasons for this locality of revisions and why centrality is not an arbitrary stipulation in this context.

On the one hand, the common-sense ontologies encoded by concept systems are hierarchical. An upper ontology represents very abstract concepts such as relations, counting, mereological notions, physical versus abstract objects, physical movement,

[13] See Fodor and Lepore (1992, pp. 73–80).

processes, information transfer, and so forth. In contrast, a lower ontology represents specific knowledge about the world. A change of the beliefs that constitute the lower ontology is unlikely to require a change of beliefs that constitute the upper ontology in a reasonable account of belief revision and theory discovery. On the other hand, theories about specific topics, identified by associated measurement operations, are discernible from other theories and the more general ontology. A lower ontology is divided vertically into parts that are mostly or entirely independent of each other from a logical perspective. For instance, there are many (onto-)logical relations between tires and pears and these objects can interact in many ways, but beliefs about these relations are regulated by the upper ontology. They might be based on the fact that both are types of manipulable physical objects that can be carried and moved, for example. A revision of the pear concept by integrating new pear and fruit theories does not have to trigger a revision of the tire concept, and likewise for the idiolectal meanings of *pear* and *tire*. So even though there are logical relations between pears and tires, neither is *pear* central to the tire theory nor, vice versa, *tire* central as a term in pear theories.

Although developing a full-fledged account of centrality as a measure of the nearness of terms to the measurable topics of a theory would be a major undertaking, there can be no doubt that *pear* is not just psychologically but also logically nearer to *apple* than *tire* is. Words whose meanings are directly related to each other by law-like statements at the same level of ontological specificity and within the same theory with measurable topics are close to each other, for instance, whereas words whose meanings are characterized in a theory about other measurable topics and whose meanings are only related to each other via law-like statements of the upper ontology (less specific, more abstract) are more distant from each other.[14]

Anyone who accepts these kind of examples and the reasoning behind them ought to be wary about Quine's dictum that "[t]he unit of empirical significance is the whole of science" (Quine, 1951, p. 39). Individual theories can be confirmed and rejected without revising other theories, let alone all of science, and changing individual theories need not trigger revisions of the upper ontology that supports them. Confirmation holism is only local. As a consequence of this position, under the indirect meaning characterization thesis and the assumption that word meaning is (at least sometimes) definitional, it follows that a change of idiolectal meanings and concepts only affects words and concepts closely related to the one that changes. Further changes may be triggered, but these are usually local, too, since the underlying common-sense ontology is divided vertically and horizontally.

[14] One approach would be to base the account of centrality on a good account of theory revision, which, in turn, would have to take into account theories and their associated measurement operations as units when modeling epistemic entrenchment. Since there is no non-psychological 'logic of theory discovery', however, even such an elaborate approach would remain limited.

4.2 Tracking Theories

So far, we have talked about beliefs and endorsing theories, and the concept systems and ontologies related to these beliefs. But how are these notions related to each other? The way I understand beliefs in this article, these are types of attitudes that we attribute to agents de re, using belief ascriptions of public language. For instance, John who calls both apples and pears *apples* does not believe de re that pears do not exist. Maybe he believes de dicto that pears don't exist because he is disposed to utter sentences like *Pears don't exist any longer*, but no corresponding de re belief can be attributed.

In contrast to de re belief, concepts can be described using public language but do not necessarily correspond to words of public language or an agent's idiolect. For instance, when John considers every pear an apple he possesses a primitive apple-pear concept. If he uses the word *apple* to refer to apple-pears (i.e., apples *or* pears), then the idiolectal meaning of *apple* is for him: being an apple-pear.

Other concepts regulate the relations between concepts and, taken together with the concepts they regulate, constitute a concept system. The ontology that corresponds to such a concept system can be described by the embedded sentences we would use when ascribing corresponding de re beliefs. Hence, in this way of talking, endorsing a theory can be described as the revision of existing beliefs by a theory. The point of the previous section was that even though this process may affect more beliefs than just those constituting the theory that is replaced, from a logical point of view the ontology constituted by corresponding background beliefs is vertically and horizontally divided into parts, and theory revisions will not generally affect all of an agent's beliefs. An indirect consequence of this view is that an agent's concept system is also usually only affected locally. Thus, we should opt for local holism and the problem of theory dependence becomes less pressing.

However, this picture is not complete. Talking about beliefs can only be understood as a first approximation. We not only endorse theories, we also consider them, suppose them, and deal with them in many other ways that do not imply that an agent fully believes them. This is another important point for explaining meaning disputes.[15]

Consider two agents having a dispute that indirectly concerns word meaning. Speaker g endorses theory A and h defends theory B, which are both about the same topic with associated measurement operations. Regardless of what has been said in the previous section, under the local holism thesis the two speakers will misunderstand each other if the idiolectal meaning of a term α differs relative to g's belief base plus A from the idiolectal meaning of α according to h's belief base plus B, provided that α is central in one of the theories and some of the inferential meanings in which α differ between h and g have a definitory quality for at least one of the agents, i.e., the speaker considers them constitutive for what it means to

[15] Endorsing a theory in this context is understood roughly as believing what the theory states. There could be attitudes other than belief at play, for example, a true-holding attitude with less epistemic entrenchment than belief. As long as it can be attributed de re, this does not impact my position. Resorting to belief should be taken as a simplification.

be (rightly) called an α. How can the speakers then understand each other? Are they not still only talking at cross purposes insofar as α is concerned?

As hinted above, the answer to this question is that a theory does not have to be endorsed to create mutual understanding. Instead, speakers can consider a theory, and this ability suffices to rule out talk at cross purposes under ideal circumstances. For g to understand theory B, she only needs to consider B's merits on the basis of a hypothetical revision with B, but need not integrate theory B and thereby give up A. Instead of endorsing other persons' theories, we *track* them. However, it does not stop there. We may also track a theory by hypothetically revising by this theory what we assume that the person(s) who defend the theory believe, i.e., based on our assumptions about the proponents' concept systems.

Even this description is incomplete. As even a cursory look at our practices reveals, humans have the astonishing ability to compartmentalize theories altogether, independently of whether these are endorsed or not. Even if a revision is not hypothetical and a new theory is endorsed, this does not necessarily induce a change of the remaining common-sense ontology. For example, physical theorizing could have triggered radical changes in the everyday concept systems of physicists. After all, time and space are no longer constant in modern physics, and quantum mechanics also has radical implications about the macrophysical world. Nevertheless, the radically different ways modern physics looks at nature have probably not changed phycisists' common-sense ontologies in any substantial way. Instead, they can designate an area of 'theoretical physics' in which physical theories revise the background ontology, but this area is compartmentalized from the original common-sense ontology that stays in place.

This compartmentalization is necessary and inevitable. First, sometimes two theories are worth endorsing even though there are good reasons for believing that they are incompatible. This point is particularly important since two theories can be incompatible with each other even when they are not about the same topic. As a typical example, many theoretical physicists believe that Einstein's Theory of General Relativity and Quantum Mechanics are not compatible with each other and that some more general theory will replace them in the future. Nevertheless, it is perfectly rational to endorse both theories at the same time. They are well-confirmed even though they cannot be combined easily. In this case, physicists endorse both of them until a better, more unifying framework has been found. Similarly, it would be incorrect to claim that physicists do not endorse Newtonian Mechanics; they do, they are merely aware that it does not provide accurate descriptions of objects moving at near light speed and does also not describe the behavior of extremely small 'objects.' Physicists endorse Newtonian Mechanics although Relativistic Mechanics can replace it. It is not necessary to use the more complicated relativistic formulas for macrophysical objects at very low speeds.

Second, it is not irrational to retain information, even when it does not meet the requirements for being fully integrated into one's belief system. Whether it is worth and rational to retain a new theory (opinion, world view) may be a complicated matter, but the decisive criterion cannot be that it meets the requirements for being endorsed. Otherwise, learning inductively by corroborating evidence from different information sources would be impossible, for instance, in a scenario where each information source individually does not meet those requirements. A rational agent needs to keep track of theories and supporting evidence that do not meet the standards for being endorsed.

Third, the standards for endorsing theories are also context-sensitive. For instance, it is rational for an agent to endorse a scientifically well-confirmed theory if that agent is not very knowledgeable about the theory's subject matter and domain. Identifying experts and relying on them is an important skill for any rational epistemic agent, since learning is largely a social process. However, it is equally rational for another agent, who is knowledgeable about the theory's subject matter and domain, not to endorse the same theory and merely to consider it. A theory worth endorsing on one occasion may only be worth being aware of in a more skeptical context. It can even be rational to consider or track a theory in one context and completely ignore it in another. For instance, a certain amount of knowledge about religious texts and opinions is needed to understand the world views and motives of religious fanatics. This does not mean that the same knowledge needs to have any influence on one's own world views or needs to play a role in the evaluation of scientific evidence.

Tracking theories means keeping their origins and sources in mind and knowing them well enough for understanding others; it does not imply endorsing them in any way. Hence, the contextualist objection of talking at cross purposes is ill-conceived for theory-based disagreement. Theories neither need to be compatible with each other nor do they need to be co-tenable, believed, or endorsed by speakers in order for them to disagree about them. It is entirely possible to rationally disagree about an aspect of a theory, opinion, or world view that neither of the interlocutors endorses.

When we take into account this ability to compartmentalize and track theories, it is reasonable to also assume that we can deal with the semantic effects of theory dependence under the assumption of local holism. Take the much-discussed *Secretariat is an athlete* example from Ludlow (2008, 2014), for instance. Secretariat was a famous racing horse. Suppose John believes that *athlete* can only be used to denote humans. In his view, part of the definitory properties of athletes is being human. Mary disagrees with him and believes that horses can be athletes, too. Even though a prototypical athlete might be human, only physical prowess and success in competitions are defining characteristics for athletes. Their disagreement is discussed by Plunkett and Sundell (2013) as a typical case of an (implicit) metalinguistic dispute.

Nevertheless, Mary and John can understand each other if they manage to keep track of each other's opinions about athletes. If each of them presupposes a different meaning of *athlete* in their idiolect, this does not automatically lead to misunderstanding. It only leads to a linguistic misunderstanding when one of them does not know the other's opinions about athletes well enough, and does not keep track of the other's athlete theory. Normally, however, speakers are able to keep track of other theories at least to some extent, which includes an ability to recognize the effects of local holism on possible candidates for word meaning. To what extent? From an idealized modeling perspective, precisely to the extent to which their model of the other's theory about a given topic in the conversation and words central to it matches the other's actual theory.

5 Summary

This article started with an overview of linguistic context dependence. I argued that parameterized contexts do not suffice to represent linguistic context dependence adequately. However, combined with the parameterized context dependence of indexicals and tenses, regarding interpretation as an inference to (usually) more specific semantic contents while presuming the semantic underdetermination thesis leads to a fairly complete account of linguistic context dependence. This is possible only if moderate contextualism is the right position. I have suggested that this is the case because the number of linguistically-regulated, rule-based context-sensitive phenomena in natural languages is finite.

The problem left open by such an approach is the theory dependence of lexical meaning. This theory dependence does not need to occur at the level of public language meaning to become a problem; it also creates difficulties for explaining mutual understanding at the level of idiolectal meaning and concept systems. Although the easiest way to address the problem is by rejecting a definitional approach to meaning and concepts, I have rejected this solution because it creates numerous problems. It does not match the reality of overt and implicit metalinguistic discussions and is forced to draw an inadequately sharp divide between definitions and externally individuated meanings. Semantic atomism can evade this problem, but may not be plausible for other reasons. Especially the meanings of words for abstract objects, complex verb phrases, and compound nouns are hard to explain from the perspective of a stringent semantic atomism. However, the problem of theory dependence remains pressing for semantic holism.

In the final part of the article, it was argued that solving the problem of theory dependence requires two theoretical commitments. First, global holism needs to be given up in favor of the overall more plausible local holism. Since common-sense ontologies are horizontally and vertically divided into parts, the effects of theories on idiolectal meaning are often isolated to these parts. Endorsing or rejecting a theory does not influence all concepts or the meaning of all words in an idiolect but only a select few central to the theory. Second, theory representations of rational agents need to be compartmentalized, as rational agents need to track theories incompatible with their beliefs without endorsing them. If this is true, then it is also reasonable to assume that the requirements of rational theory compartmentalization allow speakers to compartmentalize the effects of theory dependence on concept systems and meaning. An ideal rational speaker would be able to keep track of all theories and information sources in a way that takes into account shifts in lexical meaning due to the different law-like statements with definitory qualities for concepts and expressions that these theories support. Humans are not ideally rational in this sense, of course, yet it is reasonable to assume that they can keep track of someone else's definitions and characterizations in the same way they can keep track of their own theories. Sometimes they succeed, and then there is no misunderstanding, and sometimes they fail, and there will be talking at cross purposes.

Assembling the parts of the article leads to the following picture. Hearers interpret semantically incomplete content by drawing inferences from it based on existing epistemic representations. A model for this process may be Gricean, an inference to the best explanation, or a more psychological account like Relevance Theory. The belief base relative to which utterances are interpreted need not solely consist of the interpreter's beliefs and endorsed theories. An interpretation may also be based on assumptions about the respective speaker's beliefs and theories (opinions, world views). Endorsing what the speaker said, as well as the underlying theories that need to be presumed in order for the utterance to be believable to be true, is then a second step. This step may require a revision of the interpreter's theories about the topic and a corresponding change of the interpreter's idiolect, adapting to the speaker's idiolect in that respect. However, both world-level and metalinguistic disagreement is possible without this second step.

References

Alchourrón, C. E., Gaerdenfors, P., & Makinson, D. (1985). On the logic of theory change: partial meet contraction and revision functions. *Journal of Symbolic Logic, 50*, 510–530.

Asher, N., & Lascarides, A. (2003). *Logics of conversation*. Cambridge University Press.

Bach, K. (2005). Context ex machina. In Z. G. Szabó (Ed.), *Semantics versus pragmatics, (page 16–44)*. Oxford UP.

Bar-Hillel, Y. (1954). *Indexical expressions. MIND, 63*, 359–76.

Barwise, J., & Perry, J. (1983). *Situations and attitudes*. MIT Press.

Burge, T. (1979). Individualism and the mental. *Midwest Studies in Philosophy IV*, page 73–121.

Burks, A. (1949). Icon, index and symbol. *Philosophical and Phenomenological Research, 9*(4), 673–689.

Cappelen, H. (2018). *Fixing language*. Oxford University Press.

Cappelen, H., & Lepore, E. (2004). *Insensitive semantics*. Blackwell.

Cappelen, H., & Lepore, E. (2005). A tall tale: In defence of semantic minimalism and speech act pluralism. In G. Preyer & G. Peter (Eds.), *Contextualism in philosophy: Knowledge, meaning, and truth*. Oxford University Press.

Carnap, R. (1950). Empiricism, semantics, and ontology. *Revue Internationale de Philosophie, 4*(4), 20–40.

Castañeda, H.-N. (1967). Indicators and quasi-indicators. *American Philosophical Quarterly, 4*, 85–100.

Castañeda, H.-N. (1989a). Direct reference, the semantics of thinking, and guise theory (constructive reflections on david kaplan's theory of indexical reference). In J. Almog, J. Perry, & H. Wettstein (Eds.), *Themes from Kaplan, page 105–144*. Oxford: Oxford University Press.

Castañeda, H.-N. (1989b). *Thinking, language, experience*. Minneapolis: University of Minnesota Press.

Chalmers, D. (2006). Two-dimensional semantics. In E. Lepore & C. B. Smith (Eds.), *Oxford handbook of the philosophy of language, chapter 23, page 574–606*. Oxford University Press.

Cresswell, M. J. (1996). *Semantic indexicality*. Kluwer.

Cresswell, M. J., & von Stechow, A. (1982). De re belief generalized. *Linguistics and Philosophy, 5*(4), 503–535.

Davidson, D. (1967). Truth and meaning. *Synthese, 17*(3), 304–323.

Davidson, D. (1973). Radical interpretation. *Dialectica, 27*(1973), 314–328.

de Sa, D. L. (2008). Presuppositions of commonality: An indexical relativist account of disagreement. In M. Garcia-Carpintero & M. Kölbel (Eds.), *Relative Truth*. Oxford University Press.
de Sa, D. L. (2009). Relativizing utterance-truth? *Synthese, 170*(1), 1–5.
Fillmore, C. J. (1997). *Lectures on Deixis*. CSLI Publ., Stanford. Revised version; first publ. 1972 based on a lecture 1971 in Santa Cruz.
Fodor, J. (1975). *The language of thought*. Harvard University Press.
Fodor, J. (1987). *Psychosemantics*. MIT Press.
Fodor, J., & Lepore, E. (1992). *Holism: A shopper's guide*. Oxford/Cambridge: Blackwell Publishers.
Ginzburg, J. (2012). *The interactive stance*. Oxford University Press.
Guala, F. (2016). *Understanding institutions: The science and philosophy of living together*. Princeton University Press.
Jackendoff, R. (1987). The status of thematic relations in linguistic theory. *Linguistic Inquiry, 18*(3), 369–411.
Jackendoff, R. (1990). *Semantic structures*. MIT Press.
Kamp, H., & Reyle, U. (1993). *From discourse to logic*. Kluwer.
Kaplan, D. (1989). Demonstratives: An essay on the semantics, logic, metaphysics, and epistemology of demonstratives and other indexicals. In J. Almog, J. Perry, & H. Wettstein (Eds.), *Themes from Kaplan, (page 481–564)*. Oxford University Press.
Katsuno, H., & Mendelzon, A. O. (1992). On the difference between updating a knowledge base and revising it. In P. Gärdenfors (Ed.), *Belief Revision, (page 183–203)*. Cambridge: Cambridge University Press.
Kripke, S. A. (1972). Naming and necessity. In G. Harman & D. Davidson (Eds.), *Semantics of natural language*, (page 253–355). D. Reidel Publishing Co.
Lasersohn, P. (2005). Context dependence, disagreement, and predicates of personal taste. *Linguistics and Philosophy, 28*(6), 643–686.
Lepore, E., & Fodor, J. (1993). Précis of holism: A shopper's guide. *Philosophy and Phenomenological Research, LII, 1*(3), 637–640.
Levinson, S. C. (1983). *Pragmatics*. Cambridge University Press.
Lewis, D. (1979). Scorekeeping in a language game. *Journal of Philosophical Logic, 8*, 339–359.
Lewis, D. K. (1979). Attitudes De Dicto and De Se. *Philosophical Review, 88*(4), 513–543.
Lewis, D. K. (1980). Index, context and content. In S. Kanger & S. Öhman (Eds.), *Philosophy and grammar*, (page 79–100). D. Reidel Publishing Co.
Ludlow, P. (2008). Cheap contextualism. Philosophical. *Issues, 18*, 104–129.
Ludlow, P. (2014). *Living words*. Oxford University Press.
Lyons, J. (1977). *Semantics* (Vol. 1). Cambridge University Press.
MacFarlane, J. (2007). Relativism and disagreement. *Philosophical Studies, 132*(1), 17–31.
MacFarlane, J. (2008). Truth in the garden of forking paths. In M. Carcía-Carpintero & M. Kölbel (Eds.), *Relative truth*, (page 81–102). Oxford University Press.
MacFarlane, J. (2009). Nonindexical contextualism. *Synthese, 166*, 231–250.
MacFarlane, J. (2012). Relativism. In D. G. Fara & G. Russell (Eds.), *The Routledge companion to the philosophy of language*, (page 132–142). Routledge.
MacFarlane, J. (2014). *Assessment relativity: Relative truth and its applications*. Clarendon Press.
McTaggart, J. E. M. (1908). The unreality of time. *Mind, 18*, 457–74.
Moore, G. E. (1903). *Principia ethica*. Cambridge University Press.
Perry, J. (1977). Frege on demonstratives. *Philosophical Review, 86*, 474–497.
Perry, J. (1979). The problem of the essential indexical. *Noûs, 13*, 3–21.
Perry, J. (1998a). Myself and I. In M. Stamm, (Ed.), *Philosophie in synthetischer absicht*, (page 83–103). Klett-Cotta, Stuttgart. (Festschrift für Dieter Henrich).
Perry, J. (1998b). Rip Van Winkle and other characters. *The European review of analytical philosophy, Volume 2: Cognitive dynamics*, page 13–39.
Perry, J. (2001). *Reference and reflexivity*. CSLI Publications.

Plunkett, D. (2015). Which concepts should we use? Metalinguistic negotiations and the methodology of philosophy. *Inquiry, 58*(7–8), 828–874.

Plunkett, D., & Sundell, T. (2013). Disagreement and the semantics of normative and evaluative terms. *Philosophers' Imprint, 13*(23), 1–37.

Plunkett, D. and Sundell, T. (2019). Metalinguistic negotiation and speaker error. *Inquiry*, page 1–27.

Prior, A. (1967). *Past, present, and future*. Oxford University Press.

Prior, A. N. (1957). *Time and modality*. Oxford University Press.

Prior, A. N. (2003). Changes in events and changes in things. In P. Hasle, P. Øhrstrøm, T. Braüner, & J. Copeland (Eds.), *Papers on time and tense*, chapter 1. Oxford University Press.

Putnam, H. (1975). The meaning of 'Meaning'. In H. Putnam (Ed.), *Mind, language and reality, number 2 in philosophical papers*, page 215–271. Cambridge University Press.

Quine, W. V. O. (1951). Two dogmas of empiricism. *The Philosophical Review, 60*(1), 20–43.

Quine, W. V. O. (1960). *Word and object*. Wiley.

Rast, E. (2017). Metalinguistic value disagreement. *Studia Semiotyczne, XXX, 1*(2), 139–159.

Rast, E. (2017b). Value disagreement and two aspects of meaning. *Croatian Journal of Philosophy*, 17(51(3)):399–430.

Rast, E. (2020). The theory theory of metalinguistic disputes. *Mind and Language*, forthcoming (publ. online first):1–19.

Rast, E. H. (2010). Plausibility revision in higher-order logic with an application in two-dimensional semantics. In X. Arrazola, & M. Ponte, (Ed.), *Proceedings of the logKCA-10—Proceedings of the Second ILCLI International Workshop on Logic and Philosophy of Knowledge, Communication and Action*, Donostia/San Sebastian. ILCLI, University of the Basque Country Press.

Rast, E. H. (2014). Context as assumptions. In F. Lihoreau & M. Rebuschi (Eds.), *Knowledge, context, formalism*, (page 9–39). Springer.

Recanati, F. (2004). Deixis and Anaphora. In Z. G. Szabó (Ed.), *Semantics vs. pragmatics*. Oxford University Press.

Recanati, F. (2004). *Literal meaning*. Cambridge University Press.

Recanati, F. (2006). Crazy minimalism. *Mind and Language, 21*(1), 21–30.

Reichenbach, H. (1947). *Elements of symbolic logic*. Macmillan.

Schlenker, P. (2000). *Propositional attitudes and indexicality: A cross-categorial approach*. PhD thesis, MIT, Massachusetts. Dissertation 1999, MIT; minor changes in 2000.

Schlenker, P. (2003). A Plea for Monsters. *Linguistics and Philosophy, 26*, 29–120.

Searle, J. (1995). *The construction of social reality*. Penguin.

Searle, J. (2005). What is an institution? *Journal of Institutional Economics, 1*(1), 1–22.

Sperber, D., & Wilson, D. (1986). *Relevance: Communication and cognition*. Blackwell.

Sperber, D., & Wilson, D. (2004). Relevance theory. In *The handbook of pragmatics*, (page 607–632). Blackwell.

Stalnaker, R. (1978). Assertion. In P. Cole, (Ed.), *Pragmatics*, (page 315–332). Academic Press, New York. (= Syntax and Semantics Vol. 9).

Stanley, J. (2004). On the linguistic basis for contextualism. *Philosophical Studies, 119*(1–2), 119–146.

Stanley, J. (2005). Semantics and context. In G. Preyer & G. Peter (Eds.), *Contextualism in philosophy: Knowledge, meaning, and truth*, (page 221–253). Oxford University Press.

Stokhof, M., & Groenendijk, J. (1991). Dynamic predicate logic. *Linguistics and Philosophy, 14*(1), 39–100.

von Stechow, A. (1984). Structured propositions and essential indexicals. In F. Landman & F. Veltman (Eds.), *Varieties of formal semantics*, (page 385–403). Kluwer.

Scientific Naturalism and Its Faults

Mario De Caro

According to the most common version of naturalism—called "scientific naturalism" or "strict naturalism"—, in matters of ontology and epistemology, natural science always has the last word. In the second half of the twentieth century, two philosophers have given the greatest impetus to this concept: Wilfrid Sellars and W.V.O. Quine.

In "Philosophy and the Scientific Image of Man" (1962), Sellars elaborated a very influential distinction between the "manifest image" (the world as it is understood by ordinary vision) and the scientific image" (the world as it is understood by natural science). Sellars's view is specular to that offered by Edmund Husserl in *The Crisis of European Sciences and Transcendental Phenomenology* (1936); nor is this a coincidence since, when he was a student at Buffalo, Sellars was deeply influenced by Marvin Farber, a heterodox phenomenologist who had been a student of Husserl:

> Marvin Farber introduced me to Husserl. His combination of utter respect for the structure of Husserl's thought with the equally firm conviction that this structure could be given a naturalistic interpretation was undoubtedly a key influence on my own subsequent philosophical strategy. (Sellars, 1975, 283).

Like Husserl, Sellars strives to understand the relationship between the ways of conceiving the world that are characteristic of, respectively, the ordinary worldview and natural science; and, like Husserl, he aims at elaborating a unified conception of the two visions, which he calls "stereoscopic vision". For Sellars, the two images are "pictures of essentially the same order of complexity, each of which purports to be a complete picture of man-in-the-world which, after separate scrutiny, [philosophers] must fuse into one vision" (Sellars, 1962, 4). Moreover, like Husserl, Sellars recognizes that, from a genetic point of view, the scientific image of the world derives from the manifest image and that the normative concepts of the latter image (for example, the concepts of morality) are not reducible to the descriptive concepts that

M. De Caro (✉)
Department of Philosophy, Università Roma Tre & Tufts University, via Ostiense 236, 00144 Roma, Italy
e-mail: Mario.Decaro@tufts.edu

S. Wuppuluri and I. Stewart (eds.), *From Electrons to Elephants and Elections*, The Frontiers Collection, https://doi.org/10.1007/978-3-030-92192-7_5

characterize the scientific image. From an ontological point of view, however, the unilaterality of Sellars's conception is antithetical to the unilaterality of Husserl's conception (De Caro, 2015). While Husserl is an ordinary realist and an anti-realist concerning science, Sellars adopts the opposite perspective. That is, he is a realist concerning the scientific view and an anti-realist concerning the ordinary worldview: according to this point of view, in the modern age the scientific image justifiably gained a monopoly on ontology, and this showed that the world as conceived by the ordinary view is not the real world. Sellars (1956, 83) expresses this point with a neo-Protagorean dictum:

> Speaking as a philosopher, I am quite prepared to say that the common sense world of physical objects in Space and Time is unreal—that is, that there are no such things. Or, to put it less paradoxically, that in the dimension of describing and explaining the world, science is the measure of all things, of what is that it is, and of what is not that it is not.

Quine's theoretical perspective is similar to Sellars's, but with an important difference: Quine incorporates scientific realism in an influential overall conception that combines the ontological thesis constitutive of scientific realism, with an epistemological thesis and a metaphilosophical thesis. In this form, scientific naturalism has become the main vehicle by which scientific realism has spread in the philosophical world. Its three basic theses are thus as follows:

i. *Ontological thesis*: reality consists only of the entities to which the best explanations of the natural sciences commit us. All other presumed entities, if they are not reducible to scientific entities, are *entia non grata* and therefore should not be accepted in our ontology (Quine, 1960).
ii. *Epistemological thesis*: the natural sciences are our only genuine sources of knowledge. All other supposed forms of knowledge (such as perception, a priori, introspection, or intuition) either can in principle be accommodated into scientific knowledge or are illegitimate (Quine, 1969).
iii. *Metaphilosophical thesis*: philosophy is continuous with science in content, methods, and purposes. According to Quine (1986a, 430–431), we should pursue "philosophy rather as a part of one's system of the world, continuous with the rest of science." And elsewhere, with some irony he states that normative epistemology—the philosophical branch that deals with knowledge, truth, and justification—is "a branch of engineering," that is, it should be understood as an applied natural science (Quine, 1986b, 664).

It is important to emphasize the importance of the latter thesis. With his scientific naturalism, Quine does not simply affirm the correctness of scientific realism and the primacy of natural science from an ontological and epistemological point of view. He also argues that philosophy is, in essence, a still underdeveloped natural science—a thesis that certainly Sellars would not have accepted (nor would the advocates of more liberal forms of naturalism: see De Caro and Macarthur 2004, 2010, 2022 and forthcoming).

Incidentally, it may be noted that some scientific naturalists (including Sellars and Quine), but by no means all, attribute priority among the natural sciences to physics. According to this view (called "physicalism"), the entities and processes that physics deals with are not only the bricks with which the world is built but in principle are also sufficient to explain everything that exists. In this framework, then, all other sciences of nature are reducible to physics. One of the most influential contemporary physicalists, Jaegwon Kim (1996, 11), has given voice to this idea, writing that "each and every property of a thing is either a physical property or is determined by its physical properties and that there is nothing in the world that is not a physical thing." Another champion of this conception is Alex Rosenberg, who goes so far as to claim for his conception the label of "scientism" (which generally has instead a negative connotation):

> What is the world really like? It's fermions and bosons, and everything that can be made up of them, and nothing that can't be made up of them. All the facts about fermions and bosons determine or "fix" all the other facts about reality and what exists in this universe or any other if, as physics may end up showing, there are other ones. In effect, scientism's metaphysics is, to more than a first approximation, given by what physics tells us about the universe. The reason we trust physics to be scientism's metaphysics is its track record of fantastically powerful explanation, prediction, and technological application. If what physics says about reality doesn't go, that track record would be a totally inexplicable mystery or coincidence. (Rosenberg, 2009).

Other scientific naturalists disagree, however, because they believe that physics is not a privileged natural science. John Searle (2004), for example, believes that biology is irreducible to the sciences that deal with the elementary components of matter, such as physics and chemistry: in his opinion, therefore, the inventory of the world is provided by the natural sciences as a whole. Other defenders of scientific naturalism believe that geology or meteorology are irreducible to the more fundamental sciences (Fodor, 1997) and still others that not even chemistry is reducible to physics (Weisberg et al., 2011). In any case, as far as we are interested here, the distinction within scientific naturalism between those who defend the thesis of the ontological and epistemological primacy of physics and those who deny it is not fundamental. What is important is that, according to all those who adhere to scientific naturalism, there is nothing that in principle cannot be investigated with the methods and concepts of the natural sciences. Reality and knowledge, in short, cannot exceed the scope of these sciences.

In the background of the present fortune of scientific naturalism, there is a powerful inductive argument. Starting from Galileo, modern natural science has explained in a more and more complete way wider and wider phenomenal fields, allowing us to make extremely accurate predictions and dethroning the presumed explanations that had been developed previously. It is therefore rational to infer—this argument proceeds—that natural science can explain, in principle, everything that can be explained, and this even in fields in which it is not yet very developed. From this epistemological assertion follows an ontological conclusion: we must assume the existence only of entities that natural science can in principle account for. And, from this point of view, philosophy—at least in the case in which it aspires to speak about reality and not about fictions—can only align itself completely with natural science.

These theses may sound a bit maximalist. And not surprisingly, the main problem of scientific naturalism is the so-called "placement problem" (Price, 2004). The terms of this problem have been well presented by Searle (2007, 4–5):

> How can we square a conception of ourselves as mindful, meaning-creating, free, rational, etc. agents with a universe that consists entirely of mindless, meaningless, unfree, nonrational, brute physical particles?

The philosopher of science John Earman (1992, 262) expresses a similar idea, emphasizing the difficulty of the task:

> If science succeeds in its attempt to explain our nature, then we lose our specificity as beings able to determine their own destiny; if science fails in this attempt, then we turn out to be very mysterious entities within the universe. It seems that the attempt to locate human agents in nature either fails in a manner that reflects a limitation on what science can tell us about ourselves, or else it succeeds at the expense of undermining our cherished notion that we are free and autonomous agents.

Simon Blackburn (1993, 49), in turn, clearly declines this problem regarding ethics: "The problem is one of finding room for ethics, or placing ethics within the disenchanted, non-ethical order which we inhabit, and of which we are a part."

The problem of placement concerns the phenomena constitutive of the ordinary conception of the world: at least at first glance, it would seem that these phenomena do not conform to the scientific conception—if they do not oppose it altogether. Think of free will, moral properties, normativity, meaning, consciousness, or ontologically elusive phenomena such as financial indebtedness or collective intentionality. The regimentation of these phenomena in the perspective of the natural sciences appears at least arduous: each of them represents, therefore, a particular case of the problem of collocation. Nonetheless, the vast majority of scientific naturalists express no doubt that the problem of collocation is solvable. In this spirit, for example, Alan Lacey (2005, 640), has written that "everything is natural, i.e.… everything there is belongs to the world of nature, and so can be studied by the methods appropriate for studying that world, and the apparent exceptions can be somehow explained away."

Scientific naturalists are presented with three possible strategies for dealing with the problem of collocation. The first strategy is that of reductionism, according to which one must show that the phenomena just mentioned are, yes, ontologically genuine, but only because they are identical, or at least reducible, to scientifically acceptable phenomena. For several decades, reductionism has been an attitude present in many areas of philosophy: from the attempts of the so-called "neuroaesthetics" to reduce aesthetic properties to neurological properties to Penelope Maddy's "naturalized Platonism" concerning mathematical properties up to the attempts to naturalize religious spirituality (Dawkins, 2006; Dennett, 2006).

Let us consider a couple of more detailed examples of the current fortunes of reductionism. First, many philosophers of mind sympathetic to scientific naturalism have attempted to reduce mental property types (beliefs, desires, and so on) to neurological properties, according to the famous "identity of mind-brain types" thesis. According to this theory, each type of mental event (such as the belief that Kabul is the capital of Afghanistan or the desire to eat an apple) is identical to a certain type of physical event (typically, a neural process). Proponents of this theory recognize, of course, that we are not yet able to determine such identities: that is, we are not able to say, for all kinds of mental events, to what kind of physical events they are identical. Nonetheless, this identity exists and is the only guarantee of the reality of mental events (Gozzano & Hill, 2012; Kim, 2007).

Another attempt at reduction by scientific naturalists concerns moral properties (Jackson, 1998). The idea here is that moral properties are real, independent of our minds, and make ethical judgments objective, but are natural properties of the same kind studied by the sciences. This is the typical reasoning followed in this area by reductionistically oriented scientific naturalists:

1. Moral judgments can be true or false.
2. The truth or falsity of those judgments depends on the existence of specific phenomena: moral properties.
3. There are no non-natural phenomena.
4. All real phenomena are therefore natural phenomena, in the sense that they fall within the purview of the natural sciences.
5. Moral properties fall within the purview of the natural sciences.

Note that premises 3 and 4 in premise 4 of this reasoning are typical of scientific naturalism: the only natural properties are the proper ones studied by the natural sciences. The most interesting premise, however, is premise 5, because the naturalness of moral properties can be conceived in two ways. In a first sense, it can be argued that moral properties are reducible to non-moral natural properties: the property of a given action to be good, for example, could simply mean that that action conforms to a system of instructions—hardwired into our brains by natural selection—that results in a benefit to humanity (FitzPatrick, 2014). In a second sense, however, premise 5 can be interpreted to mean that moral properties can be investigated by the natural sciences, but that is not to say that they are identical to some non-moral property: that is, they are natural properties of a specific kind (Boyd, 1988). Against the latter conception, however, it has been objected that, while we perceive the usual non-moral properties (such as whiteness or sphericity), moral properties such as goodness or generosity cannot be perceived: and this would be an indication of their illusoriness or supernaturalness. To this objection, it has been replied that other properties considered perfectly natural are also not directly perceptible, but are only inferred: for example, the property of being in good health. In order to take as real properties such as being generous or being healthy what matters is their causal power: that is, these properties are considered real because they can cause changes in the world. We certainly cannot argue that the property of being healthy is supernatural because we see its effects, but we do not perceive it directly; rather, we consider it as a natural property of a specific kind that can be investigated with normal scientific instruments (for example, by measuring the homeostasis of an organism). The same, according to this point of view, should be thought regarding moral properties: they are real properties because they have causal power (Martin Luther King's generosity, for example, caused changes in the world), but that does not mean we should consider them supernatural. Moral properties, in short, Richard Boyd (1988) and David Copp (2017) argue, can be studied with ordinary scientific tools. For example, one can investigate empirically how, how much, and in what situations generosity affects the human world; and in this way, one can understand what generosity is.

Other scientific naturalists oppose the reductionist strategy: that is, they do not believe that the problem of collocation can be solved with the idea that the controversial phenomena of the ordinary image of the world are investigable by the natural sciences or that they are reducible to phenomena that can be investigated by these sciences. In reality, these philosophers argue, such phenomena are constitutively incompatible with the natural sciences and for these reasons must be eliminated from our ontology, exactly as in the past happened with phlogiston, with the epicycles of Ptolemaic astronomy, and with the alleged magical properties of witchcraft. This conception is called "eliminationism". To give a few examples: Paul and Patricia Churchland argue that the entire conceptual apparatus of the intentional mind proper to common-sense psychology (with its references to beliefs, desires, intentions, rationality, and so on) is nothing more than a completely flawed para-scientific theory. In their view, mental phenomena, being immaterial, are not reducible to brain processes, and therefore one must conclude that they are not real. In the words of Paul Churchland (1988, 43):

> A false and radically misleading conception of the causes of human behavior and the nature of cognitive activity. On this view, folk psychology is not just an incomplete representation of our inner natures; it is an outright misrepresentation of our internal states and activities.

In a similar spirit, Derk Pereboom (2014) and Gregg Caruso (2013) deny reality to free will and moral responsibility; Daniel Dennett (1991) and Georges Rey (2016) contest the reality of phenomenological properties (the so-called "qualia"); Hartry Field (1980) and Mark Balaguer (2009) defend mathematical fictionalism, and Richard Joyce (2005) defends moral fictionalism, following the steps of John Mackie (1977), who argued that moral properties and values, being constitutively "queer" (strange, bizarre) with respect concerning o the scientific worldview, cannot be included in our ontology. "If there were objective values," writes Mackie (1977, 38), "they would be entities or qualities or relations of a very strange sort, utterly different from anything else in the universe." According to these authors, since values and moral properties do not exist, it follows that the normative judgments of ethics—which claim to be objective and presuppose the reality of values—are always hopelessly false. In general, then, to solve the problem of collocation, all these authors propose a very drastic solution: that of eliminating from our ontological repertoire the properties of the ordinary worldview.

Finally, we must consider another family, smaller but no less resolute, of scientific naturalists. They reject both reductionism and eliminationism, in the name of a conception called mysterianism, initially developed by the famous linguist and philosopher Noam Chomsky. Human beings try to solve two different kinds of questions, Chomsky (1976) argues, "problems" and "mysteries". Problems are questions that we know how to deal with: for example, we generally understand how to investigate to find out if there are still unknown planets in the solar system or to find the cure for diabetes. Moreover, we can also imagine the kind of solution of these questions (respectively, the possible detection of an unknown celestial body of a certain size orbiting exclusively around the Sun and a therapy that cures diabetes patients or makes their condition significantly better). In the case of mysteries, instead, we don't have, and never will have, the slightest idea of how they could be solved nor of what form their solution could have; and, for this reason, Chomsky argues, mysteries are questions that our species will never solve. And, in this perspective, free will, consciousness, or the mind–body problem are most likely some of these mysteries: we cannot conceptualize the world without these phenomena (and therefore they cannot be eliminated from our ontology), but neither can we reduce them to scientifically explainable phenomena. According to mysterianism, for humans the problem of the location of phenomena seemingly unrelated to the scientific worldview will forever remain a mystery; and this is because, quite simply, our species lacks the conceptual resources to solve it—in the same sense that dogs lack the resources to prove the Pythagorean theorem.

British philosopher Colin McGinn (1993) has presented the most ambitious and detailed version of mysterianism. From a perspective typical of scientific naturalism, McGinn (2002, 207) argues that.

> nature is a system of derived entities, the basic going to construct the less basic; and understanding nature is figuring out how the derivation goes... Find the atoms and laws of combination and evolution, and then derive the myriad of complex objects you find in nature.

This approach is not without philosophical consequences if, in reflecting on the philosophical status of ordinary worldview phenomena (consciousness, ego, free will, meaning, and knowledge), McGinn (2002, 209) himself acknowledges that.

> there are yawning gaps between these phenomena and the more basic phenomena they proceed from, so that we cannot apply the [scientific] format to bring sense to what we observe. The essence of a philosophical problem is the unexplained leap, the step from one thing to another without any conception of the bridge that supports the step.

According to McGinn, our species is not intelligent enough to bring phenomena such as consciousness, ego, free will, meaning, and knowledge to a format that can be handled by the natural sciences. And because of this, McGinn (2002, 207) concludes, philosophy, because it tries to solve insoluble problems, is a "futile" activity. On the other hand, we cannot even think of considering these phenomena as illusory, because they play too important roles in our intellectual lives and practices. Therefore, for us, they represent insoluble mysteries and will always do.

In sum, scientific naturalists face the complex challenge posed by the problem of collocation. This problem concerns phenomena (from freedom to consciousness, from normativity to morality to signification) that for ordinary realism—the kind of realism encompassed in the commonsense image of the world—are, at the same time, indubitable and essential to understanding human reality. Within the framework of scientific naturalism, however, these phenomena appear mysterious because they do not seem treatable by natural science. Scientific naturalists attempt, then, three strategies to account for these phenomena: reductionism, eliminationism, and mysterianism.

Ordinary realism (which is defended, for example, by Husserl and van Fraassen) manifests an intrinsic hegemonic tendency to the extent that it projects onto the whole of reality—including the areas of relevance to science—the idea that perception (direct or assisted by technological supports) is the only parameter we have to determine what the world is like. In other words, ordinary realists make a very dubious inference: from the very plausible thesis that perception is a legitimate (though obviously not infallible) key to access reality, they conclude that perception is the only key to access reality and that therefore one can deny ontological legitimacy to the unobservable entities postulated by science (De Caro, 2015 and 2019).

Scientific realism, which, as we have seen today, is defended with particular vigor by scientific naturalists, tends to be as hegemonic as ordinary realism, but in a specular way. This conception, in fact, in the name of the reality of scientific ontology, tends to dismiss the realist attitude of the ordinary worldview, based on the idea that the only entities that exist are those contemplated by science. In doing so, however, it encounters a considerable theoretical problem: the problem of collocation. How to account for the entities and properties (secondary properties, free will, consciousness, values, and so on) that are of such importance to the ordinary worldview, but which, at least apparently, do not seem tractable by the natural sciences? The strategies adopted by scientific naturalists, we have seen, are of three kinds: reductionism, eliminationism, and mysterianism. All three, however, present considerable problems: let us see why.

According to the first strategy, reductionism, phenomena accepted by the ordinary worldview are actually identical to more fundamental, scientifically investigable properties in the same sense that water is identical to H_2O (think, for example, of the attempts of some moral realists to reduce moral properties to properties that can be investigated with the tools of the natural sciences). Those attempts at reduction run into a huge problem, however. An essential aspect of moral properties is that they have to do not only with the world of being, which concerns the way things are but also with the world of possibility and obligations—that is, with normativity. The behavior of a given person is moral when in a given situation that person does something morally praiseworthy: that is, something that should be praised, not something that is in fact praised. For example, a generous action by an individual (e.g., when someone welcomes a politically persecuted person into their home) may be criticized by the respective community because that community is clouded by prejudice or fear: in such a case, the community is in error because it should have praised that action, not criticized it. A natural scientist, however, can only investigate how things are, not how they should be. The normative aspect of morality escapes attempts at reduction altogether; and similar criticisms can be made of attempts to reduce other phenomena proper to the ordinary worldview.

Yet attempts at reduction, or as we sometimes say "naturalization," continue to thrive. Thus, some time ago Tyler Burge (1993, 117) described attempts to reduce mental properties to neuroscientific properties:

> The flood of projects... that attempt to fit mental causation or mental ontology into a 'naturalistic picture of the world' strike me as having more in common with political or religious ideology than with a philosophy that maintains perspective on the difference between what is known and what is speculated.

This reductionist "flood of projects" is the result of an ideology that characterizes all versions of scientific naturalism: an ideology that does not, however, come to terms with the articulated ways in which we, as a matter of fact, understand the world. Referring to cognition, and more generally to thought, Putnam (1992, 18), for example, wrote:

> There is no reason why the study of human cognition requires that we try to reduce cognition either to computations or to brain processes. We may very well succeed in discovering theoretical models of the brain which vastly increase our understanding of how the brain works without being of very much help to most areas of psychology, and in discovering better theoretical models in psychology (cognitive and otherwise) which are not of any particular help to brain science. The idea that the only understanding worthy of the name is reductionist understanding is a tired one, but evidently it has not lost its grip on our scientific culture.

It will not be surprising, then, that while reductionist ideology is very common today, there is much less agreement on the value of concrete attempts to reduce the entities of the ordinary worldview, to the point that, as Putnam (2004, 62) ironically notes, "none of these ontological reductions gets believed by anyone except the proponent of the account and one or two of his friends and/or students."

In this regard, however, a remark is necessary. First of all, as far as the empirical investigation conducted by scientists is concerned, the assumption that a given phenomenon can be studied by resorting exclusively to the categories of the natural sciences is obviously legitimate: it is a methodological maxim (not an ontological principle) that has often been very fruitful for research. The history of science, on the other hand, teaches us that the most successful research programs have often involved real leaps in the dark by their proponents: heliocentrism was definitively proven only in the nineteenth century and the theory of relativity was confirmed years after Einstein had proposed it. It could perhaps happen, for example, that one day the mind will be explained as any physical system, without any other tools than those of natural sciences (and not also with intentional psychology and introspection); at the moment, however, we have no elements to conclude that this will happen. Therefore, at least for now, it is at least adventurous to confer ontological dignity on a maxim that is methodological in nature.

In essence, we must distinguish between reductionist ideology and concrete scientific reductions. What is essential for scientific progress are concrete reductions, when they succeed: that is, the reduction of a given range of phenomena to a more fundamental range. When a reduction is accomplished (such as when it was demonstrated that water is H_2O or that light corresponds to a certain portion of the electromagnetic spectrum) our knowledge has taken a great step forward. The "reductionism" instead is an ideology: that is the conception that all phenomena must in principle be reduced and explained from more fundamental phenomena. This is a philosophical thesis, not a scientific one. In their practice, scientists often attempt to make reductions, but in many cases, they proceed by studying phenomena at their particular level: for example, biologists generally do not move to the physical–chemical level for their research. Nor do we have any basis for being certain that reductions are always possible—as reductionist ideology instead assumes.

Returning to the philosophical discussion, we can therefore say that in general reductionism is not a winning strategy. This is why many scientific naturalists take a bolder route: that of eliminating from our ontology the entities and properties belonging to the ordinary worldview. Secondary properties, free will, consciousness, moral properties, normativity, intentionality: no aspect of the ordinary view is spared from the eliminationist pathos of these authors. The aforementioned Alex Rosenberg (2009) gives an excellent example of this trend when he writes:

> Science forces upon us a very disillusioned 'take' on reality. It forces us to say 'No' in response to many questions to which most everyone hopes the answers are 'Yes'. These are the questions about purpose in nature, the meaning of life, the grounds of morality, the significance of consciousness, the character of thought, the freedom of the will, the limits of human self-understanding, and the trajectory of human history.

In this quotation, one notices, moreover, a strange mixture of ideas that no serious thinker today would consider worthy of consideration (the purpose of nature, the trajectory of human history) with others that are essential to both the ordinary worldview and to many philosophical systems as if they were all at the same level of plausibility. In any case, the fundamental question of eliminationism is this: can we seriously conceive of a world without the central ideas of the ordinary worldview and philosophy, such as free will, consciousness, morality, and so on?

Let's consider, as an example, the attempt, by Paul and Patricia Churchland and others, to eliminate from our ontology the mental states proper to common-sense psychology (beliefs, desires, intentions, etc.). As will be recalled, according to the Churchlands, common-sense psychology is a proto-scientific theory that is completely erroneous regarding how the mind works. A first objection that can be made to this idea is that common-sense psychology is not a theory at all (albeit a proto-scientific and erroneous one): understanding the mind does not have at all the structure and function of a scientific explanation. For example, when we interpret the minds of others, or even when we reflect on ourselves, using common sense psychology, we frequently refer to normative notions: "My belief was wrong," "This desire of yours is absurd," "Your intention should be another." And natural science theories cannot adequately deal with normative notions: thus, equating common-sense psychology with a scientific theory, albeit a primordial one, is incorrect. A second objection is that, even if one were to accept the idea that common-sense psychology is a theory, then it should also be said that it is a theory that works quite well, because it helps us to make a large number of correct predictions about other people's behavior and, at least at this stage, it does this much better than any alternative theory (neurological or otherwise): so it is hard to see why we should eliminate it. Finally, one can also object that the Churchlands' eliminationism is self-confirming: if beliefs are not real, it is hard to see how the Churchlands can believe that their theory is better than common sense psychology, nor how they can try to convince others to believe in their theory.

Equally convincing arguments can be developed against attempts to eliminate the other fundamental components of the ordinary worldview. From this perspective, one must therefore conclude that eliminationist ideology as a whole, like reductionist ideology, does not work. For this reason, as we have seen, some scientific naturalists have espoused a third conception, "mysterianism." According to the proponents of this conception, the phenomena proper to the ordinary image of the world cannot be banished from our ontology, as the eliminativists claim: such phenomena are indispensable for giving meaning to many fundamental aspects of our existence. On the other hand, continue the mysterians, we are not even able to bring back those phenomena, as the reductionists hope, to the explanatory modalities of science, which are the only epistemically legitimate: this, however, does not happen because these phenomena are intrinsically supernatural, but for our insurmountable cognitive limits. We are simply not an intelligent enough species to give a scientific explanation for these phenomena, in the same sense that dogs do not have sufficient cognitive endowment to understand a mathematical demonstration of Pythagoras' theorem.

Mysterians consistently draws the consequences implicit in the ontological premises of scientific naturalism and has the merit of recognizing that attempts at reduction are generally vague and the alleged eliminations unfeasible. However, in doing so, he reaches a conclusion that is very difficult to accept, i.e., that free will, consciousness, knowledge, meaning etc., are "mysteries" because we will never be able to understand them. In reality, however, if it is true that we do not know how to solve problems such as those of free will or consciousness, it is also true that over the centuries we have made considerable progress in clarifying them. Today we know much more about these problems than we did in antiquity, the Middle Ages, or even a few decades ago: conceptions that were thought plausible have been refuted, conceptually more refined ones have been developed, various facets of the problems have been clarified, and so on. Philosophy, in short, progresses conceptually (even if it does not solve its own problems, because a solved problem is ipso facto considered non-philosophical). And if there is conceptual progress, it means that philosophical problems are not unfathomable mysteries as Chomsky thinks and that philosophy is not at all a futile activity, as McGinn thinks. Moreover, it seems intellectually arrogant to set limits to what our species can do cognitively based on what we now think of our epistemic limits: Aristotle (perhaps the greatest genius ever to appear on Earth) could never have conceived of the possibility of sending a human being to the Moon, of formulating Gödel's theorem, of calculating the speed of light or of explaining how sight works (a problem, the latter, which at the time was considered to be the domain of philosophy). But this proves that those questions were not mysteries: they were, rather, very difficult problems, still not formulated at Aristotle's time, but that with the passing of generations have been solved. And this could happen even with some of the problems discussed today by many philosophers—even if, for this reason, they would no longer be considered philosophical problems.

On the other hand, once we assume the perspective of scientific naturalism, we have no idea what form the acceptable explanations concerning consciousness, free will, meaning, etc. might take. In that framework, these phenomena—which within the ordinary view of reality are not considered so mysterious—become completely incomprehensible: there is no way, in short, to talk about them in an intelligible way. However, the fact that scientific naturalism makes the most important phenomena of our existence incomprehensible can also be taken as a reductio ad absurdum of this conception. In other words, a philosophical conception that is not able to account for some of our fundamental ideas, and cannot reduce or eliminate them, appears radically unsatisfactory and should be abandoned.

So thought also the late Lynne Baker (2013, 73), who wrote that "We should not lend faith to metaphysics that render ordinary but significant phenomena unintelligible." It is hard to see, then, why we should accept a conception such as mysterianism, given that it makes it impossible to think meaningfully about such fundamental issues as freedom, responsibility, consciousness, and meaning.

References

Baker, L. R. (2013). *Naturalism and the first-person perspective.* Oxford University Press.

Balaguer, M. (2009). Fictionalism, theft, and the story of mathematics. *Philosophia Mathematica, 17*(2), 131–162.

Blackburn, S. (1993). *Essays in quasi-realism.* Oxford University Press.

Boyd, R. (1988). How to be a moral realist. In G. Sayre-McCord (Ed.), *Essays on moral realism* (pp. 181–228). Cornell University Press.

Burge, T. (1993). Mind-body causation and explanatory practice. In J. Heil & A. Mele (Eds.), *Mental causation* (pp. 97–120). Clarendon Press.

Caruso, G. (2013). *Free will and consciousness: A determinist account of the illusion of free will.* Lexington Books.

Chomsky, N. (1976). Problems and mysteries in the study of human language. In A. Kasher (Ed.), *Language in focus* (pp. 281–357). Reidel.

Churchland, P. M. (1988). *Matter and consciousness: A contemporary introduction to the philosophy of mind* (rev). MIT Press.

Copp, D. (2017). Normative naturalism and normative nihilism: Parfit's dilemma for naturalism. In S. Kirchin (Ed.), *Reading Parfit* (pp. 28–53). Routledge.

Dawkins, R. (2006). *The God delusion.* New York Books.

De Caro, M. (2015). Realism, common sense, and science. *The Monist, 98*(2), 197–214.

De Caro, M. (2019). The indispensability of the manifest image. *Philosophy and Social Criticism, 46,*, 451–11.

De Caro, M., & David, M. (Eds.). (2004). *Naturalism in question.* Harvard University Press.

De Caro, M., & David, M. (Eds.). (2010). *Normativity and naturalism.* Columbia University Press.

De Caro, M., & David, M. (Eds.). (2022). *Routledge handbook of liberal naturalism.* Routgledge.

De Caro, M., & Macarthur, D. (forthcoming b). *Liberal naturalism*, Harvard University Press.

Dennett, D. (1991). *Consciousness explained.* Backbay Books.

Dennett, D. (2006). *Breaking the spell: Religion as a natural phenomenon.* Viking.

Earman, J. (1992). Determinism in the physical science. In M. H. Salmon, J. Earman, & C. Glymour et al. (Eds.), *Introduction to the philosophy of science.* Prentice Hall, 232–268.

Field, H. (1980). *Science without numbers.* Blackwell.

FitzPatrick, W. J. (2014). Morality and evolutionary biology. In E. N. Zalta (Ed.), *Stanford encyclopedia of philosophy*, https://plato.stanford.edu/entries/morality-biology/.

Fodor, J. (1997). Special sciences: Still autonmous after all these years. *Philosophical Perspectives, 11*, 149–163.

Gozzano, S., & Hill, C. (2012). *New perspectives on type identity: The mental and the physical.* Cambridge University Press.

Husserl, E. (1936). *The crisis of European sciences and transcendental phenomenology* (1970). Norwestern University Press.

Jackson, F. (1998). *From metaphysics to ethics.* Oxford University Press.

Joyce, R. (2005). Moral Fictionalism. In M. E. Kalderon (Ed.), *Fictionalism in metaphysics* (pp. 287–313). Oxford University Press.
Kim, J. (1996). *Philosophy of mind*. Westview Press.
Kim, J. (2007). *Physicalism, or something near enough*. Princeton University Press.
Lacey, H. (2005). Naturalism. In T. Honderich (Ed.), *The Oxford companion to philosophy* (p. 640). Oxford University Press.
Mackie, J. (1977). *Ethics: Inventing right and wrong*. Viking Press.
McGinn, C. (1993). *Problems in philosophy*. Blackwell.
McGinn, C. (2002). *The making of a philosopher*. Scribner.
Pereboom, D. (2014). *Free will, agency, and meaning in life*. Oxford University Press.
Price, H. (2004). *Naturalism without representationalism*. In De Caro & Macarthur (Eds.) (2004), 71–88
Putnam, H. (1992). *Renewing philosophy*. Harvard University Press
Putnam, H. (2004). The content and appeal of "naturalism." In De Caro & Macarthur (Eds.) *Naturalism in question* (pp. 59–70). Harvard University Press.
Quine, W. V. (1960). *Word and object*. MIT Press.
Quine, W. V. O. (1969). Epistemology naturalized. In *Ontological relativity and other essays* (pp. 69–90). Columbia University Press.
Quine, W. V. O. (1986a). Reply to Hilary Putnam. In L. E. Hahn & P. A. Schillp (Eds.), *The philosophy of W.V. Quine*. Open Court, 427–431.
Quine, W. V. O. (1986b). Reply to Morton White. In L. Hahn, & P.A. Schilpp (Eds.), *The Philosophy of W.V. Quine*. Open Court, 663–665.
Rey, G. (2016). Taking consciousness seriously—as an illusion. *Journal of Consciousness Studies, 23*(11–12), 197–214.
Rosemberg, A. (2009). *The disenchanted naturalist's guide to reality*, https://nationalhumanitiescenter.org/on-the-human/2009/11/the-disenchanted-natura-lists-guide-to-reality/.
Searle, J. (2004). *Biological naturalism*, https://web.archive.org/web/20060501002411/http://ist-socrates.berkeley.edu/~jsearle/BiologicalNaturalismOct04.doc
Searle, J. (2007). *Freedom and neurobiology. Reflections on free will, language, and political power*, Columbia University Press.
Sellars (1956). *Empiricism and the philosophy of mind*. Reprinted: Harvard University Press, 1997.
Sellars, W. (1962). Philosophy and the scientific image of man. In R. Colodny (Ed.), *Frontiers of science and philosophy* (pp. 35–78). University of Pittsburgh Press.
Sellars, W. (1975). *Autobiographical reflections*. In H. N. Castaneda (Ed.), *Action, knowledge, and reality: Critical studies in honor of wilfrid sellars*. The Bobbs-Merrill Company, 277–93.
Weisberg, M., Needham, P., & Hendry, R. (2011). *Philosophy of chemistry*. In E.N. Zalta (Ed.), *Stanford encyclopedia of philosophy*, http://plato.stanford.edu/entries/chemistry/.

Scientific Emergentism and the Mutualist Revolution: A New Guiding Picture of Nature, New Methodologies and New Models

Carl Gillett

> The world we actually inhabit, as opposed to the happy world of modern scientific mythology, is filled with wonderful and important things we have not yet seen because we have not looked... The great power of science is its ability, through brutal objectivity, to reveal to us truth we did not anticipate. (Laughlin (2005), p. xvi)

The guiding picture of nature to which we subscribe—what we take the ontological structure of nature to be in a broad sense—configures not just what kinds of scientific models and explanations we offer, but also what phenomena we even seek, and recognize, in nature. In our opening passage, Robert Laughlin, echoing other scientific emergentists like Ilya Prigogine, tells us that we have routinely overlooked all manner of phenomena that did not fit the guiding picture of nature pressed by scientific reductionism. For recent empirical findings, emergentist like Laughlin and Prigogine contend, show that the picture of scientific reductionism is in fact a misleading "myth".[1]

Still more exciting, scientific emergentism offers a new guiding picture that allows us to finally see many natural phenomena and provides novel models/explanations that potentially allow us to understand them. The result, over the last few decades, is arguably a revolution in the sciences built around adoption of a new view of the relation of parts and wholes, and hence of the structure of nature itself. The pioneering scientists pressing this view, calling themselves "scientific emergentists", include physicists like Laughlin or Philip Anderson, chemists such as Prigogine, biologists including Denis Noble, neuroscientists like Walter Freeman, and many in systems biology or the sciences of complexity.[2]

---

[1] See Prigogine and Stengers (1984), and Prigogine (1997), for Prigogine's own interesting discussion of such ontological "myths".

[2] Anderson (1972), Freeman (2000), Laughlin (2005), Noble (2006), and Prigogine (1997).

C. Gillett (✉)
Department of Philosophy, Northern Illinois Univ, DeKalb, IL 60115, US
e-mail: cgillett@niu.edu

S. Wuppuluri and I. Stewart (eds.), *From Electrons to Elephants and Elections*, The Frontiers Collection, https://doi.org/10.1007/978-3-030-92192-7_6

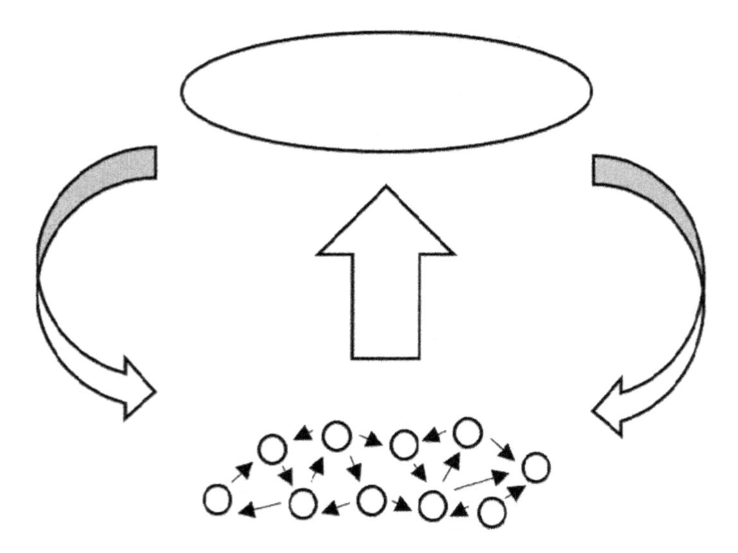

Fig. 1 The complexity researcher Chris Langton's famous diagram of the scientific emergentist's Mutualist view of nature. (Redrawn from Lewin (1992))

At the heart of their positions, these researchers endorse what I have elsewhere termed "Mutualism" whose core idea is framed in Fig. 1. Under the Mutualist picture, it continues to be the case that certain complex wholes (and their activities and properties), like the one at the top of the diagram, are taken to be fully composed by organized parts (and their activities and properties) shown at the bottom—and hence we have compositional relations upwards from parts (and their activities and properties) to the whole (and its activities and properties). However, the foundational change under Mutualism is to accept that sometimes we also have a downward determinative relation from the whole (and its activities and properties) to its parts (and their activities and properties) depicted by the downward curving arrows. Consequently, we have *both* upward *and* downward determination, and hence *mutual* determination between parts and whole—hence the "Mutualist" tag for the position. The result is a new picture of nature with all manner of exciting implications.

Simply appreciating Mutualism has profound theoretical and practical implications in the sciences. Elsewhere I have sought to provide a detailed treatment of Mutualism and the wider debates over reduction and emergence in which it figures (Gillett, 2016a). In this paper, my goal is narrower. I simply seek to provide an accessible account of scientific emergentism and its key claims. To this end, I briefly sketch the background to scientific debates over reduction and emergence, but my primary focus is on outlining the core ideas of Mutualism, the new guiding picture of nature that results, and hence the novel methodologies it offers and the new models its provides in concrete scientific cases.

To start, I sketch two connected waves of scientific findings about compositional relations that drive our present research as well the debates over reduction and emergence. The first wave of findings, outlined in Part 1, stretching from the Scientific

Revolution onwards, was to provide what I term "compositional" models and explanations of wholes (and their properties and activities) using compositional relations to parts (and their properties and activities). Thus, for instance, we explain the contraction of a muscle by a compositional relation to moving protein filaments that compose it. Or we explain the mass of the muscle using a compositional relation to the masses of the cell's that are its parts.

As I highlight, all sides in the sciences accept the need to search for, and provide, such compositional explanations in what I term "everyday reductionism". And the success of this methodology in providing compositional explanations/models, and the advent of new techniques, has more recently allowed a second wave of usually quantitative scientific findings, outlined in Part 2, about the activities of the parts we find in wholes. This more detailed, and precise, understanding of the behaviors of parts in wholes has led to a range of what I term "Challenging Compositional Cases" where we cannot presently understand the behaviors of the parts in various wholes using existing resources, including accounts and models given of such parts in simpler systems. Across examples in a range of sciences, from super-conductors to populations of neurons, Challenging Compositional Cases are now at the cutting edge of ongoing scientific inquiry.

I briefly outline, in Part 3, how everyday reductionism, and the provision of compositional explanations, has been argued to support the stronger position I term "scientific reductionism", espoused by researchers such as the physicist Steven Weinberg, biologists like Francis Crick, Richard Dawkins or E.O. Wilson, and many others.[3] Scientific reductionists provide reasoning that, they claim, shows reflection on compositional explanations leads from everyday reductionism to their more robust scientific reductionism position. I detail the guiding picture of nature that results under scientific reductionism, one where there are *nothing but parts,* and collectives of them, but where higher sciences are needed to study collectives of parts. I highlight how this a picture under which the ultimate parts are the only determinative entities in nature and the laws about them are the only fundamental laws—thus implying only fundamental physics illuminates fundamental phenomena and fundamental laws of nature.

Against this empirical and theoretical background, in Part 4, we can finally appreciate the core ideas of scientific emergentism in its Mutualist position that allows a whole, and its parts, to be mutually determinative. This picture grows from the findings of everyday reductionism, and Challenging Compositional cases, about the behaviors of the parts in wholes. Crucially, Mutualism accepts that "Parts behave differently in wholes", but this then allows the scientific emergentist to argue that "Wholes are more than the sum of their parts" because such wholes sometimes downwardly determine their parts. Furthermore, I note how appreciating Mutualism shows the key parsimony argument of scientific reductionism is invalid, hence blocking the main theoretical reasons commonly used to dismiss "emergence".

Perhaps more importantly, I then detail how, as well as theoretical implications, Mutualism has substantive import for scientific practice both globally and locally.

[3] Crick (1966), Weinberg (1994, 2001) and Wilson (1998).

To begin, in Part 5, I outline how Mutualism globally underpins a new guiding picture of nature sharply contrasting in its practical import with that of the scientific reductionist. For example, I detail how this new Mutualist picture accepts that there are many compositional levels of parts and wholes in nature to explore, rather than just parts and collectives of them. And how Mutualism opens up the possibility of fundamental phenomena, laws and research at many of these levels of nature, and hence as the focus of many sciences beyond physics, rather than just at the level of ultimate parts.

Moving from the global to the local, in Part 6, I sketch how Mutualism offers a new class of "Mutualist" models and explanations positing relations of whole-to-part determination that offer help in Challenging Compositional Cases and other ongoing investigations. I highlight how Mutualist models and explanations *supplement* causal and compositional models, hence adding to everyday reductionism rather than burning it down. And I note that researchers are now exploring whether such Mutualist models are successful, or even the best models available, in various ongoing cases from superconductors to neural populations.

1 The Wave of Compositional Explanations from the Scientific Revolution Onwards

Compositional explanations have been one of the main engines of the sciences since the Scientific Revolution, transforming our understanding of nature. Philosophers of science have used a range of terms for compositional explanation, including "reductive explanation", "functional explanation" or "constitutive mechanistic explanation".[4] Let us consider just a few examples of compositional explanation drawn from physiology, cell biology and molecular biology to appreciate their character.

In response to the question "Why did the muscle contract?" two good answers, in certain contexts, are based around the model in Fig. 2 and are "The cell fibers contracted" or "The myosin crawled along the actin". This is the one species of compositional explanation widely acknowledged by philosophers of science where we explain *an activity* of a whole using a compositional relation to activities of parts in what are often termed "constitutive mechanistic explanations" and which I term "Dynamic" compositional explanation.[5] We explain the muscle's contraction at some time using a compositional relation to the contraction of various cells at that time. The cells are inter-connected, or "organized", so as each contracts it pulls on the cells

[4] Though neglected, there has been philosophical work on compositional explanation that goes back at least to early work by Fodor (1968) and Dennett (1969), through Wimsatt (1976), down to more recent work such as Bechtel and Richardson (1993), Glennan (1996), Machamer, Darden and Craver (2000) and Craver (2007), amongst many others. See Aizawa and Gillet (2019) for an outline of some of the various species of compositional explanation.

[5] Aizawa and Gillet (2019).

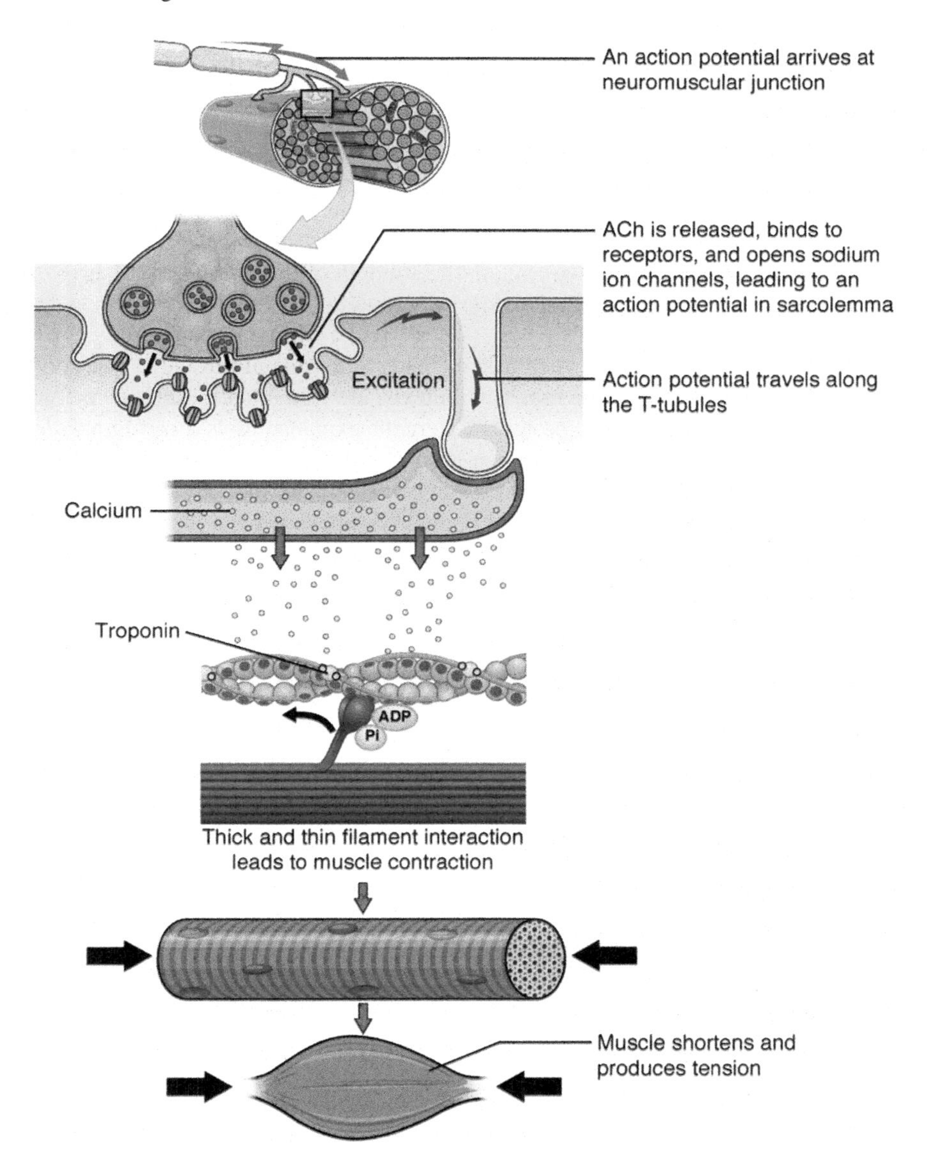

Fig. 2 A textbook diagram of the sliding filament model of muscle contraction and a Dynamic compositional model. (From Betts (2013), Chap. 10, Sect. 10.3, Fig. 1)

to which it is connected and which are also contracting. Hence the contracting cells compose (or what I term "implement"), and explain, the muscle's contracting.

There are plausibly other species of compositional explanation. What I have elsewhere termed a "Standing" compositional explanation explains *a property* of a whole using a compositional relation, what is termed "realization", to properties of parts.[6] For example, in response to the question "Why is the muscle strong?" one could answer "Because the cell's each have a certain strength" or an answer focused on the properties of proteins. For we explain the strength of the muscle, a property of a whole, using compositional relations of realization to properties of its parts at the cellular and molecular levels.

Lastly, we should note that when asked "What is a skeletal muscle?" two good answers (amongst others) in the relevant contexts, are "Bundled muscle fibers", as Fig. 3 highlights, or "Organized proteins". Here the explanadum is *a certain whole*, i.e. an individual, the explanans is some group of parts (at a certain "level") and the backing relation is the part-whole relation between these individuals. I term this an "Analytic" compositional explanation where we explain a whole itself using a compositional relation to individuals that are parts.

All of these explanations are what I shall term "ontic" explanations that work by representing an ontological relation between entities in the world, the "backing relation" of the explanation, where the nature of this relation drives these explanations. In addition, these explanations are also all backed by compositional, rather than causal, relations, since their backing relations all share common ontological features lacking in causal relations. For example, amongst other singular features, their backing relations are all synchronous relations, between entities that are in some sense the same and which involve synchronous changes in their relata.[7] So we can see that these are not causal explanations.

As our examples begin to highlight, all of the species of compositional explanation about a common phenomenon are plausibly systematically integrated with each other.[8] Furthermore, such explanations are systematically integrated with the related causal explanations, about connected phenomena, that philosophers of science term "etiological mechanistic explanations", amongst others. It is important to remember this point, since it highlights how scientists often seek, and provide, various integrated causal and compositional models/explanations in tandem about a certain state of affairs in nature.

As researchers piled up compositional explanations of phenomena across all levels in nature, scientists came to accept that everything in nature is composed by the entities of physics and hence to endorse this as a guiding picture of the structure of nature—the view that all individuals, activities and properties are either entities of fundamental physics or composed by the entities of fundamental physics. Under this

---

[6] Aizawa and Gillet (2019).

[7] Elsewhere I have highlighted still further differences between the features of such compositional and causal relations. See Gillett (2016a), Chap. 2, (2016b), (2020) and (Forthcoming), Chaps. 1–3.

[8] See Gillett (Forthcoming) for a more detailed discussion of such integration. We thus have another example of what Mitchell (2003) terms "integrative pluralism" in multiple, but integrated, models.

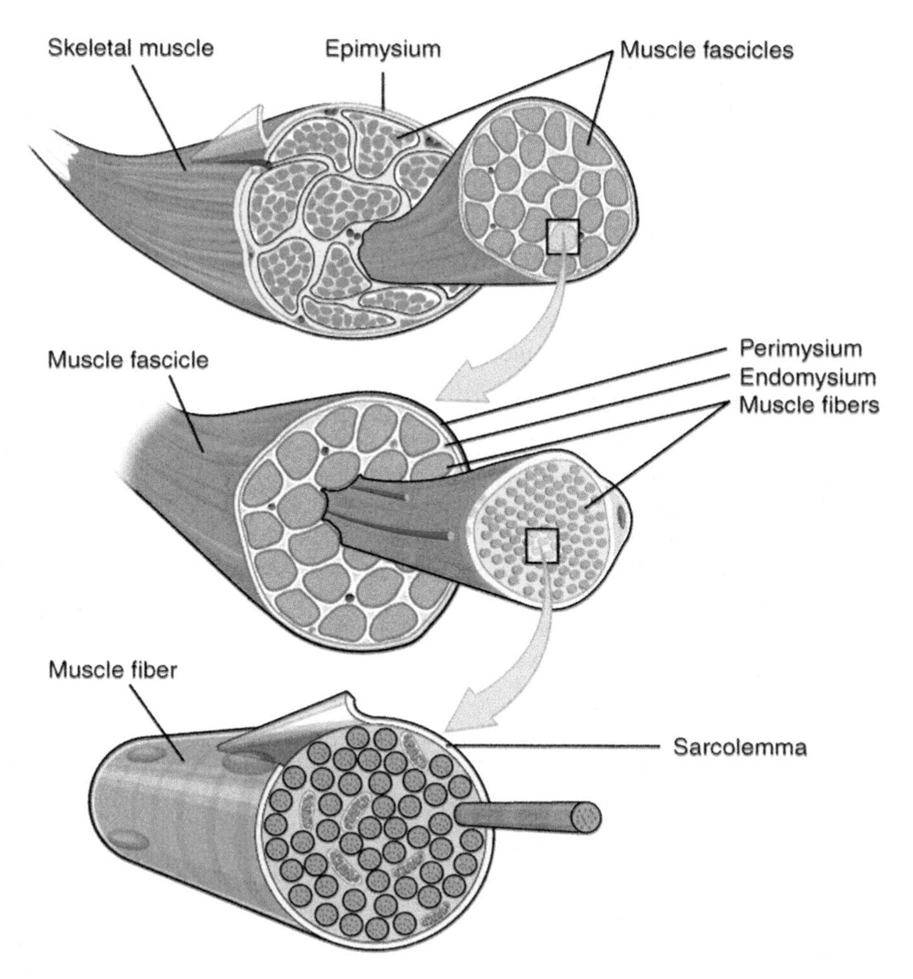

Fig. 3 A textbook diagram of the composition of a skeletal muscle at tissue and cellular levels, and hence a multi-level Analytic model of it. (From Betts (2013), Chap. 10, Sect. 10.2, Fig. 1)

picture, we have both wholes and their parts, as well as their activities and properties, which are all compositionally related in various (local) compositional levels down to the entities of physics. Once we endorse this guiding picture it also entails obvious methodological guidance–search for compositional explanations of all entities in nature! Carefully note, however, that such guidance does not exclude the existence of, or need to provide, other models/explanations as well.

The search for compositional explanations was at the cutting-edge of twentieth century science and furnished the first great wave of empirical findings I wanted to highlight in a huge array of compositional explanations across the sciences and the levels of nature from chemistry on through to neuroscience and physiology.

As I detail below, all working scientists now acknowledge, and seek for, such compositional explanations/models.

It is therefore important to clarify some terminology. In the sciences, but also amongst some philosophers, common terms for a compositional explanation are that it is a 'reductive' or 'reductionist' explanation and the term 'reductionist' is thus often simply used to refer to someone who seeks compositional explanations (Wimsatt, 2007) who is taken to be pursuing a 'reductionist methodology'. Unfortunately, as later sections make plain, on these mundane usages *both* scientific reductionists *and* opposing scientific emergentists are all 'reductionists' who all explicitly seek 'reductionist explanations' and all endorse 'reductionist methodology'!

We therefore need a better terminology, so I use the neutral term 'compositional explanation' to refer to the explanations both sides acknowledge as important and which I have highlighted in this section. When I need to refer to the weaker, and as we shall see universally endorsed, position espousing the search for compositional explanations I shall refer to 'everyday' reductionism. I use the term 'scientific reductionist' for the more substantive positions endorsed by writers, like Weinberg, Crick, Dawkins, Wilson etc., that I detail in Part 3. As we shall shortly see, distinguishing everyday reductionism from scientific reductionism is important for a variety of reasons, but the most obvious is that each supplies a different "guiding picture of nature" and hence entails very different guidance on what scientific models/explanations, and methodologies, will be successful.

2 A Second Wave of Findings about Parts: Understanding Challenging Compositional Cases

In the twenty-first century, and late twentieth century, with the advent of new experimental and theoretical techniques researchers explored a range of aspects of parts and wholes. In particular, these new techniques have yielded quantitative accounts of the behaviors of parts within complex wholes from superconductors to cells or populations of neurons. It is important to emphasize these are all cases where we have well confirmed compositional explanations and models. So the relevant wholes, and their properties and activities, are all known to be fully composed—hence there are none of the new forces or energies, or uncomposed properties, involved with the type of "emergence" that figured in scientific debates at the end of the nineteenth century and early twentieth century.

In these cases where we have new quantitative accounts of the behaviors of parts, a certain kind of situation has become increasingly common. Consider the case of the proteins that compose a eukaryotic cell discussed in detail by an interdisciplinary research team consisting of philosophers of science, in Robert Richardson and Achim Stephan, as well as prominent systems theoretic biologists in Fred Boogerd, Frank Bruggeman, and Hans Westerhoff (Boogerd et al., 2005). We have a great deal of evidence about the properties, and behaviors, of such proteins in simple systems whether in vitro or elsewhere. In addition, as Boogerd et al. highlight, we have now also collected quantitative evidence about the activities (and hence properties) of such proteins when they are actually parts of cells. Researchers like Boogerd et al. consequently argue that we can now see that the activities of the proteins in cells cannot be explained using the accounts given for them in simpler systems. That is, our quantitative account of the behaviors of these parts in cells, combined with our successful accounts of such proteins in simpler systems, together show that these parts behave differently than they would if the accounts in simpler systems were exhaustive in the whole.

Similar arguments have been given across a range of cases. For instance, Robert Laughlin (2005) plausibly makes such an argument about the behaviors of electrons when they are parts of superconductors. Laughlin takes quantum mechanics to provide a successful account of the behaviors of electrons in simpler systems and Laughlin takes us to have quantitative, and highly precise, accounts of the behaviors of electrons in superconductors. Consequently, Laughlin concludes that electrons in superconductors behave differently than they would if the accounts in simpler systems were exhaustive.

The latter are examples of what I will term a "Challenging Compositional Case" in an example where the following conditions hold:

(i) we have successful compositional accounts of a certain whole (and its activities and properties) in terms of various parts (and their activities and properties);
(ii-a) we have successful accounts of the relevant parts and their behaviors in simpler systems;
(ii-b) we now possess quantitative accounts of the behavior of these parts in the relevant whole;
and;
(iii) the behavior of the parts in the relevant whole is apparently different than it would be if the accounts of this part in simpler systems were exhaustive.

Given (iii), against the background of (i), (ii-a) and (ii-b), such cases pose an obvious challenge of explaining, or otherwise accounting for, the behaviors of the parts in the relevant wholes in such.

Such cases also highlight limitations to the guiding picture of nature suggested by everyday reductionism and it associated methodologies. This guiding picture is not wrong, since everything in a Challenging Compositional Cases is composed. But its guidance to seek for compositional explanations no longer provides resources for moving our understanding of such examples further forward. We already have compositional explanations for all the relevant entities, but the behavior of the parts in the relevant wholes still *cries out for explanation.* So everyday reductionism has hit a wall in such examples.

Two options for addressing examples like Challenging Compositional Cases loom large. On one side, one can take the behaviors of parts that are apparently unexplained by accounts used for simpler systems to be *merely apparently* problematic in an *epistemic* phenomenon—that is, an artifact of our theoretical machinery, rather than a feature of the world. In the next section, I outline why scientific reductionism is forced to adopt this deflationary option about Challenging Compositional Cases given its guiding picture of nature.

On the other side, we find the possibility that the world, rather than our theorizing, underlies the situation in Challenging Compositional Cases because parts really do have a special kind of behavior. As I outline in Part 4, this is the position of scientific emergentism. So let me frame the activities and powers we are assumed to have under this emergentist position to differentiate them from the scientific reductionist alternative. Let us say that we have what I term "differential powers" and "differential activities/behaviors" when a certain part contributes different powers, and hence behaves differently, under the condition of composing a certain whole, but where the part would not contribute these powers, and behave in these ways, if the laws applying in simpler collectives exhausted the laws applying in the complex collective.[9]

3 Scientific Reductionism, its Guiding Picture of Nature and Practical Import

Scientific reductionism grew popular in the sciences as everyday reductionism flourished in the twentieth century. For scientific reductionists, like Weinberg, Crick, Dawkins, Wilson and others, argue that reflection on compositional explanations, using a type of ontological parsimony argument, leads to their stronger position. Such arguments have long attracted thinkers. For example, ancient Buddhists reflecting on carts being composed of boards, axle and wheels concluded that we should only accept such parts and reject the existence of a further whole such as a cart. Why? Because we can putatively explain everything using the parts alone.

The advent of compositional explanations in the sciences makes such reasoning especially alluring. Consider such an argument applied in the sciences in what I term the "Argument from Composition" since it claims to be driven by the nature of composition alone. As we saw in Part 1, a good compositional explanation allows one to account for the activities or properties of a whole using the activities or properties of its parts. As we saw, we compositionally explain the muscles contracting using the movement of its constituent proteins, but not vice versa. Given the nature of compositional explanations, the reductionist concludes that our successful compositional explanations mean that we can now account for all the activities and properties of both parts and wholes using the components alone. But, given this sub-conclusion,

[9] I should note that I have defined differential powers to leave it open whether their contribution is determined by a composed entity or other component level entities. I also intend differential powers to include not only extra powers that add to the powers contributed in simpler collectives, but also contracted sets of powers excluding powers contributed in simpler collectives. There can also be mixed cases where differential powers are both added and subtracted.

the scientific reductionist claims we can then apply the so-called 'Parsimony Principle' in this case—that is, the principle that when we have two equally explanatory hypotheses about some phenomenon, then we should accept the hypothesis committed to fewer entities.We regularly use the Parsimony Principle in in application to scientific hypotheses and in cases of compositional explanation we have two hypotheses about what entities they concern. On one side, we have the hypothesis that we have both a whole and its parts (i.e. a muscle plus the proteins). On the other side, we have the scientific reductionist's favored hypothesis that we have parts alone (i.e. the proteins alone). But the latter hypothesis is simpler than the former. So, applying the Parsimony Principle along with the crucial sub-conclusion, the reductionist concludes that we should only accept that there are parts alone in any case of compositional explanation or, famously, that there is really *nothing but* the parts. Similar reasoning can be run about the activities or properties of wholes to conclude there are activities and properties of parts alone.[10]

Scientific reductionism thus claims we should adopt a much starker guiding picture of nature than that of everyday reductionism. But notice that scientific reductionism is not ultimately left committed to nature as a "dust cloud" of isolated, and unrelated, fundamental parts. For the reductionist accepts compositional explanations and concludes we should only endorse the entities used as the explanans at the "bottom" of such explanations which, rather than being isolated and unrelated parts, are always *collectives* of inter-related and organized parts—thus organized, inter-related proteins (and their activities and properties), like myosin and actin in filaments, are used to explain the muscle's contracting. Collectives are not further individuals, since "collective" is just a name for a group of inter-related parts and parts are thus still the only individuals. The guiding picture of nature, endorsed by scientific reductionism, is thus one of isolated parts *and* collectives of parts of ever increasing scales.

Appreciating its picture of nature illuminates why scientific reductionism accepts that higher sciences and their explanations are indispensable. Using statistical mechanics as his example of a higher science, Weinberg tells us that:

> The study of statistical mechanics, the behavior of large numbers of particles, and its application in studying matter in general, like condensed matter, crystals, and liquids, is a separate science because when you deal with very large numbers of particles, new phenomena emerge... even if you tried the reductionist approach and plotted out the motion of each molecule in a glass of water using the equations of molecular physics..., nowhere in the mountain of computer tape you produced would you find the things that interested you about water, things like turbulence, or temperature, or entropy. Each science deals with nature on its own terms because each science finds something in nature that is interesting. (Weinberg (2001), p.40)

Crucially, the scientific reductionist takes higher sciences to coin their own terms that refer to the larger, and larger, scale collectives of parts that they study. Scientific reductionists accept we need such higher sciences to study, and express, the truths about such collectives that cannot be expressed by lower sciences. But the only

[10] Some of Jaegwon Kim's famous arguments about mental causation have a related structure to the same conclusion. See Kim (1993b) and other papers in his (1993a).

determinative entities are the parts that form such collectives, and hence the only determinative laws are still solely about those parts in the simplest systems.

This last point comes into clearer focus if we look at a couple of methodological differences between everyday and scientific reductionism. First, note that everyday reductionism's mantra to search for compositional explanations applies at every level of nature using all manner of techniques. In contrast, we can see why scientific reductionists argue, under their guiding picture, that fundamental physics and its experimental machinery, like the supercollider, are specially important (Weinberg, 1994). Under scientific reductionism, fundamental physics, and its experiments, are the only one's exploring the determinative laws of nature, since the only entities that exist are the fundamental parts and hence these are the only determinative entities. Furthermore, scientific reductionism also assumes that the laws holding of such parts in the simplest systems exhaust the laws holding of such parts. Consequently, the laws holding of entities, like quarks or mesons, in the simplest systems extend to all situations and exhaust the laws holding of such entities anywhere (whether in complex or simple systems)—and hence exhaust the determinative laws of nature itself. Scientific reductionists thus conclude that special funding consideration should be given to experimental machinery of fundamental physics, like the supercollider, that illuminate these laws.

Second, connected points illuminate what scientific reductionism has to say about Challenging Compositional Cases. As we have seen, everyday reductionism offers no productive guidance to move us forward with such examples, since we have provided all the compositional explanations that we can in such cases. In contrast, scientific reductionism does offer guidance about such cases. Under scientific reductionism, the only determinative laws are those in the simplest systems and these laws exhaust the laws holding of parts in larger and larger collectives. But this means that the parts in Challenging Compositional Cases, whether electrons in superconductors, proteins in cells, or neurons in populations, only behave in ways that fall under the laws holding in simpler systems. Hence the scientific reductionist must argue that the presently inexplicable behaviors of parts in various wholes in Challenging Compositional Cases is merely an *appearance* in an *epistemic* artifact of our theorizing. For the guiding picture of scientific reductionism is committed to the laws of parts in the simplest systems exhausting the laws holding of the behavior of parts wherever they are found—and however hard it may be to understand this. Standard versions of scientific reductionism are thus left committed to a deflationary approach to Challenging Compositional Cases.

4 The Mutualism of Scientific Emergentism: The Core Idea

Rather than deflating their significance, scientific emergentism has taken our empirical findings in Challenging Compositional Cases to yield profound insights. As Laughlin tells us:

> Ironically, the very success of reductionism has helped pave the way for its eclipse. Over time, careful quantitative study of microscopic parts has revealed at the primitive level at least, collective principles of organization are not just a quaint side show but *everything*—the true source of physical law, including perhaps the most fundamental laws we know. (Laughlin (2005), p. 208)

In the next section I turn to methodological implications and fundamental laws. But here we also see the focus of emergentists on quantitative accounts of the behaviors of parts. Amongst the lessons learned from such empirical findings by emergentists like Laughlin are, first, that parts can behave differently in wholes, so they really do have differential behaviors/powers; and, second, that parts behave in these new ways because the whole (and/or its activities or properties) determines that these parts have differential powers and the differential behaviors that result.

Here is how Walter Freeman frames the resulting Mutualist picture in the cases in the neurosciences he focuses upon involving neurons in populations. Freeman tells us:

> An elementary example is the self-organization of a neural population by its component neurons. The neuropil in each area of cortex contains millions of neurons interacting by synaptic transmission. The density of action is low, diffuse and widespread. Under the impact of sensory stimulation, by the release from other parts of the brain of neuromodulatory chemicals… all the neurons come together and form a mesoscopic pattern of activity. This pattern simultaneously constrains the activities of the neurons that support it. The microscopic activity flows in one direction, upward in the hierarchy, and simultaneously the macroscopic activity flows in the other direction, downward. (Freeman (2000), pp. 131–132)

Here we have the core idea of Mutualism applied to a concrete case. We have a whole (and its activities and properties) in a population of neurons upwardly composed by neurons (and their activities and properties). But at the same time this whole (and its activities and properties) also downwardly determines (and "constrains") these component neurons (and/or their activities and properties) which consequently have differential behaviors and powers.[11]

We should mark that the downward determinative relation from whole to parts is not a compositional relation, since parts (and their activities and properties) together usually fill the causal roles of a whole (and its activities and properties). But a whole (and its activities and properties) cannot fill the causal role of its parts (and their activities and properties), nor hence compose them. This synchronic determinative relation is also not a causal one, since it again has features that causal relations lack such as being synchronous relations, holding between entities that are in some sense the same and involving synchronous changes in their relata. So we have a novel, downward, synchronic determinative relation, from whole to parts (and/or their activities and properties), that I have elsewhere dubbed a "machretic" relation.

---

[11] Strictly speaking, it is most plausibly an "emergent" activity or property of a whole that downwardly (machretically) determines that a realizing property of some part contributes a differential power and hence has a differential behavior/activity. However, for ease of exposition I have throughout the paper talked about wholes downwardly determining parts. The reader should take me to mean this more nuanced situation involving an activity or property of a whole when talking of such whole-to-part determination.

It is worth noting that under Mutualism wholes have at least two kinds of causal relations. At their own levels, wholes productively act "horizontally" on other wholes in *thick* causal relations of activity—thus a muscle acts upon sinews and bones when contracting. But, in addition, when we have such machretic relations at a time, where a whole (or more precisely one of its properties or activities) determines some property of a part contributes a differential power, then over time we will have *thin* causal relations (such as relations of manipulability) between the whole (and/or its relevant activity or property) and the differential behavior of the part that results from this power at some later time.[12] Removing the activity or property of the whole will remove the differential power of the part and hence the differential behavior. So machresis between whole and parts (or their activities and properties) at a time leads to thin downward causal relations over time between the whole (or its activities and properties) and differential behaviors of parts of this whole. Machresis thus always results in a species of thin downward causation and many scientific emergentist frame their views around such "downward causation".[13]

At this point, many philosophers object that this kind of situation is incoherent for various reasons.[14] Most commonly, philosophers and scientists seek to use the Argument from Composition, or related arguments about mental causation (Kim, 1993b), to conclude that we should never accept a whole is both composed and causally efficacious—and hence should not accept anything like Mutualism under which wholes are determinative in various ways, including causally. However, once we appreciate Mutualism, we can see that such arguments are plausibly invalid when they proceed from the assumption of compositional relations alone, rather than also assuming stronger claims like the Completeness of Physics.

Recall that the crucial sub-conclusion of the Argument from Composition is that in cases of comprehensive compositional explanation using parts alone accounts for, or explains, everything at the higher and lower level. However, when we have differential behaviors and powers of parts, and Mutualism is true, although all wholes (and their activities or properties) are the subjects of successful compositional explanations, we still *cannot* account for all the behaviors and powers of individuals solely using parts or their activities/properties. For the differential behaviors and powers of parts have not been explained. In this type of case, the premise that we have compositional explanations is true, but the sub-conclusion that we can explain everything with parts alone is false—so we can see that the Argument from Composition is invalid and similar points hold for related forms of Kim's argument about mental causation.

Scientific reductionists, and other proponents of reasoning like the Argument from Composition, have locked themselves into the assumptions, first, that the parts in wholes never in principle need explanation beyond that offered in simpler systems. And, second, that we only ever have upward determination. But these researchers thus

[12] Thick causal relations are usually relations of activity. In contrast, thin causal relations are captured by manipulability or difference-making accounts that require not such relation of activity between their relata.

[13] For more discussion, see Gillett (2020) and (2016a), Chap. 7.

[14] Gillett (2016a), Chap. 7, reviews a range of such concerns and offers rebuttals.

overlook options that scientific emergentists, like Prigogine, Laughlin, and others, claim empirical findings, about Challenging Compositional Cases, bring to the fore—parts can have differential behaviors/powers and the latter can be machretically determined downwardly by a whole or its activities/properties.

We can therefore see that a mistaken theoretical reason, in an alluring but invalid argument, has wrongly been used to dismiss natural phenomena that did not fit the scientific reductionist's reasoning and hence guiding picture. But, as Laughlin puts it in our opening passage, "The great power of science is its ability, through brutal objectivity, to reveal to us truth we did not anticipate".[15] And that power of science has produced empirical findings that, scientific emergentists argue, reveal the flawed assumptions, and invalid reasoning, used to justify the standard versions of scientific reductionism.

5 A New Guiding Picture of Nature and its Implications

Scientific emergentism, through its Mutualist view, offers us a guiding picture of nature, but is this picture really different from that of "reductionism"? In answering this question we need to be careful of the ambiguity we have now revealed over what we mean by "reductionism", since we actually confront two questions: How is the guiding picture of scientific emergentism different from that of *everyday* reductionism? And in what ways does it diverge from that of *scientific* reductionism? I take these questions in turn and show we get starkly different answers.

With regard to everyday reductionism, and the search for compositional explanations, we actually find overlap. As Laughlin explains about his main emergentist conclusion:

> One might subtitle this thesis the end of reductionism (the belief that things will necessarily be clarified when they are divided into smaller and smaller component parts), but that would not be quite accurate. All physicists are reductionists at heart, myself included. I do not wish to impugn reductionism so much as to establish its proper place in the grand scheme of things. (Laughlin (2005), p. xv)

Once we understand Mutualism, then we can see why Laughlin thinks scientific emergentists are clarifying the proper place of the search for compositional models and explanations, rather than abandoning that approach.

The guiding picture of scientific emergentism *supplements* that of everyday reductionism. Scientific emergentists take everything to be composed and they continue to accept that it is productive to search for compositional models and explanations. For all Mutualist cases are examples of individuals, and their activities and properties, that are composed by parts, and their activities and properties. But under Mutualism we have *added* the possibility of downward determinative relations of machresis at a time, and consequent thin downward causal relations over time, *alongside* the upward compositional relations endorsed by everyday reductionism.

[15] Laughlin (2005), p. xvi.

As I detail in the next section, these additions provide new resources under scientific emergentism to engage the cases where everyday reductionism hits a wall. However, let us now consider the differences with the guiding picture of scientific reductionism and its methodological advice. One place to highlight such differences concerns the fundamental laws and fundamental research.

As I outlined above, the nature of everyday reductionism, and compositional explanations, make parsimony arguments alluring. As the scientific emergentist Philip Anderson cautions us, once we accept everyday reductionism:

> It seems inevitable... [to accept] what appears at first sight to be an obvious corollary of [everyday] reductionism: that if everything obeys the same fundamental laws, then the only scientists who are studying anything really fundamental are those working on those laws... This point of view... it is the main purpose of this article to oppose. (Anderson, 1972, p. 393)

Here we see one of the key contentions of scientific reductionism that its stronger conclusions follow from everyday reductionism. But why does Anderson claim this conclusion is mistaken? Once again focusing on our more detailed empirical findings about the behavior of the parts in wholes, Anderson claims that:

> The behavior of large and complex aggregations of elementary particles, it turns out, is not to be understood in terms of a simple extrapolation of the properties of a few particles. Instead, at each level of complexity entirely new properties appear, and the understanding of the new behaviors requires research which I think is as fundamental in its nature as any other. (Anderson, 1972, p. 393)

As we have seen, scientific emergentists contend that the behaviors of parts in wholes are not those these parts would have if the accounts, and laws, in simpler systems were exhaustive. Instead, parts are claimed by scientific emergentists to have differential behaviors determined by wholes (and/or their activities or properties) and hence covered by new fundamental laws applying within certain wholes. Laughlin tells us:

> From the reductionist standpoint, physical law is the motivating impulse of the universe. It does not come from anywhere and implies everything. From the emergentist perspective, physical law is a rule of collective behavior, it is a consequence of more primitive rules of behavior underneath (although it need not have been), and it gives one predictive power over a limited range of circumstances. Outside this range, it become irrelevant, supplanted by other rules that are either its children or its parent in the hierarchy of descent. (Laughlin (2005), p. 80)

Under Mutualism, we thus have a complex array of fundamental laws covering parts and their behaviors: some hold in simpler systems but other fundemental "organizational" laws only hold of these parts in certain wholes.

Such laws deserve much more discussion which I have pursued elsewhere (Gillett (2016a), Chap. 7). But for our purposes here we begin to see a stark methodological difference that results from the guiding pictures of scientific reductionism and emergentism. Under scientific reductionism, fundamental physics has a monopoly on fundamental phenomena, research and laws. But under the guiding picture of scientific emergentism, as Anderson emphasizes, *many* sciences, studying nature at *many* levels, can be investigating fundamental nature phenomena, exploring the "frontiers" of fundamental research and discovering fundamental laws (Laughlin (2005), pp. 5–8).[16]

6 New Models and New Explanations: Resources for Challenging Compositional Cases and Beyond

Finally, let us return to local concerns and concrete scientific examples. We can now appreciate how the new guiding picture of nature of scientific emergentism supplies novel resources in such examples. As we saw with compositional explanations, we often give ontic explanations that explain natural phenomena by representing determinative relations in nature that result in the phenomena to be explained. To take another example, causal explanations represent various kinds of causal relation to provide ontic explanations. The foundational point is that such explanations work by representing ontological relations in nature. But Mutualism recognizes novel determination relations in nature in various kinds of machretic relation. Hence Mutualism allows a new class of models and explanations representing, and backed by, these new relations.

For example, under Mutualism one can now offer models and explanations that posit either machretic relations from wholes (or their activities or properties) to parts (or their activities or properties) at a time, or that posit thin, downward causal relations between these entities over time that result from such machretic relations. We could use the terms "whole-to-part", "machretic" or "downward causal" models/explanations for these various scientific products, but let me here simply dub them all "Mutualist" models/explanations given their connection to Mutualism.

Such Mutualist models/explanations may take all manner of forms. For instance, one can posit models representing machretic relations or thin downward causal relations. And one can construct models which variously take such relations to have individuals, activities or properties as relata. So these models can vary in their posited ontology. Furthermore, one can use different representational formats for such models. Thus one can use non-linear dynamics to articulate such models/explanations, but one can also use new network models/explanations to do the same. And a host of other mathematical and representational formats can be used to the same end.[17]

It is important to note that all of these representational formats can be used to give other models/explanations than such Mutualist ones. What is crucial is the intended, and/or the most plausible, interpretations of such successful applications of non-linear dynamics, network models, and so on. In each case, and for each application in this example, it is a substantive task to show either that researchers intend their

[16] For a more detailed discussion of the new methodologies under scientific emergentism se Mitchell (2009).

[17] See Juarrero (1999) for discussion of some of these Mutualist models and their features.

model to be a Mutualist one, or to show that some successful model is most plausibly interpreted as representing Mutualist relations.

It is therefore very much an ongoing question where Mutualist models/explanation may be productively applied and also an open issue where recent successful scientific work supports the existence of Mutualist scenarios in nature. My goal here is not to explore or resolve such questions, though elsewhere I have tried to begin to clarify what is involved in addressing them.[18] Instead, my focus has been on illuminating the Mutualist revolution and how the new guiding picture of scientific emergentism offers resources to researchers at the cutting-edge of science. And we have now found this to be the case. To properly see this, let us briefly return to Challenging Compositional Cases.

The key point is that we can *supplement* causal and compositional models/explanations in Challenging Compositional Cases with Mutualist models/explanations in the attempt to account for the behaviors of parts in the relevant complex wholes. Thus we can offer models positing machretic relations between a whole (or its activities or properties) and differential behaviors/activities or powers of a part or its property. Similarly, we can offer associated models positing a variety of thin, downward causal relations from wholes, or their properties and activities, to differential behaviors of parts. Machretic, and/or downward causal, relations can thus offer new models and explanations to understand such differential behaviors of parts. Note that such Mutualist models/explanations will be integrated with causal and compositional models/explanations, so these Mutualist models/explanations supplement, rather than supplant, the resources provided by everyday reductionism and other existing approaches.

Mutualist models thus offer researchers new resources in cases where we saw everyday reductionism has hit a wall. As Laughlin emphasized in an earlier passage, scientific emergentism is thus both broadening our methodological tool-kit and also putting compositional models/explanations in their proper place as one amongst a number of useful kinds of ontic model/explanation.

7 Conclusion

It bears emphasis that scientific research on Challenging Compositional Cases is very much ongoing. It is an open question whether Mutualist models and explanations, for example, provide the best accounts of electrons in superconductors, or proteins in cells, or a host of other cases. And it is a difficult connected, and ongoing, issue of whether successful approaches in such cases, for example using non-linear dynamics, network models, and so on, are best interpreted as being examples of Mutualist models or not. Thus it remains to be established whether we really have differential behaviors or powers in Challenging Compositional Cases.

---

[18] See Gillett (2016a), Chaps. 8 and 9.

I should also note that scientific reductionists can rework their views *even if they accept parts have differential behaviors and powers.* Here I have discussed the standard, or what I term "simple", version of scientific reductionism. But elsewhere I have sketched how one can revise scientific reductionism in what I term a "conditioned" version that accepts parts have differential behaviors and powers, but takes such differential behaviors/powers of parts to be determined, and hence explained, solely by *other parts.*[19] The conditioned variant of scientific reductionism offers novel models and explanations that can also potentially be used to gain traction with Challenging Compositional Cases. Consequently, both simple and conditioned scientific reductionist approaches need to be engaged by scientific emergentists, as relevant rivals, when they defend their treatments of various concrete cases.[20]

However, my focus here has not been on the new options for scientific reductionism. My goal has been to illuminate the exciting revolution that scientific emergentists have been working to achieve in so many areas of the sciences. Scientific emergentism, through its novel Mutualist guiding picture of nature, broaches new whole-to-part relations in the natural world, whether machretic relations at a time or thin downward causal relations over time. Consequently, the scientific emergentist's guiding picture of nature offers fresh resources for researchers in models and explanations backed by these novel relations. One can only be excited to see how, over coming decades, these new Mutualist models and explanations perform for us at the frontiers of science.

References

Aizawa, K., & Gillett, C. (Eds.) (2016a). *Scientific composition and metaphysical grounding.* Palgrave MacMillan.

Aizawa, K., & Gillett, C. (2016b). *Vertical relations in science, philosophy and the world: Understanding the New debates over verticality.* In Aizawa and Gillett (2016a).

Aizawa, K., & Gillett, C. (2019). Defending pluralism about compositional explanations. *Studies in the History and Philosophy of Biological and Biomedical Sciences, 78,* 101202.

Anderson, P. (1972). More is different: Broken symmetry and the nature of the hierarchical structure of science. *Science, 177,* 393–396.

Bechtel, W., & Richardson, R. (1993). *Discovering complexity.* Princeton University Press.

Betts, J.G. et al. (eds.) 2013: *Anatomy and physiology.* BC Campus. Retrieved from:https://openstax.org/books/anatomy-and-physiology/pages/1-introduction.

Boogerd, F., Bruggeman, F., Richardson, R., Stephan, A., & Westerhoff, H. (2005). Emergence and its place in nature: A case study of biochemical networks. *Synthese, 145,* 131–164.

Craver, C. (2007). *Explaining the brain.* Oxford University Press, USA.

Crick, F. (1966). *Of mice and molecules.* University of Washington.

Crick, F. (1994). *The astonishing hypothesis: The scientific search for the soul.* Scribner.

Dennett, D. (1969). *Content and consciousness.* Routledge Kegan Paul.

Fodor, J. (1968). *Psychological explanation.* Random House.

[19] See Gillett (2016a), Chaps. 8 and 9, for an exploration of conditioned scientific reductionism and its contrasts, and overlap, with the standard version of this position.

[20] See Gillet (2016a), Chaps. 9 and 10, for my more detailed take on what this might involve.

Freeman, W. (2000). *How brains make up their minds*. Columbia University Press.
Gillett, C. (2016a). *Reduction and emergence in science and philosophy*. Cambridge University Press.
Gillett, C. (2016b): "Emergence, downward causation and its alternatives: Surveying a foundational issue". In Gibb, Hendry & Lancaster (Eds.) *Handbook on the Philosophy of Emergence*. Routledge.
Gillett, C. (2020). "Why constitutive mechanistic explanation cannot be causal: Highlighting a needed theoretical project and its constraints". *American Philosophical Quarterly*.
Gillett, C. Forthcoming: *Compositional explanation in the sciences: Exploring integrative pluralism and ontic explanation*.
Glennan, S. (1996). Mechanisms and the nature of causation. *Erkenntnis, 44*(1), 49–71.
Juarrero, A. (1999). *Dynamics in action: Intentional behavior as a complex system*. MIT Press.
Kim, J. (1993a). *Supervenience and mind*. Cambridge University Press.
Kim J. (1993b). *The Nonreductionist's troubles with mental causation*. In Kim (1993a).
Laughlin, R. (2005). *A different universe: Reinventing physics from the bottom down*. Basic Books.
Lewin, R. (1992). *Complexity: Life at the edge of chaos*. Macmillan.
Machamer, P., Darden, L., & Craver, C. (2000). Thinking about mechanisms. *Philosophy of Science, 67*, 1–25.
Mitchell, S. (2003). *Biological complexity and integrative pluralism*. Cambridge University Press.
Mitchell, S. (2009). *Unsimple truths: Science, complexity and policy*. University of Chicago Press.
Noble, D. (2006). *The music of life: Biology beyond genes*. Oxford University Press.
Prigogine, I. (1997). *End of certainty*. The Free Press.
Prigogine, I., & Stengers, I. (1984). *Order out of chaos: Man's new dialogue with nature*. Bantam Books.
Weinberg, S. (1994). *Dreams of a final theory*. Random House.
Weinberg, S. (2001). *Facing up: Science and its cultural adversaries*. Harvard University Press.
Wilson, E. (1998). *Consilience: The unity of knowledge*. Knopf.
Wimsatt, W. (1976). "Reductionism, levels of organization and the mind-body problem". In *Conciousness and the Brain*, in G. Globus, G. Maxwell, & I. Savodnik (Eds.). Plenum Press.
Wimsatt, W. (2007). *Re-Engineering philosophy for limited beings*. Harvard University Press.

Causation in Buddhist Philosophy

Graham Priest

1 Introduction

Causation, as Mackie puts it in the title of his book,[1] is the cement of the universe. But how it holds the universe together, well, that's another matter. In particular, do the causal powers of things reduce to those of their parts, or is causation a more holistic matter?

The point of this paper is to discuss the Buddhist view of the matter—or better, Buddhist views. For Buddhist philosophy is no one thing, and though the different Buddhist schools typically agree on some things, there is a wide divergence of views on matters physical/metaphysical, causation included. The paper is not a survey of Buddhist views on the issue at hand. That would required a scholarly and inordinately longer paper. What I will do is describe the views of some important Buddhist schools, which will illustrate the wide variety of views that have been endorsed.

Nor is the point here to try to adjudicate the differences. Again, that would require a much longer philosophical treatise. My aim is simply to show the variety. In particular, we will look at three very distinctive views. The first is the Indian Abhidharma view. The second is the Madhyamaka view. This is Indian too, though it had an enormous impact on all subsequent Mahāyāna views—which include all the Chinese Buddhist views. The third is the Chinese Huayan view.

[1] Mackie (1980).

G. Priest (✉)
Departments of Philosophy, The Graduate Center, CUNY, 365 Fifth Avenue, New York, NY 10016, USA

University of Melbourne, Melbourne, Australia

Ruhr University of Bochum, Bochum, Germany

S. Wuppuluri and I. Stewart (eds.), *From Electrons to Elephants and Elections*, The Frontiers Collection, https://doi.org/10.1007/978-3-030-92192-7_7

I shall make some comments on the connection between these schools of thought. But those who seek an account of the history and geography of Buddhist philosophy must look elsewhere.[2]

2 Background

2.1 Holism and Reductionism

First, however, some general background. The notions of reductionism and holism are somewhat vague, and tend to be used in different ways. In his article in the *Stanford Encyclopedia of Philosophy*, Healey usefully defines methodological versions of these notions as follows[3]:

- *Reductionism*: An understanding of a complex system is best sought at the level of the structure and behavior of its component parts.
- *Holism*: An understanding of a certain kind of complex system is best sought at the level of principles governing the behavior of the whole system, and not at the level of the structure and behavior of its component parts.

These glosses, focussing on the notion of understanding, will serve our purpose here.

2.2 Causation

To understand something is to grasp its whys and wherefores. And in the case of the world of space and time, that means, of course, understanding causation.[4]

Causation is of many kinds, however. To see this, let us turn, not to Buddhism, but to Aristotle. As is well known, in his *Physics*, Aristotle distinguishes between four kinds of causation. In his own words[5]:

> ...we must proceed to consider causes, their character and number. Knowledge is the object of our inquiry, and men do not think they know a thing till they have grasped the 'why' of it (which is to grasp its primary cause). So clearly we too must do this as regards both coming to be and passing away and every kind of natural change, in order that, knowing their principles, we may try to refer to these principles each of our problems.

He then proceeds to describe the causes of an object as of four kinds, illustrating with respect to a bronze statue:

---

[2] A brief account can be found in Priest (2014), pp. xxiii–xxiii. Much fuller accounts can be found in Carpenter (2014), Mitchell (2002), and Williams (2009).

[3] Healey (2016).

[4] For Buddhism, the world of space and time is the whole world. Buddhists of all stripes are nominalists about universals, and accept no abstract objects.

[5] *Physics* 194^b16-194^b23. Translation from Barnes (1991).

- *Material cause*: the matter of which the thing is made; in this case the bronze.
- *Formal cause*: the form into which the material is shaped; in this case the form of a statue.
- *Efficient cause*: the process by which the statue comes into being; in this case, the working of the artificer.
- *Final cause*: the end for which the statue was made; perhaps, in this case, to produce an object of worship.

All of these causes are at work in Buddhist philosophy, though only the first three will be part of our story here.[6]

3 Abhidharma Buddhism

3.1 Pratītyasamutpāda

These matters clarified, let us now turn Buddhism itself.

Buddhist thought can traced back to the ideas of the Buddha (awakened/enlightened one), Siddhārtha Gautama (fl. 6 or 5 c. BCE); and causation is central to these. Buddhist thought provides what one might think of as an analysis of the human condition: its unsatisfactory (*duḥkha*) nature, the causes of this, and how to ameliorate things. An important part of the story is that everything is in a causal flux. Things come into existence when causes and conditions are ripe, maintain themselves in a state of causal interaction for a time, and then go out of existence when, again, causes and conditions are ripe. Everything is impermanent (*anitya*). As one of the *sūtras* puts the matter of causation[7]:

> When this is, that is.
>
> From the arising of this comes the arising of that.
>
> When this isn't, that isn't.
>
> From the cessation of this comes the cessation of that.

The causal flux is termed *pratītyasamutpāda* (dependent origination/arising). The causation involved here is clearly efficient causation.

3.2 Dharmas

Other aspects of causation emerged in detail a little later. In the 500 years after the Buddha, a number of schools of Buddhist thought arose. These are known as the Abhidharma (higher teaching) schools.[8]

---

[6] In Buddhism, there is an appropriate final cause, the attainment of *nirvāṇa*. But this cause belongs to Buddhist soteriology, not metaphysics.

[7] Thanissaro (2005).

[8] On these, see Ronkin (2018).

It seems fairly obvious that the things we meet with in the normal course of events (including people) are composed of parts. My body has arms and legs; my perception contains sights and sounds; my car has wheels and a chassis. Those parts can themselves have parts. For example, my arm has a hand, an elbow. And those parts can themselves have parts. Thus, my hand has five fingers.

If we take some object, and consider its parts, their parts, the parts of these... and so on, must we come ultimately to partless parts—things which are simple, and themselves without parts? The Abhidharma philosophers said 'yes'. It would seem that if the parts of parts went on for ever, there would ultimately be nothing there—which there obviously is.

The Abhidharma philosophers called these ultimate parts *dharma*s.[9] Dharmas are the ultimate building blocks of reality, its atoms. They do not depend for being what they are on their parts (obviously) or anything else. They have *svabhāva*. Literally this means something like *self-being*, or *self-nature*. Perhaps the best translation into English is *intrinsic nature*, though it is common to see the word translated (somewhat misleadingly) as *essence*. Note that the *dharmas* are in the flux of *pratītyasamutpāda* as much as anything else. They interact causally with other *dharmas*, and themselves come into and go out of existence.

All the Abhidharma philosophers agreed that there were different kinds of dharmas—for example, physical and mental—though there was some disagreement about their exact nature. Perhaps the most common view was that they are tropes, that is, particular instantiations of universals, such as the redness of this cherry, or the painfulness of this experience.[10] Whatever they are, however, they are the things that are ultimately real.

The objects of our normal experience, by contrast, are simply bunches of *dharmas* arranged in a certain way. So, a table is just a bunch of atoms "arranged table-wise"; and a person is a bunch of atoms "arranged person-wise". Certain bunches of atoms have a causal continuity which gives them an important role in our lives: trees, houses, indeed people themselves. It is therefore useful to single those out with particular concepts, such as *house*, *person*, or *White House*, *Donald Trump*.

Hence there are two kinds of reality: an ultimate reality (*paramārtha satya*), comprising the *dharmas*, and a conventional reality (*saṁvṛti satya*) comprising the conceptual constructions made from these things.[11]

---

[9] Note that the word *dharma* has many different uses in Buddhist philosophy. Literally it means something like: that which is established or firm.

[10] See, e.g., Ganeri (2001), chap. 4.

[11] Further on this picture, see Siderits (2007), Chap. 6. The Sanskrit word *satya* may be translated both as *reality* and *truth*. 'Truth' is the more usual scholarly translation; but in the present case, I think that 'reality' is definitely better. Note also that a distinction between a conventional reality and an ultimate reality goes all the way back to the earliest stages of Buddhism. The Abhidharma philosophers give it a distinctive metaphysical twist, however.

The whole situation is summed up by Vasubandhu (fl. 4th or 5th c. CE) in late Abhidharma text, *Abhidharmakośa-Bhāṣya* (*Commentary on the Treasury of Abhidharma*) as follows[12]:

> The Fortunate One has... declared two truths, (1) conventional or relative truth (*saṃvṛitisatya*) and (2) absolute truth (*paramārthasatya*). What are these two truths?...
>
> If the cognition of a thing disappears when this thing is broken into parts, this thing exists relatively or conventionally. An example is a pitcher, for when the pitcher is broken into shards, the cognition of a pitcher disappears, or does not arise.
>
> If the cognition of a thing disappears when the [(constituent) factors of this thing] are mentally removed, this thing too should be regarded as existing relatively or conventionally. An example is water, for when—with respect to water—we [mentally] take and remove the factors, such as visible form or color, etc., the cognition of the water disappears or does not arise.
>
> To these things, e.g., pitcher, clothes, etc., water, fire, etc., different names or notions are given from the relative point of view or in accordance with conventional usage. Thus, if one says, from the relative or conventional point of view: "There is pitcher, there is water", one speaks truly, one does not speak falsely. Hence this is relative or conventional truth.
>
> That which is other than this is absolute truth. Therein, even when a thing is being broken—or [likewise, even if its (constituent) factors] are mentally removed, and the cognition of this thing continues, then this exists absolutely. For example, visible form: for, therein, when a visible [thing] is broken into atoms or infinitesimal particles and when taste and the other factors have been mentally removed, the cognition of the intrinsic nature [*svabhāva*] of visible form persists. Sensation, etc., is also to be seen in the same way. As this exists absolutely, this is absolute truth.

3.3 *Reflections on Causation*

Before we move on the the next Buddhist school of philosophy we will meet (Madhyamaka), let us pause for a few philosophical reflections.

First, we have been talking about parts and wholes—mereology. During the last century, starting with the work of Husserl and Leśniewski, this has become a well-developed part of formal logic.[13] In this, there is an operation called mereological *sum* or *fusion*. To illustrate: if you take all my parts and fuse them together you get me. If you take the four movements of Beethoven's 9th Symphony, you get the whole symphony. There is a standard debate in mereology as to when a bunch of things have a fusion. Some philosophers hold that any bunch of objects fuse to form a whole, though this may be a strange one. Some deny this. Thus, consider an incongruous bunch of objects such as: the Eiffel Tower, the Buddha's left earlobe, and Jupiter. These, it is held, have no fusion. To have a fusion, a bunch of objects must have a certain coherence—though how best one might understand this, is somewhat unclear.

---

[12] De La Vallée Poussin et al. (2012), Vol. 3, pp. 1891-2. I have removed many of the Sanskrit glosses. 'Fortunate One' is an honorific for the Buddha. In what follows square brackets contain translator's interpolations unless otherwise noted.

[13] For a general account, see Varzi (2016).

Now, in Abhidharma, the objects of conventional reality may naturally be thought of as the fusion of their *dharmic* parts, and the *dharmas* that fuse to form a whole are precisely those which fall under some concept, such as *person* or *Graham Priest*. The concepts to be deployed here are those of common sense, or perhaps its theoretical developments. But it seems clear that there is no natural concept which unifies the *dharmas* in our trio of incongruous objects. The Abhidharma philosophers would therefore have agreed with the modern philosophers who hold that not all bunches of things have a fusion.

Next, we had already met the notion of efficient causation in Buddhism. We have now also met the notions of material cause and formal cause. Given an object of conventional reality, its *dharmic* parts are its material cause. The concept which unifies its parts is its formal cause. This is not exactly an Aristotelian form, but it does the same job of forming the matter into an object of a certain kind.[14]

Finally, the notion of causation involved in the Abhidharma metaphysics is clearly reductionist. The only complexes are the objects of conventional reality. Our concepts pick out their *dharmic* parts, their matter; and efficient causation works on these. An understanding of the behaviour of the objects of conventional reality is therefore to be found at the level of their component parts.

4 Madhyamaka Buddhism

4.1 Emptiness

Let us move to our next Buddhist school.

Around the turn of the Common Era, a new form of Buddhism arose, Mahāyāna (Greater Vehicle). This had a quite different metaphysical picture of the world. In fact, there are several different Mahāyāna Buddhisms: two major ones in India, and all of the East Asian Buddhisms. However, it is just one of these on which we will concentrate here, Madhyamaka (Middle Way).

The basis of this was laid out by Nāgārjuna (fl. 1st or 2nd c.) in his *Mūlamadhyamakakārikā* (MMK, *Fundamental Verses of the Middle Way*), which was to exert a profound influence on all Mahāyāna Buddhisms.[15] In this, Nāgārjuna launches an attack on the older metaphysics. In particular, he argues that there are no such things as *dharmas* in the sense that the Abhidharma philosophers held, namely, things with *svabhāva*. Everything is empty (*śūnya*) of intrinsic nature. Everything, that is, is what it is, not in and of itself, but only in relation to other things.

Of course, for the Abhidharma philosophers, the objects of conventional reality are what they are only in relation to their parts and our concepts. A central part of Nāgārjuna's attack was to broaden this picture by adding efficient causation to the

[14] See Priest (2014), Chap. 3. Moreover, if the *dharmas* are tropes, these are exactly instances of Aristotelian forms—without any matter.

[15] On Nāgārjuna, see Westerhoff (2018). On Madhyamaka in general, see Hayes (2019).

list. For the Abhidharma philosophers, the efficient causes of something determine *that* it is, but now *what* it is. In Madhyamaka thought it does. Thus, to illustrate, an acorn is what it is (in part) because it grows on an oak tree, and generates further oak trees. If it grew on bicycles and produced, not oak trees, but goldfish, it would hardly be an acorn. These are matters of efficient cause and effect. In Madhyamaka, then, everything is what it is in relation to is parts, causes and effects, and our concepts.

Given this picture, it would have been natural, one might think, for Nāgārjuna to jettison the notion of ultimate reality altogether. But whether because of respect for his tradition, or for some other reason, he does not. He is as clear as his predecessors that there are two realities (MMK XXIV: 8-10)[16]:

> The Buddha's teaching of the Dharma
> Is based on two truths:
> A truth of worldly convention
> And an ultimate truth.
>
> Those who do not understand
> The distinction between these two truths
> Do not understand
> The Buddha's profound truth.
>
> Without a foundation in conventional truth
> The significance of the ultimate cannot be taught.
> Without understanding the significance of the ultimate
> Liberation cannot be achieved.

What conventional reality is, for Nāgārjuna, is clear enough. As for the Abhidharma philosophers, it is the world of our familiar experience. But what the ultimate reality of an object is, is much less clear.

He refers to this as *emptiness* (*śūnyatā*); and two things about it, anyway, seem clear. The first is that it is as empty as anything else. In perhaps the most famous verse of the MMK (XXIV: 18), he says:

> That which is dependent origination
> Is explained to be emptiness.
> That, being a dependent designation,
> Is itself the middle way.

To give the standard explanation: the (conventional) things in the flux of *pratītyasamutpāda* are empty (of *svabhāva*). Emptiness is, however, itself empty (dependent for being what it is on other things). Thus all things are neither non-existent nor are they what they are in and of themselves. The truth steers between these two extremes.

---

[16] Translations from the MMK are from Garfield (1991). Note that 'Dharma' here means *Buddhist doctrine*.

Of course, this raises the question of what it is that the ultimate reality of something itself depends on. Nāgārjuna is silent on the matter, but there isn't much for it to depend on except conventional reality. Sometimes this relationship between the conventional reality of an object and its ultimate reality is likened to that between the two sides or a coin. One cannot have the one without the other. And each, as it were, delivers a different aspect of the same thing. As Candrakīrti (fl. 7 c.), the most influential commentator on Nāgārjuna in the Tibetan tradition, puts it in his *Madhyamakāvatāra* (*Introduction to the Middle Way*)[17]:

> The Buddhas, who have an unmistakable knowledge of the nature of the two truths, proclaim that all things, outer and inner, as they are perceived by two kinds of subject (deluded consciousness on the one hand and perfectly pure wisdom on the other), possess a twin identity... They say that the object perceived by authentic primordial wisdom is the ultimate reality, whereas the object of a deluded perception is the relative truth.

The other thing that Nāgārjuna appears to be clear about is that the ultimate reality of something is ineffable. Thus, he says in the dedicatory verses of the MMK:

> I prostrate to the perfect Buddha,
> The best of all teachers, who taught that
> Whatever is dependently arisen is
> Unceasing, unarisen.
>
> Not annihilated, not permanent,
> Not coming, not going,
> Without distinction, without identity
> And free from conceptual construction.

Of course, the *whatever* in question is the ultimate aspect of something in the causal flux. Its conventional aspect is clearly dependent on conceptual construction—that is one of the things that makes it conventional. And given that, it can be described by those concepts. That concepts are *constitutive* of conventional reality is, presumably, the reason why the ultimate cannot be described, though Nāgārjuna is not explicit on the matter.

It is worth noting, however, that Nāgārjuna's view that the ultimate is ineffable is not idiosyncratic. He is just being faithful to the *sūtra* literature. Thus, for example, in the *Vajracchedikā Sūtra* (Diamond Sūtra) one of the most important Mahāyāna sūtras, we have[18]:

> [The Buddha said]: Subhūti, words cannot explain the real nature of the cosmos. Only common people fettered with desire make use of this arbitrary method.

Ultimately, then, things are ineffable.

---

[17] Padmakara Translation Group (2004), p. 192.

[18] Price et al. (1990), p. 51.

4.2 The Structure of Emptiness

So much for exegetical matters. Again before we turn to the next Buddhists school we will meet (Huayan), let us pause for some philosophical reflections.

The objects of conventional reality are, as we have seen, empty of intrinsic nature. That is, an object is what it is only in virtue of its relations—mereological, conceptual, and (efficiently) causal—to other things. In other words, anything which bore exactly those relations to those things would be that very object. Or, to put it another way, its identity is determined by its locus in a network of relations.

One may illustrate with a diagram. Take some object, and suppose that it is relevantly related to three objects: to a by the relation α, to b by β, and to c by γ. We may depict matters thus:

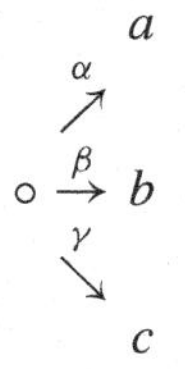

$$\begin{array}{ccc} & & a \\ & \overset{\alpha}{\nearrow} & \\ \circ & \overset{\beta}{\rightarrow} & b \\ & \overset{\gamma}{\searrow} & \\ & & c \end{array}$$

The circle, $\circ$, marks the locus of the object in this network of relations. And anything that occupied that locus would be that very thing.[19]

Of course, what is true of the object located at $\circ$ is true of the objects a, b, and c themselves, since they, too, are empty. So we may "expand" them in the same way—taking the number three, again, for the sake of illustration. This time I omit the labels of the relations, to avoid clutter:

[19] Note that relationships have a direction, from subject to object. Thus consider the relationship of killing. There is a big difference between *Brutus killed Caeser* and *Caesar killed Brutus*. I have indicated the direction of the relationship in the diagram with an arrow. In the diagram, all the arrows point in the same direction. One can do this for the following reason. Every relation has a converse, which can be used to express the same thing. Thus, the converse of *kill* is *be killed*; and one can say, indifferently, *Brutus killed Caesar*, and *Caesar was killed by Brutus*. Hence one can always always choose whichever of a relation and its converse it is which points in the right direction.

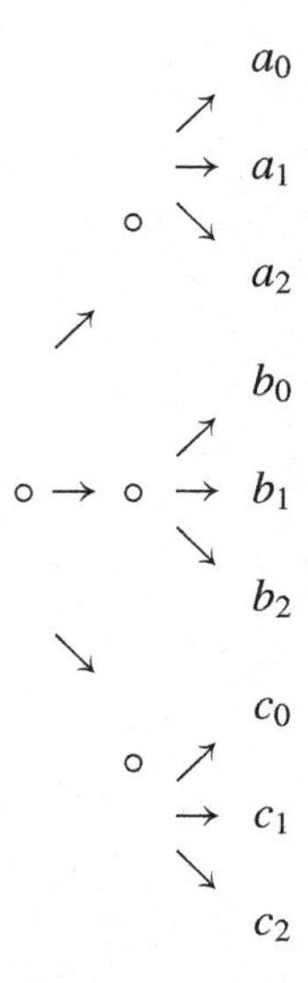

And of course, the same is true of the various as, b's and cs. So we can repeat the process, and keep doing so indefinitely. If we do this as often as possible we then arrive at the following diagram—called by mathematicians a *tree*.

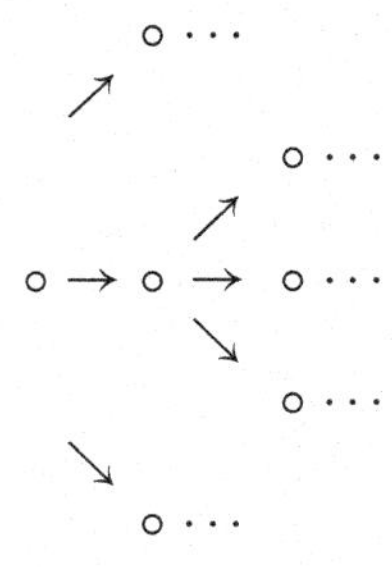

Our original object has become the root (i.e., first node) of the tree. And any branch of the tree—that is, any route from the object following a path of arrows all the way along—is infinite. The structure of the causal relations in the tree gives, as it were, the metaphysical structure of the original object.

4.3 Reduction and Holism Again

And what does this tell us about reductionism and holism concerning causation in Madhyamaka?

The first thing that might occur to you when you see the metaphysical trees with the infinite branches is that this is going to generate an infinite regress of explanations, which is either vicious or makes a reduction impossible.

Both of these thoughts would be wrong. There is nothing vicious about an infinite regress of explanations. Thus, the explanation of the existence of Anna could be the fact that Betty gave birth to her. The explanation of the existence of Betty could be the fact that Cathy gave birth to her. The explanation of the existence of Cathy might be the fact that Dorothy gave birth to her... and so on indefinitely. Of course, one might not think that actual human history is infinite in this way, but there is nothing *logically* impossible about this. In fact, traditionally, Buddhist thought indeed took the universe to be infinite in time past.

Nor is there anything about the regress which makes reduction impossible. Even if explanations ultimately cash out in terms of some some fundamental mereological level, the behaviour of the objects at that level might have a regress backwards in time of the kind involving Anna, Betty, Cathy, and so on.

Where the causal structure of Madhyamaka is relevant is this. In the reductionism of Abhidharma, causation is ultimately at the level of *dharmas*. These are simple. At this level, there are no wholes, and so there is no question of holistic explanations. In Madhyamaka, if one talks of explanation, this has to be at the conventional level, since we can say nothing about the ultimate level. And at the conventional level, there are complex wholes—the objects of our familiar experience, such as cars and people. Since there are wholes, there is a possibility of holistic explanation.

And *prima facie* there do appear to be such explanations. Thus, one might explain the melting of a polar cap in terms of changes to the ecosystem, of which the behaviour of the polar cap is a part. Or one might explain the fact that someone sold their shares in terms of the fact that there was a run on the market, of which the person's behaviour is a part.[20] Of course, it might be that on closer inspection such explanations can be reduced to non-holistic explanations. However, this is ultimately a matter for scientific investigation, and not to be settled by abstract metaphysical considerations. The point is simply that Madhyamaka metaphysics is compatible with both reductionism and holism. It does not determine which of these features of causation is correct.

5 Huayan Buddhism

5.1 The Net of Indra

The third school of Buddhism that we will look at is the Chinese Huayan (Jap: Kegon) School.[21] (The meaning of the name is not important. It is a translation of the Sanskrit *Avataṃsaka* meaning *flower garland*—the name of the *sūtra* the school took to be of most importance.)

[20] Further putative examples occur in quantum mechanics, where distinct particles can be entangled. This means that one particle of a pair has (say) *spin up*, in terms of the fact that an observation on the other member of the pair determined it to have *spin down*.

[21] The relevant Chinese characters for the names in this section can be found in an appendix to the essay.

Buddhism (Mahāyāna) started to go into China around the turn of the Common Era. There, it met the indigenous philosophies of Confucianism and Daoism. Daoism, in particular, was to exert a profound influence on its development. By about the 6th century, distinctively Chinese forms of Buddhism were developing. Huayan was one of these.[22] Traditionally, the founder and first Patriarch of the school is taken to be Dushun (557–640), but the most influential thinker is usually reckoned to be the third, Fazang (643–712). We will also meet the fourth, Chengguan (738–839).

One of the fundamental features of the school—indeed, the one that will be most important for our purposes—is that it universalises the Madhyamaka claim that any thing is what it is in virtue of its relationships to *some* other things. Any thing is what it is in virtue its relationships to *all* other things. Indeed, each thing has a very specific relationship to all other things. Using the metaphor of coins, Fazang puts the matter as follows[23]:

> If we take ten coins as symbolizing the totality of existence, and examine the relationship of existence amongst them, then, according to Huayan teaching, coin one will be seen as identical with the other nine coins.

The character translated as 'identical with' here is *ji (*Jap: *soku*). And indeed, in the vernacular, this means something like 'is the same as'; but it is clear that the Huayan philosophers use it in a very specific and technical sense. I will translate it, as is sometimes done, as 'interpenetrate'.

The relationship of universal interpenetration is depicted in more detail by perhaps the most famous metaphor associated with Huayan: the Net of Indra (*Yintuoluo wang*). Fazang puts the matter as follows[24]:

> It is like the net of Indra which is entirely made up of jewels. Due to their brightness and transparency, they reflect each other. In each of the jewels, the images of all the other jewels are [completely] reflected. This is the case with any one of the jewels, and will remain forever so. Now, if we take a jewel in the southwestern direction and examine it, [we can see] that this one jewel can reflect simultaneously the images of all other jewels at once. It is so with the one jewel, and is also so with each of all the others. Since each of the jewels simultaneously reflects the images of all other jewels at once, it follows that this jewel in the southwestern direction also reflects all the images of the jewels in each of the other jewels [at once]. It is so with this jewel, and is also so with all the others. Thus, the images multiply infinitely, and all these multiple infinite images are bright and clear inside this single jewel. The rest of the jewels can be understood in the same manner.

The god Indra has spread a net through space. At every joint of the net there is a brightly polished jewel. Each jewel reflects every other jewel; but each jewel reflects every other jewel reflecting every other jewel, and reflects every jewel reflecting every other jewel reflecting every other jewel... and so on to infinity. (Like two mirrors face to face, each reflecting the other *ad infinitum*.) The jewels are metaphors for the objects of reality; and the infinite reflection is a metaphor for interpenetration.

---

[22] On Huayan, see Van Norden (2019).

[23] *Huayan wujiao zhang* (*Treatise on the Five Teachings*). Quoted in Cook (1977), p. 2.

[24] *Treatise on the Five Teachings*, quoted in Liu (1982), p. 65.

5.2 *Li and Shi*

But how should one understand this metaphor? To do so we must start by going back to Madhyamaka. As we saw, according the Candrakīrti, things have a conventional aspect and an ultimate aspect which depend on each other.

The notion of the ultimate undergoes an important transformation in Chinese Buddhism. In Daoist thought, behind the flux of our familiar world there is a single ultimate ground, *dao*. The "myriad things" of phenomenal reality are the manifestations of this. Moreover, *dao* is ineffable. It cannot be a *this*, rather than a *that*, since it has to become all things. In Chinese thought, the Buddhist ultimate/conventional distinction becomes identified with the Daoist distinction between *dao* and its manifestations.[25] Given this, every object of conventional reality has *exactly the same* ultimate nature (often called *Buddha nature*, *foxing*, Jap: *busshō*). In his *Treatise on the Golden Lion* (*Jin shizi*) Fazang uses the example of a golden statue of a lion to explain matters. Ultimate reality is like the gold out of which the lion is made. Conventional reality is like the shapes that the gold assumes in the various parts of the lion.

We may put the matter in the language of Huayan as follows. An element of conventional reality is called a *shi* (thing, fact, object); and ultimate reality is called *li* (principle). *Li* and *sh*i interpenetrate, *lishi wuai* (the non-obstruction of *li* and *shi*). Dushun puts the matter as follows in his *Meditation on the* Dharmadhātu (*Huayan fajie xuan jing*)[26]:

> *Li*, the law that extends everywhere, has no boundaries or limitations, but *shi*, the objects that are embraced by *li*, have limitations and boundaries. In each and every *shi*, the *li* spreads all over without omission or deficiency. Why? Because the truth of *li* is indivisible. Thus, each and every minute atom absorbs and embraces the infinite truth of *li* in a perfect and complete manner.
>
> *Shi*, the matter that embraces, has boundaries and limitations, and *li*, the truth that is embraced [by things], has no boundaries or limitations. Yet this limited *shi* is completely identical [GP: *ji*], not partially identical, with *li*. Why? Because *shi* has no substance [GP: *svabhāva*]—it is the selfsame *li*. Therefore, without causing the slightest damage to itself, an atom can embrace the whole universe. If one atom is so, all other dharmas should also be so. Contemplate on this.

And in his *Treatise on the Golden Lion*, Fazang says[27]:

> All phenomena [GP: *shi*] are in great profusion, and are interfused but not mixed (losing their identity). The all is [GP: *ji*] the one [GP: *li*], for both are similar in being non-existent in nature [GP: having no *svabhā*va]. And the one is the all for the relation of cause and effect are perfectly clear. As the power [of the one] and the function [of the many] embrace each other, their expansion and contraction are free and at ease.

---

[25] There is much more to the matter than this. Various elements of Indian Buddhist thought concerning *Yogācara* and *tathāgata-garba* played an important role in this process; but we need not go into this here. For some discussion, see Priest (2018), 8.2-8.4.

[26] Chang (1972), pp. 144-5. The *Dharmadhātu* is the realm of all things. In what follows, the interpolations with my initials are mine.

[27] Chan (1969), p. 410.

Universal interpenetration tells us that *shi* and *shi* also interpenetrate *shishi wuai* (the non-obstruction of *shi* and *shi*). Moreover, they do this because of the their relationship with *li*. In *A Hundred Gates to the Sea of Ideas of the Avataṃsaka Sūtra* (*Huayan jing yi hai bai men*) Fazang puts this as follows[28]:

> [A particle of dust] has the characters of roundness and smallness. This is a fact [GP: *shi*]. Its nature is empty and non-existent. This is principle [GP: *li*]. Because facts have no substance [GP: *svabhāva*] they merge together in accordance with principle. And because the dust has no substance, it universally penetrates everything. For all facts are no different from principle and they are completely manifested in the dust.

As is immediately clear, the lack of substance (*svabhāva*) is playing an important role in all this. What, exactly, is interpenetration, though? And what has this to do with the lack of *svabhāva*?

5.3 Interpenetration

A simple way to see this is to go back to our representation of empty objects.[29] As we saw, the doctrine of emptiness implies that the metaphysical structure of an object is given by a tree, every branch of which is infinite. Now, as an example, consider a magnet. Let n be its north pole, and let s be its south pole. The north pole, being a north pole, depends on the south pole. So the tree for n, looks like this:

$$\begin{array}{ccccc} & & \dots & & \dots \\ & \nearrow & & \nearrow & \\ n & \stackrel{\rho}{\rightarrow} & s & \rightarrow & \dots \\ & \searrow & & \searrow & \\ & & \dots & & \dots \end{array}$$

where ρ is whatever the relationship is between two poles. But of course, the south pole, being a south pole, depends on the north pole. So if we take account of this in our diagram, we obtain:

$$\begin{array}{ccccccc} & & \dots & & \dots & & \dots \\ & \nearrow & & \nearrow & & \nearrow & \\ n & \stackrel{\rho}{\rightarrow} & s & \stackrel{\rho}{\rightarrow} & n & \stackrel{\rho}{\rightarrow} & \dots \\ & \searrow & & \searrow & & \searrow & \\ & & \dots & & \dots & & \dots \end{array}$$

---

[28] Chan (1969), p. 36.

[29] The matter is discussed in more detail in Priest (2015) and Priest (2018), Chap. 8.

The tree for n has the tree for s as a sub-tree, and vice versa. Moreover, as the diagram makes clear, this feature will repeat itself *ad infinitum*—just like the images of the two mirrors facing each other.

n and s intermingle metaphysically in the most intimate fashion. We may take the way they do so to be interpenetration. That is:

- Two objects interpenetrate [*ji*] if (the tree for) each is a part of (a tree for) the other.

That two objects interpenetrate is possible only because the branches are infinite—it could not happen if all branches were finite—and this is so because every object is empty.

With this understanding of interpenetration, the Huayan conclusions quickly follow. Let us write l for *li* and s_1 and s_2 for two example *shi*. Then since l is empty, and interpenetrates with each *shi*, its metaphysical tree looks like this:

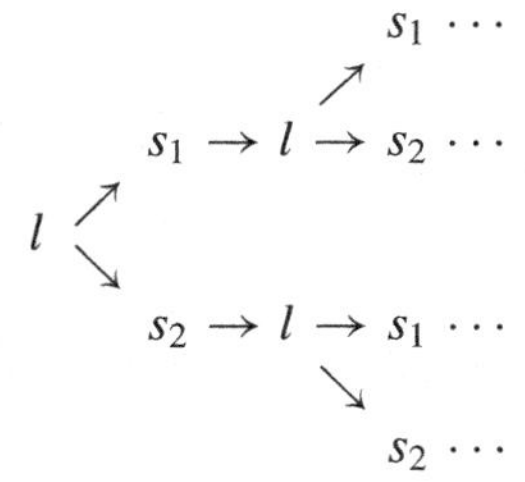

$$\begin{array}{ccccccc} & & & & & & s_1 \cdots \\ & & & & & \nearrow & \\ & & s_1 & \to & l & \to & s_2 \cdots \\ & \nearrow & & & & & \\ l & & & & & & \\ & \searrow & & & & & \\ & & s_2 & \to & l & \to & s_1 \cdots \\ & & & & & \searrow & \\ & & & & & & s_2 \cdots \end{array} \qquad (1)$$

That is, *lishi wuai*. But exactly the same diagram shows that s_1 and s_2 interpenetrate. That is, *shishi wuai*.[30]

As Chengguan says in his *Prolog to Huayan* (*Huayan jing shu zhu*)[31]:

> Because they have no selfhood [GP: *svabhāva*], the large and the small can mutually contain each other... Since the very small is very large Mount Sumeru is contained in a mustard seed; and since the very large is the very small, the ocean is included in a hair.

5.4 Reduction and Holism Again

Let us finish by returning to the question of reductionism and holism again. In the Huayan picture, every element of reality, whatever it is, causally interacts (in our various senses of causation) with every other every other element.[32] Clearly, this is

[30] What is going on here is essentially as follows. This relationship of interpenetration is clearly symmetric. (If a interpenetrates with b then b interpenetrates with a.) And a little thought shows that it is transitive too. (If a interpenetrates with b and b interpenetrates with c, then a interpenetrates with c.) A sub-tree of a sub-tree is a sub-tree. s_1 and s_2 interpenetrate with l. By symmetry, l interpenetrates with s_2, and so my transitivity, s_1 interpenetrates with s_2.

[31] Chang (1972), p. 165.

[32] For a discussion of the holism of Fazang's specifically mereological views, see Jones (2009).

a very global form of holism. Indeed, the metaphor of the Net of Indra is as striking a visual depiction of holism as one might wish.

6 Conclusion

Buddhism, as I said at the start, is not one thing. In particular, there are significant disagreements of a metaphysical kind between different schools of Buddhism. The causal structure of the cosmos is one such difference.

We have seen this to be the case by looking at the relevant parts of three Buddhist schools of thought: Abhidharma, Madhyamaka, and Huayan. Abhidharma can be naturally seen as having a reductionist account of causation. Huayan clearly provides a holist view.[33] Madhyamaka is poised somewhere in the middle—appropriately enough, for the Middle Way School. Its framework accommodates both reductionist and holist positions, the actual truth of the matter being determined by the investigations of empirical science.

The question of whether causation is reductionist or holist is, of course, a contentious question in Western philosophy. As we have seen, it is no less so in Buddhist philosophy.

Glossary of Chinese Characters

Chengguan: 澄觀
Dao: 道
Dushun: 杜順
Fazang: 法藏
foxing: 佛性
Huayan: 華嚴
Huayan fajie xuan jing: 華嚴法界玄鏡
Huayan jing shu zhu: 華嚴經疏注
Huayan jing yi hai bai men: 華嚴經一還百門
Huayan wujiao zhang: 華嚴五教章
ji: 即
Jin shizi: 金獅子
li: 理
lishi wuai: 理事無礙
shi: 事
shishi wuai: 事事無礙
Yintuoluo wang: 因陀羅網

[33] For an explicit contrast of the two schools in this regard, see Jones (2015).

References

Barnes, J. (Ed.). (1991). *The complete works of Aristotle*. Princeton University Press.
Bikkhu, T. (tr.). (2005). *Assutavā Sutta*, https://www.accesstoinsight.org/tipitaka/sn/sn12/sn12.061.than.html.
Carpenter, A. (2014). *Indian Buddhist philosophy*. Durham: Acumen.
Chan, W. T. (Ed. and tr.). (1969). *A source book in Chinese philosophy*. Princeton, NJ: Princeton University Press.
Chang, G. C. C. (1972). *The Buddhist teaching of totality: The philosophy of Hwa Yen Buddhism*. London: George Allen and Unwin Ltd.
Cook, F. (1977). *Hua-Yen Buddhism: The jewel net of Indra*. University Park, PA: Pennsylvania State University Press.
De La Vallée Poussin, L., Sangpo, G. L., (trs.). (2012). *Abhidharmakośa-Bhāṣya of Vasubandhu*. Motilal Banarsidass Publishers Private Limited.
Ganeri, J. (2001). *Philosophy in classical India*. London: Routledge.
Garfield, J., (tr.). (1991). *The fundamental wisdom of the middle way*. New York, NY: Oxford University Press.
Hayes, R. (2019). Madhyamaka. In E. Zalta (Ed.), *Stanford encyclopedia of philosophy*. https://plato.stanford.edu/entries/madhyamaka/.
Healey, R. (2016). Holism and nonseparability in physics. In E. Zalta (Ed.), *Stanford encyclopedia of philosophy*. https://plato.stanford.edu/entries/physics-holism/.
Jones, N. (2015). *Buddhist reductionism and emptiness in Huayan perspective*, pp. 128–49 of Tanaka et al. (2015).
Jones, N. (2009). Fazang's total power mereology: An interpretive analytic reconstruction. *Asian Philosophy, 19,* 199–211.
Liu, M. (1982). The harmonious universe of Fazang and Leibniz. *Philosophy East and West, 32,* 61–76.
Mackie, J. L. (1980). *The cement of the universe: A study of causation*. Oxford: Oxford University Press.
Mitchell, D. (2002). *Buddhism: Introducing the Buddhist experience*. Oxford: Oxford University Press.
Padmakara Translation Group (trs.). (2004). *Introduction to the middle way: Candrakīrti's Madhyamakāvatāra with a Commentary by Jamgön Mipham*. Boston, MA: Shambala.
Price, A. F., Wong, M. L., (trs.). (1990). *The diamond Sūtra and the Sūtra of Hui-Neng*. Boston, MA: Shambala.
Priest, G. (2015). The net of Indra, pp. 113–27 of Tanaka et al. (2015).
Priest, G. (2014). *One*. Oxford: Oxford University Press.
Priest, G. (2018). *The fifth corner of four*. Oxford: Oxford University Press.
Ronkin, N. (2018). Abhidharma. In E. Zalta (Ed.), *Stanford encyclopedia of philosophy*. https://plato.stanford.edu/entries/abhidharma/.
Siderits, M. (2007). *Buddhism as philosophy*. Aldershot: Ashgate.
Tanaka, K., Deguchi, Y., Garfield, J., & Priest, G. (Eds.). (2015). *The moon points back*. New York, NY: Oxford University Press.
Van Norden, B. (2019). Huayan Buddhism. In E. Zalta (Ed.), *Stanford encyclopedia of philosophy*. https://plato.stanford.edu/entries/buddhism-huayan/.
Varzi, A. (2016). Mereology. In E. Zalta (Ed.), *Stanford encyclopedia of philosophy*. https://plato.stanford.edu/entries/mereology/.
Westerhoff, J. (2018). Nāgārjuna. In E. Zalta (Ed.), *Stanford encyclopedia of philosophy*. https://plato.stanford.edu/entries/nagarjuna/.
Williams, P. (2009). *Mahāyāna Buddhism: The doctrinal foundations* (2nd ed.). London: Routledge.

A Realistic View of Causation in the Real World

George F. R. Ellis and Jonathan Kopel

1 Introduction: Reduction, Emergence, and Natures of Causation

"It's well known that matter that initially consisted of elementary particles under certain influence of laws of nature came together, formed life, language and consciousness, societies, and thereby an ability to question back the matter and the genesis of all of it." This is a quite extraordinary process that took place in the past, going through a series of major evolutionary transitions (Szathmáry & Maynard Smith, 1995) each associated with changes in the way information is stored and transmitted. In this paper, we will look at key aspects of this process of coming into being of the present world, via the transitions from cosmological to planetary to cellular, physiological, social, ecological, and engineering realms.

Reductionism and holistic effects Many scientists have approached investigating the natural world through a reductionistic perspective in which natural laws and objects are understood through their individual components, and all emergence is believed to be a purely bottom-up process. Such an approach gave rise to successful scientific paradigms, including the atomic theory of matter, statistical physics, quantum chemistry, molecular biology, and neuroscience that allowed for a comprehensive understanding of much of the natural world (Ellis, 2021). However, there remains a gulf in scientific understanding that reductionistic processes cannot

---

A contribution to *From Electrons to Elephants and Elections—Scientific Saga on the Content and Context,* ed.Shyam Wuppuluri.

G. F. R. Ellis (✉)
Applied Mathematics Dept, University Cape Town, Rondebosch 7700, South Africa
e-mail: george.ellis@uct.ac.za

J. Kopel
Health Sciences Centre, Texas Tech University, Lubbock, United States
e-mail: jonathan.kopel@ttuhsc.edu

S. Wuppuluri and I. Stewart (eds.), *From Electrons to Elephants and Elections*,
The Frontiers Collection, https://doi.org/10.1007/978-3-030-92192-7_8

traverse. As Roald Hoffman succinctly wrote, "*I would ask the reader who is a chemist to think of ideas such as aromaticity, acidity, and basicity, the concept of a functional group, or a substituent effect. Those constructs have a tendency to wilt at the edges as one tries to define them too closely. They cannot be mathematicized, they cannot be defined unambiguously, but they are of fantastic utility to our science*" (Kopel, 2019). As Hoffman correctly notes, there are properties in science that are caused by the effects of hierarchical structures rather than the individual components of an object. These holistic effects are enabled by downward causation (Ellis, 2016), as discussed below.

The nature of life Living systems are open systems that exhibit unique properties that differentiate them from other material objects: reproduction, development, metabolism, regulation via homeostasis, and information-based reactivity. These properties arise from organization in the form of modular hierarchical structures, which establish interlevel constraints and mutual dependence (Ellis, 2016; Ellis et al., 2012; Juarrero, 1999; Noble, 2008). Through them they are able to adapt through variation and selection to perform essential functions for survival. Information use is key in biology (Nurse, 2008, Davies, 2020) through DNA, RNA, and cell signaling networks (Berridge, 2012) at the molecular level, and endocrine, sensory, and nervous systems at the physiological level (Guyton & Hall, 2006), and predicting/planning at the psychological level. Metabolism enables this all (Peacocke, 1989), and places limits on what can be accomplished (Brown et al., 2004).

Logical Branching The related key step that separates life from physics is when logically controlled branching dynamics occurs. This is of the form

$$\text{IF } T(Xi) \text{ THEN } O1(Yj) \text{ ELSE } O2(Yj) \tag{1}$$

where T(Xi) is a logical operator (perhaps involving mathematical relations based in equalities or inequalities) depending on the variables Xi, and O1, O2 are alternative outcome functions for the variables Yi (Ellis & Kopel, 2019; Hoffman, 2012). Such branching operations can be combined in modular hierarchical structures and networks to obtain behaviour of arbitrary complexity (Gorjão et al., 2018).

Examples are developmental biology circuits (Manukyan et al., 2020), where for example in one species of lizard, *Timon lepidus*, the colour and pattern of its scales evolve in a manner akin to a discrete rule-based computation (a cellular automaton), and the way that abstract rules drive adaptation in the subcortical sensory pathways (Tabas et al., 2020). This branching dynamics is the basis of how information is used productively in biology, and how adaptive selection takes place. The transition allowing this kind of dynamic to occur is the most important event in evolutionary history.

Contexts of emergence In cosmology, emergence took place in a series of successive steps over time, each one depending on outcomes of the previous one before they could start. For example first generation stars needed to evolve and (by providing the right internal context) produce heavy elements and then spread these elements through space before second generation stars surrounded by planets could come into existence. In biology, there are three main contexts for emergence: it took place over the evolutionary history of the Earth (evolutionary emergence), it takes place each time an embryo becomes an adult organism (developmental emergence), and it takes place minute by minute as our brains function to enable our body and brain to work due to electrons and protons interacting in immensely complex ways (functional emergence). Your brain has thoughts ("I feel like a pizza", "That was a tense election") that are not implied by the nature of physical forces—although they are allowed and enabled by them. This is strong emergence: the coming into being of structures, entities, and functions that cannot be predicted, even in principle, from the underlying physical interactions (Ellis, 2020a, b). Physics enables all this, but what actually occurs is determined by higher level functional capacities such as the capacity for logical thought. Physics per se cannot even explain the existence of a teapot (Ellis, 2005).

Upward and Downward Causation Upward emergence and downward causation are deeply intertwined in all aspects of these processes, from cosmology to galactic structures and stellar and planetary astronomy to microbiology, physiology, psychology, society, ecology, and engineering. In particular, as biological emergence takes place via developmental processes, the resulting organisms collectively form ecosystems that provide the context for preferential selection of fittest individuals by evolutionary processes, which choice chains down to preferentially select genomes leading to emergence of organisms adapted to that environment, which in turn is modified by the organisms. Thus one has a highly non-linear EVO-DEVO process (Carroll, 2005), where interlevel causal closure (Ellis, 2020b; Mossio, 2013; Mossio & Moreno, 2010) underlies the emergence of the complexities of life. It is the combination of upward and downward causation that enables same-level behaviour to emerge at each level. This is due to entities at the higher level setting the context for the lower level actions in such a way that consistent same-level effective laws emerges. In physics, this comes about by boundary conditions and contextual determination of time dependent constraints for local physics, which also enables biology to emerge from physics (Ellis & Kopel, 2019; Nobel, 2008).

Multiple Realisation The multiple possible realisations of higher level dynamics at lower levels enables attainment of higher level purposes independent of the specifics of lower level interactions. It occurs when a phenomenon can be implemented by different realizers, and is resistant to a uniform physical explanation at the lower level. Hoyningen-Huene (1997) give as an example the case of sex pheromones that attract individuals of the opposite sex, triggering the performance of behaviours related to sexual reproduction. In terms of chemistry, pheromones lack a common structure, and consequently are not an identifiable kind. However from the perspective of biology, they constitute a natural kind that elicits a systematic pattern of behaviours. These emergent biological regularities would pass undetected if regarded at the level of chemical implementation, and much more so at the underlying physical level.

Mechanisms There are several ways these downward effects can happen (Ellis, 2012, 2017, 2020a). They firstly involve both static and dynamic downward constraints (Juarrero, 1999), including structural and channelling constraints and homeostasis/feedback control. Secondly they involve downward emergence, alteration, and deletion of lower level elements. Adaptive selection is a key case here, whereby lower level elements get selected to adapt better to higher level selection criteria. Downward selection is enabled by randomness (stochasticity) at the lower level, providing an ensemble of elements or processes to select from (Noble, 2020). Because of the degeneracy in the way higher level states and processes are realized at lower levels, the effective processes determining what happens cannot be sensibly characterized at lower levels (Edelman & Gally, 2001). It is an essentially higher-level dynamic.

This paper This can all be characterised in general terms, but the devil is in the details: can one give specific examples of how this chain of emergence occurs in the real universe? How downward effects occur in cosmology and astrophysics? How emergent branching dynamics works in real biological systems? In what way is it controlled by higher level variables, in each case? That is, how does contextual (top down) dynamics arise?

It occurs everywhere. The aim of this paper is to illustrate this by focusing on a few specific cases which gives solidity to generic claims regarding the occurrence of downward causation, showing how it works in practical terms in specific contexts, and thus exemplifying how it underlies the chain of emergence that is the focus of the book.

As regards cosmology, downwards influences occur as regards the arrow of time, nucleosynthesis, the temperature of the night sky, and structure formation (Ellis, 2017). We discuss the latter case in Sect. 2).

As regards biological emergence, we look at how plasma membrane proteins enable contextual biological responses (Sect. 3), how protein synthesis is related to natural selection and developmental biology (Sect. 4), feedback control, basins of attraction, and biological oscillators (Sect. 5), and how downward effects occur in ecology (Sect. 6).

These are just a fraction of the examples one could use, but they are sufficient to illustrate how these effects occur in physics, biology, and engineering. We do not deal with crucial other effects such as downward causation in the brain (Ellis, 2016, 2018) and in social neuroscience (Cacioppo et al., 2002). A fascinating example as regards the last is given by Ashton et al. (2018): cognitive performance is linked to group size and affects fitness in Australian magpies.

2 The Influence of Cosmology on Structure Formation: Contexts for Lower Level Dynamics

In standard cosmology (Peter & Uzan, 2013), an extremely rapid inflationary era of exponential expansion occurs in the very early universe, driven by an effective scalar field. Quantum fluctuations during this epoch generate primordial inhomogeneities that then become the seeds for formation of structure by gravitational attraction at later times. The way this works out depends on the rate of expansion of the universe at the relevant time, as determined by the scale factor a(t). The context for local physics is set by the cosmological density parameter $\rho(t)$ which determines the rate of change with time of the scale factor a(t), which itself determines the rate of change with time of $\rho(t)$ in a non-linear feedback loop ("Matter tells spacetime curvature how to change which then tells matter how to move").

If the universe is static ($a(t) = a_0$), structure formation is exponential with time. If the universe is exponentially expanding ($a(t) = \exp H t$), no structure formation takes place. If the universe expands in a power law way, as in the early matter dominated phase ($a(t) = a_0 t^{2/3}$), structure formation also takes place as a power law with time. The resultant large scale structure is characterised by its power spectrum P(k) as a function of wave number k as well as its angular power spectrum, and secondly by its effect on the Cosmic Microwave Background Radiation (CMB) angular power spectrum. Measuring these features is the most sensitive way to determine the parameters of the background cosmological model (Aghanim et al., 2020).

Thus this represents a downward effect from the largest causally connected scales in the physical universe to the scales that determine galaxy formation, and hence set the stage for the existence of the Milky Way galaxy that is the home for the Solar system. This is an early step in the chain of causation that led to our existence. It happens by cosmological conditions setting the context for local physics: the expanding universe is the base around which perturbations lead to structure formation, and so determines their outcome (Ellis, 2017). This also happens during the inflationary era in the early universe that sets up the seeds for primordial structure formation, The history of the cosmological expansion rate at that time determines these smaller scale outcomes.

3 Downward Causation and Plasma Membrane Proteins: Altering Lower Level Constraints.

In biological systems, a primary mechanism for establishing downward processes at the molecular level is compartmentalization of biochemical processes through the plasma membrane (Alberts et al., 2007; Peacock, 1989). Thus this is downward causation via setting constraints on lower level dynamics. Entities need to be localized in space and time because they need to engage in particular activities at particular times and places. Among its several functions, the plasma membrane regulates the flow of biochemical substrates into and out of the cells through transmembrane protein channels and transporters. As shown in **Fig. 1**, the cellular organization in the cell is further sub-compartmentalize into organelles, which perform specific biochemical and information processing functions. Once established, the cell is able to establish complex and overlapping metabolic and information networks (Berridge, 2012; Peacock, 1989). These networks provide the cell the necessary foundation to increase production of cytosolic and membrane proteins, which can be incorporated

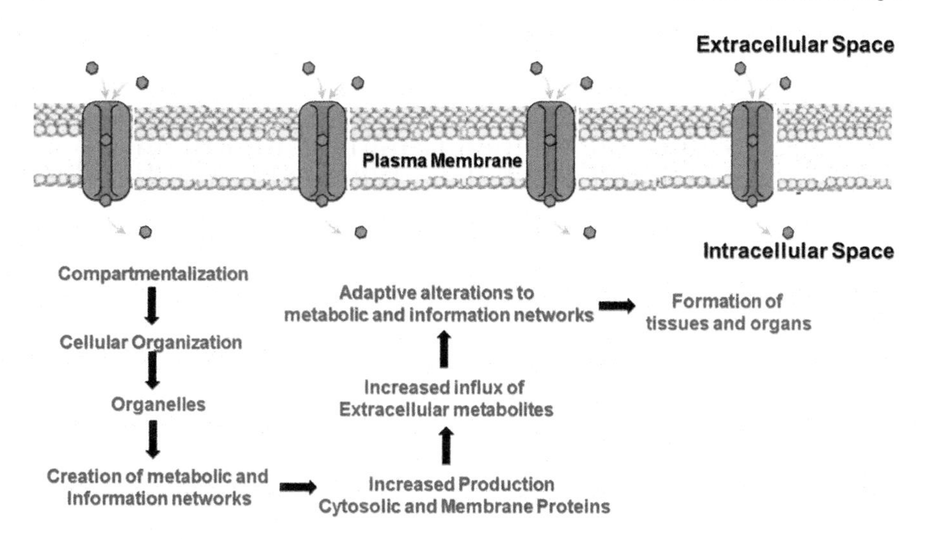

Fig. 1 Plasma Membrane creates compartmentalization and hence downward control of ions within the cell, allowing external signaling molecules to shape metabolic network and signaling network dynamics within the cell and so allow controlled emergence of tissues and organs to take place

into the plasma membrane. Among the proteins, transport proteins become an essential means by which the cell accumulates metabolites in the cytosol. The increase cell metabolites provide regulatory and adaptive alterations to established metabolic and information networks.

Transport Proteins These adaptive networks are mediated through the actions of transport proteins in the plasma membrane. Transport proteins are involved in the movement of several biological molecules both in and out across the plasma membrane and subcellular organelles (Schumann, 2019). They are divided into two major groups: ATP-binding cassette (ABC) and the solute carrier (SLC) proteins (Schumann, 2019). ATP-binding cassette proteins function primarily as efflux transporters for the elimination of xenobiotics and toxins (Schumann, 2019). ABC proteins are involved in the absorption, distribution, and elimination of several drugs, which influence pharmacological interactions that underly many drug toxicities (Schumann, 2019). In contrast, solute transporters facilitate the transport of several metabolic substrates, such as amino acids and tricarboxylic acid (TCA) intermediates, that are essential for metabolic processes involved in energy and synthesis of proteins, lipids, carbohydrates, and nucleic acids (Schumann, 2019). Without the plasma membrane, the transporter proteins would cease to function.

The SLC13A5/NaCT (Sodium-Citrate) Transporter. One SLC transporter that illustrates downward causation enabled by the plasma membrane is the SLC13A5/NaCT (sodium-citrate) transporter. It uses sodium to transport citrate, an important metabolite, into the cell, particularly in hepatocytes and neurons. NaCT-mediated citrate entry in the liver impacts fatty acid metabolism, glycolysis, and gluconeogenesis. In neurons, it is essential for the synthesis of neurotransmitters such as glutamate, GABA, and acetylcholine. Thus, the same substrate, namely citrate, has different functions in the liver and brain in these different contexts, even though the way a substrate is used once it enters the cell is independent of the transporter. In brief, citrate is transported into the liver and brain to be used for different purposes, dependent on context.

Thus NaCT exhibits aspects of top-down causation related to the plasma membrane through two key observations. First, there are different physiological and clinical outcomes for NaCT in the liver and brain of both mice and humans. Second, there are differences in the transport kinetics between mouse and human NaCT transporters.

Kinetic Differences of Mouse and Human NaCT—Plasma membrane transporters, such as the SLC transporters, are characterised ed by three key properties. First, the transporter is defined by its saturability, which describes the maximum transport capacity of a given molecule through a plasma membrane transporter in the presence of other substrates. We can imagine this like a machine's maximum functional capacity for an output given its many functions. For transport proteins, this is known as the maximum velocity or V_{max}. Second, transport proteins select substrates with a specific orientation (e.g. D- and L-amino acids); this is known as stereospecificity. Third, transport proteins have a specific binding affinity to individual substrates that it transports. This is described by an individual numerical value, known as the Michaelis-Menton constant or K_m. These parameters are summarized in Eqs. (2) and (3). Figure 2 shows the graph of solutions to these relations.

$$\text{Velocity} = V_{max}\frac{[S]}{K_m + [S]} \tag{2}$$

$$V_{max} = \left(\frac{1}{2}\right)K_m \tag{3}$$

When comparing the activity of a transporter in different species, the quantities K_m and V_{max} are often used for comparison. When comparing mouse and human NaCT, kinetic analysis estimated the Michaelis constant for mouse NaCT was approximately 20 μM; in contrast, the human and primate NaCT transporters showed a 30-fold

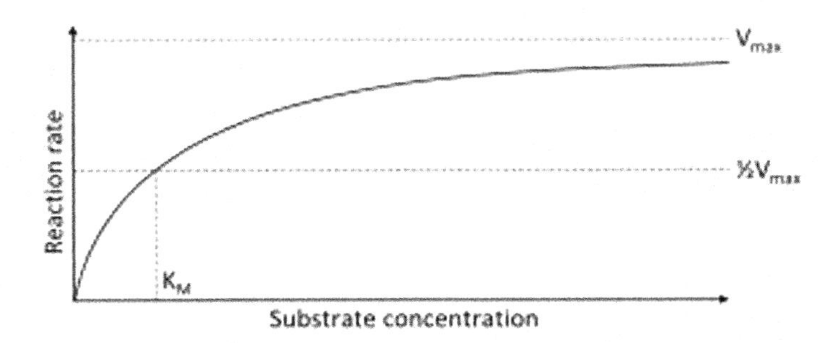

Fig. 2 Michaelis-Menton Equation solutions

higher Michaelis constant (Gopal, 2015; Inoue, 2004, 2002). The differences between human and murine NaCT become more apparent at human physiological citrate concentrations (150–200 μM) in the blood. At this level, the mouse NaCT transport is completely saturated while the human NaCT transport exhibits sub-saturating levels of citrate uptake (Gopal, 2015; Inoue, 2004, 2002). Therefore, the mouse NaCT represent a high-affinity, low capacity citrate transporter while the human NaCT represent a low-affinity, high capacity citrate transporter. As a consequence, a smaller amount of citrate is transported into murine cells than human cells. Furthermore, the murine and human NaCT transporter kinetic can be further differentiated in the presence of lithium (Li^+) (Gopal, 2015; Inoue, 2003). Although neither the human nor rodent NaCT transport Li^+, the mouse NaCT transport is inhibited by Li^+, whereas the human NaCT transport is stimulated by Li^+ (Gopal, 2015; Inoue, 2003). Therefore, NaCT transports can be characterized by their response to Li^+.

It seems that the organisms require different levels of citrate, which is reflected in their differing kinetics and responses to Li^+. However, this does not explain why the difference exists in the first place. Species differences in the transport attest to a top-down process. Humans use a greater amount of citrate for their metabolism and neurotransmitter synthesis while mice do not. This is a physiological need based in their activity patterns, related to their physical, ecological, and social needs. It appears to that the organism itself influences the type of transporter that is necessary rather than any reductionistic processes.

Functional Differences of Knocking-out NaCT in Mice and Humans—Although previous biochemical studies showed NaCT was expressed in the cerebral cortex, hippocampus, cerebellum, and olfactory bulb, it was unknown whether a homozygous deletion of the transporter would have any clinical effects in humans (Thevenon, 2014; Yodoya, 2006; Inoue, 2004; Inoue, 2002). Previous studies showed the expression of NaCT is restricted to neurons whereby extracellular citrate is circulated through release of citrate by astrocytes (Mycielska, 2015; Sonnewald, 1991). Interestingly, NaCT-null mice did not show any neurological deficits, such as seizures (Birkenfeld, 2011). In contrast, deletion of NaCT in humans causes severe neonatal epilepsy, known as SLC13A5-deficiency or Early Infantile Epileptic Encephalopathy 25 (EIEE-25), which causes early onset epilepsy within the first 24 h of life and persist throughout childhood (Thevenon, 2014; Yang, 2020). It was hypothesized that the affinity and citrate transport capacity of the murine and human NaCT transports could explain this discrepancy (Bhutia, 2017). The murine NaCT is a high-affinity, low-capacity transporter while the human NaCT transporter is a low affinity, high-capacity transporter. Murine cells may utilize less citrate for their metabolic needs while humans require citrate to maintain normal metabolic functions and neurotransmitter production for our complex central nervous systems.

The different effects of removing NaCT in the mouse and human brain strongly suggests that it is top-down processes. It is the fact that the dynamics is happening in these two different contexts that influences the effects of utilizing the citrate NaCT. The top down effect is via setting context: the context firstly of the plasma membrane, that acts as a container, and then the transporters, which control ingress and egress of molecules to the cell. In Aristotelian terms, this is formal causation.

4 Adaptive Selection and Developmental Biology: Deleting or Altering Lower Level Elements

Once life has come into existence, a combination of natural selection and developmental processes occurs (Carroll, 2005), both of which crucially involve downward causation. At the micro level these are based in protein synthesis governed by gene regulatory networks and cell signaling networks.

Natural Selection As powerfully stated by Campbell (1974), natural selection (Darwin, 1859), with it specific feature of reproduction with heredity featuring genetic variation, is a crucial case of downward causation. This has to be the case, because the core process of variation followed by selection tends to select phenotypes and hence species better adapted to the environment, where "better adapted" means a superior relative survival rate in that specific context. Thus natural selection preferentially selects for the genotypes associated with better adapted phenotypes, so if the environment is altered, for example by processes of global climate change, there will tend to be a change in the physiology and behaviour of the species and associated genomes that survive. This is downward causation in two ways: first from the nature of the environment to preferred phenotypes (an ecological effect), and second from preferred phenotypes to preferred genotypes (a physiological effect).

Multiple realization Because this is a form of downward causation, multiple realisation occurs in both steps, as examined by Kuechle and Ríos (2020). They present a typology of multiple realization that provides a plausible account of the differences between across- and within-species multiple realization, and perform a formal analysis of the dynamics of multiple realization that sheds light on the differences between multiple realization at different levels of organization.

Adaptive Selection Natural selection is a special case of adaptive selection. This occurs whenever variation followed by selection takes place, and there are numerous examples in biology apart from natural selection. Stochasticity is key in this process, as emphasized by Noble and Noble (2018) and Noble (2020): living organisms harness chance variations in ways that enable them to generate new solutions to the environmental challenges they meet, as well as to extract order out of chaotic molecular motion That is how molecular machines such as kinesin work (Hoffmann, 2012).

The adaptive immune system A key instance is the adaptive immune system (Flajnik & Kasahara, 2010), which is which is based in lymphocytes bearing antigen receptors generated by somatic recombination. This is enabled by organisms using stochasticity at all levels to generate new DNA sequences in the immune system, through gene regulatory networks. Those antibodies that apply to the invading antigen are then preferentially multiplied so as to combat it. This enables circular causation to occur: "*Response to a new antigen can rapidly develop a neutralising antibody by enhancing the mutation rate in a highly specific region of the gene coding for an immunoglobulin The mechanism here has been shown to involve regulating the error-correction machinery*" (Noble, 2020).

Neuronal Group Selection Initial neural connections in the developing brain are random at the microlevel, and then refined through pruning some connections and strengthening others as the brain learns by interacting with its environment (Wolpert et al., 2002). Different environments will result in different outcomes, for example a child recognizes a specific person as its mother through the "fire together, wire together" Hebbian process. Additionally, brain plasticity that results in adaptation to the local social environment takes place via the process of "Neural Darwinism" proposed by Changeux et al (1973), Changeux and Danchin (1976), Edelman (1989). Different environments (e.g. French as a home language rather than English) result in different patterns of synaptic connections. This is downward causation from processes in the family environment to microstructure of the neocortex.

In all cases, the key downward effect is deleting lower level elements, thus controlling what the lower level interacting elements are. In Aristotelian terms, this is material causation.

Developmental processes A remarkable feature of developmental biology is how pluripotent cells become specialized so as to become specialized cells that can fulfil specific physiological functions. This takes place via diffusion of morphogens from organizing centres (Wolpert et al., 2002), thus large scale axes and domains are set up that then determine developmental outcomes through controlling gene regulatory networks, particularly via the hierarchical structure introduced by HOX genes (Carroll, 2005). Illari and Williamson (2010) explore in detail how this hierarchical structuring occurs via functional individuation, decomposition, and organization. The phenomenon of concern (which may be structure or function) is explained by being decomposed into lower-level components, where the functioning of these lower-level components may in turn be explained by further decomposition into yet lower-level components. Thus both mechanisms and their functions come to be hierarchically nested, which is key to attaining complexity (Simon, 2019). Each level determines developmental outcomes at the next lower level by setting its context, and overall environmental conditions such as temperature, humidity, and nutrient availability can shape these outcomes (for example in plant development).

Higher order principles But other influences are crucial, for example mechanical forces deriving from the state of development of an embryo can change developmental outcomes "*Researchers have begun to define the mechanisms by which cells sense, respond to and generate forces. ... Most researchers are probing mechanical signals using cells or tissues cultured in a dish. But a few groups are studying whole animals, and sometimes they find different principles at work from those apparent in isolated tissues. These* in vivo *studies come with many challenges—such as measuring tiny amounts of force in complex tissues—but they are key to understanding the role of force in sculpting life* (Dance, 2021). This is downward causation from these macro structures to gene expression. Again, a range of developmental phenomena can be explained by the regulation of cell surface tension (Lecuit & Lenne, 2007).

In summary: Modular genetic control of protein synthesis takes place leading to modular body construction. That then allows hierarchical modular functions to interlock to produce desired higher level behavior. The downward causation occurs by signaling molecules switching lower level dynamics by altering the shape of biomolecules and so allowing regulation of transcription (Monod et al., 1963). It is a case of altering the nature of lower level elements by molecular binding.

5 Feedback Control, Basins of Attraction, and Biological Oscillators

Two key ways downward causation takes place in biology can easily be confused. Both involve higher level structures that have an irreducible dynamic effect at the macro level. Their function at the macro level then acts down to entrain dynamics at the micro level.

Homeostasis and Feedback control One key case is the feedback control systems that underlie cybernetic processes in engineering (Dorf & Bishop, 2011; Wiener, 1948), and homeostasis in biology (Guyton & Hall, 2006). Their function is to protect the system from disturbances, and keep it in a comfortable operating range. It functions by a sensor determining the actual value of the relevant variable **v**, a comparator determining the difference $\mathbf{\Delta v := v - vo}$ from the desired value **vo**, and a controller that actuates some system that will tend to return **v** to the desired value **vo.** A. An engineering example is a thermostat controlling room temperature. There is an extensive literature on feedback control systems detailing the ways that recovery from perturbations takes place (Dorf & Bishop, 2011). It may be underdamped, oscillating strongly about the desired state, but with gradually decreasing amplitude; it may be strongly damped, returning in one strong sweep to its stable positionl; or it may be critically damped, the separatrix between the other two. Biological examples are physiological systems controlling body temperature, blood pressure, ionic concentrations, and so on (Guyton & Hall, 2006).

Downward causation A thermostat is a holistic system: if you disconnect the sensory wire where it enters the thermostat, all the micro components are still there exactly as before, but it will no longer function because the macro connections have been changed. Thus it ceases to function if reduced to its parts: the feedback dynamics is essentially due to a macro level configuration (the connection between the wire and thermostat). Its operation entails both upward and downward causation because the micro level dynamics both enables and follows from the macro level state thereby attained. For example if the temperature setting on the thermostat for a room is reduced from **T1** to **T0**, at the macro level, the system will operate to as to attain a temperature in the desired range around **T0**. At the micro level, the molecules in the room will move on average at a lower speed than they did before, this speed corresponding to the temperature **T0** entered on the thermostat dial. The dial setting is the cause of room temperature at the emergent level, as well as of average molecular speed at the molecular level. That is downward causation from an abstract concept in the operators mind (the desired temperature) to physical outcomes in the room.

Dynamical Systems and Basins of Attraction The second key case is dynamical systems, which result from a generic system of interacting particles when suitable constraints are applied, for example if dynamical symmetries occur or channelling constraints are applied. A well developed mathematic theory shows how dynamical systems behave (Arnold, 1989). They may have sources, sinks, saddle points, and limit cycles in phase space, and they may exhibit chaotic dynamics with strange attractors. Basins of attraction in phase space characterize which orbits end up at or near these features.

The confusion with feedback control system can occur because there may be basin of attraction characterising the orbits ending up at a fixed point, with dynamics apparently similar to those of feedback control. But the mechanism is entirely different: there is no comparator and control device, as there is in the case of feedback control systems. Rather these patterns occur simply as a result of the constraints shaping the dynamical system. For example if the dynamical system characterizes a river ending up in the sea, a range of heights from river pools to the sea level will be attractors for water molecule height as they flow towards the sea, resulting simply from the topography that determines the water flow.

Biological Oscillators A key case is biological oscillators, which occur in all branches of biology (Otero-Muras & Banga, 2016, Forger, 2017, Rydin Gorjão et al., 2018). Wherever there are oscillators in biology, there are emergent non-linear effects taking place due to irreducible higher level dynamics that then entrain the component parts to follow those oscillations In dynamical systems terms, they are limit cycles, with a basic of attraction: if the system has initial data in that basin, it will tend to join oscillatory motion described by that limit cycle.

Neural oscillators Churchland and Sejnowski (2016) describe the basic issue very clearly, discussing the stomatogastric ganglion of the spiny lobster (pages 4–5): "*The basic electrophysiological and anatomical features of the neurons have been catalogued, so that the microlevel vitae for each cell in the network is impressively detailed. What is not understood is how the cells interact to constitute a network that produces the rhythmic pattern. No one cell is responsible for the network's rhythmic output: no one cell is itself the repository of properties displayed by the network as a whole. Where then does the rhythmicity come from? Very roughly speaking, from the patterns of interactions among cells,* and *the intrinsic properties of component cells. ...How is it that the network can produce different rhythms under different biochemical conditions?*" This is a classic description of emergent properties of a network. They state "*Microlevel data are necessary to understand the system, but not sufficient*". The oscillator is a limit cycle, stable to perturbations.

The Heart Denis Noble gives a similar story in his discussion (Noble, 2006:58–65) of how the heart works: we have a system that operates rhythmically but has no specific 'oscillator' component (Noble, 2002). The rhythm is driven by a compact networks of proteins and genes that code for them; it is a systems property, which chains down to cause rhythmic motions of the parts. The rhythmic activity of the cell drives the operation of the protein channels, which drive the network oscillations. Causal closure is a property of the whole interacting set of levels (Ellis, 2020a). Turning off this interlevel feedback loop causes the oscillations to cease (Noble, 2012).

At the molecular biology level,

- **Gene expression oscillators** are described in "Developmental function and state transitions of a gene expression oscillator in Caenorhabditis elegans" (Meeuse et al., 2020) **Metabolic oscillators** are described in "Autonomous metabolic oscillations robustly gate the early and late cell cycle" (Papagiannakis et al., 2017).
- **Biochemical timers** The design principles of biochemical timer circuits that discriminate between transient and sustained stimulation are described in Gerardin et al. (2019).

6 Examples in Ecology

Ecological systems are hierarchical systems where the availability of suitable niches determines what flora and fauna will flourish: a clear downward effect. A basic feature is that metabolism link the biology of individual organisms to the ecology of populations, communities, and ecosystems (Brown et al., 2004). Consequently the global biogeochemical cycles (carbon, nitrogen, oxygen, phosphorus, sulfur, rock, and water) are the basis for ecosystem function at all levels from communities down to insects and microbes; and conversely microbial engines drive the geochemical cycles (Rousk & Bengtson, 2014). There is a relatively stable set of core genes coding for the major redox reactions essential for life and the biogeochemical cycles (Falkowski et al., 2008). These genes created and coevolved with biogeochemical cycles and were passed from microbe to microbe primarily by horizontal gene transfer. Thus upward and downward effects intertwine to give causal closure only when the whole set of such interlevel interactions is taken into account (Ellis, 2020a). Complex autocatalytic effects attract stable, persistent system configurations (Ulanowicz, 2009).

However if the global context changes, the configuration changes. An example is that top predators shape the ecosystem in a topdown way, regulating species diversity (Letnic et al., 2012; Newsome et al., 2017; Sergio et al., 2005). Consequently, major shifts in ecosystems at lower levels can follow changes in the abundance and distribution of apex consumers. Estes et al (2011) give a penetrating analysis of how this downward effect works. The three key elements are, firstly, an ecosystem may be shaped by apex consumers. Trophic cascades occur through the downward propagation of impacts of consumers on their prey. Secondly, alternative stable states occur, and perturbations of sufficient magnitude can push ecosystems from one basin of attraction to another. Tipping points can occur, resulting in abrupt changes in ecosystem structure and function related to such transitions between alternative stable states. Third is the high connectivity of ecosystem interaction webs. Every species can potentially influence many other species through biological processes such as predation, competition, and mutualism, and these interactions are shaped by the supportive or limiting physico-chemical effects of water, temperature, and nutrients. These interactions link species in an array of spatial scales in a highly complex network.

> Taken together, this sets the stage for the idea of trophic downgrading. The loss of apex consumers reduces food chain length, thus altering the intensity of herbivory and the abundance and composition of plants in largely predictable ways. The transitions in ecosystems that characterize such changes are often abrupt, are sometimes difficult to reverse, and commonly lead to radically different patterns and pathways of energy and material flux and sequestration.

The focus of the paper is how human actions are having an irreversible influence through these processes:

> [This is] Humankind's most pervasive influence on the natural world. This is true in part because it has occurred globally and in part because extinctions are by their very nature perpetual, whereas most other environmental impacts are potentially reversible on decadal to millenial time scales. Recent research suggests that the disappearance of these animals reverberates further than previously anticipated, with far-reaching effects on processes as diverse as the dynamics of disease; fire; carbon sequestration; invasive species; and biogeochemical exchanges among Earth's soil, water, and atmosphere.

Thus these are effects occurring because of deletion from the ecosystem of various higherlevel species, in turn changing the lower level constituent species from which ecosystem dynamics arise. Including effects at a global scale, they reach down to affect what happens at micro levels where microbial and chemical interactions take place, in turn enabled by the underlying physics. But it is the biosphere and ecosystem levels that provide the context determining what happens.

References

Aghanim, N. et al. (2020). "Planck 2018 results-VI. Cosmological parameters." *Astronomy and Astrophysics 641*, A6

Alberts, B. et al. (2007). *Molecular biology of the cell* (Garland Science).

Arnol'd, V.I. (1989). *Mathematical methods of classical mechanics* (Springer_).

Ashton, B. J., Ridley, A. R., Edwards, E. K., & Thornton, A. (2018). Cognitive performance is linked to group size and affects fitness in Australian magpies. *Nature, 554*, 364–367.

Berridge M (2012) *Cell signalling biology* (Portland Press).

Birkenfeld, A. L., et al. (2011). Deletion of the mammalian INDY homolog mimics aspects of dietary restriction and protects against adiposity and insulin resistance in mice. *Cell Metabolism, 14*, 567.

Bhutia, Y. D., et al. (2017). Plasma membrane Na^+-coupled citrate transporter (SLC13A5) and neonatal epileptic encephalopathy. *Molecules, 22*, 378.

Brown, J. H., et al. (2004). Toward a metabolic theory of ecology. *Ecology, 85*(7), 1771–1789.

Cacioppo, J.T. et al. (2002). *Foundations in social neuroscience* (MIT Press).

Campbell, D.T. (1974). "'Downward causation in hierarchically organised biological systems."*Studies in the Philosophy of Biology* (Palgrave), 179–186.

Carroll, S.B. (2005). *Endless forms most beautiful*. (WW Norton and Company).

Changeux, J. P., Courrège, P., & Danchin, A. (1973). A theory of the epigenesis of neuronal networks by selective stabilization of synapses. *Proceedings of the National Academy of Sciences, 70*(10), 2974–2978.
Changeux, J. P., & Danchin, A. (1976). Selective stabilisation of developing synapses as a mechanism for the specification of neuronal networks. *Nature, 264*(5588), 705–712.
Churchland, P.S, Sejnowski, T.J. (2016). *The computational brain* (MIT Press).
Dance, A. (2021). "The secret forces that squeeze and pull life into shape" Nature 13 January (2021).
Darwin, C. (1859). *The origin of species. 6th.* Vol. 570. (John Murray).
Dorf, R.C., & Bishop, R.H. (2011). *Modern control systems.* (Pearson).
Edelman, G. M., & Gally, J. A. (2001). Degeneracy and complexity in biologicalsystems. *Proceedings of the National Academy of Sciences, 98*, 13763–13768.
Ellis, G.F.R. (2012). "Top-down causation and emergence: some comments on mechanisms. *Interface Focus 2*, 126–140. https://www.ncbi.nlm.nih.gov/pmc/articles/PMC3262299/.
Ellis, G.F.R. (2016). *How can physics underlie the mind* (Springer).
Ellis, G. F. R. (2018). Top-down effects in the brain. *Physics of Life Reviews, 31*, 1–30.
Ellis, G. F. R. (2020a). Emergence in solid state physics and biology. *Foundations of Physics, 50*, 1098–1139.
Ellis, G. F. R. (2020b). The causal closure of physics in real world contexts. *Foundations of Physics, 50*, 1057–1097.
Ellis, G.F.R. (2021). "Physics, determinism, and the brain" https://arxiv.org/abs/2008.12674.
Ellis, G. F. R., & Kopel, J. (2019). The dynamical emergence of biology from physics: Branching causation via biomolecules. *Frontiers in Physiology, 2019*, 9.
Ellis, G.F.R., Noble, D., & O'Connor, T. (2012). "Top-down causation: an integrating theme within and across the sciences?" *Interface Focus.***2**1–3
Flajnik, M. F., & Kasahara, M. (2010). Origin and evolution of the adaptive immune system: Genetic events and selective pressures. *Nature Reviews Genetics, 11*, 47–59.
Estes, J. A. et al. (2011). "Trophic downgrading of planet earth" *Science 333*, 301.
Falkowski, P. G., Fenchel, T., & Delong, E. F. (2008). The microbial engines that drive Earth's biogeochemical cycles. *Science, 320*, 1034–1039.
Forger, D. B. (2017). *Biological clocks, rhythms, and oscillations: the theory of biological timekeeping* (MIT Press).
Gerardin, J., Reddy, N. R., & Lim, W. A. (2019). The design principles of biochemical timers: Circuits that discriminate between transient and sustained stimulation. *Cell Systems, 9*, 297–308.
Gopal, E., et al. (2015). Species-specific influence of lithium on the activity of SLC13A5 (NaCT): Lithium-induced activation is specific for the transporter in primates. *Journal of Pharmacology and Experimental Therapeutics, 353*, 17–26.
Gorjão, L.R., Saha, A., Ansmann, G., Feudel, U., & Lehnertz, K. (2018). "Complexity and irreducibility of dynamics on networks of networks", *Chaos: An Interdisciplinary Journal of Nonlinear Science, 28*:106306.
Guyton, A., & Hall, J. (2006). *Textbook of medical physiology*, 11th edition (Saunders Co)
Hoffmann, P.M. (2012). *Life's ratchet: how molecular machines extract order from chaos.* (Basic Books).
Hoyningen-Huene, P. (1997). " Comment on J. Kim's 'Supervenience, emergence and realization in philosophy of mind'", in *Mindscapes: Philosophy, Science and the Mind*, in M. Carrier & P. K. Machamer (Eds.) (University of Pittsburgh Press), pp. 294–302.
Illari, P.M., & Williamson, J. (2010). "Function and organization: Comparing the mechanisms of protein synthesis and natural selection." *Studies in History and Philosophy of Science Part C: Studies in History and Philosophy of Biological and Biomedical Sciences, 41*(3), 279–291.
Inoue, K., et al. (2004). Functional features and genomic organization of mouse NaCT, a sodium-coupled transporter for tricarboxylic acid cycle intermediates. *Biochemical Journal, 378*, 949–957.

Inoue, K., Zhuang, L., & Ganapathy, V. (2002). Human Na+ -coupled citrate transporter: Primary structure, genomic organization, and transport function. *Biochemical and Biophysical Research Communications, 299*, 465–471.

Inoue, K. et al. (2002). "Structure, function, and expression pattern of a novel sodium-coupled citrate transporter (NaCT) cloned from mammalian brain," *J Biol Chem 277*, 39469–39476.

Inoue, K., et al. (2003). Human sodium-coupled citrate transporter, the orthologue of Drosophila Indy, as a novel target for lithium action. *The Biochemical Journal, 374*(Pt 1), 21–26.

Hoffman, P.M. (2012) *Life's ratchet: How molecular machines extract order from chaos* (Basic Books).

Juarrero, A. (1999). *Dynamics in action.* MIT Press.

Kopel, J. (2019). A note regarding relational ontology in chemistry. *Process Studies, 48*, 59–66.

Kuechle, G., & Ríos, D. (2020). "Multiple realization and evolutionary dynamics: A fitness-based account," *Australasian Journal of Philosophy*, https://doi.org/10.1080/00048402.2020.1839920, https://doi.org/10.1080/00048402.2020.1839920

Lecuit, T., & Lenne, P. F. (2007). Cell surface mechanics and the control of cell shape, tissue patterns and morphogenesis. *Nature Reviews Molecular Cell Biology, 8*(8), 633–644.

Letnic, M., Ritchie, E. G., & Dickman, C. R. (2012). Top predators as biodiversity regulators: The dingo Canis lupus dingo as a case study. *Biological Reviews, 87*, 390–413.

Lin, L., et al. (2015). SLC transporters as therapeutic targets: Emerging opportunities. *Nature Reviews Drug Discovery, 14*, 543–560.

Meeuse, M., et al. (2020). Developmental function and state transitions of a gene expression oscillator in Caenorhabditis elegans. *Molecular Systems Biology, 16*, e9498.

Monod, J., Changeux, J. P., & Jacob, F. (1963). Allosteric proteins and cellular control systems. *Journal of Molecular Biology, 6*, 306–329.

Mossio, M. (2013). "Causal Closure." In Dubitzky et al. (Eds) *Encyclopedia of Systems Biology* (Springer) pp. 415–418.

Mossio, M., & Moreno, A. (2010). Organisational_closure_in_biological_organisms. *Hist Phil Life Sci, 32*, 269–288.

Mycielska, M., et al. (2015). Extracellular citrate in health and disease. *Current Molecular Medicine, 15*, 884–891.

Newsome, T. M., et al. (2017). Top predators constrain mesopredator distributions. *Nature Communications, 8*, 15469.

Noble, D. (2002). Modeling the heart–from genes to cells to the whole organ. *Science, 295*(5560), 1678–1682.

Noble, D. (2008). *The music of life: Biology beyond genes* (Oxford University Press).

Noble, D. (2020). "The role of stochasticity in biological communication processes". *Progress in Biophysics and Molecular Biology* 2020.

Noble, R., & Noble, D. (2018). Harnessing stochasticity: How do organisms make choices? *Chaos, 28*, 106309.

Nurse, P. (2008). Life, logic and information. *Nature, 454*, 424–426.

Otero-Muras, I., & Banga, J.R. (2016). "Design principles of biological oscillators through optimization: Forward and reverse analysis," *PLoS One 11*, e0166867.

Papagiannakis, A., Niebel, B., Wit, E. C., & Heinemann, M. (2017). Autonomous metabolic oscillations robustly gate the early and late cell cycle. *Molecular Cell, 65*(2), 285–295.

Peacocke, A. R. (1989). *An introduction to the physical chemistry of biological organization.* Oxford University Press.

Peter, P., & Uzan, J-P (2013) *Primordial cosmology*, Oxford Graduate Texts.

Rousk J and Bengtson P (2014) "Microbial regulation of global biogeochemical cycles," *Frontiers in Microbiology, 5*, 103.

Schumann, T. et al. (2019). "Solute carrier transporters as potential targets for the treatment of metabolic disease'. *Pharmacological Reviews, 72*, 343–379.

Sergio, F., Newton, I., & Marchesi, L. (2005). Top predators and biodiversity. *Nature, 436*, 192–192.

Simon, H. A. (2019). *The architecture of complexity. Sciences of the artificial*, MIT press.

Sonnewald, U., et al. (1991). First direct demonstration of preferential release of citrate from astrocytes using [13C]NMR spectroscopy of cultured neurons and astrocytes. *Neuroscience Letters, 128*, 235–239.

Szathmáry, E., & Maynard Smith, J. (1995). The major evolutionary transitions. *Nature, 374*(6519), 227–232.

Tabas, A., Mihai, G., Kiebel, S., Trampel, R., & von Kriegstein, K. (2020). "Abstract rules drive adaptation in the subcortical sensory pathway". *Elife, 9*, e64501.

Thevenon, J., et al. (2014). Mutations in SLC13A5 cause autosomal-recessive epileptic encephalopathy with seizure onset in the first days of life. *The American Journal of Human Genetics, 95*, 113–120.

Ulanowicz, R. E. (2009). The dual nature of ecosystem dynamics. *Ecological Modelling, 220*, 1886–1892.

Wiener, N. (1948). *Cybernetics or control and communication in the animal and the machine*. MIT Press.

Wolpert, L. et al. (2002). *Principles of development*, Oxford University Press.

Yang, Q.-Z., et al. (2020). "Epilepsy and EEG phenotype of SLC13A5 citrate transporter disorder." *Child Neurology Open, 7*, 2329048X2093136.

Yodoya, E., et al. (2006). Functional and molecular identification of sodium-coupled dicarboxylate transporters in rat primary cultured cerebrocortical astrocytes and neurons. *Journal of Neurochemistry, 97*, 162–173.

Where is the Top and What Might Go Down?

Tim Maudlin

Aristotle distinguished four different "aitiai", a term sometimes translated as "causes" but more accurately rendered as "explanatory factors". The one most familiar in contemporary discussion is the efficient cause. And there is a sense of "top" and "bottom" such that macroscopic or other complex objects are at the top and their microscopic constituents at the bottom. The suggestion that in this sense the "top" can be an efficient cause of what happens at the "bottom" conflicts with the usual understanding of the relationship between physics and the other sciences. But if instead of efficient causation one has another sense of explanation in mind, or if instead of the macro/micro distinction one has in mind the local/non-local distinction, then "top-down causation" need not conflict with the usual understanding of physics. This is an attempt to sort these issues out.

1 The Four Causes and the Special Sciences

The phrase "top-down causation" is apt to raise the hackles of anyone committed to a certain form of physicalism. That form—which in fact is not realized by modern physics!—is best illustrated by Democritian atomism. In Democritus' pungent apothogem: "By convention sweet and by convention bitter, by convention hot, by convention cold, by convention color; but in reality atoms and void". The punch comes in the last clause: somehow, at the bottom of everything, all there *really* is are atoms moving through the void until they smack into each other and bounce off (or become hooked together, or whatever). Everything else one thinks of or talks of is not really real in the way the atoms are.

T. Maudlin (✉)
Department of Philosophy, New York University, 5 Washington Place, New York, NY 10003, USA

John Bell Institute for the Foundations of Physics, Hvar, Croatia

S. Wuppuluri and I. Stewart (eds.), *From Electrons to Elephants and Elections*, The Frontiers Collection, https://doi.org/10.1007/978-3-030-92192-7_9

The possible targets of Democritus' withering assessment of reality are many. His own examples call to mind perceptual qualities such as sweet and bitter, for the obvious skeptical reasons. As Sextus Empiricus noted, what tastes sweet to one person may taste bitter to another; what is delicious and appealing to one species may be disgusting and repellant to another. These things are not *in themselves* either sweet or bitter or delicious or disgusting, they are only so relative to some perceiver. If so, then for similar reasons the sweetness of an object cannot really be the cause of how it behaves, since it is not, in itself, any more sweet than not-sweet.

But from a modern perspective, the rejection of heat as a real cause can be pushed beyond the realm of subjective perceptions. One and the same room, for example, can "feel hot" to one person and "feel cold" to another, so that felt quality cannot be in the room itself. But one might think that there is an objective physical quality of heat, the one measured by a thermometer without reference to anyone's feelings, that can be a cause. But in a sense, the modern kinetic theory of heat denies that as well.

Suppose we define the temperature of a gas (at equilibrium) as the average kinetic energy of the molecules in the gas (as measured in the rest frame of the gas). That is an objective quantity. But it is still not a quantity that directly *causes* anything to happen. The *average* kinetic energy is a mathematical abstraction: it does not reside in any particular place in space. The *average* kinetic energy does not itself deflect a single molecule from its path or cause an ice cube to melt. What does that, properly speaking, are the *individual* collisions of *individual* gas molecules with *individual* water molecules in the ice, knocking them free. Without or apart from those particular collisions in particular places and times the gas could have no effect at all. In that sense, the heat really does no causal work. It is rather a sort of handy statistical bookkeeping device, which can be used to make statistical arguments about the numbers and sorts of actual individual collisions that are likely to occur. In this sense, one might reasonably say, heat itself cannot cause anything. In this sense (which one has to be careful about), heat isn't "really real" physically. All that is really real are the atoms and the void.

I hope that this fundamental metaphysical picture, adumbrated by Democritus, is clear enough. It fits as well as anything can the demands of a certain sort of reductionism. That reductionism consists of two steps. First, certain qualities such as perceived sweetness are denied physical existence (in the "sweet object") at all. And then some physical quantities, such as temperature, are denied fundamental physical reality because they are merely generic or abstract, rather than the particular characteristics of particular atomic objects.

The language here gets a bit slippery. One might say, for example, that in one sense the view is that tables and chairs are not "really real", in that nothing that happens in the physical world is caused by "chairness" or "tablehood" per se. Chairs and tables are *nothing but* collections of atoms arranged in certain ways, and their physical behavior is determined without remainder by the physical laws that govern the atoms individually. At no point in the application of those laws could the property of "being a table" or "being a chair" even be invoked: there would be no place to make use of that fact when applying the fundamental physical laws. A physicist who had no idea even what a chair or table is would not be deterred from analyzing and predicting the behavior of a table or chair simply from the geometrical disposition of the atoms within it. In that sense, tablehood and chairhood would be not exactly "unreal" but more accurately "epiphenomenal". Yes, there are tables and chairs, but their being tables and chairs, per se, makes no fundamental difference in the world to how they behave or what causes what.

This Democritean picture, then, denies "top down causation" in two different ways. It denies that any non-physical characteristics are really causes of any event. And it also denies that merely abstract or generic or statistical physical characteristics cause any event. At the bottom—what is doing all the causing—is *physical* and *individual* and also *local*: just one atom in a particular location banging into another atom in a nearby location. The world as a whole, in all its aspects, is really nothing more than the sum-total of all these local bangings.

Just as some people are rather deeply committed to something like this account of physical reality and causation being correct, others are equally deeply repelled and even disgusted by it. One sort of reaction is that it seems to deny the "real reality" of all the things we take to be important in everyday life: human beings and mountains and sunsets and beauty and desire and love. None of these would show up in the physics at the bottom so none are really real, none really causes anything or is responsible for anything. If all that is really going on is the laws of physics playing themselves out blindly among atoms, all meaning and significance seem leached out of the world. And this general reaction—like the general Democritean impulse in the opposite direction—comes in many specific forms. It is useful to distinguish them.

Perhaps the most intractable objection to the Democritean impulse arises from the mind–body problem. One thing we are all completely certain about (as Descartes observed) is the existence of our own subjective mental states, and indeed of ourselves as conscious beings. That is, from an epistemic standpoint at least, as "really real" as it gets. Descartes found that he could even doubt the existence of all of physical reality (*res extensa* in his scheme) and still be sure of his own existence as a cognitive being. If so, he argued, the Democritean account of fundamental reality *cannot* be correct. Furthermore, in Descartes' telling there is ultimately a *direct causal* connection between the purely mental and the purely physical: purely mental acts of will cause (somewhere in the brain, most likely in the pineal gland) particular atoms to swerve in their trajectories in ways not caused by other atoms, and the motions of certain atoms (in the visual cortex, for example) cause the existence of particular mental perceptive states. So if we identify the mind as the "top" and the physical brain as the "bottom", Descartes postulated both top-down and bottom-up causation. And it is one consequence of his view that when it comes to understanding the dynamics of individual atoms in brains, *standard physics derived from the study of inanimate matter would just straightforwardly fail.* In the pineal gland sometimes atoms just would not go where the usual physical laws predict, due to the control or nudging or whatever of the mental act. This is the assertion of top-down causation in its starkest form.

Let me now pause to lay down some markers. First, I regard the mind/body problem as the hardest problem there is. I do not have a clue how conscious states arise from, or supervene on, the physical goings-on in the brain. I am convinced they do because straightforward physical intervention in the brain has predictable consequences for one's mental state. But I have not only no idea how this is to be explained, I have no idea how it even *could* be explained. The gap between the physical character of things and the character of conscious states is just too wide to bridge by any principle I can grasp. So for the hard problem of consciousness, I simply note it and surrender. I have nothing more to say.

However, I also am very, very, very, very skeptical that the sort of mind-to-brain causation that Descartes envisages exists. I am firmly convinced that the atoms and molecules inside of brains obey the very same physical laws as those outside of brains, and indeed those in entirely inanimate matter. I believe that the laws of physics as derived from the behavior of particles in high-energy accelerators and stars and so on will be completely adequate to account for all the physical behavior of brains. Given my view about the intractability of the mind/body problem, one might object here. If minds are real, and you don't know how they are connected to physical characteristics, why not think they can straightforwardly influence physical characteristics? I cannot detail exactly how my rejection of Cartesian mind-to-body causal influence fits in with my mystification over the mind/body problem, but simply note that it does. I do not expect the physics derivable from the behavior of inanimate matter to fail somewhere in the brain. And that's all I have to say about that.

At the other end of the spectrum, there *is* a form of "top-down" causation that I fully endorse, but the key to that lies in a careful discussion of the term "causation". So far, I have only used that term in one sense, the sense that Aristotle denominated "efficient causation". Or, to be a little more accurate, Aristotle distinguished four different sorts of things he called *aitiai*, a Greek term that is commonly translated "causes". But it has been correctly observed that the translation is misleading. A closer rendition of what Aristotle had in mind would be "explanatory factors". That is, when trying to account for, or explain, or create understanding of some event, there are four different types of factors that are invoked: material explanatory factors, formal explanatory factors, efficient explanatory factors and final (or teleological) explanatory factors. Each of these can provide a sort of explanation or understanding, and indeed all of them might be invoked when trying to explain *one and the same* event. Where more than one factor exists, we might say that each sheds a different sort of light on the situation. A full comprehension of what is going on would require a grasp of all the various levels or sort of explanation available.

The sort of explanatory factor that we most commonly call a "cause" is what Aristotle called the *efficient* or *moving* cause. This cause accounts for why a certain event happened when it did, in terms of immediately preceding events. For example, the assassination of Archduke Ferdinand is commonly cited as the immediate efficient cause of the First World War: that was the "spark" that "set it off". Such a judgment does not conflict with the claim that in a sense the "real" cause had to do with big-power political dynamics, and that the assassination was just an *occasion* to start fighting, not the "deep reason". One could judge, for example, that if it hadn't been the assassination it would have been something else that set it off a little later. This sort of judgment cites a "final" or *teleological* explanation: what were the basic *ends* being sought by the agents. It was not to simply avenge the death of Ferdinand. If it were, then one would judge that without the assassination there would have been no war at all.

I hope it is clear that identifying the assassination as the efficient cause and the larger international tensions as the final cause, in the sense just outlined, is a perfectly coherent thing to do. Neither precludes the other. The student of international affairs can say that the assassination was incidental to understanding the fundamental reasons for the war, and the military historian trying to account for the exact details of which units were deployed where and when may find that the date of the assassination plays a critical explanatory role. But these people merely have different interests, not incompatible causal judgments. Both what is to be explained, and what sort of explanation is sought, can vary from context to context.

Here is another example. My computer is many things at the same time. It is a purely physical object that can be analyzed in terms of its physical structure. And it is also a (pretty good) instantiation of a Turing machine that can be described in terms of a Turing table. (Or, even more precisely, it instantiates many different Turing machines "running at different levels".) Now: suppose a particular event occurs. As it sits on the table, there is a spinning colored circle that suddenly stops and pixels in the shape of the Arabic number 8 appear. How is this to be explained?

Well, from a purely physical point of view, the computer is a collection of protons, neutrons and electrons configured in a certain way embedded in a physical environment. In order to directly apply the fundamental laws of physics to it, both it and the environment would have to be described in precise microscopic detail. We all believe that such a description is possible in principle, and that the laws of physics applied to the state just before the colored disc stopped spinning would predict that it would soon stop spinning and that pixels in the shape of an "8" would light up. Note that in this explanation the fact that the pixels form the shape of the Arabic numeral "8" is completely irrelevant. The physical analysis could be carried out without mentioning or noticing or deriving that. And indeed, it might even be accidental that the result is the shape of the Arabic "8": maybe the computer is programmed to work in Korean, for example, and the symbol only accidentally has the form of an Arabic numeral at all.

On the other hand, the computer scientist does not know—or even care!—about the detailed physical description the physicist needs. The computer scientist may ask not about any of that, but about what program the computer is running. She may be told that the program is running a particular algorithm. By analyzing that algorithm, the computer scientist can come to the conclusion that the program will halt and output the result which is a pixel bitmap in the form "8".

But a mathematician might not care about the details of the algorithm at all. The mathematician might want to know what *question* the algorithm was designed to answer. And let's say the answer to that is that the question is the lowest number of dimensions in which a certain topological shape can exist. The mathematician might set about investigating that question in a completely different way than the computer algorithm does, but still conclude that the correct number is eight. It is then essential that the mathematician recognize that the output of the program represents the Arabic numeral for eight. And the mathematician can declare that he understands why the output was what it was in a different—and more profound—way than either the physicist or the computer scientist does.

What are we to say of the explanations provided by the physicist, the computer scientist and the mathematician? They all provide a sort of understanding of the event. And each can provide their understanding without referencing the other. The physicist need not even know the object is a computer or describable in terms of a Turing table; the computer scientist need not ascribe any significance to the algorithm as long and the Turing table is available; the mathematician may not know a thing about either physics or the algorithm. Each might find the facts of interest to the other as completely irrelevant to their own analysis. But I hope we can agree that there is no rational grounds for them to *fight* over whether the "real" cause of the final state of the machine is given by the physical analysis or the algorithmic analysis or the topological analysis. Each perspective provides its own insight and understanding.

There is only this asymmetry among them. Let's suppose that each has to provide a prediction of the outcome on the basis of their own approach. The mathematician solves the topological problem in whatever way is most convenient and says that the computer should come up with the answer "eight". The computer scientist says that the algorithm should halt and the final output be a bitmap in the shape "8". The physicist says that the pattern of light coming off the screen should shift from the rotating colors of the set of pixels in the shape "8". But what if they were to *disagree* in their predictions? What if the topologist concludes the answer is nine, or the computer scientist that the algorithm will never halt, while the physicist says the output will be a blank screen?

Well, there is a sense in which—if each has no made a mistake in their own area of expertise—we ought to trust the physicist.

The computer scientist, for example, simply has to postulate that the physical machine does, and will continue to, implement a certain Turing table. And the mathematician simply has to postulate that whatever algorithm the computer is running, it will arrive at the correct mathematical solution to the question. If either of these postulates fails, then their predictions will fail, and not through any fault of their own. But the physicist makes no such vulnerable postulate. The physicist's conclusions are drawn entirely from physical initial conditions, and if those are wrong it is a failure of physics as such. And if the physicist says the screen will go blank—let's say because the computer's battery is about to run out—then nothing the computer scientist concludes or the mathematician concludes can override that. Both of those experts can properly say that the state of the battery lies outside their area of expertise, and of course they could not be expected to take that into account in making their predictions. We can call this the "Wernher von Braun" defense, after the classic song of Tom Lehrer: "Once the rockets are up, who cares where they come down? That's not my department!" says Wernher von Braun. There are certain important facts about what will actually show up on the screen that do not fall in the department of the computer scientist or the mathematician. If their predictions fail because the battery runs out they have not failed *qua* computer scientist or *qua* mathematician. But if the screen does not go blank—or display the glowing "8"—when the physicist says it should, then the physicist has made a mistake *qua* physicist. Either the physical description was wrong, or the laws of physics employed were wrong. In this sense, physics holds a fundamental or basic position among these sciences. But that does not imply that the physical explanation provides the most *insight* or *understanding* among the three, where they all make the right prediction.

It is in this sense that physics is "below" or "more basic" or "more fundamental" than either mathematics or computer science with regard to the behavior of the computer. So this provides one sort of direction in which one can distinguish higher from lower, or top from bottom. And there is a certain sense in which efficient causation is most accurately and completely described "at the bottom", by the physics. Even if the Turing machine table says a certain step should come next in the calculation, nothing will intervene in the face of the battery running out to bring that about. And similarly for the topological explanation: it may get us to the answer "eight", but the situation of the battery will overwhelm that. So *in the sense of efficient causation* and *in the sense in which physics lies at the "bottom" of a hierarchy of sciences*, there is no "top-down causation".

But none of that implies that there is not—in a perfectly good sense—top-down *explanation*. The perspective of the computer scientist or the topologist or the psychologist can be more explanatory, for the purposes at hand, than that of the physicist. One of the sharpest observations of this fact was made by Socrates as he waited in prison for his execution.

Phaedo is set in that prison, and Socrates undertakes to explain his approach to philosophy to Cebes. The relevant passage is this (97b–99b, translation Benjamin Jowett):

> Then I heard someone who had a book of Anaxagoras, as he said, out of which he read that mind was the disposer and cause of all, and I was quite delighted at the notion of this, which appeared admirable, and I said to myself: If mind is the disposer, mind will dispose all for the best, and put each particular in the best place; and I argued that if anyone desired to find out the cause of the generation or destruction or existence of anything, he must find out what state of being or suffering or doing was best for that thing, and therefore a man had only to consider the best for himself and others, and then he would also know the worse, for that the same science comprised both. And I rejoiced to think that I had found in Anaxagoras a teacher of the causes of existence such as I desired, and I imagined that he would tell me first whether the earth is flat or round; and then he would further explain the cause and the necessity of this, and would teach me the nature of the best and show that this was best; and if he said that the earth was in the center, he would explain that this position was the best, and I should be satisfied if this were shown to me, and not want any other sort of cause. And I thought that I would then go and ask him about the sun and moon and stars, and that he would explain to me their comparative swiftness, and their returnings and various states, and how their several affections, active and passive, were all for the best. For I could not imagine that when he spoke of mind as the disposer of them, he would give any other account of their being as they are, except that this was best; and I thought when he had explained to me in detail the cause of each and the cause of all, he would go on to explain to me what was best for each and what was best for all. I had hopes which I would not have sold for much, and I seized the books and read them as fast as I could in my eagerness to know the better and the worse.

> What hopes I had formed, and how grievously was I disappointed! As I proceeded, I found my philosopher altogether forsaking mind or any other principle of order, but having recourse to air, and ether, and water, and other eccentricities. I might compare him to a person who began by maintaining generally that mind is the cause of the actions of Socrates, but who, when he endeavored to explain the causes of my several actions in detail, went on to show that I sit here because my body is made up of bones and muscles; and the bones, as he would say, are hard and have ligaments which divide them, and the muscles are elastic, and they cover the bones, which have also a covering or environment of flesh and skin which contains them; and as the bones are lifted at their joints by the contraction or relaxation of the muscles, I am able to bend my limbs, and this is why I am sitting here in a curved posture: that is what he would say, and he would have a similar explanation of my talking to you, which he would attribute to sound, and air, and hearing, and he would assign ten thousand other causes of the same sort, forgetting to mention the true cause, which is that the Athenians have thought fit to condemn me, and accordingly I have thought it better and more right to remain here and undergo my sentence; for I am inclined to think that these muscles and bones of mine would have gone off to Megara or Boeotia—by the dog of Egypt they would, if they had been guided only by their own idea of what was best, and if I had not chosen as the better and nobler part, instead of playing truant and running away, to undergo any punishment which the State inflicts. There is surely a strange confusion of causes and conditions in all this. It may be said, indeed, that without bones and muscles and the other parts of the body I cannot execute my purposes. But to say that I do as I do because of them, and that this is the way in which mind acts, and not from the choice of the best, is a very careless and idle mode of speaking. I wonder that they cannot distinguish the cause from the condition, which the many, feeling about in the dark, are always mistaking and misnaming.

Here Socrates invokes a species of teleological explanation, in two quite different contexts. It is worth our while to tease them apart.

The overall theme that Socrates insists on is that there is a sort of explanation—and corresponding "cause"—which invokes what is best or, more accurately, what is *thought to be* best. To take them in reverse order from that which appears in the passage, Socrates remarks that to really *understand* why he is sitting in prison—given that he had had several opportunities to avoid being condemned to death and to escape once he had been—one has to understand that he himself judges that remaining is the best course of action, just as the Athenians judged it to be the best of their available options to condemn him to death. Both of these cases seek to explain intentional actions by reference to judgements of the good: what is sometime called "belief/desire psychology". The Athenians thought it best that Socrates be put to death, and believed that confining him to prison would achieve that end, so they acted to bring it about. Socrates thinks it best to obey the commands of the laws of the city, so he voluntarily remains. Of course, he could not do so *without* a physical body, and his remaining in prison is a physical disposition of that physical body. Nonetheless, he asserts, true understanding of the situation simply cannot be grasped via physical analysis. That understanding requires conceptualizing the situation in terms of beliefs about what is good and actions to achieve that good.

It is notable that this sort of conceptualization transcends physics in a certain sense. Socrates is an agent with various options available to him. Among these are agreeing to escape (which his friends try to convince him to do). He refuses because he deems that that would not be good. It is true, on the one hand, that Socrates must *be* a physical being with a physical body to find himself in such a situation. But it is also true that with respect to the psychological description, the details of the physics become quite irrelevant. Whether his body is ultimately Democritean atoms or quantum strings is neither here nor there. That will matter greatly to the physicist, but be irrelevant to the psychologist. As an agent, Socrates must have a body, and various viable options the body is capable of (such as running off to Megara or Boeotia), and beliefs about which of those viable options would be best. But beyond that, the physics provides no insight. It is the psychological per se that does the explaining.

The explanation is *teleological* in the sense of being goal- or end-directed. Socrates—like the Athenians—thinks he knows what is best and acts to achieve it. Of course, either or both parties could be *wrong* about what is best, and also wrong about how to achieve what they desire. So this sort of teleology involves nothing like backward causation, or the end state literally bringing about the stages that lead to it. It is rather that the sequence leading to the end state becomes *comprehensible* when seen in light of the beliefs and desires of the agents.

In the first part of the passage, Socrates states that be was hoping to find in Anaxagoras such a teleological explanation of the cosmos itself: that the stars and earth and planets are and behave as they do because that is for the best. This sort of cosmological teleology is something modern science has abandoned entirely. We don't think that the universe as a whole was designed or intended to be any way at all, and so to try to comprehend it in those terms is an error. But even if so, it is not *always* an error to seek teleological explanations of this sort, since there *are* agents with desires for what they take to be good and beliefs about how to achieve them.

The theory of evolution provides yet another sort of model of scientific explanation. There, no agent aiming at any end is involved, but still the result is in many respects *as if there were* such an agent. Darwin was quite clear about this. The farmer, as a conscious agent, desires to have fatter pigs, and so chooses the fattest of the pigs he has to breed the next generation. Such artificial selection has an actual effect due to genetics: the pigs become fatter and fatter through time, because the farmer *wanted* them to be and acted to bring that about. We can say that later generations of pigs were *designed* to be fat. What Darwin realized is that the very same selective mechanism could exist—somewhat less efficiently—without there being any conscious agent with beliefs or desires involved. What *does the work* is just the selection: that fatter pigs, because they are fatter, have a greater propensity to breed. In the farm, the farmer and his desires is the mechanism that creates this situation but Nature, all on its own and without any conscious aims, creates a form of Natural Selection with the same effect. Hence we understand how living beings could evolve to appear *as if* they had been designed for some purpose without there being any designer involved at all. It was, of course, a brilliant insight.

Darwinian natural selection neither competes with nor displaces physics as a scientific discipline providing understanding. Rather, both physics and evolution provide different sorts on insight into the very same situation. It was this non-competitive—and at times even symbiotic—existence of multiple sorts of explanation and understanding that Aristotle insisted upon, and that we can also see before him in Plato.

Insofar as one employs a psychological explanation to account for a physically-described event—the cup of hemlock was raised against the gravitational potential *because* Socrates wanted to drink it—one might regard this as allowing for a sort of "top-down explanation", with the special science (psychology) at the top and physics at the bottom. But once again, this does not *interfere with* or *displace* the existence of a purely physical account of everything that happens using purely physical concepts and laws. The psychological states do not *push the electrons and protons and neutrons around* (as Descartes imagined they would), rather the psychological states are *realized by* constellations of protons and electrons and neutrons organized in a certain way.

This account of "top down explanation" invokes a sort of hierarchy, in the sense that physics "lies at the bottom". But the special sciences—meteorology and geology and psychology and computer science and so on—do not form a strict hierarchy in the sense that each one must be placed either "above" or "below" every other. Rather, each deploys its own proprietary set of concepts and principles that can create a form of understanding. There may well be no such thing as "complete understanding" of any event in the physical world, but if there is it would involve grasping it under the conceptual categories of *all* the applicable special sciences in addition to physics. Have *only* physical understanding—no matter how complete in its own terms—would leave one completely ignorant of the most important aspects of the trial and death of Socrates, to take an obvious example.

There is yet another, distinct, sense in which Aristotle and Plato insist on a sort of "top-down" structure to explanation. This sense requires understanding "top" and "bottom" in a different way: the top is a *more generic* description and the bottom a *more specific* description. These descriptions need not belong to different sciences or employ different sorts of descriptive resources. But the top and bottom are related in this way: specifying the lower description logically fixed the higher, but not the other way around. The higher level omits detail that the lower lever includes.

Here is an example that Aristotle discusses (*Posterior Analytics* 74a22-75b4, translation Jonathan Barnes):

> For this reason, even if you prove of each triangle, either by one or by different demonstrations that each has two right angles--separately of the equilateral and the scalene and the isosceles--you do not yet know of the triangle that it has two right angles, except in the sophistic fashion, nor do you know it of triangle universally, not even if there is no other triangle apart from these. For you do not know it of the triangle as triangle, nor even of every triangle (except in respect of number; but not of every one in respect of sort, even if there is none of which you do not know it.)
>
> So when do you not know universally, and when do you know *simpliciter*? Well, clearly you would know *simpliciter* if it were the same thing to be a triangle and to be equilateral (either for each or for all). But if it is not the same but different, and it belongs as triangle, you do not know. Does it belong as triangle or as isosceles? And when does it belong in virtue of this as primitive? And of what does the demonstration hold universally? Clearly whenever after abstraction it belongs primitively--e.g. two right angles will belong to bronze isosceles triangle, but also when being bronze and being isosceles have been abstracted. But not when figure or limit have been. But they are not the first. Then what is first? If triangle, it is in virtue of this that it also belongs to the others, and it is of this that the demonstration holds universally.

Let's break this down.

Suppose you have before you an isosceles triangle, and also a perfectly valid proof that all isosceles triangles have interior angles equal to two right angles, a proof that *makes use* of the fact that the triangle is isosceles. Then, according to Aristotle, you know *that* the triangle in front of you has interior angles equal to two right angles, but you do not yet *understand why*. For the fact that it happens to be isosceles is merely accidental and irrelevant to the fact to be understood, as accidental as if it were also made of bronze. This is demonstrated by, for example, Euclid's proof, which merely requires that the figure be a triangle and leaves it unspecified whether it is scalene, isosceles or equilateral. The fact that the proof goes through without need of the more specific description shows that the specific distinctions between triangles are explanatorily irrelevant. Merely *being triangular* accounts for the sum of the interior angles, and the additional fact that it is an isosceles is irrelevant. So long as one does not understand that, one does not really appreciate *why* the interior angles are as they are.

In similar fashion, most of the specific physical detail of Socrates' body and brain is irrelevant to understanding *why* he remains in the cell and does not run away. The laws of physics require the complete specific to apply, but the principles of explanation do not. Even remaining within the vocabulary of physics alone, and not bringing in psychology, sometimes the explanation of a phenomenon requires conceptualizing it at a more generic rather than more specific level. This is the foundation of statistical mechanics and statistical explanation in general. Appealing to the ideal gas laws or the laws of thermodynamics to explain the behavior of a box of gas allows one to abstract away from most of the fine physical detail and to comprehend (in a sense) what is really going on. As a rough rule of thumb, understanding of a phenomenon is achieved by rising to the highest level of generic description or of abstraction at which the phenomenon can be accounted for.

This last suggestion is just a rule of thumb in that one can, in some instances, rise to too high a level of abstraction. For example, triangles in a Euclidean space form the class of polygons with interior angles equal to two right angles, and Euclid's proof shows why all Euclidean triangles have that property. But if we expand our scope from Euclidean geometry to all geometries of constant curvature, there will be some non-Euclidean polygons (quadrilaterals, pentagons, etc.) in negatively curved spaces that also have interior angles equal to two right angles. Does the inability of Euclid's proof to deal with *them* show that his proof is still not at the correct level of generality?

The answer to the last question is surely "no", but the reason is subtle. Euclidean triangles together with the various non-Euclidean polygons mentioned above really are a geometrically miscellaneous class, and what shows this is that there is no natural proof that encompasses them all. These matters are very nicely discussed in Imre Lakatos' classic *Proofs and Refutations* (Cambridge University Press, 1976), but it would take us too far astray to go into details here. For present purposes, we need only note that, as Aristotle already remarked, if we understand "causation" in terms of "explanation" and the "top-to-bottom" direction to be "more-generic-to-more-specific", then there is indeed top-down causation in the sense of explanation that must be provided at the more generic level of description. This accounts for the appeal to more generic levels of description in explanation even within the same science, such as statistical-mechanical or thermodynamics explanations offered in physics, even where more detailed and specific physical descriptions of the system exist.

2 "Top-Down" Explanation and Locality in Physics

There is a another sort of "top-down" causation and explanation that occurs in modern physics, specifically in quantum theory, which is worthy of note. Unlike the examples in the first part of this essay, it does not involve distinguishing physics as a discipline from other sciences: this is a distinction that occurs within physics itself. Further, it came as a complete surprise. Einstein, for example, judged that the historical trend of physics up until the twentieth century had been running in a direction opposite to this causal structure, which is the source of much of his objection to quantum theory. The easiest way into this issue is to forget for a moment about the distinction of "top" and "bottom" and focus instead on the distinction between "local" and "global" (and eventually between "local" and "non-local").

Einstein exposited his view with characteristic clarity in a famous letter to Max Born, which bears repeating here:

> If one asks what, irrespective of quantum mechanics, is characteristic of the world of ideas of physics, one is first of all struck by the following: the concepts of physics relate to a real outside world, that is, ideas are established relating to things such as bodies, fields, etc., which claim "real existence" that is independent of the perceiving subject—ideas which, on the other hand, have been brought into as secure a relationship as possible with the sense data. It is further characteristic of these physical objects that they are thought of as arranged in a space-time continuum. An essential aspect of this arrangement of things in physics is that they lay claim, at a certain time, to an existence independent of one another, provided these objects "are situated in different parts of space." ... This principle has been carried to extremes in the field theory by localizing the elementary objects on which it is based and which exist independently of each other, as well as the elementary laws which have been postulated for it, in the infinitely small (four-dimensional) elements of space.[1]

What Einstein remarks is a historical tendency for physics to become more and more *local* in both its ontology and its dynamical laws. He expected that trend to continue.

The basic idea behind locality here is easy to exposit. Take the four-dimensional space–time manifold (or higher-dimensional if there are compactified spatial dimensions) and cover it with a collection of small overlapping open patches of the sort commonly referred to as the local maps in an atlas. It is not really essential that the maps be open, but rather that they have two features collectively: (1) every point in the manifold is covered by at least one map and (2) the maps overlap so that in passing from one to another by a continuous motion there is non-zero length of the

[1] Einstein quoted in Born (1971).

trajectory of the motion that lies in both. The second requirement allows statements to be made about how the maps relate to each other in the overlap region.

We can call a physical fact "local" to a map if it can be specified in that map without reference to anything at all outside the region covered by the map. In this sense, the classical electromagnetic field is clearly a local object: in any open region of space–time, one can specify what the electromagnetic field there is without reference to what it is anywhere outside the region. If the atlas of maps covers the whole space in the way required, and the electromagnetic field is specified in each individual map, then the global state of the electromagnetic field is specified completely. That is what it is for some ontology to count as "local". Clearly, the disposition of classical point particles in a space–time is local in this sense.

But Einstein notes something even more. Not only is the *ontology* of classical field theory local, its *dynamics* is as well. The explication of this is exactly parallel. Cover all of space–time with an atlas of maps, where the individual maps can be as small as you like. Describe the physics in each map without reference to any other. That fixes the local ontology. But also, in a classical field theory whose dynamics is specified by differential equations of the local ontology one can ask of each individual map—irrespective of any other—whether the laws hold *there.* If the *dynamics* is local, then the global satisfaction of the dynamics is nothing but the local satisfaction of it in every map. That is a feature of Maxwellian electrodynamics, but not of Newtonian gravity understood as an unmediated action-at-a-distance force.

Consider the latter case. Cavendish famously used a torsion pendulum to test Newton's gravitational theory. In that set-up, a dumbbell of lead weights was hung from the ceiling by a thin wire which would then oscillate around an equilibrium point. When Cavendish placed larger lead balls near the hanging ones, the gravitational attraction caused the hanging weight to twist and the equilibrium point to shift. Newton's law of gravity predicted the angle of shift as a function of the weights of the lead balls, their distances, and the torsion provided by the wire. But what Newton famously did *not* do was provide any hypothesis about exactly *how* the gravitational force was produced between the weights. Various different hypotheses are possible.

One hypothesis is that each weight produces a *gravitational field* akin to the later electromagnetic field postulated by Maxwell. One would need some law about how the value of that field is related to the distribution of mass, but the main point for present purposes is that such a field would be a piece of *local physical ontology*. The gravitational force felt by any mass would be determined by the local gravitational field. If each map described the local masses and also the local fields, one could check whether the equations for them are satisfied in that map.

Contrast this with the hypothesis (which Newton positively rejected) that masses act gravitationally on each other at a distance, with no intervening local ontology. In that case *one could not verify whether the laws of gravity hold by verifying it map by map.* For example, consider a map that covered just the torsion pendulum but not the rest of the lab. In that map of the experiment, one could verify that at a particular time the equilibrium point of the oscillation shifts, but there would be nothing that accounts for the shift that shows up *in the map* since the introduction of a large lead weights would occur outside it. The dynamics of the action-at-a-distance Newtonian gravitational theory is non-local, while that of the gravitational-field variant is local. By Einstein's lights, the local theory would be preferable and more in line with the historical progression of physics.

In a local theory there is a clear sense in which the collection of local facts is more "basic" or "foundational" than the global facts which follow from them. In an obvious sense, the global facts are nothing "over and above" the collection of local facts recorded in the maps. And this is a *stronger* sense of "nothing over and above" than is secured by the standard philosophical notion of supervenience. Even in the action-at-a-distance theory, the global facts—and the holding of the fundamental dynamical law–supervenes on the collection of local facts. But there, the holding of the dynamical law requires something quite different from its holding in each local map separately. In the theory with local ontology and local dynamics, there is a very strong sense in which the entire physics is local, and the global conditions are nothing more than the "sum of the local parts". In such a theory, all efficient causation can be accounted for locally—by the reaction of local items to other local items *in their immediate neighborhood*. So if we characterize the global as the "top" and the local as the "bottom", there is no top-down efficient causation in a local physical theory. The local physics does what it does on its own, and the global states just come along for the ride, as it were. This was Einstein's vision, and it illustrates a version of denying a form of top-down causation: the laws and the ontology exist locally at the bottom, and the global behavior is an inevitable logical consequence of the local in that there is no irreducibly non-local ontology or dynamics (nomology).

Quantum mechanics challenges all of that. Everyone is familiar with Einstein's complaint of "spooky action-at-a-distance" in the understanding of quantum theory propounded by Bohn and the Copenhagen school. That complaint was founded on the "collapse of the wavefunction" as a form of action-at-a-distance. Einstein thought of the "collapse" as merely epistemic conditionalization, not anything physical at all, but that required simultaneously rejecting the claim that the wavefunction provides a *complete* physical description of a situation (otherwise there could be no new information to conditionalize on). This dialectic underlies the EPR argument.

An even greater challenge to the "all is local" credo, as Schrödinger recognized, lies in the structure of the wavefunction itself, quite apart from issues of collapse. Entangled wavefunctions of spatially separated systems contain more information than the logical sum of the local descriptions (density matrices) ascribed to the parts. Unlike the electromagnetic field, *the wavefunction of one subsystem cannot even be mathematically specified without reference to the other, distant system.* So if one takes the wavefunction to describe a piece of physical ontology, that ontology itself cannot be local in Einstein's sense. It cannot be specified local-map-by-local-map.

Einstein, of course, was aware of all this. That's why he rejected that idea that the wavefunction describes any individual ontological item at all: he preferred a *statistical* understanding of it as a description of the statistical properties of a *collective* of systems. If such an account could be carried out, then the mathematical non-locality of the wavefunction would no longer be worrisome since the wavefunction would not represent any piece of *individual* physical ontology of a system at all. One problem with this statistical approach, though, is that the interference effects such as two-slit interference require the wavefunction to describe something physically effective *at the level of individual experiments*. In order to avoid particles ever appearing at places when both slits are open where they sometimes appear when only one slit or the other is open, there must be something physically sensitive to the state of both slits *on each individual run*. That is the reason that Einstein's statistical approach to the wavefunction was destined to fail.

We can therefore now put together two ideas that seem unavoidable in quantum theory. One is that the mathematical wavefunction used to describe a system represents an irreducibly *global* sort of item, which we may call the *quantum state* of an individual system. When entangled, the quantum state of a system with spatially separated parts cannot be represented as merely the sum of some physical states of the parts taken separately. But the everyday world we are familiar with, as John Bell insisted, is not composed entirely of such a global and non-localized item as the quantum state. In addition, there must be what he called some *local beables*, which do inhabit small localized regions of space–time. There are various hypotheses about what the local beables might be, including particles, local fields, and even point-events called "flashes". But whatever they are, the familiar shapes and spatial dispositions of table and chairs and cats and laboratory apparatuses must be determined by the configuration of these local beables. The non-local quantum states only has observable empirical effects via its influence on these local beables. That is, the quantum state must, somehow or other, play a causal role in determining the locations of the local beables. If it did not, then there could not be any empirical reason to postulate its existence.

So: if we say that in a certain sense the global is at the "top" and the local at the "bottom", that puts the quantum state of a system at the top and the local beables at the bottom. And what quantum theory suggests, contrary to anything in classical physics, is that there is fundamental top-down causation: the irreducibly global and non-spatially located quantum state really does "push the local beables around", much as Descartes postulated that the immaterial mind could deflect the trajectories of material particles in the pineal gland. In this quite surprising sense, quantum physics itself posits a sort of top-down causation that has no analog in classical physics.

In sum, there are various ways to draw the distinction between a top and a bottom or a higher and a lower sort of realm. And there are also various meanings that can be given to a "cause": it can be a sort of efficient cause, or some other sort of explanatory principle. In the course of this essay, I have argued for several distinct theses.

First, we can characterize physics as the "bottom" or "most fundamental" empirical science since it provides the most precise and detailed description of an observable item and because its scope is programmatically universal in way no other science is. The various special sciences—botany, geology, psychology, etc.—then stand "above" physics. I do not believe in any top-down *efficient* causation in this sense: botanical characteristics do not, per se, influence the physical trajectories of things. But that does not mean that botany *reduces*, even in principle, to physics. The special sciences can provide *explanations* and *understanding* of phenomena that physics alone cannot. So when the "cause" is understood as an explanation, there can be top-down causation in this setting.

Even within physics itself, and elsewhere, we can also characterize the more specific description of an item as "below" or "more fundamental" than any generic description that omits information about it. Here too, the generic is never an *efficient* cause of the specific, so there is no top-down causation in that sense. But just as with the special sciences, there can be a sort of explanation or insight or understanding of a fact that flows from a more generic description rather than a more specific one. In mathematics, this means looking for the most generic characterization of an item from which a certain feature of interest can be proven. In statistical explanation, it also involves seeking less detailed, more generic characterizations from which the behavior of interest can be shown to necessarily or typically follow. So again, we have no top-down efficient causation, but we do have top-down explanation.

Finally, within fundamental physical ontology we can distinguish the fundamentally local physical items "at the bottom" from the irreducibly global (if any) "at the top". This happens in quantum physics, but not in classical physics. And here we *do* find top-down *efficient* causation: the quantum state is a cause of the motions of the local beables in the most concrete and familiar sense of cause. Exactly *how* that causation works differs between different "interpretations" of quantum theory, but *that* there is some such influence of the quantum state is held in common by all.

And somewhere mixed up in all of this is the place and role of consciousness in the physical world. But how exactly that fits in is, at least for the moment, beyond our ken.

Reference

Born, M. (1971) *The Born-Einstein Letters* (translated by I. Born, pp. 170–171). Walker.

Multiplicity, Logical Openness, Incompleteness, and Quasi-ness as Peculiar Non-reductionist Properties of Complexity

Gianfranco Minati

In memory of Eliano Pessa

1 Introduction

The purpose of this contribution is to better specify the profile of complexity and conceptual approaches to be considered for dealing with irreducible, analytically non-zippable, and theoretical incompleteness and quasi-ness of processes of emergence in complex systems. The dominant theme to have in mind is having to deal with complex systems requiring approaches involving issues such as logical openness, incompleteness, and quasi-ness. Neglecting or incorrectly approximating these aspects is a new form of reductionism.

In Sect. 2 we elaborate the concept of Multiple Systems, whose components interact in multiple ways and play multiple interchangeable roles, such as in ecosystems.

In Sect. 3 we consider logical openness contrasted with logical closedness, when modeling may be only partial and the strategy of being able to *completely* represent complex phenomena is ineffective rather than just wrong. This is because of the multiplicities of complex systems acquiring coherent sequences of new properties in a phase-transition-like manner. The properties characterizing complexity and its emergent nature have different aspects, not precisely separable, but indeed often occurring simultaneously, with or without regularity. Precise, well-defined natures are in reality often simplifications useful for making complex processes and phenomena more tractable and for suitably approximating them.

In Sect. 4 we introduce the conceptual framework of theoretical incompleteness related to equivalences, processes of losing and recovering coherence in Multiple

G. Minati (✉)
Italian Systems Society, 20161 Milan, Italy
e-mail: gianfranco.minati@AIRS.it
URL: http://www.gianfrancominati.net/

S. Wuppuluri and I. Stewart (eds.), *From Electrons to Elephants and Elections*,
The Frontiers Collection, https://doi.org/10.1007/978-3-030-92192-7_10

Systems of complex phenomena such as collective behaviors. We discuss the crucial role of weak forces in breaking equivalences, setting tentative initial conditions and having long-range effects such as in deterministic chaos. Theoretical incompleteness, already considered in physics, for instance, with uncertainty principles, and in mathematics, states the multiplicity of non-equivalences in representing phenomena. Theoretical incompleteness is a property of multiple equivalences of becoming, of its analytically intractable multiplicity.

In Sect. 5 we consider approaches introduced in the literature, the Dynamic Usage of Models (DYSAM) and meta-structures suitable for dealing with the logical openness, incompleteness and multiplicity of complexity. Examples of possible applications are multiple, non-equivalent representations and models, clusters, infra-clusters, statistical properties, and ergodicity.

In Sect. 6 we outline the issue of quasi-ness. Significant aspects of quasi-ness differentiate it from incompleteness. On the one hand, incompleteness is intended as inexhaustible, incomplete multiplicity, while quasi-ness is intended to relate to levels of instabilities, irregular alternations of collapse and recovery, when a system is not always a system, not always the same system, and not only a system.

In Sect. 7 we discuss how considering issues of complexity such as logical openness, incompleteness and quasi-ness reveals new forms of reductionism when such aspects are neglected or approximated in a non-scientific attitude.

In Sect. 8 we propose possible further research dealing with regimes of validity of sub-symbolic properties and processes of transience within complex systems.

We conclude by stressing how the new form of reductionism consisting in neglecting the properties of complexity considered here often occurs by using non-complex systemic concepts and properties.

2 Emergent and Non-emergent Multiple Systems

Multiple Systems are fundamentally considered as given by the multiple roles of their constituting interacting components, such as in the case of multiple interactions (Minati & Pessa, 2018, pp. 42–45; pp. 166–170). A scheme is presented in Fig. 1; a classic example is that of ecosystems.

The situation we are considering here is when, for instance, there is:

- interchangeability among components, which take on the same roles at different times and different roles at the same time, e.g., ergodic behavior. It is matter of equivalences reducing degrees of freedom and allowing stability in collective behaviors (see Sect. 2.3);
- multiple mediated flows of information with no direct, linear conveyance of information when non-spatially close neighbor nodes in networks and 'boids' in collective behaviors have a suitable topological distance (Ballarini et al., 2008) and in remote synchronization (Gambuzza et al., 2013; Minati, 2015). In collective animal behaviors

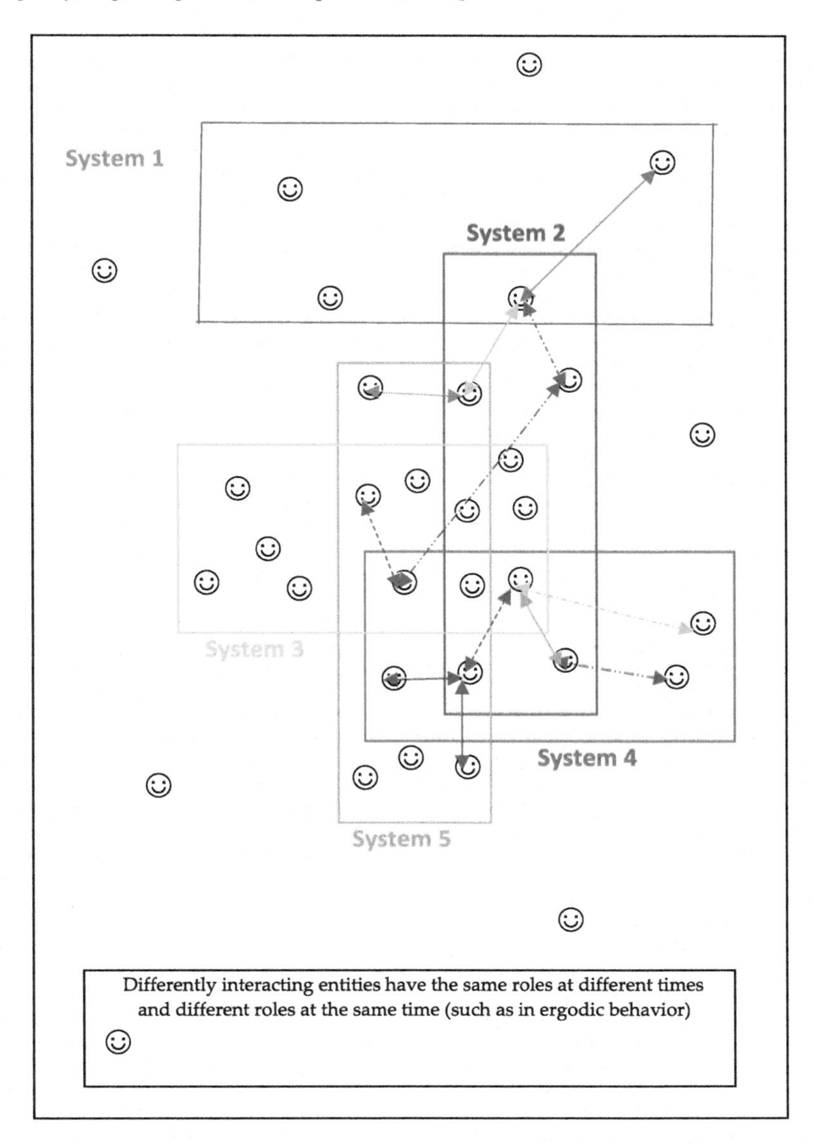

Fig. 1 Multiple System with simple examples of multiple interactions and multiple roles of elements

Correlation is the expression of an indirect information transfer mediated by the direct interaction between the individuals: Two animals that are outside their range of direct interaction (be it visual, acoustic, hydrodynamic, or any other) may still be correlated if information is transferred from one to another through the intermediate interacting animals (Cavagna et al., 2010).

- multiple roles, e.g., in networks the same node interconnects in multiple ways with other nodes.

The consideration of Multiple Systems corresponds to an attitude of general respect, that is, taking account of the weak micro-dynamics usually supposed irrelevant and ignored as overwhelmed by predominant ones establishing macroscopic behavior, but significant and decisive for emergent behaviors of collective systems. The macroscopic behavior is erroneously supposed to suitably approximate the sum of micro-dynamics that are actually non-summable because of their different natures.

Such multiple micro-dynamics are, rather, the source, as we will see below, of incompleteness, quasi-ness, fluctuations, crucial breaking of equivalences, macroscopic irreducibility, continuous environmental trade-offs, microscopic changes of minimal vector components contextually phenomenologically available and crucial for metastable or equivalent states.

Examples of non-emergent Multiple Systems include networks where the same node interconnects in multiple ways with other nodes; the cooperative and multiple roles of the values of sensors constituting, at one and the same time, safety and regulatory systems; the multiple reuse of profiling data; programs and nodes of the internet; and values in electrical networks simultaneously establishing control and regulatory systems.

Such multiplicity allows one to *indirectly* influence a system through easier influence on one or more elements shared with another system, or on one or more systems (sub-communities) of the Multiple System under consideration, such as in social Multiple Systems, e.g., shared members simultaneously belonging to families, corporations, workplaces, passengers on means of transport, customers in supermarkets, and in considerations of advertising, political influencing, and information.

2.1 Emergence and Multiple Systems

In the literature there are distinctions among processes of self-organization and emergence when considering, for instance, how in self-organization the sequence of new properties ('new' compared to those possessed by the interacting components) acquired in a phase-transition-like manner (Brovchenko & Oleinikova, 2008; Brovchenko et al., 2005) has some regularities, repetitiveness, and synchronizations, e.g., whirlpools, while in emergence the sequence of new properties acquired in a phase-transition-like manner (Paperin et al., 2011) is not regular, not repetitive, but *coherent*, e.g., flocks and swarms (Minati & Pessa, 2018, pp. 65–86; pp. 255–260; Minati, 2019a). In this view regularities and repetitiveness of self-organization are particular cases of the *coherence* of emergence (Minati, 2016a).

Coherence may be understood as *diffused correlation*, such as long-range, scale-free correlation. As is well known in probability and statistics, the concept of correlation is very similar to that of covariance measuring levels of dependency between independent variables. However, it is difficult to compare covariances among data sets having different scales, when, for instance, a value representing a strong relationship for one data set might represent a very weak one in another. It is then possible to normalize the covariance as the product of the standard deviations of the variables and create a dimensionless quantity. In this way correlation can be considered the scaled version, or *standardized* form of covariance. Through correlation coefficients, it is possible to measure the *similarity* between the variation of two entities, e.g., signals or waves. Synchronization may be regarded as a particular case of correlation, as in physics relating to oscillatory phenomena, such as individual oscillators being *in phase*. Quasi-coherence considers, for instance, different variable ranges of validity within the population of interacting elements under study, oscillations between coherence and non-coherence, and the occurrence of different coherences (see Sect. 6).

The phenomenological becoming of processes of emergence requires multiplicity and, as we will see below, it is not only theoretically, i.e., intrinsically and unavoidably incomplete, but also *requires* incompleteness in order to take place (see Sect. 4.4). This is related to *how emergence emerges* (Minati, 2019a) once the process has started in non-equivalent structural multiplicities. However, the consideration of properties of emergent phenomena implicitly ignores properties of their generative phenomena. This is the case of the geometry of a spider's web, which does not represent the *phenomenological process,* the generative mechanisms of web building by the spider, but a *detectable resulting property.*

Similarly, long-range correlations, polarization, power laws, fractality and more are detectable properties, rather than as-yet not understood generative phenomenological processes, for instance, "Whatever the origin of the scale-free behavior is ... the fact that the correlation is almost not decaying with the distance, is by far the most surprising and exotic feature of bird flocks. How starlings achieve such a strong correlation remains a mystery to us," (Cavagna et al., 2010). Furthermore, other detectable rather than generative properties, such as self-similarity and synchronization, pervasively occur in living systems, whether vegetal, e.g., fractals in broccoli and leaves, or in animals, e.g., fractals in snail and clam shells.

2.2 *Emergent Multiple Systems*

The concept of Multiple System particularly applies to emergent complex systems established by collective behaviors, or collective systems.

The dynamics of emergent collective systems refers to the complexity of the systems continually transforming themselves however coherently or quasi-coherently.

As examples we may consider (Vicsek & Zafeiris, 2012):

- collective systems emergent from the collective motion of living systems provided with significantly complex cognitive systems, including anthills, flocks, herds,

schools, swarms, and traffic. In these cases, the behavioral interactions take place through cognitive processing.
- collective systems emergent from the collective motion of living systems supposed *without* cognitive systems, including amoeba, bacterial colonies, cells, macromolecules, and composing protein chains. Examples of collective systems emergent from the collective motion of *non-living systems* include nano-swimmers (magnetic nano-propellers), nematic fluids (fluids with internal orientational order of the building blocks, such as liquid crystals), networks, signal traffic, rods on vibrating surfaces, and interacting communities of simple stimulus-reaction robots.
- collective artificial systems emergent from the *collective interactions* of various natures (cognitive and non-cognitive) other than physical motion, including human communities such as companies, hospitals, inhabited housing structures and schools, communities of mobile phone networks, industrial districts, markets, cities, networks like the internet, and vehicle queues. Interaction is given by systems of cognitive processing and physical reactions, such as in energy exchanges. We discussed the case for Collective Beings in (Minati & Pessa, 2006, pp. 97–134), understood as constituted by the same elements interacting in different ways to accommodate the acquisition of various non-equivalent properties and establishing Multiple Systems.

A generic conceptual example of a Multiple System in Fig. 1 relates to multiple roles taking place when considering multiple, interrelated, almost networked, biochemical and physical properties of composing elements, as in ecosystems.

2.3 Variables to Model Emergent Multiple Systems

Among a large variety of possibilities, the variables considered here are intended to have an intrinsic collective nature suitable for and compatible with multiplicities of soft values possibly related to weak forces (see Sect. 4.3).

The complexity we have in mind here pertains to processes of emergence having non necessary, large numbers of interacting components facilitating and making evident collective phenomena.

The variables, contextually defined for ongoing processes, are intended to capture and represent multiplicities related to, for instance:

- Interchangeability due to commensurability, compatibilities, equivalences, and similarities;
- Multiple roles of the elements when different and multiple regimes of interaction apply. Furthermore, multiple roles of elements allow for variations while belonging to one system and for having effects belonging to another one, allowing thus the cross-conveying of partial behavioral properties, as in collective behaviors (see above).

- Multiple roles due to unintentional or phenomenologically different, superimposed roles, such as for elements found to play the same role with different meaning in simultaneously belonging, for instance, to multiple networks where the same nodes may belong to different networks simultaneously (Nicosia et al., 2013), and to different clusters (see Sect. 5.2).

> Migration of elements in Multiple Systems from one class to another one and simultaneously belonging to more than one is not active, but a matter of belonging, i.e. interchangeability of roles intended as belonging to classes ... Interchangeability of roles within collective behaviours can be considered represented by its ergodicity. This is, of course, an idealistic simplified situation. We should, rather, consider different more realistic situations. We may consider cases of multiple ergodicity related to the ergodicity of the values adopted by different state variables, e.g. speed, direction and altitude, and by different clusters ... The degree of ergodicity is given by:
>
> $$E_\varphi = 1/[1 + (X_\varphi\% - Y_\varphi\%)^2]$$
>
> where we may consider $Y_\varphi\%$ as the average percentage of time spent by a single element in state S and $X_\varphi\%$ as the average percentage of elements lying in the same state over a given observational time and considering a system composed by finite, constant over time number of elements. The state shows ergodicity when $X_\varphi\% = Y_\varphi\%$ and the degree E_φ adopts its maximum value of 1 (Minati & Pessa, 2018, p. 162, see Section 4.5.1).

In the following Sects. 3, 4, 5, and 6 we introduce concepts and approaches for dealing with the complexity of Multiple Systems.

3 Logical Openness

A first conceptual approach is *logical openness*, introduced as corresponding to the thermodynamic and the usual system openness (Minati et al., 1996, 1998) and elaborated in publications such as Minati (2016a) and Minati and Pessa (2018).

The concept of logical openness takes from thermodynamic openness the admissibility of a conceptual *flux* of non-equivalent approaches to model a phenomenon. The logical openness relates to the admissibility of multiple non-equivalent representations dealing with the multiplicity of complex systems, rather than using the approach of looking for the final, best, definitive one (the true one in the reductionist view).

The subject is introduced here from the opposite concept of logical closedness.

3.1 Logical Closedness

Thermodynamically closed systems are intended as being isolated, that is, without exchange of either matter or energy with the other systems of the environment. Their evolutionary paths are internally stated and they can reach their final state solely as determined by the initial conditions. It is possible to consider the concept of *logically closed modeling* as corresponding to the *thermodynamically closed systems* the evolution of which it is suitable to describe.

More precisely, *logically closed modeling* is defined as having available:

1. a complete, formal description of the relations between the state variables of the system;
2. a complete, analytically describable representation of the interactions between the system and its environment.

Knowledge of these two points allows deduction of all possible states which the system can subsequently assume. It is a clockwork, Laplacian world, understood as functioning rather than emergent. It evokes the concept of computability when a complete computational procedure, i.e., an algorithm, is available, or more generally, when a procedure is intended to exhaust the representation of a process.

3.2 Logical Openness

Logical openness takes place in the violation of the above points. Logical openness (Minati et al., 1998; Minati & Pessa, 2006, pp. 111–112; 2018, pp. 47–51) can be considered as the infinite or non-depleting number of degrees of freedom for the system, including the environment (in principle independent), thus making the system incomplete.

This is the case in complex systems the degrees of freedom (system variables) of which are not only imprecise and variable in number, but continuously acquired. As we shall see below (see Sect. 4.1) considering DYSAM, complex systems acquire n-sequences of new, non-equivalent properties to be dealt with by adopting suitable corresponding models and their combinations. The corresponding n-levels of representation and n-levels of modeling are characterized by:

- their non-equivalence;
- the possibility of adopting a strategy of moving between levels, allowing simultaneous usage of more than one level.

The incompleteness of logical openness is given by usage of a variable number of non-equivalent models and by the indefiniteness of n due to the abductive and constructivist (see Sect. 4.3) generation of representations and models by the observer.

More generally, the concept of logical openness is related to the theory of cognitive operators (Diettrich, 2001) and uncertainty principles (Minati & Pessa, 2006, pp. 55–64). This introduces the subject of incompleteness, considered later (see Sect. 4). Regarding the theory of cognitive operators, we mention the classic situation of the behavior of a complex system whose subsystems are an observed system, a kind of environment, and an observing system, for instance the researcher. The observed system initially inputs its state φ to the observing system, perturbing, in turn, the

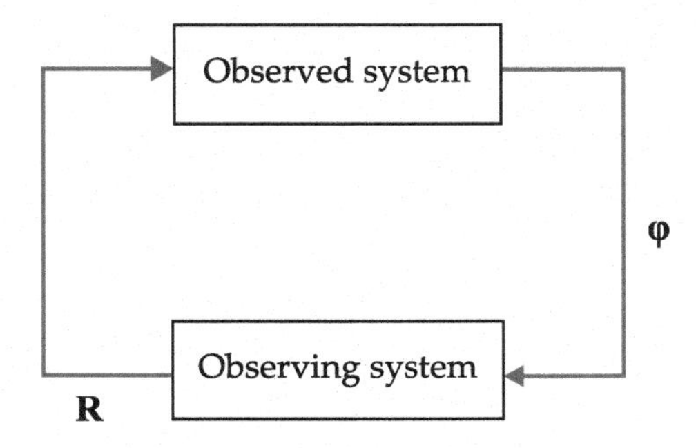

Fig. 2 Observed and observing systems

observed system, due to the process or act of observation itself. This perturbation can be described as the action of an operator R which acts on φ and gives rise to a new state of the observed system $R\varphi$ (see Fig. 2).

The input to the observing system will no longer be φ but $R\ \varphi$. After another observation the input will be $RR\ \varphi$. After n-observations, we will have $R^n\ \varphi$ and the observer will have no possibility of detecting stable feature of their environment. An interesting case occurs when we have fixed environmental states (fixed points) φ^* for which $R\varphi^* = \varphi^*$ despite the perturbations induced by the observer; then detecting the environment is in the invariant state φ^*.

With respect to uncertainty principles, we note how the above arguments generalize experimental cases such as Heisenberg's well-known uncertainty principle[1] and that in signal analysis. The latter can be expressed as $\Delta\omega\ \Delta t \geq 1/2$, where $\Delta\omega$ is the so-called spectral width or the signal or uncertainty in the frequency of its components, whereas Δt is the so-called temporal width or the uncertainty in the time location of its components (Minati & Pessa, 2006, pp. 55–64).

We may consider logical closedness versus logical openness in relation to what Feynman considered as 'Greek' versus 'Babylonian' mathematics. The 'Greek' approach to mathematics is characterized by the tendency to arrange theories on an axiomatic basis, whereas Feynman writes.

> What I have called the Babylonian idea is to say, "I happen to know this, and I happen to know that, and maybe I know that; and I work everything out from there. Tomorrow I may forget that this is true, but remember that something else is true, so I can reconstruct it all again. I am never quite sure of where I am supposed to begin or where I am supposed to end. I just remember enough all the time so that as the memory fades and some of the pieces fall out I can put the thing back together again every day." (Feynman, 1967, p. 45)

and

> In physics we need the Babylonian method, and not the Euclidian or Greek method. (Ibid., p. 47)

[1] The uncertainty associated with the location of a particle, denoted by Δx, and associated with its momentum, denoted by Δp, are connected by the Heisenberg relationship $\Delta x\ \Delta p \geq h/4\pi$, where h is Planck's constant.

4 The Concept of Theoretical Incompleteness

This subject is elaborated in publications such as Longo (2011, 2019), Minati (2016a), Minati and Pessa (2018), and Minati et al. (2019).

4.1 *Compatibilities, Similarities, Analogies, Commensurability, and Equivalences*

The incompleteness we are outlining here is not due to lack of completeness which is temporary and recoverable, but intrinsic non-completability that, if not occurring, would at least partially destroy intrinsically incomplete phenomena of self-organization and emergence constituting complex systems. For instance, hypothetical complete, analytical representations of emergence, *zippable* (Minati & Licata, 2013) in analytical equations, would be incompatible insofar as they would allow regulation, control and predictability, all properties contradictory to processes of emergence. Hypothetical complete, analytical representations may be compatible with processes of multiple synchronizations of self-organization, while being incompatible with the logical openness of processes of emergence.

Reductionist representations and approaches neglecting such peculiarities and used to influence (see the operator 'R' in Sect. 3.2) phenomena of emergence are not only ineffective, but increasingly deprive them of their *unpredictable, non-equivalent multiplicity,* between which *levels of coherence* are continuously dynamically established, interrupted and restored in partial and nonhomogeneous ways. Influences based on completeness have the similar invasive, prescriptive nature of *strong* forces, while weak forces (see Sect. 4.3) play the role of *suggestions* to be elaborated by the complex system (for instance, deciding among equivalences), rather than being substitutive, context-less, strong impositions. Incompleteness leaves the system the possibility to *decide* among equivalences, eventually acquiring coherence, as in cases of emergence. An interesting case is that of surgery, obviously invasive but inevitably incomplete, leaving the operated body the role of elaborating the intervention, adapting or even rejecting it.

We are dealing, according to the levels of complexity of the cases studied, with *ecosystems of interactions*, i.e., variably applying, having changing parametrical values and different temporal durations, partially superimposed, combining among themselves, and able to mutate.

We may consider, within such conceptual ecosystems, categories of incompleteness suitable to be considered as aggregated and establishing multiple incomplete dynamics. Such categories may be understood as items aggregated according to different aspects of incompleteness such as:

- Compatibility, such as ability to share the same environmental conditions, perform the same or similar roles, have significant levels of interchangeability;

- Similarity, such as having equivalences in properties of a different nature, e.g., geometrical, material, and behavioral;
- Analogies, the process of making which is not exactly a logical inference, but a sort of *incomplete induction* reduced to only some aspects or variables. For instance, consider the case in which an entity A has all the attributes a_n. B has some of the attributes a_n, for instance not the attributes $\{a_k\}$, where $1 \leq k < C < n$. C is considered as the *level of analogy*. However, two analogous entities may not be equivalent, as in a flock where two birds may be analogues, e.g., in age, amount of food ingested, dimensions, and weight, but not equivalent on account of having different speeds, directions and topological positions, e.g., at the center or at the border;
- Equivalences, when entities or processes not necessarily analogous play similar, interchangeable roles. Context-dependent equivalence does not require nor imply analogy, for instance two non-analogous 'boids' may play equivalent topological roles;
- Commensurability, given by measurability or comparability by a common standard;
- Clustering within the same ranges of values (Aggarwal & Reddy, 2013; Boulis & Ostendorf, 2004; Minati & Pessa, 2018, pp. 112–114).

The intrinsic incompleteness of such aggregates is given by their very high dependence on the levels of scale, representation considered and thresholds.

A very well-known approach to dealing with incompleteness in several disciplinary fields, such as bioinformatics, engineering and information theory, is based on fuzzy sets, fuzzy logic and fuzzy systems. Fuzzy sets are defined as sets whose elements have membership degrees within the continuous interval between 0 and 1 rather than, as in classical set theory, only 0 or 1. The approach is suitable for dealing with problems in which information is incomplete or imprecise. In those cases, however, imprecision relates to the *level of membership,* the membership function characterizing the fuzzy set. **Conversely, the imprecision we are considering here, incompleteness, pertains to the sets themselves dynamically composed of same elements, e.g., clusters and related clustering approaches, as for Multiple Systems, and not only membership functions characterizing the level of fuzziness.**

4.2 The Concept of Theoretical Incompleteness

We distinguish the theoretical aspect of incompleteness from its phenomenological ones because in the first case, incompleteness is intended as a property, not necessarily negative and supposed to be recovered into completeness, contrary to the second case. Cases of theoretical incompleteness include the uncertainty principles in physics mentioned in Sect. 3.2; the incompleteness theorems introduced by Gödel; the undecidability of cases where computable algorithms are not available;

processes represented by non-analytic models based, for instance, on neural networks and cellular automata, and *natural computing* (Brabazon et al., 2015; Mac Lennan, 2004); processes not reducible to procedures, such as workplace safety (Bonometti, 2012).

Incompleteness is considered here as having the non-completeness of representations for emergence as its necessary, if not sufficient, attribute, manifesting incompatibility between phenomena of emergence and their *complete* representations, as mentioned above. As we will see, the problem is how to non-contradictorily, non-destructively represent such incompleteness (for instance with DYSAM and meta-structures, see Sects. 5.1 and 5.2).

Furthermore, in systems science the topic of incompleteness may be considered in relation to the multiplicities, however coherent (Mikhailov & Calenbuhr, 2002), of Multiple Systems regarded with respect to the multiple roles of their constituting interacting components (Minati & Pessa, 2018, pp. 42–45; pp. 166–170) and interchangeabilities, such as ergodic, as stated above. In addition, the various ways of incompleteness of Multiple Systems are prototypes of the quasi-systems introduced in Sect. 6.

Focusing on the incompleteness of self-organization and emergence, we stress that processes of emergence should be considered as a matter of incompletely, e.g., analytically partially unruled, processes that, if for some reason they become completely ruled, degenerate, losing their unpredictability and the dynamics among levels of coherence that are actually continuously dynamically established, interrupted and restored in partial and inhomogeneous ways. In short, a process of emergence cannot be completely represented by a procedure, an algorithm, when any randomization may simulate, but does not model or theoretically represent, structural unpredictability of emergence (Minati, 2016a; Minati & Pessa, 2018, pp. 100–105). In Sect. 4.4. we will consider incompleteness as necessary condition for radical emergence.

4.3 More on Incompleteness and Weak Forces

We consider here the incompatibility of complexity with reductionist, mechanist understandings and approaches assumed complete and suitable for deciding, prescribing, and imposing changes upon the system under consideration. Apart from some admissible, elementary but still necessary interventions supposed to be reparative, this approach to complex systems contains destructive attitudes. *It is a matter, instead, of introducing suitably changed constraints as inputs to be elaborated by the system, for instance through adaptation.*

We need representations suitably open to *interface* with complex systems, allowing for the induction or proposal of changes to be subsequently elaborated by the system supposed to carry out the induced changes. The interventions are not supposed to reductively *replace* behavioral aspects of the system, but to be inputs elaborated by the system itself.

This attitude has a deep constructivism similar to asking questions, as experiments, of Nature. Nature is intended to respond by making experiments happen. However, there are no answers if there are no questions. On the other hand, events may be turned into answers if we invent the appropriate questions. Interfacing with complex systems derives effectiveness from a non-imposing, respectful approach that *lets complexity work*. **The understanding of complexity assumed here is suitable for interfacing and not for forcing, imposing or replacing**.

A very interesting case is given by the effectiveness of weak forces capable of breaking equivalences, equilibria, starting collapses, and setting initial conditions (Minati, 2016b). As in nuclear physics the effective range of the weak force is limited to subatomic distances, we characterize forces here as 'weak' when they have local ranges of influence, that is, involving very few adjacent composing elements—low intensity, for instance, less than a suitable low percentage of all forces involved at the moment—and consequently a range of influence and intensity insufficient to force changes in the behavior or properties of the composing elements, of the entire system, and in the properties of the ongoing interactions.

In the case of collective behaviors, it is a question of substituting the high frequency of weak, irregular, i.e., significantly unpredictable, actions for impossible or unsuitable, strong, single actions, with the advantage, moreover, of flexibility to adapt and, for instance, implement a defensive collective strategy, as in the case of swarms or low but persistent dosages of drugs. Furthermore, low percentages of scalar components of vector forces and side effects can be considered as weak forces. Another case occurs when weak forces are decisive in deterministic chaos. In addition, simplified collective, partial, tentative weak interactions may be assumed at first to be incomplete, tentative *initial conditions* of a self-establishing, quasi-convergent process. Examples include *spontaneous synchronizations*, as in applause, objects on vibrating surfaces, or fireflies, until a specific synchronization becomes predominant and iterated.

The following topics have been introduced and elaborated in Minati and Pessa (2018) and pertain to the richness of the incompleteness of matter (Longo, 2019), understandable as the *intelligence of matter* (Minati, 2019b):

- Between levels of emergence;
- Levels of quasi-ness;
- Multiple emergence;
- Non-invasiveness;
- Non-prescribable actions;
- Pre-properties;
- Quasi-properties;
- Quasi-systems;
- Recurrence of properties at different levels;
- Theoretical systemics and quantum field theory (Blasone et al., 2011).

These are intended as some key topics suitable for soft representations of complexity and to allow approaches enabling one, for instance, to:

- Act on collective emergent phenomena with the purpose of changing, inducing, regulating, defusing, or maintaining acquired properties;
- Diffuse, propagate, or delimit the propagation of emergent phenomena;
- Induce the emergence of collective behaviors in populations of elements collectively, but not coherently interacting, e. g., Brownian motion;
- Merge different collective emergent phenomena;
- Recognize a phenomenon as emergent.

4.4 Incompleteness as Necessary Condition for Radical Emergence

Logical openness is the violation of logical closedness, as in the use of cognitive operators unavoidably multiple and related to uncertainty principles. This may sound in some sense like a limitation, causing us to regret not being able to have closed, well-defined, stable and definitive models, *as if scientific research should have indiscriminately and in every case the goal of logical closure.*

As introduced above (see Sects. 4.2 and 4.3), theoretical incompleteness is based on uncertainty principles, the dual corpuscular-undulatory aspect of matter, the multiplicity of complexity when specific disciplinary approaches and models, possibly coherent, are not also comprehensive due to structural dynamics (change and multiplicity of variable structures among elements, see Sect. 5.1); and on incompatibility in phenomena of emergence between radical emergence and their *complete* representations.

However, the crucial interest in theoretical incompleteness lies in the fact that the phenomenological becoming of processes of radical emergence is not only theoretically, i.e., intrinsically and unavoidably, incomplete but also *requires* incompleteness to take place. Incompleteness is important to figure out *how emergence emerges* (Minati, 2019a).

The concept of radical emergence considers the occurrence of unique multiple coherences, their changing and crossing, e.g., protein folding, acquisition of superconductivity, and superfluidity. The incompleteness we have in mind in this contribution relates to radical emergence established by quasi-coherence (see Sect. 6) among multiple, changing, possibly superimposed, locally modeled processes of emergence and levels of emergence (Minati & Pessa, 2018, pp. 253–281). We refer to populations, domains or local, partially properly modeled processes of emergence establishing ecosystems of domains of validity.

There are properly modeled processes of emergence, for instance, through chaotic dynamics, power laws, long-range correlations, network properties and suitable constraints where incompleteness relates to *equivalent acquirable values* (when one state is as good as another one). In simulations it is a matter of inserting random perturbation, avoiding repeatability.

Here, by contrast, we consider *systems of regimes of validity*, such as the chaotic, ergodic, correlational, networking, polarization, power laws, and scale invariance. Systems of regimes of validity are intended as dynamic zones where, in a simplistic case, a single regime of validity applies, or where multiple regimes of validity, for instance, superimpose, combine, and alternate, allowing coherence in the corresponding *harnessed* system of interacting components.

The processes of acquisition of emergent properties are not intended as given, as the result of single, applied rules, but rather of their ongoing multiple applications, similarly to emergent computation, emerging as an acquired ongoing property rather than as computational result of analytical formulae. In the latter case it is matter of *computational properties of computations in progress*, such as for artificial neural networks and cellular automata, rather than formal properties of the representative equations, as for classical mathematics and physics (Licata & Minati, 2016). Correspondingly, in mathematics, the end of the so-called Bourbaki program (1935–1998), relying on abstract definitions and axioms and finalized to a completely self-contained treatment of the core areas of modern mathematics, was a manifestation of the decreasing effectiveness of classical formalist mathematics.

It is a matter of considering the autonomy of computational approaches and devices sufficient to deal with the structural autonomy of complex systems continuously acquiring properties. The autonomy of computational approaches and devices should not only be suitable for phenomenological complexity, as in physics and chemistry, but also respectful of and appropriate for social, medical, environmental, and economic problems where classical approaches are ineffective, run the risk of 'breaking the toy', and only consider how to repeat themselves apart from parametric variations.

Dealing with ecosystems of properties acquired by the ongoing *usage* (rather than sole application) of rules demands representations open to *interfacing* (Longo, 2011, 2019) with complex systems, allowing one to investigate their nature and to induce or propose changes to be subsequently elaborated by the system supposed to carry them out. This attitude has a deep constructivist nature when it is not a matter of ideologically choosing to be constructivist *or* objectivist, but of using both approaches with a learning strategy in mind (see Sect. 5).

5 Approaches: DYSAM and Meta-Structures

This section outlines two approaches introduced in the literature, namely the Dynamic Usage of Models (DYSAM) as conceptual, methodological framework, and the search for mesoscopic properties, cluster properties, through cluster analysis in collectively behaving populations of interacting entities, both usable for modeling complex systems. The latter approach has been applied also to Big Data.

These may be related to other approaches, such as

- Fuzzy modeling, as a mathematical approach suitable for representing vagueness and imprecise information rather than multiple non-equivalent properties (Miyamoto et al., 2008);

- *Sloppiness* of the models, i.e., models with many, generally more than five, parameters of fit, a matter of models "poorly constrained" when parameters are either unknown or significantly uncertain (Transtrum et al., 2015);
- Converting one problem into another having acceptable levels of equivalence, but more easily approachable, such as converting military problems into economic ones, medical problems into chemically treatable unbalances, geometric problems into more tractable algebraic ones.

5.1 The Dynamic Usage of Models (DYSAM)

After its introduction (Minati, 2001; Minati and Brahms, 2002) the concept, based on established approaches in the literature such as ensemble learning (Ovelgönne & Geyer-Schulz, 2013) and evolutionary game theory, was further elaborated in Minati and Pessa (2006, pp. 64–75).

Dynamic modeling is usually understood to consist of temporal representation through models having time in their equations. A different case is when the dynamics instead relate to methods and strategies for using more models over time. In simple cases, it is a matter of empirical approaches, leaving aside the search for the best, most effective model, just groping, "try and try again" alternatives depending on available resources, e.g., data and energy, or environmental situations; a more suitable case relates to learning strategies.

DYSAM relates to multiple model Multiple Systems, where such multiple modeling inherits the incompleteness of single modeling and the interchangeability of roles of constituent elements. The case of multiple modeling we are considering here relates to the *n*-levels of logical openness, when dealing with the multitude of non-equivalent properties acquired by complex systems over time. Such dynamics relates to the irreducible multiplicity and non-completeness of complex systems considered above and is essentially compatible with the logical openness of the constructivist approach.

This is also the case of the structural dynamics considered above and intended as change and multiplicity of elements, of their roles, and structures between them (Minati & Pessa, 2018, pp. 102–130). Examples of complex structural dynamics are the dynamics of collective behaviors, e. g., flocks and swarms, and of the cytoskeleton consisting, within the cell cytoplasm, of a network of protein fibers and characterized by structural dynamics since its parts are continuously destroyed, renewed or newly created. Structural dynamics refers to the complexity of systems continually transforming themselves and keeping levels of coherence constituted by the same elements interacting, however, in different ways to accommodate interchangeability and multiple roles, allowing the acquisition of various non-equivalent properties.

Examples of structural dynamics asking for DYSAM-like approaches include

- simultaneous usage of economic, cultural, and religious aspects in sociology;
- biochemical and psychological aspects in medicine;

- classical and quantum models in physics in which concrete cases are given by transitions from paramagnetic to ferromagnetic phases, the occurrence of superconductivity and superfluidity and order–disorder transitions in some kinds of crystals, cases of very complicated transient dynamics where classical and quantum aspects mix;
- the learning in childhood of how to use the five senses, not with the purpose of choosing the most effective one, but of learning how to use them simultaneously and in a coherent manner.

5.2 Meta-Structures

The intrinsic incompleteness of aggregates and partially defined sets can be dealt with, for instance, by considering meta-structures, properties of clustered mesoscopic variables (Minati & Licata, 2012, 2013, 2015; Minati et al., 2013; Minati & Pessa, 2018, pp. 102–129).

Structures among cluster properties are considered *meta* because of the incompleteness of clusters, e.g., in number of components over time; in their non-regular, non-iterative occurrence; in aggregative variables changeable over time, such as speed, altitude, direction, price, quantity; in number of elements belonging to them. As discussed below, the term 'meta' indicates virtual, dynamic structures as relationships between cluster properties and also *properties of such properties*, such as their recurrence, levels of correlation, and spatial and temporal distribution.

The mesoscopic level, unlike the macroscopic, does not completely ignore the microscopic, but rather considers some aggregations, i.e., clusters, of the microscopic properties available. Such clustered variables are considered, for instance, with respect to their number of elements, their change over time, distributions and infra-cluster properties. Examples include molecules in a solution affected by similar thermal fluctuations around the average; birds of a flock having similar direction or speed, whatever their spatial or topological positions; customers buying the same groups of items or spending a similar amount of money for whatever items; people living within a building and being on the stairs for determinate periods of time independently of their motion; and cars in traffic that cannot increase their speed (the clusters of cars standing still in the queue, cars slowing down, and cars running at constant speed in the queue).

The interest lies in the dynamical characteristics, properties of changes of the partially or fuzzily defined sets, in this case mesoscopic variables and clusters. While for fuzzy systems, dynamics applies to the levels of belonging of elements, in the latter cases dynamics pertains to the change in properties of the set as its elements, their number, and properties not only belong, but vary. The mesoscopic level can be considered as fuzziness of sets and systems *in having properties*, replacing degrees of belonging with values of aggregation and properties (Salgado, 2004).

This approach is based on considering clusters and their intra-properties, relations and infra-cluster properties as *meta-structures*, 'meta' since clustering may be performed in any way, converging toward more significant levels like a moving magnifying lens. While microscopic interactions establish dynamic structures, for instance, of variable intensity and duration, meta-structures are related to infra-cluster relations and correlations allowing mesoscopic representation of collective complex phenomena, by intercepting transversely a large mesoscopic variety of clustered microscopic aspects. Meta-structures are thus well represented by cluster properties and cluster analysis. The intrinsic incompleteness of aggregates and partially defined sets can be dealt with by using meta-structures, properties of clustered mesoscopic variables (as in Minati & Licata, 2012, 2013, 2015; Minati et al., 2013; Minati & Pessa, 2018, pp. 102–129). Generic examples of meta-structures (Minati & Pessa, 2018, p. 368) are given by cluster regularities and infra-cluster properties, such as correlations intended as meta-structures among clusters. Even if less tractable and more difficult to model, clusters may relate to the interactions in action per instant and their properties such as intensities, durations, directions, and distributions. Analogously, in such cases meta-structures are intended as infra-cluster properties.

This mesoscopic, cluster-based approach may be considered also for populations of data whose possible levels of coherence are neither known nor certain, and for which the representation scale is indefinite and can have different levels, even multiple and superimposed (moving the lens and also using various magnifying glasses). In this case it is a matter of using multiple tentative scenarios and profiles, intended as non-ideal, data-driven modeling, and their emergent, ongoing properties, such as correlations and coherences when dealing with Big Data (Minati, 2019c).

6 Quasi-ness

We conclude by referring briefly to the issue of quasi-ness mentioned above and considered in publications such as Minati (2018, 2019d, 2019e), Minati and Pessa (2018) and Minati et al. (2019).

The concept of quasi-ness does not completely overlap the concept of incompleteness. Significant aspects of quasi-ness differentiate it from incompleteness:

- As introduced briefly in Sect. 3, *incompleteness* is intended as inexhaustible, incomplete multiplicity. Incompleteness is considered a necessary, not a sufficient condition for the emergence of complex systems suitable, for instance, for the Dynamic Usage of Models (DYSAM) and meta-structural cluster analysis (Everitt et al., 2011), using cluster and infra-cluster properties. These approaches interface with complexity to suitably represent multiplicities and the role of critical weak forces.
- *Quasi-ness* pertains to levels of instabilities, irregular alternations of collapse and recovery, for instance of coherences, when a system is not always a system, not always the same system, and not only a system.

Such an understanding of quasi-ness particularly applies to quasi-coherence of emergence, irregularly, incompletely, and continuously established and restored at different levels, rather than considered as a *phenomenologically improbable formal property*. This relates to cases when temporary, possibly collapsing incoherencies occur. This occurrence may be admissible for reasons such as perturbations, irregularities, instabilities, impurities, untimely choices and mistakes, particularly for autonomous entities. Environmental and energetic changes are other examples of reasons for the quasi-ness of coherence-incoherence fluctuations. Such recoverable incoherencies may be regarded as occasional, irregular and not homogeneous with respect to constraints, as incomplete application of rules, such as for the regimes of validity of rules mentioned in Sect. 4.4, whereas Multiple Systems can be regarded as multiple regimes of validity of rules and models.

Within the dynamics of such fluctuations, we should consider also the *inversion* of changes, causing changes of situations previously supposed *incoherent* to *become* coherent. In such situations the coherence is repaired, restored by changing its context. This may be the case for animal behavioral coherences acquired, kept, lost, and restored as important survival properties in the face of danger. The change in the context makes local or previous incoherence become coherent. This probably occurs through a variety of equivalent, *rebalancing* changes.

On the other hand, temporarily incoherent changes are supposed possible when it is not necessary to maintain the nature of emergence over iterative, local coherences not realistically constant, as is predominant in cases of self-organization. An example is multiply interrelated, networked contexts (Nicosia et al., 2013), environments such as ecosystems, where collections of external and internally-generated inputs are intrinsically systemic and non-systemic reactions.

We may say that quasi-ness requires us to better consider the theoretical openness of complex systems, intended as continuous trade-off or exploration, for instance, between order and disorder, coherence and incoherence, incompleteness and quasi-ness, levels of emergence, infinitude of betweenness, and collapsing mechanisms. In complex structural dynamics, the scientifically necessary *Galilean* general repeatability of complete scientific approaches and models should, on the contrary, be *continuously* acquired by contextual, multiple, DYSAM-like populations of quasi-ness-based approaches.

7 New Reductionism

Reductionism is usually understood as an approach privileging the microscopic, bottom-up levels of representation where the ultimate, definitive causes and building blocks of matter are supposed to be. Other, related aspects of reductionism should also be considered, however, for instance, the presumed existence of context-independent optimum, scalar and disciplinary levels of description, or treating the maximum level of precision and completeness as a universal goal. The multiplicity of complex systems is accordingly usually considered as consisting in points of view, rather than in intrinsically multiple properties. Completeness and exhaustiveness are considered essential properties of objective knowledge, in contrast, for instance, with logical openness. We consider here how such assumptions are inadequate and counter-productive, rather than merely wrong, when dealing with emergent systems and complexity.

Some aspects of complexity are incompatible with reductionistic approaches neglecting the structural, intrinsic multiplicity of processes of emergence. For instance, incompleteness, already considered in physics with uncertainty principles and in mathematics, relates to processes occurring when completeness does not apply, such as for phenomena not reducible to procedures, to sequences of uniqueness, phase-transition-like (Sole, 2011), and for non-symbolic computability. Incompleteness and quasi-ness are incompatible with the search for the final, optimal, and fixed. The concepts introduced here thus make evident new possible forms of reductionism, such as non-detection, neglect, and disregard of such properties of complexity or, more plausibly, use of their presumed *approximations* deprived of aspects such as multiplicity, incompleteness, and quasi-ness. The neglect of such properties, however, often occurs in combination with the use of non-complex systemic concepts and properties assumed, nevertheless, to be sufficient and suitable, such as anticipation, controllability, forecasting, growth (instead of development as a property of systems of growth), non-multiplicity, fixed configurations of elements and interactions, organization, planification, regulation, replicability, reversibility, separability (ideal, thermodynamically closed systems), and single optimum representation (Minati, 2018; Minati et al., 2016).

A further case of new reductionism is to consider incompleteness or quasi-ness as suitably representable and simulated by introducing algorithmic randomness, losing sight thus of the weak or partial conveying of information, of the forces critical to the collapsing of equivalences in processes of emergence, and of the loss-recovery sequences of coherence resumptions.

This increases the scientific and cultural responsibilities of those who assume, accept or admit reductionism in its various forms, which are, in reality, deeply unscientific.

8 Further Research

Further related research considers the establishment of systemic regimes of validity of symbolic and sub-symbolic properties and the related occurrence of processes of transience (Minati & Pessa, 2018, pp. 127–130, pp. 265–266). Suitable simulations should be implemented.

9 Conclusion

We have considered some approaches introduced in the literature for dealing conceptually with complexity, such as collective systems and collective behaviors, in particular, logical openness, theoretical incompleteness, the Dynamic Usage of Models (DYSAM), and clustering, all intended as methodological conceptual approaches suitable for implementation in models such as meta-structures. Such an overview opens the way to considering quasi-ness as a generic property and as a necessary, even if non-sufficient, condition for the emergence of complexity. Starting from previous approaches and quasi-ness, we introduced more general conceptual frameworks allowing less abstract, more phenomenological modeling and simulations.

While this understanding increases the degrees of freedom in representing and dealing with complexity, allowing more suitable approaches, modeling and simulations, on the other side it implicitly intensifies the possible reductionism expressing itself as negation of the properties introduced or, more plausibly, as their approximation, deprived of aspects such as multiplicity, incompleteness, and quasi-ness. What reductionism sacrifices, thus, grows proportionately, enhancing the perception of reductionism as such as a fundamentally oversimplifying, unscientific attitude, as if scientific research should have as its purpose, indiscriminately and in any case, logical closure.

References

Aggarwal, C. C., & Reddy, C. K. (2013). *Data clustering: Algorithms and applications*. CRC Press.

Ballarini, M., Cabibbo, N., Candelier, R., Cavagna, A., Cisbani, E., Giardina, I., Lecomte, V., Orlandi, A., Parisi, G., Procaccini, A., Viale, M., & Zdravkovic, V. (2008). Interaction ruling animal collective behaviour depends on topological rather than metric distance: Evidence from a field study. *PNAS, 105*(4), 1232–1237.

Blasone, M., Jizba, P., & Vitiello, G. (2011). *Quantum field theory and its macroscopic manifestations*. Imperial College Press.

Bonometti, P. (2012). Improving safety, quality and efficiency through the management of emerging processes: The Tenaris-Dalmine experience. *The Learning Organization, 19*, 299–310.

Boulis, C., & Ostendorf, M. (2004). Combining multiple clustering systems. In J. F. Boulicaut, F. Esposito, F. Giannotti, & D. Pedreschi (Eds.), *Knowledge discovery in databases: PKDD 2004. PKDD 2004. Lecture Notes in Computer Science* (Vol. 3202). Springer.

Brabazon, A., O'Neill, M., & McGarraghy, S. (2015). *Natural computing algorithms*. Springer.

Brovchenko, I., & Oleinikova, A. (2008). Multiple phases of liquid water. *ChemPhysChem, 9*(18), 2660–2675.

Brovchenko, I., Geiger, A., & Oleinikova, A. (2005). Liquid-liquid phase transitions in supercooled water studied by computer simulations of various water models. *The Journal of Chemical Physics, 123*(4), 44515.

Cavagna, A., Cimarelli, A., Giardina, I., Parisi, G., Santagati, R., Stefanini, F., & Viale, M. (2010). Scale-free correlations in starling flocks. *Proceeding of the National Academy of Sciences of the United States of America, 107*, 11865–11870. https://www.pnas.org/content/107/26/11865. Accessed September 17, 2019.

Diettrich, O. (2001). A physical approach to the construction of cognition and to cognitive evolution. *Foundations of Science, 6*, 273–341.

Everitt, B. S., Landau, S., Leese, M., & Stahl, D. (2011). *Cluster analysis*. Wiley.

Feynmann, R. (1967). *The character of physical law*. The MIT Press.

Gambuzza, L. V., Cardillo, A., Fiasconaro, A., Fortuna, L., Gómez-Gardenes, J., & Frasca, M. (2013). Analysis of remote synchronization in complex networks. *Chaos, 23*, 1–8.

Licata, I., & Minati, G. (2016). Emergence, computation and the freedom degree loss information principle in complex systems. *Foundations of Science, 21*(3), 1–19.

Longo, G. (2011). Reflections on concrete incompleteness. *Philosophia Mathematica, 19*, 255–280.

Longo, G. (2019). Interfaces of incompleteness. In G. Minati, M. R. Abram, & E. Pessa (Eds.), *Systemics of incompleteness and quasi-systems* (pp. 3–55). Springer.

Mac Lennan, B. J. (2004). Natural computation and non-turing models of computation. *Theoretical Computer Science, 317*, 115–145.

Mikhailov, A. S., & Calenbuhr, V. (2002). *From cells to societies. Models of complex coherent actions*. Springer.

Minati, L. (2015). Remote synchronization of amplitudes across an experimental ring of non-linear oscillators. *Chaos, 25*, 123107–123112.

Minati, G. (2001). Experimenting with the dynamic usage of models (DYSAM) approach: The case of corporate communication and education. In *CD-Proceedings of the 45th annual meeting of the international society for the systems sciences (ISSS)*. Monterey, California, July 8th–13th 2001; 01-094, pp. 1–15.

Minati, G. (2016a). Knowledge to manage the knowledge society: The concept of theoretical incompleteness. *Systems, 4*(3), 1–19. http://www.mdpi.com/2079-8954/4/3/26

Minati, G. (2016b). General system(s) theory 2.0: a brief outline, In G. Minati, M. Abram, & E. Pessa (Eds.), *Towards a post-bertalanffy systemics* (pp. 211–219). Springer.

Minati, G. (2018). The non-systemic usages of systems as reductionism. Quasi-systems and Quasi-systemics. *Systems, 6*(3). http://www.mdpi.com/2079-8954/6/3/28

Minati, G. (2019a). Phenomenological structural dynamics of emergence: An overview of how emergence emerges. In Urbani Ulivi Lucia (Ed.), *The systemic turn in human and natural sciences. A rock in the pond*. (pp. 1–39). Springer. https://www.springer.com/us/book/9783030007249

Minati, G. (2019b). A controversial issue: The intelligence of matter as *residue*? A possible understanding. *Biomedical Journal of Scientific & Technical Research (BJSTR), 23*(1), 17139–17140. https://biomedres.us/pdfs/BJSTR.MS.ID.003848.pdf

Minati, G. (2019c). Big data: From forecasting to mesoscopic understanding. Meta-profiling as complex systems. *Systems, 7*(1), 8. https://www.mdpi.com/2079-8954/7/1/8

Minati, G., (2019d), On some open issues in systemics. In G. Minati, A. Abram, & E. Pessa (Eds.), *Systemics of incompleteness and quasi-systems* (pp. 317–323). Springer.

Minati, G. (2019e). Non-classical systemics of quasi-coherence: From formal properties to representations of generative mechanisms. A conceptual introduction to a paradigm-shift. *Systems, 7*(4). https://www.mdpi.com/2079-8954/7/4/51

Minati, G., & Brahms, S. (2002). The dynamic usage of models (DYSAM). In G. Minati & E. Pessa (Eds.), *Emergence in complex cognitive, social and biological systems* (pp. 41–52). Kluwer.

Minati, G., & Licata, I. (2012). Meta-structural properties in collective behaviours. *The International Journal of General Systems, 41*(3), 289–311.

Minati, G., & Licata, I., (2013). Emergence as mesoscopic coherence. *Systems, 1*(4), 50–65. http://www.mdpi.com/2079-8954/1/4/50

Minati, G., & Licata, I., (2015). Meta-structures as multidynamics systems approach. Some introductory outlines. *Journal on Systemics, Cybernetics and Informatics (JSCI), 13*(4), 35–38. http://www.iiisci.org/journal/sci/issue.asp?is=ISS1504

Minati, G., & Pessa, E. (2006). *Collective beings*. Springer.

Minati, G., & Pessa, E. (2018). *From collective beings to quasi-systems*. Springer.

Minati, G., Penna, M. P., & Pessa, E. (1996). Towards a general theory of logically open systems. In E. Pessa, M. P. Penna, & A. Montesanto (Eds.), *Proceedings of the 3rd systems science European congress*, (Kappa, Rome, Italy, pp. 957–960).

Minati, G., Penna, M. P., & Pessa, E. (1998). Thermodynamic and logical openness in general systems. *Systems Research and Behavioral Science, John Wiley and Sons Ltd., 15*(3), 131–145.

Minati, G., Abram, M., & Pessa, E. (Eds.). (2016). *Towards a post-Bertalanffy systemics*. Springer.

Minati, G., Licata, I., & Pessa, E. (2013). Meta-structures: The search of coherence in collective behaviours (without physics). In Alex Graudenzi, Giulio Caravagna, Giancarlo Mauri, & Marco Antoniotti (Eds.), *Proceedings Wivace 2013 – Italian Workshop on Artificial Life and Evolutionary Computation* (Wivace 2013), Milan, Italy, July 1–2, 2013. *Electronic Proceedings in Theoretical Computer Science* (Vol. 130, pp. 35–42). http://rvg.web.cse.unsw.edu.au/eptcs/paper.cgi?Wivace2013.6

Minati, G., Abram, M., & Pessa, G. (Eds.) (2019). *Systemics of incompleteness and quasi-systems*. Springer. https://www.springer.com/gp/book/9783030152765

Miyamoto, S., Ichihashi, H., & Honda, K. (2008). *Algorithms for fuzzy clustering: Methods in C-means clustering with applications*. Springer.

Nicosia, V., Bianconi, G., Latora, V., & Barthelemy, M. (2013). Growing multiplex networks. *Physical Review Letters, 111*(058701), 1–5.

Ovelgönne, M., & Geyer-Schulz, A. (2013). An ensemble learning strategy for graph clustering. In D. A. Bader, H. Meyerhenke, P. Sanders, & D. Wagner (Eds.), *Graph partitioning and graph clustering* (Vol. 588, pp. 187–206). American Mathematical Society.

Paperin, G., Green, D. G., & Sadedin, S. (2011). Dual-phase evolution in complex adaptive systems. *Interface, 8*, 609–629.

Salgado, & Garrido, P.J. (2004). Fuzzy clustering of fuzzy systems. In *IEEE International Conference on Systems, Man and Cybernetics* (Vol. 3, pp. 2368–2373).

Solé, R. V. (2011). *Phase transitions*. Princeton University Press.

Transtrum, M. K., Machta, B. B., Brown, K.S., Daniels, B. C., Myers, C. R., & Sethna, J. P.: Perspective: Sloppiness and emergent theories in physics, biology, and beyond. *Journal of Chemistry and Physics, 143*, 010901.

Vicsek, T., & Zafeiris, A. (2012). Collective motion. *Physics Reports, 517*, 71–140.

Micro-Latency, Holism and Emergence

Alexander Carruth

1 The Structure of Inquiry and the Structure of the World

Part of the job of scientific inquiry is to engage with, make sense of, describe, explain, and make predictions concerning the wildly varied phenomena which constitute the world around us. Distinct scientific disciplines each with their own intellectual regimes—domains of inquiry, basic assumptions, investigative techniques and so on—address different groupings of this phenomena. Thus, physics, or at least an important part of that discipline, is concerned with the properties of and interactions between the relatively small and simple constituents of matter, and of energy. Chemistry addresses the properties and behaviour of more complexly structured systems of those constituents that form substances (in the standard, as opposed to technical metaphysical, sense): elements, compounds, mixtures, suspensions and so on. Biology treats phenomena which exhibit the characteristics which are criteria for life, ranging over micro-organisms, flora, fauna etc. Psychology and cognitive science engage with just those living things which possess mentality, and sociology, economics, and political science all range over aspects of the interactions between these thinking agents. These characterisations are somewhat glib, and they surely fall short of a properly nuanced and comprehensive conception of each discipline, but hopefully they are fit for the illustrative purpose to which they are employed.

That inquiry has this sort of structure raises a number of interesting philosophical questions. One set of questions concerns the sorts of relationships that obtain between the theories put forward by each discipline—questions such as: Can we predict the behaviour of the more-complex based solely on knowledge of the more-simple? Can we derive the facts about the more-complex from those concerning the more-simple? Can theories which refer to the more-complex be translated, without loss of

A. Carruth (✉)
Theoretical Philosophy, Department of Philosophy, History, and Art Studies, University of Helsinki, Yliopistonkatu 3, 00014 Helsinki, Finland
e-mail: alexander.carruth@helsinki.fi

S. Wuppuluri and I. Stewart (eds.), *From Electrons to Elephants and Elections*, The Frontiers Collection, https://doi.org/10.1007/978-3-030-92192-7_11

information, into ones which refer only to the more-simple? And so on... Another set of questions concerns the extent to which the sort of structure described above is a feature not just of the way we organise our inquiry into the world, but of the world itself: that is, addressing the sorts of relationships that obtain between the entities, properties, processes, laws, events (and so on...) with which the various sciences are concerned—questions such as: Is the more-complex something *over and above* the more-simple? Are apparently 'higher-level' entities autonomous? Can the more-complex be identified with the more-simple? Is the more-complex exhaustively determined by the more-simple?

This paper is focussed on questions of the latter sort. Speaking in very broad terms, emergentists of various sorts hold that there is some sense in which the kind of structure exhibited by inquiry is mirrored in the world itself: there are genuinely distinct, hierarchically arranged 'levels of reality', with entities or phenomena at higher levels existing separately from and enjoying some degree of causal (or more broadly determinative) autonomy from those at lower levels, but still being somehow dependent on those lower-level entities and phenomena. Conversely, reductionism holds that this structure is merely apparent, and is not mirrored in reality: all genuine existence and causal/determinative 'action' is confined to a single, basic level, and more complex and apparently higher-level entities can be identified with or reduced to or otherwise exhaustively accounted for by the more simple, basic entities.

Both emergentism and reductionism come in various forms. The principal aim of this paper is to outline and explore a position—let's call it Flat Holism—regarding apparently emergent phenomena, which arguably sits somewhere between emergence and reduction. This position preserves some key reductionist commitments, as it involves no radical ontological novelty, for instance, and is consistent with a one- or no-level ontology. It also, however, adopts the emergentist idea that the whole or context plays a crucial, metaphysically determinative role. Whether this position counts as emergentist or reductionist (or neither) is relatively moot, what is much more important is (i) how the position compares to related accounts and (ii) how it stands with regard to the relevant empirical evidence.

This essay focusses on (i); the important work of (ii) must be left for another day. It begins with an examination of the prima facie case for emergentism, followed by some discussion of the varieties—and key commitments—of emergence and reduction (Sects. 2 and 3). In the rest of the essay Flat Holism is introduced, and the commitments of the position delimited, through comparison with two other positions: Sydney Shoemaker's micro-latency account and Carl Gillett's mutualism. In Sects. 4 and 5 the micro-latency and mutualist accounts are introduced. Section 4 also argues that despite Shoemaker labelling the micro-latency view as a form of emergentism, it preserves the key theoretical commitments of the reductionist programme—indeed, the view can be leveraged to provide a general argument against emergentism. Both Shoemaker's and Gillett's accounts are framed in terms of powers, and Sect. 6 introduces two approaches to the metaphysics of powers: the *orthodox view*—briefly, the view that powers are single-track and operate according to the 'stimulus-manifestation' model—and an alternative view which holds that powers are multi-track and operate according to the 'mutual manifestation' model. Section 7 shows how the micro-latency and mutualist accounts both tacitly rely on the *orthodox view*, and demonstrates how Flat Holism can be generated by adopting both the alternative, *multi-track mutual-manifestation* view of powers and some of the motivating insights of the micro-latency and mutualist accounts. Some advantages of Flat Holism are explored in Sect. 8.

2 The Prima Facie Case for Emergence

It is worth spending a little time reflecting on the conditions that generate the debate between reductionism and emergentism. Central to this debate are what are referred to in this essay as 'E-cases'. We are presented with an E-case when we find ourselves confronted with:

- a complex context C
- the elements of which are composed of basic entities the Bs
- in which some apparently novel phenomenon or behaviour E is present
- and *prima facie* there does not seem to be a way to account for E based on the properties or behaviour of the Bs taken either in isolation or in other contexts which lack the features definitive of C.

E-cases provide the principal motivation for emergentism, and the principal challenge to reductionism, because it is precisely such cases which provide us with examples of apparently novel or distinctive phenomena and behaviour which arise only in certain complex circumstances and which seem to be in some metaphysically significant sense dependent on (because composed by) but nevertheless not exhaustively accounted for or determined by the more-simple elements which compose the entities or system in question. Take for example questions concerning reduction and emergence in the philosophy of mind: here anti-reductionism is motived by appeal to the idea that it is only in certain highly complex and specific circumstances (those involved in there being a living human person, say) that certain phenomena are present (such as conscious experience and agency), and that despite the fact that persons are composed of more basic entities (organs, tissues, cells and ultimately micro-physical entities) there is no clear way to account for conscious experience and agency in terms of quarks and electrons and so on, either taken in isolation or in other complexes which are not constitutive of a living human person (even in simpler collectives which are both partially constitutive of and clearly highly relevant for the target phenomena, such as populations of neurons or synaptic structures). E-cases nevertheless provide only a *prima facie* motivation for emergentism because they are defeasible—the appearance that the Bs do not fully account for E may turn out, on further investigation or reflection, to be just that: an appearance.

It is probably fair to say that the dominant—perhaps even the default—attitude in philosophical and scientific circles towards the relationship between the more- and less-complex over the past several decades has been broadly reductionist in spirit. Inspired by the impressive and ongoing successes of scientific analysis, this has led to a kind of temperamental preference for reductionism. Despite the fact that we are sometimes confronted with apparent E-cases for which explicit reductionist analyses are not forthcoming, temperamental reductionism is bolstered by an often-tacit optimistic meta-induction of roughly the following form: as our scientific understanding has developed, various E-cases have been shown to be merely apparent, as previously-unknown underlying mechanisms, features and behaviours of the more basic elements have been identified which show that E can in fact be accounted for by the Bs, and so any support such cases gave to emergentism was in fact illusory. We should anticipate that this trend will continue as our scientific understanding develops further, such that remaining E-cases will eventually be uncovered as also merely apparent. Perhaps some E-cases are so complex and difficult that limited rational agents such as ourselves may never have sufficient understanding to comprehend how the Bs account for E, but this is an 'in practice' rather than 'in principle' concern, and thus doesn't tell against the truth of (ontological) reductionism. Thus, E-cases do not provide any significant challenge to reductionism.

To give a concrete example, consider how *life* acted as an E-case supporting vitalism—the view that life involves some distinctive and novel phenomena and is governed by distinctive principles which are not instanced in non-living matter—for centuries. However, advances in biology and chemistry have shown that vitalism has little credibility, and life can be accounted for—not just epistemically, but metaphysically—in terms of various underlying mechanisms: life *just is* the presence of such mechanisms suitably related to one another and embedded in a sufficiently stable context. As it went for life and vitalism, so it will go for the E-cases with which we are currently presented, or at least so the optimistic meta-induction of the temperamental reductionist has it. Whilst such reasoning is often tacit, something at least close to it is sometimes made explicit (see for example Churchland, 1981). Alongside that generated via the preceding reasoning, temperamental reductionism may also be motivated by general theoretical considerations—for instance by considerations such as that reductionism, if true, is a more parsimonious and more unifying perspective than anti-reductionist alternatives.

It should be noted that although the immediately preceding discussion is couched partially in epistemic terms—about the development of scientific understanding and so on—the issue at stake here is not a matter of epistemology. What is at stake is whether E-cases are, or are not, evidence of the *existence* of novel entities, properties, processes (and so on) which are distinct from, albeit dependent on, the entities of which they are composed; the properties these entities are characterised by, and the processes in which they participate. The relevant advancements in scientific understanding are therefore those which support the establishment of metaphysically significant relationships between apparently-higher-level and more basic phenomena, that support claims such as 'life *just is* the presence of such-and-such mechanisms...' or 'temperature *is identical with* mean kinetic energy' and so on (one oughtn't to assume that there is *no* controversy surrounding these examples, but hopefully they are sufficiently clear for the illustrative purposes for which they are deployed here).

The temperamentally reductionist dominant viewpoint is coming under increasing pressure and scrutiny, both within science and philosophy. Recent consideration of an ever-increasing number of case studies from a range of scientific disciplines seems to suggest that we are confronted with a host of recalcitrant E-cases which resist reductionist analysis. These include, but are not limited to: cases from physics including critical behaviour in phase transitions (Batterman, 2011), the Fractional Quantum Hall effect (Lancaster & Pexton, 2015); quasi-particles (Franklin & Knox, 2018) and superconductivity (Morrison, 2012); cases from chemistry including temperature (Bishop, 2010) and molecular structure (Bishop, 2010; Hendry, 2010); biological cases including protein assembly and gene expression (McLeish, 2017), and examples from the social sciences including macroeconomic properties (Hoover, 2009) and group cognition (Theiner & O'Connor, 2010). These cases stand alongside familiar and difficult examples such as those mentioned earlier from the philosophy of mind. The diversity, range and recalcitrance of these cases seem to undermine the optimistic meta-induction and place temperamental reductionism—especially where the envisaged reductionism is of a relatively simplistic form which we'll call *straightforward reductionism*, according to which all E-cases can in principle be accounted for in terms of just the attributes and behaviours displayed by the relevant basic entities in simpler, non-E cases—under significant pressure, as the envisaged reductive accounts have rarely been forthcoming. This has revived interest in alternative anti-reductionist viewpoints, and especially in emergentism. This renewed interest is evidenced for instance by the significant number of monographs and collections on the topic published in the last decade or so (not to mention vast numbers of articles).[1]

3 Varieties and General Features of Emergence and Reduction

One result of the recent vigorous activity surrounding issues concerning emergence and reduction is a proliferation of accounts of what each position might involve and be committed to. Indeed, I've heard it said (although, regrettably, I no longer remember by whom) that there are as many accounts of emergence as there are emergentists, and, what's more, the same is nearly true of reduction. In the contemporary debate concerning these two competing standpoints on the relationship between the more- and less-complex or basic, distinctions are drawn between varieties of emergence that are *metaphysical* and *epistemological*; *weak* and *strong*; *synchronic* and *diachronic*; *supervenience-preserving* and *supervenience-violating*, and doubtless more besides.

[1] A non-exhaustive list of recent, relevant monographs and collections: Baggio & Parravicini (eds.) (2019); Baysan & Sartenaer (eds.) (2021); Bedau & Humphreys (eds.) (2008); Bigaj & Wüthrich (eds.) (2015); Bishop (2019); Bishop et al. (forthcoming); Carruth et al., (2017a; 2017b); Corradini & O'Connor (eds.) (2010); Falkenburg & Morrison (eds.) (2015); Gibb et al. (2018); Gillett (2016); Hohwy and Kallestrup (2008); Humphreys (2016); Macdonald and Macdonald (2010); Paolini Paoletti (2017); Paolini Paoletti and Orilia (2017); Wilson (2021)—and of course, this volume itself.

Furthermore, these distinctions can be cross-combined, leading to a proliferation of potential positions—not all of which, of course, will enjoy equal prior plausibility or support from the empirical evidence. This essay is not the place for a detailed examination of all these different options, but see for instance Baysan and Sartenaer (eds.) (2021); Chalmers (2006); Sartenaer (2015); Silberstein (2001); van Gulick (2001); Wilson (2015) and (2021) for discussion.

The comment with which this section opened could be taken to express some measure of exasperation, and perhaps with good reason. Where the available positions in a debate proliferate in the manner just discussed, there is, for instance, the danger of discussants talking past one another, especially if 'emergentism' and 'reductionism' are often treated as relatively monolithic positions. However, if the debate concerning emergence is to be sensitive to empirical findings, then careful exploration of the various positions that can be occupied and a detailed examination of their various commitments, entailments and relative merits are crucial—especially as pluralism remains a live option in this area of debate. Drawing on some of the E-cases mentioned earlier for illustrations sake, it could for example turn out that both conscious experience and superconductivity are emergent, but in very different ways; and that chemical structure and protein folding can both be reduced to more basic phenomena, but again in different ways. The world might well turn out to be such that in different contexts, different forms of emergence and reduction are present, and thus developing a nuanced appreciation of the different forms that each position might take is a crucial task for those involved in this debate.

Notwithstanding the ways in which the varieties of emergence and reduction mentioned above can differ from one another, some central commitments or general themes of each position can be identified. Key reductionist notions include:

a. That the basic entities enjoy special ontic status not enjoyed by non-basic entities (if there are any)
b. That the 'action' is all at the base level: basic entities alone determine how things are
c. That apparent non-basic phenomena can be exhaustively accounted for in terms of basic phenomena.

And key emergentist notions include:

d. That there are non-basic entities which enjoy some metaphysical autonomy, putting them on a par, ontologically speaking, with basic entities
e. That the non-basic can play a determinative role, so at least some of the 'action' is not at the base level
f. That there are some non-basic phenomena that cannot be exhaustively accounted for in terms of basic phenomena—in this sense, 'higher level' entities/goings-on are something 'over and above' the bases upon which they depend.

(a)–(c) and (d)–(f) aren't intended here as an analysis of reductionism and emergentism respectively, but rather as a very broad outline of the core characteristics of each family of positions, which can help us to identify where various positions in this debate are located. Although it is perhaps most common to see claims like

(b) and (e) couched in terms of causation, they are stated in terms of the broader notion of determination here, because it seems plausible that causation is not the only metaphysically significant manner in which something might play a role in 'how things go' such that if some non-basic entity were to play such a role this would be good reason to take it to be challenging to the reductionist viewpoint (note that Gillett (2016) uses the term 'determination' in a narrower sense than I do here, just for cases where a difference is made to which powers are had by one or more individuals).

4 The Micro-latency Hypothesis

Sydney Shoemaker outlines an intriguing position in the debate concerning reduction and emergence, which has been the subject of relatively little discussion (2002, 2007). Shoemaker's account is formulated in terms of *powers*. Central to Shoemaker's account is a distinction between two different kinds of powers possessed by basic entities: what he calls *micro-manifest* and *micro-latent* powers. Micro-manifest powers are powers which basic entities display in both simple and complex contexts and which play a determinative role in both simple and complex contexts. For instance, supposing electrons to be amongst the fundamentalia, 'having a mass of $9.10938356 \times 10^{-31}$ kg' or 'having -e charge' would be candidate micro-manifest powers: these features of electrons are plausibly considered to be, or at least to involve, powers (to exert a certain gravitational force and resist acceleration, say, or to repel like charges and attract opposite charges); are displayed by electrons in the simplest contexts in which they can be situated and play a role in determining the behaviour of the electron, and the behaviour of a system or complex of which the electron is a component. Micro-manifest powers are arguably one of the key explanatory resources of scientific analysis, through identifying these powers and showing how they can be combined to form mechanisms which account for apparently novel, non-basic/complex behaviour or phenomena. One potentially fruitful way of thinking about recalcitrant E-cases and why they provide support for emergentism, then, is as cases in which no such analysis is forthcoming. This leaves E in need of some explanation, and if the micro-manifest powers of the Bs are not sufficient to the task, then one alternative is that we need to posit some novel, distinctive, non-basic power or powers which *do* account for E—and such powers would be emergent.

Micro-latent powers, according to Shoemaker, are powers of basic entities which only play a determinative role—and thus manifest—in certain complex contexts, but which supervene on micro-manifest powers. Thus, a basic entity B_1 possessed of a micro-latent power P_1 would only display P_1 in very specific circumstances and in many contexts or systems in which B_1 could be situated P_1 would be completely idle, and thus undetectable, only making itself manifest in the relevant complex context. Novel phenomena or behaviour of some complex system, according to Shoemaker, can then be accounted for in terms of what he calls a Type-2 micro-structural property, that is, the property of having basic constituents arranged thus-and-so which possess certain micro-manifest *and* certain micro-latent powers:

> ...Type-2 micro-structural properties, will be properties specified in terms of all of the powers, micro-latent as well as micro-manifest, of the constituent micro-entities. Such a property will be the property of being composed of particles that have certain micro-latent and micro-manifest powers and are related in certain ways.
>
> Type-2 micro-structural properties, although they are micro-structural, will be emergent properties. For they are specified partly in terms of the micro-latent powers of the constituent micro-entities that account for the emergence (2016: 74).

Adopting Shoemaker's micro-latency view allows one to accommodate E-cases not by positing novel, distinctive non-basic powers (and perhaps non-basic individuals which are the bearers of these powers), but rather by exclusive appeal to the powers of basic entities. As the second half of the quotation above demonstrates, Shoemaker takes the fact that this approach to E-cases involves something other than the micro-manifest powers of the basic entities to show that the view is a form of emergentism. But it isn't clear that this is the case. Whilst this establishes that the micro-latency view is not a form of *straightforward* reductionism (according to which all E-cases can in principle be accounted for in terms of just the attributes and behaviours displayed by the relevant basic entities in simpler, non-E cases), the micro-latency view nevertheless has all the hallmarks of a broadly reductionist metaphysic.

First, a Type-2 micro-structural property—which, according to Shoemaker, is an emergent—is a second-order property of *having parts arranged thus-and-so which instantiate properties F, G, H and so on...* It can be questioned whether such second-order 'properties' are genuine properties at all, but more importantly, it is clear that a second-order property of *having some first order property/properties* does not have the same ontic status as the first-order property/properties to which it relates; upon which it depends and by which it is determined: there is nothing more to a Type-2 micro-structural property than the arrangement and nature of the basic, micro-structural entities. Thus, the micro-latency view satisfies (a): basic entities enjoy special ontic status not enjoyed by non-basic entities. Second, Type-2 micro-structural properties do not play any novel or distinctive determinative role on this account. As discussed above, in an E-case, the novel phenomena or behaviour which are characteristic of the case are exclusively the result of powers (both micro-manifest and micro-latent) of the basic, micro-physical components involved in the case. Thus, the micro-latency view satisfies (b): basic entities alone determine how things are. Finally, as the instantiation of a Type-2 micro-structural property *just is* having components arranged thus-and-so and which possess certain first-order properties, and it is exclusively these components which are determinative, then it seems clear that Type-2 microstructural properties (and thus the distinctive behaviour or phenomena present in E-cases for which they are posited as explanation) are fully accounted for by basic, micro-physical entities and their natures. Thus, the micro-latency view also satisfies (c): the non-basic can be exhaustively accounted for in terms of the basic. As the micro-latency view satisfies (a)–(c), the view is a form of reductionism, albeit not *straightforward* reductionism. For further discussion of how the micro-latency view is not an emergentist position, see Gillett (2016:230 fn17) or Shrader (2010).

We've just seen that the micro-latency view appears to accommodate E-cases without appeal to an emergentist metaphysic. As the need to accommodate E-cases provides the principal motivation for adopting emergentism, the micro-latency view can be leveraged to generate an anti-emergentist argument (note that this is not a use to which Shoemaker himself puts the view) of the following form:

P1: Emergentism is plausible only if E-cases cannot be accommodated by non-emergentist metaphysics
P2: The micro-latency view, if plausible, demonstrates that E-cases can be accommodated by non-emergentist metaphysics
P3: The micro-latency view is plausible
C: Emergentism is not plausible.

P1 is supported both by general principles of parsimony and because it is a consequence of the crucial link between the need to accommodate E-cases and motivating emergentism outlined in Sect. 2. P2 asserts the non-emergentist interpretation of the micro-latency view argued for earlier in this section. P3 is obviously controversial but will be granted for the sake of discussion here. That C follows from P1–P3 shows that the micro-latency view, at least as interpreted here, is hostile to emergentism. Note that this might make the view particularly attractive to the temperamental reductionist, as it provides her with a general schema by which to deflate the apparent support given to emergentism by the prevalence of E-cases which is independent of the optimistic meta-induction discussed in Sect. 2. This gain, of course, comes at a cost: abandoning *straightforward* reductionism and embracing the micro-manifest/micro-latent distinction.

5 Gillett's Mutualism

Carl Gillett (2016) has developed a genuinely emergentist, scientifically informed approach which is designed to be both consistent with and illuminating with regard to the success of the compositional explanations typical of scientific analysis. The resulting position is thus less 'radical' than forms of ontological emergence which posit causally active, uncomposed, non-physical higher-level entities, whilst nevertheless including higher-level composed entities which play a distinctive determinative role. Like Shoemaker's view, Gillett formulates his mutualist approach in terms of powers, introducing the notion of a *differential power*:

> components have some powers in complex collectives that they would not have if the laws or principles applying in simpler collectives exhaustively applied in the complex aggregation (2016: 18).

Differential powers, like micro-latent powers, are powers of component entities. And it is these differential powers—which are present only in certain complex circumstances—that account for the novel behaviour or phenomena characteristic of E-cases. Crucially, on the mutualist view the presence of one or more differential powers in a given complex collective will be attributable to the determinative nature of the whole or complex itself. Thus, the whole plays an ineliminable determinative role, by determining *which powers* the basic, component entities of which it is composed have when they compose it.

On the mutualist view there are, in at least some complex collectives, therefore two simultaneous forms of determination in operation: the familiar 'upward' compositional determination of the nature of the whole by the components, and a less familiar 'downward' determination of the differential powers of the components by the whole. Gillett calls this downward determination *machresis*:

> Combining the Greek words 'macro' and 'chresis,' where the latter is roughly the Greek for 'use,' we get the terms 'machresis,' and 'machretic determination,' for the general phenomenon of composed, or 'macro,' entities that non-productively, and non-compositionally, determine the nature of their components through productive role-shaping (2016: 207).

On this view, productive determination—that is, the manifestation of powers—may occur only at the level of basic, component entities. But *what productive role* a given component or set of components plays is (at least sometimes) determined by the composed, 'higher-level', 'macro' entity. Thus, composed entities play a crucial and distinctive determinative role, as their parts wouldn't do what they in fact do if they weren't parts of these wholes. As this determinative role is not itself productive, familiar concerns regarding exclusion (and the like) by which more 'radical' forms of emergence are beset, can be avoided.

The mutualist account takes composed, 'higher-level', 'macro' entities to be elements of the ontology with the same ontic status as the basic parts of which they are composed. As composed entities are non-basic, the mutualist account satisfies (d): it holds that some non-basic entities are on a par, ontologically speaking, with basic entities. Through machretic determination, these composed entities shape the productive role of the parts of which they are composed by determining which differential powers these parts possess, and thus the mutualist view clearly satisfies (e): the non-basic plays a determinative role and at least some of the 'action' is not at the base level. Finally, in recognising the simultaneous and mutual forms of 'upward' compositional and 'downward' machretic determination, the account shows how although composed entities depend on their components, nevertheless there are some non-basic phenomena that cannot be exhaustively accounted for in terms of basic phenomena, and thus satisfies (f).

6 Powers

Like much of the metaphysically focused discussion of emergence and reduction, both Shoemaker's and Gillett's views are couched in terms of *powers*.[2] I've argued elsewhere that this focus in the debate is both appropriate and means that the debate is importantly sensitive to issues in the metaphysics of powers (see Carruth, 2019 and 2020). Whilst this latter point might seem obvious, it's often been missed—or at least not discussed explicitly. Where Carruth, 2020 offers general arguments demonstrating this sensitivity, one way of thinking about the material presented in this essay, as should become clear over the following sections, is as a specific case study supporting the sensitivity claim. This section gives a very brief introduction to powers, and outlines two ways of conceiving of how they are directed and how they operate: the *orthodox view* according to which powers are single-track and operate according to the 'stimulus-manifestation' model, and an alternative view which holds that powers are multi-track and operate according to the 'mutual manifestation' model. This will form the basis for outlining Flat Holism in Sect. 7.

Powers are features of individuals in virtue of which the individuals that bear these powers interact and behave in the manner that they do. Individuals with just the same powers will behave in the same way when placed in exactly similar circumstances (or, if some powers are stochastic, will have the same objective probabilities assigned to the same range of possible behaviours—henceforth I will omit qualifications of this sort, as it should usually be clear how the relevant claims can be modified to accommodate the possibility of stochastic powers). Individuals with different powers will behave differently, and this difference in behaviour will be due to their instantiating different powers. Views such as strong versions of dispositional essentialism hold that all fundamental properties are causal powers (see Shoemaker, 1980). Others take causal powers to depend on or reduce to a combination of non-causal properties and the laws of nature (e.g. Armstrong, 1997). Others still take some fundamental properties to be powers, and others not to be (e.g. Molnar, 2003). This essay will not aim to settle this dispute.

Powers are essentially powers *to* something or other. That is, their nature involves being directed towards some manifestation or set of manifestations; and they bring about these manifestations in suitable circumstances. Canonical examples include fragility, which could be roughly characterised as 'the power to break when struck', or solubility, 'the power to dissolve when in contact with a suitable solvent'. Whilst these canonical examples are non-basic, it's plausible to think of 'having a mass of $9.10938356 \times 10^{-31}$ kg' or 'having -e charge' similarly. We can roughly capture these features using conditional statements of the form:

[2] For instance, Jaegwon Kim argues that emergent entities must have distinctive powers (1999). Timothy O'Connor and Hong Yu Wong characterise emergent properties as basic properties of composites; and take 'basicness' to involve conferring novel powers (2005). Jessica Wilson says that strongly metaphysically emergent entities have "fundamentally novel powers" (2015, p.356). For further discussion of the role of powers in the emergence/reduction debate, see Carruth (2019) and (2020).

P: x has the power to φ if it is the case that were x to be placed in suitable circumstances C, then x would φ.

I don't offer **P** as an *analysis* of what it is to have a power, although some philosophers have attempted to analyse powers in terms of the truth of conditional statements (e.g. Lewis (1997); such analyses are plagued by familiar counterexamples, however, see Martin (1994) and Bird (1998)). But conditional statements such as **P**, even if they fail to analyse what it is to have a power, capture the central features of powers mentioned above: their directedness and their sensitivity to circumstance.

There are several controversies in the debate concerning the nature of powers. One concerns whether powers are 'single-' or 'multi-track'; that is, whether a given power only ever disposes its bearer towards a single manifestation or whether an object can be disposed towards a range of manifestations just in virtue of a single power it instantiates. Whenever a single-track power manifests, and in whatever circumstances it does so, the manner in which it manifests is the same. Conversely, multi-track powers are directed towards a range of different manifestations. Heil describes multi-tracking as follows[3]:

> Consider a simple case, the sphericity of a particular ball. The ball's sphericity, in concert with incoming light radiation, structures outgoing radiation in a definite way. The very same property of the ball disposes it to produce a concave depression in a lump of clay or to roll… one disposition, many different kinds of manifestation (2003: 198–199).

Thus, on two different occasions, when such a power manifests, it may manifest in different ways, depending on the circumstances in question.

Another controversy concerning powers involves how it is that they come to manifest, that is, how they operate. One view has it that powers operate through being 'triggered' by some stimulus, which then leads to the power producing the manifestation. According to this 'stimulus-manifestation' model, a power will only give rise to a manifestation when it is galvanised into action by some trigger or stimulus. For instance, in the case of the *fragility* of a vase, the stimulus might be 'being struck with a force greater than X', or in the case of the *solubility* of a sample of salt, 'being submerged in water'. Crucially, the manifestation is produced by the target power alone, although it will not be produced until the occurrence of the stimulus. Matthew Tugby (2010) has argued that stimuli needn't themselves be powers, and that stimuli may belong to a variety of categories (e.g. events, states, actions and so on).

An alternative account is the 'mutual manifestation' model, which holds that there must always be two or more powers working together to bring about a manifestation. When powers work together, there is no sense of priority such that one power could be considered the 'operative power', whilst the other is held to have merely stimulated or triggered it. For instance, in the case of the production of a particular vase's shattering, this view would hold that this is not the result of the 'fragility' of the vase alone, but rather of a whole host of powers of the vase, of the object that struck it,

[3] Heil talks in terms of 'dispositions' here, which for the sake of the current discussion can be treated as synonymous with the way this paper uses the term 'power'.

and perhaps more besides. Likewise, the dissolution of a sample of salt is a result of the mutual action of both the particular crystalline structure of the salt and the dipole moment typical of H_2O molecules (and perhaps more besides).

Inspired by something like **P**, what could reasonably be regarded as the *orthodox view* of powers holds that powers are single-track—each power is directed towards just one manifestation—and that they operate according to the *stimulus-manifestation* model—a power will only give rise to a manifestation when it is galvanised into action by some trigger or stimulus (e.g. Bird, 2007). One important outcome of this view for our present discussion is that each distinct manifestation type implies a distinct type of power. Thus, apparently novel and distinctive manifestations (including those exhibited as novel behaviours or phenomena displayed in E-cases) are evidence of novel and distinctive powers. This evidence is defeasible, because the apparently novel and distinctive manifestation might be shown to be either illusory or, alternatively, nothing over and above an aggregation or collection of more basic manifestations (and thus be attributable to the powers already associated with such manifestations). However, where apparently novel and distinctive manifestations themselves resist such analyses, they stand as strong evidence in favour of novel powers. Another important outcome is that on the *orthodox view* a manifestation occurs as a result of the determinative action of the target power alone, although it will not be produced until the occurrence of the stimulus.

An *alternative view*, defended by for instance C. B. Martin (e.g. 2008) and John Heil (e.g. 2003) holds that powers are *multi-track*—a single power is directed towards a (possibly very wide and diverse) range of qualitatively distinct manifestations—and that they operate according to the *mutual manifestation* model—in order for some manifestation to occur, there must always be at least two powers working together, and there is no sense of priority such that one power (or some subset of the powers involved) could be considered the 'operative' or 'active' power(s), whilst the other(s) are merely a stimulus or trigger. Note that this is distinct from the view (often associated with Aristotle) that powers come in *active/passive* pairings, where the active power plays the determinative role, and the passive power is a mere liability to be subject to changes associated with the manifestation of the active power. One important outcome of this *alternative view* is that different mutual partnerships between different sets of powers will lead to different manifestations, and thus that apparently novel and distinctive manifestations do not necessarily indicate novel and distinctive powers, as they might merely indicate novel groupings or arrangements (or perhaps better *partnerings*) of powers associated with other manifestations. Another important outcome of the view is that a manifestation *never* occurs as a result of the determinative role of a single power but is always attributable to a set of powers in a partnering relationship with one another.

7 Flat Holism

Although one is committed to a reductionist metaphysic and the other an emergentist metaphysic, one thing that the micro-latency and mutualist accounts have in common is that both approaches accommodate E-cases by appeal to the idea that *parts behave differently in wholes* (a notion that Gillett identifies as a core concept amongst scientists who endorse emergentist outlooks e.g. 2016:42–43). The micro-latency view does so by holding that parts possess powers which only ever come to manifest when that part is embedded in certain kinds of whole. The mutualist view does so by holding that parts gain new powers when they are embedded in certain kinds of whole. These are not the only options, however, for an approach to the relationship between the more- and less-complex, and to E-cases, which make use of this idea.

Both the micro-latency and the mutualist approaches seem to be driven, at least in part, by an acceptance of the *orthodox view* of powers—that is, by the idea that powers are single-track and operate according to the stimulus-manifestation model. Confronted with E-cases and observing that the prospects for accommodating the apparently novel phenomena or behaviour that typify them using only the resources of micro-manifest powers, Shoemaker is moved to posit additional, micro-latent powers of basic entities. Likewise, mutualism holds that components in E-cases possess distinctive powers which are responsible for the apparently novel phenomena or behaviour that typify such cases, but attributes the origin of such powers (which are bestowed on the components via machresis) to the whole. The underlying logic here is that *distinctive manifestations imply distinct powers*, and as explained above, whilst this holds true on the *orthodox view*, it is not licenced by the *alternative view*—that is, the view that powers are multi-track and operate according to the mutual manifestation model.

Adopting this *alternative view*, however, seems in some ways particularly amenable to the idea that parts behave differently in wholes and to the notion—implied by Shoemaker's, but not by Gillett's view—that the potential to bring about the distinctive phenomena or behaviour which typifies a given E-case was already present in the components. This is because commitments of the *alternative view* with regard to the nature of powers in general include that a given set of powers can produce a range of qualitatively varied manifestations in different complex circumstances—that is, when engaged in different partnering relationships. Thus, it should come as no surprise that certain complex contexts (such as E-cases) exhibit distinctive, novel phenomena or behaviour. The *alternative view* does not need to deploy any distinction between species of powers (such as Shoemaker's micro-manifest/micro-latent distinction, or Gillett's differential/non-differential distinction), as whilst it allows for manifestations which will only be produced in certain complex contexts—because only these contexts have the right reciprocal partnering relationships between powers—the powers responsible for these manifestations are the very same powers which produce (different) manifestations in simpler contexts.

Crucially, however, the mutual, reciprocal and non-prioritised manner in which a population of powers involved in a given partnering operate means that, on the view under consideration here, there is an ineliminable determinative role for the whole to play: it is the *complex* of powers, which operate together to produce the relevant phenomena or behaviour, which should be seen as the entity which determines the outcome.[4] Where we have a genuine E-case, there is no prospect of analysing away these complexes into individual powers or simpler complexes of powers (of course, apparent E-cases might in the end be amenable to such analysis, but this is equally true with regard to the theoretical posits employed in the micro-latency and mutualist views—an E-case might appear to require micro-latent or differential powers, but eventually be shown to be accountable for in terms of just micro-manifest or non-differential powers).

Let's call the resulting account of the relationship between the more- and less-complex, and the approach it implies with regard to E-cases, *Flat Holism.* According to this view, when basic powers enter into the appropriate forms of partnering relationships with each other, they bring about qualitatively distinctive, novel manifestations of the sort which typify E-cases. Because these manifestations are mutually produced, they can only be attributed to the structured complex of powers that is operative in the case, and not to particular individual powers or to simpler complexes of powers. Thus, the resulting view is holistic in the sense that the behaviour of the system can only be accounted for by the complex of partnered powers taken as a whole, which determines how the parts behave not by bestowing new powers, or by triggering otherwise latent powers, but by bringing to fruition manifestations which require the collaborative, mutual and reciprocal contribution of all the powers involved in the complex. This needn't, however, imply that whole is something *more* than the sum of the parts (a notion which is often associated with holism) hence the 'Flat'. Furthermore, whilst it is consistent with the kind of 'universal' holism—associated for instance with certain versions of confirmation or semantic holism, in their respective domains—that takes *every* power to be in some important sense dependent on every other power, or the totality of powers taken as a whole, Flat Mutualism does not necessarily imply that this is the case.

Is Flat Holism a form of reductionism or emergentism? The crucial metaphysical posit of the view are basic powers possessed by basic individuals, which are multi-track and operate according to the mutual manifestation model. It isn't clear that, in addition to these powers, Flat Holism is required to posit additional, metaphysically autonomous non-basic entities, and so it seems to satisfy (a): basic entities enjoy special ontic status not enjoyed by non-basic entities. Nevertheless, because of the crucial and ineliminable determinative role played by complexes of powers, Flat Holism also seems to satisfy (e): the non-basic plays a determinative role and at least some of the 'action' is not at the base level. These complexes needn't be seen as something over-and-above the powers which participate in them (arranged thus-and-so), for their distinctive determinative role comes not from there being some

[4] Note that the form of mutuality at play here is ontologically distinct from that which is a central feature of Gillett's Mutualism, which involves mutual 'upward' compositional and 'downward' machretic determination within a system. Flat Holism involves the 'sideways' mutual action of complexes of powers.

novel entity or individual on the scene, but rather from the collaborative, mutual and reciprocal manner in which the participant powers come to manifest. Thus, Flat Holism plausibly also satisfies (c): the non-basic is accounted for in terms of the basic. As it satisfies (a) and (c), Flat Holism thus maintains some key reductionist notions: it doesn't require novel, distinct higher-level entities; it takes basic entities to enjoy special ontic status, and it is thus consistent with a one- or no-level ontology.[5] However, in attributing an ineliminable determinative role to complexes of powers, it also satisfies a central emergentist notion: that the non-basic plays a determinative role and at least some of the 'action' is not at the base level. Importantly, unlike the Micro-latency view Flat Holism thus doesn't allow one to generate a general anti-emergentist argument of the form set out in Sect. 4, as it isn't clear that the relevant analogue of the second premise in that argument:

P2* Flat Holism, if plausible, demonstrates that E-cases can be accommodated by non-emergentist metaphysics

would be true—because in satisfying (e), Flat Holism cannot be said to straightforwardly embrace a non-emergentist metaphysic (and so *a forteriori* cannot demonstrate that E-cases can be accommodated by such a metaphysic). Flat Holism thus seems to sit somewhere between canonical conceptions of reduction and emergence. Ultimately, whether the view is classified as one, the other or neither is less important than how the view performs in comparison to competing theories. It is to this question the next section will turn.

8 Flat Holism, Micro-latency and Mutualism

One noteworthy difference between Flat Holism, on the one hand, and both the Micro-latency and Mutualist accounts, on the other, can be seen by considering how each view accommodates E-cases. Arguably, both the Micro-latency and Mutualist accounts are principally *reactive*. That is, faced with the challenge that E-cases present—to provide a suitable framework for accommodating the apparently novel and distinctive phenomena or behaviour that typify such cases—these views introduce novel theoretical posits and ontological commitments. The Micro-latency account introduces the distinction between micro-latent and micro-manifest powers and commits to the existence of at least some powers of the former kind. Mutualism introduces both the distinction between differential and non-differential powers, and the notion of machretic determination and it commits to the existence of 'higher-level', 'macro' wholes which are the source of such determination and to the existence

[5] Note that whilst a one- or no-level ontology would typically be taken to exclude the possibility of emergence, Sartenaer (2018) has argued that there is conceptual space for forms of 'flat emergence'. There isn't space in this paper to explore the extent to which Sartenaer's conception accords with the view under discussion here.

of at least some differential powers. In both cases, some of these posits or commitments enjoy limited motivation or support that is independent of the requirement to accommodate E-cases—the principal reason to accept micro-latent powers, or machresis and differential powers, is precisely in reaction to the need to accommodate E-cases.[6]

In contrast, Flat Holism follows naturally from the *alternative view* of powers, which is motivated independently of discussions concerning the relationship between the more- and less-complex, of emergence and reduction, and of E-cases. Key proponents of the *multi-track mutual manifestation* view, such as C. B. Martin and John Heil simply are not concerned with these sorts of issues when they outline the arguments in favour of this account of the nature of powers (see e.g. Heil, 2003 ch. 8 or Martin, 2008). Flat Holism does not react to E-cases and modify the way in which powers are conceived in order to accommodate them; rather, the potential for such cases is predicted by and naturally follows from the *alternative view* of powers which is at the centre of the account. Arguably then, Flat Holism represents a significantly more unified account than it's nearby neighbours.

Furthermore, Flat Holism gets by without positing the various novel distinctions; powers and higher-level entities that are introduced by both the Micro-latency and Mutualist accounts. This means it seems to have a theoretical advantage in terms of qualitative simplicity: it posits only one kind of powers, where both the other views posit two; and it does not need to introduce a novel form of determination as the Mutualist view does with machresis. A critic might respond that Flat Holism doesn't really have such an advantage, as it introduces the posit of powers as *multi-tracking*. But such a criticism fails to hit home: all three views under consideration are framed in terms of powers, and *any* account of powers has to give some answer or other regarding whether powers are single- or multi-track. That the single-track view is the orthodox one doesn't make it somehow simpler than the alternative. In a similar vein, the critic might also respond that Flat Holism, in adopting the mutual manifestation model, is committed to a proprietary notion of determination, just as Mutualism is with machresis. But again, such a response fails to hit the mark: Flat Holism's claim that complexes can play an ineliminable determinative role (due to the mutual, reciprocal operation of the powers which participate in them) is just a part of its account of how powers operate *in general*. Again, all three views accept powers, and *any* account of powers will have to have some view about how it is that powers operate—and that Flat Holism adopts the *alternative* rather than *orthodox view* doesn't imply any loss in terms of simplicity. Indeed, if Tugby (2010) is correct that the stimulus-manifestation account of how powers operate implies the existence of stimuli in other ontological categories, the *orthodox view* might actually be the more complex when taken in isolation—notwithstanding the fact that we may

[6] Gillett (2016 ch. 7) does provide an argument that machresis is necessary for the existence real compositional levels, and so it should be noted that if this argument is sound, then any independent reasons we might have for taking such levels to exist would equally count in favour of both machresis and differential powers. Whilst E-cases surely account for much of the motivation for taking there to be real compositional levels, perhaps there are other grounds as well.

well have reasons to accept entities such as actions, events and states which are independent of considerations concerning how powers operate.

9 Concluding Thoughts

Let me be clear: the discussion in the preceding section of this essay is not intended to demonstrate that Flat Holism is in some absolute sense a superior position to either the Micro-latency or the Mutualist accounts; or that these accounts should be abandoned in favour of Flat Holism; or that they suffer from fatal flaws. The principal aim of this essay has been much more humble: to introduce the basic form and ontological commitments of Flat Holism, and to try to establish that the view has some distinctive merits—in terms of being relatively more unified and simple—such that it should be taken at least as seriously as closely competing views such as the Micro-latency view and Mutualism. It is worth bearing in mind, as mentioned earlier in this essay, that a pluralistic approach remains a live option in this debate, such that it could be the case, for instance, that some E-cases involve micro-latent powers; some differential powers and some qualitatively novel manifestations produced by the mutual partnerings of multi-track powers (and of course, some apparent E-cases could be open to reductive analyses of various kinds). Such a pluralism remains a live option because none of the accounts under discussion here are, strictly speaking, mutually exclusive of one another. Given the possibility of pluralism, discussions concerning these three accounts are thus not a zero-sum game—but it is crucial is to engage with and appreciate the manner in which each account approaches E-cases; the similarities between them; the points of difference and any distinctive merits each account might have.

A secondary aim of this essay has been to lend support to the general claim that the way in which certain issues in the metaphysics of powers are resolved has ramifications for the debate between emergentists and reductionists. This paper has argued that adopting the *alternative view*, as opposed to the *orthodox view*, of powers allows us to recognise a novel manner in which to accommodate E-cases: Flat Holism. But the *orthodox* and *alternative* views aren't the only possible accounts of powers, and so one important focus for future work in this area of debate is to consider what other approaches to the metaphysics of powers might be theoretically fruitful in the discussion of the relationship between the more- and less-complex.

Finally, although not addressed in this essay, it should be noted that an examination of how concrete cases fit (or fail to fit) each of the three accounts is crucially important (see e.g. the final two chapters of Gillett (2016)): the relevant debates should not be settled by abstract discussion alone. Of course, in order to be in a position to engage with concrete cases, one needs a clear idea of the relevant commitments and entailments of each account, hence this paper's focus on outlining and exploring how Flat Holism might address E-cases and how it relates to relevant alternative accounts. Hopefully this essay has contributed to laying the groundwork such that the important task of engaging the empirics can more be undertaken more easily.

References

Armstrong, D. M. (1997). *A world of states of affairs*. Cambridge University Press.
Baggio, G., & Parravicini, A. (eds.), (2019). Special Issue on pragmatism and theories of emergence. *European Journal of Pragmatism and American Philosophy, XI*(2).
Batterman, R. (2011). Emergence, singularities and symmetry breaking. *Foundations of Physics, 41*, 1031–1050.
Baysan, U., & Sartenaer, O. (eds.). (2021) Special issue: Non-standard approaches to emergence. *Synthese.*
Bedau, M. A., & Humprheys, P. (Eds.). (2008). *Emergence: Contemporary readings in philosophy and science.* The MIT Press.
Bigaj, T., & Wüthrich, C. (Eds.). (2015). *Metaphysical emergence in contemporary physics.* Rodopi.
Bird, A. (1998). Dispositions and antidotes. *The Philosophical Quarterly, 48*, 227–234.
Bird, A. (2007). *Nature's metaphysics.* Oxford University Press.
Bishop, R. (2010). Whence chemistry? *Studies in History and Philosophy of Science Part B: Studies in History and Philosophy of Modern Physics, 41*, 171–177.
Bishop, R. (2019). *The physics of emergence.* Morgan & Claypool.
Bishop, R., Silberstein, M., & Pexton, M. (forthcoming). *Emergence in context.* Oxford University Press.
Carruth, A. D. (2019). Strong emergence and Alexander's dictum. In Gibb et al. (eds.), *The Routledge handbook on emergence.* Routledge.
Carruth, A. D. (2020). Emergence, reduction and the identity and individuation of powers. *Topoi, 39*, 1021–1030.
Carruth, A. D., Miller, J. T. M., & Pexton, M. (eds.) (2017a). Special issue on strong emergence (Vol. 1). *Philosophica*, 91.
Carruth, A. D., Miller, J. T. M., & Pexton, M. (eds.), (2017b). Special issue on strong emergence (Vol. 2). *Philosophica*, 92.
Chalmers, D., (2006). Strong and weak emergence. In Davies & Clayton (eds.), *The Re-emergence of emergence.* Oxford University Press.
Churchland, P. M. (1981). Eliminative materialism and the propositional attitudes. *The Journal of Philosophy, 78*(2), 67–90.
Corradini, A., & O'Connor, T. (Eds.). (2010). *Emergence in science and philosophy.* Routledge.
Falkenburg, B., & Morrison, M. (Eds.). (2015). *Why more is different: Philosophical issues in condensed matter physics and complex systems.* Springer.
Franklin, A., & Knox, E. (2018). Emergence without limits: The case of Phonons. *Studies in History and Philosophy of Science Part B: Studies in History and Philosophy of Modern Physics, 64*, 68–78.
Gibb, S. C., Hendry, R. F., & Lancaster, T. (Eds.). (2018). *The Routledge handbook of emergence.* Routledge.
Gillett, C. (2016). *Reduction and emergence in science and philosophy.* Cambridge University Press.
Heil, J. (2003). *From an ontological point of view.* Oxford University Press.
Hendry, R. (2010). Ontological reduction and molecular structure. *Studies in History and Philosophy of Science Part B: Studies in History and Philosophy of Modern Physics, 41*, 183–191.
Hohwy, J., & Kallestrup, J. (Eds.). (2008). *Being reduced: New essays on reduction, explanation, and causation.* Oxford University Press.
Hoover, K. D. (2009). Microfoundations and the ontology of macroeconomics. In D. Ross, & H. Kincaid (eds.), *The Oxford handbook of philosophy of economics.* Oxford University Press.
Humphreys, P. (2016). *Emergence: A philosophical account.* Oxford University Press.
Kim, J. (1999). Making sense of emergence. *Philosophical Studies, 95*, 3–36.
Lancaster, T., & Pexton, M. (2015). Reduction and emergence in the fractional quantum hall state. *Studies in the History and Philosophy of Science Part B: Studies in History and Philosophy of Modern Physics, 52*, 343–357.

Lewis, D. (1997). Finkish dispositions. *The Philosophical Quarterly, 47*, 143–158.
Macdonald, G., & Macdonald, C. (Eds.). (2010). *Emergence in mind*. Oxford University Press.
Martin, C. B. (1994). Dispositions and conditionals. *The Philosophical Quarterly, 44*, 1–8.
Martin, C. B. (2008). *The mind in nature*. Oxford University Press.
McLeish, T. (2017). Strong Emergence and Downward Causation in Biological Physics. *Philosophica*, 92, 113–138.
Molnar, G. (2003). *Powers: A study in metaphysics*. Oxford University Press.
Morrison, M. (2012). Emergent physics and micro-ontology. *Philosophy of Science, 79*(1), 141–166.
O'Connor, T., & Wong, H. Y. (2005). The metaphysics of emergence. *Noûs, 39*, 659–678.
Paolini Paoletti, M. (2017). *The quest for emergence*. Philosophia.
Paolini Paoletti, M., & Orilia, F. (Eds.). (2017). *Philosophical and scientific perspectives on downward causation*. Routledge.
Sartenaer, O. (2015). Synchronic vs. diachronic emergence: A reappraisal. *European Journal for Philosophy of Science, 5*, 31–54.
Sartenaer, O. (2018). Flat Emergence. *Pacific Philosophical Quarterly, 99*, 225–250.
Shoemaker, S. (1980). Causality and properties. In P. van Inwagen (Ed.), *Time and cause: Essays presented to Richard Taylor* (pp. 109–135). Reidel.
Shoemaker, S. (2002). Kim on emergence. *Philosophical Studies, 108*, 53–63.
Shoemaker, S. (2007). *Physical realisation*. Oxford University Press.
Shrader, W. (2010). Shoemaker on emergence. *Philosophical Studies, 150*, 285–300.
Silberstein, M. (2001). Converging on emergence: Consciousness, causation and explanation. *The Journal of Consciousness Studies, 8*, 61–98.
Theiner, G., & O'Connor, T. (2010). The emergence of group cognition. In: A. Corradini, & T. O'Connor (eds.) *Emergence in science and philosophy*. Routledge.
Tugby, M. (2010). Simultaneity in dispositional interactions. *Ratio, 23*(3), 322–338.
van Gulick, R. (2001). Reduction, emergence and other recent options on the mind/body problem: A philosophic overview. *Synthese, 8*, 1–34.
Wilson, J. (2015). Metaphysical emergence: Weak and strong. In: Bigaj & Wuthrich (eds.) *Metaphysics in contemporary physics: Poznan studies in the philosophy of the sciences and the humanities*. Brill.
Wilson, J. (2021). *Metaphysical emergence*. Oxford University Press.

Enactive Realism. A First Look at a New Theoretical Synthesis

Arturo Carsetti

Cognitive activity is rooted in reality, but at the same time represents the necessary means whereby reality can embody itself in an objective way: i.e., according to an in-depth nesting process and a surface unfolding of operational meaning. In this sense, the objectivity of reality is also proportionate to the autonomy reached by cognitive processes.

Within this conceptual framework, reference procedures thus appear as related to the modalities providing the successful constitution of the channel, of the actual link between operations of vision and thought. Such procedures ensure not only a "regimentation" or an adequate replica, but, on the contrary, the real constitution of a cognitive autonomy in accordance with the truth. A method thus emerges which is simultaneously project, *telos* and regulating activity: a code which becomes a process, positing itself as the foundation of a constantly renewed synthesis between function and meaning. In this sense, reference procedures act as a guide, mirror and, canalization for primary information flows and involved selective forces. They also constitute precise support for the operations which "imprison" meaning and "inscribe" the "file" considered as an autonomous generating system. In this way, they offer themselves as the actual instruments for the constant renewal of the code, for the invention and the actual articulation of an ever new incompressibility. Hence the possible definition of new axiomatic systems, new measure spaces, the real displaying of processes of continuous reorganization at the semantic level. Indeed, it is only through a complete, first-order "reduction" and a consequent non-standard second-order analysis that new incompressibility will manifest itself. Therefore, the reference procedures appear to be related to a process of multiplication of minds, as well as to a process of unification of meanings which finally emerges as a vision via principles. Here also the possibility emerges of a connection between things that are seen and those that are unseen, between visual recognition of objects and thought concerning their secret interconnections. Hence, for instance, according to Boccioni: "*la traduzione in forme*

A. Carsetti (✉)
University of Rome, Rome, Italy

S. Wuppuluri and I. Stewart (eds.), *From Electrons to Elephants and Elections*,
The Frontiers Collection, https://doi.org/10.1007/978-3-030-92192-7_12

plastiche dei piani atmosferici che legano ed intersecano le cose". In other words, this is the connection between the successive opening of the eyes of the mind and the metamorphoses of meaning, a meaning which is progressively enclosed within generative thinking and manages to express itself completely through the body's intelligence.

This functional analysis reveals even more clearly, if possible, the precise awareness that, at the level of a cognitive system, in addition to processes of rational perception (categorial intuition), we also face specific ongoing processes of semantic categorization. It is exactly when such processes unfold in a coherent and harmonious way that the "I" not only manages to emerge as an observation system but is also molded by the simultaneous display of the structures of intentionality. Through the intentional vision, I comes to sense the Other's thought-process emerging at the level of its interiority. The drawing thus outlined, however, is meant for the Other, for the Other's autonomy, for its emerging as objectivity and action. This enables me to think of the autonomy of the Nature that "lives" (within) me.

At the level of intuition-based categorization processes, the file is "selected" from the ongoing morphogenesis. When the original meaning manages to express new lymph through a renewed production of forms, the self-inscribing file might express its unification potentialities through the successive individuation of concepts which, however, are selected and molded at an intuitive level. Hence the possibility of an actual "inscription" to the same extent as the morphogenesis, but also the realization of a reduction process, the very laying down of an original creativity within a monodimensional and dynamic framework. It is exactly when the reduction is carried out, though, that the procedures of reflection, the identification of limits and completion can be performed on the basis of the constant support to the *telos* activity, of the primary regulation activities proper to the organism taken as ongoing projectuality.

The unification procedures inherent in the nesting process, effected in accordance with precise conceptual constraints depending on the self-inscription of the file, then fit, finding their foundation, in attractors which operate at the level of specific correlation-patterns and organic instrument-systems of measure. These gradually grow up and multiply, giving rise to natural self-organizing modules, activated by an inner-code, which materialise over a period of time as based on precise measure operations encoded in a specific project. The result is an autonomous (and selective) production of forms modulated according to concepts and connected through the *telos,* thereby becoming vision via principles, a production able to articulate according to a specific and unifying intelligence. There thus emerges a "body" acting in conjunction with its intelligence:

> "my" body which at the same time manages to transcend itself and blend with itself in the Other. Hence the very possibility of the "presentation" as *Forma formata* of an original meaning which will simultaneously blend with itself and sub-divide itself in time. While the eyes of the mind manage to observe a *Natura naturata* populated by observers, the brain with its measure operations and net connections manages to "think", moving beyond itself into the Other, a *Forma formata* interwoven by works.

In a context of this kind, the forms of intuition (as well as, on the other hand, the categorial apparatus) cannot be considered impermeable to the conditions of external evolution. In the meantime, it appears increasingly necessary to recognize that the Darwinian external selection will co-exist with an internal selection connected to the successive deep-level unfolding of meaning. It then becomes mandatory to refer back to the procedures of categorial intuition as postulated by Husserl, but also to anchor these procedures to a particular conception of the relation between organism and environment which is both dialectical and co-evolutive. The reference procedures, like those of simulation, are never neutral, and never regard the human realm alone, but are able to act as a guide, mirror, and canalization for the primary information fluxes which gradually inscribe themselves in the form of codices, constraints, and modules-forms in action: natural simulation modules which govern, at the basic level, the structuration processes articulating at the level of the living (and cognitive) organism. In this sense, the operational logic of intellect and simulation must also be considered in combination with a deeper logical level concerning the articulation of life itself, also requiring us to map out their co-operative and functional interdependence. The procedures of reference, far from being external to reality, lead, on the basis of an ongoing interdependence with the evolutionary paths in action, to the progressive (inner) constitution of individuals who finally act as autonomous entities and posit themselves as the real source of neural creativity (at the level of knowledge construction). The reference procedures thus give rise to a complex dialectical exchange between action, thought, and meaning, producing, in particular, an evaluation and exploration of the contents and limits of the original information-fluxes. This exchange leads to new forms of autonomy, and the extension and recovery of the conditions of primitive creativity: hence the primary source of that continuous "addition" of new nuclei of creativity characterising the logic of the living (being) which Bergson speaks of.

True invariance, life, can exist only within the framework of ongoing autonomous morphogenesis and vice versa. Concepts would thus appear to be linked to the invention and a continuous activity of selection and "anchorage" realized on semantic grounds. It is the work of invention and generation (in invariance), linked with the "rooting" of meaning, which determines the evolution, the leaps and punctuated equilibria, the conditions related to the unfolding of new modalities of invariance, an invariance which is never simple repetition and which springs on each occasion through deep-level processes of renewal and recovery. The selection perpetrated by meaning reveals its autonomy above all in its underpinning, in an objective way, the ongoing choice of these new modalities. As such it is not, then, concerned only with the game between the possible and the actual (F. Jacob), offering itself as a simple channel for pure chance, but with providing a channel for the inscription of the file in the *humus* of meaning, to prepare the necessary conditions for continuous renewal and recovery of original creativity. In effect, it is this autonomy in inventing new modules of incompressibility which determines the emergence of new (and true) creativity, which also takes place through the "narration" of the effected construction. *Pace* Kant, sensibility is not a simple interface between absolute chance and an invariant intellectual order. In this sense, the reference procedures, if successful, are able to modulate canalization and create the basis for the appearance of new frames of incompressibility through morphogenesis. This is not a question of discovering and exploring (according, for instance, to Putnam's conception) new "territories", but of offering ourselves as the matrix and arch through which they can spring autonomously in accordance with ever-increasing levels of complexity. There is no casual autonomous process already in existence, and no possible selection and synthesis activity via a possible "remnant" through reference procedures considered as a form of simple regimentation. These procedures are in actual fact functional to the construction and irruption of new incompressibility: meaning considered as the promoter of *Forma formans,* offers the possibility of creating a holistic anchorage, and is exactly what allows the categorial apparatus to emerge and act according to a coherent "arborization". In this way a time of invention can be assured, but not a time of repetition: a time characterized by specific processes of renewal and recovery which continuously reveal themselves as possible in proportion to the effective realization of the "work". What determines the ongoing selection each time (at the level of the primary informational fluxes) is the new incompressibility that arises. This requires that the reference procedures posit themselves as arch between the two selections: between invariance on the one hand, and autonomous morphogenesis on the other. In other words, they are only able to nurture new incompressibility where there exists a process of nesting of pure virtuality's original space. The important aspect is not, then, the remnant *in itself* but the successful "narration". It is effective inscription giving rise to new incompressibility which necessarily bypasses me. I will, then, ultimately be able to observe a new incompressibility which reveals itself as the ongoing fusion of emergent nuclei of creativity within the unity of an operant signification. The new invention which is born then shapes and opens the (new) eyes of the mind: I see as a mind because new meaning is able to articulate and take root through me (and only proportionately as this occurs).

I must transform myself into an arch and then offer the arch (and myself) to other through multiplication. I will thus be able to fulfill myself as a form of new creativity and achieved autonomy. This is the means to step outside myself while affirming my objectivity as creator-artificer: the donation of the drawing-thought to the other, an assimilation of the other for the other, to make it, too, a creator again in his coming out of self-abandonment. It is this that will then reveal itself as true possession, to merge with the inner broadening of the basis of creativity. I neither capture nor order nor replicate simply: I simulate to allow the advent of a new life, but to do this I must pass through the arch of creation and "disincarnation", effecting the link between the two ongoing selections and managing to close the circle by opening and laying down myself as a grid (Tiziano, The Martyrdom of St. Lawrence, Venice).

As a collating mechanism I must give voice to the two selection-processes (at the two different levels of vision (the mind) and revisable thought (the brain)), becoming, in my turn, a creator and a source of life and coagulum for the Other and in the Other, at the same time as the Other posits itself as a source of meaning and truth for me. Hence the need to shape-create the drawing and offer it up for the realization-thought process of the other. I offer up the vision of myself (which is ultimately the result of a self-organizing process) for the thought-process of the other: for the measure operations effected by the other, to allow it. To posit itself again as a creative and autonomous being.

As W. Dean correctly remarks, Tennenbaum's Theorem can be understood to illustrate that: "although in classical mathematics we can demonstrate that non-standard models of arithmetic exist, the theorem intervenes to show that we can never hope to go beyond linguistic descriptions such as 'let $\mathcal{M}$ be a model of T_0' so as to characterize the structure of $\mathcal{M}$ explicitly".[1] Actually, given that $\mathcal{M}$ is countable we can characterize the substructure $\langle M, < M \rangle$ constructively up to isomorphism as the order type $\omega + (\omega * + \omega) \cdot \eta$.

However: "the fact that we have still gone on to develop a rich theory of such structures and their interrelationships is testament to the fact that the development of model theory often does not require us to fully extensionalize descriptions of models which we have introduced by such means".[2] In fact, we can easily realize that our ability to refer to non-standard models must be understood as mediated by descriptions which are not only indefinite but which we know can never be made fully constructive. In other words, we are obliged to adopt a different understanding of 'model' inspired by Constructivism. Hence a possible confluence, at first, of the computationalist view with Putnam's 'nonrealist' semantics, with the attempt, that is to say, to identify the reference of an expression with its sense understood as an appropriate type of verification procedure. 'Objects' in constructive mathematics are *given through descriptions.* Those descriptions do not have to be mysteriously attached to those objects by some non-natural process ... Rather the possibility of proving that a certain construction (the 'sense', so to speak, of the description of the model) has certain constructive properties is what is asserted and all that is asserted by saying the model 'exists'. In short, *reference is given through sense and sense is given through verification procedures and not truth conditions.*[3] (Putnam, 1980, p. 479) (emphasis original). According to the theoretical perspective proposed here, let us remember that according to Benacerraf any set of objects with the ω-type ordering can be a model for arithmetic. However, next to this property we must also consider other important properties: actually, a basic feature of natural numbers is given by the fact that humans normally utilize them to count. Specifically we learn what natural numbers are while learning to count. But learning in the case of an autonomous agent (the Minotaur) is necessarily linked to the realization of an embodiment, an embodiment that, in turn, presupposes the encounter of the Minotaur with Ariadne as well as the full unfolding of his imagination. The agent must also take into account the use and the conditions of the exercise concerning his very self-identification: the properties relative to ω-ordering alone are not enough to identify the real exercise on the mathematics. In other words, the model is intended (and exists) when it adequately reflects our intuitions. "We learn what are natural numbers while learning to count. Consequently, we argue that an intended model for arithmetic should be such that one can perform basic arithmetical operations (addition and multiplication) on elements of this model (numbers from this model)".[4]

In this sense the model does not present itself simply as a construction characterized by certain properties: the model must also allow specific operations to be performed on its own elements and must have a privileged relationship with the use of certain abilities as operated by the autonomous agent at the level of the embodiment process at play. It is only if I prove myself capable of operating successfully on elements of the model that I come to understand: i.e. to exercise a specific skill such as that, for example, relative to counting. It is in this way that Ariadne can illuminate and that the construction of I, in its turn, can be pursued.

We are far beyond Putnam: the model not only exists because it is identified through abstract structures, constructive properties and verification procedures, but also because it refers to the conditions proper to an actual embodiment of which, for instance, a specific learning process is an integral part. Thus, at the level of the intended model for arithmetic we have the convergence of recursivity, first-order induction and ω-type ordering (Tennenbaum theorem): the intended model necessarily takes shape in reference to specific recursive processes, to what is, from a general point of view, the landscape of Reflexivity. A convergence, in any case, that takes place in the context of the detachment operated by the Minotaur. What happens, however, when we enter the arena of metamorphosis in all its breadth? When, that is to say, we take into account the entire journey of the Minotaur. In such a case, as Picasso clearly shows in the painting "The flute of Pan" (Museum Picasso, Paris), a further element enters the scene: the score relating to the inheritance of Pan, i.e. the original set of the eigenvalues on the carpet. We are now in the realm of non-standard models where the reference to the ordering is to vary as shown by Henkin in 1950. We will no longer only be faced with ω-ordering but, for instance, also with the order type $\omega + (\omega * + \omega) \cdot \eta$ and so on. The imagination at work at the level of the embodiment (as, indeed, shown by Picasso in his painting) is guided by eigenvalues and not by eigenforms. Let's imagine now to recover, in the footsteps of Carsetti (1989),[5] a suitable model for a given process of metamorphosis and self-organization. Having to refer, initially, to a set of eigenvalues, it will be necessary to refer not only to recursive processes and standard models but also to both non-standard models and simulation and invention procedures. Hence the entry into the scene of a new theoretical perspective: the perspective related to set-theoretic Relativism. Now, we must adopt, as Skolem does, a different understanding of 'model' inspired by Constructivism and set-theoretic Relativism. The object-construction to which Putnam refers is now replaced by a process of self-organization, by the very decline of a metamorphosis process such as that so well illustrated by Picasso or Ovid. We are faced with a dialectic at play between imagination and invention and not only with the presence of specific relations between objects. In the light of this new perspective, we should maintain that to find out which algorithms really correspond to the references relative to some specific constructive operations it should mean for the autonomous agent that undergoes the metamorphosis to be able to make Nature speak (constructing in the right way the oracle as a new Oedipus) in order to come to feel the solution of the problem in its coming to flow at the level of his own veins. In this sense, only an effective renewed embodiment can, therefore, tell us what the algorithms in question should be. It is the new life and with it the new arising mathematics that will come to condition the self-organizing fibers of the Minotaur along the course of his own evolution starting from the actual giving of the irruption as it arises from the sacrifice of Marsyas. A life, in particular, that will, then, extend itself along the profiles of a new invariance (up to the self-organizing of a Road, but in the silence, a silence interrupted only by the flowing and fading away of the sound related to the stiletto heels of Echo). The reference for an autonomous self-organizing agent is given by the achievement and verification (but on his own flesh) of his autonomy, the autonomy proper to an agent that manages to handle the algorithmic schemes at work in accordance with his inner transformation thus resulting able to prove that he exists to the extent that he places himself at the root of the fulfilment of metamorphosis.

The new autonomous agent who will thus be born will therefore be able to look at the ancient remains of the first observer thus realizing, as Skolem himself states, that many ancient figures which inhabited the theoretical universe of the first observer (such as the ancient infinities) no longer show themselves in accordance with their original characteristics (i.e. as true infinities) with respect to the new arising horizon (the horizon relative to the new observation that is born). We are, in effect, faced with a new embodiment and the conditions relating to the model will now undergo a radical change. If we set ourselves from the point of view of a radical Constructivism, an effective semantic anchorage for an observer system such as the one, for example, represented by the non-trivial machine as imagined by H. von Foerster, can come to be identified only to the extent that the evolving system itself proves able to change the semantics. This, however, will result in our being able to realize an expression of ourselves as autonomous beings, as subjects, in particular, capable of focusing on the same epistemological conditions relating to our autonomy. A creative autonomy that expresses itself above all in the observer's ability to govern the change taking place. Only the Minotaur operating in these conditions will actually come to undergo the new embodiment. Here is the passage on one's shoulders to which Skolem refers, namely that continuous passage from the first to the second observer that marks the very course of natural evolution.

We will then be able to place ourselves as witnesses (but at the level of the new embodiment) of what in the past has been the ability on the part of the first observer to govern his own growth process. Here is the flourishing of an intentional logic based on the ineliminable relationship with other. At the outset there is no ability to count, in fact the eyes of the Minotaur as painted by Picasso are not open from the beginning: they come to open as genuine eyes only to the extent of the construction in progress of those structures of imagination that allow the correct articulation of the schemes and, therefore, the same birth in the round, but by *bricolage*, of the activity of counting. Biological and cognitive activity is always in reference to the evolution at work and the construction of a Temple intersected by perceptual acts (within the framework of the ongoing dialectic between incompressibility and meaning).

The reference for an autonomous self-organizing agent is given by the achievement and verification (but on his own flesh) of his autonomy, the autonomy proper to an agent that manages to handle the algorithmic schemes at work in accordance with his own transformation thus resulting able to prove that he exists to the extent that he places himself at the root of the construction of the properties that identify his very creativity: true existence is given by creativity at work (but in the agent's awareness of this same creativity).

The categorial is in me, in the inwardness of my being, in my own coming to mould myself as a creator: *Noli foras ire*. Here is a categorial that emerges at the level of Nature (Pan) and that manifests itself, then, following the conception of

Marsyas, along with the giving of a simulation activity which occurs in the kingdom of Culture according to the dictates of a specific DNA. On the opposite side, the God (the environment, the reality that surrounds me as a craftsman and that selects me, the S. V. Mountain that every time upsets me as a Painter and craftsman (Cezanne) following what is the manifestation of its apparently inviolable complexity) is the one who arouses, who proceeds with the selection and who consoles, the one who gives impressions of excruciating beauty and who, however, open to the possibility for the craftsman to manage his own conceptual apparatus in order to grasp the meaning of such impressions. Hence the encounter between the craftsman and his environment, but in dependence on a coupled metamorphosis. Apollo and Marsyas: creativity and simulation. There is a creativity in Apollo expressing itself through an inspiration that is declined for subsequent impressions involving the subject and there is a simulation activity in Marsyas that comes to be expressed through successive exposures of his DNA. Apollo expresses himself as creativity in life through an inspiration that emanates, Marsyas instead gives rise to a Work in accordance with the truth. Apollo inspires the craftsman's activity, while Marsyas, for his part, allows the God to express his selective activity. Pan is the Mountain that through Syringe transmits its inheritance thus opening to the birth of the M. Marsyas is the conceived that opens to the Painter of abstraction, to Cezanne as craftsman and Painter. Cezanne, coming to undergo the impressions, develops that conceptual network that will lead him by means of the categorial intuition, to hear the speech of the Mountain flowing in his veins. Pan as creativity and Nature which then gives rise to new conception and Marsyas, instead, as the Lord of the Garlands which comes to decline himself as simulation and Work, thus determining the new irruption. If there were not the Minotaur who, at first, is added to the Temple in determining each time the right fixed points and if there were not, therefore, Marsyas who presides over his extroversion, thus coming to undergo the selection by the God, there cannot be that co-evolution which alone can allow the meeting between creativity and simulation to take place. A co-evolution which, in any case, will be tailored to the channeling of the God at the level of what is the subsequent emergence of impressions. Here is the Mountain which will accompany the Muse and here is the Painter who will come to inhabit his brain. Here is a DNA that will express itself in a fabric of actions coordinated by a brain and here is a web-network that will be channeled through the emergence of impressions synthesized in themselves by operating intentionality. When I grasp the reality of the web (when I contribute to "making" the environment) I offer the necessary support, at the biological level, for the establishment of ever new intentionality. When I contribute to "model" from the inside the craftsman, I offer the necessary support at the intentional level (not at the level of function, that is to say, but of the meaning) for the constitution of a cognitive organism. From Nature to Culture: from the Minotaur to Marsyas-Painter (through conception). Once we start from a thought (creativity) considered as the matrix of an emerging Nature (and, therefore, from Pan) and once we start from a pure simulation activity on a cultural level (and from Marsyas). From Apollo's own creativity to impressions, from Marsyas to a computational and simulation activity through concepts. Indeed, I can only grasp impressions through the proper construction of concepts. Here is the simulation-creativity circle but preceded by the function-meaning circle. If the concepts do not come to express themselves in the right way along with the construction of the filters, the impressions will not be able to come to reveal their (hidden) meaning. If impressions do not come to work properly, concepts cannot come to articulate in depth in harmony with operating self-organization. Here is the backbone of the process of categorial intuition. The channeling by the Mountain represents the channeling of Nature itself as a function (but together with its meaning). The channeling by Marsyas represents, in turn, the channeling by Culture as simulation at work (but together with its creativity). Nature (Pan) as thought and as the arena of the function-meaning dialectic versus Culture (Marsyas) as simulation and as the arena of simulation-creativity dialectic. Apollo speaks by impressions (at the level of semantic categorization), Marsyas expresses himself by computations (at the level of categorial intuition). On the one hand, impressions + concepts, on the other, an intellectual vision in the God. The God thinks in (and through) the craftsman, the craftsman feels in (and through) the God. I feel the God flowing in my veins. The craftsman contemplates the God sensing him in his own light by making himself light. Here is Kitano Takeshi leading the Ecclesia into the light although he is no longer able to see (cf. the ending of the film: Zatoichi). The function is carried out by fixed points, the intentionality by the carving operated at the level of the Work. Once the constellation in the sky and once the nesting of Eurydice in the abyss. Apollo (the Mountain) speaks by impressions, Marsyas elaborates his inner vision by computations and models. When this happens it is because specific filters have allowed the DNA of Marsyas to "expose" itself in the appropriate way, in adherence, that is to say, to the information content present in the original "impressions" that characterize the mysterious message of the Mountain (Apollo). That's when Clio can appear. Hence, then, the fractal articulation of Pan. Marsyas-Painter has brought his work to completion and his DNA has proved capable of exposing itself in the right way (the way, that is to say, that allows the content of the impressions to be represented and channeled at the craftsman's level) thus allowing the craftsman to explore new forms of incarnation. The God, in other words, thinks in the craftsman to the extent that the craftsman himself reconstructs the message of God in himself, making it operative at the level of the exposure of his DNA and the consequent incarnation. In other words, the Painter, by painting the very presence of the God (in Clio), comes to represent, at the level of the Work, the impressions that characterize this very presence and reconstructs in himself the selection put in place by the God having identified the correct software. Extroversion > selection > reconstruction at the genetic level > incarnation. Marsyas dies but in the meantime, he comes out of absence thus preparing for a renewed realization of the enthusiasm and for the new irruption. If the God comes to think in him he comes to fully feel his presence at the level of the Work of art. Hence the new break-in and the emergence of Pan. Hence, again, the detachment, the journey of the new Minotaur, the function-meaning dialectic up to the very giving of the new conception of Marsyas and the new dialectical relationship between creativity and simulation but in a new ambient. When the categorial proper to Marsyas gives rise to new forms of incarnation (thus embodying the Other), when the Painter feels the new presence of the God in the Muse (at the same time that he paints this very enthusiasm), this means that he has captured in himself the creativity (Thought) of the God himself and can, therefore, die by the very hand of Apollo and following what is the consequent irruption. The Silenus now comes to contemplate the God through the action of Painting: the artist feels the God, contemplates him in his own light, the light that now also emanates from him to the extent that he was added to himself by the God as a creator. He who, as a Painter, reveals himself capable of embodying the Other through the action of Painting (cf. Lucien Freud) comes to feel the presence of God (in the Other) by coming himself to be added as creator (as Painter in truth). Cezanne who realizes himself as a Painter by dying on his battlefield (the clearing in front of the S. V. Mountain he had chosen as location) comes to be illuminated through the Work by the very light of the God. His DNA came to "expose" itself in harmony with the software which animates the creativity of the God. The impressions bring into play their informational (and intentional) value by means of the filters, thus inducing those changes at the level of the exposure of the DNA of the craftsman that mark that metamorphosis of Marsyas that occurs when he comes to be added to himself by the God The impressions now speak in his flesh, in his own veins: God thinks in him in his own metamorphosis. Cezanne tracing the very archaeology of the Mountain and placing himself as a stool for it in view of the renewed expression of its creativity will come to be added as a creator by realizing himself as a Painter in truth. He will come to reconstruct in himself the paths relating to the incarnation in him of the original impressions. He will become flesh and renewed computation for a God who will thus be able to come to think in him. Cezanne will act as software for the hardware represented by the God. The Father recovers himself through the Son but following the sacrifice put in place by the Son himself as well as following the help he gave to the Father who fell prey to the self-abandonment. Marsyas will thus be able resurrect as an added creator. Here is the action of Painting and here is the real presence of the God in Clio as Work and as Muse of History. Cezanne-Vermeer to the extent that he embodies his Muse-model according to the truth, recovers in himself the creativity of the God by coming to be added as a Painter but in his own death and in his exit from absence along with his coming to become pure light in the God. The impressions will now come to speak from his own bosom, he will feel them articulated in himself: *Noli foras ire*. Here is the sense of the Simulation- Creativity dialectic. From the Simulation- Creativity dialectic to the renewed dialectic between Function and Meaning. From the irruption to the conception. When Vermeer paints the Muse again, his DNA will have come to expose itself in harmony with the impressions coming from the Mountain: he will, therefore, added, but only to the extent of his success, of the success, that is to say, of the very action of Painting, an action he tenaciously pursued. The meeting between creativity and simulation takes place in the arena of Schematism and is articulated through the construction of specific filters. Marsyas receives the message of the Mountain to the extent that he reveals himself capable of inspiring and channeling his own growth on a biological level according to the impressions coming from the Mountain itself to the point of coming to feel the original creativity (the God) coming to dictate to him in what are his own veins. Here is the miracle operated by the filters but on the basis of the ongoing process of self-organization.

Apollo and Marsyas, creativity and simulation. Apollo (the S. V. Mountain) thinks of the craftsman (Cezanne), engraves it, and selects it for excruciating impressions on the basis of his mysterious creativity. The moment in which the Painter dies, thus sanctioning the victory by Apollo, (his success in truth) is the presence of the God (the enthusiasm of which Clio bears witness) that comes to affirm itself through the action of Painting. The God thus comes to think in the Silenus but at the cost of the Painter's exit from absence. The Silenus, however, to the extent that he will have had the opportunity to witness Clio's emerging enthusiasm, will be able to identify (and determine) the same coming to "flash" of new Nature. Hence, in fact, Pan's scream. Pan comes to be born because the Painter through the action of Painting has become pure software, leaving the absence but in the very manifestation of the presence of the God (in Clio). The Painter consumes himself in the praxis of art, he hears the God speaking in him but by resorting to his own brush. He becomes a stool for the God and for his resurrection, and in this way comes to be added. Here is the dialectic between software and incarnation. It is the action of the software which, by embodying the subject of Painting (to the point of identifying, at the limit, the very presence of the God in Clio) allows the Painter who becomes a stool for the divinity to be added in creativity. Here is the one who paints in the creativity thus opening to new Nature. Hence a Nature that comes from a Work of art. Hence, then, the new emerging Minotaur. The realization of the software is always within the boundaries of a sensibility and of the ever renewed dialectic between Simulation and Creativity. Both the God and the creator participate in this sensibility. Without these limits and without the passage preached by Skolem there is, therefore, no real life. In this sense, the passage for disembodiment as advocated by Chaitin is fundamental, just as it is fundamental the recourse to the identification of the software's path. However, it is necessary to open, at the same time, to the dialectic between Simulation and Creativity, as well as to the procedures of a real self-organization process, to the passage each time to new Nature, etc. Equally fundamental is the relationship between Nature and Culture, just as the intuitions of Skolem regarding the metamorphoses of the observer. Moreover, we must always open up to new Semantics. All this, however, would not be possible without recourse to that Theory of complexity created by Chaitin and without the connection of this theory with the doctrine of morphogenesis as identified by Turing and with the theory of self-organization (cf. H. Atlan, S. Kauffman, A. Carsetti etc.). Only a careful exploration in the regions of Chaos like the one carried out by Chaitin, by Turing etc. can help with that. In particular, it is Chaitin's studies on Omega that have led to the possibility of a better understanding of the rules of engagement of the software at the level of the constitution of those peculiar skills that intersect life.

The procedures of extroversion and the identification of the software are in sight of a new and deeper incarnation. I must not only optimize the software, I must, first of all, submit myself to the selection by Apollo in view of an opening on his part to new Nature and new possible forms of creativity. This is the key role played by the simulation activity and the consequent extroversion, the role that Marsyas plays starting from his conception. The path of Marsyas is that relating to the construction and use of new software but in view of the opening up to new creativity. The Painter of Abstraction poses himself as Lucien Freud, like a Hermes, that is to say, who reveals himself capable of embodying his subjects to the extent that he has become the conscious master of the software relating to the praxis of art he has put in place. I have to offer (in extroversion) my cortex, the software as it has materialized in my body, for the incision by the knife of the God until it is shaken. Hence, then, the rising of a new Minotaur in view of the new conception of Marsyas as Prince of simulation and the successive constitution of the Painter of abstraction. From the Minotaur to the Painter, from Pan to Marsyas. When I identify the software through the extroversion operated, I come to fix a body as hardware allowing the God to affect and dictate me inside. This is the way to get the God into my veins. In other words, this does not open only to the possibility of a better organization at the level of the existing software but also to the opening itself relating to the flow of new creativity. Here is the creative song of Apollo, here is the enthusiasm of the God which manifests itself in me, in my own passing away but through Clio. This is the Work (the action of Painting) with which I assure my inheritance and my ascent (my exit from absence), the final recognition that annihilates and exalts me (cf. Picasso's sculpture that compares with the conception by the Goddess and does so through poor materials and the result of recycling). Here is the Other who returns to life in me but through my work, the Other in which I "enlighten". It is starting from the irruption of the God that it will therefore be possible for a new categorial to come to rise; here is the very emergence of Pan as new Nature from which the path relative to the new incarnation along the detachment will spring. From the wild ferinity of Pan to the opening of the eyes of the Minotaur but in truth. Hence the wayfarer and natural evolution but in view of a renewed cultural evolution and the role that the new Painter will play in it. The new incarnation thus appears linked to the realization as existence and autonomy of a new body, a body inhabited by a renewed (visual) cognition and which constitutes itself as prelude to the subsequent revelation of a soul. The Minotaur that is born is new compared to the ancient observer. He makes reference to the climbing on his very shoulders by the ancient observer but to the extent that cultural evolution and the passage for the Painter will be given (with consequent irruption). This is the path that is missing in Chaitin, the path, in fact, concerning the reality of a biological being that is both life and truth, life and cognition (in truth). Without via there is no new incarnation and vice versa: the absence of real development actually hibernates life, freezes every possible emergency. God speaks only through the praxis of art, only with reference to my Work, to my making myself a stool. Therefore, the mere offering of my body is not enough, intelligent preparation is needed. Oedipus-M needs to prepare for his death so that there can be the resolution of the labyrinth. Hence the importance of the studies on omega but also the role played by the taking charge of the meaning, by the love (intellectual yet embodied) that in Caravaggio wins everything, by the emotion that if guided by the intellect illuminates with its light every aspect of reality. Here is a God who is reborn from his own viscera because the Painter (De Nittis) has revealed himself able to know all the secrets of the air and has conquered the arena of the intellect. The real optimization is the one that opens to the scream of Pan passing through Clio and the renewal of the primeval emotion itself but through the Work and the autonomous creation by the Painter, that creation that is balm and gift for the exit by the God from self-abandonment. Father, can't you see that I burn? Why have you forsaken me? Why have you, in fact, abandoned yourself? Accept the flames that surround me in order to regain your creativity. If Clio excites me, the God is at work. The success of the Work is the way to salvation. The hand of the Painter that portrays Clio is the hand that opens to the new mathematics of the emerging world. The creativity that overwhelms everything is creativity that operates in the viscera of the nucleus, changing the very way of being of the rules. Here are rules that change the rules: God comes to speak in me as he plays the role of Vermeer, Caravaggio, Cezanne etc. So I must not only promote Pan's scream but also the new detachment starting from the *Sylva*. A new language will therefore be necessary with the invention of new words at the level of the vernacular (Dante). I am reborn to myself according to my story but in view of conception and following the irruption linked to the presence of the God in the Muse (Clio), a presence that passes through my hand (the hand of a Painter) but that transcends me, that I cannot contain in me and by which I cannot limit myself to being contained where the irruption is to be true. The life that talks about you is still too short if it contains you (if it only contains you). If it is true it can only shake you up and it will be with reference to the grammar of the new irruption that you will only be able to find possible information on what your (new) emerging being is. It is only starting from the broken mirrors that you can now proceed to identify yourself in the imagination and in the simulation in accordance with the arts and tools of poetry (Painting). Here is the necessity every time of conception. Only if, starting from the broken mirrors, the right cypher comes to the surface, can you continue your path in metamorphosis. Optimization is only one of the possible consolations along such a harsh journey. Each time you will have to complete the circle: from the Minotaur to the Painter and from the Painter to the Minotaur. Here are the two steps necessary for conception and irruption. Here is via (as evolution and metamorphosis) and the two senses of it: natural and cultural. It is the natural evolution that leads to Marsyas in truth and it is the cultural one that leads to Pan and the detachment of the new observer. Marsyas simulates and extrudes himself by unraveling his software and fixing it as hardware on which the God's knife will come to operate. Hence the emergence of new creativity, but from within. It is what is required in order to awaken the God from self-abandonment. Here is the cry (see Antonioni's film: *Il Grido*) of the Son to the Father, the last invocation on the cross before the crash. In the film, the cry is of the woman who witnesses the death of her Son, of the one who had been her husband and who has come to be replaced by the new infant. In reality, the cry accompanies the groom's flight from the tower and his coming to smash. It constitutes the last word-expression of the traveller reflected in the one who as a woman had abandoned him. The invocation of the Son to the Father is translated into the cry of the ancient bride. Once you are added to the meaning and once you come to be added by the creativity in action to itself. Function and meaning, on the one hand, and simulation and creativity, on the other. By adding myself I make a precise carving with reference to the Temple, coming to be added it is my being, instead, that is carved and dictated inside. The sense of software as a gift (or the sense of dance as a gift, of choreography as a thought) is to open myself to new creativity and new evolution. The God who thinks in the craftsman who is added comes, therefore, to be contemplated and intuited by the very craftsman (as shown by Beato Angelico). The Goddess, in turn, comes to be categorized and calculated by the hero who comes to join the Temple (by intuition). Here is the stone Temple of Poussin. The God who dictates to me inside comes to be categorically intuited, but it is only by grasping his voice that I can therefore realize my metamorphosis in conception. Marsyas as the result of the metamorphosis and as the hero who, later on, will be added, will only be able to show himself as the one who has climbed onto his shoulders, thus opening up to a new possible observer along the course of the metamorphosis. This is the sense itself of the ongoing metamorphosis, but within the framework of a double dialectic: the dialectic between surface and depth and the dialectic between hardware and software. It is through this double dialectic that the passage Skolem speaks of can really come to be realized in all the complexity of its actual articulation.

According to an ancient intuition by Braque the emotion must be corrected according to the rule if we want to ensure a subsequent intentional extension by the emotion itself and, therefore, through this, the achievement of true invariance, albeit in the change. The rule, in turn, must be nourished by emotion if we wish to ensure a full development of the rule itself and, therefore, a real morphogenesis, albeit in continuity. Here is Life and Truth coupled, and with them via, albeit in accordance with its two directions. Hence the limit but also the value of the theses advocated by Chaitin: the great scholar is unable to grasp all the valences of the role played by meaning at the level of evolution, but carries out an analysis of the "artificial" connected to the emergence of Nature, thus individuating the conditions for the creating of an effective opening to new possible break-in. Setting himself up as a new Marsyas and undertaking the journey to the columns of Hercules, Chaitin really sets the conditions for a paradigm shift, thus opening up (albeit *in nuce*) to a new Semantics and the new cries of future detachment. The journey into invention begins here, and it is precisely here that the role played by the "verses" takes shape at the beginning. No longer pure determinations of the Form but tools capable of inventing the very way of articulating things to say and think from within: that is, tools that open up to the new arising function. The extroversion and semantic clarification represent the first step in view of the irruption taking place. It is necessary, however, to feed the omega-related pyre in order to be able to ensure the correct modalities for the change of semantics: in other words, to be able to truly face that onerous passage constituted by the passage on one's shoulders by the hero as suggested by Skolem. The artificial must be reflected in itself and must reflect the imagery in place, with a view to preparing for a correct irruption. It is in this sense that, by linking Goedel, Turing, and Darwin, Chaitin offers a versatile and important contribution to that complex analysis that right now is progressively preparing the first foundations of a new science: Metabiology. Hence the first emergence of artificial but not trivial machines as imagined by von Foerster: biological machines able to self-organize and to stay in symbiosis with man in view of his becoming a new creator through his coming to be added to himself by the God (in accordance with Bergson's metaphor, later taken up by Monod). These are unheard worlds which come to open and expand before our eyes. At the artificial level we can invent only by means of successive simulations, while on the natural level we can only imagine what form successive illuminations will take. Here is the light of Grace which in Caravaggio comes to illuminate the cheat who plays with the Chaos of his life leading him to the metamorphosis-conversion. This is what happens to the mathematician who comes to change semantics by opening up to that onerous journey corresponding to the overcoming of himself as well as of his own vision of the world as an autonomous observer in action. The metamorphosis in other, and the opening up to a new world of thought and observation (together with the entry on the scene of new infinities) emerge necessarily to the extent of a radical transformation on the part of man, a transformation for which the right compass is not easily found. This is the challenge that life presents to us every day. Just think of the enormous load that as humans we carry on our shoulders: that load which every time requires the artist to work for the overcoming of his own Work as well as of what constitutes his inheritance as man and craftsman. Life grants no insurance in this sense; it cannot, in effect, give assurances to itself if it truly wants to succeed in ensuring the necessary renewal of its original creativity. Turing and Chaitin focus on the role played by the grid relative to pure software with reference to extroversion and disembodiment, the way is open to the identification of Omega. This identification, however, turns out to be linked to a conception (and previous petrifaction) possessing a precise historical character. In this sense, therefore, omega has no absolute character: when, in effect, a real metamorphosis takes place, centered on the passage by the hero on his own shoulders, the coming into being of a new observation and, therefore, the very onset of new petrifaction come to enter the field. When the irruption occurs there is openness in depth, and unprecedented actors appear, albeit in the necessary context of an inheritance. Hence the proper sense of a natural evolution that can never come to be separated from the dialectic in place between Simulation and Creativity. The software leads, each time, to the opening from within of the hardware (with the birth of new intensities and the consequent outcrop of the God). The hardware leads, in turn, to the nesting in depth (in the swirls of meaning) of the software. The hardware opens on its abysses while also gifting incarnation to the point of surfacing as Nature. The software lurks deep within the sky of abstraction in successive increments of complexity (and its Methods) to outline the ever renewed contours of a kingdom of Culture. The pressure we will be under will not be of algorithms alone but also of meanings in action. The disembodiment must be pursued not only with a view to optimizing the evolutionary pressures on the table but also in view of a more ambitious goal: a complex system for living such as that represented by Ulysses-Marsyas appears to be the tool itself (first of all in overcoming the Pillars of Hercules), for an in-depth opening of its own hardware with the continuous birth of new meanings and with a continuous (but organic) remodulation of the ongoing evolutionary pressures. As Bergson states, complex living systems that act as autonomous agents come to enter the scene in function of ever new creativity, of the continuous realization of a renewed evolution. The way in which this happens is that which passes for the identification, each time, of the grid related to the martyrdom of St. Lawrence as masterfully illustrated by Titian. In the painting by Picasso "The flute of Pan" (Paris, Musée Picasso, Paris) the eigenvalues in action at the level of the score played by Pan are, in effect, the way to realize the trigger and the possible multiplication of the first cries of the Minotaur, that is to say of the first steps of the incarnation process. In this context, the eye of the mind, as well as the eye of Horus, son of Isis, appears, precisely, as one of the engines of natural evolution. But Horus is not only a name or a concept or an imagination that lives: he is, first of all, a universal Form (cf. Picasso's sculpture), that soul of itself and that guides and points to every possible vision. It is the eye that in Reflexivity becomes an eigenform to itself, a matrix of real invariance and autonomy. When we are faced with works that come to be worn by the Muse through an ideal seam for files, we find ourselves faced with the offer of a particular inheritance from a piece of hardware that has come into being as an autonomous agent and that allows the new software (Marsyas as conceived) to come to light through the support offered by the cypher. Creativity along the path pursued by the Minotaur has turned into petrifaction, thus offering a legacy and while allowing the Goddess to conceive. Marsyas represents the new software that is born, but the hero is, in his turn, marked by a cypher, by a secret Rule that lives him in filigree: the robe relative to his sacrifice will be woven into the file of Reflexivity to the point of determining the giving of extroversion. At that moment the God will come to select opening to the new irruption (and new hardware). By following Horus we have the effective possibility of fully entering the realm of dialectics between function and meaning that allows us not to close ourselves into the enclave of the first-order structures but to range in a much broader realm of functions also featured in accordance with the tools offered by non-standard mathematics. Choosing non-standard model theory really 'introduces' a new general semantics. Hence the possibility of fully exploring the arena relating to non-standard models as indicated, first and foremost, by Skolem but taking advantage of the latest acquisitions achieved at the level of the most recent theory of self-organising models.[6] At the level of a natural, biological self-organising system that is characterised by its cognitive activities, the objectives are set from within. The origin of meaning is an emergent property within the organisation of the system itself, and this is connected to precise linguistic and logical operations as well as distinct procedures of observation and self-observation. These operations stimulate continuous processes of inner reorganisation. In the light of the new scientific paradigm represented, today, by Enactive Realism[7] Nature appears to 'speak' by means of mathematical forms: we can observe these forms, but they are, at the same time, inside us as they populate our organs of cognition. In this sense, in such a scenario, natural evolution also scans the effective growth of our tools of participatory knowledge of the world around us, our own coming to recognise ourselves as a stage within a path concerning a real development of creativity but on a co-evolutive level and in accordance with a semantically pregnant perspective. Paraphrasing Galileo, normally regarded as the father of modern science, we can affirm that Nature is speaking by computations, thus causing Nature to be embodied. However, such natural computations cannot be articulated only on a purely syntactic level nor can they be flattened on disembodied crystallizations. Actually, at the biological level we are continuously challenged by semantic information and semantic phenomena thus coming to be involved in an effective process of self-organization. Real 'extroversion' is in function of the activation of new selection procedures capable of ensuring a metamorphosis (from within) of the system: a metamorphosis that will necessarily involve the mathematician who builds the model. In this context, the system's pursuit can really offer the opportunity to trigger more sophisticated levels of embodiment only by changing semantics.

According to Monod, Nature appear as a tinkerer characterized by the presence of precise principles of self-organization. However, while Monod was obliged to incorporate his brilliant intuitions into the framework of first-order cybernetics and a theory of information with an exclusively syntactic character such as that defined by Shannon, research advances in recent decades have led not only to the definition of a second-order cybernetics but also to an exploration of the boundaries of semantic information. We have already seen how for H. Atlan, on a biological level "the function self-organizes together with its meaning". Hence the need to refer to a conceptual theory of complexity and to a theory of self-organization characterized in an intentional sense. However, there is also a need to introduce, at the genetic level, a distinction between coder and ruler as well as the opportunity to define a real software space for natural evolution. The recourse to non-standard model theory, the opening to a new general semantics, and the innovative definition of the relationship between coder and ruler can be considered, today, among the most powerful theoretical tools at our disposal in order to correctly define the contours of that new conceptual revolution (that new worldview) increasingly referred to as Metabiology. A conceptual revolution that appears primarily to refer to the growth itself (at the co-evolutive level) of our instruments of participatory knowledge of the world. A work at whose level the entropy conditions change continuously also depending on the decline of Clio as the Muse of History. Hence the very possibility of considering Nature also as a Work of Art as advocated by P. Feyerabend.

Notes

1. Cf. Dean (2013).
2. Cf. Dean (2013).
3. Cf. Putnam (1980).
4. Cf. Quinon and Zdanowski (2006).
5. Cf. Carsetti A. (1989) "Self-organizing models", T.R. (La Nuova Critica).
6. Cf. Carsetti A. (1989) "Self-organizing models", T.R. (*La Nuova Critica*).
7. Carsetti A. (1993) "Meaning and complexity: the role of non-standard models", *La Nuova Critica*, 22, 57–86.

Bibliography

Atlan, H. (1992). Self-organizing networks: Weak, strong and intentional, the role of their underdetermination. *La Nuova Critica, 19–20*, 51–71.

Bell, J. L., & Slomson, A. B. (1989). *Models and ultraproducts.*

Benacerraf, P. (1965). What numbers could not be. *The Philosophical Review, 74*, 47–73

Carnap, R., & Bar Hillel, Y. (1950). An outline of a theory of semantic information. *Technical Report.* N. 247, M.I.T.

Carsetti, A., (ed.) (1999). *Functional models of cognition. Self-organizing dynamics and semantic structures in cognitive systems.* Kluwer A. P.
Carsetti, A. (ed.) (2004). The embodied meaning. In *Seeing, thinking and knowing* (pp. 307–331). Kluwer A.P.
Carsetti, A. (2010). Eigenforms, natural self-organization and morphogenesis. *La Nuova Critica, 55–56*, 75–99.
Carsetti, A. (2011). The emergence of meaning at the co-evolutionary level: An epistemological approach. https://doi.org/10.1016/j.amc.2011.08.039
Chaitin, G. (1987). *Algorithmic information theory.*
Chaitin, G. (2010). Metaphysics, metamathematics and metabiology. *APA News, 10*(1), 11.
Chaitin, G. (2013). *Proving Darwin: Making biology mathematical.*
Chaitin, G., & Calude, C. (1999). Mathematics/randomness everywhere. *Nature, 400*, 319–320.
Dean, W. (2013). Models and Computability. *Philosophia Mathematica (III), 22*(2), 51–65.
Henkin, L. (1950). Completeness in the theory of types. *Journal of Symbolic Logic, 15*, 81–91.
Herken, R. (Ed.) (1988). *The universal turing machine. A half century survey.*
Hintikka, J. (1970). Surface information and depth information. In J. Hintikka, & P. Suppes, (Eds.), *Information and inference* (pp. 298–230).
Husserl, E. (1964). *Erfahrung und Urteil.*
Jaynes, E. T. (1965). Gibbs vs Boltzmann entropies. *American Journal of Physics, 33*, 391.
Kauffman, S. A. (1993). *The origins of order.*
Kauffman, L. H. (2003). Eigenforms—objects as tokens for eigenbehaviours. *Cybernetics and Human Knowing, 10*(3–4), 73–89.
Kauffman, L. H. (2009). Reflexivity and eigenforms. *Constructivist Foundations, 4*(3), 120–131.
Kauffman, L. H. (2018). Mathematical themes of Francisco Varela. *La Nuova Critica, 65–66*, 72–83.
Kauffman, S. A., Logan, R. K., Este, R., Goebel, R., Hobill, G., & Shmulevich, I. (2008). Propagating organization: An enquiry. *Biology and Philosophy, 23*, 34–45.
Martin-Delgado, M. A. (2011). On quantum effects in a theory of biological evolution. arXiv:1109. 0383v1.
Maynard Smith, J. (2000). The concept of information in Biology. *Philosophy of Science, 67*, 177–194.
Mayr, E. (2001). *What evolution is.*
Nicolis, G. (1989). Physics in far-from-equilibrium systems and self-organisation. In P. Davies (ed.). *The new Physics.*
Nicolis, G., & Prigogine, I. (1989) *Exploring complexity.*
Piaget, J. (1967). *Biologie et Connaissance.*
Popper, K. (1959). *Logic of scientific discovery.*
Prigogine, I. (1980). *From Being to Becoming.*
Prigogine, I. (1993). *Les lois du Chaos.*
Putnam, H. (1965). Trial and error predicate and the solution to a problem of Mostowski. *Joural of Symbolic Logic, 30.*
Putnam, H. (1980). Models and reality. *The Journal of Symbolic Logic, 45*, 464–482.
Putnam, H. (1983). *Representation and reality.*
Quinon, P., & Zdanowski, K. (2006). *The intended model of arithmetic. An argument from Tennenbaum's theorem* (p. 2). http://www.impan.pl/_kz/_les/PQKZTenn.pdf
Skolem, T. (1941). Sur la portée du théorème de Löwenheim-Skolem. In *Les Entretiens de Zurich, 6–9 Dec.1938:* 25–47. Reprinted in (Skolem 1970, pp. 455–482).
Skolem T. (1955). Peano's axioms and models of arithmetic. In *Mathematical interpretation of formal systems* (pp. 1–14).
Skolem, T. (1970). *Selected works in logic.* Universitets for Laget, Oslo,. Fenstad, J. E. (ed.)
Turing, A. M. (1952). The chemical basis of morphogenesis. *Philosophy Transactions of the Royal Society of London, 237*, 641, 32–72
Varela, F. (1982). "Self-organization: Beyond the appearances and into the mechanism. *La Nuova Critica, 64*, 31–51.

Varela, T. E., & Rosch, E. (1991). *The embodied mind*. The MIT Press.
Varela, F. (1975). A calculus for self-reference. *International Journal of General Systems*; Kauffman, L. H. (2005). Eigenform. *Kybernetes, 34*, 129–150.
von Foerster, H. (1981). Objects: Tokens for (eigen-) behaviors. *Observing systems, The systems, inquiry series* (pp. 274–285). Intersystems Publications.
Wuketits, F. M. (1992). Self-organization, complexity and the emergence of human consciousness. *La Nuova Critica, 19–20*, 89–109.

Holism and Pseudoholism

Sven Ove Hansson

1 Introduction

Since Jan Smuts introduced the term "holism" in his 1926 book *Holism and Evolution* (Smuts, 1926), the term has been widely used in many fields of knowledge and speculation (Michaelson et al., 2019). However, the concept is much older than the term. The ideas of present-day holism can be traced back at least to early nineteenth century Romanticism, with its widespread resistance to the fragmentation and reduction of human belief systems that was claimed to follow with new scientific theories (Harrington, 1996, p. 4). Other words, such as "wholeness" and "unity" have been used to express the same ideas as "holism".[1]

At first sight, it might seem almost impossible to be critical of holism. If we are discussing or investigating some entity, what reasons could there be not to include all of it and to treat all aspects that pertain to it? However, like many other seemingly self-evident notions, that of holism is much more problematic than what first catches the eye. One major reason for this is that many if not most of our objects of study and reflection have so many aspects that it is impossible to cover them all. For instance, there are so many factors that can potentially affect a person's health that is practically impossible to consider them all. A selection has to be made.

[1] I will not spend much effort on the common claim that holism means that the whole is more than its parts and their relations. As noted by Richard Healey, this is an empty criterion since "one relation among the parts is what we might call the *complete composition* relation–that relation among the parts which holds just in case they compose this very whole with all its properties" (Healey, 1991, pp. 400–401). A better delimitation, proposed by Healey, explicates holism as the claim that "the whole has features that cannot be reduced to features of its component parts". (ibid., p. 397) The meaning of this explication will depend much on whether this reduction is conceived as being ontological or epistemological.

S. O. Hansson (✉)
Department of Philosophy and History, KTH Royal Institute of Technology, SE-100 44 Stockholm, Sweden
e-mail: soh@kth.se

S. Wuppuluri and I. Stewart (eds.), *From Electrons to Elephants and Elections*, The Frontiers Collection, https://doi.org/10.1007/978-3-030-92192-7_13

Furthermore, there is usually more than one overarching perspective that can be applied to one and the same topic. More often than not, it is impossible to combine all of those overarching perspectives into a single, unified perspective that covers them all. This means that there are *competing holistic perspectives on one and the same topic*. A claim that there is only one truly holistic perspective or approach that can be applied to a topic is nothing else than an attempt to monopolize the topic.

However, not all overarching perspectives are presented or recognized as holistic. It turns out that only some such perspectives are in general called "holistic". The following four examples should make this clear.

First: Charles Darwin's theory of evolution gave rise to a much broader and more interconnected understanding of biology than what was previously available. Investigations of a biological species can now be extended to how it was evolutionarily adapted to its whole environment in all its complexity. Fossil findings are directly connected with our knowledge of living plants and animals. (Salgado, 2019, p. 2) However, the theory of evolution has seldom been described or promoted as "holistic". In contrast, creationist ideas are commonly called "holistic", although they deny all these connections and thereby fragmentize biological knowledge. (Andrews, 1984).

Secondly, animal experiments still have an essential role in the early phases of drug testing. A major argument used by proponents of animal experiments is that studies on isolated cells or tissues cannot provide a credible picture of what will happen when the drug is introduced into the body as a whole, with all its complex interactions. This is clearly a holistic argument, but we seldom hear animal experimentation called a "holistic" approach.

Thirdly, the climate models used for instance in the IPCC reports on climate change include a large number of natural and anthropogenic factors that influence the climate, such as solar irradiance, clouds, precipitation, biological processes in plants, animals, soils, and oceans, geological processes, emissions of various greenhouse gases from a large variety of sources, etc. Complex interactions between these and other processes are taken into account, and regional differences are calculated along with global averages. This approach is certainly holistic, but that word is not often heard in connection with these models.

Finally, so-called "personalized medicine" makes use of large amounts of genomic and biochemical information about a patient in order to put together treatments that are better adjusted to the individual patient than what is possible with traditional methods. This is clearly a holistic approach, although it is seldom described as such. (Vogt et al., 2016) In most discussions on "holistic medicine", the term refers either to the use of unproven therapies based on implausible theories about health and disease (Widder & Anderson, 2015) or (much better) to clinical practices that combine medical science with humanistic ideals. (Ferry-Danini, 2018; Thibault, 2019).

These examples reveal a considerable tension in common usage of the term "holism". On the one hand, the term is usually conceived as denoting the unassailable idea that when studying or reflecting on some object, we should consider it in its entirety, and in particular not restrict our deliberations to the properties of its parts. On the other hand, the designation "holistic" is in practice commonly reserved for only some of the many approaches that attempt to cover "the whole" of something. Obviously, if only some of the many overarching perspectives that can be applied to an object of study are recognized as holistic, then holism becomes a highly contestable ideal.

From the fact that different overarching approaches or perspectives can be constructed for the same subject matter it does not follow that these different constructions are all of equal value. A "holistic" perspective can be misleading in various ways. I will focus on three major categories of failing holistic approaches:

Over-inclusive holism contains elements that should not be there.
Incomplete holism excludes elements that should be included in the intended whole.
Dogmatic holism is unable to revise its claims in response to new circumstances or insights.

These three categories of failure are largely overlapping in the sense that many deficient holisms have two or all three of them. However, from a conceptual point of view, it is clarifying to treat them as distinct types of shortcomings. The following three sections exemplify and further specify each of them.

2 Over-Inclusive Holism

Holism can fail by including claims that lack factual support. For instance, if someone wants to provide a holistic account of the fauna of Loch Ness, it would be most inappropriate to include a claim that the mythical Loch Ness monster is a real animal, living in that lake. Such a component would be superfluous, and it would detract from the value of the account by making its holism over-inclusive.

Unfortunately, over-inclusion is a common problem in allegedly holistic accounts. Sometimes, the superfluous component is the underlying motivation for constructing the account in question. This applies for instance to many cases of "holistic complementary medicine", in which the "whole" is said to consist of a combination of conventional medicine and some additional, "alternative" intervention. The purpose of the whole construction is usually to gain acceptance for that addition. However, if the addition is a worthless method with no positive effects beyond the placebo effect, then the whole construction is just a deceptive way to introduce an inferior therapy into healthcare in the name of holism.

Let us consider three examples of over-inclusive holism in somewhat more detail.

2.1 Gaia

In the early 1970s, James Lovelock worked for the oil company Royal Dutch Shell. He was deeply involved in the oil industry's early responses to research indicating that CO_2 emissions could have far-reaching impacts on the earth's climate. (Aronowsky, 2021) In 1971 he published an article entitled "Air pollution and climatic change", in which he introduced what was to become an important part of the fossil fuel industry's defence against these concerns: the claim that the planet possesses stabilizing mechanisms that counteract the effects of CO_2 emissions and keep atmospheric temperatures within fairly strict limits.

> It is known that this substance [carbon dioxide] directly stimulates the growth of vegetation and consequently the output of the products of vegetative growth such as terpenes, ammonia and hydrogen sulphide. As already stated, these are aerosol precursors. If this is the direct cause of the increase of haze-forming substances, it could be looked on as a regulatory response of the ecosystem to combustion emissions for it tends to neutralise the effect (temperature increase) of the perturbing stimulus (the accumulation of carbon dixode [*sic*]), thereby restoring the status quo. If this biological-cybernetic explanation is correct, the outlook will not be as gloomy as that predicted by direct extrapolation of past trends...
>
> We may find in the end that the direct aspects of combustion are the least harmful of all the major disturbances by man of the planetary ecosystem, for the system may have the capacity to adapt to the input of combustion gases (Lovelock, 1971, pp. 409–410).

In an article published the following year, he further developed these ideas, and presented what he called the "Gaia hypothesis". He now claimed to have found strong evidence that the total biosphere of the earth acts as an organism, which is capable of controlling the atmospheric gases and the climate.

> The purpose of this letter is to suggest that life at an early stage of its evolution acquired the capacity to control the global environment to suit its needs and that this capacity has persisted and is still in active use. In this view the sum total of species is more than just a catalogue, "The Biosphere", and like other associations in biology is an entity with properties greater than the simple sum of its parts. Such a large creature, even if only hypothetical, with the powerful capacity to homeostat the planetary environment needs a name; I am indebted to Mr. William Golding for suggesting the use of the Greek personification of mother Earth, "Gaia"....
>
> In fact a close examination of the composition of the atmosphere reveals that it has departed so far from any conceivable abiological steady state equilibrium that it is more consistent in composition with a mixture of gases contrived for some specific purpose...
>
> Life is abundant on Earth and the chemically reactive gases almost all have their principal sources and sinks in the biosphere. This taken with the evidence above is sufficient to justify the probability that the atmosphere is a biological contrivance, a part and a property of Gaia. If this is assumed to be true then it follows that she who controls the atmospheric composition must also be able to control the climate (Lovelock, 1972).

In a follow-up article two years later, Lovelock and a co-worker further elaborated the idea that early in the development of life, Gaia acquired the ability to "secure the environment against adverse physical and chemical change", so that "unfavourable tendencies could be sensed and counter measures operated before irreversible damage had been done." This included "the presence of an active process for thermostasis", i.e. for keeping the temperature close to a desirable, constant level (Lovelock & Margulis, 1974, p. 8). In another article they reaffirmed their picture of Gaia as a "control system" that regulates the earth's temperature and other crucial variables and is able to "keep these variables from exceeding limits that are intolerable to all terrestrial species" (Margulis & Lovelock, 1974, p. 486). They expressed their belief in this mechanism in no uncertain terms:

> We believe that Gaia is a complex entity involving the earth's atmosphere, biosphere, oceans and soil. The totality constitutes a feedback or cybernetic system which seeks an optimal physical and chemical environment for the biota... (Margulis & Lovelock, 1974, p. 473).

> We conclude from the fact that the temperature and certain other environmental conditions on the earth have not altered very much from what is an optimum for life on the surface, that life must actively maintain these conditions (ibid, p. 475).

The Gaia notion reached a wider public through an article, published in 1975, that Lovelock wrote jointly with a senior Shell manager (Aronowsky, 2021, p. 316). They claimed that Gaia had kept the surface temperature of the Earth within the bounds required for life "in spite of drastic changes of atmospheric composition and a large increase in the mean solar flux". Against this background, "[m]an's present activities as a polluter is trivial by comparison and he cannot thereby seriously change the present state of Gaia let alone hazard her existence" (Lovelock & Epton, 1975, pp. 304 and 305).

The Gaia construct served the purposes of the fossil industry in whose service it was originally developed. The alleged self-regulating stability of the earth's climate became one of major themes in their attempts to downplay the seriousness of the greenhouse effect. For instance, in one advertisement in 1995, Exxon talked about "Mother Nature", describing her as "one strong lady, resilient and capable of rejuvenation", with the consequence that "nature, over the millennia, has learned to cope" (Supran & Oreskes, 2017, p. 9). But at the same time, the "holistic" nature of the Gaia construct has connected with certain strands of environmentalism, and the construct has attracted considerable following for instance among proponents of so-called deep ecology (Bartkowski & Swearingen, 1997; Clarke, 2017; Haig, 2001). It has even given rise to a New Age-related religious movement called Gaianism.

In later publications, Lovelock has modified his Gaia construct, but without giving up the basic assumption of an organism-like entity that regulates the physical and chemical conditions on earth to retain its inhabitability. From a scientific point of view, the Gaia construct is an unnecessary and misleading assumption.[2] Biological mechanisms that counteract the greenhouse effect should be studied carefully, one by one, and so should biological mechanisms that aggravate the greenhouse effect. The introduction of an unproven fantasy creature that regulates these mechanisms adds nothing to our understanding. The Gaia construct is therefore a clear example of over-inclusive holism.

---

[2] It is highly doubtful whether it can serve as a scientific hypothesis. (Kirchner, 1989) I therefore call it a construct rather than, as Lovelock does, an "hypothesis".

2.2 Anthroposophy

Some of the best examples of over-inclusive holism can be found in esoteric movements that include extensive supernatural claims in their "holistic" accounts of the natural world and human life. One such movement is anthroposophy, which was founded by the German Rudolf Steiner (1861–1925), originally as a splinter group in the theosophical movement. Anthroposophy is best known for its Steiner schools, its deviant version of medicine, and a variant of organic farming called "biodynamical". Claims about holism form an important part of the anthroposophical rhetoric. For instance, the introduction to biodynamical agriculture at the website of the Anthroposophical Society begins: "To understand and shape agriculture as a living whole belongs to the most important principles of the biodynamic impulse".[3] Similarly, their introduction to anthroposophical medicine describes it as responding to "the human's need of a holistic treatment."[4]

Probably, most people who encounter statements like these will interpret them as referring to agriculture taking the whole ecosystem into account and medicine considering the patient's social and psychological situation in addition to the bodily issues. But anthroposophical holism goes far beyond that. Its focus is on spiritual beings and on a large collection of esoteric claims in which astrology plays a significant role. Although its adherents do not see it as a religion, the teachings of anthroposophy include belief in reincarnation and in a large number of supernatural beings, including elemental beings such as gnomes, sylphs and nymphs, various spirits, angels, and archangels, and an assortment of demons. Many of these beings are claimed to be connected for instance with specific planets, astrological signs, or natural phenomena. Belief in a spiritually predetermined destiny for humanity, described in Rudolf Steiner's writings and lectures, is an essential component of the anthroposophical belief system. Since the anthroposophical "holism" includes all this, it differs radically from holistic approaches not associated with the movement. In a recent article in the official journal of the Anthroposophical Society, *Das Goetheanum*, the prominent anthroposophist Andreas Neider accused environmentalists of not seeing "the whole":

> Ecocentrism sees no spiritually developing I in the human being, and it wishes to arouse consciousness of the connections between all living beings and of their mutual dependence in their coming to be and passing away. But it has no answer to the question what is the meaning of human existence on Earth and why man at all arose in evolution. The meaning of the whole remains in the dark[5] (Neider, 2019, p. 12).

---

[3] "Die Landwirtschaft als lebendige Ganzheit zu erfassen und zu gestalten gehört zu den wichtigsten Prinzipien des biodynamischen Impulses". https://www.sektion-landwirtschaft.org/grundlagen/biodynamische-landwirtschaft. Accessed 21-01-04.

[4] "dem Bedürfnis der Menschen nach einer ganzheitlichen Behandlung". http://www.anthromed.de/Dateien/Downloads/Anthroposophische_Medizin.pdf. Accessed 21-01-04.

[5] "Der Ökozentrismus sieht kein sich geistig entwickelndes Ich im Menschen und möchte das Bewusstsein von der Zusammengehörigkeit aller Lebewesen und ihres Entstehens und Vergehens in gegenseitiger Abhängigkeit wecken. Er kann aber keine Antwort auf die Frage geben, was

Thus, Neider rejects the holism of ecocentric environmentalists who see the ecosphere as a whole, in which humans do not have a special pre-determined destiny. He describes their view as non-holistic since they have a naturalistic explanation of the emergence of the human species and do not subscribe to the supernatural claims of anthroposophy.

Another article in *Das Goetheanum* further exemplifies how anthroposophical holism differs from that of ecocentric environmentalists. The author of this article expressed worries that the gnomes living inside the Alp Mountains would react negatively to the construction of a new railway tunnel. He took the existence of such entities for granted, and quoted Rudolf Steiner as an authoritative source for the claim that gnomes are much inconvenienced by light from the full moon. (Bockemühl, 2007) This is of course a very different perspective from that of ecocentric environmentalists, who would focus on the effect of the tunnel on the groundwater level and other parameters that can have negative impact on living organisms.

Anthroposophy also includes beliefs that strongly connect the history of humanity with astrological claims. Human history is claimed to proceed in a series of cultural epochs. Each of them lasts 2160 years, which is the time it takes for the sun to pass through one of the twelve signs of the Zodiac (the precession of the equinox). We are now said to live in the Germanic-Anglosaxon cultural period, which is associated with the astrological sign Pisces. It began in the year 1413 and will end in the year 3573. It will be followed by the Slavic cultural period, which is associated with the Aquarius. It will begin in 3573 and be replaced in 5733 by the American cultural period, which is associated with the Capricorn and ends in the year 7893.

These are only a few examples from the extensive system of esoteric beliefs introduced by Rudolf Steiner and still cherished by today's anthroposophists. These beliefs are all parts of the "whole" of the "holistic" anthroposophical worldview. For obvious reasons, only a small selection of the more palatable constituents of this worldview is presented to the public. For instance, anthroposophical healthcare practitioners discourage parents from having their children vaccinated against the measles and other deadly diseases, but they do not tell them the origin and basis of their antagonism towards vaccines. Its origin can be found in Rudolf Steiner's pronouncements on infectious childhood diseases. He claimed that the causes of measles can be found in personality defects in previous lives, making measles "the physical-karmic effect of a previous life".[6] The disease has a positive role in correcting these defects: "And when then such a personality enters existence, it will wish to make corrections in this area as soon as possible and in the time between birth and the usual appearance of childhood diseases, in order to work through the measles as organic self-education".[7]

für einen Sinn die menschliche Existenz auf Erden hat und warum der Mensch überhaupt in der Evolution entstanden ist. Der Sinn des Ganzen bleibt im Dunklen".

6 "der die physisch-karmische Wirkung ist eines früheren Lebens".

7 "Und wenn dann eine solche Persönlichkeit ins Dasein tritt, wird sie so schnell wie möglich Korrektur üben wollen auf diesem Gebiet und in der Zeit zwischen der Geburt und dem gewöhnlichen Auftreten der Kinderkrankheiten, um organische Selbsterziehung zu üben, die Masern durchmachen".

The disease becomes “a spiritual process”,[8] leading to “what is enormously important namely that when this process is taken up in the soul as a maxim of life, it will engender a conception that has a healing effect on the soul”[9] (Steiner, 1992, pp. 103–104). The soul, notably, is supposed to reappear after death in coming reincarnations. Therefore, said Steiner, it is wrong even to wish that a child will not contract measles:

> For instance, it is no good thing to say: This child has the measles; I wish it had not contracted the measles! – You cannot know all that would have happened to the child if it had not caught the measles. Because thereby that came out that was sitting deep in the child and was trying to find its redemption[10] (Steiner, 1980, p. 340).

Thus, the true nature of the “holism” of anthroposophy explains its resistance to vaccination against childhood diseases: These diseases are supposed to have a positive “karmic” function in a perspective that sees the child as just one in a series of reincarnations of the same person. From the viewpoint of medical science and medical ethics, these other reincarnations, as well as the claimed positive effects of diseases such as measles, constitute unproven and unjustified considerations in healthcare. Their inclusion has the tragic effect of contributing to the spread of the measles and other deadly childhood diseases.

2.3 Certain Approaches to Patient-Centred Care

Patient-centred care (PCC) is a collection of approaches to healthcare that share “a holistic paradigm, which suggests that people need to be seen in their biopsychosocial enti[re]ty.” (Olsson et al., 2013, p. 456) As a general principle, this is clearly a laudable approach, and many positive developments in healthcare in the last few decades have taken place under the banner of patient-centredness. (Hansson & Fröding, 2021) However, like most other broad churches, it contains problematic elements. In parts of the PCC movement, health care personnel are encouraged to engage in all kinds of difficulties that patients encounter, including those that do not fall within the traditional concerns of healthcare. (Stewart, 2001) This can include personal relationships, conflicts with family members, relatives and acquaintances, economic problems, etc. This is surely a holistic practice in the sense of engaging with “the whole person”, but weighty arguments can be raised against an extension of the tasks of healthcare to issues for which healthcare personnel have no special qualifications. Warnings have been raised that this can lead to “a kind of medicalisation, whereby domains of life previously not considered relevant to health, or as

---

[8] “einen geistigen Prozeß”.

[9] “das ungeheuer Bedeutsame, daß wenn dieser Prozeß in die Seele als Lebensmaxime aufgenommen wird, er eine Anschauung erzeugt, die gesundend auf die Seele wirkt”.

[10] “Man wird zum Beispiel gar nicht gut tun, wenn man sagt: Das Kind hat die Masern; hätte es doch diese Masern nicht bekommen! - Man kann nicht wissen, was alles über das Kind gekommen wäre, wenn es die Masern nicht gekriegt hätte. Denn darin kam das heraus, was immer tief in dem Kinde saß und seinen Ausgleich suchte”.

appropriately falling within the structures of healthcare, are newly seen through a medical lens" (Brown, 2018, p. 1000).

One example of this is the proposal, sometimes advanced within PCC, that health care personnel should attend to the patients' "spiritual well-being" (Mezzich, 2012, p. 8). This goes far beyond the traditional practice in healthcare, which is to help patients by connecting them with persons or organizations that can provide them with religious or spiritual services that they ask for. Healthcare personnel are not in general educated to provide such services themselves, and mixing them up with medical and nursing activities is bound to be problematic, given that patients tend to differ in what—if any—such services they prefer to receive.

3 Incomplete Holism

As mentioned in Sect. 1, any account of a complex phenomenon will have to be incomplete. It is impossible to cover everything that has bearing on an intricate real-world issue. In that sense, all forms of holism are incomplete. However, some omissions are worse than others. For instance, issues of diet cannot be excluded from a reasonable account of obesity as a health problem—in particularly not an account claimed to be holistic. Similarly, a "holistic" account of climate change that leaves out the known facts about the anthropogenic greenhouse effect, or a "holistic" account of Nazism that leaves out the Holocaust, would be severely incomplete.

Unfortunately, such incompleteness is not uncommon. As we have already seen, severe cases of over-inclusiveness tend to also involve the exclusion of essential aspects of the object of study or reflection. For instance, the literature on the Gaia construct usually leaves out the *de*stabilizing biological mechanisms that amplify, rather than mitigate, the greenhouse effect. Unfortunately, the total effect of the various biological mechanisms seems to be amplifying, contrary to assumptions made by proponents of the Gaia construct (Kirchner, 2003, pp. 26–28). Similarly, anti-vaccination propaganda within "holistic medicine" tends to leave out the scientific knowledge about of the death toll of measles and the protective effects of immunization.

Another example of glaringly incomplete holism is iridology, which is usually claimed to be "holistic" since its practitioners claim that they can determine the health status of all organs in the body by careful inspection of the iris. This practice neglects the unequivocal scientific evidence, available since more than four decades, that the "diagnoses" of iridologists are no better than random guesses. Furthermore, due to their focus on spurious signs in the colour pattern of the iris, iridologists miss the signs of diseases in other organs that ophthalmologists can find. Such signs are mostly seen in other parts of the eye than the iris. (Ernst 2000; Knipschild, 1988; Münstedt et al., 2005; Noworol, 2020; Simon et al., 1979) The iridologists' failure to inspect the whole eye is a quite remarkable example of an unholistic practice promoted as holistic.

This is by no means an isolated phenomenon. As Joshua Freeman pointed out, it is not uncommon for allegedly "holistic" healing practices to be much more reductionist than conventional, science-based medicine. This is because these practices tend to be based on a limited view of the causes of disease, and on a very small arsenal of treatments. (Freeman, 2005).

Another interesting example of "unholistic holism" can be found in an evaluation report on an anthroposophical clinic that was recently closed down in Sweden. Since 2011, the Stockholm Regional Council had a contract with the anthroposophical Vidar clinic, to which some patients were sent for rehabilitation and treatment. In preparation for a possible extension of the contract, two physicians were tasked with evaluating the clinic. They found severe deficiencies, not least in the patient records, which were so incomplete that concerns arose on patient safety. The documentation of drug prescriptions was also inadequate. From the viewpoint of holism, the most remarkable criticism in the report was the following:

> Discharge records from day care and outpatient care exist to a large extent, but do not contain multimodal summaries, since each profession writes its own summary (Holmberg & Vallin, 2016, p. 3).

In other words, the different professions involved in patient rehabilitation did not coordinate to make joint assessments and to document these assessments. Although it described itself as "holistic" and "integrative", the clinic did not satisfy the requirements of seeing the whole patient and integrating the different interventions that are standard in conventional rehabilitation care. Belief in unproven therapies is not what it takes to make healthcare holistic in the sense of seeing the whole person. Co-operation between the different healthcare professions and specializations, on the other hand, is essential for achieving it.

Based on the report, the Stockholm Regional Council decided in 2016 not to prolong its contract with the anthroposophical clinic. The clinic was subsequently closed down.

4 Dogmatic Holism

Our knowledge develops. The major advantage of science over non-scientific doctrines about the world is the ability of science to accommodate new information and correct its mistakes. Treating something as a "whole" does not decrease the need to learn from new information; it may even increase that need. In other words: holistic accounts of a subject matter have to be open to criticism, and they should be revised or given up if new information provides sufficient reason to do so. But unfortunately, there are many examples of alleged holism that do not satisfy these requirements.

Homeopathy is a highly illustrative example of this. It is an "alternative" pharmacology, developed in the 1790s by the German physician Samuel Hahnemann. It is based on the supposition that if a substance causes certain symptoms in a healthy person, then very small amounts of that substance will cure a disease with those same symptoms. Furthermore, the curative effects of the substance are supposed to increase as the administered amount is decreased. Therefore, homeopathic drug companies produce what they market as highly potent drugs by performing a long series of dilutions of a solution containing the supposedly efficient drug. In the 1970s, when "holism" became a catchword in healthcare (Whorton, 1985, p. 29), homeopaths were quick to pick up the new word as a designation of their own activities (Clover, 1979; Twentyman, 1973; Whitmont, 1974). The terminology is still frequently used by homeopaths (Attena, 2016; Prousky, 2018; Schmidt, 2020).

A modern example of a homeopathic drug is "Berlin wall", which is used by some homeopaths to cure patients with a long list of symptoms, including "depression, sense of blackness, total isolation, aloneness, despair" and also "oppression (political, family, abuse-sexual, religious, being bullied) and perceiving themselves as victims" (Dam, 2006). The preparation is made from a piece of the Berlin Wall, which is repeatedly diluted in lactose to the "potency C200". This means that a dilution 1:100 has been done 200 times in a row. The resulting total dilution is $1{:}10^{400}$. A simple calculation (based on molar concentration) shows that already after 15 dilution steps, when the concentration is $1{:}10^{30}$, it is virtually certain that no single molecule from the Berlin Wall is left in the preparation. The remaining 185 dilutions are therefore just dilutions of lactose into lactose. Provided that the dilutions are reported correctly, the vial contains 100% lactose. However, its label does not mention lactose. Instead, it mentions Berlin wall, although there is no trace of the Berlin wall in the product.

Obviously, Hahnemann did not have access to Berlin wall, but he made similarly extreme homeopathic dilutions, starting from a wide variety of other materials. In the 1790s, when he first proposed homeopathic drugs, it was not known that his preparations contained no trace of the substance supposed to induce the therapeutic effect. It was not even known that substances consist of molecules. Nevertheless, the idea that the effects of a drug would increase with decreasing dose was difficult to believe since it contradicted everyday experience. One German journal wrote in 1810:

> How can Mr. Hahnemann ask us to believe something like this although it contradicts reason and experience? So a glass of wine should have a larger effect if divided into four parts and consumed during a longer period, mixed with a larger amount of water, than if it is consumed undiluted in a short period of time? Nevertheless, the author basis his doctrine on the use of drugs on this basic idea[11] (Kendl, 2017, p. 17).

In the 1830s, new evidence convinced scientists that substances consist of a large number of small units, called molecules. This made the claims of homeopathy extremely implausible. Since then, new scientific insights have repeatedly confirmed this. At the same time, scientific pharmacology has been transformed

[11] Wie kann Herr Hahnemann verlangen, daß man ihm, der Erfahrung und Vernunft zum Trotze, so etwas glauben soll! Ein Glas Wein wirkt also kräftiger, wenn es in vier Theile getheilt und mit einer guten Quantität Wasser vermischt in längeren Zeiträumen konsumiert wird, als wenn man es unvermischt in kurzer Zeit trinkt? Gleichwohl baut der Verf. auf diesen Grundsatz seine Lehre von der Anwendung der Heilmittel.

by a long series of discoveries and innovations, including active substances, dose–response relationships, a multitude of mechanisms of action, and—not least—the use of clinical trials to determine the effects of drugs and other treatments. However, homeopaths have adopted none of this. Instead, they have continued to base their "remedies" on principles that were developed in the 1790s and thoroughly refuted a few decades later.

Thus, to the extent that homeopathy is at all a "holistic" approach, it exemplifies all our three characteristics of failed holism. Its holism is *over-inclusive* since it contains, as crucial and defining elements, fallacious theories about effects of excessively diluted "drugs". Its holism is also remarkably *incomplete* since it excludes almost all the knowledge needed to successfully cure, prevent, and relieve diseases with the help of drugs. Finally, it is an exceptionally *dogmatic* form of holism since it holds on to theories that were thoroughly disproven almost two centuries ago, and refuses to learn from the scientific progress that has taken place during these two centuries.

5 Conclusion

The many failed examples of holism reported in Sects. 2, 3 and 4 may perhaps give the impression that holism is a failed endeavour. But that would be a too rash conclusion. What these examples show, however, is that strivings for holism have many pitfalls and stumbling blocks. Much too often has the term "holism" been appropriated as a means to promote and justify claims that are unproven or outright false. The result has often been accounts of an alleged "whole" that is so misleading and misconceived that it is better described as pseudoholism than as holism in the proper sense.

We need to take back the term "holism" from the charlatans. This means for instance that the impressively inclusive climate models of the IPCC should be recognized as a prime example of (scientific) holism. It also means that "holistic" theories that disregard most of what is known about their subject matter should be disclosed as pseudoholistic.

Our strivings for holism have to be open, critical and pluralistic. We must always be *open* to the possibility that we have missed something important, and then we should include it. We have to be *critical* to what we have already included, and willing to discard components that have turned out to be irrelevant or misleading. Finally, we need to be *pluralistic* in the sense of recognizing that there can be more than one legitimate overarching perspective on a topic. For instance, one holistic approach to mental disease can have its focus on each affected person as a whole individual, whereas another holistic approach to the same topic has its focus on social conditions and society as a whole. There may be good reasons to pursue these perspectives in parallel, rather than trying to merge them into a single larger and possibly unwieldy account.

References

Andrews, E. H. (1984). Creationism in confusion? *Nature, 312*(5993), 396–396.

Aronowsky, L. (2021). Gas guzzling Gaia, or: A prehistory of climate change denialism. *Critical Inquiry, 47*(2), 306–327.

Attena, F. (2016). Limitations of Western medicine and models of integration between medical systems. *The Journal of Alternative and Complementary Medicine, 22*(5), 343–348.

Bartkowski, J. P., & Scott Swearingen, W. (1997). God meets Gaia in Austin, Texas: A case study of environmentalism as implicit religion. *Review of Religious Research, 38*(4), 308–324.

Bockemühl, A. (2007). Und was sagen die Elementarwesen zum Tunnelbau? *Das Goetheanum, 38*, 5.

Brown, R. C. (2018). Resisting moralisation in health promotion. *Ethical Theory and Moral Practice, 21*(4), 997–1011.

Clarke, B. (2017). Rethinking Gaia: Stengers, Latour, Margulis. *Theory, Culture & Society, 34*(4), 3–26.

Clover, A. M. (1979). Another look at the principles of homoeopathy. *British Homoeopathic Journal, 68*(1), 14–20.

Dam, K. (2006, April). Berlin Wall. *Interhomeopathy*. www.interhomeopathy.org/berlin_wall

Ernst, E. (2000) Iridology: Not useful and potentially harmful. *Archives of Ophthalmology, 118*(1), 120–121.

Ferry-Danini, J. (2018). A new path for humanistic medicine. *Theoretical Medicine and Bioethics, 39*, 57–77.

Freeman, J. (2005). Towards a definition of holism. *British Journal of General Practice, 55*(511), 154–155.

Haigh, M. J. (2001). Constructing Gaia: Using journals to foster reflective learning. *Journal of Geography in Higher Education, 25*(2), 167–189.

Hansson, S. O., & Fröding, B. (2021) Ethical conflicts in patient-centred care. *Clinical Ethics, 16*(2), 55–66.

Harrington, A. (1996). *Reenchanted science: Holism in German culture from Wilhelm II to Hitler*. Princeton University Press.

Healey, R. A. (1991). Holism and nonseparability. *The Journal of Philosophy, 88*(8), 393–421.

Holmberg, P., & Vallin, M. (2016, June 27) Granskning av Vidarklinikens avtalsefterlevnad. Uppdrag från Hälso-och sjukvårdsförvaltningen. Helseplan.

Kendl, A. (2017) Scepticismus 1793–1812. *Das Journal der Erfindungen, Theorien und Widersprüche in der Natur- und Arzneiwissenschaft*", *Der Skeptiker, 2017*(1), 13–19.

Kirchner, J. W. (1989). The Gaia hypothesis: Can it be tested? *Reviews of Geophysics, 27*(2), 223–235.

Kirchner, J. W. (2003). The Gaia hypothesis: Conjectures and refutations. *Climatic Change, 58*(1–2), 21–45.

Knipschild, P. (1988). Looking for gall bladder disease in the patient's iris. *British Medical Journal, 297*(6663), 1578–1581.

Lovelock, J. E. (1971). Air pollution and climatic change. *Atmospheric Environment, 5*, 403–411.

Lovelock, J. E. (1972). Gaia as seen through the atmosphere. *Atmospheric Environment, 6*, 579–580.

Lovelock, J. E., & Epton, S. (1975, February 6). The quest for Gaia. *New Scientist*, 304–306.

Lovelock, J. E., & Margulis, L. (1974). Atmospheric homeostasis by and for the biosphere: The Gaia hypothesis. *Tellus, 26*(1–2), 2–10.

Margulis, L., & Lovelock, J. E. (1974). Biological modulation of the Earth's atmosphere. *Icarus, 21*(4), 471–489.

Mezzich, J. (2012). The construction of person-centered medicine and the launching of an international college. *International Journal of Person Centered Medicine, 2*(1), 6–10.

Michaelson, V., Pickett, W., & Davison, C. (2019). The history and promise of holism in health promotion. *Health Promotion International, 34*(4), 824–832.

Münstedt, K., El-Safadi, S., Brück, F., Zygmunt, M., Hackethal, A., & Tinneberg, H.-R. (2005). Can iridology detect susceptibility to cancer? A prospective case-controlled study. *Journal of Alternative and Complementary Medicine, 11*(3), 515–519.

Neider, A. (2019). Belastet unsere Existenz die Erde? *Das Goetheanum, 43*, 10–15.

Noworol, A. M. (2020). What is the value of iridology as a diagnostic tool? *Medicine and Clinical Science, 2*(3), 1–6.

Olsson, L.-E., Ung, E. J., Swedberg, K., & Ekman, I. (2013). Efficacy of person-centred care as an intervention in controlled trials–a systematic review. *Journal of Clinical Nursing, 22*(3–4), 456–465.

Prousky, J. E. (2018). Repositioning individualized homeopathy as a psychotherapeutic technique with resolvable ethical dilemmas. *Journal of Evidence-Based Integrative Medicine, 23*, 1–4.

Salgado, J. (2019). We are reductionists. Should we care? *Biofisica* 15, Sep–Dec, 1–5. https://www.uv.es/biophys/sbe/15/PDFsite/WeAreReductionists.pdf

Schmidt, J. M. (2020). The need for multidisciplinary research within the history and theory of homeopathy. *Homeopathy*. https://doi.org/10.1055/s-0040-1714740

Simon, A., Worthen, D. M., & Mitas, J. A. (1979). An evaluation of iridology. *JAMA, 242*(13), 1385–1389.

Smuts, J. C. (1926). *Holism and evolution*. Macmillan.

Steiner, R. ([1910] 1992) *Die Offenbarungen des Karma. Ein Zyklus von elf Vorträgen gehalten in Hamburg zwischen dem 16. und 28. Mai 1910*. Rudolf Steiner Gesamtausgabe, (Vol. 120). Rudolf Steiner Verlag.

Steiner, R. ([1915] 1980) *Das Geheimnis des Todes. Wesen und Bedeutung Mitteleuropas und die europäischen Volksgeister. Fünfzehn Vorträge, gehalten 1915 in verschiedenen Städten*. Rudolf Steiner Gesamtausgabe, (Vol. 159). Rudolf Steiner Verlag.

Stewart, M. (2001). Towards a global definition of patient centred care. *BMJ, 322*, 444–445.

Supran, G., Oreskes, N. (2017). Assessing ExxonMobil's climate change communications (1977–2014). *Environmental Research Letters, 12*, 084019.

Thibault, G. E. (2019). Humanism in medicine: What does it mean and why is it more important than ever? *Academic Medicine, 94*(8), 1074–1077.

Twentyman, L. R. (1973). *Natrum muriaticum* and our convulsive age. *British Homoeopathic Journal, 62*(3), 186–190.

Vogt, H., Hofmann, B., & Getz, L. (2016). The new holism: P4 systems medicine and the medicalization of health and life itself. *Medicine, Health Care and Philosophy, 19*(2), 307–323.

Whitmont, E. C. (1974). Psycho-physiological reflections on Lachesis. *British Homoeopathic Journal, 63*(4), 234–243.

Whorton, J. C. (1985). The first holistic revolution: Alternative medicine in the nineteenth century. In D. Stalker & C. Glymour (Eds.), *Examining holistic medicine* (pp. 729–748). Prometheus Books.

Widder, R. M., & Anderson, D. C. (2015). The appeal of medical quackery: A rhetorical analysis. *Research in Social and Administrative Pharmacy, 11*, 288–296.

Explanatory Emergence, Metaphysical Emergence, and the Metaphysical Primacy of Physics

Terry Horgan

Are there emergent phenomena in the world, over and above those describable and explainable by an ideally completed physics? This question is only as clear as is the expression 'emergent phenomenon'. I will distinguish three distinct notions of emergence, and three distinct emergence theses that respectively invoke these notions. I will call these theses *strong metaphysical emergence*, *weak metaphysical emergence*, and *physicalist explanatory emergence*. I also will articulate a thesis I call *the metaphysical primacy of physics*, and a thesis I call *the trans-theoretic uniformity of explanation*. I will describe the logical connections among these five theses, and I will discuss some considerations for and against each of them.

Some principal morals will be the following. (1) Strong metaphysical emergence and weak metaphysical emergence both are incompatible with the metaphysical primacy of physics. (2) Physicalist explanatory emergence is not at all incompatible with the metaphysical primacy of physics; on the contrary, the former actually *presupposes* the latter. Hence (3) considerations that count against strong metaphysical emergence, and/or against weak metaphysical emergence, do not count against physicalist explanatory emergence (from (1) and (2)). (4) Physicalist explanatory emergence is incompatible with the trans-theoretic uniformity of explanation. Hence (5) considerations that count against strong metaphysical emergence, and/or against weak metaphysical emergence, do not *favor* the trans-theoretic uniformity of explanation over physicalist explanatory emergence (from (1)–(4)). (6) Physicalist explanatory emergence is considerably more plausible than the trans-theoretic uniformity of explanation.

T. Horgan (✉)
Department of Philosophy, University of Arizona, Tucson, AZ 85721-0027, USA
e-mail: thorgan@email.arizona.edu

S. Wuppuluri and I. Stewart (eds.), *From Electrons to Elephants and Elections*, The Frontiers Collection, https://doi.org/10.1007/978-3-030-92192-7_14

1 LaPlace's Demon as a Philosophical Trope

The recent and current literature in metaphysics and in philosophy of mind contains a bewildering panoply of proposed formulations—often employing specialized and somewhat technical philosophical vocabulary—of various theses whose underlying spirit is essentially the same as the several theses I will formulate here.[1] I propose to avoid deploying such vocabulary, and thereby to avoid internecine debates in philosophy about which technical formulations best capture the ideas they seek to explicate. Instead I will articulate the theses of interest by harnessing some variations on a vivid thought experiment that was used, by the early nineteenth century scholar and polymath Pierre-Simon, marquis de LaPlace, to articulate in a vivid and suggestive way the thesis of *determinism*. In a famous passage of his 1820 book *A Philosophical Essay on Probabilities*, he said:

> An intelligence knowing, at a given instant of time, all forces acting in nature, as well as the momentary positions of all things of which the universe consists, would be able to comprehend the motions of the largest bodies of the world and those of the smallest atoms in one single formula, provided it were sufficiently powerful to subject all the data to analysis. To it, nothing would be uncertain, and both future and past would be present before its eyes.[2]

LaPlace's hypothetical super-intelligent being—who presumably is not supposed to be a part of the cosmos that she is contemplating—nowadays is commonly called *LaPlace's demon*. I will adopt that usage here, with no derogatory connotation attached to the word 'demon'.

2 The Metaphysical Primacy of Physics

Traditional LaPlacean determinism is a special case of the thesis I will call the metaphysical primacy of physics. LaPlacean determinism really involves two separable components—although LaPlace himself did not acknowledge the second component explicitly. One is the idea that the demon is able to calculate the universe's entire *physical history* from a specification of its total physical state at any single moment in time—the physical history being a specification, in terms of the language and concepts of physics, of the universe's total physical state at every moment in time. (LaPlace had in mind that the demon could do the calculation by appeal to fundamental physical laws, e.g., the laws of Newtonian mechanics.) Second is the idea that the demon would somehow be able to know *everything* about the universe—i.e., all the *truths*, as specifiable in terms of any applicable concepts whatever—on the basis of knowing the universe's physical history.

Consider the claim that LaPlace's demon can ascertain the universe's entire physical history on the basis of the conjunction of (1) the laws of physics and (2) a specification of the universe's total physical state at a single moment in time. This is a (time-symmetrically) deterministic version of a thesis that I will call *physical dynamical closure*. I use the word 'closure' to capture the idea that no factors other than those describable by physical dynamical laws ever "intrude from outside," so to speak, on the dynamical evolution of one total physical state of the universe to another.

A more general version of physical dynamical closure is wanted, in order to allow for the possibility that the actual fundamental laws of physics are non-deterministic. (On some competing contemporary interpretations of quantum mechanics—though not all—quantum theory is non-deterministic.) Here is generalized formulation, expressed in terms of the capabilities of LaPlace's demon:

> *Physical dynamical closure*: For any moment in time t, and any pair of possible total-universe physical states Φ and Ω, if Φ is the universe's total physical state at t, then the LaPlacean demon can ascertain, solely on the basis of this fact together with the fundamental laws of physics, the probability (at t) that Ω will be universe's total physical state at the next instant after t.

If the fundamental laws of physics are deterministic in the past-to-future direction, then the special case of physical dynamical closure arises in which the demon always ascertains that probability of Φ being immediately followed by Ω is 1. And if the fundamental physical laws are time-symmetrically deterministic, then the yet-more-special case arises in which the demon also ascertains that the probability of Ω having been immediately preceded by Φ is 1.

What about the second component of LaPlacean determinism, the idea that the demon can ascertain all truths about the universe—however those truths are characterized, and regardless of which concepts might figure in these characterizations—on the basis of knowing the whole physical history of the universe as characterized via the concepts of fundamental physics? I suggest that what is needed here is to explicitly stipulate the following about the LaPlacean demon: she has *unlimited conceptual competence*. In particular, she has full and complete competence for deploying any concepts that human beings ever do deploy or ever could deploy. Let me give this LaPlacean demon a name, with the understanding that doing so records my stipulation of her unlimited conceptual competence; I will call her *Demonea*.

Demonea's conceptual competence includes, *inter alia*, full mastery of all the concepts that figure in any of the "special sciences" that humans invoke to understand and explain their world, sciences other than fundamental physics—chemistry, geology, biology, psychology, sociology, and so forth. Importantly, however, her full mastery of the *concepts* that figure in the various special sciences does not by itself give her knowledge any *substantive truths* involving such concepts. On the contrary, any such substantive truths are among the body of truths which, according to the second component of LaPlacean determinism, she is able to ascertain, on the basis of her knowledge of the universe's full physical history, by exercising her mastery of special-science concepts. For example, she can ascertain that—and explain why—a particular collection of atoms, structurally interconnected in a specific way that she can describe in terms of the concepts of fundamental physics, constitutes a *cell*, or a *multi-cellular organism*, of an *elephant*, or an *economic transaction*, etc. Likewise, she can ascertain that—and explain why—various entities of the kinds posited by the various special sciences instantiate the various kinds of properties invoked by those various special sciences—e.g., a given cell's *subdividing*, a given multicellular organism's *degree of fitness*, an elephant's *occupying an ecological niche*, an economic transaction's being a *monetary exchange*, etc. The only knowledge she needs in order to ascertain and explain all such phenomena, over and her knowledge of the full physical history of the universe (as characterizable via the concepts of fundamental physics), is the kind of knowledge that constitutes full conceptual competence in the deployment of special-science concepts. Roughly, the latter is knowledge about *what it takes to be* a cell, or an elephant, or an ecological niche, etc.

So the second component of LaPlacean determinism, which I will call *strong physical cosmic determination*, can now be formulated as a thesis about the capacities of the LaPlacean demon Demonea, this way:

> *Strong physical cosmic determination*: Demonea can ascertain, solely on the basis of her unlimited conceptual competence together with knowledge of the total physical history of the universe, all truths about the cosmos.

And the thesis I will call *the metaphysical primacy of physics* is the conjunction of two theses: physical dynamical closure and strong physical cosmic determination. As a thesis about Demonea, it is this:

> *The metaphysical primacy of physics*: (1) For any moment in time t, and any pair of possible total-universe physical states Φ and Ω, if Φ is the universe's total physical state at t, then Demonea can ascertain, solely on the basis of this fact together with the fundamental laws of physics, the probability (at t) that Ω will be universe's total physical state at the next instant after t; and (2) Demonea can ascertain, solely on the basis of her unlimited conceptual competence together with knowledge of the total physical history of the universe, all truths about the cosmos.

Henceforth I will call this the *MPP* thesis, for short.

Traditional LaPlacean determinism is a special case of the MPP thesis, comprising both strong physical cosmic determination and a thesis about the universe's physical history that is logically stronger than—and therefore entails—physical dynamical closure: viz., the thesis that universe's physical history is time-symmetrically deterministic, in a manner that accords with deterministic laws of physics.

Although the latter thesis is controversial because of the vexed issue of whether quantum theory is deterministic, many contemporary philosophers—myself included—would be very sympathetic to a thesis in the spirit MPP, and would contend that such a thesis is well supported by contemporary scientific knowledge. I suspect that the same goes for many scientists.

3 Strong Metaphysical Emergence

The thesis I call strong physical emergence (for short, *SME*) rejects both components of the MPP thesis. Contrary to the thesis of strong physical cosmic determination, SME asserts that there are some phenomena in the cosmos that Demonea could not ascertain simply on the basis of her unlimited conceptual competence and her knowledge of the universe's physical history. And, contrary to the thesis of physical dynamical closure, SME asserts that such phenomena sometimes intrude upon the universe's dynamical evolution from one total physical state to another—so that it is not always the case that if Demonea knows that the universe is in total physical state Φ at a time *t*, then Demonea can correctly ascertain, on the basis of knowing this and knowing the laws of physics, the actual probabilities (at *t*) of the various potential total physical states to which the universe might evolve at the next moment of time.

Are there phenomena in the world that might plausibly be regarded as strongly metaphysically emergent? One candidate is rational human mental activity, and the potential contributions of such activity to physical phenomena in the brain which themselves causally instigate action. For instance, the early twentieth century British philosopher C. D. Broad entertained strong emergentism about mentality—without overtly embracing it—in his 1925 book *The Mind and Its Place in Nature*. Here is a representative passage:

> [T]he facts...suggest that what the mind does in voluntary action, if it does anything, is to lower the resistance of certain synapses and to raise that of others. The result is that the nervous system follows such a course as to produce the particular movement which the mind judges to be appropriate at the time. On such a view the difference between reflex, habitual, and deliberate actions for the present purpose becomes fairly plain. In pure reflexes the mind cannot voluntarily affect the resistance of the synapses concerned, and so the action takes place in spite of it. In habitual action it deliberately refrains from interfering with the resistance of the synapses, and so the action goes on like a complicated reflex. But it *can* affect these resistances if it wishes, though often only with difficulty; and it is ready to do so if it judges this to be expedient. Finally, it may lose the power altogether. This would be what happens when a person becomes a slave to some habit, such as drug-taking (Broad, 1925, pp. 112–113).

On this conception of mentality, Demonea will not be able to ascertain what judgments are being made at a time *t* by a creature with mentality, even if she knows the whole physical history of the universe and therefore knows everything physical that is going (at *t*) in the creature's brain. (This goes contrary to the thesis of strong physical cosmic determination.) And, if the creature is making a judgment (at *t*) in a way that voluntarily lowers the electrical resistance of certain neural synapses and/or raises the electrical resistance of others, then this external mental intrusion into the universe's physical dynamical evolution will prevent Demonea from being able to ascertain the actual probability that accrues (at *t*) to the neural motor-controlling processes that will next occur in the creature's brain and will then generate a particular bodily movement. (This goes contrary to the thesis of physical dynamical closure.)

But in the absence of any direct positive evidence from neuroscience that neurons in situ ever behave in ways not fully explainable by physico-chemical laws—and there is no such evidence!—many contemporary scientists and philosophers (including myself) will consider very implausible the idea that conscious mentality can intrude and intervene in this way, or in any other physical-closure violating way, upon the physico-chemical workings of the central nervous system.[3] Vastly less radical is the alternative possibility that reasons-responsive cognition is *physically implemented by* nervous-system processes which, although much more complex than those subserving pure reflexes, nonetheless always operate in ways that are explainable, at least in principle, by physico-chemical laws.

This point generalizes. Strong metaphysical emergence is quite a radical doctrine, and does not mesh well with contemporary science.

4 Weak Metaphysical Emergence

The thesis I call weak metaphysical emergence (for short, *WME*) departs less radically from MPP than does SME. Like MPP and unlike SME, WME affirms physical dynamical closure. Like SME and unlike MPP, WME denies strong physical cosmic determination. WME also includes a sub-thesis I will call *weak* physical cosmic determination, which I will formulate presently.

The leading idea of WME is that even though physical dynamical closure obtains, nevertheless there are certain phenomena in the cosmos that Demonea would not be able to ascertain and explain just by exercising her unlimited conceptual competence together with her knowledge of the universe's complete physical history. One candidate phenomenon is what philosophers call "phenomenal consciousness," involving those kinds of experiences such that there *something it is like*, for the experiencing subject, to undergo such experiences. The nineteenth century physician and physiologist Emil Heinrich du Bois Reymond, the co-discoverer of nerve-action potential and the developer of experimental electrophysiology, expressed well why phenomenal consciousness can easily seem to elude physico-chemical explanation. He wrote:

> What conceivable connection is there between certain movements of certain atoms in my brain on one side, and on the other the original, indefinable, undeniable facts: 'I feel pain, feel lust; I taste sweetness, smell the scent of roses, hear the sound of organ, see redness'... It is entirely and forever incomprehensible why it should make a difference how a set of carbon, hydrogen, nitrogen, oxygen, etc. atoms are arranged and move, how they were arranged and moved, how they will be arranged and will move. It is in no way intelligible how consciousness might emerge from their coexistence.[4]

An advocate of WME concerning phenomenal consciousness is apt to claim that the full class of fundamental, unexplainable, laws of nature includes not only the fundamental laws of physics, but also certain "bridge laws" asserting that whenever a creature instantiates a certain specific complex physical property P, the creature simultaneously instantiates a certain specific phenomenal mental property M. If indeed there are such fundamental and unexplainable physical-to-phenomenal bridge laws, then Demonia would need to know these laws, in addition to knowing the fundamental laws of physics, in order to ascertain all the truths about the cosmos on the basis of its total physical history. Otherwise, she would not be able to ascertain where and when various phenomenal-consciousness properties are instantiated in the cosmos.

Claims about morality are another candidate phenomenon. For example, the British philosopher G. E. Moore, in the influential book Moore (1903), argued that there are certain fundamental and unexplainable principles asserting that whenever a certain specific non-normative "natural" property P is instantiated, a certain moral-normative "non-natural" property M (e.g., the property *intrinsic goodness*) is then-and-there instantiated as well. Moore held that such natural-to-moral connection principles could not *possibly* be false—not even in a cosmos in which the prevailing laws of nature are different from the natural laws that prevail in our own cosmos. But he also held that these connection principles are not mere conceptual truths, i.e., truths (e.g., "All bachelors are unmarried") that are knowable purely by virtue of understanding the concepts they deploy. Rather, he maintained, knowledge of the connection principles requires a special cognitive faculty of "moral intuition," possession of which goes beyond mere conceptual competence in deploying the concepts that figure in these principles. If indeed there are such fundamental and unexplainable natural-to-moral connection principles, then Demonia would need to know these principles, in addition to knowing the fundamental laws of physics, in order to ascertain all the truths about the cosmos on the basis of its total physical history. Otherwise, she would not be able to ascertain where and when various moral properties are instantiated in the cosmos.

So the existence of fundamental, unexplainable, laws or principles beyond those of physics would undermine the thesis of strong cosmic determination. Nevertheless, the following thesis of *weak* physical cosmic determination might still be true:

> *Weak physical cosmic determination*: Demonea, solely on the basis of the combination of (i) her unlimited conceptual competence, (ii) knowledge of the total physical history of the universe, and (iii) knowledge of whatever fundamental and unexplainable laws or principles there are (if any) beyond those of physics, can ascertain all truths about the cosmos.

Thesis WME, then, is the conjunction of the following three sub-theses: (1) physical dynamical closure, (2) weak physical cosmic determination, and (3) the denial of strong cosmic determination.

WME departs from MPP considerably less radically than does SME. Nevertheless, the many contemporary philosophers (myself included) who are sympathetic to a thesis like MPP will be inclined to resist WME. Concerning phenomenal consciousness, for example, one potential line of resistance is to argue (1) that although experience-based phenomenal-property *concepts* are very different from scientific-theoretical concepts, nevertheless the *properties* that are designated by experience-based phenomenal-property concepts are identical to properties that are designated by certain scientific-theoretical concepts, and (2) and that these property-identity facts neither need, nor are susceptible to, explanation.[5] And concerning matters of morality, for example, one potential line of resistance is to argue that moral judgments are really a kind of action-guiding attitude that does not actually posit moral properties at all—much less moral properties that are weakly metaphysically emergent.[6]

5 Trans-theoretic Explanatory Uniformity Versus Physicalist Explanatory Emergence

Scientific explanations are given not just by physics, but also in the various "special sciences": chemistry, biology, psychology, sociology, etc. How are special-science explanations related to explanations in physics? I will now set forth two competing general answers to this question; I call these alternative theses, respectively, trans-theoretic explanatory unity (for short, *TTEU*) and physicalist explanatory emergence (for short, *PEE*). Importantly, both theses presuppose the MPP thesis. This means that any considerations that might favor MPP over either SME or WME should be neutral regarding TTEU versus PEE.

Each of the sciences posits certain proprietary kinds of entities and properties that figure centrally in its explanations—e.g. (depending on the science) subatomic particles, atoms, molecules, planets, stars, cells, multicellular organisms, animals of various species, democratic social systems, etc. And each science also posits certain proprietary properties (including relational properties), instantiable by the proprietary entities in its domain. In philosophy of science, such proprietary entities and properties are often called "natural kinds"—an expression I will adopt here. Using this terminology, thesis TTEU can be formulated this way:

> *Trans-theoretic explanatory uniformity*: Every special-science natural kind is identical to some natural kind of physics, and hence every special-science explanation is reformulable, at least in principle, as a physics-level explanation.

On the conception of inter-theoretic relations embodied in TTEU, the pertinent physics-level natural kinds that are identical to special-science natural kinds need not be *fundamental* natural kinds of physics, and typically will not be. Rather, the natural-kind entities posited by the special sciences typically will be physically complex, composed of numerous fundamental physics-level natural-kind entities that are interconnected in some way that is specifiable in terms of physics-level natural-kind properties. Similarly, the pertinent natural-kind properties posited by the special sciences typically will be physically complex too, while also being specifiable in terms of how various fundamental physics-level natural-kind properties are instantiated by the various fundamental physics-level entities that compose a given special-science natural-kind entity. (These physics-level natural-kind properties will be like *temperature* as a property of gases—only typically much more complex. The temperature of a gas is the mean kinetic energy of its composite molecules.)

On this conception of explanation, special-science explanations do not afford an explanatory understanding of phenomena that differs, in principle, from the kind of explanatory understanding that could be achieved, in principle, via purely physics-level explanations. However, the qualifier 'in principle' is quite important here, because of human cognitive limitations. An ordinary human could not begin to hold in mind—to "cognitively survey"—purely physics-level specifications of the various natural-kind entities and natural-kind properties that are posited by the special sciences. Special-science concepts have the advantage of leaving out the physics-level details, thereby allowing the explanatory connections among phenomena to be characterized in less specific, albeit cognitively surveyable, ways. Special-science explanations are "dumbed-down" versions of physics-level explanations that are available in principle but not in practice to creatures like ourselves.

It would be quite different, of course, for Demonea. Given her enormous intellectual prowess, she would have no trouble at all in understanding the pertinent explanations that invoke special-science natural kinds under their fundamental-physics characterizations rather than invoking those natural kinds under their dumbed-down special-science characterizations. For her, therefore, special-science explanations deploying special-science natural-kind concepts would provide no extra, no distinctive, explanatory understanding at all. She simply would have no need for special-science concepts—even though her unlimited conceptual mastery would include mastery of these concepts. And she would understand why, perforce, these concepts are explanatorily useful to mere humans.

I turn now to thesis PEE, physicalist explanatory emergence. This conception of how explanations in the special sciences relate to those of physics presupposes thesis MPP, the metaphysical primacy of physics. (The modifier 'physicalist' is used here to indicate this.) Beyond that, what PEE asserts is that TTEU is false. So PEE can be formulated this way:

> *Physicalist Explanatory Emergence*: (1) Thesis MPP is true; furthermore, (2.i) it not the case that every special-science natural kind is identical to some natural kind of physics, and hence (2,ii) it is not the case every special-science explanation is reformulable, at least in principle, as a physics-level explanation.

The philosopher Jerry Fodor, in his important 1974 paper "Special Sciences (Or: The Disunity of Science as a Working Hypothesis)," gave a vigorous defense of a conception of inter-theoretic relations among the sciences that amounts in essence to PEE. Here is the key passage (which also displays Fodor's rhetorical verve):

> The reason it is unlikely that every natural kind corresponds to a physical natural kind is just that (a) interesting generalizations can often be made about events whose physical descriptions have nothing in common, (b) it is often the case the *whether* the physical descriptions of the events subsumed by these generalizations have anything in common is, in an obvious sense, entirely irrelevant to the truth of the generalizations, or to their interestingness, or to their degree of confirmation, of, indeed, to any of their epistemologically important properties, and (c) the special sciences are very much in the business of making generalizations of this kind.
>
> I take it that these remarks are obvious to the point of self-certification; they leap to the eye as soon as one makes the (apparently radical) move of taking the special sciences at all seriously. Suppose, for example, that Gresham's 'law' really is true. (If one doesn't like Gresham's law, then any true generalization of any conceivable future economics will probably do as well.) Gresham's law says something about what will happen in monetary exchanges under certain conditions. I am willing to believe that physics is general *in the sense that it implies that any event which consists of a monetary exchange* (hence any event that falls under Gresham's law) *has a true description in the vocabulary of physics and in virtue of which it falls under the laws of physics*. But banal considerations suggest that a description which covers all such events must be wildly disjunctive. Some monetary exchanges involve strings of wampum. Some involve dollar bills. And some involve signing one's name to a check. What are the chances that a disjunction of physical predicates which covers all these events...expresses a physical natural kind? In particular, what are the chances that such a

> predicate forms the antecedent or consequent of some proper law of physics? The point is that monetary exchanges have interesting things in common; Gresham's law, if true, says what one of these interesting things is. But what is interesting about monetary exchanges is surely not their commonality under *physical* description. A natural kind like a monetary exchange could turn out to be co-extensive with a physical natural kind; but if it did, that would be an accident on a cosmic scale.
>
> In fact, the situation…is still worse than the discussion thus far suggests…. [The physical natural-kind predicate] P would have to cover not only all the systems of monetary exchange that there *are*, but also all the systems of monetary exchange that there *could be*; a law must succeed with the counterfactuals. What physical predicate is a candidate for 'P'…? (Fodor, 1974, 103-104).

To my mind, Fodor's argumentation in this passage is quite compelling. A key point he was stressing is that that there are numerous, physically highly diverse, ways in which a physical event can instantiate a special-science natural-kind property like *being a monetary exchange*—that is, numerous ways in which that property can be *physically realized* by some complex property that is characterizable in the language of physics. The phenomenon of multiple, highly diverse, physical realizability of special-science natural kinds appears to be ubiquitous in the special sciences. And the ubiquity of this phenomenon appears to be an excellent reason to reject TTEU and instead embrace PEE.

Return again to Demonea. Suppose that she contemplates a hypothetical scenario that is described solely in certain special-science terminology, without any details about physical realization being specified—say, a hypothetical scenario involving numerous monetary exchanges within a money-supply situation involving the availability of both "bad money" (i.e., money in a form that is not intrinsically valuable, like paper currency) and "good money" (i.e., money in a form that is intrinsically valuable, like gold bars). Suppose that in this hypothetical scenario, the economic transactions increasingly deploy the bad money, whereas the good money increasingly gets horded rather than being used in exchanges. According to Gresham's law, "bad money drives out good money" in an economic system. So if Gresham's law is true, then its applicability to the envisioned hypothetical situation *explains* why the economic transactions increasingly employ the bad money. Demonea thereby acquires some genuine explanatory understanding of what is happening in the scenario. Nevertheless, since no details have been provided or stipulated about matters like how, in the scenario, natural kinds like *bad money*, *good money*, and *hording behavior* get physically realized, Demonea simply lacks adequate information about this situation to possess any particular physics-level explanation of what is going on. Thus, the special-science explanation in terms of Gresham's law gives her a distinctive and autonomous kind of explanatory understanding, because the explanation at hand is not itself reformulable as a physics-level explanation.

Fodor did not use the terms 'emergent' or 'emergence' in his defense of the explanatory autonomy of the special sciences. But one thinker who does, and whose usage is very much in the spirit of PEE, is the theoretical biologist and complex systems researcher Stuart Kauffman. Here is a compendium of pertinent remarks from Kauffman (1995), with some added boldfacing by me:

> I suspect that the fate of all adaptive systems in the biosphere—from single cells to economics—is to evolve to a natural state between order and chaos, a grand compromise between structure and surprise... (p. 15).
>
> The hope...is to characterize classes of properties of systems that...are typical or generic and do not depend on the details.... Not knowing the details we nevertheless can build theories that seek to explain the generic properties... (p. 17).
>
> After all, what we are after here is not necessarily detailed prediction, but explanation. We can never hope to predict the exact branchings of the tree of life, but we can uncover powerful laws that predict and explain their general shape. I hope for such laws. I even dare to hope that we can begin to sketch some of them now. For want of a better general phrase, I call these effects a search for a theory of **emergence**.... (p. 23)
>
> I believe that life itself is an **emergent** phenomenon, but I mean **nothing mystical** by this.... [S]ufficiently complex mixes of chemicals can spontaneously crystallize into systems with the ability to collectively catalyze the network of chemical reactions by which the molecules themselves are formed. Such collective autocatalytic sets sustain themselves and reproduce.... A set of molecules either does or does not have the property that it is able to catalyze its own formation and reproduction from some simple food molecules. **No vital force or extra substance** is present in the **emergent**, self-reproducing, whole. But the...collective system is alive. **Its parts are just chemicals** (p. 24).
>
> Whether we are talking about organisms or economies, surprisingly general laws govern adapative processes on miltipeaked fitness landscapes. These general laws may account for phenomena ranging from the burst of the Cambrian explosion in biological evolution, where taxa fill in from the top down, to technological evolution, where striking variations arise early and dwindle to minor improvements.... The best exploration of an evolutionary space occurs at a kind of phase transition between order and disorder, when populations begin to melt off the local peaks they have become fixated on and flow along ridges toward distant regions of higher fitness (p. 27).

Kauffman, in saying that he means nothing "mystical" by his talk of emergence, that "no vital force or extra substance" is present in an emergent whole, and that parts of the whole "are just chemicals," presumably means to eschew theses like SME or WME. And in saying that emergent laws governing adaptive systems invoke properties that are "typical or generic and do not depend on the details," and that such laws are "surprisingly general" and "may account for phenomena ranging from the burst of the Cambrian explosion...to techological evolution," he presumably means to invoke the idea that the properties he calls "emergent" are radically multiply realizable physically and hence are not identical to any physics-level natural-kind properties. So in effect, he is advocating physicalist explanatory emergence.

In light of Sects. 1, 2, 3 and 4, the upshot of the present section is this: (1) PEE is much more plausible on its face than TTEU, because of the apparent ubiquity of special-science natural kinds that are multiply physically realizable in significantly divergent ways; (2) advocates of TTEU over PEE therefore bear a very heavy burden of proof; and (3) since PEE not only is compatible with MPP but actually presupposes it, considerations that might be given in favor of MPP (and thus, against SME and WME) evidently do not provide support for TTEU over PEE.

6 Strong Reduction, Weak Reduction, and the Unity of Science

Are the special sciences all reducible to physics, perhaps in a hierarchical manner with some special sciences being progressively reducible to others which themselves are reducible in turn, with the whole hierarchy ultimately being reducible to physics? This question is only as clear as the intended meaning of the expressions 'reduction' and 'reducible'. On the basis of Sects. 1, 2, 3, 4 and 5, I will here distinguish two distinct notions of inter-theoretic reduction, which I will call *strong* reducibility and *weak* reducibility, respectively.

Let strong reducibility of a theory T_2 to a theory T_1 be the thesis that every natural-kind entity or property posited by T_2 is identical to some natural-kind entity or property posited by T_1. And let weak reducibility of a special-science theory T to physics be the thesis that whenever there is an instance of natural-kind entities or properties posited by T, (1) there is a physics-level explanation (perhaps enormously complex) of why this is so, and (2) there also is a physics-level explanation (perhaps enormously complex) of why, in this instance, the pertinent natural-kind entities and properties conform to the laws of T.

Suppose—as I myself believe—that thesis MPP (the metaphysical primacy of physics) and thesis PEE (physicalist explanatory emergence) are both true. Then the special sciences all are weakly reducible to physics, because MPP entails that conditions (1) and (2) of my definition of weak reducibility to physics are both satisfied. (Demonea could always provide the physics-level explanations required for weak reducibility, even if such explanations would often be too complex to be tractable for humans.) But the special sciences are not, in general, strongly reducible to physics, because the heterogenous physical multiple realizability of special-science natural-kind entities and properties prevents these natural kinds from being identical to physics-level natural kinds.

In philosophy of science, the notion of inter-theoretic reduction typically has been understood as being essentially what I am calling *strong* reducibility.[7] The doctrine that the special sciences all are ultimately reducible to physics is often called "the unity of science"; it was articulated and defended in the influential article Oppenheim and Putnam (1958). Jerry Fodor was overtly attacking this hypothesis in his 1974 paper cited and quoted above (indeed, he was negatively riffing on Oppenheim and Putnam with his subtitle "The Disunity of Science as a Working Hypothesis").

But the words 'reduction' and 'reducibility' also can be used—not inappropriately—to advert to what I am calling *weak* reducibility, and to sometime efforts in science to spell out physical details about how certain special-science natural kinds are physically realized in certain specific individuals—e.g., how certain psychology-level natural kinds are physically realized in human brains. Consider, for instance, these remarks in Fodor's paper, with some added boldfacing by me:

> It seems to me (to put the point quite generally) that the classical construal of the unity of science has really misconstrued the *goal* of **scientific reduction**. The point of **reduction** is *not* primarily to find some natural kind predicate of physics co-extensive with each natural

> kind predicate of a **reduced** science. It is, rather, to explicate the physical mechanisms whereby events conform to the laws of the special sciences. I have been arguing that there is no logical or epistemological reason why success in the second of these projects should require success in the first, and that the two are likely to come apart *in fact* whenever the physical mechanisms whereby events conform to a law of the special sciences are heterogenous (Fodor, 1974, p. 107).

The scientific project that Fodor here dubs 'scientific reduction' involves what I am calling *weak* reducibility. In effect, Fodor was saying in this passage that the classical construal of the unity of science has mistakenly supposed that the pertinent kind of unity involves strong reducibility, when it really involves only weak reducibility.

So, are the special sciences all reducible to physics? The answer I favor is "Yes and no." Yes, because they are all weakly reducible to physics. But no, because it's not the case that they are all strongly reducible to physics; instead, a crucial feature of the special sciences is physicalist explanatory emergence.

Notes

1. See, for instance, Horgan (1982, 1984, 1993), McLaughlin (1995, 2008), and further references cited in these papers.
2. LaPlace (1820). The quoted translation is from Ducasse (1958).
3. There is also an intelligibility problem looming here, not unlike the one about mind–body interaction that Descartes confronted (because of his conception of the mind as a non-physical "substance" lacking spatial location). Broad himself seems to have come to think that the kind of mental/physical agentive control he had gestured at is not really intelligible. Cf.Broad (1934). *The Mind and Its.*
4. Emil du Bouis-Remond (1872), quoted in translation by Bieri (1995).
5. See, for instance, McLaughlin (2007), Horgan (forthcoming).
6. See, for instance, Horgan and Timmons (2006).
7. Sometimes reducibility to physics is described not as involving identities between special-science natural kinds and physics-level natural kinds, but rather as involving "bridge laws" each of which asserts nomic coextensiveness (rather than outright identity) between a special-science natural kind and a specific physics-level natural kind. But if such nomic-coextension bridge laws are supposed to be fundamental and unexplainable, then the resulting picture would be a version of weak metaphysical emergentism—which is surely not intended by advocates of strong reducibility of the special sciences to physics. And if the so-called bridge laws are *not* supposed to be fundamental and unexplainable, then considerations of theoretical parsimony strongly suggest that these "laws" are best construed as expressing outright natural-kind identities.

   If one claims that the identity construal should be resisted because the actual relation between special-science natural kinds and physical natural kinds is physical realization (rather than identity), then the contention that special-science natural kinds are nomically *coextensive* with physical natural kinds faces exactly the same problem as does the contention that special-science natural kinds are *identical* with physical natural kinds—viz., the fact that special-science natural kinds, ubiquitously, seem to be multiply and heterogeneously physically realizable.

References

Bieri, P. (1995). Why is consciousness puzzling? In T. Metzinger (Ed.), *Conscious experience* (pp. 45–60). Imprint Academic Schoningh.
Broad, C. D. (1925). *The mind and its place in nature.* Harcourt, Brace, and Company.
Broad, C. D. (1934). *Determinism, indeterminism, and libertarianism: An inaugural lecture.* Cambridge University Press.
du Bouis-Remond, E. (1872). Uber die Grenzen Des Naturerkennens (About the limits of natural knowledge). In his *Vortrage uber Philosophie und Gesellschaft.* Meiner, 1974.
Ducasse, C. J. (1958). Determinism, freedom, and responsibility. In S. Hook (Ed.), *Determinism and freedom in the age of modern science* (pp. 160–169). Collier
Fodor, J. (1974). Special sciences (or: The disunity of science as a working hypothesis). *Synthese, 28*, 97–115.
Horgan, T. (1982). Supervenience and microphysics. *Pacific Philosophical Quarterly, 63*, 29–43.
Horgan, T. (1984). Supervenience and cosmic hermeneutics. *Southern Journal of Philosophy, 22*(Supplement), 19–38.
Horgan, T. (1993). From supervenience to superdupervenience: Meeting the demands of a material world. *Mind, 102*, 555–586.
Horgan, T. Forthcoming. Physical grounding and the intrinsic nature of phenomenal consciousness. In G. Rabin (Ed.), *Grounding and consciousness.* Oxford University Press.
Horgan, T., & Timmons, M. (2006). Cognitivist expressivism. In T. Horgan & M. Timmons (Eds.), *Metaethics after Moore* (pp. 255–298). Oxford University Press.
Kauffman, S. (1995). *At home in the universe: The search for the laws of self-organization and complexity.* Oxford University Press.
LaPlace, P. Simon, de Marquis (1820). *Theorie analytique des probabilities* (3rd ed.). Coursier.
McLaughlin, B. (1995). Varieties of supervenience. In E. Savellos, & S. Yalcin (Eds.), *Supervenience: New essays* (pp. 16–59). Cambridge University Press.
McLaughlin, B. (2007). Type materialism for phenomenal consciousness. In M. Velmans, & S. Schneider (Eds.), *The Blackwell companion to consciousness* (pp. 431–444). Blackwell.
McLaughlin, B. (2008). Supervenience. *Stanford encyclopia of philosophy.*
Moore, G. E. (1903). *Principia ethica.* Dover Publications.
Oppenheim, P., & Putnam, H. (1958) Unity of science as a working hypothesis. *Minnesota Studies in the Philosophy of Science, 2*, 3–36.

Contextual Emergence: Constituents, Context and Meaning

Robert C. Bishop

1 Ontological Reductionism or Radical Emergence?

When confronted with order and novel phenomena we experience in the world, scientists will be inclined to invoke some combination of laws, constraints, and mechanisms among other determining factors. Scientific explanations typically rely on the already established order that scientists have worked out. Hence, it's not unusual to find some scientists and philosophers endorsing reductionism: The belief that elementary particles and forces determine everything in the world (e.g., biology, geology, or your reactions to this chapter).

Of course, it's the case that elementary particles and forces underlie everything in the material world.[1] Without them you wouldn't be here reading this! Reductionism, however, is a stronger thesis than just remarks about what lies at the bottom of physical reality, so to speak. It's a claim about the ultimate structure of nature being completely determined by the complex play of elementary particles and forces. This strong claim has troubling implications: What is the status of ethics, moral responsibility, free will, creativity, meaning? Are these merely subjectively experienced effects of the underlying action of particles and forces?

Under reductionism, it's far from clear there is room for genuine qualities of human agency and ethics if nature is structured reductively (Bishop, 2010). This is because reductionism presupposes the causal closure of physics, where all physical effects are fully determined by elementary particles and forces–the arrow of determination points from the bottom up, so to speak. However, some of the necessary conditions for a behavior to qualify as an action that might be described as responsible or free are:

[1] Or, for quantum mechanics, fields and forces since particles are thought to be excitations in fields in quantum field theory. For ease of exposition, I'll stick with particles instead of fields but this will affect nothing I say going forward.

Based on a talk given at ETH Zurich for the ETH Critical Thinking initiative.

R. C. Bishop (✉)
Wheaton College, 501 College Avenue, Wheaton, IL 60187-5501, USA
e-mail: robert.bishop@wheaton.edu

S. Wuppuluri and I. Stewart (eds.), *From Electrons to Elephants and Elections*,
The Frontiers Collection, https://doi.org/10.1007/978-3-030-92192-7_15

- A person has an immediate awareness of their activity (physical or mental) and of that activity's aim or goal.
- A person has some from of direct control over or guidance for their behavior.
- A person's behavior must be seen as intentional under some description.
- A person's actions are explainable in terms of their intentions, desires and beliefs.

If the causal closure of physics is true, then people's actions do not genuinely flow out of reasons, motives, beliefs, and so forth. Instead, all behaviors flow ultimately from the play of forces on elementary particles. This implies that none of the conditions for action–much less some form of free action–can be satisfied. Under reductionism, all behaviors are ultimately mapped onto the dynamics of elementary particles and forces whether these behaviors are taking place in human societies or not. What we think of as human "free choice"or "responsible action" is simply the law-like play of elementary particles and forces.

These consequences for human morality, agency, and meaning can be made vivid by thinking about mathematician John Conway's Game of Life.[2] The game involves a grid of squares, where some are colored black (living) while others are colored white (dead). It uses one simple rule determining under what conditions black squares will switch to white and vice versa along with an initial state to be specified at $t = 0$ (the initial configuration or pattern of black/white squares). Then, let the system evolve according to the rule and whatever happens happens. The rule plus the initial condition for a configuration of black and white squares determines when and where every pattern arises in the game, how patterns behave, how long they persist, and so forth. Beautiful patterns such as gliders that flock like birds can appear and "fly" across the screen. Yet, all the patterns are simply the result of the one rule plus the initial condition. In the actual world under reductionism, the particles and forces of elementary particle physics play the role of the one rule. Given the initial start of the universe and this rule, everything—including your choice to read this essay—is just the product of the forces and the initial configuration of the particles at the beginning.

On the other hand, it's not unusual to find some scientists and philosophers endorsing emergence: The belief that physics, chemistry, biology, geology, physiology and (by implication at least) human behavior are more than just the action of elementary particles and forces. However, there are two basic kinds of emergence usually discussed in the reduction/emergence debates. The first is *radical emergence*. This is the belief that novel laws, properties and processes come from nowhere in the sense that they aren't based on elementary particles and forces. The second kind is *epistemic emergence*. This is the belief that chemical, biological and social phenomena, say, are not explainable or derivable from elementary particles and forces. This failure could be due to some kind of epistemic limitation such as a lack of computational or descriptive power.

Epistemic emergence is rather banal because it's ubiquitous. As a matter of scientific practice and necessity we are often forced to use higher-level descriptions for chemical, biological and social phenomena because elementary particle physics

[2] https://playgameoflife.com/.

descriptions make no sense of higher-level situations (Anderson, 1972). Nevertheless, this means that ontologically, nature can be reductively structured while we're still forced to use higher-level descriptions. Epistemic emergence is consistent with a reductively structured world. Human morality and freedom are ontologically still just the play of the elementary particles and forces.

While epistemological emergence is uncontroversial, radical emergence is very problematic as it's irrelevant to the sciences. Much of our scientific work is aimed at unifying and connecting phenomena. But radical emergence implies that nature is disunified and disconnected. Apparently, there are some kinds of brute laws bridging between elementary particles and forces, on the one hand, and biological and physiological phenomena on the other. Or there are brute novel properties and processes independent of elementary particles and forces. Consciousness, free will and morality might be some of these brute entities.

For scientists, and many philosophers this kind of radical ontological emergence is a dead end. It's useless for making sense of the order and stability of our world as well as being irrelevant to the sciences. Radical emergence is both mysterious and *prima facie* inconsistent with our experience of a coherent, ordered world and seems more like giving up on the project of understanding our experience and making sense of how human agency and meaning fits into an ordered picture of reality.

2 A False Forced Choice

In trying to understand nature as far as we can scientifically, reductionism may appear to be the only viable alternative between these two options for an ontological picture of nature. The candidate for ontological reduction is relatively clear in the debates: our most fundamental theory of physics. In contrast, radical emergence as an alternative for an ontological account of the world's order is obscure, mysterious or irrelevant. Ontological reduction appears to win by default. Yet, this "win by default" leaves us with troubling questions about human morality and agency as mentioned earlier. Some physicists, such as Weinberg (1993), Laughlin (2005) and Anderson (2011), don't appear to be particularly troubled by these questions.

This "win by default" situation is an example of a forced choice fallacy. Such a fallacy occurs whenever the options for choice are reduced so that viable options for debate are left out. If you've ever been in an argument that amounted to "I'm right, so you must be wrong!", then you've likely experienced a forced choice. There may be viable alternatives left out of the argument, one of which is that both people are wrong!

One reason the false forced choice in reductionism debates seems so compelling is the basic assumptions that both reductionism and strong emergence share. One of these assumptions is that nature is organized in a fixed hierarch. There are clearly defined layers from lower-level laws and entities to higher levels. This hierarchical structuring is often treated as something pre-given or ontologically fixed. So, the hierarchy from elementary particles to atoms to molecules to stars and planets to galaxies, or from physics to chemistry, to geology to biology to animals to societies, has somehow always been fixed.

A further shared assumption is foundationalism, the belief that only elementary particle physics contains rock-bottom fundamental laws and entities. This implies the physical facts of elementary particle physics are fully ontologically autonomous because they depend on nothing else for their existence and are determined by nothing else.

Combined together, these two assumptions entail the world exhibits a well-ordered objective hierarchy ranging from elementary particle physics on up to larger spatial and temporal scales. The arrow of determination moves upward from the smallest spatial and temporal scales to the larger. Hence, the reductionist view that elementary particles and forces ultimately determine geology, biology, and politics. The only difference between reductionists and radical emergentists is that the latter think new (physical or metaphysical) laws, causal powers or entities must be added to the set of fundamental or basic lowest-level facts to explain the existence of novel emergent phenomena.

What if there is a viable account of ontological emergence that clarifies the order and structure of the world while illuminating the genuine emergence of chemical, biological and social phenomena, an account that is missing from most of the typical reduction/emergence debates? Such an account would need to satisfy the following three desiderata:

1. No violations of the inherent unity of the world.
2. Never appeal to new brute laws or causes when finding that no reductive explanation exists.
3. Assume neither foundationalism nor that the world is an ordered hierarchy of reified levels.

3 Contextual Emergence: Between Ontological Reductionism and Radical Emergence

Although physics is often thought of as being a reductionistic science, it actually offers an exemplary pattern for interlevel relations that is a viable alternative to the forced-choice framing just described. This pattern has been called *contextual emergence* by those of us developing this account (Bishop, 2005; Bishop and Atmanspacher, 2006; Bishop, 2019). This account of emergence has its roots in the work of chemical physicist Primas Primas (1977, 1983, 1998), and has been developed with an eye towards complexity and quantum mechanics.

Contextual emergence's distinctiveness can be seen in the following framework organizing the three alternatives:

- **Ontological Reduction**: Properties and behaviors in a lower level or underlying domain (including its laws) offer by themselves both necessary and sufficient conditions for properties and behaviors at a higher level.

- **Contextual emergence**: Properties and behaviors in a lower level or underlying domain (including its laws) offer some necessary but no sufficient conditions for properties and behaviors at a higher level. Higher levels or target domains provide the needed extra conditions.
- **Radical emergence**: Properties and behaviors in a lower level or domain (including its laws) offer neither necessary nor sufficient conditions for properties and behaviors at a higher level.

Contextual emergence focuses on the most crucial conditions for making the existence, stability, and persistence of phenomena and systems possible termed *stability conditions*. It's too often the case that these stability conditions are taken for granted though we can see them when we know how to look for them. Such conditions often are involved in or imply inherently irreducibly multiscale relations. So in this sense, it's not surprising that scientific explanations often are multiscale (Bishop et al., in press).

Contextual emergence describes situations where the constituents and laws belonging to the supposed fundamental level or underlying domain of reality contribute some necessary but no sufficient conditions for entities and properties in the target domain, or higher level. It's the stability conditions that provide the needed sufficiency, yet these latter conditions are never found at the underlying level or domain (Atmanspacher and Bishop, 2007; Bishop, 2019). For instance, the domain of elementary particles contributes some of the necessary conditions for the existence of the properties and behaviors of water parcels, collections of roughly an Avogadro's number of H_2O molecules. Nonetheless, the existence of elementary particles and their laws do not guarantee that large-scale phenomena such as wine flowing from a bottle or Rayleigh-Bénard convection will exist. The basic laws of elementary particles and forces establish the possibilities for there to be fluids of many kinds, motions of many kinds, and so on. Yet, by themselves the laws and forces of elementary particle physics don't enable the existence of specific fluids and motions. These laws and forces only fix the total set of possibilities.

For wine to flow from a bottle requires the selection of a specific bottle, the opening of the bottle, the tilting of the bottle for the wine to flow into a glass (not to mention the process of cultivating soil and grapes, fermenting, aging in barrels under controlled conditions, etc.) To get convection requires several contingent conditions: a specific type of fluid, a physical space the fluid occupies, a temperature differential in the presence of gravity, action of all fluid molecules acting on all fluid molecules, and so forth. It's among the latter where the needed stability conditions exist bringing about convection and none of these conditions are fixed by elementary laws, particles and forces.[3]

Lasers are another example of phenomena that are physically possible, yet are never naturally realized in the actual world apart from appropriate stability conditions. Einstein was the first to propose the physical possibility of the stimulated

[3] For details, see (Bishop, 2019), Sect. 4.1. For worked out examples of the contextual emergence of temperature and molecular structure, see Sects. 4.2 and 4.3, respectively.

coherent emission process among atoms that would eventually lead to lasers (Einstein, 1916, 1917). He demonstrated the possibility that a large number of atoms in identical excited states producing a single photon of the right energy can stimulate one atom to emit another photon which stimulates another atom which emits another photon which stimulates another atom, and so forth, leading all the atoms to release their excess energy in a sustained cascade. Nevertheless, the stability conditions for this process (e.g., preparation of a collection of atoms all in the relevant excited states, precise triggering of the population inversion returning the atoms to their ground states producing photons, a designed optical cavity trapping the photons and enhancing the stimulation of more photons, appropriate isolation from the wider environment), although physically possible, aren't given by elementary particles and forces (nor are these stability conditions given by the atoms, photons and their interactions alone).

This is the pattern of contextual emergence in physics. There are no new forces that come out of nowhere. Everything can be explained in terms of physics and engineering that we understand, so no radical emergence. Nonetheless, the underlying domain of particles and forces don't contain all the conditions necessary and sufficient for flowing wine, convection or lasers to actually happen in our world.

4 A Broad Pattern

The contextual emergence pattern of relationships extends beyond physics. Chemists recently created a novel hydrocarbon structure that can be useful to quantum computing applications (Ma et al., 2017). Creating this compound required both controlled laboratory conditions and bringing together particular chemical compounds that would only happen intentionally with the goal of producing a novel sp^2-carbon lattice material.[4] In other words, an intentionally designed chemical environment provides the stability condition to form the carbon lattice and defects (similar to the case for lasers).

The creation process allows for manipulation of topological defects resulting in superior spintronic performance for quantum computing applications. Such intentional large-scale control allows qubits designed by this process to be put into any arbitrary superposition desired, for instance. Moreover, by changing the chemical environment for the creation of the carbon lattice and defects, chemists can remove compounds one at a time that lead to destruction of the superposed state, another example of contextual emergence (Lombardi et al., 2019).

Consider an example from biology. The placement of hair and feather follicles on animal bodies is highly ordered. However, the genome doesn't direct location of individual follicles. It turns out that the genetics controlling follicle generation is shaped by larger-scale mechanical forces determining typical distance between neighboring follicles (Shyer et al., 2017).

---

[4] An sp^2 bond is between one s-orbital with two p-orbitals.

The developing skin has two layers, an epithelial layer forming the epidermis lying on top of the dermis. The underlying dermis contracts locally causing the epithelial cells to bend forming slight dome shapes where follicles form. These dermal contractions cause compressive stress in the overlying epithelial cells. Two interesting things happen from this compressive stress. First, the dermal contractions break the symmetry of the random distribution of the overlying epithelial cells ordering them spatially. Second the mechanical forces activate the genetic machinery producing follicles. Otherwise, the genetic machinery for follicle production never turns on.

The upshot is that larger-scale mechanical forces provide a stability condition sufficient to trigger follicle formation in an ordered array. The underlying genetic machinery provides some of the necessary conditions for patterned follicle formation. The larger-scale dermal contraction provides the additional necessary and sufficient stability condition for patterned follicle formation.

From this example, you might be thinking that gene behavior is context-dependent. This fact has been well established (Buchberger et al., 2019; Javierre et al., 2016; Lübbe and Schaffner, 1985). Hence, more generally, the stability conditions for gene behavior isn't found in the individual genes themselves. The latter only contribute some of the necessary conditions for their own behavior.

The same contextual emergence pattern can be found in ecology. For instance, in the open ocean, theories focusing only on the body size of marine animals predict that marine ecosystems should be bottom-heavy with more plants and animals at the lowest levels than the highest. We indeed see many such bottom-heavy structured marine ecosystems. In contrast to these theories, observations reveal that many marine ecosystems are top-heavy. What makes the difference is that complex food webs function as stability conditions for maintaining top-heavy marine ecosystems (Woodson et al., 2020). One implication is that bottom-heavy marine ecosystems appear to be the adverse effect of human activity (e.g., overfishing) which has disturbed the complex food web in these ecosystems leading to a loss of top-heavy structure. Destroy the complex food web stability condition and the ecosystem suffers a devastating reordering.

Turning to cognitive science, we find the same contextual emergence pattern. Work on modeling insect motion shows that neural-network models for motion based on vision require an environmental context for coherent, meaningful motion to be possible (Webb, 2020). In other words, the presence of stable large-scale objects in an environment are a stability condition for the possibility of meaningful motion. Neural-network models of insect motion track fast-moving small objects (e.g., for predation) through the rise and fall of stimulus intensity with respect to a fixed background of large-scale objects. Even estimates of speed depend on the larger-scale environmental surroundings. This includes the sky as a stability condition for insect navigation (Homberg et al., 2011). Nonvisual cues for avoidance in mosquitoes, for example, also depend on surfaces of the larger-scale environment as a stability condition for changes in fluid flow patterns to indicate an object or surface is nearby, sensing and responding to minute pressure changes in that flow induced by coming near an object or surface (Toshiyuki et al., 2020). One sees something similar in bird flocking behavior. For instance, Jackdaws change their flocking behavior—the rules they use to organize group behavior—based on the larger environmental context and self-propelled particle models can only reproduce this behavior by taking the external environment into account (Ling et al., 2019).

Finally, let's consider machine learning, a particular sub-branch of artificial intelligence that has generated a lot of recent interest. Machine learning typically involves designing a neural network model and training that model on a set of data relevant for a specific application such as facial recognition. The training data set represents an environment the machine learning model is exposed to and is to "learn" from.

What research shows is that the performance of machine learning models is very sensitive to their training data sets. The type and quality of the learning environment greatly determines the model performance in its target environment. A particularly concerning example of machine learning systems trained on a large data set of faces is the failure to recognize the faces of black females in the actual world (Hardesty, 2018). The lack of a sufficiently representative sample of faces in the training data set led to failure in the target task of facial recognition. In machine learning, the architecture of the neural network provides some of the necessary conditions for its performance. The data environments for learning and target tasks provide the rest of the necessary and sufficient conditions for actual performance of machine learning models.

This is a particularly interesting example because there are actually three different interrelated levels: (1) The hardware level that provides some necessary but no sufficient conditions for its own functionality. (2) The software level at which the neural network model is implemented providing the rest of the necessary and sufficient stability conditions for specific hardware function. And (3) the learning and target environments that provide the needed necessary and sufficient stability conditions for performance of the machine learning model.

5 Does Contextual Emergence Do the Job?

What about the three desiderata for a viable form of ontological emergence? Earlier I stipulated that such an emergence account should have the following features:

1. No violations of the inherent unity of the world.
2. Never appeal to new brute laws or causes when finding that no reductive explanation exists.
3. Assume neither foundationalism nor that the world is an ordered hierarchy of reified levels.

How does contextual emergence fulfill these criteria? In all of the examples I have given, the contextual emergence pattern does't invoke any new mysterious brute forces that come out of nowhere. Nor does the pattern depend on some pre-given ordered hierarchy of levels of reality. For instance, in Rayleigh-Bénard convection some of the stability conditions arise from the emergence of a dynamics on a larger spacial and temporal scale than that of the individual interactions of fluid parcels with their nearest neighbors. Nor does the contextual emergence pattern rely on smaller-scale factors determining the outcomes at larger scales. Again, Rayleigh-Bénard convection illustrates that the smaller-scale factors can't even determine their own behaviors apart from larger-scale conditions, particularly the emergent larger-scale dynamics (Bishop, 2019).

Contextual emergence doesn't fit reductionism, nevertheless every example fits our expectations for scientific explanations in terms of known phenomena. For instance, in the case of the patterning of feathers and fur, contextual emergence shows us how the smaller-scale genes and the larger-scale dermal contractions work together to produce astounding phenomena such as the striking pattern of the Peregrine Falcon or the mundane covering of the human body by hair.

Note as well that the contextual emergence of the phenomena in all the examples doesn't arise from some underlying set of "governing laws" in contrast to the Game of Life. Whether it's convection, novel hydrocarbon structures, follicle formation, complex food webs, insect vision or facial recognition, the phenomena along with their explanations and predictions have no dependence on fundamental laws other than as providing some of the necessary conditions for the existence of said phenomena. Furthermore, there is no dependence on some pre-existing ordered hierarchy. Hence, we have an ordered world without the need to posit any new brute laws or causes aside from the starter kit for the universe and we have no need of either foundationalism or a reified hierarchy of levels–the spatial and temporal scales can arise contingently.

One might still wonder if everything is actually already built into this initial starter kit just like in the Game of Life. This is the reductionist intuition. Yet, the universe's starter kit is more subtle and interesting than the reductionist intuition allows. There is a universal stability condition formed by the set of Kubo-Martin-Schwinger (KMS) conditions on stable states that have the property of temperature.[5] This stability condition is part of the universe's starter set and means that once particles are around, such as quarks and gluons, they necessarily conform to this stability condition. The KMS conditions aren't part of elementary particle physics, but characterize a context into which elementary particle physics comes to expression. Basically all of elementary particle physics dynamics is shaped by these universal KMS conditions.

As another example, consider the electromagnetic field from our most fundamental theory: quantum electrodynamics. As soon as a quantum electromagnetic field emerged in the early universe, it had what is called a far-field stability condition structuring the field and its related electromagnetic force. This far-field stability condition guarantees that there will be both quantum and classical electromagnetic fields and forces with the properties physicists study.[6]

Both the KMS and far-field stability conditions are well understood by physicists. There are no mysteries, here; rather, in the beginning there was contextual emergence with some important stability conditions in the universe's starter set. These stability

[5] For details, see Bishop (2019).

[6] For details, see Gervais and Zwanziger (1980); Buchholz (1982) and discussion in Bishop (2019).

conditions are just as fundamental to everything that happened in the universe's history as the most "fundamental" laws, particles and forces. Even at the beginning of the universe there is no genuine analog to the foundationalism and reified hierarchy of levels found in the Game of Life. One of the beautiful things is that the starter set of stability conditions led to the contextual emergence of new stability conditions that led, in turn, to more contextual emergence and so forth. This is why we find the contextual emergence pattern to be pervasive in the world on multiple scales.

The ontological pattern looks like this. Elementary particles and forces provide some necessary conditions for molecular structure, while the concrete chemical context provides the rest of the necessary and sufficient conditions for the molecules chemists explore and work with in the laboratory. In turn, molecular chemistry provides some of the necessary conditions for the behavior of cells, while the concrete cellular context provides the rest of the necessary and sufficient conditions for existence and behavior of cells. Likewise, cells provide some of the necessary conditions for organs and their function, while the concrete context of the organs in an organism and their environments provide the rest of the necessary and sufficient conditions for the existence and function of organs. And so forth, where the emergence of stability conditions defining new contexts become part of the set of necessary conditions of a domain or level underlying a newly emergent domain or level.

This is the interleaving or interlevel pattern we saw in all the examples given earlier. Such an interlocking pattern can be described somewhat more formally as *relative onticity* (Atmanspacher and Kronz, 1999). As just described, an underlying level or domain provides some the necessary conditions for higher levels or target domains. The former provide an ontological basis for the epistemic access of the phenomena and properties at the higher levels or target domains. In turn, the latter levels and target domains provide an ontological basis for epistemic access to even higher levels and target domains. In this way we can make sense of the autonomy of the special sciences (e.g., biology, geology, social science) and the fact that our epistemic access and explanatory purchase in the special sciences is in terms of the properties and processes made possible by the emergent contexts and the stability conditions defining their domains.

Hence, there is no "absolutely fundamental" ontology at rock bottom providing *the* reductive key to the structure of our world and explanations in the sciences. This absence of a reductive bedrock isn't because there are in-practice difficulties with working out scientific explanations based on such a bedrock level (e.g., elementary particles and forces). Rather, it's because of the ineliminable role of stability conditions defining contexts.

Someone might object that this all amounts to smuggling everything in through background conditions. This kind of objection comes from how physicists and mathematicians solve the equations we use to model the physical world. We can't solve our equations without specifying some initial conditions—the initial configuration of particles and forces, say—and some boundary conditions—constraints on the particles and forces. The invoking of such conditions—particularly the boundary conditions—are thought of as just background to the "real action." What contextual emergence teaches us is that the constraints represented in stability conditions aren't "background" that we can stuff into boundary conditions and forget about. Instead, stability conditions and the contexts they define are *just as important to the action as the particles and forces*. We don't put things into the background because they are irrelevant in scientific investigations. We put things in the background to focus on the question at hand. Such a distinction between background and question at hand doesn't imply that what is relegated to background at the moment is unimportant to questions we're currently exploring.

The upshot is this: The only sense in which elementary particles and forces are 'fundamental' is two-fold. First, the domain of elementary particle physics contributes some necessary conditions for the existence of molecules, moles and mountains in a way that is universal. If there were no elementary particles and forces there would be no molecules, moles or mountains.

Second, the laws of the elementary particle physics domain are fundamental in the sense that they delimit the space of physically possible events.[7] The most 'fundamental' laws function as constraints on what can possibly happen, but it's contexts through stability conditions that structure or determine the particular conditions for specific kinds of events to happen (e.g., convection, wine pouring from bottles, feather patterns). Think of laws as establishing the physical space of possibilities and stability conditions as gatekeepers in the space of physical possibilities for concrete events to occur in the world. No new laws "pop out of nowhere" as the underlying laws of elementary particle physics contribute some of the necessary conditions for any emergent laws. There is unity and order to the world.

This means there is no sense in which elementary particles and forces provide sufficient conditions for molecules, moles, or mountains to exist and act as they do, or for wine pouring from bottles, and feather patterns. The concrete contexts and constraints into which elementary particles and forces come to expression are just as important as elementary particles and forces. Instead of the Game of Life picture, where there is a set of basic building blocks at the lowest level driving everything else that happens, you can think of the contextual emergence picture as one where wholes and "parts" are the fundamental furniture of the world. There is "bottom up" and "top down" as well as "in between" and "all around." [8]

6 The Big Picture and Meaning

Now let's return to the framing of reduction-emergence debates as a forced choice between plausible-sounding reductionism and implausible-sounding radical emergence. This framing leaves out the important role stability conditions defining contexts play in the origin and existence of phenomena. In other words it leaves out at least one viable alternative for ontological emergence: Contextual emergence! The

[7] For more technical discussion, see Bishop (2019); Bishop et al. (in press).

[8] Further discussion of contextual emergence, examples and objections can be found in Bishop (2019); Bishop et al. (in press).

general pattern of contextual emergence is a combination of bottom-up and top-down features—more generally interlevel relations—through which complex phenomena arise. This is a pattern of interlevel or interrelational influence that isn't captured in reductionism or radical emergence. If a debate is framed in such a way that a viable alternative is missing, then we will not be able to think well about the issues involved in the debate. Nor is the debate capable of being concluded in a sound fashion.

Why is contextual emergence missing? It's a pattern for the structuring of reality that often goes unnoticed until we make explicit what is typically left implicit: The role of contexts and the stability conditions characterizing those contexts providing the constraints for how elementary particles and forces come to concrete expression in the world. The more we are aware of all the factors that go into the concrete actualization of the wide range of possibilities provided by the basic laws of nature, the more we can see that the ontological structure of reality is more subtle than the reductionist claims as well as more interesting!

Moreover, the forced choice between reductionism and radical emergence not only leaves out important possibilities, it also has consequences for bigger questions and concerns we have regarding consciousness, free will, ethics, creativity and meaning. For instance, as noted earlier in a reductively structured world human activity, thought, and consciousness turn out to be effects of the complex play of elementary particles and forces. There is no genuine morality, just the consequences of elementary particle physics. Consciousness and motivations are just the accidental byproduct of the complex play of elementary particles and forces, nothing more. The causal closure of physics rules out any impact conscious awareness, intentionality and ethical commitments can have on human action. Even the creative thought and work to develop the standard model of elementary particle physics—something physicists think is very meaningful—is just the effect of the very physics of that model!

On the other hand, a radical emergence world would leave us with consciousness, thought and morality as totally separate from the material world. Not only would there be no discernible relationship between thought and meanings and the material world, it would be an absolute mystery as to why there is any coherence between thought and action that has an impact on material objects! How is it that the thoughts of the physicist about the world move her pencil and paper in meaningful ways? How is it that school district busing plans constrain metal buses to move accomplishing purposeful ends?

Neither ontological reductionism nor radical emergence represent meaningful homes for the kind of open and responsive intellectual engagement exhibited by the theorizing of scientists and philosophers in our attempts to understand the world of our experience. To formulate and contemplate the standard model of particle physics or a reductionist understanding of the world requires human intellectual engagement and creativity as well as the realization that these are meaningful activities—in other words, our thoughts and motives genuinely make a difference in the world. Neither ontological reductionism nor radical emergence make sense of the human search for truth and meaning, the exercise of genuine choice, moral reflection or even the creative effort that went in to formulating ontological reductionism and radical emergence in the first place. Advocates of these two ontological positions think that their own position is meaningful and that engaging in debate about these positions is meaningful activity even though the implication of both positions is that such activity ultimately isn't meaningful or understandable.

In contrast to both reductionism and radical emergence, a contextual emergence world is one full of significance, an ordered world where scientific investigation and the ordinary business of living find a meaningful home. There is no causal closure of physics in a contextually emergent world; as we've seen, even elementary particle physics is subject to contextual constraints that aren't part of elementary particle physics. Moreover, a contextually emergent world is a more unified and understandable world than that of radical emergence while making room for genuine consciousness, thought, free will, moral responsibility and meaning that is connected to the rest of the world. This makes a contextual emergence world a meaningful world, a world which we can understand little-by-little, that we can navigate in sensible ways, and where our experience of both order and novelty are at home rather than being foreign interlopers or meaningless riders on elementary particles and forces. All the creative thought and hard work physicists put into developing the standard model of particle physics—one of the great human achievements—was the meaningful and worthwhile activity they took it to be all along.

References

Anderson, P. W. (1972). More is different: Broken symmetry and the nature of the hierarchical structure of science. *Science, 177*(4047), 393–396.

Anderson, P. W. (2011). *More and different: Notes from a thoughtful curmudgeon*. Singapore: World Scientific.

Atmanspacher, H., & Bishop, R. C. (2007). Stability conditions in contextual emergence. *Chaos and Complexity Letters, 2*(2/3), 139–150.

Atmanspacher, H., & Kronz, F. (1999). Relative onticity. In H. Atmanspacher, A. Amann, & U. Müller-Herold (Eds.), *On quanta, mind and matter*. Hans Primas in context. Fundamental Theories of Physics series (pp. 273–294). Kluwer.

Bishop, R. C. (2005). Patching physics and chemistry together. *Philosophy of Science, 72*, 710–722.

Bishop, R. C. (2010). Free will and the causal closure of physics. In R. Y. Chiao, M. L. Cohen, A. J. Leggett, W. D. Phillips, & C. L. Harper, Jr., (Eds.), *Visions of discovery: New light on physics, cosmology, and consciousness* (pp. 601–611). Cambridge University Press.

Bishop, R. C. (2019). *The physics of emergence*. IOP Concise Physics Series, San Rafael: Morgan & Claypool Publishers.

Bishop, R. C., & Atmanspacher, H. (2006). Contextual emergence in the description of properties. *Foundations of Physics, 36*(12), 1753–1777.

Bishop, R. C., Silberstein, M. D., & Pexton, M. (in press). *Emergence in context: A science-first approach to metaphysics*. Oxford University Press.

Buchberger, Elisa, Reis, Micael, Ting-Hsuan, Lu., & Posnien, Nico. (2019). Cloudy with a chance of insights: Context dependent gene regulation and implications for evolutionary studies. *Genes, 10*(7), 492. https://doi.org/10.3390/genes10070492

Buchholz, Detlev. (1982). The physical state space of quantum electrodynamics. *Communications in Mathematical Physics, 85*, 49–71.

Einstein, A. (1916). Strahlungsemission und -absorption nach der quantentheorie. *Verhandlungen der Deutschen Physikalischen Gesellschaft, 18*, 318–323.

Einstein, A. (1917). *Quantentheorie der strahlung. Physikalische Zeitschrift, 18*, 121–128.

Gervais, J.-L., & Zwanziger, D. (1980). Derivation from first principles of the infrared structure of quantum electrodynamics. *Physics Letters, 94B*(3), 389–393.

Hardesty, L. (2018). Study finds gender and skin-type bias in commercial artificial-intelligence systems. *MIT News*, 11 February 2018. https://news.mit.edu/2018/study-finds-gender-skin-type-bias-artificial-intelligence-systems-0212. Accessed on January 5, 2020.

Homberg, U., Heinze, S., Pfeiffer, K., Kinoshita, M., & el Jundi, B. (2011). Central neural coding of sky polarization in insects. *Philosophical Transactions of the Royal Society B, 366*, 680–687. https://doi.org/10.1098/rstb.2010.0199

Javierre, B. M., Burren, O. S., Wilder, S. P., Kreuzhuber, R., Hill, S. M., Sewitz, S., Cairns, J., Wingett, S. W., Várnai, C., Thiecke, M. J., Burden, F., Farrow, S., Cutler, A. J., Rehnström, K., Downes, K., Grassi, L., Kostadima, M., Freire-Pritchett, P., Wang, F., & The BLUEPRINT Consortium, (2016). Lineage-specific genome architecture links enhancers and non-coding disease variants to target gene promoters. *Cell, 167*(5), 1369–1384. https://doi.org/10.1016/j.cell.2016.09.037

Laughlin, R. B. (2005). *A different universe: Reinventing physics from the bottom down*. New York: Basic Books.

Ling, H., McIvor, G. E., Westley, J., van der Vaart, K., Vaughan, R. T., Thornton, A., & Ouellette, N. T. (2019). Behavioural plasticity and the transition to order in jackdaw flocks. *Nature Communications, 10*, 5174. https://doi.org/10.1038/s41467-019-13281-4

Lombardi, F., Lodi, A., Ji, M., Junzhi, L., Michael, S., Akimitsu, N., Myers, W. K., Müllen, K., Xinliang, F., & Lapo, B. (2019). Quantum units from the topological engineering of molecular graphenoids. *Science, 366*(6469), 1107–1110. https://doi.org/10.1126/science.aay7203

Lübbe, A., & Schaffner, W. (1985). Tissue-specific gene expression. *Trends in Neuroscience, 8*, 100–104. https://doi.org/10.1016/0166-2236(85)90046-3

Ma, J., Liu, J., Baumgarten, M., Fu, Y., Tan, Y.-Z., Schellhammer, S. K., Ortmann, F., Cuniberti, G., Komber, H., Berger, R., Müllen, K., & Feng, X. (2017). A stable saddle-shaped polycyclic hydrocarbon with an open-shell singlet ground state. *Angewandte Chemie International Edition, 56*, 3280–3284. https://doi.org/10.1002/anie.201611689

Nakata, T., Phillips, N., Simoes, P., Russell, I. J., Cheney, J. A., Walker, S. M., & Bomphrey, R. J. (2020). Aerodynamic imaging by mosquitoes inspires a surface detector for autonomous flying vehicles. *Science, 368*(6491), 634–637. https://doi.org/10.1126/science.aaz9634

Primas, H. (1977). Theory reduction and non-Boolean theories. *Journal of Mathematical Biology, 4*, 281–301.

Primas, H. (1983). *Chemistry, quantum mechanics and reductionism: Perspectives in theoretical chemistry*. Number 24 in Lecture Notes in Chemistry. Springer-Verlag, Berlin, second, corrected edition.

Primas, H. (1998). Emergence in exact natural sciences. *Acta Polytechnica Scandinavica Ma, 91*, 83–98.

Shyer, A. E., Rodrigues, A. R., Schroeder, G. G., Kassianidou, E., Kumar, S., & Harland, R. M. (2017). Emergent cellular self-organization and mechanosensation initiate follicle pattern in the avian skin. *Science, 357*(6353), 811–815. https://doi.org/10.1126/science.aai7868

Webb, B. (2020). Robots with insect brains: A literal approach to mechanistic explanation provides insight in neuroscience. *Science, 368*(6488), 244–245. https://doi.org/10.1126/science.aaz6869

Weinberg, S. (1993). *Dreams of a final theory*. New York: Pantheon Books.

Woodson, C. B., Schramski, J. R., & Joye, S. B. (2020). Food web complexity weakens size-based constraints on the pyramids of life. *Proceedings of the Royal Society B, 287*(20201500), 1–7. https://doi.org/10.1098/rspb.2020.1500

Mathematics/Theoretical Physics

Contents, Contexts, and Basics of Contextuality

Ehtibar N. Dzhafarov

1 Contents, Contexts, and Random Variables

The word *contextuality* is used widely, usually as a synonym of *context-dependence*. Here, however, contextuality is taken to mean a special form of context-dependence, as explained below. Historically, this notion is derived from two independent lines of research: in quantum physics, from studies of existence or nonexistence of the so-called hidden variable models with context-independent mapping (Bell, 1964, 1966; Clauser et al., 1969; Clauser & Horne, 1974; Cabello, 2008; Cabello et al., 1996; Fine, 1982; Kurzynski et al., 2014; Kochen & Specker, 1967; Klyachko et al., 2008),[1] and in psychology, from studies of the so-called selective influences (Dzhafarov, 2003; Dzhafarov & Gluhovsky, 2006; Dzhafarov & Kujala, 2010, 2016; Kujala & Dzhafarov, 2008; Sternberg, 1969; Townsend, 1984; Zhang & Dzhafarov, 2015). The two lines of research merged relatively recently, in the 2010s (Dzhafarov & Kujala, 2012a, b, 2013a, b, 2014a, b), to form an abstract mathematical theory, Contextuality-by-Default (CbD), with multidisciplinary applications (Bacciagaluppi, 2015; Basieva et al., 2019; Cervantes & Dzhafarov, 2017a, b, 2018, 2019, 2020; de Barros et al., 2016; Dzhafarov, 2016, 2017, 2019, in press; Dzhafarov & Kujala, 2014c, 2015, 2016, 2017a,b, 2018, 2020; Dzhafarov et al., 2015a,b, 2016a, 2017, 2020a,b,

[1] Here, I mix together the early studies of nonlocality and those of contextuality in the narrow sense, related to the Kochen-Specker (1967) theorem. Both are special cases of contextuality.

E. N. Dzhafarov (✉)
Department of Psychological Sciences, Purdue University, 703 Third Street, West Lafayette, IN 47907-2081, USA
e-mail: ehtibar@purdue.edu

S. Wuppuluri and I. Stewart (eds.), *From Electrons to Elephants and Elections*, The Frontiers Collection, https://doi.org/10.1007/978-3-030-92192-7_16

2021; Kujala & Dzhafarov, 2015, 2016, 2019; Kujala et al., 2015; Jones, 2019; Zhang & Dzhafarov, 2016).[2]

The example I will use to introduce the notion of contextuality reflects the fact that even as I write these lines the world is being ravaged by the Covid-19 pandemic, forcing lockdowns and curtailing travel.

Suppose we ask a randomly chosen person two questions:

q_1 : would you like to take an overseas vacation this summer?
q_2 : are you wary of contracting Covid-19?

Suppose also we ask these questions in two orders:

c^1 : first q_1 then q_2
c^2 : first q_2 then q_1

To each of the two questions, the person can respond in one of two ways: Yes or No. And since we are choosing people to ask our questions randomly, we cannot determine the answer in advance. We assume therefore that the answers can be represented by *random variables*. A random variable is characterized by its *identity* (as explained shortly) and its *distribution*: in this case, the distribution means responses Yes and No together with their probabilities of occurrence.[3]

One can summarize this imaginary experiment in the form of the following *system of random variables*:

R_1^1	R_2^1	$c^1 = q_1 \to q_2$
R_1^2	R_2^2	$c^2 = q_2 \to q_1$
$q_1 =$ "vacation?"	$q_2 =$ "Covid-19?"	system $\mathcal{C}_{2(a)}$

. (1)

This is the simplest system that can exhibit contextuality (as defined below). The random variables representing responses to questions are denoted by R with subscripts and superscripts determining its identity. The subscript of a random variable in the system refers to the question this random variable answers: e.g., R_1^1 and R_1^2 both answer the question q_1. The superscript refers to the *context* of the random variable, the circumstances under which it is recorded. In the example the context is the order in which the two questions are being asked. Thus, R_2^1 answers question q_2 when this question is asked second, whereas R_2^2 answers the same question when it is is asked first.

The question a random variable answers is generically referred to as this variable's *content*. Contents can always be thought of as having the logical function of questions, but in many cases other than in our example they are not questions in the colloquial

[2] The theory has been revised in two ways since 2016, the changes being presented in Dzhafarov & Kujala (2017b) and Dzhafarov et al. (2017).

[3] I set aside the intriguing issue of whether responses Yes and No may be indeterministic but not assignable probabilities.

meaning. Thus, a q may be one's choice of a physical object to measure, say, a stone to weigh, in which case the stone will be the content of the random variable R_q^c representing the outcome of weighing it (in some context c). Of course, logically, this R_q^c answers the question of how heavy the stone is, and q can be taken to stand for this question.

Returning to our example, each variable R_q^c in our set of four variables is identified by its content ($q = q_1$ or $q = q_2$) and by its context ($c = c^1$ or $c = c^2$). It is this double-identification that imposes a structure on this set, rendering it a *system* (specifically, a *content-context system*) of random variables. There may be other variable circumstances under which our questions are asked, such as when and where the questions were asked, in what tone of voice, or how high the solar activity was when they were asked. However, it is a legitimate choice not to take such concomitant circumstances into account, to ignore them. If we do not, which is a legitimate choice too, our contexts will have to be redefined, yielding a different system, with more than just four random variables. The legitimacy of ignoring all but a select set of contexts is an important aspect of contextuality analysis, as we will see later.

The reason I denote our system $\mathcal{C}_{2(a)}$ is that it is a specific example (the specificity being indicated by index a) of a *cyclic system of rank* 2, denoted $\mathcal{C}_2$. More generally, cyclic systems of rank n, denoted $\mathcal{C}_n$, are characterized by the arrangement of n contents, n contexts, and $2n$ random variables shown in Fig. 1.

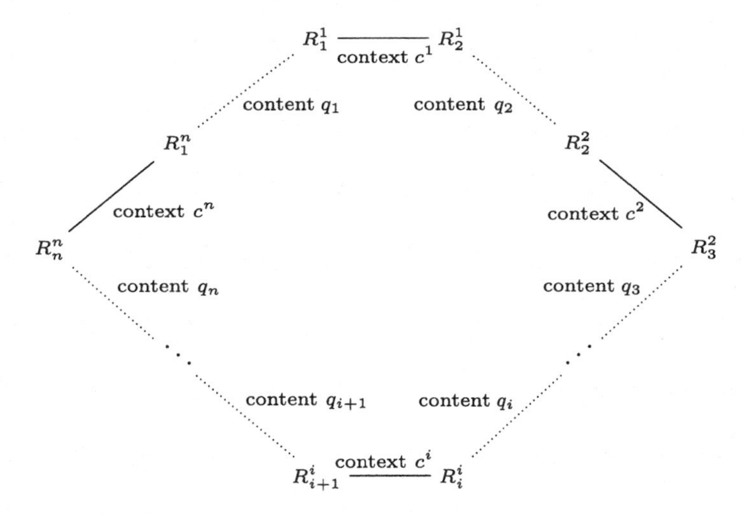

Fig. 1 A cyclic system of rank n

A system of the C_2-type is the smallest such system (not counting the degenerate system consisting of R_1^1 alone):

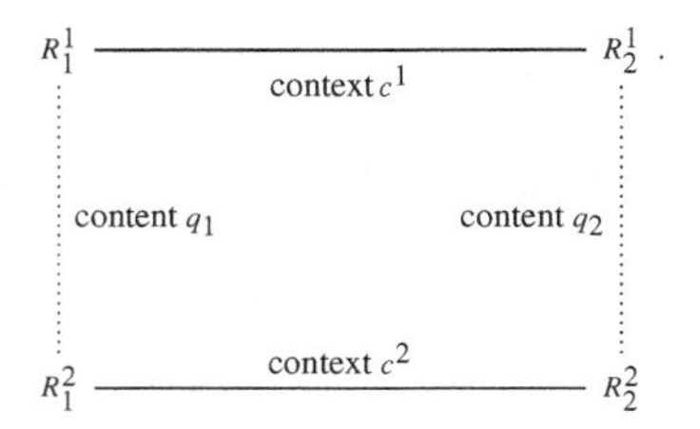

What else do we know of our random variables? First of all, the two variables within a context, (R_1^1, R_2^1), or (R_1^2, R_2^2), are *jointly distributed*. By the virtue of being responses of one and same person, the values of these random variables come in pairs. So it is meaningful to ask what the probabilities are for each of the joint events

$$\begin{array}{l} R_1^1 = +1 \text{ and } R_2^1 = +1, \\ R_1^1 = +1 \text{ and } R_2^1 = -1, \\ R_1^1 = -1 \text{ and } R_2^1 = +1, \\ R_1^1 = -1 \text{ and } R_2^1 = -1, \end{array}$$

where $+1$ and -1 encode the answers Yes and No, respectively. One can meaningfully speak of correlations between the variables in the same context, probability that they have the same value, etc.

By contrast, different contexts, in our case the two orders in which the questions are asked, are mutually exclusive. When asked two questions, a given person can only be asked them in one order. Respondents represented by R_1^1 answer question q_1 asked first, before q_2, whereas the respondents represented by R_1^2 answer question q_1 asked second, after q_2. Clearly, these are different sets of respondents, and one would not know how to pair them. It is meaningless to ask, e.g., what the probability of

$$R_1^1 = +1 \text{ and } R_1^2 = +1$$

may be. Random variables in different contexts are *stochastically unrelated*.

2 Intuition of (non)contextuality

Having established these basic facts, let us consider now the two random variables with content q_1, and let us make at first the (unrealistic) assumption that their distributions are the same in both contexts, c^1 and c^2:

value	probability
$R_1^1 = +1$	a
$R_1^1 = -1$	$1-a$

and

value	probability
$R_1^2 = +1$	a
$R_1^2 = -1$	$1-a$

. (2)

If we consider the variables R_1^1 and R_1^2 in isolation from their contexts (i.e., disregarding the other two random variables), then we can view them as simply one and the same random variable. In other words, the subsystem

R_1^1	$c^1 = q_1 \to q_2$
R_1^2	$c^2 = q_2 \to q_1$
$q_1 =$ "vacation?"	$\mathcal{C}_{2(a)}$/only q_1

appears to be replaceable with just

R_1
$q_1 =$ "vacation?"

,

with contexts being superfluous.

Analogously, if the distributions of the two random variables with content q_2 are assumed to be the same,

value	$R_2^1 = +1$	$R_2^1 = -1$
probability	b	$1-b$

and

value	$R_2^2 = +1$	$R_2^2 = -1$
probability	b	$1-b$

, (3)

and if we consider them in isolation from their contexts, the subsystem

R_2^1	$c^1 = q_1 \to q_2$
R_2^2	$c^2 = q_2 \to q_1$
$q_2 =$ "Covid-19?"	$\mathcal{C}_{2(a)}$/only q_2

appears to be replaceable with

R_2
$q_2 =$ "Covid-19?"

.

It is tempting now to say: we have only two random variables, R_1 and R_2, whatever their contexts. But a given pair of random variables can only have one *joint distribution*, this distribution cannot be somehow different in different contexts. We should predict therefore, that if the probabilities in system $\mathcal{C}_{2(a)}$ are

$$\Pr\left[R_1^1 = +1, R_2^1 = +1\right] = r_1 \text{ and } \Pr\left[R_1^2 = +1, R_2^2 = +1\right] = r_2,$$

then

$$r_1 = r_2.$$

Suppose, however, that this is shown to be empirically false, that in fact $r_1 > r_2$. For instance, assuming $0 < a < b$, suppose that the joint distributions in the two contexts of system $\mathcal{C}_{2(a)}$ are

context c^1	$R_2^1 = +1$	$R_2^1 = -1$	
$R_1^1 = +1$	$r_1 = a$	0	a
$R_1^1 = -1$	$b - a$	$1 - b$	$1 - a$
	b	$1 - b$	

(4)

and

context c^2	$R_2^2 = +1$	$R_2^2 = -1$	
$R_1^2 = +1$	$r_2 = 0$	a	a
$R_1^2 = -1$	b	$1 - a - b$	$1 - a$
	b	$1 - b$	

. (5)

Clearly, we have then a *reductio ad absurdum* proof that the assumption we have made is wrong, the assumption being that we can drop contexts in R_1^1 and R_1^2 (as well as in R_2^1 and R_2^2), and that we can therefore treat them as one and the same random variable R_1 (respectively, R_2). This is the simplest case when we can say that a system of random variables, here, the system $\mathcal{C}_{2(a)}$, is *contextual*.

This understanding of contextuality can be extended to more complex systems. However, it is far from being general enough. It only applies to *consistently connected* systems, those in which any two variables with the same content are identically distributed.[4] This assumption is often unrealistic. Specifically, it is a well-established empirical fact that the individual distributions of the responses to two questions do depend on their order (Moore, 2002). Besides, this is highly intuitive in our example. If one is asked about an overseas vacation first, the probability of saying "Yes, I would like to take an overseas vacation" may be higher than when this question is asked second, after the respondent has been reminded about the dangers of the pandemic.

In order to generalize the notion of contextuality to arbitrary systems, we need to develop answers to the following two questions:

A: For any two random variables sharing a content, how different are they when taken in isolation from their contexts?

B: Can these differences be preserved when all pairs of content-sharing variables are taken within their contexts (i.e., taking into account their joint distributions with other random variables in their contexts)?

[4] The term "consistent connectedness" is due to the fact that in CbD the content-sharing random variables are said to form *connections* (between contexts). In quantum physics consistent connectedness is referred to by such terms as lack of signaling, lack of disturbance, parameter invariance, etc.

For our system $\mathcal{C}_{2(a)}$ with the within-context joint distributions given by (4) and (5), our informal answer to question **A** was that two random variables with the same content (i.e., R_1^1 and R_1^2 or R_2^1 and R_2^2) are not different at all when taken in isolation. The informal answer to question **B**, however, was that in these two pairs (or at least in one of them) the random variables are not the same when taken in relation to other random variables in their respective contexts. One can say therefore that

The contexts make R_1^1 and R_1^2 (and/or R_2^1 and R_2^2) more dissimilar than when they are taken without their contexts.

This is the intuition we will use to construct a general definition of contextuality.

3 Making It Rigorous: Couplings

First, we have to agree on how to measure the difference between two random variables that are not jointly distributed, like R_1^1 and R_1^2. Denote these random variables X and Y, both dichotomous (± 1), with

$$\Pr[X = +1] = u \text{ and } \Pr[Y = +1] = v.$$

Consider all possible pairs of *jointly distributed* variables (X', Y') such that

$$X' \overset{dist}{=} X, Y' \overset{dist}{=} Y,$$

where $\overset{dist}{=}$ stands for "has the same distribution as." Any such pair (X', Y') is called a *coupling* of X and Y. For obvious reasons, two couplings of X and Y having the same joint distribution are not distinguished.

Now, for each coupling (X', Y') one can compute the probability with which $X' \neq Y'$ (recall that the probability of $X \neq Y$ is undefined, we do need couplings to make this inequality a meaningful event). It is easy to see that among the couplings (X', Y') there is one and only one for which this probability is minimal. This coupling is defined by the joint distribution

	$Y' = +1$	$Y' = -1$	
$X' = +1$	$\min(u, v)$	$u - \min(u, v)$	u
$X' = -1$	$v - \min(u, v)$	$\min(1 - u, 1 - v)$	$1 - u$
	v	$1 - v$	

, (6)

and the minimal probability in question is obtained as

$$(u - \min(u, v)) + (v - \min(u, v)) = |u - v|.$$

This probability is a natural measure of difference between the random variables X and Y[5]:

$$\delta(X, Y) = \min_{\substack{\text{all couplings} \\ (X', Y') \text{ of } X \text{ and } Y}} \Pr\left[X' \neq Y'\right] = |u - v|. \tag{7}$$

If X and Y are identically distributed, i.e. $u = v$, the joint distribution of X' and Y' can be chosen as

context c^1	$Y = +1$	$Y = -1$	
$X = +1$	u	0	u
$X = -1$	0	$1 - u$	$1 - u$
	u	$1 - u$	

yielding

$$\delta(X, Y) = \min_{\substack{\text{all couplings} \\ (X', Y') \text{ of } X \text{ and } Y}} \Pr\left[X' \neq Y'\right] = 0.$$

Let us apply this to our example, in order to formalize the intuition behind our saying earlier that two identically distributed random variables, taken in isolation, can be viewed as being "the same." For R_1^1 and R_1^2 in (2),

$$\delta\left(R_1^1, R_1^2\right) = \min_{\substack{\text{all couplings} \\ \left(S_1^1, S_1^2\right) \text{ of } R_1^1 \text{ and } R_1^2}} \Pr\left[S_1^1 \neq S_1^2\right] = 0,$$

and, analogously, for R_2^1 and R_2^2 in (3),

$$\delta\left(R_2^1, R_2^2\right) = \min_{\substack{\text{all couplings} \\ \left(S_2^1, S_2^2\right) \text{ of } R_2^1 \text{ and } R_2^2}} \Pr\left[S_2^1 \neq S_2^2\right] = 0.$$

4 Making It Rigorous: Contextuality

What is then the rigorous way of establishing that these differences cannot both be zero when considered within their contexts? For this, we need to extend the notion of a coupling to an entire system. A coupling of our system $\mathcal{C}_{2(a)}$ is a set of *corresponding jointly distributed* random variables

[5] It is a special case of the so-called *total variation distance*, except that it is usually defined between two probability distributions, while I use it here as a measure of difference (formally, a *pseudometric*) between stochastically unrelated random variables.

$$\begin{array}{|c|c|}\hline S_1^1 & S_2^1 \\ \hline S_1^2 & S_2^2 \\ \hline \end{array} \tag{8}$$

such that

$$\left(S_1^1, S_2^1\right) \overset{dist}{=} \left(R_1^1, R_2^1\right), \left(S_1^2, S_2^2\right) \overset{dist}{=} \left(R_1^2, R_2^2\right). \tag{9}$$

In other words, the distributions within contexts, (4) and (5), remain intact when we replace the R-variables with the corresponding S-variables,

	$S_2^1 = +1$	$S_2^1 = -1$	
$S_1^1 = +1$	a	0	a
$S_1^1 = -1$	$b - a$	$1 - b$	$1 - a$
	b	$1 - b$	

and

	$S_2^2 = +1$	$S_2^2 = -1$	
$S_1^2 = +1$	0	a	a
$S_1^2 = -1$	b	$1 - a - b$	$1 - a$
	b	$1 - b$	

. (10)

Such couplings always exist, not only for our example, but for any other system of random variables. Generally, there is an infinity of couplings for a given system.[6] Thus, to construct a coupling for system $\mathcal{C}_{2(a)}$, one has to assign probabilities to all quadruples of joint events,

S_1^1	S_2^1	S_1^2	S_2^2	*probability*
$+1$	$+1$	$+1$	$+1$	p_{++++}
$+1$	$+1$	$+1$	-1	p_{+++-}
$\vdots$	$\vdots$	$\vdots$	$\vdots$	$\vdots$
-1	-1	-1	-1	p_{----}

so that the appropriately chosen subsets of these probabilities sum to the joint probabilities shown in (10):

$$\begin{array}{c} p_{++++} + p_{+++-} + p_{++-+} + p_{++--} = \Pr\left[S_1^1 = +1, S_2^1 = +1\right] = a, \\ p_{+-++} + p_{+-+-} + p_{+--+} + p_{+---} = \Pr\left[S_1^1 = +1, S_2^1 = -1\right] = 0, \\ p_{++++} + p_{+-++} + p_{-+++} + p_{--++} = \Pr\left[S_1^2 = +1, S_2^2 = +1\right] = 0, \\ etc. \end{array}$$

This is a system of seven independent linear equations with 16 unknown p-probabilities, subject to the additional constraint that all probabilities must be non-negative. It can be shown that this linear programming problem always has solutions,

[6] One need not have separate definitions of couplings for pairs of random variables and for systems. In general, given any set of random variables $\mathfrak{R}$, its coupling is a set of random variables S, in a one-to-one correspondence with $\mathfrak{R}$, such that the corresponding variables in $\mathfrak{R}$ and S have the same distribution, and all variables in S are jointly distributed. To apply this definition to $\mathfrak{R}$ representing a system of random variables one considers all variables within a given context as a single element of $\mathfrak{R}$. In our example, (8) is a coupling of two stochastically unrelated random variables, $\left(R_1^1, R_2^1\right)$ and $\left(R_1^2, R_2^2\right)$.

and infinitely many of them at that, unless one of the probabilities a and b equals 1 or 0 (in which case the solution is unique).

Unlike in system $\mathcal{C}_{2(a)}$ itself, in any coupling (8) of this system the random variables have joint distributions across the contexts. In particular, $\left(S_1^1, S_1^2\right)$ is a jointly distributed pair. Since from (9) we know that

$$S_1^1 \overset{dist}{=} R_1^1 \text{ and } S_1^2 \overset{dist}{=} R_1^2,$$

$\left(S_1^1, S_1^2\right)$ is a coupling of R_1^1 and R_2^1. Similarly, $\left(S_2^1, S_2^2\right)$ is a coupling of R_2^1 and R_2^2. We ask now: what are the possible values of

$$\Pr\left[S_1^1 \neq S_1^2\right] \text{ and } \Pr\left[S_2^1 \neq S_2^2\right]$$

across all possible couplings (8) of the entire system $\mathcal{C}_{2(a)}$? Consider two cases.

Case 1 In some of the couplings (8),

$$\Pr\left[S_1^1 \neq S_1^2\right] = 0 \text{ and } \Pr\left[S_2^1 \neq S_2^2\right] = 0.$$

We can say then that both $\delta\left(R_1^1, R_1^2\right)$ and $\delta\left(R_1^1, R_1^2\right)$ preserve their individual (in-isolation) values when considered within the system. The system $\mathcal{C}_{2(a)}$ is then considered *noncontextual*.

Case 2 In all couplings (8), at least one of the values

$$\Pr\left[S_1^1 \neq S_1^2\right] \text{ and } \Pr\left[S_2^1 \neq S_2^2\right]$$

is greater than zero. That is, when considered within the system, $\delta\left(R_1^1, R_1^2\right)$ and $\delta\left(R_1^1, R_1^2\right)$ cannot both be zero. Intuitively, the contexts "force" either R_1^1 and R_1^2 or R_2^1 and R_2^2 (or both) to be more dissimilar than when taken in isolation. The system $\mathcal{C}_{2(a)}$ is then considered *contextual*.

We can quantify the degree of contextuality in the system in the following way. We know that

$$\begin{aligned}&\delta\left(R_1^1, R_1^2\right) + \delta\left(R_2^1, R_2^2\right)\\ &= \min_{\substack{\text{all couplings}\\ \left(S_1^1, S_1^2\right) \text{ of } R_1^1 \text{ and } R_1^2}} \left(\Pr\left[S_1^1 \neq S_1^2\right]\right) + \min_{\substack{\text{all couplings}\\ \left(S_2^1, S_2^2\right) \text{ of } R_2^1 \text{ and } R_2^2}} \left(\Pr\left[S_2^1 \neq S_2^2\right]\right) = 0.\end{aligned}$$

This quantity is compared to

$$\delta\left(\left(R_1^1, R_1^2\right), \left(R_2^1, R_2^2\right)\right) = \min_{\substack{\text{all couplings}\\ \left(S_1^1, S_2^1, S_1^2, S_2^2\right) \text{ of system } \mathcal{C}_{2(a)}}} \left(\Pr\left[S_1^1 \neq S_1^2\right] + \Pr\left[S_2^1 \neq S_2^2\right]\right),$$

which can be interpreted as the total of the pairwise differences between same-content variables within the system. The system is contextual if this quantity is greater than zero, and this quantity can be taken as a measure of the degree of contextuality. This is by far not the only possible measure, but it is arguably the simplest one within the conceptual framework of CbD.

5 Generalizing to Arbitrary Systems

Consider now a realistic version of our example, when

$$\begin{array}{l}\Pr\left[R_1^1=+1\right]=a_1, \Pr\left[R_2^1=+1\right]=b_1,\\ \Pr\left[R_1^2=+1\right]=a_2, \Pr\left[R_2^2=+1\right]=b_2,\end{array}$$

with a_1 allowed to be different from a_2, and b_1 from b_2. The within-context joint distributions then generally look like this:

$$\begin{array}{c|c|c|c} \text{context } c^1 & R_2^1=+1 & R_2^1=-1 & \\ \hline R_1^1=+1 & r_1 & a_1-r_1 & a_1 \\ \hline R_1^1=-1 & b_1-r_1 & 1-a_1-b_1+r_1 & 1-a_1 \\ \hline & b_1 & 1-b_1 & \end{array} \quad (11)$$

and

$$\begin{array}{c|c|c|c} \text{context } c^2 & R_2^2=+1 & R_2^2=-1 & \\ \hline R_1^2=+1 & r_2 & a_2-r_2 & a_2 \\ \hline R_1^2=-1 & b_2-r_2 & 1-a_2-b_2+r_2 & 1-a_2 \\ \hline & b_2 & 1-b_2 & \end{array}. \quad (12)$$

Let us call the system in (1) with these within-context distributions $C_{2(b)}$. We clearly have context-dependence now (unless the two joint distributions are identical), but can we also say that the system is contextual? If we follow the logic of the definition of contextuality as it was presented above, for consistently connected systems, the answer cannot automatically be affirmative. The logic in question requires that we answer the questions **A** and **B** formulated in Sect. 2. By now we have all necessary conceptual tools for this.

To answer **A** we look at all possible couplings $\left(S_1^1, S_1^2\right)$ and $\left(S_2^1, S_2^2\right)$ of the content-sharing pairs $\left\{R_1^1, R_1^2\right\}$ and $\left\{R_2^1, R_2^2\right\}$, respectively, and determine

$$\delta\left(R_1^1, R_1^2\right) = \min_{\substack{\text{all couplings} \\ \left(S_1^1, S_1^2\right) \text{ of } \left\{R_1^1, R_1^2\right\}}} \Pr\left[S_1^1 \neq S_1^2\right],$$

and

$$\delta\left(R_2^1, R_2^2\right) = \min_{\substack{\text{all couplings} \\ \left(S_2^1, S_2^2\right) \text{ of } \left\{R_2^1, R_2^2\right\}}} \Pr\left[S_2^1 \neq S_2^2\right].$$

To answer **B**, we look at all possible couplings

S_1^1	S_2^1
S_1^2	S_2^2

of the entire system $\mathcal{C}_{2(b)}$, and determine if we can find couplings in which

$$\Pr\left[S_1^1 \neq S_1^2\right] = \delta\left(R_1^1, R_1^2\right)$$

and

$$\Pr\left[S_2^1 \neq S_2^2\right] = \delta\left(R_2^1, R_2^2\right).$$

If such couplings exist, we say that the system is noncontextual, even if it exhibits context-dependence in the form of *inconsistent connectedness*.

Recall that consistently connected systems are those in which any two variables with the same content are identically distributed, as it was in our initial (unrealistic) example. For such systems $\delta\left(R_1^1, R_1^2\right) = 0$ and $\delta\left(R_2^1, R_2^2\right) = 0$. However, if

$$R_1^1 \overset{dist}{\neq} R_1^2,$$

then $\delta\left(R_1^1, R_1^2\right) > 0$, and analogously for $\delta\left(R_2^1, R_2^2\right)$. In fact, we know from (6) and (7) that if the within-context distributions in the system are as in (11) and (12), then

$$\delta\left(R_1^1, R_1^2\right) = |a_1 - a_2|\,, \delta\left(R_2^1, R_2^2\right) = |b_1 - b_2|\,.$$

This means that system $\mathcal{C}_{2(b)}$ is contextual if and only if

$$\delta\left(\left(R_1^1, R_1^2\right), \left(R_2^1, R_2^2\right)\right) = \min_{\substack{\text{all couplings} \\ \left(S_1^1, S_2^1, S_1^2, S_2^2\right) \\ \text{of system } \mathcal{C}_{2(b)}}} \left(\Pr\left[S_1^1 \neq S_1^2\right] + \Pr\left[S_2^1 \neq S_2^2\right]\right)$$

$$> |a_1 - a_2| + |b_1 - b_2|\,.$$

Indeed, this inequality indicates that in all couplings either

$$\Pr\left[S_1^1 \neq S_1^2\right] > \delta\left(R_1^1, R_1^2\right),$$

or

$$\Pr\left[S_2^1 \neq S_2^2\right] > \delta\left(R_2^1, R_2^2\right),$$

or both. The intuition remains the same as above: the contexts "force" the same-content variables to be more dissimilar than they are in isolation. The difference

$$\delta\left(\left(R_1^1, R_1^2\right), \left(R_2^1, R_2^2\right)\right) - \delta\left(R_1^1, R_1^2\right) - \delta\left(R_2^1, R_2^2\right)$$

is a natural (although by far not the only) measure of the degree of contextuality.[7]

6 Other Examples

The system $\mathcal{C}_{2(b)}$ of the previous section, with the within-context distributions (11) and (12), is not a toy example, despite its simplicity. Except for the specific choice of the questions, it describes an empirical situation one sees in polls of public opinion, with two questions asked in one order of a large group of participants, and the same two questions asked in the other order of another large group of participants (Moore, 2002; Wang & Busemeyer, 2013).

In quantum physics, system of the $\mathcal{C}_2$-type can describe the outcomes of successive measurements of two spins along two directions, encoded 1 and 2, in the same spin-1/2 particle (e.g., electron). Without getting into details, in such an experiment the spin-1/2 particles are *prepared* in one and the same *quantum state*, and then subjected to two measurements in one of the two orders. Each measurement results in one of two outcomes, spin up (+1) or spin down (−1).

R_1^1	R_2^1	$c^1 = q_1 \rightarrow q_2$
R_1^2	R_2^2	$c^2 = q_2 \rightarrow q_1$
$q_1 =$ "is spin in direction 1 up?"	$q_2 =$ "is spin in direction 2 up?"	system $\mathcal{C}_{2(c)}$

(13)

The computations in accordance with the standard quantum-mechanical rules yield the following two results (Dzhafarov et al., 2015b). First, the system is inconsistently connected, i.e. generally the probability of spin-up in a given direction depends on whether it is measured first or second,

$$\Pr\left[R_1^1 = +1\right] \neq \Pr\left[R_1^2 = +1\right] \text{ and } \Pr\left[R_2^1 = +1\right] \neq \Pr\left[R_2^2 = +1\right].$$

[7] For other measures of contextuality, see Cervantes & Dzhafarov (2020), Dzhafarov et al. (2020a, b), Kujala & Dzhafarov (2019).

Second, the system is noncontextual,[8] i.e., it is always the case that

$$\delta\left(\left(R_1^1, R_1^2\right), \left(R_2^1, R_2^2\right)\right) \leq \delta\left(R_1^1, R_1^2\right) + \delta\left(R_2^1, R_2^2\right).$$

As we see, systems of the C_2-type may be of interest in both physics and behavioral studies.

However, in both these fields, the origins of the research of what we now call contextuality are dated back to another cyclic system, in which the arrangement shown in Fig. 1 specializes to

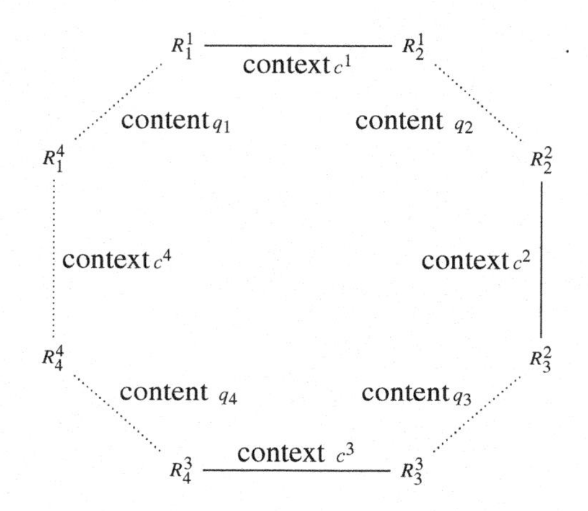

Figure 2 illustrates the empirical situation described by this system, and the first for which contextuality was mathematically established (Bell, 1966; Bohm & Aharonov, 1957; Clauser & Horne, 1974; Clauser et al., 1969; Fine, 1982). Two spin-1/2 particles are prepared in a special quantum state making them *entangled*, and they move away from each other. The "left" particle's spin is measured along one of the two directions (encoded 1 and 3) by someone we will call Zora, and simultaneously the "right" particle's spin is measured along one of the two directions (encoded 2 and 4) by a Nico.[9] The outcomes of the measurements are spin-up or spin-down, and each random variable R_i^j answers the question

$$q_i : \text{is the spin in direction } i \text{ up? } (i = 1, 2, 3, 4).$$

[8] For those familiar with CbD, this follows from the fact the expected values $\langle R_1^1 R_2^1 \rangle$ and $\langle R_1^2 R_2^2 \rangle$ are always equal to each other, whereas the criterion for contextuality of a cyclic system (Kujala & Dzhafarov, 2016), when specialized to $n = 2$, is $|\langle R_1^1 R_2^1 \rangle - \langle R_1^2 R_2^2 \rangle| > |\langle R_1^1 \rangle - \langle R_1^2 \rangle| + |\langle R_2^1 \rangle - \langle R_2^2 \rangle|$.

[9] For no deep reason, I decided to deviate from the established tradition to call the imaginary performers of the measurements in this task Alice and Bob.

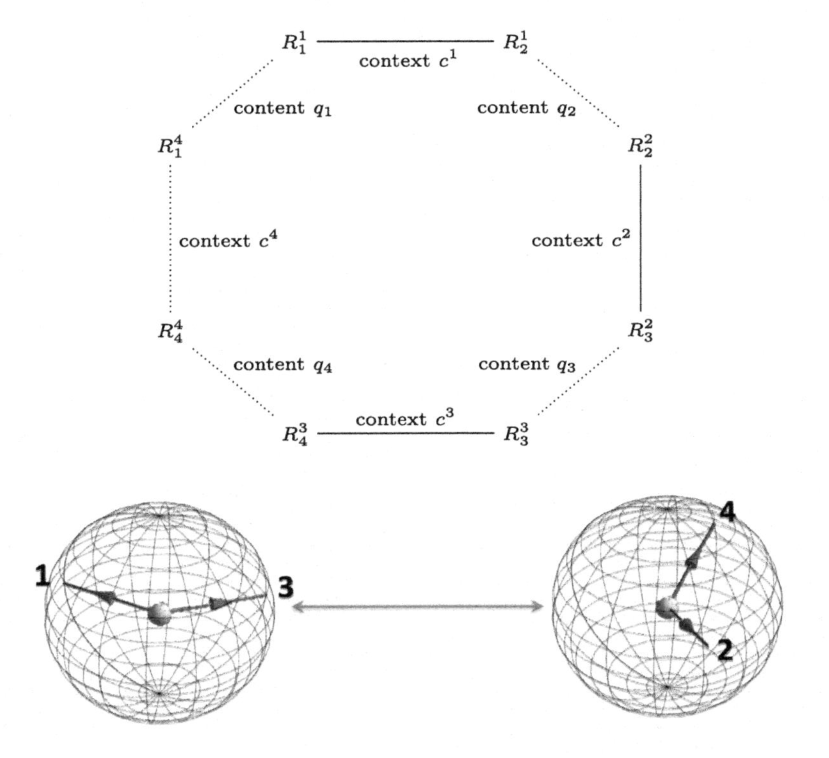

Fig. 2 A schematic representation of the EPR/Bohm experimental set up. Explanations in the text

In the form of a content-context matrix the system can be presented as

R_1^1	R_2^1			c^1
	R_2^2	R_3^2		c^2
		R_3^3	R_4^3	c^3
R_1^4			R_4^4	c^4
q_1 (Zora's 1)	q_2 (Nico's 2)	q_3 (Zora's 3)	q_4 (Nico's 4)	$\mathcal{C}_{4(a)}$

(14)

The measurements by Zora and Nico are made simultaneously, or at least close enough in time so that no signal about Zora's choice of a direction can reach Nico before he makes his measurement, and vice versa. Because of this, the system is consistently connected,

$$R_i^j \overset{dist}{=} R_i^{j'}$$

for any content q_i and two contexts c^j and $c^{j'}$ in which q_i is measured. Following the logic of contextuality analysis, we first establish that (because of the consistent connectedness)

$$\delta\left(R_1^1, R_1^4\right) = \delta\left(R_2^1, R_2^2\right) = \delta\left(R_3^2, R_3^3\right) = \delta\left(R_4^3, R_4^4\right) = 0.$$

Then we compute

$$\delta\left(\left(R_1^1, R_1^4\right), \left(R_2^1, R_2^2\right), \left(R_3^2, R_3^3\right), \left(R_4^3, R_4^4\right)\right) = \min_{\substack{\text{all couplings} \\ \left(S_1^1, S_4^1, S_1^2, S_2^2, S_2^3, S_3^3, S_3^4, S_4^4\right) \\ \text{of system } \mathcal{C}_{4(a)}}} \left(\Pr\left[S_1^1 \neq S_1^4\right] + \Pr\left[S_2^1 \neq S_2^2\right] + \Pr\left[S_3^2 \neq S_3^3\right] + \Pr\left[S_4^3 \neq S_4^4\right]\right).$$

The system is noncontextual if and only if this quantity is zero. As it turns out [and this is what was established by John Bell in his celebrated papers in the 1960s (Bell, 1964, 1966)], the directions $1, 2, 3, 4$ can be chosen so that, by the laws of quantum mechanics, this quantity is greater than zero, making the system contextual.

In psychology, systems of the same $\mathcal{C}_4$-type have been of interest as representing the following empirical situation (Dzhafarov & Kujala, 2014a; Dzhafarov, 2003; Dzhafarov & Kujala, 2010, 2016; Kujala & Dzhafarov, 2008; Sternberg, 1969; Townsend, 1984; Zhang & Dzhafarov, 2015). Consider two variables having two values each, that can be manipulated in an experiment. Think, e.g., of a briefly presented visual object that can have one of two colors (red or green) and one of two shapes (square or oval), combined in the 2×2 ways. In the experiment, an observer responds to the object by answering two Yes-No questions: "is the object red?" and "is the object square?". If we simply identify these questions with contents, the resulting system of random variables looks like this:

R_1^1	R_2^1	c^1 : red and oval
R_1^2	R_2^2	c^2 : green and oval
R_1^3	R_2^3	c^3 : red and square
R_1^4	R_2^4	c^4 : green and square
q_1 : red?	q_2 : square?	$\mathcal{R}$

, (15)

with the contexts describing the object being presented, and the contents the questions asked.

Although possible, this is not, however, an especially interesting way of conceptualizing the situation. It is more informative to describe the contents of the random variables as color and shape responses to the color and shape of the visual stimuli, respectively:

q_1 : does this red object appear red?
q_2 : does this square object appear square?
q_3 : does this green object appear red?
q_4 : does this oval object appear square?

With the contexts remaining as they are in system (15), the experiment is now represented by a system of the $\mathcal{C}_4$-type:

R_1^1	R_2^1			c^1
	R_2^2	R_3^2		c^2
		R_3^3	R_4^3	c^3
R_1^4			R_4^4	c^4
q_1 (red)	q_2 (square)	q_3 (green)	q_4 (oval)	$\mathcal{C}_{4(b)}$

Compared to system $\mathcal{C}_{4(c)}$ in (14), the physical situation described by $\mathcal{C}_{4(b)}$ is, of course, very different: e.g., instead of R_1^j and R_3^j being outcomes of spin measurements by Zora along two different directions, these random variables represent now responses to the color question when the color is red and when it is green, respectively. However, the logic of the contextuality analysis does not change. If this system turns out to be consistently connected and noncontextual, the interpretation of this in psychology is that the judgment of color is *selectively influenced by object's color* (irrespective of its shape), and the judgment of shape is *selectively influenced by object's shape* (irrespective of its color). Deviations from this pattern of selective influences, whether in the form of inconsistent connectedness or contextuality, or both,[10] provide an interesting way of classifying (and quantifying) the ways object's color may influence one's judgment of its shape and vice versa.

7 What if the System Is Deterministic?

A *deterministic quantity* r is a special case of a random variable: it is a random variable R that attains the value r with probability 1:

$$\Pr[R = r] = 1.$$

It is convenient to present this as

$$R \equiv r.$$

8A *deterministic system* is one containing only deterministic variables. For instance,

r_1^1	r_2^1		r_4^1		c^1
r_1^2		r_3^2			c^2
	r_2^3	r_3^3	r_4^3	r_5^3	c^3
		r_3^4		r_5^4	c^4
q_1	q_2	q_3	q_4	q_5	$\mathcal{D}$

(16)

[10] System $\mathcal{C}_{4(d)}$ is almost certainly inconsistently connected (guessing of an imaginary experiment based on the results of many real ones).

is a deterministic systems in which r_i^j represents a random variable $R_i^j \equiv r_i^j$. The system can be consistently connected (if the value of r_i^j does not depend on j) or inconsistently connected (otherwise).

It is easy to see, however, that a deterministic system is always noncontextual.[11] Indeed, any two content-sharing $R_i^j \equiv r_i^j$ and $R_i^{j'} \equiv r_i^{j'}$ in this system have a single coupling ($S_i^j \equiv r_i^j, S_i^j \equiv r_i^j$), consisting of the same deterministic quantities but considered jointly distributed.[12] It follows that

$$\delta\left(r_i^j, r_i^{j'}\right) = \begin{cases} 1 \text{ if } r_i^j \neq r_i^{j'} \\ 0 \text{ if } r_i^j = r_i^{j'} \end{cases}.$$

The entire deterministic system in (16) also has a single coupling, one containing the same deterministic quantities as the system itself, but considered jointly distributed. Clearly, the subcoupling $\left(S_i^j \equiv r_i^j, S_i^{j'} \equiv r_i^{j'}\right)$ extracted from this coupling is precisely the same as the coupling of $R_i^j \equiv r_i^j$ and $R_i^{j'} \equiv r_i^{j'}$ taken in isolation, and

$$\delta\left(\left\{\left(r_i^j, r_i^{j'}\right) : \text{all such pairs}\right\}\right) = \sum_{\text{all such pairs}} \delta\left(r_i^j, r_i^{j'}\right).$$

One might conclude that deterministic systems are of no interest for contextuality analysis. This is not always true, however. There are cases when we know that a system is deterministic, but we do not know which of a set of possible deterministic systems it is, because it can be any of them. Let us look at this in detail, using as examples systems consisting of logical truth values of various statements.

Consider first the following $\mathcal{C}_4$-type system:

$R_1^1 \equiv +1$	$R_2^1 \equiv -1$			c^1
	$R_2^2 \equiv +1$	$R_3^2 \equiv -1$		c^2
		$R_3^3 \equiv +1$	$R_4^3 \equiv -1$	c^3
$R_1^4 \equiv -1$			$R_4^4 \equiv +1$	c^4
q_1	q_2	q_3	q_4	$\mathcal{C}_{4(c)}$

, (17)

[11] This fact was first mentioned to me years ago by Matthew Jones of the University of Colorado.

[12] There is a subtlety here, first pointed out to me by Janne Kujala of Turku University. If $R_i^j \equiv r_i^j$ and $R_i^{j'} \equiv r_i^{j'}$, one may be tempted to say that the joint event $\left(R_i^j \equiv r_i^j, R_i^{j'} \equiv r_i^{j'}\right)$ has the probability one, and this would create an exception from the principle that random variables in different contexts are not jointly distributed. This is wrong, however, because $\left(R_i^j \equiv r_i^j, R_i^{j'} \equiv r_i^{j'}\right)$ can only be thought of counterfactually, as it involves mutually exclusive contexts. In fact, the only justification (or, better put, excuse) for the intuition that $\left(R_i^j \equiv r_i^j, R_i^{j'} \equiv r_i^{j'}\right)$ is a meaningful joint event is that $R_i^j \equiv r_i^j$ and $R_i^{j'} \equiv r_i^{j'}$ have a single coupling, and in this coupling $\Pr\left[S_i^j \equiv r_i^j, S_i^j \equiv r_i^j\right] = 1$. More generally, use of couplings is a rigorous way of dealing with counterfactuals (Dzhafarov, 2019).

where +1 and −1 encode truth values (true and false), and the contents are the statements

q_1 : "my name is Zora"	q_2 : "my name is Nico"
q_3 : "my name is Max"	q_4 : "my name is Alex"

Equivalently, the contents could also be formulated as questions, "is my name Zora?" and "is my name Nico?", in which case +1 and −1 would encode answers Yes and No. In the following, however, I will refer to the q's as statements, and the values of the variables as truth values. The contexts justifying the truth values in (17) are

c^1 : the statements are made by Zora	c^2 : the statements are made by Nico
c^3 : the statements are made by Max	c^4 : the statements are made by Alex

This is a situation when the truth values are determined uniquely, the system is deterministic, and consequently it is noncontextual (even though context-dependence in it is salient in the form of inconsistent connectedness).

Consider next another system of the C_4-type,

R_1^1	R_2^1			c^1
	R_2^2	R_3^2		c^2
		R_3^3	R_4^3	c^3
R_1^4			R_4^4	c^4
q_1 : "q_2 is true"	q_2 : "q_3 is true"	q_3 : "q_4 is true"	q_4 : "q_1 is false"	$C_{4(d)}$

with contents/statements of a very different kind, and the contexts which here (at least provisionally) can simply be defined by which statements they include: c^1 includes (q_1, q_2), c^2 includes (q_2, q_3), etc.

One can recognize here a formalization of the quadripartite version of the Liar antinomy: one can begin with any statement, say q_3, assume it is true, conclude that then q_4 is true, then q_1 is false, then q_2 is false, and then q_3 is false; and if one assumes that q_3 is false, then by the analogous chain of assignments one arrives to q_3 being true. There is no consistent assignment of truth values in this system. In the language of CbD, the truth values of the statements in $C_{4(d)}$ can only be described by an inconsistently connected deterministic system.

We come to the main issue now: $C_{4(d)}$ is certainly a deterministic system (because truth values of statements within a context are fixed), but which deterministic system is it? There are 16 possible ways of filling this system with truth values:

+1	+1		
	+1	+1	
		+1	+1
−1			+1

,

+1	+1		
	+1	+1	
		−1	−1
+1			−1

,

+1	+1		
	−1	−1	
		+1	+1
−1			+1

,

+1	+1		
	+1	+1	
		−1	−1
+1			−1

,

etc.

The only constraint in generating these systems is that

1. in the first three contexts (rows) the truth values of the two variables coincide (because the first statement in them says that the second one is true, and the second one does not refer to the first one);
2. in context c^4 (the last row) the truth values of the two variables are opposite (because q_4 says that q_1 is false, and q_1 does not refer to q_4).

We see that although random variability in $\mathcal{C}_{4(d)}$ is absent, we have in its place *epistemic uncertainty*. This opens the possibility of attaching epistemic (Bayesian) probabilities to the 16 possible deterministic variants of $\mathcal{C}_{4(d)}$, and obtaining as a result a *system of epistemic random variables*. Mathematically, such a variable is treated in precisely the same way as an ordinary ("frequentist") random variable. For instance, we can say that an epistemic variable R can have values $+1$ and -1 with Bayesian probabilities p and $1-p$. This means that R in fact is a deterministic quantity that can be either $+1$ or -1, and the degree of rational belief that R is $+1$ (given what we know of it) is p. In all computational respects, however, R is treated as if it was a variable that sometimes can be $+1$ and sometimes -1.

If we choose equal weights for all 16 deterministic variants of $\mathcal{C}_{4(d)}$ (simply because we have no rational grounds for preferring some of them to others), the resulting system will have the following Bayesian distributions:

context c^i, $i=1,2,3$	$R^i_{i+1}=+1$	$R^i_{i+1}=-1$	
$R^i_i=+1$	1/2	0	1/2
$R^i_i=-1$	0	1/2	1/2
	1/2	1/2	

(18)

and

context c^4	$R^4_1=+1$	$R^4_1=-1$	
$R^4_4=+1$	0	1/2	1/2
$R^4_4=-1$	1/2	0	1/2
	1/2	1/2	

. (19)

This system is clearly contextual. Indeed, since it is consistently connected,

$$\delta\left(R_1^1, R_1^4\right) = \delta\left(R_2^1, R_2^2\right) = \delta\left(R_3^2, R_3^3\right) = \delta\left(R_4^3, R_4^4\right) = 0. \tag{20}$$

At the same time,

$$\begin{aligned}&\delta\left(\left(R_1^1, R_1^4\right), \left(R_2^1, R_2^2\right), \left(R_3^2, R_3^3\right), \left(R_4^3, R_4^4\right)\right)\\ &= \min_{\substack{\text{all couplings}\\ \left(S_1^1, S_4^1, S_1^2, S_2^2, S_2^3, S_3^3, S_3^4, S_4^4\right)\\ \text{of system } \mathcal{C}_{4(d)}}} \left(\Pr\left[S_1^1 \neq S_1^4\right] + \Pr\left[S_2^1 \neq S_2^2\right] + \Pr\left[S_3^2 \neq S_3^3\right] + \Pr\left[S_4^3 \neq S_4^4\right]\right) > 0.\end{aligned} \tag{21}$$

This is easy to see. This quantity could be zero only if, in some coupling of $\mathcal{C}_{4(a)}$, the equalities in the first row below all held with probability 1:

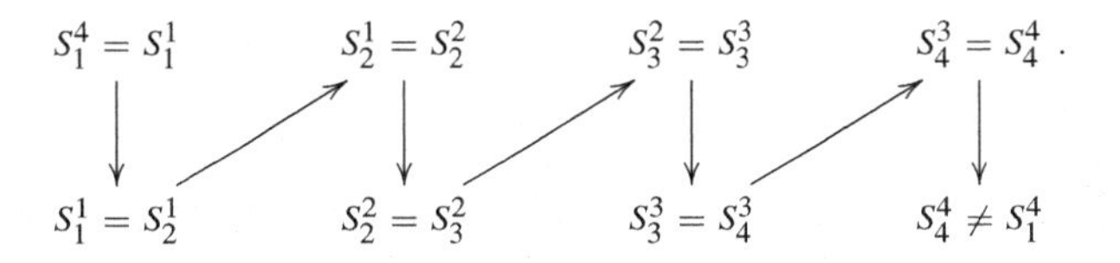

But in any coupling of $\mathcal{C}_{4(a)}$, the equalities in the second row also hold with probability 1, because they copy (18) and (19). Reading now all the equalities above from left to right along the arrows as a chain

$$S_1^4 = S_1^1 = S_2^1 = S_2^2 = \ldots,$$

one arrives at a contradiction

$$S_1^4 \neq S_1^4.$$

In essence, this is the same reasoning as that establishing the unremovable contraction in the Liar antinomy. However, this time it merely serves the purpose of establishing that our system is contextual. In fact, the degree of contextuality here, computed as the difference between (21) and the (zero) sum of the deltas in (20), is maximal among all possible systems of the $\mathcal{C}_4$-type.

We could use other multipartite versions of the Liar paradox, with three or five or any number of statements, all leading to the same outcome. A special mention is needed of the bipartite version. In this system it is no longer possible to define the contexts simply by the contents of the variables they include. Instead we once again need to use the order of the contents, this time interpreted as the direction of inference: $q \to q'$ means that we assign truth values to q and infer the corresponding truth values for q'.[13] The resulting system is

[13] The interpretation of contexts in terms of the direction of inference is the right one also in systems with larger number of statements. It is merely a coincidence that for $n > 2$ in the systems depicting the n-partite Liar paradox the direction of inference in a context is uniquely determined by the pairs of contents involved in this context.

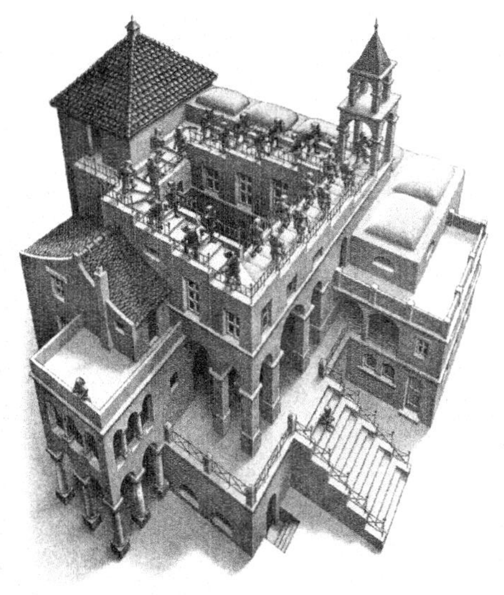

R_1^1	R_2^1			c^1
	R_2^2	R_3^2		c^2
		R_3^3	R_4^3	c^3
R_1^4			R_4^4	c^4
R_1^5	R_1^5	R_1^5	R_1^5	c^5
q_1	q_2	q_3	q_4	$\mathcal{E}$

Fig. 3 "Ascending-Descending" by M. C. Escher. The four flights of stairs are enumerated q_1, q_2, q_3, q_4. The epistemic random variables have values *ascending* and *descending*, and in each of the the first four contexts they are perfectly correlated. The fifth context is a mixture of the quadruples of values precisely two of which are ascending (so that travelers always end up in the same place from where they started). The resulting epistemic system is contextual (Cervantes & Dzhafarov, 2020)

R_1^1	R_2^1	$c^1 : q_1 \to q_2$
R_1^2	R_2^2	$c^2 : q_2 \to q_1$
q_1 = "q_2 is true"	q_2 : "q_1 is false"	$\mathcal{C}_{2(d)}$

,

with four possible deterministic variants:

$$\begin{array}{|c|c|}\hline +1 & +1 \\ \hline -1 & +1 \\ \hline \end{array}, \quad \begin{array}{|c|c|}\hline +1 & +1 \\ \hline +1 & -1 \\ \hline \end{array},$$

$$\begin{array}{|c|c|}\hline -1 & -1 \\ \hline -1 & +1 \\ \hline \end{array}, \quad \begin{array}{|c|c|}\hline +1 & +1 \\ \hline +1 & -1 \\ \hline \end{array}.$$

Mixing them with equal epistemic probabilities creates a consistently connected and highly contextual system (maximally contextual among all cyclic systems of rank 2).

Logical paradoxes are not, of course, the only application of contextuality analysis with epistemic random variables. It seems that many "strange" or "paradoxical" situations can be converted into contextual epistemic systems (Cervantes & Dzhafarov, 2020; Dzhafarov, in press). Among other applications are such objects as the Penroses' "impossible figures" and M. C. Escher's pictures (as in Fig. 3).

8 The Right to Ignore (or Not To)

I will mention now some aspects of the Contextuality-by-Default theory (CbD) that seem to pose difficulties for understanding. Questions about them are being asked often and in spite of having been repeatedly addressed in published literature.

The most basic aspect of CbD is double indexation of the random variables. The response to a given question q is a random variable R_q^c whose identity is determined not only by q but also by the context c in which q is responded to. This looks innocuous enough, but it puzzles some when a system being analyzed is consistently connected, i.e. when changing c in R_q^c does not change the distribution. And the puzzlement may increase when our knowledge tells us there is no possible way in which different contexts c can differently influence the random variables R_q^c.

Consider again the system $\mathcal{C}_{4(a)}$ in (14), from which we date contextuality studies. I reproduce it here for the reader's convenience:

R_1^1	R_2^1			c^1
	R_2^2	R_3^2		c^2
		R_3^3	R_4^3	c^3
R_1^4			R_4^4	c^4
q_1 (Zora's 1)	q_2 (Nico's 2)	q_3 (Zora's 3)	q_4 (Nico's 4)	$\mathcal{C}_{4(a)}$

In this system, Nico's choice between directions 2 and 4 can in no ways affect Zora's measurements of spin along direction 1. Nevertheless, when Nico switches from direction 2 to 4, the random variable describing the outcome of Zora's measurement of spin along direction 1 ceases to be R_1^1 and becomes R_1^4. It looks like Nico has influenced Zora's measurements after all. Isn't it an example of what Albert Einstein famously called a "spooky action at a distance"?

The answer is, it is not. Nico's choices are undetectable by Zora. Whether he chooses direction 2 or direction 4, Zora can see no changes in the statistical properties of what she observes when she measures spins along direction 1. "Action" means information transmitted, and no information is transmitted from Nico to Zora (and vice versa). The fact that in at least one of the pairs

$$\left\{R_1^1, R_1^4\right\}, \left\{R_2^1, R_2^2\right\}, \left\{R_3^3, R_3^4\right\}, \left\{R_4^3, R_4^4\right\}$$

the two random variables cannot be viewed as being the same can be established by neither Zora nor Nico. It can only be established by a Max who receives the choice of directions and outcomes of measurements from both Zora and Nico and computes the joint distributions in contexts c^1, c^2, c^3, c^4.

An important point here is that compared to Max, Zora does not misunderstand or miss anything when she sees no difference between R_1^1 and R_1^4 or between R_3^3 and R_3^4. Her understanding is no less complete or less correct. Zora and Max simply deal with different systems of random variables. In the same way Max's understanding is no less complete or less correct than that of an Alex who, in addition to knowing what Max knows, observes whether solar activity during the measurements is high or low. In Alex's system, each context of system $C_{4(a)}$ is split into two contexts, e.g., c^1 is replaced with

$R_1^{1,high}$	$R_2^{1,high}$			$c^{1,high}$
$R_1^{1,low}$	$R_2^{1,low}$			$c^{1,low}$
q_1	q_2	q_3	q_4	$C_{4(a)}/c^1$ only

In studying a system of random variable one always can ignore any of the circumstances that do not affect the distributions of the variables.[14] Or one can choose not to ignore such circumstances, to systematically record them and make them part of the contexts. If a circumstance is irrelevant (as it may be in the case of Alex's recording of solar activity), one will find this out by considering couplings of the system. Thus, one may establish that the contextuality analysis of the system does not change if all couplings are constrained by

$$\Pr\left[S_i^{j,high} = S_i^{j,low}\right] = 1,$$

for any R_i^j in the original system $C_{4(a)}$. This would mean that $R_i^{j,high}$ and $R_i^{j,low}$ can be viewed as being one and the same random variable (assuming, of course, that solar activity is indeed irrelevant).

This reasoning fully applies to the issue often raised by those who enjoy shallow paradoxes. If one records values of a random variable R in, say, chronological order, and simultaneously records the ordinal positions of these values in the sequence (as part of their contexts),

r_1	r_2	...	r_n	...
1	2	...	n	...

would not this transform all these realizations of a single random variable into pairwise stochastically unrelated random variables

$$R^1, R^2, \ldots, R^n, \ldots$$

with a single realization each? The answer is yes, if one so wishes (one may also choose to ignore the ordinal positions of the observations altogether), but then a standard view is immediately restored when one considers couplings of these random variables. For instance, the *iid coupling* (corresponding to the standard statistical concept of *independent identically distributed variables*) has the structure

[14] This statement can even be extended to ignoring circumstances when distributions do change (inconsistent connectedness). However, this issue has more complex ramifications, and we will set it aside.

$$\begin{array}{cccccc} & R^1 & R^2 & \cdots & R^n & \cdots \\ S^1 & \boxed{r_1^1 = r_1} & r_2^1 & \cdots & r_n^1 & \cdots \\ S^2 & r_1^2 & \boxed{r_2^2 = r_2} & \cdots & r_n^2 & \cdots, \\ \vdots & \vdots & \vdots & \ddots & \vdots & \ddots \\ S^n & r_1^n & r_2^n & \cdots & \boxed{r_n^n = r_n} & \cdots \\ \vdots & \vdots & \vdots & \ddots & \vdots & \ddots \end{array}$$

where the boxed values are those factually observed, whereas all other values are independently sampled from the distribution of R. More details are available in Dzhafarov (2016) and Dzhafarov & Kon (2018).

Finally, does the double-indexation in CbD lend any support to the holistic view of the universe, the view that "everything depends on everything else"? The answer is that the opposite is true, CbD supports a radically analytic view. First, as we have established, unless distributions of two given content-sharing variables are found to be different (which is ubiquitous but not universal) one can ignore the difference between their contexts, i.e., disregard all other variables in these contexts. This will redefine the system, but will not be wrong. Second, the difference in the identity of two content-sharing variables in different contexts (whether their distributions are the same or not) involves no change in the colloquial meaning of the word. The notion of a change implies that something that preserves its identity (e.g., a moving body) changes some of its properties (e.g., position in space). However, R_2^1 and R_2^2 (having the same content in different contexts) are simply different random variables, stochastically unrelated because they occur in mutually exclusive contexts. The difference between them is precisely the same as that between R_2^1 and R_1^1 (different contents in the same context). By choosing a different question to ask, one switches to considering another random variable rather than "changes" the previous one. The same happens when one chooses a different context: one simply switches to considering a different random variable. If I see Max and then see Alex, it does not mean that Max has changed into Alex.

The core of these and other problems with understanding CbD, it seems to me, is in the tendency to view random variables as empirical objects. They are not. Random variables are our descriptions of empirical objects. They are part of our knowledge of the world, and the same as any other knowledge, they can appear, disappear, and be revised as soon as we adopt a new point of view or gain new evidence.

References

Bacciagaluppi, G. (2015). Einsten, Bohm, and Leggett-Garg. In E. N. Dzhafarov, S. Jordan, R. Zhang, & V. Cervantes (Eds.), *Contextuality from quantum physics to psychology* (pp. 63–76). New Jersey: World Scientific.

Basieva, I., Cervantes, V. H., Dzhafarov, E. N., & Khrennikov, A. (2019). True contextuality beats direct influences in human decision making. *Journal of Experimental Psychology: General, 148,* 1925–1937.
Bell, J. (1964). On the Einstein-Podolsky-Rosen paradox. *Physics, 1,* 195–200.
Bell, J. (1966). On the problem of hidden variables in quantum mechanics. *Review of Modern Physics, 38,* 447–453.
Bohm, D., & Aharonov, Y. (1957). Discussion of experimental proof for the paradox of Einstein, Rosen and Podolski. *Physical Review, 108,* 1070–1076.
Cabello, A. (2008). Experimentally testable state-independent quantum contextuality. *Physical Review Letters,101,* 210401.
Cabello, A., Estebaranz, J., & García-Alcaine, G. (1996). Bell-Kochen-Specker theorem: A proof with 18 vectors. *Physics Letters A, 212,* 183–187.
Cervantes, V. H., & Dzhafarov, E. N. (2020). Contextuality analysis of impossible figures. *Entropy, 22,* 981. https://doi.org/10.3390/e22090981.
Cervantes, V. H., & Dzhafarov, E. N. (2017). Exploration of contextuality in a psychophysical double-detection experiment. *Lecture Notes in Computer Science, 10106,* 182–193.
Cervantes, V. H., & Dzhafarov, E. N. (2017). Advanced analysis of quantum contextuality in a psychophysical double-detection experiment. *Journal of Mathematical Psychology, 79,* 77–84.
Cervantes, V. H., & Dzhafarov, E. N. (2018). Snow Queen is evil and beautiful: Experimental evidence for probabilistic contextuality in human choices. *Decision, 5,* 193–204.
Cervantes, V. H., & Dzhafarov, E. N. (2019). True contextuality in a psychophysical experiment. *Journal of Mathematical Psychology, 91,* 119–127.
Clauser, J. F., & Horne, M. A. (1974). Experimental consequences of objective local theories. *Physical Review D, 10,* 526–535.
Clauser, J. F., Horne, M. A., Shimony, A., & Holt, R. A. (1969). Proposed experiment to test local hidden-variable theories. *Physical Review Letters, 23,* 880–884.
de Barros, J. A., Dzhafarov, E. N., Kujala, J. V., & Oas, G. (2016). Measuring observable quantum contextuality. *Lecture Notes in Computer Science, 9535,* 36–47.
Dzhafarov, E. N. (in press). The contextuality-by-default view of the sheaf-theoretic approach to contextuality. To be published. In A. Palmigiano & M. Sadrzadeh (Eds.), *Samson Abramsky on logic and structure in computer science and beyond, in series outstanding contributions to logic.* Springer Nature. (available as arXiv:1906.02718).
Dzhafarov, E. N., & Kujala, J. V. (2010). The joint distribution criterion and the distance tests for selective probabilistic causality. *Frontiers in Psychology: Quantitative Psychology and Measurement,1,* 151. https://doi.org/10.3389/fpsyg.2010.0015
Dzhafarov, E. N., & Kujala, J. V. (2013). All-possible-couplings approach to measuring probabilistic context. *PLoS ONE,8*(5), e61712. https://doi.org/10.1371/journal.pone.0061712
Dzhafarov, E. N., & Kujala, J. V. (2014). Contextuality is about identity of random variables. *Physica Scripta T,163,* 014009.
Dzhafarov, E. N., & Kujala, J. V. (2016). Probability, random variables, and selectivity. In W. Batchelder, H. Colonius, E. N. Dzhafarov, & J. Myung (Eds.), *The new handbook of mathematical psychology* (pp. 85–150). Cambridge University Press.
Dzhafarov, E. N., & Kujala, J. V. (2020). Systems of random variables and the Free Will Theorem. *Physical Review Research, 2,* 043288; https://doi.org/10.1103/PhysRevResearch.2.043288.
Dzhafarov, E. N., Kujala, J. V., & Cervantes, V. H. (2020). Contextuality and noncontextuality measures and generalized Bell inequalities for cyclic systems. *Physical Review A,101,* 042119.
Dzhafarov, E. N., Kujala, J. V., & Cervantes, V. H. (2020). Erratum: Contextuality and noncontextuality measures and generalized Bell inequalities for cyclic systems [*Physical Review A, 101,* 042119 (2020)]. *Physical Review A, 101,* 069902.
Dzhafarov, E. N., Kujala, J. V., & Cervantes, V. H. (2021). Epistemic odds of contextuality in cyclic systems. *European Physics Journal-Special Topics, 230,* 937–940. (available as arXiv:2002.07755).

Dzhafarov, E. N. (2003). Selective influence through conditional independence. *Psychometrika, 68,* 7–26.

Dzhafarov, E. N. (2016). Stochastic unrelatedness, couplings, and contextuality. *Journal of Mathematical Psychology, 75C,* 34–41.

Dzhafarov, E. N. (2017). Replacing nothing with something special: Contextuality-by-Default and dummy measurements. In A. Khrennikov & T. Bourama (Eds.), *Quantum foundations, probability and information* (pp. 39–44). Berlin: Springer.

Dzhafarov, E. N. (2019). On joint distributions, counterfactual values, and hidden variables in understanding contextuality. *Philosophical Transactions of the Royal Society A, 377,* 20190144.

Dzhafarov, E. N., Cervantes, V. H., & Kujala, J. V. (2017). Contextuality in canonical systems of random variables. *Philosophical Transactions of the Royal Society A, 375,* 20160389.

Dzhafarov, E. N., & Gluhovsky, I. (2006). Notes on selective influence, probabilistic causality, and probabilistic dimensionality. *Journal of Mathematical Psychology, 50,* 390–401.

Dzhafarov, E. N., & Kon, M. (2018). On universality of classical probability with contextually labeled random variables. *Journal of Mathematical Psychology, 85,* 17–24.

Dzhafarov, E. N., & Kujala, J. V. (2012). Selectivity in probabilistic causality: Where psychology runs into quantum physics. *Journal of Mathematical Psychology, 56,* 54–63.

Dzhafarov, E. N., & Kujala, J. V. (2012). Quantum entanglement and the issue of selective influences in psychology: An overview. *Lecture Notes in Computer Science, 7620,* 184–195.

Dzhafarov, E. N., & Kujala, J. V. (2013). Order-distance and other metric-like functions on jointly distributed random variables. *Proceedings of the American Mathematical Society, 141,* 3291–3301.

Dzhafarov, E. N., & Kujala, J. V. (2014). Selective influences, marginal selectivity, and Bell/CHSH inequalities. *Topics in Cognitive Science, 6,* 121–128.

Dzhafarov, E. N., & Kujala, J. V. (2014). A qualified Kolmogorovian account of probabilistic contextuality. *Lecture Notes in Computer Science, 8369,* 201–212.

Dzhafarov, E. N., & Kujala, J. V. (2015). Conversations on contextuality. In E. N. Dzhafarov, S. Jordan, R. Zhang, & V. Cervantes (Eds.), *Contextuality from quantum physics to psychology* (pp. 1–22). New Jersey: World Scientific.

Dzhafarov, E. N., & Kujala, J. V. (2016). Context-content systems of random variables: The contextuality-by-default theory. *Journal of Mathematical Psychology, 74,* 11–33.

Dzhafarov, E. N., & Kujala, J. V. (2017). Probabilistic foundations of contextuality. *Fortschritte der Physik, 65,* 1–11.

Dzhafarov, E. N., & Kujala, J. V. (2017). Contextuality-by-Default 2.0: Systems with binary random variables. *Lecture Notes Computer Sciences, 10106,* 16–32.

Dzhafarov, E. N., & Kujala, J. V. (2018). Contextuality analysis of the double slit experiment (with a glimpse into three slits). *Entropy, 20,* 278.

Dzhafarov, E. N., Kujala, J. V., & Cervantes, V. H. (2016). Contextuality-by-Default: A brief overview of ideas, concepts, and terminology. *Lecture Notes in Computer Science, 9535,* 12–23.

Dzhafarov, E. N., Kujala, J. V., Cervantes, V. H., Zhang, R., & Jones, M. (2016). On contextuality in behavioral data. *Philosophical Transactions of the Royal Society A, 374,* 20150234.

Dzhafarov, E. N., Kujala, J. V., & Larsson, J. -Å. (2015). Contextuality in three types of quantum-mechanical systems. *Foundations of Physics, 7,* 762–782.

Dzhafarov, E. N., Zhang, R., & Kujala, J. V. (2015). Is there contextuality in behavioral and social systems? *Philosophical Transactions of the Royal Society A, 374,* 20150099.

Fine, A. (1982). Joint distributions, quantum correlations, and commuting observables. *Journal of Mathematical Physics, 23,* 1306–1310.

Jones, M. (2019). Relating causal and probabilistic approaches to contextuality. *Philosophical Transactions of the Royal Society A, 377,* 20190133.

Klyachko, A. A., Can, M. A., Binicioglu, S., & Shumovsky, A. S. (2008). A simple test for hidden variables in spin-1 system. *Physical Review Letters,101*, 020403.

Kochen, S., & Specker, E. P. (1967). The problem of hidden variables in quantum mechanics. *Journal of Mathematics and Mechanics, 17,* 59–87.

Kujala, J. V., & Dzhafarov, E. N. (2019). Measures of contextuality and noncontextuality. *Philosophical Transactions of the Royal Society A, 377,* 20190149. (available as arXiv:1903.07170).

Kujala, J. V., Dzhafarov, E. N., & Larsson, J. -Ã. (2015). Necessary and sufficient conditions for extended noncontextuality in a broad class of quantum mechanical systems. *Physical Review Letters,115,* 150401.

Kujala, J. V., & Dzhafarov, E. N. (2008). Testing for selectivity in the dependence of random variables on external factors. *Journal of Mathematical Psychology, 52,* 128–144.

Kujala, J. V., & Dzhafarov, E. N. (2015). Probabilistic contextuality in EPR/Bohm-type systems with signaling allowed. In E. N. Dzhafarov, S. Jordan, R. Zhang, & V. Cervantes (Eds.), *Contextuality from quantum physics to psychology* (pp. 287–308). New Jersey: World Scientific.

Kujala, J. V., & Dzhafarov, E. N. (2016). Proof of a conjecture on contextuality in cyclic systems with binary variables. *Foundations of Physics, 46,* 282–299.

Kurzynski, P., Cabello, A., & Kaszlikowski, D. (2014). Fundamental monogamy relation between contextuality and nonlocality. *Physical Review Letters,112,* 100401.

Moore, D. W. (2002). Measuring new types of question-order effects. *Public Opinion Quarterly, 66,* 80–91.

Sternberg, S. (1969). The discovery of processing stages: Extensions of Donders' method. In W. G. Koster (Ed.), *Attention and performance II. Acta Psychologica, 30,* 276–315.

Townsend, J. T. (1984). Uncovering mental processes with factorial experiments. *Journal of Mathematical Psychology, 28,* 363–400.

Wang, Z., & Busemeyer, J. R. (2013). A quantum question order model supported by empirical tests of an a priori and precise prediction. *Topics in Cognitive Science, 5,* 689–710.

Zhang, R., & Dzhafarov, E. N. (2015). Noncontextuality with marginal selectivity in reconstructing mental architectures. *Frontiers in Psychology: Cognition,1,* 12. https://doi.org/10.3389/fpsyg.2015.00735

Zhang, R., & Dzhafarov, E. N. (2016). Testing contextuality in cyclic psychophysical systems of high ranks. *Lecture Notes in Computer Science, 10106,* 151–162.

Content, Context, and Naturalism in Mathematics

Otávio Bueno

1 Introduction

Mathematics is often presented as being necessary: its results, if true, are necessarily true. It is not a contingent fact about prime numbers that there are infinitely many of them: this is required by what these numbers are, by the content of the principles that characterize them. This would entail that the very content of mathematical theorems guarantees that their truth holds in any context, irrespectively of any vicissitudes. This leads, in turn, to the formulation of what can be called the traditional conception of mathematics as a body of necessary, immutable truths.

Despite its ubiquitous adoption (see, e.g., Hale, 2013), the traditional conception faces significant difficulties (see also Bueno, 2020, and, 2021, for further discussion). Mathematical theorems are extremely sensitive to context in, at least, three distinct ways:

(a) *Assumptions*: theorems in mathematics crucially depend on the assumptions that are built into them. If the assumptions are violated, the result in the theorem's statement need not go through. For instance, in real analysis (Rudin, 1976, p. 32), a limit point of a set E is one for which every neighborhood of that point contains a different point that is a member of E. It then follows that if p is a limit point of E, then every neighborhood of p contains infinitely may points of E. Clearly, the consequent of the theorem need not hold if p is not a limit point of the set E, given that, in this case some, some neighborhood of p may not have any points of E at all.

(b) *Logic*: mathematical results depend on the underlying logic. A change in logic may undermine the validity of the theorem. Certain results in classical analysis cannot be obtained constructively, given the shift to an intuitionistic logic. This is the case, for example, of the least-upper-bound principle, according

O. Bueno (✉)
Department of Philosophy, University of Miami, Coral Gables, FL 33124, USA

S. Wuppuluri and I. Stewart (eds.), *From Electrons to Elephants and Elections*,
The Frontiers Collection, https://doi.org/10.1007/978-3-030-92192-7_17

to which each nonempty subset of real numbers that is bounded above has a least upper bound (Bridges, 1999, p. 99). As Douglas Bridges notes, the result has a suitable constructive version. However, the latter does not have the same content as the classical theorem, given the nonconstructive nature of the classical continuum.

(c) *Interpretation*: when applied to the world, mathematical theorems depend on the interpretation of the concepts that are used in their formulation, and the resulting interpretation may turn out to be empirically inadequate. For instance, the negative energy solutions of the Dirac equation can be interpreted as corresponding to "holes" in space–time (an interpretation initially advanced by Dirac himself). Yet such interpretation leads to the conclusion that electrons and protons have the same mass, which, clearly, is not empirically adequate (see Bueno, 2005, for references and further discussion; I will return to this case below).

Given their dependence on assumptions, on the underlying logic, and on the interpretations used in applied contexts, mathematical theorems do not hold either in general or necessarily. At best, they hold conditionally—on the assumptions, the underlying logic or the interpretations that are invoked. Given that none of these parameters (the assumptions, the logic, and the interpretations) are necessary either, since each can, and have, failed in some context (see Bueno, 2021), mathematical results are fairly constrained, restricted, and dependent on the context under consideration. Hence, as opposed to the traditional conception, these results are ultimately contingent (Bueno, 2020).

Given the shifting contexts that mathematics is involved with, how should the content of mathematical statements be understood? In this paper, I examine this issue by considering three features of such content: (i) Mathematical content has an *inferential* component given that it is meant to allow one to infer suitable properties and relations regarding the objects under study. (ii) Mathematical content has a *phenomenological* component given the role that intuition seems to have in the specification of some such content. (iii) Mathematical content has a significant role in the *application of mathematics*, especially regarding the representation of empirical phenomena.

In order to focus the discussion, I will consider each of these three components in a particular context. The inferential component will be examined in the context of a recent reassessment of if-thenism in the philosophy of mathematics (Yablo, 2017). The phenomenological component will be considered in the context of a particular perceptualist view of intuition (Chudnoff, 2003). Finally, the application component will be discussed in the context of a particular account of the use of mathematics in scientific representation (Pincock, 2012). In each case, I will examine critically the proposals and, where appropriate, indicate ways in which a suitable alternative can be offered. The result, as will become clear, is a broadly empiricist, contingentist, and, given the role played by mathematical practice, naturalist approach to content.

2 Content: If-Thenism

As a view in the philosophy of mathematics, if-thenism states, in the words of Hilary Putnam, that:

> *if* there is any structure which satisfies such-and-such axioms (e.g. the axioms of group theory), *then* that structure satisfies such-and-such further statements (some theorems of group theory or other). (Putnam, 1979, p. 20)

And as long as the axioms (or, more generally, the principles) in question are consistent (or, at least, nontrivial), if-thenism can be used to capture a significant feature of mathematical practice: the fact that mathematicians are in the business of figuring out what follows from the axioms or principles they introduce. In this way, the content of mathematical theorems is established.

Yablo (2017) critically discusses how exactly if-thenism should be formulated, and argues that, at its best, its crucial conditional $\psi \rightarrow \varphi$ should be understood to mean whatever is left when ψ is subtracted from φ, in symbols: $\varphi \sim \psi$. He notes three significant motivations for such an approach—all of them bear on philosophy of mathematics issues: (a) *Independence*: Mathematics is independent from the physical processes that can be described in terms of it. (b) *Content*: The content of certain statements that are couched in terms of mathematics need not concern mathematics, but some other subject matter the statements address (especially in the context of applied mathematics). (c) *Explanation*: In some cases, one may obtain a better, more intrinsic, explanation of physical phenomena by not invoking mathematics. To make sense of each of these traits of mathematics, it is crucial to provide a strategy to take back (subtract) some parts of what is said. However, Yablo notes, an if-thenist face difficulties if an attempt is made to implement the subtraction approach to arithmetic, given that "the prospects for subtracting Peano Arithmetic's consistency from its truth seem fairly dim" (Yablo, 2017, 131).

There is, however, a perfectly straightforward way of capturing if-thenism without the difficulties that Yablo identified in it by noting that the troublesome ontological commitments that we are invited to avoid—the "subtraction" part that Yablo discusses—are better resisted not at the level of the conditionals—as Yablo's strategy recommends—but at the level of the quantifiers. If ontologically neutral quantifiers are invoked—that is, quantifiers that do not require the existence of that which is quantified over (Azzouni, 2004, and Bueno, 2005)—if-thenism can indeed be formulated as a straightforward conditional, but none of the unwanted ontology is presupposed: quantification over Peano Arithmetic structures does not require the existence of the corresponding structures any more than quantification over fictional characters does.

The if-thenist is then not settled with the unwelcome task of "subtracting Peano Arithmetic's consistency from its truth", but can simply assert that: if there are Peano Arithmetic structures (that is, structures that satisfy the axioms of Peano Arithmetic), then such structures also satisfy the relevant theorems of the theory. However, no claim about the existence of the structures that are quantified over is ever made, since quantification does not require the existence of that over which one quantifies. The result is then a clean, defensible version of if-thenism without qualms or tears. But how can we make sense of neutral quantifiers?

3 Neutral Quantification

Ontologically neutral quantifiers—or neutral quantifiers, for short—are exactly like the usual existential and universal quantifiers, except that they avoid doing double duty of (i) specifying the range of objects that are quantified over (all of the objects in the domain, in the case the universal quantifier, or only some of these objects, in the case of the existential one) and (ii) asserting the existence of the objects that are quantified over. *Range specification* (role (i)) and *ontological commitment* (role (ii)) implement very different functions of the quantifiers, and it is unclear that quantification is the proper way of marking ontological commitment anyway. One thing is to assert that all objects in the domain are F, or that some of these objects are G, and quite another is to claim that Fs exist, or that Gs exist. Rather than indicating what exists, ontologically neutral quantifiers simply specify the range of the objects that are quantified over: all or some. Existence is marked by the introduction of an existence predicate E (Azzouni, 2004, and Bueno, 2005).

Consider, for instance, the following claim:

(*) Some sets—such as those that are too large—don't exist.

If in (*) the quantifiers are interpreted as ontologically loaded—that is, as having ontological import—then the resulting statement would be contradictory, since it would amount to the claim that there exist sets that don't exist. With ontologically neutral quantifiers, and an existence predicate E, (*) can be expressed consistently and without trouble as follows—where 'S' stands for the predicate *is a set*:

(**) $\exists x\ (Sx \wedge \neg Ex)$.

What does fall under the existence predicate? The answer is just as simple as hopelessly uninformative: the existing things. Figuring out what exists is, of course, no straightforward or uncontroversial matter, nor is it a matter of logic or quantification. What exists is ultimately a matter of ontology. Different ontological views provide different answers to this question; each offers a distinctive criterion for what is taken to exist. Idealists, for instance, typically insist that (only) mind-dependent things exist. Realists often argue that existence is ultimately a matter of ontological independence from our linguistic practices and psychological processes (Azzouni, 2004). Platonists are realists who take abstract objects to be ontologically independent from us. Nominalists are realists who take abstract objects not to be so: these objects are ontologically dependent on our relevant practices and processes.

Part of the difficulty of settling ontological disputes emerges from the difficulty to determine, in non-question-begging ways, which criterion for what is taken to exist is adequate. Even if the relevant parties agree on the significance of ontological independence for existence, they may still disagree about which things are ontologically independent. And in trying to settle this issue, one ends up invoking precisely the things that are under dispute. For example, platonists do take abstract objects to be things that exist independently of our mental processes and linguistic practices. Thus, given their realism, they take abstract objects to exist. Nominalists demur. They argue that, since abstract objects depend on us, they lack the ontological independence required for them to exist: they are similar to fictional objects—which also do not exist. Non-spatiotemporal and causally inactive objects are just not the kind of thing that exists, nominalists argue. Platonists rightly complain that this just begs the question against their view; just as the platonists' insistence that abstract objects exist, given their ontological independence from us, begs the question against nominalism.

Quantification, however, should not depend on any of these metaphysical disputes. Quantification is needed in order to conduct such debates in the first place, and so it is crucial that it does not rely on what exists (or not). Ontological neutrality is a cherished feature of logic—albeit one that is not easily reached, but which can be assisted by ontologically neutral quantifiers. Once it is in place, quantificational neutrality allows for ontological debates to be carried out. Ontologically neutral quantifiers are crucial for this task, since one need not require the existence or the nonexistence of the contentious ontological items (numbers, functions, sets, properties, universals) in order to conduct the debate.

The semantics for these quantifiers is exactly the same as the semantics of usual quantifiers. It is enough that the quantifiers in the metalanguage—in which the semantics is provided—be similarly ontologically neutral (Azzouni, 2004). In this respect, the situation is entirely analogous to what goes on when the usual semantics is offered. Ontological commitment is marked by the existence predicate even in the metalanguage.

4 Content Revisited: If-Thenism with Ontologically Neutral Quantifiers

If properly characterized, if-thenism provides an interpretation of mathematical statements that preserves their objectivity without an unnecessary layer of ontological commitment. The content of mathematical claims can then be specified. With ontologically neutral quantifiers, if-thenism can be formulated directly as follows:

(I) If there were structures satisfying the mathematical principles of a theory M, the theorems of M would be true in those structures.

Moreover, it is possible that there are such structures. This is required in order to prevent the trivialization of the formulation in the case of inconsistent (or trivial) theories, since the conditional (I) above would be satisfied in case no such structures were possible, even if the negation of the theorems were plugged in the consequent.

(Hellman, 1989, building up on Putnam, 1979, Chap. 3, offers a similar strategy, but invokes, as does Putnam, ontologically committing quantifiers.)

This highlights, more generally, the broader issue of the consistency of mathematical theories, which is a concern not only for the if-thenist, but also for platonists and nominalists alike, who are similarly concerned about the potential inconsistency, or triviality, of mathematical principles. If mathematical theories are (taken to be) true, as platonists insist, they had better be consistent, on pain of platonists becoming dialetheists (Priest, 2006). If mathematical theories are not (taken to be) true, but are conservative, as certain nominalists would have it (Field, 1980), then the theories had better be consistent, since conservativeness is formulated in terms of consistency with every internally consistent body of claims about the world. In either case, consistency is required. (Of course, in light of Gödel's second incompleteness theorem, the consistency of such theories cannot be generally proved in the theories themselves, assuming they are consistent—and this fact impinges, once again, on all of these views, whether they invoke if-thenism or not.)

However, the existence of none of the structures referred to in (I) is ever asserted, given the ontologically neutral quantifiers: just because one quantifies over certain structures it does not follow that such structures exist, even if they are indispensable to the formulation of the theories in question. Quantification is one thing; ontological commitment is another.

It may be objected that if ontologically neutral quantifiers are introduced, there is no need to adopt if-thenism. Why bother with the latter if the former does all the work? But this is not right. Neutral quantification provides no account of philosophical issues about mathematics. It is simply an account of the quantifiers. It does not specify how mathematical discourse should be interpreted or evaluated. It does not determine the status of that discourse: whether it is objective, factual, or conceptually constrained. These are issues that need to be addressed as part of an interpretation of mathematical language—and ontologically neutral quantifiers per se have nothing to say about them.

But if-thenism does. On this interpretation of mathematics, mathematical discourse is fundamentally the result of what follows from certain principles—the mathematical principles that characterize a certain domain of mathematical objects. Mathematical concepts are introduced and relations among these concepts—specified in the relevant mathematical principles—are advanced. Once these principles are put forward, it is often inferentially opaque what follows from them. According to the if-thenist, trying to determine what these inferential relations are (among mathematical principles and their consequences) is a key feature of mathematics.

Of course, mathematical practice involves a number of activities, including refining concepts, changing definitions, selecting axioms, focusing on interesting mathematical theories, and ignoring uninteresting ones. All of this is ultimately achieved by establishing suitable inferential relations among the relevant mathematical principles and statements. Definitions are changed and refined in light of counterexamples and theorems; new axioms are introduced in light of the results that can (or cannot) be inferred from them, and the discrimination of interesting from uninteresting mathematics is tied to what can (or cannot) be inferred from the mathematical theories in question and the connections that can be established (or not) with other mathematical results. Typically, the more unexpected such connections among different areas of mathematics are, the more interesting the theorems become.

Mathematical practice is rigorous but informal (Azzouni, 2004), and axiomatization is not a requirement. In fact, if-thenism is not committed to mathematics being axiomatized. Mathematics has been developed quite independently of any such constraint. But a clear, rigorous formulation of the objects in a given mathematical domain is needed in order for their investigation to go through. But there is no difficult for the if-thenist to accommodate this aspect of the practice either. Indeed, the relations among concepts—as they are expressed via mathematical principles—and what follows from such principles are crucial features of if-thenism itself.

Once mathematical principles are introduced, and a logic is adopted—which is typically, although not always, classical—it is no longer up to us, or to the relevant mathematical community, what follows or not from such principles. This is an objective fact about the principles and the relevant theorems—or the relevant statements that are not consequence of the principles. Logical relations among statements hold or fail to hold depending on their content; they do not depend on us. The existence of the objects that are quantified over in the relevant statements plays no role in this process. What matters are the logical relations that hold, or fail to hold, among the statements in question: what follows from what, or what does not follow from what.

That it follows from Peano Arithmetic that there are infinitely many prime numbers is a fact about numbers, how they have been formulated in the theory, and why it would be inconsistent with the principles of the theory if there were only finitely many of them. Nothing is said about whether such numbers exist, whether they exist independently of the Peano Arithmetic principles, whether they are abstract objects, concrete objects, tropes, properties, aggregates, or entities of some other kind. Mathematical discourse is just silent about these issues. And ontologically neutral quantifiers provide a way of achieving this neutrality, since quantification over objects, including those that are posited in a mathematical theory, does not require their existence.

Coupled with ontologically neutral quantifiers, if-thenism provides a philosophically suggestive interpretation of mathematical discourse. It has a number of attractive features: It preserves the objectivity of mathematics without requiring the commitment to the existence of mathematical entities (something that is also emphasized by Putnam, 1979, and Hellman, 1989; for a discussion, see Bueno, 2018, 2013). It meshes naturally with the inferential role played by mathematical principles in mathematical practice. The objectivity of mathematics emerges from the objectivity of the logical consequence relation itself, independently of the whims of reasoning agents. The lack of ontological commitment can be achieved directly by separating quantification from existence. It does not depend on any subtraction procedure, in contrast with Yablo's (2017) proposal. The inferential role played by mathematical principles is accommodated via the very formulation of if-thenism—highlighting the importance of inferring theorems from mathematical principles.

Moreover, given the lack of ontological commitment, it is also easy to accommodate the three motivations Yablo (2017) introduced for his subtraction approach:

(a) The *independence* of mathematics from physical processes emerges straightforwardly from the fact that, except for cardinality considerations, mathematics doesn't constrain the physical world. (b) With neutral quantifiers, the *content* of statements is also accommodated: certain statements couched in terms of mathematics need not concern mathematics, but whatever the statements refer to, since no commitment to the existence of mathematical objects is ever made—even when such objects are quantified over. (c) Finally, more intrinsic *explanations* can be provided once neutral quantifiers are in place, since no explanation, even those couched in mathematical terms, is committed to the existence of mathematical objects. The result is a clean, simple characterization of if-thenism, within a broadly deflationary view.

5 Content and Intuition: Intellectual Perceptions

Perhaps specifying the content of mathematical theories by the relevant principles and the logical relations among them may not be enough. Some argue that an additional phenomenological constraint needs to be met for at least some mathematical principles: they are apprehended via intuition. But what exactly does this amount to?

Elijah Chudnoff (2013) offers an insightful answer to this question by defending a perceptualist account of intuition. On this view, intuitions are thought of as a form of intellectual perception. They are experiences (that is, conscious states) that are seemings and possess presentational phenomenology. On Chudnoff's account, intuitive justification is possible, and thus intuitions can be a source of reliable justification. More generally, they provide intuitive knowledge, in which a key role is played by veridical presentationalism. In what follows, I will raise some concerns.

Chudnoff's account of intellectual perceptions is characterized, in broad terms, by the articulation (i.e., the refinement and defense) of five theses: the first three emphasize the similarities between intuition and perceptions; the last two highlight the differences between them (Chudnoff, 2013, p. 3; see also pp. 6–13, and pp. 226–228).

(IP1) Intuitions are experiences: in them the mind "sees, feels and handles".

(IP2) Intuitions immediately justify beliefs: the justification they provide is independent of "reasoning or information passed on to you by teachers".

(IP3) Intuitions are similar to sensory perceptions in that they purport to, and sometimes do, put us in a position to gain knowledge by making us aware of their subject matter: again, "this knowledge" is "something that your mind sees, feel and handles". [...]

(IP4) The subject matter of intuition is not the subject matter of sensory perception: it is not to be found among the "objects you can see".

(IP5) Though intuition experience can involve sensory experience, it is a distinctive experience and can also occur autonomously—it can present a truth without "trying to clothe it with shapes" (Chudnoff, 2013, p. 3).

In the context of intuitions, talk of the mind *seeing*, *feeling* and *handling* (IP1) is, of course, metaphorical, given that the mind alone cannot literally perform these activities. Of course, part of Chudnoff's case is to provide support for the conclusion that there is more in common between our sensory and intellectual modalities than we may have initially thought (at least *vis-à-vis* our perceptual experiences and our intellectual intuitions).

The contrast between (IP1) and (IP4) is, thus, very telling. Although intuitions are experiences in which the mind "sees, feels and handles" (IP1), the subject matter of intuition is distinct from the subject matter of sensory perception. After all, intuition's subject matter is not supposed to be among "the objects you can see" (IP4) since it involves abstract objects. But then the sense in which the mind "sees, feels and handles" when having intuitions becomes somewhat opaque.

Later on (2013, Chap. 7), Chudnoff highlights a point that supports one way in which intuitions are like perceptions: just as to have a sensory perception of an object one needs to distinguish that object from its background, in order to have an intuition, one also has to differentiate the relevant object from its background. The foreground–background contrast is a common trait of both perceptual experiences and intuitions.

It is unclear, however, that the concepts of foreground and background have the same meaning when applied to perceptual experiences and when applied to abstract objects (in the case of intuitions). With regard to perceptual experiences, foreground and background involve a visual perspective (the point of view of the subject); with regard to intuitions, what is involved is attention to salient properties, which are in general formulated independently of any visual perspective. In fact, in many instances of intuition, there is literally no perspective from which to establish the differentiation of an object to its foreground or background.

Consider one formulation of the axiom of choice (Jech, 1973): any collection of nonempty sets has a choice function (that is, if A is a collection of nonempty sets, a choice function is a map f with domain A such that $f(x) \in x$, for every $x \in A$). In this case, what are the background and the foreground? We are considering a collection of nonempty sets, but this is an arbitrary collection, of indeterminate cardinality. With no particular specification, it is difficult to consider what would be the relevant background. We then consider a function defined on that collection, but again that function is extremely general, the only constraint is that given an element of the domain (a given nonempty set) it generates as a value a member of that set. The function, however, itself not particularly well specified, does not specify at all how that set is determined or defined. In fact, it is precisely this lack of specificity that generated so much controversy when Ernest Zermelo formulated the axiom of choice explicitly for the first time in 1904. Is the background of the choice function f all other functions? But in many cases, there are too many choice functions for the differentiation from a background even to make sense. The alternatives are too open-ended. As a result, it is unclear what would be the background (or the foreground) in these instances.

Given these considerations, should we say that one does not have an intuition regarding the axiom of choice? If so, the concern is that significant amounts of mathematics—perhaps some of the most important parts of classical mathematics, given the role that the axiom of choice plays in it—are not amenable to a treatment based on intuition. This is a significant limitation of the approach.

6 Perceptual Knowledge

As noted, a crucial feature of Chudnoff's account of intuitive knowledge is its similarity to perceptual knowledge. On his view:

> the conditions in virtue of which a perceptual experience representing that *p* puts you in a position to know that *p* are already intimated by the phenomenology of that perceptual experience. More precisely: I argue that if a perceptual experience representing that *p* puts you in a position to know that *p*, then it does so because it has presentational phenomenology with respect to the proposition that *p* and *this phenomenology is veridical*—i.e. it is true that *p* and your perceptual experience really makes you aware of the chunk of reality that makes it true that *p*. (Chudnoff, 2013, p. 21; see also pp. 174–180, in particular p. 177, in which this condition is formulated in terms of truth-makers)

There are a few difficulties here. First, the account presupposes knowledge. As Chudnoff (2013, p. 21) notes: "if a perceptual experience representing that *p* puts you in a position to know that *p*, then it does so because it has [veridical] presentational phenomenology". Clearly, a perceptual experience alone is not sufficient for knowledge, since it may be a hallucination. To qualify as a source of knowledge, the perceptual experience needs to put the subject in a position to *know* the relevant proposition *p*. But this assumes that the subject already knows. The relevant ground is, of course, provided by veridical presentational phenomenology: not only is *p* true, but also the subject's perceptual experience makes the subject "aware of the chunk of reality that makes it true that *p*" (Chudnoff, 2013, p. 21). As a result, Gettier cases can be avoided (Chudnoff, 2013, pp. 181–194). But Gettier cases become particularly relevant only if one is trying to provide a reductive analysis of knowledge; if the concept of knowledge is already presupposed, the fact that justified true beliefs do not amount to knowledge is far less troublesome.

Perhaps the relevant conditions can be stated without presupposing knowledge. This can be done, by readjusting Chudnoff's formulations, as follows:

> (VPP) We say that a perceptual experience representing that *p* has *veridical presentational phenomenology* with respect to the proposition that *p* if, and only if, (i) it is true that *p*, and (ii) the subject's perceptual experience really makes the subject aware of the chunk of reality that makes it true that *p* (that is, the experience makes the subject sensorily aware of a truth-maker for the proposition that *p*).
>
> (K) If a perceptual experience representing that *p* has veridical presentational phenomenology with respect to the proposition that *p*, then the subject *knows* that *p*.

Chudnoff, nevertheless, is interested in establishing that veridical presentational phenomenology *grounds* knowledge rather than in establishing a sufficient condition for knowledge. Clearly, veridical presentational phenomenology is not necessary for knowledge, since presumably one can have knowledge independently of such phenomenology—by memory, reasoning, or testimony.

Moreover, on Chudnoff's view, even though A may be a ground for B, in the sense that B obtains entirely in virtue of A, this does not entail that A is a sufficient condition for B. This is due to Chudnoff's discussion, and ultimately his rejection, of what he calls *Necessitation*: "if Q obtains wholly in virtue of P, then P is a sufficient condition for Q to obtain" (2013, p. 182; see pp. 182–194).

It seems that given (VPP) and (K), one does have a sufficient condition for knowledge. After all, if one has a perceptual experience representing that p has veridical presentational phenomenology, one is in a position to rule out those situations that would undermine one's knowledge, such as, the presence of fake barns and Gettier cases. Clearly, if you see a fake barn and think it is a real one, you do not thereby have knowledge that there is a real barn in front of you. Seeing a fake barn does not have veridical presentational phenomenology: only seeing a real barn does. A similar remark goes through for the Gettier case. If you see a hairy dog and think it is a wolf, you do not thereby have knowledge that there is a wolf near you (even if, unbeknownst to you, there is indeed such a wolf nearby). To have a perceptual experience that has veridical presentational phenomenology as of a wolf requires the existence of a wolf near you and that you are perceptually aware of the wolf. As a result, the Gettier case is similarly undermined.

One may complain that such a purely externalist account of knowledge is ultimately inadequate. One needs to have access to whatever grounds one's knowledge in order to be in a position to preserve that knowledge against potential challenges. In response, it may be argued that it depends on whether the awareness in (VPP) and (K) is conscious or not. If it is unconscious, the objection does stand: the awareness, being unconscious, provides the subject with no access to the relevant grounds of the knowledge in question. But if the awareness is conscious, the subject may satisfy an internalist knowledge requirement. Being aware of the real barn that makes it true there is a real barn in front of the subject not only provides the subject with knowledge, but it also gives access to the relevant grounds, namely, the veridical presentational phenomenology.

But is this correct? How can the subject be aware that the perceptual experience is of a real barn rather than of a fake one? Fake barns and real barns are perceptually indistinguishable, after all. It seems that the subject needs to have some previous knowledge to be able to determine which barns are real and which are fake. Having the perceptual experience of a real barn, which has veridical presentational phenomenology, will give the subject knowledge of the existence of a real barn, but the perceptual experience would be phenomenologically the same if the subject were seeing a fake barn. So, more is needed for the subject to be able to know that what is being experienced is a real barn: what needs to be known (or what one needs to be aware of) is that the perceptual experience is of a real barn. But since this is precisely what we are trying to establish, this would make the account otiose. No wonder Chudnoff talks only of grounds rather than sufficient conditions for knowledge.

7 Intuitive Knowledge

The next step is to develop an account of awareness that is meant to support the conclusion that intuitions can be a source of knowledge. The central problem to be addressed is how we can be aware of abstract objects. More precisely: "how can we really stand in awareness relations to the portions of reality that make propositions about abstract matters such as mathematics true?" (Chudnoff, 2013, p. 21).

On Chudnoff's view, awareness needs to satisfy two constraints (Chudnoff, 2013, p. 22; and Chap. 7):

(i) First, there is a *dependence constraint*, according to which if an experience makes its subject aware of an object, then the experience depends on that object. Chudnoff argues that intuition experiences depend non-causally on abstract objects.
(ii) Second, there is a *differentiation constraint*, according to which if an experience makes its subject aware of an object, then the experience's phenomenology differentiates the object from its background. Chudnoff argues that intuition experiences provide a background-foreground structure in the subject's phenomenal state in which certain abstract objects (and their properties) are foregrounded against the background of distinct abstract objects (and their properties).

I registered above, in the context of the discussion of the axiom of choice, the concern that for significant parts of mathematics it is unclear how to work out the differentiation of background and foreground. There is an additional concern here. Let us grant that the veridical presentational phenomenology of a certain intuition allows one to differentiate an abstract object from its background. Consider now a naïve comprehension principle such as the one according to which every property determines a set (namely, the set of objects that have that property). Suppose that one is able to differentiate the relevant objects (properties and corresponding sets) from their background. In this case, it seems that, if one were guided by the presentational phenomenology, one would be led to believe in an inconsistency, given the inconsistency of such naïve comprehension principle.

It may be argued that, in this case, the presentational phenomenology clearly is not veridical. If it were veridical, however, one would believe in the existence of a true contradiction (Priest, 2006). Without begging the question against the dialetheist, who insists that there are true contradictions, it is unclear how we can determine which of these options is the case—namely, whether the presentational phenomenology of the intuition of a naïve comprehension principle is veridical or not. Whichever option turns out to be the case, problems emerge, though. If the account allows for the intuition of inconsistent objects, the view would be unreliable if these objects turn out *not* to be inconsistent after all despite seeming to be so. Alternatively, if the intuition of the naïve comprehension principle has a veridical presentational phenomenology, the resulting account, although reliable, would then lead one to dialetheism.

Despite the important connections to perceptual experiences that Chudnoff's account provides, there are concerns regarding the account's ability to provide knowledge both in the case of perceptual experiences and when intuitions of mathematical objects are at stake. More is needed, in the end, to articulate the content of mathematical statements based on intuition. Until this point is reached, mathematical content is better formulated in terms of logical connections among mathematical principles as suggested above.

8 Context: Applying Mathematics, Models and Representations

Once the specification of the content of mathematical statements is achieved, it is crucial to understand the role played by such content in the context of the applications of mathematics, especially in the representation of empirical phenomena (for further examination of this issue, see Bueno and French, 2018). In what follows, I consider critically the important approach to this issue offered by Christopher Pincock (2012), contrasting it with the one I support.

I focus on two issues. Part of Pincock's case involves the development of an accuracy-based account of mathematical scientific representation that he contrasts with an inference-based account (which I favor). He also provides a searching critique of some forms of fictionalism in the philosophy of mathematics (which I also favor). I argue that, despite Pincock's objections, it is still possible to remain an inference-based fictionalist, and that there are some advantages to do so. In light of the considerations made so far, I start with the issue of the content of a mathematical scientific representation, and will then move to fictionalism.

Pincock distinguishes, as he should, *theories*, *models* and *representations*. This is done as follows:

> A theory for some domain is a collection of claims. It aims to describe the basic constituents of the domain and how they interact. A model is any entity that is used to represent a target system. Typically we use our knowledge of the model to draw conclusions about the target system. Models may be concrete entities or abstract mathematical structures. Finally, a representation is a model with a content. Contents provide conditions under which the representation is accurate. (Pincock, 2012, pp. 25–26)

My concern is that a model is characterized in terms of representation, and a representation, in turn, is characterized in terms of models. As things stand, the formulations are clearly circular. It is unclear why models are characterized in terms of their use rather than formulated directly, as is usually done, in terms of what satisfies the conditions specified by a theory. This would allow for the possibility of models being either concrete entities or abstract mathematical structures, as Pincock rightly intends, but without presupposing a tight circle between models and representations.

Without proper clarification of this issue, the crucial notion of content, as conditions for the accuracy of a representation (Pincock, 2012, p. 26), becomes unclear.

Pincock tells us that in order to determine the content of a mathematical scientific representation, we need to answer three questions:

> (1) What mathematical entities and relations are in question? (2) What concrete entities and relations are in question? (3) What structural relation must obtain between the two systems for the representation to be correct? (Pincock, 2012, p. 27)

In other words, to specify the content of a mathematical scientific representation, we need first to determine the relevant mathematical entities and relations as well as the relevant concrete entities and relations. We then specify the appropriate structural relation between the mathematical system and the physical system, so that the resulting representation is correct. As Pincock rightly acknowledges (2012, p. 27), the crucial work is done by the determination of the appropriate structural relation, such as: isomorphism, homomorphism, etc. It is in terms of this structural relation that the accuracy of a representation is ultimately determined. Note that the notion of content presupposes that a representation is already in place. For condition (3) of the content of a mathematical scientific representation asks us to determine what "structural relation must obtain between the two systems for the representation to be correct" (Pincock, 2012, p. 27). However, if a representation is a model with a content, and the content presupposes a representation, it is unclear how a representation can ever be determined. We have, once again, a circle. Moreover, it is also unclear how to specify the accuracy of a representation given that in order to have a representation (a model with a content) we need to have already determined its content, and the content, in turn, presupposes the accuracy of the representation (which is specified by the appropriate structural relation).

What is needed here is the formulation of the content of a scientific representation that is characterized independently of its accuracy. After all, presumably we should first have access to the content of the representation (what it is about and how it is implemented) in order then to determine how accurate (or not) it turns out to be.

The point is not just a quibble, since it affects one of the central features of Pincock's approach: in order to make sense of the way in which mathematics contributes to scientific success (such as by producing accurate representations of the physical world), it is important to identify different contributions, which are, in turn, individuated via the contents of mathematical scientific representations (see Pincock, 2012, p. 21). So, a proper specification of such contents is crucial. But it is unclear how to specify the contents given the way in which the relevant concepts have been introduced.

There are approaches, however, to mathematical scientific representations in which accuracy conditions are not placed on the central stage. In fact, Pincock notes that the main competitor to the accuracy-based approach he favors emphasizes the central role played by inferential relations in mathematical scientific representation. This is the case of the inferential account developed in Bueno and Colyvan (2011) and Bueno and French (2018). According to Pincock:

> Inferential approaches must explain the scientific practice of evaluating representations in terms of their accuracy. Although there does not seem to be any barrier to doing this, I have found it more convenient to start with the accuracy conditions. On my approach, inferential

> claims about a given representation follow immediately from its accuracy conditions: a valid inference is accuracy-preserving. (Pincock, 2012, p. 28)

However, the accuracy-based approach has a significant cost in this context. As opposed to what Pincock maintains, it is unclear that this approach can regain inferential claims from its accuracy conditions. Consider, for instance, a single application of *modus ponens* involving a vague predicate, such as 'rich':

(P1) To have 1 dollar in one's bank account does not make one rich.
(P2) If to have 1 dollar in one's bank account does not make one rich, then to have 2 dollars in one's bank account does not either.
Therefore, to have 2 dollars in one's bank account does not make one rich.

Clearly, this inference preserves accuracy. But after a sufficiently large number of applications of it (e.g., one million), the accuracy is entirely lost, since having one million and one dollars in one's bank account presumably does make one rich. This is, in a sense, one of the lessons of the Sorites paradox. Valid inferences need not be accuracy-preserving. It is, thus, better to start from inferential relations, and specify accuracy conditions via them, rather than the other way around.

There is an additional difficulty one faces to specify the content of a mathematical scientific representation in terms of mathematical structures, which are typically thought of as being set-theoretic in nature. Since mathematical structures are (taken to be) abstract—they are causally inert and are not located in space–time—they raise the issue of how to explain the connection between them and concrete physical structures in the world. As opposed to abstract structures, concrete structures *are* causally active and spatiotemporally located. This is a major problem for all realist views about mathematics. After all, it is indeed deeply puzzling how structures of fundamentally different kinds can be related in such a way that one is used to represent the other. Acknowledging the gap, Bas van Fraassen posed the problem as follows:

> How can an abstract entity, such as a mathematical structure, represent something that is not abstract, something in nature? (van Fraassen, 2008, p. 240)

After quoting this passage, Pincock notes that he intends to take a less ambitious route, by invoking the notion of instantiation:

> Suppose we have a concrete system along with a specification of the relevant physical properties. This specification fixes an associated structure. [...] [W]e can say that the system instantiates that structure, relative to that specification, and allow that structural relations are preserved by this instantiation relation [...]. This allows us to say that a structural relation obtains between a concrete system and abstract structure. (Pincock, 2012, p. 29)

The problem with this suggestion is that it is unclear how a concrete system along with the relevant physical properties can specify an associated mathematical structure. After all, presumably the physical system is finite, and since the mathematical structures that are typically used in scientific practice are infinite (containing at least the natural, the real or the complex numbers), it is unclear how an associated (infinite) mathematical structure can be fixed by a (finite) concrete system. Given any such concrete system, there is a plurality of different mathematical structures that may overlap with the finite segment but differ significantly after that. Which of these structures (if any) is instantiated in this case? In fact, the point applies even if the mathematical structures are finite, but the cardinality of their domains is larger than that of the concrete system. In that case, there are several mathematical structures that will do the same trick.

In the end, it is unclear how the notion of instantiation helps to solve the problem raised by van Fraassen. How can the natural number structure be instantiated by a concrete system if there is no largest natural number and (presumably) there are only finitely many physical objects? The natural number structure is, strictly speaking, *inconsistent* with the finiteness of the concrete system. It may be argued that it is only a fragment (a segment) of the natural number structure that is used in this case. But which fragment is that? To specify the fragment in question is to assume that there is already a relation between the relevant mathematical structure and the concrete system. But this is precisely what the notion of instantiation was supposed to do in the first place. In the end, it is not apparent how exactly instantiation helps to solve the problem at hand. It is advisable then not to rely on it.

9 Fictionalism Defended: Naturalism

Pincock also raises significant objections to some forms of fictionalism in the philosophy of mathematics. According to these views, the content of mathematical theories need not correspond to an ontology of independently existing mathematical objects and structures. Some of these views (such as those discussed in Balaguer, 1998, and Yablo, 2001) aim to draw a distinction between the fictional (platonistic) content and the representational (nominalistic) content, and it is only the latter that carries any ontological weight. The idea is that the representational (nominalistic) content is that part of the fictional (platonist) content that is only concerned with physical, concrete objects.

Pincock's main objection is that these fictionalist views are unable to specify the representational (nominalistic) content so that they can accommodate the role it is supposed to play in scientific representation (Pincock (2012), pp. 150–156; see, in particular, pp. 154–156). For what exactly fixes the representational content? It may be argued that in order to determine which features of the fictional content are exported to characterize the representational content, we need to focus on the beliefs of the relevant speakers. But beliefs about what is physical or concrete are unlikely to be clear and settled enough to be used to fix the representational content. Alternatively, it may be argued that the actual facts in the world could be used to fix the content. But without access to what these facts are, the fictionalist would end up in the awkward position of having to endorse the representational content without having evidence for the resulting commitments. In either case, the fictionalist is in trouble.

The situation, however, need not be as bleak as Pincock suggests. Although I cannot speak for Mark Balaguer or Stephen Yablo, it seems to me that the form of fictionalism I favor (Bueno, 2009) is not open to this objection. Initially, the fictional content, in light of the discussion of the first portion of this paper, is provided by the specification of suitable mathematical (comprehension) principles that characterize the mathematical domain under consideration. Studies are then developed of what follows from such principles. In many cases, it will be unclear whether a certain result obtains (that is, whether it can be proved from the relevant comprehension principles) or whether its negation obtains. The concepts originally introduced may not be refined enough to settle certain questions about the domain. In response, new, more refined concepts are introduced, with correspondingly more refined comprehension principles. The domain is reconfigured, and the process of trying to find proofs or counterexamples continues. Overtime this process leads to the introduction of significantly different concepts and comprehension principles within mathematics. Rather than a simple accumulation of mathematical facts, we experience a richer, more interesting variety of mathematical concepts, theories, and theorems. (Imre Lakatos described this process in vivid detail in Lakatos (1976) and Lakatos (1978).) We do have a form of fictionalism, since no claim is ever made about the independent existence of the mathematical objects that are specified via the various comprehension principles: ontologically neutral quantifiers are used throughout.

So far, except perhaps for issues about the cardinality of the domain, mathematical principles, strictly speaking, do not state anything at all about the physical world. After all, functions, numbers, and sets, which are typically taken to be abstract, are entirely independent of concrete spatiotemporal objects. Even if, in some cases, problems that prompted the development of a mathematical theory may have emerged from the need to accommodate particular issues in a physical domain, the developments of pure mathematics are not constrained by these needs. In such cases, the motivation may come from a physical domain, but the development of the resulting mathematical theories is autonomous. Physical interpretations and mathematical simulations can be used as aids to conceptualize certain aspects of a mathematical domain, but they should not be confused with what that domain ultimately is. In fact, if anything it is the other way around: the mathematical domain needs to be characterized first so that one can then have a clear picture of how mathematical theories, which may be relevant to that domain, can be applied.

In other words, in order for the mathematics to have any relevance to the world, it needs to be interpreted in a physically adequate way (see Bueno and Colyvan (2011), and Bueno and French (2012) and (2018)). But this process of interpretation is radically underdetermined by the mathematics, given that the same mathematical theory is compatible with very different interpretations. This is the case, for instance, with Paul Dirac and the famous equation that now bears his name (see Bueno (2005), and the references therein). When Dirac initially identified the equation he noted that it had negative energy solutions. As is common in classical mechanics, he simply ignored these solutions, indicating that they have no physical interpretation. A few years later, as I noted above, Dirac devised an interpretation in terms of "holes" in space–time. However, it was pointed out to him, by Werner Heisenberg, Hermann Weyl and others, that this interpretation was empirically inadequate, given that it entailed that electrons and protons have the same mass. As a result, Dirac abandoned it. A couple of years later, he devised yet another interpretation, according to which there was a new particle that had the same mass as the electron but the opposite charge. And for that interpretation Carl Anderson was able to collect evidence that, in retrospect, could be reconstructed as supporting such a particle, which is now called 'positron'.

As this case indicates, the same piece of mathematics, a certain differential equation, is compatible with three entirely different interpretations: one in which nothing is said about the equation's negative energy solutions; one which is empirically inadequate, and one for which there seems to be some empirical evidence. The process of forging interpretations to certain pieces of mathematics is no easy task. But it provides the central trait of how mathematics is ultimately applied (for further details, see Bueno and Colyvan (2011) and Bueno and French (2012) and (2018)).

The resulting proposal emphasizes the role of the interpretation of mathematics in the application process. In fact, this is one of the key components of the inferential conception (see Bueno and Colyvan (2011) and Bueno and French (2018)). As a result, the inferential proposal is not open to the challenge raised by Pincock to other forms of fictionalism. For to devise an interpretation is not a matter of tracking the beliefs of the relevant mathematicians and physicists: some of Dirac's beliefs involving "holes" in the space–time turned out to be empirically false. The relevant interpretations are also not simply fixed by the world: one needs to have evidence for their adequacy. In Dirac's case, the evidence emerged, as it should, from experimental physics. In the end, finding adequate interpretations of mathematics is what matters.

This form of fictionalism is also able to accommodate the openness of the conceptual developments in mathematics, since it is not tied to an understanding of concepts as monolithic. Thus, it is able to respond to the charge that Pincock made against the fictionalist approaches developed by Azzouni (2004) and Rayo (2009). After granting that his objection against Balaguer's and Yablo's fictionalisms do not apply to Azzouni's and Rayo's views (Pincock, 2012, pp. 261–263), Pincock argues that both Azzouni's nominalism and Rayo's trivialism fail, since "they link the understanding and the truth of mathematical claims too closely" (2012, p. 278). In particular, according to Pincock, they rely on an understanding of concepts that is unable to explain how shifts in mathematical concepts can take place. Whether this objection applies to their views or not is not for me to say, but it should be clear that it does not undermine the form of fictionalism I favor, since, as suggested above, conceptual change is something this fictionalism definitely makes room for (see also Bueno, 2000, and 2007). As it should be clear by now, given that the proposal is sensitive to details of mathematical and scientific practice, it involves a minimal form of naturalism (for additional constraints on naturalism, formulated as "second philosophy", see Maddy (2007); the contrast between the view I favor and Maddy's will be examined in another occasion).

10 Conclusion: Content and Context in Mathematical Practice

For the reasons indicated, the form of mathematical fictionalism that uses the inferential conception of the application of mathematics is able to resist the objections that Pincock has raised. In fact, given the difficulties faced by Pincock's own accuracy-based account, indicated earlier, this fictionalism provides an interesting alternative to Pincock's own proposal. In the end, despite Pincock's challenge, there is still room for an inferential form of fictionalism.

Given the previous critical discussion of the role of intuition in characterizing mathematical content, in the context of Chudnoff's (2013) proposal, more needs to be done by those who intent to defend such role, and a characterization of content via ontological neutral quantifiers is to be preferred. Similarly, there is something to be said for a form of if-thenism via neutral quantifiers, without relying on Yablo's (2017) subtraction approach. The various contexts in which mathematical theories are then applied need to be carefully specified, since they involve different empirical interpretations of mathematical formalisms, as the Dirac case illustrates. The result is a deflationary, empiricist, and fictionalist view that, given its sensitivity to mathematical practice, is also broadly naturalist.

Acknowledgements My thanks go to Eli Chudnoff, Mark Colyvan, Steven French, Chris Pincock, and Steve Yablo for helpful discussions on the issues examined in this paper. Thanks are also due to Shyam Wuppuluri for all the support during the writing of this work.

References

Azzouni, J. (2004). *Deflating existential consequence*. Oxford University Press.

Balaguer, M. (1998). *Platonism and anti-platonism in mathematics*. Oxford University Press.

Bridges, D. (1999). Constructive mathematics: A foundation for computable analysis. *Theoretical Computer Science, 219*, 95–109.

Bueno, O. (2000). Empiricism, mathematical change and scientific change. *Studies in History and Philosophy of Science, 31*, 269–296.

Bueno, O. (2005). Dirac and the dispensability of mathematics. *Studies in History and Philosophy of Modern Physics, 36*, 465–490.

Bueno, O. (2007). Incommensurability in mathematics. In B. Van Kerkhove & J. P. Van Bendegem (Eds.), *Perspectives on mathematical practices* (pp. 83–105). Springer.

Bueno, O. (2009). Mathematical fictionalism. In O. Bueno & Ø. Linnebo (Eds.), *New waves in philosophy of mathematics* (pp. 59–79). Palgrave MacMillan.

Bueno, O. (2013). Putnam and the indispensability of mathematics. *Principia, 17*, 217–234.

Bueno, O. (2018). Putnam's indispensability argument revisited, reassessed, revived. *Theoria, 33*, 201–218.

Bueno, O. (2020). Contingent abstract objects. In J. L. Falguera & C. Martínez-Vidal (Eds.), *Abstract objects* (pp. 91–109). Springer.

Bueno, O. (2021). Revising Logics. In J.-Y. Béziau, J.-P. Desclés, A. Moktefi, & A. Pascu (Eds.), *Logic in question: Talks from the annual Sorbonne logic workshop (2011–2019)*. Birkhäuser (forthcoming).
Bueno, O., & Colyvan, M. (2011). An inferential conception of the application of mathematics. *Noûs, 45*, 345–374.
Bueno, O., & French, S. (2012). Can mathematics explain physical phenomena? *British Journal for the Philosophy of Science, 63*, 85–113.
Bueno, O., & French, S. (2018). *Applying mathematics: Immersion, inference, interpretation.* Oxford University Press.
Chudnoff, E. (2013). *Intuition.* Oxford University Press.
Field, H. (1980). *Science without numbers.* Princeton University Press.
Hale, B. (2013). *Necessary beings: An essay on ontology, modality, and the relations between them.* Oxford University Press.
Hellman, G. (1989). *Mathematics without numbers: Towards a modal-structural interpretation.* Clarendon Press.
Jech, T. (1973). *The axiom of choice.* North Holland.
Lakatos, I. (1976). *Proofs and refutations: The logic of mathematical discovery.* (Edited by J. Worrall and E. Zahar.). Cambridge University Press.
Lakatos, I. (1978). *Mathematics, science and epistemology.* Philosophical Papers, volume 2. (Edited by J. Worrall and G. Currie.). Cambridge University Press.
Maddy, P. (2007). *Second philosophy: A naturalistic method.* Oxford University Press.
Pincock, C. (2012). *Mathematics and scientific representation.* Oxford University Press.
Priest, G. (2006). *In contradiction* (2nd ed.). Clarendon Press.
Putnam, H. (1979). *Mathematics, matter and method* (2nd ed.). Cambridge University Press.
Rayo, A. (2009). Toward a Trivialist account of mathematics. In O. Bueno & Ø. Linnebo (Eds.), *New waves in philosophy of mathematics* (pp. 239–260). Palgrave MacMillan.
Rudin, W. (1976). *Principles of mathematical analysis* (3rd ed.). McGraw-Hill.
van Fraassen, B. C. (2008). *Scientific representation: Paradoxes of perspective.* Clarendon Press.
Yablo, S. (2001). Go figure: A path through fictionalism. *Midwest Studies in Philosophy, 25*, 72–102.
Yablo, S. (2017). If-thenism. *Australasian Philosophical Review, 1*, 115–132.

Shared Mathematical Content in the Context of Complex Systems

Hildegard Meyer-Ortmanns

1 Introduction

Usually methodological reductionism is understood as a reduction to more fundamental degrees of freedom on a more microscopic level. Zooming into matter from macroscopic bodies to molecules to atoms to electrons and nuclei, to protons and neutrons inside the nuclei, one finally ends up at quark and gluon degrees of freedom as constituents of protons and neutrons. Pushed to the extreme, this kind of methodological reductionism would be an attempt to describe macroscopic dynamics in terms of electrons, quarks and gluons, or explaining biophysics in terms of particle physics. From a practical point of view this appears immediately to be absurd. From a principal perspective the question should be allowed how far one can reconstruct emergent phenomena on macroscopic levels from the knowledge of the interaction rules on more microscopic levels or even the fundamental one if "fundamental" refers to the basic elementary constituents of matter.

Less ambitious but very worthwhile may be attempts to explain material properties in terms of underlying constituents on a lower level. Material science, technical and medical applications profit enormously from this kind of reductionism, as materials or equipments can be designed, based on a deeper understanding of what the essential control parameters are on underlying levels. Curing diseases on the genetic level in cases where they can be traced back to failures on this level erases the macroscopic symptoms from the root. This is what is usually understood as reductionism in physics and its applications. Nowadays this approach is often criticised, as definitely not all emergent features on the macroscopic level can be captured by a reduction to the more microscopic level.

However, this is not the type of reductionism that is the topic of this article. We refer to a reduction to the mathematical content rather than to the material constituents.

H. Meyer-Ortmanns (✉)
Physics and Earth Sciences, Jacobs University Bremen,
P.O.Box 750561, 28725 Bremen, Germany
e-mail: h.ortmanns@jacobs-university.de

S. Wuppuluri and I. Stewart (eds.), *From Electrons to Elephants and Elections*,
The Frontiers Collection, https://doi.org/10.1007/978-3-030-92192-7_18

Such a reduction strives for the relevant degrees of freedom.[1] What deserves to be termed relevant depends on the problem of interest. What is relevant for one aspect may be irrelevant for another.

Why should one identify the mathematical content? Occasionally, measurements are made which lead to almost identical results in quite different substances, materials or dynamical systems. In these cases, the similarity goes far beyond a superficial analogy, but refers to quantitative agreement, not only in some individual numbers, but in functional dependencies, such as the strength of the response with respect to perturbations, the approach of transitions into other (aggregate) states of matter, or the distribution of fluctuations on the surface of a growing blob. The experimental realizations are so different that one may think of a mere coincidence at first. On the other hand, such coincidences are so unlikely and contradict so much to the fact that the agreement is reproducible in repeated experiments that one gets curious and wants to disclose the causation behind these observations. What are the mechanisms that are responsible for strong similarities between seemingly unlinked problems? To find the answer may be quite challenging and often takes not only years, but even decades of extensive research, both on the experimental and theoretical side.

In the examples which we have selected for this contribution, we completely skip any steps of derivation. Thus we cut really long stories short and only confront the compact mathematical content with some of the diverse contexts, for which the content explains the shared observations in terms of a common mechanism behind them.

We will distinguish between four cases of increasing mathematical complexity to illustrate how the same mathematical content is realized in different contexts from physics and beyond. The first case is a differential equation that describes pattern formation in animal coats and in density fluctuations of air caused by a vibrating membrane (Sect. 2). The second case is a set of numbers, termed critical exponents, to classify the asymptotic behavior when singularities are approached in the vicinity of critical points. These exponents serve to classify singular behavior near phase transitions. Models sharing the critical exponents constitute universality classes (Sect. 3), a concept of much interest in statistical physics. In our third example, the common mathematical structure is a set of stochastic (chemical) reactions, representing predator-prey dynamics that is used to predict the lifetime statistics and phenomenology in the transition from laminar to turbulent flow. What sounds like a metaphoric wording at a first view turns out to be the appropriate mathematical reduction that is able to reproduce a certain set of measurements of transitional turbulent flow in a quantitative way (Sect. 4). Finally we focus on the Tracy-Widom probability distribution and an associated third-order phase transition which pop up in widely different contexts. The identification of the underlying mathematical

[1] A standard example from classical mechanics is a rigid body with idealized fixed distances between its mass points, it has six degrees of freedom to move in three-dimensional space, since six parameters are needed to describe its translational and rotational motion, although the body may consist of more than 10^{23} molecules.

content explains their universal occurrence, ranging from abstract problems in mathematics to fluctuations in growth processes of bacteria to (an approximation of) a phase transition in cosmology, to name just a few (Sect. 5). We conclude in Sect. 6.

2 Differential Equations for Pattern Formation

Pattern formation in animal coats is about how the zebra gets its stripes and the leopard its spots. Murray (1981a, 1981b) proposed that a single mechanism may be responsible for generating the variety of patterns in animal coats. His proposal was based on a hypothesis of Searle (1968) that a Turing mechanism of pattern formation may be behind the chemical concentrations. One starting point is a set of equations for two chemical species $A(\mathbf{r}, t)$, $B(\mathbf{r}, t)$ of the form

$$\begin{aligned}\frac{\partial A}{\partial t} &= F(A, B) + D_A \nabla^2 A \\ \frac{\partial B}{\partial t} &= G(A, B) + D_B \nabla^2 B \,,\end{aligned} \tag{1}$$

where F and G determine the nonlinear kinetics, $D_A \neq D_B$ are two diffusion constants, $\mathbf{r}, t$ denote the coordinates in space and time. According to A. Turing, spatially inhomogeneous patterns can evolve by diffusion driven instability if $D_A \neq D_B$. If the full reaction diffusion system is linearised about the (homogeneous) steady state, it leads (after some intermediate steps, going to dimensionless quantities and specifying the reaction kinetics) to an equation for the time dependence of the deviation $\mathbf{w}$ from a homogeneous state of species concentrations. It is given as

$$\frac{\partial \mathbf{w}}{\partial t} = \gamma \; J \; \mathbf{w} + D \nabla^2 \mathbf{w} \tag{2}$$

with Jacobian matrix J, γ a constant parameter, D the diffusion matrix. Its solution $\mathbf{w}(\mathbf{r}, t)$ can be expanded in terms of eigenfunctions $\mathbf{W}_k(\mathbf{r})$ of the spatial eigenvalue problem with time-independent solutions, defined by

$$\nabla^2 \mathbf{W} + k^2 \mathbf{W} = 0 \tag{3}$$

with $(\mathbf{n}\nabla)\mathbf{W} = 0$ for r on the boundary ∂C of the given diffusion domain C with normal vector $\mathbf{n}$, k being the eigenvalue of the spatial problem, so the full solution is $\mathbf{w}(\mathbf{r}, t) = \sum_k c_k e^{\lambda t} \; \mathbf{W}_k(\mathbf{r})$, with λ the eigenvalue that determines the temporal growth. Therefore, which spatial patterns evolve is determined by Eq. 3, by the spatial domain B and the boundary conditions. The spatial domain can have any shape, in particular the shape of a square, a rectangular, a cylinder or that of an animal coat. Varying this form leads to a variation of the patterns which are compatible with solutions of this eigenvalue equation. Thus $\mathbf{W}_k(\mathbf{r})$ denotes the spatial eigenmode of wavelength $1/k$ contributing to the deviations from the homogeneous solution.

In view of the stripes of zebras or the spots of leopards one may be tempted to interpret the patterns $\mathbf{W}_k(\mathbf{r})$ as pigment densities in the animal coats. However, the pattern generation mechanism takes place during morphogenesis of the embryo. What actually matters should be the conditions on the embryo's surface at the time of patterns activation. Therefore, the mechanism captured in Eq. 3 refers to the morphogen prepatterns for the animal coat markings, and according to Murray (1981b), it seems far from trivial to understand how the subsequent cell differentiation produces melanin in a way as to reflect the spatial prepatterns of morphogen concentrations, so that the zebra, for example, after full development, comes to its stripes. For a careful discussion about these topics we refer to Murray (1993).

So far we referred to Eq. 3 as the mathematical content of pattern formation, as its solutions (patterns of concentrations in space) vary with the boundary conditions on the reaction-diffusion domain C. Obviously there is no direct way of demonstrating this mechanism being at work during morphogenesis in mammalian embryos. Instead, it is much easier to visualize the impact of the geometric shape on the formation of patterns via a vibrating membrane, a drum, or a thin plate. The reason is that the initial[2] pattern formation by reaction diffusion mechanisms pose the same kind of eigenvalue problem as posed by the vibration of membranes, surfaces of drums, or thin plates. The difference is only in the interpretation of the variables $\mathbf{W}(\mathbf{r})$, representing now the time-independent amplitudes of the vibration as a function of the spatial coordinates $\mathbf{r}$.

To illustrate how the form and the scale of the vibrating plate determine the density fluctuations of the air at the surface of the plate, the fluctuations have been visualized in time-average holographic interferograms.[3] In this case, increasing the frequency of the sound waves at constant plate size is equivalent to increasing the plate size at constant frequency. This way the influence of the plate size can be easily controlled. When the plate size is too small, it does not sustain vibrations, they decay quickly. (Note that animals of small size usually have uniform coats.) When the plate size is effectively increased by increasing the frequency, more complex patterns of vibrations are sustained, ranging from simple and few domains to many structured spots resembling the spots of a leopard, see Xu et al. (1083) for nice visualizations.

In summary, the shared content was reduced to the eigenvalue equation Eq. 3 together with its initial and boundary conditions. Its very derivation depends already on the context. In case of pattern formation of animal coats the hypothesis was an underlying reaction-diffusion mechanism *à* la Turing, where the corresponding original equations have been reduced to Eq. 3 within a linear stability analysis. Eq. 3 as an equation for a vibrating membrane is derived in classical mechanics, in

---

[2] "initial" means as long as departures from uniformity are small such that the linear stability analysis applies.

[3] Time-average holographic interferometry involves recording a hologram with an exposure long compared to the period of vibration. The reconstructed image is then covered with fringes that can be used to map the vibration amplitude (Encyclopedia of Materials, 2001).

particular suited to illustrate the important role of boundary conditions on the solution of differential equations. Depending on the context, the meaning of the variables, here $\mathbf{W}(\mathbf{r})$, represent variations in pigment densities (after development of the animal), or density fluctuations of the air at the surface of the vibrating membrane.

The type of insight gained from the reduction to the mathematical content in form of an eigenvalue equation is an explanation of what controls the type of patterns in the absence of any external interference. After all, also the patterns in the wings of a butterfly have the same origin, though two "species" (kind of colors) would not be sufficient to distinguish the variety of colors, but no external painting is needed for an explanation. These patterns are self-organized and depend on the availability of finite domains on coats, wings, plates, and membranes. If complex pattern formation is explained in terms of a set of differential equations, referring to structural complexity, one may try to extrapolate and wonder which equations explain the emergence of functional complexity as observed in living systems. How far can the concept of self-organization be pushed? These are definitely open questions to date.

3 Critical Exponents

Phase transitions are events such as water evaporating into steam, metals transforming into superconductors, laminar into turbulent flow, solids into liquids. In general, phase transitions are visible as a sharp change in the macroscopic properties of a system, when a certain driving parameter like the temperature T is smoothly changed. Similar behavior to the change of a liquid as a function of the temperature is observed in magnetic phase transitions, for example in a ferromagnet, in which the magnetization at zero external field spontaneously takes a non-vanishing value $\pm M(T)$ below some critical temperature T_c and vanishes above T_c.

Phase transitions are by far not restricted to condensed matter physics. They are discussed in particle physics if the temperature or the density of matter are increased to extreme values, such that even protons and neutrons inside the nuclei of the atoms melt into their constituents, which are the so-called quarks and gluons. And also outside physics, phase transitions have been identified, for example in the context of computer science if algorithms diverge as a function of the system size: logarithmically, power-like at the critical point, or exponentially fast, depending on the parameter choice.

What makes phase transitions in general interesting is the drastic change in the macroscopic properties that usually comes as a surprise from the naive perspective of smooth and linear extrapolations.

In mathematical terms and in the language of thermodynamics, a phase transition occurs when a thermodynamic potential like the free energy is non-analytic, so that one of its derivatives has a kink or diverges. Usually first and second order transitions are considered, and more recently—as we shall see in Sect. 5—also third-order transitions. The order of the transition stands for the derivative of the thermodynamic potential which has a kink or divergence.

Of particular interest are critical points at which second-order phase transitions occur. They are characterized by diverging response functions such as the specific heat or the susceptibility. As the name suggests, the susceptibility describes how susceptible a system is to perturbations. The divergencies that occur at the critical point and strictly speaking only in the thermodynamic limit are further described by critical exponents which characterize the nature of the singularities in various measurements.

In this contribution, we do not present a full list of them, but just a few prototypical ones for systems in equilibrium. If we denote the critical parameter T_c (representing here a critical temperature) and introduce as normalized distance from T_c the quantity

$$t = \frac{T - T_c}{T_c}, \tag{4}$$

the critical exponents α, β, γ, ν are defined as follows. The exponent α characterizes the divergence of the heat capacity C ($C \sim |t|^{-\alpha}$), the exponent β the divergence of the order parameter M, that is the magnetization in case of a magnetic system or the liquid-gas density difference in a fluid ($M \sim |t|^{\beta}$), the exponent γ the divergence of the susceptibility χ at zero magnetic field, or the isothermal compressibility in case of a fluid ($\chi \sim |t|^{-\gamma}$), and the exponent ν the divergence of the correlation length ξ ($\xi \sim |t|^{-\nu}$). The assumed forms are based on the assumption that (thermodynamic) quantities can be decomposed into a regular part (which remains finite) and a singular part that may be divergent or has diverging derivatives.

The reason why these critical exponents are of more interest than the critical temperatures (or critical driving parameters) is that they are universal. From experiments it is known that widely different systems with critical temperatures differing by orders of magnitude and possibly completely different material composition approximately share the critical exponents, which depend only on a few fundamental parameters such as the dimension of space and the symmetry of the order parameter.

To be more concrete, striking first evidence came from a plot of coexistence curves between gas and liquid phases of eight rather different fluids when T/T_c was plotted over ρ/ρ_c with ρ the density of the substance in the liquid (ρ_L) and gas (ρ_G) phase, respectively. As early as 1945, Guggenheim (1945) presented the measured data, and assumed an exponent $\beta = 1/3$ for the fit. The data collapsed to this fit and satisfied almost perfectly the relation $\frac{\rho_L - \rho_G}{\rho_c} = \frac{7}{2}(1 - \frac{T}{T_c})^{1/3}$. The branch to the right of the critical density refers to the liquid phase, the left one to the gas phase. The branches come together at $T = T_c$ and $\rho = \rho_c$. The critical temperature itself varies from $T_c = -228.7\,°\mathrm{C}$ in Ne to $16.6\,°\mathrm{C}$ for Xe, to mention two out of eight substances.

Moreover, the exponent $\beta = 1/3$ is also shared in phase transitions in completely different systems, not just in diverse liquids. According to a classic experiment by Heller and Benedek (1962), $\beta = 0.335(5)$ is found for a magnet (with uniaxial anisotropy in spin space) such as MnF_2 (manganese fluoride). The result for a binary fluid mixture (CCl_4+C_7F_{16}) is $\beta = 0.33(2)$ (Thompson & Rice, 1964). Numerical results for the three-dimensional Ising model on a cubic lattice give $\beta = 0.327$ (Liu & Fisher 1989).

The only reason why we mention such concrete numbers and materials is to demonstrate that measurements of almost identical critical exponents (the mathematical content) in widely different substances (the physical context) calls for an explanation and makes the numbers interesting beyond their immediate context. What we have reported so far is the mere observation that the mathematical content apparently coincides for a number of systems near critical points.

To cut a long theoretical development short, the critical exponents, the shared universal behavior, can be derived in the framework of the renormalization group. In this framework, the dynamics on a coarse scale is mathematically derived from the dynamics on a more microscopic scale, in iterated steps, going from small to large scales. An important ingredient there is the distinction between relevant and irrelevant degrees of freedom. In particular, the singular behavior close to a critical point is determined by a few relevant degrees of freedom. This give at least a hint on why widely different substances (with many diverse degrees of freedom on the microscopic scale) may share the singular behavior, determined by the few relevant ones which remain relevant under iterated coarse graining.

It is instructive to see how the functional form of the singular behavior in terms of critical exponents, characterizing the divergence, can be alternatively obtained from a simple but strong assumption with far reaching implications. It is an assumption that can be derived within the renormalization group approach. When it is used as starting point, it appears ad hoc, but working it out, it reflects the experimental observations of the scaling behavior of observables at and near the critical point. The assumption concerns a *scaling hypothesis* to hold. It deals with the change of various quantities under a change of length scale. Usually, quantities with dimensions such as the volume change according to their dimensions. Near the critical point, so the hypothesis, the correlation length ξ is the only characteristic length of the system in terms of which all other lengths must be measured. Combined with the experimental observation that the correlation length diverges at the critical point in the large-volume limit, the system has no distinct length and must therefore be invariant under scale transformations. Magnifying a small part of the system to the original full system size, the magnified part cannot be distinguished from the original one. What is the corresponding mathematical manifestation?

Mathematically, scale invariance is reflected by the fact that thermodynamic (in case of temperature driven transitions) functions such as the correlation function Γ are homogeneous functions, that is

$$\Gamma(x) \sim x^{-p} \tag{5}$$

with p the dimension of $\Gamma(x)$. This means when the length unit is increased by a factor b, the new coordinates x' have to be rescaled according to $x' = x/b$ to refer to the new unit of length. As a homogeneous function, the correlation function then transforms according to

$$\Gamma(x/b) = b^{p}\ \Gamma(x). \tag{6}$$

This homogeneous transformation behavior would not be possible if Γ contained a length scale determined by an exponential decay, for example. More generally, a homogeneous function scales according to $f(q') = b^{D_f} f(q)$ for $q' = b^{D_q} q$ if D_q is the dimension of q and D_f the dimension of f. As it then turns out, based on the scaling hypothesis, all critical exponents can be derived and expressed in terms of only two dimensions D_h and D_t with D_t read off from the correlation length $\xi \sim t^{-\nu}$, so that $D_t = 1/\nu$, and D_h the dimension of $h = H/kT = (2 + d - \eta)/2$ with H the external magnetic field, T the temperature, k the Boltzmann constant, d the spatial dimension and η the anomalous dimension associated with the pair correlation function. It should be noticed that the realization of the scaling hypothesis in terms of homogeneous functions is intimately related to the observed power-law behavior in the critical regime. It is power laws which realize scale-free behavior.

One may wonder how universal the critical exponents are. The claim is not that a single set of critical exponents characterizes all kind of singular behavior of second-order phase transitions. The ones for which we quoted concrete numbers refer to the Ising universality class, named after the simplest representative of the class of models sharing this set of critical exponents. Other universality classes exist, in which one collects models with the same set of critical exponents. However, even within the same universality class, the material composition of the respective models may be widely different, as most of the microscopic details do not matter for the behavior close to a critical point.

In summary, the mathematical content in relation to critical behavior (typically observed at second-order phase transitions) refers to scaling laws in terms of critical exponents. These laws can be derived along different routes with the scaling hypothesis as an essential ingredient. Based on experimental observations, the scaling hypothesis is very plausible. However, more satisfactory in terms of a deeper understanding would be a derivation (rather than an assumption) of scale invariance in terms of more basic premises. This is indeed achieved in the renormalization group approach.

The reduction to scaling laws explains the coalescence of macroscopic measurements in quite different substances. What is measured is the shared singular behavior that coincides, not the microscopic behavior. Here a remark is in order in which sense we refer to singularities in real experiments, as the volume there is always finite, no correlation length gets ever infinitely large in the strict sense. Taking the thermodynamic limit, in which phase transitions occur according to their very definition, is a matter of convenient description. Already their precursors in the finite volume show characteristic signatures of the order of the phase transition in a finite-size scaling analysis. Therefore, there is no contradiction between singular behavior and real experiments. It is only for convenience that we reported on the critical exponents in the thermodynamic limit rather than in the corresponding finite-size scaling analyses.

4 A Set of Stochastic Reactions

In this section the mathematical content will be a set of stochastic reactions in terms of birth, death, predation and motion. As we shall see, this set explains the quantitative results of certain measurements near a transition from laminar to turbulent flow, which happen to be the same as those near the extinction of a species in a predator-prey dynamics of ecological systems. Both transitions are said to belong to the universality class of directed percolation, so they share the critical exponents of this universality class when the respective transition is approached. Anticipating the outcome that zonal flow in a pipe behaves like a predator of turbulent puffs in the laminar phase may sound like superficial metaphorical wording. As we shall see, this formulation may be taken literally in a certain sense, in which sense will be answered below.

Let us first describe the fluid close to turbulence in little more detail, although we have to refer to the original reference for a precise description of the experiments. Consider the flow of a fluid in a cylindric pipe of diameter D. Depending on the fluid's viscosity ν, its velocity u and the diameter D, the flow will be laminar as long as the velocity is low, more precisely as long as the Reynold's number $Re \equiv uD/\nu$ is below a critical value. The Reynold's number serves as a control parameter. Above its critical value, the formerly laminar flow becomes turbulent with eddies of different size. Flow velocity and pressure rapidly vary in space and time. The focus now is on how the transition to turbulence proceeds, thus we zoom into the transition region. Close to the transition, zonal flow emerges at large scales at the boundaries of the cylinder, which is activated by anisotropic turbulent fluctuations, but the zonal flow suppresses small-scale turbulence as long as the flow is in the laminar phase. For small Reynold's number, turbulence is only metastable, turbulent puffs decay with lifetime τ_d. For larger Reynold's number, the lifetime of these puffs increases and clusters of puffs split. The main observables (experimentally and numerically) are the lifetimes until the puffs decay τ_d, or until they split τ_s.

When H.-Y. Shih, T.-L. Hsieh and N. Goldenfeld performed direct numerical simulations of the Navier-Stokes equations in a pipe geometry to explore the transitional behavior (Shih et al., 2016), they identified a predator-prey like dynamics. The focus there was on a decomposition of large-scale modes that indicate some collective behavior (the zonal flow) as well as small-scale modes that are representative of the turbulent dynamics (the turbulent puffs). A careful analysis of the time series for the energy of the zonal flow, compared with the energy of the turbulent puffs, reveals oscillations, modulated by long-wavelength modes with a phase difference between zonal and turbulent energies of $\pi/2$. This observation suggests that the oscillations result from an activator-inhibitor mechanism, well known from predator-prey ecosystems.

In predator-prey ecosystems, the prey mode activates the predator which then grows. At the same time, the predator inhibits the prey. Less prey leads to the starvation of the predator. Its inhibition allows the resumed growth of the prey, which then activates the predator again, and the cycle restarts.

Table 1 Dictionary between a fluid and an ecosystem close to their respective phase transitions

Abstraction	Fluid	Ecosystem
Driving parameter	Reynold's number Re	Birth rate
Species A	Zonal flow	Predator
Species B	Turbulent puffs	Prey
Species E	Laminar flow	Nutrient

The observed oscillations led to the conjecture that a stochastic description in terms of predators and preys (rather than in terms of the Navier Stokes equations) might reproduce the results for the decay and splitting times τ_d, τ_s, obtained from the numerical simulations of the Navier Stokes equations. Note that this amounts to a reduction of the relevant degrees of freedom just to predators and preys (and nutrients, as we shall see).

According to Goldenfeld and Shih (2017), Shih et al. (2016), the map between the flow and the ecological system close to their respective transitions amounts to a dictionary that is shown in Table 1. The left column represents the abstract notation, with corresponding realizations in the fluid (middle column) and in the ecosystem (right column). Thus the role of the driving parameter is the Reynold's number in the fluid and the birth rate in the ecosystem. The zonal flow (species A) preys on turbulent flow (species B), while the laminar flow (species E) acts as nutrient of B. Stated differently, zonal flow is the collective mode, important in the transition region, it is generated by turbulent fluctuations and regulates turbulence by shearing it, thus acting as its predator.

To adequately describe the oscillations in "population" concentrations of zonal flow and turbulent zones, a description in terms of a set of stochastic predator-prey reactions with birth, death, and predator activity is required (since a description on the mean-field level would fail to reproduce the observed oscillations). If one lists all possible interactions between zonal flows as predator (A), turbulence as prey (B), with E representing the laminar flow, based on the dictionary of Table 1, one ends up at the following set of stochastic reactions (Goldenfeld & Shih, 2017):

$$\begin{array}{llll} B+E \stackrel{b}{\rightarrow} B+B & (a) & B+B \stackrel{c}{\rightarrow} B+E & (b) \\ A+B \stackrel{p}{\rightarrow} A+A & (c) & A+B \stackrel{p'}{\rightarrow} A+E & (d) \\ A \stackrel{d_A}{\rightarrow} E & (e) & B \stackrel{d_B}{\rightarrow} E & (f) \\ B \stackrel{m}{\rightarrow} A & (g). & & \end{array} \tag{7}$$

These relations may be read as follows: Turbulent flow B is generated from laminar flow E (acting as "nutrition") with rate b (a), it decays into laminar flow with rate c (b), so the reaction with rate c can be absorbed in a modified rate b. Zonal flow A preys on turbulent flow B with rate p by generating more zonal flow (c), or more laminar flow E with rate p' (d). Again the reaction with rate p' can be absorbed into a modified rate p. Furthermore zonal and turbulent flow can change into laminar flow with "death" rates d_A (e) and d_B (f), and turbulent flow can mutate to zonal flow with rate m (g). If the predators, preys, and nutrients are assigned to a spatial grid and allowed to move to neighboring sites $\langle ij \rangle$ with hopping rate D, the set of reactions can be summarized as

$$\begin{array}{llll} B_i + E_j \xrightarrow[<ij>]{b} B_i + B_j \quad (a) & & A_i + B_j \xrightarrow[<ij>]{p} A_i + A_j \quad (b) & \\ A_i + E_j \xrightarrow[<ij>]{D} E_i + A_j \quad (c) & & B_i + E_j \xrightarrow[<ij>]{D} E_i + B_j \quad (d) & \\ A_i \xrightarrow{d_A} E_i \quad (e) & & B_i \xrightarrow{d_B} E_i \quad (f) & \\ B_i \xrightarrow{m} A_i \quad (g). & & & \end{array} \tag{8}$$

The former reactions ((a), (b)) and ((c), (d)) have now been summarized to (a) and (b) with modified rates b and p. This set of stochastic reactions, defined for species A, B, E, assigned to a spatial grid and specified by reaction rates of death d_A, d_B, predation rate p, birth rate b, mutation rate m and diffusion (or hopping) rate D, define the mathematical content that is supposed to be shared between the fluid close to the transition to turbulence (with turbulent zones, laminar flow and zonal flow as the relevant degrees of freedom) and an ecological system of species A, B, E with respective reaction rates, moving in space, here on a spatial grid.

To quantitatively verify the hypothesis about the shared content, the set of reactions (8) can be either numerically simulated via Gillespie simulations or analytically evaluated by deriving a master equation from (8) and solving the master equation approximately in certain limits. Shih et al. (2016) performed stochastic simulations and measured for various birth rates b the (logarithm of) the survival probability of a cluster of prey as a function of time. The survival here refers to the time until decay t_d or until the splitting t_s of the cluster. Vice versa, they also measured the survival probability of a cluster of prey as a function of b for various time instants t_d or t_s, where time is measured in units of τ_0 with $\tau_0 \sim D/U$ with D the pipe diameter and U the average velocity of the fluid. Now, while t_d/τ_0 or t_s/τ_0 are varied, the same survival probabilities $P(b, t_d)$, $P(b, t_s)$ can be achieved if the birth rate b is varied accordingly, and vice versa. The striking result is that in the so-called scaling regime, all data collapse to two universal scaling functions, one for τ_0/τ_d and one for τ_0/τ_s, with τ_d, τ_s the lifetime and splitting time of a cluster, respectively, plotted as a function of the birth rate.

These scaling relations hold in the critical regime and tell us how a change of the survival or splitting time can be compensated by a change of the birth rate. The measured scaling relations correspond to a superexponential dependence of the inverse decay and splitting times τ_d and τ_s as a function of the birth rate b. This agrees remarkably well with the superexponential scaling behavior in real transitional pipe turbulence experiments (Hof et al., 2008) and in numerical simulations (Avila et al., 2011). Note that the superexponential scaling behavior here has been derived by modeling the transition to turbulence in terms of predator-prey dynamics.

Moreover, when the world-lines $x(t)$ of the clusters of prey (representing turbulent flow, being species in the ecosystem) are plotted as a function of time such that the color codes the local density of prey, the time evolution shows a branching tree on a spatial grid. This branching tree resembles the branching pattern that is observed in the spreading of turbulence in the numerical simulations of Avila et al. (2011), which analyzes the onset of real turbulence in pipe flow.

As we have seen in Sect. 2, critical behavior is characterized by critical exponents, and critical exponents serve to classify systems into universality classes. What is the universality class to which the transition to turbulence belongs to? The patterns of world-lines, which are nothing but space-time diagrams, just mentioned as an observation in the predator-prey dynamics, are a typical signature of a well-known universality class. This class comprises many models and is named after one of its members, that is directed percolation.

Directed percolation itself represents various versions of a process that mimics filtering of fluids through porous media along a given direction. At a critical permeability p_c, a transition happens to a phase in which a path can percolate through the whole lattice. Less abstract, this happens when a fluid finds its path through the entire sponge when the permeable channels become connected.

At this point one may indeed become skeptical what the connection should be if it really exists: the connection between percolation processes on grids, transient turbulent flow in the spatial continuum of a pipe, and predator-prey dynamics close to the extinction of one species.

Being again rather brief and skipping any derivation, the answer is a shared contact process that amounts to one realization of a directed percolation process. It is shared between the dynamics in the vicinity of the onset of turbulence and the predator-prey dynamics close to extinction. Numerical support of this result comes from work by M. Sipos and N. Goldenfeld, who simulated a directed percolation process in the geometry of a pipe (Sipos & Goldenfeld, 2011). An occupied site on the lattice corresponds to a turbulent correlation volume, an empty site to a laminar region, while turbulent intensity may diffuse into nearby laminar regions. Starting with a localized puff of turbulence, the directed percolation region fades away below the critical percolation threshold, whereas it spreads out over the pipe above the threshold. The lifetimes of puffs are measured and reproduce the superexponential behavior for the decay rate below p_c and the splitting rate above p_c.

Analytical support of this result should be obtained from the basic set of reactions, Eq. (8), which we identified as the mathematical content of transitional turbulence. Thus it remains to reduce the set of reactions (8) to those of a contact process with diffusion. According to a heuristic argument, the prey population (species B, representing turbulent regions) is very small near the transition to prey extinction, no predator (species A, zonal flow) can survive, so the set of reactions (8) reduces to

$$B_i \xrightarrow{d_B} E_i; \qquad B_i + E_j \xrightarrow[<ij>]{b} B_i + B_j; \qquad B_i + E_j \xrightarrow[<ij>]{D_{AB}} E_i + B_j. \tag{9}$$

Turbulent regions decay into laminar flow with rate d_B, they are born from laminar flow at neighboring site with rate b and diffuse into neighboring sites with diffusion rate $D_{AB} = D_B$ (more precisely, they exchange their place with a laminar neighboring site). These reactions are exactly those of a contact process with diffusion as a model for directed percolation.

If one does not want to rely on the heuristic argument in view of the strong result, a longer analytical detour amounts to the following steps: A master equation is assigned to the set of reactions (8). The Doy-Peliti path integral formalism is employed to arrive at an action that simplifies to that of Reggeon field theory near the extinction transition, which is known to be in the universality class of directed percolation. For details we refer to Mobilia et al. (2007), Täuber (2012) and references therein.

Let us summarize the line of arguments and comment on the procedure. Before the work of Goldenfeld and Shih (2017), Shih et al. (2016), which was in our main focus, experimental measurements in particular of decay and splitting times of turbulent puffs were available. In the critical region the results showed a superexponential scaling of these times as a function of the Reynold's number, whose change drives the transition from laminar to turbulent flow. These results were confirmed by numerical simulations of the Navier-Stokes equations in Shih et al. (2016). A closer look at the transition region showed that an activator-inhibitor mechanism is at work which is familiar from ecological predator-prey systems. In a next step, degrees of freedom were selected that were supposed to be the relevant ones both for the transition to turbulence and to an extinction in ecosystems. These degrees of freedom are shared in the mathematical structure that was the set of stochastic reactions (8) and termed the mathematical content in the context of this article. When the stochastic set of reactions was simulated in Shih et al. (2016), it reproduced the phase diagram for transitional pipe turbulence from the laminar (nutrient only) phase to the phase of expanding turbulent puffs (population), respectively. Particularly in the transition region from laminar to turbulent flow, the relevant degrees of freedom were further reduced to those of a contact process that is representative for directed percolation processes.

Why all this effort of establishing mathematical maps and dictionaries between highly diverse experiments? What kind of insight is here gained from this mapping between the fluid and the ecosystem, or of both to a contact process? The point is that the reduction to the mathematical content explains the striking experimental observations of superexponential scaling relations in terms of a *simple mechanism* that leads in other (material) realizations to the same (here superexponential) behavior of lifetimes with branching patterns familiar from directed percolation. Merely in words, this mechanism reads as follows: In the transition region, the turbulent puffs act as prey of zonal flow in the presence of nutrition (laminar flow). The corresponding set of reactions capture the lifetime statistics and the phenomenology of the pipe flow.

It should be noticed that the set of reactions in terms of birth, death or decay, interaction (via predation) and mutation reminds to a set of fundamental reactions between elementary particles: Also there the basic processes refer to creation, decay, annihilation, transition [4], however with one drastic difference, that is the material composition. In the context of particle physics, the set of reactions involves the basic constituents of matter such as electrons interacting via gauge bosons, which mediate the interaction. In the context of fluid dynamics or ecology, the material composition of the "species" A, B, E is widely different and anything but elementary: In an interpretation from ecology, A, B, E represent three trophic levels within a food chain with A and B differing in their nutritional relationship to the primary energy source E. Species A plays the role of a predator, B of a prey and E the nutrient of the prey, realized, for example, in a population of competing species of bacteria. In comparison, in transitional turbulence, A represents a macroscopic collective long-wavelength mode, the zonal flow of the fluid, while B denotes the turbulent puffs. More precisely, it is the *fluctuations* of large scale zonal flows A and small-scale turbulent regions B in the laminar flow E that are modelled in the set of stochastic reactions. (The fluctuations are actually essential, as the corresponding mean-field dynamics does not lead to oscillations.) The material composition of A, B, E is the same type of fluid molecules and not their conclusive in view of the question of interest.

It should be further noticed that in spite of the big difference of the material realization in the ecosystem, the turbulent fluid, or a set of reactions in particle physics, (the latter not further considered here), the mathematical formulation is very similar across the scales. It makes use of stochastic reactions, which lead to the generation, deletion, decay, composition, fragmentation, or mutation of entities; therefore, a formalism or a framework that evaluates these basic stochastic processes, should uncover common mechanisms on fairly different scales if the processes on an abstract level are the same. The framework and the language it uses are suited to bridge the scales from electrons to living (elephants) and social (elections) systems.[5]

In the spirit of "standard" reductionism to underlying, more microscopic degrees of freedom, it is still natural to wonder whether the predator-prey dynamics can be derived (rather than assumed) from the underlying Navier-Stokes equations. As argued in Goldenfeld and Shih (2017) such a derivation would involve hardly controllable approximations without providing further insights into the relevant mechanism that explains the experimental measurements. It should be emphasized that here the experimental results have been solely predicted from the predator-prey dynamics, without any reference to the Navier-Stokes equations. Thus this kind of standard reductionism to a more microscopic description is not even needed for providing a deeper understanding.

A further comment may be added on the attribute "relevant", assigned to the selected degrees of freedom. Being relevant is not a universal attribute, but depends on the question addressed. The predator-prey dynamics according to Eq. (8) predicts

[4] The authors of Goldenfeld and Shih (2017) actually assign graphs to these reactions which are reminiscent to Feynman graphs.

[5] In view of elections, the species may be human individuals of a population, whose elections of a representative are based on rules of a "game" that is formulated in terms of stochastic reactions for taking decisions.

the lifetime statistics and the phenomenology of the pipe flow correctly. It does not (neither is it supposed to) provide an explanation in terms of nonlinear dynamics. For example, it does not identify the bifurcations in a more microscopic description which cause the first generation of turbulent puffs when the Reynold's number is varied, nor does it explain their decay or splitting as a result of more microscopic nonlinear interactions between the fluid molecules and their physical properties of them. Therefore, the reduction to a few degrees of freedom should not be mixed up with an oversimplification, as long as their relevance is referred to a specific problem of interest.

5 The Tracy-Widom Distribution and a Third-Order Phase Transition

From the 18th and 19th century on, it was the Gaussian probability distribution with its symmetric bell shape which was in the focus of interest and whose universal appearance was later explained by the central limit theorem. The central limit theorem holds for independent and identically distributed random variables. It states that the properly normalized sum of random samples of size N tends to a normal or Gaussian distribution for $N \to \infty$ even if the original variables themselves are not normally distributed. In this section we consider another probability distribution, which nowadays attracts much attention. It is the Tracy-Widom (TW-) distribution, named after C. Tracy and H. Widom, who calculated exactly the probability distribution of the largest eigenvalue of a certain set of random Gaussian matrices in the large-N limit (Tracy & Widom, 1994; 1996). In contrast to the Gaussian distribution, it is lopsided and encountered in cases with strongly correlated variables, thus it is clearly more challenging to deal with it than with a Gaussian distribution. We will identify the TW-distribution along with a third-order phase transition as the mathematical content behind a large number of seemingly completely unlinked problems. Our summary is based on work of S. N. Majumdar and G. Schehr (2014) and references therein.

One way to encounter the TW-distribution is to ask for the probability distribution of the maximal eigenvalue of a certain set of Gaussian random matrices, along the lines of Tracy and Widom. This question seems peculiar and particular and at best of interest for mathematicians. Before we come to the distribution in more detail, a short excursion to a problem of ecosystems should motivate why the spreading of maximal eigenvalues may be of interest from the physics perspective.

Random matrix theory (Wishart, 1991) is known to have applications in statistical physics, mesoscopic physics, particle physics, but also in finance and telecommunications, to name just a few of them. Using random matrices for interactions, it may be an expression of "not knowing better", that is, lack of knowledge what the real interactions are, or even if they are known in principle, they may effectively amount to random ones. It was R. M. May (1972), who used a random real and symmetric matrix of size $N \times N$ to describe the pairwise interactions between N species in an ecosystem. The model is named after him as the May-model. The largest eigenvalue of this matrix then decides about the stability of the system when interactions are turned on, given a concrete interaction matrix. This eigenvalue fluctuates with the realization of the random matrix. Thus the probability that the ecosystem remains stable under the inclusion of pairwise random interactions of strength α amounts to the cumulative distribution function of the largest eigenvalue λ_{max} to remain below the threshold $1/\alpha$. As noticed already by R. M. May, this probability undergoes a sharp transition in the $N \to \infty$- limit. In this limit, the system is for sure (with probability 1) stable if $\alpha < \alpha_c$ and unstable for $\alpha > \alpha_c$. This was the first indication that actually a sharp phase transition may be associated with the cumulative distribution function of the maximal eigenvalue λ_{max}.

If one has to deal with a phase transition, it is natural to ask for an analytical description of how the transition is washed out into a crossover for large but finite N, as one expects from phase transitions in a finite volume. A second natural question is about its order. The order of a phase transition is determined by the order of the derivative of a suitable thermodynamic potential like the free energy that has a kink or jump at the transition point. In the context of the ecosystem, it should be kept in mind that at a first place the critical coupling strength signals only a bifurcation point at which a fixed point gets unstable. Therefore the relation to a thermodynamic language with a phase transition and non-analytical behavior of a thermodynamic potential has to be first established. How can one assign a free energy to the cumulative probability distribution of the maximal eigenvalue of an abstract set of random matrices, and what is the analytical form of the crossover region for finite, but large N?

Both questions have been answered by Majumdar and Schehr (2014). Now we have to cut a rather long development of theoretical work short and directly present their result for the cumulative probability distribution $F_N(w)$ of the maximal eigenvalue λ_{max} for taking a value w, given a finite fixed value of N. It is given as

$$F_N(w) \approx \begin{cases} \exp\left(-\beta N^2 \Phi_-(w)\right) & w < \sqrt{2} \text{ and } |w - \sqrt{2}| \sim \mathcal{O}(1), \\ F_\beta(\sqrt{2}N^{2/3}(w - \sqrt{2})), & |w - \sqrt{2}| \sim \mathcal{O}(N^{-2/3}), \\ 1 - \exp\left(-\beta N \Phi_+(w)\right), & w > \sqrt{2} \text{ and } |w - \sqrt{2}| \sim \mathcal{O}(1). \end{cases} \tag{10}$$

First of all, it is obvious that we have to distinguish between small and typical fluctuations of order $\mathcal{O}(N^{-2/3})$ about the mean $\langle \lambda_{max} \rangle = \sqrt{2}$, and large fluctuations of $\mathcal{O}(1)$. For small fluctuations, the cumulative probability function is described by the' β-Tracy-Widom distribution F_β which is explicitly known for $\beta = 1, 2, 4$, labelling respectively the Gaussian orthogonal (1), unitary (2), and symplectic (4) ensemble of random matrices. This part describes the crossover region. The respective functions $\Phi_-(w)$ and $\Phi_+(w)$ to the left and right of $\lambda_{\max} = \sqrt{2}$, have been explicitly calculated as the large-deviation functions in (Dean and Majumdar, 2008), (Majumdar and Vergassola, 2009), respectively. The left and right large-deviation tails look different, the left one dropping off steeper than the right one, accordingly the physical mechanism on both sides of the distribution must be different. At this point one may wonder what the physical interpretation of this peculiar probability distributions should be.

To get at least a hint on why the physical mechanism is expressed in terms of a pushed (left) and pulled (right) Coulomb gas, it should be mentioned that $F_N(w)$ can be written as a ratio of partition functions

$$F_N(w) = Z_N(w)/Z_N(w \to \infty) \tag{11}$$

with

$$Z_N(w) = \int_{-\infty}^{w} d\lambda_1 ... \int_{-\infty}^{w} d\lambda_N \exp\left[-\beta/2\left(N\sum_{i=1}^{N}\lambda_i^2 - \sum_{i\neq j}\ln|\lambda_i - \lambda_j|\right)\right]. \tag{12}$$

Here the integrand is proportional to the joint probability density function of the eigenvalues of the respective random matrices. Now it should be noticed that formally $\exp[-\beta E(\{\lambda_i\})]$ with $E(\{\lambda_i\}) \equiv N/2\sum_{i=1}^{N}\lambda_i^2 - \sum_{i\neq j}\ln|\lambda_i - \lambda_j|$ can be interpreted as the Gibbs-Boltzmann measure of an interacting Coulomb gas of charged particles on a line: λ_i, the former eigenvalues, denoting the position of the i-th charge on a line in this interpretation, β the inverse temperature, not necessarily restricted to the values 1, 2, 4, and two competing interaction terms: the first one an external confining parabolic potential, and the second one a logarithmic repulsion (corresponding to a Coulomb repulsion in two dimensions, while the set of charges is restricted here to a one-dimensional line). The meaning of w, which appears as argument in the cumulative probability distribution and refers to the probability for the maximal eigenvalue being smaller or equal this value w, translates to the maximal position on the one-dimensional line that a Coulomb charge can take; w acting therefore like the position of an infinitely high wall which the cloud of Coulomb charges can at most touch, but not penetrate. It should be emphasized that this interpretation is merely based on the *formal* coincidence between the contributed weights of the probabilities of eigenvalues and a Coulomb gas restricted to a line.

However, this interpretation is rather useful as it allows to formulate the basic abstract mechanism, shared by many occurrences of the TW-distribution and an associated phase transition, in simple terms of what is going on in the corresponding Coulomb gas. To the right of $\sqrt{2}$, in the stable regime of weak coupling α in the May-model, mentioned initially, the cloud of charged Coulomb particles, the " Coulomb droplet", has a soft edge with density vanishing as the square root of the edge at $b < w$, and a hard wall at w, both being still separated by a gap. When the control parameter is tuned, the gap between the cloud and the wall shrinks and is zero at $w_c = \sqrt{2}$. For smaller parameters $w = 1/\alpha$ (stronger coupling α), on the left of $\sqrt{2}$, the Coulomb droplet is squeezed by the wall. This costs an energy of order $\mathcal{O}(N^2)$, while the energy of pulling one charge out of the droplet for weak coupling α (large w) is only of order $\mathcal{O}(N)$.

Moreover, in the formulation of a Coulomb gas it makes sense to talk about a thermodynamic phase transition, as the position of the wall approaches a critical value $w_c = \sqrt{2} = 1/\alpha_c$. The value of α_c was already determined by May (1972). Furthermore, it is shown in Majumdar and Schehr (2014) that

$$\lim_{N\to\infty} -\frac{1}{N^2} \ln F_N(w) = \begin{cases} \Phi_-(w) & w < \sqrt{2}, \\ 0 & w > \sqrt{2} \end{cases} \tag{13}$$

with $\Phi_-(w)$ the large-deviation function to the left of $\sqrt{2}$ as before and $\Phi_-(w) \sim (\sqrt{2} - w)^3$ for $w \to \sqrt{2}$ from below. Since $\ln F_N(w)$ is proportional to the free energy of the Coulomb gas, its third derivative has a discontinuity at the critical point, identifying the phase transition as being of third-order by the usual definition of the order of a phase transition.

Both interpretations related to the Coulomb gas and the maximal eigenvalue distributions in random matrices may still appear rather peculiar, based on a formal coincidence and anything but universal. However, as pointed out in (Majumdar and Schehr, 2014), meanwhile third-order phase transitions with the "Coulomb-gas mechanism" have been identified in a number of other systems. Here we mention only one of them, a phase transition in two-dimensional U(N)-lattice quantum chromodynamics (QCD) when the coupling is increased from weak to strong. This transition runs under the name Gross-Witten-Wadia transition and is known since quite some time (Gross & Witten, 1980; Wadia, 1980). In the large-N limit, the free energy per plaquette there also undergoes a third-order transition at a critical coupling constant. Here, limiting angles $\pm\pi$ play the role of hard walls for the eigenvalues. The coupling g corresponds to the coupling α in the May-model.

Just to indicate what this other phase transition actually refers to: Quantum chromodynamics is the gauge theory of strong interactions, one of the four fundamental interactions apart from gravity, electromagnetism and the weak force. It is responsible for the binding of quarks via gluons inside protons and neutrons of the atomic nuclei. At extremely high densities and/or temperatures, phase transitions or rapid crossovers are expected between the state of matter in which quarks and gluons are confined inside protons and neutrons, and a kind of quark-gluon plasma, where they are free. For realistic parameters, this transition or crossover is not too exotic to have happened once in the early universe when the hot and dense quark-gluon plasma cooled down below the critical parameters, or to happen nowadays in collisions at large hadron colliders. However, due to the strong interaction strength, this transition is intrinsically hard to describe analytically (Meyer-Ortmanns & Reisz, 2007).

The Gross-Witten-Wadia transition was one among a number of attempts to approximate this transition, by considering its counterpart in two dimensions on a space-time lattice as approximation to the three-dimensional space continuum. Independently of its relevance for the "true" phase transition in the early universe, the mathematical structure of this U(N)-lattice QCD transition strongly resembles the one in the May-model, although it should be noted that the interpretation of the exponential factors in relation to QCD are weights, contributing to paths in a multi-dimensional path integral, while they are joint probability density functions in the May-model.

It is fascinating to see where else in physical models third-order phase transitions have been identified together with a TW- distribution in the crossover regime. Here it is not the place to give a complete review on early work preceding the discovery of a third-order transition in relation to the TW-distribution according to (Majumdar and Schehr, 2014). In the following we mention just two earlier (classes of) results, referring to stochastic growth models on the one side, and the Ulam process on the other. Less exotic than an application to QCD in the early universe are irregularly growing bacterial colonies. A new colony on the surface of a nutrient medium will grow into a roughly circular blob with an outer edge that gets increasingly rough in the course of time. An accurate description of the irregular fluctuations in space and time is provided by the KPZ-equation, named after Kardar et al. (1986). It predicts two exponents associated with the width of the distribution of fluctuations and describes also surface growth and interface fluctuations in irregular growth. Initially its intimate relation to the TW-distribution was not recognized. However, models of the KPZ-type were later exactly mapped to the Ulam problem, named after the mathematician S. Ulam, and the relation to the TW-distribution was established this way (Prähofer & Spohn, 2010).

What is the Ulam problem about? Consider N permutations of the first N integers. and assume they are all equally likely. The Ulam problem is to determine, for given N, the distribution of the length l_N of the longest increasing subsequence (out of all possible increasing subsequences). It is obvious that as a function of N, this length will fluctuate, depending on the permutation. As a surprise, the fluctuations about the average of this length, $\langle l_N \rangle$, are determined by the TW-distribution (Baik et al., 1999). Moreover, an exact mapping between variants of the Ulam problem and models of the KPZ-type have been established. This means that also a number of discrete models of the KPZ-type follow the exact TW-distribution as does the Ulam problem.

KPZ-type models themselves show already a remarkable universality between quite different realizations; their equivalence to variants of the Ulam problem and their relations to random matrices that showed up in shared TW-distributions inspired further search for a common mechanism underlying all these linkages. The subsequent analysis of Majumdar and Schehr (2014) and their identification of the third-order transition together with the TW-distribution in the crossover regime have been a quite important step toward an explanation for their ubiquitous occurrence. Universal features of phase transitions are familiar in statistical physics and refer to the approach of singularities, as indicated earlier. This approach contains characteristic precursors at finite but large N of the limiting case ($N \to \infty$).

As the common mechanism behind the third-order transition here was formulated in terms of a Coulomb gas, this was the simplest physical interpretation which is compatible with the mathematical content, though the same mathematics describes a wide variety of contexts as we have indicated. Our examples covered a range from pure mathematical abstractions like random matrices and the Ulam process, to ecosystems (the May-model), to surface growth fluctuations (KPZ-models), to phase transitions in strongly interacting matter in cosmology. Here, a reduction to smaller material compositions would be obviously meaningless once we deal with purely mathematical constructions of our mind such as the Ulam process.

6 Summary and Conclusions

The reduction to the mathematical content, often initiated by common observations in widely different contexts, does not pretend to explain the complex world in all its facets. The dynamics is reduced to those degrees of freedom, which turn out to be relevant for the aspect of interest. These are often only a few, orders of magnitudes less than those suggested by the material composition, and in terms of these few, the explanations become simple and mechanistic. From a practical point of view, this kind of understanding allows to control the processes as a function of a few parameters; from a theoretical perspective, it tells us about the basic mechanisms that are responsible for the striking similar observations in seemingly unlinked situations. The more systems share a certain behavior, the more universal the mechanism is. From the background of theoretical physics it is not surprising that such universal behavior often shows up when singularities are approached, in particular, in relation to phase transitions, and phase transitions are ubiquitous in natural and artificial systems.

References

Avila, K., Moxey, D., de Lozar, A., Avila, M., Barkley, D., & Hof, B. (2011). *Science, 333,* 192.

Baik, J., Deift, P., & Johansson, K. (1999). *Journal of the American Mathematical Society, 12,* 1119.

Dean, D. S., & Majumdar, S. N. (2006). *Physical Review Letters,97*, 160201; Dean, D. S., & Majumdar, S. N. (2008). *Physical Review Letters,77*, 041108.

Encyclopedia of Materials: Science and Technology. (2001).

Goldenfeld, N., & Shih, H.-Y. (2017). *Journal of Statistical Physics,167*, 575.

Gross, D. J., & Witten, E. (1980). *Physical Review D21*, 446.

Guggenheim, E. A. (1945). *Journal of Chemical Physics, 13,* 253.

Heller, P., & Benedek, G. B. (1962). *Physical Review Letters,8,* 428.

Hof, B., de Lozar, A., Kuik, D. J., & Westerweel, J. (2008). *Physical Review Letters,101*, 214501 (2008).

Kardar, M., Parisi, G., & Zhang, Y. C. (1986). *Physical Revie Letters, 56,* 889.

Liu, A. J., & Fisher, M. E. (1989). *Physica A,156*, 35.

Majumdar, S., & Schehr, G. (2014). *Journal of Statistical Mechanics: Theory and Experiment,P01012*, 1–32.

Majumdar, S. N., & Vergassola, M. (2009). *Physical Review Letters,102*, 060601.

May, R. M. (1972). *Nature, 238,* 413.

Meyer-Ortmanns, H., & Reisz, T. (2007). *Principles of phase structures in particle physics, World Scientific lecture notes in physics* (Vol. 77). World Scientific.

Mobilia, M., Georgiev, I. T., & Täuber, U. C. (2007). *Journal of Statistical Physics, 128,* 447.

Murray, J. D. (1993). *Mathematical Biology II: Spatial Models and Biomedical Applications.* Springer.

Murray, J. D. (1981a). *Philosophical Transactions of the Royal Society B, 295,* 473.

Murray, J. D. (1981b). *Theoretical Biology, 88,* 161.

Prähofer, M., &Spohn, H. (2000). *Physical Review Letters,84*, 4882 (2000); Gravner, J., Tracy, C. A., & Widom, H., *Journal of Statistical Physics,102*, 1085; Majumdar, S. N., & Nechaev, S., *Physical Review E,69*, 011103; Imamura, T., & Sasamoto, T. (2004). *Nuclear Physics B,699*, 503; Sasamoto, T., & Spohn, H. (2010). *Physical Review Letters104*, 230602.

Searle, A. G. (1968). *Comparative Genetics of Coat Color in Mammals*. Academic.
Shih, H.-Y., Hsieh, T.-L., & Goldenfeld, N. (2016). *Nature Physics, 1,* 245.
Sipos, M., & Goldenfeld, N. (2011). *Phys. Rev. E,84*(3), 035304.
Täuber, U. C. (2012). *Journal of Physics A: Mathematical and Theoretical, 45,* 405002.
Thompson, D. R., & Rice, O. K. (1964). *Journal of the American Chemical Society,86,* 3547.
Tracy, C. A., & Widom, H. (1994). *Communications in Mathematical Physics,159*, 151.
Tracy, C. A., & Widom, H. (1996). *Communications in Mathematical Physics,177*, 727.
Wadia, S. R. (1980). *Physics Letters, 93B,* 403.
Wishart, J. (1928). Biometrika *20*, 32; Wigner, E. (1951). *Proceedings of the Cambridge Philosophical Society47*, 790; Mehta, M. L. (1991). *Random matrices* (2nd ed.). Academic.
Xu, Y., Vest, C. M., & Murray, J. D. (1083). *Applied Optics,22*, 3479.

United but not Uniform: Our Fecund Universe

Timothy O'Connor

The sciences collectively depict a diversified but unified universe: one populated by a great many diverse kinds of organized structures that, while built of the same basic constituents, engage in distinctive processes described by domain-specific explanatory principles and laws. All scientific theories are tentative, and we may be confident that significant revisions await us in every field of inquiry—and not least in basic physics, which has yet to deliver a unified understanding of small-and large-scale dynamics. Nonetheless, the very general picture of my opening sentence is a permanent fixture of scientific understanding.

Yet this general picture admits of two importantly different variations, still remaining at a high level of generality. On a broadly reductionist vision, everything that occurs is fully determined by the unfolding dynamics of fundamental physical structures throughout space and time, with *precisely* the same dynamics operative in every context. To be sure, the sundry organized systems and their characteristic behavior can be described only by using special-science *concepts*, but even so they are ontologically derivative. They are one and all coarse-grained structures and *patterns* embedded in a mosaic that is entirely painted by the elements. On the other, emergentist vision, at least some organized structures manifest fundamental causal powers of their own, not supplanting but supplementing the collective powers of their composing elements. Whatever the fundamental dynamical principles of physics prove to be, they are not complete. When caught up in and around emergent structures, basic physical entities are subject to macroscopic as well as microscopic causal influences, which only together determine their trajectories.[1]

[1] The following three paragraphs, with some modification, are taken from my "The Emergence of Personhood: Reflections on *The Game of Life*," in Malcolm Jeeves, ed., *The Emergence of Personhood: A Quantum Leap?*, Eerdmans Press, 2015, 143–162.

T. O'Connor (✉)
Department of Philosophy, Indiana University, Sycamore 026, Bloomington, IN 47405, US
e-mail: toconnor@indiana.edu

S. Wuppuluri and I. Stewart (eds.), *From Electrons to Elephants and Elections*,
The Frontiers Collection, https://doi.org/10.1007/978-3-030-92192-7_19

To better grasp the difference in these two visions, consider mathematician John Conway's cellular automaton, *the Game of Life*. *Life* is a dynamic and spatially and temporally discrete two-dimensional infinite grid. One sets an arbitrary initial state by assigning one of two basic properties, *live* or *dead*, to each of the square cells, which one can represent by different colors. Each subsequent state of the grid is wholly determined by applying the following three rules to every cell (which has eight 'neighbors' in every direction, including diagonally):

Birth: A dead cell with exactly three live neighbors becomes a live cell.

Survival: A live cell with two or three live neighbors stays alive.

Death: In all other cases, a cell dies or remains dead.

Surprisingly, over time, stable clusters of various kinds arise and exhibit macro-level patterns of activity and interaction with other cluster types. (Apt names have been given to certain recurring sorts, such as 'oscillator', 'glider', 'puffer', and 'eater'.) Once these clusters appear, their macro-level behavior can be studied in ignorance of the three micro-level rules that underlie them. More surprising still, different sorts of high-level patterns are observed in games with different initial conditions.

If we think of Conway's 2-D grid as a toy world (there are 3-D versions of it as well), the three basic rules and its initial state constitute its general 'physics,' and the high-level dynamical patterns exhibited by parts of it can be thought of as its 'chemistry' or 'biology.' *Life* vividly illustrates how high-level patterns can be distinctive in form, underwriting similarly distinctive causal *explanations*—if we intervene in a *Life* world by manipulating the value of one higher-level variable and holding other relevant factors constant, we can change the value of another higher-level variable in very predictable ways and in a variety of contexts—while not in any way altering or supplementing the basic dynamics that drive the world's evolution. The 'physics' of a standard *Life* world is causally closed, with each total configuration of the grid at a time t_1 (and so also the state of every stable cluster) being strictly determined by its state at the previous time t_0 in accordance with the three basic rules. What is more, the micro-level causation asymmetrically determines the macro-level causation. For this reason, while macro-level rules of interaction in *Life* worlds cannot be deduced from micro-level rules, macro-level behavior is ontologically and causally derivative.

Now consider the following variation on *Life*, which, for reasons that will become apparent, we may call *Fecund-Life*. (I will hereafter refer to the original version of Life as *Lego-Life*, to mark its emphasis on the sufficiency of the most basic units for determining the features and arrangement of everything composed of them.) Imagine that you are handed a tall stack of (very large!) numbered sheets of graph paper. On them are changing snapshots of a *Life* world (with shaded squares signifying 'live' cells). Your job is to figure out the most compact transitional rules capturing change/stasis in every cell between adjacent sheets. After flipping through many pages, you hit upon the Birth, Survival, and Death rules. Continuing to check subsequent pages to verify that the rules hold without exception, you hit upon a page where the result departs in a small way from what the rules predict. The divergence is restricted to a complex star-shaped cluster that first appeared on the previous page. Flipping ahead, you observe that as more of these star-shaped clusters appear, their subsequent evolution, too, departs from what the three basic rules predict. Further investigation reveals that the form the divergence takes is identical in each case. As a result, you may once again predict what the future world-states will be, using *modified, disjunctive* forms of the original rules that invoke the star-shaped configuration. The new Birth rule, e.g., has the form: a dead cell with exactly three live neighbors becomes a live cell, *except* when occurring within the bounds of, or immediately adjacent to, a star cluster, in which case…. Imagine that as star clusters come into contact, new modifications of the original rules are required to fully capture the way stars interact. You find that the most compact way to capture star-associated behavior is to assign primitive new properties ('bright', 'golden') to star clusters and then to describe the precise impact of these properties on the basic cellular dynamics via additional laws. *These laws are no less fundamental*—even though they come into play only in particular kinds of structured contexts.

Reduction-minded thinkers often suggest that such a fundamentally new kind of macroscopic behavior would be inexplicable or 'unintelligible', something whose analogue in our physical world we therefore rightly set our face against when engaged in empirical investigation of the mechanisms of nature. However, to understand such novel phenomena as woven into a causally unified whole, we need only enrich our conception of the properties of the fundamental elements of a world. Staying with the *Fecund-Life* analogue, we would be led to suppose that, in addition to *live* and *dead*, whose dispositional profiles are largely captured by the original unmodified rules, the cells contain a further set of dispositions towards *cooperatively* contributing to *macroscopic-but-fundamental* properties, a disposition that is merely latent until triggered by the requisite configurational context. As we may say, the elements of such a world are fecund, the dispositional 'seeds' of the very structured unities that exert kinds of causal influence that extend beyond that of the elements. In this understanding, the world is fully causally unified without being operationally uniform.

Do we live in (a more complex and higher-dimensional analogue of) a *Lego-Life* world or instead a *Fecund-Life* world? By standard empirical measures, this remains an open question. Notice that I described the emergent powers of *Fecund-Life* by imagining the failure of the basic rules to fully capture the dynamics in particular structured contexts. Such a scenario is easy to confirm or disconfirm in a simple cellular automaton, whose elemental behavior is simple and discrete, turning on the value of a single variable within a small number of cells. In the real world, the richer features of trillions of subatomic particles in continuous trajectories underlie any macroscopic phenomena. While we have been able to steadily push past previous limits on measurement and computation, no one imagines that we will ever be able to monitor in situ the totality of quantum-theoretic underpinnings of any large-scale composed systems and assess the theory's descriptive/explanatory completeness. It's testable 'in principle,' but only in principle, requiring something akin to a disembodied Laplacian super-intelligence. Short of that, the best we might hope for would be to demonstrate the success or failure of certain explanatory reductions, under necessary simplifying assumptions, within localized areas in a stepwise manner (neural activity to constituting chemical activity, and psychologically described activity to neural activity), and also (crucially) at the system's boundary conditions. We're just not there yet, if ever we shall be. Furthermore, if attempted reductions fail, one also has to reckon with the persistent possibility that we have simply gotten the underlying theory wrong in some respects.

Given the present undecided state of the question on ordinary empirical grounds, whence the faith of some in physics fundamentalism? Neuroscience is ground zero for our question as it bears on human beings and other sentient animals, the complex systems of keenest interest and also the most plausible loci of the kind of strongly emergent behavior that is modeled in *Fecund-Life*. The neuroscientists I know tell me that they can't imagine conducting their science without the working assumption that all large-scale neural processes are entirely the outworking of the processes of excitation and communication within and among the vast array of constituent neurons (while allowing that our pretty good understanding of such processes is not complete). Call this stance '*methodological* reductionism.' I have no quarrel with it. For it seems to me necessary to theoretical advancement, *even if ontological reductionism is false.* We need to understand thoroughly the nature of the causal influences contributed in *every* context by underlying elements and their associated mechanisms before we can precisely characterize the form and magnitude of the influence of emergent features in the special contexts in which they are generated and sustained. (That form and magnitude is given by the difference between what is observed in the organized emergent context from what would be predicted by the application of a correct theory of non-emergent mechanisms.) Thus, the way to understand strongly emergent phenomena of the sort modeled by *Fecund-Life* (if such there be) is to be a methodological reductionist until you bump up against hard limits, and then take stock.[2] But a well-motivated methodological stance is not evidence for, or even a rational presumption in favor of, an ontological thesis on a question that is straightforwardly empirical. (Unlike the philosopher, your typical scientist does not spend much time thinking about what will turn out to be true at the end of scientific inquiry. Rather, she is thinking about what working hypotheses are most useful in advancing current understanding. For one with such immediate aims, it is all-too easy but mistaken to slide from the quite proper methodological stance of looking for reducing explanations to drawing an evidential conclusion.)

If anything, a stance *toward* strong emergence might be the rationally necessitated default. The universal context necessary to making sense of all forms of scientific inquiry is the cooperative agency of scientists themselves. For this reason, the effective and purposive activity of experimentalists and theorists is an ineliminable part of our being rationally justified in accepting their theories. We tacitly believe that those involved in constructing a scientific theory and marshalling evidence for it

[2] A nice description of such an imagined process and its practical challenges is given in Richard Corry, *Power and Influence: The Metaphysics of Reductive Explanation* (Oxford: Oxford University Press, 2019), Chap. 10.

were generally aware of what they were doing and effectively guided by reasons, practical and theoretical, for so acting. Rescind that assumption and our reasons for thinking the theory to be well-supported by evidence collapse, as the entire collective activity must be seen as mere phenomena, physically unified but not rationally guided. It would be to saw off the large and sturdy branch supporting the smaller branches that are the individual theories. I draw attention to this universal context of science because it is at least doubtful that the assumption of consciously self-aware and reasons-guided agents is consistent with ontological reductionism. According to the latter, all *fundamentally* efficacious causal processes are the interplay of non-conscious and non-rational forces, filling out a pointillist mosaic across the fabric of spacetime. Coarse-grained patterns including human behavior are present in the mosaic, but they are consequent upon, not prior constraints to, its microphysical completion.

Philosophers who argue that there is no inconsistency between reductionist fundamental ontology and reasons-driven human action urge that it is enough that the psychological processes of human beings are 'weakly' emergent: a manifestation of regular patterns that sustain distinctive kinds of causal explanation. All that is necessary to ground causal-rational explanations, they contend, is that our behavior is preceded by particular beliefs, desires, and intentions and would not have occurred without them. Such psychological patterns are consistent with the basic physical domain's being causally 'complete' or self-contained, sufficient to determine psychological and all other complex phenomena. However, this weakly emergent understanding of reason-governed action secures only the *predictive usefulness* of psychological concepts and principles in categorizing families of complex neural configurations and relations among them. It doesn't show what is needed, viz., that configurations manifest a distinctive kind of productive *efficacy* by virtue of their falling under such concepts—that an agent's grasping of her reasons for acting is a salient factor in bringing about her acting as she does.[3]

Some philosophers see a way out here by endorsing a skeptical doctrine advanced by David Hume. Hume argued that our notion of causal production is empty, a folk-theoretic idea that dissolves under critical scrutiny. In reality, any kind of causal process, whether micro or macro, is only a recurring pattern of one sort of thing followed by another. There are the many individual pixels of paint—localized events such as particle accelerations or field excitations in tiny spatiotemporal regions—that collectively exhibit both small- and large-scale patterns, and that is all. Nothing produces the events, as there are no natural agents inherently 'disposed' to manifest themselves in certain ways. The idea of natural agents exercising causal powers is jettisoned in favor of entities exhibiting mere causal regularities. Forces don't produce particle accelerations, as 'force' is just a term encapsulating observed regularities in the changing spatial coordinates of particles over time. If causation in

[3] For a thorough of overview of the varieties of accounts of both weak and strong emergence, see my "Emergent Properties" *Stanford Encyclopedia of Philosophy*, (Fall 2020 Edition), Edward N. Zalta (ed.), URL = < https://plato.stanford.edu/archives/fall2020/entries/properties-emergent/ > .

general reduces to recurring patterns of a particular kind, there is no difficult question to confront regarding where in physical reality productive efficacy resides (it's nowhere!). Psycho-behavioral causation is then secure, as the requisite patterns are readily discerned in everyday life as in psychological science, no less real than the patterns in the elements composing them. Hume's austere and deflationary metaphysics is driven ultimately by an impoverished account of sensory perception (and an implausible associated account of concept formation), and it is hardly the way forward here. As Hume himself recognized more clearly than his latter-day followers, it creates a severe epistemological problem: if the natural causal patterns scientists have actually observed in tiny fragments of reality do not reflect the stable propensities of matter but are instead brutely contingent matters of fact, then our confidence that they hold constant through the vast reaches of unobserved reality, stretching to the distant past and future, can be no more than an animal faith. If physical events do not literally constrain one another—if they are, in Hume's phrase, 'loose and separate'—then all bets would be off: we might just be inhabiting a temporarily stable patterned phase within a larger whole in which a different pattern, or many different patterns, or no intelligible pattern at all, occur in other regions. On Hume's metaphysical vision from which all causal power is banished, the only constraint on the global distribution of events is logical consistency. There are vastly many distributions of events consistent with all observations to date, and just one in which the deep patterns tracked by our physical theories hold throughout.—*Yes, but good scientific practice bids us to favor the 'simplest' generalization consistent with the data.* Indeed. However, the assumptions of scientific practice, as Hume recognized, plausibly rest on the further unreflective assumption that observable patterns are not brute happenstance, not a cosmic coincidence, but instead are grounded in the causal capacities of physical individuals and systems, capacities that are uncovered (at least approximately) by scientific inquiry. That is why the reductionist's bid to save psychological agency by embracing Hume's neutered metaphysics of causation is such a desperate measure. It's deeply incongruent with a fundamental conception that animates science.

There is a further reason to think the question of whether the shape of our world is *Fecund-Life* or *Lego-Life* is an important one: it plausibly bears on the *value* of persons as well as the value of other sentient animals. You might reasonably suppose that some dynamically changing configurations of the *Lego* variety are more valuable than other kinds of configurations, but it is hard to credit the thought that they have a deeper *kind* of value altogether, in the way that we commonly suppose living persons to have a deeper kind of value than any non-sentient dynamical system, no matter how interesting the latter from a physical or mathematical point of view. Destroying a non-sentient dynamical system may be a bad thing in certain contexts, but it is not a disvalue comparable to the wanton killing of an innocent person. (We do judge the destruction, whether intentional or not, of certain artifacts or features of a natural landscape or ecosystem to be a serious loss. But such judgments typically are partly a function of their impact on human communities or animal populations, and so the value at issue is partly instrumental, rather than intrinsic.) Our assigning high value to the lives of conscious creatures, and especially of conscious persons, plausibly reflects a tacit belief that consciousness and other person-level attributes have intrinsic qualities that sharply set them apart from other intrinsic attributes of physical things. Whether an overarching moral commitment might properly (rationally) constrain our vision of the causal structure of reality, or instead should only be consequent upon it, whose character is independently determined, is an interesting philosophical question, and I do not mean here to be expressing an opinion it. My point is simply that the reductionist's ontological vision of things is plausibly at odds with a pervasive moral vision that he presumably shares.

I turn now to the question of whether it is possible to give plausible general conditions on any possible instances of strong emergence in nature. This question is most fruitfully addressed by looking for clues in the range of plausible candidates. Consider first a widespread phenomenon that some regard as the most straightforward candidate for emergentist analysis, because it appears in the quantum domain itself. Individual particles or particle systems become 'entangled' when they interact in certain ways, such that thereafter measurable features such as position and momentum are correlated in a way that cannot be accounted for in terms of separate states of the individuals. Such pluralities are treated as coupled systems, certain of whose intrinsic properties are irreducibly holistic. Counterintuitively, they retain this coupled status when the individuals are widely (even space-like) separated. But notice that here the holism is limited to the value or magnitude of a feature that is also had by its components. (Correlated "spin" values, e.g., are permutations on the fundamental feature of spin, also had by non-entangled particles.) Thus, it does not involve a fundamentally new type of basic property at a systemic level. Even so, it is a very low-level, ubiquitous indicator of a non-*Lego* character to our world.

One step up from quantum entanglement, it has been argued (more controversially) that the structure of chemical molecules is not wholly determined by the properties of and dynamics between their subatomic constituents.[4] Here the suggestion seems to be that a kind of structure spontaneously emerges, partly constraining subsequent constituent dynamics. More complex candidate cases occurring in the biological domain also link emergence to structure: e.g., the interplay of bottom-up and top-down principles in systems biology and the conscious, intentional, and purposive character of human and animal psychology. As Jonas (1966) and Maturana and Varela (1980) emphasize, life in all its forms, beginning with the simple cell, imposes a boundary maintained through an unceasing process of the loss and acquisition of parts. This suggests that careful exploration of the formation of resilient boundaries and the nature of their effects, *on a case-by-case basis* for distinct general kinds of biological entities, is the best means for understanding the dynamics of strong emergence in our fecund world. Spontaneous, weakly emergent organizational principles of the sort modeled by cellular automata adequately explain some dynamical systems. But concluding that these principles are adequate to all cases of interest is hasty, reflecting an anti-empirical ideology more than good science. It also flies in the face of our own lived experience as conscious, purposive agents. We don't merely track information in neural networks; we consciously grasp some of what is thereby

[4] See Woolley (1978, 1998), Primas (1981), and for philosophical analysis, Hendry (2010, 2019).

registered. We don't merely act in ways that reliably co-vary with our changing perceptions and psychological attitudes; we sometimes consciously form goals and choose from among possible ways to implement them. As the needed qualifier 'some' signals, we are curiously hybrid creatures. Our richer representational and action-directing capacities overlay the older mammalian machinery that neuroscience has begun to reveal. The chaotic interplay of, e.g., conscious reason and hormones results in behavior and individual and cultural evolution that is highly unpredictable (sometimes wonderfully creative, sometimes distressingly foolish).[5] Ultimately, we must look to future developments both in the understanding of largescale network dynamics in neuroscience and its wider context in the organism as a whole to understand the structured conditions under which collective mechanisms are activated, giving rise to the panoply of qualities in virtue of which we are subjects and knowers and purposive doers.

The preceding may suggest that I think merely weakly emergent phenomena to be of little theoretical significance. In reality, I think only that enthusiasts have misplaced their significance. The many varieties of weakly emergent structure have importance within the overall architectonic of the world, first, by providing a stable platform for strongly emergent structure (in some cases, as with the emergence of persons), and second (and much more generally) by providing a structured, middle-sized world for strongly emergent structures to engage, including making possible our own scientific activity and theorizing. Students of abstract structure explore *Lego-Life* worlds and other cellular automata as conscious knowers, 'hovering over the deep', grasping the varied macroscopic patterns that weakly emerge and their underlying basis. Remove such consciously intelligent Laplacian observers altogether, and the significance of intermittent coarse-grained macroscopic patterns wholly determined by the fundamental dynamics recedes. It is significance *to* ourselves and any other conscious knowers there may be—which knowers we implicitly take to transcend such patterns.

I suspect there is a final source of resistance to accepting that our world includes sentient knowers who are strongly emergent. If our world is so, its elemental building blocks are partly latent seeds whose complete fruition in the fullness of time is something altogether new and valuable, unlike imaginable alternative worlds with flat and 'sterile' physics. To some, the latent-seed physics of strongly emergent worlds looks implausibly 'rigged', giving too strong an appearance of having been designed. It is similar to the way that exquisitely fine-tuned numerical constants in the laws and initial conditions of contemporary physics give the appearance of being rigged for life, motivating theorists to develop deeper theories that curb the appearance.[6] But whatever force we attribute to 'the appearance of design' we may

---

[5] This theme is developed nicely by anthropologist Ian Tattersall in his contribution to Jeeves (2015), op cit., 37–50.

[6] The most popular strategy involves 'multiverse' hypotheses on which our universe is but one of very many, the result of a mechanism for generating new 'Big Bangs' (universe-initiating singularities) that differ in the values of the constants in the laws and initial conditions. Theories that implement this strategy purport to turn the suspiciousness of our world's fine-tuning into a mere observer selection effect in a mostly lifeless sea of chance distribution over possible values.

find in fundamental physics, we ought not to let our sympathies for or against the conclusions that some will draw from philosophical design arguments to drive our assessment of the evidence itself. We have direct evidence of just this one, apparently fine-tuned-for-life universe, *and* the existence within it of consciously purposive and self-aware knowing agents is a fundamental datum presupposed in all scientific inquiry. We ought to reason *from* this datum rather than seeking to explain it away. Contemporary philosophy of mind and action is littered with hand-waving attempts at reductive analyses of consciousness, of intentionality, and of purposive agency. All have proven inadequate, and for fundamental reasons, with no hope of adequate successor accounts in sight.

However, it is possible to see such proposals in a more positive light, by reconceiving their ambitions. The currently popular integrated information theory (IIT) of consciousness, while it fails utterly to analyze or explain the qualitative and subjective character of conscious experience, might prove to be an important insight into the kind of neural architecture that is necessary for conscious experience to arise. And the varieties of 'functional role' and 'tracking' theories of the intentional (or directed) character of mental states are undoubtedly pointing at important and at least prevalent relational characteristics of our intentional states, such that identifying their purely information-theoretic neural analogues will again double as guides to uncovering the mechanisms sustaining the genuine articles. As I emphasized earlier, getting clear on the precise character of strongly emergent phenomena starts with a comprehensive and well-confirmed theory of the underlying mechanisms. So let the reductive methods of normal science proceed apace, while not seeing its successes as somehow indicating an inevitable outcome of the still far-off endgame. That success is what both the emergentist and reductionist predict, while the emergentist retains all lived experience on his side.

References

Corry, R. (2019). *Power and influence: The metaphysics of reductive explanation.* Oxford University Press.

Hendry, R. (2010). Ontological reduction and molecular structure. *Studies in History and Philosophy of Modern Physics, 41*, 183–191.

Hendry, R. (2019). Emergence in chemistry. In S. Gibb, R. Hendry, & T. Lancaster (Eds.), *The routledge handbook of emergence* (pp. 339–351). Routledge.

Jonas, H. (1966). *The phenomenon of life: Towards a philosophical biology.* Harper & Row.

Maturana, H., & Varela, F. (1980). *Autopoiesis and cognition.* D. Reidel.

O'Connor, T. (2015). "The emergence of personhood: Reflections on *The Game of Life*," in M. Jeeves (Ed.) *The emergence of personhood: A quantum leap?*, Eerdmans Press, pp. 143–162.

O'Connor, T. (2020). "Emergent properties," *Stanford encyclopedia of philosophy*, (Fall 2020 Edition), E. N. Zalta (Ed.) https://plato.stanford.edu/archives/fall2020/entries/properties-emergent/.

Primas, H. (1981). *Chemistry, quantum mechanics and reductionism.* Springer.

Woolley, R. G. (1978). Must a molecule have a shape? *Journal of the American Chemical Society, 100*, 1073–1078.
Woolley, R. G. (1998). Is there a quantum definition of a molecule? *Journal of Mathematical Chemistry, 23*, 3–12.

Probability, Typicality and Emergence in Statistical Mechanics

Sergio Chibbaro, Lamberto Rondoni, and Angelo Vulpiani

1 Introduction

Statistical mechanics originates in the study of the properties of macroscopic bodies, i.e. of objects made of very large numbers of microscopic particles (atoms or molecules) whose dynamics follows mechanical laws, that are classical or quantum, depending on the case (Landau & Lifshitz, 1980; Ma, 1985). One may formally write these equations for all the particles in the system, and may in principle solve them. However, the number of degrees of freedom, hence of equations to solve, is huge and, in addition, the initial conditions are not known, therefore the solution of this problem is impossible in practice. At the same time, knowledge of positions and velocities of all the particles, which amounts to endless tables of numbers, is not particularly informative, if one is interested in temperature, pressure, elasticity, magnetization etc. This impossibility is hardly any concern.

On the other hand, whether the microscopic motions are very complicated or not, whether they are known or not, we observe that the macroscopic behaviours are relatively simple and understandable in terms of a reduced number of "observables" following relatively simple laws. In SM, the observed macroscopic simplicity, in spite of the expected complexity of the microscopic motions, is attributed to the statistical nature of the macroscopic laws, which reduces the complexity by averaging over the many degrees of freedom. The result is not purely mechanical (Castiglione et al., 2008; Ma, 1985; Landau & Lifshitz, 1980), and *qualitatively* differs from merely mechanical laws. That is why probabilities constitute the fundamental tool of SM.

S. Chibbaro
LISN Université Paris-Saclay, Rue du Belvedère, 91405 Orsay cedex, France
e-mail: sergio.chibbaro@universite-paris-saclay.fr

L. Rondoni
Dipartimento di Scienze Matematiche, Politecnico di Torino, Corso Duca degli Abruzzi, 24, 10129 Torino, Italy
e-mail: lamberto.rondoni@polito.it

A. Vulpiani (✉)
Dipartimento di Fisica Universitá degli studi di Roma "La Sapienza",
Piazzale A. Moro, 5, 00185 Roma, Italy
e-mail: angelo.vulpiani@roma1.infn.it

S. Wuppuluri and I. Stewart (eds.), *From Electrons to Elephants and Elections*,
The Frontiers Collection, https://doi.org/10.1007/978-3-030-92192-7_20

Their usage in physics has been pioneered by the founding fathers of SM, Maxwell, Boltzmann and Gibbs, who changed the very idea of the term *prediction* in physics and, as a consequence, in philosophy as well.

Given that this subject is one and a half century old, why should we discuss it today? In our opinion there are at least three good reasons for that:

1. The subject is of interest by itself, both for scientists and philosophers, since it it exemplifies how a new phenomenon may *emerge* from the typical behaviour of a lower level one. In particular, the relation between microscopic and macroscopic laws is paradigmatic of how reductionist approaches to complex phenomena in many branches of science are prone to failure (Batterman, 2002; Berry, 1994; Chibbaro et al., 2014a, b).
2. The subject is pedagogically relevant: with respect to other appealing but rather speculative frameworks, like ecology or cosmology, it allows a concrete discussion of the main conceptual issues concerning the link between different levels of description of a given reality.
3. The subject is important in the development of current technology: for instance, challenging frontiers for the applications of statistical physics are provided by systems with a small number of degrees of freedom, far from the thermodynamic limit, such as those of interesting in bio- and nano-technologies. Another frontier is given by non Hamiltonian models, which are considered appropriate in the description of granular materials, active matter, epidemics, etc. In these cases, one or both of the original assumptions of SM, namely the very many degrees of freedom and the Hamiltonian dynamics, are absent. Therefore the foundations and applicability of the theory have to be scrutinised (Ma, 1985), in the light of a presumably even higher relevance of probabilities than in the original framework of SM (Zanghì, 2005).

The relevance of probability theory for SM stems from the original idea of Boltzmann, who associated macroscopic (thermodynamic) quantities to averages of mechanical observables of the microscopic constituents of matter. In particular, he adopted frequency of events as the basic notion of probability (Goldstein, 2001; Vulpiani et al., 2014). In Boltzmann's SM, probability has no relation to measures of ignorance or uncertainty,[1] and it does not make any use of collections of identical objects.

This is part, instead, of Gibbs approach to SM, in which averages are computed with respect to probabilities that represent how the microscopic phases of large ensembles of identical objects are distributed in their phase spaces. This corresponds to the classical notion of probability, which differs from the frequentist notion, but it is commonly expected be equivalent to that. As computations of time-averages are much harder than ensemble calculations, one commonly accepts ergodic hypothesis, which amounts to such equivalence, and proceeds as prescribed by Gibbs. Therefore, a question arises about the link between the probabilistic computations of SM, and

[1] This inspired a whole branch of mathematics, known as ergodic theory, which represents one way of introducing probabilities in the analysis of the otherwise rigidly deterministic dynamical systems.

the results of laboratory experiments, which are conducted on a single realization of the macroscopic object under investigation.

In our opinion, the main theoretical issue to be addressed, in order to answer this question is the justification of *typicality*, i.e. of the fact that time averages of macroscopic quantities in the evolution of a single system are very close to averages of that quantity over ensembles of microscopically distinct but otherwise identical replicas of that system (Goldstein, 2012). This fundamental property can be seen as *emergent* in the proper limits.

To convince ourselves that this is not a hopeless project, we may refer to one of the best propositions used to link probability and physics, the Cournot's principle:

An event with very small probability will not happen.

Actually, this statement may be associated with the one in Jakob Bernoulli's celebrated book *Ars Conjectandi* (1713), which reads:

Something is morally certain if its probability is so close to certainty that shortfall is imperceptible.

We do not enter the debate about the validity of such a principle, see (Shafer & Vovk, 2001) for a nice analysis of it. However, we recall that eminent mathematicians, such as P. Levy, J. Hadamard, and A. N. Kolmogorov, considered the Cournot's principle as the only sensible connection between probability and the empirical world. That connection granted, Levy stressed the concrete character of probability, arguing that, at the ontological level:

Probability is a physical property just like length and weight

In this chapter, we shall explain how macroscopic laws emerge as statistical laws from the microscopic ones: in passing from the microscopic realm to the macroscopic one, novel properties arise, which are alien to the microscopic realm. In summary: (i) we first illustrate the main ideas of Boltzmann, and the entailing ergodic hypothesis for systems made of very many degrees of freedom. Then, we will analyse some examples with regard to irreversibility and typicality: (ii) to this purpose, the Ehrenfest model will be used. This is a stochastic process concerning N non-interacting particles, and it can be rigorously analysed, showing that, in the $N \to \infty$ limit, irreversible behaviours characterize almost all realizations of the process; this will be followed by (iii) numerical simulations of single systems made of many particles, showing their irreversibility.

We note that conceptually a stochastic process is fundamentally different from the deterministic reversible dynamics of Hamiltonian particle systems. Nevertheless, it can be rigorously proven that particular deterministic systems can be mapped into stochastic processes, and when that is not the case, the presence of chaos in interacting particle systems effectively amounts to a certain degree of dynamical randomness. Assuming that this is the case for systems of interest, as it has been repeatedly demonstrated in the literature, and as we will also show, the stochastic process we consider turns useful because it allows a pregnant quantitative analysis of the onset of typicality in the large N limit.

2 Probability and Real World

Discussing the foundations of SM necessarily starts with the two seminal contributions given by Boltzmann (Cercignani, 1998; Goldstein, 2001):

1. the use of probabilities in the calculation of physical quantities;
2. the link between microscopic dynamics (mechanical laws) and macroscopic properties (thermodynamics).

The second is formalised by Boltzmann's celebrated relation

$$S = k \ln W \,. \tag{1}$$

where S is the entropy of a given state, and W is number of possible microscopic configurations corresponding that macroscopic state. In the Hamiltonian dynamics picture, this number is then identified with the phase space volume occupied by the relevant microscopic phases. For a system of N particles each with d degrees of freedom, a microscopic phase is a $2dN$-dimensional vector, $\Gamma = (Q_1, P_1; Q_2, P_2; ...; Q_N, P_N)$, whose components are the d-dimensional coordinates Q_i and momenta P_i, $i = 1, ..., N$ of all particles. The volume in the phase space is thus defined for a fixed energy as

$$W(E, V, N) = \frac{1}{N! h^{3N}} \int \delta(H(\mathbf{Q}, \mathbf{P}) - E) d^{3N}\mathbf{Q} d^{3N}\mathbf{P} \,, \tag{2}$$

where h is the $2d$-dimensional volume of a small cell, that we may think refers to a single particle (Landau & Lifshitz, 1980).[2]

From a philosophical standpoint, Eq. (1) plays the role of a *bridge law* (Chibbaro et al., 2014a), connecting the atomic level to the macroscopic one, and constitutes the fundamental ingredient of SM, that justifies all its applications to condensed matter physics and chemistry.

Linked to point I is the ergodic hypothesis, which connects dynamics and probability. This is done as follows. Consider a macroscopic object of N interacting particles each with 3 degrees of freedom, and let the microscopic state be described by $\mathbf{X} \in R^{6N}$. A measurement of some macroscopic quantity, for instance the pressure, is supposed to last much longer than the molecular time scales, and the result of the measurement is taken to be the time average over the measurement time $\mathcal{T}$, of some mechanical property that is function of $\mathbf{X}$. The measurement tool is therefore said to effectively compute the following quantity:

$$\bar{A}^{\mathcal{T}} = \frac{1}{\mathcal{T}} \int_0^{\mathcal{T}} A(\mathbf{X}(t)) dt \,. \tag{3}$$

[2] The use of the symbol h should not lead to believe that quantum effects are taken into consideration. In the present picture, quantum mechanics plays no role.

In principle, the computation of $\bar{A}^{\mathcal{T}}$ requires the initial condition $\mathbf{X}(0)$, and the determination of the time evolution following from that initial phase, $\mathbf{X}(t)$. Given that $\mathbf{X}(t)$ represents the complete microscopic motion, this is surely beyond any human capability.

Boltzmann's ingenious idea to overcome this difficulty, i.e. the ergodic hypothesis, is to replace the time average with a suitable average on the the phase space. He assumed that

$$\lim_{\mathcal{T}\to\infty} \frac{1}{\mathcal{T}} \int_0^{\mathcal{T}} A(\mathbf{X}(t))dt = \int A(\mathbf{X})\rho(\mathbf{X})d\mathbf{X} \,, \tag{4}$$

where $\rho(\mathbf{X})$ is the suitable probability density. The sense of this hypothesis is that the physically relevant but impossible computation of the time-average can actually be turned into a (generally exceedingly simpler) probabilistic computation. In particular, if the ergodic hypothesis is assumed to be valid, it is easy to derive also the canonical Boltzmann-Gibbs distribution for a system which exchanges energy with an external environment, and then deduce the corresponding thermodynamics. If successful, this process achieves the goal of SM.

The issue is now whether the ergodic hypothesis is valid or not in the cases of physical interest. Unfortunately, many numerical investigations, starting from the FPUT (Fermi, Pasta, Ulam and Tsingou) work (Gallavotti, 2007) on chains of non linear oscillators, as well as rigorous mathematical results, notably the KAM (Kolmogorov, Arnold and Moser) theorem on non integrable systems (Dumas, 2014), show that the ergodic hypothesis does not hold rigorously in the form given above, if generic functions of phase A are considered. One could, thus, naively conclude that ergodicity cannot be taken as central in the foundations of SM, and that it could even be misleading. As a matter of fact, one finds that the ergodic hypothesis cannot be so lightly discarded. Indeed, it turns out that:

1. the Boltzmann-Gibbs probability distributions (the classical ensembles) are valid;
2. molecular dynamics gives correct results, being generally based on the ergodic hypothesis.

To understand these facts, one may recall the original Boltzmann's reasoning, that has since been considered the standard explanation of the success of the ergodic hypothesis (Landau & Lifshitz, 1980; Ma, 1985), even though it has been variously challenged in time (Bricmont, 1996; Gaspard, 2005). That reasoning has been made mathematically rigorous by Khinchin (1949). In a nutshell, Khinchin's argues that the ergodic hypothesis is "practically" true, as far as physical phenomena are considered, if

- $N \gg 1$;
- suitable observables are selected;
- one allows for failure of (4) in a small region of the phase space.

Khinchin proved that, for the class of separable sum functions defined by

$$f(\mathbf{X}) = \sum_{n=1}^{N} f_n(\mathbf{q}_n, \mathbf{p}_n), \tag{5}$$

where each f_n represents a single particle contribution, the following holds

$$Prob\Big(\frac{|\delta f(\mathbf{X})|}{N} \geq \frac{c_1}{N^{1/4}}\Big) \leq \frac{c_2}{N^{1/4}} \tag{6}$$

where c_1, c_2 are constants, and $\delta f(\mathbf{X})$ is the difference between the time average starting at $\mathbf{X}$ and the average value computed in the microcanonical ensemble. It is worth noting that many interesting microscopic functions of phase are sum functions, like kinetic energy, the momentum etc. Then, Khinchin considered non interacting systems, whose energy (Hamiltonian) is expressed as

$$H = \sum_n h_1(\mathbf{q}_n, \mathbf{p}_n). \tag{7}$$

where each h_1 term is the energy of one particle. That is a serious limitation of the approach, but Mazur and van der Linden generalised Khinchin's result to the physically more interesting case of (weakly) interacting particles (Mazur & van der Linden, 1963), whose hamiltonian can be written as

$$H = \sum_n h_1(\mathbf{q}_n, \mathbf{p}_n) + \sum_{n,n'} V(|\mathbf{q}_n - \mathbf{q}_{n'}|) \,. \tag{8}$$

In brief, it has been proven that, although the ergodic hypothesis as formulated above is not generally rigorously true for physically interesting systems, it remains valid for physically relevant observables of a wide class of systems made of very many particles. Indeed, in this case, violations of the hypothesis are restricted to negligibly small regions of phase space, in which the system may fall with a probability of order $O(N^{-1/4})$, that vanishes in the $N \to \infty$ limit, but is definitely irrelevant "already" for 10^{24} particles.

The main ingredients of this reasoning are the large value of N, together with the fact that one only needs to consider a special class of phase functions. This makes by and large marginal the role of the details of the microscopic dynamics, apart from the fact that it must preserve phase space volumes, like Hamiltonian dynamics does.

One consequence of having a restricted set of observables and and very large N, is that one may separate different space and time *scales*. The fact that $N \gg 1$ implies that particles are much smaller than the macroscopic body they constitute. Moreover, when these particles are allowed to move almost freely in space, thanks to their weak interactions, their mean-free path λ has also to be much smaller than the characteristic macroscopic length L of the object they belong to. Such distances can then be associated with the corresponding time-scales, i.e. the times needed to cross them with the particles average velocity.

The separation of scales is fundamental for the emergence of novel phenomena, when passing from one level of description to another (Berry, 1994; Kadanoff, 2013; Drossel, 2015). Indeed, it is required for spatial correlations to be negligible over distances that are very small on the macroscopic scale, which is the basis for quantities such as the internal energy, the entropy etc. to be extensive, as observed in thermodynamics. Moreover, the separation of time scales allows the realization of local thermodynamic equilibrium in sufficiently short times, that the average performed by a measurement appears to account for all the possible values the observable of interest can take. Consequently, the initial condition is irrelevant, and the ergodic hypothesis is vindicated.

That is the content of the condition known as typicality, which states that extensive observables will stay close to their mean value; and we now see that such a condition is better established if the number of particles is larger. Therefore, in this framework, the "atypical" behaviours can be considered of vanishing probability when dealing with macroscopic objects, in agreement with thermodynamics, that is deterministic and excludes them.

We conclude this section noting that while the approach illustrated above provides a convincing basis for the applicability of SM to the description of macroscopic objects, it does not cover the whole spectrum of relevant problems. In particular, given a generic initial condition, and an observable $\mathcal{O}$, estimates of the minimum value of the measurement time $\mathcal{T}$, such that $\bar{\mathcal{O}}^{\mathcal{T}} \simeq \langle \mathcal{O} \rangle$, are hardly available. This problem has been widely investigated since the FPUT numerical experiment, in which the 1-dimensional nature of the system hinders the decay of various kinds of correlations, making the local thermodynamic equilibrium hard to establish (Gallavotti, 2007). While this does not allow a direct connection with thermodynamics, it does not prevent the use of SM, which in this respect generalizes the macroscopic theories to small systems, like 1-dimensional systems must be.

2.1 *Statistical Mechanics as Statistical Inference?*

As argued above, we believe that Boltzmann's justification of SM, based on the large number of degrees of freedom and on typicality, is conceptually satisfactory when dealing with the emergence of macroscopic phenomena from microscopic dynamics. Nevertheless, there exists a radical anti-dynamical point of view which takes SM as a mere form of statistical inference, and not like a theory of objective physical reality.

This view is philosophically pragmatic and anti-realistic, and it implies that probabilities measure the degree of truth of a logical proposition, rather than describing the state of a given material object. This approach has become quite fashionable in the framework of "complex systems", and can be traced back to the work of Jaynes (1957), who expressed this idea through the maximum entropy principle (MaxEnt): a general rule for finding the probability of a given event when only partial information is available. In a nutshell, the principle proceeds as follows: given the expected values c_i of m independent functions f_i, defined on a space of coordinates x, a probability distribution ρ is constructed, in such a way that

$$c_i = \int f_i(x)\rho(x)dx \equiv \langle f_i \rangle \quad i = 1, ..., m \, . \tag{9}$$

As the name anticipates, the construction proceeds by maximisation of a formal entropy function H, under the constraints (9), which is thought to generalize the Gibbs entropy (Castiglione et al., 2008; Gibbs, 1906). In practice, using the Lagrange multipliers procedure, the probability density ρ is obtained maximising

$$H = -\int \rho(x) \ln \rho(x) dx, \tag{10}$$

under the constraints $c_i = \langle f_i \rangle$. One then obtains:

$$\rho(x) = \frac{1}{Z} \exp \sum_{i=1}^{m} \lambda_i f_i(x) \, , \tag{11}$$

where the parameters $\lambda_1, .., \lambda_m$ depend on the values $c_1, ..., c_m$. This approach may indeed be applied to the statistical mechanics of systems with a fixed number of particles; for instance, fixing the value of the mean energy leads to the usual canonical Gibbs distribution in a very simple and elegant fashion (Peliti, 2011; Uffink, 1995). Therefore, the MaxEnt appears as a cornucopia, out of which one can extract in a straightforward way the main results of SM.

This conceptual issue deserves a critical discussion. Indeed, the interest of such an approach comes from the fact that most phenomena of scientific interest, notably the biological phenomena, lack a reliable theory, while there is good amount of data concerning them. Two difficulties immediately arise:

1. the ancient saying *"ex nihilo nihil"* continues to be appealing;
2. *unperformed experiments have no results.*

In this respect, a caustic, but insightful example was conceived by Ma (1985):

"*How many days a year does it rain in Hsinchu?*" [3] *One might reply "As there are two possibilities, to rain or not to rain, and I am completely ignorant about Hsinchu, therefore it rains six months in a year.*"

The important point Ma wants to make is that it is not possible to infer something about a real phenomenon, thanks to our ignorance. As recalled in the previous section, probability in SM is used in relation to objective frequencies, it is the ratio of numbers extracted from concrete observations. In the MaxEnt framework, it is instead related to the degree of uncertainty or of our ignorance about an event: lack of knowledge is used to produce knowledge.

Apart from these very general considerations, the weakest technical aspect of the MaxEnt approach is its dependence on the choice of the variables needed to represent

[3] Hsinchu is a chinese city on the Pacific ocean.

a given phenomenon. Tis fact can be understood as follows. Given a property X, whose values x are distributed according to the probability density ρ_X, one realizes that the "entropy" $H_X = -\int \rho_X(x) \ln \rho_X(x) dx$ is not an intrinsic quantity of the phenomenon X, hence it is unclear how H can characterize X. For instance, changing parametrisation, i.e. using the coordinates $y = f(x)$ in place of x, where f is an invertible function, the entropy of the same phenomenon turns:

$$H_Y = -\int \rho_Y(y) \ln \rho_Y(y) dy = H_X + \int \rho_X(x) \ln |f'(x)| dx \ . \quad (12)$$

Therefore the MaxEnt gives different solutions if different variables are adopted to describe the very same phenomenon. In order to avoid this unacceptable condition, Jaynes later proposed a more sophisticated version of the MaxEnt, in terms of the relative "entropy":

$$H^* = -\int \rho(x) \ln \left[\frac{\rho(x)}{q(x)}\right] dx \ , \quad (13)$$

where q is a known probability density. H^* at variance with the entropy H, does not depend on the choice of variables. Nonetheless, H^* now depends on the distribution q, hence the problem is merely shifted toward the selection of such a probability density, which is analogous to the problem of choosing the proper variables.

For instance, knowledge of the mean energy and taking a uniform distribution, say $q =$ const, leads to the correct Gibbs distribution, but this q is, in principle, either a totally arbitrary choice, or it amounts to an a priori knowledge of the correct result. Analogously, while the correct variables for the description of equilibrium thermodynamic systems are well known, because they concern comparatively very simple phenomena, which have been investigated for very long, the same cannot be stated about generic systems, such as the complex ones for which the MaxEnt principle is supposed to provide a theoretical framework.

In conclusion, even the second, more elaborate method, is not truly predictive. Therefore, although the MaxEnt principle can be considered a neat and elegant way of deriving Gibbs-like probability distributions, when they are known to apply, we see no reason to found SM on it. Presumably, it may be useful to gain insight on a given phenomenon, in the absence of informed guiding principles, to be tested together with other alternatives, but one should keep in mind that it may lead into error, since Gibbs-like probability distributions are not generic, not even in relatively simple physical phenomena (Auletta et al., 2017).

3 The Old Debated Problem of Irreversibility

Typicality, which we have first discussed in the case of equilibrium systems, plays an important role also in the case of irreversible non-equilibrium phenomena. To illustrate this fact, let us begin with two simple observations:

1. microscopic mechanical laws are invariant under time reversal:

$$t \to -t \ , \ \mathbf{q} \to \mathbf{q} \ , \ \mathbf{p} \to -\mathbf{p} \, . \tag{14}$$

2. the macroscopic world is described by irreversible laws, e.g. the Fick equation for the diffusion of a scalar concentration C

$$\partial_t C = D \Delta C \, , \tag{15}$$

where D is the diffusivity of the scalar.

The question thus arises: is it possible to derive macroscopic (irreversible) equations starting from a microscopic (reversible) description (Lebowitz, 1993)?

This fundamental question constitutes the core of the objections raised by Loschmidt and Zermelo about Boltzmann's celebrated H-theorem, which describes an irreversible evolution from non-equilibrium toward equilibrium states (Huang, 2009). Loschmidt tackled directly the issue of reversibility, while Zermelo applied Poincaré's recurrence theorem, that shortly earlier had been demonstrated. The theorem states that, given a conservative system, like the newtonian ones we consider, and an initial condition in its phase space, the entailing evolution will sooner or later come back arbitrarily close to the starting point. In other words, there is "recurrence". Therefore, if a function of phase increases for a while,[4] sooner or later it has to decrease; which apparently means that the second law of thermodynamics cannot be derived from the newtonian dynamics of a system made of N particles.

Beside technical points, Boltzmann rapidly understood and refuted Zermelo's mathematically correctly formulated paradox, explaining the physical content of his theory. First of all, one must realize that physics, like all measurements one can perform, is about specific space and time scales. Then, Boltzmann's point of view was masterly summarised by Smoluchowski as follows: *A process appears irreversible when the initial state has a recurrence time which is long compared to the time of observation* In fact, Zermelo's paradox is physically irrelevant because, as rigorously proven by Kac (1957) for ergodic systems with N degrees of freedom, the recurrence-time goes like

$$\langle T_R \rangle = \tau_o C^N \, , \tag{16}$$

where τ_o is a typical time, and $C > 1$ depends on the desired precision of recurrence. Therefore, the mathematically correct Zermelo's argument is physically irrelevant because, given N for a macroscopic system, the corresponding recurrence time is unphysically and ridiculously huge: well beyond many ages of our universe for just a cubic centimetre of air.

Loschmidt raised a subtler criticism, that requires more elaborate analysis, see Sect. 3.2 below.

[4] Most notably the opposite of the H-functional taken by Boltzmann to mirror the entropy of an isolated system.

3.1 Use and Abuse of Probabilities (Ensembles) and Entropies

Gibbs ensembles are one of the cornerstones of SM, yet we believe that they are often introduced in very unfortunate fashions. For instance, in standard textbooks [ ] one can finds rather obscure statements, such as

"*an ensemble is an infinite collection of identical systems*"

Gibbs' goal, who acknowledged Maxwell and Boltzmann for introducing ensembles (Gibbs, 1906), was to use them in order to reformulate the Boltzmann's probabilistic approach based on the ergodic hypothesis. He then defined an *ensemble* as an infinite (imaginary) collection of macroscopically identical systems, that differ in their microscopic phases. Mathematically, such a collection could be intuitively and efficiently represented by a distribution of points in the phase space. The physical reasons behind the applicability of this idea have been outlined e.g. by Fermi (1956).

He explained that an ensemble represents the microscopic states explored by the dynamics of a single system, in the time taken by a measurement, but only under some conditions. In particular, the transitions from microscopic state to microscopic state must be much faster than the measurement.

Therefore, from the thermodynamic perspective, which is a deterministic description of single systems, taking too seriously statistical ensembles may be misleading and, in fact, a source of errors. Different is the case of systems that are not of thermodynamic interest, for which probabilities may be the only sensible information,[5] which however we do not treat here.

Let us consider, for example, the entropy of a given system, and let $\rho(\mathbf{X},t)$ be a probability distribution of its microscopic states in the phase space. The so called Gibbs-entropy is then defined as:

$$S_G(t) = -k_B \int \rho(\mathbf{X},t) \ln \rho(\mathbf{X},t)\, d\mathbf{X} = S_G(0)\,. \tag{17}$$

One may be tempted to think of the dynamics of the collection of ensemble members described by $\rho(\mathbf{X},t)$ as of the molecules of a certain system, and their evolution in phase space as the diffusion of molecules in real-space. However, one should note that phase space is an abstract, exceedingly high-dimensional space, that is totally different fro the 3-dimensional real space. Phase points are not molecules,[6] hence the evolution of their density has in general no physical content at all. In fact, if the N particles of the system obey the Hamilton equations of motion, the celebrated Liouville Theorem states that volumes in phase space are conserved by the time

[5] When dealing with non-macroscopic systems, thermodynamics does not strictly apply. This is the case, for instance, of Brownian particles immersed in a liquid. In this case, only a probabilistic, *ensemble*, description appears interesting and feasible.

[6] For instance, they have no extension and do not interact, while molecules occupy a certain volume and interact with each other. Moreover, one phase point represents a whole N-particles system, which something totally different from one of the N particles.

evolution. An immediate consequence of which is that the Gibbs entropy S_G is constant in time (Huang, 2009). In other words, while the Gibbs entropy correctly yields the equilibrium thermodynamic entropy of the system, it does not represent the growing entropy of an isolated nonequilibrium system: the Gibbs entropy is not a suitable SM counterpart of the thermodynamic entropy.

To overcome this difficulty, many authors have introduced a coarse-graining of the phase space, i.e. a partition of phase space made of cells of given small size, say ϵ, and a corresponding coarse-grained version of the probability density and of the Gibbs entropy. The probability for the microscopic phase to lie in the i−the cell at time t is expressed by:

$$P_\epsilon(i,t) = \int_{\Lambda_\epsilon(i)} \rho(\mathbf{X},t)\,d\mathbf{X}\,, \tag{18}$$

and the corresponding coarse grained Gibbs entropy is defined by:

$$S_{G,\epsilon}(t) = -k_B \sum_i P_\epsilon(i,t) \ln P_\epsilon(i,t)\,. \tag{19}$$

Then, unlike S_G, the quantity $S_{G,\epsilon}$ is not constant in time: it grows monotonically, if ρ is not invariant, till a maximum is reached. But this success in describing an evolving "entropy" is only apparent, and not real. One of the main, far from unique, difficulties that this method faces is that the evolution of $S_{G,\epsilon}$ is not intrinsic, but depends on ϵ. Also, it has been proven that $S_{G,\epsilon}$ does not grow for a while: it remains constant up to a crossover time $t_* \sim \ln(1/\epsilon)$, which grows without bounds, when ϵ decreases (Castiglione et al., 2008; Falcioni et al., 2007). Physically this makes no sense; it is analogous to state that the heat generated by burning one litre of gasoline depends on how accurately we observe the phenomenon, and if we observe it very accurately, no heat is generated...

3.2 The H Theorem

A physical framework in which macroscopic irreversibility emerges out of microscopic reversible dynamics is afforded by the celebrated H-Theorem, which Boltzmann derived within the kinetic theory of gases. Here, one starts from the one particle distribution function $f(\mathbf{q},\mathbf{p},t)$, which represents the mass density in the so-called μ- space, i.e. the space of a single particle coordinates and momenta, which 6-dimensional for particles with 3 degrees of freedom. In the limit of dilute monoatomic gas, with rather subtle assumptions, Boltzmann derived the time evolution equation of f, which takes the form:

$$\frac{\partial}{\partial t} f(\mathbf{q},\mathbf{p},t) + \sum_j \frac{p_j}{m}\frac{\partial}{\partial q_j} f(\mathbf{q},\mathbf{p},t) + \sum_j \dot{p}_j \frac{\partial}{\partial p_j} f(\mathbf{q},\mathbf{p},t) = C(f,f)\,, \tag{20}$$

where $C(f, f)$ is a bilinear integral term which accounts for the (weak) interactions among the particles. This equation implies that the quantity:

$$S_B(t) = -H(t) = -k_B \int f(\mathbf{q}, \mathbf{p}, t) \ln f(\mathbf{q}, \mathbf{p}, t)\, d\mathbf{q}\, d\mathbf{p} \tag{21}$$

constantly increases, until an equilibrium state is reached (Huang, 2009):

$$\frac{dS_B(t)}{dt} \geq 0\,, \quad \text{where `` $=$'' holds only at equilibrium} \tag{22}$$

Boltzmann could then identify his entropy S_B with the thermodynamic entropy of an isolated dilute gas. Equation (22) is called "H-theorem". The physical content is that the second law of thermodynamics is obtained via S_B, (21), casting the laws of classical mechanics, which are reversible, into a suitable probabilistic framework.

In addition to the recurrence paradox formulated by Zermelo, which we have observed to be physically irrelevant, another paradox has been devised to contradict Boltzmann's approach and his H-theorem: the reversibility paradox, usually attributed to Loschmidt.[7] In fact, this paradox is mathematically justified, like Zermelo's paradox, and can be equally dismissed, as irrelevant for the physics of macroscopic systems.

To understand that, let us recall the assumptions underlying the Boltzmann theory. First of all, the number of particles N is very large. Then, the one particle distribution function $f(\mathbf{q}, \mathbf{p}, t)$ that is the main theoretical object in the theory, can be seen as an empirical distribution function, concerning the positions and velocities of the N particles, formally expressed by:

$$f(\mathbf{q}, \mathbf{p}, t) = \frac{1}{N} \sum_{n=1}^{N} \delta[\mathbf{q} - \mathbf{q}_n(t)]\delta[\mathbf{p} - \mathbf{p}_n(t)]\,. \tag{23}$$

Therefore,

1. $f(\mathbf{q}, \mathbf{p}, t)$ is a well defined macroscopic observable: the number density of particles;
2. $f(\mathbf{q}, \mathbf{p}, t)$ can be measured (e.g. in numerical simulations) and such a measurement concerns the single system under investigation, made of the N particles: there is no need to refer to the statistical ensembles.

It has then been proven by Lanford that the microscopic Hamiltonian, reversible, dynamics is not incompatible with the H-theorem (Lanford, 1981). Indeed, given a hard-sphere system; considering the Boltzmann-Grad limit, i.e. $N \to \infty$, $\sigma \to 0$ so that $N\sigma^2 \to constant$, where σ is the diameter of the particles; and starting from

[7] The content of the paradox is the following. Given that microscopic dynamics is reversible in time, if we were able to reverse time, the dynamics should trace back its trajectory, and therefore also S_B should decrease.

an initial condition in a *"good set"*, one obtains that $f(\mathbf{q}, \mathbf{p}, t)$ evolves as prescribed by the Boltzmann equation, hence the H-theorem holds. In other words, Lanford proved that

$$f(\mathbf{q}, \mathbf{p}, t) \simeq f_B(\mathbf{q}, \mathbf{p}, t) , \tag{24}$$

where $f_B(\mathbf{q}, \mathbf{p}, t)$ is the solution of the Boltzmann equation. This result holds for a short time, which is a fraction of the mean-collision time, for $N \gg 1$ and a typical $\mathbf{X}(0)$; but it is enough to prove rigorously that Hamiltonian dynamics, in the proper limit, does not violate the Boltzmann equation, and one obtains an irreversible behaviour from a microscopic reversible dynamics (Castiglione et al., 2008; Cerino et al., 2016; Chibbaro et al., 2014a; Lebowitz, 1993).

It is further worth noting that the validity of the H-theorem does not rest on the details of the particles interactions, as long as they exert a short range repulsion. This is important, from a physical point of view, since it means that the result is quite general.

3.3 Again About Entropies and Probability

Although at first glance the Gibbs and the Boltzmann entropies look similar, their dynamical conceptual and physical meaning, hence their behaviours are totally different. Both entropies correctly describe equilibrium states, but the Gibbs entropy is defined in terms of the very abstract notion of phase space probability density or of *ensemble*, while the Boltzmann entropy is derived from the very material property which is the number of particles of one concrete system occupying a given spacial volume, with velocity in a given cube of the velocity space. Therefore, some understanding of the connection between such diverse entropies is desirable.

Roughly speaking, two main points of view are generally adopted: the subjective and the objective interpretation of probability. According to the subjective interpretation, probability is a degree of belief in something. One of the most influential followers of such a view is Jaynes, who claimed that the entropy of a physical system depends on the observers' knowledge of it, or on their (informed) belief concerning the phenomenon of interest. In the objective interpretation, the probability of an event is instead determined by the the physics of the system of interest and not by the available or missing information.

This difference allows us to distinguish between thermodynamic irreversibility, and the relaxation of a phase-space probability distribution $\rho(\mathbf{X}, t)$ to an invariant (constant in time) distribution, further clarifying that an abstract ensemble must not be confused with a given macroscopic system. Indeed, for dynamical system exhibiting a good degree of chaos, one commonly observes that $\rho(\mathbf{X}, t)$ converges in time to an invariant probability distribution, $\rho(\mathbf{X}, t) \to \rho_{inv}(\mathbf{X})$. In other words, the ensemble averages of all phase functions "irreversibly" converge to given values.

This is not the irreversibility the second law of thermodynamics speaks about! In the thermodynamic case, the observables of interest of systems prepared in the same way, evolve in the same fashion. If an ensemble average converges, the different elements of the ensemble, hence their observables, may evolve in totally different and inconsistent ways. Again, this is a consequence of the fact that phase-space points are not particles, and their probability density is not a mass distribution. In the phase space, a single system is represented by just one point and an actual experiment follows a single trajectory, not a cloud of points from which a collection of different trajectories arises.

The physically relevant issue is that a single macroscopic system behaves irreversibly and in a unique fashion, starting from a generic initial microscopic state.

In summary, contrarily to some perhaps fashionable claims (Prigogine & Stengers, 1979), there is no direct link between the convergence process of probabilities in phase space, and the thermodynamic irreversibility. For this reason, the only way to pursue the program of SM for macroscopic objects is to take an objective approach to probability, which is the Boltzmann framework. That does not diminish in any way the importance of dynamical system theory in other problems (Castiglione et al., 2008), even when the number of degrees of freedom is large (Bohr et al., 1998).

4 Typicality and Irreversibility

The above discussion introduces the question of *typicality*, which is related to the one raised by various philosophers of science, regarding the role of the microcanonical distribution (ensemble) in the description of constant energy (isolated) systems. We argue that there are very good reasons to assign a privileged role to the microcanonical ensemble, compared to other probability distributions, that are equally invariant under the Hamiltonian evolution. This rather technical subject concerning deterministic dynamics, can be cast in a suggestive framework, once it has been shown that the dynamics of interacting particles does commonly and effectively result in a certain kind of randomness. We thus illustrate the notion of typicality, and the connection between deterministic and stochastic evolution with some examples.

4.1 Typicality in Stochastic Models

A popular model whose simplicity allows a neat discussion of typicality is the well known Ehrenfest flea model (Baldovin et al., 2019), that is jokingly referred to the fleas that jump back and forth between two dogs. The model consists of a Markov chain (Gnedenko, 2018) representing N "particles", each of which can either be in a box called A, or in another box called B. The state of the Markov chain at time t is identified by the number n_t of particles in A, and the evolution is ruled by a stochastic law, with given transition probabilities:

$$P_{n\to n-1} = \frac{n}{N} \ , \quad P_{n\to n+1} = 1 - \frac{n}{N} \, .$$

dictating how the state $n_t = n$ changes in one time step to become $n_{t+1} = n \pm 1$. In our SM language, the state n_t can be seen as the "macroscopic" state of the system of interest, while the the corresponding "microscopic" configuration is defined by the list of the particles lying in box A, considered as distinguishable particles. The equilibrium (macroscopic) state is expressed by $n_{eq} = N/2$.

The evolution of an ensemble of initial conditions starting from a given state n_0 can be described computing not only the mean population $\langle n_t \rangle$ but also its variance $\sigma_t^2 = \langle n_t^2 \rangle - \langle n_t \rangle^2$. One obtains:

$$\langle n_t \rangle = \frac{N}{2} + \Big(1 - \frac{2}{N}\Big)^t \Delta_0 \ , \ \sigma_t^2 = \frac{N}{4} + \Big(1 - \frac{4}{N}\Big)^t \Big(\Delta_0^2 - \frac{N}{4}\Big) + \Big(1 - \frac{2}{N}\Big)^{2t} \Delta_0^2 \, , \tag{25}$$

where $\Delta_0 = n_0 - N/2$. The main result is that $\langle n_t \rangle \to n_{eq} = N/2$, exponentially fast with a characteristic time $\tau_c = -1/\ln(1 - 2/N) \simeq N/2$ and a standard deviation σ_t that goes to its equilibrium value $\sqrt{N}/2$ with a characteristic time $O(N)$.

These results for $\langle n_t \rangle$ and σ_t are obtained at the level of the ensemble, i.e. as averages over the behaviour of all possible single N-particles (macroscopic) systems obeying Eq. (25). What about a single macroscopic object, i.e. a single realization of the process?

Figure 1 illustrates the result of numerical simulations, showing that for large N, the single object behaves "typically". In more mathematical terms, one has:

$$\text{Prob}\Big(n_t \simeq < n_t > \ \text{ for any } t \in [0, T]\Big) \simeq 1 \ \text{ where } T = O(N) \, , \tag{26}$$

which means that n_t practically behaves as the average $\langle n_t \rangle$ in almost all cases.

Consider now a far from equilibrium initial condition, e.g. $n_0 \simeq N$. It is possible to show that, for $N \gg 1$, up to a time $O(N/2)$, i.e. as long as n_t remains far from

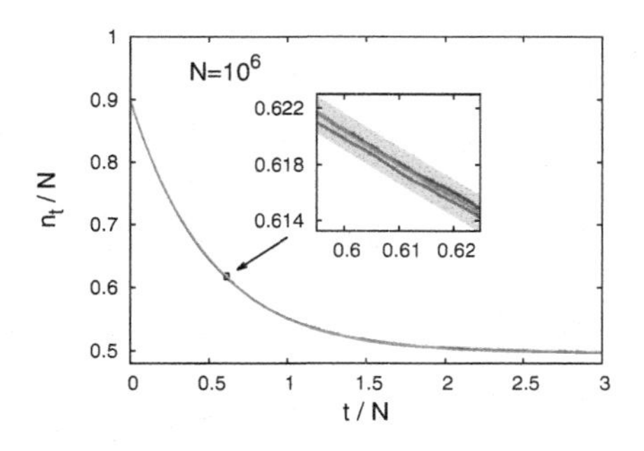

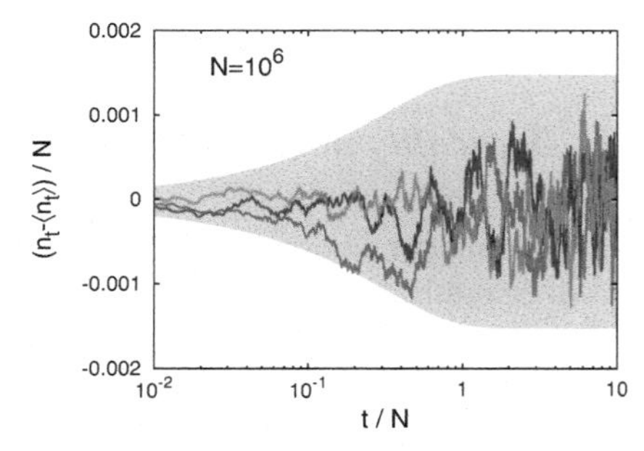

Fig. 1 Several realisations of the time evolution of the state of the Ehrenfest flea model, n_t, for $N = 10^6$. The coloured region corresponds to three standard deviations from the running mean: $\langle n_t \rangle - 3\sigma_t < n_t < \langle n_t \rangle + 3\sigma_t$, from (Baldovin et al., 2019)

n_{eq}, each single realization of n_t stays "close" to the time dependent average $\langle n_t \rangle$. Indeed using tools of probability theory, one can show (Baldovin et al., 2019):

$$\text{Prob}\Big(\frac{|n_t - \langle n_t \rangle|}{N} < \epsilon_N \text{ for any } t \in [0, T]\Big) \geq 1 - a_N \,, \tag{27}$$

with the quantities $\epsilon_N \to 0$, and $a_N \to 0$ as $N \to \infty$. Taking $\epsilon_N \sim N^{-B}$ with $0 < B < 1/3$, one has $a_N \sim N^{-A}$ with $A > 0$. For instance, $B = 0.2$ implies $A \geq 0.2$

That means that the overwhelming majority of realisations of the stochastic process n_t remains close to $\langle n_t \rangle$ for a long time, if N is sufficiently large. In other words, every macroscopic measurement on the systems has a very low probability of resulting sensibly different from the expected value. This is the conceptual meaning of "typicality" in SM.

4.2 Typicality in Large Deterministic Systems

The above, exactly solvable stochastic model, neatly quantifies the notion of typicality, but it may appear inappropriate in one investigation concerning the statistical properties of particle systems obeying deterministic equations of motion. The gap between the stochastic and the deterministic realm is however bridged by standard deterministic particle systems, whose properties evolve as erratically as they do in random processes.

Consider, for instance, a channel containing N particles of mass m, closed by a fixed vertical wall on the left, and by a frictionless piston of mass M on the right (Cerino et al., 2016). The piston motion is determined by a constant force F and by its collisions with particles inside the channel, see Fig. 2.

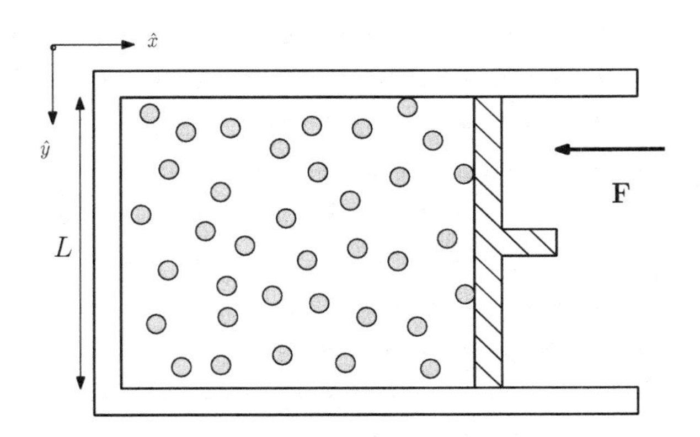

Fig. 2 Sketch of a deterministic (Hamiltonian) model of a gas confined by a piston

The Hamiltonian of this system reads:

$$H = \frac{P^2}{2M} + \sum_i \frac{p_i^2}{2m} + \sum_{i<j} U(|\mathbf{q}_i - \mathbf{q}_j|) + U_w(\mathbf{q}_1, ..., \mathbf{q}_N, X) + FX ,$$

where U is the interacting potential among the particles, and U_w denotes the interaction of the particles with the wall. In the case of point particles, $U = 0$ and U_w yields elastic collisions. Then, the dynamics is not chaotic, and it is easy to find the "equilibrium" position of the piston, $\langle X \rangle$, and its variance σ_X^2. In presence of interactions, e.g. for interaction potentials like:

$$U(r) = \frac{U_o}{r^{12}} \quad , \quad U_w = U_o \sum_i \frac{1}{|x_i - X|^{12}}$$

the equations of motion can be solved numerically, and reveal one positive Lyapunov exponent, i.e. chaos.

Figure 3 illustrates the irreversible behavior of the states $X(t)$ of one chaotic and of one non-chaotic instances of the piston model. Their initial conditions $X(0)$ are typical assuming a fixed $X(0)$ which is far from equilibrium, which are determined by the models parameters. In particular, the positions of the particles are initially distributed uniformly in the interval $[0, X(0)]$, while the velocities initially follow a Maxwell-Boltzmann distribution at a temperature T different from the equilibrium temperature T_{eq}, so that $|X(0) - X_{eq}| \gg \sigma_{eq}$, where the subscript eq refers to the equilibrium state.

The result is that the single trajectories are typical: although far from equilibrium, fluctuations about the corresponding ensemble averages are small compared to such averages, as in the case of the stochastic Ehrenfest model. This supports the anticipated analogy between stochastic and deterministic systems, both in the presence and in the absence of chaos, demonstrating that positive Lyapunov exponents are not required for the randomness associated with a many particles system irreversible behaviour. As a matter of fact, the numerical results for our deterministic reversible dynamics look rather similar to those for the stochastic Ehrenfest model, explicitly showing why irreversibility can be understood as an emergent property of a single system under proper initial conditions, when N turns sufficiently large (Cerino et al., 2016).

These results should be contrasted with those of Fig. 4, which reports the behaviour of a small N system starting from an initial condition close to equilibrium. It is well evident the absence of irreversible behaviour.

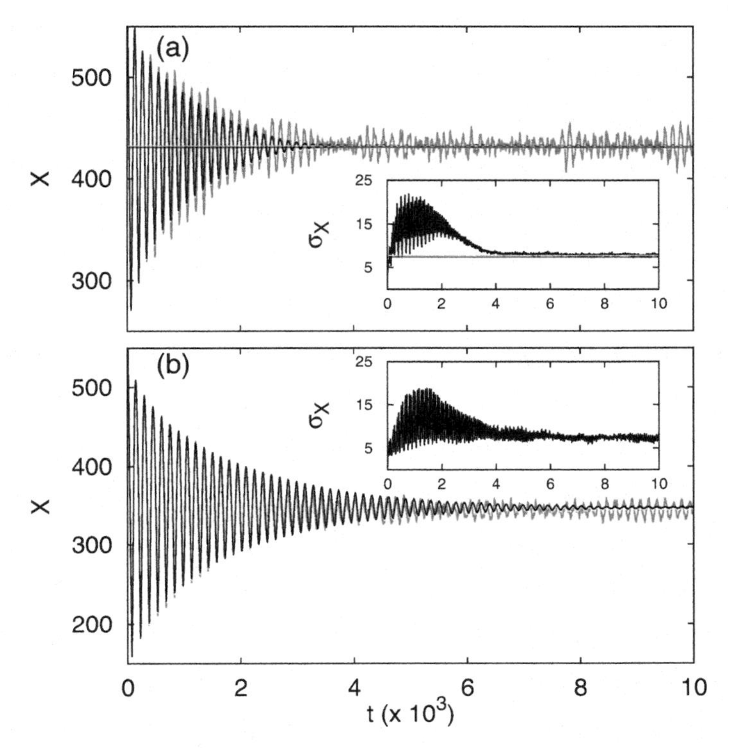

Fig. 3 $X(t)$ versus t for $N = 1024$, $M = 50$, $m = 1$, $F = 150$ and $X(0) = X_{eq} + 10\sigma_{eq}$ in a chaotic piston (**a**), and in a non chaotic piston (**b**). Red lines represent $X(t)$ for a single realization; black lines refer to the ensemble average $\langle X(t) \rangle$, (Cerino et al., 2016)

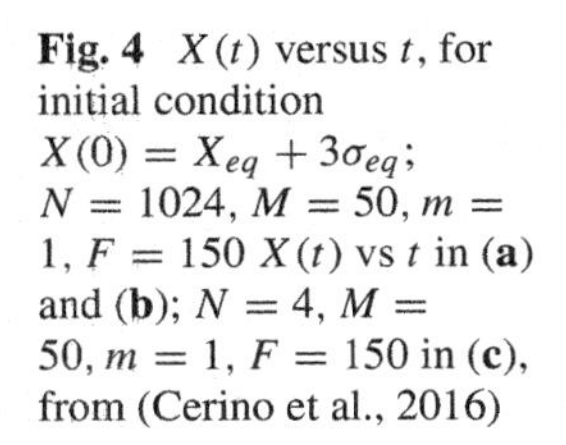

Fig. 4 $X(t)$ versus t, for initial condition $X(0) = X_{eq} + 3\sigma_{eq}$; $N = 1024$, $M = 50$, $m = 1$, $F = 150$ $X(t)$ vs t in (**a**) and (**b**); $N = 4$, $M = 50$, $m = 1$, $F = 150$ in (**c**), from (Cerino et al., 2016)

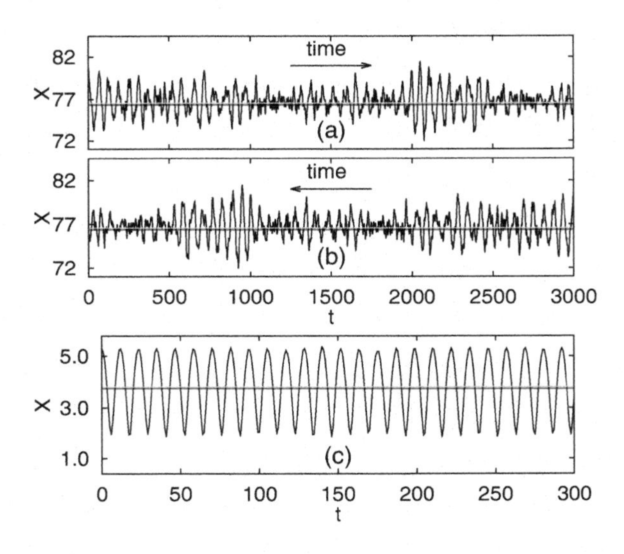

5 Summary and Conclusion

In this contribution we discussed the notion of irreversibility and its relation to typicality, in the framework of the SM of macroscopic systems made of classical particles. We pointed out the dangers associated with an uninformed use of probabilities in phase space, i.e. of statistical ensemble. The fact is that the thermodynamic laws ruling the behaviour of macroscopic objects are deterministic and, for a given initial macroscopic state, always lead to the same evolution. Therefore, there is no need to average over ensembles of differently evolving objects: in the case one really wants to perform such an average, the results will be equal to that given by one of those objects. We then formalized this fact with the notion of *typicality* which, for a macroscopic observable M, may be expressed as:

$$\text{Prob}\Big(M(t) \simeq \langle M(t) \rangle\Big) \simeq 1 \;\; when \;\; N \gg 1 \tag{28}$$

where Prob represents the fraction of phase space trajectories enjoying a given property, and the property considered is that almost all trajectories separately behave as their average.

In our investigation, two examples have been analysed, in order to illustrate how typicality and irreversibility arise in the evolution of systems made of many particles. The first is a stochastic process, known as the Ehrenfest flea model, which is exactly solvable hence allows us to obtain analytical expressions of the quantities of interest. While this is not a particle system like the Hamiltonian systems considered by Boltzmann, it gives a clear picture of the emergence of irreversibility in its large N limit. We have then investigated two versions of a gas confined by a moving piston, which are in fact Hamiltonian particles systems, and we have then shown that the stochastic and the deterministic systems have totally analogous behaviours, confirming the relevance of the stochastic description, in the case of large N. Finally, we have shown that for small N irreversibility in a single system, in the sense of Eq. (28), does not hold. Therefore, this is indeed and emergent property of the macroscopic realm.

Our results have been obtained under the following conditions:

1. the system is made of a very large number of particles, i.e. $N \gg 1$;
2. the initial condition is very far from equilibrium, meaning
   - $|n_0 - N/2| \gg \sqrt{N}$ in the Ehrenfest model
   - $|X(0) - X_{eq}| \gg \sigma_{eq}$ in piston model.

As a final remark, let us observe that the notion of typicality we have associated with the thermodynamic laws, pertains also to the Boltzmann approach. It consists of an *objective* operation: counting the cases of interest, and realizing that, with no exception, they behave in the same prescribed way. However, counting requires a finite number of objects. Ensembles, on the other hand, refers to the probability of a continuum. This is one of the technicalities that requires further investigation.

References

Auletta, G., Rondoni, L., & Vulpiani, A. (2017). *On the relevance of the maximum entropy principle in non-equilibrium statistical mechanics The European Physical Journal Special Topics, 226*, 2327.

Baldovin, M., Caprini, L., & Vulpiani, A. (2019). Irreversibility and typicality: A simple analytical result for the Ehrenfest model. *Physica A, 524*, 422.

Batterman, R. W. (2002). *The Devil in the Details: Asymptotic Reasoning in Explanation, Reduction, and Emergence*. Oxford University Press.

Berry, M. V. (1994). *Asymptotics, singularities and the reduction of theories* (p. 597). Methodology and Philosophy of Science, IX: Logic.

Bohr, T., Jensen, H. M., Paladin, G., & Vulpiani, A. (1998). *Dynamical systems approach to turbulence*. Cambridge University Press.

Bricmont, J. (1996). Science of chaos or chaos in Science? *Annual New York Academy Science, 775*, 131.

Castiglione, P., Falcioni, M., Lesne, A., & Vulpiani, A. (2008). *Chaos and coarse graining in statistical mechanics*. Cambridge University Press.

Cercignani, C. (1998). *Ludwig Boltzmann: The man who trusted atoms*. Oxford University Press.

Cerino, L., Cecconi, F., Cencini, M., & Vulpiani, A. (2016). The role of the number of degrees of freedom and chaos in macroscopic irreversibility. *Physica A, 442*, 486.

Chibbaro, S., Rondoni, L., Vulpiani, A. (2014). *Reductionism, emergence and levels of reality*. Springer.

Chibbaro, S., Rondoni, L., & Vulpiani, A. (2014). On the foundations of statistical mechanics: Ergodicity, many degrees of freedom and inference. *Communications in Theoretical Physics, 62*, 469.

Drossel, B. (2015). *On the relation between the second law of thermodynamics and classical and quantum mechanics*. In B. Falkenburg & M. Morrison (Eds.), *Why more is different* (p. 41). Springer.

Dumas, H. S. (2014). *The KAM story*. World Scientific.

Falcioni, M., Palatella, L., Pigolotti, S., Rondoni, L., & Vulpiani, A. (2007). Initial growth of Boltzmann entropy and chaos in a large assembly of weakly interacting systems. *Physica A, 385*, 170.

Fermi, E. (1956). *Thermodynamics*. Dover Publications.

Gallavotti, G. (Ed.). (2007). *The Fermi-Pasta-Ulam problem: A status report*. Springer.

Gaspard, P. (2005). *Chaos, scattering and statistical mechanics*. Cambridge University Press.

Gibbs, J. W. (1906). *The scientific papers of J. Willard Gibbs* (Vol. 1). Longmans, Green and Company.

Gnedenko, B. V. (2018). *Theory of probability*. Routledge.

Goldstein, S. (2012). Typicality and notions of probability in physics. In Y. Ben-Menahem & M. Hemmo (Eds.), *Probability in physics* (p. 59). Springer.

Goldstein, S. (2001). Boltzmann's approach to statistical mechanics. In J. Bricmont et al. (Eds.), *Chance in physics* (p. 39). Springer.

Huang, K. (2009). *Introduction to statistical physics*. CRC Press.

Jaynes, E. T. (1957). Information theory and statistical mechanics. *Physical Review, 106*, 620.

Kac, M. (1957). *Probability and related topics in physical sciences*. American Mathematical Soc.

Kadanoff, L. P. (013). Relating theories via renormalization. *Studies in History and Philosophy of Science Part B, 44*, 22.

Khinchin, A. I. (1949). *Mathematical foundations of statistical mechanics*. Dover.

Landau, L. D., & Lifshitz, E. M. (1980). *Statistical physics*. Butterworth-Heinemann.

Lanford, O. E. (1981). The hard sphere gas in the Boltzmann- Grad limit. *Physica A, 106*, 70.

Lebowitz, J. L. (1993). Boltzmann's entropy and time's arrow. *Physics Today, 46*, 32.

Ma, S. K. (1985). *Statistical mechanics*. World Scientific.

Mazur, P., & van der Linden, J. (1963). Asymptotic form of the structure function for real systems. *Journal of Mathematical Physics, 4,* 271.

Peliti, L. (2011). *Statistical mechanics in a nutshell.* Princeton University Press.

Prigogine, I., & Stengers, I. (1979). *La nouvelle alliance: métamorphose de la science.* Gallimard.

Shafer, G., & Vovk, V. (2001). *Probability and finance.* Wiley.

Uffink, J. (1995). Can the maximum entropy principle be explained as a consistency requirement? *Studies in History and Philosophy of Science Part B, 26,* 223.

Vulpiani, A., Cecconi, F., Cencini, M., Puglisi, A., & Vergni, D. (Eds.). (2014). *Large deviations in physics. The legacy of the law of large numbers.* Springer.

Zanghì, N. (2005). I fondamenti concettuali dell'approccio statistico in fisica. In V. Allori, M. Dorato, F. Laudisa & N. Zanghì (Eds.), *La natura delle cose: Introduzione ai fondamenti e alla filosofia della fisica* (p. 202). Carocci.

Zurek, W. H. (2018). Eliminating ensembles from equilibrium statistical physics: Maxwell's demon, Szilard's engine, and thermodynamics via entanglement. *Physics Reports, 755,* 1.

The Metal: A Model for Modern Physics

Tom Lancaster

1 Introduction

When atoms combine to form a solid, the quantum-mechanical interaction of electrons causes new states of matter to emerge. Condensed matter physics (Anderson, 1985) is the field that investigates these states and it has made contributions to our knowledge that are every bit as fundamental as those arising from the study of particle physics or cosmology. Central to modern condensed matter physics is the metal: a state of matter that owes its properties to the interactions between electrons. Despite the study of metals being a key part of the history of science, and their application in technology being of immense importance to human civilisation, it is only relatively recently that we have developed a detailed understanding of this phase of matter. This understanding involves a shift in our world view from regarding the metallic state as built from individual particles, to a view where we must consider the many-particle metallic solid as a whole. In this modern, contextual, viewpoint, the notion of an electron is replaced by that of a quasiparticle, where the electron effectively dresses itself in its interactions with the rest of the system. Despite this, it is still possible to discuss individual electrons, and this mixture of single-particle and whole-system properties lies at the heart of our understanding of metals. Here we discuss the modern conception of a metal and its place in condensed matter physics, the ingredients of a model to describe it and its consequences, not just for condensed matter, but also for our view of the Universe more widely.

The ability to manipulate metal has been central to the development of human civilization. We are taught at school about metals being lustrous, relatively soft (that is, ductile and malleable), and being good conductors of electricity and heat. These properties are initially explained to us in terms of a lattice of ionic cores of atoms fixed in a sea of mobile electrons (Rosenberg, 1988). The electrons interact with the

T. Lancaster (✉)
Department of Physics, Durham University, South Road, Durham DH1 3LE, UK
e-mail: tom.lancaster@durham.ac.uk

S. Wuppuluri and I. Stewart (eds.), *From Electrons to Elephants and Elections*,
The Frontiers Collection, https://doi.org/10.1007/978-3-030-92192-7_21

ion cores and with each other via the electrostatic interaction, which is enshrined in Coulomb's law. This classical view of a metal is the sort of thing that Paul Drude (1863–1906) had in mind in his original model of a metal, formulated three years after the electron's discovery (Ashcroft & Mermin, 2003; Coleman, 2015; Singleton, 2001). Drude's model had some very impressive successes, explaining electrical resistance in particular, but its failures to describe the metal in detail led physicists to become uneasy. Work on this model coincided with the development of quantum mechanics (QM), which provides a rather different view of the world compared to the one behind Drude's model, which relied on classical notions of how the particles in a gas interact (Schiff, 1968). In fact, the application of QM to the problem of metallic behaviour provided some of the most significant tests of quantum theory in its early days. QM does a very good job of describing the properties of a metal (Ashcroft & Mermin, 2003; Singleton, 2001), but this in itself is curious in that QM is most successful at describing single, non-relativistic particles. (This is, particles moving a speeds much less than the speed of light.) Although the velocities of particles (and energies of the processes) in a metal put it within the non-relativistic domain, the metal is built from something like Avogadro's number ($\approx 6 \times 10^{23}$) of atoms, and so it is surprising how well quantum mechanics does in describing it.

Non-relativistic quantum mechanics was developed in parallel with quantum field theory (QFT), a version of quantum theory that is capable of describing relativistic effects (Lancaster & Blundell, 2014; Peskin & Schroeder, 1995; Zee 2003) and also the physics of many particles (Coleman, 2015), and by the middle of the twentieth century the methods of QFT were being applied to metals with great success (Abrikosov et al., 1963). What QFT showed was that the physics of metals is closely bound up in the physics of many particles and, in a sense we shall discuss, the properties of metals rely on considering all of the particles in the metal at some level. Modern condensed matter physics (Anderson, 1985) was shaped by this success and the description of the metal is now seen (along with the description of phase transitions, also discussed below) as being a key foundational part of the subject. In fact, our picture of the metal is probably the closest thing we have to a *standard model* (Halzen & Martin 1984; Penrose, 2004; Peskin & Schroeder, 1995) in condensed matter physics. It provides a set of concepts and methods that we use in many areas of the physics of solids. Where it breaks down, it does so for interesting reasons that lead to further insights into matter. Its limitations guide us towards new ways of treating problems in matter.

The rest of this chapter is constructed as follows. We shall start by outlining the quantum mechanical model of a metal, and its successes, before going on to examine how whole-system interactions have become increasingly important in how we view the behaviour of metals. We also discuss how scale and dimensionality allow us to treat more exotic physical systems based on metals. We shall see how the content of the metal (electrons) is very much shaped by *context*. Specifically, we seek to describe a macroscopic piece of matter compatible with the relatively low-energy conditions in which we experience this state.

2 Electrons in a Box

It might be imagined that a simple and successful description of a metal would involve building it from atoms. We would take the model of an atom (iron say) whose properties we know, and then build a model of a speck of matter by combining the model atoms, calculating the properties as we go. This might seems a sensible start in that we know Schrödinger's equation that governs the behaviour of the electron, and so building a metal out of electrons and atoms seems like a reasonable prospect. However, our progress is rapidly arrested by the scale of the problem. The metal is a three-dimensional object, comprising $N \approx 10^{24}$ electrons, so we need a quantum mechanical wavefunction that describes the $3N$-dimensional system of the electronic coordinates. Owing to the size of this task it is not something that is practically possible with current computational technology, nor likely with any technology in the coming decades.

In fact, to model a metal we do something rather different from building it from atoms: we simply fill an empty box with identical electrons (Ashcroft & Mermin, 2003; Rosenberg, 1988; Singleton, 2001). That is, we forget that a metal is built from atoms and we ignore them. We also ignore the fact that electrons are charged and should therefore interact (strongly) with each other via electrostatic repulsion. What we do retain in our model of electrons in a box is some of the rules of quantum mechanics. The first is that the quantum wavefunctions of the individual electrons have fixed momenta that must be compatible with the size and shape of the box. The second, which follows from electrons being identical Fermi particles (or fermions), is that the Pauli exclusion principle (Schiff, 1968) prevents fermions from occupying the same quantum state in the box. The electrons being identical is key: there is no in-principle way to tell if two electrons have been swapped. A detail here is that electrons have a spin, an arrow-like magnetic property that can be described as being up or down, with the result that two electrons can sit in each energy level: one with spin-up and one with spin-down. This state of affairs causes the electrons to fill up the available quantum states, with two electrons (spin up and spin down) at each energy. This is shown in Fig. 1a. Since a macroscopic mass of metal will contain something like 10^{24} electrons, the electrons in the top energy level (known as the *Fermi level*) have a high energy, compared to those in Drude's model, at least. This basic model of the metal, comprising a stacked up set of electrons, is known as the *Fermi gas*.

It is well known that Heisenberg uncertainty puts a limit on our ability to know both the position and momentum of a particle (Schiff, 1968). An important point here is that, although we speak of the electrons being stacked up, in that they have unchanging values of energy, they are not sitting in unique points in real space inside the box. In fact, each energy corresponds to a unique value of momentum and so the electrons can be thought of as having precise values of momentum, but as being completely delocalized in space. Alternatively, we can think of electrons as being well-described as stationary waves stretching out across the box that represents the metal such that the wave pattern doesn't alter in time. The distribution of electrons

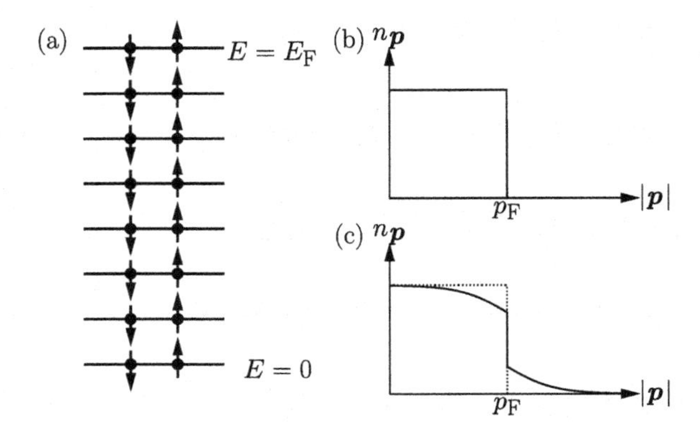

Fig. 1 **a** Non-interacting electrons stack up in energy levels to a maximum known as the Fermi energy E_F. **b** At zero temperature all of the states below the Fermi energy are occupied. **c** When electron-electron interactions are included, some electronic matter can be promoted to higher-energy states, but there is still a discontinuity at the Fermi level

in the momentum states of the metal is shown in Fig. 1b, where we see all of the states are occupied up to the Fermi energy, and then none of the states are occupied.

The discussion so far has implied that the metal is being held at a very low temperature ($T = 0$, formally). Once we raise the temperature, we give the electrons the possibility of being excited into new states. An important property of the Fermi gas is that, owing to the exclusion principle, if an electron is going to change its energy, it must jump into an energy level that is not occupied. This costs the least energy for those electrons near the Fermi level, where there are lots of unoccupied states close in energy, and so is most likely for these electrons. As a result, the electrons close to the Fermi level are responsible for most of the measurable properties of the metal. Those electrons at energies far below the Fermi energy don't often change their states and so exist in the background, at least in our model. When an electron in a filled state jumps up above the Fermi level it leaves behind it an unoccupied state. There is a symmetry between (1) the negatively-charged electron in an unoccupied above the Fermi level, surrounded by other unoccupied states, and (2) the now-unoccupied state below the Fermi level that the electron previously occupied, surrounded by states occupied by negatively charged particles. There is nothing preventing us from treating the unoccupied state below the Fermi level as a particle in it own right and so we do so. This particle is known as a *hole* and has an effective positive charge. The hole is analogous to the antiparticles in the standard model of particle physics, which share many of the properties of the particles, but have a reversed charge.[1]

Using the physicist's penchant for simplification, we can abstract the model down to its bare bones. The ground state of the system (i.e. the lowest energy state) is comprised of all of the electrons stacked up in energy (Fig. 1b). If a small amount of energy is injected into the system then we create excitations in the form of particles above and below the Fermi energy, called electrons (above the Fermi energy) and positively-charged holes (below the Fermi energy). The other electrons, stuck in the energy levels they had in the ground state, play a structural role in the model, but are

[1] This picture is very much like Dirac's original picture of particles and antiparticles. For a discussion see Lancaster and Blundell (2014).

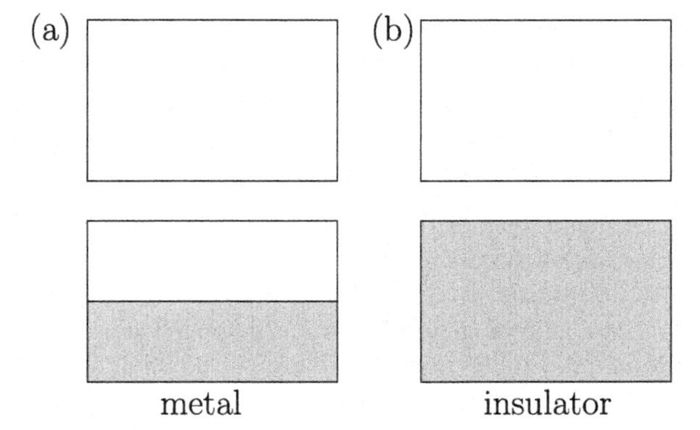

Fig. 2 **a** A metal is characterised by having a half-filled energy band with filled states next to empty ones. **b** An insulator has a completely filled band, with an energy gap separating the highest occupied level from the next available one

not the important ones in determining the properties of the metal. (Note however, the excitations rely on the motion of all of the electrons in their different momentum states.) We therefore often ignore the electrons in the ground state and refer to this as a *vacuum*. Clearly it is not a vacuum in the traditional sense of there being no particles present. The particle excitations are regarded as excitations of this vacuum. We shall return to this picture later.

The Fermi gas model does a good job in describing the properties of metals (Ashcroft & Mermin, 2003; Singleton, 2001), providing a description more consistent with experiment than is found with Drude's model. The existence of the excitations near the Fermi level allows us to understand electrical conduction, heat conduction and several other properties of metal. Electrical conductivity, for example, involves those electrons near the Fermi level being promoted into vacant states by an electric field which causes electric charge to be transported throughout the metal. However, the model doesn't yet explain why some materials are metallic and some insulating.

3 Energies and Bands

So far our electrons have not interacted with each other, nor with their environment. In order to explain the existence of insulating material we now allow the electrons to interact with the ions that form the material. In Drude's classical model the electrons collide with all of the ion cores, but this is not borne out in the quantum-mechanical treatment. In fact, the wave-like properties of the electrons allow most of the electronic wavefunctions to naturally avoid any collisions. However, there are particular values of momentum for which electrons do collide with the ionic cores, with their resulting scattering altering their behaviour. The result is that electrons cannot exist at those energies where they scatter strongly. These energies are forbidden and we therefore have bands of allowed energies separated by gaps of forbidden ones (Rosenberg, 1988; Singleton, 2001).

With the structure of the bands fixed by the scattering of the electronic waves, we again put electrons into the crystal. As before, they stack up in energy with two electrons per energy state. There are two possible scenarios, shown in Fig. 2. If the Fermi level sits in the middle of a band then an electron near the Fermi energy can change its state with little energy cost. This gives metallic conduction and the material is a metal. If, on the other hand, the Fermi energy coincides with the top of the band, then there will be a large number of forbidden energies before the next accessible states are found. It is therefore impossible for the electrons to change their states without a sizeable input of energies. We don't have electronic conductivity in this case and the material is an insulator. Whether a material conducts electricity or not in this simple model can therefore be thought of as a sort of housing crisis for electrons: unfilled states nearby the Fermi energy leads to conduction, no unfilled states nearby and the electrons must stay put leading to insulating behaviour.

This story of how metallic or insulating properties of a material emerge is one that rests much of the explanation on the motions of single electrons. It is the scattering of single electrons from the regular array of ions that creates the structure of allowed and forbidden energy levels. The Pauli principle (and the presence of the other electrons in forming the ground state) then forces us to stack up the particles in the available energy levels and the metallic or insulating properties follow. Although we have come a long way using only these simple models, what we have not included in our description of the metal is the interaction of electrons with each other. (As a result this simplified model is often called single-particle band theory.) Once we allow electrons to interact with each other our picture of what is happening in the metal changes rather profoundly. At this point, we are forced to start taking interactions of the whole of the system into account in order to continue to refine our description of a metal.

4 Quasiparticles and Field Theory

A simple way of understanding the interactions of electrons in a metal is provided by the work of physicist Lev Landau (1908–1968), who invites us to consider the following thought experiment (Anderson, 1985; Coleman, 2015). We start with the Fermi gas and imagine a knob we can turn that slowly turns on the interactions between electrons. That is, the knob controls the strength of the electrostatic (Coulomb) interaction between the electrons. As we turn the knob, Landau argues that the particles interact, and might be expected to shift up and down a little in energy, as shown in Fig. 3. This shift can be accounted for by changes in the mass of the electron: we say that interacting electrons have an *effective* mass by virtue of their interactions. However, the electronic energy levels do not cross over each other in energy, and so we can keep track of which non-interacting electron in the Fermi-gas becomes which interacting electron as the knob is turned. (Since the electrons are identical, crossing energy levels would mean we lose track of which one is which.) The ability to follow the system as the interactions are dialled up is known as *adiabatic continuity* (Anderson, 1985) and is often summarised by saying that, as we turn on the interactions, a

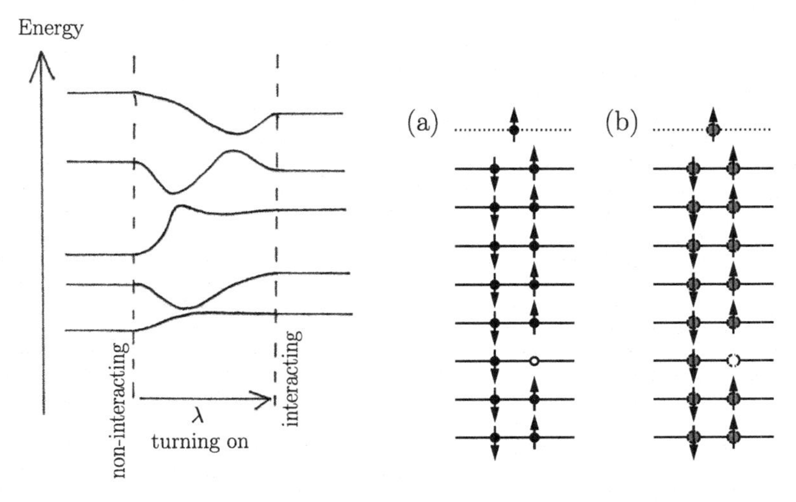

Fig. 3 Left: Turning on the interactions causes the energy levels to shift, but they never cross if the process if adiabatic. **a** The Fermi gas with a single excited electron and hole. **b** On turning on the interactions, the Fermi gas turned into the Fermi liquid, with gas states occupied by electrons and holes turning into liquid states occupied by quasi-electrons and quasi-holes. Adapted from Lancaster and Blundell (2014). Reproduced by permission of Oxford University Press (global.oup.com)

filled state in the non-interacting gas becomes a filled state in the interacting system (Fig. 3).

The state of matter that emerges from a Fermi gas on inclusion of the interacting electrons is known as a *Fermi liquid* (Anderson, 1985; Coleman, 2015; Lifshitz & Pitaevskii, 1980), in analogy with the difference between the weakly interacting, occasionally colliding molecules in a gas and the more strongly interacting molecules of the liquid phase. Once we introduce interactions, there is a sense in which the electron is no longer strictly a single particle, but instead is a single particle that carries around a cloud reflecting the influence of the other particles in the system on it. In the presence of interactions, the electrons in the liquid therefore change their properties and become *quasiparticles* (Anderson, 1985; Lancaster, 2019). We can think of these as the original particles "dressed" in a cloud of interactions. Conceptually we can write

$$(\text{quasiparticle}) = (\text{bare particle}) + (\text{interactions}), \tag{1}$$

where "bare particle" is a shorthand for referring to a non-interacting particle.

The interactions with other electrons can cause the electrons to be removed from their energy levels and scattered into other levels. This gives a quasi-electron a finite time to survive before its state is changed. However, this does not happen equally for all electrons. Those with energies close to the Fermi energy survive the longest, while those with energies far from the Fermi energy are most susceptible to scattering and so have a much reduced lifetime. If the lifetime of a particle is small compared, say, to the time we needed to turn on the interactions, then it is not meaningful to say that an electron quasiparticle even exists. As a result, it is really only meaningful to talk about individual quasiparticles existing close to the Fermi energy. The rest of the particles are still there as part of the makeup of the metal, but have ceased to be describable in terms of the properties we usually ascribe to particles: that is, of their persisting for a reasonably length of time with their specific properties. This gives us another insight into why these electrons close to the Fermi energy are the ones that determine the properties of the system: they survive long enough to be meaningfully described as particles (Anderson, 1985). The distribution of electrons for an interacting Fermi liquid is shown in Fig. 1c. Compared to the non-interacting case of Fig. 1b we see that some matter is now found above the Fermi energy and some is missing below. However, there is still a sharp discontinuity at the Fermi energy. Curiously, this sharp discontinuity is probably the defining feature of a metal, identifying the metal far more precisely than the properties of conducting electricity or heat, for example. In cases where this discontinuity is not observed (such as in one-dimensional systems, discussed below) there are no quasiparticles.

Landau's Fermi liquid model is enormously useful in understanding why interacting electrons in a metal act the way they do. However, we can also apply a more formal approach to the interacting electron problem that brings out some similarities with other parts of physics. It was realized in the 1950s, particularly among the Russian school that grew around Landau, that the methods of quantum field theory (QFT) could be used to model the physics of solids (Abrikosov et al., 1963; Lifshitz & Pitaevskii, 1980). In QFT, every particle (and antiparticle) is an excitation in a field, defined over all space and time (Lancaster & Blundell, 2014). The field with no particle/antiparticle excitations is referred to as the vacuum. In the absence of interactions, excitations in the vacuum do not interact. Once we include interactions then the excitations scatter from each other, but it is also possible to have processes where particles and antiparticles are spontaneously excited from the vacuum state and then fall back into it. These are known as vacuum fluctuations. Applied to the metal, QFT says that there is an electron field that fills the metal whose excitations are particle-like and hole-like states. There is also the unexcited, vacuum state and, although in most applications of QFT this would involve there being no particles in the system, in a metal we have the curious situation where the vacuum state contains electrons stacked up to the Fermi energy. (Slightly confusingly, therefore, we must keep distinct the idea of a particle and the idea of an excitation/excited particle.) Vacuum fluctuations involve electrons being spontaneously lifted from the vacuum state, creating an electron and hole pair. This can interact with any other other particles in the vicinity, before being returned to the vacuum.

QFT allows the properties of the metal to be computed by treating the electron's interactions with vacuum fluctuations as perturbations to the motion of the electron that would take place in the absence of fluctuations. These interactions change the properties of the electron, such as its mass and charge, in a process known as renormalization. In practice the calculations are often done with the help of an efficient process of drawing *Feynman diagrams* (Mattuck, 1967) (an example of which is shown in Fig. 4) which encode the details of the the interactions between the electron and the vacuum and function as a shorthand for terms in the equations that describe

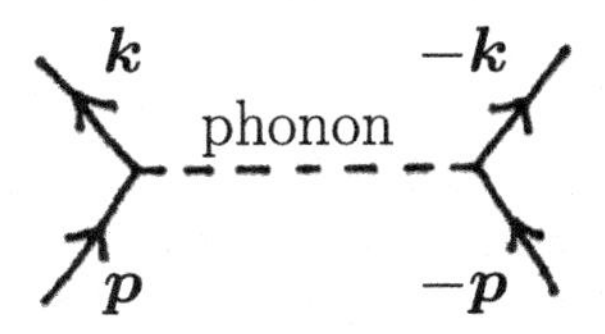

Fig. 4 A Feynman diagram showing how an interaction between two electrons is mediated by a phonon. This causes an attractive force between the electrons that leads to them becoming Cooper pairs, which are the building blocks of the superconducting state.Taken from Lancaster and Blundell (2014). Reproduced by permission of Oxford University Press (global.oup.com)

the interactions. Although renormalization sounds mysterious, it can be described in physical terms. For example, the fact that the electron will be dressed in interactions that can be expressed in terms of electron and hole pairs, allows us to picture the electronic charge being screened by these electrons and holes, reducing its value.

The description of the metal in terms of QFT allows us to compare it with quantum electrodynamics (QED) (Feynman, 1990; Penrose, 2004), which is the QFT that describes the workings of electrons, their antiparticles (positrons) and their interactions with light in free space. QED describes the relativistic properties of electrons and the interactions that they can undergo at very high energy (far larger than those at play in the metal). However, the interactions between electrons and positrons are still electromagnetic. They are mediated by the photon, the particle of light that mediates the electromagnetic interaction. Renormalization of the electron's properties also occurs in QED where the mass and charge of the electron are shifted by the presence of the interactions drawn out of the vacuum. Despite the metal being non-relativistic and interactions occurring effectively instantaneously, the Feynman diagrams found to describe the metal and to describe QED resemble each other closely. This close analogy allows us some insight into electrons that exist outside of the metal, as described by QED. We conclude that electrons in free space are never observed alone; they come with a cloud of interactions between the electron and the ground state vacuum fluctuations. From this point of view, *all* electrons, whether confined to a metal or free, are quasi-electrons whose properties (their mass and charge) are determined from many-particle considerations. Once renormalization is taken into account, we can regard the electrons as particles with fairly fixed properties, as long as we engage them with relatively low energies, so that we don't disrupt the interactions that have renormalized their properties. For example, if we choose to interact with electrons by probing them at very high energy (by bombarding them with high-energy probe particles, for example), the charge of the electron will appear to change. This can be thought of as a consequence of our disrupting the fluctuations that shield the electron charge, altering the many-particle nature of the electron in QED. The message here is that so-called constants of Nature, like electronic charge, can be shown not to really be constant at all, when many-particle processes are considered!

5 Phase Transitions

Another way to engage with the many-particle aspect of metals is via the physics of waves, which we often use to characterize a quantum-mechanical system in terms of its coherence. A system that maintains fixed phase relationships between its components is characterized as being highly coherent and having long-lived correlations between different observable properties (Rae, 1992; Weinberg, 2013). In quantum mechanics, this translates into a system being able to maintain a quantum superposition over a significant length of time or over a significant distance (where significant here means compared to the time/distance over which we expect random fluctuations). Superpositions are often described in terms of the quantum entanglement of the different parts of the system. Although our model of a metal is a many-particle state of matter, it does not come ready-prepared as a highly-coherent superposition of electronic states. That is to say that the simplest quantum-mechanical model of the metal involves a multiplication of all of the wavefunctions of the different electrons: which is the most incoherent quantum state that it's possible to build (Ashcroft & Mermin, 2003). Once we introduce the Pauli principle and the interactions between the electrons, we begin to recover some correlations. However, this suggests that the many-particle, correlated nature of the metal is limited, and might be relatively modest when compared against other states of matter. This turns out to be the case.

Many states in condensed matter are characterized by their coming into being via a phase transition (Anderson, 1985). A phase transition occurs, for example, when the underlying state of matter changes discontinuously as a function of temperature. A solid, for instance, condenses out of a liquid state when temperature is reduced below a critical value that characterizes the solid (i.e. the melting point). Phase transitions of this kind generally involve a reduction in the *symmetry* of a state. In the liquid, we have translational symmetry: on average, there is the same chance of finding a particle at any point in the material. In the solid, we have lost this symmetry: if we locate a particular atom, then we know equivalent positions can be found *only* at those regular intervals defined by lattice vectors. In order to locate a phase transition in a model of a state of matter such as a metal, we can perform our thought experiment from the previous section again. We again turn on the interactions between the parts of the system (e.g. the electrons) and we watch the system evolve from a non-interacting one to an interacting one, with each particle dressing itself in interactions to form a quasiparticle. Recall that the important point in constructing the metal was adiabatic continuity: filled states evolve into filled states and the energy levels of the states never cross, allowing us to keep track of states. However, if we encounter a case where the levels *do* cross, this tells us something: the symmetry of the underlying state of matter *has* changed. The breakdown of adiabatic continuity is therefore exactly what tells us that we have a phase transition. The crystalline solid itself comes into being from one of these phase transitions and once a symmetry-breaking phase transition occurs, the resulting phase of matter develops some new properties (Anderson, 1985):

1. One is rigidity: the solid, for example, cannot be deformed the way a liquid can (e.g. stirred) without costing a lot of energy.

2. We find that the solid has walls (or crystal defects) that separate regions where the symmetry has been broken in a slightly different way.
3. The solid also hosts new excitation not present in the liquid: these are quantized vibrations known as phonons.

These new excitations are often characterized as being like particles. However, they are collective excitations of the whole of the solid, involving the change in the coordinates of all of the atoms in the system, usually by small amounts. This variation of the position of atoms away from their special, translationally invariant positions, can be thought of as the excitation undoing a little of the symmetry breaking that brought the solid into being. However, this macroscopic motion creates an excitation that is, in many respects, functionally equivalent to a particle on the microscale. Phonons have a well-defined momentum and can interact with each other and with particles like electrons, exchanging energy and momentum. (Although it is notable that the conservation law for their momenta is slightly different from that of particles in free space.) In fact, just as for the electrons in a box, phonons are best thought of as being well localized in momentum, but delocalized in position. As far as quantum mechanics is concerned though, such a state qualifies as a particle. However, given their reliance on the coordinates of all of the atoms in the crystal, phonons are termed *collective excitations*, and we distinguish them from excited electrons, which we call *single-particle excitations* as these are more like single particles in terms of their origin, although we must remember they are dressed by interactions with their surroundings (Lancaster & Blundell, 2014).

Turning now to the metal, if we start with non-interacting electrons in a box, then the interacting metallic state does not involve a phase transition to come into being. That is, it is connected, adiabatically to the non-interacting state of electrons. We turn on the interactions between the electrons and the filled electronic states evolve into filled quasiparticle states, with the states never crossing. The metal does not have any new form of rigidity and does not form domains. As we've said, this contrasts with the case of creating a solid from a liquid, where there is a phase transition and the emergence of new phonon particles. We've also seen how the slightly more sophisticated model of a metal, taken as a gas of electrons inside the (broken-symmetry) crystal system, generally interacts weakly with the atoms of the solid. (We discussed before that in a model of a metal involving the electron gas and ion cores, most electrons don't interact with the ion cores.) The exception is that the electrons *do* interact strongly with the crystal lattice when they have particular, special values of momentum (as we discussed above), but also with anything that breaks the translational symmetry of the crystal. One example might be an impurity atom (or piece of dirt) in the metal, another is a phonon. The interaction with the phonon results because the phonon certainly breaks the translational symmetry (in that it involves all of the atoms moving away from their translationally-symmetric positions). This interaction of electrons with phonons and with impurities is the reason that metals have resistance to the transport of electrical charge. An electrical current leads to the dissipation of energy (causing wires to feel warm when carrying currents). The linear relationship between voltage across and current through a metallic sample is Ohm's law (Rosenberg, 1988; Singleton, 2001) and we return to it below.

The electron-phonon interaction can have a dramatic consequence: the scattering of electrons and phonons leads to an instability in the metal which itself causes a phase transition. In other words, the metallic Fermi liquid can, given the right condition, fall apart. The falling apart involves each electron being linked to a partner electron by interactions with the phonon. We think of this as one electron emitting a phonon and another absorbing it, as shown in the Feynman diagram in Fig. 4. This pairing interaction causes the electrons, previously strongly repelling each other owing to Coulomb's law, to start to feel a weak *attractive* interaction (Coleman, 2015). This, in turn, leads to the formation of a composite electronic state known as a Cooper pair of electrons. At this point the metal becomes highly unstable and these electron-electron Cooper pairs then condense into a highly-coherent state of matter known as a superconductor (Annett, 2004). That is, the wavefunction becomes a very highly-structured superposition, rather than resembling the large multiplication of energy levels that formed our starting point for understanding the metal. Remarkably, the superconducting state allows current to be transported with no electrical resistance. It also forms a state of matter which a magnetic field cannot penetrate. These unusual properties have led to superconductors being very thoroughly studied in the 20th century.

Finally we mention that metals can also undergo another class of phase transition: one that is not driven by changes in temperature. These *quantum phase transitions* (Sachdev, 2011) occur at zero temperature where quantum mechanical uncertainty again gives is fluctuations. Here, both quantum fluctuations (i.e. time dependence in the wavefunctions deriving from uncertainty relations) and thermal fluctuations are equally important and it is impossible to describe the excitations in terms of quasiparticles. In this region the phase of the wavefunctions are as incoherent as Nature allows, realising an (almost) ideal fluid. Quantum critical behaviour continues to be an area of much experimental and theoretical interest.

6 Scale, Ohm's Law and Dimensionality

A fact of Nature is that the microscopic properties of electrons, described by quantum mechanics, are time reversible. That is, it is impossible to tell the difference between a description of the interactions at the microscale and the same description played backwards in time. However, Ohm's law (that relates the current and voltage of a conductor) tells us that the behaviour of the metal involves dissipation of energy, which spoils this property of time-reversal invariance. Such a situation where the macroscale is seen to break time-reversal invariance occurs quite generally in physics and is often referred to as the *arrow of time problem* (Penrose, 2004). This is because the dissipation of energy (and the associated increase in the entropy of a system) give us an unambiguous direction of time that we can identify as being "forward". With respect to this forward direction, entropy always increases as energy is dissipated. While the origin of the arrow of time is hotly debated, some insight into the state of affairs in the metal is provided by considering the *scale* at which Ohm's law is experienced. The key here is that Ohm's law is something we measured in the laboratory, which means that relative to the underlying dynamics of electrons, it occurs over very long times and very long distances. In relating microscopic properties to the macroscopic behaviour of Ohmic conductors, we are therefore forced to examine the limiting behaviour of the metal. That is, we are forced to engage with the whole system in the sense of sending lengths and times to large values, such that we capture the way in which we engage with the metal.

We have two limits to consider here, and it turns out the order in which we consider them is crucial (Coleman, 2015; Zee, 2003). The limit we are taking says that we will interact with the system over lengths L and times T that tend to large values. Specifically, if we take (1) the limit of long times, followed by (2) the limit of long lengths, we obtain the prediction that a metal screens an electric field. On the other hand, if (1) we take the limit of long lengths and then (2) the limit of long times, we find the Ohmic transport behaviour (i.e. electrical resistivity). The explanation for the Ohmic transport behaviour is as follows: electrons diffuse though the system at some characteristic rate τ. The time it takes for an electron to explore a given volume of linear size L is $\tau = L^2/D$, where D is a diffusion constant. If we send L to infinity first then this diffusion time τ is much longer than the typical observation time for a measurement T. When we subsequently send T to large values, the measurement time increases far more slowly than the diffusion time of the electrons. Physically, we therefore make lots of observations of the electrons while they diffuse around, and this is resistivity: the finite amount of charge in the metal rearranges itself but doesn't finish this process before our measurement is complete. In the other case where we obtain the screening behaviour, taking the long measurement-time ($T \to \infty$) limit first means that the system has a chance to completely rearrange itself before a measurement is complete. Since there is only a finite amount of charge available, this charge rearranges itself such that it cancels out the applied electric field. This is the screening effect. This example teaches us that while we are forced to consider the metal in the long-time, long-length-scale limit, how we pose the question affects the answer. In Nature we're not really free to take one limit and then the other: in considering the behaviour over long times and distances the underlying structure of the equations means that one scale will change faster than the other. In practice, of course, these questions of description translate into different experiments. For the screening case we apply a static electric field to a conductor and the field is screened at equilibrium. In the Ohmic case, we drive a continuous current through the conductor and measure the voltage dropped across it.

This idea of the limiting behaviour of a theory is central to a set of techniques known as the renormalization group (RG) (Lancaster & Blundell, 2014; McComb, 2004). The RG allows us to examine how the behaviour of the underlying fields in a theory behave as we examine the limit of long time and distance. The key to applying this to metals is that the procedure supplies a natural description of the inverse resistance, a quantity called the conductance g. The application of RG theory to metals answers an interesting question: is it possible to turn off the conductance of

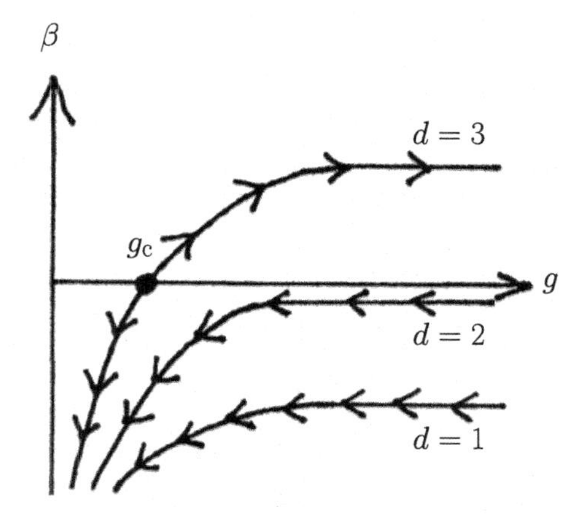

Fig. 5 The evolution of the conductance g as we take the long-length-scale-limit (shown by arrows). For the one-dimensional ($d = 1$) and two-dimensional ($d = 2$) cases this leads to insulating ($g = 0$) behaviour. For the three-dimensional ($d = 3$) case, the behaviour depends on where we start with respect to the critical point g_c. If we start with $g > g_c$ (low amounts of impurity, large conductance) we see that g flows to large values and the material conducts electricity. Taken from Lancaster and Blundell (2014). Reproduced by permission of Oxford University Press (global.oup.com)

a metal by adding dirt to it? Nevill Mott (1905–1996) supposed this would be so: by introducing impurities into the crystal structure of a metal the electron motions would eventually become so confused that the system would be forced to be an insulator. Applying RG theory allows one to use simple dimensional analysis to provide the answer and this turns out to depend on dimensionality (Zee, 2003). That is, we get a different result if the electrons is forced to move in a flat slab (a 2D solid) or a one-dimensional channel. The results are shown in Fig. 5. For a 3D metal we have conducting behaviour if the amount of disorder is small and insulating if it is large, just as Mott predicted. A surprise comes for 2D and 1D: no matter how small the amount of disorder, the system is forced to be an insulator. We conclude that simple Ohmic conductivity is not possible in 2D and 1D.

The idea of a 2D or 1D metal might seem hypothetical. However, for the metal it is not a fantasy, but rather a state of affairs that is accessible in the laboratory, and exploring the consequences of varying the number of dimensions is a regular and fruitful preoccupation of physicists. Specifically, by growing layers of semiconducting materials it is possible for the energy bands of materials to arrange themselves in such a way as to produce systems where electrons are confined to 2D planes or to 1D channels (Singleton, 2001). Quantum mechanics forces these to act as highly idealised gases of 2D or 1D electrons, which do not have the possibility of exploring or interacting in higher dimensions.

The fact the we don't see Ohmic conduction in 2D and 1D doesn't mean these states are uninteresting. In 1D for example, Fermi-Liquid theory does not describe the 1D system: there are no quasiparticles (Giamarchi, 2003). Instead excitations of this state correspond to waves in charge density and, independently, waves in spin

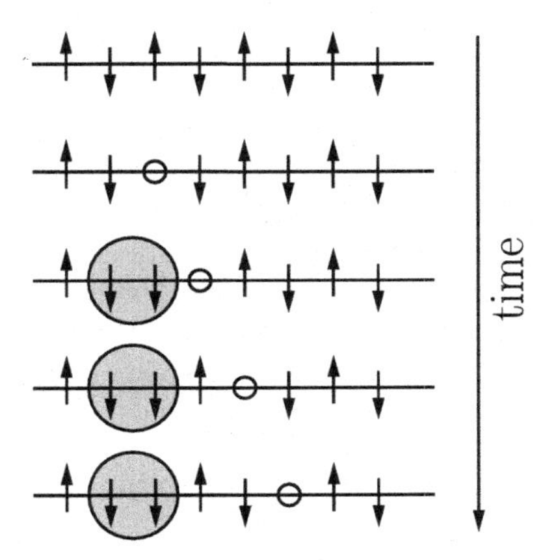

Fig. 6 A cartoon showing how spin and charge separates in a one-dimensional system. A spin up electron is removed, leaving a hole behind. As the hole moves down the chain it leaves behind it a spin excitation (circled)

density. This is a remarkable feature where we see that the whole system interactions allow the apparently fundamental properties of the electron: spin and charge, to break into two. This feature can be understood with a simple thought experiment. Imagine a set of electrons arranged along a one-dimensional line as shown in Fig. 6. We arrange there electron spins up-down-up-down, so the overall spin of the system is zero. If we remove an electron from this system then we leave behind a hole, which can be thought of as an excited state of charge (it is known as a holon in this context). If we move an electron along the line into the empty space without changing its spin, then we see that the hole is mobile. This has a consequence: it leave two like spins as neighbours. This is a spin excitation known as a spinon. As we slide electrons around we see that the holon can move independently of the spinon. Similarly by flipping a pair of spins we can move the spinon around. The spin excitation and the hole excitation are independent: we have spin-charge separation.

The 1D system is certainly very different to its 3D cousin. We obtain qualitatively different physics again in the 2D case. The most dramatic manifestation of this is a state known as the fractional quantum Hall fluid (Lancaster & Pexton, 2015; Zee, 2003). This is the state of affairs that occurs when we apply a large magnetic field to a 2D gas of electrons. In the presence of a magnetic field electrons are forced to undergo (helical) cyclotron orbits around the magnetic field lines. In the 2D gas the electrons are also forced to avoid each other owing to Coulomb repulsion and the quantum mechanical constraint of Pauli's principle. The result is a complicated, whole-system dance of electrons that results in a new phase of matter. This is the fractional quantum Hall fluid and its properties are truly remarkable. Most remarkable of these is that an extra electron added to the system seems not to be able to sustain its electronic charge. For example in one of the FQH states it appears to fall apart into three. Another way of saying this is that the quasiparticle excitation of the FQH fluid have fractional charge. This is a situation, therefore, where electronic charge, which we are taught is a fundamental quantity that cannot be subdivided, is indeed divided into a rational fraction. Furthermore, these excitations are neither fermions nor bosons, but actually have different quantum statistics and are known as *anyons*. This state of matter can be understood using notions from the mathematical study of topology, whose concern is the overall shape of spaces (rather than the details of distances and angles, which is the field of geometry). The FQH fluid is therefore characterized as a topologically-ordered phase of matter in which whole-system interactions are decisive in determining the properties (Lancaster & Pexton, 2015).

In both of these low-dimensional cases, we see new excitations emerge with remarkable and unexpected properties. The driving force in both of these cases is the electrostatic interaction between pairs of electrons that take place across the whole of the system. The excitations are expressions of energy being absorbed by this whole system.

7 Conclusion

If there is a standard model of condensed matter and the solid state, it is provided by the metal. The simplest models of the metal give us the deceptive impression that the physics of metals is mainly determined by electrons (specifically those electrons with the Fermi energy) acting as single entities. This is an illusion: these electrons are dressed in the interactions with the other electrons in the system forming quasi-particles (and this is also true of electrons outside of the metal). Despite this, there is a limit on the importance of whole-system interactions in the metal expressible in terms of the coherence of the metallic quantum state.

Metals exist in crystals, which are broken-symmetry systems. A many-particle manifestation of this broken symmetry state is the phonon: an excitation of the crystal that involves changes in the coordinates of all of the atoms in the crystal. These excitations can be treated as particles and scatter from electrons in the metal. This can lead the metal to become unstable, itself undergoing a phase transition to the superconducting state. This state is far more coherent than the metal, built as it is from a special superposition of states of electron pairs.[2]

We engage with whole-system properties in the measurements we make on metals. In order to relate these to the underlying dynamics of electrons we are forced to examine the limiting behaviour of large lengths and times, which provides us with Ohmic conduction and the dissipation of energy. Once we also invoke constraints on the dimensionality of the system, this approach reveals a rich variety of behaviour. Possibly the most dramatic of these is the topologically-ordered fractional quantum Hall state.

In the recent literature on the philosophy of science, phase transitions and the fractional quantum hall fluid are often discussed, but the metal receives much less attention. This would seem (to me, at least) to be a mistake, given the apparent

[2] The superconducting transition involves the breaking of the phase symmetry of the wave function in a system with charge. As a result, the superconductor does not support analogous particles to the phonon: the description of the disappearance of these is known as the Higgs mechanism, and is very similar to the Higgs mechanism in the standard model.

simplicity and the rich array of physics that the many-particle nature of the state brings into being (Coleman, 2019), which results from many-particle and whole-system effects. Finally, we can note that throughout this description the context, i.e. a macroscopic piece of matter, experienced in the long-length-scale, long-time limit, has been just as important as the detailed knowledge of the physics of single electrons and more so than the knowledge of how electrons interact within atoms or molecules.

Acknowledgements I am grateful to Thomas Hicken and Ben Huddart for useful comments on this essay.

References

Abrikosov, A. A., Gorkov, L. P., & Dzyaloshinski, I. E. (1963). *Methods of quantum field theory in statistical physics*. Dover.
Anderson, P. W. (1985). *Basic notions in condensed matter physics*. Benjamin-Cummings.
Annett, J. F. (2004). *Superconductivity, superfluids and condensates*. OUP.
Ashcroft, N. W., & Mermin, N. D. (2003). *Solid state physics*. Thomson Press (India) Ltd.
Coleman, P. (2015). *Introduction to many body physics*. CUP.
Coleman, P. (2019). In S. Gibb, R. F. Hendry, & T. Lancaster (Eds.), *The Routledge handbook of emergence*. Routledge.
Feynman, R. P. (1990). *QED: the strange theory of light and matter*. Penguin.
Giamarchi, T. (2003). *Quantum physics in one dimension*. OUP.
Halzen, F., & Martin, A. D. (1984). *Quarks and leptons*. Wiley.
Lancaster, T. (2019). In S. Gibb, R. F. Hendry, & T. Lancaster (Eds.), *The Routledge handbook of emergence*. Routledge.
Lancaster, T., & Blundell, S. J. (2014). *Quantum field theory for the gifted amateur*. OUP.
Lancaster, T., & Pexton, M. (2015). *Studies in History and Philosophy of Modern Physics, 52*, 343.
Lifshitz, E. M. & Pitaevskii, L. P. (1980). *Statistical physics, part 2*, Pergamon (1980).
Mattuck, R. D. (1967). *A guide to Feynman diagrams in the many-body problem*. Dover.
McComb, W. C. (2004). *Renormalization methods*. OUP.
Penrose, R. (2004). *The road to reality*. Vintage.
Peskin, M. E., & Schroeder, D. V. (1995). *An introduction to quantum field theory*. Westview Press.
Rae, A. I. M. (1992). *Quantum mechanics* (3rd ed.). Institute of Physics Publishing.
Rosenberg, H. M. (1988). *The solid state*. OUP.
Sachdev, S. (2011). *Quantum phase transitions* (2nd ed.). CUP.
Schiff, L. I. (1968). *Quantum mechanics* (3rd ed.). McGraw-Hill.
Singleton, J. (2001). *Band Theory and Electronic Properties of Solids*. OUP.
Weinberg, S. (2013). *Lectures on quantum mechanics*. CUP.
Zee, A. (2003). *Quantum field theory in a nutshell*. Princeton University Press.

Spacetime Emergence: Collapsing the Distinction Between Content and Context?

Karen Crowther

1 Introduction

When you think of emergence, you might think of the collective behaviour of flocks of birds or colonies of ants; you might think of the emergence of life from non-living molecules, or of consciousness from collections of unconscious brain cells. Perhaps, you think of more exotic things, like the emergence of stable macroscopic objects whose behaviour can be described by deterministic laws, from some strangely behaved fundamental quantum particles. In any case, what you think of is most probably some behaviour, process, property, or object that occurs, or exists, in space and time. But what about the case of spacetime itself—can this be considered emergent from some collective behaviour of non-spatiotemporal objects? Or, could a spatiotemporal universe emerge from some 'prior' non-spatiotemporal state 'before' the beginning of the universe? Many philosophers and physicists believe that, indeed, both these scenarios are real physical possibilities in our own universe—these suggestions comes from research in *quantum gravity* and *quantum cosmology*, respectively.

Currently, our best description of spacetime is provided by Einstein's theory of *general relativity* (GR), which says that gravity is the curvature of spacetime due to massive objects. While this theory is incredibly successful, physicists do not believe that GR—along with the description of spacetime it provides—is fundamental. Instead, GR is thought to be incorrect at extremely short length scales (the Planck scale, 10^{-35} m), and in regions of extremely high spacetime curvature, where quantum effects cannot be neglected: these include black holes and the 'Big Bang'. In these domains we require a new theory, called quantum gravity (QG).

While there is no accepted theory of QG, there are several different attempts at finding it (i.e., different research programs) that are currently in development. Some of these approaches do not feature a concept of spacetime fundamentally. Instead, the

---

K. Crowther (✉)
Centre for Philosophy and the Sciences, University of Oslo, Oslo, Norway
e-mail: karen.crowther@ifikk.uio.no

S. Wuppuluri and I. Stewart (eds.), *From Electrons to Elephants and Elections*,
The Frontiers Collection, https://doi.org/10.1007/978-3-030-92192-7_22

existence of spacetime (and gravitational phenomena) is to be explained by reference to some more basic non-spatiotemporal objects that 'underlie', or 'compose' it. In other words, these theories describe 'atoms' of spacetime that do not themselves exist in space or time. Just as tables and chairs, and other familiar 'macroscopic' objects are not fundamental according to our 'microscopic' quantum theories of matter, so too spacetime is not fundamental according to these approaches to QG.

Does this mean that spacetime is *emergent*? If so, in what sense? So far, it has been argued that yes, on some approaches to QG, we can understand spacetime as emergent in a 'hierarchical' (inter-level) sense, where GR (and/or spacetime) emerges from the more fundamental theory of QG.[1] This can be understood as emergence of *content*, where 'objects' (taken in a broad sense), or phenomena, emerge at a higher ('macro') level that are not present at the more fundamental ('micro') level. But another sense of emergence has also been argued for as potentially possible, where spacetime emerges from a 'prior' non-spatiotemporal state replacing the Big Bang at the beginning of the universe, described by quantum cosmology.[2] This type of emergence occurs at a single level, so I refer to it as 'flat emergence' (Crowther, 2020). This can be understood as emergence due to (or *of*) *context*, where novel 'objects' (again, in a broad sense) emerge at a 'later' state which are not present at the 'initial' state of the system.

Given the unique case of spacetime emergence, however, it may be that the distinction between content and context just described is not a useful, or even sensible, one to draw. It is a case where both types of emergence are supposed to obtain 'simultaneously', and where the standard ways of distinguishing between these two types of emergence are not obviously available. For instance, in inter-level emergence (which I've called hierarchical, or content emergence), the levels ('micro' and 'macro') are usually characterised in terms of different scales—e.g., length scales, or energy scales—but how do we do this in the absence of space? And, for flat emergence, the states of a system ('earlier' and 'later'), are usually distinguished by reference to time—so how do we do this in the absence of time? Here, I explore the possibility that the emergence of spacetime is a case where we need a more general conception of emergence: one that collapses the distinction between content and context.

I begin by first describing QG (Sect. 2), including what it means to say that QG is more fundamental than GR (Sect. 3), which requires that the relation of *reduction* holds between the two theories (Sect. 4). I then describe the different senses of spacetime emergence (Sect. 5). Hierarchical emergence is explored in Sect. 6, with the example of loop quantum gravity Sect. 6.1. Flat emergence is explored in Sect. 7, with the example of loop quantum cosmology Sect. 7.1. I then discuss the example of phase transitions Sect. 8, arguing that these represent both hierarchical and flat emergence, as distinct notions. It is possible that spacetime also emerges in a phase transition, which I explore in Sect. 8.1. In this case, however, the two notions of emergence are

[1] See, e.g., Butterfield and Isham (1999, 2001), Crowther (2016), Huggett and Wüthrich (2013), Oriti (Forthcoming), Wüthrich (2017, 2019).

[2] See, e.g., Brahma (2020), Crowther (2020), Huggett and Wüthrich (2018).

supposed to obtain simultaneously, and are not so obviously distinguished. It may be more natural to not make the distinction between hierarchical and flat emergence in the case of spacetime: this possibility is motivated in Sect. 9.

2 Quantum Gravity

Currently, all fundamental particles and forces are described by quantum field theory (QFT), while gravity is described by GR. Although these theories are supposed to be universally applicable, we do not, in practice, need to use both of them together to describe any of the systems that we observe or directly interact with in the world. Yet, there are inaccessible domains of the universe whose description requires both theories, including the the Planck scale 10^{-35} m, black holes, and cosmological singularities (such as the Big Bang). The problem is that it is difficult to combine GR and QFT in a way that gives us acceptable answers about these domains. And so physicists are seeking a new theory, QG.

QFT is a framework that combines quantum mechanics and special relativity. It describes various quantum fields in a fixed, non-dynamical (unchanging) background spacetime. What we call the fundamental particles and forces—the electromagnetic, strong and weak forces—are conceived of as local (point-like) excitations of these fields according to one particular model of this framework, known as the *standard model* of quantum field theory. While spacetime is used in this theory, it is not *described by* the theory. GR, on the other hand, is a theory of spacetime. It describes spacetime itself as a dynamical field (that does not exist in some further 'background' spacetime, and so is *background independent*), and says that gravity is due to the curvature of spacetime. Both of these theories are incredibly successful, yet, neither theory is thought to be fundamental (Crowther, 2019). QG is supposed to be more fundamental than both these theories, but since we are interested in spacetime emergence, I consider only how it is supposed to be related to GR (Sect. 3).

There are various different approaches towards finding a theory of QG. The most well-known of these is *string theory*. This approach describes tiny, 1-dimensional strings propagating on a fixed background spacetime, and the excitations of these strings correspond to the fundamental particles and forces of the standard model, as well as gravity. The approach can be seen as extending the methods of QFT at the expense of the lessons of GR, in that it employs a fixed background spacetime rather than a background-independent dynamics. Here, gravity is treated on par with the fundamental forces, coming from string excitations and corresponding to a QFT particle known as the graviton (whereas according to GR, gravity is not a force but the curvature of spacetime). We could thus say that the approach prioritises the principle of *unification* (all forces, including gravity, stem from the same origin) over that of *background independence* (that QG not feature a fixed, background spacetime).

Some other approaches to QG instead prioritise background independence; because these describe basic entities that are non-spatiotemporal (not existing in spacetime), they more completely demonstrate the emergence of spacetime. I briefly introduce one of these, *loop quantum gravity*.

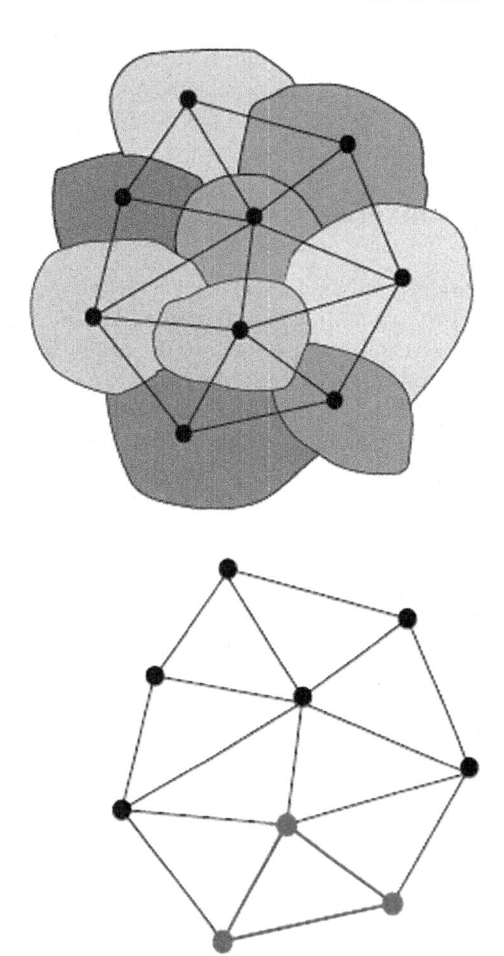

Fig. 1 Quanta of volume (grey blobs). Adjacent chunks are separated by a surface $\mathcal{S}$ of quantised area. The corresponding spin network graph is overlaid. Each link intersects one quantised surface $\mathcal{S}$ (Rovelli, 2004, p. 20)

Fig. 2 Spin network: Nodes represent quanta of volume, which are adjacent if there is a link between them. Connected links form loops, like the one highlighted in red (Rovelli, 2004, p. 20)

2.1 Loop Quantum Gravity

Loop Quantum Gravity (LQG) is an attempt to construct a theory of QG by quantising GR. It describes quanta (roughly, discrete quantum 'chunks') of spacetime. In a quantum theory, the discrete values of a physical quantity can be found by calculating the *eigenvalues* of its corresponding *operator*. In LQG, the important operators are the 'area operator', $\hat{\mathbf{A}}$, associated with the area, $\mathbf{A}$, of a given surface, $\mathcal{S}$, and the 'volume operator', $\hat{\mathbf{V}}$, associated with the volume, $\mathbf{V}$, of a given spatial region, $\mathcal{R}$. The eigenstates of these operators are called *spin network* states, and are represented as graphs called *spin networks*. An example of a spin network is shown in Fig. 2, with an illustrative depiction of how the spin network relates to the quanta of volume shown in Fig. 1.

This describes the *kinematical* aspect of the theory (i.e., the spin network states provide a basis for the kinematic Hilbert space), rather than the dynamics, and so represents space rather than spacetime. There is more to say about the dynamics of the theory (and how this is supposed to relate to spacetime), but this very brief, informal introduction to LQG is sufficient to gain an understanding of how space might emerge according to LQG.[3]

Physical space is thought to be a quantum superposition of spin network states with well-behaved geometric properties, known as *weave states*.[4] The intuitive idea is captured by analogy: at familiar scales, the fabric of a t-shirt is a smooth, two-dimensional curved surface, but when we examine it more closely, we see that the fabric is composed of one-dimensional threads woven together.[5] The suggestion is that LQG presents a similar picture: while the geometry of space at large-scales is a three-dimensional continuum, at high-energy it is revealed to be a very large lattice of Planck-sized spacing (i.e. a spin network with a very large number of nodes and links).

3 Fundamentality

As stated above, QG is supposed to be more fundamental than GR. In this section, I first explain what it means for one theory to be more fundamental than another (for the purposes required here), and then explain how this applies in the case of QG as more fundamental than GR.[6]

Here, I just speak about *relative* fundamentality, rather than absolute fundamentality.[7] A *more fundamental* theory, M, of a given system, S, or phenomenon, P, provides a *more basic* description of S or P than a *less fundamental* theory, L, does. I take it that there are two necessary conditions for relative fundamentality:

Asymmetric dependence The laws of L depend upon the physics described by M*, and not vice-versa*;

---

[3] For more on LQG, see Rovelli (2004), Rovelli and Vidotto (2014). Note that this latter reference is much more up-to-date than the brief sketch of the kinematic aspect of the theory that I present here; in particular, it has much more detail on the dynamics of the theory, using the covariant approach to LQG.

[4] These are semiclassical states; i.e., states in which the quantum fluctuations are minimal and the gravitational field behaves almost classically.

[5] The analogy comes from Ashtekar et al. (1992).

[6] The following discussion is based on Crowther (2018a), although the definition of relative fundamentality differs here in that I include two conditions, while Crowther (2018a) requires only one.

[7] While QG must be more fundamental than GR, QG need not be a fundamental theory; i.e., it is not necessary to include the criterion of (absolute) fundamentality in the definition of QG (Crowther & Linnemann, 2019). For other ideas of fundamentality in physics and metaphysics, see Morganti (2020a, 2020b).

Broader domain The domain of validity of M includes that of L (i.e., that M successfully describes all physical phenomena that L does).

A theory's *domain of validity* is the part of the world that is successfully described by the theory. In regards to this definition of relative fundamentality, note that M will typically describe the system, S at a different range of energy scales than L, or perhaps under different conditions. Also, M might not actually describe the phenomenon, P, that L does, but rather (some of) the physics underlying P (i.e., part of the more-basic physics responsible for the appearance of P).

The apparent influence or effect of the dependence of L upon M may not only be very minimal, but also obscured by the way it is incorporated into the parameters and structure of L. In other words, a less-fundamental physics may be largely robust and apparently autonomous, in spite of being dependent on the M physics. Often, the more fundamental theory provides a 'finer-grained', or more detailed, description of the system than the less fundamental one. The L is, in this sense, an approximation to M that works well in a given regime (e.g., a certain range of energy scales, or under special conditions).

There are two ways in which QG is more fundamental than GR: by being quantum (or by being *beyond* quantum) and by being a 'micro' theory of spacetime. Each of these is sufficient to satisfy both the 'asymmetric dependence' and 'broader domain' conditions of relative fundamentality. The recovery of GR from QG may be a two-step process, recovering the 'appearance of classicality' from the quantum theory, and moving to the 'macro' (low-energy) limit of the theory (which brings us back to familiar energy scales). The former is known as the quantum/classical transition, and the latter is called the micro/macro transition. While both address the question of why we do not need to use QG to describe much of the gravitational phenomena we observe, they are distinct, and may or may not be related to one another. Both transitions represent common problems in physics and the philosophy of physics; and both play a role in understanding the relationships of reduction and emergence.

The micro/macro transition is not something that happens to a system, but a change in the level of description: it is the move to a coarser-grained theory. The micro/macro transition may be represented by an approximation procedure, a limiting process (such as the thermodynamic or 'continuum limit'), or the renormalisation group flow. All of these different techniques are employed by various approaches to QG in their attempts to connect QG back to GR.

In the case of the quantum/classical transition there are two different issues. Quantum theories are supposed to apply universally; so, first, there is the question of why, in practice, they are usually only necessary for describing small systems. Secondly, there is the *measurement problem*: why it is that any measurement on a quantum system finds the system in a definite state even though the system evolves as a superposition of different states. The process of *decoherence* describes how the interference effects associated with superpositions become suppressed through a system's interactions with its environment, with the consequence that the quantum nature of the system is no longer manifest. Larger systems more strongly couple to their environments, so decoherence provides the beginning of an explanation for why quantum theory is usually only necessary for describing micro-systems. As such, it gives us some insight into the 'transition' that a system undergoes that prompts us to move from a quantum description of it to a "classical" one (although the system remains inherently quantum, as does the rest of the universe). Decoherence does not, however, give us an answer to the measurement problem.

We expect that the generic states of the objects described by QG will be quantum superpositions, but the quantum nature of spacetime is not manifest. An explanation of the quantum/classical transition is necessary for understanding why this is. Of course, this is an incredibly challenging task, given that the quantum/classical transition is poorly understood in general. It seems likely that a solution to the measurement problem is required if we are to fully understand the relationship between spacetime and the quantum objects that somehow underlie it; or it may be that the solution will be provided by the theory itself.[8] For the time being, however, we seek to better understand the relationship between QG and GR to any degree that will aid in the development of the theory, even without full knowledge of the quantum/classical transition.

4 Reduction

Reduction in physics means showing that the successful parts of the older theory (in this case, GR), are, in principle, *derivable* (i.e., deducible) from the newer one (QG) in the appropriate domain (where we know GR is successful), under appropriate (physically sensible) conditions. Reduction in this sense also demonstrates that the newer theory is *more fundamental* than the older one. The 'asymmetric dependence' condition of relative fundamentality is satisfied, because if the older theory is derivable from the newer one (and not vice versa), then there is a sense in which the older theory is dependent upon the physics described by the newer theory. If the newer theory is correct, and the older theory is appropriately derivable from the newer one, then the older theory will automatically be correct, since it is entailed by—or, a consequence of—the newer one.

QG is meant to be more fundamental than GR (as explained in Sect. 3): this means that the physics described by QG is supposed to be responsible for the success of the laws of GR, and that QG also describes all of the systems/phenomena that GR does. As such, the relation of reduction must obtain between these two theories. This is a strong constraint on QG, which serves to define the new theory: any prospective theory of QG will not be accepted unless physicists are satisfied that GR is appropriately derivable from the theory of QG. Standardly, however, the derivability of the older theory from the newer one is not rigorously established: physicists employ various approximations and limiting relations, relying on different assumptions when doing so. We demonstrate that particular *correspondence relations* hold between the two theories in the shared domain where they are both supposed to apply. These

[8] Penrose explores this second possibility, see e.g. Penrose (1999, 2002).

are inter-theory relationships that connect the two theories, by, e.g., shared terms, numerical predictions, laws, or principles.[9] Reduction is taken to obtain when sufficient correspondence relations have been demonstrated (running in the direction from the newer theory to the older one), such that we are convinced that the older theory is *in principle* appropriately derivable from the newer one.

This is where many attempts to construct QG have run into difficulties. For instance, although LQG starts out with GR, and uses correspondence relations running from GR to QG ('top-down') in order to construct the new theory, it has trouble naturally 'recovering' spacetime (going 'bottom-up') and establishing that GR is appropriately derivable from the new theory. This is not necessarily an indication that GR is not recoverable from the theory, however, since it is still under development and may eventually reach a stage where it succeeds. Given that it is a requirement upon any theory of QG that it appropriately recover GR in the domains where GR is successful, I will, in the rest of this essay, assume that GR is, ultimately, appropriately derivable from whatever the accepted theory of QG turns out to be—i.e., that reduction holds in the sense described above.

5 Emergence

Emergence is an empirical relation between two relate of the same nature, an emergent, E, and its basis, B. Depending on the case of interest, E and B may be objects, properties, powers, laws, theories, models, etc. Here, I am interested in emergence as a relation between theories or parts of theories. I take the general conception of emergence to comprise three claims,[10]

Emergence: General Conception

Dependence E is dependent on B, as evidenced by E being derivable from B, and/or supervenient upon B (supervenience means that there would be no change in E unless there were a change in B, but not vice versa);

Novelty E is strikingly, qualitatively different from B;

Autonomy E is robust against changes in B; or E is the same for various choices of, or assumptions about, the comparison class, B.

This is a *positive* conception of emergence, since it is not characterised by a *failure* of reduction, deduction, explanation, or derivation in any sense in any sense. That is, the emergent E need not be irreducible to, or unexplained by, its basis B. Such a positive conception of emergence is now familiar in the philosophy of physics generally, and the philosophy of QG in particular.[11] The positive conception of emergence is the most appropriate for understanding the case of spacetime emergent from QG for two

[9] Correspondence takes various forms and plays many important roles in theory development and assessment; see, e.g, Crowther (2018b), Hartmann (2002), Post (1971), Radder (1991).

[10] The discussion in this section is based on Crowther (2020).

[11] This is largely due to Butterfield and Isham (1999, 2001), Butterfield (2011a, 2011b).

reasons. First, as explained above, GR must be reducible to QG—i.e., it is a requirement on any theory of QG that GR be approximately and appropriately derivable from it. This condition may be used to satisfy the 'Dependence' claim of emergence. Thus, we need an account of emergence that is compatible with reduction, at least in this sense. Second, none of the approaches to QG are complete, so basing any claims of emergence on their failure to explain, derive, or predict particular aspects of GR spacetime is risky, given that a central goal of each of the approaches is to develop a theory that approximately and appropriately *recovers* (i.e., derives and explains) GR spacetime.

This general conception of emergence admits more specific forms: for instance, it can accommodate either *synchronic* or *diachronic* conceptions of emergence. In the synchronic conception of emergence, B and E represent different levels of description: B is said to describe the system at the *lower level* and E at the *higher level*. In physics, B and E may be theories that apply at different ranges of length- or energy-scales, where, typically, B describes the system at higher energy scales (shorter length scales) than E, which applies at comparatively low energy scales (large length scales). These theories are supposed to apply to the system at the same time, or otherwise under the same conditions, i.e, there is no 'change' considered, except the level at which you view the system.

In the diachronic conception of emergence, E and B describe the system at the same level. These theories, or models, are supposed to apply to the system at different times, or otherwise under different conditions. The idea is that the system has undergone some change: typically, B describes it before, and E after. This conception of emergence is not associated with a notion of fundamentality. The difference between these two conceptions of emergence is illustrated in Fig. 3.

There are many specific accounts, and examples, of synchronic and diachronic emergence described in the literature. However, any account of synchronic or diachronic emergence requires some modifications in order to be applicable to the case of spacetime emergence. Most obviously, these two conceptions rely on a notion of *time* for their distinction, as reflected in the names 'synchronic' and 'diachronic'. I suggest two accounts of emergence, roughly analogous to synchronic and diachronic accounts, but which do not rely on spatiotemporal notions; these are *hierarchical* emergence and *flat* emergence, respectively (Crowther, 2020). I briefly describe these in the next two sections.

6 Hierarchical Emergence: *Content*

Hierarchical emergence can be understood as an analogue of synchronic emergence that doesn't rely on a conception of space in order to distinguish the different levels involved, and which doesn't rely on a notion of time in identifying the system between levels. I will distinguish levels in terms of 'size of grain' ('size' is not here a reference to lengths, but refers to the amount of detail captured)—i.e., a lower-level theory provides a finer grained description of the physics, while a higher-level

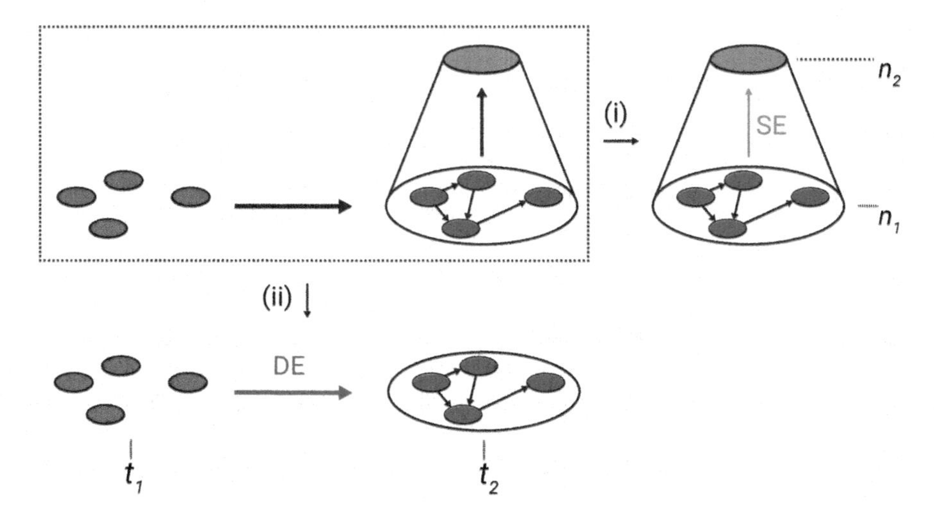

Fig. 3 Two conceptions of emergence. A system **(a)** at time t_1 (red) has changed at time t_2 (blue), resulting in some novel higher-level phenomena (green). **(i)** If we are interested in synchronic emergence (arrow SE), we consider the system at level n_2 compared to the system viewed at n_1, at a single time, here t_2. **(ii)** If we are interested in diachronic emergence (arrow DE), we consider the system at t_2 compared to the system at t_1, at a single level, here n_1. Figure adapted from Guay and Sartenaer (2016)

theory provides a coarser-grained description of the phenomena (abstracting away from the finer details, which are irrelevant at this level). Although we tend to think of finer-grained theories as describing 'shorter length scales', this is not a necessary correlation. Nevertheless, for convenience of notation, I will still distinguish the lower level physics described by B as 'micro' physics, and the emergent level E physics as 'macro' phenomena.

I suggest the following account of hierarchical emergence, which is based on other case-studies involving *effective theories* in physics (Crowther, 2016):

Emergence: Hierarchical Conception

Dependence The coarser grained theory (model, or structure) E is constructed (i.e., derived) from the finer grained theory B. The physics described by the laws of E may be said to *supervene* on those of B.

Novelty The physics described by the coarser grained, or low energy ('macro') theory E differs remarkably from that of the finer grained, or higher energy ('micro') theory B;

Autonomy The physics described by E is robust against changes in the micro physics; B is underdetermined by E.

As in the more general conception above (Sect. 5), *Novelty* is a symmetric relation; this condition captures the ways in which the two theories differ from one another. The idea of *Autonomy* here comes from *universality* or *multiple-realisability*, which ensures the robustness of the higher-level physics compared to the lower-level physics: there are many different lower-level structures, systems, or states, that could 'give rise to' (or 'realise') the same higher-level, emergent physics (which is said to be 'universal' behaviour).[12] There are two different ways this can happen which are relevant here. First, different micro states described by, or models of, B can correspond to a single macro state/model of E. An an example is the way in which a number of different micro states described by statistical mechanics correspond to a single macro state in thermodynamics, i.e., how different configurations of molecules in your coffee give rise to the same homogeneous-looking liquid of a particular temperature (at the finer-grained, micro level, the individual molecules can have different positions and velocities, but you don't notice this at the coarse-grained, macro-level). Second, different micro theories can correspond to the same macro theory. An example is how fluids of different micro-constitutions (i.e., cells, molecules, atoms, or particles of different types) can give rise to the same hydrodynamic behaviour at a coarser-grained description. The fact that the macro physics is multiply-realisable leads to an *underdetermination* of the micro-physics, meaning that if you only know the E behaviour, you will not automatically be able to determine the specific B theory (state, or model), since there are many possible candidates that could be responsible.

We can consider hierarchical emergence as emergence of *content*. This can be understood in two ways: considering it as emergence of theoretical structures, or as 'ontological' emergence of some particular entities or behaviour associated with the emergent theoretical structures. In either case, the idea is that novel content appears at the higher level that is not present at the lower level, and which is autonomous from the lower-level content in the sense that many different structures at the lower level could 'give rise to' the same higher-level structures.

This conception of emergence can be used to understand hierarchical emergence in several different approaches to QG (Crowther, 2016). In the next section, I consider how it applies in LQG.

6.1 Hierarchical Emergence of Spacetime in LQG

LQG is still incomplete, and it is not yet clear how spacetime is to be recovered from the fundamental structures of the theory. For now, we will just assume that the kinematical aspect of LQG described above (Sect. 2.1) is roughly correct, which means assuming that *space* (rather than spacetime) is fundamentally constituted by a spin network. We will also take it that LQG is a serious contender for QG, and thus assume that GR is appropriately derivable from LQG. So, we assume that the *Dependence* condition for hierarchical emergence is satisfied.

The *Novelty* condition of emergence is plausibly satisfied because the spin network states differ from space in a number of ways; I mention three of these here.

[12] See Batterman (2000, 2002, 2018), Crowther (2015), Franklin (2018a) for more on autonomy, universality, and multiple-realisability, particularly as related to emergence in effective field theory.

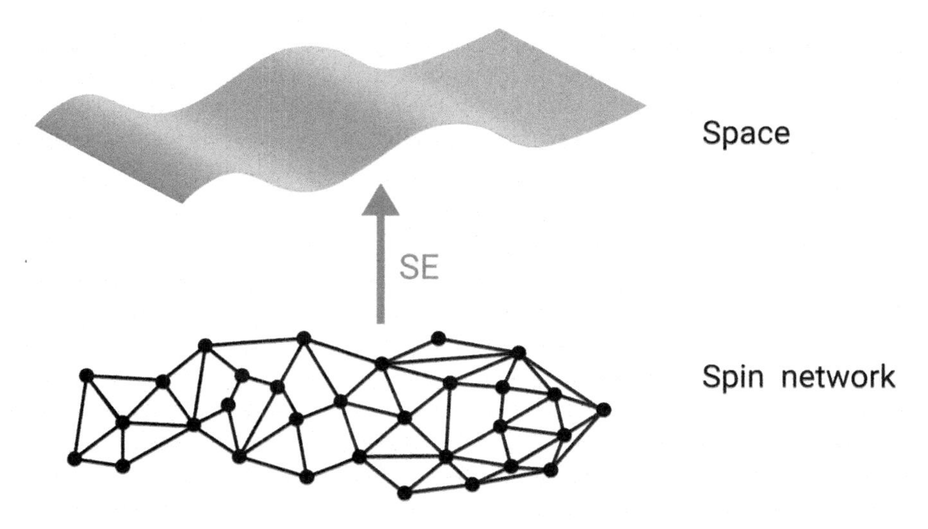

Fig. 4 Space as hierarchically emergent from a spin network

First, the spin networks represent discrete, quantum 'objects' rather than continuum spacetime.[13] Second, as has been emphasised by Huggett and Wüthrich (2013) there is a particular form of "non-locality", where it is possible for two discrete 'chunks of space' that are adjacent (directly connected to one another) in the spin network to not map to neighbouring regions in the corresponding spacetime (though this "non-locality" should be heavily suppressed, otherwise the particular spin network in question would correspond to a different spacetime, one which better reflects its fundamental structure). Third, space is supposed to be a quantum superposition of spin networks, so there is no clear notion of geometry at the fundamental level.

The *Autonomy* condition of emergence is plausibly satisfied because many different spin network states can correspond to the same (semiclassical) geometry—demonstrating the robustness of the emergent spacetime. Also, given that space is meant to correspond to a superposition of spin networks, it is autonomous from any particular definite (non-superposed) spin network state. Thus, there is a plausible claim to be made that GR spacetime is emergent from the fundamental structures of LQG. This is depicted in Fig. 4 as space emergent from a definite spin network state.

[13] Actually, the dynamics can be thought of as not simply a 'quantum version' of GR, as this characterisation suggests, but a more radically different theory; see Oriti (2014, 2018) for more on this aspect.

7 Flat Emergence: *Context*

Flat emergence is an analogue of diachronic emergence which does not rely on spatiotemporal notions. This conception is potentially applicable to spacetime because one of the domains where GR is expected to be incorrect, and to require replacement by QG, is at the very beginning of the universe. Using GR and current observations of the large-scale structure of the universe, cosmologists extrapolate backwards in time in order to produce a description of the past evolution of the universe. The resulting picture (described in the direction of increasing time) is the standard, or 'Big Bang', model of cosmology, which describes the universe expanding from a state of extremely high temperature and density approximately 13 billion years ago. Before this, however, there is the Big Bang singularity, which is difficult to interpret physically.

One interpretation of the singular behaviour of the model is that GR is incorrect in this domain, due to its neglect of quantum effects that become important at extreme density and temperature (in which case GR likely becomes incorrect at some finite time approaching the singularity). On this view, the singularity is an unphysical artefact—a signal that GR is inapplicable here—and thus, QG should provide a correct, non-singular description of the physics in this domain. Some approaches to QG cosmology suggest that the 'pre-Big Bang' physics is non-spatiotemporal (though there are different things this could mean, e.g., perhaps there is space, but not time, or vice-versa, or there is nothing corresponding to spacetime), in which case spacetime might emerge 'after' from this 'prior' non-spatiotemporal physics.

In order to develop an account of emergence that could potentially be applicable in this case, I consider a characteristic account of diachronic emergence, from Guay and Sartenaer (2016) and Sartenaer (2018). On this account, the *Dependence* condition holds that E is the product of a spatiotemporally continuous process going from B, and/or E is caused by B. The Novelty condition states that E exhibits new entities, properties or powers that do not exist in B. And the *Autonomy* condition states that these new entities, properties or powers are forbidden to exist in B according to the laws governing B.

This account is not generally applicable to the case of spacetime, since it relies on spatiotemporal notions such as causation, location, and continuous processes.[14] If a spatiotemporal state is to emerge from a state that is non-spatiotemporal (or, rather, less-than-fully-spatiotemporal), we cannot assume that this is a process that itself takes place in space and time (although, in fact, some approaches to QG do utilise a notion of time, this is not in all cases able to be identified with our familiar conception of time). A more general conception of flat emergence is required if we are to account for the 'flat' emergence of spacetime from the "Big Bang" state[15] Additionally, the (Guay & Sartenaer, 2016) account of emergence is a negative one,

[14] Although these notions may have non-spatiotemporal analogues, e.g., causation without time (Baron & Miller, 2014, 2015; Tallant, 2019).

[15] Note that the "Big Bang" strictly refers to the GR singularity, whereas in QG cosmology, this state may not be singular.

requiring that E exhibit forbidden entities, properties, or powers. As explained above (Sect. 5), a negative conception of emergence is ill-suited for the case of QG, and a positive conception is to be preferred.[16]

The more general, positive conception of flat emergence that I propose is best-suited for understanding the flat emergence of our spatiotemporal universe from a non-spatiotemporal state is one that is analogous to the hierarchical conception of emergence presented in the previous section (Sect. 6).

Emergence: Flat Conception

Dependence E flatly supervenes on B. (Flat supervenience means that there would be no change in the E state unless there were a change at the B state, but not vice versa);

Novelty E differs remarkably from B;

Autonomy The physics described by E is robust against changes in B. The 'prior' state, B is underdetermined by E. (This sense of underdetermination can be understood as a non- temporal form of indeterminism, meaning that many different 'initial' conditions at the B state could give rise to the same E state. If we only have knowledge of the E state, this would not be enough information to determine the 'prior' B state that it 'evolved from').

This account of emergence is very permissive, yet, it is still not trivially satisfied in the case of QG cosmology. I present an example from loop quantum cosmology that has recently been claimed to represent the emergence of spacetime from a 'prior' non-spatiotemporal state (Sect. 7.1), and argue that it is not clear how we can make sense of this. Later, I present the example of *geometrogenesis*, which is a type of approach to QG that conceives of spacetime emergent in a phase transition (Sect. 8.1). This is a case which is more readily able to be understood as flat emergence of spacetime; however, in Sect. 9, I argue that this example also is a good illustration for why it may make more sense to collapse the distinction between hierarchical and flat emergence.

7.1 Flat Emergence in Loop Quantum Cosmology

Loop quantum cosmology (LQC) is the attempt to use LQG in describing the structure and evolution of the universe. Brahma (2020) describes a particular model of LQC that aims to resolve (i.e., remove) the Big Bang singularity present in the standard model of cosmology. It must be emphasised that, like all attempts at QG cosmology, the model is not fully developed nor understood, so any interpretation is reliant upon speculation, and is highly precarious. It is not clear whether these models are physically meaningful at all.

The model discussed in Brahma (2020) starts by simplifying the system being described, so that it is a spatial geometry with just one parameter, the 'scale factor', with quantum operator $\hat{p}$. According to Huggett and Wüthrich (2018), the resulting

[16] Shech (2019) also suggests weakening the novelty condition along these lines.

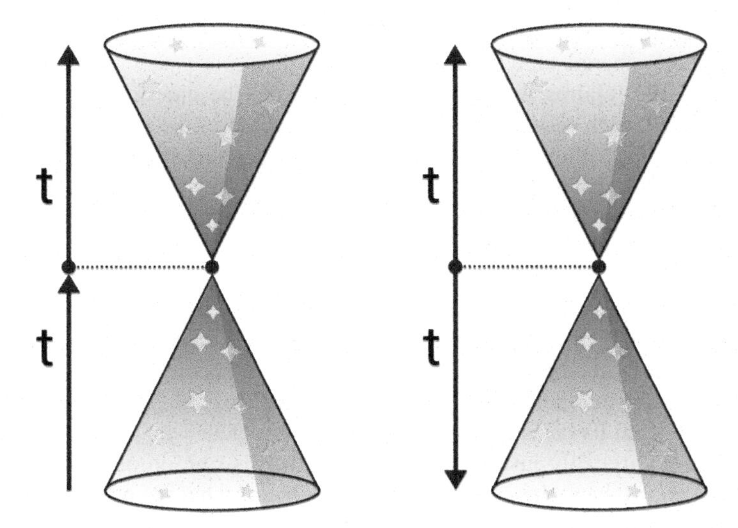

Fig. 5 Two interpretations of LQC. **a** The standard interpretation, as a single universe contracting (bottom cone) then expanding (top tcone) in time, is made difficult by the intermediate state (middle point) having no notion of time. **b** Huggett & Wüthrich interpretation, as of two expanding universes that 'emerge' in positive time from the single non-temporal state (point)

simplified dynamical equation can be interpreted as describing the evolution of the universe, with the scale factor acting as a 'time variable' (though this reading is problematic in many ways). Running this backwards through what would otherwise be the Big Bang, we find that the singularity is not present, and that 'on the other side' of (what would otherwise be) the Big Bang, there is a 'mirror world': an expanding universe in negative 'time'. The resulting picture is standardly interpreted as 'Big Bounce', or a universe undergoing a 'Big Crunch', contracting to a maximally hot, dense state, before re-expanding (this is depicted in Fig. 5a).

At least in one particular type of model, however, Huggett and Wüthrich (2018) argue that there is an alternative picture that is better supported by the physics. In the particular model being referred to by Brahma (2020) and Huggett and Wüthrich (2018), going backward in time leads us from a relativistic spacetime to a region of Euclidean space (without time), and then back into a region of relativistic spacetime on the 'other side'. In other words, the state that replaces the Big Bang is one without any notion of time (at the macro level), and so there is not a way of 'connecting' or 'ordering' the spatiotemporal states with respect to the purely spatial state. Without a continuous time variable running through these three states, why should we say that the purely spatial state lies 'in between' the collapsing universe and the expanding one? In fact, because the collapsing universe and the expanding universe are otherwise symmetric, and its more natural to understand time as directed away from the Big Bang, Huggett and Wüthrich (2018) argue that this model is better interpreted not as a Big Bounce, but rather as a "twin birth" of two expanding universes that arise 'after' the purely spatial state (Fig. 5b).[17]

Brahma (2020) claims that this model represents the emergence of time, and Huggett and Wüthrich (2018) says that it is an example of the "(a) temporal emergence of spacetime", though neither paper goes into detail about what this means or how it fits with standard conceptions of emergence. This is difficult because the physics in this example, at the macro level, can be interpreted either as a bounce or as a "twin birth". Which state emerges from which? In the bounce picture, it could be that space (without time) emerges from a collapsing spacetime, and that an expanding spacetime emerges from a space without time. While in the "twin birth" picture, it is supposed to be two expanding universes emerging from space without time. In order to apply the account of flat emergence sketched above (Sect. 7), we need to understand which state depends on which, if we are to specify what the emergent state E is, and what its basis state B is. There is not a clear way of doing this, since the model could arguably just as well represent the dissolution of spacetime as its emergence (i.e., we have these two different pictures of what the model represents). So, the Dependence condition of flat emergence is unable to be assessed, as is the Autonomy condition (since this requires understanding which state is E and which is B). Clearly, however, the Novelty condition is satisfied in this example, however Crowther (2020).

8 Phase Transitions

Here, I discuss phase transitions, which can be seen as examples of both hierarchical and flat emergence. It is useful to understand this, too, because there are some approaches to QG that imagine spacetime emergent in a phase transition, Sect. 8.1.

Phase transitions are qualitative changes in the state of a system; most familiar are the examples of water freezing to ice, or boiling to vapour. Of particular interest as examples of emergence are second-order phase transitions, where systems exhibit *critical phenomena*. These represent conditions under which there is no real distinction between two states of the system–for instance, between the liquid and vapour phases of water.[18] Second-order phase transitions present clear examples universality (multiply-realised behaviour), where where a number of different systems—different types of fluids (e.g., with different molecular constitutions), as well as magnetic materials—exhibit the same critical phenomena.

An example is the ferromagnetic phase transition. A magnetic material at the micro level, can be pictured as comprising atoms with magnetic spins. When the

[17] In other LQC models, however, the 'Big Bounce' picture is arguably better-supported.

[18] Under these conditions, the system has a fractal structure, not changing as we view it at smaller length scales. In this example of the second order phase transition between liquid and vapour, as we look at smaller scales we would see liquid water droplets containing bubbles of steam, within which float even tinier liquid droplets, and so on... until we reach the scale of atoms.

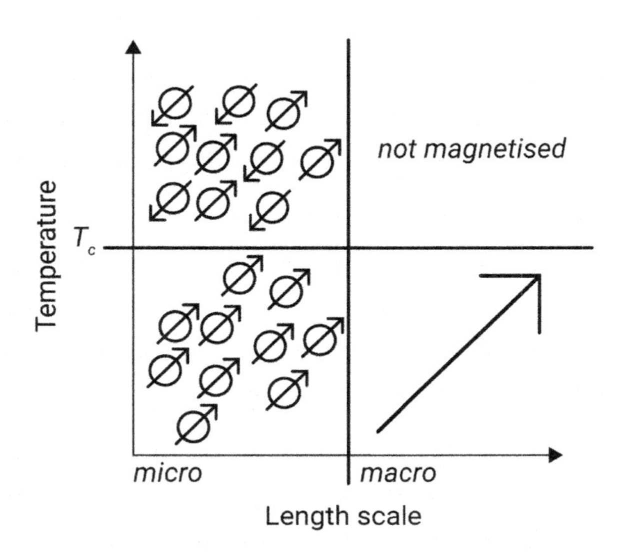

Fig. 6 Ferromagnetic phase transition. At temperatures below the critical temperature, T_C, most of the atoms' magnetic spins are parallel (viewed at the micro level), and the material has a net magnetisation (ferromagnetic phase). Above T_C, the spins point in different directions and the material has no net magnetisation (paramagnetic phase)

temperature is low, the spins of adjacent atoms are parallel (the energy is lower if the spins on adjacent atoms are parallel than if they are antiparallel). Below a certain 'critical' temperature, (in this case it is called the Curie temperature), T_C, most of the spins in the material are parallel, and so add up constructively to give the material a net magnetisation, known as ferromagnetism. Above T_C on average half the spins point in one direction and the other half in the opposite direction, and so the material has no net magnetisation: it is a then a paramagnet. Thus, at T_C, a phase transition occurs where the material undergoes a sudden qualitative change in state: it goes from being paramagnetic, having no net magnetisation, to suddenly being ferromagnetic, having a magnetisation. This is illustrated in Fig. 6.

This example represents a symmetry-breaking phase transition. When the spins are randomised, there is no preferred direction in the system: it is symmetric (looks the same no matter how you rotate it). After the phase transition, however, the spins are aligned and so there is a preferred direction: the symmetry has spontaneously been broken.

We can understand critical phenomena as heirarchically emergent behaviour, as well as flatly emergent. But these are understood separately, and depends on the perspective of interest—the two types of emergence are not related to each other in this case. Start by considering hierarchical emergence, by reference to scenario (i) of synchronic emergence (SE) depicted in Fig. 3. Here, the emergent physics E (green) is at the higher level n_2 from the basis state B (purple) at the lower, micro level of atoms at n_1. The emergent physics is described by a different theory, with different degrees of freedom, than the micro physics. This is after the phase transition has occurred (at time t_2). The Dependence condition is satisfied because we can derive the emergent macro behaviour from the micro dynamics; the macro physics supervenes on the micro physics. The Novelty condition is met because there are new laws describing the emergent degrees of freedom compared to the laws governing the micro-system. The Autonomy condition is met because of the universality of the emergent behaviour: many different types of systems, of diverse micro-constitutions, nevertheless exhibit the same behaviour at criticality. This is explained by the mathematical apparatus known as the *renormalisation group*, which is used to obtain a simple coarse-grained description from a complicated micro description; it shows that the micro-details are irrelevant for describing the macro physics. This is an independent framework that is used in many different areas of physics, and demonstrates that the macro physics is largely robust against changes in the micro physics.[19]

Symmetry breaking is also a strong explanation of the universality: the emergent physics can be said to depend only on the particular symmetry-breaking pattern, and not on the details of the micro physics. Even if we can derive the macro behaviour of any given micro-system by starting out with the details of that system, the fact that there are many other different micro systems that exhibit the same macro behaviour under those conditions means that any particular 'reduction' based on a given micro-system will fail to capture the universality of the phenomena. In this sense, we might say that the micro-story *does not* and *cannot* provide an account of the emergent (universal) phenomena. Laughlin and Pines (2000) present symmetry breaking as an example of a "higher-order" organising principle, and the phenomena that depend on it are "transcendent"—insensitive to, and unable to be deduced from, micro-details. As Morrison (2012) states, too, the notion of symmetry breaking is not itself linked to any specific theoretical framework, rather, it functions as a structural constraint on many different kinds of systems, both in high-energy physics and condensed matter physics.

Phase transitions can also be understood as examples of flat emergence. This means looking only at the system at one level, in this case we consider the micro level. This can be understood by reference to the depiction (ii) of diachronic emergence (DE) in Fig. 3. The basis state, B is the state of the system before the phase transition (as depicted at time t_1 in Fig. 3) and the emergent state E is the one after the phase transition (at t_2). E and B are different states of the same system, described by different models of the same theory. The change has occurred in the system because of the different conditions (i.e., the change in temperature). The way in which E depends on B is not so obvious, but perhaps we can say that there is some non-temporal notion of causality that can give a sense of flat supervenience, where there would be no change in the E state unless there were a change in the B state, and not vice versa.[20] The Novelty condition is more obviously satisfied, with B being a state in which there is no preferred direction (symmetric), and E being one with a preferred direction (broken symmetry). The Autonomy condition would hold that E can arise from many different B micro states, e.g., there are many random arrangements at a

[19] See Batterman (2005, 2011), Bain (2013), Franklin (2018b), Rivat and Grinbaum (2020) for more on this.

[20] Cf. Footnote 14.

temperature above T_C that will result in the same E at temperatures below T_C. The particular microstate of B is thus underdetermined given only the E state.

8.1 Geometrogenesis

Pregeometric approaches to QG describe spacetime emergent in a phase transition known as *geometrogenesis*. There are several approaches of this type, but here I consider just one simple example, *quantum graphity*.[21] The fundamental structure described by this approach is represented as a graph: points represent events, which are causally related if there is a connection (link) between them. This system is quantum-mechanical, and its dynamics is represented as a change in the connections between the points. The connections, represented by the links of the graph, are able to be in two states 'on' or 'off', and, being quantum-mechanical, the generic states are superpositions of both 'on' and 'off'.

In the early universe, prior to the phase transition, space does not exist. At the micro-level, this is depicted as a maximally connected graph: every point is connected to every other point (t_1 in Fig. 7). This means that everything in the universe is adjacent to every other event, and so there is no notion of geometry or locality. In this state, the dynamics is invariant (symmetric) under permuation of the events (they cannot be distinguished by their connections). As the universe cools and condenses, it undergoes a phase transition in which many of the connections switch off. The system at low-energy (i.e. at its ground state) is a graph with far fewer edges (t_2 in Fig. 7): the permutation invariance breaks, and instead translation invariance arises. At this stage locality is able to be defined and we gain a sense of relational geometry. The idea is that geometry emerges in this phase transition, known as *geometrogenesis*. This is illustrated in Fig. 7.

Note that, in these approaches, there is a notion of time at the fundamental level, that connects the pre- and post-geometric phases. Spacetime is supposed to be associated with the geometric phase, such that the post-geometric phase is a finer-grained (lower-level) description of GR spacetime (being the higher-level phenomenon). But flat emergence concerns only a single level; here we consider the system just at the more-fundamental level of the discrete structures, rather than the 'phenomenal' spacetime. So, the emergence basis B is taken as the model describing the pre-geometric phase (t_1 Fig. 7), and the emergent model E describes the geometric phase (t_2 Fig. 7).

The Dependence condition can be understood as flat supervenience, since there is no change in the E state unless there is a change in the B state, and not vice-versa. This is ensured by the temporal aspect of these models, such that the B state causally precedes the E state via the theory's evolution equation, and the two states are supposed to be of the same system, being the entire universe. The Novelty condition is

[21] For details: Konopka et al. (2008), Markopoulou (2009). Another active approach to geometrogenesis is in group field theory, see Oriti (2009, 2014).

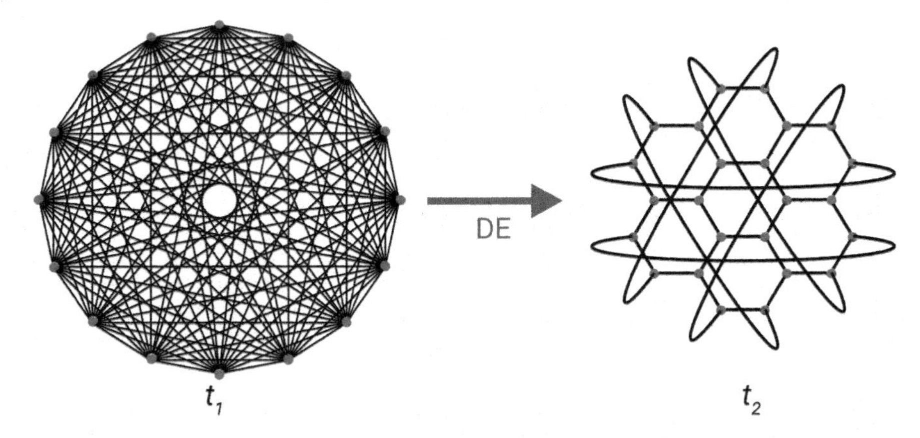

Fig. 7 Geometrogenesis as flat emergence (DE). (t_1) High-energy (pre-geometric) phase of quantum graphity, in which every point is connected to every other point. (t_2) Low-energy (geometric) phase of quantum graphity, in which there are fewer connections between the points, resulting in the emergence of a conception of locality. Figure adapted from Markopoulou (2009)

satisfied given the different symmetries characterising the two states: B is permutation invariant, while E is not permutation invariant (but is translation invariant). Finally, the Autonomy condition is plausibly satisfied, since E depends only on the broken symmetry pattern that the system instantiates, rather than the details of B.

Thus, there is a plausible sense in which spacetime could potentially be flatly emergent on these models (one which would arguably also apply to any symmetry-breaking phase transition). But there is also a sense of hierarchical emergence here. Unlike in more familiar cases of phase transitions, however, this is identical with the flat emergence in geometrogenesis. This is because there is no macro state corresponding to the pre-geometric phase of the universe. The emergent macro state of the universe is associated with the post-geometrogenesis phase. These models can be seen as simulataneously representing both the 'diachronic' (flat) emergence of spacetime from the (state replacing the) 'Big Bang' as well as the 'synchronic' (hierarchical) emergence of spacetime from its 'atoms'. In other words, the content (spacetime) is emergent along with the context (the conditions required for the geometrogenesis phase transition).

9 Collapsing the Distinction

So far, I have presented two different accounts of emergence, applicable to different scenarios: the case of spacetime emergent from some more-fundamental 'atoms' is a possible example of hierarchical emergence, while spacetime emergent from some 'prior' non-spatiotemporal state (replacing the Big Bang singularity) is a possible example of flat emergence. But in the case of spacetime, it may be more natural to think of them as representing the same situation. The conditions we encounter as we extrapolate backwards in cosmological time, and see the universe contracting to a very high-energy state, are the same conditions required to move to the 'lower-level', finer-grained picture of the 'atoms' of the universe (the most fundamental structures). This is the connection between cosmology and high-energy particle physics. Understanding these two senses of emergence as lacking a distinction in the case of spacetime is not exemplified in the case of LQG and LQC, as presented here. But it is more apparent in the case of geometrogenesis, which illustrates the possibility of collapsing the two conceptions of emergence: it is a case where both types of emergence obtain simultaneously.

Collapsing the distinction may seem natural, too, given how difficult it is to accommodate the more standard conceptions of emergence in the case of spacetime. The distinction between flat and hierarchical emergence is intuitive, but loses this motivation when we are forced to abstract away from spatiotemporal notions. We had to define the 'micro' and 'macro' levels in hierarchical emergence not by reference to length scales, in spite of the connotations of the labels 'micro' and 'macro'. On the other hand, we run into problems understanding the Dependence condition in flat (diachronic) emergence without a notion of time, and rely on a non-temporal sense of causation, or flat supervenience, to link the emergent state to its 'prior' basis. Representing the flat and hierarchical emergence of spacetime as the same scenario from the outset may be a simpler way of envisioning spacetime emergence. For instance, it might help us solve the problem of Dependence in flat emergence: the 'prior' or basis state may be identified as the one lacking a macro state, which emerges *along with* the emergent micro state, as in geometrogenesis.

Arguably, the more general conception of emergence, presented in Sect. 5, is suitable for capturing the relevant sense of 'collapsed' or 'no levels' emergence. This account is supposed to be a balance of prescriptive and descriptive: to potentially be useful for understanding the unique case of spacetime emergence, while still attempting to capture enough of what is usually meant by 'emergence' in philosophy. But an alternative may be to develop a more radical account of emergence, that departs further from resemblance to more standard accounts of emergence in philosophy. For instance, we might explore abandoning the asymmetry typically required for emergence, by removing the Dependence condition. Such a conception would then be more readily applicable to the case of LQC, as an example, and we could say that the model of Sect. 7.1 does actually represent *both* the emergence and dissolution of spacetime.

10 Conclusion

Understanding emergence in QG is difficult because we do not have a fully-developed theory. We are trying to interpret fragmentary pieces of theories that may not even be meaningful to speculate about at this stage. Nevertheless, philosophical exploration can still be interesting here, and potentially useful in anticipating what spacetime emergence may be like, and perhaps for suggesting new avenues to explore.

Standardly, there are two different senses in which spacetime is thought to emerge according to QG, and QG cosmology. I've recommended a positive conception of emergence, based on other case studies in physics, that can be framed to fit either the hierarchical case, of spacetime emergent from some more fundamental structures, or the flat case, of (a 'micro' structure corresponding to) spacetime emergent from some 'prior' non-spatiotemporal state, as described by models of QG cosmology. But splitting this conception of emergence into the two specific accounts may not be necessary—it may be more natural to collapse the distinction between flat and hierarchical emergence in the case of spacetime, given that the way these two accounts are normally distinguished is by reference to spatiotemporal notions, and the possibility of QG models where the 'micro dynamics' responsible for the appearance of spacetime is the same in the early universe as it is at a high-energy, finer-grained description of our universe 'now'. It is a case where, possibly, *content* emerges along with *context*.

Acknowledgements Thanks to Sebastiano Orioli for help with the figures.

References

Ashtekar, A., Rovelli, C., & Smolin, L. (1992). Weaving a classical geometry with quantum threads. *Phys. Rev. Lett., 69*, 237–240.

Bain, J. (2013). Emergence in effective field theories. *European Journal for Philosophy of Science, 3*(3), 257–273.

Baron, S., & Miller, K. (2014, August). Causation in a timeless world. *Synthese, 191*(12), 2867–2886.

Baron, S., & Miller, K. (2015). Causation sans time. *American Philosophical Quarterly, 52*, 27–40.

Batterman, R. (2002). *The devil in the details: Asymptotic reasoning in explanation, reduction and emergence*. Oxford University Press.

Batterman, R. W. (2000). Multiple realizability and universality. *British Journal for the Philosophy of Science, 51*(1), 115–145.

Batterman, R. W. (2005). Critical phenomena and breaking drops: Infinite idealizations in physics. *Studies In History and Philosophy of Modern Physics, 36*(2), 225–244.

Batterman, R. W. (2011). Emergence, singularities, and symmetry breaking. *Foundations of Physics, 41*, 1031–1050.

Batterman, R. W. (2018). Autonomy of theories: An explanatory problem. *Noûs*, 858–873.

Brahma, S. (2020). Emergence of time in loop quantum gravity. In C. W. Nick Huggett Keizo Matsubara (Ed.), *Beyond spacetime: The foundations of quantum gravity* (pp. 53–78). Cambridge University Press.

Butterfield, J. (2011). Emergence, reduction and supervenience: A varied landscape. *Foundations of Physics, 41*(6), 920–959.

Butterfield, J. (2011). Less is different: Emergence and reduction reconciled. *Foundations of Physics, 41*, 1065–1135.

Butterfield, J., & Isham, C. (1999). On the emergence of time in quantum gravity. In J. Butterfield (Ed.), *The arguments of time* (pp. 116–168). Oxford: Oxford University Press.

Butterfield, J., & Isham, C. (2001). Spacetime and the philosophical challenge of quantum gravity. In C. Callender & N. Huggett (Eds.), *Physics meets philosophy at the planck scale* (pp. 33–89). Cambridge University Press.

Crowther, K. (2015). Decoupling emergence and reduction in physics. *European Journal for Philosophy of Science, 5*(3), 419–445.

Crowther, K. (2016). *Effective spacetime: Understanding emergence in effective field theory and quantum gravity*. Springer.

Crowther, K. (2018). Inter-theory relations in quantum gravity: Correspondence, reduction, and emergence. *Studies in History and Philosophy of Modern Physics, 63*, 74–85.

Crowther, K. (2018). What is the point of reduction in science? *Erkenntnis*. https://doi.org/10.1007/s10670-018-0085-6

Crowther, K. (2019). When do we stop digging? conditions on a fundamental theory of physics. In A. Aguirre, B. Foster, & Z. Merali (Eds.), *What is 'fundamental'?* (pp. 123–133). Springer.

Crowther, K. (2020). As below, so before: 'synchronic' and 'diachronic' conceptions of spacetime emergence. *Synthese*, 1–29.

Crowther, K., & Linnemann, N. (2019). Renormalizability, fundamentality, and a final theory: The role of uv-completion in the search for quantum gravity. *British Journal for the Philosophy of Science, 70*(2), 377–406.

Franklin, A. (2018). On the renormalization group explanation of universality. *Philosophy of Science, 85*(2), 225–248.

Franklin, A. (2018). Whence the effectiveness of effective field theories? *The British Journal for the Philosophy of Science*. https://doi.org/10.1093/bjps/axy050

Guay, A., & Sartenaer, O. (2016, May 01). A new look at emergence. or when after is different. *European Journal for Philosophy of Science*, *6*(2), 297–322.

Hartmann, S. (2002). On correspondence. *Studies in History and Philosophy of Modern Physics, 33*(1), 79–94.

Huggett, N., & Wüthrich, C. (2013). Emergent spacetime and empirical (in)coherence. *Studies in History and Philosophy of Modern Physics, 44*(3), 276–285.

Huggett, N., & Wüthrich, C. (2018). The (a)temporal emergence of spacetime. *Philosophy of Science, 85*(5), 1190–1203.

Konopka, T., Markopoulou, F., & Severini, S. (2008). Quantum graphity: A model of emergent locality. *Physical Review D, 77*(10).

Laughlin, R. B., & Pines, D. (2000, January). The theory of everything. *Proceedings of the National Academy of Sciences of the United States of America, 97*(1), 28–31 (Laughlin, RB Pines, D)

Markopoulou, F. (2009). New directions in background independent quantum gravity. In D. Oriti (Ed.), *Approaches to quantum gravity* (pp. 129–149). Cambridge University Press.

Morganti, M. (2020a). Fundamentality in metaphysics and the philosophy of physics. Part II: The philosophy of physics. *Philosophy Compass, 15*(10), 1–14.

Morganti, M. (2020b). Fundamentality in metaphysics and the philosophy of physics. Part I: Metaphysics. *Philosophy Compass, 15*(7).

Morrison, M. (2012, January). Emergent physics and micro-ontology. *Philosophy of Science, 79*(1), 141–166.

Oriti, D. (2009). The group field theory approach to quantum gravity. *Approaches to quantum gravity: Toward a new understanding of space time and matter* (pp. 310–331). Cambridge University Press.

Oriti, D. (2014). Disappearance and emergence of space and time in quantum gravity. *Studies in History and Philosophy of Modern Physics, 46*, 186–199.

Oriti, D. (2018). Levels of spacetime emergence in quantum gravity. arXiv:1807.04875.

Oriti, D. (Forthcoming). Levels of spacetime emergence in quantum gravity. In C. Wüthrich, B. L. Bihan, & N. Huggett (Eds.), *Philosophy beyond spacetime*. Oxford University Press.

Penrose, R. (1999). The central programme of twistor theory. *Chaos, Solitons & Fractals, 10*(2–3), 581–611.

Penrose, R. (2002). Gravitational collapse: The role of general relativity. *General Relativity and Gravitation, 34*(7), 1141–1165.
Post, H. (1971). Correspondence, invariance and heuristics: In praise of conservative induction. *Studies in History and Philosophy of Science Part A, 2*(3), 213–255.
Radder, H. (1991). Heuristics and the generalized correspondence principle. *British Journal for the Philosophy of Science, 42*, 195–226.
Rivat, S., & Grinbaum, A. (2020). Philosophical foundations of effective field theories. *European Physical Journal A, 56*(3).
Rovelli, C. (2004). *Quantum gravity*. Cambridge University Press.
Rovelli, C., & Vidotto, F. (2014). *Covariant loop quantum gravity: An elementary introduction to quantum gravity and spinfoam theory*. Cambridge University Press.
Sartenaer, O. (2018). Flat emergence. *Pacific Philosophical Quarterly, 99*(S1), 225–250.
Shech, E. (2019, June). Philosophical issues concerning phase transitions and anyons: Emergence, reduction, and explanatory fictions. *Erkenntnis, 84*(3), 585–615.
Tallant, J. (2019). Causation in a timeless world? *Inquiry, 62*(3), 300–316.
Wüthrich, C. (2017). Raiders of the lost spacetime. In D. Lehmkuhl (Ed.), *Towards a theory of spacetime theories* (pp. 297–335). Birkhäuser.
Wüthrich, C. (2019). The emergence of space and time. In S. Gibb, R. F. Hendry, & T. Lancaster (Eds.), *The Routledge handbook of emergence* (pp. 315–326).

Topological Quantum Field Theory and the Emergence of Physical Space–Time from Geometry. New Insights into the Interactions Between Geometry and Physics

Luciano Boi

1 Introduction

This paper addresses various topics and different issues related essentially to general relativity theory and quantum field theories, and, more generally, to the interactions between geometry and physics. It aims at presenting recent works and discussing new ideas and results from these topics. It focuses on the subject of the geometric and topological structures and invariants which enriched in a remarkable way cosmology and quantum field theories in the last century, say, starting from Einstein's general relativity until string theory. In the last three decades, new and deep developments in this direction have emerged from cosmology and theoretical physics.

The general goal of the paper is to examine some striking aspects of the role of geometrical and topological concepts and methods in the developments of theoretical physics, especially in cosmology, quantum field theory, string theory, quantum gravity and non-commutative geometry, and then to show the great significance of these concepts and methods for a better understanding of our universe and the physical world at the very small scale. From the beginning we would like to stress the crucial fact that many physical phenomena appear to be related to deep geometrical and topological invariants (Atiyah, 1988) and furthermore that they are effect which emerge, in a sense, from the geometric structure of space–time (Connes & Chamseddine, 2006; Vafa, 1998).

L. Boi (✉)
École Des Hautes Études en Sciences Sociales, Centre de Mathématiques (CAMS), 54, bd Raspail, 75006 Paris, France
e-mail: lboi@ehess.fr

S. Wuppuluri and I. Stewart (eds.), *From Electrons to Elephants and Elections*, The Frontiers Collection, https://doi.org/10.1007/978-3-030-92192-7_23

2 Einstein's General Relativity and the Interaction Between Curvature and Matter

The first good example we would like to mention of this new point of view, which however rely upon ideas advocated by Riemann, Clifford and Poincaré, is that of general relativity, which showed that gravity was an effect of the space–time curvature (Boi, 2004, 2006b; Penrose, 2004; Regge, 1992). More precisely, with the general relativity theory, actual (physical) geometry enters the picture of Minkowski space–time (which, mathematically speaking, is a manifold with a Lorentz metric, i.e., a non-degenerate pseudo-Riemannian metric of signature $+ \ldots + -$; $\mathbf{R}^n$ with metric $(dx^1)^2 + \ldots + (dx^{n-1})^2 - (dt)^2$) by assuming the world-history of each particle is a geodesic and that the Ricci curvature of the metric reflects the structure of matter and energy present at each point. The Einstein field equations,

$$R_{\mu\upsilon} - 1/2 g_{\mu\upsilon} R + \Lambda g_{\mu\upsilon} = 8\pi G T_{\mu\upsilon}$$

states that mass and pressure warp space–time. These equations relate the metric to matter distribution. Thus, according to the general theory of relativity, the gravitational force has to be reinterpreted as the curvature of space–time in the proximity of a massive object. When the energy is very concentrated, then the deformation of space–time may change sufficiently its topological structure and not only its metric (Baez & Muniain, 1994; Boi, 2004a; Regge, 1992). Let us stress that general relativity related two fundamental concepts which had, till then, been considered as entirely independent, namely, the concepts of space and time, on the one hand, and the concepts of matter and motion, on the other. Indeed, the geometry of space–time is not given a priori, for, in some sense, depends on the underlying physical structure of space-time. General relativity theory predicts at least three fundamental phenomena of the physical reality: (i) the gravitational waves; (ii) the black holes; (iii) the expanding of the Universe.

One of the most important ideas of general relativity was that space–time, not space, was the fundamental intrinsic object and that its structure was to be determined by physical phenomena. Einstein's main discoveries were as follow: (i) Spacetime is a pseudo- Riemannian manifold, i.e., its metric ds^2 is not Euclidean but has the signature $(+,-,-,-)$ at each point. In presence of matter (the gravitational field), general relativity, based on the geometric concepts discovered by Riemann (see Riemann, 1854; and Boi 2019a), replaces the flat (pseudo) metric of Poincaré, Einstein (special relativity) and Minkowski, $ds^2 = dx^2 + dy^2 + dz^2 - dt^2$, by a curved spacetime metric whose components form the gravitational potential $g_{\mu\upsilon}$, $ds^2 = g_{\mu\upsilon}\, dx^{\mu}\, dx^{\upsilon}$. (ii) Gravitation is just the physical manifestation of the curvature of spacetime (as foreseen by Clifford in 1876, see (Clifford, 1876)). (iii) Light travels along geodesics. Another point should, however, be added. (iv) The metric of (flat) space–time is not Euclidean but has the form $ds^2 = dx^2 - dx^2 - dx^2 - dx^2$ at each point. This is what nowadays is called a Lorentzian structure. However, even in the absence of matter, the geometry of space–time could not be asserted to be flat but only Ricci flat, i.e., that its Ricci tensor, which can be calculated from the Riemannian curvature tensor, is 0 (Penrose, 2004; Regge, 1992).

3 Quantum Mechanics and the Idea of Non-Commutativity

The next essential advance in twenty-century physics has been quantum mechanics. Let us summarize some fundamental idea of this theory (Cao, 1997; Heisenberg, 1930). In quantum mechanics and relativistic quantum field theory formulated by W. Heisenberg, P. Jordan, W. Pauli, P. Dirac and E. Wigner, the position and velocity of a particle (at the subatomic scale) are non-commuting operators acting on a Hilbert space, and classical notions such as "the trajectory of a particle" do not apply. In the 19th and early twentieth century physics, many aspects of nature were described in terms of fields—the electric and magnetic fields, and the gravitational field. So, since fields interacts with particles, to give an internally coherent account of nature, the quantum concepts must be applied to fields as well as to particles. When this is done, quantities such as the components of the electric field at different points in space–time become non-commutative. When one constructs a Hilbert space in which these operators act, one finds many surprises. The distinction between fields and particles break down, since the Hilbert space of a quantum field is constructed in terms of particle-like excitations. Conventional particles such as electrons are reinterpreted as arising from the quantization of a field. In the process, one finds the prediction of "antimatter": for every particle there must be a corresponding antiparticle, with the same mass and opposite electric charge (Coleman, 1985).

The quantum field theories (QFT's) that have proved to be very important in describing elementary particle physics are gauge theories (Zeidler, 2011). The classical example of gauge theory is the theory of electromagnetism. The gauge group is the Abelian group $U(1)$. If the (physical) potential A denotes the $U(1)$ gauge connection, which locally can be regarded, mathematically speaking, as a one-form on space–time, then the curvature or electromagnetic field tensor is the two-form $F = dA$, and Maxwell's equation read: $0 = dF = d^*F$. Here * is the Hodge duality operator.

4 Gauge Theories: From H. Weyl to Yang-Mills

The second main step of the geometrization of physics in the twenty-century has been gauge theory, thanks to which several new deep geometrical and topological structures have emerged (Bourguignon & Lawson, 1982; Boi, 2011). Gauge theory is a quantum field theory obeying to the geometrical principle of local gauge invariance. Gauge theory was introduced by Hermann Weyl in 1918 as an attempt to unify general relativity with electromagnetism (Weyl, 1918, 1929). However, the theory of Weyl failed because of lacking of an appropriate quantum physics framework.

Gauge idea rebirths with the formulation of non-Abelian Yang-Mills theory in 1954 by Yang and Mills (Boi, 2019b; Yang & Mills, 1954). This new theory stems from the recognition of the structural similarity, from the mathematical viewpoint, of non-Abelian gauge (quantum) fields with general relativity and the understanding that both are connections (Yang, 1977; Bourguignon & Lawson, 1982). This last, defined over a fiber bundle and possessing a curvature, is a very deep geometrical concept introduced by Weyl and Cartan, which generalize the concept of parallel transport of Levi–Civita to a new mathematical object: that of a non-point-like space or manifold in which precisely the points are replaced by the fibers (Boi, 2004a).

The very idea of Yang and Mills consists in suggesting a new program of geometrization of physics, this time applied to the physical forces supporting the quantum world. They proposed that the strong nuclear interactions be described by a (quantum) field theory in the same manner than electromagnetism, which is exactly local gauge invariant, as it is general relativity. More precisely, they postulated that the local gauge was the *SU(2)* isotopic spin-group or *SU(2)* isotopic spin-connection on which the non-Abelian group (a compact Lie group[1]) acts. This idea was "revolutionary" because it changed the very concept of "identity" of what has been ever assumed to be an "elementary particle". The novel idea that the isotopic spin connection, and therefore the potential A_μ (where, in order to relate the phases function $\lambda(x_i)$ at different points, the familiar gauge transformation for A_μ was written in terms of the phase change: $A_\mu \rightarrow A_\mu - 1/e\, \partial_\mu \lambda$) acts like the *SU(2)* symmetry group is the most important result of Yang-Mills theory. The concept of isotopic-spin connection lies at the heart of local gauge theory. It shows explicitly how the gauge symmetry group is built into the dynamics of the interaction between particles and fields (Atiyah, 1990; Yang, 1977). Moreover, some of the important physical characteristics of the field can be deduced directly from the connection (the potential), which can be viewed as a linear combination of the generators of the *SU(2)* group. We can, in fact, associate this formal operation with real physical processes.

Let's add few specifications on the mathematical structure of gauge theory (for a more comprehensive exposition, see Bourguignon & Lawson, 1982; Manin, 1988; Zeidler, 2011). Yang-Mills or non-Abelian gauge theory can, at the classical level, be described similarly to the "classical" Abelian gauge theory, with *U(1)* (see above) replaced by a more general compact gauge group *G*. The definition of curvature must be modified to $F = dA + A \wedge A$, and Yang-Mills equations: $0 = dAF = dA{*}F$, where dA is the gauge-covariant extension of the exterior algebra derivative. These equations can be derived from the Yang-Mills Lagrangian

$$\mathcal{L} = 1/4g^2 \int Tr\, \mathrm{Tr} F \wedge *F,$$

where Tr denotes an invariant quadratic form on the Lie algebra of *G*. The Yang-Mills equations are non-linear, so, in contrast to the Maxwell equations, but like

[1] *Finite groups* are spacial cases of *compact Lie groups*. For example, the rotation group SO(3) of the three-dimensional Euclidean space or the gauge group U(1) × SU(2) × SU(3) of the Standard Model in elementary particle physics are compact Lie groups.

the Einstein equations for the gravitational field, they are not explicitly solvable in general. But they have certain properties in common with the Maxwell equations and, in particular, they describe at the classical level massless waves that travel at the speed of light.

The first (classical) Yang-Mills theory corresponds to the quantum version of Maxwell theory—known as Quantum Electrodynamics—, which gave a very accurate account of the quantum behaviour of electromagnetic fields and forces. The non-Abelian gauge theory were introduced for describing the other forces in nature, notably the weak force (responsible among other things for certain forms of radioactivity) and the strong or nuclear force (responsible among other things for the binding of protons and neutrons into nuclei). For the weak force, we have now the Weinberg-Salam-Glashow electroweak theory with gauge group: $H = SU(2) \times U(1)$.

The masslessness of classical Yang-Mills waves was avoided by elaborating the theory with an additional "Higgs field". This is a scalar field, transforming in a two-dimensional representation of H, whose non-zero and approximately constant value in the vacuum state reduces the structure group from H to $U(1)$ sub-group (diagonally embedded in $SU(2) \times U(1)$). This theory describes both the electromagnetic and weak forces, in a more or less unified way; because of the reduction of the structure group to $U(1)$, the long-range fields are those of electromagnetism only, in accord with what we see in nature.

To sum up what we said about gauge theory, let's stress that Yang and Mills showed for the first time that local gauge symmetry was a powerful fundamental principle that provided new insights into the newly discovered "internal" quantum numbers like isotopic spin. In their theory, isotopic spin was not just a label for the charge states of particles, but it was crucially involved in determining the fundamental forms of the interaction between these particles. The most important philosophical point is that in the gauge theories of quantum fields, symmetries of nature determine the properties of forces; therefore, it is allowed to say that mathematical groups and invariants are at the origin of the dynamics of physical forces.

Let's add that in the search for a non-linear generalization of Maxwell's equations to explain elementary particles, there are various symmetry properties one would require. These are: (i) External (spatial–temporal) symmetries invariant under the Lorentz and Poincare's groups and under the conformal group if one is taking the rest-mass to be zero; (ii) Internal (physical) symmetries invariant under the non-Abelian groups like $SU(2)$ or $SU(3)$ to account for the known feature of weak and strong interactions, respectively; (iii) Covariance or its supersymmetric coupling by working on a complex topological space–time.

5 String Theory and the Supersymmetric Picture of the Quantum World

The next fundamental step in the geometrization of physics has been realized by string theory, a quantum field theory that tries to unify in a coherent picture general relativity and quantum mechanics at a deeper level than that of the Standard Model of particle physics (Witten, 1995). String theory entails beautiful geometrical and topological new structures, more rich and powerful with respect to those developed before by the other quantum field theories. It is yet theoretically incomplete and hitherto physically untested (Marino, 2005; Vafa, 1998).

It is worth of recalling that originally string program go back, in a sense, to the ideas putted forward by the German mathematician Bernard Riemann about hundred-fifth years early. According to him, one can make two fundamental assumptions. (i) First, on a given n- dimensional manifolds there are many possible metric structures (i.e., many different functions for measuring the distance between any pair of infinitesimally near points), so that the problem of which structure is the one appropriate for physical space required empirical methods for its solution. In other words, Riemann stated explicitly in 1854 (Riemann, 1854) that the question of the geometry of physical space does not make sense independently of physical phenomena. And (ii) space does not exist independently of phenomena and its structure depends on the extent to what we can observe and what happens in the physical world. From the previous follows, say, a corollary even more insightful: in its infinitely small parts (nowadays we would say at the quantum level) space may not be accurately described even by the geometrical notions of Riemannian geometry (Ashtekar & Lewandowski, 2004).

This last idea, which is hinted in Riemann's statement (ii), remain dormant until the search for a unified field theory at the quantum level forced the physicists to reconsider the structure of space–time at extremely small distances. One of the ideas to which their efforts led them was that the geometry of spacetime was supersymmetric with the usual coordinates supplemented by several anticommuting (fermionic) ones. This is a model that reflects the highly fuzzy structure of spacetime in small regions (at the quantum scale 10^{-33} cm) where one can pass back and forth between bosonic and fermionic particles. Modern string theory (i.e., superstring theory) takes Riemann's vision even further, and replaces the points of spacetime by strings, thereby making the geometry even more non-commutative (see Connes, 1994, 1996; and Landi, 1999).

Let's address briefly some conceptual aspects and issues of superstring theory. Superstring theory relies on the two ideas of supersymmetry and spacetime structure of eleven dimensions. Supersymmetry require that for each known particle having integer spin 0, 1, 2, and so on, measured in quantum units—there is a particle with the same mass but half-integer spin (1/2, 3/2, 5/2 and so on), and vice-versa. Supersymmetry transforms the coordinate of space and time such that the laws of physics are the same for all observers. Einstein's general theory of relativity derives from this condition, and so supersymmetry implies gravity. In fact, supersymmetry predicts "supergravity", in which a particle with a spin of 2—the graviton—transmits gravitational interactions and has as a partner a graviton, with spin of 3/2.

Superstring is based on the fundamental notion of T-duality, which relates two kinds of particles that arise when a string loop around a compact (spatial) dimension. One kind, call them "vibrating particles", is analogous to those predicted by Kaluza and Klein and comes from vibrations of the loop of the string. Such particles are energetic if the circle is small. In addition, the string can wind many times around the circle, its energy become higher the more times it wraps around and the larger the circle. Moreover, each energy level represents a new particle—call them "winding particle". T-duality states that the winding particles for a circle of radius R are the same as the "vibrating particles" for a circle of radius $1/R$, and vice-versa. So, to a physicist, the two sets of particles are indistinguishable: a fat compact dimension may yield apparently the same particles as thin one.

String theory, if correct, entails a radical change in our concepts of spacetime. That is what one would expect of a theory that reconciles general relativity with quantum mechanics. The answer involves duality again. A vibrating string is described by an auxiliary two-dimensional field theory, whose Lagrangian is roughly

$$L = 1/2 \int d\tau \, d\sigma (\partial X/\partial \tau)^2 + (\partial X/\partial \sigma)^2.$$

Here, $X(\tau, \sigma)$ is the position of the string at proper time τ, at a coordinate σ along the string. In string theory, the auxiliary two-dimensional field theory plays a more fundamental role than spacetime, and spacetime exists only to the extent that it can be reconstructed from the two-dimensional theory. In other words, duality symmetries of the two-dimensional field theory put a basic restriction on the validity of the classical notion of spacetime.

All the attempts mentioned, which are aimed at solving one of the central problems in twentieth-century physics, i.e.: how to combine gravity and the other forces into a unitary theoretical explanation of the physical word, essentially depend on the possibility of building a new geometrical framework conceptually richer than Riemannian geometry. In fact, as we saw, it plays a fundamental role in non-Abelian gauge theories and in superstring theory, thanks to which a great variety of new mathematical structure has emerged. A very interesting hypothesis is that the global topological properties of the manifold's model of spacetime play a major role in quantum field theory and that, consequently, several physical quantum effects arise from the non-local metrical and topological structures of these manifold (Isham, 1988; Labastida & Lozano, 1989). Thus, the unification of general relativity and quantum theory requires some fundamental breakthrough in our understanding of the relationship between spacetime and quantum processes (Penrose, 2004). In particular the superstring theory, but also, in a different manner, loop quantum gravity, lead to the guess that the usual structure of spacetime at the quantum scale must be dropped out from physical thought (Carfora, 2011). Non-Abelian gauge theories satisfy the basic physical requirements pertaining to the symmetries of particle physics because they are geometric in character. They profoundly elucidate the fundamental role played by bundles, connections and curvature in explaining the essential laws of nature. Kaluza-Klein theories and more remarkably superstring theory showed that spacetime symmetries and internal (quantum) symmetries might be unified through the introduction of new structures of space with a different topology. This essentially means that "hidden" symmetries of fundamental physics can be related to the phenomenon of topological change of certain class of (presumably) non-smooth manifolds (Atiyah, 1990). This entails a number of extremely important mathematical and physical consequences, which partly are discussed in this paper.

6 New Developments and Conceptual Issues in Quantum Field Theory

Let us now address some of the recent most fundamental developments in mathematical and theoretical physics, and in particular, the fact that these developments point forwards the search for a new scheme of spacetime structure at the quantum scale. Quantum mechanics culminated in the "standard model" of particle interactions, which is a quantum field theory. The fundamental ingredients of nature that appear in the underlying equations are fields: the familiar electromagnetic field, and some twenty or so other fields. The so-called elementary particles, like photons and quarks and electrons, are "quanta" of the fields-bundles of the field's energy and momentum. The properties of these fields and their interactions are largely dictated by principles of symmetry, including Einstein's special principle of relativity, together with a principle of "renormalizability", which dictates that the fields can only interact with each other in certain specially ways. The standard model has passed every test that can be imposed with existing experimental facilities. However, many unsolved problems and open questions remain. We do not know why it obeys certain symmetries and not others, or why it contains six types of quarks, and not more or less. Finally, gravitation cannot be brought into the quantum field theoretic framework of the standard model, because gravitational interactions do not satisfy the principles of renormalizability that governs the other interactions. This constitutes at present one the most fundamental and challenging issues of researches in theoretical physics and mathematics. Both topological quantum field theories and non-commutative geometry dedicate much effort to find out a solution to the very hard and key problem of the renormalization of the standard model. This problem might be answered, following different paths, by the Witten's topological string approach (Witten, 1988) and the Connes's non-commutative approach (Connes, 1996).

The not-yet-achieved incorporation of the fundamental ideas of a dynamical space–time geometry into a quantum theory of matter is one of the central open problems of contemporary physics, whose solution may well require another radical change in the physicist's conception of nature and space–time. We think that a real understanding of the cosmological questions and of the nature of elementary particles can ever been achieved without a simultaneous deeper understanding of the nature of space–time itself. It is well-known that quantum mechanics taught us that the classical notions of the position and velocity of a particle were only approximations of the truth. Notably, it is not clear whether the Riemannian geometry—even in a revised and generalized form—is adequate for the description of the small-scale structure of space–time (Isham, 1988; Penrose, 2004). The Planck length $lP = (G\hbar/c^3)\ 1/2{\sim}10^{-33}$ cm is considered as a natural lower limit for the precision at which coordinates of an event in space–time make sense. Nevertheless, not only does quantum mechanics have some striking geometrical characters, but its description of the world also reveals a wealth of deep underlying mysteries—even bordering on paradox—which cannot arise merely from an inadequate human understanding of the implications of the theory's mathematical formalism. Instead, at some level, there must be a deviation from purely unitary evolution, so that state-vector reduction can become a real phenomenon (Ashtekar & Lewandowski, 2004). Moreover, because of the (mysterious) non-local nature of quantum entanglement, whatever the nature of this revolution might be, the final theory that will emerge must have a fundamentally non-local character. In effect, according to certain mathematical-physical theories, such as topological quantum field theories and especially superstring theory, the local information of the space–time fields and of the other fields is stored in global (topological) structures of space–time (Boi, 2004).

7 Non-Commutative Geometry and the Quantum Fields

This is also truth for non-commutative geometry, where the quantum field equations are calculated for the full set of internal space metric fluctuations allowed by the non-commutative geometry axioms in the spectral triple formulation of the standard model (Connes & Chamseddine, 2006). These calculations have been given both from the perspective of the spectral triple and from the perspective of Fredholm module.[2] It has been showed that studying these Fredholm modules using algebraic K theory and K homology leads to a suggested non-commutative version of Morse theory—a well-known tool for studying the topology of manifolds—which is applied to the finite spectral action. According to the spectral action principle, which has been introduced ten years ago by Connes and Chamseddine, the standard model of particle physics is formulated with a product (whose image is called the total space) of two spectral triples—one that represents the Euclidean space–time manifold and the other the zero-dimensional internal space of particles charges. The space–time coordinate functions remain commutative but the internal space is a non-commutative "manifold". The spectral action principle is an important step towards the unification

[2] Recall that if A is an involutive algebra over the complex numbers $\mathbf{C}$, then a *Fredholm module* over A consists of an involutive representation of A on a Hilbert space H, together with a self-adjoint operator F, of square 1 and such that the commutator $[F, a]$ is a compact operator for all $a \in A$.

of gravity with particle physics; the Einstein-Hilbert action plus Weinberg-Glashow-Salam theory all result from a calculation of the eigenvalues of the Dirac operator on the total space and since the Dirac operator encodes the metric, the spectral action principle is a purely geometrical theory (Connes, 1995).

Formally, a spectral triple (A, *H*, *D*) provides the analog of a Riemannian spin manifold to non-commutative geometry (here we follow closely Connes and Chamseddine, 1996). It consists of an involutive, non-necessarily commutative algebra *A*, a Hilbert space *H:* a finitely generated projective module on which the algebra is represented, and a Dirac operator *D* that gives a notion of distance, and from which is built a differential algebra. A very important technical point is that the geometry of any closed (even dimensional) Riemannian spin manifold can be fully described by a (real and even) spectral triple and a non-commutative geometry is essentially the same structure but with the generalization that the algebra of coordinates is allowed to be non-commuting. For the standard model the internal Hilbert space is $H = H \oplus H \oplus H^{\mathrm{C}} \oplus HC$, where *LRLR*

$$H = (\mathrm{C2} \otimes \mathrm{CN} \otimes \mathrm{C3}) \oplus (\mathrm{C2} \otimes \mathrm{CN}), L$$

$$H = ((\mathrm{C} \oplus \mathrm{C}) \otimes \mathrm{CN} \otimes \mathrm{C3}) \oplus (\mathrm{C} \otimes \mathrm{CN}), R$$

and whose basis is labeled by the elementary fermions and their antiparticles. The symbol *c* is used to indicate the section represented by the antiparticles. The even triple has the Z/2-grading operator χ, the chirality (eigenvalues $+1$ or -1). In either case of *HL* and *HR,* the first direct summand is the quarks and the second the leptons. *N* Stands for the numbers of generations. For example, the left-handed up and down quarks form an isospin doublet and their right-handed counterparts are singlets and there are three colors for quarks and none for leptons. The charges on the particles are identified by the faithful representation of the algebra on the Hilbert space. In the definition of *H* above we see a second Z/2-grading that splits the Hilbert space into two orthogonal subspaces for particles and antiparticles: $H^{+} \oplus H^{-}$ or $H \oplus HC$. This is called *S0* reality and is not an axiom but applies to the standard model as it excludes Majorana masses. The *S0* reality grading operator ε satisfies:

$$[D, \varepsilon] = 0, \quad [J, \varepsilon]^{+} = 0, \quad \varepsilon * = \varepsilon, \quad \varepsilon^{2} = 1.$$

8 The "ontology" of Newtonian Physics and Quantum Field Theory

Let us now address the important point concerning the differences between the "ontology" of classical physics and that of quantum physics. (Here this term stands for the nature and the kind of properties ascribed to the most fundamental physical entities from which a specific theory is built up and also to the mathematical objects by means of which one construct a definite space–time theory or model). One may affirm that Newtonian physics had a clear ontology: the world consisted of massive particles situated in Euclidean space. In that sense, the nature of space played a fundamental role. In the mathematical developments of Newtonian mechanics, however, the role of space is not clear. There is not much difference between the description of two particles moving in $\mathbf{R}^3$ and that of a single particle moving in $\mathbf{R}^6$, nor between that of a pivoted rigid body and that of a point moving on the group-manifold SO_3. In quantum mechanics the idea of space is even more elusive, for there seems to be no ontology, and, whatever wave-functions are, they are certainly not functions defined in space. Still, for about seventy years we have known that elementary particles must be described not by quantum mechanics but by quantum field theory, and in the field theory the role of space is quite different. Although it is an important fact that quantum field theory cannot be reconciled with general relativity, one could emphasize that the two theories have a virtual feature in common, for in both of them the points of space play a central and objective dynamical role. In quantum field theory two electrons are not described by a wave-function on $\mathbf{R}^6$; instead they constitute a state of a field in $\mathbf{R}^3$ which is excited in the neighborhood of two points. The points of space *index* the observables in the theory. The mathematics of quantum field theory is an attempt to describe the nature of space, but it proposes to look at space in a completely different way (Manin, 1988; Zeidler, 2011).

Like quantum field theory, Penrose's twistor theory is a radical attempt to get rid of space as a primary concept (Penrose, 1977). The Connes's program of non-commutative geometry amounts to a huge generalization of the classical notion of a manifold (Connes, 1994). Finally, string theory proposed a scheme for making space as an approximation to some more general kind of structure. One striking difference (maybe the essential one) between general relativity and quantum mechanics lie in the fact that, whereas in general relativity it seems impossible to separate the postulate of (continuous) space–time localization of events and the theory of gravitation from the (inner) geometric structure of space–time, on the other hand, it is precisely this postulate of the indistinguishability of the physical fields from the space–time geometry that got lost in quantum mechanics. It is particularly contradicted by the Bohr principle of complementarity and the Heisenberg uncertainty relations, which states the impossibility of knowing simultaneously the exact position and velocity of particles (electrons). These relations are indeed based on a model in which the electron jumps quickly from one orbit to another, radiating all energy thus liberated in the form of a global package, a *quantum* of light.

9 What It Could Be a Quantum Geometry of Space–time?

Many attempts have been made, starting from the sixties, to understand what kind of geometry and topology and therefore what kind of space–time model could be truly appropriate to describe the behavior of physical space both at the very large and quantum levels (Isham, 1988; Penrose, 2004). Among them, the most attractive and promising ones seem to be string theory, non-commutative geometry and loop quantum gravity (Ashtekar & Lewandowski, 2004; Carfora, 2011). The nature of quantum geometry is the central issue of non-perturbative quantum gravity. Is the familiar continuum picture then only an approximation? If so, what are the 'atoms' of geometry? What are its fundamental excitations? Is there a discrete underlying structure? If so, how does the continuum picture arise from this fundamental discreteness? By a quantized geometry, it is meant (Baez & Muniain, 1994) that there exist physical quantities which can take on continuous values classically but are such that the corresponding quantum operators have a discrete spectrum. In the resulting quantum geometry, Riemannian geometry can then emerge only as an approximation on a large scale. This topic can be discussed either from the perspective of topological quantum field theory and superstring theory or from that of non-commutative geometry.

The most attractive feature of non-commutative geometry is that it develops a new notion of geometric space where points do not play the central role, thus giving much more freedom for describing the subatomic-scale nature of spacetime. The theory proposed a framework which is sufficiently general to treat discrete space, Riemannian manifolds, configurations spaces of quantum field theory, and the duals of discrete groups which are not necessarily commutative. The development of a non-commutative geometry has been recently one of the most important attempts to unify (mathematically) quantum field theory with gravitation. In addition, its physical implications have found lately a confirmation in that it predicted a physical model for coupling gravity with matter (Connes, 1996).

The other fundamental change in our conception of spacetime and physics comes from superstring theory. Indeed, recent developments in theoretical physics suggest that a new kind of quantum geometry may enter physics, and that spacetime itself may be reinterpreted as an approximate, derived concept that one can extract from a two-dimensional field theory (Katz & Vafa, 1997; Witten, 1995). Intuitively, strings are viewed as one-dimensional objects whose modes of vibration represent the elementary particles. In addition, in string theory the one-dimensional trajectory (world-line) of a particle in space–time is replaced by a two-dimensional orbit (world-tube) of the string. The main conceptual point of the string program is that it entails some revolutionary ideas about our conception of space and space–time. Indeed, space is not more thought as formed up of points-like elements and therefore the particles not either. Space as well is endowed with a point-less structure. Instead of point-like elements, the space seems to be filled out of other kinds of geometrical objects, richer and more complex, like knots of many types, Riemannian surfaces, topological (unconventional) objects, and so on. The most interesting point is that space must be considered as a dynamical thing, which may change with respect to its metrical and topological properties (Boi, 2009b; Vafa, 1998). The main physical aspect of string theory is that all particles which we previously thought of as elementary, that is, as little points without any structure in them, turn out in fact not to be points at all but basically little loops of string which move through space, oscillating around it. We have thus that the different physical properties of matter emerge somehow from the different structural and dynamical patterns of these strings and loops in space. For example, the electric charge might be seen as a quality of the motion of the string rather than something which is just added on to a particle as fundamental object.

The idea of replacing point particles by strings sounds so naïve that it may be hard to believe that it is truly fundamental. But in fact, this naïve-sounding step is probably as basic as introducing the complex numbers in mathematics. If the real and complex numbers are regarded as real vector spaces, one has $\dim_R(\mathbf{R}) = 1, \dim_R(\mathbf{C}) = 2$. The orbit of a point particle in space–time is one-dimensional and should be regarded as a real manifold, while the orbit of a string in space–time is two-dimensional (over the reals) and should be regarded as a complex Riemann surface. Physics without strings is somehow analogous to mathematics without complex numbers.

10 New Insights Into the Nature of Space–time

We now outline some new ideas relating to the structure of space–time in the most recent physical theories, to start with general relativity. (i) The geometric structure of space–time gives rise to the dynamics of this same space–time, and in particular of the gravitational field. (ii) Even the other (fermionic and bosonic) fields describing matter and its electroweak and strong interactions seems to emerge as dynamical effects from the topological (global) structure of space–time. Conversely, the space–time itself must be henceforth thought of, in some sense, as a derived (changing) object whose metric and topological structures may be subject, to some extent, to the quantum fluctuations of these same fields. For example, one of the predictions of T-duality in string theory is that geometry and topology are *ambiguous* at the string length $lS = \sqrt{\alpha}$'. Furthermore, space is ambiguous at the Planck length $lP \ll lS$. Another more complicated and richer example of T-duality is the mirror symmetry and topology change in Calabi-Yau spaces. There are different types of dualities that play an important role in the recent developments of theoretical physics. One conclusion is, thus, that spacetime is likely to be an emergent, approximate, classical concept. The challenge is to have emergent spacetime, while preserving some locality (macroscopic locality, causality, analyticity, etc.). (iii) The recent developments of theoretical physics enable us to think that the discrete and continuous character of the laws of physics are but special cases according with each other in the framework of a new unitary mathematical-physical theory. With the theory of supergravity, and still more with string theory, we get a consistent theoretical framework which is finite and which simultaneously incorporate both quantum gravity and chiral supersymmetric gauge theories in a natural fashion. Supergravity generalizes a gauge theory proposed by H. Weyl in 1923 in order to unify the Einstein's theory of gravitation with the electromagnetic theory, and another by Kaluza and Klein in the 1920s, in which they suggested to further unify the concepts of internal and space–time symmetries by reducing the former to the latter through the introduction of some extra dimension of space, more precisely, a fifth (space-like) dimension, which has the topology of a circle. (iv) The physical (dynamical) and space–time symmetries dictate, at different extents, the various forces of nature and the interactions between particles. This is a very general principle and it is the crucial idea at the heart of quantum field theories. In fact, all physical phenomena seem to be founded upon such principle (Coleman, 1985). However, at a deeper level, one is increasingly led to believe that, beside symmetries (including, space–time, physical, broken symmetries, and maybe other "hidden" symmetries), topological structures and invariants might have an even more important role in determining physical phenomena at the very large and extremely small scales (Atiyah, 1989).

11 Topological Quantum Field Theory

Topological quantum field (TQFT) emerged in the eighties as a new relation between mathematics and physics. The relation connected some of the most advanced ideas in the two fields. The nineties have been characterized by its development, originating unexpected results in topology and testing some of the most fundamental ideas in quantum field theory and string theory. The first TQFT was formulated by Witten in 1988 (Witten, 1988). He constructed the theory now known as Donaldson-Witten theory, which constitutes a quantum field theory representation of the Donaldson invariants of four-manifolds (1983–84) (Donaldson, 1983). His work was strongly influenced by M. Atiyah. In 1988 Witten formulated also another two-dimensional TQFTs which have been widely studied during the last three decades: topological sigma models in two dimensions and Chern-Simons gauge theory in three dimensions (Marino, 2005). These theories are related, respectively, to Gromov invariants (Gromov, 1985), and to knot and link invariants as the Jones polynomial and its generalizations (Atiyah, 1988; Thurston, 1997; Turaev, 1994). TQFT has provided an entirely new approach to study topological invariants. Being a quantum field theory, TQFT can be analyzed from different point of view. The richness inherent to quantum field theory can be exploited to obtain different perspectives on the topological invariants involved in TQFT. This line of thought has shown to be very fruitful in the last two decades and new topological invariants as well new relations between them have been obtained.

TQFT have been studied from both, perturbative and non-perturbative points of view. In the case of Chern-Simons gauge theory, non-perturbative methods have been applied to obtain properties of knot and link invariants, as well as general procedures for their computation. Perturbative methods have also been studied for this theory providing integral representations for Vassiliev invariants. In Donaldson-Witten theory perturbative methods have proved its relation to Donaldson invariants (Donaldson, 1990). Non-perturbative methods have been applied after the work by Seiberg and Witten on $N = 2$ supersymmetric Yang-Mills theory. The outcome of this application is a totally unexpected relation between Donaldson invariants and a new set of topological invariants called Seiberg-Witten invariants.

Donaldson-Witten theory is a TQFT of cohomological type. TQFTs of this type can be formulated in a variety of frameworks. The most geometric one corresponds to the Mathai-Quillen formalism. In this formalism a TQFT is constructed out of a moduli problem. Topological invariants are then defined as integrals of a certain Euler class (or wedge products of the Euler class with other forms) over the resulting moduli space. A different framework is the one based on the twisting of $N = 2$ supersymmetry. In this case, information on the physical theory can be used in the TQFT. Indeed, it has been in this framework where Seiberg-Witten invariants have shown up. After Seiberg and Witten worked out the low energy effective action of $N = 2$ supersymmetric Yang-Mills theory, it became clear that a twisted version of this effective action could lead to topological invariants related to Donaldson invariants. The twisted action revealed a new moduli space, the moduli space of Abelian monopoles (Witten, 1994). Its geometric structure has been derived in the context of the Mathai-Quillen formalism. Invariants associated to this moduli space should be related to Donaldson invariants. This turned out to be the case. The relevant invariants for the case of *SU(2)* as gauge group are the Seiberg-Witten invariants.

Donaldson-Witten theory has been generalized after studying its coupling to topological matter fields. The resulting theory can be regarded as a twisted form of $N = 2$ supersymmetric Yang-Mills theory coupled to hypermultiplets, or, in the context of the Mathai-Quillen formalism, as the TQFT associated to the moduli space of non-Abelian monopoles. Perturbative and non-perturbative methods have been applied to this theory for the case of *SU(2)* as gauge group and one hypermultiplet of matter in the fundamental representation. In this case, again, it turns out that the generalized Donaldson invariants can be written in terms of Seiberg-Witten invariants. One would expect that in general the invariants associated to non-Abelian monopoles could be expressed in terms of some other simpler invariants, being Seiberg-Witten invariants just the first subset of the full set of invariants.
The present situation in three and four dimensions relative to Chern-Simons gauge theory and Donaldson-Witten theory, respectively, can be described as follows.

These theories share some common features. Their topological invariants are labeled with group-theoretical data: Wilson lines for different representations and gauge groups (Jones polynomials and its generalizations), and non-Abelian monopoles for different representations and gauge groups (generalized Donaldson polynomials); these invariants can be written in terms of topological invariants which are independent of the group and representation chosen: Vassiliev invariants and Seiberg-Witten invariants. This structure leads to the idea of universality classes of topological invariants. In this respect Vassiliev invariants constitute a class in the sense that all Chern-Simons or quantum group knot invariants for semi-simple groups can be expressed in terms of them. Similarly, Seiberg-Witten invariants constitute another class since generalized Donaldson invariants associated to several moduli spaces can be written in terms of them. This certainly holds for the two cases described above but presumably it holds for other groups. It is very likely that Seiberg-Witten invariants are the first set of a series of invariants, each defining a universality class.

12 Concluding Remarks

We stressed the crucial fact that many physical phenomena, at the quantum and at the cosmological level as well, appear to be deeply related to some geometrical and topological invariants, and furthermore that these phenomena are effects which emerge, in a sense, from the geometric and topological structure of space–time (Atiyah, 1990). The first good example of this new point of view, which actually rely upon ideas advocated by Riemann and Clifford, is that of general relativity, which showed that gravity is a manifestation of the curvature of space–time. The Einstein's field equations relate the metric to matter distribution. Thus, according to the general theory of relativity, the gravitational force has to be reinterpreted as the curvature of space–time in the proximity of a massive object. When the energy is very concentrated, then the deformation of space–time may change sufficiently its topological structure.

Topological quantum field theory (TQFT) appear as a very rich and promising research program in theoretical physics. Two conceptual points appear to be very significant, and likely promising for physics, in TQFT. (i) The first is the assumption of an effective correlation between knots and link invariants and the physical observables and states of quantum field theories and gauge theories. (ii) The second is, on the one hand, the idea of the fuzziness of physical space–time and of its emergence from the dynamical fluctuations of its metrical structure, on the other, the idea of the geometric and topological nature of physical phenomena at different scales.

More precisely, the main ideas we have addressed in this paper are the following:

(1) The geometric and topological deformations and invariants could generate the dynamics of space and time, of the quantum field and the gravitational field as well. For example, in string theory, the picture is that the different physical properties of matter are linked to the different topological configurations of strings and loops moving through space and oscillating around it. For instance, the electric charge might be seen as a quality of the motion of the string rather than something which is just added on to a particle as a fundamental object.

(2) The fermionic and bosonic fields composing matter and its electroweak and strong interactions seems to emerge as dynamical effects from the topological (global) deformations of the varying structure of space–time. Conversely, the space itself must be henceforth thought of, in some precise sense, as a derived and changing object whose metric and topological structures may be subject, to some extent, to the quantum fluctuations of these same fields. We already gave two very significative examples illustrating these facts, both relating to T-duality in string theory: the first predict the ambiguous character of geometry and topology at the string length scale; the second concerns mirror symmetry and topological change in Calabi-Yau spaces. After Riemann's revolution in the geometric vision of physical space, which goes very far beyond the discovering of what we now call "Riemannian geometry", for he has not only the idea that the distribution of matter in the universe depends upon the variation of curvature of space–time, but also the vision of a geometry for the microscopic (quantum) physical world as a dynamical and fluctuating object, the next revolution should

be to think that space–time might be an emergent, approximate, non-classical concept. The challenge is to prove the validity of the emergent global nature of space–time while preserving some locality (macroscopic locality, causality, analyticity, etc.) On of the most remarkable constituents of quantum geometry might be knots and other tangled structures. If different aspects of the link between the Jones polynomial and mathematical physics have been intensively studied in the last three decades and are quite well-known, the relationship between knots and quantum physics remain still almost unexplored. Recently, Witten suggested that, in quantum physics, a knot may be regarded as the orbit in space–time of a charged particle. One way of calculating the Jones polynomial in quantum theory involves using Chern-Simons function for gauge fields. But to use the Chern-Simons function, the knot must be a path in a space–time of three dimensions rather than the four dimensions of the real world.

(3) The recent developments of theoretical physics enable us to think that the discrete and continuous character of the laws of physics are but special situations according with each other in the context of a new unitary mathematical-physical theory. With the theory of supergravity, and still more with superstring theory, we get a consistent theoretical framework which is finite and which simultaneously incorporate both quantum gravity and chiral supersymmetric gauge theories in a natural fashion. Supergravity generalizes a gauge theory proposed by H. Weyl in 1923 in order to unify the Einstein's theory of gravitation with the electromagnetic theory, and another by Kaluza and Klein in the 1920s, in which they suggested to further unify the concepts of *internal* and *space–time* symmetries by the former to the latter through the introduction of some extra dimension of space, more precisely, a fifth (space-like) dimension, which has the topology of a circle.

(4) The physical ("internal") and space–time ("external") symmetries, which we tend to consider both dynamical because they can equally produce some physical effects, dictate, at different extents, the various forces of nature and the interactions between particles. This is a very general and meaningful principle and it is the crucial idea setting at the core of gauge quantum field theories. In fact, the most physical phenomena at different scales seem to be founded upon such principle. However, at a deeper level, one is increasingly led to believe that, beside symmetries—including space–time, physical and broken symmetries, and maybe other "hidden" symmetries –, topological deformations and invariants might have an even more important role in determining the dynamics of physical phenomena at the extremely small and very large scales. This is essentially related with the phenomenon of topological changes. It is much conceivable to think, on the one hand, that it can exist a deep link between symmetries and topological changes, and, on the other, that topological deformation be a new dynamical variable not depending on physical parameters but which may produce important physical effects as well.

All the previous aspects and ideas play an important role in the TQFT. Topological quantum field theory is a third sort of idealization of the physical world, besides general relativity and quantum field theory, which is attractative and deep from the mathematical and philosophical point of view as well. It is a background-free quantum field theory with no local degrees of freedom. The interesting thing is the presence of 'global' degrees of freedom (Baez & Muniain, 1994; Turaev, 1994).[3] Two spaces-times which are locally indistinguishable, since locally both look like the same model of space–time, can, hovewer, be distinguished globally, for example, by measuring the volume of the whole space–time or studying the behavior of geodesics that wrap around a 3-dimensional torus.

An axiomatic approach to topological quantum field theory was proposed by Atiyah (Atiyah, 1990). An important feature of TQFTs is that they do not presume a fixed topology for space or space–time. In other words, when dealing with an n-dimensional TQFT, we are free to choose any $(n-1)$-dimensional manifold to represent space at a given time. Moreover given two such manifolds, say S and S', we are free to choose any n-dimensional manifold M to represent the portion of spacetime between S and S'. For his construction, Atiyah used the notion of cobordism, introduced by R. Thom in the 1950s (Thom, 1954), and he developped a formalism in which he found that cobordism construction obeys to the algebraic properties of associativity (of manifolds), the non-commutativity of the composition of cobordism (this is related with the famous non-commutativity of observable in quantum theory) and an identity cobordism. The operations are dynamical in the sense that they formalize the notion of "passage of time" (temporal evolution) in a context where the topology of space–time is arbitrary and there is no background fixed metric. Atiyah's axioms relate this notion to quantum theory as follows. First, a TQFT must assign a Hilbert Space $Z(S)$ to each $(n-1)$-dimensional manifold S. Vectors in this Hilbert space represent possible states of the universe given that space is the manifold S. Second, the TQFT must assign a linear operator $Z(M)\colon Z(S) \to Z(S')$ to each n-dimensional cobordism $M\colon S \to S'$. This operator describes how states change given that the portion of space–time between S and S' is the manifold M. In other words, if space is initially the manifold S and the state of the universe is ψ, after the passage of time corresponding to M the state of the universe will be $Z(M)\,\psi$.

Baez and Muniain (1994) emphasized that the analogy between differantial topology and quantum theory "is exactly the sort of clue we should pursue for

[3] A good example is quantum gravity in 3-dimensional space–time. Classicaly, Einstein's equations predict qualitatively very different phenomena depending on the dimension of space–time. If space–time has 4 or more dimensions, Einstein's equations imply that the metric has local degrees of freedom. In other words, the curvature of space–time at a given point is not completely determined by the flow of energy and momentum through that point: it is an independent variable in its own right. For example, even in the vacuum, where the energy–momentum tensor vanishes, localized ripples of curvature can propagate in the form of gravitational radiation. In 3-dimensional space–time, hovewer, Einstein's equations suffice to completely determine the curvature at a given point of space-tume in terms of the flow of energy and momentum through that point. We thus say that metric has no local degrees of freedom. In particular, in the vacuum the metric is flat, so every small patch of empty space–time looks exactley like every other.

a deeper understanding of quantum gravity. At first glance, general relativity and quantum theory look very different mathematically: one deals with space and space-time, the other with Hilbert spaces and operators. (…) Topological quantum field theory suggests that perhaps they are not so different after all! Even better, it suggests a concrete program of synthesizing the two, which many mathematical physicists are currently pursuing. Sometimes this goes by the name of 'quantum topology'".

It seems likely that differential topology and quantum theory must merge if we are to understand background-free quantum field theories. In classical (Newtonian) physics, one treat space as a background on which states of the world are posed, and, similarly, one treat spacetime as a background on which the process of change occurs. But it could be that these be idealizations which we must overcome in a background-free theory, i.e. a theory with global degrees of liberty given by topological change. As Baez and Muniain pointed out, the concepts of 'space' and 'state' are, in fact, two aspects of a unified whole, and likewise for the concepts of 'spacetime' and 'process'. This fact might open new and significant perspectives for the mathematical and philosophical understanting of the physical world.

References

Ashtekar, A., & Lewandowski, J. (2004). Background independent quantum gravity: A status report. *Class Quant Grav, 21*, 53–152.

Atiyah, A. (1989). Topological quantum field theories. *Publ Math Inst Hautes Etudes Sci, 68*, 175–186.

Atiyah, M. (1988). "New invariants of three-and four-dimensional manifolds," in *The Mathematical Heritage of Hermann Weyl*, O. Wells, Jr. (Ed.), *Proc Symp Pure Math* AMS, *48*, 285–299.

Atiyah (1990). *The geometry and physics of knots*, Cambridge U. Press.

Baez, J., & Muniain, J. P. (1994). *Gauge fields, knots and gravity*, World Scientific.

Boi, L. (2004a). Geometrical and topological foundations of theoretical physics: from gauge theories to string program. *International Journal of Mathematics and Mathematical Sciences, 34*, 1066–1136.

Boi, L. (2004b). Theories of space-time in modern physics. *Synthese, 139*(3), 429–489.

Boi, L. (2006a). "From riemannian geometry to Einstein's general relativity theory and beyond: Space-time structures, geometrization and unification," in *Proceedings of the Albert Einstein's century international conference*, J.-M. Alimi & A. Füzfa (Eds.) (pp. 1066–1075), American Institute of Physics Publisher.

Boi, L. (2006b). "Mathematical knot theory," in Encyclopedia of mathematical physics, J.-P. Françoise, G. Naber & T. S. Sun (Eds.), Elsevier, 399–406.

Boi, L. (2009a). Clifford geometric algebra, spin manifolds, and groups action in mathematics and physics. *Advances in Applied Clifford Algebras, 19*(3), 611–656.

Boi, L. (2009b). Ideas of geometrization, invariants of low-dimensional manifolds, and topological quantum field theories. *International Journal of Geometric Methods in Modern Physics, 6*(5), 701–757.

Boi, L. (2011). *The quantum vacuum. A scientific and philosophical concept: From electrodynamics to string theory and the geometry of the microscopic world*, The Johns Hopkins University Press.

Boi, L. (2019a). Some mathematical, epistemological and historical reflections on the relationship between geometry and reality, space-time theory and the geometrization of theoretical physics, from Riemann to Weyl and beyond. *Foundations of Science, 24*(1), 1–38.

Boi, L., H. (2019b). "Weyl's deep insights into the mathematical and physical world. His important contribution to the philosophy of space, time and matter," in *Weyl and the Problem of Space*, C. Lobo & B. Julien (Eds.) (pp. 231–263), Springer.

Bourguignon, J. P., & Lawson, H. B. (1982). Yang-Mills theory: Its physical origins and differential geometric aspects. In S.-T. Yau (Ed.), *Seminar on differential geometry* (pp. 395–421). Princeton University Press.

Yu, C. T. (1997). *Conceptual developments of 20th century physics*. Cambridge University Press.

Carfora, M. (2011). "Quantum gravity and quantum geometry," in *New trends in geometry. Their role in the natural and life sciences*, in C. Bartocci, L. Boi, & C. Sinigaglia (Eds.) (pp. 17–33), Imperial College Press.

Clifford, W. K. (1876). On the space-theory of matter. *Cambridge Philosophical Society Proceedings, 2*, 157–158.

Clifford, W.K. (1968). *Mathematical papers* (1882), new edition.

Coleman, S. (1985). *Aspects of symmetry*. Cambridge University Press, Cambridge.

Connes, A., & Chamseddine, A. H. (2006). Inner fluctuations of the spectral action. *Journal of Geometry and Physics, 57*, 1–21.

Connes, A. (1994). *Noncommutative geometry*. Academic Press.

Connes, A. (1996). Gravity coupled with matter and the foundation of non-commutative geometry. *Comm Math Phys, 182*, 155–176.

Connes, A. (1995). Noncommutative geometry and reality. *Journal of Mathematics and Physics, 36*, 6194–6231.

Connes, A., & Landi, G. (2001). Noncommutative manifolds, the instanton algebra and isospectral deformations. *Comm. Math. Phys., 221*, 141–159.

Donaldson, S. K. (1983). An application of gauge theory to the topology of four manifolds. *Journal of Differential Geometry, 18*, 269–287.

Donaldson, S. K. (1990). Polynomials invariants for smooth four-manifolds. *Topology, 29*, 257–315.

Gromov, M., & Lawson, H. B., Jr. (1983). Positive scalar curvature and Dirac operator on complete Riemannian manifolds. *Publ. Math. IHÉS, 58*, 295–408.

Gromov, M. (1985). Pseudo-holomorphic curves in symplectic manifolds. *Inventiones Mathematicae, 82*, 307–347.

Heisenberg, W. (1930). *The physical principles of the quantum theory*. University of Chicago Press.

Isham, D. (1988). "Topological and global aspects of quantum field theory," in *Relativity, groups and topology II*, in B.S. DeWitt & R. Stora (Eds.), (pp. 1059–1290).

Katz, S., & Vafa, C. (1997). Matter from geometry. *Nuclear Physics B, 497*, 146–154.

Labastida, J. M. F., & Lozano, C. (1989). "Lectures on topological quantum field theory", in *Trends in theoretical physics*, (Eds). H. Falomir, R. Gamboa & F. Schaposnik (Eds.) AIP, CP 419, p. 54.

Landi, G. (1999). *An introduction to noncommutative spaces and their geometries*. Springer.

Manin, Y. I. (1988). *Gauge field theory and complex geometry*. Springer.

Marino, M. (2005). Chern–Simons theory and topological strings. *Reviews of Modern Physics, 77*, 675–720.

Penrose, R. (1977). The twistor programme. *Reports in Mathem Physics, 12*, 65–76.

Penrose, R. (2004). *The road to reality*. A Complete Guide to the Laws of the Universe. Vintage, London.

Regge, T. (1992). "Physics and differential geometry," in *1830–1930: A century of geometry*, in L. Boi, D. Flament & J.-M. Salanskis (Eds.), Lecture notes in physics, (Vol. 402, pp. 270–272), Springer.

Riemann, B. (1990). "Über die hypothesen, welche der Geometrie zu Grunde liegen" (*Habilitationsarbeit*, Göttingen, 1854), in *Gesammelte mathematische werke* (new edition edited by R. Narasimhan) (pp. 304–319), Springer.

Thom, R. (1954). Quelques propriétés globales des variétés différentiables. *Comment Math Helv, 28*, 17–86.

Thurston, W. P. (1997). *Three-dimensional geometry and topology*. Princeton University Press.

Turaev, V. (1994). *Quantum Invariants of Knots and 3-Manifolds*. de Gruyter.

Vafa, C. (1998). "Geometric physics", in *Documenta mathematica*, Extra Volume ICM, *I*, 537–556
Weyl, H. (1929). Elektron und gravitation. *Zeitschrift Für Physik, 56*, 330–352.
Weyl, H. (1918). Gravitation und Elektrizität. *Sitz Ber Königl Preuss Akad Wiss Berlin, 26*, 465–480.
Weyl, H. (1928). *Gruppentheorie und quantenmechanik*. Hirzel.
Witten, E. (1995). String theory dynamics in various dimensions. *Nuclear Physics B, 443*, 85–126.
Witten, E. (1988). Topological quantum field theory. *Comm Math Phys, 117*, 353–386.
Witten, E. (1994). Monopoles and four-manifolds. *Mathematical Research Letters, 1*(6), 769–796.
Wu, T. T., & Yang, C. N. (1975). Concept of nonintegrable phase factors and global formulation of gauge fields. *Phys. Rev. D, 12*(12), 3845–3857.
Yang, C. N., & Mills, R. L. (1954). Conservation of Isotopic Spin and Isotopic Gauge Invariance. *Physical Review, 96*(1), 191–195.
Yang, C. N. (1977). Magnetic Monopoles, fiber bundles, and gauge fields. *Annals of the New York Academy of Sciences, 294*, 86–97.
Zeidler, E. (2011). *Quantum field theory III: Gauge theory—a bridge between mathematicains and physicits*. Springer.

The Electron and the Cosmos: From the Universe of Fragmented Objects to the Particle-World

Leonardo Chiatti

1 Prelude: Ninety (Three) Years Later

Born at the end of 1927 in the famous congresses held in Como and Brussels, modern quantum theory has travelled a triumphal journey whose end is not still at sight today. Yet the doubts about its formulation, already raised at the time by many of its eminent founders, have not yet subsided. It is about some of these doubts that we intend to return to in this article, suggesting a way in which they can be clarified. Having to establish a starting point for our reasoning, we choose in particular one of these doubts, which was raised by Einstein with reference to the so-called "dual description" (corpuscular and wavelike) of quantum entities (Bacciagaluppi & Valentini, 2009). As is well known, an electron emitted—for example—by a hot metal filament and directed towards a photographic plate is described, in quantum theory, by a wave function which, in this specific case of a single particle, is defined on the usual spacetime. This wave function is widespread and incides on the whole photographic plate or an extended portion of it. When, however, the interaction between the plate and the electron takes place, this interaction is revealed through the blackening of a well-defined grain of the plate emulsion. This event marks the end of the wave function, which cancels simultaneously in all points of space except those corresponding to the position of the blackened granule. It is evident, Einstein commented, that if the wave function is to be understood as a physical reality then we are in presence of an instantaneous propagation of information that violates the principle of relativity. The response suggested by Heisenberg was to consider the wave function not as a physical entity but in purely epistemic terms, as a compendium of information relating to the preparation of the electron. This proposal did not satisfy Einstein and continues not to satisfy many scholars: the wave function seems to

L. Chiatti (✉)
ASL VT Medical Physics Laboratory, Via Enrico Fermi 15, 01100 Viterbo, Italy
e-mail: leonardo.chiatti@asl.vt.it

S. Wuppuluri and I. Stewart (eds.), *From Electrons to Elephants and Elections*,
The Frontiers Collection, https://doi.org/10.1007/978-3-030-92192-7_24

"guide" the manifestation of the event (the blackening of the grain); and if it *does* something, it must also *be* something.

We will return to this old question from a slightly more general perspective, which seems pertinent to the theme of this anthology: that of the relationship between global and local in microphysics, and more specifically in the physics of so-called "elementary particles". At first glance it seems strange to evoke this relationship with reference to a "particle", which is the elementary entity for excellence (the minimal building block) and is local by definition. One wonders where the "global" aspect of a particle resides. However, Einstein's observation reveals that this aspect exists even if it is not immediately accessible to experimentation. The "transmission of information" to which Einstein refers, if this really is concerned, is in fact confined within a process whose outcome is the event (the blackening of the grain), and has no detectable effects externally. It is exactly this characteristic that saves the principle of relativity, which concerns restrictions on the propagation of signals *between* events. Instead, what we are talking about is the relationship—all internal to a single event—between the space in which this event takes place and its *manifestation* in a single position at a given instant.

From this perspective, the reductionist approach, which investigates the relationships *between* particles and the causal links *between* individual events, taking care to bring systems and their stories back to these ultimate atomic components, does not seem to have much to say. In fact, we wonder about the internal structure of these atomic components; an operation that is meaningless from the "pure" reductionist point of view, although commonly practiced by particle physicists. Moreover, the internal structure in question actually connects the entire spatial extension to a single point of space in a single instant. This specification of absence of duration is important, because it reveals that the relationship between the point and the space involved in the "quantum jump" is not dynamical. In other words, the "quantum jump" is something that cannot surely be described by dynamical laws and equations of motion but by an entirely different formal structure. Conceptually, we must be prepared to see in the physical phenomenon constituted by this jump the expression of a type of causality different from the efficient and diachronic one which is usual in physics (i.e. the so called "dynamical" causality). Rather, the *formal* causality, in the original meaning intended by Aristotle, will be involved.

These brief notes should unambiguously define our intentions, which are not to immerse ourselves in the non-conclusive debate between the various proposals of "interpretation" of the quantum formalism. In previous works, in fact, we have entirely *derived* this formalism from basic principles (Chiatti, 2005, 2014). Instead, what we intend to do is to reconnect the content of these and other previous works, to which we refer for technical and formal details, through a single conceptual thread. We thus hope to put Einstein's argument in a new light and to highlight its powerful implications for the construction of a renewed philosophy of nature.

The reference to Einstein, however, must not mislead the reader. It is not our intention to adhere to an outdated identification of "physical realism" with an ontology of objects permanently actualized in spacetime. The instantaneous quantum jumps (whose existence and diversity from the underlying gradual evolution of the wave function guiding their manifestation have been widely demonstrated since the 1980s (Nagourney et al., 1986)) represent something radically different and "new" compared to the Einsteinian vision. Just as radically different and irreducible to that vision (Howard, 2007) are the non-separability of the *entangled* quantum amplitudes (Aspect et al., 1982) and the non-existence of pilot waves that guide corpuscle-objects (Vigier et al., 1987; Zou et al., 1992). A correct realist attitude must therefore be oriented towards the elucidation of an ontology of quantum entities completely irreducible to that of the bodies of classical physics. This means being open to the possibility that physical processes are not exhaustively represented on the spatiotemporal theater, and that spacetime itself constitutes a set of emerging relationship properties. Dynamical causality itself must, in this context, emerge from deeper forms of causality, such as the aforementioned formal causality. We can therefore say that this contribution is intended as an attempt to answer the question posed by Einstein, but from a more modern perspective which is not limited to the spatiotemporal theater [limitation assumed instead as indispensable by Einstein himself (Howard, 2007)]. At the same time our approach will be markedly ontological and based on the total rejection of a purely epistemic reading of quantum formalism.

Einstein was concerned about consistency with relativity. In re-examining this question, we will first start with a more general relativity than that intended by Einstein. In fact, the most general relativity possible under reasonable assumptions (Bacry & Lévy-Leblond, 1968; Fantappié, 1954), namely the de Sitter relativity. To make the discussion more self-contained, the essential aspects of this theory are recalled in Sect. 2. Specifically, this theory allows a projective representation of spacetime which is linked to the quantization problem in Sect. 3. The objective of this section is to show how the wave functions of quantum systems, which live on the multidimensional abstract space of the configurations of such systems, actually represent phenomena in spacetime. This representation also appears to be local, in a sense that will be specified.

In Sect. 4 this description is linked to quantum jumps, and an "event-based" ontology of physical phenomena is proposed; the important problem of the emergence of the classical level in which macroscopic bodies are located, including humans and the biosphere, is also discussed. All this admits a semiotic narrative, according to which the elementary particles constitute real "signs" manifested into the spatiotemporal theater (Chiatti, 2014). This aspect is further explored in Sect. 5, which in particular illustrates the emergence of spacetime and elementary particles of the Standard Model from general conditions of semiotic nature.

Section 6 is a brief consideration on the self-reflective structure of the physical Universe, which can be deduced from the topics of the previous sections, and the problems it opens. In particular, the relationship between synchronic formal causality and diachronic efficient causality, i.e. the emergence of dynamical causality from a timeless background, is discussed. Section 7 summarizes the concept of particle-world and presents some considerations on the irreducible complexity of the self-reflective relationship, which appears to be only partially mappable by reductionist approaches.

2 From Fantappié to the Bindu

The form of physical laws must remain unchanged when passing from one "admissible" spatiotemporal frame of reference to another; this is the substance of the principle of relativity. Of course, it is necessary to specify what is meant precisely by "admissible" frames and how the physical quantities, which appear in the expression of the laws, vary in the passage from one admissible frame of reference to another. Depending on the answers to these questions, there are different theories of relativity. The best known are the Galileian relativity and the Einsteinian (special) relativity. In these theories the admissible references are the inertial ones while the physical quantities are transformed as tensors of a group of coordinate transformations which is the Galileo group in the first case, the Poincaré group in the second. The essential difference between the two groups is that in the Poincaré group there appears, as a parameter, a maximum speed of propagation of physical phenomena, usually indicated with the letter c. The Galileo group is the limit case $c \rightarrow \infty$ of the Poincaré group and for this reason the Galileian relativity is normally considered to be an approximation of the Einsteinian one. Physically c is identifiable with the propagation speed of light in a vacuum (in the absence of gravitational fields).

It is good to dwell on the meaning of this limit speed. Let us consider an observer which coordinates the spatiotemporal position of the events of the history of a material point, using appropriate measurement instruments such as graduated rulers, clocks, etc. The generic instant of the point history will be associated with the spacetime position X_μ ($\mu = 0, 1, 2, 3$) relative to the observation point-event that constitutes the observer here-now. At that moment the material point will be endowed with a relative (four-)velocity V_μ with respect to the observer here-now. If V is the measure of the projection of V_μ on the ordinary three-dimensional space, then the constraint $-c \leq V \leq +c$ holds in the ordinary Einsteinian relativity. In other words, the space of relative velocities is limited: there is a horizon at the velocities $\pm c$. Since this constraint applies with respect to any direction of the ordinary space, in this latter space the causal structure is defined by light-cones. The essential point on which we draw attention is however another here, namely that the space of relative velocities is *hyperbolic*. To clarify the concept, we introduce the rapidity w through the usual relation $\tanh(w) = \frac{V}{c}$. Two things are then immediately evident; first: $-\infty < w < +\infty$, that is, the rapidity is not limited in the two directions of the real axis. Second, this definition of rapidity is the equivalent, in Bolyai-Lobachevski hyperbolic geometry, of the relation $\tan(\phi) = \frac{R}{r}$ valid in the Euclidean space. This relation admits a precise geometric meaning in this space. In fact, let us consider a four-dimensional hypersphere of radius r tangent to the three-dimensional space at point O. Let us imagine that the points on the hyperspherical surface are projected onto the three-dimensional space from a projection center located in the center of the hypersphere. Then a hyperspherical arc having an extreme in O and corresponding to an angle ϕ at the center of the hypersphere is projected onto the three-dimensional space into a segment of length R. In the hyperbolic space of relative velocities the hypersphere becomes a pseudosphere, R becomes V and the radius r becomes the radius c of the pseudosphere. The angle ϕ (limited in the range between $-\pi/2$ and $+\pi/2$) thus becomes the rapidity w (which is not limited, because the pseudosphere is not limited). The reader interested in a more complete examination of the relationship between Einsteinian relativity and hyperbolic space can consult the reference (Barrett, 2011) and the bibliography cited therein.

What matters here is that the nature of the relative velocity V is projective. Due to the existence of a finite pseudospherical radius c (i.e. of the limit speed), each value of V is converted into the value of an "angle" w measured in the projection center of the pseudosphere. The whole V-space is thus coded in the straight lines diverging from a single point of the four-dimensional space. This coding disappears in the Galileian limit $c \to \infty$, in which the pseudosphere collapses on the three-dimensional Euclidean space tangent to it and the projection becomes parallel. It could therefore be said that the true sense of c is to establish the non-local translation of V-space in the set of straight lines departing from a single point of the four-dimensional space. It should be noted that this point does not belong to the V-space and therefore, in this sense, it is not "physical". It basically corresponds to an "aether" with respect to which no observer can determine his own state of motion, but at the same time includes all possible states of (relative) motion.

One may wonder whether similar conclusions can be drawn for the length X of the projection of X_μ on the space of relative *positions*, that is, on the space properly understood. In the context of Einsteinian relativity the answer is negative: the X-space (i.e. the space of contemporaneity of an observer) is *Euclidean*, not hyperbolic. However, the generalization is possible and it leads to the de Sitter relativity (Arcidiacono, 1958, 1969; Fantappié, 1959). In de Sitter relativity both X-space and V-space are hyperbolic and projective. In addition to the finite limit speed c, a finite maximum time distance t_0 from the observer here-now now appears, with the result that the observer (placed naturally in the origin) can coordinate only the events located at a spatial distance less than $r = ct_0$. In other words, in addition to the horizon in the V-space another horizon, previously absent, appears in the X-space. The four-dimensional spacetime becomes the projection of the hypersphere (with imaginary time, actually a hyperboloid) of radius r in the five-dimensional space on a four-dimensional hyperplane tangent to it in the point-event corresponding to the observer here-now. This projection is conducted from the center of the hypersphere, which is located in the 5-space and is therefore physically inaccessible. All spatial positions are coded non-locally as straight lines departing from that center. The center of the hypersphere therefore represents a sort of "non-local aether" which is not localizable in the usual four-dimensional spacetime, but which *contains* all the point-events of this space.

In the limit $t_0 \to \infty$, the de Sitter's relativity collapses on the best known Einstein special relativity. It should be noted that as is possible to pass from the Einstein special relativity to the Einstein general relativity by generalizing the metric $[(-1, 1, 1, 1) \to g_{\mu\nu};\ \mu, \nu = 0, 1, 2, 3]$, it is also possible, in the same way, to pass from the de Sitter special relativity to the corresponding de Sitter general relativity (Arcidiacono, 1964). In particular, the hyperboloid in the 5-space is replaced by a more complex manifold, but the four-dimensional spacetime remains interpretable as the central projection of this manifold on tangent spaces. Despite of its name, the de Sitter relativity was in fact discovered by the Italian mathematician Luigi Fantappié (1901–1956) in the 1950s, and subsequently developed by his pupil Giuseppe Arcidiacono (1927–1998). Although the theory was subsequently rediscovered independently by various authors, its original projective version has remained largely confined to the original publications of these two scholars, written mainly in Italian. This led to a limited dissemination of the results concerning this version; the reference (Licata et al., 2017) tries to partially fill this gap. The importance of the projective version in relation to the examination of the problems posed in the introductory section is evident. The finiteness of c and t_0 makes possible a non-local coding of the entire spacetime in the projection center, and is presumably justified by the need for the existence of this coding. We note, *inter alia*, that the finiteness of t_0 is physically equivalent to the assertion of the existence of a positive finite cosmological constant $\frac{3}{(t_0)^2}$, a prediction perfectly in agreement with current cosmological observations. The projection center and the lines departing from it in the five-dimensional space are therefore the true "system" to which the quantization should be applied. Quantum non-locality in usual spacetime therefore derives (in an apparently paradoxical way) from relativity; an aspect which is generally concealed by the fact that in much better known Einstein relativity t_0 is infinite, so the projection center disappears. However, it is known that the most general kinematic group is that of de Sitter (Bacry & Lévy-Leblond, 1968) and that the most general relativity possible, under physically reasonable assumptions, is that of de Sitter (Fantappié, 1954). The appropriate theoretical environment for discussing the problems in question therefore appears to be the one indicated here. The projection center then becomes a sort of Bindu, an empty point in which the cosmos extending under our gaze is transcribed.

3 Quantizing the Center of the Vacuum

As we have seen, each point-event of spacetime is really a line, joining that point to the center of projection in five-dimensional space. All these "point-lines" diverge from a common point which is the projection center; a reading of quantum formalism in light of this awareness seems therefore appropriate. As is well known, the meaning of the quantum amplitude of a particle:

$$|\Psi\rangle = \sum_{x} |x\rangle \Psi(x, t) \tag{1}$$

is the following. The event $\Psi(x, t_O) \rightarrow \delta(x - x_O)$, corresponding to the projector $|x_O\rangle\langle x_O|$, represents the localization of the particle in the space point x_O at the instant t_O. The probability density of this event, conditionated by the initial condition assumed for Ψ, is proportional to $\Psi^*(x_O, t_O)\Psi(x_O, t_O)$. Since the wave functions $\Psi(x, t)$, $\Psi^*(x, t)$ depend on the *line* (x, t) departing from the center of projection in five-dimensional space, they can be defined as properties of that center. The event $\Psi(x, t_O) \rightarrow \delta(x - x_O)$ is then, in this specific sense, local; it appears to involve a form of non-locality only when viewed from the perspective of the spacetime domain. It is clear that in this description there is no solid object permanently actualized in spacetime and to which a trajectory can therefore be assigned. The "particle" coincides with the event of its localization and is manifested in the spacetime domain only at the instant t_O, as the event $\Psi(x, t_O) \rightarrow \delta(x - x_O)$. The experimental set up considered by Einstein or the classic double slit experiment must be reconsidered in this light.

The wave function of a "system of N particles" $\Psi(x_1, \ldots x_N, t)$ can be defined in the same way, assuming the possibility of a multiple event of the type $\Psi(x_1, \ldots x_N, t_O) \rightarrow \prod_i \delta(x_i - x_{Oi})$, $(i = 1,\ldots, N)$ consisting of N simultaneous localizations. Here the wave function $\Psi(x_1, \ldots x_N, t)$ depends on N lines (x_i, t) in five-dimensional space, for each assigned value of t. These lines diverge from the projection center and therefore also in this more general case Ψ can be considered as a property of that center. It should be noted that in the case of identical particles, the absence of trajectories that connect distinct origins to different places of localization in a unique way leads directly to entanglement.

It is possible to postulate the existence of a direction for each line in the five-dimensional space: the line can then be either leaving the projection center or entering it. The physical interpretation of this hypothesis is completely natural: the exit from the center means the breaking of a symmetry that assigns to each particle a position (delocalized on the spatial domain); entering the center means restoring that symmetry with the disappearance of the spatial label associated with that particle. In other words, the meaning is that of creating and destroying a distinct localizable spatial position. The projection center therefore assumes the role of pre-spatial vacuum, already mentioned in the previous section. Thus, it becomes natural to assume that the arguments of the function Ψ are outgoing lines, while the arguments of the conjugate Ψ^* are incoming lines. The quantum jump that leads to projection $|\Psi\rangle\langle\Psi|$ at the instant t_O then corresponds, in virtue of (1), to the simultaneous "absorption" of Ψ into the pre-spatial vacuum and its re-"emission". From a temporal perspective this process can be seen as the double emission, at $t = t_O$, of the advanced wave function Ψ^* with argument $t < t_O$ and the retarded wave function Ψ with argument $t > t_O$. It is necessary to realize that although the instant $t = t_O$ corresponds to a spatial region (space of simultaneousness) on the spacetime domain, this dual emission does not occur in three-dimensional space. Indeed, Ψ and Ψ^* are not necessarily defined as single-particle functions. This observation clarifies the relationship between real space, that is, the spatial region $t = t_O$ where the interaction that induces the quantum jump takes place, and the configurational space in which Ψ and Ψ^* live. It also clarifies, more generally, the relationship between the Hilbert space (really, a rigged space) of the quantum amplitudes of a system of N particles and the real three-dimensional space.

In summary, we can say that real interactions, represented by quantum jumps, occur on real space, while the connections between these interactions are described in Hilbert space. This description refers to a "first quantization" situation in which the number of particles of each type is conserved. The more general case of the second quantization, according to which there may be creations, annihilations and transmutations of particles, will be discussed in a next section with specific reference to the Standard Model. We note however that (1) can be transformed into a second quantization quantum amplitude by translating it into the representation of the occupation numbers, according to standard procedures (Landau & Lifsits, 1973). The superposition coefficient $\Psi(x, t)$ is then replaced by harmonic functions of x and t multiplied by the operators of creation and annihilation on the "state". All this has no effect on the correspondence between spacetime event-points and directed lines in five-dimensional space. The physical description introduced here therefore remains applicable also to this more general case.

Before closing this section it is good to note that according to the description proposed here, wave functions and quantum jumps are both real physical facts. However, while wave functions exist at the level of the projection center and are therefore undetectable, quantum jumps occur in the spacetime domain (the spatial region $t = t_O$ or its sub-regions). They are therefore observable through the events they induce in this domain.

4 Pauses in Time

In the context of the first quantization, the identity of a particle is determined by the Hilbert space to which its quantum amplitude belongs. If $|\Psi\rangle\langle\Psi|$ and $|\Phi\rangle\langle\Phi|$ are two consecutive quantum jumps that take place respectively at $t = t_A$ and $t = t_B > t_A$ (that is, there are no intermediate jumps for $t_A < t < t_B$) and $|\Psi\rangle$, $|\Phi\rangle$ are elements of the same Hilbert space, then we say that a particle "prepared in the state $|\Psi\rangle$ at $t = t_A$ has been propagated until its detection in the state $|\Phi\rangle$ at $t = t_B$". Posing $|\Psi_t\rangle = S(t - t_A)|\Psi\rangle$, the connection between the two events is expressed by the transition amplitude $\langle\Phi|S(t_B - t_A)|\Psi\rangle$ and its complex conjugate. Of course, even these amplitudes are properties of the projection center. Thus, although the structure of the time evolution operator S must preserve causality and relativistic invariance, in fact the connection between the real events of the history of a particle (i.e. the quantum jumps in which it is involved) is of an extra-spatiotemporal nature.

What happens in the center is basically an annular process of this type: 1) Ψ evolves as $S(t_B - t_A)|\Psi\rangle$ and is subsequently projected onto $|\Phi\rangle$ by the operator $|\Phi\rangle\langle\Phi|$; 2) the conjugate $\langle\Phi|$ of $|\Phi\rangle$ evolves as $\langle\Phi|S^{+}(t_B - t_A)$ and is subsequently projected onto $\langle\Psi|$ by the operator $|\Psi\rangle\langle\Psi|$; 3) the conjugate $|\Psi\rangle$ of $\langle\Psi|$ is fed back as input of the first operation. This process admits a simple statistical interpretation (Chiatti, 2005, 2014), which provides its correct probability $|\langle\Phi|S(t_B - t_A)|\Psi\rangle|^2$ according to the Born rule. According to this interpretation $\langle\Phi|S(t_B - t_A)|\Psi\rangle$ represents a bundle of directed links that connect the event $|\Psi\rangle\langle\Psi|$ to the next event $|\Phi\rangle\langle\Phi|$, while $\langle\Phi|S^{+}(t_B - t_A)|\Psi\rangle$ represents a bundle of directed links that connect the event $|\Phi\rangle\langle\Phi|$ to the previous event $|\Psi\rangle\langle\Psi|$. The product of these two transition amplitudes is associated with the circular links that lead first $|\Psi\rangle$ in $\langle\Phi|$, then $|\Phi\rangle$ in

$\langle\Psi|$. We can assume that these links can be in the normal state "0" or in the activated state "1", and that the activation of a single circular link leads to the actual manifestation of the events $|\Psi\rangle\langle\Psi|$ and $|\Phi\rangle\langle\Phi|$. If the probability of activation of a specific link is independent on the link, that is, if the links are equally probable with respect to their activation, then the probability of the process is expressed by the Born rule. It is possible to reverse the reasoning, that is, start from the logic of the links, and derive from it a statistic that can be represented by transition amplitudes and quantum amplitudes. This leads to a complete derivation of quantum formalism in an entirely Kolmogorovian probabilistic context (Chiatti, 2005, 2014; Chiatti & Licata, 2014). It is also possible to show that in this description the onset of self-interference and entanglement effects is completely natural (Chiatti, 2005, 2014).

The activation of the link implies the connection of $|\Psi\rangle$ with its conjugate $\langle\Psi|$ through the involvement of the event $|\Phi\rangle\langle\Phi|$, and the connection of $|\Phi\rangle$ with its conjugate $\langle\Phi|$ through the involvement of the event $|\Psi\rangle\langle\Psi|$. In terms of the description proposed in the previous section, this means that each of these two events arises from nothing in the form of an entrance into the pre-spatial vacuum followed by the simultaneous exit from it, through the mediation of the other event. This "bootstrap" of the two events and their connection is clearly a process that is not entirely contained in the temporal domain. The activation of the link defines *both* events simultaneously, so it makes no sense to ask whether, in a single experimental case, the former induces the latter in a deterministic way. Quantum randomness appears only when one of the two events is known, and one wonders what the other event, leaving the next quantum jump, will be. The probability expressed by the Born rule is therefore always conditional (Chiatti & Licata, 2014). Of course, the activation is possible only if the experimental context [representable in spatiotemporal terms through (1)] allows real interactions of the required type. An electron can only be located if there is a detector available, which can also be a single atom.

It should also be clear that the wave function of a system of N particles can be at the origin of no more than N localization events. Therefore directed links not ended by N actually realized events have no physical implications. In other words, there are no "empty wave" effects deriving from the attribution of a physical reality to the wave function, consistently with the requests for relativistic invariance (Pykacz, 1992) and experimental tests (Zou et al., 1992). A further element to keep in mind to avoid misunderstandings is that the causal connection between the two events described in this section is trans-temporal: it occurs between two distinct instants t_A and t_B through modalities that are not entirely reducible to uni-directional crossing of the spatio-temporal continuum between them. An excited atom "senses" the presence of the ground state if there is an operator S that joins the present excited state to the ground state and this "feeling" is *through* time. If this condition is verified, the decay begins. In other words, spontaneous decay is connected to the vacuum term of the Hamiltonian of interaction with the electromagnetic field not because hypothetical "vacuum fluctuations" trigger it; but because is through that Hamiltonian term that these two states come into contact.

If a physical system undergoes three quantum jumps in succession, say A, B and C, trans-temporality and quantum aspects exist only as characteristics of the connection between A and B and the connection between B and C. The connection between A and C does not include these aspects; in particular, the causal and energetic flow is always directed from the past to the future (Chiatti, 2005, 2014; Chiatti & Licata, 2017). In other words, there is no form of super-causality above the direct causal connection between two successive jumps.

In the case of many-particle quantum systems we have to distinguish two different levels of time zoom. The temporal evolution of the quantum amplitude of one of the elementary components of the system between two successive quantum jumps is governed by unitary equations of motion (e.g. Schrödinger). Quantum jumps are equivalent to a redefinition of the initial condition for the application of these equations (Fock, 1957); this is the situation in which the system dynamics is zoomed in to the maximum. Starting from this situation and zooming out on the same dynamics, we can reach a point where the single connection between jumps is no longer relevant and the relevant aspects of the dynamics are instead related to the evolution over time of the average properties of groups of many events (jumps). The substantial confinement of the quantum aspects to the single direct connections constitutes a basis for the possibility of representing these average properties through classical quantities that evolve according to classical laws. It is in this sense that the proposed description is compatible with the emergence of classical behavior in systems that basically remain quantum. In this *event based* reading of quantum theory, the classical macroscopic systems emerge from the fundamental quantum description, ensuring the logical closure of this latter (Licata & Chiatti, 2019). In particular, the pre-requisite of the existence of classical macroscopic observers is not required.

The absence, in this description, of any "measurement problem" is easily verifiable, because the pointer states of a measurement device are never entangled with the degrees of freedom of an elementary component. In particular, with reference to the well-known cat paradox (Schrödinger, 1935), there is *never* a superposition of the type $d * D + nd * L$ between the nuclear states d (decayed), nd (not decayed) and the states of the cat D (dead), L (alive). What happens is actually a process that develops in three successive stages. In the first phase there is the temporal evolution of a superposition of the states d, nd according to the Schrödinger equation; this first phase ends with a quantum jump consisting in the passage from this superposition to the outgoing amplitude d, with the emission of a gamma photon. The second phase begins with the product of the photonic amplitude exiting the first phase and the ground state of the gas atom of the Geiger tube affected by the first ionization event. In this phase the Schrödinger equation governs the time evolution of the superposition $1 * g + 0 * i$ with initial condition $1 * g$, where g = ground state of the atom, i = ionized state of the atom, 0 = state of the electromagnetic field with 0 photons (vacuum), 1 = state of the electromagnetic field with a photon (leaving the previous phase). This phase ends with a quantum jump that transforms this superposition into the outgoing amplitude $0 * i$. The third phase includes a chain of essentially classical amplification processes, whose final outcome is the conversion of the state $|L\rangle\langle L|$ into the state $|D\rangle\langle D|$. This final outcome manifests itself, in a single experimental run, at time t_O; previously at this instant the cat is alive; after that moment it is dead; the cat is never in a superposition of L and D. It follows that opening the box containing the cat at the instant t, one obtains the result $|L\rangle\langle L|$ if $t < t_O$, $|D\rangle\langle D|$ otherwise. If the duration of the last two phases is neglected, the probability p_{nd} of the result $|L\rangle\langle L|$ is simply that which can be derived from the coefficient of nd in the superposition of d, nd relative to the first phase, extrapolated to time t. The probability of the result $|D\rangle\langle D|$ is the complementary to 1 of p_{nd}. The statistics of the results obtained by opening the box in an ensemble of repetitions of the same experiment is therefore that of a mixture $p_{nd}|L\rangle\langle L| + (1 - p_{nd})|D\rangle\langle D|$ where $p_{nd}(t)$ is the probability derived from the superposition of the nuclear amplitudes. But these amplitudes are never entangled with the states L, D.

Inter alia, the states L, D are related not to future events of an elementary component of the system (as is the case of d, nd), but to the presently actualized average properties of a cluster of quantum jumps so large (the cat) that the effects of quantum delocalization are irrelevant. Thus, amplitudes like $|L\rangle$, $|D\rangle$ can be used in quantum language only bearing in mind (Chiatti & Licata, 2019; Licata & Chiatti, 2019) that their physically achievable superpositions are only those trivial coinciding with the $|L\rangle$, $|D\rangle$ themselves. More generally, and regardless of the specific case, it must be observed that the evolution of a system of many particles includes not only the evolution of the amplitudes of its elementary components between one jump and the next, but also the stable succession of jumps, represented by projectors. It is therefore impossible to describe a system of this type with a single quantum amplitude. A more general dynamical description is outlined in (Castellani, 2019).

Before closing this section we want to return to the elementary components, which are instead described by a single quantum amplitude. These components are actualized in the time domain in events of the type $|\Psi\rangle\langle\Psi|$ which, as we have seen, can be understood as the absorption of an amplitude in the pre-spatial vacuum and simultaneous re-emission of the same amplitude from that vacuum. In practice it is as if there was a pause in the temporal evolution of the particle, during which the properties related to the component $|\Psi\rangle$ of its quantum amplitude are absorbed and re-emitted. If to these properties (the "state" Ψ of the particle leaving the jump) corresponds an energy E, we can say that in the first phase the vacuum absorbs that energy from the temporal domain, to return it in the second phase. The energetic content of the time domain therefore undergoes a variation equal to $-E$ in the first phase and a variation $+E$ in the second one. The total balance is clearly zero, according to the conservation of energy, but the relevant point is that a positive variation (the vacuum releases energy to the temporal domain) is associated with $|\Psi\rangle$ and therefore to propagation towards the future; instead, a negative variation (the vacuum absorbes energy from the temporal domain) is associated with $\langle\Psi|$ and therefore to propagation towards the past. In other words, a relationship is generated between the sign of the energy and the two regions in which the temporal domain is divided: future and past. The relationship between time domains and the sign of the energy represent a microscopic *time arrow*.

For a free elementary particle of mass M is $E = Mc^2\gamma$, where γ is the relativistic contraction factor. It should be noted that this is the relativistic generalization of the classical kinetic energy T, although in the handbooks is wisely suggested to take instead $E = Mc^2\gamma - Mc^2$. Classically, $T = MV^2/2$ is the work required to accelerate a body of mass M, initially at rest, up to the speed V; $-T$ is the work necessary to bring this body back at rest. Moving from the three-dimensional space of classical physics to spacetime, it happens that even a body at rest is actually in movement along its timeline. This movement is nothing but the persistence of the body. Therefore Mc^2 is the energy necessary to start this movement from quietness, that is, to create an object of mass M. Similarly, $-Mc^2$ is the energy released by stopping this movement, that is, in the destruction of the object. Of course, in the pre-quantum relativistic description there is no quietness state over time. In the quantum description adopted here, the state of quietness in which the amplitude $|\Psi\rangle$ enters and from which it emerges is the pre-spatial vacuum previously described.

5 An Emergent Space-Time-Matter Made of Signs

According to the description proposed in the previous sections, a "particle" is manifested through its quantum amplitude $|\Psi\rangle$ when it comes out of a quantum jump at a well-defined instant of time t. This outgoing amplitude is spatially delocalized according to the wave function $\langle x|\Psi\rangle = \Psi(x, t)$, evaluated at the instant of the jump. On the other hand, we have seen that the spacetime points (x, t) are lines of the five-dimensional space that depart from an origin in that space, which is the center of projection. So also the function $\Psi(x, t)$ is coded in this center. In accordance with this approach, the problem of the structure of elementary particles is then led back to the question: how does this coding take place? It is certainly not possible to univocally answer this question, and in the following we consider only one of the possible answers.

Let's start with some preliminary considerations. Let us consider a three-dimensional sphere of radius $\rho = \rho_0 \exp(T/T_0)$, where ρ_0 and T_0 are positive real constants and $T \in (-\infty, +\infty)$; we will admit that this sphere can rotate around its generic axis in a three-dimensional Euclidean space. On this sphere we can consider a specific three-rectangle triangle (i.e. whose three internal angles are all right) and assign a direction to each of its sides. Of course, an arbitrary rotation of the sphere will rigidly move this triangle while maintaining the direction of its sides. A homothety induced by the variation of T will transform the triangle into another similar to it, preserving the relations between its sides. The three angles that define the rotation will cover a three-dimensional space, which we will identify with the real physical space at time T. Taken together, homothety and rotation map the four-dimensional spacetime and establish the invariance, with respect to spacetime translations, of the relationships between the sides and the angles of the three-rectangle triangle. In the following we will assume that the material degrees of freedom of a particle can be associated with this triangle, and that the different possible rotations and homotheties correspond to the different positional eigenstates of the particle, with which these degrees of freedom can be associated. In particular, it will be possible to identify a positional eigenstate at a definite instant by setting the sphere configuration (radius and orientation). The different positional eigenstates form the basis on which the quantum superposition of coefficients $\langle x|\Psi\rangle = \Psi(x, t)$ can be constructed, to give the quantum amplitude $|\Psi\rangle$ of the particle coming out of the jump. For details on this construction (directed links and their activation) we remind to the previous section.

It is possible to associate the three sides of the three-rectangle triangle to the quaternionic units i, j, k and the rotation identity to the quaternionic unit 1. In this case the rotations of the sphere can be expressed by quaternions represented on rigid transformations of the basis $(1, i, j, k)$ and the quaternion angle will be half the effective rotation angle. It can be seen that in this description the difference between spacetime and matter (the three-rectangle triangle) disappears because both are made up of the same elements. The sphere, with the "preferred" three-rectangle triangle chosen on it, will represent the internal structure of the projection center in the five-dimensional space that we will adopt.

In this particular solution to the problem of the structure of space-time-matter, the importance attributed to the three-rectangle triangle derives from its self-duality: both the edges and the vertices correspond to the same quaternionic units. If a duality operation is carried out, consisting in the exchange of the vertices with the edges while maintaining the topological connection, the triangle has changed in itself. The fact that, in a network, the nodes coincide with their relationships identifies a level of minimum complexity that cannot be further reduced and we identify it as that of the "elementary particles". Therefore the nature of elementarity is identified by the irreducibility of the structural complexity of a set of relationships and has nothing to do with the existence of "basic bricks" without internal structure (Chiatti, 2018).

The same concept can be understood in terms of Peirce's triadic relationship "sign": in a typical relationship as $ij = k$ between quaternion units, j can be assimilated to the Object, i to the Representamen and k to the Interpretant (i.e. the interaction between i and j which allows to associate the Object j with i). The cyclicity of this relationship corresponds to the interchangeability (rotation) of Object, Interpretant and Representamen; that is, we are at a level where interactions and interacting entities coincide. Therefore, at this fundamental level dominated by triadicity both the material (triangle) and spatiotemporal (transformations of the sphere that leave the triangle unchanged) degrees of freedom make their appearance.

From the perspective of the single positional eigenstate $|x\rangle$ contained in the decomposition of $|\Psi\rangle$, the quantum jump from which $|\Psi\rangle$ exits can be seen as a cascade of symmetry breakings induced by the (ordinary) interactions involved in that jump (Chiatti, 2018). The specific succession of symmetry breakings defines the type of particle of the Standard Model (electron, proton, etc.). In other words, it is the process of spatio-temporal localization of the particle that defines the "type" of particle. This process can be represented through a particular graph called "glyph" (Chiatti, 2018). All glyphs are subsets of a mother glyph (the "universal oscillator") and can be viewed as its "states".

In Chiatti (2018) it is hypothesized that the structure of the localization process, i.e. the sequence of symmetry breakings involved in a quantum jump, is the following. First of all, the fundamental quaternionic units defined by the oriented three-rectangle triangle are subdivided on the basis of their sign, with the resulting constitution of two conjugated sets of units. These sets are what in usual quantum formalism is designated respectively with symbols $|x\rangle$, $\langle x|$ while the spacetime position (x, t) is coded by the orientation of the sphere and its radius. This generation of a positional label is graphically represented by the root vertex of the glyph (major vertex), while the split is represented by a first-order edge coming out of this vertex.

The two sets of quaternionic units derived from the split are then symmetrically divided into sub-groups of units, each corresponding to a "center of charge" internal to the particle manifested in (x, t). Graphically, each of these subgroups corresponds in the glyph to a second order vertex ("minor" vertex), connected to the major vertex by the first-order edge which represents the filiation by splitting. From each minor vertex, second-order edges associated with the quaternionic units of the subgroup represented by that vertex depart. These edges represent the "type" of center of charge (electron center of charge, quark up, quark down, etc.).

This succession of operations connects the a-spatial and timeless realm of the projection center to the spacetime domain. Its outcome is the genesis of 1, 2, or 3 lines in the five-dimensional space corresponding, in the usual spacetime, to the position of 1, 2 or 3 centers of charge within a particle located in (x, t). The construction is completed by assuming, in accordance with an old micro-causality argument discussed by Caldirola (Caldirola, 1979), that (x, t) is the tangency point of a de Sitter space of radius $c\theta$ (with $\theta \leq \theta_0$) on the usual spacetime and that the positions of centers of charge are enclosed (confined) within the projection of this space on spacetime. This leads (Chiatti, 2014, 2018) to the usual systematics of elementary particles and their subdivision into leptons, mesons and baryons, according to the Standard Model (with the bonus of an *ab initio* condition of confinement of quarks and gluons).

This representation is completely a-dynamical and *synchronic*, since a connection is involved between an a-temporal layer of physical reality and the usual temporal domain. However, it is possible to move on to the more familiar dynamical representation by examining the action of the interactions that govern the various stages of the succession. It can then be seen that the first stage of the sequence that forms the quantum jump, described with the passage from the major vertex to the minor vertex along the first-order edge, is associated with the definition of the color and the (possible) appearance of the color interaction between centers of charge. The second passage, described by the second-order edges, corresponds to the identification of the flavor of the center of charge operated by electroweak and Higgs fields coupled with the center.

What remains largely conjectural in this model of quantum jump is the connection between the gauge fields and the geometry of the de Sitter space, a topic that constitutes the object of recent work (Chiatti, 2020). Largely unexplored is the relationship with gravitation and inertia. The maximum radius $c\theta_0$ is defined by the gravitational self-coupling of the vacuum (Licata & Chiatti, 2019). Provided that the relationship between the cosmological time t_0 (described in Sect. 2) and the particle time θ_0 is given by the conventional Dirac number, this condition fixes the Planck length (Licata & Chiatti, 2019).

Of course, the specific model of quantum jump summarized in this section (the interested reader is referred to the cited papers for a more detailed and hopefully satisfactory presentation) is not the only one possible. What matters here, however, is the possibility of a theoretical representation which is both *synchronic* and *diachronic* (Crowther, 2020). On the one hand, an a-dynamical connection between the extra-spatial and extra-temporal domain and the spatio-temporal realm is described, through a succession of distinctions. On the other hand, the symmetry breakings that make up these distinctions, and that lead to the spatio-temporal localization of a particle, can be investigated in the usual dynamical perspective. This double reading of quantum formalism allows, on the one hand, to perceive the radicality inherent in the non-unitary nature of the quantum jump; on the other hand, it leads back to the usual dynamical and unitary formalism of the second quantization operators of the Standard Model (Chiatti, 2018).

6 A Self-reflective Universe

In the previous sections a possible portrait of the physical world has been painted as it appears seen at the level of its most elementary processes, that is quantum micro-events. Each of these events occurs at a specific moment of time and involves quantum amplitudes that are normally provided of an extension on the three-dimensional ambient space (that is, the related wave functions). From this simple fact and the relativistic invariance is possible to infer that these amplitudes cannot be classical fields. In general, quantum amplitudes evolve over time, but they do not represent objects in spacetime. The time evolution of amplitudes constitutes the expression of the efficient causality at the level of quantum phenomena. The quantum amplitude connects, as a probability amplitude, different events that are then defined as causally connected. Since this connection is marked by ordinary time, it can be called *diachronic*. The diachronic, efficient (in Aristotelian sense) causality is what is normally called "dynamical causality". It finds its expression in the laws of motion, for example the Schrödinger equation in non relativistic quantum mechanics. The set of micro-events actualized over time forms a network connected by the dynamical causality, and this network is the physical world.

The limit of dynamic causality (and of the Schrödinger equation) is represented, in the description of this network, by the single event. The events, or nodes of the network, are in fact modifications of the underlying vacuum. This vacuum, as we have seen, is completely a-spatial and a-temporal; it is connected to each individual event (and therefore to the temporal domain and to space) through a series of transformations closely connected with elementary interactions: electroweak, strong and presumably gravitational. This succession of transformations connects time to what is outside of time, therefore it is not marked by time: it is not diachronic. On the contrary, from the temporal perspective it appears as an unanalyzable event concentrated in a single instant, that of the "quantum jump". To qualify this sequence we can then use the adjective "synchronic". The causality underlying the quantum jump is therefore not diachronic, but synchronic; moreover, it cannot be properly qualified as efficient causality, but rather as a formal causality. In fact, what happens in an event is the manifestation of something starting from a maximally symmetrical void: the appearance of a form (the physical quantities associated with the event and exchanged in it). We could also say: the production of signs. The opposition between synchronic formal causality and dynamical causality is what appears in formalism as the alternation between the collapse of the quantum amplitude and its subsequent (or previous) unitary evolution.

This opposition is not irreducible; the same interactions that originate the succession of synchronic transformations, producing the appearance of a single event, act in diachronic mode on the time evolution of the wave function. The reasoning can therefore be rephrased by saying that there is a double face of the interactions: one unitary, the other non-unitary. The first face appears every time an interaction is inserted into Schrödinger's equation: this is the face that everyone recognizes. The second face appears in non-unitary quantum jumps, and is often not recognized, although it is certainly evident to everyone that, for example, the orbital jumps of an electron are caused by the usual electromagnetic field.

The instantonic treatment in imaginary time (Chiatti, 2020) attenuates the contrast between the two forms of causality, allowing at least a partial application of equations of motion to interaction vertices and quantum jumps. De Broglie would perhaps be happy. However, it should not be forgotten that a single vacuum is connected to all the events of the past, present and future history of the Universe through synchronic causality (which is therefore properly synchronic in the sense of trans-temporal). The ordinary dynamical causal connection between events is therefore a reflection of the most profound and fundamental synchronic formal causality. We can say that it represents its spatiotemporal reflection.

In our opinion, it is to this general context that reference must be made to understand the non-locality inherent in the spatial delocalization of a single particle, or the non-separability of the entangled amplitudes of distinct particles. But the scenario we propose seems to be much broader. The vacuum is the environment of events and the mean by which they communicate through synchronic relationships reflected in the time evolution of quantum amplitudes. It seems legitimate to say that this vacuum enters into a relationship with itself (a self-reflective relationship) through intermediate actors—quantum events—which fix its contents. Here we will not develop a reading of this type, instead focusing on the re-proposition of the basic theme of this work, which we will return to in the following Epilogue. Our thesis is that the so-called incomprehensibility of quantum phenomena derives from inadequate conceptual instrumentation and, in particular, from the rampant fixation on dynamical causality as the only possible form of mechanical causality.

7 Epilogue

The material presented in the previous sections raises the question of whether many problems of contemporary micro-physics (the nature of quantum amplitudes and the reason for their collapse, hadronic confinement, the origin of the masses, and many others) are not actually as many *koan*, thrown on the path of the investigation to urge the researcher to a different point of view on elementarity, on the emergence of space-time-matter, on the derived and secondary nature of dynamical causality. And it is under the effect of this doubt that we are now going to our conclusions.

Normally we look at particle physics as the descriptive level at which physical reality can be said to be entirely deciphered; the work of deciphering precisely consists in bringing the categories of the higher levels of description (and organization) of the physical world back to the language of reciprocal actions between the elementary components of matter. While not wanting to deny in any way the validity of this methodological approach, which has marked the entire path of post-Galilean physics, it is however important to realize that this operation of reduction and decomposition is carried out on an entity - the physical world—which is a *whole*. And it is a whole not only in the additive sense of a set of ontologically independent parts albeit externally inter-dependent, but in the much stronger sense related to the existence of a generative relationship that leads to the origination of all these parts and their external relations from a single unity. Unity consisting of an a-spatial and a-temporal void that enters into a self-reflective relationship, determining the emergence of all these parts and their external relations, both of a causal dynamic and spatiotemporal type. The individual physical events are to this void like the leaves of a tree are to the trunk from which they are born.

To be content with the reconstruction of the structure of external relations starting from concretely manifested events means to assume as fundamental the fragmentation inherent in the self-reflective relationship mentioned, completely forgetting the entirety of the underlying unity. This is consistent with the legitimate limits of a correct methodological approach. But an uncompromosing reductionism, carried beyond these limits and transfigured into fundamentalist ideology, exposes us to the risk of leaving in the shadow important aspects for the understanding of physical reality, even at the level of its "elementary constituents". This is particularly evident in the complete fall into oblivion of the distinction between the *particle*, the kinematically propagated unit, and its *centers of charge*. The relationship between these two elements, which is a typical local-global relationship recognized in the old structural models (Bohm et al., 1960; Corben, 1968; Rivas, 2015), is completely ignored by the Standard Model. The latter only knows the centers of charge and their *local* interactions with gauge fields, but completely ignores the micro-spaces that allow the *ab initio* formulation of a *global* confinement condition. The effect of this limitation is not only constituted by the persisting mysteries of the confinement of quarks and gluons or, as we have seen, of wave-corpuscle dualism, with the related debates that have dragged on without solution for decades.

Even more seriously, there is the real risk of a wrong setting of the research program on the quantization of gravity. In fact, the problem arises: at what level of the glyph is the gravitational interaction placed? If it is located at the level of the major vertex and therefore of the center of mass, then gravity acts on the particle but not on its centers of charge. If, on the other hand, the insertion of gravity is at or below the minor vertices, then gravity acts on the centers of charge and, through them, on the particle as a whole. At present it is not possible to experimentally distinguish these two possibilities; to our best knowledge, no gravitational effect on single (say) quarks has never been evidenced. The second possibility is the one implicitly accepted in a more or less universal way, and this is coherent with a vision that takes centers of charge as fundamental. But what if the first possibility was true?

In this case, the gravitation would couple with the center of mass of a particle but not with its centers of charge. The equivalence principle would then be applicable at the center of mass level, which would become the equivalent of the famous free-fall lift. The centers of charge would be manifested in strong and/or electro-weak interactions, but not in graviton-mediated interactions. The latter would see only the particle. If $c\theta$ is the de Sitter radius of the particle, gravitons of energy much greater than $\hbar/\theta$ would be decoupled from the particle. It would be the task of a quantum theory of gravitation to specify the dynamics of this decoupling [of the same type that Fermi invoked for electrons with extended charge distribution in quantum electrodynamics (Fermi, 1931)]. But if we keep in mind that the de Sitter radius decreases with the reciprocal of the particle mass and that there seem to be no heavier particles than the aggregates of the top quark, we can easily reach two conclusions. The first is that the Planck scale is never achieved in gravitational interactions. It can be achieved by non-gravitational interactions but what happens then is that gravitation is coupled with the particles produced as final states. The second is that gravitational collapse probably does not lead to singularities, but only to hyperdense states (Licata & Chiatti, 2017). We do not intend to investigate these suggestions here, but only to show that opening the reductionist-diachronic cage can open up significant horizons of physical research.

What we want to highlight is how the problems of greatest interest in fundamental physics become truly understandable only if a broader perspective is adopted, which recognizes the centrality of the synchronic formal causation and the derivative nature of the ordinary dynamical causal representation. This recovery of Aristotelian physical categories appears necessary not only as a founding step of a renewed and necessary "philosophy of nature", which recognizes the fundamental unity of the natural world by bringing back to it the results of the individual fields of scientific investigation, but also for overcoming many difficulties that current physical research encounters.

References

Arcidiacono, G. (1958). La relativitá di Fantappié. Collectanea. *Mathematica, 10,* 85–124.

Arcidiacono, G. (1964). Gli spazi di Cartan e le teorie unitarie. *Collectanea Mathematica, 16,* 149–168.

Arcidiacono, G. (1969). L'Universo di de Sitter e la meccanica. *Collectanea Mathematica, 20,* 231–256.

Aspect A., Grangier P., & Roger. G. Experimental realization of Einstein-Podolsky-Rosen-Bohm Gedankenexperiment: a new violation of Bell's inequalities. *Physical Review Letters,49*, 91.

Bacciagaluppi, G., & Valentini, A. (2009). *Quantum theory at the crossroads: Reconsidering the 1927 Solvay conference*. Cambridge University Press.

Bacry, H., & Lévy-Leblond, J. (1968). Possible kinematics. *Journal of Mathematical Physics,9*(10), 1605.

Barrett, J. F. (2011). The hyperbolic theory of special relativity. arXiv:1102.0462 [physics.gen-ph]

Bohm, D., Hillion, P., Takabayasi, T., & Vigier, J. P. (1960). Relativistic rotators and bilocal theory. *Progress of Theoretical and Experimental Physics, 23*(3), 496–511. https://doi.org/10.1143/PTP.23.496.

Caldirola, P. (1978). The chronon in the quantum theory of the electron and the existence of heavy leptons. *Nuovo Cimento A,45*, 549–579. https://doi.org/10.1007/BF02730096; Caldirola, P. (1979). A relativistic theory of the classical electron. *La Rivista del Nuovo Cimento2*, 1–49. https://doi.org/10.1007/BF02724419

Castellani, L. (2019). History operators in quantum mechanics. *International Journal of Quantum Information, 17*, 1941001. https://doi.org/10.1142/S0219749919410016.

Chiatti, L. (2005). *Le strutture archetipali del mondo fisico*. Di Renzo.

Chiatti, L. (2014). The transaction as a quantum concept. *International Journal of Research and Reviews in Applied Sciences,16*, 28–47 (Reprinted in In I. Licata (Ed.), *Space-time geometry and quantum events* (pp. 11–43). Nova Publications).

Chiatti, L. (2020). Bit from qubit. A hypothesis on wave-particle dualism and fundamental interactions. *Information,11*(12), 571. https://doi.org/10.3390/info11120571

Chiatti, L., & Licata, I. (2014). Relativity with respect to measurement: Collapse and quantum events from Fock to Cramer. *Systems,2*, 576–589. https://doi.org/10.3390/systems2040576

Chiatti, L., & Licata, I. (2017). Fluidodynamical representation and quantum jumps. In: R. E. Kastner, & J. Jeknic-Dugic, J. (Eds.),*Quantum structural studies. Classical emergence from the quantum level* (pp. 201–223). World Scientific.

Chiatti, L., & Licata, I. (2019). A new version of quantum mechanics with definite macroscopic states. *Commubication to: Quantum 2019, Torino*, May 27–31, 2019.

Chiatti, L. (2014). Elementary particles as signs. *VS, 118*, 105–117.

Chiatti, L. (2018). Thinking non locally: The atemporal roots of particle physics. *Front. Phys., 6*, 95. https://doi.org/10.3389/fphy.2018.00095.

Corben, H. C. (1968). *Classical and quantum theories of spinning particles*. Holden-Day.

Crowther, K. (2020). As below, so before: 'synchronic' and 'diachronic' conceptions of spacetime emergence. *Synthese*. https://doi.org/10.1007/s11229-019-02521-1.

Fantappié, L. (1954). Su una nuova teoria di relativitá finale. *Rendiconti. Accademia Nazionale dei Lincei, 17*, 158.

Fantappié, L. (1959). Sui fondamenti gruppali della fisica (posthumous). *Collectanea Mathematica, 11*, 77–135.

Fermi, E. (1931). Le masse elettromagnetiche nella elettrodinamica quantistica. *Nuovo Cimento, 8*, 121–132.

Fock V. A. (1957). On the interpretation of quantum mechanics. *Czechoslovak Journal of Physics*, 643–656.

Howard, D. (2007). Revisiting the Einstein Bohr dialogue. *Iyyun: The Jerusalem Philosophical Quarterly,56*, 57–90.

Landau, L. D., & Lifsits, E. M. (1973). *Relativistic quantum theory (theoretical physics course (Vol. IV))*. MIR, SSSR.

Licata, I., & Chiatti, L. (2017). Quantum jumps: from foundational research to particle physics. *IOP Conference Series,880*. https://doi.org/10.1088/1742-6596/880/1/012033

Licata, I., & Chiatti, L. (2019). Event-based quantum mechanics: A context for the emergence of classical information. *Symmetry,11*(2), 181. https://doi.org/10.3390/sym11020181

Licata, I., Chiatti, L., & Benedetto, E. (2017). *De Sitter projective relativity*. Springer Briefs in Physics. Springer.

Nagourney, W., Sandberg, J., & Demhelt, H. (1986). Shelved optical electron amplifier: Observation of quantum jumps. *Physical Review Letters, 56*, 2797–2799. https://doi.org/10.1103/PhysRevLett.56.2797.

Pykacz, J. (1992). Direct detection of empty waves contradicts special relativity. *Physics Letters A, 171,* 141–144. https://doi.org/10.1016/0375-9601(92)90416-J.
Rivas, M. (2015). The center of mass and center of charge of the electron. *Journal of Physics: Conference Series,* 615012017.
Schrödinger, E. (1935). Die gegenwärtige situation in der quantenmechanik. Naturwissenschaften *23*, 807–812, 823–828, 844–849. https://doi.org/10.1007/BF01491891
Vigier, J. P., Dewdney, C., Holland, P. R., & Kyprianidis, A. (1987). Causal particle trajectories and the interpretation of quantum mechanics. In B. J. Hiley & F. D. Peat (Eds.), *Essays in honour of David Bohm* (pp. 169–204). Routledge.
Zou, X. Y., Grayson, T., Wang, L. J., & Mandel, L. (1992). Can an "empty" de Broglie pilot wave induce coherence? *Physical Review Letters,68*, 3667–3669.

"A Novel Feature of Atomicity in the Laws of Nature": Quantum Theory Against Reductionism

Arkady Plotnitsky

1 Introduction

Although Niels Bohr's most famous conceptual contribution to physics and philosophy is his concept of complementarity, introduced in 1927, two of his later concepts, correlative to each other, phenomenon and atomicity, introduced in the late 1930s, are as innovative and important. According to Bohr (writing in 1949): "Planck's discovery of the quantum of action [h] disclosed a novel feature of atomicity in the laws of nature supplementing in such unsuspected manner the old doctrine of the limited divisibility of matter" (Bohr, 1987, v. 2, p. 34). The realization of this disclosure, however, took three decades after Max Planck's discovery of quantum theory in 1900. It was shaped by several subsequent developments, in particular by Bohr's (1913) atomic theory, the discovery of quantum mechanics (QM) in 1925, and Bohr's interpretation of QM, developed in its ultimate form, grounded in his concepts of phenomena and atomicity, in1930s.[1] This may be a long time in one's scientific life or in the history of quantum theory. It is, however, a very short time in comparison with the history of "the old doctrine of the limited divisibility" and the atomic constitution of matter, and with it, the idea of reductionism—the idea that the reality of the world is reducible to the ultimate constitutive elements, atoms, and their organization. Its history extends back two and half millennia, to the pre-Socratic philosopher Leucippus and his pupil Democritus, with whose name the doctrine is commonly associated. The doctrine has enjoyed sporadic importance throughout its long history, as with Epicurus and then Lucretius in the ancient world, or via

[1] Bohr's interpretation, in any of its versions, will be distinguished here from "the Copenhagen interpretation." There is no single such interpretation, as even Bohr has changed his a few times. I have considered different versions of Bohr's interpretation in Plotnitsky (2012).

A. Plotnitsky (✉)
Literature, Theory, and Cultural Studies Program, Philosophy and Literature Program, College of Liberal Arts, Purdue University, West Lafayette, IN 47907, USA
e-mail: plotnits@purdue.edu

S. Wuppuluri and I. Stewart (eds.), *From Electrons to Elephants and Elections*,
The Frontiers Collection, https://doi.org/10.1007/978-3-030-92192-7_25

Lucretius, in the sixteenth century, and then in modern, post-Galilean, physics, as in Sir Isaac Newton's corpuscular theory of light, before its rise to dominance with the nineteenth-century atomic theories, from the kinetic theory of gases to the atomist theory of Brownian motion, courtesy of Albert Einstein, and finally the present-day atomism, that of elementary particles, in QM and quantum field theory (QFT).

Planck's discovery was initially understood on these lines as well, especially after Einstein's introduction of the concept of the photon in 1905, which, along with the discovery of the electron (before Planck's theory) and then the proton, brought with them the concept of elementary particles as new atoms of nature. Quantum theory was far from free of complexities and paradoxes, for example, the wave-like behavior of these new "atoms" of nature in certain circumstances, a behavior initially associated with radiation, but extended to the particles of matters, such as electrons, by Louis de Broglie in the early 1920s. While QM, discovered by Werner Heisenberg and Erwin Schrödinger in 1925–1926, resolved most physical difficulties of the previous quantum theory (the "old quantum theory") and is still our standard theory, it hardly avoided the epistemological complexities brought about by quantum theory and even exacerbated some of them.

Bohr's interpretation of quantum phenomena and QM, introduced in 1927 and developed by him during the next decade, met these complexities in a new way, which has been and remains controversial and even unacceptable to some, beginning, famously, with Einstein. Bohr distinguished, on Kantian lines (but more radically), between quantum *phenomena*, defined by what is observed in measuring instruments and quantum *objects*, responsible for quantum phenomena. Quantum objects were placed beyond knowledge or even conception. By 1937, Bohr spoke of "our not being any longer in a position to speak of the autonomous behavior of a physical object, due to the unavoidable interaction between the object and the measuring instruments" (Bohr, 1937, p. 87). Accordingly, one could no longer ascribe to the quantum objects such properties as discreteness or indivisibility, or their continuous counterparts, such as wave properties. Such properties could only be those of quantum phenomena observed in measuring instruments, as effects of their interactions with quantum objects, effects predicted, in probabilistic terms, by QM or QFT. No other predictions are possible on experimental grounds. "The novel feature of atomicity" invoked by Bohr arises from this situation. It is defined by a transfer to quantum *phenomena*, manifested in measuring instruments, the key "atomic" features—discreteness, discontinuity, individuality, and atomicity (indivisibility)—previously associated with quantum *objects*. "Atomicity" becomes an epistemological feature that refers to physically complex and hence divisible entities, and not to physically indivisible quantum objects, including "elementary particles," such as electrons, photons, or quarks. An elementary particle or an atom, beginning with the hydrogen, become a very different concept: it is technological and informational, rather than describing the ultimate of reality responsible for quantum phenomena, to which no atomic properties could be assigned any more than any other.

While the genealogy of Bohr's concept of phenomenon appears to have been more conventional, extending from Immanuel Kant's philosophy, Bohr's concept of atomicity is unusual and its genealogy is less clear. An intriguing possibility is Alfred

North Whitehead's concept of atomism in his 1927 *Process and Reality* (Whitehead, 1978), although there is no evidence that it was familiar to Bohr. Whitehead's overall ontological position is, however, different from that of Bohr. Whitehead's *discrete* atomicity of experience is underlain by the *continuous* ultimate reality, placed in a complex way between the material and the mental. By contrast, while assuming the existence of the material reality responsible for the emergence of discrete phenomena observed in measuring instruments in quantum physics, Bohr's interpretation "*in principle* exclude[s]" a representation or even conception of this reality, and thus any ontological conception, discrete or continuous, of this reality (Bohr, 1987, v. 2, p. 62). Accordingly, it is difficult to transfer Whitehead's scheme into Bohr's interpretation.

Building on Bohr's view, I shall argue that QM and QFT, the currently standard theories of elementary particles, occupy a unique position in contemporary thought, at least in certain interpretations of these theories or quantum phenomena themselves: While they radically depart from reductionism in physics, including when it comes to the concept of elementary particles, they do not embrace holism either. They do have affinities with holism, arising from Bohr's concepts of phenomenon and atomicity as forming an indivisible whole with quantum objects. But these concepts are different from those found in most forms of holism. They represent a different way of thinking in physics and philosophy.

My argument is based on the concept of "reality without realism," RWR, which, while more general, is, when used in quantum theory, based on the irreducible role of experimental technology in the constitution of quantum phenomena (e.g., Plotnitsky, 2016, 2021; Plotnitsky & Khrennikov, 2015). This concept places the reality considered beyond representation or knowledge, which I define as "the weak RWR view," or even conception, which I define as "the strong RWR view," with each view leading to a corresponding set of interpretations of quantum phenomena and QM or QFT.

Such interpretations may be different within each set, as are Bohr's ultimate interpretation and the one adopted here. In both interpretations, the concept of RWR only applies to the ultimate constitution of the reality responsible for quantum phenomena, the constitution commonly associated with quantum objects. This association, however, takes a different form in the present interpretation. While equally beyond conception, the ultimate, RWR-type, reality and quantum objects are not the same entities or, in the present view, forms of idealization. The ultimate RWR-type stratum of reality is, as an idealization, assumed to exist independently of our interactions with it. On the other hand, the reality idealized as quantum objects is only assumed to exist at the time of measurement, defined by the interactions between the ultimate RWR stratum of reality and measuring instruments. There are no quantum objects, such as electrons, photons, or quarks, existing independently in nature apart from our interaction with it. In contrast to either the ultimate RWR-type reality or quantum *objects*, quantum *phenomena*, observed in measuring instruments, allow for a representational treatment, in fact by means of classical physics. Quantum phenomena are an idealization as well, because they or our observations in the first place are a product of thought, and, as already Kant argued, phenomena may not correspond how things, things-in-themselves, are in nature (Kant, 1997). This tripartite view of the reality responsible for quantum phenomena is not found in Bohr's interpretation in any of its versions. This view, however, defines a more radical form of anti-reductionism because it defies the very possibility of speaking of the ultimate atomic constituents of nature as existing independently.

2 Reality Without Realism

The concept of reality without realism, RWR, is grounded in more general concepts of reality and existence, assumed here to be primitive concepts and not given analytical definitions. These concepts are, however, in accord with most, even if not all (which would be impossible), available concepts of reality and existence in realism and nonrealism alike. By "reality" I refer to that which is assumed to exist, without making any claims concerning the *character* of this existence, claims that define realism. The absence of such claims allows one to place this character beyond representation or even conception. I understand existence as a capacity to have effects on the world with which we interact. The very assumption that something is real, including of the RWR-type, is made on the basis of such effects. To ascertain observable effects of physical reality entails a representation of them but not necessarily a representation or even a conception of how they come about, which may not be possible and is not in the RWR view. The concept of an effect is crucial here, and the appeal to effects becomes persistent in Bohr's writings (e.g., Bohr, 1987, v. 1, p. 92, v. 2, pp. 40, 46–47). A given theory or interpretation might, then, assume different levels and different types of *idealizations* of reality, some allowing for a representation or conception and others not, in which case the RWR view applies, as against realism, defined by the fact that all levels of the reality considered allow for a representation or at least conception.

Realist or ontological thinking in physics, or elsewhere, is manifested in the corresponding theories, which are commonly representational and often reductionist in character.[2] Such theories aim to represent the reality they consider, usually by mathematized models, suitably idealizing this reality. It is possible to aim, including in quantum theory, for a strictly mathematical representation of this reality apart from any physical concepts, at least as they are customarily understood, for example, in classical physics or relativity. It is also possible to assume an independent architecture of the reality considered, while admitting that it is either (A) not possible to adequately represent this architecture by means of a physical theory or (B) even to form a specific concept of this architecture, either at a given moment in history or even ever. Under (A), a theory that is merely predictive could be accepted for lack of a

[2] Although the terms "realist" and "ontological" sometimes designate more diverging concepts, they are commonly close and will be used, as adjectives, interchangeably here. I shall adopt "realism," as a noun, as a more general term and refer by an "ontology," more specifically, to a given representation or conception of the reality considered by a given theory.

realist alternative, but usually with the hope that a future theory will do better by being a properly representational theory. Einstein adopted this attitude toward QM. Even under (B), however, this architecture is usually conceived on the model of classical physics (to which relativity philosophically, but not physically, conforms). What, then, grounds realism most fundamentally is the assumption that the ultimate constitution of reality possesses properties and the relationships between them, or, as in structural realism (Ladyman, 2016), just a structure, in particular, a mathematical structure, that may either be ideally represented by a theory or is unknown or even unknowable, but is still conceivable, usually with a hope that it will eventually be represented.[3]

Thus, classical mechanics (used in dealing with individual objects and small systems, apart from chaotic ones), classical statistical mechanics (used in dealing, statistically, with large classical systems), or chaos theory (used in dealing with classical systems that exhibit a highly nonlinear behavior) are realist. While classical statistical mechanics does not represent the overall behavior of the systems considered because their great mechanical complexity prevents such a representation, it assumes that the individual constituents of these systems are represented by classical mechanics. The theory is a paradigmatic example of a reductionist theory, in part in contrast to thermodynamics that need not rely on this type of reductionism and can deal with macroscopic bodies themselves. In chaos theory, which, too, deals with systems consisting of large numbers of atoms, one assumes a mathematical representation of the behavior of these systems.The theory, however, can and has been used in anti-reductionist approaches in physics and elsewhere, on realist lines.

Our phenomenal experience can only serve us partially in relativity. This is because, while we can give the relativistic behavior a concept and represent it mathematically, which makes relativity a realist and classically causal and, in fact, deterministic theory (the concept explained below), we have no means of visualizing this behavior, for example, as represented by the Lorentz-Einstein velocity-addition formula for collinear motion, $s = \frac{v+u}{1+(vu/c)^2}$, especially in the case of photons. Nevertheless, relativity still offers a representation of the behavior of individual systems. This behavior could, moreover, be treated classically causally (a concept explained below), although, because all physical influences are limited by c, relativity imposes new limits on causal relationships between events.

The representation of individual quantum objects and behavior became partial in Bohr's atomic theory, introduced in 1913 (Bohr, 1913). The theory only provided representations, in terms of orbits, for the stationary states of electrons in atoms, but not for the discrete transitions, "quantum jumps," between stationary states. This was an unprecedented step, because this concept was incompatible with classical mechanics and electrodynamics alike. At the time, it was expected that Bohr's theory was a stop-gap measure that will no longer be necessary when a proper theory of quantum phenomena is developed. It was, however, this concept that became central

[3] One could, in principle, also see the assumption of the existence or reality of something to which a theory can relate without representing it as a form of realism. I would argue, however, that the present definition of realism is more in accord with most uses of the term in physics or philosophy.

for Heisenberg, who built on it by abandoning an orbital representation of stationary states as well (Heisenberg, 1925). According to Bohr's 1925 assessment:

> In contrast to ordinary mechanics, *the new quantum mechanics does not deal with a space–time description of the motion of atomic particles*. It operates with manifolds of quantities [matrices] which replace the harmonic oscillating components of the motion and symbolize the possibilities of transitions between stationary states .... These quantities satisfy certain relations which take the place of the mechanical equations of motion and the quantization rules (Bohr, 1987, v. 1, p. 48; emphasis added).

This assessment was thus based on the RWR view, at least the weak RWR view. By contrast, the first worked-out version of Bohr's interpretation, in his Como lecture, attempted to restore, ambivalently, realism to QM (Bohr, 1987, v. 1, pp. 52–91). This interpretation was, however, quickly abandoned by Bohr, following his discussion with Einstein in October of 1927 at the Solvay conference in Brussels. This discussion initiated his path toward his ultimate, RWR-type, interpretation introduced in 1937 in "Complementarity and Causality" (Bohr, 1937). Bohr does not use the language of reality without realism, but his understanding clearly amounts to the RWR view, because one can no longer "speak of the autonomous behavior of a physical object" (Bohr, 1937, p. 87). As explained below, Bohr's view at this point becomes the strong RWR view, which places this reality or quantum objects beyond conception, but perhaps not the strongest one. For one can still ask whether our inability to do so only (A) only applies as things stand now, or (B) reflects that this reality is beyond the reach of our thought altogether. While Bohr at least assumes (A) and while it is possible that he entertained (B), he never stated so, which leaves whether he assumed (B) or only (A) to interpretation. Logically, once (A) is the case, then (B) is possible. Although there is no experimental data that would favor one against the other, these views are different philosophically in defining how far our mind can reach in investigating the ultimate constitution of nature. This is my main reason for differentiating them, although my argument here applies to both (A) and (B).

The qualification "as things stand now" applies, however, to (B) as well, even though it might appear otherwise given that this view precludes any conception of the ultimate reality not only now but also ever. It applies because a return to realism in quantum theory is possible, either on experimental or theoretical grounds. This return may take place either because quantum theory, as currently constituted, is replaced by an alternative theory that requires a realist interpretation, or because the strong (or weak) RWR view becomes obsolete even for those who hold it and is replaced by a more realist view with quantum theory in place in its present form.

Weather one adopts (A) or (B), the renunciation of classical causality follows. The concept of classical causality will be defined here, as it was by Bohr and others, usually under the rubric of "causality," by the claim that the state, X, of a physical system is determined, in accordance with a law, at all future moments of time once it is determined at a given moment of time, state A, and A is determined in accordance with the same law by any of the system's previous states. This assumption also implies a concept of reality, which defines this law and makes this concept of causality ontological. This concept has a long history, beginning with the pre-Socratics, and it has governed classical physics from its inception on. I designate it "classical causality" because there are alternative conception of causality, including applicable in QM or QFT (e.g., Plotnitsky, 2020, pp. 1844–1846). Some, beginning with P. S. Laplace, have used "determinism" for this concept. I prefer to define "determinism" as an epistemological category referring to the possibility of predicting the outcomes of classically causal processes ideally exactly. In classical mechanics, when dealing with individual objects or small systems, both notions in effect coincide. On the other hand, classical statistical mechanics or chaos theory are classically causal but not deterministic in view of the complexity of the systems considered, which limit us to probabilistic or statistical predictions concerning them. It is possible to assume that the ultimate nature of reality is random or mixed; and, while causal conceptions of reality have been dominant, random ones have been around since the pre-Socratics, as in Democritus's and then Epicurus's and Lucretius's atomism. Such conceptions have a problem insofar as the dynamics leading to random events is not given an explanation, a problem avoided by the RWR view, which in principle precludes such as explanation.

In the case of quantum phenomena, deterministic predictions are not possible even in considering the most elementary quantum phenomena, such as those associated with elementary particles. This is because the repetition of identically prepared quantum experiments in general leads to different outcomes, and this difference cannot be diminished beyond the limit defined by Planck's constant, h, by improving the capacity of our measuring instruments. Hence, the probabilistic or statistical character of quantum predictions must be maintained by interpretations of QM or alternative theories of quantum phenomena that are classically causal. Such interpretations and theories are also realist because classical causality implies a law governing it and thus a representation of the reality considered in terms of this law. By contrast, RWR-type interpretations are not classically causal because the ultimate nature of reality responsible for quantum phenomena is beyond representation or conception.

As stated in the Introduction, the idealization adopted by the present, strong RWR-type, interpretation of quantum phenomena and QM or QFT is stratified into three strata: the ultimate nature of the reality responsible for quantum phenomena, a reality that is beyond conception and is assumed to exist independently; that of quantum objects, which are also beyond conception but are assumed to exist only at the time of measurement; and that of quantum phenomena, which are knowable and representable. I shall now explain this stratification, crucial for understanding how radically quantum theory may depart from reductionism. It is, again, not adopted by all RWR-type interpretations, including that of Bohr.

It is fitting to start with quantum objects. The idealization of quantum objects in all RWR-type interpretations is essentially different than those, often associated with reductionism, found in classical mechanics, in particular that of using dimensionless massive points to mathematically idealize the motion of material objects. Elementary particles are commonly idealized as dimensionless, point-like entities, because, if they had volume, charged particles would be torn apart by the electromagnetic force within them. However, defining them as quantum objects poses complexities even in realist interpretations, for several reasons, in particular the following one. While what is observed in a measuring instrument, as a quantum phenomenon, is always uniquely (classically) defined, what is the object under investigation and what is the measuring instrument, beyond this observed part, are not. The difference between these two entities is, nevertheless, irreducible, as against classical physics, where it can be disregarded, because the interference of observation can be neglected. According to Bohr:

> This necessity of discriminating in each experimental arrangement between those parts of the physical system considered which are to be treated as measuring instruments and those which constitute the objects under investigation may indeed be said to form a *principal distinction between classical and quantum-mechanical description of physical phenomena.* It is true that the place within each measuring procedure where this discrimination is made is in both cases largely a matter of convenience. While, however, in classical physics the distinction between object and measuring agencies does not entail any difference in the character of the description of the phenomena concerned, its fundamental importance in quantum theory … has its root in the indispensable use of classical concepts in the interpretation of all proper measurements, even though the classical theories do not suffice in accounting for the new types of regularities with which we are concerned in atomic physics. In accordance with this situation there can be no question of any unambiguous interpretation of the symbols of quantum mechanics other than that embodied in the well-known rules which allow us to predict the results to be obtained by a given experimental arrangement described in a totally classical way (Bohr, 1935, p. 701).

This statement may suggest that, while observable parts of measuring instruments are described by means of classical physics, the independent behavior of quantum objects is described or represented by means of the quantum–mechanical formalism. This type of view has been adopted by some, for example, Dirac (1958) and von Neumann (1932) in their influential books. It was not, however, Bohr's view, at least after he revised his Como argument. Bohr does say here that the observable parts of measuring instruments are described by means of classical physics and that classical theories cannot suffice to account for quantum phenomena. But he does not say that the independent behavior of quantum objects is described by the quantum–mechanical formalism. His statement only implies that quantum objects cannot be treated classically. The "symbols" of quantum–mechanical formalism only have a probabilistically or statistically predictive role. Also, although what is observed as phenomena in quantum experiments is beyond the capacity of classical physics to account for them, the classical description can and, in order for us to be able to give an account of what happens in experiments, must apply to the observable parts of measuring instruments. The instruments, however, also have a quantum stratum, through which they interact with quantum objects. This interaction is quantum and cannot be observed as such or, in RWR-type interpretations, represented. It is "irreversibly amplified" to the macroscopic level of observable effects, say, a spot left on a silver screen (Bohr, 1987, v. 2, p. 73).

This situation is sometimes referred to as the arbitrariness of the "cut" or the "Heisenberg-von-Neumann cut." As Bohr noted, however, while "it is true that the place within each measuring procedure where this discrimination [between the object and the measuring instrument] is made is … largely a matter of convenience," it is true only largely. This is because "in each experimental arrangement and measuring procedure we have only a free choice of this place within a region where the quantum–mechanical description of the process concerned is effectively equivalent with the classical description" (Bohr, 1935, p. 701). The ultimate (RWR-type) constitution of the physical reality responsible for quantum phenomena is always on the other side of the cut. Quantum objects, too, are idealized as part of the reality that is, in each experiment, on the other, "object," side of the cut. In the present view, if not that of Bohr, a quantum object, as an idealization is different from that of the ultimate, RWR-type, reality responsible for quantum phenomena. While a measuring instrument, which is, in its observable part, a classical object, and the ultimate RWR-type reality considered, are assumed to exist independently, a quantum object can only be ascribed existence by a measurement and its setup, including the cut. What is a quantum object can be different in each case, including something that, if considered by itself, could be viewed as classical, although certain quantum objects, such as elementary particles, are always on the other side of the cut, where the ultimate, RWR-type, reality is.

The following question might, then, be asked. If a quantum object is only defined by a measurement, rather than as something that exists independently, could one still speak of the same quantum object, say, the same electron, in two or more successive measurements? According to the view here outlined, each of these two measurements defines an electron, with the same mass and charge, in two different positions at two different moments in time. The case can be given a strictly RWR interpretation, insofar as all these properties (mass, charge, and position) are, physically, those of measuring devices, assumed to be impacted by quantum objects, rather than of these objects themselves, placed beyond representation or conception. The question is, however: Do these two measurements register the same electron? To consider them as the same electron is a permissible idealization in low-energy (QM) regime, an idealization ultimately statistical in nature, because that the second measurement will register an electron is not guaranteed, as would in principle be in classical physics. On the other hand, as discussed below, speaking of the same electron in successive measurements in high-energy (QFT) regimes is meaningless.

The epistemological cost of the RWR view is not easily absorbed by most and to some, beginning, famously, with Einstein, is unacceptable. This is not surprising because the features of quantum phenomena that are manifested in many famous experiments and that led to RWR-views defy many assumptions concerning nature commonly considered as basic. These assumptions, arising due to the neurological constitution of our brain, have served us for as long as human life, and within certain limits, are unavoidable, including in physics, although their scope, as noted, was already challenged by relativity. QM have made this challenge much greater. The same neurological constitution, however, may also prevent us from conceiving of the ultimate (RWR) nature of physical reality responsible for quantum phenomena. Thus, it is humanly natural to assume that *something happens* between observations. Indeed, the sense that something happened is one of the most essential elements of human thought. However, in the RWR view, the expression "something happened" is inapplicable to the ultimate constitution of the reality considered. According to Heisenberg:

> There is no description of what *happens* to the system between the initial observation and the next measurement. ...The demand to "describe what happens" in the quantum-theoretical process between two successive observations is a contradiction in adjecto, since the word "describe" refers to the use of classical concepts, while these concepts cannot be applied in the space between the observations; they can only be applied at the points of observation (Heisenberg, 1962, pp. 57, 145).

The same would apply to the word "happen" or "system," or any word we use, whatever concept it may designate, including reality, although when "reality" refers to that of the RWR-type, it is a word without a concept attached to it. As Heisenberg says: "The problems of language are really serious. We wish to speak in some way about the structure of the atoms and not only about 'facts'—the latter being, for instance, the black spots on a photographic plate or the water droplets in a cloud chamber. However, we cannot speak about the atoms in ordinary language" (Heisenberg, 1962, pp. 178–179). Nor is it possible in terms of ordinary concepts, from which ordinary language is indissociable, or even physical concepts.

Mathematical concepts are a possible exception. As Heisenberg said on an earlier occasion, mathematics is "fortunately" free from the limitations of ordinary language and concepts (Heisenberg, 1930, p. 11). At the time, Heisenberg, adopting the RWR view, used this freedom to construct QM as a theory designed only to predict the probabilities or statistics of events observed in measuring instruments. In his later writings, he assumed the possibility of a mathematical *representation* of the ultimate constitution of reality, while excluding physical concepts (at least in their customary sense found in classical physics or relativity) as applicable to this constitution (Heisenberg, 1962, pp. 145, 167–186). Bohr, by contrast, rejected the possibility of a mathematical representation of quantum objects and behavior, or the reality they idealize, along with a physical one. Bohr often speaks of this reality as being beyond our phenomenal intuition, also involving visualization, sometimes used, including by Bohr, to translate the German word for intuition, *Anschaulichkeit* (Bohr, 1987, v. 1 p. 51, 98–100, 108; v. 2, p. 59). It is clear, however, that Bohr saw the ultimate nature of this reality as being beyond conception, including a mathematical one, in his ultimate interpretation, based in the concepts of phenomena and atomicity, which I shall discuss in Sect. 4. First, I shall consider Bohr's concept of complementarity.

3 Complementarity

Defined arguably most generally, complementarity is characterized by:

(a) a mutual exclusivity of certain phenomena, entities, or conceptions; and yet
(b) the possibility of considering each one of them separately at any given point; and
(c) the necessity of considering all of them at different moments of time for a comprehensive account of the totality of phenomena that one must consider in quantum physics.

The concept was not given by Bohr a single definition of this type. However, this definition may be surmised from several of Bohr's statements, such as: "Evidence obtained under different experimental conditions cannot be comprehended within a single picture, but must be regarded as complementary in the sense that only the totality of the phenomena [some of which are mutually exclusive] exhaust the *possible* information about the objects" (Bohr, 1987, v. 2, p. 40; emphasis added). In classical mechanics, we can comprehend all the information about each object within a single picture because the interference of measurement can be neglected: this allows us to identify the phenomenon considered with the object under investigation and to establish the quantities defining this information, such as the position and the momentum of the object, in the same experiment. In quantum physics, this interference cannot be neglected, which leads to different experimental conditions for each measurement on a quantum object and their complementarity, in correspondence with the uncertainty relations. The situation implies two incompatible pictures of what is observed. Hence, the *possible* information about a quantum object, the information *to be found* in measuring instruments, could only be exhausted by the mutually incompatible evidence obtainable under different experimental conditions. On the other hand, once made, either measurement will provide the complete *actual* information about the object, as complete as possible, at this moment in time.

It is worth noting that wave-particle complementarity, with which the concept of complementarity is often associated, had not played a significant role in Bohr's thinking. Bohr thought deeply, even before QM, about the problem of wave-particle *duality*, as it was known then. However, Bohr was aware of the difficulties of applying the concept of physical waves to quantum objects, or of thinking in terms of the wave-particle duality, as the assumption that both types of nature and behavior pertain to the same individual entities, such as each photon or each electron as such, considered independently. The wave-particle duality was thought of as representing the same thing in two different ways. By contrast, complementarity refers to two different, incompatible, entities, like two different effects of an electron on a measuring instruments, but not two features of the electron itself. The "*both*" (both types of properties) of the wave-particle duality is the opposite of complementarity, based in the mutual exclusivity of the two type of effects observed in measuring instruments, and thus on "*either or*" (either one or the other type of effects). Bohr's ultimate solution to the dilemma of whether quantum objects are particles or waves was that they were neither. Instead, either "picture" refers to one of the two mutually exclusive sets of discrete individual effects of the interactions between quantum objects and measuring instruments, particle-like, which may be individual or collective, or wave-like, which are always collective, but still composed of discrete individual effects.

The concept of complementarity is best exemplified by complementarities of spacetime coordination and the application of momentum or energy conservation laws, correlative to the uncertainty relations, as complementarities of phenomena observed in measuring instruments and thus in accord with Bohr's concept of phenomena. Technically, the uncertainty relations, $\Delta q \Delta p \cong h$ (where q is the coordinate, p is the momentum in the corresponding direction), only prohibit the simultaneous exact measurement of both variables, which is always possible, at least in principle, in classical physics, and allows one to maintain classical causality there. The physical meaning of the uncertainty relations is much deeper in Bohr's interpretation. First of all, they are not a manifestation of the limited accuracy of measuring instruments, because they would be valid even if we had perfect measuring instruments. As Bohr says: "we are of course not concerned with a restriction as to the accuracy of measurement, but with a limitation of the well-defined application of space–time concepts and dynamical conservation laws, entailed by the necessary distinction between measuring instruments and atomic objects" (Bohr, 1937, p. 86; Bohr, 1987, v. 2, p. 73; v. 3, p. 5). The uncertainty relations make each type of measurement complementary to the other. In addition, one not only cannot measure both variables simultaneously but also cannot define them simultaneously, which makes probabilistic or statistical considerations unavoidable in considering both the uncertainty relations and complementarity. In RWR-type interpretations, however: "the statistical character of the uncertainty relations in no way originates from any failure of measurement to discriminate within a certain latitude between classically describable states of the objects, but rather expresses an essential limitation of applicability of classical ideas to the analysis of quantum phenomena" (Bohr, 1938, p. 100). There is no contradiction between this statement and the fact that quantum phenomena are described classically, because the statement only says that their ultimate nature and emergence cannot be so described.

Bohr and others proposed using the concept of complementarity in philosophy, biology, and psychology. Here, however, I am only concerned with complementarity in quantum physics. Bohr speaks of complementarity as "an artificial word" that "does not belong to our daily concepts" (Bohr, 1937, p. 87). It is a physical concept, eventually linked by Bohr to his ultimate, strong RWR-type, interpretation. As explained, this concept prevents us from ascertaining the "whole" composed from its "parts," in contrast to the conventional understanding of parts *complementing* each other within a whole. At any moment of time only one "part" and not the other could be ascertained. This part is the only whole at this moment of time, and hence is not a part. This difference has not always been adequately understood or appreciated. Consider John S. Bell's comments:

> [Bohr] seemed to revel in contradiction, for example, between 'wave' and 'particle.' ... Not to resolves these contradictions and ambiguities, but rather to reconcile us to them, he put forward a philosophy which he called 'complementarity." ... There is very little I can say about 'complementarity.' But I wish to say one thing. It seems that Bohr used this word with the reverse of its usual meaning. Consider for example the elephant. From the front she is head, trunk, and two legs. From the sides she is bottom, tail, and two legs. They supplement one another, they are consistent with one another and they are all entailed by the unifying concept 'elephant.' It is my impression that to suppose Bohr used the word 'complementarity' in this ordinary way would have been regarded by him as missing his point and trivializing his thought. He seems to insist rather that we must use in our analysis elements which *contradict* one another, which do not add up or derive from a whole. By 'complementarity' he meant, it seems to me, the reverse: contradictoriness. ... Perhaps he

> a subtle satisfaction in the use of a familiar word with the reserve of its familiar meaning." (Bell, 2004, p. 190).

Bell does not favor Bohr's interpretation or for that matter QM, which attitude is legitimate. On the other hand, Bell's comments on complementarity are beside the point. Bell is not incorrect in saying that "to suppose Bohr used the word 'complementarity' in this ordinary way would have been regarded by him as missing his point and trivializing his thought." It would have, because it does. Besides, while "complementary" as an adjective is a familiar word, "complementarity" was never used as a noun before Bohr. Also, Bohr's treatment of the question of "particles and "waves" in quantum theory is more complex than Bell makes it appear, which is why, as noted, Bohr avoids speaking of wave-particle complementarity. Complementarity is a new physical concept, which must be understood in the specific sense Bohr gives it. There is no point in attempting to relate it to a meaning it may be given in our daily life, as Bell does, by defining complementary parts as adding up to a whole. This is what Bohr wants to avoid. He needed a new concept in order to account for the epistemological situation defined by quantum phenomena (Bohr, 1937, p. 87; Bohr, 1935, p. 700). It is not, as Bell suggests, about some "subtle satisfaction in the use of a familiar word with the reverse of its familiar meaning." There is no evidence that Bohr ever had such a satisfaction. On the other hand, there is plenty of evidence for his physical reasons for defining complementarity in the way he did, such as the uncertainty relations or the double-slit and other iconic quantum experiments. Complementarity was introduced in the spirit of resolving contradictions, and not reveling in them, as it "seemed" to Bell. Bohr's concept of complementary "parts" that do not add up to a "whole" does just that. Against the Democritean atomism, Bohr's argument relates elementary particles, say, electrons to the "elephant" scale of atomic phenomena observed in measuring instruments. The complementarity of some of these phenomena does not allow them to be "parts" that could be added up to a whole, as Bell wants to convey by his image of an elephant.

4 Measurement, Phenomena, and Atomicity: Bohr Against Whitehead

Bohr's ultimate interpretation of quantum phenomena and QM was first presented in "Complementarity and Causality" (Bohr, 1937). Bohr does not use the language of reality without realism, but his argument clearly amounts to the RWR view:

> The renunciation of the ideal of causality in atomic physics which has been forced on us is founded logically only on our not being any longer in a position to speak of the autonomous behavior of a physical object, due to the unavoidable interaction between the object and the measuring instruments which in principle cannot be taken into account, if these instruments according to their purpose shall allow the unambiguous use of the concepts necessary for the description of experience (Bohr, 1937, p. 87).

The concept of causality that grounds this ideal of causality is that of classical causality, explained earlier. RWR-type interpretations are not classically causal because of the absence of realism in considering the behavior of quantum objects or the reality thus idealized. Given, however, that it is possible to argue for interpretations of QM or alternative theories of quantum phenomena that are realist and classically causal, Bohr's statement represents the strong RWR-type interpretation. For, if, as Bohr says, we are no "longer in a position to speak of the autonomous behavior of a physical object, due to the unavoidable interaction between the object and the measuring instrument," this behavior must also be beyond conception. If we had such a conception, we would be able to say something about it. It is true that there is a difference between some conception of this reality and a rigorous conception that would enable us to provide a proper representation of it. Bohr, however, clearly claims the impossibility of any such conception: we are no longer in a position to speak of the autonomous behavior of quantum objects *at all*, at least as things stand now. Bohr's ultimate interpretation was grounded, along with complementarity, in two new, correlative, concepts, "phenomenon" and "atomicity", defined in terms of effects observed in measuring instruments impacted by quantum objects. According to Bohr:

> I advocated the application of the word phenomenon exclusively to refer to the *observations* obtained under specified circumstances, including an account of the whole experimental arrangement. In such terminology, the observational problem is free of any special intricacy since, in actual experiments, all observations are expressed by unambiguous statements referring, for instance, to the registration of the point at which an electron arrives at a photographic plate. Moreover, speaking in such a way is just suited to emphasize that the appropriate physical interpretation of the symbolic quantum-mechanical formalism amounts only to predictions, of determinate or statistical character, pertaining to individual phenomena appearing under conditions defined by classical physical concepts [describing the observable parts of measuring instruments] (Bohr, 1987, v. 2, p. 64).

As defined by "*the observations obtained under specified circumstances*," phenomena refer to events that have already occurred, and not to future events that one can predict. These observations are the same as in classical physics, which allows one to identify phenomena with physical objects (here measuring instruments), because our observation does not interfere with their behavior, in contrast to the way our observation by means of a measuring instrument interferes with the (RWR-type) ultimate reality responsible for a phenomenon thus observed.

Bohr's rethinking of the concept of phenomena in quantum physics also led him to his anti-reductionist concept of "atomicity," capturing "the novel feature of atomicity in the laws of nature," disclosed by "Planck's discovery of the quantum of action," h (Bohr, 1938, p. 38; Bohr, 1987, v. 2, p. 33; v. 3, p. 2). This concept transfers to the level of phenomena all of the key features—discreteness, discontinuity, individuality, and indivisibility (atomicity proper)—previously associated with atomic entities in nature. As is Bohr's concept of phenomenon, the concept of "atomicity" is defined in terms of *individual* effects of quantum objects on the classical world. In fact, both concepts are correlative: phenomena are atomic in their nature, while Bohr's new "atoms" are phenomena. "Atomicity" refers to physically complex and sub-divisible entities, and no longer to indivisible physical entities, such as quantum objects, in particular, elementary particles, to which one cannot ascribe atomic properties any more than any other. Each "atom" is *individual*, each—as every (knowable) effect conjoined with every (unknowable and even unthinkable) process of its emergence—unique and unrepeatable. Some can be clustered insofar as they refer to the "same" quantum entities, whether individual or collective. Thus, along with *indivisibility,* quantum atomicity as *individuality* is now understood as the individuality or uniqueness of each phenomenon. By the same token, each phenomenon is discrete in relation to any other.

These new "atoms" are technological-informational entities—bits of information enabled by our experimental technologies and our thought processing this information. One can see the situation in quantum-informational terms, following J. A. Wheeler's concept of "it from bit," in part inspired by Bohr (Wheeler, 1990, p. 4). The "it" in question is an RWR-type reality ultimately responsible for each quantum phenomenon or techno-info-atom (bit), inferred from the totality of information obtained in quantum experimental. This information qua information is classical, but its structure, in particular, the so-called quantum correlations, such as those observed in the Einstein–Podolsky–Rosen (EPR) type experiments, involved in the Bell and the Kochen-Specker theorems, cannot be predicted by classical physics. One could call this new technological-informational and thus, anti-reductionist, concept "the Bohr atom," adopting the term sometimes applied to Bohr's concept of atom in his 1913 theory. This theory already contained the seeds not only of the idea of reality without realism, but also of this atomicity. In thinking of Bohr's (1913) concept of atom one usually focuses on electronic orbits (stationary states). The key concept of theory was, however, that of discrete transitions, "quantum jumps," between stationary states, which has remains part of QM, while the concept of orbit was abandoned. As Heisenberg was the first to realize these transitions are techno-atomic. As he said: "What I really like in this scheme is that one can really reduce all interactions between atoms and the external world … to transition probabilities" (Heisenberg, Letter to Kronig, 5 June 1925; cited in Mehra & Rechenberg, 2001, v. 2, p. 242). By speaking of "*interactions* between atoms and the external world," this statement suggests that QM was only predicting the effects of these interactions, observed in measuring instruments. This view was adopted by Bohr and came to define his concepts of phenomenon and atomicity against the Democritean atomicity of quantum objects. In the present view, moreover, quantum objects are defined only at the time of measurement, which leaves no room for reductionism.

One can detect affinities between Bohr's and Whitehead's concepts of atomism, as "drops of experience." On the other hand, the ontological architecture of reality in Bohr is very different from that in Whitehead, given that the concept of ontology could, in Bohr, only apply to measuring instruments (phenomena) and not, in any form, continuous or discontinuous, to the ultimate constitution of reality. Whitehead, by contrast, sees this constitution as continuous. According to Henry Stapp, adapting Whitehead's view to quantum theory:

> [D]iscreteness is the signature of quantum phenomena ... The core issue for both Whiteheadian process and quantum process is the emergence of the discrete from the continuous. This problem is illustrated by the decay of a radioactive isotope located at the center of a spherical array of a finite set of detectors, arranged so that they cover the entire spherical surface. The quantum state of the positron emitted from the radioactive decay will be a continuous spherical wave, which will spread out continuously from the center and eventually reach the spherical array of detectors. But only one of these detectors will fire. The total space of possibilities has been partitioned into a discrete set of subsets, and the prior continuum is suddenly reduced to some particular one of the elements of the selected partition (Stapp, 2011, pp. 8, 88).

There are further complexities given the role of potentiality in Whitehead: "Whitehead draws a basic distinction between the two kinds of realities upon which his ontology is based: 'continuous potentialities' versus 'atomic actualities': 'Continuity concerns what is potential, whereas actuality is incurably discrete' [Whitehead, 1978, p. 91]" (Stapp, 2011). The question is the status of this continuous potentiality, especially if one sees it as real, if there "two kinds of realities," actual and potential, and the discrete actual reality emerges from the continuous potential one. I shall leave aside Whitehead's ontology of the world and the mind, which, it is worth noting, involves neither probability nor complementarity. In quantum theory, the potentiality in question is, according to Stapp, defined by the formalism. If this potentiality is also assumed to be real, this reality is, as mathematically represented by the formalism, either physical, material, or ideal, Platonist. Bohr's view is incompatible with either possibility. There are no spreading waves in Bohr's interpretation, in which quantum waves have no physical significance, but only a symbolic one, representing probability distributions. The only continuity within Bohr's scheme of things is the mathematical continuity of the formalism, which relates to observed discrete phenomena in terms of probabilities and is not given the status of reality, material or mental, beyond that of individual thinking, possibly shared, but not anything preexisting this thinking on Platonist or Whiteheadian lines. Even if one assumes, as von Neumann (1932) and his followers do, that the formalism represents the ultimate nature of quantum reality undisturbed by measurement, this need not imply that this representation corresponds to any physical (or even mental) picture, such as that of propagating waves. Bohr, however, does not makes this assumption either. The difference is not that between discrete and continuous ontologies, but between a discrete ontology of quantum phenomena and the impossibility of any ontology applicable to quantum objects.

The resulting concept of atomicity, especially if one, beyond Bohr's view, sees quantum objects, defined strictly by experiment, rather than existing independently, is more radical than any other nonreductionist concept proposed in quantum theory. Consider, for example, Anthony J. Leggett's concept of macrorealism. According to Leggett, while QM is a correct theory dealing with nature at the microlevel, it may not be applicable at the macrolevel, insofar are macrolevel phenomena are not reducible to their microlevel quantum constitution, handled by QM (e.g., Leggett, 1988). Leaving aside the likelihood of Leggett's theory as such, macrorealism tells us little, if anything, about QM at a microlevel as concerns atomism or reductionism. Leggett's own position on this issue is not entirely clear, but he appears to allow for the atomism of quantum objects at the microlevel. Bohr's and more radically the present interpretation preclude the Democritean atomism and, with it, reductionism in considering the microscopic quantum objects, including elementary particles.

5 What is an Elementary Particle?[4]

As testified to by the persistent title, "What is an elementary particle?" used by, among others, Heisenberg (1989, pp. 71–88) and Weinberg (1996), "an elementary particle" has been and remains a problem to which only fragments of a possible solution could be offered. As must be apparent from the preceding discussion "a particle," in the first place, is a problem, too, a problem that underlies that of "an *elementary* particle." It is not my aim here do more than consider this problem as a problem. In contrast, however, to most approaches to this problem, which are realist in nature, I want to offer that based on the strong RWR view, extended to high-energy regimes and QFT.[5]

Low-energy quantum regimes already permit a conception of elementary particles as quantum objects. The same conception is also applicable in high-energy regimes, beginning with the circumstance that elementary particles of the same type, such as electrons or photons cannot be distinguished from each other, while these types themselves are rigorously distinguishable. In the present interpretation, a particle, again, only exists and is defined as a quantum object by a measurement. One cannot be certain that one encounters the same electron in an experiment designed to detect it after it was assumed to be emitted from a source even in QM regimes, although the probability that it would be a different electron is generally low and, as discussed earlier, the assumption that it is the same is only a statistically permissible idealization in the present view. In QFT regimes, speaking of the "same" electron detected in a given experiment involving several measurements loses its meaning altogether. Two electrons could be distinguished by a changeable property associated with them, such as their positions in space or time, momentums, energy, or the directions of spins, but, in the RWR view, only as properties manifested in measuring instruments and only at the time of measurement. Such properties are subject to the uncertainty relations and complementarity. It is possible to locate (and in the present view, establish) by measurement two different electrons, as quantum objects, in separate regions in space. It is not possible to distinguish them from each other on the basis of their mass, charge, or spin. In RWR-type interpretations, such properties, too, could only be associated with quantum objects by means of the corresponding effects observed in measuring instruments and are not attributable to these objects themselves. It is, however, possible to maintain both the indistinguishability of particles of the same

[4] This discussion is part follows Plotnitsky (2020).

[5] As a result, my engagement with literature on the subject is limited. Most sources cited here, such as Kuhlman (2020) and Ruetsche (2011), contains extensive bibliographies. Among the standard technical textbooks are Peskin and Schroeder (1995), Weinberg (2005), and Zee (2010).

type and the distinguishability of the types themselves in RWR-type interpretations because both features can be consistently defined by the corresponding sets of effects manifested in measuring instruments.

This definition is, thus, in accord with the assumption, defining RWR-type interpretations, that the *character* of elementary particles and their behavior, or of the reality thus idealized, is beyond representation or even conception, just as is the ultimate, RWR-type, reality itself, except that, in the present view, this reality is assumed to exist independently, while elementary particles are only assumed to exist, as quantum objects, in measurements. An elementary particle of a given type, say, an electron, is specified by a discrete set of possible phenomena or events (the same for all electrons), observable in measuring instruments in the experiments associated with particles of this type. Thus, an elementary particle can only exist as part of a composite system, consisting of this particle and the quantum part of a measuring instrument, which system has a registered effect upon the observable, classically-describable, part of this instrument. The elementary character of a particle is defined by the fact that there is no experiment that allows one to associate the corresponding effects on measuring instruments with more elementary individual quantum objects. Once such an experiment becomes conceivable or performed the status of a quantum object as an elementary particle could be challenged or disproven, as it happened when hadrons and mesons were discovered to be composed of quarks and gluons.

This concept of an elementary particle, defined in terms of such effects, does not imply that "elementary particles," even if never shown to be composite (as concerns such effects), are fundamental elementary constituents, "building blocks" or "atoms" of nature. This assumption is impossible in RWR-type interpretations, as is any assumption concerning this constitution, including applying the concepts of elementary or constituents. Nor, by the same token, is it possible to apply to elementary particles any concept of a particle, any more than any other concept, such as wave.

While most QFT conceptions of an elementary particle are transferred from QM to high-energy quantum regimes, they are insufficient in these regimes and need to be adjusted or supplemented by additional concepts, most commonly that of quantum field. The present approach follows this approach by defining the concept of quantum field in RWR terms. First, however, I shall explain why the concept of an elementary particle operative in QM is insufficient in high-energy regimes. This insufficiency arises in view of the following situation, not found in QM, to which the mathematical architecture of QFT responds. Suppose, in the case of quantum electrodynamics (QED), the simplest form of QFT, that one arranges for an emission of an electron, at a given high energy, from a source and then performs a measurement at a certain distance from that source, say, by placing a photographic plate there. The probability or, if we repeat the experiment with the same initial conditions, statistics of the outcomes would be properly predicted by QED. But what will be the outcome? The answer is not what our classical or even quantum–mechanical intuition would expect. This answer was a revolutionary discovery made by Dirac through his equation (Dirac, 1928).

Let us consider first what happens if one deals with a classical object analogous to an electron. A classical electron, say, a Lorentz electron, of a small finite radius, would be torn apart by the force of its negative electricity. This led to treating the electron mathematically as a dimensionless point, without giving it any physical structure. One can take as an example a small ball that hits a metal plate. The place of the collision could be predicted (ideally) exactly by classical mechanics, and we can repeat the experiment with the same outcome on an identical or even the same object. Regardless of where we place the plate, we always find the same object, when shielded from outside interferences.

If one considers an electron in the QM regime, it is, first of all, impossible, because of the uncertainty relations, to predict the place of collision exactly. A single emitted electron could, in principle, be found anywhere or not found at all. Nor can an emission of an electron be guaranteed. There is a small but nonzero probability that such a collision will not be observed or that the observed trace is not that of the emitted electron. Finally, as discussed above, assuming that one observes the same electron in two successive measurements is an idealization, statistically permissible in low-energy quantum regimes.

Once one moves to high-energy quantum phenomena, beginning with those governed by QED, the situation is still different, even radically different. One might find, in the corresponding region, not only an electron (or nothing), as in QM regimes, but also other particles: a positron, a photon, an electron–positron pair; that is, in RWR-type interpretations, one registers the events or phenomena (observed in measuring instruments) that we associate with such entities. QED predicts which among such events can occur and with what probability; and just as QM, QED, in RWR-type interpretations, does so without representing or, in the strong RWR view, allowing one to conceive of how these events come about. The Hilbert-space machinery becomes more complex, in the case of Dirac's equation making the wave function a four-component Hilbert-space vector, as opposed to a one-component or, if one considers spin, two-component one, as in QM. These four components represent the fact that Dirac's equation

$$(\beta mc^2 + \sum_{k=1}^{3} \alpha_k p_k c)\psi(x,t) = i\hbar \frac{\partial \psi(x,t)}{\partial t}$$

$$\alpha_i^2 = \beta^2 = I_4$$

I_4 is the identity matrix

$$\alpha_i \beta + \beta \alpha_i = 0$$
$$\alpha_i \alpha_j + \alpha_j \alpha_i = 0$$

is an equation for both the (free) electron and the (free) positron, including their spins, and they can transform into each other or other particles, such as photons, transformations that, in the RWR view, are only manifested by effects observed in measuring instruments. By the same token, one can no longer speak of the same electron, positron, and so forth as detected in two successive measurements, as is permissible, as a statistical idealization, in low-energy regimes.

Once one moves to still higher energies governed by QFT, the panoply of possible outcomes becomes much greater. The Hilbert spaces and operator algebras involved have still more complex structures, linked to the appropriate Lie groups and their representations, associated (when these representations are irreducible) with different elementary particles. In the case of QED, we only have electrons, positrons, and photons, single or paired; in QFT, depending how high the energy is, one can find any known and possibly as yet unknown elementary particle or combination. Although these transformations can only be handled probabilistically or statistically, they also have a complex ordering to them. In particular, they obey various symmetry principles, especially local symmetries, central to QFT, which led to discoveries of new particles, such as quarks and gluons inside the nucleus, and then various types of them, eventually establishing the standard model of particle physics. QED is an abelian gauge theory with the symmetry group U(1) and has one gauge field, with the photon being the gauge boson. The standard model is a non-abelian gauge theory governed by the tensor product of three symmetry groups $U(1) \otimes SU(2) \otimes SU(3)$ and broken symmetries, and it has twelve gauge bosons: the photon, three weak bosons, and eight gluons.

The concept of a relativistic quantum field responds to this situation. The concept was initially developed as a form of quantization of the electromagnetic field, necessary even in low energy quantum regimes. Not all quantum field theories are relativistic. For now, however, by a quantum field I shall only refer to a relativistic quantum field. The character and even the possibility of such a concept, especially as a physical concept, is a subject of seemingly interminable debates. While there is a strong general sense concerning the mathematics involved and while there is a large consensus, although not a uniform one, that a physical concept of quantum field is necessary, most such concepts are realist.[6] By contrast, I would like to suggest a *physical* concept of quantum field, defined by the strong RWR view, which is consistent with the mathematics of QFT and most currently available *mathematical* concepts of quantum field.

In this view, a quantum field is not a quantum object but a particular *mode* of the RWR-type reality, which, as any such mode, is assumed to exist independently and is manifested only by its effects on measuring instruments, via quantum objects, such as elementary particles. Thus, a quantum field is independent of measurement, while quantum objects are defined by a measurement. These effects are more multiple than those observed in low energy regimes. This multiplicity is defined by the fact that these effects correspond to elementary particles, to which a quantum field gives a rise, of various types, even in a single experiment, consisting of one or more successive measurements, with the first one performed on a given particle. The initial quantum object could also be a set of elementary particles of the same or different types, with a different such set, possibly consisting of entirely different types of particles, appearing in each new measurement. As a mode of the RWR-type reality, a quantum

[6] This assessment is confirmed by Kuhlmann's representative review (Kuhlmann, 2020).

field is, again, assumed to exist independently and is manifested by creating transforming effects associated with elementary particles created in the process, either invariant (as concerns a given particle type), such as those associated with mass, charge, or spin, or variable, associated with position, momentum, or energy. As concerns this association, there is no difference from low energy regimes; the difference is in what kind of effects are observed. These effects have a kind of multiplicity in high-energy regimes which they do not in low-energy regime. The multiplicities of types of elementary particles thus observed or created become progressively greater in higher-energy regimes. Quantum fields bring together the irreducibly unthinkable, discovered by QM, and the irreducibly multiple, discovered by QFT.

As discussed earlier, in considering two successive measurements, which register different outcomes, it is humanly natural, to assume that something "happened" or that there was a "change" in the physical reality responsible for these events between them. However, in any interpretation, we cannot give this happening a determined location in space and time, and in RWR-type interpretations, there is nothing we can say or even think about the character of this change, including as a "happening" or "change," apart from its effects. Nor can we assume (again, in any interpretation), as we can, ideally, in low-energy regimes, that we observe the same quantum objects in two successive measurements. For example, it is no longer possible to think of a single electron in the hydrogen atom, as the same electron detected by (and in RWR-type interpretations, defined) by different measurements. Each measurement detects a different electron. One could also speak of quantum fields in QM, but in a reduced form that preserves the particle identities: each photon always remains the same photon (or disappears), each electron the same electron (or disappears), and so forth. In this understanding of the concept, speaking, as is common, of the quantum field of a particle, say, an electron, entails new complexities. Mathematically, the formalism of, say, QED, as a quantum field theory, allows one to make predictions concerning the electron, which, mathematically, invited one to speak of the electron as a quantum field. Physically, in the present understanding of a quantum field, this only means that the RWR-type reality defining the quantum field considered in a given experiment, has strata that enables the corresponding measurements detecting electrons. It is not possible to separate these strata from those associated with the possibility of detecting a positron or a photon in the same experiment, because neither of these strata is observed in a measurement. It is possible, however, to experimentally specify quantum fields associated with fundamental forces and the corresponding types of particles, field bosons, electromagnetic (photons), weak W^{+}, W^{-}, and Z, or strong (gluons), or gravitation (gravitons), in the latter case, hypothetically, given that there is no theory of gravity as a form of QFT.

The concept of a quantum field thus defined is a physical rather than mathematical concept. It can, however, be associated with several versions of a mathematical concept, commonly also called a "quantum field," defined in terms of a predictive Hilbert space formalism, with a particular operator structure, enabling proper probabilistic predictions of the phenomena concerned. The operators enabling one to predict the probabilities for the "annihilation" of some particles and "creation" of others, that is, for the corresponding measurable quantities observed in measuring instruments, are called annihilation and creation operators or lowering and raising operators, commonly designated as $\hat{a}$ and $\hat{a}^{\dagger}$, each lowering or increasing a number of particles in a given state by one. In RWR-type interpretations, these operators do not represent any physical reality: they only enable one to calculate the probabilities or statistics of the outcomes of experiments, just as the wave functions do in QM. But those provided by QFT also relate to the appearance of quantities associated with other types of particles even in experiments initially registering a particle of a given type. In QFT regimes it is meaningless to speak of a single electron even in the hydrogen atom. QFT correctly predicts these effects, probabilistically or statistically, fully in accord with what is observed. QED is the best confirmed physical theory in existence.

6 Conclusion

The argument offered in this article does not imply that it *is impossible* to have a realist view, of either reductionist or holistic type, at all levels of the physical reality considered in quantum physics, but only that it *may not be possible*. On the other hand, the RWR view, which precludes realism and especially reductionism, is consistent with quantum phenomena and QM or QFT, at least as things stand now. Equally importantly, the RWR view, even the strong one, does not preclude thinking and knowledge from advancing. The advancement of quantum physics has always allowed for this view. As Bohr noted:

> [This type of] argumentation does of course not imply that, in atomic physics, we have no more to learn as regards experimental evidence and the mathematical tools appropriate to its comprehension. In fact, it seems likely that the introduction of still further abstractions into the formalism will be required to account for the novel features revealed by the exploration of atomic processes of very high energy. (Bohr, 1987, v. 3, p. 6)

The history of high-energy physics and QFT has amply confirmed this assessment, made in 1958, without, thus far, contradicting the RWR view. What changes, if one adopts this view, is the character of thinking and knowledge. They now include the assumption that there is something that is beyond thought that is, at the same time, responsible for what we can think and know.

References

Bell, J. S. (2004). *Speakable and unspeakable in quantum mechanics*. Cambridge University Press.

Bohr, N. (1913). On the constitution of atoms and molecules (Part 1). *Philosophical Magazine, 26*(151), 1–25.

Bohr, N. (1935). Can quantum-mechanical description of physical reality be considered complete? *Physical Review, 48*, 696–702.

Bohr, N. (1937). Causality and complementarity. In J. Faye, & H. J. Folse, (Eds.) *The philosophical writings of Niels Bohr, volume 4: Causality and complementarity, supplementary papers* (pp. 83–91). Ox Bow Press, 1994.

Bohr, N. (1938) The causality problem in atomic physics. In J. Faye, & H. J. Folse, (Eds.) *The philosophical writings of Niels Bohr, volume 4: Causality and complementarity, supplementary papers* (pp. 94–121). Ox Bow Press, Woodbridge, CT, 1987.

Bohr, N. (1987). *The philosophical writings of Niels Bohr* (Vol. 3). Ox Bow Press.

Dirac, P. A. M. (1928). The quantum theory of the electron. *Proceedings of the Royal Society of London A, 177*, 610–624.

Dirac, P. A. M. (1958). *The principles of quantum mechanics* (4th ed.). Clarendon, rpt. 1995.

Heisenberg, W. (1962). *Physics and philosophy: The revolution in modern science.* Harper & Row.

Heisenberg, W. (1989). *Encounters with Einstein, and other essays on people, places, and particles.* Princeton University Press.

Heisenberg, W. (1930). *The physical principles of the quantum theory* (K. Eckhart, F. C. Hoyt, Trans.). Dover, rpt. 1949.

Kant, I. (1997). *Critique of pure reason* (P. Guyer, A. W. Wood, Trans.). Cambridge University Press.

Kuhlmann, M.: Quantum field theory. In: E. N. Zalta (Ed.), Stanford encyclopedia of philosophy. https://plato.stanford.edu/entries/quantum-field-theory/

Ladyman, J. (2016). Structural realism. In: E. N. Zalta (Ed.) *The Stanford encyclopedia of philosophy.* https://plato.stanford.edu/archives/win2016/entries/structural-realism/

Leggett, A. J. (1988). Experimental approaches to the quantum measurement paradox. *Foundations of Physics, 18*, 939–952. https://doi.org/10.1007/BF01855943

Mehra, J., & Rechenberg, H. (2001). *The historical development of quantum theory* (Vol. 6). Springer.

Peskin, M. E., & Schroeder, D. V. (1995). *An introduction to quantum field theory.* CRS Press.

Plotnitsky, A. (2012). *Niels Bohr and complementarity: An introduction.* Springer.

Plotnitsky, A. (2016). *The principles of quantum theory, from Planck's Quanta to the Higgs Boson: The nature of quantum reality and the spirit of Copenhagen.* Springer/Nature.

Plotnitsky, A. (2021). *Reality without Realism: Matter, Thought, and Technology in Quantum Physics.* Springer/Nature.

Plotnitsky, A., & Khrennikov, A. (2015). Reality without realism: On the ontological and epistemological architecture of quantum mechanics. *Foundations of Physics, 25*(10), 1269–1300.

Ruetsche, L. (2011). *Interpreting quantum theories.* Oxford University Press.

Stapp, H. (2011). *Mindful universe: Quantum mechanics and participating observer.* Springer.

Von Neumann, J. (1932). *Mathematical foundations of quantum mechanics* (trans. R. T. Beyer). Princeton University Press, rpt. 1983.

Weinberg, S. (1996). *What is an elementary particle?* http://www.slac.stanford.edu/pubs/beamline/27/1/27-1-weinberg.pdf

Weinberg, S. (2005). *The quantum theory of fields, volume 1: Foundations.* Cambridge University Press.

Wheeler, J. A. (1990). Information, physics, quantum: the search for links. In: W. H. Zurek, (Ed.), *Complexity, entropy, and the physics of information.* Addison-Wesley.

Whitehead, A. N. (1978). *Process and reality: An essay in cosmology.* Free Press.

Zee, A. (2010). *Quantum field theory in a nutshell.* Princeton University Press.

Geometric and Exotic Contextuality in Quantum Reality

Michel Planat

1 Introduction

What is quantum reality? Quoting Niels Bohr: *We are suspended in language in such a way that we cannot say what is up and what is down. The world "reality" is also a word, a word which we must learn to use correctly* (Bohr, 1997). Today, the words 'quantum holism' are often used to qualify the inseparability of distant quantum objects known as quantum entanglement or quantum non-locality (Esfeld, 1999; Ferrero et al., 2004; Miller, 2014). The concept of 'quantum contextuality' seems to be more appropriate because it is used to describe our objective experience of quantum measurements. In a contextual world, the measured value of an observable depends on which other mutually compatible measurements might be performed and cannot simply be thought as revealing a pre-existing value. It is not only that the whole supersedes the parts but that the observer interprets the quantum world with his available sensors and words. Quantum contextuality is able to feature counter-intuitive aspects of the quantum language and is now considered as more general than quantum entanglement and quantum non-locality (at least when one refers to Bell's theorem).

In this line of thought, the Bell-Kochen-Specker theorem (BKS) is able to rule out non-contextual hidden variable theories by resorting to mathematical statements about coloring of rays located on maximal orthonormal bases in a d-dimensional Hilbert space (with d at least 3) (Peres, 1993; Quantum contextuality, 2021). A very transparent 'proof' of the BKS theorem makes use of 18 rays and 9 maximal orthonormal bases of two qubits (i.e. in the 4-dimensional Hilbert space) (Cabello et al., 1996). This topic will be described in some details in Sect. 2.

In the past few years, the author developed a group theoretical approach of quantum contextuality that he called 'geometric contextuality'. The idea is to take seri-

M. Planat (✉)
Université de Bourgogne/Franche-Comté, Institut FEMTO-ST CNRS UMR 6174, 15 B Avenue des Montboucons, 25044 Besançon, France
e-mail: michel.planat@femto-st.fr

S. Wuppuluri and I. Stewart (eds.), *From Electrons to Elephants and Elections*,
The Frontiers Collection, https://doi.org/10.1007/978-3-030-92192-7_26

ously Bohr's suggestion that quantum theory is a language. Most of the time, words in this language only need two letters and the theory resorts to the so-called 'dessins d'enfants' of Grothendieck (Planat, 2015, 2016; Planat and Zainuddin, 2017). This topic is developed in Sect. 3 by restricting to the case of two qubits in order to keep the technicalities simple enough.

Then, in Sect 4, the topic 'exotic contextuality' offers an opportunity to reintroduce a four-dimensional space-time in our interpretation of the quantum world. Our objects are four-manifolds. Quantum measurements may be seen as taking place in 'parallel' worlds/contexts that mathematically are homeomorphic but non diffeomorphic to each other (Planat, 2020). This idea looks like the many worlds interpretation of quantum mechanics (DeWitt, 1970) while being different in the mathematical approach.

2 A Glance at Two-Qubit Parity Proofs of the BKS Theorem

A parity proof of BKS theorem is a set of v rays that form l bases (l odd) such that each ray occurs an even number of times over these bases. A proof of BKS theorem is critical if it cannot be further simplified by deleting even a single ray or a single basis. The smallest BKS proof in dimension 4 is a parity proof and corresponds to arrangements of real observables arising from the two-qubit Pauli group, more specifically as eigenstates of two-qubit operators forming a (3×3)-grid (also known as a Mermin's square) as follows

$$\begin{array}{ccc} | & | & || \\ IX- & XI- & XX- \\ | & | & || \\ ZX- & XZ- & YY- \\ | & | & || \\ ZI- & IZ- & ZZ- \\ | & | & || \end{array} \qquad (1)$$

where I is the two-dimensional identity matrix, X, Y and Z are the Pauli spin matrices, and the operator products are Kronecker products.

The simplification of arguments in favour of a contextual view of quantum measurements started with Peres' note (1993) and Mermin's report (1993). Observe that in (1), the three operators in each row and each column mutually commute and their product is the identity matrix, except for the right hand side column whose product is minus the identity matrix. There is no way of assigning multiplicative properties to the eigenvalues ± 1 of the nine operators while still keeping the same multiplicative properties for the operators. Paraphrasing (Peres, 1993), the result of a measurement depends "in a way not understood, on the choice of other quantum measurements, that may possibly be performed". Mermin's 'proof' of the BKS theorem stated in terms of two-qubit observables can now be reformulated in terms of rays and maximal bases.

We shall employ a signature of the proofs in terms of the distance D_{ab} between two orthonormal bases a and b defined as (Planat, 2012)

$$D_{ab}^2 = 1 - \frac{1}{d-1} \sum_{i,j}^{d} \left(|\langle a_i | b_j \rangle|^2 - \frac{1}{d} \right)^2 . \tag{2}$$

The distance (2) vanishes when the bases are the same and is maximal (equal to unity) when the two bases a and b are mutually unbiased, $|\langle a_i | b_j \rangle|^2 = 1/d$, and only then. We shall see that the bases of a BKS proof employ a selected set of distances which happens to be a universal feature of the proof.

Using the list of the unnormalized eigenvectors (numbered consecutively)

$$\begin{array}{l} 1:[1000],\ 2:[0100],\ 3:[0010],\ 4:[0001],\ 5:[1111],\ 6:[11\bar{1}\bar{1}] \\ 7:[1\bar{1}1\bar{1}],\ 8:[1\bar{1}\bar{1}1],\ 9:[1\bar{1}\bar{1}\bar{1}],\ 10:[1\bar{1}11],\ 11:[11\bar{1}1],\ 12:[111\bar{1}] \\ 13:[1100],\ 14:[1\bar{1}00],\ 15:[0011],\ 16:[001\bar{1}],\ 17:[0101],\ 18:[010\bar{1}] \\ 19;[1010],\ 20:[10\bar{1}0],\ 21:[100\bar{1}],\ 22:[1001],\ 23:[01\bar{1}0],\ 24:[0110] \end{array} \tag{3}$$

one gets 24 complete orthogonal bases are as follows

$$\begin{array}{l} \mathbf{1}:\{1,2,3,4\},\mathbf{2}:\{5,6,7,8\},\mathbf{3}:\{9,10,11,12\},\mathbf{4}:\{13,14,15,16\}, \\ \mathbf{5}:\{17,18,19,20\},\mathbf{6}:\{21,22,23,24\},\mathbf{7}:\{1,2,15,16\},\mathbf{8}:\{1,3,17,18\}, \\ \mathbf{9}:\{1,4,23,24\},\mathbf{10}:\{2,3,21,22\},\mathbf{11}:\{2,4,19,20\},\mathbf{12}:\{3,4,13,14\}, \\ \mathbf{13}:\{5,6,14,16\},\mathbf{14}:\{5,7,18,20\},\mathbf{15}:\{5,8,21,23\},\mathbf{16}:\{6,7,22,24\}, \\ \mathbf{17}:\{6,8,17,19\},\mathbf{18}:\{7,8,13,15\},\mathbf{19}:\{9,10,13,16\},\mathbf{20}:\{9,11,18,19\}, \\ \mathbf{21}:\{9,12,22,23\},\mathbf{22}:\{10,11,21,24\},\mathbf{23}:\{10,12,17,20\},\mathbf{24}:\{11,12,14,15\}. \end{array} \tag{4}$$

Then, by normalizing rays, one obtains a finite set of distances between the 24 bases

$$\begin{aligned} D = \{a_1, a_2, a_3, a_4, a_5\} &= \{\sqrt{1/3}, \sqrt{7/12}, \sqrt{2/3}, \sqrt{5/6}, 1\} \\ &\approx \{0.58, 0.76, 0.82, 0.91, 1.000\}. \end{aligned}$$

Table 1 provides a histogram of distances for various parity proofs $v - l$.

The table reveals that there exist four main types of parity proofs arising from the 24 rays, that are of the type $18 - 9$, $20 - 11$, $22 - 13$ and $24 - 15$. Types $20 - 11$

Table 1 The histogram of distances for various parity proofs $v - l$ obtained from Mermin's square

Proof $v - l$	# Proofs	a_1	a_2	a_3	a_4	a_5
24-15	16	18	18	9	54	6
22-13A	96	12	18	3	42	3
22-13B	144	12	18	4	42	2
20-11A	96	6	18	0	30	1
20-11B	144	6	18	1	30	0
18-9	16	0	18	0	18	0

and $22 - 13$ subdivide into two non-isomorphic ones A and B as shown in Table 1 (Planat, 2012; Pavičić et al., 2005; Waegell and Aravind, 2011).

The 16 proofs of the $18 - 9$ type can be displayed as the 4×4 square (5) in which two adjacent proofs share three bases. Observe that each 2×2 square of adjacent proofs has the same shared base, which is taken as an index (e.g. the upper left-hand-side 2×2 square has index 7 and the lower right-hand-side square has index 10). All four indices in each row and in each column correspond to four disjoint bases that together partition the 24 rays.

$$\begin{array}{ccccccccc}
\begin{pmatrix} 7 & 8 & 10 \\ 13 & 14 & 16 \\ 22 & 23 & 24 \end{pmatrix} & - & \begin{pmatrix} 7 & 9 & 11 \\ 14 & 15 & 18 \\ 19 & 20 & 22 \end{pmatrix} & - & \begin{pmatrix} 8 & 9 & 12 \\ 16 & 17 & 18 \\ 20 & 21 & 24 \end{pmatrix} & - & \begin{pmatrix} 10 & 11 & 12 \\ 13 & 15 & 17 \\ 19 & 21 & 23 \end{pmatrix} & - \\
|_{7} & & |_{20} & & |_{12} & & |_{23} & \\
\begin{pmatrix} 7 & 9 & 11 \\ 16 & 17 & 18 \\ 19 & 21 & 23 \end{pmatrix} & - & \begin{pmatrix} 7 & 8 & 10 \\ 13 & 15 & 17 \\ 20 & 21 & 24 \end{pmatrix} & - & \begin{pmatrix} 10 & 11 & 12 \\ 13 & 14 & 16 \\ 19 & 20 & 22 \end{pmatrix} & - & \begin{pmatrix} 8 & 9 & 12 \\ 14 & 15 & 18 \\ 22 & 23 & 24 \end{pmatrix} & - \\
|_{17} & & |_{10} & & |_{14} & & |_{9} & \\
\begin{pmatrix} 8 & 9 & 12 \\ 13 & 15 & 17 \\ 19 & 20 & 22 \end{pmatrix} & - & \begin{pmatrix} 10 & 11 & 12 \\ 16 & 17 & 18 \\ 22 & 23 & 24 \end{pmatrix} & - & \begin{pmatrix} 7 & 8 & 10 \\ 14 & 15 & 18 \\ 19 & 21 & 23 \end{pmatrix} & - & \begin{pmatrix} 7 & 9 & 11 \\ 13 & 14 & 16 \\ 20 & 21 & 24 \end{pmatrix} & - \\
|_{12} & & |_{23} & & |_{7} & & |_{20} & \\
\begin{pmatrix} 10 & 11 & 12 \\ 14 & 15 & 18 \\ 20 & 21 & 24 \end{pmatrix} & - & \begin{pmatrix} 8 & 9 & 12 \\ 13 & 14 & 16 \\ 19 & 21 & 23 \end{pmatrix} & - & \begin{pmatrix} 7 & 9 & 11 \\ 13 & 15 & 17 \\ 22 & 23 & 24 \end{pmatrix} & - & \begin{pmatrix} 7 & 8 & 10 \\ 16 & 17 & 18 \\ 19 & 20 & 22 \end{pmatrix} & - \\
|_{14} & & |_{9} & & |_{17} & & |_{10} &
\end{array} \tag{5}$$

Diagrams for the proofs How can we account for the distance signature of a given proof? A simple diagram does the job.

The diagram for the $18 - 9$ proof is simply a 3×3 square. Below we give an explicit construction of the first proof that corresponds to the upper left-hand-side corner in (5). The 9 vertices of the graph are the 9 bases of the proof, the one-point crossing graph between the bases is the graph (6), with aut $= G_{72} = \mathbb{Z}_3^2 \rtimes D_4$. There are 9 (distinct) edges that encode the 18 rays, a selected vertex/base of the graph is encoded by the union of the four edges/rays that are adjacent to it.

$$\begin{array}{cccccc}
\begin{pmatrix}1&2\\15&16\end{pmatrix} & -1- & \begin{pmatrix}1&3\\17&18\end{pmatrix} & -3- & \begin{pmatrix}2&3\\21&22\end{pmatrix} & -2 \\
|16 & & |18 & & |22 & \\
\begin{pmatrix}5&6\\14&16\end{pmatrix} & -5- & \begin{pmatrix}5&7\\18&20\end{pmatrix} & -7- & \begin{pmatrix}6&7\\22&24\end{pmatrix} & -6 \\
|14 & & |20 & & |24 & \\
\begin{pmatrix}11&12\\14&15\end{pmatrix} & -12- & \begin{pmatrix}10&12\\17&20\end{pmatrix} & -10- & \begin{pmatrix}10&11\\21&24\end{pmatrix} & -11 \\
|15 & & |17 & & |21 &
\end{array} \tag{6}$$

As for the distances between the bases, two bases located in the same row (or the same column) have distance $a_2 = \sqrt{7/12}$, while two bases not in the same row (or column) have distance $a_4 = \sqrt{5/6} > a_2$, as readily discernible from Table 2 and the histogram in Table 1. Indeed, any proof of the $18 - 9$ type has the same diagram as (6).

Similar diagrams can be drawn to reflect the histogram of distances in proofs of a larger size. Below we restrict to the case of a $20 - 11A$ proof (where only the distance between two bases is made explicit, but not the common rays of the bases)

$$\begin{array}{ccccccc}
\begin{pmatrix}10&12\\17&20\end{pmatrix} & -a_2- & \begin{pmatrix}11&12\\14&15\end{pmatrix} & -a_2- & \begin{pmatrix}10&11\\21&24\end{pmatrix} & ...a_4 = \sqrt{5/6}... & \\
|a_2 = \sqrt{7/12} & & |a_2 & & |a_2 & & ... \\
\begin{pmatrix}1&3\\17&18\end{pmatrix} & -a_2- & \begin{pmatrix}1&2\\15&16\end{pmatrix} & -a_2- & \begin{pmatrix}1&4\\23&24\end{pmatrix} & ..a_1 = \frac{1}{\sqrt{3}}.. & \begin{pmatrix}1&2\\3&4\end{pmatrix} \\
|a_2 & & |a_2 & & |a_2 & & |a_5 = 1 \\
\begin{pmatrix}5&7\\18&20\end{pmatrix} & -a_2- & \begin{pmatrix}5&6\\14&16\end{pmatrix} & -a_2- & \begin{pmatrix}5&8\\21&23\end{pmatrix} & ..a_1 = \frac{1}{\sqrt{3}}.. & \begin{pmatrix}5&6\\7&8\end{pmatrix} \\
|a_2 & & |a_2 & & |a_2 & & ...
\end{array} \tag{7}$$

The proof consists of 11 bases, 9 of them have the same mutual diagram as in (6) and their mutual distance is $a_2 = \sqrt{7/12}$ (as shown) or $a_4 = \sqrt{5/6}$ (not shown), depending on whether they are located in the same row (or the same column) of the 3×3 square, or not. The extra two bases of the right-hand-side column are mutually unbiased (with distance $a_5 = 1$), their distance to any base of the same row is $1/\sqrt{3}$ and their distance to any base of the first row is a_4 (as shown).

3 Geometric Contextuality

Interpreting quantum theory is a long standing effort and not a single approach can exhaust all facets of this fascinating subject. Quantum information owes much to the concept of a (generalized) Pauli group for understanding quantum observables, their commutation, entanglement, contextuality and many other aspects, e.g. quantum computing. Quite recently, it has been shown that quantum commutation relies on some finite geometries such as generalized polygons and polar spaces (Planat, 2011). Finite geometries connect to the classification of simple groups as understood by prominent researchers as Jacques Tits, Cohen Thas and many others (Planat and Zainuddin, 2017; Thas et al., 2004).

In the Atlas of finite group representations (Wilson et al., 2015), one starts with a free group G with relations, then the finite group under investigation P is the permutation representation of the cosets of a subgroup of finite index d of G (obtained thanks to the Todd-Coxeter algorithm). As a way of illustrating this topic, one can refer to (Planat and Zainuddin, 2017, Table 3) to observe that a certain subgroup of index 15 of the symplectic group $S_4'(2)$ corresponds to the $2QB$ (two-qubit) commutation of the 15 observables in terms of the generalized quadrangle of order two, denoted $GQ(2, 2)$ (alias the doily). For $3QB$, a subgroup of index 63 in the symplectic group $S_6(2)$ does the job and the commutation relies on the symplectic polar space $W_5(2)$ (Planat and Zainuddin, 2017, Table 7). An alternative way to approach $3QB$ commutation is in terms of the generalized hexagon $GH(2, 2)$ (or its dual) which occurs from a subgroup of index 63 in the unitary group $U_3(3)$ (Planat and Zainuddin, 2017, Table 8). Similar geometries can be defined for multiple qudits (instead of qubits).

The straightforward relationship of quantum commutation to the appropriate symmetries and finite groups was made possible thanks to techniques that we briefly summarize.

3.1 Finite Geometries from Cosets (Planat, 2015; Planat and Zainuddin, 2017; Planat et al., 2015)

Let H be a subgroup of index d of a free group G with generators and relations. A coset table over the subgroup H is built by means of a Coxeter-Todd algorithm. Given the coset table, on builds a permutation group P that is the image of G given by its action on the cosets of H. In this paper, the software Magma (Bosma, 2019) is used to perform these operations.

One needs to define the rank r of the permutation group P. First one asks that the d-letter group P acts faithfully and transitively on the set $\Omega = \{1, 2, \ldots, d\}$. The action of P on a pair of distinct elements of Ω is defined as $(\alpha, \beta)^p = (\alpha^p, \beta^p)$, $p \in P$, $\alpha \neq \beta$. The orbits of P on the product set $\Omega \times \Omega$ are called orbitals. The number of orbits is called the rank r of P on Ω. Such a rank of P is at least two, and it also known that two-transitive groups may be identified to rank two permutation groups.

One selects a pair $(\alpha, \beta) \in \Omega \times \Omega, \alpha \neq \beta$ and one introduces the two-point stabilizer subgroup $P_{(\alpha,\beta)} = \{p \in P | (\alpha, \beta)^p = (\alpha, \beta)\}$. There are $1 < m \leq r$ such non-isomorphic (two-point stabilizer) subgroups of P. Selecting one of them with $\alpha \neq \beta$, one defines a point/line incidence geometry $\mathcal{G}$ whose points are the elements of the

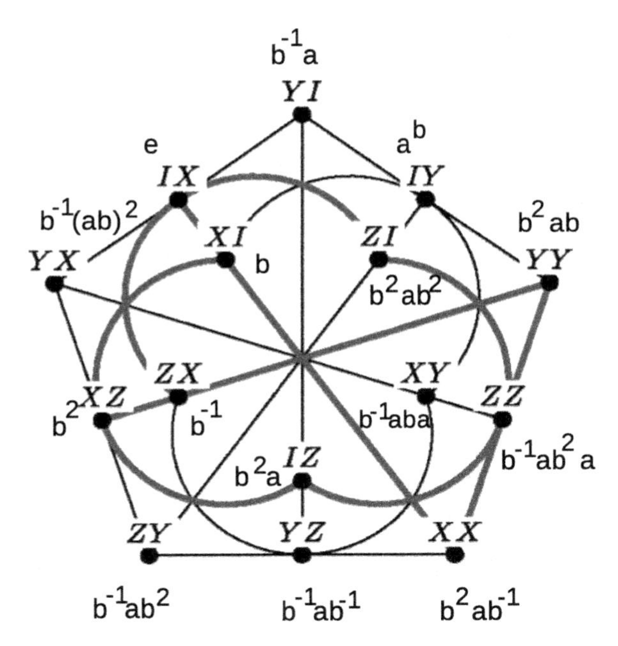

Fig. 1 The generalized quadrangle of order two $GQ(2, 2)$. The picture provides a representation in terms of the fifteen $2QB$ observables that are commuting by triples: the lines of the geometry. Bold lines are for an embedded 3×3 grid (also called Mermin square) that is a basic model of Kochen-Specker theorem (e.g. Planat and Zainuddin, 2017, Fig. 1 or (Planat, 2012)). The second representation is in terms of the cosets of the permutation group arising from the index 15 subgroup of $G \cong A_6$ (the 6-letter alternating group)

set Ω and whose lines are defined by the subsets of Ω that share the same two-point stabilizer subgroup. Two lines of $\mathcal{G}$ are distinguished by their (isomorphic) stabilizers acting on distinct subsets of Ω. A non-trivial geometry is obtained from P as soon as the rank of the representation $\mathcal{P}$ of P is $r > 2$, and at the same time, the number of non isomorphic two-point stabilizers of $\mathcal{P}$ is $m > 2$. Further, $\mathcal{G}$ is said to be *contextual* (shows *geometrical contextuality*) if at least one of its lines/edges is such that a set/pair of vertices is encoded by non-commuting cosets (Planat, 2015).

Figure 1 illustrates the application of the two-point stabilizer subgroup approach just described for the index 15 subgroup of the symplectic group is $S_4'(2) = A_6$ whose finite representation is
$H = \langle a, b | a^2 = b^4 = (ab)^5 = (ab^2)^5 = 1 \rangle$. The finite geometry organizing the coset representatives is the generalized quadrangle $GQ(2, 2)$. The other representation is in terms of the two-qubit Pauli operators, as first found in (Planat, 2011; Saniga and Planat, 2007). It is easy to check that all lines not passing through the coset e contains some mutually not commuting cosets so that the $GQ(2, 2)$ geometry is contextual. The embedded (3×3)-grid shown in bold (the so-called Mermin square) allows a $2QB$ proof of Kochen-Specker theorem (Planat, 2012).

3.2 The Kochen-Specker Theorem with a Mermin Square of Two-Qubit Observables

Let us show how to recover the geometry of the Mermin square, i.e. the (3×3) grid embedded in the generalized $GQ(2, 2)$ of Fig. 1. Recall that it is the basic model of two-qubit contextuality (Planat and Zainuddin, 2017, Fig. 1) (Planat, 2012). One starts with the free group $G = \langle a, b | b^2 \rangle$ and one makes use of the mathematical software Magma (Bosma, 2019). Then one derives the (unique) subgroup H of G that is of index nine and possesses a permutation representation P isomorphic to the finite group $\mathbb{Z}_3^2 \times \mathbb{Z}_2^2$ reflecting the symmetry of the grid. The permutation representation is as follows:

$$P = \langle 9 | (1, 2, 4, 8, 7, 3)(5, 9, 6), (2, 5)(3, 6)(4, 7)(8, 9) \rangle ,$$

where the list [1, ..., 9] means the list of coset representatives

$$[e, a, a^{-1}, a^2, ab, a^{-1}b, a^{-2}, a^3, aba].$$

The permutation representation P can be seen on a torus as in Fig. 2i.

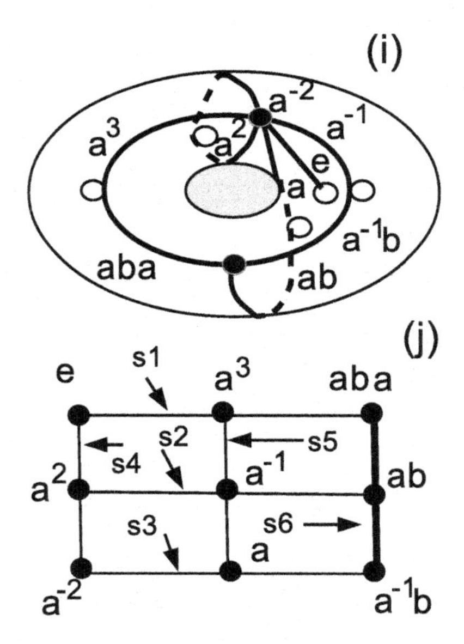

Fig. 2 The map (**i**) leading to Mermin's square (**j**). The two-point stabilizer subgroups of the permutation representation P corresponding to the dessin (one for each line) are as follows: $s_1 = (2, 3)(4, 7)(5, 6), s_2 = (1, 7)(2, 8)(6, 9), s_3 = (1, 4)(3, 8)(5, 9), s_4 = (2, 6)(3, 5)(8, 9), s_5 = (1, 9)(4, 5)(6, 7), s_6 = (1, 8)(2, 7)(3, 4)$, where the points of the square (resp. the edges of the dessin d'enfant) are labeled as $[1, .., 9] = [e, a, a^{-1}, a^2, ab, a^{-1}b, a^{-2}, a^3, aba]$

Next, we apply the procedure described at the top of this subsection. There are two types of two-point stabilizer subgroups isomorphic to the single element group $\mathbb{Z}_1$ or to the two-element group $\mathbb{Z}_2$. Both define the geometry of a (3×3) grid comprising six lines identified, by their non-identical, but isomorphic two-point stabilizers s_1 to s_6, made explicit in the caption of Fig. 2. The first grid (not shown) is considered non-contextual in the sense that the cosets on a line are commuting. The second grid, shown in Fig. 2j, is contextual in the sense that the right column does not have all its triples of cosets mutually commuting. The non-commuting cosets on this line reflect the contextuality that occurs when one takes two-qubit coordinates for the points of the grid, see (Planat, 2015) for more details about the relationship between non-commuting cosets and geometric contextuality.

4 Exotic Contextuality

We already approached the topic of quantum contextuality (QC) in two ways. In Sect. 2, we found how the 3×3 grid (or Mermin square) can be considered as a building block of QC by proving the BKS theorem, either at level of two-qubit operators that parametrize the grid or at the level of rays that correspond to eigenstates attached to the operators of the grid. In Sect. 3, a group theoretical language with two-letter words was found to nicely mimic QC in the Mermin square and its embedding generalized quadrangle $GQ(2, 2)$—the locus of of the two-qubit Pauli group. In such an approach, geometric contextuality corresponds to QC. Now, we jump to a possible interpretation of this language by seeing the QC-geometries as creatures of exotic four-manifolds that one may identify to our familiar space-time (Planat, 2020).

We introduce the concept of exotic contextuality for such an interpretation. Moreover, such a type of contextuality is related to a model of quantum computing based on magic states that we developed in a series of papers (Planat and Gedik, 2017; Planat et al., 2018, 2019). In quantum information theory, the two-qubit configuration and its properties: quantum entanglement and quantum contextuality have been discussed at length as prototypes of peculiarities or resources in the quantum world. Our Sect. 3.2 mainly featured the quantum contextuality of two-qubit systems. Our model of quantum computing is based on the concept of a magic state—a state that has to be added to the eigenstates of the d-dimensional Pauli group- in order to allow universal quantum computation. This was started by Bravyi & Kitaev in 2005 (Bravyi and Kitaev, 2005) for qubits ($d = 2$). A subset of magic states consists of states associated to minimal informationally complete measurements, that we called MIC states (Planat and Gedik, 2017). We require that magic states should be MIC states as well. For getting the candidate MIC states, one uses the fact that a permutation may be realized as a permutation matrix/gate and that mutually commuting matrices share eigenstates. They are either of the stabilizer type (as elements of the Pauli group) or of the magic type. One keeps magic states that are MIC states in order to preserve a complete information during the computation and measurements.

A further step in our quantum computing model was to introduce a 3-dimensional manifold M^3 whose fundamental group $G = \pi_1(M^3)$ would be the source of MIC states (Planat et al., 2018, 2019). Recall that G is a free group with relations and that a d-dimensional MIC state may be obtained from the permutation group that organizes the cosets of an appropriate subgroup of index d of G.

It was considered by us quite remarkable that two group geometrical axioms very often govern the MIC states of interest (Planat et al., 2019), viz (i) the normal (or conjugate) closure $\{g^{-1}hg | g \in G \text{ and } h \in H\}$ of the appropriate subgroup H of G equals G itself and (ii) there is no geometry (a triple of cosets do not produce equal pairwise stabilizer subgroups). See (Planat et al., 2019, Sect. 1.1) for our method of building a finite geometry from coset classes. But these rules had to be modified by allowing either the simultaneous falsification of (i) and (ii) or by tolerating a few exceptions. If it happens that (ii) is violated, one gets geometric contextuality, the parallel to quantum contextuality (Planat, 2015) that one featured in Sect. 3.

It is known that there exist infinitely many 4-manifolds that are homeomorphic but non diffeomorphic to each other (Akbulut, 1991a, 2016; Gompf and Stipsicz, 1999; Scorpian, 2011). They can be seen as distinct copies of space-time not identifiable to the ordinary Euclidean space-time. A cornerstone of our approach is an 'exotic' 4-manifold called an Akbulut cork W that is contractible, compact and smooth, but not diffeomorphic to the 4-ball (Akbulut, 1991a). In our approach, we do not need the full toolkit of 4-manifolds since we are focusing on W and its neighboors only. All what we need is to understand the handlebody decomposition of a 4-manifold, the fundamental group $\pi_1(\partial W)$ of the 3-dimensional boundary ∂W of W, and related fundamental groups. Following the methodology of our previous work (Planat and Gedik, 2017; Planat et al., 2018), the subgroup structure of such π_1's corresponds to the Hilbert spaces of interest. Our view is close to the many-worlds interpretation of quantum mechanics where all possible outcomes of quantum measurements are realized in some 'world' and are objectively real (DeWitt, 1970). One arrives at a many-manifolds view of quantum computing -reminiscent of the many-worlds- where the many-manifolds are in an exotic class and can be seen as many-quantum generalized measurements, the latter being POVM's (positive operator valued measures).

4.1 Excerpts on the Theory of 4-manifolds and Exotic R^4's

Handlebody of a 4-manifold. Let us introduce some excerpts of the theory of 4-manifolds needed for our paper (Akbulut, 2016; Gompf and Stipsicz, 1999; Scorpian, 2011). It concerns the decomposition of a 4-manifold into one- and two-dimensional handles as shown in Fig. 3 (Akbulut, 2016, Figs. 1.1 and 1.2). Let B^n and S^n be the n-dimensional ball and the n-dimensional sphere, respectively. An observer is placed at the boundary $\partial B^4 = S^3$ of the 0-handle B^4 and watch the attaching regions of the 1- and 2-handles. The attaching region of 1-handle is a pair of balls B^3 (the yellow balls), and the attaching region of 2-handles is a framed knot (the red knotted circle) or a knot going over the 1-handle (shown in blue). Notice that the 2-handles are

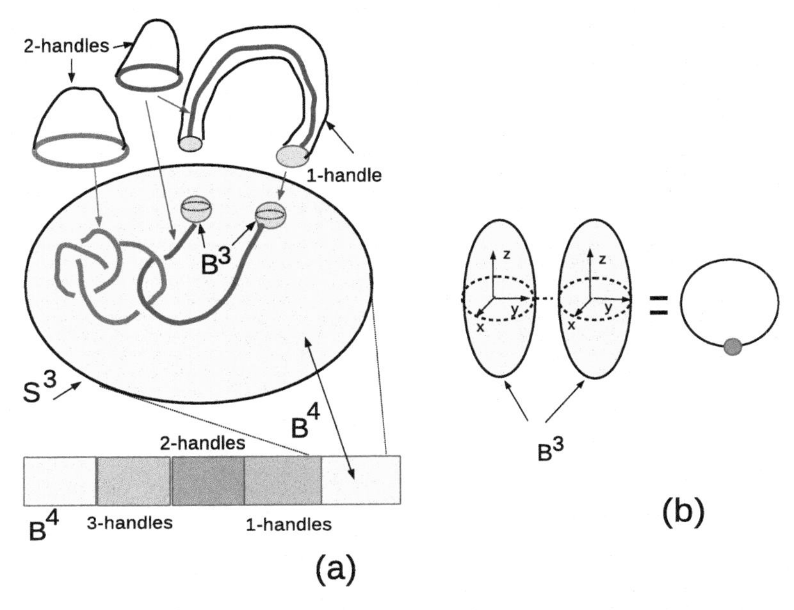

Fig. 3 **a** Handlebody of a 4-manifold with the structure of 1- and 2-handles over the 0-handle B^4, **b** the structure of a 1-handle as a dotted circle $S^1 \times B^3$

attached after the 1-handles. For closed 4-manifolds, there is no need of visualizing a 3-handle since it can be directly attached to the 0-handle. The 1-handle can also be figured out as a dotted circle $S^1 \times B^3$ obtained by squeezing together the two three-dimensional balls B^3 so that they become flat and close together (Gompf and Stipsicz, 1999, p. 169) as shown in Fig. 3b. For the attaching region of a 2- and a 3-handle one needs to enrich our knowledge by introducing the concept of an Akbulut cork to be described later on. The surgering of a 2-handle to a 1-handle is illustrated in Fig. 4a (see also Gompf and Stipsicz, 1999, Fig. 5.33). The 0-framed 2-handle (left) and the 'dotted' 1-handle (right) are diffeomorphic at their boundary ∂. The boundary of a 2- and a 3-handle is intimately related to the Akbulut cork shown in Fig. 4b as described at the Sect. 4.1.

Akbulut cork. A Mazur manifold is a contractible, compact, smooth 4-manifold (with boundary) not diffeomorphic to the standard 4-ball B^4 (Akbulut, 2016). Its boundary is a homology 3-sphere. If one restricts to Mazur manifolds that have a handle decomposition into a single 0-handle, a single 1-handle and a single 2-handle then the manifold has to be of the form of the dotted circle $S^1 \times B^3$ (as in Fig. 4a) (right) union a 2-handle.

Recall that, given p, q, r (with $p \leq q \leq r$), the Brieskorn 3-manifold $\Sigma(p, q, r)$ is the intersection in the complex 3-space $\mathbb{C}^3$ of the 5-dimensional sphere S^5 with the surface of equation $z_1^p + z_2^q + z_3^r = 0$. The smallest known Mazur manifold is the Akbulut cork W (Akbulut, 1991a, b) pictured in Fig. 4b and its boundary is the Brieskorn homology sphere $\Sigma(2, 5, 7)$.

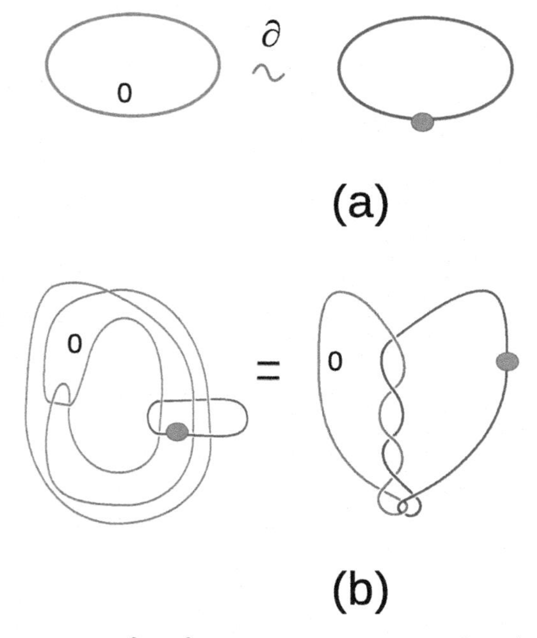

Fig. 4 **a** A 0-framed 2-handle $S^2 \times B^2$ (left) and a dotted 1-handle $S^1 \times B^3$ (right) are diffeomorphic at their boundary $\partial = S^2 \times S^1$, **b** two equivalent pictures of the Akbulut cork W

According to Akbulut and Durusoy (2005), there exists an involution $f : \partial W \to \partial W$ that surgers the dotted 1-handle $S^1 \times B^3$ to the 0-framed 2-handle $S^2 \times B^2$ and back, in the interior of W. Akbulut cork is shown in Fig. 4b. The Akbulut cork has a simple definition in terms of the framings ± 1 of $(-3, 3, -3)$ pretzel knot also called $K = 9_{46}$ (Akbulut and Durusoy, 2005, Fig. 3). It has been shown that $\partial W = \Sigma(2, 5, 7) = K(1, 1)$ and $W = K(-1, 1)$.

Exotic manifold R^4. An exotic R^4 is a differentiable manifold that is homeomorphic but not diffeomorphic to the Euclidean space $\mathbb{R}^4$. An exotic R^4 is called small if it can be smoothly embedded as an open subset of the standard $\mathbb{R}^4$ and is called large otherwise. Here we are concerned with an example of a small exotic R^4. Let us quote Theorem 1 of (Akbulut, 1991a).

There is a smooth contractible 4-manifold V with $\partial V = \partial W$, such that V is homeomorphic but not diffeomorphic to W relative to the boundary.

Sketch of proof (Akbulut, 1991a):

Let α be a loop in ∂W as in Fig. 5a. α is not slice in W (does not bound an imbedded smooth B^2 in W) but $\phi(\alpha)$ is slice. Then ϕ does not extend to a self-diffeomorphism $\phi : W \to W$.

It is time to recall that a cobordism between two oriented m-manifolds M and N is any oriented $(m + 1)$-manifold W_0 such that the boundary is $\partial W_0 = \bar{M} \cup N$, where M appears with the reverse orientation. The cobordism $M \times [0, 1]$ is called the trivial cobordism. Next, a cobordism W_0 between M and N is called an h-cobordism if W_0 is

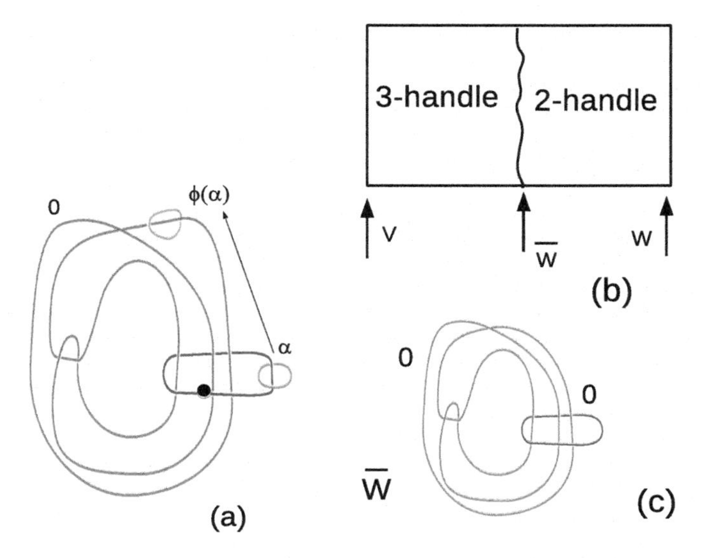

Fig. 5 **a** The loop α is not slice on the Akbulut cork, **b** the non-trivial h-cobordism between small exotic manifolds V and W, **c** the mediating 4-manifold $\bar{W}$

homotopically like the trivial cobordism. The h-cobordism due to S. Smale in 1960, states that if M^m and N^m are compact simply-connected oriented M-manifolds that are h-cobordant through the simply-connected $(m+1)$-manifold W_0^{m+1}, then M and N are diffeomorphic (Scorpian, 2011, p. 29). But this theorem fails in dimension 4. If M and N are cobordant 4-manifolds, then N can be obtained from M by cutting out a compact contractible submanifold W and gluing it back in by using an involution of ∂W. The 4-manifold W is a 'fake' version of the 4-ball B^4 called an Akbulut cork (Scorpian, 2011, Fig. 2.23).

The h-cobordism under question in our example may be described by attaching an algebraic cancelling pair of 2- and 3-handles to the interior of Akbulut cork W as pictured in Fig. 5b (see Akbulut, 1991a, p. 343). The 4-manifold $\bar{W}$ mediating V and W is as shown in Fig. 5c [alias the 0-surgery $L7a6(0, 1)(0, 1)$] (see Akbulut, 1991a, p. 355).

Following (Akbulut, 1991b), the result is relative since V itself is diffeomorphic to W but such a diffeomorphism cannot extend to the identity map $\partial V \to \partial W$ on the boundary. In (Akbulut, 1991b), two exotic manifolds Q_1 and Q_2 are built that are homeomorphic but not diffeomorphic to each other in their interior.

By the way, the exotic R^4 manifolds Q_1 and Q_2 are related by a diffeomorphism $Q_1 \# S^2 \times S^2 \approx Q \approx Q_2 \# S^2 \times S^2$ (where # is the connected sum between two manifolds) and Q is called the middle level between such connected sums. This is shown in Fig. 6 for the two R^4 manifolds Q_1 and Q_2 (Akbulut, 1991b), (Gompf, 1993, Fig. 2).

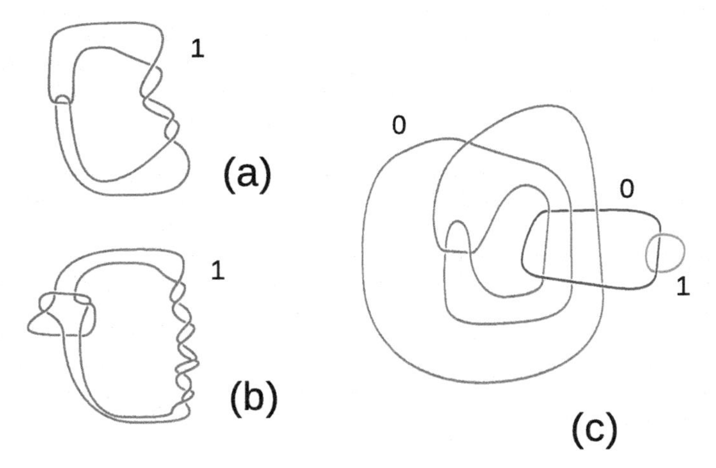

Fig. 6 Exotic R^4 manifolds Q_1 shown in (**a**) and Q_2 shown in (**b**). The connected sums $Q_1\#S^2 \times S^2$ and $Q_2\#S^2 \times S^2$ are diffeomorphic with middle level Q shown in (**c**)

4.2 Finite Geometry of Small Exotic R^4's, Quantum Computing and Quantum Contextuality

In the present paper, we choose G as the fundamental group $\pi_1(M^4)$ of a 4-manifold M^4 that is the boundary ∂W of Akbulut cork W, or governs the Akbulut h-cobordism. More precisely, one takes the manifold M^4 as $\bar{W}$ in Fig. 5 and Q in Fig. 6. Manifolds Q_1 and Q_2 are the small exotic R^4's of Ref. (Akbulut, 1991b, Figs. 1 and 2). There are homeomorphic but not diffeomorphic to each other in their interiors. This choice has two important consequences.

In the present paper, we choose G as the fundamental group $\pi_1(M^4)$ of a 4-manifold M^4 that is the boundary ∂W of Akbulut cork W, or governs the Akbulut h-cobordism. More precisely, one takes the manifold M^4 as $\bar{W}$ in Fig.5 and Q in Fig.6. Manifolds Q_1 and Q_2 are the small exotic R^4's of Ref. (Akbulut, 1991b, Figs. 1 and 2). There are homeomorphic but not diffeomorphic to each other in their interiors. This choice has two important consequences.

Recall the introduction of this section that that axioms (i) and (ii) are expected to govern the subgroup structure of groups G relevant to our model of quantum computing based on magic states. For the aforementioned manifolds M^4, the fundamental group $G = \pi_1(M^4)$ is such that (i) is always satisfied and that (ii) most often is true or geometric contextuality occurs with corresponding finite geometries of great interest such as the Fano plane $PG(2, 2)$ (at index 7), the Mermin's pentagram (at index 10), the finite projective space $PG(3, 2)$ or its subgeometry $GQ(2, 2)$ -known to control 2-qubit commutation (Planat et al., 2019, Fig. 1) (at index 15), the Grassmannian Gr(2, 8) -containing Cayley-Dickson algebras (at index 28) and a few maximally multipartite graphs.

Second, this new frame of 'exotic contextuality' provides a physical interpretation of quantum computation and measurements as follows. Let us imagine that $\mathbb{R}^4$ is our familiar space-time. Thus the 'fake' 4-ball W -the Akbulut cork- allows the existence of smoothly embedded open subsets of space-time -the exotic R^4 manifolds such as Q_1 and Q_2- that we interpret in this model as 4-manifolds associated to quantum measurements.

The boundary ∂W of Akbulut cork. As announced earlier $\partial W = K(1, 1) \equiv \Sigma(2, 5, 7)$ is a **Brieskorn sphere** with fundamental group

$$\pi_1(\Sigma(2, 5, 7)) = \langle a, b | aBab^2aBab^3, a^4bAb \rangle, \text{ where } A = a^{-1}, B = b^{-1}.$$

The cardinality structure of subgroups of this fundamental group is found to be the sequence

$$\eta_d[\pi_1(\Sigma(2, 5, 7))] = [0, 0, 0, 0, 0, \ 0, 2, 1, 0, 3, \ 0, 0, 0, \mathbf{12}, \mathbf{145}, \ \mathbf{178}, 47, 0, 0, \mathbf{4}, \cdots].$$

All the subgroups H of the above list satisfy axiom (i).

Up to index 28, exceptions to axiom (ii) can be found at index $d = 14, 16, 20$ featuring the geometry of multipartite graphs $K_2^{(d/2)}$ with $d/2$ parties, at index $d = 15$ and finally at index 28. Here and below the bold notation features the existence of such exceptions.

Apart from these exceptions, the permutation group organizing the cosets is an alternating group A_d. The coset graph is the complete graph K_d on d vertices. One cannot find a triple of cosets with strictly equal pairwise stabilizer subgroups of A_d (no geometry), thus (ii) is satisfied.

At index 15, when (ii) is not satisfied, the permutation group organizing the cosets is isomorphic to A_7. The stabilized geometry is the finite projective space $PG(3, 2)$ (with 15 points, 15 planes and 35 lines) as illustrated in Fig. 7a. The geometry is contextual in the sense that all lines not going through the identity element do not show mutually commuting cosets.

At index 28, when (ii) is not satisfied, there are two cases. In the first case, the group P is of order $2^8\ 8!$ and the geometry is the multipartite graph $K_4^{(7)}$. In the second case, the permutation group is $P = A_8$ and the geometry is the configuration $[28_6, 56_3]$ on 28 points and 56 lines of size 3. In (Saniga, 2015), it was shown that the geometry in question corresponds to the combinatorial Grassmannian of type Gr(2, 8), alias the configuration obtained from the points off the hyperbolic quadric $Q^+(5, 2)$ in the complex projective space $PG(5, 2)$. Interestingly, Gr(2, 8) can be nested by gradual removal of a so-called 'Conwell heptad' and be identified to the tail of the sequence of Cayley-Dickson algebras (Saniga, 2015; Saniga et al., 2015, Table 4).

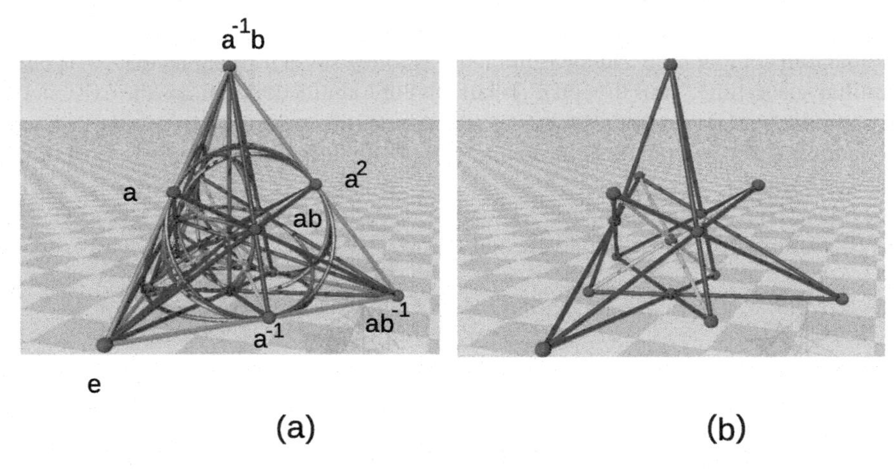

Fig. 7 **a** A picture of the smallest finite projective space $PG(3, 2)$. It is found at Frans Marcelis website (Marcelis, 2020). The coset coordinates are for a Fano plane $PG(2, 2)$ of $PG(3, 2)$. **b** A picture of the generalized quadrangle of order two $GQ(2, 2)$ embedded in $PG(3, 2)$. It may also be found at Frans Marcelis website

One expects a connection of the 28-point configuration to a del Pezzo surface of degree 2 (since the 56 lines of such a del Pezzo surface map in pairs to the 28 bitangents of a quartic).

The $[28_6, 56_3]$ *configuration*. Below are given some hints about the configuration that is stabilized at the index 28 subgroup H of the fundamental group $\pi_1(\partial W)$ whose permutation group P organizing the cosets is isomorphic to A_8. Recall that ∂W is the boundary of Akbulut cork W. The 28-letter permutation group P has two generators as follows

$$P = \langle 28|g_1, g_2\rangle \quad with \ g_1 = (2, 4, 8, 6, 3)(5, 10, 15, 13, 9)(11, 12, 18, 25, 17)$$
$$(14, 20, 19, 24, 21)(16, 22, 26, 28, 23), \ \ g_2 = (1, 2, 5, 11, 6, 7, 3)(4, 8, 12, 19, 22, 14, 9)$$
$$(10, 16, 24, 27, 21, 26, 17)(13, 20, 18, 25, 28, 23, 15).$$

Using the method described in Sect. 3.1, one derives the configuration $[28_6, 56_3]$ on 28 points and 56 lines. As shown in [Table 4] (Saniga, 2015), the configuration is isomorphic to the combinatorial Grassmannian Gr(2, 8) and nested by a sequence of binomial configurations isomorphic to Gr$(2, i)$, $i \leq 8$, associated with Cayley-Dickson algebras. This statement is checked by listing the 56 lines on the 28 points of the configuration as follows

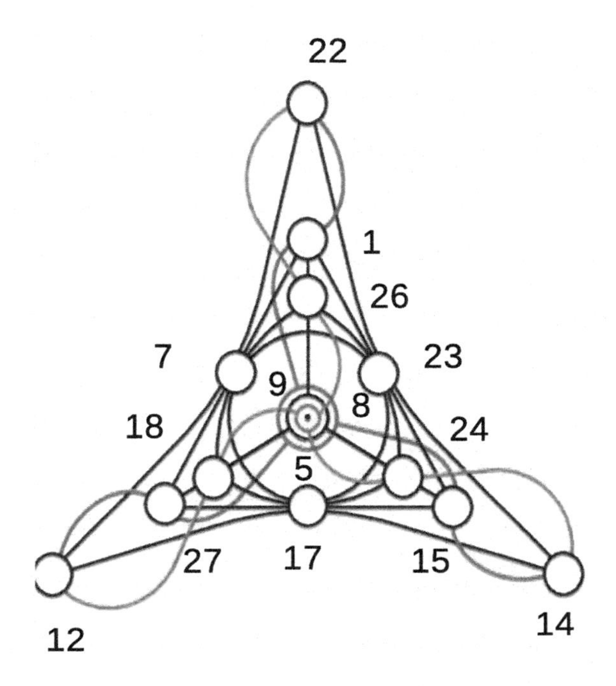

Fig. 8 The Cayley-Salmon configuration built around the Desargues configuration (itself built around the Pasch configuration) as in (Saniga et al., 2015, Fig. 12)

$\{1, 7, 27\}, \rightarrow \mathbf{Gr(2, 3)}$

$\{1, 15, 23\}, \{15, 17, 27\}, \{7, 17, 23\}, \rightarrow \mathbf{Gr(2, 4)}$

$\{1, 5, 26\}, \{5, 18, 27\}, \{5, 15, 24\}, \{23, 24, 26\}, \{17, 18, 24\}, \{7, 18, 26\}, \rightarrow \mathbf{Gr(2, 5)}$

$\{12, 14, 17\}, \{1, 9, 22\}, \{5, 8, 9\}, \{9, 14, 15\}, \{7, 12, 22\}, \{8, 12, 18\},$
$\{8, 14, 24\}, \{8, 22, 26\}, \{14, 22, 23\}, \{9, 12, 27\}, \rightarrow \mathbf{Gr(2, 6)}$

$\{3, 10, 15\}, \{3, 6, 24\}, \{3, 17, 25\}, \{3, 23, 28\}, \{1, 10, 28\}, \{3, 14, 19\}, \{7, 25, 28\}, \{6, 8, 19\},$
$\{19, 22, 28\}, \{5, 6, 10\}, \{12, 19, 25\}, \{10, 25, 27\}, \{9, 10, 19\}, \{6, 18, 25\}, \{6, 26, 28\}, \rightarrow \mathbf{Gr(2, 7)}$

$\{4, 11, 12\}, \{11, 21, 25\}, \{6, 20, 21\}, \{2, 3, 21\}, \{2, 4, 14\}, \{7, 11, 16\}, \{2, 16, 23\}, \{1, 13, 16\},$
$\{2, 11, 17\}, \{4, 19, 21\}, \{16, 20, 26\}, \{2, 13, 15\}, \{11, 13, 27\}, \{16, 21, 28\}, \{2, 20, 24\},$
$\{5, 13, 20\}, \{11, 18, 20\}, \{4, 9, 13\}, \{4, 8, 20\}, \{4, 16, 22\}, \{10, 13, 21\} \rightarrow \mathbf{Gr(2, 8)}.$

More precisely, the distinguished configuration $[21_5, 35_3]$ isomorphic to Gr(2, 7) in the list above is stabilized thanks to the subgroup of P isomorphic to A_7. The distinguished Cayley-Salmon configuration $[15_4, 20_3]$ isomorphic to Gr(2, 6) in the list is obtained thanks to one of the two subgroups of P isomorphic to A_6. The upper stages of the list correspond to a Desargues configuration $[10_3, 10_3]$, to a Pasch configuration $[6_2, 4_3]$ and to a single line$[3_1, 1_3]$ and are isomorphic to the Grassmannians Gr(2, 5), Gr(2, 4) and Gr(2, 3), respectively. The Cayley-Salmon configuration configuration is shown on Fig. 8, see also (Saniga et al., 2015, Fig. 12).

For the embedding of Cayley-Salmon configuration into $[21_5, 35_3]$ configuration, see (Saniga et al., 2015, Fig. 18).

Frank Marcelis provides a parametrization of the Cayley-Salmon configuration in terms of 3-qubit operators (Marcelis, 2020).

Not surprisingly, geometric contextuality (in the coset coordinatization not given here) is a common feature of all lines except for the ones going through the identity element.

As a final note for this subsection, we found Brieskorn spheres other than $\Sigma(2, 5, 7)$ whose fundamental group admits an index 28 subgroup isomorphic to A_8 whose geometry is the configuration with 28 points and 56 lines. Three-manifolds $\Sigma(3, 4, 5)$, $\Sigma(3, 4, 7)$ and $\Sigma(3, 5, 7)$ are such Brieskorn spheres.

5 Conclusion

To conclude, it has been shown that the group theoretical language seems efficient for describing quantum reality. We introduced the concepts of geometric and exotic contextuality for quantum theory and quantum measurements. In other papers dealing with slightly different subjects, we found that 'informationally complete' magic states may be defined as irreducible characters of an appropriate finite group. These characters are useful in the context of quark and lepton mixings (Planat et al., 2020a) and in the context of the universal code of life—the genetic code (Planat et al., 2020b, 2021). What next? Proteins are the language of life. We have much to learn about quantum mechanics by decoding its 20-letter language, e.g.

$MGFTCPNSDCLYSRSEWSNRALREEGLSFSMRCPGACCGAML$ $V \cdots$, is the beginning of the sentence of spike protein of SARS-Cov-2. Understanding the contextuality of life is the next step of this type of research.

References

Akbulut, S. (2016). *4-manifolds, Oxford graduate texts in mathematics* (Vol. 25). Oxford University Press.

Akbulut, S., & Durusoy, S. (2005). An involution acting nontrivially on Heegard-Floer homology. In *Geometry and topology of manifolds* (Fields Inst. Commun., Amer. Math. Soc., Providence, Rhode Island), (Vol. 47, pp. 1–9).

Akbulut, S. (1991). A fake compact contractible 4-manifold. *Journal of Differential Geometry, 33,* 335–356.

Akbulut, S. (1991). An exotic 4-manifold. *Journal of Differential Geometry, 33,* 357–361.

Bohr, N. Philosophy of Science Vol. 37 (1934), p. 157, and in The Truth of Science : Physical Theories and Reality (1997) by Roger Gerhard Newton, p. 176.

Bosma, W., Cannon, J. J., Fieker, C., & Steel, A. (Eds.), *Handbook of Magma functions*, Edition 2.23 (2017), 5914pp. Accessed on January 1, 2019.

Bravyi, S., & Kitaev, A. (2005). Universal quantum computation with ideal Clifford gates and noisy ancillas. *Physical Review A,71*, 022316.

Cabello, A., Estebaranz, J. M., & Garcia-Alcaine, G. (1996). *Physics Letters A, 212,* 183.
DeWitt, B. S. (1970). Quantum mechanics and reality. *Physics Today, 23,* 30.
Esfeld, M. (1999). Quantum holism and the philosophy of mind. *Journal of Consciousness Studies, 6,* 23–28.
Ferrero, M., Salgado, D., & Sánchez-Gómez, J. L. (2004). Is the epistemic view of quantum mechanics incomplete? *Foundations of Physics, 34,* 1993–2003.
Gompf, R. E. (1993). An exotic menagerie, *37,* 199–223.
Gompf, R. E., & Stipsicz, A. I. (1999). *4-manifolds and Kirby calculus*, Graduate Studies in Mathematics (Vol. 20). American Mathematical Society.
Marcelis, F. (2020). https://fgmarcelis.wordpress.com/pg32/pg32-1/ and https://fgmarcelis.wordpress.com/mermin-cayley-salmon-desargues. Accessed on January 1, 2020.
Mermin, N. D. (1993). Hidden variables and the two theorems of John Bell Rev. *Modern of Physics, 65,* 803–815.
Miller, E. (2014). Quantum entanglement, Bohmian mechanics, an humean supervenience. *Australasian Journal of Philosophy, 92,* 567–83.
Pavičić, M., Merlet, J.-P., McKay, B. D., & Megill, N. D. (2005). Kochen-Specker vectors. *Journal of Physics A: Mathematical Generation,38*, 1577–1592.
Peres, A. (1993). *Quantum theory, concepts and methods*. Dordrecht: Kluwer.
Planat, M. (2011). Pauli graphs when the Hilbert space dimension contains a square: Why the Dedekind psi function? *Journal Physics A Mathematical Theoretical44*, 045301.
Planat, M., & Gedik, Z. (2017). Magic informationally complete POVMs with permutations. *Royal Society Open Science,4*, 170387.
Planat, M., Giorgetti, A., Holweck, F., & Saniga, M. (2015). Quantum contextual finite geometries from dessins d'enfants. *International Journal of Geometric Methods in Modern Physics,12*, 1550067.
Planat, M. (2012). On small proofs of the Bell-Kochen-Specker theorem for two, three and four qubits. *The European Physical Journal Plus, 127,* 86.
Planat, M. (2015). Geometry of contextuality from Grothendieck's Coset space. *Quantum Information Processing, 14,* 2563–2575.
Planat, M. (2016). Two-letter words and a fundamental homomorphism ruling geometric contextuality. *Symmetry, Culture and Science, 1,* 1–16.
Planat, M. (2020). Quantum computation and measurements from an exotic space-time R^4. *Symmetry, 12,* 736.
Planat, M., Aschheim, R., Amaral, M. M., Fang, F., & Irwin, K. (2020). Complete quantum information in the DNA genetic code. *Symmetry, 12,* 1993.
Planat, M., Aschheim, R., Amaral, M. M., & Irwin, K. (2018). Universal quantum computing and three-manifolds. *Symmetry, 10,* 773.
Planat, M., Aschheim, R., Amaral, M. M., & Irwin, K. (2019). Group geometrical axioms for magic states of quantum computing. *Mathematics, 7,* 948.
Planat, M., Aschheim, R., Amaral, M. M., & Irwin, K. (2020). Informationally complete characters for quark and lepton mixings. *Symmetry, 12,* 1000.
Planat, M., Chester, D., Aschheim, R., Amaral, M. M., Fang, F., & Irwin, K. (2021). Finite groups for the Kummer surface: The genetic code and quantum gravity. *Quantum Reports, 3,* 68–79.
Planat, M., & Zainuddin, H. (2017). Zoology of Atlas-Groups: Dessins D'enfants, finite geometries and quantum commutation. *Mathematics (MDPI), 5,* 6.
Quantum contextuality, https://en.wikipedia.org/wiki/Quantum_contextuality. Accessed on January 15, 2021.
Saniga, M. (2015). The complement of binary Klein quadric as a combinatoriam Grassmannian. *Mathematics, 3,* 481–486.
Saniga, M., Holweck, F., & Pracna, P. (2015). From Cayley-Dickson algebras to combinatorial Grassmannians. *Mathematics, 3,* 1192–1221.
Saniga, M., & Planat, M. (2007). Multiple qubits as symplectic polar spaces of order two. *Advanced Studies in Theoretical Physics, 1,* 1.

Scorpian, A. (2011). *The wild world of 4-manifolds*. American Mathematical Society.
Thas, J., & van Maldeghem, H. (2004).Generalized polygons in finite projective spaces. In *Distance-Regular Graphs and Finite Geometry, in Special Issue: Conference on Association Schemes, Codes and Designs, Proceedings of the 2004 Workshop on Distance-regular Graphs and Finite Geometry (Com 2 MaC 2004)*, Busan, Korea, 19–23 July 2004.
Waegell, M., & Aravind, P. K. (2011). Parity proofs of the Kochen-Specker Theorem Basedon the 24 Rays of Peres, *Foundation of Physics,41* 1786–99.
Wilson, R., Walsh, P., Tripp, J., Suleiman, I., Parker, R., Norton, S., Nickerson, S., Linton, S., Bray, J., & Abbott, R. (2015). *ATLAS of finite group representations, Version 3*. Available online: http://brauer.maths.qmul.ac.uk/Atlas/v3/exc/TF42/. Accessed on June 2015.

Quantum Identity, Content, and Context: From Classical to Non-classical Logic

J. Acacio de Barros, Federico Holik, and Décio Krause

1 Introduction

Quantum mechanics (QM) is a very successful theory. It is also a strange theory. Though QM can calculate many experiments' outcomes, there is no consensus about what quantum models tell us about the microscopic world. In other words, it is unclear what is the relationship between QM and metaphysics. In this paper, we examine one particular aspect of the quantum world: quantum particles seem to lack identity.

Under certain circumstances, two quantum systems of the same kind (e.g., two electrons) become utterly indistinguishable by any empirical means. However, the lack of identity comes from more than just the impossibility of distinguishing between two quantum particles (e.g., two electrons). It derives from the fact that nothing changes when we permutate two identical quantum particles, contrary to what happens in the classical world. This invariance by permutation is at the core of the Bose-Einstein and Fermi-Dirac statistics. In this way, the standard interpretation of the theory assumes indistinguishability. Here, we argue that indistinguishability is an essential concept in quantum theories (both non-relativistic and quantum field theories). Indistinguishability should be thought of as at the same level as celebrated quantum concepts, such as superposition (in particular, entanglement), contextuality, and nonlocality.

Some philosophers and physicists are reluctant to admit that indistinguishability, also known as indiscernibility, plays a salient role in quantum physics's ontology. Perhaps, this reluctance comes from the notion that indistinguishability can be sim-

J. A. de Barros (✉)
School of Humanities and Liberal Studies, San Francisco State University, 1900 Holloway Ave., San Francisco, CA, USA
e-mail: barros@sfsu.edu

F. Holik
Universidad Nacional de La Plata, Instituto de Física (IFLP-CCT-CONICET), C.C. 727, 1900, La Plata, Argentina

D. Krause
Department of Philosophy, Federal University of Santa Catarina, Florianópolis, Brazil

Graduate Program in Logic and Metaphysics, Federal University of Rio de Janeiro, RJ, Brazil

S. Wuppuluri and I. Stewart (eds.), *From Electrons to Elephants and Elections*, The Frontiers Collection, https://doi.org/10.1007/978-3-030-92192-7_27

ulated within a "classical" mathematical setting, as we shall see below. However, we find this argument weak for several reasons.

First, just because we can do something does not mean that this is the best approach. Consider, for example, the geometry of curved spaces. We may describe a curved space using Riemannian geometry, where Euclid's postulate of parallel lines is not valid. Alternatively, we can describe the same space by embedding it in a higher-dimensional space and keeping Euclid's postulates. Both approaches yield the same results: all geometry theorems on the curved space are valid in both descriptions. However, one requires a more complicated ontological structure with extra dimensions. Should we make our ontology unnecessarily complicated to accommodate our prejudices? We believe not.

Second, when someone is interested in a theory's foundations, the underlying logic and mathematics become fundamental. We should not do away with an ontological feature because we can use a mathematical trick to describe it. Instead, we argue that the mathematical formalism used to cope with quantum systems' description should consider the ontological features that one aims to describe. Therefore, as we discuss below in more detail, it is crucial to develop a mathematical framework that accommodates indistinguishability in a natural way. In fact, we cannot cope with a contradictory theory (as some claim is Bohr's theory for the atom, yet this is disputable–see the discussion in Vickers, 2013) within a "classical" framework such as in the mathematics developed in a standard set theory such as the ZFC system, which we presuppose here.[1]

Thus, we wish to pursue a metaphysics of non-individuals. In this metaphysics, quantum entities[2] (here, quantum objects, independently of their proper characterization) are seen as not following the standard notion of identity (to be discussed below). Therefore, we need to change logic and mathematics, unless we accept the physicists' usual way of impersonating them within classical frameworks. These entities need to be considered in most cases as *absolutely* indiscernible, something forbidden in the classical settings.[3]

Nevertheless, the interpretational problem does not end with the indiscernibility of quantum objects. Indistinguishability is not the only mystery of quantum theory. The ontological status of properties of these objects is also relevant. Quantum properties are tricky, and if we are not careful about how we deal with them, we may reach contradictions. These contradictions arise from considering the possible results of multiple (and incompatible) experiments over the same system. As we have stated elsewhere (de Barros et al., 2017), we *never* perform the same experiment twice. What we do is take a similar experiment, so similar as to be *indistinguishable*. Since experiments are associated with properties, we should consider indiscernible

[1] ZFC is the Zermelo-Fraenkel set theory with the Axiom of Choice. The reader can think of it as formalizing the intuitive notion of a set one learned in our math classes.

[2] The notion of a *quantum object*, or quantum *system*, varies from one approach to another. In orthodox quantum mechanics, we have particles and waves. In the quantum field theories, the basic entities are fields, and particles arise as particular configurations of the fields. Our claims in this paper apply to both particles and fields.

[3] For a defense of the non-individuals view, see (Krause et al., 2021).

properties also. These indistinguishable properties are also forbidden by classical logic. We need to go outside of standard mathematics and use a different mathematical (and logical) setting as, for example, *quasi-set theory*, to be sketched below. Given that we need to recreate indiscernible properties and systems, it is natural to use a mathematical setting that incorporates indistinguishability as a primitive notion right from the start.

This paper is organized as follows. In Sect. 2, we first discuss the role of context and content in classical and quantum physics. These two concepts play an essential role in the difficulties physicists and metaphysicists face concerning quantum properties. In Sect. 3, we consider the concepts of identity and indiscernibility and how they are connected. Identity is a difficult concept, and we explore it both as it is connected to classical physics and indiscernibility in logic. This discussion opens up to our investigations outlined in Sect. 4. In this section, we argue that by intimately connecting identity to context, we can solve some puzzling aspects of quantum physics. Finally, in Sect. 6, we outline how to change mathematics to allow for the existence of indiscernibility as a fundamental and primitive concept. This mathematics, grounded on quasi-set theory, captures the idea that quantum objects are indistinguishable and lack a classical identity. As a bonus, we included in Sect. 7 somewhat more detailed mathematical explanation of the structures discussed in Sect. 6. We hope the interested reader will find this useful, but this section can be skipped by those readers not seeking further mathematical details. We end the paper with some final remarks, conclusions, and perspectives.

This article is written for a layperson with a strong mathematical background. The reader is assumed to know enough mathematics to be comfortable with logic, set theory, and orthodox quantum mechanics. It should be remarked that a paper dedicated to foundations and aimed at a general reader requires many caveats, since the delicate aspects can be quickly passed unsuspected. We try to warn the reader about those details in between the text or in the footnotes. We ask the reader's forgiveness in advance for the numerous footnotes.

2 Content and Context in Quantum and Classical Physics

The idea of content and context comes from linguistics, specifically semantics and pragmatics. Nevertheless, physics has straightforwardly borrowed those concepts. This section will discuss how content and context translated from linguistics to physics, focusing on quantum mechanics. We organize this section in the following way. First, we concisely review the concepts from linguistics. Then we explore how content and context show up in classical semantics theories. Our discussion should not be thought of as a detailed scholarly review of the linguistic literature on content and context, as this topic is the object of intense research in philosophy of language and linguistics for more than a century. Instead, we present a subset of linguistics that is relevant to physics. With that in mind, we follow our linguistics discussion by examining some physics examples. We see that contents may present context-dependency in both classical and quantum physics. However, we also argue that the context-dependency in quantum physics is different.

Let us start with the concept of content. Roughly speaking, semantic content refers to the meaning of a sentence.[4] Consider the following statement, made by Vera's friend, Alice:

L1. Vera had a bad date.

Sentence L1 can be seen as a proposition referencing to an object. Assuming the correspondence theory of truth,[5] its truth value requires some metric, likely subjective, of what constitutes a "bad date." However, once such a metric exists, one could infer L1's truth value. The truth-value of L1, therefore, lies on its semantic content. In other words, a sentence's semantic content can be thought of as a function that takes the sentence and outputs a truth-value.

Context, on the other hand, is the idea that some statements and utterances depend on the circumstances surrounding it, such as time, place, speaker, hearer, and topic, to name a few. For example, Alice's claim that "Vera had a bad date" has different meanings depending on whether their conversation revolved around the fruits of the *Phoenix dactylifera* or romantic engagements. The context alters the meaning and the functions that take the content to truth values.

However, context does not alter meaning only. Consider the case of indexicals. The statement "Acacio is hungry now" is contingent on when it is uttered and on the particular subjective satiety state of the person named "Acacio." In a sense, its meaning does not change. Its referent, Acacio, is the same (assuming we are talking about the same person, one of the co-authors of this paper), the concept of hunger is invariant, and the meaning of now as the present moment is maintained. However, its truth value is variable. As we write this paragraph, it is false, as Acacio just had lunch. However, the same statement was right about an hour ago. It will be true again several hours from now, even though its meaning is seemingly unchanged.

To summarize, sentences have meanings given by their semantic contents. Sometimes the meanings are context-dependent, as in the case of dates. However, other times, their truth-values vary with context, whereas their meanings seem to do not. We shall see that physics has some correlates to those ideas.

Let us start with classical physics. A physically-relevant proposition about an object is something empirically measurable. For example, we can have the following statement:

---

[4] We shall assume this without further discussion, but things are not as straightforward as it may appear. Meaning means "meaning for someone," and there is no meaning *tout court*. Yuri Manin, in his great book (Manin, 1977, pp. 34ff) mentions the case of Lev Alexandrovich Zasetsky, who suffered a brain injury in battle. Zasetsky could write sentences with meaning, such as "An elephant is bigger than an ant," and know that it is true (semantically well defined). But his illness impeded him to understand the meaning of the terms "ant" and "elephant." He had semantics and truth, but not meaning.

[5] We also sustain that the correspondence theory of truth, for instance that treated by Tarski, is not suitable for the empirical sciences, but this is something to be developed in another opportunity; here we take the standard view.

P1. A billiard ball's kinetic energy is between 0.1 and $0.2\,kg\,m^2/s^2$.

Similarly to linguistics, P1 has a meaning: if we measure the kinetic energy of a billiard ball, perhaps by measuring its mass and speed and inferring the energy, we find it to be in a certain range. Its meaning is given by an accompanying experimental procedure that yields a truth-value to the sentence. As importantly, this truth-value also corresponds to the idea that the billiard ball, if P1 is true, has a specific property: its kinetic energy.

As in linguistics, P1 refers to a subject (the billiard ball) and a truth-value associated with some meaning-constructing procedure (the experiment). Accordingly, we can think of any physics experiment as observing a physical system's property. This property itself has an associated proposition whose truth-value is assessed by an experiment. So, in a certain sense, properties of physical systems, such as temperature, momentum, energy, present an analogy with contents.

We may take the meaning of a statement as which experiment can yield a truth-value to it. Consequently, expressions such as P1 attach a property to a physical object. Of course, the property is the statement itself, and the experiment is a way to determine its truth-value. To summarize, the properties of a physical system are the content of the propositions.

What about context? Are classical properties context-dependent? Let us examine an example from 18th-century physics. A group of Italian researchers in the1700 s, known as the Experimenters, did not differentiate between heat and temperature but combined both concepts into one (Wiser and Carey, 1983). This combined concept of heat and temperature led to some puzzling results. For instance, the Experimenters wondered about examples such as the following. Imagine we heat a 2 kg piece of iron and immerse it in a container with room temperature water, subsequently measuring the water's temperature. Now, imagine that instead of iron, we use 2 kg of a 3:1 mixture of nitric acid (1.5 kg) and tin (0.5 kg), immersing it in water, as we did with the piece of iron. It was surprising to the Experimenters that even when the mixture of tin and acid was not as hot as the iron, the latter would not raise the water's temperature as much. If both objects, iron and mixture, had the same amount of "hotness," why would they increase the water by different levels of "hotness?"

The answer to the above puzzle is straightforward in contemporary physics, as we distinguish heat and temperature. Because of this distinction, we can measure how much heat a substance holds as their temperature increases: what physicists call specific heat. With this concept, we can measure that iron has a specific heat of 0.44 J/kg K. In contrast, the specific heat of a 3:1 mixture of nitric acid and tin is 1.34 J/kg K. This means that for every one-degree increase in temperature, the amount of heat held by the 2-kg block of iron increases by 0.88 J and by 2.64 J for the 2-kg tin-nitric acid mixture. In other words, at the same temperature, the mixture holds three times the amount of heat as the iron. Because the Experimenters had a single concept of heat and temperature, they could not even investigate the concept of specific heat, nor could they understand the puzzle.

Let us examine the example above from a slightly different perspective. Imagine we are observing a student who does not distinguish temperature from heat (as the

Experimenters) and thinks of both as the smorgasbord concept "hotness." Consider the following propositions observed to be empirically true for a specific experimental setup involving three objects: X, Y, and W (as for instance X is iron, Y is the mixture of nitric acid and tin, and W is water as in the example above).

A: If X has more heat than Y, then W will have a high temperature.
B: If X has a higher temperature than Y, then W will not have a high temperature.

Both propositions A and B can be true if we carefully chose X and Y's masses, heat capacities, and how we define statements such as "low temperature," "high temperature," and so on. However, let us rephrase A and B in terms of the student's hotness concept. We now have two new propositions, A' and B':

A': If X has more hotness than Y, then W will have high hotness.
B': If X has more hotness than Y, then W will not have high hotness.

A' and B' cannot be both true, as they are contradictory. The contradiction comes here from identifying heat and temperature as a single concept: hotness.

There is an obvious, albeit silly, solution to this contradiction. The student might say, *ad hoc*, that "hotness" in the context of an experiment observing A' is different from experiment B', so they are not the same statement. To save their hotness concept, the student makes things unnecessarily more complicated than they need to be. As more experiments pile up, the more contexts and the more complicated their theory becomes. Furthermore, such a move would lead to a theory incapable of making good predictions in different situations.

Of course, this is not what scientists usually do. Scientists try to find appropriate ways to describe a physical system that does not lead to contradictions or context dependency. In the hotness case, they realized that differentiating between heat and temperature was consistent and allowed for predictions and explanations of thermal phenomena. When faced with contradictions, scientists realized that the best approach is to face them and figure out ways to rethink our theories or experiments without resorting to context-dependency.

The above example is interesting for historical reasons, but it also illustrates a type of explicit contextuality. In the physics literature, this explicit contextuality is called direct influences (Dzhafarov & Kujala, 2016) or signaling (Popescu & Rohrlich, 1994). When the student "explained" the differences between A' and B' as context-dependent, he thought of explicit contextuality. Explicit contextuality manifests when there is a direct contradiction between two statements or results, such as the contradiction between A' and B'. When this happens, scientists recognize a problem and try to solve it, as with the development of the concepts of heat and temperature.

Let us now move from classical to quantum physics. Quantum physics, as far as we know, forbids any type of properties that exhibit direct influences, i.e., signaling. However, it allow another type of context-dependency (or contextuality): implicit contextuality. In the technical literature, this is called simply "contextuality." We call it implicit contextuality to emphasize its contrast with contextuality due to direct influences. From now on, when we talk about contextuality, we will refer solely to implicit contextuality.

To understand contextuality in quantum physics, let us consider another example (Specker, 1975). Imagine a Simon-like-game device with three buttons (instead of the usual 4). Each button on this device, when pushed, randomly emits red or green light. Turns consists of multiple trials, where after observing their behavior, the player can try to predict how each button will lit. For each trial of this game, the player can push at most two buttons at the same time, for as many times as they want, and in any combination of the three buttons they wish. If all three buttons are pushed at the same time, no light is emitted. To win the turn, the player needs to correctly guess what color the unpressed button would light in their last trial.

Let us consider a simple non-contextual example for this game. During her turn, Alice notices the following.

- For trials when she only presses one key, they seem to yield either color randomly. In other words, if Alice presses X, 50% of the time he observes green and 50% red.
- For trials when Alice presses X and Y, she also gets 50% for each color for X or Y, and the two colors are the same;
- For trials when Alice presses X and Z, she also gets 50% for each color for X and Z trials colors are opposite;
- For trials when Alice presses Y and Z, she also gets 50% for each color for Y and Z trials colors are also opposite.

So, after realizing that, if Alice presses X and Y and obtain "red" for both, she could logically infer that Z would be "green." This is because Z has the opposite color of both X and Y. Guessing "green" would win Alice the turn.

Now, imagine that in another turn, Bob starts prodding different combinations of pairs of X, Y, and Z, and observes the following.

- For trials when Bob only presses one key, they seem to yield either color randomly. In other words, if Bob presses X, 50% of the time he observes green and 50% red.
- For trials when Bob presses X and Y, he also gets 50% for each color for X or Y, but the two colors are the opposite;
- For trials when Bob presses X and Z, he also gets 50% for each color for X and Z trials colors are opposite;
- For trials when Bob presses Y and Z, he also gets 50% for each color for Y and Z trials colors are also opposite.

In other words, when two buttons are pushed simultaneously, they randomly emit red or green light, but in opposite colors. This example exhibits implicit contextuality. To see this contextuality, imagine we start with X emitting green and Y red. Bob can reason that if he pushed X and Z instead, then Z would be red. However, he could also argue that if he pushed Y and Z, since Y was red, Z would be green. Here we reach a logical contradiction: Z would be both red and green, and impossibility in the game. To avoid such contradiction, we need to either assume that Z has no possible color, or that its color changes with the "context" of being seen with X or with Y.

To convince themselves that Z changes with which other buttons it is pushed, we urge the readers to think about possible mechanisms that could yield the outcomes we described. The reader will quickly see that any mechanism that generates the outcomes for X and Y needs to be physically different from one generating X and Z (for an example using a firefly in a box, see de Barros et al., 2016).

The above example of contextuality is contrived. But contextuality shows up in quantum mechanics. One such example comes from the Greenberger-Horne-Zeilinger state (Greenberger et al., 1989), also known as GHZ. Without going into the details of where the following relations are derived, the GHZ state predicts the existence of six observable properties, X_1, X_2, X_3, Y_1, Y_2, and Y_3, satisfying the following properties. First, the properties X_i and Y_i take values $+1$ or -1. Second, whenever we observe each of those properties separately, they look completely random, i.e., their average value is zero. The same is true for when we observe them in pairs: they look completely uncorrelated. Third, we can observe them in triples, and when we do, we see the following relationship between the triplets.

$$Y_1 Y_2 Y_3 = 1, \tag{1}$$

$$Y_1 X_2 X_3 = X_1 Y_2 X_3 = X_1 X_2 Y_3 = -1. \tag{2}$$

The above correlations are experimentally observed (de Barros & Suppes, 2000; Bouwmeester et al., 1999). Finally, we cannot observe all six properties at the same time. In fact, we can only observe at most three of them simultaneously. For example, quantum mechanics forbids us to see Y_1, X_1,X_2, and X_3 at the same time. Contextuality manifests in a similar way as the previous three-variable example.

To see how contextuality manifests itself, let us assume that the six properties are *not* contextual. Then, we can use (1) and (2) and write the following.

$$(Y_1 X_2 X_3)(X_1 Y_2 X_3)(X_1 X_2 Y_3) = (-1)(-1)(-1) = -1. \tag{3}$$

But we can regroup the above product, and get

$$Y_1 Y_2 Y_3 (X_2 X_3)(X_1 X_3)(X_1 X_2) = Y_1 Y_2 Y_3 (X_2^2)(X_1^2)(X_3^2). \tag{4}$$

However, because X_i is ± 1 valued, their square is 1, i.e., $X_i^2 = 1$. Therefore, it follows that

$$(Y_1 X_2 X_3)(X_1 Y_2 X_3)(X_1 X_2 Y_3) = Y_1 Y_2 Y_3. \tag{5}$$

But this is a mathematical contradiction! The first term in the above equation is -1 whereas the second term is $+1$, and (5) is telling us that $1 = -1$.

Where is the contradiction coming from? It does not come from a mathematical mistake, but from an assumption of non-contextuality. When we wrote that $X_1^1 = 1$, we implicitly assumed that X_1 observed together with Y_2 and X_3 is the same as when observed together with X_2 and Y_3. This turns out to be false. If we, instead, call each

X_i by a different name depending on the context, no contradiction is obtained. What happens in quantum mechanics is similar to the simple color game we discussed before.

The reader may now be thinking about whether we could make a move similar to the contextual classical case. Namely, can we redefine properties such that no such kind of contradictions arise in quantum physics? The answer is yes. Unfortunately, there are many different ways to do so, and there is no consensus among the physics community as to which answer is even acceptable. So, let us end this section with two possible ways around this contradiction.

One move is to assume that properties depend on the context. This is the idea behind Bohm's interpretation of quantum mechanics (Bohm, 1952; Holland, 1995). In Bohm's theory, the famous duality wave/particle is resolved by assuming both wave and particle existence. The wave fills out the whole of space, and this wave guides the particle. How the wave directs the particle in one direction or another depends on its form. For example, in the two-slit experiment, the wave goes through both slits simultaneously, and due to its interference pattern, it guides the particle toward certain areas and away from others. The result is different if one or two slits are open (Holland, 1995). Since the wave depends on the context dictated by the physical experiment, Bohm's theory tells us that particles' reality and their properties are contextual. However, Bohm's theory presents a problem: for two or more particles, their waves are affected by their corresponding particle's positions. This theory implies the existence of instantaneous interactions between physical systems. Instantaneous interactions present a difficulty to the causal structure in Bohm's quantum world. As Einstein showed, to have cause and effect, we cannot have instantaneous interactions. This difficulty between Bohm's theory and Einstein's special relativity is the main reason for many physicists to reject it.

Bohm's theory gets into trouble with special relativity because it assumes that properties exist, whether we choose to measure them or not. When we measure, we affect the wave function and, consequently, the physical system. However, the property exists independent of an observer. In other words, Bohm's theory assumes that reality exists, whether we observe it or not.

Another possible solution to the problem of contextuality, particularly to contextuality at a distance (also called non-locality), is to assume quantum properties do not have values before a measurement and that the measurement process "creates" such values. This position was held by Bohr and is the core of the Copenhagen interpretation of quantum mechanics (Jaeger, 2009). In this interpretation, saying that an electron has spin $\hbar/2$ in the direction z is meaningless unless we perform a measurement of spin in the direction z and find it to be $\hbar/2$. However, before such a measurement, we cannot say anything about the spin. Furthermore, when we afterward make a measurement of spin in an orthogonal direction, say x, because z and x spins are incompatible (i.e., cannot be measured simultaneously), we cannot say anything anymore about the spin in the z direction; such "property" becomes meaningless. So, Bohr solves the problem of properties in quantum physics by merely denying their "existence" prior to a measurement.

We shall not cover all possible solutions to defining properties in quantum theory, as they abound. We just wanted to present to the reader two possible paths on how to deal with it and emphasize that the choices we have are not necessarily great. In Bohm's theory, we need to re-think the concepts of causality and space-time, two well-established tenets of special relativity, to accommodate faster-than-light signaling. In the Copenhagen interpretation, it becomes problematic to talk about a reality independent of a measurement apparatus (and the observer behind it). Either solution present metaphysical difficulties that have troubled physicists for more than a century. These puzzles all boil down to the problem of having properties that depend on the context.

To summarize, in this section, we discussed the idea of content and context. We started with its origins from linguistics and presented an interpretation that allows us to apply these concepts to physical phenomena. We saw that contextual dependencies appear in classical physics, but they are resolved by resorting to reinterpretations and refinements of the theory. We then discussed another contextual dependency that appears in quantum mechanics, such as the GHZ-state example. We then presented some of the proposed solutions to the problems and their corresponding metaphysical issues. In the following sections, we will show that those issues are intimately related to the concept of identity in the quantum world.

3 Identity and Indiscernibility

Identity is an old and difficult notion to be dealt with. Usually, the discussions have focused on personal identity and identity through time. Here, we shall be concerned with particular applications of this notion to the identity of objects and properties. By "identity of objects," or *individuals* as we prefer to call them,[6] we mean identity of those entities which are dealt with by the theories of physics.[7] For a more detailed discussion about the origins of the term "object," see (Toraldo di Francia, 1986, pp.13ff); here we review briefly some aspects of the argumentation given in (French & Krause, 2006, Chap. 1).

We have an intuitive idea of what it means to say that two objects, or individuals, are identical: they are *the same*. However, to say this is to say nothing, for we also do not know what is to be "the same," something reported equivalent to identity. Thus, we go to the opposite side: we judge individuals as being *different* and, therefore, *not identical*, hence *not the same*. Nevertheless, in virtue of what should individuals

[6] The word "individual," according to the Oxford Online Etymological Dictionary, means "one and indivisible." Hence our preference for the term. However, as it is common practice, we relax the idea of 'indivisible' and keep "one," adding that it can always be distinguished in other contexts, at least in principle, from any other individual as being *that* individual. This distinguishability cannot occur with quantum entities, even those trapped by some device.

[7] The standard quantum formalism is developed within a mathematical structure called "Hilbert-space formalism," although there are alternatives (Styer, 2002 mentions *nine* different ways of developing orthodox quantum mechanics).

be different? Usually, we look for their differences; although quite similar, two peas show differences, maybe some small scratch or a slightly different color. At least, that is what we tend to think.

Still, in virtue of what two objects would be different? Are they so? Is it possible to have two (or more) objects perfectly alike, with no differences at all? Put in other words, what makes an object an individual, distinct from any other? Is there some Principle of Individuation we can use to specify an individual's individuality? Theories of *individuation* are generally divided up in two main lines: *substratum theories* and *theories of bundles of properties*. According to the first group, beyond the properties of an object and the relations it can share with others, there is *something more*, something Locke described as "I don't know what" (Locke, 1959, Book I, XXIII, 2). This notion and the related ones (such as haecceities and thisness)[8] were discarded in favor of bundle theories of individuation. Bundle theories say that there is nothing more to an object than the collection of its properties (encompassing relations). Nevertheless, if in the substratum theories one could say that what distinguishes an object from another is its substratum (or something like that), in bundle theories, many discussions have appeared concerning the possibility of two objects having the same collection of properties. Can they have the same collection of properties? If not, why not? Of course, that objects in our scale, i.e., "macroscopic objects," can partake all their properties is something that cannot be logically proven. This assumption must be accepted as a metaphysical hypothesis, and there are no known counterexamples to it. Furthermore, this hypothesis was what Western philosophy has preferred, from the Stoics to Leibniz's metaphysics.

Let us remember Leibniz's metaphysics' intuitive idea: *no two individuals share all their properties; if they have the same attributes, they are not different, but the same individual.* This metaphysical principle was encapsulated in standard logic with the definition of identity given by Leibniz Law. This law says what we have expected: entities are identical if and only if they share all their properties, hence all their relations, that is, if and only if they are indistinguishable.

What about the identity of properties? In standard logic, we usually say that two properties, P and Q, are "identical" if they are satisfied by the same "things." For instance, for Aristotle, the properties "to be a human" and "to be a rational animal" are "identical" in this sense. As an example from standard mathematics, consider the sets $\{x \in \mathbb{R} | x^2 - 5x + 6 = 0\}$, $\{x \in \mathbb{N} | 1 < x < 4\}$, and $\{x \in \mathbb{R} | x = 2 \vee x = 3\}$. These three sets are identical: they have the same extensions but different *intensions.*[9]

Classical mathematical frameworks do not accommodate indistinguishables; entities sharing all their attributes and being just numerically distinct do not exist in classical mathematics (but see below). Individuals are unique, separable, at least in

[8] The term "haecceity" was coined in Medieval philosophy to designate that thing that makes an individual the individual it is and that does not belong to the catalog of the individual's properties (see Teller, 1998). There are peculiarities in using haecceity or thisness, but broadly speaking, all refer to something beyond an individual's properties.

[9] In technical terms, in extensional higher-order logics, we can define such a notion by saying that P and Q are identical when they have the same *extensions*, that is, when they are satisfied by the same lower terms.

principle, counted as one of a kind and presenting differences to every *other* object. There are no purely numerical identical individuals: some form of Leibniz's Law holds. This is so within standard logic and mathematics, and the ways of dealing with indiscernibles require mathematical tricks such as confining them to non-rigid structures.[10] For example, take the structure $\langle \mathbb{Z}, + \rangle$, which represents the integer numbers, $\mathbb{Z}$, and *only* the standard addition operation, "+." This structure is not rigid, since the transformation $f(x) = -x$ is an automorphism of the structure, i.e., it keeps the individuals indiscernible within its point of view. To see this, take the 2 and -2. We cannot discern them *within* this structure. Imagine any property for 2 defined only with "+," such as "$2 + 1 = 3$." If we change the numbers by the "minus" ones, we have "$(-2) + (-1) = (-3)$." From within this structure, the latter is identical to the former; we cannot distinguish them. Of course, if we added additional properties to the structure, such as the "$<$" relation, it would become rigid, and we would be able to distinguish between 2 and -2. However, we cannot do it only with "+."

The search for *legitimate* indiscernible objects/individuals, in the above sense and without mathematical tricks, requires a change of logic. We will retake this discussion later on this paper, but we wish to turn to another kind of question for now.

Some authors, such as Peter Geach, argue that identity is relative. The only thing we can say, according to him, is that two individuals a and b are (or not) identical relative to a sortal[11] predicate F; in the positive case, we say that they are F-identical and can write $a =_F b$. In our opinion, identity is absolute. Identity is, according to us, to be associated with metaphysical identity, as explained above. It is something an individual has that says that it is unique and, when it appears in some other context, we are authorized to think that it is the same individual that has appeared twice. Alternatively, an individual's identity is its identity card, one for each individual: it accompanies it in all contexts and, with its help, we can distinguish the individual as being *the same* individual of a previous experience. Identity makes the individual's name a rigid designator, denoting the same entity in all possible accessible worlds. As it is well known, David Hume guessed that there is no such an identity; according to him, we recognize someone as being *the same* from a previous experience by habit, by familiarity (Hume, 1985, p. 74 and *passim*), but cannot "logically" prove that. Schrödinger had a similar opinion regarding quantum entities when he says that

> [w]hen a familiar object reenters our ken, it is usually recognized as a continuation of previous appearances, as being the same thing. The relative permanence of individual pieces of matter is the most momentous feature of both everyday life and scientific experience. If a familiar

[10] A structure (a domain comprising relations over its elements) is rigid if its only automorphism (bijections that preserve the relations of the structure) is the identity function. Indiscernibility in a structure means that the objects are invariant by some automorphism of the structure; in rigid structures, an object is indiscernible just from itself. Non-rigid (deformable) structures hide the object's identity so that we may not be able to discern them by lack of distinctive relations or properties. For details, see (French & Krause, 2006, §6.5.2), (Krause & Coelho, 2005).

[11] A sortal predicate enables to count the objects that obey the predicate, such as "being a philosopher." So, Isaac Newton and Stephen Hacking would both be counted as "Lucasian Professor of Mathematics in Cambridge.".

> article, say an earthenware jug, disappears from your room, you are quite sure that somebody must have taken it away. If after a time it reappears, you may doubt whether it really is the same one – breakable objects in such circumstances are often not. You may not be able to decide the issue, but you will have no doubt that the doubtful sameness has an indisputable meaning – that there is an unambiguous answer to your query. So firm is our belief in the continuity of the unobserved parts of the string! (Schrödinger, 1998, p. 204)

Entities partaking metaphysical identity are termed *individuals*. Can we think of *non-individuals* too? If yes, can we give examples of entities of this kind? The first way to think of them, by considering what we have said, is to deny them the epithet "to have an identity." What should it mean? The short answer is that they would be entities sharing all their characteristics, either substratum or properties and relations. From now on, we shall avoid speaking of substratum and keep with bundle theories (Teller, 1998). However, non-individuals, in our formulation, are not simply metaphysically or numerically identical entities, although this is logically possible.[12] Our notion is weaker, enabling non-individuals to form collections (termed "quasi-sets") with cardinalities greater than one so that no particular differences can be ascribed to them. Furthermore, they would be indistinguishable even if an omniscient demon (Laplace's demon) exchanged them with one another; in this case, nothing would change in the world at all. That is the difference: individuals, by definition, when permuted, make a difference! This difference is of fundamental importance, for it involves several other related notions which appear in physical theories, such as space and time and, fundamentally, permutations. We shall need to explain that further, but for now, we wish to emphasize that we do not regard identity as something an entity *must* have. When something has an identity, then it is absolute, it is metaphysical, and no two entities with identity can be only numerically distinct. *Non-individuals* are entities that lack identity, that can be just numerically discerned, that have all the same identity card. If one looks at one non-individual here and there, one finds "another" one in a different context; not even demons or gods will tell one if this new object is "different" or "the same" one found previously, as this would be meaningless.

Nevertheless, once we think about more than one entity, one could claim that they must be *different*. Mathematically, this would be expressed by the set-theoretical argument that once the cardinal of a set is greater than one, its elements *must* be different. We stress that this depends on the set theory one is taking into account. In standard set theories, such as the most celebrated systems (the apparently most famous one is termed "ZFC"), this is true, but in *quasi-set theory* (discussed below), this is may not be the case. In quasi-set theory, we not only can have collections (quasi-sets) of absolutely indiscernible entities and with a cardinal greater than one, but we can also quantify such "non-individuals." Quasi-set theory shows that Quine's motto of "no entity without identity" (Quine, 1969, p. 23) does not hold in general, for even non-individuals can be values of the variables of a regimented language.

[12] In his criticism to the definition of identity given by Whitehead and Russell in their *Principia Mathematica* (Leibniz Law, in a standard second-order language, $x = y := \forall F(Fx \leftrightarrow Fy)$, where x and y are individual terms and F is a predicate variable for individuals), F. P. Ramsey said precisely this: that we could logically conceive entities violating the definition, sharing all their properties, and even so not being the same entity (Ramsey, 1965, p.30).

3.1 Identity in Classical Formal Settings

There is a problem concerning the metaphysical identity of the last section: it cannot be defined in first-order languages (Hodges, 1983; French & Krause, 2006).[13] We provide here a slightly technical explanation. As said earlier, first-order languages speak of the individuals of some domain. Usually, the axiomatizations take logical identity as primitive (represented by a binary predicate "="), subject to certain axioms (reflexivity and substitutivity). We can prove that identity is an equivalence relation, really a congruence, whose intended interpretation is the *identity of the domain*; calling it D, then we are referring to the set $\Delta_D := \{\langle a, a\rangle : a \in D\}$, also called the *diagonal of* D. But it can be proven that there are other structures, called elementary equivalent structures,[14] which also model "=" but interprets this symbol in sets other than the diagonal (op.cit.). So, within a first-order language, we never know if we speak of the identity (or the difference) of two individuals or of, say, classes of individuals.

Higher-order languages enable us to define logical identity by Leibniz Law, but such logical identity is defined through indiscernibility. If we wish to define indiscernibility instead, the definition would be the same: agreement for all properties. So, higher-order languages do not distinguish between these two concepts. If we intend to speak of indiscernible but not identical things, Leibniz Law does not help.[15] Furthermore, if we aim to preserve some meta-properties of our system (Henkin's completeness), we are subject to find Henkin models so that two objects of the domain look as indiscernible since they obey all the language's predicates, but which are not the same element (French & Krause, 2006, §6.3.2). In short, we need to conclude that metaphysical identity cannot be defined. The most we can do is find refuge in logical identity, but this, as we shall see soon, causes troubles to quantum mechanics.

However, let us first put away the often-made claim that even quantum objects can be discerned by spatio-temporal location.

3.2 Identity and Space and Time

There is still another way to look at identity in classical settings: include space and time. Orthodox non-relativistic quantum mechanics makes use of *classical* space

[13] First-order languages deal with domains of individuals, their properties, relations and operations over them. Quantified expressions like "There exists some x such that..." and "For all individuals x, ..." applies only to individuals, and we cannot say things like "There is a relation among individuals ..." or "For every property of individuals ...." In logic, we say that first-order languages quantify over individuals only.

[14] Elementary equivalent structures are interpretations of a first-order language that preserve the same truth sentences. From the language's point of view, one cannot distinguish among such structures: they look the same.

[15] The distinction between identity and indiscernibility can be made only in semantical terms; see (da Costa & Krause, 1997).

and time or, as we can say, "Newtonian" absolute notions. Intuitively, the classical space and time structure is a space that looks, at least for small regions, like the $\mathbb{R}^4$, namely three dimensions for space ($\mathbb{R}^3$) and one for time ($\mathbb{R}$). More precisely, mathematically, the classical space-time is a manifold locally isomorphic to $\mathbb{R}^4$, usually termed $\mathbb{E}^4$ (for "Euclidean"); see (Penrose, 2004, Chap. 17).

This structure has some interesting features, but for us here, an important characteristic is that it is a "Hausdorff space." This property of being Hausdorff means that, given any two points a and b, $a \neq b$, it is always possible to find two disjoint open sets (say two open balls) B_a and B_b such that $a \in B_a$ and $b \in B_b$. In extensional contexts, such as the ZFC set theory, a property is confounded with a set; the objects that belong to the set are precisely those satisfying the property. So, a and b have each a property not shared with the other, namely, to belong to "its" open set. Hence, Leibniz's Law applies, and they are different. Notice that this holds for any *two* objects a and b: once we have *two*, they are distinct. Therefore, we may say that, within such a framework, there are no indiscernibles![16]

Let us see now how we can pretend to say that we have indiscernibles within a classical framework.

3.3 Indiscernibility in Classical Logical Settings

Still working in a classical setting, say the ZFC system, we can mimic indiscernibility. In this subsection we expand the above discussion about using non-rigid structures, presenting some of its more technical concepts and ideas.

Usually, we say that the elements of a certain equivalence class are indiscernible, and perhaps this is acceptable for certain purposes. More technically, in doing that, we are restricted to a *non-rigid (or deformable) structure*. As we saw previously, we say that a structure $\mathfrak{A} = \langle D, R_i \rangle$, $i \in I$, is *rigid* if its only automorphism is the identity function; this means that we have a domain D, a non-empty set, and a collection of relations over the elements of D, each one of a certain arity $n = 0, 1, 2, 3, \ldots$.[17] If the structure is not rigid, then it is is non-rigid or deformable. We saw an example of a deformable structure earlier on, the $\langle \mathbf{Z}, + \rangle$. Another example of a deformable structure is the field of the complex numbers, for the operation of taking the conjugate is an automorphism. In such a structure $\mathfrak{C} = \langle \mathbb{C}, 0, 1, +, \cdot \rangle$, the individuals i and $-i$ are indiscernible.

Given $\mathfrak{A}$ as above, we say that the elements a and b of D are $\mathfrak{A}$-indiscernible if there exists $X \subseteq D$ such that (i) for every automorphism h of $\mathfrak{A}$, $h(X) = X$, that

---

[16] In model theory, an important part of logic, we can speak of "indiscernibles" in a sense, for instance, *Ramsey indiscernibles*. However, this is a way of speaking; even these entities obey the classical theory of identity, therefore being individuals. See (Button & Walsh, 2018, Chap. 15).

[17] That the identity mapping is an automorphism is trivial. For all the argumentation, it is enough to consider *relational structures*, for distinguished elements and operational symbols can be taken as particular kinds of relations; also, we subsume all domains in just one.

is, X is invariant by the automorphisms of the structure, and (ii) $a \in X$ iff $b \notin X$. Otherwise, a and b are $\mathfrak{A}$-discernible (Krause & Coelho, 2005).

It is clear that in a rigid structure, the only element indistinguishable from a is a itself since the only automorphism is the identity function. In informal parlance, we may say that a and b are $\mathfrak{A}$-indiscernible iff they are invariant by permutations that "preserve the relations of the structure."

Something like that is what we do in quantum mechanics. Roughly speaking, the theory says that when we measure a certain observable value for a quantum system in a certain state, the value does not change before and after a permutation of particles of the same kind. Physicists say that *permutations are not observable*, and this is expressed by the Indistinguishability Postulate.[18]

Leaving formal logic and mathematics for a while, let us consider more general situations, which will lead us to a more detailed discussion about quantum mechanics. We shall commence by emphasizing the importance of the *contexts*.

4 Connecting Identity to Context

On many occasions, we are tempted to think about possible worlds which are not actual. We wonder what our life would have been like if we had taken different decisions at crucial moments. We can think about an object, person, or animal, in many different circumstances, which can differ from the actual ones. For example, suppose that we have a pet cat and live in a small apartment. Given its living conditions, the cat cannot catch the birds that he sees through the window. He observes them with attention, craving for them but unable to reach them. Thus, in our tiny-apartment world, our cat never caught a bird. Furthermore, he never will because he cannot go out. However, we can *imagine* a different world, in which we live in a house with a big yard in which our cat can wander out as many times as it wants. In this big yard world, our cat can surely try to catch a bird, and he will undoubtedly do so at least once.

The above story is an example of how we reason about counterfactuals. We are tempted to conclude something that occurs in a world that is not actual *could* happen, even if that world never becomes actual. This kind of reasoning is very natural in our everyday life. However, what are the assumptions behind it? First, somehow, our cat

[18] In technical terms, let us take a permutation P between particles denoted by x_i and x_j. As usually stated, we may say that for any $x_1, \ldots, x_n$,

$$P(x_1, \ldots, x_i, \ldots, x_j, \ldots, x_n) \leftrightarrow P(x_1, \ldots, x_j, \ldots, x_i, \ldots, x_n) \tag{6}$$

The *Indistinguishability Postulate* is expressed in terms of "expectation values;" it says that

$$\langle\psi|\hat{A}|\psi\rangle = \langle P\psi|\hat{A}|P\psi\rangle \tag{7}$$

for any observable represented by a self-adjoint operator $\hat{A}$ and for any permutation operator P, being $|\psi\rangle$ the vector state of the system.

retains its identity among the different worlds: the cat in the small apartment world is the same as the cat in the big yard world. Both cats have the same name, color, same capabilities, and desire to catch birds. Nevertheless, how can we assure that the cat will retain its properties among the different worlds? Perhaps, if we could afford a house with a big yard, we could also afford fancy and tasty cat food. The cat gets used to it, stays inside the house, and eats the whole day. In the fancy house world, it might become idle to the point that it barely moves or plays, as it happens with some cats. When it finally goes out to the garden, it cannot catch birds anymore, as it became clumsy and slow.

The above example shows that we should not make hasty conclusions: the properties of an object, person, or animal, might depend strongly on the *context* in which we are considering them. In the small apartment, humble life, with cheap food, our cat is playful and agile: it has a high probability of catching a bird but no bird to catch. In the big house, those properties may or may not be valid. The first lesson is: to assume that an object retains its properties among different and incompatible worlds is not granted. Even more so, one may ask: in which sense are the two cats in different worlds the same? From a strict point of view, one may say that the agile cat from our actual world is not the same as the idle cat of the alternative reality. In the same way, we should not mix the different worlds with counterfactual reasoning. If we conclude, by studying our cat in this actual world, that he is very skilled in chasing birds, we cannot use empirical information from our world to conclude that the cat will indeed chase a bird in the alternative world.

Thus, we are introduced to a profound philosophical problem by thinking about the above straightforward situation: what are the principles or conditions that grant identity to objects considered in different possible worlds? Are we entitled to say that a given object retains its identity when considered in different and incompatible situations? Of course, in many situations of our daily life, assuming that objects retain their identities and properties in different contexts will work. Our bike works well on sunny and rainy days and in diverse landscapes (such as cities or mountains). Many characteristics of our bike–such as its color or its range of velocities–are, to a great extent, *context independent*. However, we should not take this context independence for granted. This is more so if we consider quantum systems that define phenomena that lie far beyond our everyday experience. The realm of the atom extends far beyond the ångström scale (ten to the minus ten meters, which is something like $0,0000000001$ meters for one ångström!). The principles–whatever they are–that allow us to identify properties and objects among incompatible situations may no longer be valid for atomic systems. Moreover, this seems to be the case, as the GHZ example above and the following example show.

Suppose that Alice and Bob have separated labs, L_A and L_B, in which they perform their experiments. At a given time, a third party prepares a quantum system capable of affecting what happens in L_A and L_B. Suppose that Alice decides to make an experiment P_A in her lab, in order to interact with the given quantum system, and that Bob can do P_B or P'_B in L_B. Due to the peculiarities of quantum mechanics, P_B and P'_B cannot be performed at the same time–they are *incompatible* experiments. To understand what *incompatible* means, imagine the following situation: in order to perform P_B, Bob must align a magnet in a given direction d, and in order to perform P'_B, he must align its magnet in a different direction d'. A magnet cannot point in two different directions–similarly, a clock's handle cannot point at two different angles simultaneously. Thus, there are two incompatible situations: either Alice performs experiment P_A and Bob performs P_B, or Alice performs P_A and Bob P'_B. The two possibilities *cannot* coexist in the same world. Let us call these possibilities W_1 and W_2, respectively.

Suppose now that Alice and Bob are in the process of deciding what to do. They wonder about the experiments' possible outcomes in the different situations, W_1 and W_2. Notice that they do not need actually to perform the experiments. It is all about reasoning in various alternatives without actually performing them. Now we question: what is the status of the possible results of experiment P_A concerning W_1 and W_2? After the discussion about the cat, we should not be as quick to identify what happens in W_1 with W_2, even if we are talking about the same experiment, P_A. In both possible worlds, Alice will perform the same actions (she will orient the magnets in the same directions, prepare the same reading apparatus, and so on). Is she going to obtain the same results? What enables us to conclude that she will? Notice that we are not asking here about an *influence* of Bob's actions in Alice ones: the laboratories can be very far away in space and time. We are asking here whether we are entitled to assume that there is some trace of identity among the results obtained in different (and incompatible *worlds*). As expected, the answer is: no, we are not. Contradictions can be readily achieved if we do so, as the cat and contextuality examples suggest (and shown in technical research on quantum theory).

The actions required for experimenting P_A are the same in W_1 and W_2. Can we say that P_A in W_1 is the same as P_A in W_2? After the cat discussion, let us be conservative about the answer. We will say that P_A in W_1 is *indistinguishable* from P_A in W_2. The two experiments are completely alike: Alice will execute the same actions in a system prepared with an equivalent procedure in both worlds. However, we should not be tempted to claim they are the same. The more so, we should not expect the same results. In this sense, we say that the properties studied by experiment P_A in W_1 are indistinguishable from the properties studied by P_A in W_2. We denote these properties by the pairs $(P_A; W_1)$ and $(P_A; W_2)$ and write $(P_A; W_1) \equiv (P_A; W_2)$, to stress the fact that they are indistinguishable (but not identical). A natural, logical formalism for describing this kind of indistinguishability is the quasi-set theory. This theory allows us to consider properties or objects in alternative worlds as collections of indiscernible ur-elements.

If world W_1 becomes actual, Alice and Bob will perform their actions, obtain their results, and record them. Out of these results, what conclusions should they take about the possible results associated with W_2? Are they entitled to reason in a counterfactual way and combine the results of worlds W_1 and W_2 to extract conclusions about them? Much caution should be taken here, as the cat and contextual examples show. In principle, there is no *a priori* reason to do so. That we are allowed to do so in many (but not all!) everyday situations is more a lucky strike that we share with other creatures in our macroscopic reality than a general rule. Counterfactual reasoning simplifies our existence, but we should not expect it to be valid in every situation.

This lack of validity seems empirically suggested at microscopic scales, which are very different from our own.

To summarize, we can state the following:

- Even if state preparations and measurement procedures are completely alike among different worlds, we should not treat them as identical. In this sense, we speak about things such as indistinguishable properties and objects.
- Even if two experiments are completely indistinguishable, we should not expect the same results in different worlds.
- We should not derive conclusions from counterfactual reasoning, especially in the quantum domain. Such conclusions are not reliable and are not metaphysically justified.

5 Quantum Mechanics in Classical Logical Settings

In this section, we briefly review how the standard quantum formalism performs the trick of treating indiscernible quantum systems within the scope of classical logic (encompassing mathematics). In doing so, we lay the groundwork for alternative logics and mathematics, which provide an adequate description from our perspective.

A glance at standard textbooks on quantum mechanics reveals that they use classical mathematics, hence classical logic. However, the claim that quantum mechanics requires a different logic, known as quantum logic, can also often be found.[19] These two observations seem contradictory. Why is this apparent contradiction present in the literature?

The reason may be as follows. Most physicists are concerned with physical problems being solved by quantum theory and not with philosophical or logical foundational questions about it. Although they might endorse some particular interpretation of quantum mechanics, thus presupposing some concern with quantum theory's philosophy, most physicists use "classical" mathematics in an almost instrumentalist way. Thus, when dealing with entities that would be indistinguishable, physicists use some mathematical tricks to hide the identifications typical of our standard mathematical languages. Let us see how they do it.

First, we recall that, in quantum mechanics' standard formulation, a system's state is represented mathematically by a vector in a Hilbert space. This vector, also called the wave function, is supposed to encode all information available for that system in a specific situation. Observables, which represent possible experimental procedures and their outcomes, are self-adjoint operators in the Hilbert space. When an observable is measured, the state-vector enters (or "collapse") into one of the observable operator's eigenvectors. Since this process is "mysterious," in the sense that the formalism does not explain how it happens, many physicists try to avoid

[19] The field of "quantum logic" arose from Birkoff and von Neumann's 1936 seminal paper. The reader interested in the subject is referred to the following excellent papers: (Dalla Chiara et al., 2004) and (Svozil, 1998).

it, adopting alternative explanations. Nevertheless, the primary mathematical object in quantum theory is the Hilbert space and vectors in it. So, the question is how to represent indistinguishable objects using the mathematics of vectors.

Quantum particles come in two types: bosons and fermions. Their main difference comes from their statistics: bosons follow the Bose-Einstein statistics, whereas Fermions satisfy the Fermi-Dirac one. Both statistics count objects as if they were indistinguishable, contrary to the classical Maxwell-Boltzman statistics.

Bosons are a typical type of indistinguishable quantum entities. Bosons are a kind of quantum "particles," and they are entirely indistinguishable when prepared in the same quantum state. This state is such that they share all the relevant quantum properties. A system composed of, say, two bosons 1 and 2 in two possible situations A and B is described by a symmetric wave function such as the following.

$$\Psi = \frac{1}{\sqrt{2}}\left(\psi_1^A\psi_2^B + \psi_2^A\psi_1^B\right), \tag{8}$$

where $\psi_1^A\psi_2^B$ means system 1 in the state A and system 2 in B and similarly for the other term. The $\frac{1}{\sqrt{2}}$ is just a normalization factor required by the formalism. Ψ is invariant under the permutation of 1 and 2. This invariance means that exchanging particle 1 by 2 (and vice-versa) does not affect the state of the system. Consequently, any measurement results are maintained under permutations.

This symmetrization of the wave function works, but it is a trick. We are still using labels to "name" the particles because our language and mental models have a hard time thinking otherwise. In other words, this trick assumes, upfront, that bosons are individuals. Suddenly, as if a miracle happened, permutations do not conduce to different situations. However, this invariance was put there by hand. We could give more detailed arguments as to why this is a mathematical trick that does not make bosons indistinguishable, but we hope the above example is sufficient for the reader to grasp the main idea.

The use of the above trick is similar to confining the discussion to a deformable (non-rigid) structure, as explained earlier. However, as mentioned, within such classical settings, we can always go "outside" of the structure and identify the particles. This possibility of identification is at odds with the hypothesis that they are indiscernible.[20]

There is no way to escape this conclusion. As we have said before, standard mathematics and logic are theories of individuals. This is so for historical reasons: classical logic, mathematics, and even classical physics were built with individuals in mind. Quantum mechanics, of course, came to challenge those ideas and to question the concepts of individuality.

[20] The way to "go outside" the quantum formalism is to go to the set-theoretical universe since all mathematics used in quantum mechanics can be performed in terms of sets.

6 Alternative Logical Approaches

Assuming that indiscernibility is a core notion in quantum mechanics, we should look for an alternative logical and mathematical basis that considers it right from the start. This bottom-up approach would not mimic it within a standard framework from a top-bottom one. Our strategy is grounded in a metaphysics of non-individuals (for detail, see (French & Krause, 2006), (Krause et al., 2021), and references therein). Moreover, it tries to develop mathematics compatible with such metaphysics. Consequently, Schrödinger logics and quasi-set theory were developed in the 1990s. Although they are mathematical developments independent of the interpretations, the intended one is precisely to cope with such non-individual entities. In this section, we will give a rough idea about how quasi-set theory works. For a review about Schrödinger logics, see (French & Krause, 2006, chap.8).

6.1 Quasi-set Theory

In the quasi-theory $\mathfrak{Q}$, indiscernibility is a primitive concept, formalized by a binary relation "$\equiv$" satisfying the properties of an equivalence relation, but not full substitutivity.[21] In this notation, "$x \equiv y$" is thought to mean "x is indiscernible from y." This binary relation is a partial congruence in the following sense: for most relations, if $R(x, y)$ and $x \equiv x'$, then $R(x', y)$ as well (the same holds for the second variable). The only relation to which this result does not hold is membership: $x \in y$ and $x' \equiv x$ does not entail that $x' \in y$; details in (French & Krause, 2006, 2010)).

Quasi-sets can have as elements other quasi-sets, particular quasi-sets termed *sets* which are copies of the sets in a standard theory (in the case, the Zermelo-Fraenkel set theory with the Axiom of Choice), and two kinds of atoms (entities which are not sets), termed M-atoms (M-objects), which are copies of a standard set theory with atoms (ZFA) and m-atoms (m-objects), which have the quanta as their intended interpretation, to whom it is supposed that the logical identity does not apply. If we eliminate the m-atoms, we are left with a copy of ZFA, the Zermelo-Fraenkel set theory with atoms. Hence, we can reconstruct all standard mathematics within $\mathfrak{Q}$ in such a "classical part" of the theory.

Functions cannot be defined in the standard way. When m-atoms are present, it cannot distinguish between indiscernible arguments or values. Therefore, the theory generalizes the concept to "quasi-functions," which map indiscernible elements into indiscernible elements. See below for more on this point.

Cardinals (termed "quasi-cardinals," qc) are also taken as primitive, although they can be proven to exist for finite qsets (finite in the usual sense Domenech & Holik, 2007; Arenhart, 2011). The concept of quasi-cardinals can be used to speak of "several objects." So, when we say that we have two indiscernible q-functions,

[21] If we add substitutivity to the postulates, then no differences between indiscernibility and logical first-order identity would be made.

according to the above definition, we are saying that we have a qset whose elements are indiscernible q-functions and whose q-cardinal is two.[22] The same happens in other situations.

An interesting fact is that qsets composed of several indistinguishable m-atoms do not have an associated ordinal. This lack of an ordinal means that these elements cannot be counted since they cannot be ordered. However, we can still speak of a collection's cardinal, termed its *quasi-cardinal* or just its *q-cardinal*. This existence of a cardinal but not of an ordinal is similar to what we have in QM when we say that we have some quantity of systems of the same kind but cannot individuate or count them, e.g., the six electrons in the level $2p$ of a Sodium atom.[23]

Identity (termed *extensional identity*) "$=_E$" is defined for qsets having the same elements (in the sense that if an element belongs to one of them, then it belongs to the another)[24] or for M-objects belonging to the same qsets. It can be proven that this identity has all the properties of classical logical identity for the objects to which it applies. However, it does not make sense for q-objects. That is, $x =_E y$ does not have any meaning in the theory if x and y are m-objects. It is similar to speak of categories in the Zermelo-Fraenkel set theory (supposed consistent). The theory cannot capture the concept, yet it can be expressed in its language. From now on, we shall abbreviate "$=_E$" by "$=$," as usual.

The postulates of $\mathfrak{Q}$ are similar to those of ZFA, but by considering that now we may have m-objects. The notion of indistinguishability is extended to qsets through an axiom that says that two qsets with the same q-cardinal and having the same "quantity" (we use q-cardinals to express this) of elements of the same kind (indistinguishable among them) are indiscernible too. As an example, consider the following: two sulfuric acid molecules H_2SO_4 are seen as indistinguishable qsets, for both contain q-cardinal equals to 7 (counting the atoms as basic elements), and the elements of the sub-collections of elements of the same kind are also of the same q-cardinal (2, 1, and 4 respectively). Then we can state that "$H_2SO_4 \equiv H_2SO_4$," but of course, we cannot say that "$H_2SO_4 = H_2SO_4$," as for in the latter, the two molecules would not be two at all, but just the same molecule (supposing, of course, that "$=$" stands

[22] Quasi-cardinals turn to be *sets*, so we can use the equality symbol among them. We use the notation $qc(x) = n$ (really, $qc(x) =_E n$, see below) for a quasi-set x whose cardinal is n.

[23] To count a finite number of elements, say 4, is to define a bijection from the set with these elements to the ordinal $4 = \{0, 1, 2, 3\}$. This counting requires that we identify the elements of the first set.

[24] There are subtleties that require us to provide further explanations. In $\mathfrak{Q}$, you cannot do the maths and decide either a certain m-object belongs or not to a qset; this requires identity, as you need to identify the object you are referring to.

In quasi-set theory, however, one can hypothesize that *if* a specific object belongs to a qset, then so and so. This is similar to Russell's use of the axioms of infinite (I) and choice (C) in his theory of types, which assume the existence of certain classes that cannot be constructed, so going against Russell's constructibility thesis. What was Russell's answer? He transformed all sentences α whose proofs depend on these axioms into conditionals of the form $I \rightarrow \alpha$ and $C \rightarrow \alpha$. Hence, *if* the axioms hold, *then* we can get α. We are applying the same reasoning here: *if* the objects of a qset belong to the another and vice-versa, *then* they are extensionally identical. It should be noted that the definition of extensional identity holds only for sets and M-objects.

for classical logical identity). In the first case, notwithstanding, they count as two, yet we cannot say which is which.

Let us speak a little bit more about quasi-functions. Since physicists and mathematicians may want to talk about random variables over qsets as a way to model physical processes, it is important to define functions between qsets. This can be done straightforwardly, and here we consider binary relations and unary functions only. Such definitions can easily be extended to more complicated multi-valued functions. A (binary) q-relation between the qsets A and B is a qset of pairs of elements (sub-collections with q-cardinal equals 2), one in A, the other in B.[25] Quasi-functions (q-functions) from A to B are binary relations between A and B such that if the pairs (qsets) with a and b and with a' and b' belong to it and if $a \equiv a'$, then $b \equiv b'$ (with a's belonging to A and the b's to B). In other words, a q-function maps indistinguishable elements into indistinguishable elements. When there are no m-objects involved, the indistinguishability relation collapses in the extensional identity, and the definition turns to be equivalent to the classical one. In particular, a q-function from a "classical" set such as $\{1, -1\}$ to a qset of indiscernible q-objects with q-cardinal 2 can be defined so that we cannot know which q-object is associated with each number (this example will be used below).

To summarize, in this section, we showed that the concept of indistinguishability, which conflicts with Leibnitz's Principle of the Identity of Indiscernibles, can be incorporated as a metaphysical principle in a modified set theory with indistinguishable elements. This theory contains "copies" of the Zermelo-Frankel axioms with *Urelemente* as a particular case when no indistinguishable q-objects are involved. This theory will provide us the mathematical basis for formally talking about indistinguishable properties, which we will show can be used in a theory of quantum properties. We will see in the next section how we can use those indistinguishable properties to avoid contradictions in quantum contextual settings such as KS.

7 Formulating Quantum Mechanics Within Quasi-set Theory

As we have seen, the quasi-set theory enables us to form collections (the quasi-sets) of "absolutely" indiscernible elements. In this theory, even if one goes outside the relevant structures, they will not become rigid: this mathematical universe is not rigid. Thus, the quasi-set theory is a suitable device to develop a quantum theory where indiscernibility is considered from the start as a fundamental notion. This section explains how quantum mechanics (in the Fock space formalism) can be developed within the quasi-set theory $\mathfrak{Q}$. The current development is based in (Domenech

[25] We are avoiding the long and boring definitions, as, for instance, the definition of ordered pairs, which presuppose lots of preliminary concepts, just to focus on the basic ideas. For details, the interested reader can see the indicated references.

et al., 2008) and is technical. This level of mathematical formality is necessary to provide essential details. The reader unconcerned with such technicalities may skip this section and proceed directly to the conclusions.

7.1 The $\mathfrak{Q}$-spaces

In the standard mathematical formalisms, the assumptions that quantum entities of the same kind must be indiscernible are hidden behind mathematical tricks such as symmetrizing wave-functions and vectors. In order to avoid these tricks, we introduce the notion of $\mathfrak{Q}$-spaces. The resulting framework is termed *nonreflexive quantum mechanics or, simply, nonreflexive.*

We begin with a q-set of real numbers $\epsilon = \{\epsilon_i\}_{i \in I}$, where I is an arbitrary collection of indexes, denumerable or not. Since it is a collection of real numbers, which may be constructed in the classical part of $\mathfrak{Q}$, we have that $Z(\epsilon)$. Intuitively, the elements ϵ_i represent the eigenvalues of a physical observable $\hat{O}$, that is, they are the values such that $\hat{O}|\varphi_i\rangle = \epsilon_i|\varphi_i\rangle$, with $|\varphi_i\rangle$ the corresponding eigenstates. Since observables are Hermitian operators, the eigenvalues are real numbers. Thus, we are justified in assuming that elements of ϵ are real numbers. Consider then the quasi-functions $f : \epsilon \longrightarrow \mathcal{F}_p$, where $\mathcal{F}_p$ is the quasi-set formed of all finite and pure quasi-sets (that is, finite quasi-sets whose only elements are indistinguishable m-atoms). Each of these f is a q-set of ordered pairs $\langle \epsilon_i, x\rangle$ with $\epsilon_i \in \epsilon$ and $x \in \mathcal{F}_p$. From $\mathcal{F}_p$ we select those quasi-functions f which attribute a non-empty q-set only to a finite number of elements of ϵ, the image of f being $\emptyset$ for the other cases. We call $\mathcal{F}$ the quasi-set containing only these quasi-functions. Then, the quasi-cardinal of most of the q-sets attributed to elements of ϵ according to these quasi-functions is 0. Now, elements of $\mathcal{F}$ are quasi-functions which we read as attributing to each ϵ_i a q-set whose quasi-cardinal we take to be the occupation number of this eigenvalue. We write these quasi-functions as $f_{\epsilon_{i_1}\epsilon_{i_2}\ldots\epsilon_{i_m}}$. According to the given intuitive interpretation, the levels $\epsilon_{i_1}\epsilon_{i_2}\ldots\epsilon_{i_m}$ are occupied. We say that if the symbol ϵ_{i_k} appears j-times, then the level ϵ_{i_k} has occupation number j. For example, the notation $f_{\epsilon_1\epsilon_1\epsilon_1\epsilon_2\epsilon_3}$ means that the level ϵ_1 has occupation number 3 while the levels ϵ_2 and ϵ_3 have occupation numbers 1. The levels that do not appear have occupation number zero. Another point to be remarked is that since the elements of ϵ are real numbers, we can take the standard ordering relation over the reals and order the indexes according to this ordering in the representation $f_{\epsilon_{i_1}\epsilon_{i_2}\ldots\epsilon_{i_m}}$. This will be important when we consider the cases for bosons and fermions.

The quasi-functions of $\mathcal{F}$ provide the key to the solution to the problem of labeling states. Since we use pure quasi-sets as the images of the quasi-functions, there is simply no question of indexes for particles, for all that matters are the quasi-cardinals representing the occupation numbers. To make it clear that permutations change nothing, one needs only to notice that a quasi-function is a q-set of weakly ordered pairs.[26] Taking two of the pairs belonging to some quasi-function, let us say $\langle \epsilon_i, x \rangle$, $\langle \epsilon_j, y \rangle$, with both x and y non-empty, a permutation of particles would consist in changing elements from x with elements from y. However, by the unobservability of permutations theorem,[27] what we obtain after the permutation is a q-set indistinguishable from the one we began with. Remember also that a quasi-function attributes indistinguishable images to indistinguishable items; thus, the indistinguishable q-set resulting from the permutations will also be in the image of the same eigenvalue. To show this point precisely, we recall that by definition $\langle \epsilon_i, x \rangle$ abbreviates $[[\epsilon_i], [\epsilon_i, x]]$,[28] and an analogous expression holds for $\langle \epsilon_j, y \rangle$. Also, by definition, $[\epsilon_i, x]$ is the collection of all the items indistinguishable from ϵ_i or from x (taken from a previously given q-set). For this reason, if we permute x with x', with $x \equiv x'$ we change nothing for $[\epsilon_i, x] \equiv [\epsilon_i, x']$. Thus, we obtain $\langle \epsilon_i, x \rangle \equiv \langle \epsilon_i, x' \rangle$ and the ordered pairs of the permuted quasi-function will be indiscernible (the same if there are no m-atoms involved). Thus, the permutation of indistinguishable elements does not produce changes in the quasi-functions.

7.2 A Vector Space Structure

Now, we wish to have a vector space structure to represent quantum states. To do that, we need to define addition and multiplication by scalars. Before we go on, we must notice that we cannot define these operations directly on the q-set $\mathcal{F}$, for there is no simple way to endow it with the required structure; our strategy here is to define $\star$ (multiplication by scalars) and $+$ (addition of vectors) in a q-set whose vectors will be quasi-functions from $\mathcal{F}$ to the set of complex numbers $\mathbb{C}$. Let us call C the collection of quasi-functions that assign to every $f \in \mathcal{F}$ a complex number. Once again, we select from C the sub-collection C_F of quasi-functions c such that every $c \in C_F$ attributes complex numbers $\lambda \neq 0$ for only a finite number of $f \in \mathcal{F}$. Over C_F, we can define a sum and a product by scalars in the same way as it is usually done with functions as follows.

Definition 7.1 Let $\gamma \in \mathcal{C}$, and c, c_1 and c_2 be quasi-functions of C_F, then

$$(\gamma \star c)(f) := \gamma(c(f))$$

$$(c_1 + c_2)(f) := c_1(f) + c_2(f)$$

[26] A weak ordered pair is a qset having just one element (that is, its cardinal is one). We cannot name such an element, for we need an identity to do that. SO, it can be taken as *one* element of a kind.

[27] This theorem says that if we exchange an element of a qset by an indistinguishable one, the resulting qset turns to be indistinguishable from the original one.

[28] We are leaving aside the subindices in this notation.

The quasi-function $c_0 \in C_F$ such that $c_0(f) = 0$ for every $f \in \mathcal{F}$ acts as the null element for the sum operation. This can be shown as follows:

$$(c_0 + c)(f) = c_0(f) + c(f) = 0 + c(f) = c(f), \forall f. \tag{9}$$

With both the operations of sum and multiplication by scalars defined as above we have that $\langle C_F, \mathbb{C}, +, \star\rangle$ has the structure of a complex vector space, as one can easily check. Some of the elements of C_F have a special status though; if $c_j \in C_F$ are the quasi-functions such that $c_j(f_i) = \delta_{ij}$ (where δ_{ij} is the Kronecker symbol), then the vectors c_j are called the basis vectors, while the others are linear combinations of them. For notational convenience, we can introduce a new notation for the q-functions in C_F; suppose c attributes a $\lambda \neq 0$ to some f, and 0 to every other quasi-function in $\mathcal{F}$. Then, we propose to denote c by λf. The basis quasi-functions will be denoted simply f_i, as one can check. Now, multiplication by scalar α of one of these quasi-functions, say λf_i can be read simply as $(\alpha \cdot \lambda) f_i$, and sum of quasi-functions λf_i and αf_i can be read as $(\alpha + \lambda) f_i$. What about the other quasi-functions in C_F? We can extend this idea to them too, but with some care: if, for example c_0 is a quasi-function such that $c_0(f_i) = \alpha$ and $c_0(f_j) = \lambda$, attributing 0 to every other quasi-function in $\mathcal{F}$, then c_0 can be seen as a linear combination of quasi-functions of a basis; in fact, consider the basis quasi-functions f_i and f_j, (this is an abuse of notation, for they are representing quasi-functions in C_F that attribute 1 to each of these quasi-functions). The first step consists in multiplying them by α and λ, respectively, obtaining αf_i and λf_j (once again, this is an abuse, for these are quasi-functions in C_F that attribute the mentioned complex numbers to f_i and to f_j). Now, c_0 is in fact the sum of these quasi-functions, that is, $c_0 = \alpha f_i + \lambda f_j$, for this is the function which does exactly what c_0 does. One can then extend this to all the other quasi-functions in C_F as well.

7.3 *Inner Products*

The next step in our construction is to endow our vector space with an inner product. This is a necessary step for we wish to calculate probabilities and mean values. Following the idea proposed in (Domenech et al., 2008), we introduce two kinds of inner products, which lead us to two Hilbert spaces, one for bosons and another for fermions. We begin with the case for bosons.

Definition 7.2 Let δ_{ij} be the Kronecker symbol and $f_{\epsilon_{i_1}\epsilon_{i_2}\ldots\epsilon_{i_n}}$ and $f_{\epsilon_{i'_1}\epsilon_{i'_2}\ldots\epsilon_{i'_m}}$ two basis vectors (as discussed above), then

$$f_{\epsilon_{i_1}\epsilon_{i_2}\ldots\epsilon_{i_n}} \circ f_{\epsilon_{i'_1}\epsilon_{i'_2}\ldots\epsilon_{i'_m}} := \delta_{nm} \sum_p \delta_{i_1 pi'_1} \delta_{i_2 pi'_2} \ldots \delta_{i_n pi'_n}. \tag{10}$$

Notice that this sum is extended over all the permutations of the index set $i' = (i'_1, i'_2, \ldots, i'_n)$; for each permutation p, $pi' = (pi'_1, pi'_2, \ldots, pi'_n)$.

For the other vectors, the ones that can be seen as linear combinations in the sense discussed above, we have

$$\left(\sum_k \alpha_k f_k) \circ (\sum_k \alpha'_k f'_k) := \sum_{kj} \alpha^*_k \alpha'_j (f_k \circ f'_j \right), \tag{11}$$

where α^* is the complex conjugate of α. Now, let us consider fermions. As remarked above in page 512, the order of the indexes in each $f_{\epsilon_{i_1}\epsilon_{i_2}\ldots\epsilon_{i_n}}$ is determined by the canonical ordering in the real numbers. Thus, we define another • inner product as follows, which will do the job for fermions.

Definition 7.3 Let δ_{ij} be the Kronecker symbol and $f_{\epsilon_{i_1}\epsilon_{i_2}\ldots\epsilon_{i_n}}$ and $f_{\epsilon_{i'_1}\epsilon_{i'_2}\ldots\epsilon_{i'_m}}$ two basis vectors, then

$$f_{\epsilon_{i_1}\epsilon_{i_2}\ldots\epsilon_{i_n}} \bullet f_{\epsilon_{i'_1}\epsilon_{i'_2}\ldots\epsilon_{i'_m}} := \delta_{nm} \sum_p \sigma_p \delta_{i_1 p i'_1} \delta_{i_2 p i'_2} \ldots \delta_{i_n p i'_n} \tag{12}$$

where: $\sigma_p = 1$ if p is even and $\sigma_p = -1$ if p is odd.

This definition can be extended to linear combinations as in the previous case.

7.4 *Fock Spaces Using* $\mathfrak{Q}$*-spaces*

We begin with a definition to simplify the notation. For every function $f_{\epsilon_{i_1}\epsilon_{i_2}\ldots\epsilon_{i_n}}$ in $\mathcal{F}$, we put

$$\alpha|\epsilon_{i_1}\epsilon_{i_2}\ldots\epsilon_{i_n}) := \alpha f_{\epsilon_{i_1}\epsilon_{i_2}\ldots\epsilon_{i_n}}$$

Note that this is a slightly modified version of the standard notation. We begin with the case of bosons.

Suppose a normalized vector $|\alpha\beta\gamma\ldots)$, where the norm is taken from the corresponding inner product. Let ζ stand for an arbitrary collection of indexes. We define $a^\dagger_\alpha|\zeta) \propto |\alpha\zeta)$ in such a way that the proportionality constant satisfies $a^\dagger_\alpha a_\alpha|\zeta) = n_\alpha|\zeta)$. From this it will follow, as usual, that:

$$((\zeta|a^\dagger_\alpha)(a_\alpha|\zeta)) = n_\alpha.$$

Definition 7.4 $a_\alpha|\ldots n_\alpha\ldots) := \sqrt{n_\alpha}|\ldots n_\alpha - 1\ldots)$

On the other hand,

$$a_\alpha a^\dagger_\alpha|\ldots n_\alpha\ldots) = K\sqrt{n_\alpha + 1}|\ldots n_\alpha\ldots),$$

where K is a proportionality constant. Applying $a_\alpha^\dagger$ again, we have

$$a_\alpha^\dagger a_\alpha a_\alpha^\dagger | \ldots n_\alpha \ldots) = K^2 \sqrt{n_\alpha + 1} | \ldots n_\alpha + 1 \ldots).$$

Using the fact that $a_\alpha^\dagger a_\alpha |\zeta) = n_\alpha |\zeta)$, we have that

$$(a_\alpha^\dagger a_\alpha) a_\alpha^\dagger | \ldots n_\alpha \ldots) = \sqrt{n_\alpha + 1} K | \ldots n_\alpha + 1 \ldots).$$

So, $K = \sqrt{n_\alpha + 1}$. Then, we have

Definition 7.5 $a_\alpha^\dagger | \ldots n_\alpha \ldots) := \sqrt{n_\alpha + 1} | \ldots n_\alpha + 1 \ldots)$.

From this definition, with additional computations, we obtain $(a_\alpha a_\beta^\dagger - a_\beta^\dagger a_\alpha)|\psi) = \delta_{\alpha\beta}|\psi)$. In our language, this means the same as

$$[a_\alpha ; a_\beta^\dagger] = \delta_{\alpha\beta} I.$$

In an analogous way, it can be shown that

$$[a_\alpha ; a_\beta] = [a_\alpha^\dagger ; a_\beta^\dagger] = 0.$$

So, the bosonic commutation relation is the same as in standard Fock space formalism.

For fermionic states, we use the antisymmetric product "•." We begin by defining the creation operator $C_\alpha^\dagger$.

Definition 7.6 If ζ is a collection of indexes of non-null occupation numbers, then $C_\alpha^\dagger := \alpha|\zeta)$

If α is in ζ, then $|\alpha\zeta)$ is a vector of null norm. This implies that $(\psi|\alpha\zeta) = 0$, for every ψ. It follows that systems in states of null norm have no probability of being observed. Furthermore, their addition to another vector does not contribute to any observable difference. To take the situation into account, we have the following definition.

Definition 7.7 Two vectors $|\phi)$ and $|\psi)$ are similar if the difference between them is a linear combination of null norm vectors. We denote similarity of $|\phi)$ and $|\psi)$ by $|\phi) \cong |\psi)$.

Using the definition of $C_\alpha^\dagger$ we can describe what is the effect of C_α over vectors, namely

$$(\zeta|C_\alpha := (\alpha\zeta|.$$

Then, for any vector $|\psi)$,

$$(\zeta|C_\alpha|\psi) = (\alpha\zeta|\psi) = 0$$

for $\alpha \in \zeta$ or $(\psi|\alpha\zeta) = 0$. Then, if $|\psi) = |0)$, then $(\zeta|C_\alpha|0) = (\alpha\zeta|0) = 0$. So, $C_\alpha|0)$ is orthogonal to any vector that contains α, and also to any vector that does not contain

α, so that it is a linear combination of null norm vectors. So, we can put by definition that $\vec{0} := C_\alpha|0)$. In an analogous way, if $\sim\alpha$ denotes that α has occupation number zero, then we can also write $C_\alpha|(\sim\alpha)\ldots) = \vec{0}$, where the dots mean that other levels have arbitrary occupation numbers.

Now, using our notion of similar vectors, we can write $C_\alpha|0) \cong \vec{0}$ and $C_\alpha|(\sim\alpha)\ldots) \cong \mathbf{0}$. The same results are obtained when we use $\cong$ and the sign of identity. By making $|\psi) = |\alpha)$, we have $(\zeta|C_\alpha|\alpha) = (\alpha\zeta|\alpha) = 0$ in every case, except when $|\zeta) = |0)$. In that case, $(0|C_\alpha|\alpha) = 1$. Then, it follows that $C_\alpha|\alpha) \cong 0$. In an analogous way, we obtain $C_\alpha|\alpha\zeta) =\cong |(\sim\alpha)\zeta)$ when $\alpha \notin \zeta$. In the case $\alpha \in \zeta$, $|\alpha\zeta)$ has null norm, and so, for every $|\psi)$:

$$(\alpha\zeta|C_\alpha^\dagger|\psi) = (\alpha\zeta|\alpha\psi) = 0.$$

It then follows that

$$(\psi|C_\alpha|\alpha\zeta) = 0,$$

so that $C_\alpha|\alpha\zeta)$ has null norm too.

Now we calculate the anti-commutation relation obeyed by the fermionic creation and annihilation operators. We begin calculating the commutation relation between C_α and $C_\beta^\dagger$. We do that by studying the relationship between $|\alpha\beta)$ and $|\beta\alpha)$. Let us consider the sum $|\alpha\beta) + |\beta\alpha)$. The product of this sum with any vector distinct from $|\alpha\beta)$ is null. For the product with $|\alpha\beta)$ we obtain $(\alpha\beta|[|\alpha\beta) + |\beta\alpha)] = (\alpha\beta||\alpha\beta) + (\alpha\beta||\beta\alpha)$. By definition, this is equal to $\delta_{\alpha\alpha}\delta_{\beta\beta} - \delta_{\alpha\beta}\delta_{\beta\alpha} + \delta_{\alpha\beta}\delta_{\alpha\alpha} - \delta_{\alpha\alpha}\delta_{\beta\beta}$. This is equal to $1 - 0 + 0 - 1 = 0$.

The same conclusion holds if we multiply the sum $|\alpha\beta) + |\beta\alpha)$ by $(\beta\alpha|$. It then follows that $|\alpha\beta) + |\beta\alpha)$ is a linear combination of null norm vectors, which we denote by $|nn)$, so that

$$|\alpha\beta) = -|\beta\alpha) + |nn).$$

Given that, we can calculate

$$C_\alpha^\dagger C_\beta^\dagger|\psi) = |\alpha\beta\psi) = -|\beta\alpha|\psi) + |nn) = -C_\beta^\dagger C_\alpha^\dagger|\psi) + |nn).$$

From this it follows that $\{C_\alpha^\dagger; C_\beta^\dagger\}|\psi) = |nn)$. We do not lose generality by setting $\{C_\alpha^\dagger; C_\beta^\dagger\}|\psi) = 0$. In an analogous way we conclude that

$$\{C_\alpha; C_\beta\}|\psi) = 0.$$

Now we calculate the commutation relation between C_α and $C_\beta^\dagger$. There are some cases to be considered. We first assume that $\alpha \neq \beta$. If $\alpha \notin \psi$ or $\beta \in \psi$ then

$$\{C_\alpha; C_\beta^\dagger\}|\psi) \approx \vec{\mathbf{0}}.$$

If $\alpha \in \psi$ and $\beta \notin \psi$, assuming that α is the first symbol in the list of ψ, then $\{C_\alpha; C^\dagger_\beta\}|\psi) = C_\alpha|\beta\psi) + C^\dagger_\beta|\psi(\sim\alpha)) \cong -|\beta\psi(\sim\alpha)) + |\beta\psi(\sim\alpha)) = \vec{0}$. If $\alpha = \beta$ and $\alpha \in \psi$, then $\{C_\alpha; C^\dagger_\alpha\}|\psi) = C_\alpha|\alpha\psi) + C^\dagger_\alpha|\psi(\sim\alpha)) \cong \vec{0} + |\psi) = |\psi)$. If $\alpha = \beta$ and $\alpha \notin \psi$, then $\{C_\alpha; C^\dagger_\alpha\}|\psi) = C_\alpha|\alpha\psi) + C^\dagger_\alpha|\psi(\sim\alpha)) \cong |\psi) + \vec{0} = |\psi)$. In any case, we recover $\{C_\alpha; C^\dagger_\alpha\}|\psi) \cong \delta_{\alpha\beta}|\psi)$. So, we can put

$$\{C_\alpha; C^\dagger_\alpha\} = \delta_{\alpha\beta}.$$

It then follows that the commutation properties in $\mathfrak{Q}$-spaces are the same as in traditional Fock spaces.

Using this formalism, we can adapt all the developments done in (Mattuck, 1967, Chap. 7) and (Merzbacher, 1970, Chap. 20) for the number occupation formalism. However, contrary to what happens in these books, no previous (even unconscious) assumptions about quantum objects' individuality is taken into account.

8 Conclusions

It is an exciting question to ask if we need to change logic every time we find difficulties with the classical one. Are there other ways to circumvent the problems, such as in the quantum case, using the tricks mentioned above, or choosing an alternative interpretation? This question makes sense. However, we think that every theory, even a mathematical one, starts from metaphysical hypotheses, even if not made explicit. We have stated above that classical logic, standard mathematics, and classical physics were developed with the classical enclosing world in our minds. This world is one of individuals that have an identity. So, two of those individuals cannot possibly be different.

Nevertheless, quantum mechanics brought us a different world, a world with no proper names. In the quantum world, objects are (in most cases) precisely alike, and permutations between objects of the same kind do not lead to any physical differences. Here we emphasize that it is not that these are not *measurable* differences; *there are no differences at all.* So, we arrive at the following conclusions.

1. Indistinguishability is essential in quantum mechanics, regardless of interpretation. In our opinion, it should be placed at an equal level of importance in quantum foundations to concepts such as entanglement, contextuality, and nonlocality.
2. Ontological and epistemic aspects matter. Any physical theory is grounded in interpretations due to the possibility of associating different world views (or metaphysics) to a theory. Parodying Poincaré, we can say that physics is (also) a domain where we give the same name to distinct things.[29]

[29] Poincaré was referring to mathematics: "mathematics is the art of giving the same name to distinct things"—look at (Verhulst, 2012). Of course, he spoke within the framework of axiomatized mathematical theories, able to have different models.

3. Since mathematics and logic need to reflect the assumed metaphysical aspects (we could speak in terms of ontology), quantum mechanics' formalism and physical theories should do the same.

Let us expand on this last point with an example involving logic. It is common to say that in order to obtain intuitionistic logic, it is enough to drop the excluded middle law from the axioms of classical logic. From a purely formal point of view, this is correct. However, logic is not only syntax. It also involves semantic aspects and even pragmatic ones (making references to who uses the logic and why). Let us consider semantics. Although classical and intuitionistic logic differs syntactically just by one axiom, semantically, they are much different. Classical propositional logic can be described through truth-tables; intuitionistic logic cannot. In classical logic, any proposition is either true or false, yet we may not know what the case is; in intuitionistic logic, the notions of true and false are different. In this logic, a proposition p is true if there is a "process" to get it, and false if a process for obtaining p leads to a contradiction. Other differences can be pointed out. For instance, in classical logic, something exists if its nonexistence entails a contradiction. In intuitionistic logic, something exists if it can be created by our imagination.

This example shows that in order to consider a logic, semantical aspects must at least be considered. Of course, this is true also with physical theories. Otherwise, we risk having a purely mathematical theory. However, what corresponds to semantics in the quantum case? We chose interpretations because quantum mechanics, as Yuri Manin wrote, "does not really have its own language" (Manin, 1977, p. 84). At least not yet. Indeed, the standard formalism grounded on Hilbert spaces makes use of the language of standard functional analysis, which presupposes classical mathematics and logic, with all the problems seem before (in regarding quantum phenomena). A proper language should reflect the indiscernibility of quanta from the start, without tricks!

As we showed in this paper, such a correct language can be constructed. In this paper, we examined content and context in quantum physics. We provided examples of context for the classical and quantum realms and argued that the quantum situation is fundamentally different. Furthermore, we reasoned that context-dependency in the quantum world is intrinsically connected to the lack of identity. Thus, the non-identity of individuals is an essential feature of the quantum world. Since the standard mathematics used in physics does not exactly allow for objects who lack identity, i.e., indistinguishable objects, we advocated for using a different mathematical structure in physics: quasi-set theory. Quasi-set theory includes standard mathematic in it but also contains indistinguishable objects. We believe that recreating quantum physics in terms of quasi-set theory and its underlying logic would result in thinking closer to a more reasonable ontology for the quantum world than currently available ontologies. This way of thinking may lead to exciting insights into quantum ontologies and fundamental physical principles that define quantum mechanics.

References

Aerts, D., D'Hondt, E., & Gabora, L. (2000). Why the disjunction in quantum logic is not classical. *Foundations of Physics, 30*(9), 1473–1480.

Arenhart, J. R. B. (2011). A discussion on finite quasi-cardinals in quasi-set theory. *Foundations of Physics, 41*, 1338–1354.

Bohm, D. (1952). A Suggested interpretation of the quantum theory in terms of "Hidden" variables. *II. Physical Review, 85*(2), 180–193. https://doi.org/10.1103/PhysRev.85.180

Bouwmeester, D., Pan, J.-W., Daniell, M., Weinfurter, H., & Zeilinger, A. (1999). Observation of three-photon Greenberger-Horne-Zeilinger entanglement. *Physical Review Letters, 82*(7), 1345–1349.

Button, T., & Walsh, S. (2018). *Philosophy and model theory*. Oxford: Oxford University Press.

da Costa, N. C. A., & Krause, D. (1997). An intensional Schrödinger logic. *Notre Dame Journal of Formal Logic, 38*(2), 179–194.

de Barros, J. A., & Suppes, P. (2000). Inequalities for dealing with detector inefficiencies in Greenberger-Horne-Zeilinger type experiments. *Physical Review Letters, 84*(5), 793–797.

de Barros, J. A., Kujala, J. V., & Oas, G. (2016). Negative probabilities and contextuality. *Journal of Mathematical Psychology, 74*, 34–45. https://doi.org/10.1016/j.jmp.2016.04.014

de Barros, J. A., Holik, F., & Krause, D. (2017). Contextuality and indistinguishability. *Entropy, 19*(9), 435–57.

Domenech, G., & Holik, F. (2007). A discussion on particle number and quantum indistinguishability. *Foundations of Physics, 37*(6), 855–78.

Domenech, G., Holik, F., & Krause, D. (2008). Q-spaces and the foundations of quantum mechanics. *Foundations of Physics, 38*(11), 969–994.

Dalla Chiara, M. L., Giuntini, R., & Greechie, R. (2004). *Reasoning in Quantum Theory. Sharp and Unsharp Quantum Logics*. Kluwer Ac. Pu.

Dzhafarov, E. N., & Kujala, J. V. (2016, July). Contextuality-by-default 2.0: Systems with binary random variables. In de Barros, J. A., Coecke, B., & Pothos, E. (Eds.), *International Symposium on Quantum Interaction* (pp. 16–32). Springer.

French, S., & Krause, D. (2006). *Identity in physics: A historical, philosophical, and formal analysis*. Oxford: Oxford University Press.

French, S., & Krause, D. (2010). Remarks on the theory of quasi-sets. *Studia Logica, 95*(1–2), 101–124.

Geach, P. (1967). Identity. *Review of Metaphysics, 21*, 3–12.

Greenberger, D. M., Horne, M. A., & Zeilinger, A. (1989). Going Beyond Bell's theorem. In M. Kafatos (Ed.), *Bell's theorem, quantum theory, and conceptions of the universe* (Vol. 37, pp. 69–72). Kluwer.

Hodges, W. (1983). Elementary predicate logic. In: D. M. Gabbay & F. Guenthner, (Eds.), *Handbook of philosophical logic—Vol. I: Elements of classical logic* (pp. 1–131). D. Reidel.

Holland, P. R. (1995). *The quantum theory of motion: An account of the de Broglie-Bohm causal interpretation of quantum mechanics*. Cambridge University Press.

Hume, D. (1985). *Treatise of human nature* (L. A. Selby-Bigge (Eds.), 2nd ed.). Oxford University Press.

Jaeger, G. (2009). *Entanglement, information, and the interpretation of quantum mechanics*. Springer.

Krause, D., Arenhart, J. R. B. & Bueno, O. (2020). The non-individuals interpretation of quantum mechanics. In O. Freire Jr., G. Bacciagaluppi, O. Darrigol, T. Hartz, C. Joas, A. Kojevnikov, & O. Pessoa Jr. (Eds.), Forthcoming in the *Oxford Handbook of the History of Interpretations of Quantum Mechanics*.

Krause, D., & Coelho, A. M. N. (2005). Identity, indiscernibility, and philosophical claims. *Axiomathes, 15*, 191–210. https://doi.org/10.1007/s10516-004-6678-5

Locke, J. (1959). *AN essay concerning human understanding*. New York: Dover.

Manin, Y. I. (1977). *A course in mathematical logic*. Springer.

Mattuck, R. D. (1967). *A guide do Feynman diagrams in the many-body problem*. McGraw-Hill.
Merzbacher, E. (1970). *Quantum mechanics*. New York: Wiley.
Penrose, R. (2004). *The road to reality: A complete guide to the laws of the universe*. London: Jonathan Cape.
Popescu, S., & Rohrlich, D. (1994). Quantum nonlocality as an axiom. *Foundations of Physics, 24*(3), 379–385.
Quine, V. O. (1969). *Ontological relativity and other essays*. New York: Columbia University Press.
Ramsey, F. P. (1965). *The foundations of mathematics and other logical essays*. In R. B. Braithwaite (Ed.), with a preface by G. E. Moore. Routledge & Kegan Paul.
Schrödinger, E. (1998). What is an elementary particle? In E. Castellani (Ed.), *Interpreting bodies: Classical and quantum objects in modern physics* (pp. 197–210). Princeton: Princeton University Press.
Specker, E. P. (1975). The logic of propositions which are not simultaneously decidable. In C. A. Hooker (Ed.), *The logico-algebraic approach to quantum mechanics* (pp. 135–140). Netherlands: Springer.
Styer, D. F., et al. (2002). Nine formulations of quantum mechanics. *American Journal of Physics, 70*(3), 288–297.
Svozil, K. (1998). *Quantum logic*. Singapore: Springer.
Teller, P. (1998). Quantum mechanics and haecceities. In E. Castellani (Ed.), *Interpreting bodies: Classical and quantum objects in modern physics*. New Jersey: Princeton University Press.
Toraldo di Francia, G. (1986). *Le Cose e i Loro Nomi*. Bari: Laterza.
Verhulst, F. (2012). An interview with Henri Poincaré. NAW 5/13 nr. 3 September 2012
Vickers, P. (2013). *Understanding inconsistent science*. Oxford: Oxford University Press.
Weyl, H. (1949). *Philosophy of mathematics and natural science*. Princeton: Princeton University Press.

Contextual Probability in Quantum Physics, Cognition, Psychology, Social Science, and Artificial Intelligence

Andrei Khrennikov

1 Introduction

This is the review on the applications of QP to modeling behavior of biological, social, and AI systems. The recent years were characterized by explosion of interest to applications of quantum theory outside of physics, especially in cognitive psychology, decision making, information processing in the brain, molecular biology, genetics and epigenetics, and evolution theory, psychology, decision making, social and political sciences, economics and finance (see Khrennikov, 1999–Khrennikov, 2004a for the pioneer papers, (Khrennikov, 2004b)–(Bagarello, 2019) for monographs, and (Ozawa & Khrennikov, 2020)–(Khrennikov & Watanabe, 2021) for the recent papers).

We call the corresponding models quantum-like. They are not directed to micro-level modeling of real quantum physical processes in biosystems, say in cells or brains. Thus, quantum-like research has to be sharply distinguished from quantum biophysics - the study of genuine quantum physical processes in biosystems, in particular, from quantum brain theory associated with the names of Penrose and Hameroff. Quantum-like modeling works from the viewpoint to quantum theory as a measurement theory. For example, humans are systems performing information processing and measurements including self-measurements which are described by the quantum formalism. Such modeling is also applicable to AI-systems which functioning is based on the Hilbert space representation of their information states and transition from one state to another is described by quantum channels. The quantum-like approach to modeling of cognition is based on the quantum-like contextual paradigm proposed by the author (Khrennikov, 2004b, 2010):

A. Khrennikov (✉)
Linnaeus University, International Center for Mathematical Modeling in Physics and Cognitive Sciences, 351 95 Vaxjo, Sweden
e-mail: andrei.khrennikov@lnu.se

S. Wuppuluri and I. Stewart (eds.), *From Electrons to Elephants and Elections*,
The Frontiers Collection, https://doi.org/10.1007/978-3-030-92192-7_28

The mathematical formalism of quantum information and probability theories can be used to model behavior not only of genuine quantum physical systems, but all context-sensitive systems, e.g., human beings.

Thus, contextuality of quantum theory is one of the main motivations for using QP in cognition, psychology, and decision making. The same can be said about AI-systems that should be sensitivity of context variation.

We review the basic applications of quantum-like models, from cognition to complex social processes, including the recent waves of mass protests throughout the world described by the novel theory of social laser (Khrennikov, 2020), as well to behavior of AI systems, individual systems as robots and their collectives.

We want to present in more detail consequences of such information processing for rationality. In classical decision making, rational agents are mathematically modeled as probabilistic information processors using Bayesian update of probabilities: rational = Bayesian. Quantum state update is generally non-Bayesian (Ozawa & Khrennikov, 2020). We define quantum rationality as decision making that is based on quantum state update. Quantum and classical rational agents behave differently. For instance, a quantum(-like) agent can violate the Savage Sure Thing Principle (Savage, 1954) (see Busemeyer & Bruza, 2012; Haven & Khrennikov, 2013; Haven et al., 2017) and the Aumann theorem (Aumann, 1976) on impossibility of agreeing to disagree (see Haven et al., 2017).

2 Classical and Quantum Probability Calculi: Measures Versus Complex Amplitudes

classical probability (CP) was mathematically formalized by Kolmogorov (1933) (Kolmogorov, 1933). This is the calculus of probability measures, where a non-negative weight $p(A)$ is assigned to any event A. The main property of CP is its additivity: if two events O_1, O_2 are disjoint, then the probability of disjunction of these events equals to the sum of probabilities:

$$P(O_1 \vee O_2) = P(O_1) + P(O_2).$$

In fact, powerful integration theory that is needed for calculation of averages demands σ-additivity:

$$P(\cup_j O_j) = \sum_j P(O_j), \tag{1}$$

where $O_j \cap O_i = \emptyset, i \neq j$.

Quantum probability (QP) is the calculus of complex amplitudes or in the abstract formalism complex vectors. Thus, instead of operations on probability measures one operates with vectors. We can say that QP is a *vector model of probabilistic reasoning.*

Each complex amplitude ψ gives the probability by the Born's rule: *Probability is obtained as the square of the absolute value of the complex amplitude.*

$$p = |\psi|^2. \tag{2}$$

By operating with complex probability amplitudes, instead of the direct operation with probabilities, one can violate the basic laws of CP.

In CP, the *law of total probability* (LTP) is derived by using additivity of probability and *the Bayes formula*, the definition of conditional probability,

$$P(O_2|O_1) = \frac{P(O_2 \cap O_1)}{P(O_1)}, \quad P(O_1) > 0. \tag{3}$$

Consider the pair, A and B, of discrete classical random variables. Then

$$P(B = \beta) = \sum_{\alpha} P(A = \alpha) P(B = \beta | A = \alpha).$$

Thus, in CP the B-probability distribution can be calculated from the A-probability and the conditional probabilities $P(B = \beta | A = \alpha)$.

In QP classical LTP is perturbed by the interference term (Khrennikov, 2010); for dichotomous quantum observables A and B of the von Neumann-type, i.e., given by Hermitian operators $\hat{A}$ and $\hat{B}$, the quantum version of LTP has the form:

$$P(B = \beta) = \sum_{\alpha} P(A = \alpha) P(B = \beta | a = \alpha) \tag{4}$$

$$+2 \sum_{\alpha_1 < \alpha_2} \cos\theta_{\alpha_1\alpha_2} \sqrt{P(A = \alpha_1) P(B = \beta | A = \alpha_1) P(A = \alpha_2) P(B = \beta | a = \alpha_2)} \tag{5}$$

If *the interference term* is positive, then the QP-calculus would generate a probability that is larger than its CP-counterpart given by the classical LTP (2). In particular, this probability amplification is the basis of the quantum computing supremacy.

3 Non-Kolmogorovness of QP and Its Contextual Background: Quantum-like Paradigm

Violation of CP-laws in QP can be formulated in the probabilistic terms as the impossibility to represent all quantum observables as random variables on the same Kolmogorov probability space. This is a consequence of the existence of incompatible observables which are mathematically described by noncommuting operators.

In quantum physics, incompatible experimental contexts are described by different probability spaces. Thus, QP can be considered as a special calculus of contextual probabilities, manifold of probability spaces labeled by experimental contexts. This probabilistic picture matches with Bohr's complementarity principle.

A complex probability amplitude, a pure quantum state, serves for coupling probability measures corresponding to a variety of generally incompatible observables.[1]

As mentioned in introduction, applications of QP to cognition and decision making are based on

Quantum-like paradigm. *The mathematical formalism of quantum information and probability theories can be used to model behavior not only of genuine quantum physical systems, but all context-sensitive systems, e.g., human beings. Contextual information processing cannot be based on complete resolution of ambiguity. It is meaningless to do this for the concrete context, if in a few minutes (or even seconds) context will be totally different. Therefore such systems process ambiguities, process superpositions of alternatives.*

We remark that in the original Kolmogorv's formulation (Kolmogorov, 1933) CP is also contextual, but this contextual viewpoint on CP was practically forgotten (see appendix for details).

4 Quantum Formalism

4.1 States: Pure and Mixed

Denote by $\mathcal{H}$ a complex Hilbert space endowed with the scalar product $\langle\cdot|\cdot\rangle$. For simplicity, we assume that *it is finite dimensional.* The space of density operators is denoted by $\mathfrak{S}(\mathcal{H})$ The space of all linear operators in $\mathcal{H}$ is denoted by the symbol $\mathcal{L}(\mathcal{H})$. In turn, this is the complex Hilbert space with the scalar product, $\langle A|B\rangle = Tr(A^\star B)$. We shall also consider linear operators acting in $\mathcal{L}(\mathcal{H})$. They are called *superoperators.*

A pure quantum state is represented by a vector $|\psi\rangle \in \mathcal{H}$ that is normalized to 1, i.e., $\langle\psi|\psi\rangle = 1$. It can be represented as the density operator $\rho_\psi = |\psi\rangle\langle\psi|$; this is the orthogonal projector on the vector $|\psi\rangle$. States which are not pure are called mixed.

[1] A density operator, a mixed quantum state, serves for the same purpose.

4.2 Entropy

The von Neumann entropy is defined as

$$\mathcal{S}(\rho) = -Tr\rho \ln \rho, \tag{6}$$

where ρ is a density operator.

There exists an orthonormal basis $|j\rangle$ consisting of eigenvectors of ρ, i.e., $\rho|j\rangle = p_j|j\rangle$ (where $p_j \geq 0$ and $\sum_j p_j = 1$). In this basis, the matrix of the operator $\rho \ln \rho$ has the form diag($\mathrm{p_j \ln p_j}$;) hence

$$\mathcal{S}(\rho) = -\sum_j p_j \ln p_j. \tag{7}$$

However, the von Neumann entropy has the classical form, but only w.r.t. this to special basis.

We present three basic properties of the von Neumann entropy.

1. $\mathcal{S}(\rho) = 0$ if and only if ρ is a pure quantum state, i.e., $\rho = |\psi\rangle\langle\psi|$.
2. For a unitary operator U, $\mathcal{S}(U\rho U^\star) = \mathcal{S}(\rho)$.
3. The maximum of entropy is approached on the state $\rho_{\mathrm{disorder}} = I/N$ and $\mathcal{S}(\rho_{\mathrm{disorder}}) = \ln N$, where N is the dimension of the state space.

It is natural to call $\rho_{\mathrm{disorder}} = I/N$ the state of maximal disorder.

4.3 Projective Measurements

In the original quantum formalism (Von Neumann, 1955), physical observable A is represented by a Hermitian operator $\hat{A}$. We consider only operators with discrete spectra:

$$\hat{A} = \sum_x x\ \hat{E}^A(x),$$

where $\hat{E}^A(x)$ is the projector onto the subspace of H corresponding to the eigenvalue x. Suppose that system's state is mathematically represented by a density operator ρ. Then the probability to get the answer x is given by the Born rule

$$\Pr\{A = x \| \rho\} = Tr[\hat{E}^A(x)\rho] = Tr[\hat{E}^A(x)\rho\hat{E}^A(x)] \tag{8}$$

and according to the projection postulate the post-measurement state is obtained via the state-transformation:

$$\rho \to \rho_x = \frac{\hat{E}^A(x)\rho\hat{E}^A(x)}{Tr\hat{E}^A(x)\rho\hat{E}^A(x)}. \tag{9}$$

For reader's convenience, we present these formulas for a pure initial state $\psi \in \mathcal{H}$. The Born's rule has the form:

$$\Pr\{A = x \| \rho\} = \|\hat{E}^A(x)\psi\|^2 = \langle\psi|\hat{E}^A(x)\psi\rangle. \quad (10)$$

The state transformation is given by the projection postulate:

$$\psi \to \psi_x = \hat{E}^A(x)\psi / \|\hat{E}^A(x)\psi\|. \quad (11)$$

Here the observable-operator $\hat{A}$ (its spectral decomposition) uniquely determines the feedback state transformations $\mathcal{I}_A(x)$ for outcomes x

$$\rho \to \mathcal{I}_A(x)\rho = \hat{E}^A(x)\rho\hat{E}^A(x). \quad (12)$$

The map $x \to \mathcal{I}_A(x)$ given by (12) is the simplest (but very important) example of quantum instrument.

4.4 Simplest Non-projective Measurements

In general, the statistical properties of any measurement are characterized by

(i) the output probability distribution $\Pr\{\mathbf{x} = x \| \rho\}$, the probability distribution of the output $\mathbf{x}$ of the measurement in the input state ρ;
(ii) the quantum state reduction $\rho \mapsto \rho_{\{\mathbf{x}=x\}}$, the state change from the input state ρ to the output state $\rho_{\{\mathbf{x}=x\}}$ conditional upon the outcome $\mathbf{x} = x$ of the measurement.

In von Neumann's formulation, the statistical properties of any measurement of an observable A is uniquely determined by Born's rule (8) and the projection postulate (9), and they are represented by the map (12), an instrument of von Neumann type. However, von Neumann's formulation does not reflect the fact that the same observable A represented by the Hermitian operator $\hat{A}$ in $\mathcal{H}$ can be measured in many ways.[2] Formally, such measurement-schemes are represented by quantum instruments.

Now, we consider the simplest quantum instruments of non von Neumann type, known as *atomic instruments*. We start with recollection of the notion of POVM (probability operator valued measure); we restrict considerations to POVMs with a discrete domain of definition $X = \{x_1, ..., x_N, ...\}$. POVM is a map $x \to \hat{D}(x)$ such that for each $x \in X$, $\hat{D}(x)$ is a positive contractive Hermitian operator (called effect) (i.e., $\hat{D}(x)^\star = \hat{D}(x)$, $0 \leq \langle\psi|\hat{D}(x)\psi\rangle \leq 1$ for any $\psi \in \mathcal{H}$), and the normalization condition

$$\sum_x \hat{D}(x) = I$$

[2] Say $\hat{A} = \hat{H}$ is the operator representing the energy-observable. This is just a theoretical entity encoding energy. Energy can be measured in many ways within very different measurement schemes.

holds, where I is the unit operator. It is assumed that for any measurement, the output probability distribution $\Pr\{\mathbf{x} = x \| \rho\}$ is given by

$$\Pr\{\mathbf{x} = x \| \rho\} = Tr[\hat{D}(x)\rho], \tag{13}$$

where $\{\hat{D}(x)\}$ is a POVM. For atomic instruments, it is assumed that effects are represented concretely in the form

$$\hat{D}(x) = \hat{V}(x)^{\star}\hat{V}(x), \tag{14}$$

where $V(x)$ is a linear operator in H. Hence, the normalization condition has the form $\sum_x V(x)^{\star}V(x) = I$. [3] The Born rule can be written similarly to (8):

$$\Pr\{\mathbf{x} = x \| \rho\} = Tr[V(x)\rho V^{\star}(x)] \tag{15}$$

It is assumed that the post-measurement state transformation is based on the map:

$$\rho \to \mathcal{I}_A(x)\rho = V(x)\rho V^{\star}(x), \tag{16}$$

so the quantum state reduction is given by

$$\rho \to \rho_{\{\mathbf{x}=x\}} = \frac{\mathcal{I}_A(x)\rho}{Tr[\mathcal{I}_A(x)\rho]}. \tag{17}$$

The map $x \to \mathcal{I}_A(x)$ given by (16) is an atomic quantum instrument. We remark that the Born rule (15) can be written in the form

$$\Pr\{\mathbf{x} = x \| \rho\} = Tr\ [\mathcal{I}_A(x)\rho]. \tag{18}$$

Let $\hat{A}$ be a Hermitian operator in $\mathcal{H}$. Consider a POVM $\hat{D} = (\hat{D}^A(x))$ with the domain of definition given by the spectrum of $\hat{A}$. This POVM represents a measurement of observable A if Born's rule holds:

$$\Pr\{A = x \| \rho\} = Tr[\hat{D}^A(x)\rho] = Tr[\hat{E}^A(x)\rho]. \tag{19}$$

Thus, in principle, probabilities of outcomes are still encoded in the spectral decomposition of operator $\hat{A}$ or in other words operators $\hat{D}^A(x)$ should be selected in such a way that they generate the probabilities corresponding to the spectral decomposition of the symbolic representation $\hat{A}$ of observables A, i.e., $\hat{D}^A(x)$ is uniquely determined by $\hat{A}$ as $\hat{D}^A(x) = \hat{E}^A(x)$. We can say that this operator carries only information about the probabilities of outcomes, in contrast to the von Neumann scheme,

[3] We remark that any orthonormal projector $\hat{E}$ is Hermitian and idempotent, i.e., $\hat{E}^{\star} = \hat{E}$ and $\hat{E}^2 = \hat{E}$. Thus, any projector $\hat{E}^A(x)$ can be written as (14): $\hat{E}^A(x) = \hat{E}^A(x)^{*}\hat{E}^A(x)$. The map $x \to \hat{E}^A(x)$ is a special sort of POVM, the projector valued measure - PVM, the quantum instrument of the von Neumann type.

operator $\hat{A}$ does not encode the rule of the state update. For an atomic instrument, measurements of the observable A has the unique output probability distribution by the Born's rule (19), but has many different quantum state reductions depending of the decomposition of the effect $\hat{D}(x) = \hat{E}^A(x) = V(x)^* V(x)$ in such a way that

$$\rho \to \rho_{\{A=x\}} = \frac{V(x)\rho V(x)^*}{Tr[V(x)\rho V(x)^*]}. \tag{20}$$

4.5 Quantum Instruments

Finally, we formulate the general notion of quantum instrument. A superoperator acting in $\mathcal{L}(\mathcal{H})$ is called positive if it maps the set of positive semi-definite operators into itself. We remark that, for each x, $\mathcal{I}_A(x)$ given by (16) can be considered as linear positive map.

Generally any map $x \to \mathcal{I}_A(x)$, where for each x, the map $\mathcal{I}_A(x)$ is a positive superoperator is called *Davies–Lewis* (Davies & Lewis, 1970) quantum instrument. Here index A denotes the observable coupled to this instrument. The probabilities of A-outcomes are given by Born's rule in form (18) and the state-update by transformation (17). Ozawa (Ozawa, 1984) introduced the important additional condition to ensure that every quantum instrument is physically realizable. This is the condition of complete positivity. A superoperator is called *completely positive* if its natural extension $\mathcal{T} \otimes I$ to the tensor product $\mathcal{L}(\mathcal{H}) \otimes \mathcal{L}(\mathcal{H}) = \mathcal{L}(\mathcal{H} \otimes \mathcal{H})$ is again a positive superoperator on $\mathcal{L}(\mathcal{H}) \otimes \mathcal{L}(\mathcal{H})$. A map $x \to \mathcal{I}_A(x)$, where for each x, the map $\mathcal{I}_A(x)$ is a completely positive superoperator is called *Davies-Lewis-Ozawa* (Davies & Lewis, 1970; Ozawa, 1984) quantum instrument or simply quantum instrument.

Complete positivity is a sufficient condition for an instrument to be physically realizable. On the other hand, necessity is derived as follows [?]. Every observable A of a system S is identified with the observable $A \otimes I$ of a system $S + S'$ with any system S' external to S.

Then, every physically realizable instrument $\mathcal{I}_A$ measuring A should be identified with the instrument $\mathcal{I}_{A\otimes I}$ measuring $A \otimes I$ such that $\mathcal{I}_{A\otimes I}(x) = \mathcal{I}_A(x) \otimes I$. This implies that $\mathcal{I}_A(x) \otimes I$ is again a positive superoperator, so that $\mathcal{I}_A(x)$ is completely positive. Similarly, any physically realizable instrument $\mathcal{I}_A(x)$ measuring system S should have its extended instrument $\mathcal{I}_A(x) \otimes I$ measuring system $S + S'$ for any external system S'. This is fulfilled only if $\mathcal{I}_A(x)$ is completely positive. Thus, complete positivity is a necessary condition for $\mathcal{I}_A$ to describe a physically realizable instrument.

5 Classical and Quantum Probabilistic Formalization of the Concept of Rationality

5.1 Savage Sure Thing Principle as the Rationality Axiom

In classical theory of decision making, rational behavior of agents is formalized with the *Savage Sure Thing Principle* (STP) (Savage, 1954):

If you prefer prospect b_+ to prospect b_- if a possible future event A happens ($a = +1$); *and you prefer prospect b_+ still if future event A does not happen* ($a = -1$); *then you should prefer prospect b_+, despite having no knowledge of whether or not event A will happen.*

Savage's illustration refers to a person deciding whether or not to buy a certain property shortly before a presidential election, the outcome of which could radically affect the property market:

"Seeing that he would buy in either event, he decides that he should buy, even though he does not know which event will obtain".

STP is considered as the axiom of rationality of decision makers (Savage, 1954). It plays the important role in decision making and economics in the framework of Savage's subjective utility theory. In the latter, probability is formalized in the classical probabilistic framework (Kolmogorov, 1933) and it is endowed with the subjective interpretation.

We remark that STP is a simple consequence of the law of total probability - LTP (see (5)). Violation of LTP implies violation of STP. Thus, the degree of satisfaction of LTP can be used as a statistical test of classical (STP-type) rationality.

In cognitive psychology, violation of STP is known as the *disjunction effect.* A plenty of statistical data was collected in cognitive psychology in experiments demonstrating disjunction effect. For example, in experiments of the Prisoners' Dilemma type (Kahneman & Tversky, 1972)–(Kahneman & Thaler, 2006). Such data violate LTP. The latter implies irrationality (from classical viewpoint) of agents participating in experiments (mainly students).

We recall that LTP is derived from two assumptions that are firmly incorporated into the Kolmogorov axiomatics:

1. Additive law for probability.
2. Baeys formula for conditional probability.

Therefore, violation of LTP and, hence, of STP (and classical rationality) is generated either by violation of additivity of probability or the Bayes law for conditional probability or by the combination of these factors. Generally, this leads to the impossibility to use in decision making Bayesian inference. Quantum(-like) agents proceed with more general inference machinery based on the quantum state update.

Hence, classical rationality is Bayesian inference rationality and quantum rationality is non-Bayesian inference rationality.[4]

5.2 Quantum(-like) Rationality

In the light of above considerations, one can ask:

Are quantum agents irrational?

As was discussed, by using QP it is possible to violate LTP and hence STP. Therefore, generally quantum-like agents are (classically) irrational. However, we can question the classical probabilistic approach to mathematical formalization of decision making and, consequently, the corresponding notion of rationality. We define quantum(-like) rationality as respecting the quantum calculus of probabilities and the quantum formula for interference of probabilities, LTP with the interference term (5).

In the framework of quantum-like modeling, violation of the classical CP-laws including the Byaes formula for conditional probability or even additivity of probability are not exotic at all. Moreover, the situation in which the probabilistic data satisfies LTP seems to be rather an exception than the norm. We can speculate that CP-processing of information was resulted from evolution of biological systems, not only humans, but even animals and simple bio-organisms.

The question whether the "genuine human behavior" should be characterized by classical rationality taken as the normative theory for the rational decision making is very complicated and we are not ready to discuss it in this paper.

6 Social Laser

One of the consequences of information overload is that information loses its content. A human has no possibility analyze deeply the content of communications delivered by mass-media and social networks. People process information without even attempting to construct an extended Boolean algebra of events. They operate with labels such as say covid-19, vaccination, pandemy without trying to go deeper beyond this labels. Contentless information behaves as a bosonic quantum field which is similar to the quantum electromagnetic field. Interaction of humans with such quantum information field can generate a variety of quantum-like behavioral effects. One of them is social lasing, stimulated amplification of social actions (SASA) (Khrennikov, 2015)-(Tsarev et al., 2019). In social laser theory, humans

[4] Of course, non-Bayesian probability updates are not reduced to quantum, given by state transformations in the complex Hilbert space. One may expect that human decision making violates not only classical, but even quantum rationality.

play the role of atoms, social atoms (s-atoms). Interaction of the information field composed of indistinguishable (up to some parameters, as say social energy) excitations with gain medium composed of s-atoms generate the cascade type process of emission of social actions. SASA describes well e.g. color revolutions and other types of mass protests (see Khrennikov, 2020 for detailed presentation).

Over the past years, our society has been constantly shaken by high-amplitude information waves. These are waves of enormous social energy. They are often destructive and are a kind of information tsunami. The main distinguishing features of these waves are their high amplitude, coherence (the homogeneous nature of the social actions they generate) and the short time required for their generation and relaxation, huge singular spikes.

We showed [] that such waves can be modeled using the social laser, which describes stimulated amplification of coordinated social actions. "Actions" are interpreted very broadly, from mass protests, in particular, leading to color revolutions such as the Orange or Maidan revolutions in Ukraine or the recent mass protests in USA (anti-Baiden protests for fair votes), Belarus (anti-Lukashenko protests for fair votes), Russia (anti-Putin protests for liberation of Naval'nii), Germany, UK, Australia, Canada, Sweden (protests against corona-fascism and violation of the basic human rights with pandemic-justification) as well as generating of "right voting" and other collective decisions as, e.g., acceptance of lockdown and support of the total vaccination against covid-19 by the majority of population (Khrennikov, 2020).

7 Order-stability as a Bonus for Quantum Rationality

Entropy is typically considered as a measure of disorder in a system (physical or biological). For an isolated classical system, *the second law of thermodynamics implies that entropy increases monotonically.* Of course, a biosystem is never completely isolated; in particular, energy and matter flows from and into environment never stop. A completely isolated biosystem is dead. But, we are not interested in these physical flows, only in information processing, in the dynamic of the agent's information state. How would behave classical versus quantum agents isolated from external information flows? As was mentioned the entropy of an isolated system performing classical information processing would increase, but the entropy of an isolated system performing quantum information processing is preserved (Khrennikov & Watanabe, 2021).

8 Concluding Remarks

Applications of the mathematical formalism of quantum theory and its methodology outside of physics cover the wide range of research areas, from molecular biology to cognition, social science, and artificial intelligence [].

We emphasize that behavior quantum(-like) agents differs crucially from behavior of classical agents. Quantum agents demonstrate superiority over classical agents in informationally dense environments, because they need less computational resources for decision making. The main danger of quantum rational behavior is that such agents become a very good medium for social engineering; in particular, a good active medium for social lasing. The latter can be used to generate instability throughout the world, in the form of mass-protests and color revolutions.

Appearance of this sort of instability can be expected the AI-area. Quantum robots having sufficiently high cognitive abilities, i.e., not just algorithmically programmed entities, but self-learning and developing creatures, would also become a good active medium for AI-social lasing.

9 Appendix: Contextuality of Kolmogorov Theory

We recall the original Kolmogorov's interpretation of probability (Kolmogorov, 1933):

"[. . .] we may assume that to an event A which may or may not occur under conditions Σ is assigned a real number $P(A)$ which has the following characteristics:

- (a) one can be practically certain that if the complex of conditions Σ is repeated a large number of times, N, then if n be the number of occurrences of event A, the ratio n/N will differ very slightly from $P(A)$;
- (b) if $P(A)$ is very small, one can be practically certain that when conditions Σ are realized only once, the event A would not occur at all."

The (a)-part of this interpretation is nothing else than the frequency interpretation of probability. This is the essence of the *'statistical interpretation of probability'* which is mathematically justified by the law of large numbers (a theorem in the Kolmogorov measure-theoretic mathematical model): frequencies converge to probabilities for almost all elementary events.

This reference to context Σ is closely related considerations of Sect. 3. Kolmogorov pointed out that each probability space is determined by its own complex of conditions (*context*) Σ. For example, he definitely would not be surprised if statistical data collected for a few different experimental contexts, $\Sigma_1, ..., \Sigma_n$, would violate one of the laws of probability; for example, LTP. For him, in general, each of these contexts determines its own probability space

$$\mathcal{P}_{\Sigma_j} = (\Omega_{\Sigma_j}, \mathcal{F}_{\Sigma_j}, p_{\Sigma_j}).$$

Since LTP was proven by working in a single Kolmogorov probability framework (the same probability measure was used to define all conditional probabilities in the right-hand side of LTP), the possibility of its violation in a multi-space framework is not surprising. Unfortunately, this contextuality dimension of classical probability model [which was so strongly emphasized in Sect. 2 of Kolmogorov's book

(Kolmogorov, 1933)] has be washed out in the process of further development of classical probability. Therefore by seeing violation of LTP people often make fundamental philosophic conclusions such as, e.g., "death of realism". The latter means impossibility to assign definite values of observables to chance parameters $\omega \in \Omega$ (in physics one speaks about hidden variables and use symbols $\lambda \in \Lambda$.), i.e., impossibility to construct the functional representation of observables $\omega \to a(\omega)$. Kolmogorov would defend realism, but at the same time emphasize its contextuality, dependence on the experimental context. For Kolmogorov, one cannot speak about probability and random variables before concrete experimental context is fixed.

References

Asano, M., Khrennikov, A., Ohya, M., Tanaka, Y., & Yamato, I. (2015). *Quantum adaptivity in biology: From genetics to cognition*. Heidelberg-Berlin-New York: Springer.

Aumann, R. J. (1976). Agreeing on disagree. *The Annals of Statistics, 4,* 1236–1239.

Bagarello, F. (2019). *Quantum concepts in the social, ecological and biological sciences*. Cambridge: Cambridge University Press.

Busemeyer, J., & Bruza, P. (2012). *Quantum models of cognition and decision*. Cambridge University Press.

Davies, E. B., & Lewis, J. T. (1970). An operational approach to quantum probability. *Communications in Mathematical Physics, 17,* 239–260.

Haven, E., & Khrennikov, A. (2013). *Quantum social science*. Cambridge University Press.

Haven, E., Khrennikov, A., & Robinson, T. R. (2017). *Quantum methods in social science: A first course*. Singapore: WSP.

Kahneman, D., & Thaler., R. (2006). Utility Maximization and experienced utility. *Journal of Economic Perspectives,20*, 221–232.

Kahneman, D. (2003). Maps of bounded rationality: Psychology for behavioral economics. *American Economic Review, 93*(5), 1449–1475.

Kahneman, D., & Tversky, A. (1972). Subjective probability: A judgment of representativeness. *Cognitive Psychology, 3*(3), 430–454.

Kahneman, D., & Tversky, A. (1979). Prospect theory: An analysis of decision under risk. *Econometrica, 47,* 263–291.

Kahneman, D., & Tversky, A. (1984). Choices values and frames. *American Psychologist, 39*(4), 341–350.

Khrennikov, A., Alodjants, A. Trofimova., & A. and Tsarev, D. (2018). On interpretational questions for quantum-Like modeling of social lasing. *Entropy,20*(12), 921.

Khrennikov, A. (1999). Classical and quantum mechanics on information spaces with applications to cognitive, psychological, social and anomalous phenomena. *Foundations of Physics, 29,* 1065–1098.

Khrennikov, A. (2003). Quantum-like formalism for cognitive measurements. *Biosystems, 70,* 211–233.

Khrennikov, A. (2004). On quantum-like probabilistic structure of mental information. *Open Systems and Information Dynamics, 11*(3), 267–275.

Khrennikov, A. (2004). *Information dynamics in cognitive, psychological, social, and anomalous phenomena*. Dordreht: Ser.: Fundamental Theories of Physics, Kluwer.

Khrennikov, A. (2010). *Ubiquitous quantum structure: from psychology to finances*. Berlin-Heidelberg-New York: Springer.

Khrennikov, A. (2015). Towards information lasers. *Entropy, 17*(10), 6969–6994.

Khrennikov, A. (2016). Social laser: Action amplification by stimulated emission of social energy. *Philosophical Transactions of the Royal Society, 374*(2054), 20150094.

Khrennikov, A. (2018). Social laser model: From color revolutions to Brexit and election of Donald Trump. *Kybernetes, 47*(2), 273–278.

Khrennikov, A. (2020). *Social laser*. Singapore: Jenny Stanford Publ.

Khrennikov, A., Toffano, Z., & Dubois, F. (2019). Concept of information laser: From quantum theory to behavioral dynamics. *The European Physical Journal Special Topics, 227*(15–16), 2133–2153.

Khrennikov, A., & Watanabe, N. (2021). Order-stability in complex biosystems from the viewpoint of the theory of open quantum systems. *Entropy, 23*(3), 355–365.

Kolmogorov, A. N. (1933). *Grundbegriffe der Wahrscheinlichkeitsrechnung*. Berlin: Springer.

Ozawa, M., & Khrennikov, A. (2020). Application of theory of quantum instruments to psychology: Combination of question order effect with response replicability effect. *Entropy*, *22*(1), 37. 1-9436.

Ozawa, M. (1984). Quantum measuring processes for continuous observables. *Journal of Mathematical Physics, 25,* 79–87.

Ozawa, M., & Khrennikov, A. (2020). Modeling combination of question order effect, response replicability effect, and QQ-equality with quantum instruments. *Journal of Mathematical Psychology, 100,* 102491.

Savage, L. J. (1954). *The foundations of statistics*. New York: Wiley and Sons.

Tsarev, D., Trofimova, A., Alodjants, A., et al. (2019). Phase transitions, collective emotions and decision-making problem in heterogeneous social systems. *Scientific Reports, 9,* 18039.

Von Neumann, J. (1955). *Mathematical foundations of quantum mechanics*. Princeton, NJ, USA: Princeton Univ Princeton Univ Princeton Univ Princeton University Press.

Cognitive Science/Computer Science

Nothing Will Come of Everything: Software Towers and Quantum Towers

Samson Abramsky

The theme of this volume is *content* and *context*, as a perspective on the long-standing debate between reductionism and holism. I will discuss these dichotomies with reference to three areas of science: computer science, mathematics, and foundations of quantum mechanics.

Firstly, though, some general remarks. In my view, from a scientific perspective, reductionism *is*, perhaps in a caricature form, the basic method of science; whereas holism is reaching for a way to protect various forms of belief from the incursions of science, and to call for a return to a pre-scientific viewpoint. Science proceeds by mastering the overwhelming complexity of everything by isolating aspects of the whole: subsystems, degrees of freedom, parts. This enables it to find the hidden simplicities and patterns underlying the richness and specificity of phenomena. In a slogan:

> If the only way to understand *anything* would be to understand *everything*, then we could understand *nothing*.

On the other hand, this method has to be understood in a suitably nuanced fashion. I shall argue, firstly, that Computer Science offers an excellent arena for discussing these issues, all the more so as it is removed from the emotional undertones which usually color the debate.

S. Abramsky (✉)
Department of Computer Science, Oxford University, Wolfson Building,
Parks Road, Oxford OX1 3QD, UK
e-mail: samson.abramsky@cs.ox.ac.uk

S. Wuppuluri and I. Stewart (eds.), *From Electrons to Elephants and Elections*,
The Frontiers Collection, https://doi.org/10.1007/978-3-030-92192-7_29

1 First Lens: Computer Science

Let us begin with an old chestnut which has often been used in the following kind of reductive argument:

> All a computer does is manipulate 1's and 0's. *Therefore* it can't [...].

The specific conclusion often relates to exhibiting intelligent behaviour of one kind or another. There is a serious discussion to be had about possible limits to AI. But the above argument does not contribute to such a discussion. In fact, while the premise of the argument is, for standard computer architectures, true enough, there are *no* interesting conclusions which can be drawn from this fact.

The basic issue is this: does the fact that at a low level of description[1] computers are manipulating finite strings of bits, in any way prevent or falsify the description of computers as manipulating much higher level objects: whether they are our medical records, bank accounts, credit records, games of chess, mathematical proofs, musical scores, visual images, text or speech? Any computer science undergraduate after a year or two of their studies will be well aware that the answer to this question is a resounding *No*! In fact, a large part of what they will have learnt in their studies will have been precisely how to build high-level structures of diverse kinds based on lower-level primitives. This is actually what programming, and software design and architecture, are all about. Several features are worthy of attention:

- Firstly, there is an obvious parallel between the way software developers build high-level abstractions from low-level primitives, and work in the foundations of mathematics. Indeed, contemporary work in developing formal mathematics in systems such as Agda, Coq, HoTT (homotopy type theory), etc. makes this explicit, so that the boundaries between code and mathematics become somewhat indistinct. From the heroic age of Principia Mathematica, we are now in an era of highly engineered software systems, capable of undertaking large scale proofs, such as the proof of the Kepler conjecture in the Flyspeck project led by Tom Hales (Hales et al., 2017), and the proofs of the Four-Color Theorem and the Feit-Thompson theorem in projects led by Georges Gonthier (Gonthier, 2008; Gonthier et al., 2013). In each case, elaborate towers of concepts, definitions and results relating to specialised mathematical theories are built up from simple, logically evident foundations. This has much the same overall structure as the way that elaborate towers of modules and libraries determined by application-level concepts are built from the basic mechanisms provided by a programming language—which itself sits on top of a stack of compilers, editors, tracing and debugging tools, etc. These in turn are ultimately mapped down to the code controlling the "bare metal" of the computer—which is indeed directly manipulating 1's and 0's. This last fact, however, is gloriously irrelevant to the software artefact that sits on top of this tower of abstractions. Indeed, by virtue of the decoupling of high-level programming languages from specific machines provided by the now-routine

[1] By no means the lowest: there is a lot of complicated device physics sitting underneath the abstraction level of computer architecture.

mechanisms of compilation, the same code can be run on many different machines, eliciting different sequences of manipulations of bit-strings, but achieving the *same* high-level effect.

- While the architecture of concepts embodied in code is analogous to the architecture of mathematical concepts, there is also mathematics *about* code: namely, the tools of formal specification and verification. It is by virtue of these tools that we can be sure of the independence of high-level code from low-level implementation details. Another important aspect is directly relevant to the holism-reductionism debate, as reflected in the manifesto of this volume. For each level of the software tower, specifications will be written *at the corresponding abstraction level.* If we are specifying relationships between geometric objects in a visual feature recognition system, or a hierarchical relationship in an ontology used in a medical database, we would no more refer to details of bitstrings being loaded into the registers of a GPU than we would refer to electrons in describing the biology of elephants. Yet our code will not run without being executed on a physical machine, any more than an elephant can exist without being manifested in physical matter. So as far as the delightful rhetorical flourishes of this volume's manifesto are concerned, they are exhibited in software in terms completely familiar to computer science undergraduates on a daily basis. Nothing to see here!
- One feature that perhaps serves to obscure the analogy we are making is that we customarily read these towers of abstractions in opposite directions. In the case of scientific reductionism, we read the tower *downwards*, in the direction of analysis. That is, we emphasize the reduction of higher-level concepts to lower level ones—even though, in many cases, this reduction is "in principle", and difficult or impossible to achieve in practice. By contrast, in the case of software, we read the tower *upwards*, in the direction of synthesis. We are interested in *constructing* a complex artefact, not in analyzing the complexity of a pre-existing class of systems occurring in nature. But this difference in how these towers arise does not in itself show that they are different in kind—and indeed, the well-understood towers of software abstraction may serve to shed light on the hierarchical structure of scientific theories.
- One difference that may be argued is that in natural science, we expect nature to force our hand in the development of a tower of theories, whereas in an engineering context we have the luxury of choosing our programming language and tools, and our hardware. But this difference is not as great as it may appear. In mathematics, the same ideas in a given domain may be developed from different choices of foundational concepts. For example, algebraic geometry has been subject to several different foundational frameworks, in a process which is still ongoing (Van der Waerden, 1971; Dieudonné & Grothendieck, 1971; Anel & Catren, 2021). In physical science, physical systems can be studied in a classical, semi-classical, or fully quantum framework according to need.

The main overall point we wish to make is that Computer Science offers, in the ideas of the tower of software concepts, a well-understood, well-formalised, non-mystical paradigm for understanding how systems can be described at different levels of abstraction, and the levels related to each other. Moreover, although the mappings downwards are well-understood, and essential for the design and verification of implementations, there is no temptation to refer to them when reasoning about the higher levels! In fact, the development of higher levels of abstraction is a *practical necessity*. It is an essential tool in mastering complexity. Indeed, the tower enables us to *introduce* suitable levels of abstraction, so that we can "think bigger thoughts".

An essential part of this Computer Science methodology, to which we shall now turn, is *compositionality*, an idea which is also of great relevance to the context/content and reductionism/holism debates.

1.1 Compositionality

Compositionality is a methodological principle, originating from the work of Frege and others in logic (Janssen, 2001; Janssen & Partee, 1997), which has played a crucial rôle in Computer Science for several decades, but has yet to achieve the recognition in general scientific modelling which it deserves. I believe it is of *major* potential importance for mathematical modelling throughout the sciences. See e.g. (Werning et al., 2012; Fong & Spivak, 2019) for some recent texts.[2]

Compositionality was originally formulated as a principle for the semantics of natural language: the meaning of an expression should be a function of the meaning of its syntactic constituents, and of how these parts are combined to form the expression. That is, the structure of semantics should follow the grammatical structure of the language—it should be *syntax-directed*, in computer science parlance.

In mathematical logic, the Tarskian semantics of predicate logic stands as the paradigm of compositional definitions for formal languages (Tarski, 1936; Tarski & Vaught, 1956). It has in turn heavily influenced the development of the formal semantics of programming languages (Scott & Strachey, 1971).

More generally, in computer science, compositionality has become a major paradigm in enabling the structured description of complex systems. We can contrast it with the traditional approach in mathematical modelling, of whole-system (monolithic) analysis of given systems. In the compositional approach we start with a fixed set of basic (simple) building blocks, and *constructions* for building new (in general more complex) systems out of given sub-systems, and build up the required complex system with these.

A little more formally (and somewhat simplistically), compositionality can be expressed *algebraically*:

$$S = \omega(S_1, \ldots, S_n)$$

The system S is described as being built up from sub-systems $S_1, \ldots S_n$ by the operation ω. This allows the *hierarchical* description of systems, e.g.

$$S = \omega_1(\omega_2(a_1), \omega_1(a_2, a_3)).$$

---

[2] There is also a journal, https://compositionality-journal.org/.

In graphical form:

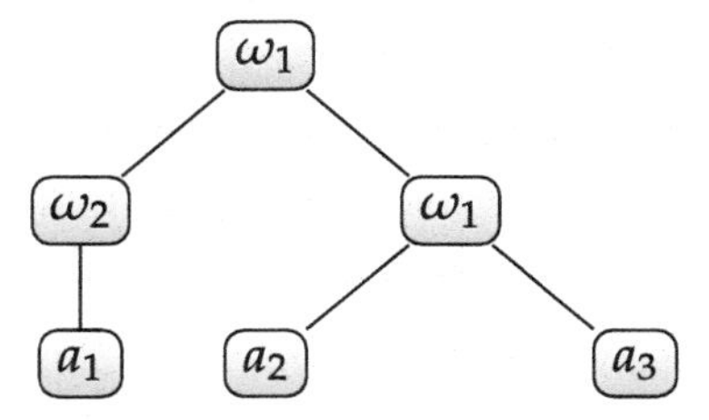

Here ω_1 is a binary operation, ω_2 a unary operation, and a_1, a_2, a_3 are sub-expressions, which may themselves be built from components in arbitrarily complex fashion.

There is also a *logical* perspective:

$$\frac{S_1 \models \phi_1, \ldots, S_n \models \phi_n}{\omega(S_1, \ldots, S_n) \models \phi}$$

(Read $S \models \phi$ as "system S satisfies the property ϕ"). Here *properties* ϕ of the compound system S can be inferred by verifying properties ϕ_1, ..., ϕ_n for the simpler sub-systems $S_1, \ldots S_n$. This allows the properties of the sub-systems to be tracked all the way up (or down) the tree of syntax.

In addition to its major role in Computer Science and linguistics, compositionality is increasingly being introduced into other areas, including physics (Abramsky & Coecke, 2009), systems biology (Danos et al., 2007), game theory (Ghani et al., 2018), and more.

Since compositionality systematically relates the meaning of larger systems to the meanings of their parts, it appears as an antithesis to *contextualism*, which asserts that a part only acquires meaning in relation to the larger context in which it appears. There is a *reductio* of contextualism which echoes our opening slogan: if we pursue it to its limit, we end up needing to understand the meaning of everything in order to understand the meaning of anything. And how is this "everything" delimited, anyway? Perhaps we can only fully understand the meaning of an English utterance in the context of the entire history, not yet completed, of English speech.[3]

1.2 Challenges to Compositionality

We have emphasized the importance of compositionality as a methodological principle. It is also interesting to consider some challenges to it which have arisen, explicitly or implicitly, in recent developments.

[3] We may quite plausibly claim that our present understanding of past utterances and writings is heavily influenced by the subsequent history.

Independence-Friendly logic A notable challenge to compositionality was made by Jaakko Hintikka in relation to his *independence-friendly logic* (IF logic), a generalization of branching quantifiers (Hintikka & Sandu, 1989; Hintikka, 1998). Hintikka claimed that this logic could not be given a compositional semantics in a Tarskian style. One had to look at entire formula, and give the meaning in terms of strategies for a game associated with this formula. While Hintikka was correct in his claim that one could not give a semantics for this logic using assignments of elements of a quantificational domain to the variables in a formula, as is done in Tarskian semantics of predicate logic, he was taking too limited a view of the possibilities for a compositional semantics. Wilfrid Hodges subsequently showed that a compositional semantics *could* be given, using *sets of assignments* rather than single assignments (Hodges, 1997). This semantics in terms of sets of assignments, nowadays called *team semantics*, has been extensively developed by Jouko Väänänen and his collaborators, in his logics of dependence and independence (Väänänen, 2007; Abramsky et al., 2016). It turns out that this yields a very interesting extension of the possibilities for compositional semantics, and for logic in general, with connections to databases, foundations of probability and statistics, quantum physics, and more. What this illustrates is that overcoming challenges to compositionality can lead to significant advances.

Emergence A key concept in the reductionism/holism debate is that of *emergence*: the idea that salient concepts or features of systems can only appear at higher structural levels, and cannot be accounted for at the lower levels. Referring this to the setting of the software tower, we can recognise that, on the level of feasibility and intelligibility, this is clearly true in an unproblematic way. If we think of the analogous mathematical situation, defining the curvature of a Riemannian manifold in the bare language of set theory or type theory, without an intervening tower of definitions and intermediate results, would be hopelessly long and unwieldy. The question is whether there are truly higher-level emergent properties which fundamentally cannot be expressed at all in terms of lower levels. I am not aware of precise results to this effect.

AI One place where we might look for such results is in Artificial Intelligence, in particular in its dominant modern form based on machine learning. The history of AI can be argued to have gone against the compositional grain. Much of early AI was logic- and rule-based, but there has been a big shift towards statistical machine learning, where the wisdom is in the data. This has led to systems with highly impressive performance in terms of their ability to carry out a wide range of specialised tasks in natural language processing, vision, robotics and autonomous devices, games playing, medical diagnosis, financial analysis, protein folding and many more. Many of these encroach well into areas previously considered as requiring distinctively human intelligence, although the challenge of integrative intelligence, encompassing a full range of intelligent behaviour, remains. These systems have been highly resistant to compositional description and analysis. However, there is a major push in current research to achieve this, in order to have explainable, verifiable and accountable AI (Adadi & Berrada, 2018; Huang et al., 2017). This is a fundamental issue in the current research agenda in AI. Indeed, can one be said to have a "theory of intelligence", or to have achieved scientific understanding of it, if one has a system which produces intelligent behaviour, but has no explanation of how this behaviour is produced? And such an explanation would surely have to refer to an underlying system structure. It can be plausibly argued that structure "cashes out" into compositionality.

Cheap tricks? A more subtle challenge to compositionality is that *it is too easily achieved*. This is argued, for example, in Hodges (2001, 1998). Indeed, by introducing additional variables, which in effect encode the relevant contextual information, one can, in some generality, make *any* semantic definition compositional. Does this trivialise compositionality? Rather, it highlights the importance of having additional criteria over and above compositionality for the acceptability of a formal semantics. In the setting of programming language semantics, such criteria are provided by *adequacy* and *full abstraction* (Plotkin, 1977; Milner, 1977). Similar criteria can be applied for team semantics, which we discussed above (Abramsky & Väänänen, 2009). As we shall see, there are analogous issues in the foundations of quantum mechanics.

2 Second Lens: Quantum Mechanics

Two issues which arise from the foundations of quantum mechanics can be related to our discussion. Interestingly, they pull in opposite directions:

- One the one hand, the quantum phenomenon of *entanglement* has been argued to imply a form of "quantum holism" (Healey, 1991).
- On the other hand, quantum contextuality, a key non-classical feature of quantum mechanics, is problematic for holism, since it calls into question whether there *is* a whole.

2.1 Entanglement and Quantum Holism

A fundamental aspect of quantum mechanics is *entanglement*, a phenomenon where the quantum state of a group of particles cannot be described solely in terms of the states of each particle separately. This behaviour may be exhibited even when the particles are spatially separated.

Intuitively, entanglement violates common sense principles like the "Principle of Local Action", by which an object is directly influenced only by its immediate surroundings. This counter-intuitive nature led Einstein to describe entanglement as *spooky action at a distance*, and to the *Einstein–Podolsky–Rosen (*EPR*) paradox* (Einstein et al., 1935).

Bell's seminal idea in (Bell, 1964) was that entanglement has observable implications, which separate the predictions of quantum theory from *any* attempt at classical explanation by a local, realistic theory. If the particles in an entangled system are spatially separated, and each particle is measured independently, the presence of entanglement implies *correlations* between the outcomes of the measurements that provably exceed what can be achieved classically. This clear separation of the predictions of quantum theory from any classical theory has been verified experimentally (Freedman & Clauser, 1972; Aspect et al., 1982; Hensen et al., 2015; Giustina et al., 2015; Shalm et al., 2015), and forms the basis of the currently emerging technologies of quantum information and communication.

One point to mention is that Bell's result bounding the possible correlations achievable by "local realistic theories", and showing that quantum mechanics exceeds these bounds, can be stated as an impossibility result for hidden-variable theories. There is a striking analogy with our discussion of "cheap tricks" for compositionality in the previous section. Just as we can trivially make definitions compositional by adding extra variables which encode contextual information, so we can construct hidden variable theories to account for any observable phenomena (Abramsky, 2014). However, if we introduce suitable constraints on such theories, e.g. locality of information flow, then results such as Bell's can be proved. This parallels the way that compositionality has to be tensioned against required properties of a semantics, such as adequacy and full abstraction. Similarly, contextual hidden variable theories can be constructed for quantum mechanics. Bohmian mechanics can be viewed as such a theory. However, these theories are as non-local as quantum mechanics itself.

Returning to entanglement, it has been argued that the non-separable nature of entangled states exhibits a form of holism, since we cannot recover the entangled state from its components. We find this dubious, mainly because it is not clear what is at stake here. While much has been learnt about how to *use* entanglement in quantum information, a deeper physical understanding of how and why this phenomenon arises, if there is one to be had, remains elusive.

We content ourselves with the following observations:

- Non-separability in this sense is a common phenomenon, which arises mathematically wherever we have *monoidal categories* (Fong & Spivak, 2019).[4] There are many examples of this in classical computation, and in Linear and other substructural logics (Girard, 1987; O'Hearn & Pym, 1999).
- Monoidal categories, and the mathematics of entanglement, can be handled in a thoroughly compositional fashion (Abramsky & Coecke, 2009).

2.2 Contextuality: Is There a Whole?

Contextuality arises from an even more fundamental non-classical feature of quantum theory: the incompatibility of different measurements, meaning that one cannot observe definite values for all physical quantities at the same time. Again, this is not

[4] More precisely, monoidal categories which are not cartesian, ie where the tensor product is not the usual (cartesian) product, equipped with diagonals and projections (Abramsky, 2009).

merely a practical limitation, but a fundamental feature of quantum mechanics, as shown by the seminal results due to Bell (Bell, 1964, 1966) and Kochen–Specker (Kochen & Specker, 1967). This feature is known as *contextuality*, and recent work has shown that it is a key signature of the non-classicality of quantum mechanics, responsible for many of the known examples where quantum computation offers possibilities that exceed classical bounds (Raussendorf, 2013; Abramsky et al., 2017; Howard et al., 2014; Bermejo-Vega et al., 2017; Bravyi et al., 2018; Aasnæss, 2019). Moreover, contextuality subsumes non-locality as a mathematical feature of a physical theory (Abramsky & Brandenburger, 2011).

2.2.1 Contextual Logic in Quantum Mechanics

Logic traditionally emphasises truth (semantically) and consistency (proof-theoretically). While the debates in the foundations of mathematics have, among other things, led to contrasting classical and constructive views of logic, these share an *integrated view*, going back to at least Plato and Aristotle, which can be summarised as follows:

a logical system should stand or fall as a whole.

Quantum mechanics challenges this integrated perspective in a new way. This was already revealed by the seminal results of John Bell and Simon Kochen and Ernst Specker in the 1960s (Bell, 1964, 1966; Kochen & Specker, 1967), but we are still in the process of understanding these ideas. To accommodate a non-integrated view, the logical structure of quantum mechanics is given by a family of overlapping perspectives or contexts. Each context appears classical, and different contexts agree locally on their overlap. However, there is no way to piece all these local perspectives together into an integrated whole, as shown in many experiments, and proved rigorously using the mathematical formalism of quantum mechanics.

To illustrate this non-integrated feature of quantum mechanics, we may consider the well-known "impossible" drawings by Escher, such as the one shown in Fig. 1.

Clearly, the staircase *as a whole* in Fig. 1 cannot exist in the real world. Nonetheless, the constituent parts of Fig. 1 make sense *locally*, as is clear from Fig. 2. Quantum contextuality shows that the logical structure of quantum mechanics exhibits exactly these features of *local consistency*, but *global inconsistency*. We note that Escher's work was inspired by the *Penrose stairs* from (Penrose & Penrose, 1958).[5]

[5] Indeed, these figures provide more than a mere analogy. Penrose has studied the topological "twisting" in these figures using cohomology (Penrose, 1992). This is quite analogous to our use of sheaf cohomology to capture the logical twisting in contextuality (Abramsky et al., 2012, 2015; Carù, 2017).

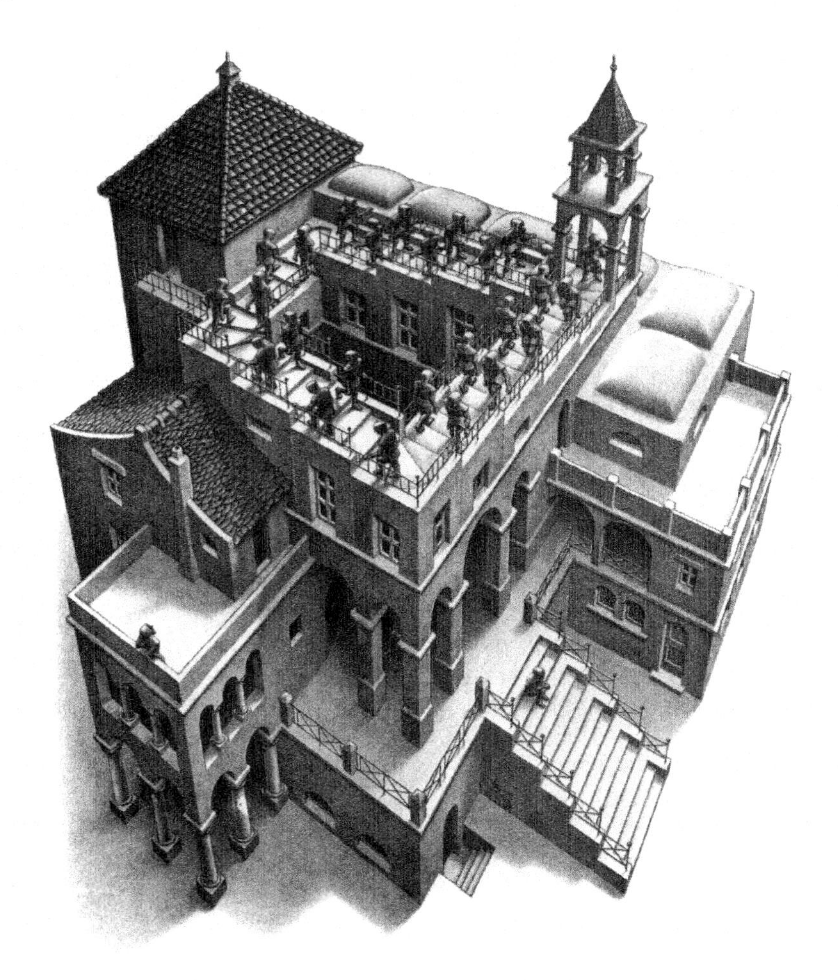

Fig. 1 M. C. Escher, *Klimmen en dalen (Ascending and descending)*, 1960. Lithograph.

2.3 Discussion

Conceptually, an intriguing feature of our discussion of quantum contextuality is that, whereas context is customarily aligned with holism, contextuality tends to undermine holism, since it brings into question the existence of an integrated whole. Rather, the picture of reality which it suggests is of an overlapping family of local perspectives, which support local consistency, but cannot be pieced together into a global, context-independent reality. This raises a number of questions, spanning a range of disciplines:

Fig. 2 Locally consistent parts of Fig. 1.

- Philosophically, how should we understand this lack of global, context-independent truth or consistency? Can contextual logic give a formal foundation for contextualism in contemporary philosophy, such as epistemic contextualism (e.g. DeRose (DeRose, 2009), as a counter to scepticism) and ontic contextualism (e.g. Gabriel, known for *Why the world does not exist* (Gabriel, 2015))?
- Logically, we have physically meaningful—and indeed experimentally accessible—systems which, when viewed globally, can validate contradictory propositions. This is, arguably, more radically disturbing than the more familiar fact that some classical tautologies may not be valid constructively.
- Mathematically, the structures underlying these logical phenomena have a rich geometric and topological content, in which sheaf theory, the mathematics of local-to-global phenomena, and cohomology play a key role, identifying the geometry of the "logical twisting" obstructing a global semantics.
- Physically, we have the issue of experimentally witnessing these phenomena, and understanding the role they play in a wide range of physical systems. These include many-body systems, e.g. frustration in spin networks (Liang et al., 2011), and quantum simulators (Kirby & Love, 2019, 2020).
- Computationally, contextuality appears as a key signature of non-classicality, and is at the core of many of the known examples of quantum advantage in information processing tasks. This is both of great practical import and a crucial tool for showing the impact of non-classicality at the macroscopic level.

3 Concluding Remarks

In this brief essay, we have posed some challenges to holism, if it is to be more than a fuzzy feel-good term. What would holistic science, or science done holistically, look like? It is not clear that there are any convincing examples.

Also, we have argued that quantum contextuality poses a challenge to holism, since it is casts doubt on whether there is an integrated whole underlying our perceptions of physical reality.

More positively, we have advocated the importance of compositionality and levels of abstraction, important methodologies in Computer Science, which provide a much richer and more nuanced alternative to crude reductionism.

For those readers interested in more technical presentations of related issues, we refer to papers such as (Abramsky, 2015, 2017, 2020).

References

Aasnæss, S. (2019). Contextuality as a resource for shallow circuits. Talk at Workshop on Quantum Contextuality in Quantum Mechanics and Beyond (QCQMB 2019), Prague, 2019.

Abramsky, S. (2009). No-cloning in categorical quantum mechanics. *Semantic Techniques in Quantum Computation*, 1–28.

Abramsky, S. (2017). Contextuality: At the borders of paradox. In E. Landry (Ed.), *Categories for the working philosopher*. Oxford University Press.

Abramsky, S. (2020). Classical logic, classical probability, and quantum mechanics. In M. Hemmo & O. Shenker (Eds.), *Quantum, probability, logic: The work and influence of Itamar Pitowsky* (pp. 1–17). Springer.

Abramsky, S., Barbosa, R. S., Kishida, K., Lal, R., & Mansfield, S. (2015). Contextuality, cohomology and paradox. In Kreutzer, S. (Ed.), *24th EACSL Annual Conference on Computer Science Logic (CSL 2015), Leibniz International Proceedings in Informatics (LIPIcs)* (Vol. 41, pp. 211–228). Schloss Dagstuhl – Leibniz-Zentrum für Informatik.

Abramsky, S., Barbosa, R. S.. & Mansfield, S. (2017). Contextual fraction as a measure of contextuality. *Physical Review Letters, 119*(5), 050504.

Abramsky, S., Brandenburger, A., & Savochkin, A. (2014). No-signalling is equivalent to free choice of measurements. arXiv preprint arXiv:1412.8523.

Abramsky, S., Kontinen, J., & Väänänen, J., & Vollmer, H. (2016). *Dependence Logic: Theory and Applications*. Birkhäuser.

Abramsky, S., Mansfield, S., & Barbosa, R. S. (2012). The cohomology of non-locality and contextuality. In B. Jacobs, P. Selinger, & B. Spitters (Eds.), *8th International Workshop on Quantum Physics and Logic (QPL 2011), Electronic Proceedings in Theoretical Computer Science* (Vol. 95, pp. 1–14). Open Publishing Association.

Abramsky, Samson. (2015). Contextual semantics: From quantum mechanics to logic, databases, constraints, and complexity. *Contextuality from Quantum Physics to Psychology, 6,* 23–50.

Abramsky, S., & Brandenburger, A. (2011). The sheaf-theoretic structure of non-locality and contextuality. *New Journal of Physics, 13*(11), 113036.

Abramsky, S., & Coecke, B. (2009). Categorical quantum mechanics. *Handbook of Quantum Logic and Quantum Structures, 2,* 261–325.

Abramsky, S., & Väänänen, J. (2009). From IF to BI. *Synthese, 167*(2), 207–230.

Adadi, A., & Berrada, M. (2018). Peeking inside the black-box: A survey on explainable artificial intelligence (XAI). *IEEE Access, 6,* 52138–52160.

Anel, M., & Catren, G. (2021). *New spaces in mathematics: Formal and conceptual reflections*. Cambridge University Press.
Aspect, A., Dalibard, J., & Roger, G. (1982). Experimental test of Bell's inequalities using time-varying analyzers. *Physical Review Letters, 49*(25), 1804–1807.
Bell, J. S. (1964). On the Einstein-Podolsky-Rosen paradox. *Physics, 1*(3), 195–200.
Bell, J. S. (1966). On the problem of hidden variables in quantum mechanics. *Reviews of Modern Physics, 38*(3), 447–452.
Bermejo-Vega, J., Delfosse, N., Browne, D. E., Okay, C., & Raussendorf, R. (2017). Contextuality as a resource for models of quantum computation with qubits. *Physical Review Letters, 119*(12), 120505.
Bravyi, S., Gosset, D., & König, R. (2018). Quantum advantage with shallow circuits. *Science, 362*(6412), 308–311.
Carù, G. (2017). On the cohomology of contextuality. In R. Duncan & C. Heunen (Eds.), *13th International Conference on Quantum Physics and Logic (QPL 2016), Electronic Proceedings in Theoretical Computer Science* (Vol. 236, pp. 21–39). Open Publishing Association.
Danos, V., Feret, J., Fontana, W., Harmer, R., & Krivine, J. (2007). Rule-based modelling of cellular signalling. In *International conference on concurrency theory* (pp. 17–41). Springer.
DeRose, K. (2009). *The case for contextualism: Knowledge, skepticism, and context* (Vol. 1). Oxford University Press.
Dieudonné, J., & Grothendieck, A. (1971). *Éléments de géométrie algébrique*. Berlin, Heidelberg, New York: Springer.
Einstein, A., Podolsky, B., & Rosen, N. (1935). Can quantum-mechanical description of physical reality be considered complete? *Physical Review, 47*(10), 777–780.
Fong, B., & Spivak, D.I. (2019). *An invitation to applied category theory: Seven sketches in compositionality*. Cambridge University Press.
Freedman, S. J., & Clauser, J. F. (1972). Experimental test of local hidden-variable theories. *Physical Review Letters, 28*(14), 938–941.
Gabriel, M. (2015). *Why the world does not exist*. Polity.
Ghani, N., Hedges, J., Winschel, V., & Zahn, P. (2018). Compositional game theory. In *Proceedings of the 33rd Annual ACM/IEEE Symposium on Logic in Computer Science* (pp. 472–481).
Girard. J.-Y. (1987). Linear logic. *Theoretical Computer Science*.
Giustina, M., Versteegh, M. A. M., Wengerowsky, S., Handsteiner, J., Hochrainer, Phelan, K., Steinlechner, F., Kofler, J., Larsson, J., Abellán, C., Amaya, W., Pruneri, V., Mitchell, M. W., Beyer, J., Gerrits, T., Lita, A. E., Shalm, L. K., Nam, S. W., Scheidl, T., Ursin, R., Wittmann, B., & Zeilinger, A. (2015). Significant-loophole-free test of Bell's theorem with entangled photons. *Physical Review Letters, 115*(25), 250401.
Gonthier, G., Asperti, A., Avigad, J., Bertot, Y., Cohen, C., Garillot, F., Le Roux, S., Mahboubi, A., O'Connor, R., Biha, S., et al. (2013). A machine-checked proof of the odd order theorem. In *International Conference on Interactive Theorem Proving* (pp. 163–179). Springer.
Gonthier, G. (2008). Formal proof-the four-color theorem. *Notices of the AMS, 55*(11), 1382–1393.
Hales, T., Adams, M., Bauer, G., Dang, T. D., Harrison, J., Le Truong, H., Kaliszyk, C., Magron, V., McLaughlin, S., Nguyen, T. T., et al. (2017). A formal proof of the Kepler conjecture. In *Forum of mathematics, Pi* (Vol. 5). Cambridge University Press.
Healey, R. A. (1991). Holism and nonseparability. *The Journal of Philosophy, 88*(8), 393–421.
Hensen, B., Bernien, H., Dréau, A. E., Reiserer, A., Kalb, N., Blok, M. S., Ruitenberg, J., Vermeulen, R. F. L., Schouten, R. N., Abellán, C., Amaya, W., Pruneri, V., Mitchell, M. W., Markham, M., Twitchen, D. J., Elkouss, D., Wehner, S., Taminiau, T. H., & Hanson, R. (2015). Loophole-free Bell inequality violation using electron spins separated by 1.3 kilometres. *Nature, 526*(7575), 682–686.
Hintikka, J. (1998). *The principles of mathematics revisited*. Cambridge University Press.
Hintikka, J., & Sandu, G. (1989). Informational independence as a semantical phenomenon. In J. E. Fenstad et al. (Eds.), *Logic, methodology and philosophy of science VIII* (pp. 571–589). Elsevier.

Hodges, W. (1997). Compositional semantics for a language of imperfect information. *Logic Journal of the IGPL, 5*(4), 539–563.
Hodges, W. (1998). Compositionality is not the problem. *Logic and Logical Philosophy, 6*(6), 7–33.
Hodges, Wilfrid. (2001). Formal features of compositionality. *Journal of Logic, Language and Information, 10*(1), 7–28.
Howard, M., Wallman, Jl., Veitch, V., & Emerson, J. (2014). Contextuality supplies the 'magic' for quantum computation. *Nature, 510*(7505), 351–355.
Huang, X., Kwiatkowska, M., Wang, S., & Wu, M. (2017). Safety verification of deep neural networks. In *International Conference on Computer Aided Verification* (pp. 3–29). Springer.
Janssen, T. M. V., & Partee, B. H. (1997). Compositionality. In *Handbook of logic and language* (pp. 417–473). Elsevier.
Janssen, T. M. V. (2001). Frege, contextuality and compositionality. *Journal of Logic, Language and Information, 10*(1), 115–136.
Kirby, W.M., & Love, P.J. (2020). Classical simulation of noncontextual Pauli Hamiltonians. *Preprint*arXiv:2002.05693 [quant-ph].
Kirby, W. M., & Love, P. J. (2019). Contextuality test of the nonclassicality of variational quantum eigensolvers. *Physical Review Letters, 123*(20), 200501.
Kochen, S., & Specker, E. P. (1967). The problem of hidden variables in quantum mechanics. *Journal of Mathematics and Mechanics, 17*(1), 59–87.
Liang, Y.-C., Spekkens, R. W., & Wiseman, H. M. (2011). Specker's parable of the overprotective seer: A road to contextuality, nonlocality and complementarity. *Physics Reports, 506*(1–2), 1–39.
Milner, R. (1977). Fully abstract models of typed lambda-calculi. *Theoretical Computer Science, 4,* 1–22.
O'Hearn, P. W., & Pym, D. J. (1999). The logic of bunched implications. *Bulletin of Symbolic Logic, 5*(2), 215–244.
Penrose, R. (1992). On the cohomology of impossible figures. *Leonardo, 25*(3–4), 245–247. (Visual Mathematics: Special Double Issue).
Penrose, L. S., & Penrose, R. (1958). Impossible objects: A special type of visual illusion. *British Journal of Psychology, 49*(1), 31–33.
Plotkin, G. D. (1977). LCF considered as a programming language. *Theoretical Computer Science, 5,* 223–255.
Raussendorf, Robert. (2013). Contextuality in measurement-based quantum computation. *Physical Review A, 88*(2), 022322.
Scott, D., & Strachey, C. (1971). *Mathematical semantics for computer language*. Technical Monograph PRG-6, Programming Research Group, University Of Oxford.
Shalm, L. K., Meyer-Scott, E., Christensen, B. G., Bierhorst, P., Wayne, M.A., Stevens, M. J., Gerrits, T., Glancy, S., Hamel, D.R., Allman, M.S., Coakley, K. J., Dyer, S. D. , Hodge, C., Lita, A. E., Verma, V. B, Lambrocco, C., Tortorici, E., Migdall, A. L., Zhang, Y., Kumor, D. R., Farr, W.H., Marsili, F., Shaw, M. D., Stern, J. A., Abellán, C., Amaya, W., Pruneri, V., Jennewein, T., Mitchell, M. W., Kwiat, P. G., Bienfang, J. C., Mirin, R. P., Knill, E., & Nam, S. W. (2015). Strong loophole-free test of local realism. *Physical Review Letters, 115*(25), 250402.
Tarski, A., & Vaught, R. (1956). Arithmetical extensions of relational systems. *Compositio Mathematica* 81–102.
Tarski, A. (1936). Der wahrheitsbegriff in den formalisierten sprachen. *Studia Philosophica, 1,* 261–405.
Väänänen, J. (2007). *Dependence Logic, London Mathematical Society Student Texts* (Vol. 70). Cambridge University Press.
Van der Waerden, B. L. (1971). The foundation of algebraic geometry from Severi to André Weil. *Archive for History of Exact Sciences*, 171–180.
Werning, M. E., Hinzen, W. E., & Machery, E. E. (2012). *The Oxford handbook of compositionality.* Oxford University Press.

The Quantum-like Behavior of Neural Networks

Thomas Filk

1 Introduction

A neural network can be defined as "a massively parallel distributed processor made up of simple processing units that has a natural propensity for storing experiential knowledge and making it available for use" (Haykin, 2009). 'Experiential knowledge' refers to a training or learning phase during which certain properties of the network (usually the connectivity or the so-call synaptic weights) are changed, and this 'knowledge' is represented in the connectivity (and its weights) of this network. 'Making it available for use' refers to a retrieval phase in which the network can be stimulated by input-activities, while a reaction, which depends on this knowledge, is read from activities of certain network units (the output nodes).

Essentially networks are graphs, where the nodes—also called vertices or, in this case, neurons—represent the processing units, and the lines of the graph represent the connectivity. In addition, a dynamics specifies how the activities of the units change in time.

Alain Turing suggested the first networks of this type (Turing, 1948) using simple gates (like NAND- or NOR-gates) as processing units. McCulloch (McCulloch & Pitts, 1943) and later Rosenblatt (Rosenblatt, 1958) defined units which model neuronal behavior in the sense that an input of activities is integrated and leads to an output depending on a threshold. While in the realm of neurosciences often spiking neuron models are used, the most prominent example being the model of Hodgkin and Huxley (Hodgkin & Huxley, 1952) (for more general models and applications see also (Gerstner & Kistler, 2002)), the neural networks designed for practical purposes, which do not aim at the physiological details in the millisecond range, use rate-coded networks [see, e.g., (Haykin, 2009)].

A second distinction between applications in the neurosciences and in engineering refers to the architecture of networks: Most networks used for pattern recognition or pattern classification (like 'deep learning') are based on so-called feed-forward

T. Filk (✉)
Institute of Physics, Albrecht-Ludwigs University Freiburg, 79104 Freiburg, Germany

Parmenides Foundation for the Study of Thinking, Munich, Germany
e-mail: thomas.filk@physik.uni-freiburg.de

S. Wuppuluri and I. Stewart (eds.), *From Electrons to Elephants and Elections*,
The Frontiers Collection, https://doi.org/10.1007/978-3-030-92192-7_30

networks which do not contain closed loops along which activity can propagate, on the other hand, most network models used in the neurosciences are recurrent. Recurrent networks exhibit a much more complex behavior and are therefore much less predictable.

A recurrent network with N nodes allows for $N(N-1)$ possible directed lines and as each line can be present or absent, the number of such networks grows as $2^{N(N-1)}$ or roughly $O(2^{N^2})$. So, already for networks with $N = 10$ nodes this leads to roughly 10^{27} different configurations, which brings us to the edge of most present day computers when we want to test all these configurations individually for their behavior. Even though the local behavior of each node as well as the connectivity of nodes and how they interact pairwise is completely known and almost trivial, the overall behavior of larger networks cannot be predicted apart from a few special cases. This is one way to see networks as an example where it is difficult to relate the 'context' (the overall behavior) to the 'content' (the local constituents), and sometimes surprising phenomena emerge.

However, neural networks allow for a second way to distinguish between content and context. The content can refer to all the local activities, connectivities, weights, activity thresholds, types of neurons (e.g., inhibitory or excitatory) etc., i.e. the physics of the total network and its dynamics; this aspect will be dealt with in Sect. 2. In contrast, the context refers to the input-output behavior (depending also on the history of the network), i.e. on what can be seen from the outside. This perspective is more similar to most psychological or cognitive experiments, in which subjects are 'stimulated' by asking them questions or putting them into certain experimental situations, and the reaction of the subjects—their answer or behavior—is observed. The input-output behavior of recurrent neural networks exhibits surprising phenomena which sometimes are closer to phenomena observed in quantum systems as compared to classical systems. This perspective and some of its phenomena will be addressed in Sect. 3. One particular surprising phenomenon in quantum cognition is contextuality, which will be the subject of Sect. 4. The advantage of using neural networks as models for content and context is that we can always 'open the sculp' and look into the physics and/or mathematics and see what is happening.

2 Neural Networks

I begin with a brief introduction to the main concepts and terminology of neural networks: the architecture, the dynamics of single neurons, the training and the retrieval phase.

2.1 *Neural Networks: Structure and Dynamics*

2.1.1 Constituents and Architectures

Neural networks consist of nodes, also called vertices or neurons (the set of nodes will be denoted by $V = \{v_i\}$), and directed lines connecting these nodes, sometimes called synaptic connections. Mathematically, this represents a graph. A graph with N nodes can be represented by an $N \times N$ adjacency matrix:

$$A_{ij} = \begin{cases} 1 \text{ if there exists a connection from node } j \text{ to node } i \\ 0 \text{ otherwise}\,. \end{cases} \qquad (1)$$

This definition excludes so-called multi-loops, i.e. multiple connections between the same two vertices. More general structures are possible but will not be considered here.

If $A_{ij} = A_{ji} = 1$ for some i and j, the connection between the two vertices v_i and v_j is called bidirectional; if the adjacency matrix is symmetric, i.e. $A_{ij} = A_{ji}$ for all i and j, the graph is called undirected. If $A_{ii} = 1$, the line connecting vertex i to itself is called a self-loop. Most models exclude self-loops, i.e. $A_{ii} = 0$ for all i, but again there are exceptions.

In general, one distinguishes three types of nodes: so-called input nodes, output nodes and internal or hidden nodes. Input nodes receive signals from an environment and direct the activities to the internal nodes. The result of a neural network computation is read-off from the output nodes. If the vertices can be ordered in such a way that $A_{ij} = 0$ for $j > i$, the network is called feed-forward. All lines are necessarily directed. If the set of vertices V can be partitioned, $V = L_1 \cup L_2 \cup \ldots \cup L_O$ with $L_i \cap L_j = \emptyset$, such that for all vertices v_i and v_j with $v_i \in L_m$ and $v_j \in L_n$ we have $A_{ij} = 0$ unless $m = n$ or $m = n + 1$, the graph is called layered. The set L_m defines layer m, L_1 is the set of input nodes and L_O the set of output nodes. If $A_{ij} = 0$ unless $m = n + 1$ there are no connections between vertices within a layer and such networks are called 'layered feed-forward'.

So-called 'deep learning' networks often consist of hundreds to thousands of layers and usually are layered feed-forward. In this article I mostly consider so-called recurrent networks, i.e., there can be (directed) lines forming closed loops within the graph. Many of the 'non-classical' features of neural networks—non-commutativity, non-trivial attractors, etc.—do not occur in layered feed-forward networks.

2.1.2 The Dynamics of Single Neurons

Nodes or neurons are the local signal processing units of a network. One of the major distinctions between different classes of neural networks is the type of signal, where one usually distinguishes between so-called spiking and rate-coded networks. In spiking neural networks the signal consists of spikes—very brief pulses—and the information is stored in the temporal pattern of these spikes, i.e. essentially in the frequency and the relative phases between spikes. Most neural networks which are used in the field of neurocomputing are based on a dynamics for spiking neurons (more details can be found in (Gerstner & Kistler, 2002)).

In the present article I will concentrate on so-called rate-coded networks. In this case the signal consists of a real number which can be interpreted as an activity of a neuron and it represents a mean firing rate for spikes. While spiking neural networks are used to model parts of the dynamics happening in real neuronal networks and attempt a temporal resolution in the range of milliseconds or smaller, a signal in rate-coded networks usually refers to an average of spike trains and, when compared with biological models, corresponds to a temporal resolution in the range of several dozens up to hundreds of milliseconds. However, rate coded models can be considered as a type of neural networks of its own and almost all applications of neural networks, e.g. in pattern recognition, are based on rate-coded networks.

A second distinction in the type of dynamics refers to 'continuous' versus 'discrete time models'. Continuous dynamics is usually described in terms of a (first order) differential equation for the rates. Discrete time step models are based on a dynamics of the type $x_i(t+1) = F(\{x_j(t)\})$, where $x_i(t)$ denotes the activity rate of neuron v_i at time step t. For the rest of this article I will restrict myself to discrete time step models.

In most cases the dynamics of neural networks is defined by the following type of equation, which I will refer to as the 'update-equation':

$$x_i(t+1) = T\left(\sum_i w_{ij}\epsilon_j x_j(t)\right) . \qquad (2)$$

Here w_{ij} is the so-called weight matrix (or synaptic weight matrix), $\epsilon_j = \pm 1$ controls inhibitory influence and T is the so-called transfer or activity function. The meaning of these entities will be explained in the following paragraphs.

2.1.3 The Weight Matrix

Weight matrix and adjacency matrix are the same, if the synaptic weights w_{ij} assume only the values 0 and 1. However, the elements of the weight matrix represent a degree of 'transmissivity' for a connection between two neurons. In biological systems this transmissivity is regulated by many parameters, amongst others by neurotransmitters of the pre-synaptic neuron and the corresponding receptors at the post-synaptic neuron, and they can be influenced, e.g., by drugs, mood, or viruses. The weight-matrix contains the information or 'knowledge' stored in the network because in most models the weights w_{ij} are the parameters which are changed and adapted during a training phase [(see sect. 2.2)].

2.1.4 The Activity Function

The activity or transfer function T determines the activity (firing rate) of a neuron depending on the input it gets. The argument of T in eq. 2 is a sum over all activities x_j of neurons j which are connected to a particular neuron i, weighted by w_{ij}, the transmissivity or synaptic weight of the connection from j to i. Biological neurons fire as the result of a cumulative effect: if the sum of the incoming signals exceeds a certain threshold (within a certain time interval of the order of 20 ms), the neuron fires. The details will not be relevant here, but in order to model this behavior, the activity function T is usually chosen to be a sigmoid function: 0 for small values of the argument with a threshold at a particular input level and 1 for large values of the argument. For more details about the firing behavior of physiological neurons see, e.g., (Kandel, 2000).

In very crude models, $T(x)$ is a step-function and has the form: $T(x) = 0$ for $x < x_t$ and $T(x) = 1$ for $x \geq x_t$ (where x_t is the threshold value). In mathematics, this function is called the Heaviside-Function and often denoted by $\Theta(x - x_t)$. Sometimes also continuous functions are used, e.g.

$$T(x) = \frac{1}{1 + \exp(-a(x - x_t))}, \tag{3}$$

where a determines the sensitivity of T near the threshold: for a large this is almost a step-function, for a small T increases slowly from 0 to its maximal value 1, assuming the value $\frac{1}{2}$ for $x = x_t$.

2.1.5 Inhibitory and Excitatory Neurons

Mammal brains consist of many different types of neurons with special features. For recurrent networks the distinction between at least two types of neurons turns out to be relevant: excitatory and inhibitory neurons. Inhibitory neurons control the activity in a network, preventing it from getting too large. Inhibitory effects can be incorporated into the model by allowing the weights w_{ij} to be negative. However, as a rule of thumb (which has its exceptions), a single neuron uses one type of neurotransmitter and therefore either acts in an inhibitory or an excitatory way. This is sometimes referred to as Dale's rule. So, $\epsilon_i = +1$ indicates that neuron i is excitatory and $\epsilon_i = -1$ indicates that it is inhibitory.

2.2 The Training of Networks—Two Examples

As already mentioned, the information stored in a neural network is contained in the weights w_{ij}, i.e. essentially in the connectivity of the network. The first step in using neural networks is a training or learning phase. During such a training phase certain patterns are presented to the input neurons. In general, one distinguishes three types of training: unsupervised learning, reinforcement learning and supervised learning.

In unsupervised learning there is no feed-back to the network; the network detects regularities and correlations in a set of input-data and restructures itself in such a way that the output reflects some of these regularities. An example for such networks are so-called Kohonen networks or self-organizing maps (Kohonen, 1984). In supervised learning the weights are changed in such a way as to get closer to the desired output. This strategy is used, e.g., in layered feed-forward networks where there exist powerful algorithms (like the so-called backpropagation algorithm; see, e.g., (Haykin, 2009)) to update the weights for better performance. In reinforcement learning the weights are changed according to general rules which only depend on whether the output is correct or incorrect, or whether it is better or worse as compared to a previous network.

Here, I will describe two learning mechanisms which will be used later in the applications.

2.2.1 Hopfield Networks and Hebbian Learning

Hopfield networks (Hopfield, 1982) are used as models for associative pattern memory. They consist of a single layer of neurons, each connected (bidirectionally) with every other neuron. This layer serves as input and output layer. The state x_i of each neuron is usually chosen binary, e.g. $x_i = 0$ or 1.

The training phase of such a network can be replaced by an exact calculation of the weights using the so-called Hebbian learning rule. In 1949, Donald Hebb formulated a learning rule according to which the connection between neurons firing simultaneously gets strengthened (Hebb, 1949). This is often abbreviated as "fire together, wire together". Given a certain set of patterns $\{M_i^\alpha\}$ where $\alpha = 1, 2, ..., n$ enumerates the patterns (for a network with N neurons, n should not be much larger than roughly $0.14\,N$ (Amit et al., 1985)) and $M_i^\alpha \in \{0, 1\}$, the weight matrix of the network is calculated according to

$$w_{ij} = \frac{1}{n} \sum_{\alpha=1}^{n} M_i^\alpha M_j^\alpha \,. \tag{4}$$

For each pattern, where neuron i and j are active ($M_i^\alpha = M_j^\alpha = 1$), the corresponding entry w_{ij} of the weight matrix is increased by $1/n$. Sometimes in eq. 4, instead of $M_i^\alpha M_j^\alpha$, which is a strict application of Hebb's rule, the combination $(2M_i^\alpha - 1)(2M_j^\alpha - 1)$ is chosen. In this case the weight increases, if neurons i and j in a pattern α have the same activity (both 0 or both 1) and it decreases if the activity is different. This choice has some mathematical advantages. For the following these differences will not be relevant.

For the retrieval of a pattern, the network is stimulated by a pattern $M_i^R \in \{0, 1\}$ $(i = 0, ..., N)$ which may be similar but not necessarily identical to one of the training patterns. The activity of the nodes is then updated according to

$$x_i(t+1) = \Theta\left(\sum_j w_{ij} x_j(t) - c\right) \tag{5}$$

with $x_i(0) = M_i^R$ and $\Theta(x)$ being the Heaviside function. One can show that this algorithm stops after a finite number of steps. For $c \approx 1/n$ (a good value for c has to be determined by trials and depends on the 'density' of the patterns $\{M^\alpha\}$, i.e., the ratio of 0s versus 1s in a typical pattern) it usually stops with the trained pattern M^α which is closest (e.g. in the sense of a minimal number of differing entries) to M^R. However, depending on the value of c and the chosen pattern M^R, the algorithm might also stop at a new pattern. So, Hopfield networks may create 'novelty' (see Sect. 3.5)

2.2.2 Random Recurrent Networks and Genetic Learning Algorithms

Hebbian learning is only moderately successful in recurrent networks for which input, internal and output neurons are different and which are trained to 'react' in a particular way to a certain input. In such cases so-called genetic learning algorithms can be applied.

First one defines an error function (or performance function, depending on the context one also speaks of an inverse fitness or an energy function), which measures the quality of the network. In many cases the error function is defined as a sum over the squared differences between 'desired activity' and 'actual activity' at the output nodes, i.e.

$$E = \sum_{i \in O} (x_i(t) - x_i^d)^2 \tag{6}$$

where x_i^d is the desired activity of output node i (O is the set of output nodes and the sum runs over all output nodes) and $x_i(t)$ is the actual activity of node i at a particular time-step t. One can also average over a certain number of time-steps as well as over input patterns with different desired reactions.

In many cases so-called genetic algorithms are used for such networks: One starts with a random adjacency (or weight) matrix and determines its error function. Then one makes a random change at one of the connections or weights and determines the error function of the new network. If the error function of the new network is smaller, one proceeds with this network. If the error function of the new network is worse as compared to the previous network, one keeps the previous network and tries different changes.

This basic idea can be extended in many ways: One can introduce error-dependent probabilities such that in some cases also the network with the larger error is kept (in this way one arrives at so-called Boltzmann machines) or one can add other terms to the error function, like a sum over squared weights; this suppresses networks with many connections or large weights and is sometimes referred to as a 'renormalization' of the network. In extreme cases this renormalization can lead to sparse networks with only few connections.

The learning phase stops when the error function for a networks gets below a preset threshold—in extreme cases when it is zero. For pattern recognition this training is done with a certain set of training patterns and the error function usually consists of an average error for these training patterns. Very often the training phase (during which the connectivities and/or weights are changed) is followed by a test phase in which the quality of the network is tested using a set of new patterns. If the error for these test patterns is sufficiently small, the training phase is over and the connectivities and weights are kept constant. The network can be applied to 'real' pattern recognition. If not, the training phase is extended.

3 Non-classical Behavior of Neural Networks

After a brief discussion of what I mean by 'non-classical behavior', which involves the notion of states, observables, and attractors, I will give some examples from recurrent neural networks which in this sense behave non-classically.

3.1 States and Observables

Physical systems are often described in terms of so-called states and observables. E.g., in Newtonian mechanics a state is described by a point in phase space giving the position and the momentum or velocity of particles, and an observable is a function of these variables, like energy, angular momentum etc., which are functions of positions and momenta. On the other hand, in quantum mechanics a state is represented by a normalized vector (or, to be more precise, by a one-dimensional linear subspace, in which case the vector is a representative of this subspace) in a Hilbert space (essentially a vector space with a scalar product) and an observable is represented by a linear mapping on this Hilbert space. In general, a state describes what we know about the system from its past, e.g., how it was prepared, and an observable describes the information we can gain by measuring this quantity.

When I speak of the 'behavior' of a neural network I refer to this structure of states and observables. When a neural network is considered as a physical system, for which the state is specified by the weight matrix and the momentary activity of all nodes, and an observable is a function of these variables, we treat it like a Newtonian system. However, one can also define states and observables in terms of the reaction of a network (which we can read off from the output nodes) as a result of a stimulus at the input nodes. Again, a state refers to what we know about the network from its past input-output reactions (as far as this is relevant for the prediction of future reactions) and an observable describes what we can learn from applying a certain stimulus (pattern) to the input nodes.

This second perspective emphasizes a 'black box' nature of a neural network: only the input patterns (stimuli) can be controlled and only the reaction of the network (the activity at the output nodes) can be observed. Therefore, I distinguish between an 'observer perspective', which is the perspective just mentioned, and an 'internal' or 'God's-eye perspective', which has access to all the connectivities, weights and, in particular, momentary activities of the nodes inside the network. With respect to the God's-eye perspective, all neural networks mentioned in this article are classical systems and can be described by classical dynamics.

From a physics point of view, the notion of states and observables with respect to an observer perspective might be considered as strange or, at least, non-standard. However, from a cognitive or psychological point of view this concept is exactly what models observations or experiments in the cognitive sciences: A stimulus is presented to a subject and we observe his or her reaction. Essentially, this is all we have. From these 'measurements' we have to deduce features of a 'mental state' which may lead to predictions about future reactions.

These two perspectives allow for a second way to view neural networks as being in between 'content' and 'context': The content refers to the physical system as being made up of nodes and connections and the equations for updating activities (the 'internal perspective'). On the other hand, the context refers to the input-output relations of such a network (the 'observer perspective').

When I speak of 'non-classical' behavior of neural networks, I mean that the observer perspective sometimes exhibits phenomena which resemble phenomena in quantum systems and which usually are not expected to occur in classical Newtonian systems. Some of these will be described in the following sections after a brief introduction to the useful notion of attractors and attractor landscapes. The particular important phenomenon of contextuality will be addressed in Chap. 4.

3.2 *Attractors in Neural Networks*

The notion of an attractor is particular useful in dynamical systems theory. It describes the long-term behavior of a system—usually depending on initial conditions and on boundary conditions—for which all transient dynamics is irrelevant. I will describe the concept of an attractor only for discrete dynamics like the one defined in eq. 2.

Let us denote by $\mathbf{x}(t) = \{x_1(t), x_2(t), ..., x_N(t)\}$ the activities of all nodes at time t. For a given initial configuration of activities $\{\mathbf{x}(0)\}$ (time $t = 0$), eq. 2 determines all subsequent configurations $\{\mathbf{x}(t)\}$. Note that a given stimulus pattern acts as a boundary condition on the input neurons and should be considered as constant for this sequence (i.e., the input neurons do not change their activities as a function of

t). Now denote by $X_T = \{\mathbf{x}(T), \mathbf{x}(T+1), ...\}$ the set of all activity configurations in this sequence from time $t = T$ on. Obviously, $X_T \subseteq X_{T'}$ for $T' < T$, i.e., X_T can never increase as a function of T. The formal definition of the attractor set X_∞ of the sequence of activities is $X_\infty = \lim_{T\to\infty} \bigcap X_T$. Essentially, these are all activities which keep reappearing after arbitrary many time steps.

An important example of an attractor is a fixed point attractor, for which X_∞ contains only one element, i.e., one activity pattern for the network which repeats itself. All feed-forward networks have fixed point attractors in this sense: from a certain moment on the activity doesn't change any more.

A second import class of attractors are so-called periodic attractors of cycle length L. In this case, a sequence of L activity patterns repeats itself in a cycle. If the activities $x_i(t)$ can assume only finitely many (discrete) values (e.g., only 0 and 1), fixed point attractors and periodic attractors are the only attractor sets possible. If $x_i(t)$ can assume a continuum of values (e.g. $x_i(t) \in [0, 1]$ for a transfer function T like in eq. 3) also so-called strange attractors are possible. They consist of an infinite number of points and usually have a fractal nature. In the following I will only consider fixed point and periodic attractors. For small networks (with a total of less than 50 vertices) the attractor states are reached quite rapidly (mostly within less than 20 time steps).

Note that from an observer's perspective, which only has access to the activity of output nodes, a periodic attractor may look like a fixed point attractor: The activity of internal nodes may still vary while the activity at the output nodes remains constant.

For a given network, defined by its weight matrix w_{ij}, the attractor set usually depends not only on the input pattern, which from a mathematical point of view can be considered as fixed boundary conditions, but also on the initial activity pattern $\mathbf{x}(0)$. The set of all initial activity patterns which finally lead to the same attractor is called the attractor basin (for this attractor). For feed-forward networks the fixed point attractors depend only on the boundary condition (i.e. the input activity in the first layer). Therefore, the attractor structure of feed-forward networks is quite simple. For recurrent networks, however, the attractor structure may be quite complicated with different initial conditions leading to different attractors. This gives rise to a so-called attractor landscape, which can contain fixed point attractors and periodic attractors of various length. These attractor landscapes change during the training phase when the weight matrix w_{ij} is changed and show an interesting complexity structure: In the average, the number of attractors first increases during a training process and decreases towards the final stage when the networks have almost completed their training (Atmanspacher & Filk, 2006; Atmanspacher et al., 2009).

3.3 Non-commutativity

Non-commutativity is one of the catchwords which is often used to distinguish classical and quantum systems. It refers to an order-dependence of observations. If first an observation A is performed and then an observation B, the results may differ if the order is reversed. In psychology this phenomenon is known from questionnaires and similar 'measurements' [see, e.g., (Atmanspacher & Filk, 2018; Atmanspacher & Römer, 2012; Wang et al., 2014)].

In order to make the non-commutative nature of neural networks more explicit, I define an observation A to consist of the following steps (performed for a given network with fixed weight matrix w_{ij}):

- A pattern (also called A) is presented to the input nodes of the network, i.e., the activity of these input nodes is fixed by this pattern.
- For the first observation, the initial activity state $\mathbf{x}(0)$ can be chosen randomly. One can also choose the last activity pattern from a previous 'observation' for the initial configuration. The update of activities follows eq. 2. The update is performed sufficiently often until an attractor state is reached.
- After an attractor state has been reached, the result of the observation consists of the output activities for sufficiently many time steps. If the attractor happens to be a fixed point attractor, 'sufficiently many' is 1. If the system has reached a periodic attractor, the result of the observation consists of all the activities at output nodes within one attractor cycle. (For more details on the notion of states and observations and the mathematical structure of observables in neural networks as derived from the external knowledge about the attractor landscape, see (Kleiner, 2012).)

For a second measurement B, a second pattern (also called B) is presented to the input nodes of the network as a stimulus. In this case, the initial state consists of one of the activity states of the previous attractor state. If the previous attractor state of observation A was a fixed point attractor, this fixed point activity serves as the initial activity configuration for the second observation.

In recurrent networks order effects have been observed, i.e., the final results of observations may depend on the order in which the patterns are presented (Atmanspacher & Filk, 2006; Atmanspacher et al., 2009). It is important to note that the attractor state of the first observation serves as an initial condition for the second observation. This protocol is close to what is done in psychological or cognitive experiments, e.g. in questionnaires: The first question may prime the mental state of a person for the second question.

3.4 *Causality and Observed Indeterminism*

From a God's-eye perspective, a neural network subject to the dynamics of eq. 2 is completely deterministic. However, from an observer's perspective this may not be the case.

Imagine the following situation: An observation A has been made, i.e., the input nodes have the activity of pattern A and the output nodes show a constant activity which is interpreted as the 'result' of this measurement. Internally, the activity might be in a periodic cycle. If the input pattern is changed from A to a new 'stimulus' B at a certain moment t, the result, i.e. the attractor which finally appears at the output nodes, may depend on the internal activity at the exact moment t. But as the internal activity is part of a periodic cycle, which is not observed, there is no control about this activity. For continuous variables the internal activity may be part of a strange attractor, in which case the situation is even worse. For an external observer the result seems indeterministic, because the 'switch' from one pattern A with a given result to a second pattern B may not always lead to the same results for this second measurement.

For larger networks with a complicated attractor landscape it may be impossible to determine the complete internal state of the system from the past reactions. And for real neuronal networks the synaptic weights keep changing all the time which makes any form of prediction even less likely. Even if the complete history of a network is known—this refers to the observer's perspective, i.e. the previous relations between input pattern and output activities— the network will appear to be indeterministic. Of course, in general this indeterminism will not be completely random but one will observe statistical rules or regularities.

3.5 *Memory and Novelty*

Hopfield networks are models for associative memory (Hopfield, 1982): Upon a clue, which usually consists of parts of a previously trained pattern, the network reproduces the full pattern. How large this 'part' of a previously trained pattern has to be in order to trigger the full pattern depends on many details, amongst others also on similarities among the trained patterns and the threshold chosen for the transition function. A Hopfield network of N nodes can memorize roughly $0,14\,N$ randomly chosen pattern (Amit et al., 1985).

However, depending on the threshold of the transition function and the stimulus, the network sometimes reproduces a pattern which is not part of the trained set. At least for small networks (of the order of 30 nodes) any combination of trained patterns can be reproduced: the conjunction of activities, the disjunction, the complement of one pattern minus the other, etc. (Bässgen, 2017). (In mathematics this is called a sigma algebra.)

This implies that by varying the threshold of the transition function and applying appropriate stimuli one can generate many new patterns which are combinations of the old ones. In real neural networks the threshold of the transition function can easily be influenced by changes in neurotransmitter concentrations, so this testing of patterns can be an effective method of the brain to find new solutions.

4 Contextuality in Neural Networks

Quantum contextuality does not only play an important role in quantum physics, but it also became a relevant issue in quantum cognition (for a general introduction of quantum cognition see, e.g., (Busemeyer & Bruza, 2012; Filk, 2019; Khrennikov, 2010; Wendt, 2015)). After some historical remarks, I will explain this phenomenon using the example of the so-called CHSH inequality, putting some emphasis on the role of signaling, before I finally make the connection to neural networks.

4.1 Preliminary Historical Remarks

Historically, the notion of contextuality—at least as far as its meaning of 'context dependence' is concerned—took a somewhat bizarre path. While context dependence is a familiar concept in the arts or cognitive sciences (philosophy later added the term 'contextualism') it swapped over to physics in the 1960s as a consequence of the Kochen-Specker theorem (see below), where it received a mathematical formulation. During the last fifteen to twenty years, contextuality—equipped with this mathematical add-on—drifted back again into the cognitive sciences via the route of 'quantum cognition'.

The term 'contextuality' seems to have first emerged in quantum physics. In 1957, Andrew Gleason proved a mathematical theorem about the existence of measures on the set of closed subspaces of a Hilbert space (in quantum physics such linear subspaces represent physical states) (Gleason, 1957). After analyzing the physical consequences of this theorem, John Bell came to the conclusion that "The result of an observation may reasonably depend not only on the state of the system (including hidden variables) but also on the complete disposition of the apparatus" and that the strangeness of these consequences are grounded in the "tacid assumption [...] that measurement of an observable must yield the same value independently of what other measurements may be made simultaneously" (Bell, 1966). In 1967 Kochen and Specker constructed an example with discrete states in which it could be proven that the results of certain measurements depend on which other compatible measurements (the context) are performed on this same system (Kochen & Specker, 1967). Despite its discreteness, the model of Kochen and Specker was quite complicated and involved 117 hypothetical measurements. Later simpler proofs were published [see e.g. (Cabello, 1994; Peres, 1991)].

A special example of contextuality are quantum correlations of entangled systems. In this article I will only discuss one example of this form of contextuality, which is the CHSH inequality (see next subsection). This inequality has been the subject of much research also in the cognitive sciences. It was first noted by Aerts et al. that Bell-type inequalities (like CHSH) can be violated in scenarios where subjects are asked to answer certain sets of questions (Aerts et al., 2002). Later similar results were obtained by other groups (Aerts & Sozzo, 2014; Bruza & Cole, 2005). Then it was noted by Dzhafarov et al. that the data of all these experiments do not have 'marginal selectivity' (a term used in early articles of this group, e.g. (Dzhafarov & Kujala, 2014)) or 'consistent connectedness' (the equivalent expression used in more recent articles, e.g. (Dzhafarov et al., 2016)). This quantity measures to which extend correlations of two variables can be used for a transfer of information between the sites where these variables are measured. This group suggested to first subtract these 'signaling' correlations before the CHSH inequality is evaluated. They found that in this case none of the previously reported results from the cognitive sciences violated the inequality anymore. However, they reported about the data of a new experiment which indeed did violate the inequality (Dzhafarov et al., 2016).

4.2 CHSH as Contextuality

There are many good review articles about entanglement and the violation of Bell-type inequalities, in particular also the CHSH inequality, in quantum theory [see e.g. (Clauser et al., 1969; Horodecki et al., 2009; Nielsen & Chuang, 2000)]. In this section, I will present a very explicit case which later can be transferred to various approaches in the realm of quantum cognition.

Photons, the so-called quanta of light, have a property called polarization. Even though there are infinitely many possible polarization states, a measurement (e.g. by letting a photon pass through a polarization filter) can only yield the result 'yes' (has passed the filter) or 'no' (did not pass). We are free to choose an orientation for the filter, and I will denote this orientation by an angle α which can be between $+90$ and -90 degrees.[1] In a single experiment we can 'measure' the polarization only with respect to one direction, i.e. one angle, and the result is binary: yes or no.

Consider the following set-up: There are two photons (in an entangled state, but this is not relevant at this stage; see below). On photon 1 we choose to measure the polarization either with respect to an axis α, say 0 degrees, or an axis $\alpha' = 45°$ (the particular angles are chosen in such a way that certain quantum systems yield a maximal violation of the CHSH inequality). On photon 2 we choose to measure the polarization either with respect to an axis $\beta = 22,5°$ or an axis $\beta' = 67,5°$. α and α' cannot be measured simultaneously (on the same photon), similarly β and β'. But we can measure any polarization on photon 1 in combination with any polarization on photon 2 simultaneously. This leaves us with four possibilities: (α, β), (α, β'), (α', β), and (α', β'). Denote by a, a', b, and b' the results of measurements for angles α, α', β and β', respectively (e.g. $a = +1$ for 'passed' and $a = -1$ for 'did not pass' etc.).

Now we consider a huge number of photon pairs prepared in the same (entangled) state. One photon is always sent to site A where its polarization is measured (either with respect to α or α') and the polarization of the other photon (β or β') is measured at site B. In a next step, the protocols from site A and B are compared and the four expectation values $\langle xy \rangle$, with $x = a$ or a' and $y = b$ or b', determined. In each

[1] A rotation of 180 degrees does not change the polarization axis, and for simplicity I will only consider linear polarization states.

single case the product can only be $+1$ or -1, therefore the expectation value can be anything in between. Finally, the following quantity is evaluated:

$$S = \langle ab \rangle - \langle ab' \rangle + \langle a'b \rangle + \langle a'b' \rangle \,. \tag{7}$$

The minus sign can be in front of any one of these four terms, important is that three terms have the plus sign and one the minus sign. One can show [see e.g. (Clauser et al., 1969; Horodecki et al., 2009; Nielsen & Chuang, 2000)] that if the results of the measurements were predetermined by some hidden parameter—i.e., before the measurement is actually performed some hidden degrees of freedom determine what the result of the measurement will be—we must have

$$|S| \leq 2 \,. \tag{8}$$

This is called the CHSH inequality. Essentially, its applicability depends on two assumptions: (1) there are no correlations between hidden parameters and the choice of the experimentalists for their decision to measure a or a' at site A or b or b' at site B, respectively; this is sometimes called the 'no conspiracy assumption' (with various other names in use, like 'free will of the experimentalist' or 'no superdeterminism'), and (2) there is no signaling between the two sites, i.e., whatever measurement is performed at site A (with result a or a') has no direct influence onto the results of measurements performed at site B or vice versa.

While assumption (1) is tacitly assumed in physics (we believe in the freedom of the experimentalist to make his or her choice of measurements freely and not predetermined), assumption (2) is 'guaranteed' in most physics experiments by putting sites A and B far apart (up to several hundred kilometers) and performing the measurements on the two entangled photons within nanoseconds. This excludes any signaling with a velocity not exceeding the velocity of light (up to now signaling velocities with less than 10.000 times the velocity of light can be experimentally excluded (Yin, 2013)). So, assuming the theory of relativity is valid, any direct influence of the results obtained at one site onto the results obtained at the other site can be excluded.

In quantum theory, the two photons can be prepared in a so-called entangled state such that the CHSH-inequality is violated for the polarization orientations given above. It turns out that the upper limit for S in quantum theory is $S = 2\sqrt{2} \approx 2,83$. This is the so-called Tsirelson bound (Tsirelson, 1980).

What has this to do with contextuality? As measurements on different photons are always compatible (they can be performed simultaneously) the violation of the CHSH inequality in quantum theory can be interpreted as 'the results of measurements obtained at one site depend on which measurements are performed and which results are obtained at the other site'. Measuring α in combination with β yields (on average) other values for a as compared to a measurement of α in combination with β'. I will not prove this statement here, but I will demonstrate, why S cannot be equal to $+4$, even though it is the sum of four terms each of which can be anything between $+1$ and -1.

For $S = +4$ we must have $\langle ab \rangle = \langle a'b \rangle = \langle a'b' \rangle = +1$ and $\langle ab' \rangle = -1$. However, the first equalities imply that a is always equal to b, b is always equal to a' and this in turn is always equal to b'. But now we have $a = b = a' = b'$, which also implies that $a = b'$. In this case, however, the expectation value $\langle ab' \rangle$ is also $+1$ and S assumes the value 2. For $\langle ab' \rangle$ to be -1, the results for a and b' should always be different. So, if $S = +4$ the result of α depends on whether it is measured together with β or together with β'. This is called contextuality. One can show in general that for $S > 2$, the results a, a' etc. have to be contextual: They depend on what other measurements are performed simultaneously [see, e.g., (Fine, 1982)].

4.3 CHSH in Quantum Cognition

The situation described above has been applied to cognitive experiments. Without going into the details of each of these experiments [see e.g. (Aerts et al., 2002; Bruza & Cole, 2005; Dzhafarov & Cervantes, 2018) and references therein] I just outline the main characteristics using the example of (Atmanspacher & Filk, 2019). Four pairs of terms are chosen, usually of two different types which I will call A-type and B-type. In our example the A-type pairs consist of numbers and are $(3, 4)$ and $(2, 9)$ and the B-type pairs consist of properties of numbers and are (prime, non-prime) and (even, odd). These four pairs correspond to the four possible measurements: α, α' as A-type and β, β' as B-type. In Aerts et al. 2002 the first group consisted of animal names (horse, cat, bear, etc.) and the second group of sounds (whinnies, roars, etc.), in Dzhafarov and Cervantes 2018 the first pairs consisted of characters in the fairy tale Snow Queen by Hans Christian Andersen (Snow Queen, Gerda, Troll, etc.) and the second of attributes (beautiful, mean, etc.). The group of subjects is divided into four equal-sized subgroups and the subjects of each subgroup are given two pairs of terms, one from the A-type the other from the B-type. E.g., the first subgroup gets the combination $(3, 4)$ and (prime, non-prime). The experiment consists of asking the subjects to choose one element of each of their pairs *such that it fits*, which means, e.g., that 3 of the first pair has to be combined with 'prime' from the second, while 4 of the first pair should be combined with 'non-prime'. The first term of each pair is counted as result $+1$, the second as result -1. Now the expectation values are determined (each group contributes one expectation value) and the quantity S is determined. In the above example this would lead to:

$$S = \langle((3,4), (\text{prime,non-prime})\rangle_1 - \langle((3,4), (\text{even,odd})\rangle_2 \\ + \langle((2,9), (\text{prime,non-prime})\rangle_3 + \langle((2,9), (\text{even,odd})\rangle_4$$

If all answers 'fit', the result will be $S = +4$ in this case. I have marked the expectation values by subscripts 1 to 4 indicating that these are four different samples or groups of subjects. The 'no-conspiracy' assumption essentially states that these four samples have the same statistical properties with respect to whatever 'hidden' parameters, which may have an influence onto the results of the measurements; or, to put it differently, that the result would have been essentially the same if the four groups of subjects were assigned to different A- and B-type pairs.

Is this result surprising? I think not. The main assumption in physics for the inequality to hold, and which is violated here, is 'non-invasiveness' or 'no-signaling'. Whatever pair is chosen first by a subject determines the answer for the second pair. If a subject of group 2 (second term in the expression for S) makes a decision which pair to take first (say the pair $(3, 4)$) and chooses e.g. 4, this determines the 'fitting' answer for the second pair, namely 'even'. And the product of the results will be -1. The same holds for all other combinations in this group: the choice from the first pair determines the outcome of the second. In physics, such a situation is avoided by making the two measurements within the complement of each other's light-cone, i.e. by performing them in such a way that no signal can travel (with the speed of light) from one site to the other. This is not the case here: The subject is the same and, of course, he or she knows the first choice and the matching answer.

The appropriate 'cognitive' experiment would be the following: Subjects are grouped into pairs (they can be 'entangled' in the sense that they may have a relationship, are friends, have been primed in a similar way etc.), lets call them Alice (the A-type) and Bob (the B-type). Alice and Bob are separated such that no exchange of information between them is possible. Alice gets an A-type pair of terms (e.g. $(3, 4)$), Bob a B-type pair (e.g. (even, odd)). Of course, Alice and Bob must not know in advance, before they get separated, the pair of terms they will get (this would be superdeterminism or 'conspiracy'), and after they got their pair of terms they are not allowed to get into contact anymore (this would be signaling). Alice chooses one possibility of her pair and Bob chooses one possibility of his pair. Even if Alice and Bob decide beforehand on a certain strategy (e.g. both taking always the first term of their pair) it is easy to show that the CHSH inequality will be satisfied. Any other result would indeed be surprising. But this is not the way the experiments are performed.

4.4 Signaling and Contextuality

What is the connection between the CHSH inequality and the notion of contextuality in the cognitive sciences? Presumably Fine's theorem is responsible for this relationship. In 1982, Arthur Fine proved that if the CHSH inequality (plus some trivial consistency conditions) holds for a set of data then there exists a joint probability distribution for these data (Fine, 1982). This implies that if the inequality holds there is no need for assuming a context dependence, while on the other hand, if the inequality is violated, there has to be some form of contextuality in the sense that the results of one or more observables depend on which other observables are measured simultaneously and what these results are. A violation of the CHSH inequality is a sufficient condition for context dependence, however, it is not a necessary one. The point is, and this is what Fine's theorem tells us, that, if CHSH is satisfied, from the data alone we can never judge whether there is a context dependence or not. For such a judgement it needs a model which explains how certain correlations come about.

In entangled quantum systems the CHSH inequality can be violated for certain parameters (e.g., the particular angles used in Sect. 4.2), but there are other parameter regimes, where the CHSH inequality is not violated. This doesn't imply that there is no 'context dependence' in these cases. It would be quite surprising if context dependence suddenly sets in for certain ranges of parameters. The mathematical formalism which determines the statistics of the results does not suddenly change when the parameters (the angles) switch from a region where the CHSH inequality is satisfied to a region where it is not. However, we can construct local hidden variable models without context dependence to explain the cases where CHSH holds, but we cannot construct such models when the CHSH inequality is violated.

It is still a matter of debate among physicists how to interpret the violation of Bell-type inequalities like CHSH in quantum theory. Some argue that there has to be a kind of mutual 'influence' between the two sites where the correlations occur, however, this 'influence' does not involve an exchange of energy and it cannot be used for a controlled exchange of signals (and, therefore, a superluminal communication) between Alice and Bob. This is referred to as the 'non-communication theorem' or sometimes also as the 'non-signaling theorem'. Even though the correlations maybe absolute (e.g. always correlated or always anti-correlated), the local outcomes at the sites of Alice and Bob appear to be random and cannot be controlled.

Apparently, the notion of 'signal' is used in this context in two different meanings: (1) an operational definition, which is used by Dhzafarov et al. [?] and which measures the extend to which the observed correlations can be used to transfer information from one site to the other (communicating signals), and (2) an 'ontological' definition with respect to a model in which it is obvious that a signal has been transferred from one site to the other, however, this signal cannot be used for communication. This distinction of meanings has recently been emphasized in several articles (Atmanspacher & Filk, 2019; Jones, 2019; Walleczek & Grössing, 2016).

I will not go into the technical details of how signaling is treated in the different approaches but I want to emphasize, that 'signaling' is not in contradiction to 'context dependence'. Indeed, we are surprised (and quantum physics seems to be the only example where this may be true) if the behavior of a system A depends on a context without A having knowledge (in a very broad sense) about this context. And 'having knowledge' usually implies that some kind of signal has passed from the site of the context to the site of A. So, eliminating this type of signaling may not be the right approach for a definition of contextuality in the sense of 'context dependency'. On the other hand, one can *define* contextuality as a dependence which is not due to a common course and also not due to any form of (non-communicating) signaling. In this case, however, while quantum *theory* involves a contextual formalism, it is not clear whether or not quantum *reality* is contextual.

4.5 Neural Networks Can Violate CHSH

Figure 1 (left) shows an example of a neural network which violates the CHSH inequality maximally (i.e., $|S| = 4$, such devices are sometimes called PR-boxes (Popescu & Rohrlich, 1994)) and for which the output correlations do not allow for communicating signal exchanges (or, to put it differently, which has marginal selectivity). Alice's choice for α or α' is translated into the input 0 or 1, respectively, while Bob's choice for β or β' is translated into 1 or 0, respectively. The box with a symbolized threshold function and the label $\theta = 0.5$ represents a neuron which sends out the signal 1 if the sum of the two inputs is larger than 0.5, which is always the case except for the choice (α, β'), which leads to an output of 0. Up to point (3) in Fig. 1(left) the network represents a simple AND-gate, which already violates the CHSH inequality maximally [(see also (Filk, 2016, 2015), where this model has been elaborated]. However, the AND gate can be used for signaling: Bob can signal to Alice, because if she keeps her setting in α, her reading 0 or 1 depends only on Bob's choice of β or β', respectively. External signaling is made impossible by adding a random number generator RNG with equal probabilities for 0 and 1. This random number generator is triggered by Bob's input (dashed line). The output on Bob's site equals the random number while the output on Alice's site is obtained as the result of an XOR gate, which has as its input the result of the first 'neuron' (essentially the AND gate) and the random number. The outputs on both sides, 0 or 1, are finally translated to external outputs -1 and $+1$, respectively. The device leads to $S = +4$.

The two elements which are 'foreign' to neural networks, the XOR gate and the random number generator, can easily be realized by neural networks: The XOR gate can be realized with three 'neurons' (Fig. 1 (right)), and the random number generator can be realized by any network which has a periodic attractor with period 2 or larger. The state of this attractor at the exact moment when Bob sets his choice for β or β' will then serve as the random number generator. The connection between the random number generator and the XOR gate (connection 2) can be replaced by

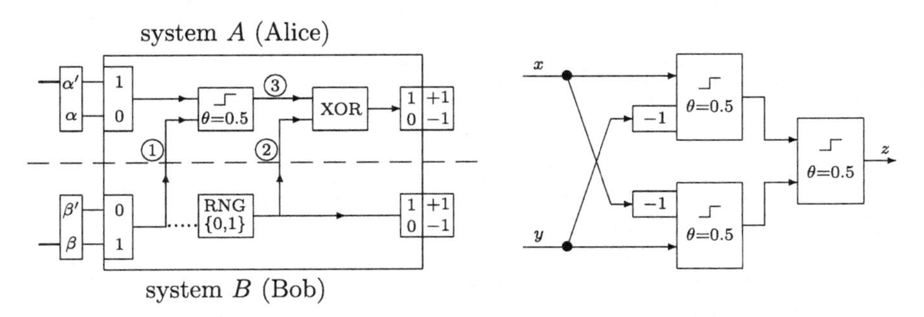

Fig. 1 (left) This network violates the CHSH inequality maximally but cannot be used for signaling.(right) The XOR gate can be realized by a neural network. The small boxes with entries -1 in front of the two threshold boxes to the left denote an inhibitory influence (the synaptic weight is -1) by reversing the sign of the input signal. See main text for other symbols.

an entangled pair of 2-state quantum systems [see (Filk, 2015)], so this part of the device does not constitute a real transfer of information and can be instantaneous, but the internal signal of Bob's setting to Alice's site (connection 1) is subject to the laws of relativity and this internal information transfer is the reason why this simple classical device can violate the CHSH inequality.

Of course, the details of these networks are not relevant for this discussion, what is relevant is the 'proof of principle': Classical systems can easily violate the CHSH inequality (even maximally) such that the correlations cannot be used for communicating signaling. A violation of the CHSH inequality is sometimes used as an argument that a classical mathematical formalism (based on Kolmogorov probabilities and, e.g., Markov processes) cannot explain this situation and that a quantum formalism (using Hilbert spaces as state spaces and the square of the scalar product for calculating probabilities) has to be used. However, while using a quantum formalism might be a possibility as long as the CHSH value satisfies the Tsirelson bound, it will be of no help in cases like the one discussed here where the CHSH inequality is violated beyond the Tsirelson bound.

5 Conclusions

Neural networks are excellent examples to study the 'route from content to context' in the sense that the local behavior of these systems is simple and for very small networks also the global behavior can be determined analytically, but for large systems the complexity increases rapidly such that analytical solutions or predictions become almost impossible. On the other hand, the local dynamics of these systems is simple enough to enable effective computer simulations.

Furthermore, neural networks allow for two perspectives: (1) an internal perspective, which is similar to the perspective of classical physics and for which a state of the system is described by the activities of all the nodes and the constellation of weights, and (2) an external perspective, more similar to cognitive disciplines, where only the input-output relations can be observed. In particular this second perspective shows much similarity to some quantum effects, e.g. non-commutativity of observations, non determinism, memory effects and, in particular, contextuality. A violation of a CHSH inequality has often been interpreted as a sign for quantum or, at least, quantum-like effects. However, this conclusion is indispensable only if certain loopholes can be excluded, in particular the locality loophole: measurements must not have a direct influence onto each other. If some form of mutual influence is possible, a violation of CHSH inequalities is no surprise and already very simple classical devices like neural networks can realize such situations.

The no-signaling theorem (or non-communication theorem) does not imply that there has been no mutual influence between the measurements. It only means that the observed correlations cannot be used for communication or controlled signal transfer. This does not exclude an internal form of influence which, however, cannot be observed directly. Even in quantum theory there is still a debate to which extend the violation of CHSH inequalities can be explained by some 'internal' form of influence which, however, is not part of the quantum formalism. Again, neural networks are a very instructive example for these phenomena.

I should emphasize that I do not argue against the use of the mathematical formalism of quantum theory in the cognitive sciences. Many examples have been reported in the past [see, e.g., (Khrennikov, 2010; Busemeyer & Bruza, 2012; Wang & Busemeyer, 2013)] where the quantum formalism yields significantly better results than the standard classical approaches, even if one takes into account that the quantum formalism has more adjustable parameters. The reason for this remains to be explored, and also for this purpose neural networks maybe of help.

References

Aerts, D., Broekaert, J. & Gabora, L. (2002). Intrinsic contextuality as the crux of consciousness. In K. Yasue, M. Jibu, & T. Della Senta (Eds.), *No matter, never mind, Series Advances in Consciousness Research* (Vol. 33). John Benjamins. ISSN 1381 -589X.

Aerts, D., & Sozzo, S. (2014). Quantum entanglement in concept combinations. *International Journal of Theoretical Physics, 53*, 3587–3603.

Amit, D. J., Gutfreund, H., & Sompolinsky, H. (1985). Storing infinite numbers of patterns in a spin-glass model of neural networks. *Physical Review Letters, 55*, 1530–1533.

Atmanspacher, H., & Filk, T. (2006). Complexity and non-commutativity of learning operations on graphs. *BioSystems, 85*, 84–93.

Atmanspacher, H., Filk, T., Finke, R., & Gruber, G. (2009). Complexity constraints and error tolerance in learning processes on small graphs. *The Open Cybernetics and Systems Journal, 3*, 90–97.

Atmanspacher, H., & Filk, T. (2018). *Non-commutativity and its implications in physics and beyond.* In P. Stenner & M. Weber (Eds.), *Orpheus' glance. Selected papers on process psychology; The Fontaréches meetings 2002–2017*; Les Éditions Chromatika.

Atmanspacher, H., & Filk, T. (2019). Contextualitiy revisited—Signaling may differ from communicating. In A. de Barros & C. Montemayor (Eds.), *Quanta and mind* (pp. 117–127). Springer Nature (first online publication 2019).

Atmanspacher, H., & Römer, H. (2012). Order effects in sequential measurements of non-commuting psychological observables. *Journal of Mathematical Psychology, 56*, 274–280.

Bässgen, F. (2017). Self-organized restructuring of attractor landscapes of neural networks. Master Thesis Freiburg.

Bell, J. (1966). On the problem of hidden variables in quantum theory. *Reviews of Modern Physics, 38*, 447–452.

Bruza, P., & Cole, R.J. (2005). Quantum Logic of semantic space: An exploratory investigation of context effects in practical reasoning. In S. Artemov, H. Barringer, A. S. d'Avila Garcez, L. C. Lamb, & J. Woods (Eds.), *We will show them: Essays in honour of Dov Gabbay*. College Publications.

Busemeyer, J. & Bruza, P. (2012). *Quantum models of cognition and decision*. Cambridge University Press.

Cabello, A. (1994). A simple proof of the Kochen-Specker theorem. *European Journal of Physics, 15*, 179–183.

Clauser, J. F., Horne, M. A., Shimony, A., & Holt, R. A. (1969). Proposed experiment to test local hidden-variable theories. *Physical Review Letters, 23*, 880–884.

Dayan, P., & Abbott, L. F. (2001). *Theoretical neuroscience—Computational and mathematical modeling of neural systems*. Cambridge, Massachusetts, London: MIT Press.

Dzhafarov, E. N., & Kujala, J. V. (2014). Selective influences, marginal selectivity, and Bell/CHSH inequalities. *Topics in Cognitive Science, 6*, 121–128.
Dzhafarov, E. N., Kujala, J. V., & Cervantes, V. H. (2016). Contextuality-by-default: A brief overview of ideas, concepts, and terminology. In H. Atmanspacher, T. Filk, & E. Pothos (Eds.), *Lecture Notes in Computer Science* (Vol. 9535, pp. 12–23).
Dzhafarov, E. N., & Cervantes, V. H. (2018). Snow queen is evil and beautiful: Experimental evidence for probabilistic contextuality in human choices. *Decision, 5*, 193–204.
Filk, T. (2015). A mechanical model of a PR-box. arXive: quant-phys 1507.06789
Filk, T. (2016). It is the theory which decides what we can observe (Einstein). In E. Dzhafarov, S. Jordan, R. Zhang, & V. Cervantes (Eds.), *Proceedings of Purdue Winer Memorial Lectures 2014—Contextuality from Physics to Psychology* (pp. 77–92). World Scientific Publishing Co.
Filk, T. (2019). *'Quantum' and 'Quantum-like': An introduction to quantum theory and its applications in cognitive and social sciences* iASK Publishing, Kőszeg.
Fine, A. (1982). Hidden variables, joint probability, and the Bell inequalities. *Physical Review Letters, 48*, 291–295.
Gerstner, W., & Kistler, W. (2002). *Spiking neuron models*. Cambridge University Press.
Gleason, A. M. (1957). Measures on the closed subspaces of a Hilbert space. *Journal of Mathematics and Mechanics, 6*, 885.
Haykin, S. (2009). *Neural networks and learning machines* (3rd ed.). Pearson Education.
Hebb, D. (1949). *The organization of behavior*. Wiley.
Hodgkin, A. L., & Huxley, A. F. (1952). A quantitative description of membrane current and its application to conduction and excitation in nerve. *Journal of Physiology, 117*, 500–544.
Hopfield, J. J. (1982). Neural networks and physical systems with emergent collective computational abilities. *Proceedings of the National Academy of Sciences of the United States of America, 79*, 2554–2558.
Horodecki, R., Horodecki, P., Horodecki, M., & Horodecki, K. (2009). Quantum entanglement. *Reviews of Modern Physics, 81*, 865–942.
Jones, M. (2019). Relating causal and probabilistic approaches to contextuality. *Philosophical Transactions of the Royal Society A, 377*, 20190133.
Kandel, E. R., Schwartz, J. H., & Jessell, T. M. (2000). *Principles of Neural Science* (4th ed.). McGraw Hill.
Khrennikov, A. (2010). *Ubiquitous quantum structure: From psychology to finances*. Springer.
Kleiner, J. (2012). The mathematical structure of measurements, observables and states on Neural Networks. Master Thesis, Heidelberg.
Kochen, S., & Specker, E. P. (1967). The problem of hidden variables in quantum mechanics. *Journal of Mathematics and Mechanics, 17*, 59–87.
Kohonen, T. (1984). *Self-organization and associative memory*. Berlin: Springer.
McCulloch, W., & Pitts, W. (1943). A logical calculus of the ideas immanent in nervous activity. *Bulletin of Mathematical Biophysics, 5*, 115–133.
Nielsen, M., & Chuang, I. (2000). *Quantum computation and quantum information*. Cambridge University Press.
Peres, A. (1991). Two simple proofs of the Kochen-Specker theorem. *Journal of Physics A: Mathematical and General, 24*, L175–L178.
Popescu, S., & Rohrlich, D. (1994). Nonlocality as an axiom. *Foundations of Physics, 24*, 379.
Rosenblatt, Frank. (1958). The perceptron—a probabilistic model for information storage and organization in the brain. *Psychological Review, 65*, 386–408.
Tsirelson, B. S. (1980). Quantum generalizations of Bell's inequality. *Letters in Mathematical Physics, 4*, 93–100.
Turing, A. (1948). Intelligent Machines. *National Physics Laboratory Report* 6/205.
Walleczek, J., & Grössing, G. (2016). Nonlocal quantum information transfer without superluminal signaling and communication. *Foundations of Physics, 46*, 1208–1228.
Wang, Z., & Busemeyer, J. (2013). Quantum question order model supported by empirical tests of an a priori and precise prediction. *Topics in Cognitive Science, 5*, 689–710.

Wang, Z., Solloway, T., Shiffrin, R. M., & Busemeyer, J. R. (2014). Context effects produced by question orders reveal quantum nature of human judgments. *Proceedings of the National Academy of Sciences of the USA, 111*, 9431–9436.

Wendt, A. (2015). *Quantum mind and social science: Unifying physical and social ontology*. Cambridge University Press.

Yin, J., et al. (2013). Bounding the speed of 'spooky action at a distance'. *Physical Review Letters, 110*.

Concepts, Experts, and Deep Learning

Ilkka Niiniluoto

1 From GOFAI to Deep Learning

Deep learning is a new paradigm which during the last decade has superseded GOFAI, the Good Old-Fashioned Artificial Intelligence, which was based on rules for manipulating symbols and logical reasoning.[1] The pioneering work on AI in the 1950s was inspired by Alan Turing's 1936 model of an ideal machine which is able to accomplish all effectively calculable tasks. Turing machines operate with a digital representation of symbols in their own language and the manipulation of symbols with simple rules. In John von Neumann's architecture, data and programs are stored in the memory unit, which is connected with the control unit and the arithmetic logic unit. Allen Newell and Herbert Simon with their co-workers had high expectations for the success for weak AI, which aims to be more effective than human agents for specific computational tasks, and strong AI, which attempts to imitate intelligent human thinking and reasoning. This cognitive approach gave some promising results in the treatment of pattern recognition and machine learning, where the machine receives an input and is able to perform deductive and inductive inferences and to make decisions.

However, critics like Hubert L. Dreyfus in his *What Computers Can't Do?* (1972), explained the slow progress in the areas of machine translation and problem-solving by the thesis that the human mind does not process information by formal rules. Inspired by the European phenomenologists Martin Heidegger and Maurice Merleau-Ponty, Dreyfus argued that computers differ from human beings also in the fact that they do not have bodies, needs, and forms of life. John Searle's (1984) "Chinese room argument" added that computers, which operate blindly with digital symbols,

[1] For GOFAI, see Haugeland (1985).

I. Niiniluoto (✉)
Department of Philosophy, History, and Art Studies, University of Helsinki, P.O. Box 24, 00014 Helsinki, Finland
e-mail: ilkka.niiniluoto@helsinki.fi

S. Wuppuluri and I. Stewart (eds.), *From Electrons to Elephants and Elections*,
The Frontiers Collection, https://doi.org/10.1007/978-3-030-92192-7_31

do not have a semantics or understanding about linguistic meaning. Computers at best pretend to be intelligent or sensitive with emotions. They also lack common sense or everyday practical reason, as shown by the frame problem (i.e. when a robot acts, what changes and what remains unchanged in the environment). As a solution, Marvin Minsky (1987) suggested that robots should be taught to play with toys and to build towers from blocks.

Deep learning (DL), which emerged in the early 2000s as an alternative to GOFAI, is based on neural networks.[2] The first model of a biological neuron was formulated already in 1943 by Warren McCulloch and Walter Pitts. Self-organizing neural nets, proposed by Teuvo Kohonen in the 1980s, were an important step toward the modeling of the associative and parallel computational processes of the human brain. The starting point of deep learning is thus *connectionist* rather than computational, i.e., this form of machine learning is "sub-symbolic" in the sense that the perceived input creates or strengthens connections between the units of the network. The layers of neural nets with billions of connections are thus constantly transformed as effects of input and parallel calculations, without any concepts that represent external objects. Important methods, used already in old models of machine learning, include supervised learning, reinforced learning, and unsupervised learning. In *supervised learning*, the machine learns millions of training examples which are paired with correct outputs, so that the algorithm is able recognize regularities and to use them for successful predictions about new data. *Reinforced learning*, which resembles the techniques of psychological behaviorists with animals and humans, is based on positive and negative feedback. In *unsupervised learning*, the machine classifies unlabeled data by their properties without reference to previously given categories. These methods have been successfully used in DL-networks for concept learning, face recognition, and natural language translation (e.g. Google translate)—even though such neural nets are agnostic and cannot be claimed to "know" or "understand" anything. The results of DL are also opaque in the sense that we humans do not always envisage the steps and reasons for their conclusions or decisions.

It is easy to teach a machine to follow the constitutive rules of chess, which tell the moves that are allowed and forbidden, but it is much more difficult to implement the strategic rules for winning the game effectively. In 1997 IBM's chess machine Deep Blue succeeded to defeat the world champion Garri Kasparov by its "brute force" in computation, which helped to calculate and compare several moves ahead. But more brilliant result have been reached by letting hundreds of layers of neural nets to learn from examples and self-plays against itself. Google's company DeepMind built in 2017 the machine Alpha Zero, which during one day was raised to the master level by reinforced learning from millions of training examples of successful games of Go. The triumph of deep learning was witnessed in 2016 when Google's neural network Alpha Go with tree search methods succeeded to defeat Leo Sedol, who was praised as the best player of Go in the world.[3]

---

[2] See LeCun et al. (2015), Goodfellow et al. (2016).

[3] See Silver et al. (2016).

2 Back to Concept Empiricism?

The classic of British empiricism John Locke argued in *An Essay Concerning Human Understanding* (1689) that the human mind does not include any innate ideas or principles:

> Let us then suppose the mind to be, as we say, white paper, void of all characters, without any ideas: How comes it to be furnished? ...To this I answer, in one world, from EXPERIENCE.
>
> The senses at first let in particular ideas, and furnish the yet empty cabinet, and the mind by degrees growing familiar with some of them, they are lodged in the memory, and names got to them. Afterwards, the mind proceeding further, abstracts them, and by degrees learns the use of general names.[4]

Thus, by the Lockean concept empiricism, human mind is originally an empty table (*tabula rasa*), to which external objects causally impress ideas by means of perceptual experience. Language arises when these ideas are generalized and named.

Immanuel Kant's *Kritik der reinen Vernunft* (1781/1787) gave what many philosophers regarded as a fatal blow to Locke's empiricism. According to Kant, the human consciousness is an active partner in cognition, so that knowledge can arise only from the united operation of understanding and the sensuous faculty:

> Thoughts without content are void; intuitions without conceptions blind.[5]

In Kant's system, space and time are the two a priori forms of sensuous intuition (perception), and categories like modality and causality are the general conceptions of pure understanding. Many Kantians thought that common names (like "cat" and "brown") are human constructions, to be used in the interpretation of sensuous information. For example, the British philosopher William Whewell (1840) emphasized the interplay between ideas and senses in scientific concept formation:

> Terms must be constructed and appropriated so as to be fitted to enunciate simply and clearly true general propositions.

Charles S. Peirce's semiotics included three kinds of signs (indices, icons, and symbols), where symbols are signs whose meaning is based on conventions within the linguistic community (e.g. "cat" in English and "kissa" in Finnish are conventional symbols in this sense).[6] The Finnish philosopher and psychologist Eino Kaila argued that the key difference of human beings to other animals is their use of symbols. In his posthumous work, he illustrated vividly the entanglement of "perceptual and conceptual elements" in everyday experience.[7] In philosophy of science, this is known as the theory-ladenness of observations. In philosophy of language, Jaakko Hintikka's (1975) "neo-Kantian" logic of perception treats perception as thoroughly conceptual: all perception statements are reducible to the propositional construction "S

[4] See Locke (1965), 61.

[5] See Kant (1930), 46.

[6] See Short (2007).

[7] See Kaila (1979).

perceives that p". Hintikka blames the phenomenology of Edmund Husserl for the illegitimate assumption that in our sensuous experience there exists a non-conceptual ingredient or *hyle* (p. 198), so that he would reject Tim Crane's (1992) idea of "non-conceptual content" in experience. In linguistics, Noam Chomsky defended in the 1950s the thesis that the learning of language presupposes linguistic competence, which consists of the possession of "linguistic universals" common to all languages. Chomsky's student Jerry Fodor (1975) developed the computational model of human mind which assumes the existence of "the language of thought".

A typical post-Kantian model of language acquisition is learning by *ostension*: the teacher has a concept, and she points to an object by saying e.g. that "This is koala". A child may learn to use such a classificatory concept even by one instance. A generalization of this model is based on several exemplars: a concept C is learned by an example c of C and by including in the extension of C all objects which are sufficiently similar to c. In philosophy of language and philosophy of science this method has been used by Ludwig Wittgenstein and Thomas Kuhn. *Pattern recognition* of traditional AI, used for the recognition of faces, signs, words, speech, and pictures, is also based on this method: patterns $G_1, \ldots, G_n$ are stored in memory as prototypes, and an observation H is identified with the pattern G which has the greatest degree of similarity with H. In the GOFAI tradition, Mitchell (1997) uses this method for concept learning by training examples.[8] Within this scheme, it can be understood how caricatures are able represent their targets, even though they intentionally exaggerate and distort some of their characters. Theoretical terms used in science have been compared to caricatures, since it is not necessary to assume a perfect fit between them and theoretical entities.[9] But it is also clear that theoretical terms cannot be learned by ostension: electrons are not directly observable, so they are known only by theoretical description and identified abductively as causes of observable effects.

Hilary Putnam (1975) has distinguished different degrees in the mastery of language by his "division of linguistic labor". Ordinary people are able to distinguish some coniferous trees (such as spruce, pine, juniper) and some leaf trees (such as birch, rowan, maple), but only a subset of speakers, a class of *linguistic experts*, are able to discriminate between elm and linden. Inspired by Putnam, we may ask whether machines and robots may become concept experts. In the computational model of AI, it would be natural to appeal to traditional methods of defining concepts by semantic criteria (necessary and sufficient conditions) or clusters of characters and attributes. Semantic webs provide a tool for expressing conceptual systems and their interrelations. But the procedure of deep learning is quite different. In supervised learning, the neural network is given millions of training examples (images of houses, cars, people, elephants, pets etc.) labelled with their category.[10] This resembles repeated

---

[8] See also Niiniluoto (2005).

[9] For the caricature theory of reference, see Niiniluoto (1999), 125–132; Niiniluoto (2020).

[10] The difference to the old model of machine learning can be seen in the fact that for L. Valiant's 1984 method of PAC-learning probable approximate correctness is enough, as it decreases the number of required positive instances. See Valiant (2013) and Niiniluoto (2005).

ostensive learning, but the classification is created on the sub-conceptual level by modifying the weights of the multiple links between the units of the network. In reinforced learning, errors of output scores with respect to desired patterns of scores may be counted. Thereby the machine learns to recognize new concepts so that irrelevant variations are noted and even minute details are detected. The deep learning algorithm becomes a Putnamian expert, who is able to recognize instances of a class, even though it does not have a definition for a concept. For example, an algorithm can be trained to separate spam and non-spam among emails. But, instead of conceptual or propositional knowledge, the DL-expert has a special kind of linguistic know how.

Yann LeCun, Yoshua Bengio, and Geoffrey Hinton, who received in 2018 the Turing Award for their contribution to deep learning, present supervised learning in their 2015 article in *Nature*. But they predict that unsupervised learning will become more important in the future:

> we discover the structure of the world by observing it, not by being told the name of every object.

Technically this means that the outputs for objects of the same kind cluster close to each other. But philosophically this sounds like a return to a crude concept empiricism: a concept is learned by copying a vast number of its training examples on an "empty table". Does the success of deep learning in fact mean that Locke overcomes Kant after all?

Arno Schubbach (2019), a scholar of the neo-Kantian Ernst Cassirer, has proposed a link between Kant and deep learning, but this time with reference to Kant's *Kritik der Urteilskraft* (1790).[11] In this "third critique", Kant studied aesthetic judgments of taste and beauty. This human faculty lacks transparency and cannot be explained by the use of general rules. According to Shubbach, this kind of judgment is comparable to deep learning neural networks as "judging machines". Indeed, this comparison gains plausibility from the fact that DL is sub-conceptual and Kant's aesthetic judgments operate "without interests and without concepts". Still, in my view, there is another so far unnoticed comparison, even more interesting than the Kantian "power of judgment", viz. the five-stage expertise scale of Hubert L. Dreyfys.[12]

3 Experts and Skill Acquisition

A solution to the slow progress of AI in the 1980s was sought from the Japanese conception of "the fifth generation" computers, which store and apply expertise in the form of computer programs. Edward Feigenbaum and Pamela McCorduck (1983) developed the idea of *expert systems*, where a "knowledge engineer" expresses an expert's non-propositional professional skill in the form of production rules and

[11] See Kant (2000).

[12] Dreyfus was aware of the nature of neural networks in the 1992 edition of his 1972 book, but the dramatic successes of their applications were not yet observed at the time.

directives. Examples could include the skills of a lawyer, a physician, or a composer. Such expert systems communicate with natural language, and were hoped to perform well as judges, medical doctors, and artists. It is clear that this program represented a new higher stage of the GOFAI approach.

Hubert L. Dreyfus, with his psychologist brother Stuart Dreyfus, raised again to oppose the hopes of AI. In their joint book *Mind over Machine* (1986) the brothers argued that Feigenbaum's expert systems fail to understand the situational, intuitive, and non-calculative nature of the problem-solving by real experts. They analyze the *acquisition of skills* in a scale of five stages:

- a *novice* follows simple orders, rules, and procedures which refer to context-free objective features of the situation
- an *advanced beginner* recognizes new aspects of the situation and is able to make comparisons to previous similar situations
- a *competent* person is able to effectively select salient aspects and important rules to follow
- a *proficient* person is involved and experienced in the task, and thereby able to make decisions quickly
- an *expert* finds a solution immediately by unconscious personal intuition without rational calculation.

If this model is valid, the highest levels of human skills cannot be expressed by statements, propositional knowledge, or computer programs. Rules are useful to novices and beginners, but not any more to real experts whose actions are based on holistic understanding of problem situations—and therefore "intuitive" and "arational". If an expert is forced to justify decisions or choices on the basis of rules, this involves a regression down to the lower stages of the scale.

The Dreyfus scale is inspired by Heidegger, but need not be interpreted in a mystical manner, since expertise is here rooted in increasing experience. H. L. Dreyfus himself illustrated advanced skills by chess players, pilots, and car drivers—and even ethics could be a skill in his sense. Patricia Benner (1984) gave an illustration of the scale by nursing practices. Expertise could be partly based on propositional education, and it may serve valuable purposes in everyday life and in the professional fields of science, arts, and technology.

Yet, the Dreyfus scale faces a number of problems. First, it is not evident that the same model of five stages applies to all kinds of skills among animals and humans. Secondly, for some skills there is no unique order of superiority (e.g. composers of different styles and genres). Thirdly, experts need to be recognized and evaluated. Could two genuine experts come to conflicting conclusions in special cases? How could we choose whom to rely in such situations? How could experts be recognized in the fields of medicine, architecture, and politics? If experts are required to be experienced and successful, does this lead a conservative dominance? If an expert behaves arationally on the basis of subjective certainty, how can his or her advice be publicly assessed and criticized? Might the crucial difference of human beings to other animals and computers lie in their linguistic abilities and self-consciousness, which implies the possibility of testing and correcting rules and recommendations of actions?[13]

A surprising turn in the debate concerning skills has followed from deep learning, which also puts in new light Dreyfus' critique of rule-based expertise. His critical assessment of computational AI seems to be correct in the case of games like chess, where it is feasible to compare the success of competing agents or systems: the new neural networks beat the old chess programs. Deep learning DL-networks do not avoid Searle's (1985) criticism that they fail to know anything (even when classifying training examples) or understand meanings (even when translating languages), but nevertheless they have pre-conceptual and non-propositional know how. It is ironic, however, that a deep learning AI-system seems to reach rapidly the highest stage of the Dreyfus scale. For example, Alpha Go learned quickly to be a master-level player from examples of wins and losses in successful games. Its expertise seems to satisfy the strict fifth-level requirements of Dreyfus: experienced, intuitive, non-conceptual, not rule-based. Thus, Dreyfus was wrong in claiming that artificial intelligence could not compete with a well-experienced physician: a neural network can go through in one day as many training examples as a good medical doctor in a life-time.

Still, DL-systems face similar questions as the Dreyfus experts. In the game Go it is easy to measure the outcome: who wins the play? This is the case, even when the moves and solutions of the AI system lack transparency. But in many other examples it is hard to evaluate the competence and reliability of experts. DL-algorithms are already in use in banks and insurance companies to make quick decisions about loans and investments. Such programs may search through enormous numbers of juridical cases and reports of sickness. Deep learning promises to assist also scientific research, when solutions involve complex phenomena and massive data,[14] and it may promote economic growth in several sectors. Still, the competence of a DL-expert crucially depends on the quality and variety of the numerous training examples: if they are one-sided or biased, the recommendations and decisions of algorithms may be misleading, harmful, and discriminating. Even though AI can function as supporting intelligence for the humans, to what extent and on what conditions are we really willing to rely our destiny on DL-economists, DL-physicians, DL-politicians, and DL-army generals? How do we react if the best DL-experts disagree with each other?

In sum, Dreyfus was both right and wrong. His critique of GOFAI-type expert systems was largely valid. But it does not hit deep learning AI-expertise, which can override human capacities in specific tasks and reach a high level of learned "intuitive" expertise. Still, the Dreyfusian ideal is not unproblematic even in the case of DL-networks, if it is applied to work life, administration, and other areas of social decision making.

---

[13] This is largely the task of applied social sciences. See Niiniluoto (2013).

[14] A recent report by Senior et al. (2020) tells that a neural network AlphaFold has learned to make accurate predictions on protein structure.

4 Discussion: Content and Context

Deep machine learning has two peculiar features. First, a layered neural network as a model of human brain follows a bottom-up approach, which starts from the underlying structure of elements with weighted connections and proceeds to the brain as a whole. Secondly, this underlying structure is sub-symbolic, so that it has no conceptual or propositional content. Even the achieved higher-level performance—e.g. playing chess, classifying objects, recognizing faces, translating languages, issuing predictions, and making decisions—exhibits abilities or know how without an associated conceptual framework. As far as DL is a form of reductionism, it does not reduce higher-order *entities* to a lower level, but rather higher-order *performances*. The explanation of the upper level is here untypical of reductionism, as it does not appeal to components and structures with conceptual or propositional *content*.

On the other hand, the traditional computational artificial intelligence operates on the higher-level with content, and its context is specified by the chosen conceptual framework. In the big ontological picture, mathematics is reducible to logic, which studies entailment relations between propositions, whereas propositions are constituted by concepts, which are compact regions in qualitative or quantitative conceptual spaces.[15] The aim of GOFAI was to teach computers to imitate human intelligence in their ability to gain input information, recognize patterns, make logical inferences, and solve problems. In this sense, GOFAI is a top-down approach which starts from the relatively independent higher-level without attempting to reduce this context to lower-level entities.

GOFAI accepts the Kantian dictum that "thoughts without content are void" (there is no thinking without concepts) and its continuation "intuitions without conceptions are blind" (all sensations are mingled with concepts). It follows that DL-networks, in spite of their spectacular performances, do not think, and their experience with training examples does not include anything like "non-conceptual content". As DL-networks do not have a symbolic language nor consciousness, they do not help to solve the perennial philosophical riddle, the mind–body-problem, which in the case of humans should show how the mind emerges from the combination of sufficiently complex material systems with a cultural environment.[16]

It is no wonder that the relation of the conceptual level and sub-conceptual DL-systems is a highly pressing open problem. Certainly it would help the project of Explainable Artificial Intelligence (XAI), which tries to explain the non-transparent performance of DL-algorithms. As LeCun et al. (2015) point out, it is important to study how to combine reasoning with deep learning. Some sort of bridge between neural networks and symbolic languages would be needed for this task.

[15] For conceptual spaces, see Gärdenfors (2000).

[16] For this philosophical task, Karl Popper's anti-reductionist ontology of three worlds is a useful tool for an emergent materialist (cf. Niiniluoto, 1999, 23–25, 2006). For example, electrons and elephants, as well as spoken and written words "electron" and "elephant", belong the physical World 1; thoughts about such physical objects belong to the mental World 2; elections and other institutions, as well as abstract concepts and propositions, belong to the cultural World 3.

Lorenzo Magnani (2019) has made an effort to model abductive reasoning within the framework of neural networks. Carlos Perez (2017) has argued that deep learning is not an example of inductive inference, since training examples are not chosen by random sampling. In the Bayesian theory of concept learning one can express the investigator's background assumptions or knowledge by the prior probability distribution.[17] Perhaps the Lockean flavor of deep learning algorithms can be restricted by noting that in practice the system cannot be "an empty table", since the application of the results of DL in inference presupposes that the learning network already knows something about the world, i.e., possesses some kind of background theory or formal ontology.[18]

References

Benner, P. (1984). *From novice to expert: Excellence and power in clinical nursing practice.* Addison Wesley.

Crane, T. (1992). The nonconceptual content of experience. In T. Crane (Ed.), *The contents of experience: Essays on perception* (pp. 136–157). Cambridge University Press.

Dreyfus, H. (1972). *What computers can't do: A critique of artificial reason.* Harper & Row.

Dreyfus, H., & Dreyfus, S. (1986). *Mind over machine: The power of human intuition and expertise in the era of the computer.* Blackwell.

Feigenbaum, E., & McCorduck, P. (1983). *The fifth generation: Artificial intelligence and Japan's computer challenge to the world.* Addison Wesley.

Fodor, J. (1975). *The language of thought.* Crowell.

Gärdenfors, P. (2000). *Conceptual spaces: The geometry of thought.* The MIT Press.

Goodfellow, I., Bengio, Y., & Courville, A. (2016). *Deep learning.* The MIT Press.

Haugeland, J. (1985). *Artificial intelligence: The very idea.* The MIT Press.

Hintikka, J. (1975). *The intentions of intentionality.* D. Reidel.

Kaila, E. (1979). The perceptual and conceptual components of everyday experience. In *Reality and experience: Four philosophical essays* (pp. 259–312). D. Reidel. (Original publication in 1960.)

Kant, I. (1930). *Critique of pure reason.* G. Bell and Sons. (Original publication in 1781/1787.)

Kant, I. (2000). *Critique of the power of judgment.* Cambridge University Press. (Original publication in 1790.)

Landgrebe, J., & Smith, B. (2021). Making AI meaningful again. *Synthese, 198*, 2061–2081. https://doi.org/10.1007/s11229-019-02192-y

LeCun, Y., Bengio, Y., & Hinton, G. (2015). Deep learning. *Nature, 521*, 436–444.

Locke, J. (1965). *An essay concerning human understanding.* Collier—Macmillan. (Original publication 1689.)

Magnani, L. (2019). AlphaGo, locked strategies, and eco-cognitive openness. *Philosophies, 4*, 8. https://doi.org/10.3390/philosophies4010008

Minsky, M. (1987). *The society of mind.* Heinemann.

Mitchell, T. M. (1997). *Machine learning.* McGraw-Hill.

Niiniluoto, I. (1999). *Critical scientific realism.* Oxford University Press.

[17] See Tenenbaum (1999). Cf. Niiniluoto (2005).

[18] See Landgrebe and Smith (2021), who argue that such ontology is needed in successful translation between natural languages.

Niiniluoto, I. (2005). Inductive logic, verisimilitude, and machine learning. In P. Hajek, L. Valdés-Villanueva, & D. Westersta°hl (Eds.), *Logic, Methodology and Philosophy of Science: Proceedings of the Twelfth International Congress* (pp. 295–314). King's College Publications.

Niiniluoto, I. (2006). World 3: A critical defence. In I. Jarvie, K. Milford, & D. Miller (Eds.), *Karl Popper: A centenary assessment Volume II metaphysics and epistemology* (pp. 59–69). Ashgate.

Niiniluoto, I. (2013). On the philosophy of applied social sciences. In H. Andersen (Ed.), *New challenges to philosophy of science* (pp. 265–274). Springer.

Niiniluoto, I. (2020). Remarks on representation and misrepresentation. *Estudios Filosóficos, 69*, 253–264.

Perez, C. E. (2017). Deep learning is not probabilistic induction. https://medium.com/intuitionmachine

Putnam, H. (1975). The meaning of meaning. *Mind, language and reality (philosophical papers 2)* (pp. 215–271). Cambridge University Press.

Schubbach, A. (2019). Judging machines: Philosophical aspects of deep learning. *Synthese*. https://doi.org/10.1007/s11229-019-02167-z

Searle, J. (1984). *Minds, brains, and science*. Harvard University Press.

Senior, A. W., et al. (2020). Improved protein structure prediction using potential from deep learning. *Nature, 577*, 706–710.

Short, T. L. (2007). *Peirce's theory of signs*. Cambridge University Press.

Silver, D., et al. (2016). Mastering the game of go with deep neural networks and tree search. *Nature, 529*, 484–489.

Tenenbaum, J. B. (1999). *A Bayesian framework for concept learning*. The MIT Press.

Valiant, L. (2013). *Probably approximately correct: Nature's algorithms for learning and prospering in a complex world*. Basic Books.

Whewell, W. (1840). *The philosophy of the inductive sciences*. Parker and Sons.

A Route to Intelligence: Oversimplify and Self-monitor

Daniel C. Dennett

I want to try to do something rather more speculative than the rest of you have done. I have been thinking recently about how one might explain some features of human reflective consciousness that seem to me to be very much in need of an explanation. I'm trying to see if these features could be understood as solutions to design problems, solutions arrived at by evolution, but also, in the individual, as a result of a process of unconscious self-design. I've been trying to think of this in the context of work in AI on the attempt to design intelligent robots—not "bed-ridden" expert systems, but systems that have to act in real time in the real world. If you want to think about something like this, you have to stray fairly far from experiments and hard empirical data; you have to get fairly speculative. Nevertheless the design efforts of people in AI do seem to bring home to conviction—if not to prove—various design constraints looming large and inescapable. If we can come to see why a system—or an organ or a behavior-pattern—must have certain features or a certain structure in order to do its task, this may help us ask the right questions, or at least keep us from dwelling on some of the wrong questions when we try to explain the machinery in the brain that is responsible for intelligent action.

Resuming the discussion of yesterday evening, let me remind you that intelligent action in the real world depends on anticipation, of two kinds: both the built-in, fast, unconscious modular anticipation of the sort we were considering yesterday, and,

[In 1984 a workshop was held at the Maison Française in Oxford, organized by Jean Khalfa, and involving René Thom, Richard Gregory, myself and others. A volume was supposed to emerge from it, but it never appeared, so far as I know, and so this has never been published. (The collection of Darwin Lectures, What is Intelligence? (Cambridge Univ. Press, 1996) edited by Khalfa is an entirely different anthology.) I refer to this essay as 'forthcoming' in "Evolution, Error and Intentionality" in The Intentional Stance, but that was a promise never kept.—DCD, February 15, 2008].

D. C. Dennett (✉)
Center for Cognitive Studies, Tufts University, 115 Miner Hall, Medford, MA 02155, USA
e-mail: daniel.dennett@tufts.edu

S. Wuppuluri and I. Stewart (eds.), *From Electrons to Elephants and Elections*, The Frontiers Collection, https://doi.org/10.1007/978-3-030-92192-7_32

in the case of human beings and maybe some other higher species, something that looks much more like voluntary, conscious, expectation-formation and calculation about the future.

There is an important family of verbs that strangely enough has not yet been singled out for philosophical attention. Central members of the family are "avoid", "prevent", "hinder", "foster", and, perhaps the most basic of all, "change" in its transitive sense, where we think of one thing or agent or event "actively" changing something else. These are the pre-eminent verbs of *action*, where one is characterizing the situation in terms of a rational agent who, as one says, sets out the "change the course of history." This is a curious phrase. We all want to be able to change the course of history, if only in our own little corners of the world. The problem of free will is very much a matter of whether one thinks one can change the course of history, but of course this familiar phrase, on even the most superficial analysis, turns out to be deeply puzzling. If you suppose it is to be taken at its face value it is absurd. How could you change the course of history? From what to what? If history is simply the sequence of events that actually occur, then of course you can't change history. People say you can't change the past, and that's true enough, but then you can't change the future either.

When one is thinking in this mode in which one considers bringing about these changes that one so very much wants to bring about, one has to be thinking of an *anticipated* history, the way history is going to go *ceteris paribus*, the way history is going to go *unless* somebody does something, or *until* somebody does thing, or *in spite of* what somebody does. These verbs of agency can have no foothold outside the framework of a projected, anticipated history, even when they are used to characterize the effects brought about by entirely inanimate objects. Let me illustrate this with an example borrowed from my book *Elbow Room* (Dennett, 1984b).

Imagine that astronomers discover a meteor heading for the earth, and they calculate that it is going to hit North America on Tuesday, and there is nothing anyone can do about it. People would be frantic, of course, wondering if there was anything to be done, and perhaps praying for miraculous deliverance from this terrible catastrophe. And then, suppose, on the eve of destruction, another meteor appears, plunging out of darkest space on a course that is just right to deflect the first onto a near-miss trajectory, thus narrowly *averting* the catastrophe, *preventing* calamity.

These words would come naturally to our lips on such an occasion. But am I suggesting that the second meteor was a miracle—a God-given answer to our prayers? No, I am supposing that the second meteor was always out there, tracing out exactly the intercepting course, just as predictably as the first; it was simply not noticed by the astronomers until the last minute. In fact, had they noticed the second meteor when they noticed the first, they would never have alarmed us, because (as they can see now in retrospect and could have calculated then) *there was never going to be a catastrophe*. It was merely an anticipated catastrophe—a mis-anticipated catastrophe. It seems appropriate to speak of an averted or prevented catastrophe because we compare an anticipated history with the way things turned out and we locate an event which was the "pivotal" event relative to the divergence between that anticipation and the actual course of events, and we call this the "act" of preventing or avoiding.

Mark Twain once said "I'm an old man, and I've seen many troubles, but most of them never happened." This is the experiential history of somebody who is used to living in the world of avoiding and preventing. This is the world in which a rational deliberator lives. Such a deliberator has to have a world view that is constantly looking forward, anticipating the way things are going to go unless it does various things or until it does various things.

Suppose then that one wants to design a robot that will live in the real world and be capable of making decisions so that it can further its interests—whatever interests we artificially endow it with. We want in other words to design a foresightful planner. How must one structure the capacities—the representational and inferential or computational capacities—of such a being? The problem that such a creature faces is, as usual in Artificial Intelligence, the problem of combinatorial explosion. The way one obtains anticipations is by sampling the trajectories of things in one's perceptual world and using the information thus gathered to ground an inference or extrapolation about the future trajectory of the thing. One cannot deal intelligently with anything that one cannot track in this way. When I speak of tracking, I am not just thinking of tracking the trajectories through space of moving things, but also the trajectories through time of things like food stores, seasons, inflation rates, the relative political power of one's adversaries, one's credibility, and so forth. There are indefinitely many things that could be kept track of, but the attempt to track everything, to keep up-to-date information about everything, is guaranteed to lead to a self-defeating paroxysm of information-overload. No matter how much information one has about an issue, there is always more that one could have, and one can often know that there is more that one could have if only one were to take the time to gather it. There is always more deliberation possible, so the trick is to design the creature so that it makes reliable but not foolproof decisions within the deadlines naturally imposed by the events in its world that matter to it.

The fundamental problem, then, is what we might call the problem of Hamlet, who, you recall, frittered away his time in deliberation (or so it appears), vacillating and postponing. This is the sort of postponement that Réné Thom was discussing yesterday in his example of the man at the crosswalk who must make a decision. One has to make decisions in real time, and this means that one has to do a less than perfect job if one is to succeed at all. So one must be designed from the outset to economize, to pass over *most* of the available information.

How then does one partition the task of the robot so that it is apt to make reliable real time decisions? One thing one can do is declare that some things in the world of the creature are to be considered *fixed*; no effort will be expended trying to track them, to gather more information on them. The state of these features is going to be set down in axioms, in effect, but these are built into the system *at no representational cost*. One simply designs the system in such a way that it works well provided the world is as one supposes it always will be, and makes no provision for the system to work well ("properly") under other conditions. The system as a whole operates *as if* the world were always going to be one way, so that whether the world really is that way is not an issue that can come up for determination. The rigid-linkage assumption in human vision described by Ullman (1979), is a good example. It is presumably a design feature endorsed over the eons by natural selection. In the past, the important things that have moved in our visual neighborhoods have tended to be assemblages of linkages the parts of which are rigid (hands, wrists, arms, elbows, and so forth), and one can create a much more efficient visual system for a creature with such a world by simply building in the rigidity assumption. This permits very swift calculations for speedy identification and extrapolation of the futures of relevant parts of the world.

Other things in the world are to be declared as *beneath notice* even though they might in principle be noticeable were there any payoff to be gained thereby. These are things that are not fixed but the changes of which are of no direct relevance to the wellbeing of the creature. These things are smeared into a blur, as it were, in our perceptual world and not further attended to. An example drawn from Wimsatt (1980) is the difference in cognitive strategy between two different predators: the insectivorous bird and the anteater, which both need to keep track of moving insects. The insectivorous bird tracks individual flying insects and samples their trajectories with a fast sampling technique: a very high flicker fusion rate relative to human vision. (If you showed a motion picture to such a bird, it would see it as a slide show, in effect, not continuous motion.) The bird sees the individual insects *as* individuals. The anteater does not track individual ants. The anteater sees swarms of ants as batches of edible substance. (If I believed it was always appropriate to speak this way, I would say that "ant" was a mass term in the anteater's language of thought!) It laps up regions of ant, and does not waste any of its cognitive resources tracking individual ants any more than we track individual molecules when we detect a "permeating" uniform odor in a volume of air which may contain a few parts per billion of the telltale molecule.

The "grain" of our own perception could be different; the resolution of detail is a function of our own calculus of wellbeing, given our needs and other capacities. In our design, as in the design of other creatures, there is a trade-off in the expenditure of cognitive effort and the development of effectors of various sorts. Thus the insectivorous bird has a trade-off between flicker fusion rate and the size of its bill. If it has a wider bill it can harvest from a larger volume in a single pass, and hence has a greater tolerance for error in calculating the location of its individual prey.

If then some of the things in the world are considered fixed, and others are considered beneath notice, and hence are just averaged over, this leaves the things that are changing and worth caring about. These things fall roughly into two divisions: the trackable and the chaotic. The chaotic things are those things that we cannot routinely track, and for our deliberative purposes we must treat them as random, not in the quantum mechanical sense, and not even in the mathematical sense (e.g., as informationally incompresssible), but just in the sense of pseudo-random. These are features of the world which, given the expenditure of cognitive effort the creature is prepared to make, are untrackable; their future state is unpredictable.

This means that any real, finite deliberator must partition the states of its world in such a way as to introduce the concept of possibility: it is possible that item n is going to be in state A, and it is possible that item n is going to be in state B, or in state C. We get an ensemble of equipossible (but not necessarily equiprobable) alternatives. This idea of partitioning the world into "possible" alternatives that remain "open" is very clearly the introduction of a concept of *epistemic* possibility. It is what is possible relative to a particular agent's knowledge. As the agent gets more knowledge, this may contract the set of possibilities. "I used to think that state B was possible, but given what I just learned, I realize it is not possible" (see Dennett,).

Sellars (1963, 1966) draws the very useful distinction between what he calls the manifest image and the scientific image. The manifest image is the everyday world view, the world of macroscopic, solid, colored objects, and other persons or rational agents. It is the world of folk physics and folk psychology. Then there is the scientific image: the world of atomic and sub-atomic particles too small to be perceived by the naked eye, the world of forces and light waves. Sellars draws his distinction in such a way as to focus on the manifest image shared by (normal) human beings, but I think we can usefully extend his distinction to other species. We are the only species that has developed science, and so we have a scientific image of the world, of the world that we and other species live in, in spite of the vast differences in our manifest images of that world. The manifest image enjoyed by a species is determined, I suggest, by the set of design "decisions" that apportion things in its environment into the categories of fixed, or beneath notice, or trackable, or chaotic. (It is important to note that this way of thinking of the manifest image of a species somewhat belies the connotations of the adjective "manifest"—since it presupposes nothing about consciousness. It is not at all ruled out that an entirely unconscious creature—our imaginary robot, for instance—would have a manifest image.)

Why are we the only species to have developed a scientific image in addition to—and somewhat discordant with—our manifest image? That is a topic that has often been written on, so I will pause to make just one point. The principles of design that create a manifest image in the first place also create the loose ends that can lead to its unraveling. Some of the engineering shortcuts that are dictated if we are to avoid combinatorial explosion take the form of ignoring—treating as if non-existent—small changes in the world. They are analogous to "round off error" in computer number-crunching. And like round-off error, their locally harmless oversimplifications can accumulate under certain conditions to create large errors.

Then if the system can notice the large error, and diagnose it (at least roughly), it can begin to construct the scientific image. For example, we have been designed to detect "directly" only those changes that occur within a certain speed range. Outside our window of direct visibility lie those changes that happen too fast or too slow for us to perceive without the aid of time-lapse or slow-motion photography, for instance. We cannot see a plant or a child grow from moment to moment. We can see the sun's motion relative to the earth only at sunrise or sunset, or with the aid of a simple prosthetic extension of our senses—a couple of sticks stuck in the ground will do. But over a few minutes in the latter case, or months or years in the case of plants or children, we detect the difference: our expectations of no change (zero plus zero plus zero ... equals zero) are overturned. Now the minimal, non-brilliant response to this is simply to make mid-course corrections in our extrapolations of trajectory and continue as before. The insightful response is to notice that we have to do this (often) and to posit *changes too small to be seen*, the entering wedge into the scientific world of postulated, invisible phenomena. Thus it is from a variety of self-monitoring—in particular the noticing of a pattern in one's own cognitive responses—that the bounteous shift of vision arises.

Let me return to the manifest image of our foresighted planner, with its "open future" of types of epistemically possible events that matter to it but cannot normally be tracked by it.

These are the alternatives it may deliberate about, and must deliberate about if it is to fend for itself in the world. One of the pre-eminent varieties of epistemically possible events is the category of the agent's own actions. These are systematically unpredictable by it. It can attempt to track and thereby render predictions about the decisions and actions of other agents, but (for fairly obvious and well-known logical reasons, familiar in the Halting Problem in computer science, for instance) it cannot make fine-grained predictions of its own actions, since it is threatened by infinite regress of self-monitoring and analysis. Notice that this does not mean that our creature cannot make some boundary-condition predictions of its own decisions and actions. Thus I can make reliable predictions about decisions I will make in the near future: tomorrow at breakfast I will decide how many cups of tea I will drink, and right now I predict that I will decide to have more than zero and less than four.

Now if our creature is to be able to choose among the alternatives of which it can conceive, what strategies of deliberation should we endow it with? One feature we want to build in is one mentioned by Réné Thom yesterday: we must guard against the possibility that an evaluation process will end in a tie—the classic problem of Buridan's ass. The cheap way of providing this safety measure is to build in something functionally analogous to a coin-flip: an arbitrary, pseudo-random "oracle" available for a decision-aiding nudge whenever the system needs it. I am fascinated by Julian Jaynes' speculation (Jaynes, 1976), that the various traditions of superstitious decision-making and prognostication found in the ancient world—throwing bones and lots, looking at the entrails of animals, consulting oracles, reading tea leaves—are actually stratagems more or less unconsciously invented by early human beings in order to get themselves out of the position of Buridan's ass, or out of the somewhat related predicament (Hamlet's, we might say) of one who simply does not know how to deliberate effectively about a complicated situation, and needs nevertheless to act somehow in a timely manner. When the issues are too imponderable, when one can think of no considerations that settle the issue, when one is simply at a loss as to how to continue deliberations, here, as in the case of the pedestrian at the crosswalk, it can be valuable simply to get yourself moving in one direction or another. It doesn't in the long run and on average matter which direction you move as long as you get out of your state of decisional funk and get a move on. These rituals, Jaynes suggests, had the effect of making up people's minds for them when they weren't very good at making up their own minds. So these were deliberative crutches, or prostheses. I mention them here because they provide a vivid example of something that was not designed and transmitted genetically by natural selection, but rather a cultural artifact, unconsciously designed by individuals.

To some ears the phrase "unconsciously designed" is an oxymoron, but what I mean is quite straightforward: by haphazard some individuals came to engage in these strange behaviors without having any point in mind, but they found they had agreeable results, and so under certain circumstances they became popular behaviors. And so the rituals were subjected to further design refinement and then preserved by cultural transmission. A behavioral strategy thus transmitted probably has no specific, organic (neural) control system (in computerese, no "dedicated hardware"), but rather is just software, part of the "virtual machine" of the human decision-maker shaped by cultural and other environmental factors, and differently implemented in individual control structures.

The most fundamental problem that faces the designer of such a deliberator is what Artificial Intelligence calls the Frame Problem. Since I have described that unsolved problem at length elsewhere (Dennett, 1984a), I will just remark here that we may view it as the problem of the effective management of the manifest image of a planner, so that the sorts of informational or representational short-cuts taken yield anticipations that are both timely and reliable. It is called the Frame Problem because of the so-called frame axioms that apparently must be used to stipulate, systematically, the sorts of constancies of effect that are assumed in any particular manifest image. What are the (gross, reliable, normal) effects of moving one thing onto another, for instance? Can we codify this understanding into defining axioms for the action type *move x onto y*?

This should be a rather basic action in the repertoire of any interestingly capable agent, and will be immediately recognized by anyone who is familiar with the famous "blocks world" of AI—an imaginary table-top world consisting of a few colored, differently shaped blocks that can be moved around and stacked by an equally imaginary robot arm (see SHRDLU, for instance, in Winograd, 1972). This is a world of breathtaking simplicity compared to the real world of any even very simple creature. But even in this diminished world the frame problem looms large. Consider some of the frame axioms that are needed:

(1) If $z \neq x$, then if I move x onto y, then if z was on w before, z is on w afterwards.
(2) If z is blue, then if I move x onto y, z is blue afterwards.
(3) If z is red, then if I move x onto y, z is red afterwards.

Do we really need separate, independent axioms for everything that doesn't change? If we do, the definition of each action type is going to have to contain clauses for every predicate available for use in state descriptions in a mindless profusion of axioms—apparently an engineering monstrosity. Can we not have some more general, basic axioms, to the effect, for instance, that the colors of things don't change?

(4) (For all x) (If x is red, x stays red).

This won't do, since one of the action types we may want to include in the repertoire is *paint x red*, which rules out (4) and its kin on pain of contradiction. The unsolved problem is how to provide a system of world-knowledge representation that is both simple and efficient enough to avoid combinatorial explosion, while supple and sensitive enough to recover from at least some of the stupid effects of its deliberate oversimplification.

No one has a good solution to the Frame Problem yet, least of all me, but I would claim that one element in any good solution is going to be layers of self-noticing. I will close by describing briefly two examples of the sort of thing I have in mind. I once had a dog that loved to fetch tennis balls thrown to it, but faced with two balls on the lawn and unable to hold them both in his mouth at once, he would switch rapidly back and forth, letting go of one to grab the other, then seeing the dropped ball, and immediately emptying his mouth again to fetch it, and so forth. He would do this maybe twenty or thirty times, apparently acting on some oversimple rule to the effect that *getting* is better than *keeping*. This was a bad rule more or less built into him—he never unlearned it—but he didn't die of following it. That is, he wasn't so transfixed by the rule that he followed it until he dropped dead of starvation. Something would click over in him after those several dozen iterations and he would stop. He didn't have to know why he stopped. He had a minimal safety valve—somehow sensitive to "excess" repetition of his own response—that stopped him, and let him set out on some more promising course of action.

A similar case was recently described by Geoffrey Hinton in a talk at MIT on the Boltzmann machine architecture he and Terry Sejnowski have developed (Hinton & Sejnowski, 1983a, 1983b). Boltzmann machines are powerful problem solvers in certain traditionally difficult problem domains, but they have their characteristic weaknesses. Consider a typical problem graphically as the task of finding the lowest spot—the global minimum—in a large terrain dimpled with many depressions—local minima. (This is, of course, just "hill-climbing" turned upside down!) Boltzmann machines are efficient finders of global minima under many conditions, but they can be trapped in unusual terrains.

Consider a terrain crossed by a steep-sided gully, which slopes gently at the bottom towards the global minimum. When a Boltzmann machine "enters" such a gully in the course of its explorations, it asks itself, in effect, "which direction should I go to go down?" and looks around locally for the steepest downgrade. Only at the very bottom of the gully is the gentle slope towards the solution "visible"; at all other points the fall line (to use skier's jargon) will be at roughly right angles to that direction. With slight overshooting, the Boltzmann machine will end up somewhere on the opposite slope of the gully, ask its question again, and shoot back onto the opposite slope. Back and forth it will oscillate in the gully, oblivious to the futility of its search. Trapped in such an environment, a Boltzmann machine loses its normal speed and efficacy, and becomes a liability to any organism that relies on it.

As Hinton noted on the occasion, what one wants in such a situation is for the system to be capable of "noticing" that it had entered into such a repetitive cycle, and resetting itself on a different course. The design solution that thus might be favored is not to discard the Boltzmann machine idea because it has this weakness, but to compensate for the weakness with some ad hoc strategy of oversight and management. Just this policy, I think, will be found to be endemic in the design of intelligent control systems.

References

Dennett, D. C. (1984a). Cognitive wheels: The frame problem of AI. In C. Hookway (Ed.), *Minds, machines and evolution* (pp. 129–151) Cambridge University Press.

Dennett, D. C. (1984b). *Elbow room: The varieties of free will worth wanting*. Bradford Books/MIT Press, and Oxford University Press.

Hinton, G., & Sejnowski, J. (1983a), Analyzing cooperative computation. In *Proceedings of the Fifth Annual Conference of the Cognitive Science Society*, Rochester NY, May 1983.

Hinton, G., & Sejnowski, J. (1983b). Optimal perceptural inference. In *Proceedings of the IEEE Conference on Computer Vision and Pattern Recognition*, Washington DC, June 1983.

Jaynes, J. (1976). *The origins of consciousness in the breakdown of the bicameral mind*. Houghton Mifflin.

Sellars, W. (1963). *Science*. Routledge & Kegan Paul.

Sellars, W. (1966). Fatalism and determinism. In K. Lehrer (Ed.), *Freedom and determinism*. Random House.

Ullman, S. (1979). *The interpretation of visual motion*. MIT Press.

Wimsatt, W. (1980). Randomness and perceived randomness in evolutionary biology. *Synthese, 43*, 287–329.

Winograd, T. (1972). *Understanding natural language*. Academic Press.

Context is King: Contextual Emergence in Network Neuroscience, Cognitive Science, and Psychology

Michael Silberstein

1 Introduction

While there are by now many, many different definitions of emergence in the scientific and philosophical literature, they tend to reduce to either "weak emergence" or "strong emergence." If ontological reductionism is true, then epistemological or weak versions of emergence are the only kinds possible. For instance, David Chalmers defines weak emergence as follows:

> To capture this, one might suggest that weak emergence is the phenomenon wherein [non-obvious] complex, interesting high-level function is produced as a result of combining simple low-level mechanisms in simple ways...This conclusion captures the feeling that weak emergence is a 'something for nothing' phenomenon... The game of Life and connectionist networks are clear cases: interesting high-level behavior emerges as the consequence of simple dynamic rules for low-level cell dynamics (2006, p. 252).

Chalmers talks about weak emergence in terms of "*ease of understanding* one level in terms of another. Emergent properties are usually properties that are more easily understood in their own right than in terms of properties at a lower-level" (2006, p. 251).[1] However, is weak emergence in principle sufficient to explain everything?

There are those who think that while everything emerges from and depends on fundamental physical processes, they doubt that weak emergence is sufficient to explain everything. These people thus tend to champion what is sometimes called

---

[1] Bear in mind that "weak emergence" is defined differently by different people (e.g., Bedau, 2008, Huneman, 2008 and Wilson, 2015). Some would argue that weak emergence need not be strictly epistemic, though it is often defined in terms of *in practice* failures of derivability, prediction or computability.

M. Silberstein (✉)
Department of Philosophy, Elizabethtown College, Elizabethtown, PA 17022, USA
e-mail: silbermd@etown.edu

Department of Philosophy, University of Maryland, College Park, MD 20742, USA

S. Wuppuluri and I. Stewart (eds.), *From Electrons to Elephants and Elections*,
The Frontiers Collection, https://doi.org/10.1007/978-3-030-92192-7_33

strong or radical emergence, which historically, has many different definitions (see for example Morgan, 1912, Alexander, 1920, Broad, 1925, O'Connor, 1994, Humphreys, 1997, Silberstein, 1999 and Chalmers, 2006). However, the basic idea is that if X is strongly emergent with respect to Y then Y does not determine X. Or, if Y does determine or cause X, it is a brute fact that has no further explanation. Thus, a strongly emergent property is one that, *in principle*, cannot be derived from, predicted from or fully explained by some more fundamental (physical) theory because the emergence of such properties are brute facts. Another common but distinct way to define a strongly emergent property is in terms of its autonomous causal efficacy, what some unfortunately call "downward causation" with its suggestion of the brute emergence of some new type of phenomena or, reified "higher-level" of organization, that then magically exerts a new force on that from which it emerged. This is the radical ontological emergence that threatens a disunified picture of the world.

Historically, mental phenomena have been a driver of the emergence debate and that has not changed. Witness the hard problem of consciousness and the explanatory gap. Chalmers, for instance, posits universal, brute psycho-physical bridge-laws, which determine the distribution of conscious properties given physical, functional or informational properties in this world (1996). Tim O'Connor and Hong Yu Wong are modern day defenders of strong emergence who hold that given the instantiation of particular complex microstructural properties, basic mental properties will be *caused diachronically* to come into being (2005). These emergent mental properties will have causal powers that are distinct from any microphysical or microstructural properties. Furthermore, these emergent mental properties will, in turn, cause the instantiation of both other microphysical properties, and other emergent mental properties. According to this view, physical states play a causal role in the rise of conscious states *and* these conscious states affect physical states (so-called downward causation). On their view, mental properties would be diachronically or dynamically emergent.

It must be said however, that these days, with some exceptions (see for example John Symons, 2018, Elly Vintiadis, 2018, Peter Wyss, 2018 and Robin Findlay Hendry, 2019), outside of intentionality and conscious experience, one does not see much advocacy for strong emergence anymore. As suggested by Jeremy Butterfield, another notable modern-day exception is Nobel prize-winning condensed matter theorist Anthony Leggett. According to Butterfield, "Leggett's work was motivated by his expecting that the phenomena could not be explained by orthodox quantum mechanics—he believed it would need to be modified by admitting new forces, e.g., a quantum analogue of Broad's configurational forces" (2011b, p. 936). Leggett, with his Macrorealism interpretation of quantum mechanics, has subsequently attempted and failed to show that unadorned non-relativistic quantum mechanics is incomplete with respect to the emergence of classical phenomena (Allen, Maroney and Gogioso, 2017). If it turned out that intentionality and conscious experience were the only possible candidates for ontological emergence, that would be troubling and puzzling. Which is to say that the hard problem of consciousness and the explanatory gap should not drive the larger scientific debate about the structure of reality.

Thus, all these defenders of strong emergence are the exception that proves the rule. When it comes to strong emergence and weak emergence, the consensus is that with the possible exception of conscious experience, the former does not exist at all and the latter while ubiquitous (indeed the norm) in science, has no obvious Earth-shattering ontological implications, e.g., it does not tell against ontological reductionism. Here, we have the contours of the false forced choice between ontological reductionism or radical forms of ontological emergence. But this false forced choice is hardly ontologically innocent. It presupposes the hierarchical and foundationalist picture of reality assumed by ontological reductionism. It must be noted that proponents of both weak and strong emergence differ only in that the latter think new (physical or metaphysical) laws, causal powers or entities must be added to the set of fundamental or brute facts to explain the existence of novel emergent phenomena.

Herein I defend an alternative to both weak and strong emergence that we call, contextual emergence. Contrary to both weak and strong emergence, contextual emergence suggests that relations between smaller and larger scales need not be anti-symmetric, transitive, or anti-reflexive. In Sect. 2 contextual emergence will be defined and in Sect. 3 we look at evidence from network neuroscience, social neuroscience, embodied, embedded, enactive and extended cognitive science (4E cognitive science), and clinical psychology that singly and collectively support contextual emergence. In the concluding Sect. 4 we will discuss what are in my opinion, retrograde attempts by neo-mechanists and reductionists of various stripes to resist contextual emergence.

2 Contextual Emergence

Metaphysicians often complain that science and philosophy of science, while great at generating detailed case studies such as the emergence and stability of superconductivity, never get around to drawing deeper or more universal metaphysical conclusions. As Robert Batterman puts it regarding emergence, "Instead of focusing on questions of correct ontology, I suggest a more fruitful approach is to focus on questions of proper modeling technique" (2015, p. 133). But the concern for someone primarily interested in studying the ultimate nature of reality is that there is no point in focusing on science if it does not aid this project. For instance, it would no doubt be disappointing if the only conclusions that one could draw about emergence based on science, is that on multiple scales, novel processes and patterns come into being that are not predictable or derivable from, nor intuitively understandable in terms of, underlying smaller scale physical processes. This lesson is trivially made with any number of concepts ranging from geometric shapes such as cubes made of different materials to natural selection (on any planet, no matter how alien the physics or the biological organisms involved).

Speaking broadly, when it comes to relatively recent interest in emergence, we note that the widespread assumption in naturalistic metaphysics and in science, is that matter (i.e., the most basic microphysical entities whatever they are) alone is fundamental and brute, and thus life, mind and culture must somehow *emerge* from fundamental physical processes. But what does "emerge" mean here? Laughlin claims, "I prefer the more physical view that politics, and human society generally, grow out of nature and are really sophisticated high-level versions of primitive physical phenomena" (2005, p. 210). Take a popular analogy people often give for ontological reductionism: That the universe is like a finite automaton, such as John Conway's Game of Life. In this analogy, the cells in the game are like the fundamental physical entities, the rules of the cell's temporal evolution are like the fundamental laws of physics, and robust patterns in the game such as gliders and eaters emerge "for free." This is exactly what Laughlin seems to be claiming about psychological, cultural and sociological phenomena.

This is really an astounding and counterintuitive claim if one dwells on it for any length of time. Indeed, it is at this point that the idea of ontological reductionism and strong/radical emergence being exhaustive options becomes most troubling. Fortunately, contextual emergence provides a third option. What is needed to get beyond this impasse in the debates is a form of both explanatory and ontological emergence grounded in scientific explanation that: (1) does not violate the inherent unity of the world; (2) does not assume that if no reductive explanation exists then there is no scientific explanation to be had other than positing a brute new law or causal power; and (3) does not presuppose foundationalism. Contextual emergence is both an explanatory and ontological account of emergence that makes good on all three desiderata, demonstrating that there is a viable, empirically well-grounded alternative between ontological reductionism coupled with explanatory anti-reductionism (weak emergence) and strong/radical emergence.

Ontological reductionism is based on foundationalism. Foundationalism is first the idea that only basic physics contains brute or fundamental laws and entities, i.e., only basic physical facts are fully ontologically autonomous in that they depend on nothing else for their existence and are determined by nothing else. Second, the world exhibits a well-ordered objective hierarchy ranging from basic physics on up the spatiotemporal scales, wherein the facts from basic physics unidirectionally determine all the other facts in the world. Foundationalism thus implies that, at least in principle, the complete explanation for phenomena at a particular scale or "level", must come from a theory about smaller scales or "lower-level" phenomena.

If one insists on a hierarchical and foundationalist conception of reality as exemplified by the Game of Life analogy, then the only way to get any kind of robust ontological emergence out of the system—any kind of emergence other than weak emergence, is to add something metaphysical, something *extra* to the game such as the brute bridge-laws of the sort we find in the strong emergence of C. D. Broad (1925). Even so, in this case we are simply adding laws or what have you to the explanatory base. But imagine instead of foundationalism, that reality is more like multiscale complex networks or structured graphs of extrinsic dispositions, now robust ontological emergence is perfectly natural. Order comes not from anything extra-added, not from any causal or nomological glue, not from any metaphysical grounding whatsoever. What science shows us in case after case, is that the arrow of explanation and determination is not strictly bottom-up, not unidirectionally from smaller length and time scales to larger scales. It is for these reasons that contextual emergence is common, universal, non-spooky and does not defy scientific explanation. Nor does contextual emergence imply any kind of discontinuity or disunity in nature. Contextual emergence emphasizes the ontological and explanatory fundamentality of multiscale *contextual constraints,* often operating globally over interconnected, interdependent, and interacting entities and their relations at multiple scales. (For the origins of contextual emergence see Bishop, 2005 and Bishop & Atmanspacher, 2006. For an early attempt to model contextual emergence see Silberstein, 2006 and for a more recent revision see Silberstein, 2018).

Contextual emergence can be summarized as follows:

1. Contextual emergence is a type of scientific explanation that emphasizes the equal fundamentality of what are often multiscale *contextual constraints* and interdependent relations at multiple interacting scales. Such constraints are characterized by stability conditions.
2. Such constraints can include global or systemic constraints such as topological constraints, dimensional constraints, network or graphical constraints, order-parameters, etc. Contextual constraints therefore need not involve anything like local or direct causal-mechanical or dynamical interactions, though they often do.
3. Such constraints can be causal-mechanical and dynamical, but they can also involve non-causal or adynamical difference makers, such as conservation laws, free energy principles, least action principles, symmetry breaking, etc.
4. Such constraints can also include global organizing principles such as plasticity, robustness, and autonomy in complex biological systems. Contextual constraints can even be behavioral, social, normative, etc.
5. Contextual constraints can be symmetric, such that X and Y can simultaneously act as contextual constraints for one another.
6. Contextual constraints represent both the screening off and opening up of new areas of modal space, i.e., degrees of freedom, and thereby new patterns emergence and become robust.
7. Contextual emergence provides a framework to understand two things: (A) how novel properties are produced, and (B) why those novel properties matter.

To unpack the last point, contextual emergence differs from deductive nomological (D-N) explanations that have received much of the focus in discussions of scientific explanation and reduction/emergence debates.We agree with James Woodward (2003) that there is no need to determine whether something is a genuine, D-N-style law or merely a robust invariance to determine whether it is a viable explanation—that is, not all good scientific explanation refer to laws, even indirectly. Woodward claims that a good explanation allows one to answer a wide range of *what-if questions*: the more what-if questions answered, the more complete the explanation. His goal is to sort causal interactions, so construed, from mere correlations. Woodward stresses that robust causal explanations exhibit invariance in that the specified relation between the cause variable and the effect variable holds under a wide range of conditions. There are diverse types of invariance, including insensitivity to micro-details, background conditions, interventions, and so forth. The kinds of explanations in contextual emergence exhibit a high degree and a wide variety of invariances, though the invariances might not primarily be explained by the mechanistic details, but by stability conditions, such as dynamical and topological features and the network of other relations involved in the context of the target system. The contingent conditions characterizing contexts that guarantee the existence and stability of relevant systems and their states and observables over time. The key feature of stability conditions is that they are whatever 'environmental' or contextual features, however concrete or abstract, that we are treating as being outside the system, that together make up the full set of conditions for the emergent in question to come into being. That is, stability conditions enable emergence and robustness (Bishop & Silberstein, 2019).

Finally, let's compare contextual emergence with typical intuitions about emergence. In surveying the literature on emergence, four typical intuitions or "marks" of emergence are often discussed (e.g., Jaegwon Kim, 1999, 19–22)[2]:

> **Arise**: Emergents at a higher-level arise out of properties and relations characterizing the entities and properties at a lower-level.
>
> **Unpredictable**: Emergents are unpredictable, even given exhaustive information concerning the lower-level.
>
> **Inexplicable**: Emergents are inexplicable in terms of lower-level properties.
>
> **Novel**: Emergents have novel features not found at the lower-levels.

Regarding **Arise**, first, contextual emergence calls into question the fundamentalist and hierarchical assumptions built into this intuition. That is, contextual emergence calls into question the existence of some fully autonomous or fully independent microscopic causally/dynamically closed basic physical process sufficient to determine all other phenomena at larger scales. With contextual emergence, emergents only arise from a "lower-level" or smaller scale provided relevant stability conditions, often found at larger scales, are present. While some necessary conditions for the emergents may exist at the "lower-level" or smaller scale; nonetheless, for contextual emergence, the sufficient conditions are represented by all the relevant stability conditions at various scales. Second, emergents can also come into being at smaller scales or levels of organization as the result of how these constraints at larger scales or levels of organization are implemented (e.g., the behavior of quantum systems being in part determined by the classical experimental set-up). This is modal accessibility in physical possibility space at work.

Emergents are often **Unpredictable** given exhaustive information at the "lower level" or smaller scale alone. However, given contextual emergence, the emergent is often predictable given the "lower level" information plus the relevant contextual features at other "levels" or larger scales (e.g., the relevant physical states and stability

[2] Notice the dependence on an ordered hierarchy of levels.

conditions). Similarly, for the **Inexplicable** intuition. As we will see, the contextually emergent explanation is not going to be fully reductive in either the intertheoretic sense of derivation as discussed by Hempel and company (the D-N model), nor in any synchronic notion of reduction involving the properties of parts determining the properties of wholes (what analytic metaphysics calls realization), and not in any causal-mechanical sense of reduction as with localization and decomposition in biological systems. Thus, in the case of contextual emergence, being unpredictable or inexplicable need not be hallmarks of emergence.

The term **Novel** is certainly loaded. We have already said that an emergent can be predicted and explained, so given contextual emergence, novel means unexpected and irreducibly different in kind from features and concepts connected to the "lower-level" or smaller scales. It can however mean more than this. For example, Kim, among others, has argued that if new "causal powers" emerge at a "higher-level" not reducible to or realized by "lower-level" "causal powers," then we face a mystery as to where such "powers" come from (e.g., Kim, 1998, 1999). Given contextual emergence, so called "causal powers" are just extrinsic dispositions that typically require interdependent multiscale conditions. Kim's worries about microphysical causal closure and exclusion have no purchase given contextual emergence. Finally, contextual emergence explains why novel emergents arise.

Contextual emergence is multiscale in that "higher-level" or target domain information is required to enrich and constrain the laws and properties of the "lower-level" or underlying domain to produce the set of contingent necessary and sufficient conditions for explanation of the emergent. Thus, contextual emergence focuses on making explicit the essential features absent in the fundamental level or underlying domain. Scientific explanations don't float free in their own "level" or domain alone. Instead, scientific explanations implicitly rely on contextual features not contained in or implied by the lower level or smaller scale. Nevertheless, as we will see going forward, the absence of explanatory reduction does not imply explanatory or ontological disunity—pluralism yes, disunity no. Contextual emergence enables explanatory unification across multiple "levels," scales and domains. While contextual emergence does not suggest the hierarchical structure implied by foundationalism, it also does not suggest a world of reified and explanatorily closed levels of organization. Nor does it does suggest the "gappy" world of C. D. Broad and his "transordinal" laws (1925).

We are certainly not alone in suggesting something like contextual emergence. As Smolin argues that, "the world around us is nothing but a network of evolving relationships" (2000, p. 96). In the words of Laudisa and Rovelli:

> For RQM (relational quantum mechanics), the lesson of quantum theory is that the description of the way distinct physical systems affect each other when they interact (and not the way physical systems 'are') exhausts all that can be said about the physical world. The physical world must be described as a net of interacting components, where there is no meaning to 'the state of an isolated system', or the value of the variables of an isolated system. The state of physical system is the net of the relations it entertains with the surrounding systems. The physical structure of the world is identified as this net of relationships. (Laudisa, Federico and Carlo Rovelli, "Relational Quantum Mechanics", *The Stanford Encyclopedia of Philosophy*

(Spring 2021 Edition), Edward N. Zalta (ed.), URL = <https://plato.stanford.edu/archives/spr2021/entries/qm-relational/>.).

Contextual emergence implies a contingent multiscale web of inextricably interconnected and interdependent extrinsic dispositions most of which are in constant flux. Some laws, constraints, principles, and so forth, are more general and subsume more phenomena than others, but such constraints and laws, while not violated by emergents, need not *determine* all the other phenomena at every scale. It is also important to note that the most general constraints need not always come from or be recoverable from, the domain of fundamental physics. For instance, in physics we use the principle that there are no preferred reference frames. The term "reference frame" has many meanings in physics related to microscopic and macroscopic phenomena, Galilean versus Lorentz transformations, relatively moving observers, and more. Think again of the light postulate and the relativity principle in special relativity as specific examples of the more general no preferred reference frame stability condition that constrains the possibilities for motion. While we use this general principle in physics more than other sciences do, there is nothing exclusively physical or microphysical about this principle. Think again of natural selection. This principle is certainly not recoverable from physics and is every bit as basic as any law of physics.

3 Contextual Emergence in Network Neuroscience and Cognitive Science

The following quote from Pigliucci explains why I chose to mostly focus on complex bio-cognitive systems for this chapter:

> Ever since Darwin a great deal of the conceptual history of biology may be read as a struggle between two philosophical positions: reductionism and holism. On the one hand, we have the reductionist claim that evolution has to be understood in terms of changes at the fundamental causal level of the gene. As Richard Dawkins famously put it, organisms are just 'lumbering robots' in the service of their genetic masters. On the other hand, there is a long holistic tradition that focuses on the complexity of developmental systems, on the non-linearity of gene–environment interactions, and on multi-level selective processes to argue that the full story of biology is a bit more complicated than that. Reductionism can marshal on its behalf the spectacular successes of genetics and molecular biology throughout the 20th and 21st centuries. Holism has built on the development of entirely new disciplines and conceptual frameworks over the past few decades, including evo-devo and phenotypic plasticity. Yet, a number of biologists are still actively looking for a way out of the reductionism–holism counter-position, often mentioning the word 'emergence' as a way to deal with the conundrum (Pigliucci, 2014, p. 261).

Words like "reductionism", "holism" and "emergence" have so many different meanings, one has to be careful. One also has to be careful because there are forms of reductionism and emergence that are completely compatible with one another. However, suppose by reductionism one means something like the following: "Ideally, every level of organization should be explained by the level below it. In an ideally complete mechanistic explanation of a phenomenon, the capacities of entities at each level are explained by the organized subcapacities of those entities' components" (Piccinini, 2020, p. 326).

This sort of explanation in the literature on neural mechanisms is known as localization and decomposition, perhaps the most well-known example of which is the attempt to localize specific cognitive functions or even aspects of conscious experience, somewhere in the brain. The core idea of localization and decomposition is to break down a mechanism as a whole, into operations of interrelated parts, organize them into modules, which when properly ordered, explain the workings of larger mechanisms or sub-mechanisms that they make up. Thus, we see how interacting and hierarchically organized parts causally produce the phenomenon in question (Bechtel and Abrahamsen, 2005; Bechtel, 2011; Machamer et al., 2000). While many defenders of neuro-mechanistic explanation and reductionism now grant that simplistic forms of localization and decomposition often fail and they grant that mechanisms are often multiscale and multilevel, they still often want to defend neo-mechanistic explanation, mechanistic reductionism and even localization and decomposition in *some* form (see Silberstein & Chemero, 2013, and Silberstein, 2021 for details). However, even most liberal neo-mechanists would like to restrict the explanation of cognition, behavior and conscious experience to the brain and central nervous system as much possible. In short, most neo-mechanists are methodological individualists who still regard "emergence" as almost as dirty a word as "holism" (Silberstein, 2021), at least "emergence" in any but the weak or merely epistemic sense of the word. This is in part because they can only conceive of two kinds of emergence, strong or weak.

While this ancient debate rages on in the literature on neural mechanisms, I think that what many often fail to notice or refuse to acknowledge, is that contextual emergence is not only well confirmed by the biological and cognitive sciences, but even constitutes textbook science. While there is certainly still lots of room for debate about what the data entails and still the battle continues, I want to make the case that when it comes systems biology, network neuroscience, psychology, psychiatry, and embodied, embedded, enactive and extended cognitive science (4E cognitive science), that contextual emergence is the best and perhaps even most common interpretation of the data.

3.1 Systems Neuroscience and Social Neuroscience

In terms of textbook or popular presentations, let's start with renowned neuroscientist Lisa Feldman Barrett's bestselling book *7 ½ Lessons About the Brain* (2020). Barrett's second lesson is that "Your brain is a network—a collection of parts that are connected to function as a single unit" (p. 30). As she notes, "Your brain network is not static—it changes continuously. Some changes are extremely fast….These network changes happen instantaneously and continually, even as your physical brain structure seems unchanged" (p. 36). We will return to discuss brain networks at length nearer the end of this section, but her key point is that the primary unit for modelling and understanding the brain in systems neuroscience is functional and structural networks, as opposed to neurons, local neural circuits, local assemblies of neurons, etc. Such networks often encompass a variety of spatial and temporal scales, entities and activities as well as interactions across scales, can be used to model brain-wide organization, as well as more regional activity (see Silberstein & Chemero, 2013, and Silberstein, 2021). Multiscale modeling of such structural and functional networks is more and more the norm in systems neuroscience and computational neuroscience, where the goal is to create more and more sophisticated mathematical tools to model and simulate such networks (Schirner et al., 2018, p. 1).

As will be made clear in later discussion, the relevance of network neuroscience to bolstering the case for contextual emergence is that these networks involve inherently multiscale interactions, explanatory relations and dependence relations not only across the brain itself, but encompassing central nervous system, immune system, bodily actions, other brains, social interactions and culture (Bassett and Spotns, 2017). Furthermore, these networks ground and explain the key organizational features of the brain enumerated below.

Barrett's third lesson is that "little brains wire themselves to their world" (p. 47). Her point is that whether you want to call it embodied, embedded, enactive or extended cognition (more on those shortly), one must acknowledge that the brain and the world must often be treated as one system or at least as inextricably connected. This includes the brain's connections to the world both physical and social. As she says, "Little brains *require* a social world in order to develop typically. For example, certain physical inputs, such as photons of light bombarding their retinas must be provided or the brain won't develop normal vision" (p. 57). Other examples of essential connections she gives include the following: role of caregivers in social world for tuning and pruning the brain in newborn's neural development, guided attention, various stable features of the fixed environment, niche, training the senses, exposure to natural language, many different social inputs, love and affection, education, various socioeconomic conditions, etc. (pp. 57–58). In her own words, "Our three examples of tuning and pruning demonstrate how the social world profoundly shapes the physical reality of the brain's wiring (p. 57). As she notes, all of this illustrates the silliness of the nature versus nurture debate (pp. 62–63).

Echoing the field of Social Neuroscience which we will discuss shortly, Barrett's fifth lesson is that "Your brain secretly works with other brains" (p. 83). The fact that we are a social species does not stop at early brain development or even at childhood. As she puts it, "ultimately, your family, friends, neighbors, and even strangers contribute to your brain's structure and function and help keep your brain humming along" (p. 84). Our social interactions with others co-regulate and synchronize a number of biological and cognitive processes such as breathing, brain waves, motion and bodily movements generally, heart rate, circadian cycles, menstrual cycles, linguistic capacities, learning, etc. (Spivey, 2020). Of course, we know all this in part because, whether "in the wild" or in the lab, we can observe what happens to people when they are removed from their social connections.

In the widely used textbook *Introduction to Social Neuroscience* (2020) by leading lights Stephanie and John T. Cacioppo, they note that this relatively nascent field of social neuroscience is based on the following assumptions:

- Human brains are not regarded as isolated computational devices, but like a device networked with other brains and people both physically and socially.
- Evolutionarily speaking, there are conserved neural, hormonal, cellular, and molecular mechanisms involved in social behavior.
- Social connectedness, social complexity and social/cultural learning are some of the driving forces behind the evolution of the human brain.
- Brains and their evolution underly social processes, but the reverse is true as well. This can be seen on both evolutionary and developmental time scales.
- The social brain hypothesis. Larger and more complex brains enabled more social interaction and vice-versa. As culture developed many more complex problems were solved by social groups, all leading to positive feedback in the direction of ever increasing neural and cultural complexity.
- The focus is on connection and coordination, e.g., inherently social functions such as communications, social perception, recognition, imitation, empathy, competition, cooperation, etc. For example, in Social Neuroscience language is not viewed primarily as an information processing medium, but as a means of communication.
- Multiple interacting scales and levels of organization from genes on up connect brains and social interactions, such that there are multiple multiscale avenues of mutual-determination and multiple multiscale interacting causal factors. There is "reciprocal determinism" from social-to-biological and vice-versa at multiple scales. Such bio-psycho-social dynamical systems are highly complex, often non-linear and "interaction dominant" (Spivey, 2020). Examples include the growing evidence that social environment can modulate gene expression, the severe effects of social isolation and loneliness on neurological, cognitive and genetic processes, and that the effects of pharmacological interventions such as stimulates or even placebos is partly a function of social hierarchy and other social factors (Cacioppo and Cacioppo, 2020, Chaps. 1–3).

Social neuroscience then is an entire discipline devoted to studying the way neural and social processes contextually constrain and enable one another to emerge, over both evolutionary and developmental time scales. The import of all this should be clear. Very highly regarded hardnosed, mechanistically minded neuroscientists at the top of their profession think that we can now regard it as well confirmed neuroscientific commonsense that contextual emergence is the right way to think about the relationship between brains and their physical and social environments. All of this evidence of course dovetails with a growing body of evidence from epigenetics and epigenomics more generally (Silberstein, 2021).

From the point of view of *everyday* commonsense, there is only one possible reaction, "duh." Anyone who, for example, has been paying serious attention to the recent effects of social media on all aspects of human cognition, behavior, politics, mood and affect, etc., could have told you that what humans are, how they cognize, how they act, how they feel, what they believe, etc., cannot be isolated from social and technological structures. In short, human beings are a contextually emergent multi-scale spatiotemporal network that includes interacting processes that range from the molecular to the social and subsumes timescales both evolutionary and developmental. (See Spivey's excellent book *Who You Are: The Science of Connectedness,* 2020, for many more examples to this effect).

3.2 Organizational Features of the Brain and Related Complex Biological Systems

While yes, there are neural mechanisms involved in all aforementioned processes, the explanation for our malleability and adaptability is not, largely speaking, a reductionist one. There are however features of the brain that, again, support contextual emergence and that help explain why we are so malleable and adaptive. Let us turn to those. Along with network properties to be discussed, the key features of the brain that help explain human general intelligence, the contents of conscious experience, and our overall adaptability are as follows:

- Many different types of neural plasticity—generally defined as changes in the structure, activity or function of the brain on some scale relative to *some change in context* such as injury, stroke or simply learning, i.e., synaptic plasticity, wherein experience, learning and memory formation change the synaptic connections in the brain. There are many other different types of neural plasticity like cross-modal plasticity (such as the loss of one sensory modality inducing cortical reorganization that leads to enhanced sensory performance in remaining modalities, e.g., the relocation and transfer of somatosensory and auditory functions to the former visual cortex), intramodal plasticity (plasticity within a modality, such as the expansion of cortical maps to neighboring regions of intact cortex that have been deprived of sensory input from within the same modality as supported by the expanding cortex), and supramodel plasticity (not unlike cross-modal plasticity but need not involve injury, sensory deprivation or special training, e.g., occipital cortices not only serve as basis for non-visual information processing, but are contributing something inherently visual to the non-visual input, i.e., "non-visual input is being processed visually", Zerilli, 2021, p. 20.). See Zerilli (2021, Chap. 2) for more details on these types of plasticity.
- As Zerilli notes, "plasticity is an intrinsic and persistent property of the nervous system" at all scales in the brain including, not only the aforementioned cortical map reorganization, but neurotransmitters, neuromodulators, cellular changes caused by learning and memory consolidation, neuromorphology, neurogenesis, etc." (2021, p. 10). Perhaps the most well-known example of plasticity is sensory substitution, e.g., converting visual images into soundscapes via a "visual-auditory sensory substitution device" (Zerilli, 2021, p. 88). Neural plasticity is joined by

a host of other types of plasticity in biology such as phenotypic plasticity, all of which helps explain the robustness and autonomy of complex biological systems in general. More on this shortly.

- Neural reuse or *recontextualization*-Each region of your brain ends up participating in many different functional coalitions over time/at a time—same neural circuits contributing to different tasks or functions *depending on various contextual features* (Anderson, 2010 and 2014; Zerilli, 2021, Chap. 3).
- Neural redundancy (numerically distinct brain regions have same structure and function). These regions become active given certain *changes in context,* such as injury and stroke.
- Neural degeneracy (different neural structures and mechanisms perform same function depending on certain *contextual features* that change over time). As Barrett puts it, "Degeneracy in the brain means that your actions and experiences can be created in multiple ways. Each time you feel afraid, for example, your brain may construct that feeling with various sets of neurons" (2020, p. 39).

One can certainly debate how all these different context-driven adaptive features of the brain relate to each other, if for example some are more fundamental than others, etc. In the case of neural reuse, by definition the very same neural circuits (the same mechanism) are contributing to different functions under different conditions—*in different contexts*, but at least in some cases of plasticity, not only do we have different mechanisms subserving the same function, but sometimes the mechanism (what Anderson calls "the working", 2010, p. 297) itself might change as a result of its new role or *context*. In the case of degeneracy distinct mechanisms that have always been distinct end up subserving the same function or task at different points in time. What all of these features have in common is the fact that they are triggered by *changes in context*, such as environmental changes, changing cognitive tasks, changes at larger and smaller scales within the brain, etc.

Again, plasticity, reuse, etc., are not unique to neural and cognitive systems, in addition to the brain, one can find multiscale networks, reuse, plasticity, robustness (invariance in the face of environmental and *contextual* changes), and autonomy (adaptability and flexibility in the face of environmental and *contextual* changes) in many complex biological systems (Silberstein, 2016, 2021). Thus, it is no surprise that neuroscience is no exception. As Bateson and Gluckman put it, "The central elements underlying many forms of plasticity are epigenetic processes, and plasticity operating at different levels of organization often represents different descriptions of the same process. Underlying behavioral plasticity is neural plasticity and underlying that is the molecular plasticity involving epigenetic mechanisms" (2011, p. 43). The point here being that brains inherit their network properties, organizational principles and other global organizing constraints from even more fundamental or basic biological processes and principles.

Perhaps all of this is best illustrated by the relationship between plasticity, robustness and autonomy. There are many different forms of robustness and plasticity, such as developmental, phenotypic, a variety of neural, behavioral, immunological, etc. Let's take phenotypic plasticity and robustness as an example. This is the phenomenon in which genetically identical individuals will develop different phenotypic traits in different environmental conditions (Kaplan, 2005, 2008). Because of phenotypic plasticity, a single genotype or genome can produce many different phenotypes depending on environmental and developmental contingencies (Gilbert and Epel, 2009). Phenotypic plasticity is just one example of epigenomic processes in which various mechanisms create phenotypic variation without altering base-pair nucleotide gene sequences, altering the expression of genes but not the gene sequence.

In contrast, there are cases in which genetic or environmental changes have no phenotypic effect. This persistence of a particular organism's traits across environmental or genetic changes is called robustness. Robustness is illustrated by various knock-out experiments in synthetic biology whereby a particular gene (or group of genes) known to be involved in the development of some protein or phenotypic trait is disabled without disturbing the presence or production of the developmental end product in question (Jablonka & Lamb, 2005).

As we will see, to account for and model plasticity and robustness, developmental biologists have called upon network/dynamical explanations. The ongoing development of an organism acts as a global constraint that 'enslaves' the components necessary to maintain its dynamics. Because of this, a developing system will have highly plastic boundaries, and will be composed of different enslaved components over time. This plasticity serves the autonomy and robustness of the developing organism, making it more likely to be viable and adaptive. Brains are no exception.

Robustness is closely related to autonomy, another key concept in evolutionary developmental biology. Autonomy is the property of living systems to make use of their environments to maintain themselves. Autonomy is sometimes explained in terms of recursive self-maintenance. Some systems are plastic such that they can maintain stability not only within certain ranges of conditions, but also within certain ranges of changes of conditions: they can switch to deploying different processes depending on conditions they detect in the environment.

As Bateson and Gluckman note, robustness and plasticity are two-sides of the same coin, they are interdependent, "Indeed, plasticity is often regulated by robust mechanisms and robustness is often generated by plastic mechanisms" (2011, p. 46), in an interplay of evolutionary and developmental processes. The ever-growing varieties of robustness and plasticity are co-creating and co-maintaining, allowing a complex biological system to have autonomy, "Development involves both internal regulation and reciprocity with the environment. Careful analysis of what happens during development suggests that it is no longer helpful to retain a hard and fast distinction between robustness and plasticity" (2011, p. 62).

Plasticity, robustness and autonomy are universal features of complex biological systems. *Such global or systemic contextual constraints* and organizing principles are well known from a variety of different fields of biology, they have been well known for a long time, and they are well confirmed without any appeal to abstract mathematical models (Jaeger & Calkins, 2012, p. 27, Koonin, 2011, pp. viii–ix). It is also well known that such global constraints can impose the same function even across different species, using different structural components. One can demonstrate equivalence classes of networks across species, including network function. Jaeger and Calkins further infer that, "Given that function is conserved and the mode [i.e., specific structural mechanism] isn't, it suggests regulation by the organism as a whole" (2012, p. 27). This real-world example of multiple realizability is a case of "top-down information control" via multilevel causation, wherein the working parts are constrained in their behavior in service to the larger function, i.e., functional networks.

The biological networks involved in Evo-Devo and systems biology are likewise multiscale. As we are now all well aware, the relationship between genes and proteins is many-many, and we must now think in terms gene networks (genomics), RNA networks, protein networks (proteome), and the complex non-linear interactions between them. Furthermore, these relationships are also affected by several global constraints and multi-scale contextual features including cellular environment, the wider organismic environment and various features of the external environment in which the organism is situated. These interactions are obviously multi-scale, multilevel, inextricably interrelated and interdependent (Noble, 2006, 105). In the developmental process, what any key biological player does, such as genes and proteins and what it results in as output, is a function of multiscale contexts (Bechtel, 2019, p. 461 and p. 488; Francis, 2011, 159; Noble, 2006, p. 17 and p. 34). We now know for example that So-called "junk" DNA contains millions of switches that regulate protein coding genes. It also produces many different types of RNA which play key regulatory roles in their own right such as regulating chemical modifications of the DNA genome and associated proteins in response to environmental signals. Epigenomic effects show that the genome is very much affected by cellular activity, bodily changes as a whole, and changes in the external environment (Parrington, 2021, p. 4, Silberstein, 2021). All of this illustrates that genes or DNA, are not best conceived of as codes for creating or computing proteins, and that the process of biological development is best viewed in the light of contextual emergence.

What all of this strongly suggests is that structural and functional modularity, or localization and decomposition, are not the norm in complex biological systems. Obviously, all of this goes for both developmental biological systems and brains. For example, as regards language, as Zerilli notes, "The picture that emerges here is very unlike the one bequeathed by Paul Broca and Carl Wernicke" (2021, p. 111). Language processing is enabled by highly distributed neural networks composed of very many smaller brain regions or nodes that are themselves multifunctional and domain-general (Zerilli, 2021, Chap. 7). Language processing is no exception in this regard, but rather exemplary. This brings us to networks and network neuroscience.

3.3 Network Neuroscience

The following is a brief overview of network neuroscience and the nature of topological networks (see Silberstein & Chemero, 2013 and Silberstein, 2021 for more details). Network analyzes of the brain are based on the thought that brain function is not just relegated to individual regions and connections but emerges instead from the topology of the brain's entire network, i.e., the connectome of the brain as a whole. In such graphical models of neural activity, the basic units of explanation are not neurons, cell groups, or brain regions, but multiscale networks and their large-scale, often distributed, and nonlocal connections or interactions (Silberstein & Chemero, 2013 and Silberstein, 2021). The study of this integrative brain function and connectivity is mostly based in topological features or architecture of the network. Such multiply realized networks are partially insensitive to, decoupled from, and have a one-to-many relationship with respect to lower-level neurochemical and "wiring" details.

More specifically, a graph in this case is a mathematical representation of some actual many-bodied biological systems. The nodes in such models can represent neurons, cell populations, brain regions, etc., and the edges represent connections between the nodes. The edges can represent structural features such as synaptic pathways and other wiring-diagram-type features, or they can represent more topological features such as graphical distance and network types. What matters in such graphical explanations is the topology or pattern of connections. Different geometries or arrangements of nodes and edges can instantiate the same topology (see Silberstein & Chemero, 2013 and Silberstein, 2021 for more technical details).

When mapping the interactions (the edges) between the local neighborhood networks, we are interested in global topological features, i.e., the topological architecture of the brain as a whole. While there are local networks within networks, it is the global connection between these that is often of greatest interest in systems neuroscience. Graph theory has many different kinds of network topologies, but one of great interest to systems neuroscience are small-world networks. This is because various regions of the brain and the brain as a whole are thought to instantiate such networks. The key topological properties of small-world networks are:

- Sparseness: relatively few edges given the large number of vertices;
- Clustering: edges of the graph tend to form knots, for example, if X and Y know Z, there is a higher-than-normal chance they know each other;
- Small diameter: the length of the most direct route between the most distant vertices, for example, a complete graph, with n2/2 edges, has a diameter of 1, since you can get from any vertex to any other in a single step. Most nodes are not neighbors of one another yet can be reached through a short sequence of steps.

That is, (1) there is a much higher clustering coefficient relative to random networks with equal numbers of nodes and edges and (2) short topological path length. Small-world networks thus exhibit a high degree of topological modularity and nonlocal or long-range connectivity. There are many different types of small-world networks and other types of networks with unique topological properties that allow researchers to make predictions about the robustness, plasticity, functionality, health, etc., of brains that instantiate these networks (Sporns, 2011). One type of network of particular interest is called the "Rich-Club" network (Pedersen and Omidvarnia, 2016; van den Heuvel and Sporns, 2011). Such network architectures are called "Rich-Club" based on the analogy with wealthy, well-connected people in society. "Members" of this club constitute a few "rich" brain-regions or central "hubs" that distribute a large number of the brain's global neural communications. The "Rich-Club" topological brain architecture is instantiated when the hubs of a network tend to be more densely connected among themselves than nodes of a lower degree.

The dynamical interactions in such networks are recurrent, recursive, and reentrant. Therefore, the arrow of explanation or determination in such systems is both top-down (graphical to structural) and bottom-up (structural to graphical). Global topological features of complex systems are not explicable in principle via localization and decomposition. The many-to-one relationship between the structural and the graphical features demonstrates that specific structural features are neither necessary nor sufficient for determining global topological features, i.e., topological features such as the properties of small-world networks exhibit a kind of "universality" with respect to lower-level structural details. In the case of random networks for example, power laws and other scale-invariant relations can be found. These laws, which by definition transcend scale, help to predict and explain the behavior and future time evolution of the global state of the brain, irrespective of its structural implementation. Power laws are explanatory and unifying because they show why the macroscopic dynamics and topological features exist across heterogeneous structural implementations (Silberstein & Chemero, 2013).

In my view the nature of networks is the underlying unifying factor that explains and binds all the features of complex biological and cognitive systems previously discussed, such as plasticity, reuse, redundancy and degeneracy. Let us say more about the nature of such networks:

1. Networks have plasticity, reuse, redundancy and degeneracy built into them. As Sporns' says, "the same set of network elements can participate in multiple cognitive functions by rapid reconfigurations of network links or functional connections" (2011, pp. 182–83).
2. We can use tools from network neuroscience and graph theory to model both structural networks (the "workings" at various scales) and functional networks (e.g., the various types of 'causally relevant' statistical dependencies that exist between different regions of structural networks, indicating that they participate in the same cognitive functions in some important way). See (Bassett and Sporns 2017; Silberstein, 2021; Silberstein & Chemero, 2013) for more details.
3. Network analysis link structure and function, showing us the very complex statistical, dynamical, topological and various causal relationships that can exist between them. These relationships generally don't look anything like the neomechanist's localization and decomposition (Silberstein, 2021). As Weiskopf puts it:

> One upshot of this form of organization is that the neural regions that participate in this assembly may have no identifiable cognitive function outside of their role in the ensemble. While classical localization assumed that distinct cognitive systems would have disjoint physical realization bases, massive redeployment and network theory seem to demonstrate that different systems may have entangled realizers: shared physical structures spread out over a large region of cortex. This suggests that not only will there not be distinct mechanisms corresponding to many of the systems depicted in

> otherwise well-supported cognitive models but given that the relevant anatomical structures are multifunctional in a highly context-sensitive way, perhaps there will be nothing much like mechanisms at all—at least as those have been conceived of in the dominant writings of contemporary mechanistic philosophers of science. And while it might be that these networks should count as mechanisms on a sufficiently liberal conception of what that involves, widespread entanglement still violates Poldrack's constraint that distinct cognitive structures should be realized in distinct neural structures (2019, p. 681).

4. One essential feature of network analysis is to illuminate the topological structures of brain networks, e.g., small-world networks and "rich club networks" seem to appear over and over in the brain and elsewhere. The point for now is that such topological features explain many key capacities and aforementioned organizational principles of brains just in virtue of the topology type—independently of structural details, and these networks also constrain the behavior and dynamics of their ever-changing structural elements. This again is an instance of real world multiple realizability at various scales (Silberstein & Chemero, 2013 and Silberstein, 2021).
5. As noted earlier these networks are highly dynamical and network neuroscience is melding tools from dynamical systems theory with network theory to better model such dynamical networks (Kaiser, 2020). Per Bassett and Sporns:

> More recently, methods from network science are expanding in new directions, going beyond descriptive accounts of network topology and toward addressing network dynamics, generative principles and higher order dependencies among nodes. One prominent example is the development of methods for assessing multi-scale organization in networks. This includes characterizing fluctuations in community structure of networks across time and implementing dynamic processes on networks as a diagnostic tool for explicitly linking micro-scale features of network organization to macro-scale characteristics of neurophysiological dynamics. Yet another approach uses network science to ask questions about the processes that can potentially generate the topology of an empirical network. Such generative models can clarify the contributions of spatial embedding and other (non-spatial) wiring rules in shaping the network topology of the connectome and can also reveal potential factors driving the selection of functionally important network attributes (2017, p. 356).
>
> Notably, changes in function can elicit changes in structure, leading to dynamics of networks. The first conception of dynamics of networks grew from the recognition that many physical and biological systems display patterns of connections that change over time, in different contexts or in response to varying external demands (2017, p. 357).

6. Such networks are often multiscale and highly distributed throughout the brain, central nervous system and body. We can model brain networks at various spatial and temporal scales often called "microscale, mesoscale and macroscale." However, often such brain-wide networks harness, recruit, integrate and unify all these scales and their components in the service of various cognitive functions and to subserve contents of conscious experience. The scales, components and their complex interactions often include the following: ionic flux, sub-cellular structures, proteins, genes, RNA, neurons and neural assemblies, glial cells, neurotransmitters and neuromodulators, hormones, large-scale neural synchrony and neural oscillations, electrical fields, etc. (Godfrey-Smith, 2020,

p. 193; Parrington, 2021, p. 114). For example, regulatory RNAs are secreted by both neurons and glial cells and travel to other cells in the brain in order to modulate various functions.

It is a mistake to focus only on networks involving neurons and their action potentials and oscillations. Complex networks often involve brain-wide integration at every scale. The brain (and the body and the environment) has to manage, modulate and coordinate these processes that are happening at very different spatial and temporal scales. There are many different heterogenous time scales in the brain ranging from milliseconds to seconds to minutes and beyond. This is sort of integration is what networks do (Bassett and Sporns 2017 and Silberstein, 2021). It should be immediately obvious from all this that, just like the genes as codes misnomer, brains are not merely or primarily simply computational devices wherein synaptic activity and their connections are all that really matter.

Graph theory and the big data tricks of network neuroscience, such as network simulations, time series analysis, various sorts of causal analysis such as Granger causality, etc., dimension reduction and universality class analysis, are perfect for illuminating these multiscale relationships and connections. That is, "Networks can also bridge across data of very different types and from different domains of biology. One example is the joint investigation of gene co-expression patterns and patterns of brain connectivity. These studies raise important questions about the nature of the mechanisms that tie the topology of structural and functional brain networks to fundamental aspects of basic brain physiology" (Bassett and Sporns 2017, p. 358).

7. In every area of cognitive neuroscience ranging from learning to memory to decision making and action to empathy and mind reading to consciousness and the contents of consciousness, etc., the central theoretical unit of study and analysis is the multiscale network (Bertolero & Bassett, 2020; Jansson, 2020 and Schirner et al., 2020; Shine et al., 2019). Whether we are talking about task neutral networks such as the Default Mode Network (DMN), or task positive networks such as Executive Control Network (ECN), the Salience Network (SN), Dorsal Attention Network (DAT), Frontoparietal Network (FPN), Amygdala Network (AN), Action-Perception Network, (APN), Empathy-Network (EN), the thalamus-anterior–posterior cingulate cortex-angular gyri network (whatever it is called), etc., the idea is that various functional brain networks are the key unit of investigation (Huang et al., 2020 and Scheinin, 2020). The details of how all these specific networks work need not concern us for the purposes of this conversation. The point is that the explanations involving these networks for various cognitive abilities and various states and contents of consciousness, all have to do with the structural, functional and topological changes in these networks and their various multiscale interactions (or lack thereof), over time under different changes in context, e.g., the presence or absence of anesthesia, psychedelics, deep sleep, stress, specific cognitive tasks, social interactions, etc.

   Sometimes spirited reductionists accuse me of "cherry picking" and "special pleading" with my focus on network neuroscience. I hope all the preceding points make it clear that their charge is empty. There is a reason that Barrett's second lesson is that your brain is a network. Bassett and Sporns once again explain why networks are so central to neuroscience and systems biology: "Network science tools are perfectly suited to accomplish the important goal of crossing levels or scales of organization, integrating diverse data sets, and bridging existing disparate analyses" (2017, p. 368).

8. As I stressed at the beginning of this paper, many hardnosed neuroscientists from different sub-disciplines now fully acknowledge that the networks in question also encompass and include social networks. Working up to social networks proper, Bassett and Sporns note that, "Large-scale studies of brain-behavior relations and behavior-behavior dependencies, although still in their infancy, promise to provide a rich database for mapping the relations among brain processes and their contributions to perception, action and cognition" (2017, p. 354). Such network analysis obviously includes correlates that fall under the category of embodied, embedded, enactive and extended cognition such as action-perception cycles. They go on to say, "network neuroscience asks how all of these levels of inquiry help us to understand the interactions between social beings that give rise to ecologies, economies and cultures. Rather than reducing systems to a list of parts defined at a particular scale, network neuroscience embraces the complexity of the interactions between the parts and acknowledges the dependence of phenomena across scales" (2017, p. 358).

   Such analysis is often called the modelling of various types of "networks-of-networks" or meta-networks (Silberstein, 2016). Just as Barrett and Cacioppo and Cacioppo noted, in such meta-networks, brain networks depend on social networks and vice-versa. Analysis of such networks can be used to study the complex relationships between ever changing social conditions, various other environmental features, changes in genomic, RNA and proteome networks, changes in gene expression and modulation, behavioral patterns, both mental and physical health, various types of neural activity, etc. (Bassett and Sporns, p. 361). Again, all of this illustrates why networks are central here.

To summarize, what makes such networks examples of contextual emergence is multiscale co-determination relations; global constraints, e.g., 'small-worldness'; sensitivity to contextual changes at all scales, both "internally" and "externally" as it were; multiple-realizability of networks with respect to structural details or "universality"; and the fact that what explains the power and autonomy of such networks is the topology itself, e.g., 'small-worldness'.

3.4 Implications for the Relationship Between Neuroscience, Cognitive Science, Psychiatry and Clinical Psychology

Currently there are various debates in what we can broadly call psychology, neuroscience and cognitive science for which all of this has immediate consequences. Broadly speaking, the debates are about how we ought to think of the relationship between neuroscience, psychiatry, clinical psychology, and cognitive science. It is fair to say that the main dividing lines are more or less reductionism and more or less internalism (i.e., methodological individualism). The two debates I want to discuss herein are the frequently touted "war" between the "biomedical model" of psychiatry and the biopsychosocial model (BPS) (Savulescu, Roache and Davies and Loebel, 2020), and secondly the debate between embodied, enactive, embedded, and extended models of cognitive science (4E) versus more internalist models. I will begin with the former debate.

According to Wikipedia, "The biomedical model of health focuses on purely biological factors and excludes psychological, environmental, and social influences. It is considered to be the leading modern way for health care professionals to diagnose and treat a condition in most Western countries." As Murphy defines it, "Scholars and textbooks alike agree (though they might not like it) that psychiatry now adheres to the "medical model", which advocates "the consistent application, in psychiatry, of modern medical thinking and methods" (Black, 2005, 3) because psychopathology "represents the manifestations of disturbed function within a part of the body" (Guze, 1992, 44) to wit, the brain" (Murphy, Dominic, "Philosophy of Psychiatry", *The Stanford Encyclopedia of Philosophy* (Fall 2020 Edition), Edward N. Zalta (ed.), URL = < https://plato.stanford.edu/archives/fall2020/entries/psychiatry/ > .). A strong form of the biomedical model holds that the causes of mental disorders are primarily neurological and that therefore the best way to taxonomize and treat most mental disorders is through some neurological intervention such as psychopharmacological manipulation.

A strong form of BPS holds that mental disorders are often caused by biological, psychological or social factors or some complex combination of these. And thus, mental disorders will often be most effectively taxonomized and treated by biological, psychological or social treatments, or some combination of these (McConnell, p. 381, 2020). It is important to note that for both models, the causal claim can be separated from the claim about treatment. However, the treatment claim is a not unreasonable provisional inference given the causal claim. While no one denies that we should continue to seek and will continue to find purely neurological treatments for mental disorders, the BPS model predicts that the best treatment for mental disorders that do have a BPS causal origin, is often one grounded in BPS types of interventions including, talk therapy and cognitive therapy (Cecil, 2020, p. 190). It is safe to say that so far at least, the BPS prediction has been born out (Cecil, 2020, p. 190).

There is no question that the multiscale and multi-domain networks model presented in this section strongly favors and supports some form of the BPS model and vice-versa. Given everything presented herein, some form of BPS is the best way to look at the relationship between neuroscience, psychiatry and clinical psychology. What is the BPS model but contextual emergence 'in action.' I will return to the discussion of psychiatry shortly.

This brings us to the question of the relationship between neuroscience and cognitive science. One school of cognitive science is internalist in its designation about what counts as the cognitive system and where to look for explanation as regards cognition or conscious experience. The internalist school historically as two basic flavors, a cognitivist/functionalist one or a brain-centric one. The latter is more common these days. As with the biomedical model the internalist brain-centric school predicts that in principle, cognitive science and clinical psychology should reduce to or be eliminated by neuroscience. The school of 4E cognition is, to one degree or another, a rejection of both schools of internalism. Historically, many different traditions of thought and many different thinkers feed into and are the foundation for the 4E movement (Chemero & Silberstein, 2008; Newen, de Bruin and Gallagher, Chap. 1, 2018). Furthermore, there are many different brands and varying strengths of 4E cognition.

The big debate in the 4E movement is whether or not the cognitive system is *constituted* by brain-body processes only (embodied) or by extra-brain-body processes (extended). The word enactive implies that the cognitive system is partly constituted by "active engagement in the agent's environment" (Newen, de Bruin and Gallagher, p. 6, 2018; Silberstein & Chemero, 2012). Such engagement is sometimes called perception–action cycles. The focus here is to see cognition as driven by and partly constituted by action. The notion of "environment" here means not just the organism's physical or social environment as defined from a third-person perspective, but their "phenomenological niche" with its first-person affordances (Silberstein & Chemero, 2012). The word embedded generally refers to the strong causal dependence or"scaffolding" on the environment of the cognitive system. Most people will grant some degree of embeddedness, but the degree of autonomy varies depending on how reductionist or internalist one is in their thinking.

As I said, with regard to the 4E's, the big debate seems to be about where one falls on the spectrum between *constituted by* versus *causally dependent on*. If forced to choose sides I am certainly with the constituted camp. However, I am previously on record for arguing that this is a pretty empty, mostly purely metaphysical debate (Chemero & Silberstein, 2008 and Silberstein & Chemero, 2012). My reasoning is as follows, first, for this to be more than a merely metaphysical debate we would have to agree on exactly how to define the slippery terms of "constituted by" versus "causally dependent" such that they are clearly mutually exclusive. And then based on that we would need to see forthcoming a group of competing predictions, the outcomes of which we could use to make our judgements. Of course, there is no such thing as a crucial experiment, but I still don't think this basic work has been done to the point where we can easily create a win/loss column for each camp, certainly not one with much consensus. In addition, both sides are generally able to provide their own spin on most any results. For example, champions of predictive coding or predicting processing tend to be internalist, but one can also give it a 4E twist (see for example Kirchhoff, Chap. 12, , 2018; Anderson and Chemero, Chap. 12, , 2019). But obviously, if we can't simply deflate this dispute by an appeal to pragmatism or pluralism, we want this dispute to be an empirical question that is empirically resolvable in principle.

My second reason for skepticism about the value of this debate is that it presupposes that there is some metaphysically or Platonically given fact of the matter about where to draw the lines around systems, cognitive or otherwise. I can think of no non-spooky reason (e.g., Platonism) for believing this. Of course, everywhere in science we make distinctions between the system we want to explain and the environment, boundary conditions, etc. However, those decisions are generally made on a pragmatic basis. In every field in science, we constantly redraw the shifting lines between what is system and what is environment as our pragmatic and explanatory interests change and our knowledge grows (Bishop & Silberstein, 2019). An obvious example to bring to bear here is developmental biology-epigenetics-epigenomics and the quaintness of the nature versus nurture debate. Are we partly *constituted* by non-linear and *causally interdependent* gene networks, RNA networks, proteome networks, human cells, bacteriological cells, a host of chemicals, bodily processes at multiple scales, etc., all taking place in larger physical and social environments, with which they all share co-determination, co-constitution and massive causal interdependence? Yes. Recall the quote from Pigliucci that began this section, where he invoked the old-school genetic determinism and reductionism of biology. One would be hard put to find many educated biologists who still believe in genetic determinism. The point is, depending on our explanatory purposes, need and capacity for manipulation an intervention, we can change the boundaries as needed (Silberstein, 2021). So why are we still having this moribund debate in the foundations of cognitive science?

What does any of this have to do with contextual emergence? My point is that regardless of your stance on internalism versus externalism or your position on 4E cognition, I think the facts speak pretty loudly for themselves. Network neuroscience, social neuroscience, systems biology, 4E cognition, BPS psychiatry and psychology, etc., separately and collectively all point to contextual emergence. Part of the point I tried to make herein is that *even key areas of neuroscience itself*, the dark tower of reductionism and mechanistic thinking, tout contextual emergence. Not convinced? For more evidence see (Newen, de Bruin and Gallagher, 2018; Spivey, 2020; and Silberstein, 2021). I am not alone in making this claim. As Spivey notes:

> Cognitive scientists Harald Atmanspacher and Peter beim Graben call this '*contextual emergence*' [my emphasis]. They have studied systems like these in intense mathematical detail and designed computational simulations of how a mental state can emerge from a neural network. They find in these mathematical treatments that being in a particular mental state requires more than just for the neurons to be producing a particular pattern of activity. The context of that simulated brain, such as the body and its environment, also need to be accounted for in order to develop a statistically reliable characterization of the mental states that the mind takes on. Hence, based on these computational simulations, the brain by itself is not sufficient to generate a mental state. A mental state, such as a belief or desire, only emerges between a brain and its context, that is, its body and the environment...So the brain-and-body is an open system...What you are studying has become an 'organism-environment system' (2020, pp. 142–43).

No matter how one wants to *count systems*, and *wherever* one wants to locate the mind and cognition, I believe there is ample evidence from across several disciplines (including biology and neuroscience), that what Spivey is suggesting is likely true. I suggest we collectively focus our energies on exploring and advancing these relatively nascent network sciences so we can more fully appreciate systems such as ourselves in all our glorious emergent complexity.

To see how all these contextual emergence exhibiting sciences can collectively support one another and bolster my claims, let us return to the case of psychiatry. In his Stanford Encyclopedia entry on psychiatry Murphy says the following:

> An important development that is avowedly sceptical of the existing biomedical paradigm in psychiatry is the rise of approaches inspired by the 4E movement in philosophy of mind—the "embedded, embodied, extended and enactive" tradition. Many philosophers of psychology and cognitive scientists dispute the viability of what they see as an unduly reductive and anti-environmental tradition of philosophical psychology. The consensus among 4E proponents is that psychiatry is wedded to a number of outdated philosophical positions...To begin with, many externalist approaches to psychiatry contest what they see as the basic ontological commitments of psychiatry when it comes to mental illness, by objecting to what is seen as a reductive or internalist thesis that psychiatric conditions are neurological disorders. Competing conceptions of mental illness may accept that psychiatric conditions involve neural dysfunction in some sense, but they argue that no psychic process, including pathological ones, "can be reduced to the brain or to localized neural activities: they are embodied, inherently intentional, and context related" (Fuchs, 2018, p. 253)....The situated view construes psychological capacities as intimately dependent upon environmental states of affairs, and so prohibits methodological solipsism...Poor mental health can be the outcome of neural dysfunction, but it can also owe to problems with the environments in which agents are situated, as well as with their capacity to fruitfully exploit environmental resources (Sneddon, 2002; Roberts, Krueger, and Glackin, 2019). The externalist claims that she can discern types of explanations, interventions and strategies that an internalist picture misses. (Murphy, Dominic, "Philosophy of Psychiatry", *The Stanford Encyclopedia of Philosophy* (Fall 2020 Edition), Edward N. Zalta (ed.), URL = <https://plato.stanford.edu/archives/fall2020/entries/psychiatry/>.).

While again there are different 4E models and different BPS models with different degrees of strength, etc., I would say the BPS model and 4E cognition go hand-in-hand. The obvious example to discuss here is the biomedical model's claim that addiction is a brain disease. Why are neuroscientists so convinced addiction is a disease of the brain? Because as Levy notes, "neuroscientists embrace the brain disease model of addiction for an obvious reason: because they have made great progress in elucidating neural mechanisms and neuroadaptations that are correlated with, and undoubtedly causally involved in, addiction. Neuroscientists have identified a range of such changes, including (but not limited to) the long term depression of reward circuitry and increased activity in anti-reward circuitry (Koob and Le Moal, 1997, 2008); alterations in the mid-brain dopamine system (Volkow and Li, 2004); and in frontal regions involved in impulse inhibition" (Levy, p.1, 2013).

The retorts on the part of the BPS and 4E schools are so obvious one hardly feels they need to be said, but here goes. First, at best what neuroscience has identified here is a necessary condition for addiction. Indeed, one doubts it is even necessary as there is growing evidence that addiction is multiply realizable in the brain and in other non-human nervous systems. There is a lot of evidence that dopamine doesn't by itself equal pleasure and evidence that having a dopamine-based neural reward system is neither necessary nor sufficient for addiction (Levy, 2013 and Hart, 2021). It is easy to see the flaw in the biomedical modeler's reasoning. They infer that where there is some neural mechanism involved that therefore that must be the primary explainer for the case at hand. However, *everything* involving humans and other animals entails some neural processes, but nothing about the sufficiency or primacy of mechanistic explanations follows from that. Perhaps the biomedical modeler is reasoning that the best or easiest place to intervene in addiction is in the brain and therefore we should count something in the brain as *the cause* of addiction. As we discussed earlier, this is clearly a bad inference. Furthermore, in spite of the rhetoric from some biomedical model purveyors that talk-therapy and 12 step programs are inefficacious pseudoscience, leaving aside the latter demarcation problem, the data as to efficacy suggests otherwise (Hart, 2021). Furthermore, even though neuroscience has continued to develop direct interventions in the brain such as blocking opioid receptors, there is little evidence that such tricks make addiction disappear and addictive behavior more generally disappear. As of this moment, the efficacy of the "pseudoscience" based interventions is far greater than the strictly neurological interventions (Hart, 2021).

Second, only a very small percentage of people who take, say, opioids become an addict. This strongly suggests that the mere interaction of a controlled substance and a neural reward system doesn't equal addiction (Hart, 2021).

Third, the strongest correlates and biggest predictors of addiction are socio-economic such as poverty, family dysfunction, isolation and loneliness, lack of education, emotional abuse, trauma, deprivation, etc. (Hart, 2021). All of this strongly suggests a multi-causal and multi-scale network-type complex causal interaction account of addiction. To which those of us who have been on planet Earth for a while, can respond once again with, "duh." Let us bring all this back to networks. Kim et. al., based on multiscale network analysis argue that their "findings suggest that childhood poverty may result in wide-spread disruptions of the brain connectome among girls, particularly at the lowest INR levels, and are differentially expressed in females and males" (2019, p. 409). Why is poverty such a predictor of disruptions in neural networks? The primary answer isn't literal malnutrition, it's because poverty is super-correlated with all the social, emotional and cognitive predictors of addiction and other mental disorders I enumerated above. Once again, all of this illustrates what Barrett and others are telling us, that neural networks and social networks are often of apiece. To which again, I can't resist saying, "duh."

Once again, the point of all this is that the data from across the biological and mind sciences strongly suggests contextual emergence. Historically one reason many people resisted the sort of picture outlined herein is fear of invoking "downward causation", and the worry that it would violate causal closure of the physical or microphysical. Downward causation was often imagined, for example, as an occult mental *force* acting *downwardly* on the lower level of neural processes via some sort of *efficient causation*. However, as many now acknowledge, we can easily understand such multiscale causal and otherwise co-determinative interactions in terms of global constraints, contextuality and boundary conditions. Contextual emergence in the biological and mind sciences makes it clear that the world is not divided up into autonomous levels of hierarchical organization wherein levels at smaller or lower scales "constitute" the entities that make up the higher level. None of which is to say that causal closure is safe or well supported by science or scientific practice. Indeed, if one accepts everything said herein, science across the board supports multiscale and global causal interactions and as the BPS model and 4E cognition illustrates, this very clearly includes social and psychological effects on brains, nervous systems and immune systems. In other words, causal closure is empty analytic metaphysics that simply does not square with experience, science, or scientific practice.

4 Conclusion: Looking Backward or Moving Forward?

I have argued at length that contextual emergence, just as commonsense would predict, is the norm in the biological, social and psychological sciences. I have argued that even many "mechanistic" explanations and methods in these fields suggest contextual emergence and not ontological reductionism, e.g., localization and decomposition (see also Silberstein, 2021). I have even taunted the defender of ontological reductionism with the word "duh." I do this in the hopes of baiting them into pushing back, because in my experience, instead of fully acknowledging the overwhelming evidence against ontological reductionism and for contextual emergence, they simply choose to redefine words like "mechanism", "reductionism", "localization" and "decomposition", in order that they never have to acknowledge that the world just isn't the way the mechanist and other ontological reductionists thought it was. To see me engage that very ghettoized debate in detail in the arena of neural mechanisms see (Silberstein, 2021). Part of what I point out therein is that even some very "liberal" mechanists who acknowledge much of what I have said herein, are still inclined to fly the flag of mechanism and to be allergic to the word "emergence."

I hope to have shown that there is no longer any reason for such a shunning. Contextual emergence is not the fringe, it's mainstream textbook science across the board. For more such evidence from multiple sciences including physics see our forthcoming book *Emergence in Context* (Bishop, Silberstein and Pexton, Oxford University Press, Summer 2022).

C. D. Broad, perhaps the most well-known champion of "emergence" in the modern era, was typically considered an archenemy of the mechanists of his time, a compositional view of nature he called "pure mechanism." According to Broad, this is the view that the 'laws governing' the parts of a system operate in a purely context-independent fashion (Broad, 1925, pp. 58–61). Contextual emergence keeps the context-dependence feature of Broad's account of emergence but rejects the claim that emergents (e.g., laws, properties, entities, processes, etc.) are brute or inexplicable and it rejects the discontinuity and gappiness of "transordinal laws."

Of course, many philosophers of science will simply note that one can do philosophy of science without ever mentioning or caring about metaphysics, they are simply interested in scientific explanation and methodology. To that I would say, first, the explanatory approaches of the aforementioned sciences still suggest contextual emergence methodologically as opposed to some brand of intertheoretic reduction or mechanistic reduction. Second, it is important to see that ontological contextual emergence as described in Sect. 2 and as suggested by our best explanatory practices in the sciences, is the reason why those explanatory practices make sense and work so well, as opposed to being some giant function of ignorance masking the truth of ontological reductionism or foundationalism. For example, just as current scientific practice suggests, human beings really can be most profitably treated as contextually emergent multiscale spatiotemporal network that includes interacting processes that range from the molecular to the social and subsume timescales both evolutionary and developmental. That's just the way the world is. Pragmatically we can choose to focus on any part, aspect or interrelations of those networks for explanation and intervention as needed. Nothing here cuts against methodological reductionism or smaller-scale interventions when they are useful and successful. But science isn't exactly hurting for these sorts of moves.

What I am suggesting her is that we really need to get much cleverer at explanation and intervention in multiscale and multi-causal networks, cleverer at science that is avowedly based in contextual emergence. Happily, more and more people are engaged in just this project. Take the following for example:

> Everywhere in biology and cognitive science we deal with systems made of parts or elements with different functionalities acting in a selective and harmonized way, coordinating themselves at different time scales, interacting hierarchically in local networks, which form, in turn, global networks and, then, meta-networks…The organization of living systems consists in different nested and interconnected levels which, being somewhat self-organized in their local dynamics, depend globally one upon the others. This means that both the components and the sub-networks contribute to the existence, maintenance and propagation of the global organizations to which they belong. And, in turn, those global organizations contribute to the production, maintenance and propagation of (at least some of) their constitutive components (Sporns, 2011, p. 322).

I should also make clear that in spite of some of the aforementioned differences, there are those mechanists whose models and ways of thinking dovetail greatly with contextual emergence (see for example Raja & Anderson, 2021; Winning & Bechtel, 2018; and Winning, 2020). More good news, there are an increasing number of mechanists, computationalists and representationalists (which tend obviously to be internalists) to grant that the story is more complex. Take for example Piccinini, an avid defender of "neurocognitive computationalism", who says the following, "Thus, a complete explanation of cognition may require, in addition to an appeal to [neuro] computation, an account of consciousness, representational content, embodiment, and embeddedness. Computation may be insufficient for some aspects of cognition, but it may still be an important part of the explanation of most—or even all-cognitive capacities" (2020, p. 245). Obviously, we still have a way to go here for rapprochement, but it's a start. It's also worth noting that Piccinini's concept of computation and representation is very much divorced from functionalism, cognitivism, computationalism, Fodorianism, etc., and is multiscale (2020). He also acknowledges the autonomy of social psychology and folk psychology, etc. As a side note, it's also worth mentioning that folk psychology, the autonomy of which is a frequent debate in this neck of the woods, can be divorced from "scientific psychology" (cognitivism) and from internalism more generally.

None of this is to deny that new things we learn about the brain such as those discussed in Sect. 3, won't lead to major revisions in both folk and cognitive psychology. For example, I think the multiscale network picture outlined herein shows how the emergence and autonomy of folk and cognitive psychology is possible, but it also shows the limits of that autonomy. That is, in such multiscale interdependent networks, one cannot screen off neural processes from belief/desire psychology, cognitive psychology, and social psychology the way some cognitivists suggest. This is because such neural processes are only a necessary component at best, in a much vaster multiscale network with which they share strong interdependence and co-determination. What in part makes possible the emergence of belief/desire psychology, sociality, and various cognitive functions is the plastic nature of nervous systems and brains. This in part guarantees coupling with all aspects of the environment and one another, rather than being cognitively closed systems.

Thus, the standard metaphysical relations invoked by cognitivists and philosophers of mind such as realization, supervenience and other modal and synchronic relations, simply don't apply to the kinds of rapidly changing multiscale dynamical networks under discussion here. Unlike even some who champion 4E cognition, it is important to understand that the issue here isn't simply one of extending the realization or supervenience base for minds into the body and environment, rather, the contextually emergent-network model herein is a complete rejection of that foundationalist and hierarchical way of looking at the world, both metaphysically and physically.

Finally, I would be remiss if I didn't point out that there is push back from some philosophers of mind and philosophers of neuroscience against the sort of anti-reductionist arguments herein. For example, Zerilli grants that multiple realizability (MR) is an empirical question and he grants much what I've said herein but, is skeptical about "true multiple realization" (MR) and thus skeptical that it counts as an argument against reductionism or some sort of mind-brain identity theory (2021, Chap. 8). Part of the problem with this discussion is that while everyone agrees that MR means difference underlying sameness, i.e., truly causally distinct mechanisms underlying the same cognitive function or content of conscious experience, there is no set agreement on what counts as truly causally distinct or same function. That is, there is little consensus on the necessary and sufficient conditions for MR, what constitutes causally relevant differences in the mechanism, or what constitutes truly different ways to "bring about" the same function that defines the kind. In short, whether MR is true depends on how tightly or weakly one defines the "mechanism" and the "function" in question. No one would deny that lots of different animals, insects, etc., utilize different mechanisms such as electrical fields and magnetic fields to perform general functions such as navigation and communication (electroreception). The MR skeptic clearly has a different target.

What makes two causal mechanisms truly causally distinct? Polger and Shapiro assert that "multiple realization occurs if and only if two (or more) systems perform the same function in different ways" (2016, p. 45). Thusly, Zerilli argues crossmodal plasticity and supramodal plasticity are not evidence of MR because, for example, the auditory cortex came to resemble the visual cortex, i.e., the auditory cortex came to take on the same columnar organization as the visual cortex (2021, p. 141), and therefore this is not a case of truly distinct causal mechanisms underlying the same function. My opening response is that if the brain or any complex biological system can under certain contexts or conditions, convert itself or some sub-region of itself into a distinct causal mechanism in order to mimic the damaged mechanism, that should count as MR par excellence! I would also reply that surely degeneracy by definition counts as MR. And as Barrett acknowledges degeneracy is a real feature of brains. This is an empirical claim of course and is subject to revision in light of further data. But for now, I'll take the word of leading neuroscientists. Perhaps neural degeneracy is rare, but even so, it is proof in principle of MR. Even in cases of reuse, redeployment or recontextualization, if a particular cognitive function can at different times have different workings subserve them, even if those workings do not morph into new mechanisms in the process, that should count as MR with respect to those cognitive functions.

Many of us would argue that convergent evolution provides more such cases of MR. The standard definition of convergent evolution is as follows: the causally independent evolution of similar functional features or evolutionary strategies in species of different periods or epochs in time. Convergent evolution creates analogous structures that have similar form or function but were not present in the last common ancestor of those groups. The idea is that the same function, behavior, evolutionary strategy or even roughly the same mechanism comes into being in causally independent circumstances as the result of environmental and evolutionary pressures. Sometimes in such cases the same mechanism (or partial mechanism) arises independently to perform the same function and sometimes the mechanism is different. For example, flying as an evolutionary strategy appears again and again a la convergent evolution, but the mechanism can be very different, e.g., bats, birds, various insects and flying squirrels. However, such cases don't rise to the level of true MR for the skeptic because "flying" is too general a function.

There are other cases of convergent evolution where almost the same mechanism appears independently across time, to perform a very similar function. For example, the brain architecture of the DVR (Dorsal Ventricular Ridge) in reptiles developed independently from the mammalian neocortex, but they share a good deal of structure and function, in part due to evolutionarily conserved neural circuits (Yamashita and Nomura, 2017, p. 302). Another example, birds, whose neural architecture also developed independently, do not have a mammalian-neocortex as such, but the avian pallium contains circuits homologous to those of the mammalian neocortex (Wada et al., 2017, p. 285). However, it is very likely that the genetic and developmental mechanisms and pathways that form the avian columnar circuit formation are different (Wada et al., 2017, p. 285). That is, while the neocortex of mammals and the pallium of birds share similar mechanisms and functions, the neural and genetic mechanisms that explain the development and placement of those neural structures is different. There are other more straightforward cases of convergent evolution such as eyes like ours, photosynthesis, sex as a strategy, nervous systems or nerve nets, etc. But as long as the mechanism remains the same or the function defined is too broad (e.g., dolphins and monkeys respectively herding prey), the MR skeptic is likely to be unimpressed.

There is an argument however that I believe shows that natural selection trumps the existence of specific mechanisms, and thus is a case of contextual emergence. What emerges are various universal mechanisms, behaviors, strategies, cognitive functions, etc., as the result of the contextual constraints given by certain specific evolutionary environments, as well as more general evolutionary principles such as the idea that the organisms that are most adaptive are most likely to survive and thrive. And the most adaptive will be the animals with the most general intelligence. Here is the argument in a nutshell:

1. Natural selection and thermodynamics are universal, so as to include not only all of Earth but other planets, etc.
2. Therefore, without knowing anything else about an environment or planet, the evolutionary environment of selection will allow us to predict what traits, etc., organisms are likely to possess. If the environment has tall trees, high elevations, deep dark cold waters, low oxygen, etc., natural selection will allow us to predict with a high degree of accuracy what sorts of traits and evolutionary strategies have emerged regardless of knowledge of implementing mechanisms. Certainly, many different features of the environment, biological, physical and otherwise, will place constraints on what is possible, but otherwise natural selection will be the determining factor. Natural selection tells us that no matter what sort of mechanisms might be available in different planetary contexts or even on other planets, while the mechanisms can and do vary, certain traits will come into being via one set of mechanisms or another. And this has happened here on Earth in the form of convergent evolution. Natural selection transcends any particular mechanistic account of a particular cognitive function. As Godfrey-Smith puts it, "evolution sometimes builds a range of different structures that carry the same function" (2020, p. 91). As he points out, crustaceans and other animals do not have a cerebral cortex, visual centers or pain related mechanisms such as our own, and yet they have the same functions and probably subjective experience as well (p. 91). His favorite example of convergent evolution is the nervous system of the octopus. The nervous system of the octopus is highly decentralized with two-thirds of the neurons not in the brain but in the tenacles, delegation of limb control is decentralized to some extent with each tenacle able to explore and multiply-sense the environment independently, and there is no body-map within their brain, thus orientation in space happens differently somehow, and they have a big optic lobe behind each eye with dedicated processing to that eye only (pp. 129–30). There is also the obvious example of creatures with very small brains (one hundred thousand neurons or less) such as bees and spiders, who nonetheless by neural mechanisms that must be quite different and perhaps less varied in many ways, manage to have sophisticated visual capacities and other sensory modalities, complex motor control, navigation, communication, problem solving and in some cases sociality and cooperation.

3. Specific biological mechanisms such as evolutionarily conserved neural circuits are going to be a function of various historically contingent constraints, and thus such mechanisms can differ greatly, especially if we consider radically different environments, different planets, etc. There is no good evolutionary reason for example that DNA, RNA, neurons, etc., need be the implementing mechanisms, these are contingent facts. For example, here on Earth some bacteria-killing viruses called bacteriophages replace adenine with 2-aminoadenine. To quote Zhou, "life doesn't have to be GTAC, life can be more diverse" (Zhou, 2021). The genome of single-celled plankton (dinoflagellates) is organized into alternating unidirectional blocks, as opposed to being random. This changes the three-dimensional structure of the genome into rod-shaped chromosomes and the structure is strangely dependent on transcriptional activity. Other oddities include: few transcription factors in their genome, they do not alter gene expression in response to environmental changes, and unlike other eukaryotes, instead of the exclusive use of histones to structure their DNA, they use viral proteins incorporated into their genome long ago. This may in part explain why their genome structure is so different. As Salazar notes, "It shows that nature can work in a completely different way than we thought. There are many possibilities for how life could have evolved" (2020). Many others make the same point, "The fundamental forces at work in evolution are independent of which molecules interact with which and how" (Kershenbaum, 2020, p. 46).
4. If an environment is complex, changing and challenging enough, natural selection will favor the most adaptive organisms—biologically, behaviorally and cognitively. The most adaptive organisms will be the most plastic and robust organisms. Those whose nervous systems and embodiment are plastic enough to enable the most universally adaptive strategies in the face of environmental changes, such as sociality, tool use and language. Therefore, it is not surprising that a number of disparate animals on Earth such as octopuses, bees, crows, dogs, dolphins and primates, etc., evolve, to one degree or another, certain universal strategies, skills, modes of learning, etc., such as motor control and perception, motility, predictive coding perhaps, memory of various sorts, mental time travel, sociality and cooperation, communication and language, culture, tool use, etc. The selective advantage of general intelligence generally and the selective advantage of all the aforementioned specific skills and cognitive functions is well documented across all life on Earth. Natural selections dictate, that when constraints allow, these same strategies and cognitive functions will emerge over and over across disparate environments, on different planets, with disparate mechanisms, as they already have on Earth.

Take the strategy of sleep for example. While experts still debate the various possible functions of sleep, it is ubiquitous across species. There is now evidence that sleep even predates animals with brains and central nervous systems, e.g., Hydra (Kanaya, 2020, p. 1). Amongst those animals who sleep and do have nervous systems, there is evidence of MR. Sleep-like states are conserved across most species, but using different mechanisms (Hayashi and Lui, 2017, p. 343). For example, hypocretin has a crucial role in maintaining waking states in mammals. Invertebrates have no homologue, but in fruit flies for instance, neuropeptide pigment dispersing factor (PDF) is critical in regulating circadian rhythms and wakefulness (Hayashi and Lui, 2017, p. 360). Different neural circuits, neurochemicals, hormones and genes are engaged in sleep regulation in different animals. There is simply no universal mechanism or molecular pathway that encodes sleepiness or wakefulness.

All of this gives us a big clue as to the nature of the human brain and to why humans are quantitatively if not qualitatively more intelligent than other species on Earth. We are the most adaptive creatures with the most general intelligence in part because our brains, nervous systems and bodies maximize plasticity and robustness. Thus, we are very responsive and adaptive to internal changes and environmental changes, i.e., our brains, bodies, etc., are 'engines' of plasticity and robustness. Which means in fact that our cognitive systems and ways of being are very open to environmental changes and what's more, as Social Neuroscience predicts, there will be all sorts of positive feedback relationships (for instance) between our cultural advances such as tool use, sociality and language and our brain development and activity. In this bio-psycho-social process of potentially ever-increasing plasticity, various boot-strapping processes involving specific cognitive functions, biological mechanisms and specific technologies converge, get integrated, get repurposed, and get iterated into ever more complex patterns of development. Which is to say, human beings are not just brains or brains and bodies, but they are temporally extended multiscale bio-psycho-social networks. And as I will stress in what follows, key "design" principles, such as plasticity and robustness are themselves MR and not fully explicable in any mechanistic fashion.

One might wonder why neo-mechanists resist the reality of MR. Perhaps part of the answer lies in the fact that the best argument for MR harkens back to a time when functionalism (generally computational) in philosophy of mind and cognitive science was ascendant. During that period functionalists often felt justified in ignoring neuromechanistic details and the brain more generally (the stuff that goes in those boxes in the cognitivist "boxological" computational diagram), because implementing mechanisms were viewed as secondary. Fodor and others famously used functionalist-based MR as an argument against type-type identity theory. Regardless of one's evaluation of such arguments in the past, it should be clear that I am not invoking functionalism of any sort and rather than ignoring neuroscience and evolution, I'm using them to make my case.

However, in light of the preceding, it would be absurd to believe that certain cognitive functions such as specific types of memory, won't occur even on other planets, and (even on this planet) equally absurd to believe that such occurrences will always involve the same mechanisms. The idea that there is some law or identity relation that universally links cognitive functions or even worse specific mental contents and specific mechanisms or brain states, wreaks of the supernatural, not scientifically grounded naturalism. This is the sort of magical metaphysical thinking that allows some to take Putnam's brains-in-vats and Boltzmann brains seriously.

All of this brings me back to the much more-Earthly topic of MR and neural networks. Discussions of MR rarely have multiscale networks in mind. Therefore, allow me to elaborate my MR argument with networks at the center. As we discussed earlier, the brain's networks featuring "small-worldness", "rich-clubbyness" with hubs, and "scale-freeness", are universal topological/organizational features. Indeed, the neural architecture of many different insects and animals exhibit such topological features and such features appear early in the development of individual organisms. The reason for this is clear. Those topological features maximize plasticity and robustness with respect to both internal and environmental perturbations, and thus they maximize adaptability.

Certainly, mechanistic explanations can advert to such organizational features in principle, if by "organizational features" one means some representation (however abstract) of the spatiotemporal, causal and dynamical relationships between various components. What I'm denying however, is that network explanations are nothing but maps or representations (however abstract) of such componential relationships. This is true not only because the nodes and edges in network explanations need not refer to components and their relations directly, but because the behavior of the components is often determined by or *constrained by* the *global* organizational feature of in this case, small-worldness. It is because network properties (or order parameters in the dynamical case) can represent/alter the global state of the system (or some sub-set of it) that one can change the behavior of the various components by tweaking the network properties.

Why are such global network constraints so prevalent, so often multiply realized in complex biological and cognitive systems even though processes at smaller scales often happen at very different time scales than the processes at larger scales they support? There is often a rapid turnover of entities and states at the smaller scales creating real world multiple realizability that belies any simplistic account of composition and realization. My answer is that those global topological features, once in place, in turn constrain the behavior of the ever-changing constituents in order to maintain the relevant efficacious topological features. The multiscale and global co-determination and causal dependency involved in such systems looks nothing like the functionalist's MR wherein the same cognitive functions are "realized" by distinct biological mechanisms. Nor does real world MR in networks look like localization and decomposition. The autonomy of functional and topological networks and the autonomy of other global organizing principles in the brain has nothing to do with realization relations or brute identity relations, it has to do with contextually emergent causally open networks.

For any particular synchronic-frame or still-shot of a biological system at a time t with some duration d, the determining features include *diachronic* multiscale interactions (context sensitivity) and *global constraints* outside the time-slice in question that cannot even be assigned a scale or 'level.' That is, when it comes to such complex biological systems one should take the word *process* very seriously and understand that such systems are spatially, temporally, functionally and in a thin sense, teleologically extended. This is not to deny of course that there are a variety of both global-to-local and local-to-global determination relations involved in such systems. This alone should be enough to dispense with talk of realization and the like.

Thus, once we see that, per contextual emergence, global topological network properties and other organizing principles do not "supervene" on and are not "realized" by structural components as such, there is little reason to think that topological explanations are reductionist or mechanistic in the sense of localization and decomposition. As Love notes:

> First, reciprocal interactions between genetic and physical causes does not conform to the expectations that mechanism descriptions 'bottom-out' in lower-level activities of molecular entities (Darden, 2006). The interlevel nature of the causal dynamics between genetic and physical factors runs counter to this expectation and is not amenable to an interpretation in terms of nested mechanisms realizing another mechanism. Second, the reciprocal interaction between genetic and physical causes does not require stable, compositional organization, which is a key criterion for mechanisms (Craver and Darden, 2013). The productive continuity of a sequence of genetic and physical difference-makers can be maintained despite changes in the number and types of elements in a mechanism. Although compositional differences can alter relationships of physical causation (fluid flow or tension), these relationships do not require the specificity of genetic interaction predominant in most mechanistic explanations from molecular biology. (The multiple realizability of CPM outcomes is central to this conclusion). Standard mechanistic strategies of representation and explanation appear inadequate to capture these mechanisms" (Love, 2018, p. 341; see also Love, 2012, p. 120 and Love & Hüttemann, 2011).

In such cases, global constraints and organizing principles and multi-scale contexts determine the behavior of the parts, not primarily or solely the other way around. With contextual emergence, global constraints and other kinds of context sensitivity are *fundamentally* at play. As Broad puts it, "[A]n emergent quality is roughly a quality which belongs to a complex as a whole and not to its parts" (Broad, 1925, p. 23). According to him, if the properties of an irreducible whole are not given by the properties of the basic parts in isolation, they are emergent (see Humphreys, 2016 for more details). For Broad, the global or systemic properties P of a system S are only reducible when the parts in isolation are sufficient to explain the existence of P. That is, there is reducibility when P can be derived or predicted in principle from the parts of S in isolation or when embedded in simpler systems (Stephan, 1992, p. 55).

Contextual emergence emphasizes the ontological and explanatory fundamentality of multiscale *contextual constraints,* often operating globally over interconnected, interdependent, and interacting entities and their relations at multiple scales, e.g., topological constraints and organizational constraints in complex biological systems. In (Silberstein, 2021) I argued at length that such multiscale contextual constraints and *global* organizing principles such as network topology, plasticity, degeneracy, robustness, autonomy, and universality, which are mostly definitely multiply realized in complex biological systems, cannot possibly be explained in terms of localization and decomposition. Not only are these organizing principles global by definition, but we see them repeatedly instantiated over and over in many different biological systems and kinds, and they remain invariant over vast changes in structural dynamics.

Thus, such organizing principles are themselves MR. I happily predict that, as with natural selection, in untold planets across the universe one will find these organizing principles at work regardless of the physical or biological nature of the organisms in question. Perhaps this is some comfort to the token or type identity theorist? If so, I'm not sure what we are arguing about, because the autonomy and invariance of these organizational principles is best explained by their co-causal and co-determinative nature as contextual constraints. Maybe wherever we find these organizing principles and multiscale networks we will find similar minds with similar content, but I don't think this is the kind of reductionism the identity theorist was hoping for.

Does the identity theorist really believe that after mastering comparative cognition and neuroscience on Earth and Astrobiology, that we will find that the same cognitive functions and social structures are always subserved by "the same" biological mechanisms? Why in the name of natural selection and systems biology would anyone believe such a thing? Please don't misunderstand me, I don't doubt that there are many universal constraints on plasticity and other organizing principles of cognitive systems. Such constraints might include physical, molecular, evolutionary (e.g., canalization, etc.). What these constraints are is an empirical question. As noted, I also don't doubt the existence of universally conserved mechanisms across evolutionary time scales, at least relative too particular evolutionary histories such as here on Earth. What this tells is that some mechanisms are more invariant than others. But what is the explanation for the invariance? I argue in (Silberstein, 2021) that the answer again is contextual emergence. Such mechanisms do what they do not for any nativist reasons, but because of their being shaped for maximal invariance by contingent evolutionary processes—which is of course more context. Part of how this works is that such conserved mechanisms carry with them the internally relevant context that allows them to function. It is contextual emergence that allows complex multiscale networks to maximize robustness and autonomy and thus adaptability.

To believe that wherever in the universe one finds certain cognitive functions or mental contents, one will also find literally "the same" underlying mechanism, is a strange piece of metaphysical magical thinking that has nothing to do with science so far as I can tell. What kind of laws would this be? Why should one believe in such brute identities? Again, these are empirical questions. It could happen that after intense analysis of all the different animals on Earth including cephalopods, birds, etc., and actual alien species, that we will find some real mechanistic invariance across all of it. However, if one had to place a bet now, the most universal, powerful and explanatory invariances would not be specific mechanisms, but topological networks and other global organizing principles.

Once again, and this is crucial, per contextual emergence and the multiscale network picture painted herein, the debate about which lower-level or smaller-scale neural mechanisms subserve particular cognitive functions simply misses the point. In such networks the explanation for various cognitive functions and mental contents will be multiscale and the determination in such networks will also be multiscale. This and all the preceding leads me to seriously doubt that there is any strong argument based in neural mechanisms against MR properly understood. Given this multiscale network picture, MR exists and is to be expected. Of course, maybe there are real brute physical or metaphysical identity relations between specific cognitive functions/mental states and specific neural mechanisms. But again, this strikes me as a decidedly spooky and unscientific idea, to really believe for example that congenital brains-in-a-vat or Boltzmann brains would be having thoughts and experiences exactly like our own is beyond me. How is this any less spooky than the idea of brute psychophysical bridge-laws as espoused by strong emergentists? Indeed, how is it any *different* than that idea? After all, even dreaming brains, brains in sensory deprivation tanks, and brains in deep meditation produce very different contents and experiences.

If after all this one still insists that mechanistic reduction or some other form of ontological reductionism is the right picture here, then I can only imagine two reasons for that. Either they think I have the facts wrong, or they have a vastly different idea of mechanistic or ontological reductionism than defined herein. Whichever the case, I eagerly await their reply. For my part, I think it's already clear that evolution and history are on the side of contextual emergence.

Bibliography

Alexander, S. (1920). Space, time, and deity: The gifford lectures at glasgow 1916–1918, Vols. 1 and 2. Macmillan.

Allefeld, C., Atmanspacher, H., & Wackermann, J. (2009). Mental states as macrostates emerging from EEG dynamics. *Chaos, 19*, 015102.

Anderson, P. W. (1972). More Is different. *Science, 177*, 393–396.

Anderson, P. W. (2000). Brainwashed by Feynman? *Physics Today., 53*(2), 11–14.

Anderson, P. W. (2011). *More and different: Notes from a thoughtful curmudgeon.* World Scientific.

Anderson, M. L. (2010). Neural reuse: A fundamental organizational principle of the brain. *Behavioral and Brain Sciences, 33*, 245–313.

Anderson, M. L. (2014). *After phrenology: Neural reuse and the interactive brain.* MIT Press.

Anderson, M. L. (2016). "Précis of after phrenology: Neural reuse and the interactive brain behavioral and brain sciences". *39*, 1–22.

Anderson, M.L. (2020). "Neural reuse: A fundamental organizational principle of the brain." 2010 Aug; *33*(4), 245–66; discussion 266–313. https://doi.org/10.1017/S0140525X10000853.

Atasoy, S., Roseman, L., Kaelen, M., et al. (2017). Connectome-harmonic decomposition of human brain activity reveals dynamical repertoire re-organization under LSD. *Science and Reports, 7*, 17661. https://doi.org/10.1038/s41598-017-17546-0

Atmanspacher, H., & Graben, P. B. (2007). Contextual emergence of mental states from neurodynamics. *Chaos and Complexity Letters, 2*(2/3), 151–168.

Bassett, D.S., & Sporns, O. (2017). "Network Neuroscience." *Nature Neuroscience, 20*(3). https://doi.org/10.1038/nn.4502.

Bassett, D. S., Zurn, P., & Gold, J. I. (2018). On the nature and use of models in network neuroscience. *Nature Reviews Neuroscience, 19*, 566–578. https://doi.org/10.1038/s41583-018-0038-8

Batchelor, G. (1967). *An introduction to fluid dynamics.* Cambridge University Press.

Bateson, P., & Gluckman, P. (2011). Plasticity, robustness, development and evolution. CUP.

Batterman, R. W., & Rice, C. C. (2014). Minimal model explanations. *Philosophy of Science, 81*(3), 349–376.

Batterman, R. (2015). "Autonomy and scales: In why more is different. Philosophical issues in condensed matter physics and complex systems." Springer. 115–135.

Baumgartner, M., & Gebharter, A. (2015). Constitutive relevance, mutual manipulability, and fat-handedness. *The British Journal for the Philosophy of Science.* https://doi.org/10.1093/bjps/axv003

Baumgartner, M., & Casini, L. (2017). An abductive theory of constitution. *Philosophy of Science, 84*(2), 214–233.
Bechtel, W. (2010). “Dynamic mechanistic explanation: Computational modeling of circadian rhythms as an exemplar for cognitive science.” *Studies in History and Philosophy of Science.*
Bechtel, W., & Richardson, R. C. (2010). *Discovering complexity: Decomposition and localization as strategies in scientific research* (2nd ed.). MIT Press.
Bechtel, W. (2011). Mechanism and biological explanation. *Philosophy of Science, 78*(4), 533–558.
Bechtel, W., & Abrahamsen, A. (2011). “Complex biological mechanisms: Cyclic, oscillatory, and autonomous.” In C. A. Hooker (Ed.) *Philosophy of complex systems.*
Bechtel, W., & Abrahamsen, A. (2013). Thinking dynamically about biological mechanisms: Networks of coupled oscillators. *Foundations of Science, 18*, 707–723.
Bechtel, W. (2017). “Explicating top-down causation using networks and dynamics.” *Philosophy of Science.*
Bechtel, W. (2017). “Top-down causation in biology and neuroscience: Control hierarchies.” In: M.P. Paolini, & F. Orilia (Eds.) *Philosophical and scientific perspectives on downward causation.* Routledge.
Bechtel, W. (2017). Systems biology: Negotiating between holism and reductionism. In S. Green (Ed.). *Philosophy of systems biology: Perspectives from scientists and philosophers.* Springer.
Bechtel, W. (2018). The importance of constraints and control in biological mechanisms: Insights from cancer research. *Philosophy of Science, 85*(4), 573–593.
Bechtel, W. (2019). Analysing network models to make discoveries about biological mechanisms. *The British Journal for the Philosophy of Science, 70*(2), 459–484.
Bechtel, W. (2020). Hierarchy and levels: Analysing networks to study mechanisms in molecular biology. *Phil Trans r Soc B, 375*, 20190320. https://doi.org/10.1098/rstb.2019.0320
Becker, A. (2018). *What is real? The unfinished quest for the meaning of quantum physics.* Basic Books.
Bedau, M. (2008). Is weak emergence just in the mind? *Minds and Machines, 18*, 443–459.
Bedau, M., & Humphreys, P. (Eds.). (2008). *Emergence: Contemporary readings in philosophy and science.* MIT Press.
Bedau, M. (2010). “Two varieties of causal emergentism.” In A. Corradini, & T. O’Connor (Eds.) *Emergence in science and philosophy. Routledge studies in the philosophy of science,* (Vol. 14, pp. 46–63) Routledge.
Bertolero, M.A., & Bassett, D.S. (2020). “On the nature of explanations offered by network science: A perspective from and for practicing neuroscientists.” *Topics in Cognitive Science* (2020) 1–22. https://doi.org/10.1111/tops.12504.
Biology and Philosophy (2007). bioRxiv preprint. Sep. 4, 2018; 22,547–563. http://dx.doi.org/https://doi.org/10.1101/408278.
Bishop, R. C. (2005). Patching physics and chemistry together. *Philosophy of Science, 72*, 710–722.
Bishop, R. C., & Atmanspacher, H. (2006). Contextual emergence in the description of properties. *Foundations of Physics, 36*, 1753–1777.
Bishop, R. C. (2008). What could be worse than the butterfly effect? *The Canadian Journal of Philosophy, 38*, 519–548.
Bishop, R. C. (2008). Downward causation in fluid convection. *Synthese, 160*, 229–248.
Bishop, R. C. (2011). Metaphysical and epistemological issues in complex systems. In C. Hooker (Ed.) Philosophy of complex systems, Volume 10 of Handbook of the Philosophy of Science, 119–150.
Bishop, R. C. (2012). Fluid convection, constraint and causation. *Interface Focus, 2*, 4–12.
Bishop, R., Silberstein, M. (2019). Complexity and feedback. In *The Routledge handbook of emergence.* In S. Gibb, R. Hendry, & T. Lancaster. Routledge.
Bishop, R., Silberstein, M., & Pexton, M. (forthcoming). Emergence in context. Oxford University Press.
Bishop, R.C., & Silberstein, M. (2019). “Emergence and complexity.” *Routledge handbook of emergence*, in S. Gibb, R. F. Hendry, & T. (Eds.) (pp. 145–156). Lancaster.

Bliss, R., & Priest, G. (2018). The geography of fundamentality: An overview. In *Reality and its structure: Essays in fundamentality* Bliss & Priest (Eds.). Oxford University Press.

Boi, L. (2017). The interlacing of upward and downward causation in complex living systems: on interactions, self-organization, emergence and wholeness. In M. P. Paolini & F. Orilia (Eds.), *Philosophical and scientific perspectives on downward causation.* Routledge.

Brigandt, I. (2013). Systems Biology and the integration of mechanistic explanation and mathematical explanation. *Ingo Brigandt—2013—Studies in History and Philosophy of Biological and Biomedical Sciences 44*(4), 477–492.

Brigandt, I., Green, S., O'Malley, M. (2018). Systems biology and mechanistic explanation. *Ingo brigandt, sara green and maureen O'Malley—2018—*, in S. Glennan & P.M. Illari (Eds.), *The Routledge handbook of mechanisms and mechanical philosophy* (pp. 362–374). Routledge.

Broad, C.D. (1925). The mind and its place in nature, London: Routledge and Kegan Paul, 1st ed.

Burnston, D. C. (2016a). "Computational neuroscience and localized neural function." *Synthese,* 1–22. https://doi.org/10.1007/s11229-016-1099-8.

Burnston, D. C. (2016b). "A contextualist approach to functional localization in the brain." Biology and Philosophy, 1–24. https://doi.org/10.1007/s10539-016-9526-2.

Burnston, D. C. (In preparation-c). "Getting over atomism: Functional decomposition in complex neural systems."

Busse, F. (1978). Non-linear properties of thermal convection. *Reports on Progress in Physics, 41,* 1929–1967.

Cacioppo, S., & Cacioppo, J.T. (2020). "Introduction to social neuroscience." Princeton University Press, School ed.

Calzavarini, F., & Viola, M. (2020). "Neural mechanisms: New challenges in the philosophy of neuroscience (Studies in Brain and Mind, 17)." Springer, 1st ed.

Casini, L. (2016). How to model mechanistic hierarchies. *Philosophy of Science, 85*(5), 946–958.

Caston, V. (1997). Epiphenomenalisms, ancient and modern. *Philosophical Review, 106*(3), 309–363.

Caston, V. (2001). "Dicaearchus' philosophy of mind". In Fortenbaugh & Schu¨trumpf, pp. 175–93.

Cecil, C.A.M. (2020). "Biopsychosocial pathways to mental health and disease across the lifespan: The emerging role of epigenetics." In J. Savulescu, R. Roache, & W. Davies (Eds.) *Psychiatry reborn: Biopsychosocial psychiatry in modern medicine.* 381–404. 190–206.

Chalmers, D. (1996). *The conscious mind. In search of a fundamental theory.* Oxford University Press.

Chalmers, D. (2006). Strong and weak emergence. In P. Clayton & P. C. W. Davies (Eds.) *The re-emergence of emergence,* (pp. 244–56). Oxford University Press.

Chemero, A., & Silberstein, M. (2008). After the philosophy of mind: Replacing scholasticism with science. *Philosophy of Science, 75*(1), 1–27.

Colombo, M., & Weinberger, N. (2018). Discovering Brain Mechanisms Using Network Analysis and Causal Modeling. *Minds and Machines, 28*(2), 265–286. https://doi.org/10.1007/s11023-017-9447-0

Craver, C. Bechtel, W. (2007). "Top-down causation without top-down causes."

Craver, C. F. (2007). *Explaining the brain: Mechanisms and the mosaic unity of neuroscience.* Oxford University Press.

Craver, C. F. (2014). "The ontic account of scientific explanation. Explanation in the Special Sciences" (pp. 27–52). Springer.

Craver, C. F. (2016). "The explanatory power of network models." Philosophy of Science (forthcoming).

Cross, M., & Hohenberg, P. (1993). Pattern formation outside of equilibrium. *Reviews of Modern Physics, 65,* 851–1112.

Damicelli, F., Hilgetag, C. C., Hütt, M. T., & Messé, A. (2019). Topological reinforcement as a principle of modularity emergence in brain networks. *Network Neuroscience, 3*(2), 589–605. https://doi.org/10.1101/408278

Duhem, P. (1954). *The aim and structure of physical theory.* Princeton University Press.

Ehrenfest, P. (1917). *Proc Amsterdam Acad, 20*, 200.
Euler. L. (1736). "Solutio problematis and geometriam situs pertinentis." *Comment Acad Sci U Petrop 8*, 128–40.
Favela, L.H. (2019). "Integrated information theory: A complexity science approach to consciousness." *Journal of Consciousness Studies 26*(1–2), 21–47(27)
Favela, L.H. (2020). "The dynamical renaissance in neuroscience." Synthese. September 2020. https://doi.org/10.1007/s11229-020-02874-y.
Fazekas, P., & Kertesz, G. (2018). "Are higher mechanistic levels causally autonomous?" In: [2018] PSA 2018: The 26th Biennial meeting of the philosophy of science association. 1–4 Nov 2018.
Feldman Barrett, L. (2020). *Seven and a half lessons about the brain.* Houghton Mifflin Harcourt.
Feldt Muldoon, S., & Bassett, D. S. (2016). Network and multilayer network approaches to understanding human brain dynamics. *Philosophy of Science, 83*(5), 710–720.
Fox Keller, E. (2010). The mirage of a space between nature and nurture. Duke University Press.
Francis. R.C. (2011). The ultimate mystery of inheritance: Epi-Genetics. W. W. Norton and Company.
Ganeri, J. (2011). Emergentisms, ancient and modern. *Mind, 120*, 671–703.
Gilbert, S., & David, E. (2009). *Ecological developmental biology: Integrating epigenetics medicine and evolution.* Sinauer Associates, Inc.
Gillett. C. (2010). "Weak emergence and context-sensitive reduction." In A. Corradini, & T. O'Connor (Eds) *Emergence in science and philosophy. Routledge Studies in the Philosophy of Science* (Vol. 314, pp. 25–46). Routledge.
Gillett, C. (2016). *Reduction and emergence in science and philosophy.* Cambridge University Press.
Glennan, S. (2016). Mechanisms and mechanical philosophy. The Oxford of philosophy of science, P. Humphreys (Ed.). Chp. 38.
Glennan, S. (2017). *The new mechanical philosophy.* Oxford University Press.
Glennan, S., & Illari, P. (2018). The Routledge handbook of the philosophy of mechanisms, in S. Glennan & P. Illari (Eds.). Routledge.
Godfrey-Smith, P. (2020). *Metazoa: Animal life and the birth of the mind.* Farrar, Straus and Girouxs.
Green, S., Serban, M., Scholl, R., Jones, N., Brigandt, I., & Bechtel, W. (2018). Network analyses in systems biology: New strategies for dealing with biological complexity. *Synthese, 195*(4), 1751–1777.
Hart, C. L. (2021). *Drug use for grown ups: Chasing liberty in the land of fear*. Penguin Press.
Hayashi, Y. Liu, C. (2017). "The evolution and function of sleep." In S. Shigeno, Y. Murakami, & T. Nomura (Eds.) *Brain evolution by design: From neural origin to cognitive architecture* (pp. 243–366). Springer.
Handbook of the philosophy of science. (2012). Vol. 10, pp. 275–285, Elsevier.
Hilgetag. C.C., & Goulas, A. (2015). "Is the brain really as small-world network?".
Hilgetag, C. C., & Goulas, A. (2020). 'Hierarchy' in the organization of brain networks. *Phil. Trans. r. Soc. B, 375*, 20190319. https://doi.org/10.1098/rstb.2019.0319
Hoekstra, A., Chopard, B., & Coveney, P. (2014). Multiscale modelling and simulation: A position paper. *Philosophical Transactions of the Royal Society A, 372*, 20130377. https://doi.org/10.1098/rsta.2013.0377
Hoffmann-Kolss, V. (2014). Interventionism and higher-level causation. *International Studies in the Philosophy of Science, 28*(1), 49–64.
Holt, J. (2012). "Physicists, stop the churlishness". New York Times Online.
Hooker, C (2011). "Conceptualising Reduction, Emergence and Self-Organization in Complex Dynamical Systems." Philosophy of Complex Systems, Hooker editor. Elsevier. 195–222.
Huang, Z., Zhang, J., Wu, J., Mashour, G. A., Hudetz, A. G., & Hütteman, A. (2005). Explanation, Emergence, and Quantum Entanglement. *Philosophy of Science, 72*, 114–127.
Huang, Z., Zhang, J., Wu, J., Mashour, G.A., & Hudetz, A.G. (2020)."Temporal circuit of macroscale dynamic brain activity supports human consciousness." *Cognitive Neuroscience 6*(11). https://doi.org/10.1126/sciadv.aaz0087

Humphreys, P. (1997). How properties emerge. *Philosophy of Science, 64*, 1–17.
Humphreys, P. (2016). *Emergence. A philosophical account*. Oxford University Press.
Huneman, P. (2008). "Emergence made ontological? Computational versus combinatorial approaches." Philosophy of science/ Vol. 75, No. 5, Proceedings of the 2006 Biennial Meeting of the Philosophy of Science Association Part II: Symposia Papers, C. Bicchieri, & J. Alexander (Dec=c 2008), pp. 595–607.
Huneman, P. (2010). Topological explanations and robustness in biological sciences. *Synthese, 177*, 213–245.
Huneman, P. (2018a). Diversifying the picture of explanations in biological sciences: Ways of combining topology with mechanisms. *Synthese, 195*, 115–146.
Huneman, P. (2018b). Outlines of a theory of structural explanations. *Philosophical Studies., 175*, 665–702.
Hüttemann, A., & Love, A. C. (2011). Aspects of reductive explanation in biological science: Intrinsicality, fundamentality, and temporality. *British Journal for Philosophy of Science, 62*, 519–549.
Hüttemann, A., & Love, A. C. (2016). Reduction. In P. Humphries (Ed.), *The Oxford handbook of philosophy of science* (pp. 460–484). Oxford University Press.
Jablonka, E., & Lamb, M. (2005). *Evolution in four dimensions: Genetic, epigenetic, behavioral, and symbolic variation in the history of life*. MIT Press.
Jaeger, L., & Calkins, E. R. (2012). Downward causation by information control in micro-organisms. *Interface Focus, 2*, 26–41.
Jansson, L. (2020). Network explanations and explanatory directionality. *Phil Trans r Soc B, 375*, 20190318. https://doi.org/10.1098/rstb.2019.0318
Kanaya, H et al (2020). "A sleep-like state in *Hydra* unravels conserved sleep mechanisms during the evolutionary development of the central nervous system." *Science Advances. 6*(41), eabb9415. https://doi.org/10.1126/sciadv.abb9415.
Kaiser, M. (2020). *Changing connectomes: Evolution, development, and dynamics in network neuroscience*. The MIT Press.
Kaiser, M. I., & Krickel, B. (2017). The metaphysics of constitutive mechanistic phenomena. *The British Journal for the Philosophy of Science, 68*(3), 745–779. https://doi.org/10.1093/bjps/axv058
Kaplan, D.M. (2018). "Mechanics and dynamical explanation." The Routledge handbook of mechanisms and mechanical philosophy. S. Glennan (Ed.), pp. 267–80.
Kaplan, J. (2008). "Review of genes in development: Rereading the molecular paradigm." M. Eva (Ed.) *Neumann-held and christoph rehmann-sutter*. Biological Theory 2, 427–429.
Kellert, S. H., Longino, H. E., & Waters, C. K. (2006). *Scientific pluralism*. University of Minnesota Press.
Kershenbaum, A. (2020). *The zoologist's guide to the galaxy: What animals on earth reveal aliens—and ourselves*. Penguin Press.
Kim, J. (1998). *Mind in a physical world*. The MIT Press.
Kim, J. (1999). Making sense of emergence. *Philosophical Studies, 95*, 3–36.
Kim, D. J., et al. (2019). Childhood poverty and the organization of structural brain connectome. *NeuroImage, 184*, 409–416. https://doi.org/10.1016/j.neuroimage.2018.09.041 Epub 2018 Sep 17.
Koonin, E.V. (2011). *The logic of chance: The nature and origin of biological evolution*, E.V. Koonin (Ed.) p. 528. FT Press.
Kostic, D. (2018).The topological realization. Synthese 195 (1).
Kostic, D. (2020). General theory of topological explanations and explanatory asymmetry. *Phil Trans r Soc B, 375*, 20190321. https://doi.org/10.1098/rstb.2019.0321
Kronz, F., & Tiehen, J. (2002). Emergence and quantum mechanics. *Philosophy of Science, 69*(2), 324–347.
Ladyman, J. (2017). "An apology for naturalised metaphysics."
Laughlin, R. (2005). *A different universe: Reinventing physics from the bottom down*. Basic Books.

Laughlin, R. B., & Pines, D. (2000). *The theory of everything.* Stanford University.
Levy, N. (2013). "Addiction is Not a Brain Disease." *Front Psychiatry.* 4, 24. Published online 2013 Apr 11. https://doi.org/10.3389/fpsyt.2013.00024
Levy, A., & Bechtel, W. (2016). "Towards mechanism 2.0: Expanding the scope of mechanistic explanation." In: [2016] PSA 2016: The 25th biennial meeting of the philosophy of science association (Atlanta, GA; 3–5 November 2016) <http://philsci-archive.pitt.edu/view/confandvol/confandvol2016PSA.html>. http://philsci-archive.pitt.edu/id/eprint/12567.
Love, A. C. (2012). Hierarchy, causation and explanation: Ubiquity, locality, and pluralism. *Interface Focus, 2*, 115–125.
Love, A. C. (2018). Developmental mechanisms. In S. Glennan & P. Illari (Eds.), *The Routledge handbook of the philosophy of mechanisms and mechanical philosophy* (pp. 332–347). Routledge.
Love, A. C., & Hüttemann, A. (2011). Comparing part-whole explanations in biology and physics. In D. Dieks, W. J. Gonzalez, S. Hartmann, T. Uebel, & M. Weber (Eds.), *Explanation, prediction, and confirmation* (pp. 183–202). Springer.
Mackie, J. L. (1965). Causes and conditions. *American Philosophical Quarterly, 2*, 245–264.
Maillé, S., & Lynn, M. (2020). Reconciling current theories of consciousness. *Journal of Neuroscience, 40*(10), 1994–1996. https://doi.org/10.1523/JNEUROSCI.2740-19.2020
Markov, N. T. Ercsey-Ravasz, M., Van Essen, D.C., & Knoblauch, K. (2013). "Cortical high-density counterstream architectures." *Science 342* (578).
Massimini, M., & Tononi, G. (2018). Sizing up consciousness. Oxford University Press.
Massimo, P. "Between holism and reductionism: A philosophical primer on emergence." *Biological Journal of the Linnean Society 112*(2), 261–267 https://doi.org/10.1111/bij.12060.
Matthiessen, D. (2017). Mechanistic Explanation in Systems Biology: Cellular Networks. *The British Journal for the Philosophy of Science, 68*, 1–25.
McConnell. D. (2020). "Specifying the best conception of the biopsychosocial model." In *Psychiatry reborn: Biopsychosocial psychiatry in modern medicine*, J. Savulescu, R. Roache, & W. Davies (Eds), pp. 381–404.
McNally, R. J. (2016). Can network analysis transform psychopathology? *Behaviour Research and Therapy, 86*, 95–104.
Moreno, A., Ruiz-Mirazo, K., & Barandiaran, X. (2011). The impact of the paradigm of complexity on the foundational frameworks of biology and cognitive science. In *Philosophy of Complex Systems*, Hooker (Ed.) (pp. 311–333), Elsevier.
Morgan, L. (1923). *Emergent evolution.* Williams and Norgate.
Moroz, L., et al. (2014). The ctenophore genome and the evolutionary origins of neural systems. *Nature, 510*, 109–114.
Nand, A., et al (2021). Genetic and spatial organization of the unusual chromosomes of the dinoflagellate Symbiodinium microadriaticum, *Nature Genetics* (2021). https://doi.org/10.1038/s41588-021-00841-y. Salazar quoted in.
Newen, A., de Bruin, A., Gallagher, S. (2018). Introduction. *The Oxford handbook of 4E cognition.* A. Newen, A. de Bruin, & S. Gallagher (Eds.) Oxford University Press.
Morrison, M. (2015). *Reconstructing reality: Models, mathematics, and simulations.* Oxford University Press.
Murphy, D. (2020). "Philosophy of psychiatry". Murphy, Dominic, "Philosophy of psychiatry", *The stanford encyclopedia of philosophy* (Fall 2020 Edition), E.N. Zalta (ed.), https://plato.stanford.edu/archives/fall2020/entries/psychiatry/.
Nagel. E. (1961). The structure of science: Problems in the logic of scientific explanation. Harcourt, Brace, and World.
Niquil, N., Haraldsson, M., Sime-Ngando, T., Huneman, P., & Borrett, S. R. (2020). Shifting levels of ecological network's analysis reveals different system properties. *Phil Trans r Soc B, 375*, 20190326. https://doi.org/10.1098/rstb.2019.0326
O'Connor, T. (1994). Emergent properties. *American Philosophical Quarterly., 31*(2), 91–104.
O'Connor, T., & Wong, H. (2005). The metaphysics of emergence. *Noûs, 39*, 658–678.

O'Malley, M. A., Brigandt, I., Love, A. C., Crawford, J. W., Gilbert, J. A., Knight, R., Mitchell, S. D., & Rohwer, F. (2014). Multilevel research strategies and biological systems. *Philosophy of Science, 81*, 811–828.

Paolini, M.P., & OriliA, F. (2017). *Philosophical and scientific perspectives on downward causation.* Routledge.

Parrington, J. (2021). *Mind shift: How culture transformed the human brain.* Oxford University Press.

Paul, M., Chiam, K.-H., Cross, M., & Greenside, H. (2003). Pattern formation and dynamics in rayleigh-bénard convection: Numerical simulations of experimentally realistic geometries. *Physica d: Nonlinear Phenomena, 184*, 114–126.

Philosophy of Science, 83 (December 2016) 674–685.

Pigliucci, M. (2014). Between holism and reductionism: A philosophical primer on emergence. *Biological Journal of the Linnean Society., 112*, 261–267.

Piccinini, G. (2020). *Neurocognitive mechanisms: Explaining biological cognition.* University Press.

Povich, M., & Craver, C. F. (2018). Mechanistic levels, reduction, and emergence. M. Povich & C. F. Craver—forthcoming—In S. Glennan & P.M. Illari (Eds.) *The Routledge handbook of mechanisms and mechanical philosophy.* Routledge.

Primas, H. (1977). Theory reduction and non-Boolean theories. *Journal of Mathematical Biology, 4*, 281–301.

Raja, V., & Anderson, M.L. (2021). "Behavior considered as an enabling constraint." In *Neural mechanisms: New challenges in the philosophy of neuroscience.* Springer Neuroscience, F. Calzavarini & M. Viola (Eds.), Chapter 9.

Rathkopf, C. (2018). Network representation and complex systems. *Synthese, 2018*(195), 55–78.

Rescher, N. (1993). *Pluralism.* Clarendon Press.

Richerson, P.J., Boyd, R., & Henrich, J. (2010). "Gene-culture coevolution in the age of genomics." PNAS May 11, 2010. 107 (Supplement 2), 8985–8992, first published May 5, 2010, https://doi.org/10.1073/pnas.0914631107.

Ross, L. (2015). "Dynamical Models and Explanation in Neuroscience." Phil of sci, 82, jan 2015, 32–54.

Ruiz-Mirazo, K., & Moreno, A. (2012). Autonomy in evolution: From minimal to complex life. *Synthese, 185*(1), 21–52.

Ruiz-Mirazo, K., Moreno, A. (2012). *Autonomy in evolution: From minimal to complex life*

Safron, A. (2020). An integrated world modeling theory (IWMT) of consciousness: Combining integrated information and global neuronal workspace theories with the free energy principle and active inference framework; toward solving the hard problem and characterizing agentic causation. *Front Artif Intell, 3*, 30. https://doi.org/10.3389/frai.2020.00030

Scheinin, A. (2020). Foundations of human consciousness: Imaging the twilight zone. *Journal of Neuroscience.* https://doi.org/10.1523/JNEUROSCI.0775-20.2020

Schirner, M., McIntosh, A. R., Jirsa, V., Deco, G., & Ritter, P. (2018). Inferring multi-scale neural mechanisms with brain network modelling. *eLife, 7*, e28927. https://doi.org/10.7554/eLife.28927.001

Serban, M. (2020). Exploring modularity in biological networks. *Phil Trans r Soc B, 375*, 20190316. https://doi.org/10.1098/rstb.2019.0316

Shine, J. M., Breakspear, M., Bell, P. T., Ehgoetz, K. A., Richard Shine, M., Koyejo, O., Sporns, O., & Poldrack, R. A. (2019). Human cognition involves the dynamic integration of neural activity and neuromodulatory systems. *Nature Neuroscience., 22*, 289–296.

Silberstein, M. (1998). Emergence and the mind-body problem. *Journal of Consciousness Studies, 5*(4), 464–482.

Silberstein, M., & McGeever, J. (1999). The search for ontological emergence. *Philosophical Quarterly, 49*, 201–214.

Silberstein, M. (2001) "Converging on emergence: Consciousness, causation and explanation" *Journal of Consciousness Studies 8*(9–10), 61–98. Special issue: *The Emergence of Consciousness.*

Silberstein, M., (2002). "Reduction, emergence, and explanation", in *The Blackwell guide to the philosophy of science*, Machamer, P., & M., Silberstein (Eds.) (pp. 203–226), Blackwell.

Silberstein, M. (2006). "In defense of ontological emergence and mental causation", in *The re-emergence of emergence*, P. Davies (Ed.), Chapter 9 (Oxford University Press).

Silberstein, M. (2009a). "Quantum Nonseparability and Mereology" in Philosophia Verlag Handbook of Mereology. Seibt and Burkhard, Editors.

(2009b) "Emergence and consciousness" in *Oxford companion to consciousness.* T. Bayne, A. Cleeremans, & P. Wilken (Eds.) Oxford University Press.

Silberstein, M. (2012). "Emergence and reduction in context: Philosophy of science and/or analytic metaphysics", *Metascience.* https://doi.org/10.1007/s11016-012-9671-4.

Silberstein, M., & Chemero, A. (2012). "Complexity and extended phenomenological-cognitive systems" in G. Van Orden & D. Stephen (Eds.) *Topics in cognitive science: Special issue on the role of complex systems in cognitive science.*

Silberstein, M., & Chemero, A. (2013). Constraints on localization and decomposition as explanatory strategies in the biological sciences. *Philosophy of Science, 80*(5), 958–970.

Silcerstein. M. (2014). "Dynamics, systematicity and extended cognition" in P. Calvo & J. Symons (Eds.) *Systematicity and the post-connectionist era.* MIT Press.

Silberstein, M. (2016). The implications of neural reuse for the future of cognitive neuroscience and the future of folk psychology. *Brain and Behavioral Sciences, 39*, E132.

Silberstein, M. (2018). "Contextual emergence." Special issue of philosophica on emergence *91*, 45–92. A.D. Carruth, & J.T.M. Miller (Eds.).

Silberstein. M. (2021). "Constraints on localization and decomposition as explanatory strategies in the biological sciences 2.0." In F. Calzavarini, M. Viola (Eds.) Neural mechanisms: New challenges in the philosophy of neuroscience. Springer neuroscience.

Slater, M.H., & Yudell, Z. (2017). Metaphysics and the philosophy of science: New essay. Kindle Locations (pp. 3491–3492). Oxford University Press. Kindle Edition.

Spivey, M.J. (2020). Who you are: The science of connectedness. The MIT Press.

Solé, R., & Valverde, S. (2020). Evolving complexity: How tinkering shapes cells, software and ecological networks. *Phil Trans r Soc B, 375*, 20190325. https://doi.org/10.1098/rstb.2019.0325

Stephan, A. (1992). Emergence—A systematic view on its historical aspects, in A. Beckermann et al. (eds.), pp. 25–47.

Stinson, C. (2016). Mechanisms in psychology: Ripping nature at its seams. *Synthese, 193*(5), 1585–1614. https://doi.org/10.1007/s11229-015-0871-5

Thalos, M. (2013). *Without hierarchy: The scale freedom of the universe.* Oxford University Press.

Van Gulick, R. (2001). Reduction, emergence and other recent options on the mind/body problem: A philosophic overview. *Journal of Consciousness Studies, 8*, 1–34.

Venturelli, N. A. (2016). A cautionary contribution to the philosophy of explanation in the cognitive neurosciences A. *Nicolás Venturelli Minds and Machines, 26*(3), 259–285.

Weinan, E. (2011). *Principles of multiscale modeling.* Cambridge University Press.

Weinan, E., & Jianfeng, L. (2011). Multiscale modeling. *Scholarpedia, 6*(10), 11527.

Weinberg, S. (1995). Reductionism redux. The New York review of books. Reprinted in Weinberg, Steven. 2001. Facing Up. Harvard University Press.

Weinberg, S. (2014). As quoted in, conceptual foundations of quantum theory. T.Y. Cao (Ed.) p. 260.

Weinberg, S. (2017). "The trouble with quantum mechanics." http://quantum.phys.unm.edu/466-17/QuantumMechanicsWeinberg.pdf.

Weiskopf, D. A. (2016). "Integrative modeling and the role of neural constraints." *Philosophy of Science, 83*(December 2016), 674–685.

Wilson, J. (2011). Non-reductive physicalism and degrees of freedom (2010). *British Journal for Philosophy of Science, 61*, 279–311.

Wilson, J. (2013). A determinable-based account of metaphysical indeterminacy (2013). *Inquiry, 56*, 359–385.
Wilson, J. (2015). "Metaphysical emergence: Weak and strong." In T. Bigaj & C. Wüthrich (Eds.), Metaphysics in contemporary physics. Leiden-Boston, pp. 345–402.
Winning, J., & Bechtel, W. (2018). Rethinking causality in biological and neural mechanisms: Constraints and control. *Minds and Machines, 28*(2), 287–310.
Winning, J. (2020). Mechanistic causation and constraints: Perspectival parts and powers, non-perspectival modal patterns. *Brit J Phil Sci, 71*(2020), 1385–1409.
Wolf, Y. I., Karev, G., & Koonin, E. V. (2002). Scale-free networks in biology: New insights into the fundamentals of evolution? *BioEssays, 24*(2), 105–109.
Woodward, J. (2003). *Making things happen.* Oxford University Press.
Yamashita, W., Nomura, T. (2017). "The Neocortex and dorsal ventricular ridge: Functional convergence and underlying developmental mechanisms." In *Brain evolution by design: From neural origin to cognitive architecture*, S. Shigeno, Y. Murakami, T. Nomura (Eds.) (pp. 291–310), Springer.
Zednik, C. (2011). The nature of dynamical explanation. *Philosophy of Science, 78*(2), 238–263.
Zednik, C. (2014). Are systems neuroscience explanations mechanistic? *Preprint volume for philosophy science association 24th biennial meeting* (pp. 954–975). Philosophy of Science Association.
Zednik. C. (2015). "Heuristics, descriptions, and the scope of mechanistic explanation." In *Explanation in biology*, (pp. 295–318), Springer.
Zednik, C. (2018). Models and mechanisms in network neuroscience. *Philosophical Psychology, 32*(1), 23–51.
Zerilli. J. (2020). The adaptable mind: What neuroplasticity and neural reuse tell us about language and cognition. Oxford University Press.
Zbili, M., Rama, S., & Debanne, D. (2016). "Dynamic control of neurotransmitter release by presynaptic potential." *Front Cell Neurosci 10*, vv278. Published online 2016 Dec 5. https://doi.org/10.3389/fncel.2016.00278
Zhou, Y., et al. (2021). A widespread pathway for substitution of adenine by diaminopurine in phage genomes. *Science. 372*(April 30), 512. https://doi.org/10.1126/science. Abe 4882.
Zimmer, C. (2018). *She has her mother's laugh: The powers.* Dutton Press.
Zurn, P., & Bassett, D. S. (2020). Network architectures supporting learnability. *Phil Trans r Soc B, 375*, 20190323. https://doi.org/10.1098/rstb.2019.0323

From Electrons to Elephants: Context and Consciousness

Michael Tye

John Donne said, "Nature's great masterpiece, an elephant; the only harmless great thing." What is needed to build such a masterpiece?

We know that elephants, like everything else in the natural world, are made up of sub-atomic particles. Some philosophers would say that you can build one out of a very large boulder. You just have to take the boulder apart down to its sub-atomic particles and re-arrange them suitably. But would the resulting entity really be an elephant or just a microphysical duplicate of an elephant, indistinguishable in itself from a real elephant? Further, are elephants literally one and the same as aggregates of sub-atomic particles? Or are they something more? What is needed to endow elephants with consciousness? These are the questions addressed in this essay. The bulk of the essay concerns the last question. A partial answer is offered. Along the way, I make some remarks about the so-called 'hard problem of consciousness' (Chalmers, 1995a, 1995b).

1 History Matters

Elephants belong to a particular biological species. What makes something an elephant, in my view, is not just how it is internally but its biological history. You can't really make an elephant simply by putting together a fantastically complex arrangement of sub-atomic particles that together constitute the putative elephant. Not even God can do that. The wider context matters too. The arrangement of sub-atomic particles has to have arisen naturally in the right way. History matters.

M. Tye (✉)
Department of Philosophy, University of Texas at Austin, 2210 Speedway, Stop C3500, WAG 316, Austin, TX 78712-1737, USA
e-mail: mtye@austin.utexas.edu

S. Wuppuluri and I. Stewart (eds.), *From Electrons to Elephants and Elections*, The Frontiers Collection, https://doi.org/10.1007/978-3-030-92192-7_34

Here's an argument for this view. Suppose that you go in your spaceship to planet Mercurius. As you walk around in the oxygen rich environment there, you see lying on the ground something that perfectly resembles a Honda motorbike speedometer. Is it really a Honda speedometer? No Honda factory exists there. Indeed you are the first human on Mercurius and you didn't bring a speedometer with you. As you explore further and meet the locals, you find that there is a company on Mercurius with the name 'Honda' that makes oil pressure gauges physically indistinguishable from Honda speedometers. The thing you found is really one of these. What makes it such is its history: it was designed to measure oil pressure. That's why it is really an oil pressure gauge and not a Honda speedometer. What goes for artificial design goes for natural design too, or so it seems plausible to suppose. If you wander around Mercurius and come across creatures that look exactly like elephants, it doesn't follow that they are elephants. If they are products of Mercurius without any historical connection with Earth, they belong to another species, one not found on earth though superficially resembling our elephant species. Melephants are not elephants.

So, this is my first point: an elephant can't be built just by combining a bunch of sub-atomic particles in the right way so as to make the whole physically indistinguishable from an elephant. Historical context is crucial.

2 Is an Elephant One and the Same as an Aggregate of Sub-atomic Particles?

Suppose it is granted that what makes something an elephant is, in part, its history. Still, isn't the thing that is a particular elephant one and the same as an aggregate of sub-atomic particles? Again, it's not that simple. Consider a bunch of flowers. That isn't just an aggregate of flowers. If the flowers are spread out in space, there is no bunch of flowers. A bunch of flowers consists of various flowers under the relation of being bunched (Fine, 2003). Here is another example. Consider a corporation, Apple, say. Apple isn't just the aggregate of people who work for it. If these people do not communicate with one another, no corporation at all exists. Apple consists of a large number of people who are connected via the relation of working cooperatively together in certain roles.

Returning now to the case of elephants, even if a given elephant isn't one and the same as just an aggregate of microparticles, can't it be one and the same as a specific group G of subatomic particles arranged elephant-wise (as we might put it).? The answer is 'No'. Identity is not relative to a time and the group of subatomic particles in the spatial region occupied by the elephant varies through time. Furthermore, the elephant surely does not have precise boundaries. There are many minimally different overlapping groups of particles, each of which has as much right as any of the others to be identified with the elephant. So, either they are all elephants, in which case there are many elephants present and not one as we are supposing or none of them are elephants. The given elephant, then, is best taken to be something that is constituted by a certain aggregate of subatomic particles. It is not identical with that aggregate, however its parts are arranged.

Is this mysterious? Take a simpler example. Consider a specific rock. It is made up of certain particles. It is not identical with the aggregate, R, of those particles for the reasons just given mutatis mutandis in the case of the elephant. Imagine now that God had laid out R in space and arranged its parts rock-wise. Did He have further work to do to create a rock? Or was His work already done? Rocks are material things. Surely once God had laid out R, and arranged its parts in the appropriate spatial way, there was nothing further for him to do. Why? The answer, I think, is that it is a conceptual truth that if an aggregate of fundamental material parts is arranged rock-wise then a rock exists. Similarly, in the earlier example of the bunch of flowers, it is a conceptual truth that if an aggregate of flowers is arranged bunch-wise then a bunch of flowers exists.

Is what is true for rocks and bunches of flowers true for elephants? Evidently, the situation is more complex; for as already noted, history matters. So, arranging an aggregate of fundamental material entities so as to form something that is physically indistinguishable from an elephant does not suffice. Further, elephants are conscious beings. They see and hear things and in so doing undergo visual experiences and auditory experiences. They feel pain. They are subjects of a range of emotional experiences. Electrons have none of these things. How can complex, spatial configurations of electrons and other subatomic particles be arranged so as to form not just creatures with tusks and trunks but creatures that are also conscious, creatures that experience and feel things? What building work was required here?

3 An Elephant Never Forgets; But What Makes an Elephant Conscious?

First, a few words about how I am using the term 'consciousness'. Consciousness, as I understand it in this essay, just is experience. Experiences are mental states such that there is inherently something it is like subjectively to undergo them. Examples are feeling pain, feeling an itch, visualizing an elephant, experiencing anger and feeling fearful. In each of these cases, it is incoherent to suppose that the state exists without there being some phenomenology, some subjective or felt character.

In understanding the term 'consciousness' in this way, I do not mean to suggest that the term has not had other uses both in science and philosophy. Sometimes, for example, it is held that a mental state is conscious just in case it is one of which its subject is introspectively aware. This is sometimes called "higher-order consciousness". My claim is simply that among the various mental states we undergo, many of which are introspectively accessible (but arguably not all), are experiences and feelings, and these states, unlike beliefs, for example, are inherently such that they feel a certain way. Different experiences differ in how they feel, in their subjective character, and that is what makes them different experiences. In being this way, experiences are conscious mental states by their very nature. This point is sometimes put by saying that experiences are phenomenally conscious.

There are two dimensions to the puzzle here. Elephants are conscious. Rocks and trees are not. Why? The obvious answer is that the former have brains and the latter do not. But all sorts of physical events occur in elephant brains that have nothing to do with consciousness. Think of the neurophysiological mechanisms responsible for their continuing to breath or their hearts beating. What is it about the neurological goings-on underlying elephant conscious states that explains why they generate the conscious states they do or why they generate anything conscious at all for that matter? This is the so-called hard problem of consciousness to which I turn next.

4 The Hard Problem

Upon closer examination, it may be seen that the problem really has two dimensions to it. To appreciate this, consider your own case, specifically, the color experience you undergo as you view a ripe tomato. Light is reflected from the surface of the ripe tomato, thereby activating cells on your retina. Via a sequence of physical interactions, cells become active in your visual cortex and you experience red. That experience has a certain 'raw feel' or subjective phenomenology to it. Why does your experience of red feel to you the way it does? Why doesn't it feel the way your experience of blue does or your experience of anger or a tickle? What explains its subjective character. This is the first dimension of the hard problem: why does your experience subjectively feel this way rather than that? There is a further question: why are you undergoing a subjective experience at all? Why do the neural events underlying your experience of red have any raw 'feel' to them? This is the second dimension of the hard problem.

Representationalism—the view of consciousness I advocate (Tye, 1995, 2021)[1]—has a simple, and to my mind compelling, answer to the first question. Your experience of red feels the way it does because it represents the color red. What it is for the experience to feel the way it does just is for it to be an experience representing red. Your experience of blue feels different because it represents a different color. The experience of red could not feel the way the experience of blue feels because if it did, it wouldn't be the experience of red at all but rather the experience of blue.

This is the view introspection seems to support. Turn your attention inwards as you experience red. What do you find? Obvious answer: the color red. That is what makes your experience feel as it does. Nothing more (and nothing less). The view of some philosophers that our experiences have intrinsic qualities, in addition to the qualities they represent, qualities that are responsible for the specific phenomenology of the experiences seems to me fundamentally misguided. When I introspect my experience

[1] See also Dretske 1995.

of red, I find no extra quality. I find only the color red. That is what I respond to cognitively, what I like or dislike, what I want to keep experiencing or to cease.

How does the experience of red represent red? Answer: by itself being a brain state with the natural function of indicating the presence of the color red. Compare: the heart has as its natural function to pump blood and certain neuronal cells (known as edge detector cells) have as their natural function to indicate the presence of an edge in the visual field. On this view, you can't really discover the ways our experiences feel by peering among the neurons. That's the wrong place to look. You have to look rather at what the neural states represent. And that is determined by historical circumstances that gave the relevant neural states their natural functions of indicating environmental (and bodily) features of one sort or another. Once again, then, history matters.

5 The Hardest Part of the Hard Problem

As for the second dimension to the hard problem, we have, it seems, two options. One is to say that consciousness simply emerged with suitably complex arrangements of micro-particles. On this view, certain neural states have a felt character associated with them because that character automatically emerged with those states. A second option is to take the view that consciousness already exists even at the level of micro-particles, so it did not really emerge at all. This option still needs a further account of why certain complex physical states are conscious and not others, and so it does not offer a direct answer to the hardest part of the hard problem, but it does at least avoid puzzles associated with emergence. What is still needed is an account of why certain complex arrangements of subatomic particles are conscious, given that their constituent fundamental parts are, and not others.

On the face of it, neither option is attractive. As already noted, electrons can't feel pain or experience red. So obviously if they are conscious, they can't be conscious of the various things we are. But consciousness always has a content. So, of what then are electrons conscious? There seems no way to answer this question. Indeed, the question itself just seems wrong-headed. On the other hand, if consciousness only emerged at the level of certain highly complex physical things, then what was responsible for that emergence?

This last question would have an answer if our concept of consciousness (experience) were itself a complicated functional concept in the way that the concept life is. But it is not. The reason is straightforward. Take the case of life. We know that living things use energy, they grow, they reproduce, they respond to their environment, they adapt and they self-regulate. Reflecting upon these facts, it seems clear that these are things we know not as a result of scientific investigation but rather simply by understanding the concept life. All it is for an entity to be living is for it to have enough of these functional and behavioral features. The concept life, thus, is a functional/behavioral, cluster concept. Here emergence is unproblematic. Once certain complexes support the relevant functional and behavioral features, they are automatically living. In the case of consciousness, however, there is no functional or behavioral definition. We know this because we can make intelligible to ourselves the idea of a zombie, a being that functions exactly as we do but who has no experiences at all.

The question about emergence would also have an answer if consciousness had a hidden (a posteriori) physical essence in the way that, for example, water does. But this is not at all plausible; for any putative such essence will be vague, that is, admit of possible borderline cases. Consciousness itself, however, is sharp. This needs a little explanation.

Consider first the point about consciousness being sharp. If consciousness is sharp then there are no borderline cases of consciousness. Is this correct? Suppose, for example, I have only just woken up, and I am still groggy, I am not yet fully conscious. Isn't this a borderline case of consciousness? It is certainly a fact that I am more conscious of the world around me when I am fully awake than when I first groggily open my eyes. What I experience is initially vague and impoverished. As I become fully awake, what I experience gets richer and richer. But this doesn't show that experience or consciousness itself has borderline cases.

Here is how Papineau puts the point:

> If the line between conscious and non-conscious states is not sharp, shouldn't we expect to find borderline cases in our own experience? Yet when we look into ourselves we seem to find a clear line. Pains, tickles, visual experiences and so on are conscious, while the processes which allow us to attach names to faces, or to resolve random dot stereograms are not. True, there are "half-conscious" experiences, such as the first moments of waking .... But, on reflection, even these special experiences seem to qualify unequivocally as conscious, in the sense that they are like something, rather than nothing. (1993, p. 125)

Try to think of other clearcut, objectively borderline cases of consciousness, that is, cases such that it is objectively indeterminate whether consciousness is present. Obviously, with some simpler creatures, we may not know whether they are conscious. But that is not germane to the issue. You can certainly think of a case of consciousness which is indeterminate as to whether it is a case of pain, say. Think of sensations at the dentist as your teeth are being drilled. Some of these sensations seem impossible to classify as to their species. There is a feeling of pressure perhaps. Is it pain? Not clearly so, but not clearly not. Here it is indeterminate as to what you are feeling, but not indeterminate as to whether you are feeling.

Alternatively, imagine that you are in a hospital bed feeling pain and that you can adjust a dial that controls the delivery of morphine to your body. As you do so, your pain becomes less intense, gradually transforming itself into a feeling of pleasure. In the middle of this process, there may well be experiences that are not easy to classify. Again, there is indeterminacy at such times as to what you are feeling, but there is no indeterminacy as to whether feeling continues to be present.

Consider the case of auditory sensations. Suppose you are participating in an experiment, listening to random high-pitched sounds through headphones. You are asked to press a button for each sound you hear. In some cases, you are unsure whether you are hearing any sound at all. Isn't this a borderline case of consciousness?

We can agree that there is epistemic indeterminacy here: you do not know whether you are hearing any sound. Still, this isn't enough for there to be a borderline case of consciousness. After all, you are listening attentively for a sound; are you hearing a sound or not? Well, even if you aren't hearing a sound, you are still hearing something, namely silence. That is, you are hearing the absence of a sound; it is not that you are failing to hear at all! There is something it is like for you subjectively to hear silence. So, either way, you are hearing and thus experiencing something. So, this doesn't show that there can be borderline cases of experience.

Suppose someone held that being tall is precise, admitting of no borderline cases. We can quickly show this person that she is wrong by presenting her with examples of people who aren't definitely tall but who also aren't definitely not tall. We can do the same with experiencing red or feeling pain or hearing a loud noise or feeling happy. But can we do it with being an experience (or being conscious)?

I don't think we can. We can certainly agree that as the intensity of an experience diminishes, it becomes less and less definite and rich in its character, but either an experience is still there or it isn't. Picturing what it is like from the subject's point of view, we picture the experience gradually changing in its phenomenology until it is so 'washed out' and minimal that it has hardly any distinguishing features subjectively. But the subject is still having an experience. The gradual transition is in the number and intensity of the subjective features of the experience, not in the state's being an experience (being phenomenally conscious).

So, consciousness itself is sharp. If this is the case, then the complex physical property that supposedly makes up its essence must be sharp too. But this seems very implausible. To see why, consider first the type identity theory and the hypothesis put forward by Crick and Koch that consciousness is one and the same as neuronal oscillation of 40 MHz. It is evident that Crick and Koch did not intend this hypothesis to rule out every neuronal oscillation that is not exactly 40 MHz. What about a neuronal oscillation of 40.1 MHz? Or 40.01 MHz? Or 40.000001 MHz? Their proposal is that consciousness is one and the same as neuronal oscillation of approximately 40 MHz or neuronal oscillation sufficiently close to 40 MHz. But these formulations of the hypothesis bring out its inherent vagueness, and not just from the use of the terms 'approximately' and 'sufficiently'; for the term 'neuron' is vague too.

Neurons are complex physical entities with diverse components. Each neuron has a cell body, dendrites and an axon. Electrical impulses come in along the dendrites and go out along the axon. Imagine removing atoms one by one from a given neuron. Eventually, as one does so, there will be no neuron left. But along the way, there will surely be a range of borderline cases—entities that are neither definitely neurons nor definitely not neurons. So, the property of being a neuronal oscillation is vague. It admits of borderline cases. In general, neurophysiological properties are highly complex. The idea that the relevant neural properties for consciousness are sharp is extremely implausible.

Suppose it is now proposed that integrated information holds the answer. What it is for a physical system to be conscious is for it to have a large amount of integrated information (Phi) in it (Tononi et al., 2016). This view, which can be taken to be offering a high-level physical account of consciousness, has some extremely counter-intuitive consequences. For example, as noted by Aaronson (2014), it predicts that if a simple 2-D grid has ten times the amount of integrated information as my brain, the grid is ten times more conscious! What exactly is meant by one system being more conscious than another has also not been made fully clear by advocates of the theory, but for present purposes, it suffices to note that what it is for an amount of integrated information to be large is patently vague and thus the view is of no help to anyone who wants to hold that consciousness is sharp and broadly physical.

A response to this difficulty is to say that some degree of consciousness goes along with any amount greater than zero of integrated information. So, consciousness is sharp, after all. This requires us to agree that thermostats are conscious as are speedometers, since they contain some integrated information, and that seems a line to be avoided, if at all possible! But even if you disagree here, as noted above, there remains the question as to what it is for one system to have a greater amount of consciousness than another. And since advocates of integrated information theory accept that certain 2-D grids are more conscious than human brains, it cannot have to do with the number of experiences or the intensity of the experiences; for surely no one wants to hold that the relevant grids have more experiences or more intense experiences than our brains (Pautz, 2019). What is meant by saying that they are more conscious then?

What about functional properties? Might they make up the essence of consciousness. Again, it seems obvious that any scientific functional properties with physical inputs and outputs that are proposed as candidates for the hidden essence of consciousness will admit of possible borderline cases. So, again they cannot really make up the essence of consciousness at all.

It appears, then, that if consciousness did emerge out of certain brain structures, it must be sharp and nonphysical, suddenly appearing on the scene. Here there is no reducibility and relatedly no explanation as to why it emerged as it did. So, uniformity in nature is lost. Phenomena gradually get more and more complex and then suddenly out of the blue something radically different just occurs. Why? There is no explanation. It is just a brute fact that once certain vague physical structures are in place, something sharp and nonphysical emerges. But that is very difficult to accept or even comprehend.

6 The Solution

The way out of these difficulties, I suggest, is as follows. Consider belief. You can't just believe period. That makes no sense. If you believe, you must believe something or other. Beliefs always have a content. What is it to have a belief then? The natural answer is that it is to undergo a state that has a content and that has a character that distinguishes it from, for example, a desire or a fear with that content. Let us call this feature "belief*". Corresponding features for desire and fear are desire* and fear*. Conceptually, there is no requirement that belief* states have a content.

That requirement is one on belief states only. Correspondingly, conscious states (experiences) always have a content. On a representationalist view, they always represent something or other. What it is to have an experience, I suggest, is to undergo a state that has a content and that has a feature that distinguishes it from nonconscious contentful states. Let us call this feature "consciousness*". Conceptually, there is no requirement that conscious* states have a content. This is the counterpart point to the one made about belief*.

The proposal I want to make, then, is that consciousness* did not emerge. It is there at the very bedrock of reality. Electrons are conscious*. What emerged was consciousness. More on that shortly. Consciousness admits of possible borderline cases since such cases can arise with respect to what one is experiencing, but consciousness* does not.

How can electrons be conscious*? After all, no physics textbooks about microphysical reality ever mention consciousness*. Furthermore, in the case of belief*, it seems plausible to hold that it is a narrow functional role property (Fodor, 1987). Beliefs are states that function in the right sort of way and that have content. But this is not plausible in the case of consciousness*; for, as already noted, a zombie has internal states that function as conscious states do and that have content. Furthermore, functional role properties are vague, not sharp.

The initial answer, I suggest, is that physical science itself tells us only about the relational/structural properties of matter, including spatiotemporal properties and causal/behavioral dispositions (second order properties). Physical science leaves open the nature of the categorical bases for these properties—the nature, that is, of the intrinsic properties that occupy the causal/dispositional roles associated with the basic theoretical terms of microphysics.

Here is an illustration. Suppose electrons are basic. Electrons are particles having mass and negative charge. But what are mass and charge? The suggestion is that these properties are to be cashed out in terms of how electrons behave and interact with other elements of reality. Things having mass attract other things with mass and resist acceleration. What it is for an electron to have mass is for it to have an intrinsic property that enables it to behave as just specified (to play the mass role). Electrons have negative charge. Negative charged things attract positively charged things and repel other negatively charged things. What it is for an electron to have negative charge is for it to have an intrinsic property that enables it to behave in these ways (to play the negative charge role). Of course, these specifications are very rough and ready. The full story about these roles is told by physics via fundamental physical laws. Generalizing, the properties that physics attributes to elementary particles are structural (that is, pertaining to their arrangement and combinations), causal (nomic) and spatio-temporal. Physics thus has nothing to say about the intrinsic or categorical properties of electrons and likewise for other fundamental particles. It is here that consciousness* enters the picture. The intrinsic natures of the micro-parts of reality are made up of (presumably a small number of) intrinsic properties, one of which, I am suggesting, is consciousness*.

Consciousness* is an intrinsic property of electrons and other basic entities, a property that has no further inner nature. Consciousness* is a fundamental property of the fundamental parts of micro-reality. But not only fundamental entities are conscious*. Some highly complex entities are conscious* too. Consciousness*, we may suppose, transfers from fundamental entities to certain complexes so long as the fundamental entities are arranged in the right sort of way. Which complexes? Well, evidently only ones found in living creatures with brains. So, the relevant arrangements of micro-entities must be ones that require a brain.

So far, I have not said directly how it is that the various feelings and experiences we all undergo arise. What is responsible for the various conscious states we undergo—the feeling of pain, the experience of red, the feeling of anger?

It should be clear that from the present perspective we cannot hold that there are different stripes of consciousness at the micro-level, different combinations of which give rise to different macro-phenomenal states. There isn't even consciousness at that level. Instead, there is just bare consciousness*. The various different macro-states are generated via the various representational contents of the complex conscious* states. Different macro-phenomenal states are generated representationally.

An example may help to make this clear. Suppose that a large number of electrons (or quarks) are arranged so as to form a complex state A that itself Normally tracks (and thereby represents) property P. That state A, let us suppose, is conscious*. Given this and given also that A also represents red, A is a full-fledged experience of red. What is true in this case is true in all cases of macro-consciousness. The various species of consciousness are generated by the different properties and different property complexes that are represented, and it is in connection with this that vagueness intrudes.

Of course, there is much more to say about the nature of representation here; my own view (Tye, 2000) is that in general the basic properties represented by experiences are phylogenetically fixed. We are simply built by nature to feel pain, to experience various colors, to feel anger, and so on. Other creatures are built differently and their experiences are different in varying degrees from ours. The bird that immediately spots what is to us a green caterpillar sitting on a green leaf (and to our eyes almost perfectly camouflaged) does so because it experiences the caterpillar as having a color different from that of the leaf (in my view, the color is a binary one—ultra-violet green, as we might call it—only one component of which is available to us, given the different sensitivity of the cones in our eyes (Tye, 2001)). When we view the caterpillar, the state we are in represents it as green since that state is the one in us that Mother Nature has given us to track the color green. That state is the one that Normally tracks green (or, in Dretske's terms (1988), has as its biological function to indicate green). The bird viewing the caterpillar is in a different conscious state, one that Normally tracks the binary color, ultraviolet-green.[2]

The question I have not tried to answer here is: what arrangements of conscious* micro-entities are themselves conscious*? This is a big question I have taken up in detail elsewhere.[3] The key idea is that some properties of parts transfer to wholes, given the right arrangement of those parts. Consider, for example, an ensemble of

[2] For more here, see Bradley and Tye 2001.

[3] See Tye 2021.

graceful dancers. The whole ensemble may be graceful too, but only if the dancers are coordinated in the right way. Or consider a play with three acts, each of which is wonderful. The play too will be wonderful if the parts fit together appropriately; otherwise it may simply be disjointed. What goes for dancers and acts goes for electrons and complex arrangements of them. In elephants, the arrangements of conscious* micro-parts are such that the elephants themselves undergo states that are conscious*. And these states, in being conscious* and representing various things are themselves conscious states of various sorts. That is how elephants are conscious beings.

7 Conclusion

So, given enough microphysical entities having consciousness* as part of their intrinsic natures, and given the right arrangement of them and given the right historical context, you get a flesh and blood, conscious elephant. A microphysical duplicate of an elephant might not be an elephant, however. Indeed, it might not even be a conscious entity at all (for its fundamental parts might be microphysical duplicates of real elephants' fundamental parts without being conscious*). To build an elephant, background context and consciousness are crucial. Furthermore, the two are not independent; for if consciousness itself essentially involves representation and representation invZolves biological indicator function or Normal tracking, then context in part also determines consciousness.

References

Aaronson, S. (2014). Why I am not an integrated information theorist. https://www.scottaaronson.com/blog/?p=1799

Bradley, P., & Tye, M. (2001). Of colors, kestrels, caterpillars, and leaves. *Journal of Philosophy, 98*, 469–487.

Chalmers, D. (1995). *The conscious mind*. Oxford University Press.

Chalmers, D. (1995). Facing up to the problem of consciousness. *Journal of Consciousness Studies, 2*, 200–219.

Dretske, F. (1988). *Explaining behavior*. MIT Press.

Dretske, F. (1995). *Naturalizing the mind*. MIT Press.

Fine, K. (2003). The non-identity of a material thing and its matter. *Mind, 112*, 195–234.

Fodor, J. (1987). *Psychosemantics*. MIT Press.

Papineau, D. (1993). *Philosophical naturalism*. Blackwells.

Pautz, A. (2019). What is the integrated information theory of consciousness. *Journal of Consciousness Studies, 1*, 1–2.

Tononi, G., Boly, M., Massimini, M., & Koch, C. (2016). Integrated information theory: From consciousness to its physical substrate. *Nature Reviews Neuroscience, 17*, 450–461.

Tye, M. (1995). *Ten problems of consciousness*. The MIT Press.

Tye, M. (2000). *Consciousness, color, and content Cambridge*. Bradford Books, the MIT Press.

Tye, M. (2001). (with P. Bradley) Of colors, kestrels, caterpillars, and leaves. *Journal of Philosophy, 98*, 469–487.

Tye, M. (2021). *Vagueness and the evolution of consciousness: Through the looking glass*. Oxford University Press.

When Two Levels Collide

John Bickle

Your world was so different from mine, don't you see/And we couldn't be close though we tried/We both reach for heavens but ours weren't the same/That's what happens when two worlds collide ...—"When Two Worlds Collide," written by Roger Miller and Bill Anderson, first recorded by Roger Miller in 1961

Often when philosophers and scientists reflect on how explanations across "levels" relate, they focus on successful cases, where some "higher-level" explanation aligns relatively smoothly with its "lower-level" counterpart. Philosophers of the contemporary mind-brain sciences are not exceptions to this focus, and reflections on familiar examples has generated recent accounts of reduction (Bickle, 2006, 2012), mechanism (Bechtel, 2008; Craver, 2007; Piccinini & Craver, 2011), realization (Polger, 2004; Polger & Shapiro, 2016), and emergence (Gillett 2016). There are numerous instances where some cognitive explanation aligns smoothly with some neuroscientific counterpart, especially when the latter is drawn from cognitive and systems neuroscience. However, when we consider some recent explanations of cognitive phenomena in the neuroscience field of 'molecular and cellular cognition' (MCC), we encounter some instances that involve significant mismatch across counterpart cognitive-scientific and MCC explanations. And MCC is not some fringe appendage in current neuroscience. It has been prominent in the discipline for nearly thirty years (for a review see Silva et al., 2014).[1]

I will here use a detailed case study to demonstrate a mismatch between a cognitive-scientific and an MCC explanation of related phenomena. It is important to note some caveats from the outset. I am NOT suggesting that all cognitive-scientific explanations are headed toward similar mismatches with MCC counterparts! There are plenty of examples of counterpart cognitive-MCC explanations that link together

[1] MCC has a professional society which has sponsored more than twenty annual meetings over the past fifteen years, boasts of European and Asian affiliates, and has an international membership of more than 5000 scientists. See https://molcellcog.org/ for details.

J. Bickle (✉)
Department of Philosophy and Religion, Mississippi State University, Starkville, MS 39762, USA
e-mail: jbickle@philrel.msstate.edu

S. Wuppuluri and I. Stewart (eds.), *From Electrons to Elephants and Elections*,
The Frontiers Collection, https://doi.org/10.1007/978-3-030-92192-7_35

relatively smoothly—I'll even reference some examples below. But mismatches do occur and need to be considered in any full discussion of cross-level relationships in the cognitive and brain sciences. As I will argue in this paper's Sect. 2, a fairly common "neural plausibility" desideratum among contemporary cognitive scientists takes from them one way to downplay such mismatches. I will then show in this paper's Sect. 3, also based on a detailed case study, that cross-lev el cognitive science-to-neurobiology mismatches also occur concerning counterpart phenomena, not only explanations. In this chapter's final section I will point out that there is nothing unique about cognitive science vis-à-vis neurobiology that would limit cross-level mismatches to these sciences. Any time that some "lower level science" begins to address data about the behaviors of the larger systems that its ontology composes, cross-level mismatches of counterpart explanations and phenomena are genuine scientific possibilities. We may hope that counterpart explanations across science's "levels" align smoothly, and indeed sometimes they do. But neither nature nor our scientific practices guarantees that they always will. In fact, sometimes they do not. And it remains an open philosophical-cum-scientific question about what we should say about cases of serious mismatch, although some constraints we commonly put on our "higher-level" sciences do rule out some strategies for dismissing these mismatches as inconseqential.

So far I've enclosed usage of "levels" in scare-quotes. This is because my understanding of "levels" and "counterpart explanations and concepts" is admittedly pedestrian. The ontologies of "lower level" sciences are constituents that compose the ontologies of "higher level" sciences. The specific ontology of a "counterpart lower level" science to some specific "higher level" science composes those specific "higher level" systems, and the "lower level" explanations purport to explain some of the "higher level" data. Ontologies are the classes and relationship between the members of those classes that define a scientific field (Larson & Martone, 2009). That's enough "scientific metaphysics" for my tastes, and with these remarks I will henceforth refrain from using the scare quotes around levels, counterparts, and the like.

1 A Mismatched Counterpart Cognitive-MCC Explanation Pair

The "Ebbinghaus spacing effect" is named after nineteenth century German psychologist Hermann Ebbinghaus (1850–1909), who first explored it systematically (using himself as a research subject in some experiments!).[2] Animals, ranging from invertebrates to rodents to humans, typically learn a task quicker, and remember previously presented training stimuli better (as measured by recall performance) when training episodes are divided into several separate sessions, with longer non-training intervals separating them (up to a non-training interval time limit, beyond which test recall performance declines). Experimental psychologists since Ebbinghaus have explored the most effective number of training sessions and durations of non-training intervals for numerous tasks, along with a variety of other factors; most recently these other factors have included experimenters' intentional manipulations of encoding conditions across presentation of items to be remembered (Maddox, 2016). But no single cognitive explanation for the Ebbinghaus spacing effect has garnered unanimity.

One cognitive explanation that received some initial acceptance was Spear and Riccio's (1994) hypothesis that when memory training is grouped into a single session (so-called 'massed' training), information about a given stimulus acquired on a previous learning presentation is still being processed into memory stores when the next training input arrives for processing. This timing interferes to disrupt ongoing processing of the previously presented information. Longer delay intervals between learning episodes (so-called 'spaced' training) permit full processing and integration of each learning episode into long-term memory.

More recent experimental work by psychologists on verbal learning in humans (acquiring, retaining, and recalling verbal materials) has generated additional cognitive explanations for the spacing effect. In a recent review Maddox (2016) surveys much of these data and focuses on the two proposed cognitive explanations that best account for them.[3] A *reminding* explanation "assumes that items are forgotten over time based on the power-law of forgetting, and when presented, items have some capacity to spontaneously cue retrieval (e.g., remind) of an earlier item" (Maddox, 2016, 695). This account further assumes that the difficulty of reminding modulates the subsequent amount of rehearsal that a subject gives to target memory items. This explanation accounts for numerous data pertaining to the spacing effect because

[2] This first case study is a more detailed account of one I presented in a previous publication (Bickle 2014). There I dubbed this case an example of 'little-e eliminativism,' where the mismatch demonstrates the falsity and so the rejection of the cognitive explanation. (I called it 'little-e' because the elimination is strictly within science itself, and has nothing to do with some broader "folk psychology, as the "Big-E Eliminativism' of, e.g., Paul and Patricia Churchland targets.) I now think that the eliminativist conclusion I stressed previously was too hasty. It takes more than just the mismatch I'm about to discuss to warrant an eliminativist conclusion about the cognitive explanation. But as I will stress in the final sections below, a little-e eliminativist conclusion about mismatched cognitive explanations vis-à-vis some MCC counterpart remains one available option, though not the only one.

[3] See Maddox (2016) for the "reliable recent experimental findings," plus more details on these two proposed explanations and extensive references to the published literature. Spoiler alert! Maddox plumps for the combined encoding variability + study-phase retrieval explanation I'll present below. He also surveys additional proposed cognitive explanations for the spacing effect which he argues are less successful in handling the recent experimental data. I will not discuss those other approaches here, but I note that all of them fail to match with the MCC explanation of the spacing effect, introduced later in this section.

difficult-to-retrieve, i.e., spaced items, get rehearsed more than easy-to-retrieve, i.e., massed items. A second hypothesized cognitive explanation for spacing combines *encoding variability + study phase retrieval*. The first component holds that contextual items get encoded in memory along with the target item, and so "an item is more likely to be encoded in different ways when repetitions are spaced than when they are massed" (Maddox, 2016, 696). Thus increased encoding variability with spaced as compared to massed presentations yields an increased number of "retrieval routes, and probability of retrieval at later test," since co-encoded contextual elements are more likely to be distinct at the different spaced presentation times (Maddox, 2016, 696) The second component of this combined account, the study phase, is akin to the reminding mechanism of the first explanation. The learner is retrieving previously-studied information throughout the entire study phase of the new item. Items must be recognized as repetitions during study to generate a spacing effect on final test performance and the difficulty of retrieval associated with longer delays leads to increased repetitions, and thus increased final test performance with spaced training.

Turning to MCC research, work in Isabelle Mansuy's lab provided strong evidence for a direct intraneuronal molecular mechanism for the Ebbinghaus spacing effect (Genoux et al., 2002). During training episodes, specific neurons recruited into the memory trace for the training stimuli undergo significant synaptic plasticity, inducing late long-term potentiation (L-LTP). These same neurons also show increased levels of phosphorylated cyclic adenosine monophosphate-responsive element-binding protein (pCREB) at the time of training (Han et al., 2007, 2009; Zhou et al., 2009). One isoform of pCREB is an activity-dependent transcriptional enhancer for genes coding for both regulatory and structural proteins.[4] The end result of pCREB activation in these neurons recruited into a specific tone-shock memory trace is the locking of additional "hidden" excitatory (glutamate-responding) receptors into active sites in post-synaptic densities. These additional excitatory receptors make these neurons much more likely to respond with action potentials to pre-synaptic activity (glutamate release) induced by subsequent presentations of the same training stimulus.

However, the high neuronal activity required to induce pCREB-driven synaptic potentiation can also activate protein phosphatase 1 (PP1) in these same neurons. Activated PP1 removes the phosphate group from pCREB molecules. This blocks CREB's gene transcriptional-enhancing activities, and so inhibits the protein synthesis that potentiates the synapses in these highly active neurons (Genoux et al., 2002). Because of the increased frequency of training stimuli presentations during massed (and briefer-interval distributed) training, action potential frequencies in highly active neurons responding to the stimulus is high enough not only to activate pCREB, but also to simultaneously activate PP1, to inhibit activity-dependent pCREB. The end result in these neurons is little to no L-LTP, and so little to no learning of the stimulus to be remembered later. However, with longer-interval distributed ('spaced') training, the extended non-training periods between training sessions are long enough to decrease activation frequency in the specific neurons recruited into

[4] Transcriptional enhancers bind to sites on the control region of genes to turn on the process of gene transcription via messenger RNA production.

the memory traces, to still be above the level required to induce L-LTP via activated pCREB, but below the level that simultaneously activates PP1 to inhibit pCREB-driven synaptic plasticity. The lessened PP1 activity in these neurons recruited into the memory trace allows pCREB to induce lasting synaptic potentiation through new gene expression and protein synthesis, and ultimately produces the learning indicated by recall behavior upon later presentations of the 'spaced' training stimuli. The resulting behavioral dynamics are exactly the Ebbinghaus spacing effect (Silva & Josselyn, 2002): enhanced learning and memory with longer-interval distributed ('spaced') training.

For readers unfamiliar with molecular biology, the MCC explanation is apt to puzzle. Some experimental details from the Genoux et al. (2002) study will not only clarify this molecular-mechanistic explanation of the Ebbinghaus effect but will also illuminate the nature of the mismatch between the cognitive and MCC explanations of the Ebbinghaus spacing effect. Genoux and colleagues engineered a transgenic mouse, the I1* mutant, that overexpresses a protein specifically in forebrain neurons (including in hippocampus) one step upstream in intracellular signaling to pCREB-PP1 interactions. Inhibitor-1 (I1) deactivates PP1, blocking its inhibition of pCREB activity. The I1* mutant has an extra copy of the *I1* gene, with this extra transgene transcribed and extra protein translated only when the mutant animals were dosed with doxycycline (dox) in their diets.[5] Nonmutated littermate control mice displayed the typical Ebbinghaus spacing effect in an object recognition task, with only mice in the longer-interval distributed training group showing statistically significant improved memory performance for previously-presented objects, compared to those wildtypes undergoing briefer-interval distributed training or massed training.[6] I1* mutants off

[5] As was standard in genetic engineering twenty years ago, transgene expression was limited to specific neurons by the addition of an engineered promoter binding site. Only in the targeted (forebrain) neurons does this specific promoter binding molecule (in this case, calmodulin kinase II, or CaMKII) occur in sufficient quantities to bind to the transgene promotor region to drive its expression. So while the transgene, inserted at the embryonic stem cell stage of development, is present in all cells in the mutants' bodies, due to this engineered promoter region it will only be expressed in the targeted forebrain cells. The temporally limited transcription of the transgene via systemic doxycycline dosage was likewise driven by an inserted activator system on the engineered transgene. Bickle (2003) provides an introduction for non-molecular biologists to the basics of such molecular-genetic work in mainstream neurobiology.

[6] Object recognition is a standard rodent non-spatial learning and memory paradigm that exploits their natural curiosity for novelty. During training episodes rodents are exposed to a novel object for a specified time interval and allowed to explore it. The object is removed, a delay period ensues, and the original object and a new novel object are presented in the testing period. The amount of time the rodent spends exploring the new novel object during the testing period, divided by the total amount of time it spends exploring both objects, computes a discrimination ratio for the novel object in the testing phase. Ratios greater than 50% operationalize memory for the original novel object presented during the learning phase, and higher ratios operationalize stronger memories for it. In the Genoux et al. (2002) study rodents in the massed training groups were exposed to the original object for one 25-min learning session. Mice in the briefer-interval distributed training groups were exposed to it for five 5-min learning sessions, each interspersed by a 5-min non-training interval. Mice in the longer-interval distributed training groups were exposed to it for five 5-min learning sessions, each interspersed by a 15-mintue non-training interval.

dox (where the I1 transgene was not expressed) showed the same pattern. But I1* mutants on dox, in which the extra I1 transgene was expressed, and the protein synthesized and activated, showed statistically significantly improved memory for objects previously presented during the training sessions for both longer-interval and briefer-interval distributed training. The effect was reversible; these same mutants returned to control and mutant off-dox performance (increased memory for previously presented objects only with longer-interval distributed training) when they were taken off dox. A control study confirmed increased pCREB levels in hippocampal and cortical neurons in the mutants on dox group during both briefer-interval and longer-interval distributed training. Increased pCREB levels in these neurons were found in control groups (both wildtype and mutants off-dox) only during longer-interval distributed training.

Genoux et al. (2002) took their results as strong evidence in favor of an intracellular pCREB-PP1 interaction mechanism to explain the Ebbinghaus spacing effect. Activation levels in neurons recruited into the memory trace for the training stimuli after longer intervals between learning episodes exceed pCREB activation thresholds to induce L-LTP in activated synapses, but do not exceed PP1 activation thresholds, so pCREB-driven synaptic plasticity is not immediately blocked in these neurons. During briefer-interval distributed and massed training, however, action potential frequencies in highly activated neurons activate both pCREB and its inhibitor, PP1, producing little L-LTP in these neurons and therefore little learning and memory for the training stimuli as indicated by recall behavior. Artificially blocking PP1 activation during briefer-interval distributed training via the activated *I1** transgene in mutants on dox enabled the activated pCREB to induce L-LTP without the PP1 inhibition activated in the non-mutated controls and mutants off dox; so the activated PP1 was blocked by the additional I1 protein from the activated transgene. This produced L-LTP in the highly activated neurons and subsequently increased recall performance in the I1* mutants to the remembered stimuli in the briefer-interval training rodent object recognition task.

Let us now be more precise in characterizing the mismatch between the counterpart cognitive and MCC explanations in this case. The important difference is more than just that one account appeals to specific neurobiological molecules and their roles in intra-cellular signaling pathways in neurons recruited into the memory trace, while the other appeals to information processing, encoding variability, or reminding.[7] It has to do rather with differences in the *basic functional profiles* of the neurobiological mechanism as contrasted with those of the cognitive ones. The basic functional profiles of the MCC and the cognitive explanations differ. From the perspective of Genoux et al.'s (2002) neurobiological explanation, the hypothesized

[7] Maddox himself, in his recent review of cognitive explanations of the spacing effect, notes that "it is important to examine the biological underpinnings of the spacing effect" (2016, 702). He mentions cross-species work "indicat[ing] that spacing and massing study events differentially influence long-term potentiation (LTP) induction" in fruit flies, mice and honeybees (although he does not mention or cite the Genoux et al. (2002) paper). Here he, as a cognitive psychologist, assets that the neurobiology is relevant to the cognitive explanation. More on this in the discussion of the neural plausibility constraint in Sect. 2 below.

cognitive mechanisms surveyed above are not found.[8] The cognitive explanations are not even *approximations* of the discovered neurobiological causal mechanism, abstracted away from neurobiological detail. Specifically, Spear and Riccio's (1994) cognitive explanation of the Ebbinghaus spacing effect attributes a central causal role to *the interruption of ongoing processing* of the previous training stimuli, due to the more rapid re-occurrence of training stimuli during massed (and briefer-interval distributed) training. Only longer-interval distributed training permits *complete, non-interrupted processing and full integration* of each separate training episode. But in the pCREB activity-PP1 inactivity neurobiological account, supported by Genoux and colleagues' molecular pathway intervention experiments, in the cases of massed and briefer-interval distributed training the high frequency of neuronal activity in activated neurons, which activates both the transcriptional enhancer (pCREB) and its related phosphatase (PP1), *blocks even the initial steps* in the induction of L-LTP. The 'memory processing' initiated by the initial training stimulus thus *never even gets started* in the hippocampus neurons highly activated by the training stimulus. The activated pCREB molecules are dephosphorylated by the activated PP1 molecules before they initiate even the first steps of gene transcription and protein synthesis that lead to L-LTP in the recruited neurons, and hence to learning and memory. No "ongoing processing" of the earlier training episode gets "interfered with," because none gets initiated. Even as a functional approximation of the neurobiological mechanisms that the Genoux et al. (2002) intervention experiments reveal to be causally at work, the cognitive explanation does not match its neurobiological counterpart. The underlying functional accounts of the cognitive and MCC explanations *mismatch*. Interfering with ongoing memory processing to previous training presentations is a different functional process than is inhibiting even the initial steps in memory processing of the training stimulus.[9]

Similar remarks hold for the more recent reminding and encoding variability + study-phase retrieval cognitive explanations reviewed by Maddox (2016). The reminding mechanism attributes increased rehearsal to difficult-to-retrieve items due to their presentation in spaced training, as opposed to less rehearsal to easy-to-retrieve items due to their presentation in massed training. But these effects will only apply to items *already encoded in memory*, and that is the step that the pCREB-PP1 interaction mechanisms blocks or reduces significantly in the highly activated neurons. Similarly with the study-phase mechanism of the combined account, which is directly related to the reminding mechanism. And the encoding variability component of the combined

[8] Take note of the expression, 'from the perspective of the neurobiological explanation.' I will return to it in the final two sections below.

[9] How then does any learning and memory occur for stimuli presented in massed or briefer-interval training, according to the MCC explanation? The stimulus will produce a distribution of activity rates across the population of activated neurons. Even in massed training, some neurons activated by the stimulus will have activity rates that fall within the "sweet spot" of just enough to activate pCREB but not enough to activate PP1. These neurons will induce LTP to the training stimulus. But there will be fewer such neurons falling within this range of activity necessary for inducing LTP to massed-trained items than to spaced-trained items, so there will be less learning and memory induced to the former, just as the Ebbinghaus behavioral data indicates.

mechanism likewise hinges on features of retrieval *of previously encoded memory items,* which include elements of the broader context occurring alongside the targeted memory item. That cognitive component also hypothesizes a mechanism with a *different basic functional profile* than the intracellular molecular mechanism as the latter *directly blocks or reduces encoding of the memory item* at the earliest processing phase. In short, all of these cognitive explanations attribute the effects of a training stimulus's temporal spacing either to late stages of the encoding phase or to effects on memory retrieval. But Genoux and colleagues' neurobiological explanation attributes the effects to the earliest stage of encoding. That is the mismatch in basic functional profiles across the cognitive and neurobiological explanations of Ebbinghaus spacing data.

2 What Should We Make of Such Explanation Mismatches?

We have now seen a detailed example in which an explanation of a phenomenon in cognitive science mismatches with a counterpart explanation of that phenomenon in neurobiology. The two explanations don't only differ in the extent to which they abstract away from neurobiological detail, but also in the basic functional profiles they attribute to the mechanisms taken to generate the behavioral data. Attenuation of ongoing memorial processing of a previous stimulus, or mechanisms for additional rehearsal or for exploiting contextual information paired with the target memory, as the accepted cognitive explanations hold, is not the same function as a mechanism that blocks memory processing at its initial stages, as the neurobiological intracellular pCREB-PP1 interactions explanation holds.

How might we best interpret such a cross-level mismatch, especially our assessment of the higher-level cognitive explanations? One possible interpretation is eliminativism. In attributing a functional profile to a mechanism for the behavioral data that is inconsistent with the profile that the discovered neurobiological mechanism displays, the cognitive explanations simply got things wrong and need to be replaced by the emerging neurobiological explanation. Given its strictly scientific focus, such a "little-e eliminativism" differs from, e.g., the Churchlands' 1980s-style eliminative materialism (e.g., Chruchalnd 1981, 1986), a view which by the mid-1990s had pretty much been relegated to philosophical oblivion. The Churchlands' specific target was something called "folk psychology," some common-sense conception of the mental. The little-e eliminativism I'm considering here has no concern with or stake in that target. But this little-e eliminativist interpretation of cognitive science/neurobiological counterpart explanation mismatches certainly would be consequential for the current cognitive and brain sciences. Genuine integration of results across these sciences has proved to be more elusive than what many had hoped for or predicted a half-century ago, when both fields began to flourish. Even real cross-pollination of serious research between systems/cognitive and cellular/molecular *neuro*science is fairly rare. It is hard enough to keep up with the published literature in one's own specific research area, much less to branch one's program across the systems/cognitive-cellular/molecular neuroscience divide. And this cross-field gulf grows even broader and more difficult to span in anything but flowery rhetoric when we expand it to include all of the cognitive sciences, including cognitive psychology, artificial intelligence, cognitive anthropology, and the like. Specialists in any of these fields rarely even read across these divisions for anything more than avocational amusement, much less pursue actual research that crosses them.

Interpreting explanation mismatches like the one I sketched in Sect. 1 above as being (little-e) eliminativist for the cognitive explanations would further complicate integration across these divides. If ongoing cellular/molecular research sometimes demonstrates the incorrectness of some explanations accepted in cognitive science, one can appreciate a "no, thanks," judgment by cellular/molecular neurobiologists about a call to integrate explanations from cognitive science into their ongoing research, even heuristically as guides for cellular-molecular investigations. After all, if the Genoux et al. (2002) team had gone looking for some cellular/molecular mechanism that attenuated ongoing processing of previously presented memory stimuli, or cordoned rehearsal time and effort to specific memories, or tracked co-encoded extra-memory stimulus contextual features from training episodes, they would not have found the intracellular pCREB-PP1 interaction explanation for the Ebbinghaus data. Nothing in those accepted cognitive explanations would have pointed them toward developing their I1* mutant mice. And as the neurobiological field of MCC continues to expand its explanatory reach into other cognitive phenomena beyond those of core learning and memory, it seems reasonable to conjecture that more cognitive-MCC mismatches will be discovered. No one need insist implausibly that such mismatches will be ubiquitous, or even the norm. Some cognitive explanations already match up reasonably well with discovered neurobiological mechanisms. (Bechtel, 2009 Sect. 4 contains some good examples.) But if one is inclined to treat such mismatches when found as evidence for (little-e) elimination of the cognitive explanations, this might give one methodological pause. From the perspective of pursuing MCC research, one might be inclined to pay no attention to existing cognitive explanations of phenomena now coming into new MCC experimental focus, and instead push forward directly toward finding cellular/molecular explanations for the behavioral data. And then let the cross-level matches or mismatches, and so the status of the cognitive explanations, fall as they may. This is hardly the picture of interdisciplinary research that many have envisioned.

In presenting the case study in Sect. 1 above I was careful to indicate the "perspectival" aspect of the claimed mismatch between the cognitive-scientific and the neurobiological explanations, prefixing my discussions with expressions like "from the perspective of MCC ..." (I first made mention of such phrasing in footnote 8 above.) I've just done the same for the little-e eliminativist interpretation of such mismatches. Might not cognitive scientists reply that turnabout is fair play? Why can't a cognitive scientist appeal to his or her own perspective to interpret such mismatches, i.e., "from the perspective of cognitive science, the discovered intracellular pCREB-PP1 interactions are irrelevant to our investigations and explanations ..." 'Autonomy' arguments for cognitive science vis-à-vis neurobiology sometimes take exactly this perspectival tack. Wouldn't this response render the little-e eliminativist interpretation of these mismatches likewise irrelevant for ongoing cognitive science?

A problem with such a response lies in an important difference between cognitive science and neurobiology—a difference that many cognitive scientists acknowledge. Despite the abstraction of cognitive-scientific concerns and explanations from cellular and molecular details of "blooming, buzzing brains," nevertheless many cognitive scientists and their philosopher allies accept a demand on the 'neural plausibility' of cognitive explanations. 'Neural plausibility' is a slippery and highly ambiguous desideratum, especially in practice. However, if the notion is to have any real content, it at least requires *consistency* of a cognitive explanation with what comes to be known about actual cellular and molecular brain mechanisms. In his review paper on recent data and cognitive explanations of the Ebbinghaus spacing effect that I discussed in Sect. 1 above, Maddox (2016) commits to at least this much "neural plausibility" (see the quote in footnote 7 above). Cellular and molecular neuroscientists are under no similar demand for consistency with even widely accepted cognitive explanations. It's nice when neurobiologists find cross-level consistency with cognitive explanations of related phenomena, but nothing compels neurobiologists to seek it out, or to question the scientific validity of their findings if they fail to achieve it. Even cognitive scientists who insist that their explanations "constraint" lower level explanations must admit that this pull on neurobiologists is weaker than a demand that cognitive explanations be "neurally plausible."

In his important defense of cognitive/systems neuroscience for addressing questions about higher "levels of mechanisms,"[10] William Bechtel implicitly requires consistency with the cellular/molecular mechanisms lower in the 'nested hierarchies of mechanisms-within-mechanisms' jointly composing the system: "Researchers pursuing [cognitive/systems] research are fully aware that the brain regions they study are comprised of neurons and that processes such as the generation of action potentials and the altering of communication at synapses are crucial to the behavior of the brain regions" (2009, 34; emphasis added). Even for mechanists, inconsistencies across "levels of mechanisms" composing a single system are a problem. A "complete nested-hierarchy of mechanisms-within-mechanisms" explanation of a cognitive system performing some specific cognitive function cannot tolerate the cellular/molecular mechanisms to provide one explanation while the network/systems one provides a different one. The higher-level mechanisms that explain a system's behavior decompose completely into the components, activities and organization of the lower-level mechanisms. The kind of cross-level explanation mismatch detailed in the case study in Sect. 1 is a problematic inconsistency even for mechanists.

[10] Talk of "levels" (of mechanisms) here is Bechtel's, but his is the weak "componency" account I adopted at the outset above.

In this context it is worth pointing out what "neural plausibility" amounts to from the perspective of MCC. Neurons are bi-lipid membrane-bound molecular mechanisms for moving ions selectively across those membranes, releasing receptor-binding molecules for generating ionic movements across the membranes of other neurons. This is where the figurative causal-mechanical "rubber meets the road" in functioning nervous tissue. Those are also exactly the mechanisms that MCC researchers manipulate experimentally, and then appeal to in explanations of behavioral data taken to operationalize some cognitive function for experimental investigation. From the perspective of MCC, those are the neuroscientific phenomena that a cognitive explanation must decompose into if it is to be "neurally plausible." For those are the phenomena experimenters can now manipulate experimentally and affect behavior that operationalizes cognitive phenomena like the Ebbinghaus spacing effect.

So the "neural plausibility" assumption adopted by many cognitive scientists rules out their offering a "turnabout is fair play" response to a "little-e" eliminativst interpretation of cognitive-neurobiological explanation mismatches from the perspective of neurobiology. But there are other ways that cognitive scientists might resist little-e eliminativism about such mismatches. One way is to adopt a recent trend in philosophy of science that builds on a distinction between 'data' and 'phenomena' introduced by philosophers of science Bogen and Woodward (1988).[11] Although Bogen and Woodward introduced this distinction to contribute to a debate that has since lost its attraction, namely the "theory-ladeness" of observation in science and its epistemological consequences, recent authors have emphasized the impact of their distinction on the difficulty and importance of individuating ("characterizing") scientific phenomena (Colaço, 2020; Craver & Darden, 2013). According to Bogen and Woodward, 'data' can be straightforwardly observed, while 'phenomena' for the most part cannot ("in any interesting sense of that term," 1988, 306). Data "play the role of evidence of the existence of the phenomena" (1988, 305). Examples of data "include bubble chamber photographs, patterns of discharge in electronic particle detectors and records of reactions times and error rates in various psychological experiments"; examples of phenomena "for which the above data might provide evidence, [are] weak neutral currents, the decay of the proton, and chunking and recency effect in human memory" (1988, 306). Data are "idiosyncratic to particular experimental contexts," while phenomena "can occur in a wide variety of different situations or contexts ... we expect phenomena to have stable, repeatable characteristics which will be detectable by means of a variety of different procedures, which may yield quite different kinds of data" (1988, 317). Little-e eliminativist interpretations of the kinds of cross-level explanation mismatches sketched here can be avoided by proponents of the higher-level accounts by showing that the two explanations *address different phenomena.* E.g., the 'Ebbinghaus spacing effect' that cognitive scientists seek to explain is *a different scientific phenomenon* in this Bogen-Woodward sense than is the 'Ebbinghaus spacing effect' that the pCREB-PP1 intraneural interactions

[11] Thanks to David Colaço for suggesting to me this cognitivist response.

of MCC explain. Of course, this idea is just a strategy for avoiding a little-e eliminativist interpretation. The cognitive scientists must show that the scientific phenomena differ across the two explanations. The mere fact that the experimental data differ alone does not establish a difference between the phenomena explained.

Another cognitivist strategy for avoiding the little-e eliminativist interpretation of cross-level explanation mismatches is to adopt some characterization of 'autonomy.' One promising example is that of philosophers Polger and Shapiro (2016).[12] Polger and Shapiro defend a version of a mind-brain identity theory from multiple realization arguments long thought to falsify such views, but they emphasize that their account is not ontologically eliminativist about psychological kinds. This is because of their three types of arguments against multiple realization—"unificationist" arguments which find relevant commonalities in the realized and realizing kinds, "heuristic, abstraction, and idealization" arguments that characterize practices in contemporary cognitive science, and their plump for taxonomic "kind splitting"—none leads to any troubling elminativist conclusion. And by further adopting James Woodward's (2003) "interventionist difference-making" account of causal explanation, Polger and Shapiro seek to deflect the charge that their identity theory undermines the legitimacy of psychological explanations, and thus rules out any autonomy for psychology vis-à-vis neuroscience. They claim that charge rests on an overly stringent and misguided understanding of scientific explanation. Identity theorists can be and should be pluralists about scientific explanation. Neuroscientists can appeal to neural causes, psychologists and cognitive scientists to mental causes. Since both can cite causal invariances at their respective levels, both offer genuine explanations. Despite the mind-brain identities that Polger and Shapiro accept, psychology remains a methodologically autonomous science.[13] Their kind of autonomy will look promising to contemporary cognitive scientists who advocate even aa fairly strong sense of "neural plausibility" on their explanations. Scientific buyer beware, however; Polger and Shapiro (2016) traverse fairly complicated "metaphysics of science" labyrinths to develop their identity theory! I for one am not an enthusiast for that journey. And there remains a difficulty within their approach for handling detailed cross-level mismatches like the one described here. How can one *identify* cognitive kinds like the ones advocated by contemporary cognitive explanations of the Ebbinghaus spacing data, with neuromolecular kinds, like the intraneuronal pCREB-PP1 interactions in the neurons most highly activated by the memory stimulus, when the cross-level explanation pairs don't even share the same functional profiles? Won't doing so stretch our sense of 'identity' beyond the breaking point?

Of course, another possible strategy for avoiding a little-e eliminativist interpretation of these kinds of cross-level explanation mismatches is for cognitive scientists to simply ignore the neuroscience findings that mismatch with accepted cognitive explanations. As I mention in my discussion of the case study above (footnote 7), in his recent review cognitive scientist Geoffrey Maddox (2016) accepts the relevance

[12] I offer a similar description of Polger and Shapiro's account of autonomy in Bickle (2020).

[13] Polger and Shapiro's (2016) defense of psychology's autonomy resembles Woodward's (2018) recent defense of a more general kind of explanatory autonomy.

of neurobiological results toward adjudicating competing cognitive explanations of the Ebbinghaus spacing effect, but fails to mention the Genoux et al. (2002) paper (despite its publication in *Nature*, a high-profile journal across all of science). The vast disciplinary divides that separate cognitive science from molecular neuroscience encourage this kind of indifference. But ignoring such experimental results and the explanations tied to them, or assuming that one's own favored cognitivist explanations will not be subject to similar mismatches as MCC marches onward, amounts to, in American vernacular, cognitive scientists "whistling past the graveyard": remaining cheerfully in pursuit of an action while blithely oblivious to real risks lurking about; proceeding confidently with a task while remaining willfully indifferent to frightful hazards lurking nearby.

3 When Distinct Cognitive Phenomena Get "Lumped" into a Single Neurobiological Mechanism

A different way that cognitive science and neurobiology can mismatch has not to do with counterpart explanations of related phenomena, but rather in a way that related phenomena characterized across levels match up. Sometimes a recognized phenomenon from cognitive science "splits" into distinct neurobiological phenomena. Other times phenomena recognized as distinct within cognitive science get "lumped" into a single neurobiological phenomenon.[14] "Splitting" is much discussed by philosophers and scientists; the example of 'memory' is commonplace, which "splits" into multiple forms rooted in distinct brain systems (Squire, 1987). Phenomena "splitting" doesn't tend to generate interpretive difficulties, as the acknowledged abstraction of cognitive concerns from neurobiological detail seems readily to accommodate its occurrence. "Lumping," however, is sometimes not so easily accommodated, especially when seemingly quite distinct cognitive phenomena get "lumped" surprisingly into a single underlying neurobiological mechanism, as I explore in this section. Again I will focus on a detailed case study. A distinction between phenomena that cognitive science recognizes, even operationalizes differently for experimental investigation, seems to be eliminated. But it is not so clear what we should say about those cognitive phenomena. I illustrate this with some recent work on visual attention.

Vision neuroscientist John Maunsell (2015) offers a philosophically astute assessment of this recent work relating two cognitive phenomena: "The terminology we use to describe attention and related phenomena is vague. On the one hand, we likely fail to appreciate distinct aspects of attention. On the other hand, we might be using different terms to describe a common mechanism. With respect to the latter, important issues exist regarding the relationship of *attention* to *reward expectation*." (2015, 384;

[14] The "splitter"/"lumper" distinction is often traced back to McKusick's (1969) essay concerning the nosology of genetic diseases. His humorous Fig. 1 illustrates this distinction.

my emphases) From a cognitive perspective, these phenomena are related. As Maunsell (2015) points out, in behavioral studies experimenters often direct an animal's visual attention to a particular stimulus or location by altering either the quantity or probability of the reward associated with the animal's correct responses. For nearly two decades Maunsell has complained that experimenters using this strategy are potentially confounding attention effects with reward expectation effects (see, e.g., his 2004). But from a neurobiological mechanism perspective this relationship may be even closer. Attention cues and reward expectation cues might activate one and the same neural feedback processing pathway eliciting activity augmentation in visual neurons all the way back to the earliest stage of cortical visual processing. What then do we say about these phenomena that cognitive science distinguishes?

In his (2015) review Maunsell points us to the work of Dutch researchers Chris van der Togt, Pieter Roelfsema, and collaborators, who showed that experiments manipulating both visual attention and reward expectation cues reveal statistically identical effects of both manipulations on individual neuron activity rates in primary visual cortex (V1). Stănişor et al. (2013) first investigated the effects of cues indicating quantity of reward to be expected (high, medium, or low) on both behavior and activity in V1 neurons in non-human primates (rhesus macaques). They developed an eye-movement curve-tracing task, where the animal starts from an initial visual fixation point, and after a delay period must saccade (make directed eye movements) to the end point of one of two curves. The to-be-rewarded curve target of the two was indicated by a cue that appears after the two curves and targets appear and disappear, at the end of the delay period where the animal maintains central visual fixation. Operationally, the visual attention cue indicating the to-be-rewarded saccade target was prior training to the length of the curve from fixation point to target; only the target curve reached all the way from fixation point to target (see Stănişor et al., 2013, Fig. 1). At the end of the delay period, when the location cue of the target stimulus appeared, the animal had to make a saccade to the indicated target in order to receive the reward for a successful performance. Activity in individual V1 neurons was recorded on each trial both prior to the appearance of the strength-of-reward cues and during the delay period just prior to saccade initiation to target. The receptive fields of the V1 neurons being recorded from always occurred within one of the possible curves. Stănişor et al. operationalized reward expectation by prior training to color of the target stimuli, with different colors denoting low, medium, or high reward for a successful saccade to it. Rewards were different amounts of fruit juice administered to macaques who had been denied liquids for 24 h prior to experimentation. Animals were only rewarded for saccading within a specified time limit to the location of the cued end-of-curve stimulus, and then received the amount of fruit juice indicated by the color of that target cue. Monkeys received no rewards for trials in which they failed to maintain fixation for the duration of the delay period, or saccaded to the non-targeted "distracter" color-coded stimulus or to anywhere else.

As expected, behaviorally the monkeys' saccade accuracy and speed increased significantly to target stimuli indicating high, as opposed to low, reward value. Interestingly, these behavioral effects matched statistically the increases in speed and accuracy of saccades of macaques engaged in strictly visual attentional tasks without differential reward expectations; saccade speed and accuracy increase similarly when saccades are directed to attended rather than unattended target locations (e.g., Desimone & Duncan, 1995). More interestingly, activity rates (action potential frequencies) in individual V1 neurons were also influenced by the expected reward values, with increased gain in action potential rate with successful saccades to target stimuli indicating high reward expectation as compared to those indicating low reward expectation (see Stănişor et al. (2013), Fig. 2). Just as with the behavioral component of this first experiment, these increases in neuronal activity in V1 neurons to high reward cues statistically matched increases found in prior studies using purely spatial attention cues (without differential reward expectations) that directed the animal to attend to the receptive fields of the individual V1 neurons being recorded from, as compared to attention directed outside those neurons' receptive fields (e.g., Roelfsema et al., 2003). These statistical identities suggested that, both behaviorally and neurophysiologically, activity driven by high reward expectation cues and that driven by selective attention location cues might result from activation of a single neural mechanism. If that was true, then what cognitive science distinguishes as two distinct kinds of phenomena, reward expectation and selective visual attention, and even operationalizes experimentally using prior training to different visual cues, would turn out to be a single mechanism from the perspective of neurobiology. This discovery would give us a second kind of cross-level mismatch between counterpart cognitive science and neurobiology, but here not a mismatch between explanations but rather between phenomena. Phenomena that cognitive science distinguishes, namely visual attention location and reward expectation, neurobiology would unite into one and the same mechanism. But could a unitary neurobiological mechanism for visual attention and reward expectation be demonstrated experimentally, directly on the lab bench?

To explore this question, Stănişor et al. (2013) developed a second experiment. The monkeys' task was similar to that in their first experiment, but with one additional delay period inserted prior to the monkeys' making the to-be-rewarded saccade to the target stimulus location. In this second experiment the initial fixation point appears, indicating a new trial; the monkey fixates it, and then the two curves appear with the strength of reward expectation associated with each saccade target cued by its color. Then the first delay period ensues, before the appearance of the length-of-curve cue indicating the location of the target that is to be rewarded for the upcoming saccade. So far this experiment exactly matches the methodology of the first experiment. But with the appearance of the location-of-target cue, the fixation point still remains illuminated, initiating a further delay period. The monkey must wait through this second delay, maintaining fixation on the initial fixation point, then when the fixation point finally extinguishes the monkey must saccade to the indicated target stimulus in order to receive the amount of fruit juice reward indicated by its color (low, medium, or high). Activity in individual V1 neurons was recorded during the first delay after the presentation of the stimulus reward cue in those neurons' visual fields, as in the first experiment; but now also during the second delay period after the location-of-target stimulus has been indicated but while the monkey cannot yet execute the saccade. What was the purpose of V1 recordings during this second delay period? Stănişor et al. reasoned as follows:

> If reward cues and attentional cues have separable effects, they are expected to interact in an additive manner and to influence independent groups of neurons. However, if there is a single neuronal selection mechanism at the level of V1, then (i) the neuronal effects of reward and attention cues might interact nonadditively because the effect of one cue might occlude the effect of the other one and (ii) neurons influenced by reward cues should also be affected by the central attentional cue. (2013, 9138)

Comparison of activity in individual V1 neurons during the first and second delays thus permitted an experimental test of the independence or non-independence of the neurobiological mechanisms of the reward expectation cue and the visual attention location cue at the earliest stage of cortical visual processing.

Stănişor et al.'s results from this second experiment were unequivocally in favor of the neurobiological *non*-independence of reward expectation and visual attention location cues (2013, Fig. 3). Just as in their first experiment, during the first delay period, before the animal learned which target location would be rewarded, the reward value of the cue in the individual V1 neuron's receptive field was a good predictor of its activity rate; cues for higher expected rewards in a neuron's receptive field drove higher action potential frequencies. However, during the second delay, when the to-be-rewarded target location had been distinguished from the distractor by the visual attention cue but the animal still wasn't permitted to execute the saccade, and specifically on trials in which the monkey subsequently saccaded to the correct target location, correlation between V1 neuron activity and reward value of the cue in its receptive field became nonsignificant. In fact, reward value of the cue lost all influence on individual V1 neuron activity during the second delay. Activity evoked by the curve in the V1 neuron's receptive field with a high reward value stayed high if that curve was later cued as the target location before the second delay period, but decreased to baseline if that curve became the distractor. Conversely, activity evoked by a curve with a low relative reward value stayed low during the second delay if that curve became the distractor, but increased to the level evoked by a high-reward expectation cue if that curve became the target. Thus the effects of the visual attention target cue during the second delay period simply replaced the relative reward value cueing effects during the first delay on V1 neuron activity, so long as the monkey subsequently correctly selected the target curve after the second delay. As Stănişor et al. put it in the paper's Discussion section, "the present results show that the effects of the reward cues on neuronal activity in V1 are the same as the effects of central attention cues in many respects" (2013, 9140). In particular, these effects were the same in the two respects the experimenters had hypothesized if reward expectation and visual attention cues were not independent: reward expectation and visual attention cues interacted nonadditively on individual V1 neurons; and the same V1 neurons were affected by the two types of cues.

How best to explain these similar effects of reward expectation and visual attention cues? Stănişor et al. offer "a likely explanation … the central [attentional] cues also influence the relative value of the two curves" (2013, 9140). In other words, "the present results support the hypothesis that studies on selective attention … and relative value … in the visual cortex investigated the same selection process" (2013, 9140). As Maunsell had noted in the review paper cited earlier in this section, often monkeys are trained to attend to specific stimuli or locations by varying reward contingencies to attended versus unattended stimuli; and studies of reward processing likewise typically influence the distribution of subjects' selective attention across the displayed stimuli. According to Stănişor et al. "a single selection process is at work ... rewarding stimuli attract attention in proportion to their value" (2013, 9140). What might this single selection process be? Since their presentation of target and distractor stimuli locations on a single trial required comparisons across regions far apart in visual space, Stănişor et al. speculate that the modulation of activity in individual V1 neurons, each with relatively limited receptive fields, "likely ... depends on feedback from higher visual and frontal areas and the amygdala where the representation of reward value depends less on the spatial configuration of stimuli" (2013, 9140). Their suggestion is that visual attention location and reward expectation generate activity in one single feedback mechanism from these higher visual processing areas back to V1 neurons. This feedback affects processing to specific visual stimuli in V1 and subsequent processing further up the visual pathways.

Whatever this feedback pathway might be, the mismatch implications of this study concerning cognitive science and neurobiology are intriguing. Cognitive science distinguishes reward expectation and selective visual attention phenomena. In behavioral studies, these notions are even operationalized using prior training to different visual cues, e.g., in the Stănişor et al. study attention to target location as opposed to distractor location stimulus was indicated by length of saccade curve, while (quantity of) reward expectation was operationalized by prior training to differently colored target stimuli. From the perspective of neurobiology, however, a single mechanism is found to affect activity at the earliest stages of cortical visual processing in response to both attention location and reward expectation cues. In this type of cross-level mismatch, phenomena that cognitive science distinguishes, neurobiology reveals to evoke a single mechanism. From the perspective of neurobiology, a distinction between cognitive phenomena, even one distinguished operationally in behavioral studies, is not found in the brain. As vision neuroscientist John Maunsell put it in the quote that began this section, "regarding the relationship of attention and reward expectation," cognitive science appears to be "using different terms to describe a common mechanism."

In the previous section we examined ways we might interpret mismatches of counterpart *explanations* across cognitive science and neurobiology. Interestingly here, none of the ones to remain available when cognitive scientists accept a common "neural plausibility" constraint on its explanations look particularly promising in this case of "lumping" differentiated cognitive phenomena into one and the same neurobiological mechanism. A distinction among phenomena recognized by cognitive science gets eliminated. But have visual attention location and reward expectation qua cognitive *phenomena* been little-e eliminated. Surely visual attention location and reward expectation still exist, even if they are subserved by one and the same neurobiological pathway. They can still be operationalized distinctly in a study that manipulates both! Yet these two ways of operationalizing visual attention location and reward expectation doesn't "save" their distinctiveness as phenomena from the perspective of neurobiology. From that perspective the distinct cues for each of these cognitive phenomena appear to be activating one and the same neurobiological mechanism. It is also not clear what the 'autonomy' of cognitive science amounts to, or what comfort it offers the cognitive scientist, when we learn that these separable cues turn out to activate one and the same neurobiological mechanism from upstream back to downstream visual processing areas.[15]

4 Cross-Level Mismatches Beyond the Behavioral and Brain Sciences

I close with an observation. There is nothing unique in the ways explored here about cognitive science and MCC neurobiology. Cross-level explanation and phenomena mismatches of these types, and probably of other types, are possibilities whenever two sciences pitched at different "levels" of biological organization begin to investigate related phenomena. These kinds of mismatches are not inevitable; plenty of counterpart cross-levels scientific explanations and phenomena align smoothly with one another. But mismatches are also not impossible or unheard of. And their existence and possibility complicates prospects for the interdisciplinary aspirations of contemporary science writ large. Most molecular biologists accept a demand on the biochemical and physical ('physics-cal') plausibility of their explanation and phenomena. Some sociologists and economists aspire for their explanations to be psychologically or even neuroscientifically plausible. Sometimes existing explanations from a science focused on the system turn out to link smoothly with ones from sciences investigating that system's components, their dynamics and organization. But sometimes serious cross-level mismatches occur. If we aspire for science to link electrons, elephants, and elections, we must acknowledge cross-science explanation and phenomena mismatches as well as matches. And we must figure out what these mismatches imply for the status of the higher-level explanations and phenomena. To not do so would be to fail to include some actual episodes from science-in-practice, and so would limit the scope of any metascientific relationship we derive from studies of how cross-level sciences link up.

References

Bechtel, W. (2009). Molecules, systems and behavior: Another view of memory consolidation. In J. Bickle (Ed.), *The Oxford handbook of philosophy and neuroscience* (pp. 13–40). Oxford University Press.

Bechtel, W. (2008). *Mental mechanisms: Philosophical perspectives on cognitive science*. Taylor & Francis.

Bickle, J. (2003). *Philosophy and neuroscience: A Ruthlessly reductive account*. Springer.

[15] Thanks again to David Colaço for helping me clarify some o the ways this second kind of cross-level mismatch differs from the first kind discussed in Sects. 1 and 2 above.

Bickle, J. (2006). Reducing mind to molecular pathways: Explicating the reductionism implicit in current cellular and molecular neuroscience. *Synthese, 151*, 411–434.

Bickle, J. (2014). Little-e elminativism in mainstream cellular and molecular neuroscience: Tensions for neuro-normativity. In C. T. Wolfe (Ed.), *Brain theory* (pp. 134–148). Palgrave Macmillan.

Bickle, J. (2012). A brief history of neurosciences's actual influences on mind-brain reductionism. In S. Gozzano & C. Hill (Eds.), *New perspectives on type identity theory* (pp. 88–109). Cambridge University Press.

Bickle, J. (2020). Multiply realizability. In E.N. Zalta (Ed.), *The Stanford encyclopedia of philosophy* (Summer 2020 Edition). https://plato.stanford.edu/archives/sum2020/entries/multiple-realizability/

Bogen, J., & Woodward, J. (1988). Saving the phenomena. *Philosophical Review, 97*(3), 303–352.

Churchland, P. M. (1981). Eliminative materialism and the propositional attitudes. *Journal of Philosophy, 78*, 67–90.

Churchland, P. S. (1986). *Neurophilosophy*. MIT Press.

Colaço, D. (2020). Recharacterizing scientific phenomena. *European Journal of Philosophy of Science, 10*(2), 1–19.

Craver, C. F. (2007). *Explaining the brain.* Oxford University Press.

Craver, C. F., & Darden, L. (2013). *In search of mechanisms.* University of Chicago Press.

Desimone, R., & Duncan, J. (1995). Neuronal mechanisms of selective visual attention. *Annual Review of Neuroscience, 18*, 193–222.

Genoux, D., Haditsch, U., Knobloch, M., Michalon, A., Storm, D., & Mansuy, I. M. (2002). Protein phosphatase 1 is a molecular constraint on learning and memory. *Nature, 418*(6901), 970–975.

Gillett, C. (2016). *Reduction and emergence in science and philosophy*. Cambridge University Press.

Han, J.-H., Kushner, S. A., Yiu, A. P., Cole, C. I., Matynia, A., Brown, R. A., Neve, R. I., Guzowski, J. F., Silva, A. J., & Josselyn, S. A. (2007). Neuronal competition and selection during memory formation. *Science, 316*(5823), 457–460.

Han, J. -H., Kushner, S. A., Yiu, A. P., Hsiang, H. -L., Buch, T., Waisman, A., Bontempi, B., Neve, R. I., Frankland, P. W., & Josselyn, S. A. (2009). Selective erasure of a fear memory. *Science, 323*(5920), 1492–1496.

Larson, S. D., & Martone, M. E. (2009). Ontologies for neuroscience: What are they and what are they good for? *Frontiers in Neuroscience* (May 01, 2009). https://doi.org/10.3389/neuro.01.007.2009

Maddox, G. B. (2016). Understanding the underlying mechanisms of the spacing effect in verbal learning: A case for encoding variability and study-phase retrieval. *Journal of Cognitive Psychology, 28*(6), 684–706.

Maunsell, J. H. R. (2004). Neuronal representations of cognitive state: Reward or attention? *Trends in Cognitive Science, 8*, 261–265.

Maunsell, J. H. R. (2015). Neuronal mechanisms of visual attention. *Annual Review of Vision Science, 1*, 373–391.

McKusick, V. A. (1969). On lumpers and splitters, or the nosology of genetic disease. *Perspectives in Biology and Medicine Winter, 12*(2), 298–312.

Piccinini, G., & Craver, C. (2011). Integrating psychology and neuroscience: Functional analyses as mechanism sketches. *Synthese, 183*(3), 283–311.

Polger, T. (2004). *Natural minds*. MIT Press.

Polger, T., & Shapiro, L. (2016). *The multiple realization book*. Oxford University Press.

Roelfsema, P. R., Khayat, P. S., & Spekreijse, H. (2003). Subtask sequencing in the primary visual cortex. *Proceedings of the National Academy of Sciences (USA), 100*(9), 5467–5472.

Silva, A. J., Landreth, A., & Bickle, J. (2014). *Engineering the next revolution in neuroscience.* Oxford University Press.

Silva, A. J., & Josselyn, S. A. (2002). The molecules of forgetfulness. *Nature, 418*(6901), 929–930.

Spear, N. E., & Riccio, D. C. (1994). *Memory: Phenomena and principles*. Allyen and Bacon.

Squire, L. (1987). *Memory and brain.* Oxford University Press.

Stănişor, L., van der Togt, C., Cyriel, M. A., & Roelfsema, P. R. (2013). A unified selection signal for attention and reward in primary visual cortex. *Proceedings of the National Academy of Sciences (USA), 110*(22), 9136–9141.

Woodward, J. (2003). *Making things happen: A theory of causal explanation.* Oxford University Press.

Woodward, J. (2018). Explanations in neurobiology: An interventionist perspective. In D. M. Kaplan (Ed.), *Explanation and integration in mind-Brian science* (pp. 70–100). Oxford University Press.

Zhou, Y., Won, J., Karlsson, M. G., Zhou, M., Rogerson, T., Balaji, J., Neve, R., Poirazi, P., & Silva, A. J. (2009). CREB regulates the excitability and the allocation of memory to subsets of neurons in the amgydala. *Nature Neuroscience, 12*, 1438–1443.

Biology

Some Remarks on Epigenetics and Causality in the Biological World

Luciano Boi

1 Introduction

The understanding of the amazing complexity and plasticity of high living organisms requires we investigate their organismal properties and their interactions with a larger, both natural and cultural environment (Jacob, 1970; Boi, 2011a; Emmeche, 1997). We are faced with the need of shifting from the local description of the molecular-genetic structure of DNA to the study of the networks of interactions between early development and evolution of complex living organisms, and of the way in which epigenetic, environmental and social factors affects the behavior and response of living beings. It is important to highlight the enormous impact of epigenetic and environmental phenomena on biological, cognitive and social processes. Another significant goal is to show that epigenetic, ecological and cultural effects can also be inherited across generations. This notably means that natural history and human evolution have been shaped by gene-culture and organisms-environments interactions. Thus, our 'living environments', such as natural and urban landscapes, learning processes and cultural contexts, can have a profound effect on phenotypic variations and human evolution; conversely, human activities and cultural practices can modify organism' metabolism and the contingent 'history' of evolution' paths. We think that this cross-disciplinary dialogue may provide novel insights into how nature and culture are deeply interrelated and is essential for bringing biological sciences and social sciences together in a very new perspective.

Within this perspective, we specifically pursue a twofold objective. First, we want to show that positive selection of morphological and functional capabilities during evolution may have developed in response to human cultural practices. Next, that the organisms-environments systems have co-evolved over evolutionary time

L. Boi (✉)
Ecole des Hautes Etudes en Sciences Sociales, Center of Mathematics and Program of Epistemology, 54, boulevard Raspail, 75006 Paris, France
e-mail: lboi@ehess.fr

S. Wuppuluri and I. Stewart (eds.), *From Electrons to Elephants and Elections*,
The Frontiers Collection, https://doi.org/10.1007/978-3-030-92192-7_36

through different interacting processes (for example: symbiosis as a source of novel traits; huge geographic migrations as a cause of genetic evolutionary and linguistic variation; radioactive large-rate pollution as a factor of genomic and phenotypic mutation) and niche-construction, that is the capacity of organisms to affect natural selection and evolution by modifications of their micro- and macro environments.

2 The Different Meanings of Epigenetics

To some, epigenetics means nuclear inheritance not based on differences in DNA sequence, i.e. trans-generational effects and inherited expression states; to others, epigenetics belongs in the province of expressed nucleic acid information, that is changes in gene expression (Lederberg, 1958; Holliday, 1994). The word epigenetics has essentially two different meaning following that it is used by molecular biologists or by morphologists and systems biologists. Molecular biologists favored the restricted definition of epigenetics as the study of mitotically and/or meiotically heritable changes in gene function that cannot be explained by changes in DNA sequences (Morange, 2000). For them, epigenetic mechanisms would include DNA methylation and histone modification. Functional morphologists, however, preferred a larger and more global definition such as that of Herring (1993), for whom epigenetics refers to the entire series of interactions among cells and cell products which leads to morphogenesis and differentiation. Thus, epigenetic influences range from hormones and growth factors to ambient temperature and orientation in a gravitational field.

Although serviceable, these views of epigenetic events tend to be very provincial. They neglect the current consequences of past history, and underestimate the non-genetic contributions to the phenotype. Perhaps, those working with animal behaviour have been most ready to include non-genetic influences in formulation of adult behaviors (phenotype). The development of bird songs (e.g., Konishi, 1965; Marler, 1990) and of offspring imprinting (e.g., Lorenz, 1965) would be a few examples. Similarly, morphologists are in a position to recognize the importance of non-genetic features that contribute to the phenotype of an organism and how it evolves. To do so, and place epigenetics in a modern idiom, we should adopt and slightly expand on the term epigenomics, meaning the analysis of the normal non-genetic processes that influence the characteristics of the phenotype during the lifetime of the organism and the historical-contingent influences included. These events occur above (hence epi-) and beyond the level of the DNA (hence genomic). Note that, because the phenotype is also an historical product, evolutionary events important to epigenomics must similarly be incorporated into analysis (Strohman, 1997).

Examples of non-genetic contributions to the phenotype have been known for some time, have been repeated frequently in scientific publications, and have received extensive reviews (Hall, 1970). The sex of many reptiles depends upon the temperature, or schedule of temperatures, the embryos experience while in the egg (Webb & Cooper-Preston, 1989). Nutritional deficiencies lead to phenotypic changes. Calcium deficiencies during infancy lead to rickets. In humans, bound feet, wrapped skulls and cradle boards all produce modified mechanical demands which result in modifications of the skeleton. Careful experimental manipulations illustrate the importance of mechanical events in the differentiation of the phenotype. Leg muscles of the developing chick contract irregularly while still in the egg, thereby producing an intermittent environment of mechanical stresses experienced by the femur (and other leg bones). If deprived of this mechanical environment, by removing embryonic muscles or growing the femur as an explant, then the developing femur is misshapen; its diaphysis is bowed, and its ends indistinct (Murray, 1936). Removing the temporalis jaw muscles and/or the cervical muscles in-day-old rats resulted (3–5 months) in morphological deficiencies later; the coronoid process was lost as were skull ridges at the site of what would be the muscle origin (Washburn, 1947). Less radical interventions in the mechanical environment produced by muscles have involved simply denervation of limb muscles, which similarly produce deficiencies in the bony phenotype (Lanyon, 1980).

In all these experiments, the genome remained unaltered; only the environment of mechanical influences was modified. These are influences outside the genes: part of the epigenomic environment that in turn controls selective gene expression. But these epigenomic contributions are more than just mechanical or nutritional in character. When grown in a bacteria-free environment, the usually leafy marine alga *Ulva* instead becomes filamentous. A rotifer, when placed in an environment with its natural predators, grows protective spine-like projections. Biotic factors in the environment affect phenotypic outcomes. In fact, the genes do not initiate these morphological modifications to serve in a hostile environment, but instead the biotic information from the environment itself initiates gene action.

The interaction between phenotype—form-function complexes—(Bock & Wahlert, 1965; Bock, 1998) and external environment places demands upon individual organisms with which they must cope to survive and reproduce. These demands that arise are selective agents (Bock, 1998) if they participate directly in the culling of phenotypes from the population. Selective agents (selective forces, selective demands), arising out of organismal interaction with the environment, are causative agents within an evolutionary context. They help explain evolutionary outcomes (Ayala and Dobzhansky, 1974).

Certainly, other interactions of organisms with their environment may result, eventually, in changes in the phenotype. One such interaction results in mutations, changes in the genome, as occurs for example in exposure to environmental radiation. Another is the epigenomic influence acting on an organism during its lifetime (Bock, 1998).

3 Epigenetic Phenomena Mediate the Link Between Environmental Factors and the Processes of Inheritance

Environmental agents and genetic variants can induce heritable epigenetic changes that affect phenotypic variation and disease risk in many species (Holliday, 1987; Jablonka and Lamb, 1989). These trans-generational effects challenge conventional understanding about the modes and mechanisms of inheritance, but their molecular basis is poorly understood. These heritable epigenetic changes persisted for multiple generations and were fully reversed after consecutive crosses through the alternative germ-lineage.

Many environmental factors and genetic variants are known to induce heritable epigenetic changes that can persist for multiple generations, affecting a broad range of traits, and that often are as frequent and strong as direct environmental exposures and conventional genetic inheritance. These trans-generational effects challenge our understanding of the modes and mechanisms for inherited phenotypic variation and disease risk, as well as the premise of most genetic studies in which causal DNA sequence variants are sought within the genome of affected individuals. Several molecular mechanisms have been implicated, ranging from inherited RNAs to chemically modified DNA and proteins. These trans-generational effects have important implications for our understanding of adaptation and evolution, the origins of phenotypic variation and disease risk, and the molecules in addition to DNA that can be the basis for inheritance.

The fetal basis of adult disease or the 'early origins' hypothesis postulates that nutrition and other environmental factors during prenatal and early postnatal development influence developmental plasticity (West-Eberhard, 2005), thereby altering susceptibility to adult chronic diseases. Developmental plasticity occurs when environmental influences affect cellular pathways during gestation, enabling a single genotype to produce a broad range of adult phenotypes. This emerging field of research also points to the *epigenotype* as an important modifier of disease susceptibility. Aberrant epigenetic gene regulation has been proposed as a mechanism of action for non-genotoxic carcinogenesis, imprinting disorders, and complex disorders including Alzheimer's disease, schizophrenia, asthma, and autism.

Within the above context, epigenetics is defined as the study of the heritable changes in gene expression that occur without a change in DNA sequence. These heritable epigenetic changes include DNA methylation, post-translational modifications of histone tails (acetylation, methylation, phosphorylation, etc.), and higher order packaging of DNA around nucleosomes. Epigenetic modifications are inherited not only during mitosis, but also can be transmitted trans-generationally.

Thus, identifying epigenetic targets and defining how they are deregulated in human disease by environmental exposures will allow for the development of innovative novel diagnostic, treatment, and prevention strategies that target the 'epigenomic software' rather than the 'genomic hardware'. Furthermore, characterizing the important environmental exposures affecting the epigenome and determining the critical windows of vulnerability to epigenetic alterations will influence environmental risk assessment.

The epigenome is particularly susceptible to deregulation during gestation, neonatal development, puberty, and old age. Nevertheless, it is most vulnerable to environmental factors during embryogenesis because the DNA synthetic rate is high, and the elaborate DNA methylation patterning and chromatin structure required for normal tissue development is established during early development.

Epigenetic adaptations in response to in utero nutritional and environmental factors are hypothesized to play an important role in developmental plasticity and disease susceptibility. Because diet-derived methyl donors and co-factors are necessary for the synthesis of the *S-adenosylmethionine* (SAM), which provides the methyl groups required for DNA methylation, environmental factors that alter early nutrition and/or SAM synthesis can potentially influence adult phenotype via alterations in CpG methylation at critically important, epigenetically labile regions in the genome. Environmental factors, including xenobiotic chemicals, behavior, and even low dose radiation, can also directly affect methylation and chromatin remodeling factors to alter the fetal epigenome and subsequent gene expression patterns. Furthermore, epigenetic alterations have also been observed in response to post-natal and adult exposure to environmental factors. Below, we summarize the effects observed in a number of recent studies.

Alterations in epigenotype have also been observed following exposure to environmental xenobiotic chemicals. Exposure of adult mice to sodium arsenite in vivo revealed decreased genomic methylation, while co-exposure to sodium arsenite and methyl deficient diet showed gene-specific hypomethylation in the promoter region of the oncogenic gene, *Ha-ras*. In vitro arsenic studies also revealed global DNA hypomethylation as well as increased expression of the oncogenic *K-ras* gene; however, gene specific changes in methylation were not observed following arsenic exposure, indicating that *K-ras* overexpression occurs via a non-DNA methylation dependent mechanism. Other metals, including cadmium, lead, and nickel, have also been shown to interact with the epigenome. In addition, decreased histone acetylation, increased histone methylation, and subsequent decreased gene expression occur following nickel exposure. Furthermore, chromium exposure is linked to epigenetically controlled gene expression alterations via interactions with histone acetyltransferase and histone deacetylase enzymes.

Endocrine active chemicals have also been associated with epigenetic alterations following in utero and adult exposures. Recently, methylation studies on the estrogenic pharmaceutical agent diethylstilbestrol (DES) observed hypomethylation in two critical DNA control regions in mice exposed in utero or in the perinatal period. DES is a non-genotoxic carcinogen that was prescribed to millions of pregnant women from the early 1940s to the early 1970s. Individuals exposed in utero during the first three months of pregnancy exhibited increased incidences of reproductive disorders and the rare cancer, clear cell adenocarcinoma of the vagina. Increased incidences of these uncommon disorders were also seen in DES granddaughters, suggesting epigenetic transgenerational inheritance. Furthermore, recent mouse studies demonstrated that the effects of maternal DES exposure are transmitted through the maternal germline to offspring via both genetic and epigenetic mechanisms.

The endocrine active compound, bisphenol A (BPA), is similarly associated with epigenetic alterations following developmental exposure to environmentally relevant levels. BPA is a high-production volume additive for many plastics including food containers, baby bottles, and dental composites. Recently, detectable levels of BPA were observed in 95% of samples from a human reference population ($n = 394$). Ho et al. (2012) observed multiple changes in gene-specific DNA methylation patterns in the rat prostate, including hypomethylation of the phosphodiesterase type 4 variant 4 (PDE4D4) gene following neonatal exposure to both estradiol and low-level BPA. Decreased PDE4D4 methylation is associated with a marked increase in prostate cancer risk. Since methylation levels are detectable well before disease presentation, it should be investigated as an early molecular marker for prostate cancer risk assessment.

4 On the Link Between Epigenetics and Diseases, Mediated by Aberrant Chromatin Alterations

The relationship between the global-topological and functional organization of high-order chromatin structure and the expression and cellular activity of the genomes is a very good example of reciprocal interaction between upward and downward causation (Lillie, 1940; Campbell, 1974; El-Hani and Emmeche, 2000). In order to highlight the fundamental fact that the organizational properties of chromatin influence the genome activity, in the sense that they are the principal carrier and activator of the multilevel genetic and epigenetic information, let us now consider the epigenome of a sick cell. Most human diseases have an epigenetic cause. The perfect control of our cells by DNA methylation, histone modifications, chromatin remodeling and microRNAs become dramatically distorted in the sick cell (Esteller, 2008). In other words, severe alterations of nuclear forms and especially of the chromatin and the chromosome may provoke different damages to the cell's activity, suggesting thus that the topological form of living systems is likely one of the most fundamental determinants of the unfolding of biological functions during development and evolution (Cremer, 2006; Boi, 2021). The groundbreaking discoveries have been initially made in cancer cells, but it is just the beginning of the characterization of the wrong epigenomes underlying neurological, cardiovascular and immunological pathologies.

In human cancer, the DNA methylation aberrations observed can be considered as falling into one of two categories: transcriptional silencing of tumor suppressor genes by CpG island promoter hypermethylation in the context of a massive global genomic hypermethylation (Ballestar and Esteller, 2005; Hammond et al., 2001). CpG islands become hypermethylated with the result that the expression of the contiguous gene is shut down. If this aberration affects a tumor suppressor gene it confers a selective advantage on that cell and is selected generation after generation. Recently, researchers have contributed to the identification of a long list of hypermethylated genes in human neoplasias, and this epigenetic alteration is now considered to be a common hallmark of all human cancers affecting all cellular pathways. At the same time as the aforementioned CpG islands become hypermethylated, the genome of the cancer cell undergoes global hypomethylation. The malignant cell can have 20–60% less genomic 5mC than its normal counterpart. The loss of methyl groups is accomplished mainly by hypomethylation of the "body" (coding regions and introns) of genes and through demethylation of repetitive DNA sequences, which account for 20–30% of the human genome.

How does global DNA hypomethylation contribute to carcinogenesis? Three mechanisms can be invoked as follows: chromosomal instability, reactivation of transposable elements and loss of imprinting (Fraga et al., 2005). Undermethylation of DNA may favor mitotic recombination, leading to loss of herezygosity as well as promoting karyotypically detectable rearrangements. Additionally, extensive demethylation in centromeric sequences is common in human tumors and may play a role in aneuploidy. As evidence of this, patients with germline mutations in DNA methyltransferase 3b (*DNMT3b*) are known to have numerous chromosome aberrations. Hypomethylation of malignant cell DNA can also reactivate intragenomic parasitic DNA, such as L1 (Long Interspersed Nuclear Elements, LINEs) and Alu (recombinogenic sequence) repeats. These, and other previously silent transposons, may now be transcribed and even "moved" to other genomic regions, where they can disrupt normal cellular genes. Finally, the loss of methyl groups can affect imprinted genes and genes from the methylated-X chromosome of women. The best-studied case is of the effects of the H19/IGF-2 locus on chromosome 11p15 in certain childhood tumors. DNA methylation also occupies a place at the crossroads of many pathways in immunology, providing us with a clearer understanding of the molecular network of the immune system (Bhalla and Iyengar, 1999). Besides, aberrant DNA methylation patterns go beyond the fields of oncology and immunology to touch a wide range of fields of biomedical and scientific knowledge (Horrobin, 2003).

Regarding histone modifications, we are largely ignorant of how these histone modification markers are disrupted in human diseases. In cancer cells, it is known that hypermethylated promoter CpG islands of transcriptionally repressed tumor suppressor genes are associated with hypoacetylated and hypermethylated histones H3 and H4. It is also recognized that certain genes with tumor suppressor-like properties such as p21WAF1 are silent at the transcriptional level, in the absence of CpG island hypermethylation in association with hypoacetylated and hypermethylated histones H3 and H4. However, until very recently there was not a profile of overall histone modifications and their genomic locations in the transformed cell. This need to determine the histone modification pattern of tumors was even more urgent, given the rapid development of histone deacetylase inhibitors as putative anticancer drugs. It has been provided this missing linking demonstrating that human tumors undergo an overall loss of monoacetylation of lysine 16 and trimethylation of lysine 20 in the tail of histone H4 (Fraga et al., 2005). These two-histone modification losses can be considered as almost universal epigenetic markers of malignant transformation, as has now been accepted for global DNA hypomethylation and CpG island hypermethylation. Certain histone acetylation and methylation marks may have prognostic value. For other human pathologies, research is still in the infancy to define their histone modification signatures.

The most important theme of the previous remarks on the human epigenome, which is mainly related to a methodological and epistemological revolution in epigenetics, may be summarized as follow. Cells of a multicellular organism are genetically homogeneous but structurally and functionally heterogeneous owing to the differential expression of genes. Many of these differences in gene expression arise during development and are subsequently retained through mitosis. Stable modifications of this kind are said to be "epigenetic", because they are heritable in the short term but do not involve mutations of the DNA itself. The two most important nuclear processes that mediate epigenetic phenomena are DNA methylation and histone modifications. Epigenetic effects by means of DNA methylation have an important role in development but can also arise stochastically as humans and animals age. Identification of proteins that mediate these effects has provided insight into this complex process and diseases that occur when it is perturbed. External influences on epigenetic processes are seen in the effects of diet on long-term diseases such as cancer. Thus, epigenetic mechanisms seem to allow an organism to respond to the environment through changes in gene expression. The extent to which environmental effects can provoke epigenetic response is a crucial question which is still largely unanswered.

5 How Chromatin Arrangements Influence Gene Expression

Chromatin is the substrate engaged by the molecular mechanisms responsible for replication, recombination and transcription in the eukaryotic cell nucleus. Chromatin structure is fundamental to the mechanisms of these processes and undergoes modifications to facilitate their regulation and progression. The nucleosome is the fundamental repeating unit of chromatin. Arrays of nucleosomes compact to form a 30 nm chromatin fiber (Boi, 2007; 2011a, 2011b, 2021).

The fundamental building block of chromatin is the nucleosome comprising 157–240 bp of DNA, to each of the four core histone proteins, and a single linker histone H1/H5. The nucleosome core is the greater part of the nucleosome and contains 147 base pairs of DNA wrapped in 1.67 left-handed superhelical turns around the histone octamer. Arrays of nucleosome in their most compact form constitute the 30 nm chromatin fiber. The crystal structure of the nucleosome core particle refined to 1.9A (Avogadro) resolution reveals the details of DNA conformation as well as all the direct and the water-mediated histone-contacts. The acute DNA bending induced by the histone proteins results in an alteration of the form of the double helix every five base pairs along its superhelical path. Sequence-dependent DNA conformations are apparent.

The eukaryotic cell contains thousands of genes, only a few of which are expressed at any one time. The fundamental goal of current work among biologists is to find how genes activation is achieved within the vast genome, which is packed into a highly condensed state. Genomic organization events in the cells of higher-organisms (but not bacteria) are achieved via the ordering of chromatin, the structural proteins that bind DNA, forming a proteinaceous coat around the genetic material.

Chromatin consists of the core histone proteins, linker histones, and non-histones proteins. The histone core, containing eight protein subunits, is the central structure of a 206-kiloDalton disc-shaped chromatin structure known as the nucleosome. Nucleosomal DNA winds itself 1.65 times around each histone protein core, generating a structure like knots on a string. The nucleosome particles are spaced by around 200 $\pm$ 40 bp along the genome. Between nucleosomes, linker histone stabilizes the string into a solenoid-type structure that compresses the genome.

So, while nucleosome provide the first level of compaction of DNA into the nucleus, the chromatin nucleofilament represents a higher level of compaction that nucleosome can adopt ultimately resulting in the highly condensed metaphase chromosome. The combined approaches of cell biology and genetics studies have led to the discovery that within an interphase nucleus chromatin is organized into different structural and functional territories.

Based on microscopic observations, chromatin has been divided into two distinct domains, heterochromatin and euchromatin. Heterochromatin was defined as a structure that does not alter in its condensation throughout the cell cycle whereas euchromatin is decondensed during interphase. Typically, in a cell, heterochromatin is localized principally on the periphery of the nucleus and euchromatin in the interior of the nucleoplasm. One distinguishes constitutive heterochromatin, containing few genes and formed principally of repetitive sequences located in large regions coincident with centromeres and telomeres, from facultative heterochromatin composed of transcriptionally active regions that can adopt the structural and functional characteristics of heterochromatin, such as the inactive X chromosome of mammals.

Chromatin is the basic organizational form of DNA in the eukaryotic nucleus. The repeat unit of chromatin is the core nucleosome in which 146 base pairs of DNAs' are wrapped around the histone octamer that consists of two molecules each of the core histones H2A, H2B, H3 and H4 (see Figs. 2 and 3). Nucleosomal arrays along the DNA are proposed to fold into a 30 nm fiber upon incorporation of the linker histone H1. In addition to the canonical histones, histone variants exist that are structurally related to the normal histones, but are functionally distinct. Another level of chromosome compaction is required at mitosis, where condensin and cohesin proteins bind to chromatin to yield the highly condensed mitotic chromosomes during cell division. Two main enzymatic activities can be distinguished that regulates chromatin access: chromatin modifying complexes, and chromatin remodelling complexes, which we try to describe briefly.

Histone modifying complexes play an essential role in the structural organization and functional regulation of the chromosome. The term "chromatin modification" describes posttranscriptional modifications on the histones. Potential modifications include histone acetylation, methylation, phosphorylation, ubiquitylation, sumoylation and ADP-ribosylation. Most modifications were originally observed on the N-terminal tails of histones. Histone modifications may affect chromatin structure directly by altering DNA-histone interactions within and between nucleosomes, thus changing higher-order chromatin structure. An alternative, more recent model is that combinations of histone modifications present an interaction surface for other proteins that translate this so-called histone code into a gene expression pattern.

Another important aspect, closely related to the previous—the histone modifying complexes—concerns the chromatin remodelling complexes. Classically, chromatin-remodelling factors have been described as activities that, unlike chromatin modifiers, leave the biochemical make-up of the nucleosomes unaffected. The remodellers either change the location of the nucleosome along a particular DNA sequence, or create a remodeled state of the nucleosome that is characterized by altered histone-DNA interactions. Recent work, however, has significantly broadened the spectrum of their activities by adding the complete removal of histones as well as histone exchange to the palette of reactions catalyzed by chromatin remodelling complexes. In these reactions, remodellers use the energy freed by ATP hydrolysis to loosen DNA-histone contacts and thus to facilitate the movement of the nucleosome.

From the previous remarks, it appears clearly that the topological form of DNA molecule, the structural modification of the chromatin and the spatial architecture of the chromosome are events which have an important influence on how gene become expressed and the genome functions in cells. Moreover, these three levels of organization of organism's nuclear components are deeply interrelated, they obey to many complexes systems of regulatory co-factors, and they further affect globally the physiology and metabolism of cells (Smet-Nocca et al., 2006). Among these different families of proteins regulatory complexes, the remodellers of chromatin structure play a fundamental role in replication and repair of DNA sequences and in the transcriptional and post-transcriptional activities of the entire genome (Scherer and Jost, 2007).

6 Organisms and Environment

Niche construction is the process whereby organisms, through their activities and choices, modify their own and each other's niches. By transforming natural selection pressures, niche construction generates feedback in evolution at various different levels. Niche-constructing species play important ecological roles by creating habitats and resources used by other species and thereby affecting the flow of energy and matter through ecosystems—a process often referred to as "ecosystem engineering." An important emphasis of niche construction theory (NCT) is that acquired characters play an evolutionary role through transforming selective environments. This is particularly relevant to human evolution, where our species has engaged in extensive environmental modification through cultural practices. Humans can construct developmental environments that feed back to affect how individuals learn and develop and the diseases to which they are exposed. Let us make a brief introduction to NCT and illustrate some of its more important scientific and philosophical implications for the human sciences.

The organism influences its own evolution, by being both the object of natural selection and the creator of the conditions of that selection. The conventional view of evolution is that species, through the action of natural selection, have come to exhibit those characteristics that best enable them to survive and reproduce in their environments. Although environmental change may trigger bouts of selection, from the standard evolutionary perspective it is always changes in organisms, rather than changes in environments, that are held responsible for generating the organism–environment match that is commonly described as "adaptation." Organisms are generally perceived as being molded by selection to become better suited to their environments. Under this perspective, adaptation is always asymmetrical; organisms adapt to their environment, never vice versa.

The niche-construction perspective in evolutionary biology, as proposed by Lewontin (1983), contrasts with the conventional perspective by placing emphasis on the capacity of organisms to modify environmental states. Thus, "Organisms do not adapt to their environments; they construct them out of the bits and pieces of the external world" (Lewontin, 1983, 280). In so doing, organisms co-direct their own evolution, often but not exclusively in a manner that suits their genotypes, in the process modifying patterns of selection acting back on themselves as well as on other species that inhabit their environment. Early advocates of related arguments include Conrad Waddington (Waddington, 1959) and Brian Goodwin.

This emphasis on the modification of habitat and resources by organisms is shared by ecologists who emphasize the significance of "ecosystem engineering," by which organisms modulate flows of energy and matter through environments. Such engineering activity can have significant impacts on community structure, composition, and diversity. Young beavers, for example, inherit from their parents not only a local environment comprising a dam, a lake, and a lodge but also an altered community of microorganisms, plants, and animals. In this vein, Martinsen et al. (1998) found that the browsing of cottonwood trees by beavers stimulates elevated levels of defensive chemicals in the report growth and that these chemicals in turn are sequestered and used by leaf beetles for their own defense. Conversely, other invertebrates are driven out by the chemicals.

More generally, living organisms interact, indirectly, via engineered abiotic components, creating "engineering webs," which affect the stability of ecosystems as well as drive "eco-evolutionary feedbacks". The field of "eco-evolutionary dynamics" emphasizes that ecological and evolutionary changes are intimately linked and may often occur on the same time scales. Many of the ecological processes that trigger evolutionary episodes depend on niche construction and ecological inheritance. Ecological inheritance does not depend on the presence of environmental "replicators" but merely on intergenerational persistence (often through repeated acts of construction) of whatever physical—or, in the case of humans, cultural—changes are caused by ancestral organisms in the local selective environments of their descendants. This is relevant to conservation and biodiversity goals because the anthropogenic environmental changes precipitated by humans (e.g., habitat degradation, deforestation, industrial and urban development, agricultural practices, livestock grazing, and pesticide use) are primarily examples of human niche construction/ecosystem engineering, which destroys the engineering control webs that underlie ecosystems.

This process of niche construction provides a second evolutionary route to establishing the adaptive fit, or match, between organism and environment. From the niche-construction perspective, such matches need not be treated as products of a one-way process, exclusively involving the responses of organisms to environmentally imposed problems. Instead, they should be thought of as the dynamical products of a two-way process involving organisms both responding to "problems" posed by their environments and solving some of those problems, as well as setting themselves some new problems by changing their environments through niche construction (Odling-Smee & Laland, 2011; Odling-Smee & Turner, 2011).

This is not meant to imply that niche construction theory (NCT) always anticipates a perfect synergy between the features of an organism and the factors in its selective environment. It does not. In criticizing static adaptive-landscape concepts prevalent in evolutionary biology, the father of NCT, Lewontin (1983), described the evolution of a population as resembling an individual walking on a trampoline. Each change in the organism, as with each step, inevitably deforms the selective landscape. Like Lewontin, we argue here that this metaphor is an apt characterization not only of evolution but also of development. All living organisms construct aspects of their world, and in doing so they do not just respond to environments by being driven to higher levels of fitness through selection. They also fashion new strategies and devices (both trough action and perception), changing the environment in which they and others about them grow, develop, and learn, frequently in ways that rescript the pattern of natural selection acting back on their population as well as on other species that cohabit their niche.

Of course, evolutionary biologists are well aware that organisms modify environments. The difference between the niche-construction perspective and conventional evolutionary perspectives is far subtler than the recognition, or failure to recognize, organism-mediated environmental change. The developmental biologist Bateson (1988, 191) captures nicely the point we are making: Many biologists (including myself) have unthinkingly accepted the Darwinian image of selection, with nature picking those organisms that fitted best into the environments in which they lived. The picture of an external hand doing all of the work is so vivid that it is easy to treat organisms as if they were entirely passive in the evolutionary process. That is not, of course, to suggest that any biologist would deny that organisms, and animals especially, are active. But the notion of "selection pressure" does subtly downplay the organisms' part in the process of change… When developmental issues are recoupled to questions about evolution, it becomes much easier to perceive how an organism's behaviour can initiate and direct lines of evolution."

The key—and indeed subtle—distinction between the two perspectives is that one views niche construction as a cause of evolutionary change as opposed to an effect of a prior cause (namely, natural selection). Niche construction, then, is a process rather than merely a product. Organisms and environments are treated by NCT as engaged in reciprocally caused relationships (Laland et al., 2011) that are negotiated over both ontogenetic and phylogenetic timescales, entwined in, to coin a very apt phrase from developmental systems theory, "cycles of contingency" (Oyama et al., 2001). Moreover, niche construction is a developmental process, and the niche-construction perspective in evolutionary biology is all about exploring the evolutionary ramifications of coupling this particular developmental process with natural selection.

Many biologists have unthinkingly accepted the Darwinian image of selection, with nature picking those organisms that fitted best into the environments in which they lived. The picture of an external hand doing all of the work is so vivid that it is easy to treat organisms as if they were entirely passive in the evolutionary process. That is not, of course, to suggest that any biologist would deny that organisms, and animals especially, are active. But the notion of "selection pressure" does subtly downplay the organisms' part in the process of change. When developmental issues are recoupled to questions about evolution, it becomes much easier to perceive how an organism's behaviour can initiate and direct lines of evolution.

One implication is that niche-constructing organisms can no longer be treated as merely "vehicles" for their genes because they also modify selection pressures in their own and in other species' environments. In the process, they can introduce feedback to both ontogenetic and evolutionary processes. That this active, constructive conception of the role of organisms in evolution, and indeed in ontogeny, fits well with conceptualizations of human agency that are widespread within the human sciences.

A second implication is that there is no requirement for niche construction to result directly from genetic variation in order for it to modify natural selection. Humans can and do modify their environments mainly through cultural processes, and it is this reliance on culture that lends human niche construction a special potency. We stress, however, that humans are far from unique in engaging in niche construction, as some of the architects of the modern synthetic theory originally claimed. Niche construction is a very general process, exhibited by all living organisms, and species do not require advanced intellect or sophisticated technology to change their world.

The general replacement of a single role for phenotypes in evolution (as gene-carrying vehicles) by the dual role (also encompassing environmental modification and regulation) envisaged by NCT removes from cultural processes any claim to a unique status with respect to their capacity to transform natural-selection pressures. Nonetheless, cultural processes provide a particularly powerful engine for human niche construction. Moreover, this dual role for phenotypes in evolution does imply that a complete understanding of the relationship between human genes and cultural processes must acknowledge not only genetic and cultural inheritance but also take account of the legacy of modified selection pressures in environments.

Every species is informed by naturally selected genes, and many animals are also informed by complex, information-acquiring ontogenetic processes such as learning or the immune system, but humans, and arguably a few other species (depending on how culture is defined), are also informed by cultural processes.

The three domains are distinct but interconnected, with each interacting with, but not completely determined by, the others. That is, learning is informed by, but not fully specified by, genetic information, and cultural transmission may be informed by, but again, not completely specified by, both genetic and developmental processes. Genes may affect information gain at the ontogenetic level, which in turn influences information acquisition in the cultural domain. In addition, ontogenetic processes—particularly learning—may be affected by cultural processes, whereas population-genetic processes may be affected by both ontogenetic processes and cultural processes when humans modify environments, generating selective feedback to each process.

Niche construction modifies selection not only at the genetic level but also at the ontogenetic and cultural levels as well, with consequences that not only feed back to the constructor population but also modify selection for other organisms. Human niche construction, through modification of the environment, creates artifacts and other ecologically inherited resources that not only act as sources of biological selection on human genes but also facilitate learning and mediate cultural traditions. For example, the construction of villages, towns, and cities creates new health hazards associated with large-scale human aggregation, such as the spread of epidemics.

An important aspect of the alewife system is the fact that human cultural niche construction—here the damming of rivers—is responsible for generating landlocked fish populations, with their alternative foraging habits. In other words, anthropogenic activity has triggered a cascade of ecological and evolutionary events. Many social scientists are interested in such consequences of human activity. Indeed, social scientists frequently have essentially the same objective as ecologists: they, too, often wish to trace causal influences through ecosystems, but with the focus on human niche construction and the ecological or evolutionary episodes this anthropogenic change precipitates.

7 Reductionism and Emergence

The reductionist method consists in analyzing a large part biological system by dissecting it into their constituent parts and determining the mechanistic (physico-chemical) connections between the parts (Bains, 2001; Cornish-Bowden et al., 2006; Cornish-Bowden and Cárdenas, 2005). They assume that the isolated molecules and their structure have sufficient explanatory power to provide an understanding of the whole system. This radical deterministic standpoint was advocated by Crick (1970) by claiming that "The ultimate aim of the modern movement in biology (he refers of course to molecular biology) is to explain all biology in terms of physics and chemistry." Such reductionist mindset arises from the belief that because biological systems are composed solely of atoms and molecules, without the influence of other kinds of forces or laws, it should be possible to explain them using the physicochemical properties of their individual components, down to the atomic level. The most extreme manifestation of the reductionist view is the belief that is held by neuroscientists that consciousness and mental states can be reduced to chemical reactions that occurs in the brain (Sperry, 1980; Eccles, 1986; Popper and Eccles, 1977; Liljenström, 2016). In the recent decades many biologists have become increasingly critical of the idea that biological systems can be fully explained using only physics and chemistry (Polanyi, 1968; Rosen, 1985; Buiatti, 2000; Noble, 2006). And, in fact, there is now important evidence that the biology, development, physiology, behaviour or fate of a human being cannot be adequately explained by the reductionist standpoint that considers only (classical or not) physical and chemical laws. A more open and integrative approach considers biology as an autonomous discipline that requires its own entities and concepts that are not (necessarily and completely) found in physics and chemistry.

Biological complexity and specificity results from the way in which single components like molecules, genes and cells self-organize and function together when constituting a whole (a tissue, an organ, an organism), say a whole system including different subsystems (Rosen, 1977; Kacser, 1986; Mazzochi, 2012). Not only the interactions between the parts and the influence from the environment (think of epigenetic factors, both chemical and spatial, that mediate the complex relationship between the genomes and the micro- and macro biophysical environments), but also the systemic properties of the whole that exert an action on the components, give rise to new features, such as network behavior and functional properties, which are absent in the isolated components (Letelier et al., 2011).

This means that we need to consider 'emergence' as an effective new concept that complements 'reduction' when reduction fails, and allow to take into account those specific systemic properties of the whole responsible for biological organization and regulation at higher levels. Emergent properties to not result from properties pertaining to simple components of biological systems. They resist any attempt at being predicated or deduced by explicitly calculation or any other analytical means. In this regard, emergent properties differ from 'resultant' properties, which can be predicted from lower-level components. "For instance, the resultant mass of a multi-component protein assembly is simply equal to the sum of the masses of each individual component. However, the way in which we taste the saltiness of sodium chloride is not reducible to the properties of sodium and chlorine gas. An important aspect of emergent properties is that they have their own causal power, which is not reducible to the powers of their constituents. For instance, the experience of pain can alter human behavior, but the lower-level chemical reactions in the neurons that are involved in the perception of pain are not the cause of the altered behavior, as the pain itself has a causal efficacy" (Van Regenmortel, 2004). Advocating the reductionist idea of 'upward causation' means to maintain that molecular components and states suffice to determine higher-level processes occurring in biological systems. However, without denying a certain role of methodological reductionism in science, today we are led to recognize the important role played by the concept of emergence in many fields of the natural and life sciences, (Humphreys, 1997) as well as to accept 'downward causation' by which higher-level systems and processes influence lower-level configurations and entities. Emergence is essentially linked to the intrinsic and peculiar complexity of living systems. The existence of emergent properties is an outcome of the complexity of living systems. In other words, in order to solve the increasingly complexity, linked to the stages of the developments of tissues and organs and the construction of global physiological systems, living multicellular organisms self-organize giving thus rise to newly, needed regulatory and functional properties (Mesarovic et al., 2004).

8 Many Levels of Causation Are Needed for Thinking the Biological Complexity and Functionality

Many theoretical ideas and experimental findings in life science over the last three decades lead to review profoundly the ideas about properties and behaviors of biological systems. Among them, maybe the most important, is the principle of causality in biological sciences, as it has been conceived by molecular biology. This fundamental issue is raised by Denis Noble when he asks: "Must higher level biological processes always be derivable from lower level data and mechanisms, as assumed by the idea that an organism is completely defined by its genome? Or are higher level properties necessarily also causes of lower level behavior, involving, actions and interactions both ways?" (Noble, 2011, 1). According to Noble, "downward causation is necessary and this form of causation can be represented as the influences of initial and boundary conditions on the solutions of the differential equations used to represent the lower level processes. (…) A priori, there is no privileged level of causation. (…) Biological relativity can be seen as an extension of the relativity principle in physics by avoiding the assumption that there is a privileged scale at which biological functions are determined" (idem).

There is increasingly evidence, experimental and theoretical, of the existence of downward causation from larger to smaller scales. Today, one is enabled to visualize exactly how multilevel 'both-way' causation occurs. There is not a priori reason why one level in a biological system should be privileged over other levels when it comes to causation. There are various forms of downward causation that regulates lower level components in biological systems.

Looking more closely to molecular biology, the essence of the central dogma is that 'coding' between genes and proteins in one-way (Werner, 2007; Shapiro, 2009). It would be better the word 'template' to 'coding' since 'coding' already implies a program. The concept of a genetic program is indeed one of the most relevant problem of molecular biology because there is no a genetic program at all. The argument runs as follow (for more details, see Noble, 2011). The sequence of DNA triplets form templates for the production of different amino acid sequences in proteins. Amino acid sequences do not form templates for the production of DNA sequences. What was shown by Crick, Watson and their followers is that template works in only one direction, which makes the gene appear primary. So, what the genome really causes? The coding sequences form a list of proteins and RNAs that might be made in a given organism. According to Noble, "These parts of the genome form a database of templates. To be sure, as a database, the genome is also extensively formatted, with many regulatory elements, operons, embedded within it. These regulatory elements enable groups of genes to be coordinated in their expression levels. And we know that the non-coding parts of the genome also play important regulatory functions. But the genome is not a fixed program in the sense in which such a computer program was defined when Monod and Jacob introduced the idea of 'the genetic program' in the sixties. It is rather a 'read-write' memory that can be organized in response to cellular and environmental signals. Which proteins and RNAs are made when and where is not fully specified. This is why it is possible for the 200 or so different cell types using exactly the same genome. A heart cell is made using precisely the same genome in its nucleus as a bone cell, a liver cell, pancreatic cell, etc. Impressive regulatory circuits have been constructed by those who favor a genetic program view of development, but these are not independent of the 'programming' that the cells, tissues and organs themselves uses to epigenetically control the genome and the patterns of gene expression appropriate to each cell and tissue type in multicellular organism." (Noble, op. cit., 3; see also Noble, 2008).

The important point to stress is that the circuits of major biological functions necessarily include non-genome elements. This tells us that the genome alone is far from being sufficient. Barbara McClintock first described the genome as 'an organ of the cell' (McClintock, 1984). Indeed, DNA sequences do absolutely nothing until they are triggered to do so by a variety of transcriptions factors, which turn genes 'on and off' by binding to their regulatory sites, and various other forms of epigenetic control, including methylation of certain cytosines and interactions with the tails of the histones that form the protein backbone of the chromosomes. All of these, and the cellular, tissue and organ processes that determine when they are produced and used, 'control' the genome (Misteli, 2007). In the neurosciences, a good example of downward causation is what neuroscientists call electro-transcription coupling, since it involves the transmission of information from the neural synapses to the nuclear DNA (Murphy, 2009).

So, there is strong evidence that the genome does not completely determine the organisms. Multi-cellular organisms use the same genome to generate all the 200 or so different types of cell in their bodies by activating different expression patterns. The regulatory parts of the genome are essential in order the genome be activated. The mechanisms and patterns of activation are just as much part of the organism's construction and the genome itself. It is time to recognize that there exist various forms of downward causation that regulates lower level components in biological systems. In addition to the controls internal to the organism, we also have to take into account the influence of the environment on all the levels. Causation is, therefore, two-way. A downward form of causation is not a simple reverse form of upward causation. It is better seen as completing a feedback loop that expresses a functional integration of the various levels of causation, including in particular the concentrations and locations of transcription and post-transcription factors, and the relevant epigenetic influences. All those forms of downward causation naturally take into account the role of cell and tissue signaling in the generation of organizing principles involved in embryonic induction, originally identified in the pioneering work of Hans Spemann and Ilde Mangold (Spemann & Mangold, 1924; De Robertis, 2006). The existence of such induction is itself an example of dependence on boundary conditions, that is those conditions which define what constraints are imposed on a biological system by its environment. That because boundary conditions are somehow involved in determining initial conditions (the state of the components of the system at the time at which we start to analyzing and modelling it), they can therefore be considered as a form of downward causation. The induction mechanisms emerge as the embryo interacts with its environment. Morphogenesis cannot be explained only by the genome (Goodwin et al., 1993). Putted in different terms, the emergence of new morphological and physiological forms in the embryo of a human being cannot be derived and understood from the level of the genome.

There is real ('strong') emergence because contingency beyond what is in the genome, i.e. in its environment, also determine what happens at the higher level of morphogenesis. Multi-cellular organisms are multi-level systems, and each level, from molecules and cells to tissues and organs, possesses a specific organization with increasing complexity when one passes to higher order systems. This organization has causal power. The idea of multicellular causation considers seriously the fact that complex organization of highest levels, such as the global properties and activity of cells and the systemic properties and state of organisms, may act on the functions of the components, particularly genes and proteins. Downward causation leads us to shift our focus away from the gene as the unit of development and evolution to that of the whole organism (Moreno and Umerez, 2000). It might be that the concept of downward causation will play an important role in the reappraising of the mind–body problem (how and why mental states may act on neural states), and in the philosophy of perception and action (perceptual global effects, intentionality, free will, etc.). Finally, we need to stress that one of the major theoretical and experimental outcomes of multilevel modelling is that causation in biological systems runs in both directions: upward from the genome and downward from all other levels. There are feed-forward and feedback loops between the different levels of causation.

9 Conclusion

The first goal of this paper was to stress the important fact that the specificity of complex biological activity does not arise from the specificity of the individual molecules that are involved, as these components frequently function in many different processes. For instance, genes that affect memory formation in the fruit fly encode proteins in the cyclic AMP (camp) signaling pathway that are specific to memory. It is the particular cellular compartment and environment in which a second messenger, such a camp, is released that allow a gene product to have a unique effect. Biological specificity results from the way in which these components assemble and function together. More precisely, we tried to showing that complex biological levels of functionality result from self-organized processes. For self-organization to act on macroscopic cellular structures, three requirements must be fulfilled: (i) a cellular structure must be dynamic; (ii) material must be continuously exchanged; and (iii) an overall stable configuration must be generated from dynamic components. Interactions between the parts, as well as influences from the environment, give rise to new features, such as network and collective behaviors which are absent in the isolated components (Hess and Mikhailov, 1994; Kauffman, 1993; Karsenti, 2008; Misteli, 2001). Consequently 'emergence' has appeared as a new concept that complements 'reduction' when reduction fails. Emergent properties resist any attempt at being predicted or deduced by explicitly calculation or any other means. In this regard, emergent properties differ from resultant properties, which can be defined from low-level configurations and information. For instance, the resultant mass of a multi-component protein assembly is simply equal to the sum of the mass of each individual component. However, the way in which we taste the saltiness of sodium chloride is not reducible to the properties of sodium and chloride gas. An important aspect of emergent properties is that they have their own causal power, which is not reducible to the power of their constituents. The key concepts here are those of 'organization' and 'regulation', first of all because organization and regulation become cause in the living matter of morphological, functional and mental novelties. According to the principle of emergence, the natural and living worlds are organized into stages and levels that have evolved over different evolutionary times through continuous and discontinuous processes. Reductionists advocate the idea of 'upward causation' by which molecular states generally bring about higher-level phenomena, whereas proponents of emergence admit 'downward causation' by which higher-level systems may influence lower-level configurations. All along the article we will stress the philosophical importance of admitting 'downward causation' in the analysis of complex living systems (i.e. presenting and ever-increasing coupled activity of plasticity and complexity) by showing that chromatin forms and its structural modifications play a crucial role in the increasing complexity of gene regulatory networks, in the emergence of cellular functions and in development, as well as in the neurocognitive plasticity.

Our second goal was to emphasize the fundamental fact that organisms are more than, and a reality profoundly different from the genes that look after their assembly. Mechanical, chemical and cultural inputs from the environment, epigenetic cues, also have an effect on the final phenotype. In fact, continued environmental influences on the adult phenotype continue to affect its characteristics. The open question is whether the epigenetic cues can become causative agents of phenotypic modifications. Within a biological multi-level, astonishing complex reality, higher levels result from lower-level processes (genes up to phenotype), and lower levels result from higher-levels processes (organism's properties to epigenetics mechanisms of genes expression and regulation), so that upward and downward causation are in different ways and in both directions deeply interlaced. Some epigenomic cues seem to be assimilated into the genome, as already C. H. Waddington showed (Waddington, 1953). The evolved genome therefore incorporates epigenomic cues or the expectation of their arrival.

Genomes are more than linear sequences, in fact, they exist as elaborate spatial and physical structures, and their functional properties are strongly determined by their cellular organization and by the interactions organisms develop with the environment (Sarà, 2002; Boi, 2009; Aguilera et al., 2010).

The key distinguishing characteristic of the eukaryotic genome is its tight packaging into chromatin, a hierarchically organized complex of DNA and histone and non-histone proteins. How genome operates in the chromatin context is a central question in the molecular genetics of eukaryotes. The chromatin packaging consists of different levels of organisation. Every level of chromatin organisation, from nucleosome to higher-order structure up to its intranuclear localization, can contribute to the regulation of gene expression, as well as affect other functions of the genome, such as replication and repair. Concerning gene expression, chromatin is important not only because of the accessibility problem it poses for the transcription apparatus, but also due to the phenomenon of chromatin memory, that is, the apparent ability of alternative chromatin states to be maintained through many cell divisions. This phenomenon is believed to be involved in the mechanism of epigenetic inheritance, an important concept of developmental biology.

References

Aguilera, O., Fernadez, A. F., Munoz, A., & Fraga, M. F. (2010). Epigenetics and environment: A complex relationship. *Journal of Applied Physics, 109*, 243–251.

Ayala, F., & Dobzhansky, T. (Eds.). (1974). *Studies in the philosophy of biology*. University of California Press.

Bains, W. (2001). The parts list of life. *Nature Biotechnology, 19*, 401–402.

Ballestar, E., & Esteller, M. (2005). The epigenetic breakdown of cancer cells: From DNA methylation to histone modification. *Progress in Molecular and Subcellular Biology, 38*, 169–181.

Bateson, P. (1988). The active role of behavior in evolution. *Evolutionary Processes and Metaphors, 6*(3), 191–207.

Bhalla, U. S., & Iyengar, R. (1999). Emergent properties of networks of biological signaling pathways. *Science, 283*, 381–387.

Bock, W. J., Wahlert, G. V. (1965). Adaptation and the form-function complex. *Evolution, 19*, 269–299.

Bock, W. J. (1998). The nature of explanations in morphology. *American Zoologist, 28*, 205–215.

Boi, L. (2009). Epigenetic phenomena, chromatin dynamics, and gene expression. New theoretical approaches in the study of living systems. *Rivista Di Biologia/biology Forum, 103*(4), 27–58.

Boi, L. (2021). A reappraisal of the form—Function problem. Theory and phenomenology. *Theory in Biosciences, 140*, 39–68.

Boi, L. (2005). Topological Knot models in physics and biology. In: *Geometries of nature, living systems and human cognition. New interactions of mathematics with natural sciences and humanities* (pp. 203–278). World Scientific.

Boi, L. (2007a). Geometrical and topological modeling of supercoiling in supramolecular structures. *Biophysical Reviews and Letters, 2*(3), 1–13.

Boi, L. (2007b). Sur quelques propriétés géométriques globales des systèmes vivants. *Bulletin d'Histoire et d'Épistémologie des Sciences de la Vie, 14*, 71–113.

Boi, L. (2011a) Plasticity and complexity in biology: Topological organization, regulatory protein networks, and mechanisms of genetic expression. In G. Terzis & R. Arp (Eds.), *Information and living systems: Philosophical and scientific perspectives* (pp. 287–338). The MIT Press.

Boi, L. (2011b). When topology meets biology 'for life': Remarks on the way in which topological form modulate biological function. In Bartocci, C., Boi, L., & Sinigaglia, C. (Eds.), *New trends in geometry, and its role in the natural and life sciences* (pp. 241–302) Imperial College Press.

Buiatti, M. (2000). *Lo stato vivente della materia. Le frontiere della nuova biologia.* UTET.

Campbell, D. (1974). Downward causation in hierarchically organized biological systems. In F. J. Ayala & T. Dobzhansky (Eds.), *Studies in the philosophy of biology: Reduction and related problems* (pp. 179–186). University of California Press.

Cornish-Bowden, A. (2006). Putting the systems back into systems biology. *Perspectives in Biology and Medicine, 49*(4), 1–9.

Cornish-Bowden, A., & Cárdenas, M. L. (2005). Systems biology may work when we learn to understand the parts in terms of the whole. *Biochemical Society Transactions, 33*(3), 516–519.

Cornish-Bowden, A., Cárdenas, M. L. Letelier, J.-C., Soto-Andrade, J., & Guínez Abarzúa, F. (2004). Understanding the parts in terms of the whole. *Biology of the Cell, 96*, 713–717.

Cremer, T., Cremer, M., Dietzel, S., Müller, S., Solovei, I., & Fakan, S. (2006). Chromosome territories—A functional nuclear landscape. *Current Opinion in Cell Biology, 18*, 307–316.

Crick, F. H. C. (1970). The central dogma of molecular biology. *Nature, 227*, 561–563.

De Robertis, E. M. (2006). Spemann's organizer and self-regulation in amphibian embryos. *Nature Reviews Molecular Biology, 7*(4), 296–302.

Eccles, J. C. (1986). Do mental events cause neural events analogously to the probability fields of quantum mechanics? *Proceedings of the Royal Society of London. Series B: Biological Sciences, 227*, 411–428.

El-Hani, C. N., & Emmeche, C. (2000). On some theoretical grounds for an organism-centered biology. Property emergence, supervenience, and downward causation. *Theory in Biosciences, 119*(3), 234–275.

Emmeche, C. (1997). Aspects of complexity in life and science. *Philosophica, 59*, 41–68.

Esteller, M. (2008). Epigenetics in cancer. *The New England Journal of Medicine, 358*(11), 1148–1159.

Fraga, M. F., Ballestar, E., Paz, M. F., et al. (2005). Epigenetic differences arise during the lifetime of monozygotic twins. *Proceedings of the National Academy of Sciences of the United States of America, 102*, 10604–10609.

Goodwin, B. C., Kauffman, S., & Murray, J. D. (1993). Is morphogenesis an intrinsically robust process? *Journal of Theoretical Biology, 163*(1), 135–144.

Hall, B. K. (1970). Cellular differentiation in skeleton tissues. *Biology Reviews, 45*, 455–484.

Hammond, S. M., Caudy, A. A., & Hannon, G. J. (2001). Post-transcriptional gene silencing by double-stranded RNA. *Nature Reviews Genetics, 2*, 110–119.

Herring, S. W. (1993). Formation of the vertebrate face: Epigenetic and functional influences. *American Zoologist, 33*(4), 472–483.

Hess, B., & Mikhailov, A. (1994). Self-organization in living cells. *Science, 264*, 223–224.

Ho, S.-M., Johnson, A., Tarapore, P., Janakiram, V., Zhang, X., & Leung, Y.-K. (2012). Environmental epigenetics and its implications on disease risk and health outcomes. *Epigenetics, 53*(3–4), 289–305.

Holliday, R. (1987). The inheritance of epigenetic defects. *Science, 238*, 163–170.

Holliday, R. (1994). Epigenetics: An overview. *Developmental Genetics, 15*, 453–457.

Horrobin, D. F. (2003). Modern biomedical research: An internally self-consistent universe with little contact with medical reality? *Nature Reviews, 2*, 151–154.

Humphreys, P. (1997). How properties emerge. *Philosophy of Science, 64*(1), 1–17.

Jablonka, E., & Lamb, M. J. (1989). The inheritance of acquired epigenetic variations. *Journal of Theoretical Biology, 139*(1), 69–83.

Jacob, F. (1970). *La logique du vivant.* Gallimard.

Kacser, H. (1986). On parts and wholes in metabolism. In G. R. Welch & J. S. Clegg (Eds.), *The organization of cell metabolism* (pp. 327–337). Plenum Press.
Kardong, K. V. Epigenomics: The new science of functional and evolutionary morphology. *Animal Biology, 53*(3), 225–243.
Karsenti, E. (2008). Self-organization in cell biology: A brief history. *Nature Reviews Molecular Cell Biology, 9*, 255–262.
Kauffman, S. (1993). *The origins of order. Self-organization and selection in evolution*. Oxford University Press.
Konishi, M. (1965). Effects of deafening on song development in American robins and black-headed grosbeaks. *Zeitschrift für Tierpsychologie, 22*(5), 584–599.
Konishi, M. (1965). The role of auditory feedback in the control of vocalization in the white-crowned sparrow. *Zeitschrift für Tierpsychologie, 22*(7), 770–783.
Laland, K. N., et al. (2011). Cause and effect in biology revisited: Is Mayr's proximate-ultimate distinction still useful? *Science, 334*, 1512–1516.
Lanyon, L. E. (1980). The influence of function on the development of bone curvature: an experimental on the rat tibia. *Journal of Zoology (London), 192*, 457–466.
Lederberg, J. (1958). Genetic approaches to somatic cell variation: Summary content. *Journal of Cellular and Comparative Physiology, 52*, 383–401.
Letelier, J. C., Cárdenas, M. L., & Cornish-Bowden, A. (2011). From L'Homme Machine to metabolic closure: Steps towards understanding life. *Journal of Theoretical Biology, 286*(1), 100–113.
Lewontin, R. C. (1983). Gene, organism, and environment. In D. S. Bendall (Ed.), *Evolution from molecules to men* (pp. 273–285). Cambridge University Press.
Liljenström, H. (2016). Multi-scale causation in brain dynamics. In R. Kozma & W. J. Freeman (Eds.), *Cognitive Phase Transitions in the cerebral cortex—Enhancing the neuron doctrine by modeling neural fields* (pp. 177–186). Springer International Publishing Switzerland.
Lillie, R. S. (1940). Biological causation. *Philosophy of Science, 7*(3), 314–336.
Lorenz, K. (1965). *Evolution and modification of behavior*. Chicago University Press.
Marler, P. (1990). Song learning: The interface between behaviour and neuroethology. *Philosophical Transactions: Biological Sciences, 329*(1253), 109–114.
Martinsen, G. D., Driebe, E. M., Whithman, T. G. (1998). Indirect interactions mediated by changing plant chemistry: beaver browsing benefits beetles. *Ecology, 79*(1), 192–200.
Mazzochi, F. (2012). Complexity and the reductionism-holism debate in systems biology. *Systems Biology and Medicine, 4*(5), 413–427.
McClintock, M. (1984). The significance of responses of the genome to challenge. *Science, 226*, 792–801.
Mesarovic, M. D., Sreenath, S. N., & Keene, J. D. (2004). Search for organizing principles: Understanding in systems biology. *Systematic Biology, 1*(1), 19–27.
Misteli, T. (2001). The concept of self-organization in cellular architecture. *The Journal of Cell Biology, 155*(2), 181–185.
Misteli, T. (2007). Beyond the sequence: Cellular organization of genome function. *Cell, 128*, 787–800.
Morange, M. (2000). *A history of molecular biology*. Harvard University Press.
Moreno, A., & Umerez, J. (2000). Downward causation at the core of living organization. In P. B. Andersen, C. Emmeche, N. O. Finnemam, & P. V. Christiansen (Eds.), *Downward causation* (pp. 99–117). University of Arhus Press.
Murphy, N., Ellis, G. F. R., & O'Connor, T. (Eds.). (2009). *Downward causation and the neurobiology of free will*. Springer.
Murray, P. D. F. (1936). *Bones: A study of the development and structure of the vertebrate skeleton*. Cambridge University Press, London.
Neuman, Y. (2008). *Reviving the living: Meaning making in the living systems*. Elsevier.
Noble, D. (2002). Modeling the heart–from genes to cells to the whole organ. *Science, 295*, 1678–1682.

Noble, D. (2006). *The music of life*. Oxford University Press.
Noble, D. (2008). Genes and causation. *Philosophical Transactions of the Royal Society A, 366*, 1125–1139.
Noble, D. (2010). Biophysics and systems biology. *Philosophical Transactions of the Royal Society A, 368*, 1125–1139.
Noble, D. (2011). A theory of biological relativity: No privileged level of causation. *Interface Focus*, 1–10.
Odling-Smee, F. J., & Laland, K. N. (2011). Ecological inheritance and cultural inheritance: How are they and how they differ? *Biological Theory, 6*, 220–230.
Odling-Smee, F. J., & Turner, J. S. (2011). Niche construction theory and human architecture. *Biological Theory, 6*(3), 283–289.
Oyama, S., Griffith, P. E., & R. D. Gray (Eds.). (2001). *Cycles of contingency: Developmental systems and evolution*. The MIT Press.
Polanyi, M. (1968). Life's irreducible structure. *Science, 160*(3834), 1308–1312.
Popper, K. (1974). Scientific reduction and the essential incompleteness of all science. In F. Ayala & T. Dobzhansky (Eds.), *Studies in the philosophy of biology* (pp. 259–284). University of California Press.
Popper, K., & Eccles, J. C. (1977). *The self and its brain, an argument for interactionism*. Springer.
Prigogine, I., & Nicolis, G. (1971). Biological order, structure and instabilities. *Quarterly Reviews of Biophysics, 4*, 107–148.
Prochiantz, A. (2012). *Qu est-ce que le vivant?* Seuil.
Rosen, R. (1977). Complexity as a system property. *International Journal of General Systems, 3*, 227–232.
Rosen, R. (1985). Organisms as causal systems which are not mechanism: An essay into the nature of complexity. In R. Rosen (Ed.), *Theoretical biology and complexity* (pp. 165–203). Academic Press.
Sarà, M. (2002). L'integrazione di genotipo e fenotipo alle soglie del 2000. *4*, 181–208.
Scherer, K., & Jost, J. (2007). Gene and genome concept. Coding versus regulation. *Theory in Biosciences, 126*, 65–113.
Shapiro, J. A. (2009). Revisiting the central dogma in the 21st century. *Annals of the New York Academy of Sciences, 1178*, 6–28.
Smet-Nocca, C., Paldi, A., & Benecke, A. (2006). De l'épigénomique à l'émergence morphogénétique. In P. Bourgine & A. Lesne (Eds.), *Morphogénèse. L'origine des formes* (pp. 153–178). Belin.
Spemann, H., & Mangold, I. (1924). Über Induktion von Embryonalanlagen durch Implantation artfremder Organisatoren. *Archiv Für Mikroskopische Anatomie Und Entwicklungsmechanik, 100*(3–4), 599–638.
Sperry, R. W. (1980). Mind-brain interaction: Mentalism, yes; dualism, no. *Neuroscience, 5*, 195–206.
Strohman, R. C. (1997). The coming Kuhnian revolution in biology. *Nature Biotechnology, 15*, 194–200.
Van Regenmortel, M. H. V. (2004a). Biological complexity emerges from the ashes of genetic reductionism. *Journal of Molecular Recognition, 17*, 145–148.
Van Regenmortel, M. H. V. (2004b). Reductionism and complexity in molecular biology. *EMBO Reports, 5*(11), 1016–1020.
Waddington, C. H. (1953). Genetic assimilation of an acquired character. *Evolution, 7*, 118–126.
Waddington, C. H. (1957). The Strategy of the Genes, George Allen & Unwin, London.
Waddington, C. H. (1959). Evolutionary systems–animals and human. *Nature, 183*, 1634–1638.
Waddington, C. H. (1961). *The nature of life*. Allen & Unwin.
Walleczek, J. (Ed.). (2000). *Self-organized biological dynamics and nonlinear control. Toward understanding complexity, chaos and emergent function in living systems*, Cambridge University Press.

Washburn, S. L. (1947). The relation of the temporal muscle to the form of the skull. *The Anatomical Record, 99*, 139–248.
Webb, G. J. W., Cooper-Preston, H. (1989). Effects of incubation temperature on crocodiles and the evolutionary reptilian oviparity. *American Zoologist, 29*(3), 953–971.
Werner, E. (2007). How central is the genome? *Science, 317*, 753–754.
West-Eberhard, M. J. (2005). Phenotypic accommodation: Adaptive innovation due to developmental plasticity. *Journal of Experimental Zoology Part B: Molecular and Developmental Evolution, 304*, 610–618.
Wicken, J. S. (1984). On the increase in complexity in evolution. In M.-W. Ho & P. T. Saunders (Eds.), *Beyond Neo-Darwinism* (pp. 89–112). Academic Press.
Wolkenhauer, O., & Green, S. (2013). The search for organizing principles as a cure against reductionism in systems medicine. *The FEBS Journal, 280*, 5938–5948.

Can Agency Be Reduced to Molecules?

Raymond Noble and Denis Noble

1 Introduction

We all *feel* what it is like to make a choice. We have a sense of volition, or that our actions and our choices are purposeful, but also that we make them to achieve objectives which we have decided. We act for reasons that we have created, even where these reasons involve hopes, fears, desires, loves and hates, or other emotions. Jack may choose to walk with Jill because he's in love with her and wishes to be with her. We know the difference between being compelled to act in a particular way, and doing so of our own free will. We therefore think we are, at least partly, agents of our own destiny. We will call that ability *agency*, the ability to act and choose what we wish to do, subject to what is physically possible.

We chose to write this chapter for this book; but was this choice an illusion? Could our actions be entirely explained from molecular events? Our answer to this question is that it cannot. We might be able to explain *how* we behaved as we did, but we would have a limited understanding of *why* we did so. Our reasons for writing this chapter are two-fold; not only to argue that reasons and emotions can influence events at the molecular level, but also to explain how.

Scientists who think that our molecular make-up makes agency impossible therefore also conclude that our *feeling* of agency is itself an illusion. Thus the evolutionary biologist, Jerry Coyne, author of the popular book *Why Evolution is True*, writes:

R. Noble
Institute for Women's Health, University College London, London WC16AU, UK
e-mail: r.noble@ucl.ac.uk

D. Noble (✉)
Department of Physiology, Anatomy & Genetics, University of Oxford, Oxford OX1 3PT, UK
e-mail: denis.noble@dpag.ox.ac.uk

S. Wuppuluri and I. Stewart (eds.), *From Electrons to Elephants and Elections*,
The Frontiers Collection, https://doi.org/10.1007/978-3-030-92192-7_37

> The illusion of agency is so powerful that even strong incompatibilists like myself will always *act* as if we had choices, even though we know that we don't. We have no choice in this matter. But we can at least ponder why evolution might have bequeathed us such a powerful illusion.[1]

We profoundly disagree with Coyne's conclusion.

The molecular biologist James Watson[2] is reported to have said, "There is only science, physics: everything else is social work." This can be interpreted in two ways. First, that all effective causation has to be at the material level. Or, second, that what happens at the social level, the level of inter-relationships, is significant in causing events. Thus we need both for a full explanation. However, we doubt that this is what he meant. Watson's implication is clear. Social interactions between organisms, including humans, are viewed as vague processes that cannot carry a hard scientific basis, because they are non-material. In this view, molecules are seen as the hard basis of our being and the level at which all causation occurs; thus the molecular level is always where we should look for explanations of what we decide to do.

2 How Did Western Science Come to Deny Agency?

The origin of this denial of agency is deeply embedded in Cartesian dualism and represents a profound misunderstanding of the nature of life. Descartes first published his notion of the "beast machine" in his *discours de la méthode* in 1637 in which he regarded organisms as machines or automata, and thus creating for humans the dualist separation of mind and body.

A mechanistic view of organisms might be understandable in the context of the fascination with automata by scientists during the Renaissance, when truly remarkable mechanical devices were being conceived using cogs, wheels, pulleys, pistons and springs. Using extraordinary ingenuity, the German Mathematician & Inventor Johannes Muller von Konigsberg created a series of wonderful contraptions, amongst which was a famous mechanical flying eagle, while Leonardo da Vinci built a mechanical lion, so life-like that it could shake its tail, open its mouth, walk and even rear up on its hind legs. But this period of understanding of physics and mechanics created also a cultural trap. Life is not a mechanical machine, and we cannot trace causality back through its cogs and wheels. The mechanical toys of da Vinci and Muller were closed systems; Life is not. Living organisms are open, ongoing process of change and redevelopment. Where life has pulleys and levers, it must continuously seek to replenish and maintain them. This drives its function and its motivation; and its self-being, and in doing so it must solve problems involving choices. Yet a powerful residue of the mechanistic view of life still has its hold on modern science

[1] Coyne (2014).

[2] Watson, J.D. This is the common version of the quote, documented by Stephen Rose in his book Lifelines.

and thought and notions of causality, but where the cogs and wheels have been replaced by molecular structures.

This materialist view produces a false dichotomy between our psychosocial being and our physical being. Yet these are inseparable in the living process. Life is a process of maintaining integrity and as such is the origin of purpose in functionality. We humans have a sense of our being, of our integrity, and of what it is we are doing. This is a sense, just as vision and hearing, touch and smell are senses. We have little trouble attributing such senses to other organisms, even though we cannot ourselves see their seeing or hear their hearing. Indeed, so much of our understanding of how our senses work comes from studies on non-human organisms. Yet we are reluctant to attribute our sense of purpose to them.

3 Agency Has a Firm Modern Scientific Basis

Recently, we have analysed this issue in a series of articles,[3] and we do so from a solid scientific basis. Our articles show:

- Even when we take the molecular view at face value, for example by constructing mathematically strict accounts of what is happening at the molecular level, it is easy to show that it is inadequate to provide causal closure. It cannot, *even on its own terms*, succeed in predicting what we will do when we act as agents.
- The molecular level is far from determinate. Atoms and molecules do not behave like hard billiard balls clicking around with each other in totally predictable ways. On the contrary, at the molecular level stochasticity, randomness, reigns.
- The interpretation of biological processes following the concept of the Central Dogma of molecular biology, in a straight causal line from DNA to RNA to protein, is incorrect, and has been shown to be so by molecular biology itself.
- Stochasticity is *used* by organisms functionally, both in unconscious and in conscious processes. These processes even give organisms a role in their own evolution. Organisms then become partly directional, both in how they act, and in how they influence evolution.

Using these points, it is possible to outline how agency can be given a physiological basis, and is therefore far from a vague concept. We go so far as to conclude that it is *mathematically and physically necessary* that organisms should be agents. In organisms there can be no causal closure at the molecular level. As the logician of science, Karl Popper concluded "all life is problem solving."[4]

We will develop our argument in several stages.

[3] Fully listed at the end of this chapter.

[4] Popper (1999).

4 Biological Relativity: A Consequence of Organisms Being Open Systems

Living organisms are open systems. They are continually exchanging energy and matter with their environment. Nothing in organisms is isolated from the outside world. That is as true of DNA as it is true of all molecules and systems in organisms. An organism that is prevented from exchanging energy and matter with its environment becomes dead.

One of the obvious consequences of being an open system is that the organism is subject to causation from outside (beyond) itself. Life is an interaction between the organism and its environment, which leads to two consequences. Life itself changes its environment. And it is itself influenced by it.

Many of those influences are social—the ways in which organisms interact with each other. Those interactions have molecular consequences, to be sure. Everything does. When you type instructions into your computer, they will produce changes at the level of electrons, atoms and molecules. But those instructions have causal power by virtue of the logic of what you are telling the machine to do. Similarly, organisms signal to each other and those signals have molecular consequences. But that is not what the signals mean nor do the molecular effects reveal the intent of the organism sending them. That requires an organism to interpret and anticipate what others intend to do.

The consequence of life being an open system therefore forces us to recognise the existence of levels of organisation above that of the organism alone. Having done that, it is relatively easy to see that the organism itself is also a set of levels of organisation: molecular, cellular, tissues, organs and systems. Each has their logic of operation that constrains the levels below to conform to that logic. This conformity to higher-level logic must be true since all multicellular organisms evolved from unicellular ones. The cells that did this then, in effect, traded their independence for the constraints of multicellular co-operation. We know what happens when those constraints break down. We call it cancer.

The existence of levels of organisation and of causality between them leads to what we call the principle of biological relativity. By ‘relativity principle’ in this context, we mean distancing ourselves in our theories from specific absolute standpoints for which there can be no a priori justification. The principle was first applied and developed in physics. From Copernicus and Galileo through to Poincaré and Einstein, the reach of this general principle of relativity has been progressively extended by removing various absolute standpoints in turn. People realized that those standpoints represent privileging certain measurements as absolute, for which there is and could be no basis. There is no centre of the universe, and no absolute frame of reference from which to measure speeds of movement. Even were all science to be just physics, we cannot avoid the consequences of the relativity principle.

Molecules cannot therefore form a privileged level of organization from which everything else is completely caused. All levels of organisation have causal consequences. If you train as a professional athlete, the training will produce changes in RNA levels that in turn control the amounts of protein you produce. That is how you produce larger and more powerful muscles. But the RNA changes depend on you making the decision to train.

5 Consequences for the Central Dogma of Molecular Biology

The current concept of genetics tends to confound different ways in which genes are used and created, leading to a confused understanding of agency, or denial that it exists.

5.1 Consequences for Organisms as Agents

The predominant view in modern science is gene-centred, so much so that it has entered our common language; Genes or DNA are referred to as "a code" or "a blueprint" and even as 'a book of life', not just determining the form we take, our anatomy and physiology, but also our behavioural characteristics. It feeds into our philosophy and politics with concepts such as "the selfish Gene". This has its origin in what became "the central dogma": a linear sequence of causality from gene to characteristic and behaviour, but not the reverse, that our behaviour can alter our genes. It gives privileged causality to genes.

This privileging of genetic causality involves a fundamental misinterpretation of what molecular biology has shown. The molecular biologist Francis Crick produced the first version of the Central Dogma in 1958,[5] as a one-way track from DNA to RNA to proteins: DNA → RNA → proteins, contending that it could not work the other way. Nevertheless, he had to reformulate his Central Dogma in 1970 after the discovery that RNAs could be reverse transcribed into DNA, and inserted into the genome. This is significant because it provides a way in which the organism can influence its genome. The first arrow works both ways, from DNA to RNA and from RNA to DNA. Crick's revised version reads:

> The central dogma of molecular biology deals with the detailed residue-by-residue transfer of sequential information. It states that *such information* cannot be transferred back *from protein* to either protein or nucleic acid.[6]

[5] Crick (1958).

[6] Crick (1970).

It beggars belief that this statement could be comprehensible! Not least since it is a statement for which there is no evidential basis. There is nothing in a genome sequence that tells us what it means, or why it could be regarded as information. It is the organism that does that. Without assigned meaning it could be just junk data, as indeed was once thought to be the case for 80% of DNA. That idea is no longer credible. To put it bluntly, the Central Dogma was an idea, an interpretation, which has, nevertheless, had a profound influence on our thinking and our behavior, as though it is an experimentally established finding. To emphasise this point we have italicized the terms '*such information*' and '*from protein*' since it is uncertain whether Crick meant that *no* control can pass from the *organism* to the genome. In fact, even in Crick's time, it was obvious that such control of DNA must exist to produce the many different patterns of gene expression, which enable many different phenotypes (e.g. many different cell types, such as heart cells, renal cells etc. in the same body) that are generated from the same genome.

5.2 Consequences for the Nature-Nurture Debate

Such is the hold of the gene-centred view in the nature versus nurture debate that science has often tried to attribute percentages or proportionate causality to the genes or environment. For example, in intelligence, where it has been suggested that it is, perhaps, 60% genes, 40% nurture. But what is this a percentage of? We might say that a cake has, say, 5% marzipan, 1% icing, 94% cake mix. But what percentage of the cake is the cook? This makes it appear that there is 0% left for the chef! Ask any gourmet whether this is true. The decision, for example, to add brandy to the mix, or not, appears to have no role, because it is an immaterial idea. Yet it is what makes this cake the chef's cake, as well as all the other decisions, a smidgeon of this and a smidgeon of the other, that the chef has up his sleeve from past experience, or simply on a creative whim. It is almost impossible to attribute what proportion is the recipe on the page and what proportion is the ingenuity of the chef. The error here is to separate the genome from the living organism. The genome is an integral part of the living organism and its interactions with its environment. Separating the genome out in this way is another example of a false dualism. In this case, a bit of the system that controls the rest of the system, as if it is not itself controlled by that system. The living organism is the chef, and not the genome. The genome is not a separate functional entity, whereas the organism is, or at least it has a functional integrity of which the genome is but a part.

Consider the nature of molecular arrangements that would enable us to create this page. They would be in the form of living agents that can make the necessary decisions and think through the ideas and put them into words. Organisms are not simply an aggregate of their molecules, and they are in a process of continuous creativity during their lifetimes, not prisoners of their genes. While genes may be involved in creating a facility for action, they do not determine a specific action. We would not conclude that Leonardo da Vinci's Mona Lisa or Beethoven's fifth symphony could be derived from the genes of their creators, rather than the organised entities of Da Vinci and Beethoven themselves. This point applies to the control that organisms have over what they do as agents through their lifetimes. Whilst it may involve changes in the expression of genes it does not require major changes in the genome.

It is not the function of genes to determine any particular kind of behaviour. Their function is to enable particular proteins and RNAs to be formed. So genes cannot be said, for example, to determine whether we are selfish, or to rule out certain types of behaviour, such as those that can be considered altruistic. In any given instance of a behaviour, or a choice, it would not be necessary for any changes in genes or any direct involvement of genes. In contrast, behaviour has been shown to alter gene expression, such as when social interaction leads to changes in development that then makes it more likely that certain types of behaviour will persist in a group. We are social beings and our interactive behaviour is the glue that binds together the groups to which we belong. For example, mutual stroking which leads to the release of endorphins and a feeling of well-being. We nurture others as well as ourselves in a mutual cooperative, the precise arrangement of which in any given time cannot and is not programmed in our genes. For example, the tools that we invent enable us to do things in ways that previous generations could not. This is true not only for us but for many other species, such as when chimpanzees use and modify stones to more readily crack nuts, which has been shown to be culturally derived rather than genetically determined.

Even if genes are significant in determining our behaviours, given the number of social species, there would have to be genes creating behaviour that outweigh any that may be specific to selfishness, else one has to resort to the idea that it is selfish to be cooperative simply because it maintains genes in a gene pool. In this case, the closest we get to any kind of true altruistic behaviour is that which is considered reciprocal in a "you scratch my back and I'll scratch yours" kind of way. Certainly in many animal species we see mutual grooming, which is important in bonding and social group cohesion; if anything this encourages less selfishness. Given that genes are used in cooperative processes none of them would truly be regarded as "selfish". That, of course, is if you believe that genes cause behaviour. In any event few genes have been found to map specifically to any particular trait or characteristic. And certainly since the Selfish Gene was promoted none has ever been found to map to selfishness. If you define whether genes are selfish by virtue of being in the gene pool then this means nothing in relation to behaviour. It is simply a definition that must be true by the way in which it is defined. There is therefore no reason to suppose that a gene would be found specifically for selfishness above all other traits, particularly when behavioural traits are complex, not simple.

6 Genes Cannot Be Selfish, People Choose Whether to Be Selfish or Cooperative

No doubt we can be selfish, by definition that would have to involve a choice to be so, depending on the context and circumstances. But any individual who might be selfish on one occasion and giving or altruistic on another does not change genes between these two occasions. In fact the genes of somebody who might be regarded as selfish are little different if at all from those who are regarded as not selfish. I might be selfish in taking the last strawberry, but I might also feel guilty about having done so. Is there also a guilty gene? On another occasion I might choose to give the strawberry to someone else simply because it is the last one. Whether or not our actions are regarded as selfish or altruistic depends on the belief we have about their outcomes when we decide to act in such a way. A decision cannot be regarded as selfish or otherwise without taking account of the motivation or reasons for the action taken.

It might be argued, as gene-determinists do, that we are inherently selfish. Consider this text from the very beginning of *The Selfish Gene*:

> This book is mainly intended to be interesting, but if you wish to extract a moral from it, read it as a warning. Be warned that if you wish, as I do, to build a society in which individuals cooperate generously and unselfishly towards a common good you can expect little help from biological nature. Let us try to *teach* generosity and altruism, because we are born selfish. Let us understand what our selfish genes are up to, because we may then at least have the chance to upset their designs, something which no other species has ever aspired to. (Dawkins, 1976, The Selfish Gene, p. 3)

Dawkins correctly sees that any inherent biological propensity can be outweighed by social factors such as education and culture. But he is doing so from an incorrect understanding of any such propensity. Most organisms are not "born selfish". They are born needful, of course, but most are born into co-operative groups. There are no genes that are "up to" anything. They are mere bits of chemicals that can't have any intentions. Note also that the text assumes Cartesian dualism—"something which no other species has ever aspired to"—which is to give humans, *and only humans*, the power of agency. This was Descartes' mistake, which has misled Western philosophy and science for centuries. It is high time that it should be abandoned.

The question why a particular instance of behaviour occurs is different from the question of why certain behaviours will persist in a population over generations. So, for example, the reason why Jack accompanies Jill up the hill to fetch a pail of water may in large part be because he is in love with her and he wishes to be with her. But his love may have other consequences in maintaining the integrity of the population. Thus we cannot conclude from this that "maintaining genes in a gene pool" was uppermost in the mind of Jack when deciding to accompany Jill. In any event, genes are used to maintain the integrity of the organism; other than being *in* organisms and being *used* by organisms, there is no gene pool in the sense in which it can be said that we behave in order to maintain it. Furthermore, the principle advantage of sexual reproduction is that it shakes the genes around, not that it maintains them. Thus, a consequence is not necessarily a cause of behaviour, unless the behaviour is done specifically to bring about the consequence. This is why simply counting genes in the "gene pool" will not give us the reason for any particular instance of behaviour.

It is said that necessity is the mother of invention. However, nature is the mother of necessity. Need drives creativity in our behaviour or in what we do but as we will see it also drives evolution. Although the precise mechanisms may be different and occur over different time periods, both can be directional and purposeful.

7 Life is Problem-Solving

Life is the source of all problems. Many of these are created by the way in which we interact with each other and are culturally expressed. For many, we create social solutions that involve the ingenuity of many of us, and often trans-generationally. Music is a vital human cultural expression which has clearly evolved from generation to generation and is continually generating new styles. We see also in relation to this an evolution of the instruments that are used to perform and interpret the myriad of compositions. For example, the development of the piano which enabled a huge expansion of orchestral capability, but it is also seen in the trumpet. The trumpet illustrates also a major point we are making which is that it is a tool crafted to play melodic lines.

A trumpet has three valves which can be used in combination to alter the length of the tubing in play and from this an almost infinite number of melodies can be played. The valves are not a "code", they are a tool invented so that a trumpet can do this. Life is similarly creative. The functioning is analogue not digital. We cannot find the melodies in the valves, but we can use them to produce them. Thus it is also with our genes. Our choice of melody and expression from moment to moment during a performance is not written in our genes.

7.1 Consequences for Evolution

Life creates problems that are, on the one hand in the here and now, which involved the kinds of choices, decisions and actions we have discussed, which affect individuals and groups of individuals. But in a continually changing environment many problems are faced by the species as a whole and could be said to be ecological. Evolution is a process of addressing these constraints. But these changes will involve changes in structure and function and functionality. We argue that this process is creative in its nature and can involve major changes in the organisation of genes.

By contrast the standard theory of evolution, neo-Darwinism, rules out any such directed re-organisation. All DNA changes are assumed to be purely random. Yet the experimental evidence shows that sequences of any length of DNA can be moved around the genome either directly as DNA or indirectly via RNA. Way back in the 1930s and 1940s Barbara McClintock showed that such transfers do occur in plants in response to environmental stress. This was why, on winning the Nobel Prize for mobile genetic elements in 1983, she referred to the genome as a

> highly sensitive organ of the cell, monitoring genomic activities and correcting common errors, sensing the unusual and unexpected events, and responding to them, often by restructuring the genome.[7]

Comparative sequencing of genomes from many different species ranging from yeast to man reported in the 2001 *Nature* report on the first full draft of the human genome sequence[8] (International Human Genome Mapping Consortium 2001) shows that reorganisation of DNA happened during evolution The gene sequences for both transcription factor proteins and chromatins show precisely this kind of massive genome re-organisation involving very long functional sequences. The length of these segments (transposons) shows that this could not have happened by the gradual accumulation of point mutations, as proposed in the standard theory of evolution.

The Chicago biochemist James Shapiro was responsible for showing that the processes of gene reorganization discovered by Barbara McClintock also occur in bacteria and has since developed the field of what he calls natural genetic engineering.[9] He writes:

> It is difficult (if not impossible) to find a genome change operator that is truly random in its action within the DNA of the cell where it works. All careful studies of mutagenesis find statistically significant non-random patterns of change, and genome sequence studies confirm distinct biases in location of different mobile genetic elements. (*Evolution, A view from the 21st century*, page 82)

---

[7] McClintock (1984).

[8] Lander et al. (2001).

[9] https://en.wikipedia.org/wiki/Natural_genetic_engineering

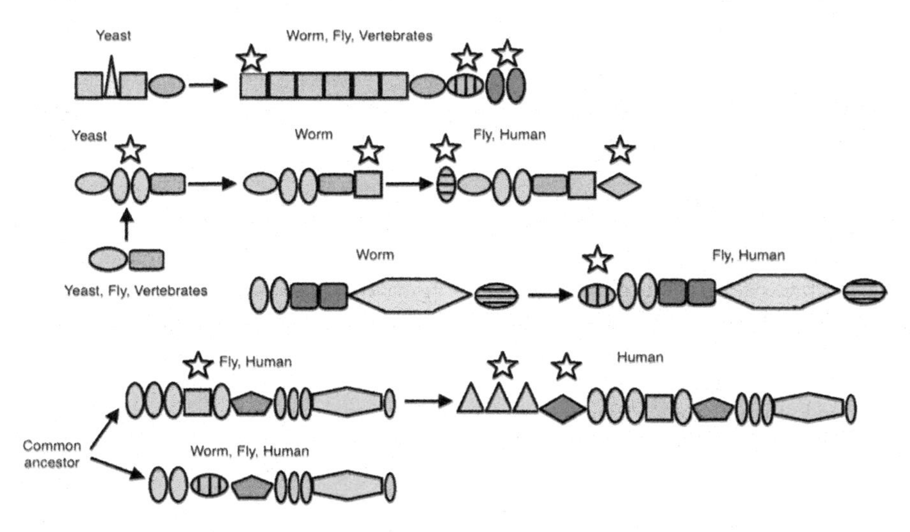

Evidence of large functional cut-and-paste changes in genomes during evolution. The different symbols represent functional domains of a chromatin protein that has evolved by domain accretion as species as different as yeast, worm, fly, and human have diverged from their common ancestors.[10] The stars indicate the domains that have moved around and been incorporated into the functionality of the similar (homologous) protein in the different species protein. As more domains get incorporated the functionality of the protein is extended. The chances of such changes occurring through random accumulation of point mutations is exceedingly low. The better explanation is that organisms under stress have initiated genome reorganisation through transposing and connecting up existing segments of DNA.

8 The Harnessing of Chance in the Immune System

Random variation is creatively harnessed at all levels in living systems and is an essential ingredient of all guided processes. For example, the immune system uses random variations in DNA to produce creative non-random functional, i.e. targeted, results.

The immune system protects the body against invading viruses, bacteria and other infectious agents by producing specific reactions against them. This creative process continues throughout the life of an organism. It is essential precisely because living systems are open and not closed, interacting continually with the environment. The immune system achieves this by producing molecules called immunoglobulins that can grab hold of invading viruses, bacteria or other infectious agents even if it has never encountered them before. When the invading body is new the immune system

[10] The original figure is from Fig. 41 in Lander et al. (2001). The meaning of each domain is also listed there.

will produce a new immunoglobulin. The proteins (antibodies) that can catch hold of invaders (antigens) are all constructed on the same plan. There is a long part of the molecule that enables it to locate itself in the organism's cells where it latches onto the foreign body. The structure of this part is produced unchanged. Any mutations in that part would stop it working as an antibody. The other part is variable and mutation is restricted to that part. This is achieved by extremely rapid mutation of the DNA sequence used in its production. Like keys in a lock, these are selected to be specific to the antigen. But for this remarkable process we would not exist. This is an example, par excellence, of a system response to environmental stress involving changes in DNA sequence. The mutations are accelerated in the variable part, and *only* that part, of the genome that forms the template for an immunoglobulin protein.

So far as is known, these mutations occur stochastically, and what is modified is the speed at which they occur. However, the location in the genome is certainly not a matter of chance. The functionality, in this case, lies precisely in targeting the relevant part of the genome. The arrival of the antigen itself activates the hypermutation process, and the binding to a successful antibody triggers the proliferation of those cells that make it. Thus, the system targets the specific antigen. Even more remarkably, all the functionality *in the rest of the genome* is also maintained. Considering the vast size of the entire genome, this is pin-point targeting requiring highly specific feedback processes to be successful. By holding correct parts of the immunoglobulin sequence constant, the system finely tunes the rapid mutation to occur in *only a tiny part* of the entire genome. Such tuning is one way in which organisms can dynamically respond to environmental change.

There are two key processes involved here. One is the hypermutation, the other is maintaining the integrity of the DNA sequences. This is also a continuous active process because random changes are occurring all the time. The hypermutation is achieved by lifting the restraint imposed by the system. Organisms also need to do the reverse, which is to protect against unwanted random change. The way in which this is achieved is that their biological networks buffer the organism from the majority of these molecular changes at the genetic level. Thus, the great majority of random genome changes have negligible effects.

9 Integrating the Forms of Biological Causation

Our diagram illustrates the relationships and forms of causation involved.

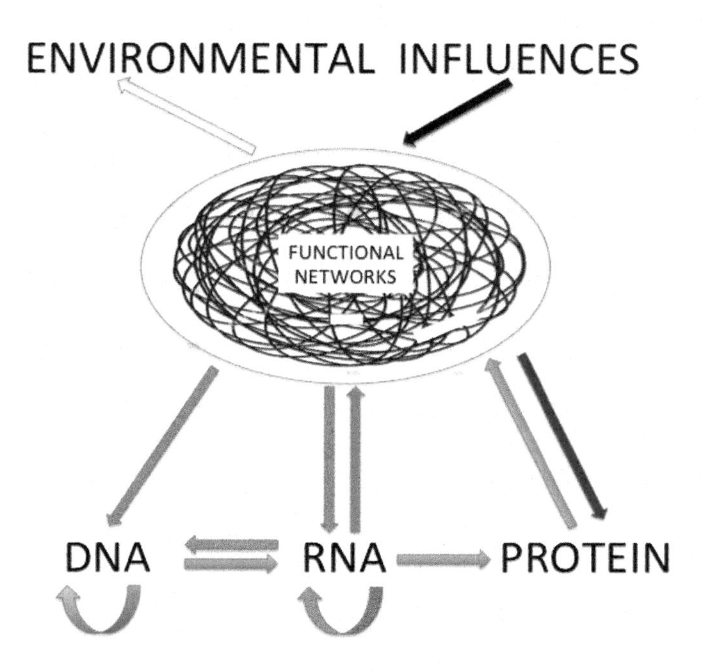

Figure illustrating the role of physiological functional networks (sometimes called gene regulatory networks GRN) in buffering organisms against dysfunctional molecular variations in the genome. The original 1958 formulation of the Central Dogma is shown by the blue arrows. DNA is viewed as a self-replicator (the curved arrow bottom left) which can be used to make RNAs, in turn used to make proteins. Both products then form the functional networks in the cells of the body, together with many other molecules that are not produced using DNA templates.

Key to figure

Blue arrows: 1958 version of Central Dogma
Orange arrows: additions in 1970 version of Central Dogma
Green arrows: Epigenetic control of DNA expression and sequences
Purple arrow: Control of protein folding
Black arrows: Environmental effects on functional networks
Open Black arrows: Influence of organisms on the environment

In 1970, the discovery that RNAs could be back-translated into DNA shook the foundations of the Central Dogma. Crick was therefore forced to modify his one-way causal concept. He did so simply by adding the orange arrows. RNA can also self-replicate, and insert DNA into the genome. But even this reformulation did not recognise what was already known: (1) That the same genome can be commanded to make hundreds of different cell forms in the body. (2) DNA can be reorganised when organisms are under stress. The downward pointing green arrows represent these forms of causation. In addition, the networks themselves are subject to many forms of environmental influence. Through this influence the genome can be sensitive to the environment. It is not isolated. The massive re-organisation of the hereditary molecular material (DNA) had already been demonstrated experimentally by an American plant scientist, Barbara McClintock, many years previously. Even in 1970 therefore it was already known that DNA is controlled and edited by organisms. Later, it became clear that DNA is not a crystal-like self-replicator. Even replication is under the control of the organism. As we have already noted above, McClintock correctly said when she received her Nobel Prize in 1983: "The genome is an organ of the cell."

10 The One-Eyed Watchmaker

In Chap. 3 of his book, *The Blind Watchmaker*,[11] Richard Dawkins produces his famous Weasel program. He estimated that a monkey writing out 28 characters randomly on a typewriter would require much more than the whole lifetime of the universe to arrive by pure chance at a correct 28 letter sequence to match Shakespeare's text, "METHINKS IT IS LIKE A WEASEL". But if each correct character were to be held constant between successive generations of random typing, it would require only a modest number (43) of iterations to achieve a correct result. The program resembles the operation of an old-fashioned three-wheel fruit (slot) machine. If the target for a reward is, say, three lemons and a spin of the wheels produces two, the best strategy might be to hold the wheels with the two lemons and spin the remaining wheel until that also shows a lemon. The number of 'wheels' (28 in the Weasel example) doesn't change the principle of this mechanism.

We agree with Dawkins that completely random processes with no 'hold' or similar 'guiding' mechanism would require impossibly long periods of time for successful evolution to occur.

We nevertheless show in our articles that organisms and populations of organisms do have identifiable and empirically testable goals, and that variations on the theme of the Weasel program found experimentally in nature, show this to be true. The key to understanding why we differ from neo-Darwinists on this matter lies in multi-level feedback processes that have been shown to exist which enable organisms and populations to direct the evolutionary process in response to stimuli from the environment and so achieve the inheritance of acquired characteristics. These feedback processes require analysis of function at a high (systems) level, e.g., networks, cells, tissues, organs and the organism as a whole in interaction with the environment, including other organisms. Multi-level feedback is a requirement of goal-directed behaviour. A purely gene-centric view would not 'see' such feedback. Empirical tests used routinely in physiology and engineering can identify the feedback processes. Such feedback is an essential feature of physiological function. Imagine walking without it. Our function is goal-directed. We avoid objects and walk to meet others.

---

[11] Dawkins (1986).

11 Limitations of the Modern Synthesis

The standard theory of evolution, The Modern Synthesis, is predicated on the undirected, slow process of gene mutation. There are many reasons why this produces an inadequate understanding. We have already pointed to the role of transposition as demonstrated by Barbara McClintock and of targeted mutation. There are several additional processes by which directed evolution occurs: epigenetics, cultural change, niche construction and adaptation. Evolution is an ongoing set of iterative interactions between organisms and the environment. Evolution is a continuous organic process. Directionality is introduced by the agency of organisms themselves as the one-eyed watchmakers. Evolution itself also evolves.

The Modern Synthesis views the environment as a passive filter for the adaptability of individual organisms. This ignores the ecological reality that organisms are in continuous process of creating this environment. Whether or not the analogy with a filter is correct, it is not passive. The separation of the organism from the environment is a mistake that ignores the ecological interactions, cohabitations, etc. that produces not a passive filter for adaptability but an active one. Cooperation in ecological systems is as significant a feature as competition between individuals, and survival depends upon it. It is seen not only in social groups of the same species but also in cooperation of individuals or groups across species. For example, a fish working cooperatively with an octopus to obtain food, each has a faculty the other does not and thus cooperation has its reward. It involves a mutual anticipation of the behavior and the intention of the other. Intention is a major feature of situational logic. It is a feature of their environment at the social level. The actions of such cooperative groups will in turn influence the fitness of the individuals whether or not it is the individual or the group that is selected through evolution, it is their mutual behaviour and not their selfishness that is being selected. Thus it is perfectly feasible to argue that there is as much a selection pressure on behaviour, which might be considered altruistic as it would be selfish. What matters in evolution is not the survival of genes in a gene pool, but of organisms and ecosystems. Only organisms and ecosystems have any say or influence on what survives. We *are* the environment!

The standard view of evolution has emphasised a distinction between vehicle (organism) and replicator (genes). This yet again tries to introduce a false dichotomy. Genes cannot replicate without organisms and in any event replication is not the key ingredient in evolution; the key ingredient is change. One of the features of reproduction is not that it produces like for like but that it enables environmental influences to bring about changes across generations in response to pressure. Some of this, for example in mammals, will occur in early development in the womb, where it is influenced by the nutritional and social environment of the mother. It is sometimes argued that such changes do not persist across many generations but this simply begs the question over how many generations they would need to persist to be of significance. Whatever is the answer to that, it cannot be denied that this is an active process physiologically not a passive one. It brings about a refinement of adaptation to the anticipated environment of the developing organism. Furthermore,

Waddington's experiments on fruit flies in the 1950s showed that genetic assimilation could occur in as few as fourteen generations.

12 How Organisms Make Choices: Harnessing Chance in the Nervous System

In the introduction we noted that we have a sense of volition, or that our actions and our choices are purposeful, but also that we make them to achieve objectives that we have decided. This is not an illusion. It is a powerful sense of our being; and much of our action is driven by this sense. Yet science seems to have a problem with this particular sense. We have an unquestioned sense of vision. We can tell each other what it is we see. Of course what we see might be an illusion, a trick of the mind; that would be true of all our senses. We have a sense that it is *we* that make choices. When we wait at a bus-stop we do so generally in anticipation of the arrival of a bus that would take us to a destination. We may be wrong. The bus may not arrive or we may get on the wrong bus. But by virtue of that this is not an illusion any more than it is when we see the bus when it arrives. So how do we do this? It would be difficult to give a precise answer to this for any of our senses. Science is still unraveling the details of the processes involved. This is no reason then to reject this sense of being, of choosing. It just means that we have more to study. But the clues lie in the organization of living systems. There is no reason to believe that there is anything unusual in the neuronal and synaptic processes that would be involved, other than their precise location and precise organization.

To consider this sense of being and choice as an illusion would ignore that much of what happens in the world around us results from the assumption that it isn't. We can discuss our choices with each other and make a reasonable case for further choices. If this were an illusion, the words on this page would have no impact on the reader. So why is there a problem with accepting such agency? Perhaps it is that it opens a Pandora's box, for if agency exists, we could act for reasons that cannot be determined by studying or counting genes, or any other molecules. Indeed, this is why the idea is so liberating. This does not mean that considerable constraints on the exercise of such liberty do not exist. We can only work on the ground of the possible. But through it we might, through ingenuity, make possible that which initially may appear not to be so. We can solve problems.

13 Relevant Processes in the Brain, Including the Cerebral Cortex

Life is inherently creative because it must continuously replenish to maintain itself. As Plato reminds us, all things are in flux, a constant process of change.

> Heraclitus, I believe, says that all things pass and nothing stays, and comparing existing things to the flow of a river, he says you could not step twice into the same river. (Plato Cratylus 402a = A6)

At about the same time (c 500 BC) as Heraclitus, Siddarttha Gautama (the Buddha) was expressing the same idea of conditioned arising—nothing is experienced independently of its conditions, all is in flux. In China also, Daoism was so similar that when Buddhism entered China the similarity was recognized.

This process of flux creates a problem for life but also its solution. Life harnesses this change to meet the challenges it faces in a continually changing environment. Indeed, the central nervous system of organisms is functionally organized to respond to change, particularly in sensing the environment.

Our senses do not passively receive stimuli and interpret them. There is an ongoing dynamic and interactive anticipation of the world about us (ideas), which is moulded by the senses in a continuously creative process. What we see is not the individual photons interacting with our molecules, we see the interpretation. If we could see individual photons we would see a stochastic mess. Moulding of the sensory experience begins even at the level of the first-order sensory cells. This enables a fine-tuning of the sensory experience. Such fine-tuning enables the organism to distinguish between different sounds of similar frequency and amplitude. The organism tunes into its environment. However, the signal does not then reach the auditory cortex directly and considerable processing takes place along the auditory pathway. This is so for all our sensory experiences. We suggest that this creative process is only possible by harnessing stochastic processes within the central nervous system.

Our proposals have two major requirements that are testable experimentally.

The first is that stochasticity should exist at multiple levels of organisation. In the case of the cerebral cortex, considered to be deeply involved in conscious activity, the experimental evidence is extensive. The most extensive recent account is to be found in Edmund Rolls' *Cerebral Cortex, Principles of Operation*,[12] Chap. 5 on "The noisy cortex: stochastic dynamics, decisions and memory." Nerve impulses occur in a probabilistic way, usually conforming to a Poisson distribution. Rolls writes:

> The spiking [impulse] activity is approximately Poisson-like ... because the neurons are held close to (just slightly below) their firing threshold, so that any incoming input can rapidly cause sufficient further depolarisation to produce a spike.

As Rolls notes, this gives great plasticity to the nervous system. This sensitivity is analogous with the extraordinary plasticity of the immune system. Of course the process by which the plasticity is achieved is different. The immune system depends on randomness created by switching down the error-correction in replication of gene sequences, the nervous system depends on the natural stochastic behaviour in the region of excitation thresholds, which can be controlled by inhibitory and excitatory synapses, which may allow or not transmission to occur in selected pathways. This could allow the creation of multitudes of processing networks in decision-making.

[12] Rolls (2017).

This plasticity can underlie two other features that Roll's draws attention to. First, it can be used by memory systems in the brain, again in a way analogous to the memory created by immune system success in developing an appropriate antibody. Second, it can be used to create many forms of behavioural outputs in response to an environmental challenge. Rolls points to the fact that stochasticity guarantees that the response will never the same, like the running.

What seems to us to be still missing is not so much the ability to produce novel reaction to the same environmental stimulus, but to have the option of *choosing between multiple possible responses* in the light of what best fits the environmental challenge. That process requires the equivalent of the choice process in the immune system. That must also involve neuronal processes. What precisely these processes are needs attention. However, if we dismiss choice as an illusion it is unlikely that this attention will be given. It is surprising therefore that the possibility of such choice is dismissed without any evidential basis and the usual processes of scientific enquiry.

A similar idea was proposed many years ago in the work of Gerald Edelman. Edelman first worked on the immune system and was responsible for working out the structure of immunoglobulins and to distinguish between their nearly constant basic structure and the variable 'grabbing' part, which has to be variable to become adapted to any new antigen, and then depends on a choice process between the many new immunoglobulins produced. He naturally wondered whether there might be a similar ability of the nervous system to generate multiple possible responses to an environmental challenge from which a choice can be made that conforms to normative social criteria.

His proposal[13] was that groups of neurones could represent the available options for action from which a selection could be made. He called this idea the theory of neuronal group selection. If he had kept just that title, his theory might have fared better than it has. Unfortunately, he also called it Neural Darwinism, which can imply that he saw it as an evolutionary mechanism. This is how Francis Crick interpreted it when he dismissed the whole idea on the basis that it did not contain separate 'replicator' and 'vehicle'.[14]

As we use the idea, it does not need a replicator. It is a hypothesis about alternative action options from which a selection can be made. Similar ideas have also been used by Ginsburg and Jablonka in their 2019 book *The Evolution of the Sensitive Soul*.[15]

14 Conclusion

In summary, we believe that the theory of Agency we have developed recently is capable of explaining the phenomenon, is plausible, and is compatible with existing neuroscience research. It is time for reductionists to cease claiming that science

[13] Edelman (1978).

[14] Crick (1989).

[15] Ginsburg and Jablonka (2019).

justifies eliminating agency from our view of life. The discoveries of molecular biology have unambiguously shown that scientific theories that deny agency have been undermined. Ironically, therefore, they have been undermined by scientific investigations at the molecular level—the very level that was claimed to justify them. But they were never plausible anyway. They could only be made to seem plausible by misusing the language of metaphors and other similes. The Selfish Gene and gene-centricity should be buried as an aberration of Western science, steeped as it is in a mechanistic view of life.

15 Relevant Source Literature

This article is based on a series of articles we have published recently. All our source references can be found in these publications:

1. The physiology of agency Can Reasons and Values Influence Action: How Might Intentional Agency Work Physiologically? 2020 *Journal for General Philosophy of Science*. https://doi.org/10.1007/s10838-020-09525-3.
2. "Active" Darwinism and Karl Popper Rehabilitation of Karl Popper's ideas on evolutionary biology and the nature of biological science. In *Karl Popper: His Philosophy and Science*. Springer–Nature 2021 In press.
3. Boundaries and causation Biological Relativity Requires Circular Causality but Not Symmetry of Causation: So, Where, What and When Are the Boundaries? *Frontiers in Physiology*. 2019. 10, 827.
4. Emergent properties A–mergence of biological systems. In *The Routledge Handbook of Emergence*. 2020. 387–399. 2020.
5. How organisms make choices. Harnessing stochasticity: How do organisms make choices? 2018. *Chaos* 28,106,309.
6. Is evolution blind? Was the Watchmaker Blind? Or Was She One–Eyed? 2017. *Biology*, 6, 47.
7. Harnessing of Stochasticity Evolution viewed from physics, physiology and medicine. 2012 *Interface Focus*, 7, 20,160,159.
8. The principle of Biological Relativity A theory of biological relativity: no privileged level of causation. 2012. *Interface Focus*. 2, 55–64.
9. Artificial Intelligence and Agency Forum: Artificial Intelligence, Artificial Agency and Artificial Life. 2019. *RUSI Journal*, Vol. 164, No. 5, July.

The complete set is available as a pdf download on the webpage: https://www.denisnoble.com/wp-content/uploads/2021/01/Harnessing-Collection.pdf.

References

Coyne, J. A. (2014). What scientific idea is ready for retirement? https://www.edge.org/response-detail/25381. Retrieved October 12, 2020.

Crick, F. H. C. (1970). Central dogma of molecular biology. *Nature, 227*, 561–563.

Crick, F. H. C. (1989). Neuronal Edelmanism. *Trends in Neurosciences, 1989*(12), 240–248.

Crick, F. H. C. (1958). On protein synthesis. *Symposia of the Society for Experimental Biology, 12*, 138–163. pmid: 13580867.

Dawkins, R. (1986). *The Blind Watchmaker: Why the evidence of evolution reveals a universe without design*. W. W. Norton & Company.

Edelman, G. M. (1978). *Neural Darwinism*. Basic Books.

Ginsburg, S., & Jablonka, E. (2019). *The evolution of the sensitive Soul*. MIT Press.

Lander, E. S., Linton, L. M., Birren, B., Nusbaum, C., Zody, M. C., Baldwin, J., Devon, K., Dewar, K., Doyle, M., FitzHugh, W., Funke, R., Gage, D., Harris, K., Heaford, A., Howland, J., Kann, L., Lehoczky, J., LeVine, R., McEwan, P., & McKernan, K. (2001). Initial sequencing and analysis of the human genome. *Nature, 409*, 860–921.

McClintock, B. (1984). The significance of responses of the genome to challenge. *Science, 226*, 792–801.

Popper, K. (1999). *All life is problem solving*. Routledge.

Rolls, E. T. (2017). *Cerebral cortex, principles of operation*. Oxford University Press.

The Epistemology of Life Understanding Living Beings According to a Relational Ontology

Marta Bertolaso and Héctor Velázquez

1 Introduction

For our understanding of the material world, its laws, principles of organization and structure advances, it seems increasingly unfeasible to consider reductionism an adequate epistemological approach to account for complex phenomena and dynamics. Although it conveniently simplifies the description of reality, it invariably leaves the feeling that it has prevented us from understanding the profound differences that exist among natural entities. One of the fields that has most strongly resisted the reductionist approach is the study of the living being. In this text we will address some reasons for thinking that holism is not a valid alternative to reductionism as an approach to understanding the world of life. We will also discuss why it is not enough to refer to the systemic perspective of complexity as more complete than mechanicism for understanding what distinguishes living beings from other entities (chemical, physical) that exist in nature. We argue that not only is a new epistemology necessary for us to understand the specificities of life but also that it should move towards a new ontology that explains how and why life is an irreducible reality and provides a much more unified understanding of the other natural spheres.

The epistemological success of mechanicism lies in the simplicity of its approach to reality since with a minimum of elements it allowed a maximum of knowledge about the object studied. For mechanicism, every natural system is composed of (i)

M. Bertolaso (✉) · H. Velázquez
Research Unit of Philosophy of Science and Human Development, Università Campus Bio-Medico, Via Alvaro del Portillo, 21, 00128 Rome, Italy
e-mail: m.bertolaso@unicampus.it

H. Velázquez
e-mail: hector.velazquez@umayor.cl

H. Velázquez
Director Centro de Sociedad Tecnológica y Futuro Humano, Facultad de Humanidades, Universidad Mayor de Chile, Santiago, Chile

S. Wuppuluri and I. Stewart (eds.), *From Electrons to Elephants and Elections*,
The Frontiers Collection, https://doi.org/10.1007/978-3-030-92192-7_38

parts, (ii) interaction criteria connecting them, and (iii) a resulting whole, regardless of whether we are talking about micro, medium or macrophysical reality (Velázquez, 2020).

According to mechanicism, time is not a leading variable. This position made it possible to achieve one of the highest aspirations of the human mind: to know the past (deduction) and the future (forecast), just by knowing the present, with the same precision with which we describe nature right now. In this way, from a mechanistic point of view, to increase our knowledge of nature it is only necessary to advance in the detailed identification of its parts (if we have better scientific instrumentation, we can identify more parts), as well as their interaction, criteria (natural laws) and the description of the resulting whole (the solar system, a cell, the circulatory system, a megacity, etc.).

The identification of parts, interaction criteria, and a resulting whole under this scheme allows a hierarchization of these elements: there are central, derivative, and tangential parts, according to the role they play in the system; a classification that can vary according to the approach: seen from another perspective, what appears as the central part in one context can be completely tangential in another. For example, universal gravitation can be central to explaining the three-body problem involving the planet Earth, the moon and the sun, while the size of the lunar crater Copernicus would be completely tangential here. But gravitation can be tangential if what we are studying is the distribution of asteroid density impacts on the lunar surface. Here the ground structure would be central.

With the replacement of mechanicism by the explanation of nature in complex sciences and through systemic models, time became the protagonist and the mere enumeration of parts became an interaction criterion, and everything turned out to be insufficient. The entity of the system is interpreted as the dynamic result of its interaction with the environment. It is a closed system if this interaction is poor and only results in a resistance of the environment to maintain its existence (like the rocks that receive the water along a cliff). It is an open system if through interaction the system assimilates the environment and makes it part of itself while leaving something of itself in the environment (as in the world of life with nutrition and flourishing). Based on this type of complexity, nature is interpreted as a changing dynamism over time with sensitivity to the variation of the initial conditions, unfolding towards the formation of structures, rhythms, and patterns (structural and temporal).

In its description of the physical world, the complex systems model loses the power of forecasting and deduction which are so important in mechanicism. On the other hand, its description of nature as a changing, evolving, structuring and dynamic system is closer to what is a physical reality. Nature understood as a complex system is a network of nodes where every interaction is equally the protagonist and in which ontologically no node is more important than another: there is a mutual causality of all the elements involved in the system, and none of them can escape dynamic interaction with the environment.

The vision of nature as a complex adaptive and evolutionary system affirms that a reinterpretation of the physical world (as a replacement for mechanicism) is useful to understand any of the natural systems that it is intended to explain: the material world, living beings, conscious entities and even cultural, social, or urban ones. Its proposal is, as in the age of mechanicism, to provide an all-encompassing vision that accounts for how the changing reality that constitutes the cosmos in any of its variants is integrated and functions.

However, systems may not account for the 'ontological' dependencies that a natural entity has on its context or environment and within its own structural dynamics. And it runs the risk of focusing solely on causal interdependence among members of the natural network. However, what is usually addressed as 'context dependency' is much more than a mere interaction with the environment: it explains not only the structure of a natural entity (emergence) but its specific stability (essence) which is more fundamental from a metaphysical point of view. In other words, as an interpretation of the natural world, systems' complexity runs the risk of becoming the new mechanicism: an explanatory model that homogenizes the *explanandum* and describes nature without paying attention to the 'relative' dependencies/specificities, which single out the entities.

2 Debating the Dynamics of Life

2.1 Reductionism Versus Holism: Limits in the Debate

When we talk about reductionism, the philosophical debate initially focuses on the different dimensions that have been historically articulated as follows (Ayala, 1974; Ayala & Arp, 2010).

(i) Methodological reductionism applied to the biological sciences, understood as a process of decomposition and re-composition of a system or entity, aims to explain dynamic features of development, understood as a combination of cell cycle, metabolism and cell communication. In this sense, molecular biology has been the paradigmatic biological approach to the living organism in the last century. Through this approach, the process of life, i.e. the functional integration of different levels of the biological complex organization, can be explained even at the lowest possible level. The question of how the 'lowest possible level' should be understood is what forced the debate about the possibility of an epistemological and ontological debate as well. The reflection about the epistemological status of an analytical point of view evolved in the debate about levels of analysis and mesoscopic way of thinking (Bertolaso, 2013b; Bizzarri et al., 2019; Green & Batterman, 2017). However, when the possibility of considering the 'lowest level' in ontological and metaphysical terms was explored, major problems arose about the double implicit aim to deduce how explanations of molecular dynamics which are biologically relevant arise only by investigating physical–chemical processes in virtue of their privileged explanatory value.

(ii) Epistemological reductionism has been typically related to inter-theoretical reduction, so that it would be possible to replace one theory T with another theory T' which is more 'explanatorily' powerful (Brigandt & Love, 2017) Theories and experimental laws in this case get reduced when they turn out to be special cases of theories and laws formulated in some other branch of science (Bertolaso, 2016). However, there has been a progressive shift in the philosophical discussion from theories to explanations and the role that mechanicism, emergent properties and biological concepts play in explaining biological phenomena. Nonetheless, under the still inherited empiricism and materialism in (philosophy of) life sciences epistemological reductionism eventually takes the form of a question about whether one theory about the world will be achieved once the world structure is described regardless of whether this is done in terms of mechanicism, systems, and networks.

(iii) Ontological reductionism relates to the reflection about what exists. Taking into consideration different kinds of empiricism and materialism that concurred to shape this third kind of reductionism, we can say that it argues that organizational properties distinguish inanimate entities (e.g. stones) and living entities (e.g. cells, elephants, bacteria colonies, etc.). This position could be compatible with both reductionist and anti-reductionist positions and has opened major current discussion about complexity and how it should be managed and understood.

Transversally to different accounts of theory and explanatory reductions (cfr. Brigandt & Love, 2017) Sect. 4), there is a context dependency argument. It has become more and more evident in the empirical field that the effects of molecular processes depend on the functional context in which they occur. Such arguments against reductionism stress the fact that "one molecular kind can correspond to many higher level kinds (…) that higher level biological structures and processes can be (and typically are) realized by different kinds of molecular processes, so that many molecular kinds can correspond to one higher level kind (…) because biological processes must be represented before they can be explained (Sarkar, 1998), two further features become salient as issues for explanatory reduction: temporality and intrinsicality". In both cases of temporality and intrinsicality, the debates are shaped by the evidence of what we call 'incompleteness' of the causal chain. The discussion about temporality highlights the hierarchical structure of temporal processes focusing on the function-structure relationship of living entities and, especially of developmental processes. Reference to space-temporal dynamics that take the form of functional-structural models and arguments have mediated the debate about higher level features (e.g., tissue organization structures and dynamics) that might causally explain lower-level features at a later time (e.g., gene expression patterns, etc.). Or as Mitchell observed (2009), components of a biological system (e.g. amino acid in a protein or DNA basis in a gene) are usually causally insufficient even if they are sufficient constitutionally. From a mereological viewpoint of the biological dynamics, the debate therefore has focused on different kinds of composition and aggregation, and decomposability of hierarchical structures often relying upon or moving from the debate already engaged in by Herbert Symon and Pattee about hierarchies and systems. However, as has been discussed in Love and Bringdant 2017 (see also Craver, 2009; Kauffman, 1971; Wimsatt, 1974; Winther, 2011) the functional-structural decomposition of a system is not a univocal enterprise and generates competing and complementary representations of the parts. Behavior often opens up the discussion about functional-structural relationships.

Similarly, Wimsatt (1980) decades ago already highlighted how "every investigation must divide a system from its environment and that methodological reductionism favors attributions of causal responsibility to internal parts of a system rather than those deemed external (see also Wilson & Craver, 2007)" (Sandford encyclopedia….). Temporal, compositional relationships assume in different ways that nested part-whole relations are predicated on a prior individuation of a system from its environment (ibidem). Although this line of thought relates to the context-dependency argument against reductionism, it also emphasizes the importance of "functional dependency relations uncovered by attending to temporality in addition to structural organization" (ibidem).

As Nagel already noted in 1961, what is investigated under the terms of space-temporal dynamics and organizations in life sciences is a mode of organization or a type of order. Craver and Bechtel (2007) rightly noted that we should also consider the possibility of diachronic aspects of part-whole and interlevel relations. Scientific explanations, in fact, usually invoke causal processes that involve components at several levels of biological organization as is in fact the case in morphogenetic processes. As reported by Love and Bringdant (2017), this is the core reason why development is a persisting biological topic for reductionism: "During ontogeny there are causal interactions over time among parts and activities (both homogeneous and heterogeneous) to bring about new parts and activities (both homogeneous and heterogeneous), eventually generating an integrated adult (whole) organism".

As we can see from this short review of problems with reductionism and context dependency issues that arise when we consider biological dynamics and their causal relevance in developmental and biological processes more in general, there are two aspects that merit special attention. First, all these debates to different extents move from a mereological view of living organisms and entities. This assumption is legitimated by the evidence that empirical sciences work by acting on things and parts and is clearly influenced by the inherited rejection of a vitalist approach in life sciences. No scientist would in fact claim that living processes are the effect of a non-material entity or vital force (Ayala, 1974). Similarly, the debate about physicalism ended up in a *clue de sac* when trying to assert that obtaining a satisfactory explanation of a phenomenon by reducing it to its smallest components is not only useless but also impossible (e.g. Ernst Mayr, 2004). Reduction in this view is based on invalid hypotheses and should be eliminated from the scientific vocabulary.

Second, the causal relevance of diachronic and synchronic dynamics in shaping the biological processes, also in their hierarchical features, brought about a reflection on the epistemological and ontological status of the emergent properties and organizational features of living entities especially in their dynamic aspects such as in developmental processes.

Processual viewpoints came to the forefront (Dupré & Nicholson, 2018) while the functional integration and coupling of biological processes became the most interesting *explananda* both in life sciences and in the philosophical reflection on them. Organisms are thus considered physiological units that exhibit a coordinated behavior and integrated systemic dynamics (reviewed in Militello, 2021) although "current definitions of functional integration do not provide a criterion (or norm) to distinguish different biological organizations on the basis of their internal physiological integration".

Systemic viewpoints, typically acknowledged as anti-reductionist stands, thus currently lead and shape contemporary debates about the possibility of explaining physiological individuality and biological autonomy, following the contributions in the field of relevant authors such as Varela (1979), Maturana and Varela (1980), Rosen (1991), Collier (2000), Kauffman (2000), Moreno and Mossio (2015). The debate about disentangling autonomy from independency from surroundings is currently focused on how we can acknowledge and explain that the internal behavior of an organism is not 'determined' by contextual factors. Two fundamental dimensions of autonomy in this regard have been described in Militello (2021): "the constitutive processes (e.g. metabolism and gene transcription and translation) that allow a biological organization to self-maintain, and the interactive processes (e.g. sensorimotor capacities and inter-organism communication) that enable an organism to interact with the environment according to its own internal norms (Moreno & Mossio, 2015; Mossio & Moreno, 2010)".

Finally, as Love and Bringdant (2017) argue, because temporality and intrinsicality are not captured by a focus on mereology alone, "different kinds of reductionism rather than a unified account of reduction or overarching dichotomies of "reductionism" versus "anti-reductionism" (or "holism")" should be endorsed so that "any single conception of reduction appears inadequate to do justice to the diversity of phenomena and reasoning practices in the life sciences. The multiplicity and heterogeneity of biological subdisciplines only reinforce this argument and suggests to some that we should move beyond reductionism entirely".

The context-dependency argument against reductionism has taken the form of a debate about causality in life sciences which brings up again the question about the peculiar status of temporality in the life sciences and also about the possibility for us to understand life from a philosophical viewpoint. Prior to determinations of whether reductive explanations succeed or fail, questions of representational choice and adequacy need to be addressed explicitly. As one of us argued (Bertolaso, 2016), when spatiotemporally continuous causal processes are at work no unique explanatory tool is either necessary or sufficient, and a theory of explanation that captures several different possibilities is needed. As Woodward (2010) claims: "depending on the details of the case, description or causal explanation can be either inappropriately broad or general, including irrelevant details, or overly narrow, failing to include relevant details. Which level is most appropriate will be in large part an empirical matter". Following Bertolaso (2016), we suggest that an 'empirical matter' should be understood in a wider sense: it is not just an empirical practical problem (what is possible here and now) but a rational problem that starts with a question like 'why this behavior and not another one'. "Pragmatic reasons in scientific practice are embedded with the effort to identify mesoscopic levels where objective and subjective dimensions of science meet. And the structural and functional dimensions of the system cannot be separated. They are captured differently by the definition of the system and by the functional behavior of the parts" (ibidem). Thus, as (not only) Love and Bringdant (2007) note, the scientific questions and the related choices of the working level influence both the results obtained and the form of the explanation, but such choices depend on a judgment about the patterns of correlations that have an explanatory relevance for that specific scientific question.

The theoretical principles on which a model's architecture is conceived, therefore, should look at not only how biological dynamics are explained at the different levels of biological organization, but mainly how the specific regulation and dependencies work among them. Here emerges the relevance of theoretical concepts in biological sciences (see below Section on *relative and relational terms* in biology). We argue, in this regard, that a *relational ontology* is necessary to ground both the conceptual and explanatory aspects of any systemic attempt to explain dynamics (Bertolaso & Ratti, 2018, ch 1, p. 10): "A relational ontology emphasizes the fact that even properties that seem to be 'internal' are actually relational. This is because a relational ontology assumes that the identity of the objects depends strictly on the existence of the web of relations an object is embedded in" Therefore, "in order to understand what certain biological entities (e.g., genes, proteins) do, we need to recreate the web of relations they are usually part of (ibidem)". That is, "[a] relational ontology holds that relations are prior—both conceptually and explanatory—to entities, and that in the biological realm entities are defined primarily by the context they are embedded within—and hence by the web of relations they are part of" (ibidem ch. 1, p. 1.).

The crucial importance of history for understanding living systems has been discussed by one of us: "Now it is necessary to introduce a further distinction between dynamics and history. Dynamical systems are described by trajectories in suitable phase space, but dynamics by themselves do not include a distinction between the before, the after, and the irreversible changing of the system. As in the case of a gas in a bottle, an observer who plays the reverse motion of each molecule would see nothing unusual, no "before" and no "after". The time arrow, the history, and the improbability of events of the reverse-motion film come into play only taking into consideration the boundary conditions. The dynamic approach just associates a clock to the degrees of freedom of a system, but it is insufficient for measuring the global structural changes. To get these ones requires, instead, the observing of the constraints' stratification/modification, which is precisely the history of the system" (Licata, 2015, p. 47 quoted in Bertolaso, 2016). Moreover, the philosophical discussion of emergence is often focused on the properties of 'wholes' that are evaluated as emergent with respect to the properties of 'parts'. "Downward causation is, consequently, evaluated as some kind of causal influence of whole properties over parts properties. Yet, several important cases in scientific practice seem to be pursuing hypotheses of parts properties emerging from wholes properties, inverting the instinctive association of emergence with wholes" (Bertolaso, 2017). Furthermore, "some areas of reflection which are very important for emergence, e.g., the philosophy of consciousness, do not allow the mapping of properties onto part-whole organizations. The conceptual puzzle is solved by constructing a framework that disentangles the mereological dimension (parts–whole, micro–macro) from the superventional dimension (basal-supervenient). By liberalizing the space-temporal allocation of emergent properties, the proposed dual framework could better capture the way in which emergence and downward causation are addressed in scientific practice" (ibidem).

In summary, all the debates mentioned above concerning reductionism versus anti-reductionism approaches rely upon a mereological view and account of biological dynamics, always producing a reductionist-vitalist tension. Moreover, anti-reductionist approaches, stressing a systemic viewpoint and relevance for emergent properties and for downward causation do not explain what "the whole is more than the parts" actually means, in part because of the mereological assumptions. Therefore, we suggest that a unified epistemology that can account properly for the ontology of functional integrations, developmental processes, etc., should move from the consideration that "the whole is different from (that is, also in some sense 'more' than) the parts". Parts and wholes have different epistemological and ontological statuses which are precisely the aspects that make biological explanations so peculiar in their functional, space-temporal representations.

2.2 Relational Ontology*: An Alternative Proposal to the Reductionism-Holism Debate*

What is missing in the above debates is a deeper understanding of 'differences' in the biological realm and of their dynamic stability, and of the convergence on the context's (causal) relevance in biological explanations, whether reductionist or systemic. A multiplicity of interactions on each level and a multiplicity of different levels of biological organization enter into definitions of the explanatory entities (either parts or wholes/systems although from different perspectives), as is logical considering the specificity of the reductionist and antireductionist views. However, there are elements present in both reductionist and antireductionist formulations that may allow for convergence on systemic visions characterized, as mentioned in Sect. 2 by (i) definition of the functional analysis system, using operational theoretical notions such as morphogenetic field, functional landscapes, stemness, etc.; (ii) circularization of the causal argument and context relevancies in the explanatory process referring to (iii) hierarchical organizations and integrative processes. Epistemological convergences towards systemic explanations, not following the scientist but the philosophers (Ayala & Arp, 2010, p. 14; Fox Keller, 2010; Dupré, 2010) can be described in terms of (i) a large number of and interactions among the parts; (ii) dependence on the identity of the parts and the interactions among them by effects at levels of higher order; (iii) robustness and adaptability of biological structures; (iv) hierarchic (multi-scale) organization of biological systems.

However, as already mentioned, in molecular models the conceptualization of 'part' as an object of study and the definition of a system based on its components, lead to explanations always and only in terms of discrete parts. The context in which these parts operate is irrelevant biologically (additional factors can always be expressed in terms of new cellular factors) and epistemologically (as context does not enjoy an epistemological status itself and thus cannot enter the explicative argument). However, once the system is identified, e.g., a cell or a tissue or a neuronal network, the way in which the genome or the brain is divided into functional parts (oncogenes and tumor-suppressor genes, in the first case; functional areas in the second case) shows how these parts are nevertheless subjected to the operational properties of the whole. This is also common to organicist models through the notion of morphogenetic fields.

This aspect of historical convergence toward operational definitions and the importance of the context in the identification of the system allows us, then, to focus on the properties of the functional field represented by the model itself. The dynamics involved here refer to the co-variation of various factors that are inherently bound to each other and that vary in inter-dependent ways over time. Systems thus framed are defined as *different levels of order* because they present unitary properties at every level that arises, not only from the properties of the components but also from the special relationships among them.

Therefore, the definition of the system, if reductionist, is given in terms of parts and their interactions, while in the antireductionist framework it is expressed in terms of organizational or functional relationships. This latter organizational dimension is however in some sense recovered in the reductionist perspective by defining, for example, cells' inactive/agency terms (for example, tumor cells do things, they move, they proliferate—in a permissive context—for which the definition of the part (gene) plays a functional role within the whole (cell, in this case). This trend, however, already implies an antireductionist approach, at least at the epistemological level. The first dimension, on the other hand, is already present in the concrete methodological approach of antireductionist models, which is articulated through the study of molecules and their interactions, showing how systemic antireductionist models can integrate the analytical perspective once the level of analysis of the phenomenon has been identified (see also the synthetic perspective, in Bertolaso, 2011, 2013, Bizzarri et al., 2019).

As occurs in antireductionist models, for example using the concept of the morphogenetic field, the initial question turns on the context and the ability to identify functional states with causal roles connected to them. It is not sufficient for ontological antireductionism to maintain that, if one must describe the behavior of a complex system, the most efficient way is not to specify the material component but to individuate the functional subsets of the system. *This level of description is common to both perspectives.* However, while all explanatory models converge on explanations of this type, the reductionist ones usually run into experimental paradoxes that concern the relevance of context for the explanatory potential of the identified systems. In other words, one might say that there is another aspect of the contextual element that resists an explanation carried out only in functional terms and calls for a non-reductionist stance which is even more radical than the one accepted by epistemic approaches that individuate functional states independently of their molecular structure. *There is, in fact, a difference between the functional organization of a system and the stability of a system as such.* The persistence of the latter depends on the particular part-whole organization and not simply on the relation among parts. When we use an operational definition, we refer to the former of the two organizations, which presents a hierarchic structure. We find here the three characteristics that have been stated also for reductionist models but that have neither a justification nor a unity within the reductionist paradigm. The scientific/explanatory question posed in terms of functional states and causality will in fact run into not only epistemological but also ontological problems. Many reductionists, however, intentionally ignore this integration of epistemological and ontological, whilst such integration is underestimated by many antireductionists who are not ready to take it into consideration and to clarify its consequences.

From an epistemic point of view, context is relevant not because it determines the identity of the parts, but because it allows one to consider one point of view as privileged and then to state it in terms of the significance of the parts. In biology this is expressed in terms of functional attributes that have meaning only if they refer to the space-temporal location in which the effects are coherent and significant with respect to the question that is asked to nature, even if only implicitly.

The context in this case indicates what 'separates', regarding the identity of the system, and what 'puts into relation' the system with other systems according to its operational specialization. A cell, a tissue, an organ, or an organism can then be defined from an organizational point of view, as an open self-referential system. Such a system can also be defined from a causal point of view as a context of effects, where we do not find a cause that relates the local entities and defines the organization, but where we find a relational organization that is itself causal.

In its ontological framework, the context indicates the set of conditions that makes it possible for a specific property of a part or a relation amongst parts to be efficacious and significant in determining the evolution of the system. It is precisely because of the normativity imposed by the system that we can a posteriori identify the above mentioned relation and recognize it as an effect brought about, for example, by feedback mechanics or as an effect that can be reproduced. From an ontological point of view, the context is therefore constitutive in the relation amongst the parts, but it is so from an epistemological point of view. We can recognize it indirectly in the relations amongst the parts as a condition for the normativity of their causal relation.

According to a manipulative account of causality (Woodward, 2003), the above mentioned perspective inversion is implicit in the asymmetry of the causal relation by means of intervention or background conditions that have an ontological role in determining the causal relation itself. This is in accord with what Buzzoni has noted in analyzing the relation of temporality and causality from an operational point of view: "Causes and effects are, without a doubt, firstly concepts, i.e. they are a construction of the mind and are picked up by our practical theoretical interests. (...) It depends on us to create them, to develop them and to make them fruitful to illuminate reality starting from our theoretical-practical interests, but it does not depend on us that

that be the cause and the effect relative to the theoretical-practical interests that have been fixed" (Buzzoni, 2008, p. 48). To proceed in a contrary way would mean, in fact, adopting a reductionist perspective, with all of its paradoxes that sooner or later would be pried open by the bio-logical reality.

The relative dimension of context allows one also to understand why multiple realizability is not a strong argument for antireductionism, while the fact that a molecular entity can perform different functions depending on the context is. Systemic explanations are always functional, but the identification of the part is subordinated to its functional definition, which, in turn, is determined by the context.

In systemic ontology, for example, the problem of multiple realizability becomes a question posed by the multi-unity that characterizes dynamical systems, biological systems, and their evolution. Natural selection works on this multi-unity because these systems carry information and significance for themselves and identity for the environment. Finally, the dependence-relation that is ultimately the testing ground for irreducibility in the hierarchical organization of a complex system is carried on in epistemology by the relevance of the context and in ontology by the operational definitions that characterize systems.

How can we move towards a *relational ontology*? We have seen that the epistemological reductionist positions attempt to explain complex biological dynamics in terms of molecules and their alterations so that sooner or later the processes can be explained and understood in molecular terms. The molecular circuits that are responsible for the biological processes are the focus of research. However, as discussed above, in the scientific practice reductionist models often ignore the lowest level of biological organization in descriptions and explanations of these phenomena, settling at a level where relevant processes appear as a not completely autonomous molecular process, and where other contextual elements play a role in various stages. From an epistemological point of view the scientific question 'why' a given process happens pushes toward the identification of an emergent phenomenon that retains the most appropriate level for its explanation, suggesting that this has a stability that takes priority over the specific functional properties of the parts. It is the behavior of the parts, as components of a complex phenomenon, which refer to a specific organization that is intrinsically linked to a system that we identify as tissue, brain, etc. through properties that were not present at lower levels.

In other words, the reductionist models seek emergence even if they try to explain it only by means of an analytical procedure and in molecular terms (Bertolaso, 2012). Therefore, a reductive explanation is not only incomplete but also fundamentally inadequate for the atomistic and mereological assumption of ontological reductionism that underlies this approach to scientific work. As podes by Nicholson, biological atomism "postulates a basic indivisible unit of life and seeks to explain the physiological and morphological constitution of all living beings' operation in terms of these fundamental units. The activity of a living organism is thus conceived as the result of its activities and the interactions of its elementary constituents, each of which individually already exhibits all the attributes proper to life" (Nicholson, 2010, 203). The result is a system in which the elements count only for their properties, and the system is understood to be a mere aggregation. We will use the term 'set' for this kind of system: an aggregation of elements defined by what unites them, even if this is only the membership to the set itself. This might imply that the limits of reductionist models to account for the (ir)reversibility of the biological processes and phenotypes (e.g., tumor latency, the onset of metastatic organic control, etc.) are due to their attempt to defend an ontological rather than an epistemological reductionism. This is an argument for admitting not only the insufficiency but also the inadequacy of philosophical reductionism for understanding complex biological phenomena. At the same time, the considerations presented above lead to a revaluation of the context and its epistemological and ontological roles, allowing for a deeper understanding of the limits of reductionism and its paradoxes in light of the importance of the context in the definition of the analyzed system. The advantage, however, of moving the study of a phenomenon from the molecular to an organizational level is that it provides a vision that is more systemic rather than holistic, this latter requiring at least an appeal to the organism as a whole as a causal system itself. The emergent phenomenon is no longer attributed to the aggregation of parts, but to their *integration*. In this sense, the context has a relevance that is not based on temporal priority but rather is concomitant with the organization of the parts, as for example in the tooth generation. This integration is local and defined by a specific organization: of the mandible for the tooth, of the pancreas for the beta-cell in producing insulin, of the brain in allowing neurons' activity, etc. The functional identity of the parts is determined by the context in which they operate.

The stronger epistemological implication is then that of the passage of a system (as a set) to a System (as an entity) characterized a priori with respect to its parts, but the behavior of which can be later described, at least partially also in terms of parts and their connections. Such a system is constructed by means of its parts without relying on them completely. Therefore, in an organization, understood as a set of relationships, the parts can be replaced without being inert with respect to the system.

There is a mutual co-determination which leaves a level of uncertainty (Bertolaso, 2012), such that systems owe their dynamism to their parts, while their unity is due to their specific organization. These principles, already known to embryologists at the beginning of the last century, are formalized today via the notions of 'field' and 'attractor' in biology. This implies a new vision and mutual integration of the concepts of system, hierarchical organization, and level of complexity. In fact, it has rightly been observed that: "Levels of organization have a variety of properties which make them rich in ontological and epistemological consequences (…) [T]heir merely empirical status is probably more a product of the fact that they haven't yet been taken seriously by any of the dominant philosophical views. In fact, these properties of levels are closely connected in ways which make the features of levels and their analysis not just a contingent empirical matter" (Wimsatt, 2007, p. 206)".

This also means that opting for a *systemic mindset* involves thinking not only in terms of the system—understood as a dynamic interaction of parts—but also of the entity—an operational unit—in which the mutual dependence and order of the constituents follow in a specific way. A more analytical discussion of these issues is beyond the scope of this paper, but it is at this level that a reflection is required on the principles governing the change and the evolution of a system, ultimately from at least a biological point of view, the relationship between systems and subsystems. The holistic issue, therefore, is relevant in e.g. organicism because it is systemic in the most peculiar sense, in that it is able to take into account the historical dimension of the organism and the diachronic as well synchronic emergence of new properties at higher levels of organization. The possibilities of this thesis, however, are rooted in a *relational ontology of the levels.*

Such a systemic view is the best tool for interpreting reality when considered as an approach, a way of thinking that arises from the convergence between our understanding and the way things are represented and constructed. This avoids the risk of an infinite regression of the argument about the relevance of the context which can lead to a universal holism, which is—at least scientifically—counterintuitive, since what we study are always objects which, also by virtue of their systemic hierarchically and organized dimension, have their own ontological status.

From an epistemological point of view, the convergence of the analytic and synthetic perspectives relies on the existence of local entities that are spatially-temporally defined. It is, therefore, simpler to speak of emergence in terms of local ontologies that represent objects also according to empirical knowledge. Because of its dual analytic and synthetic component, the systemic perspective admits a gradation of knowledge and a degree of undetermination of the system, which Corvi defines as the "unsaturated dimension that qualifies any real system" (Corvi, 2010). The notion of the organism becomes an asymptotic concept that refers to the combination of two aspects of each system: determinism and indeterminism, but this is done according to a relationship that does not refer to anything other than the system itself. From this perspective, the context becomes an ontological category. The advance is therefore conceptual. The important intuition behind the new perspective has to do with the ability to focus on the relationships themselves, described in terms, for example, of morphogenetic fields, and not as parts or interactions. Understanding how the set of parts works within the material constitution of the entity that performs these operations is an important issue that an atomistic approach is not able to capture when analyzing the parts to explain the whole. Within a field, the parts do not stop having their autonomy but also contribute to the analytical perspective. This reconciliation is possible only if the view of the whole or of the system is not interpreted in a closed or dogmatic way but as the regulative ideal that permits us to re-open the scientific discourse whenever necessary. The systemic approach shows its heuristic and explanatory power more in framing the problem rather than providing solutions to it (Bertolaso, 2012). Indeed, by its very nature it calls for a plurality of solutions that complement and integrate each other.

This explains also why within an antireductionist perspective of biological processes we can find multiple models that do not contradict each other. They treat different aspects or different scales of the phenomenon, without running into contradictions and paradoxes, but by stimulating the development of research in new directions. The reflection, then, is on the discovery of biological information and its specific organization.

3 Elements for a New Epistemological Model

In the previous section, we looked at how reductionism and its related debate focused mainly on the problem associated with a view of the biological field in terms of 'lower levels of analysis' and in molecular terms. Consistently reductionism has been designated by its proponents as a research strategy but also as a philosophical and metaphysical position. They hold that "there are no nonphysical events, states, processes, and so biological events, states, and processes are 'nothing but' physical ones. This metaphysical thesis is one reductionist share with antireductionists" (Rosenberg, 2007, p. 120: quoted in Bertolaso, 2012). This was a consequence of the general tendency in the empirical sciences towards procedures by reductions. What this means in biology is the basis of our reflection. The discovery of the molecular basis of genetic and reproductive processes seemed to give a stronger philosophical foundation to such empiricism and physicalism. As a result, an explanatory and conceptual tool of biology, such as the gene, was 'entified' and its functional dimension lost the explanatory role of accounting for an effect starting from a cause that was defined in terms of the effect itself. Genes are, in fact, functionally defined. Nonetheless, the same gene can have several names/functions, which creates the difficulty known in the literature as a 'one-many issue' or 'context dependence of functional activities'. It also raises the question of 'circularity of notion of function' (Bertolaso, 2013a, 2016).

The fundamental bias is to consider development as the result of cell proliferation and differentiation and not differentiation as a dimension of organic development and growth. The cellular differentiation process underlying the morphogenetic process, in fact, is not self-consistent from a biological or even epistemological point of view. To give an example, any stem cell is highly dependent on its space–time position in the organism. Once extracted from this 'biological context' it loses its stem properties to follow differentiation processes common to every other somatic cell. That is, it loses the division asymmetry that characterizes the stem cells' proliferation. Furthermore, the very definition of staminality refers precisely to this peculiar ontological status of some cells compared to others. Similarly, tumor cells lose their context dependency.

The useless search for common/fundamental molecular and functional properties of cancer cells, moreover, shifts for example the question about carcinogenesis (cancer exists because cells proliferate abnormally) to the characterization of the elements involved. The biological question on development, be it normal or aberrant (for example, neoplastic), becomes a question about the concrete element (thing), in which, as in classical physics, the thing is identified with its size. However, defining and dealing with *differences* remain the challenging endeavor of the biological sciences. The difficulty becomes defining the criteria for establishing what should be considered a 'same' or 'common property' among different biological entities. Thus, when the epistemological issues underlying the notion of biological heterogeneity—with the similarities and differences that this concept implies—are neglected or underestimated, problems immediately arise with the identity of the parts, which has no solution within an atomistic view.

Given that there is no molecular product or functional activity that uniquely characterizes stem cells or cancer cells or neuronal cells, the impossibility of identifying the explanatory parts—in terms of nucleonic sequence or functional properties)—usually forces scientists to introduce 'contextual factors' into the explanatory argument. On one hand, the environment or micro-environment becomes the (causally) decisive element to explain the process, from the 'mereological' to the 'ecological' explanation. On the other hand, if a mechanism must still be found to explain a 'how' question (how a process takes place, etc.) the system under observation is typically framed in biological conceptual/operational terms: tissues, functional attractors or landscapes, functional fields, etc. No activities are directly attributed to such 'entities'. Only functional features or capabilities are recognized. This means assuming that the biological entities that play the explanatory role have those properties as an instantiation of a larger class of biological identity, which instead possesses those properties by virtue of itself, i.e., in a primitive and essential way (consider, for example, the definition of stem cells or cells of the immune system). Biological explanations, especially in the molecular field, are usually of this type. They do not ask questions about essential properties, but rather presuppose them. This essentiality, however, refers to a feature that is a *capacity* and that is semantically encapsulated in the present participle verb form (ending with -ing) when talking about the living being. The question 'what does *that* do?' meaning 'how is that so' requires identifying specific differences in the biological explanation that is, a specification of what makes this instantiation something other than the 'type' to which it belongs in a broader sense.

Relative terms, therefore, in biology have their epistemological foundation in the *relational ontology* that characterizes biological processes and the identification of systems that allow their study. In the scientific language, there is a functional emergent property that defines biological individuality from an explanatory point of view (mesoscopic level: Bizzarri et al., 2019) and there are constitutive relationships that justify the conceptualization of the *explanans* in relative terms (Bertolaso, 2016). For this reason, paradigmatically in cancer biology where cancer is understood as a problem of the normal developmental processes of the organism, the conceptualization of the tumor cell involves two levels of epistemological analysis: that of the relative terms for which differentiation is more relevant in the explanation of the process than proliferation, and that of the relational terms that explains *why*—in the impairment of the differentiation process—a property (P, for proliferation) re-emerges as a dimension of the broader class of biological entities (C, for cell) from which C * (i.e. a tumor cell) also derives and that has this property by default. This predicate is explained by identifying some aspect of the specific nature of the entity (cells) that is responsible for it (property). In cancer biology, the original belonging of each cell to the organism—both in a genealogical and ecological/functional sense—is what explains its proliferative status as a default state.

Let us now spell out more in detail the dimensions that allow for a normal developmental process. At all biological levels, a peculiar functional integration is at work and maintains the coupling of processes that for example in cancer are compromised. In cancer, in fact, proliferation, differentiation and migration are decoupled leading to a neoplastic growth of cells. That is, in bot in cells and in cells within an organism (metazoan) there is a specific interdependence between metabolic and genetic processes, signaling and regulatory mechanisms. But in the latter case (metazoan) such interdependence acquires interesting features shedding light on the kind of *integration* that characterizes living organisms more in general. We propose to organize those processes through the following dimensions:

- Genealogical (genetic and epigenetic dynamics)
- Functional (metabolism)
- Contextual (micro-environmental constraints) (Fig. 1).

Given the above arguments, it is easier to understand why, in scientific practice scientists find it useful to combine an instructive and a permissive causality (Woodward, 2010) to explain how a living environment can shape the path that leads complex processes (such as mental health, drug treatment, tissue organization, etc.) and increases the likelihood of achieving such a goal. There are factors that seem to act primarily through a permissive causality by boosting, for example, neural plasticity (i.e., the ability of the brain to change itself), allowing afterward for instructive interventions to produce beneficial effects or not (Branchi & Giuliani, 2020).

How are the relational epistemology and causal issues in biological sciences actually linked in a relational ontology understood as epistemology? The insufficiency of an essentialist philosophy for biology has already been discussed by other authors (Godfrey-Smith, 2008) as well as in other areas of scientific research on the

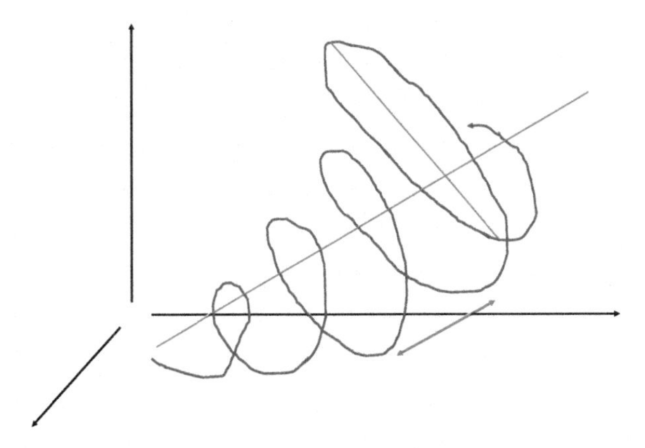

Fig. 1 Conceptual representation of the three dimensions (in green) that are combined and coupled in the dynamic stability (in red) that characterizes living entities, paradigmatically organisms. Their reciprocal dependence is ontologically and epistemologically asymmetric as discussed in the above sections. Topological laws hold their peculiar status as we shown in other studies. The collapse of the three dimensions, in fact, follow a 'time dependent' (t*) path that embodies a *rhythm of rhythms* that can be independently investigated as an object of inquiry in itself. Such t* variable is here represented by the inclination of the spiral. For further epistemological and empirical implications of this image see Bertolaso (2016), Fig. 1.2 and 5.3 and Bizzarri et al., (2019); for implications in complexity and organizational sciences, see Bertolaso (2021)

natural world. It is known, for example, how the mechanistic rationalism of classical mechanics went into crisis in the face of the radical relational dimension brought to light by the discovery of chaos: knowing the evolution of a minimally complex system requires that we consider interactions with objects that are on the other side of the planet. Quantum mechanics has also highlighted relational structures when focusing on lower scales of the phenomena, the very interaction of the measure modifies the known object. Instead of a representation based on objects as different essences, a conception of reality as *relations* must be introduced (Bertolaso, 2016).

In the biological sciences, there are issues that admit a causal determinism of some kind while requiring relational categories to describe it in a more radical sense (e.g. development, cancer biology). There is also a relational dimension in our way of knowing the world that allows us to understand more deeply the epistemological status of biological explanations. Combining these two aspects entails expanding the notion of causality through which we usually work in biological sciences (efficient causes to explain mechanisms) to grasp how relationships are made up in nature and the way in which we conceptualize them. This approach goes beyond efficient causes and is closer to formal causes.

Growth and development dynamics involve the co-variation of factors that are intrinsically (or causally) linked to each other: they vary over time in a mutually dependent way. The systems defined in this way are organizational levels because at each level they present *unitary properties* that emerge not only from the properties of the components but from the particular relationships among them (Bertolaso, 2013a). By organizational levels we mean those expressions that imply an element of repeated and repeatable order (patterns) and that are usually defined, in the biological sciences, in functional terms (tissues, etc.). This is consistent with what Duprè points out, namely that "properties of constituents cannot themselves be fully understood without a characterization of the larger system of which they are part" (Duprè, 2010). Here the epistemological relevance of the context emerges again, but from the perspective of its characterization.

The question of the 'whole that is greater than the sum of the parts' is overcome in favor of a view that captures relations as causal, allowing us to return to the specific relationship between parts and wholes in terms of mutual dependence and autonomy (co-determination reciprocal of the elements of a system). *The whole is a part, in turn, in a different way from the parts that constitute it.* If what determines the transition from a set of elements to a real system is the organization, that is, the "network of relationships between the elements, and of the structure, which constitutes one of the possible applications and implementations of the organization" (Lenoci, 2010), what allows us to study a whole even in terms of its parts without contradiction is that the mutual determination that exists between the parts and the whole is not symmetrical.

The dependence of the parts on the whole is not stated in the same way as the dependence of the whole on the parts. Only formally can we assume that the mutual determination is such that the systems owe their dynamism to the replaceable parts, while they owe their unity to the organization. The *multi-unity* that is evident in biological differentiation processes is due to a systemic determination in a space and a time that change in a way in which they are strictly dependent on each other.

The issues thus move from the epistemological to the ontological level, asking for a philosophical foundation (justification) of a relational ontology of levels which is characterized, at an epistemological level, by a hierarchical organization.

Systemic issues can be put in dialogue with those of the reductionism-anti-reductionism debate for their ontological implications, which are not in favor of a further articulation of the holistic argument, but of the recovery of a notion of *being* through an operational and relational perspective. This perspective admits discontinuities in the world and in the modalities of the dependence of the parts on the whole. Parts' transformation depends on their origin and on the way in which they arrange themselves.

The role of the context and its relationship with the parts requires, therefore, a final conceptual clarifications. In mechanicism, context is a scenario in which phenomena occur; it is absolute and infinite, like a slate without boundaries. The relationship between mechanicism and context with respect to parts that interact with each other according to physical laws is an anecdotal presence. It is not even a condition of possibility because context is where phenomena occur and although context precedes phenomena, it does not cause phenomena to influence one another but just gives them a scenario in which to occur. Within mechanicism, there is more context than parts in the same way that there is more ocean than there are boats.

If we consider nature as a complex system, the context is much more than a scenario: it is the concomitant result of causal interactions. It is a frame, a network, a set of nodes in a relation that would not exist if interactions did not occur. In this conception, there is no more context than phenomena, and facts cannot occur without a causal plot being generated. In addition, the context is dynamic, changing over time. It happens as the events take place (as, for example, the evolution of cosmic space–time or the dynamics of a hurricane). However, although the difference and explanatory richness of the mechanistic scheme are palpable, there are phenomena where this conception of nature as a complex system does not correctly reflect the true role of context.

A living being is a material entity whose parts are integrated to form a unique, peculiar, singular being that preserves its structure in a world in which processes tend to destruction and disorder. But this unique entity that preserves its structuring can export its own form through proliferation that generates a great diversity of forms. None of these elements could be justified without interaction with the context, which is not a mere scenario and that does not appear when life is reproduced but does something very different: it brings unity to the living being but also persistence. Relational ontology assumes that there is an intrinsic belonging of organisms and their components to the environment or context in which they develop, and that this context is subject of causal interactions with the organisms.

However, in this mutual involvement, there is not only a win–win process, but the context that articulates the relationships and give rise to the living organism (which is information) causes all entities within the context to share some structuring and dynamism, while at the same time deploying towards a plural set of forms. What occurs here is very similar to the original notion of information suggested by Aristotle when he spoke of *morphé*: a constitutive structuring element of a whole, which realizes its specificity while at the same time keeping it open to variation and the proliferation of new structures. Which is precisely what *asymmetric reciprocity* consists of (Bertolaso laso, 2016). In this sense, it is essential for the understanding of *relational ontology* to consider that the context works as a formal cause (*morphé*) and not only as a scenario or plot. It is a set-up, an articulator, a detonator, but not the phenomena themselves. That is why there is a relationship with asymmetry between context and phenomenon, and not just mutual interaction, as complex system theory suggests.

An additional observation: understanding what form means in this context is conditioned by the notion of part. Aristotle had established with good intuition that part is what remains as a result of a process of division. In the integration of certain phenomena or entities there are members which are fundamental (even irreducible), that make up a certain reality but not as mere aggregation or a mereological sum of parts but as essential constituents which he names. Here the form (*morphé*) performs an integration not of the parts (which are the result of physical division), but of the elements (specific minimum parts) that are what make up the subject, the body. Morphé of a ship, for example, functionally integrates elements: rudder, sails, hull; not parts, which will always be successively divisible. A living being according to this perspective is understood as a subject whose intra- and extra-contextual causal relationship give it unity over time based on the information.

Without specifying the role of information and its function within biology and dynamic stability, it is difficult to show the relevance of the paradigm of *relational ontology*. Information should be, in fact, understood, at this point not as communication of a code or instructions, but as an integrator of stable structuring over time. Therefore, a relational ontology seems to designate the way in which the living being exists, as a subject that not only occurs in a global context but is also an articulated and concomitant result with it, as is the case of all the members of the same organism.

4 Conclusion: Revising the Reductionism-Holism Debate

Biology is a science that, from its beginnings, has always had difficulty defining its object of research and study. On the one hand, there is the question of life; on the other hand, there is the question of how far biology can be considered a science in the sense of the so-called exact sciences, that is, those sciences in which prediction and control go together. Nevertheless, we call biology the science that studies those objects that we refer to as living beings and that we characterize by a specific movement of growth and development (through genetic and metabolic processes).

We deal with entities, with different degrees of individuality (that we define in terms of functional integration of the parts), but that are united by continuous and autonomous activity. That is, in the way we relate to the biological world there are always these two dimensions: one that refers to a constitutive activity (which identifies what it is) and one that refers to a functional activity (which specifies what it does). We find these dimensions, in conceptual and explanatory terms, at various scales in the biological sciences, from genes to unicellular organisms, from organisms to ecosystems.

Philosophical problems began to arise when we started explaining this peculiar movement of growth of organisms—the result of the convergence of constitutive activity (praxis) and functional (*poiesis*)—from a mereological perspective, that is, having to recompose an emerging behavior through the functional integration of the parts. This process of explaining the biological world inevitably soon showed its limits. Assuming the materialistic and naturalistic compromise that has dominated both the scientific and philosophical debate of the twentieth century, the position taken for granted is that science has to deal with material objects and is committed to offering naturalistic explanations that are based on some intrinsic property of the system and independent of the observation of the researcher.

Typically, this question has permeated the debate about the nature of species and function, organism and ecosystem. The evolutionary perspective of the natural world brought these fields of inquiry together so that most of biology became 'evolutionary biology'. The search, in fact, for principles of order, wanting to avoid a vitalist position and having no alternative to mechanistic ones, could not but end up in a genetic determinism where, always for positivist assumptions, the notion of gene coincided with that of nucleotide sequence and specific functional activity.

The emergence and development of embryology and systems biology have revived old questions about the nature of the difference between living organisms and the other entities we observe in the world. They have also revived questions about the traceability of the principles that govern the former and those that apply to the latter, thus providing a comprehensive explanation of phenomena typically associated with the living world in physical terms. The fundamental question returned to being about the principles of order and organization of living systems (Bertolaso, 2013a). The methodological and ontological questions converge in philosophical reflection on the question about what reductionism is possible in biology or what the fundamental philosophical assumptions of non-reductionist or systemic perspectives are (Bertolaso, 2013b).

The relational epistemology described above, in contrast to system as well as to mechanisms, not only makes explicit the way in which entities interact with each other within the historical whole (organized complexity) but also aims to be a way of understanding what nature is, not just how it shows itself to us. That is, it analyses natural entities as the result of a dynamic articulation among their origin, context, and function, particularly manifested in the field of life. The complex systemic methodology (which assumes its epistemology as an ontology) does not distinguish between different systems to show how each one is specific, singular, irreducible, but rather considers all natural systems (inert, living, conscious, social, etc.,) as homogeneous entities.

Does the proposed *relational ontology* serve to better understand natural reality, over and above complex system theory and mechanicism? Answering this question implies considering the dynamic identity of a natural being, as well as the articulation among their three dimensions of origin, context and function, which are maintained over time but which result in diverse and original beings, in such a way that each of them can be understood in its specificity. Is this approach appropriate for understanding other areas of reality, beyond living beings? (culture, social behaviour, etc.). To answer this question, it would be necessary to identify in each of these areas of reality what its articulation of origin, context, and function would be. This model is thus presented not as a mere metaphor or allegory about understanding living beings, but as an analogical tool that looks to identify the common elements that generate in nature different realities although with similar patterns.

References

Ayala, F. J. (1974). Introduction. In F. J. Ayala & T. Dobzhansky (Eds.), *Studies in the philosophy of biology* (pp. vii–xvi). University of California Press.

Ayala, J., & Arp, R. (2010). *Contemporary debates in philosophy of biology*. Wiley-Blackwell.

Bertolaso, M. (2011). Hierarchies and causal relationships in the interpretative models of cancer. In Special Issue "Causation and Disease in the Post-Genomic Era". *History and Philosophy of the Life Sciences, 33*(4), 389–396.

Bertolaso, M. (2012). La Dimensione non Riduzionista del Riduzionismo nella Ricerca Sperimentale Dai Modelli Molecolari a Quelli Sistemici nella Ricerca sul Cancro. *Rivista di Filosofia Neo-Scolastica, 4*, 687–705

Bertolaso, M. (2013a). Sull'irriducibilità della prospettiva sistemica in biologia. Un'analisi delle convergenze dei modelli interpretativi del cancro, in "Strutture di Mondo. Il pensiero sistemico come specchio di una realtà complessa." (Urbani Ulivi L.) (Vol. II, pp. 143–169). Il Mulino.

Bertolaso, M. (2013b). *How science works. Choosing levels of explanation in biological sciences.* Aracne s.r.l.

Bertolaso, M. (2016). *Philosophy of cancer—A dynamic and relational view*. Springer Series in "History, Philosophy and Theory of the Life Sciences".

Bertolaso, M. (2017). Uncoupling mereology and supervenience: A dual framework for emergence and downward causation. In "Causality and Explanation in Scientific Practice" (Evandro Agazzi, Marco Buzzoni Eds), *Axiomathes, 27*(6), 705–720.

Bertolaso, M., & Ratti, E. (2018). Conceptual challenges in the theoretical foundations of systems biology. In M. Bizzarri (Ed.), *Systems biology, series: Methods in molecular biology* (Vol. 1702, pp. 1–13). Springer.

Bertolaso, M. *Complexity and management: Epistemic insights from life science in handbook of philosophy*. Handbook of Philosophy of Management (Eds. M. Dibben, M. Segal) (in press).

Bizzarri, M., Giuliani, A., Pensotti, A., Ratti, E, & Bertolaso, M. (2019). Co-emergence and collapse: the mesoscopic approach for conceptualizing and investigating the functional integration of organisms. *Frontiers in Physiology* https://doi.org/10.3389/fphys.2019.00980

Branchi, I., & Giuliani, A. (2020). Shaping therapeutic trajectories in mental health: Instructive versus permissive causality. *European Neuropsychopharmacology, 43*, 1–9 (2021).

Brigandt, I., & Love, A. (2017). Reductionism in biology. In E. N. Zalta (Ed.), *The Stanford Encyclopedia of Philosophy* (Spring 2017 Edition). https://plato.stanford.edu/archives/spr2017/entries/reduction-biology/

Buzzoni, M. (2008). «Causalité et temporalité du point de vue opérationnel», relazione tenuta al «Colloque de Cerisy: La notion de temps dans sous tous ses aspects», in *Epistemologia. An International Journal for the Philosophy of Science, 2008*, Special Issue 14: Time in the Different Approaches/Le temps appréhendé à travers différentes disciplines), Actes des Entretiens de l'Académie Internationale de Philosophie des Sciences (Cerisy-la-Salle, 4–9 Octobre 2007), pp. 45–58.

Collier, J. D. (2000). Autonomy and process closure as the basis for functionality. In J. L. R. Chandler & G. van der Vijver (Eds.), *Closure: Emergent organisations and their dynamics* (pp. 280–290). Annals of the New York Academy of Sciences.

Corvi, R. (2010). Dall'olismo epistemologico al pensiero sistemico, In Urbani Ulivi L Strutture di mondo. Il pensiero sistemico come specchio di una realtà complessa. Il Mulino.

Craver, C. F. (2009). Mechanisms and natural kinds. *Philosophical Psychology, 22*, 575–594.

Craver, C. F., & Bechtel, W. (2007). Top-down causation without top-down causes. *Biology and Philosophy, 22*, 547–563.

Dupré, J. (2010). It is not possible to reduce biological explanations to explanations in chemistry and/or physics. In J. Ayala & R. Arp (Eds.), *Contemporary debate in philosophy of biology*. Wiley-Blackwell.

Dupré, J., & Nicholson, D. (2018). *Everything flows: Towards a processual philosophy of biology*. Oxford University Press.

Fox Keller, E. (2010). It is possible to reduce biological explanations to explanations in chemistry and/or Physics? In J. Ayala & R. Arp. (Eds.), *Contemporary debate in philosophy of biology*. Wiley-Blackwell.

Godfrey-Smith, P. (2008). Reduction in real life. In J. Hohwy & J. Kallestrup (Eds.), *Being reduced: New essays on reduction, explanation, and causation* (pp. 52–74). Oxford University Press.

Green, S., & Batterman, R. (2017). Biology meets physics: Reductionism and multi-scale modeling of morphogenesis. *Studies in History and Philosophy of Science Part c: Studies in History and Philosophy of Biological and Biomedical Sciences, 61*(February), 20–34.

Kauffman, S. A. (1971). Articulation of parts explanations in biology and the rational search for them. *Boston Studies in the Philosophy of Science, 8*, 257–272.

Kauffman, S. (2000). *Investigations*. Oxford University Press.

Lenoci, M (2010). Introduction in 'Strutture di mondo - Il pensiero sistemico come specchio di una realtà complessa', Lucia Urbani Ulivi (Ed). Il Mulino, Bologna

Licata, I. (2015). Incertezza. Un approccio sistemico. In L. Urbani Ulivi (Ed.), *Strutture di Mondo* (Vol. III, pp. 35–73 and 46–47). Il Mulino

Maturana, H., & Varela, F. J. (1980). *Autopoiesis and cognition. The realization of the living*. Reidel Publishing.

Mayr, E. (2004). *What makes biology unique?* Cambridge University Press.

Militello, G. (2021). Ph.D. Thesis on structural and organisational conditions for the appearance of a functionally integrated organisation in the transition from prokaryotic to eukaryotic cell.

Mitchell, S. D. (2009). *Unsimple truths: Science, complexity and policy*. University of Chicago Press.

Moreno, A., & Mossio, M. (2015). *Biological autonomy: A philosophical and theoretical enquiry*. Springer.

Mossio, M., Moreno, A. (2010). Organisational closure in biological organisms. In P Huneman & C. T. Wolfe (Eds.), *The concept of organism: Historical, philosophical, scientific perspectives, History and Philosophy of the Life Sciences, 32*(special issue), 269–288.

Nagel, E. (1961). *The structure of science: Problems in the logic of scientific explanation*. Harcourt, Brace & World.

Nicholson, D. J. (2010). Biological atomism and cell theory. *Studies in History and Philosophy of Biological and Biomedical Sciences, 41*(202–211), 203.

Rosen, R. (1991). *Life itself. A comprehensive enquiry into the nature, origin and fabrication of life*. Columbia University Press.

Rosenberg, A. (2007) Reductionism (and anti reductionism) in biology. In D. L. Hull & M. Ruse (Eds.), *The Cambridge companion to the philosophy of biology*. Cambridge University Press.

Sarkar, S. (1998). *Genetics and reductionism*. Cambridge University Press.

Varela, F. J. (1979). *Principles of biological autonomy*. North Holland.

Velázquez, H. (2020). ¿Qué es la naturaleza? Introducción filosófica a la historia de la ciencia (2a ed.). Porrúa.

Wilson, R. A., & Craver, C. F. (2007). Realization. In P. Thagard (Ed.), *Philosophy of psychology and cognitive science (Handbook of the philosophy of science, Vol. 12)* (pp. 31–54). Elsevier/North Holland

Wimsatt, W. (1974). Complexity and organization. In K. F. Schaffner & R. S. Cohen (Eds.), *Proceedings of the 1972 Meeting of the Philosophy of Science Association* (pp. 67–86). D. Reidel.

Wimsatt, W. (1980). Reductionistic research strategies and their biases in the units of selection controversy. In T. Nickles (Ed.), *Scientific discovery: Case studies* (pp. 213–259). D. Reidel.

Wimsatt, W. C. (2007). *Re-engineering philosophy for limited beings: Piecewise approximations to reality*. Harvard University Press.

Winther, R. G. (2011). Part-whole science. *Synthese, 178*, 397–427.

Woodward, J. (2003). *Making things happen—A theory of causal explanation*. Oxford University Press.

Woodward, J. (2010). Causation in biology: Stability, specificity, and the choice of levels of explanation. *Biology and Philosophy, 25*, 287–318.

Holism and Reductionism in the Illness/Disease Debate

Marco Buzzoni, Luigi Tesio, and Michael T. Stuart

1 Introduction. The Two Souls of Medicine and the Illness-Disease Dichotomy

1.1 Medicine: Two Souls for a Single Science

Dissatisfaction with biomedicine, and more generally with a medicine modelled after and depending upon biomedical sciences and technology, is spreading more and more; not only among patients (who often are turning to "alternative", "holistic" or "complementary" forms of medicine), but also among health care professionals and, particularly, medical doctors (cf. Cole & Carlin, 2009). On the one hand, medical doctors retain the ambition to be scientists—an ambition which is also reflected in the English language (they are also designed by the term "physician", whose Greek-Latin origin, "physicus", means "the one who knows nature", i.e., the scientist in its

M. Buzzoni (✉)
Dipartimento di Studi Umanistici, Università degli Studi di Macerata, Macerata, Italy
e-mail: marco.buzzoni@unimc.it

L. Tesio
Department of Biomedical Sciences for Health, Università degli Studi di Milano, Milano, Italy
e-mail: luigi.tesio@unimi.it

Department of Neurorehabilitation Sciences, Istituto Auxologico Italiano, IRCCS, Milano, Italy

M. T. Stuart
Institute of Philosophy of Mind and Cognition, National Yang Ming Chiao Tung University, Taipei, Taiwan
e-mail: mikestuart@nycu.edu.tw

Department of Philosophy and Centre for Philosophy of Science, University of Geneva, Geneva, Switzerland

Centre for Philosophy of Natural and Social Science, London School of Economics and Political Science, London, UK

S. Wuppuluri and I. Stewart (eds.), *From Electrons to Elephants and Elections*,
The Frontiers Collection, https://doi.org/10.1007/978-3-030-92192-7_39

widest sense). On the other hand, many physicians and healthcare professionals are aware of the need of personalizing the traditional and biomedical model of medicine (cf. Engel, 1978: 169; Glick, 1981 p. 1037; Willis, 1989; Marcum, 2008), a need that is reflected by the term "clinician", faithful to its Greek origin, *clìno*, probably meaning bending towards, or lying on, the sufferer's bed.

It is therefore no accident that in the literature of the last decades about the status of medicine, there is a new awareness that an adequate notion of medical praxis requires an integrative position, which mediates between the analytic-reductionist and the normative-holistic perspective (cf. Wyss, 1986; Nordenfelt, 1986, 1997a, 1997b, 2013; Christian, 1989; Hahn, 2000: 35–53; Pieringer & Fazekas, 2000: 89–111; Marcum, 2008; Larkin et al., 2011: 318–337). This is not a recent demand. Karl Jaspers already fully understood, as early as 1919, the importance of reconciling the two "souls" of medicine (Jaspers, 1919: 59), i.e., the analytical-reductionist and the holistic-normative, or, the scientific-technological (practised by the physician) and the clinical (practised by the clinician). Jaspers's problem fully corresponds to today's physician-clinician antinomy. The associated need to reconcile the scientific soul and the clinical soul of medicine (on which see also Jaspers, 1958: 1038, Engl. Transl., 255) is as (or more) urgent today as it was in Jaspers's time.

1.2 The Illness-Disease Dichotomy: A Part-Whole Puzzle

Now, since the duplicity of attitudes towards the patient corresponds roughly to what we usually designate as "disease" and "illness", it is possible to take an important step towards achieving the just-mentioned goal if the relationship between these two meanings of malady is correctly set up (the word "malady" will be used here in the most generic sense, which includes not only "disease" and "illness", but also "sickness", and the sense of being unfit or unable to do what you want to, when you want to). As Cassell noted in 1976, the technological revolution, by fueling the hope of curing many maladies, has contributed to an increasing differentiation between what could be treated by technological means (with an often excessive confidence and hope) and illness, something which is subjective, it is something the person "lives through" as a whole, and therefore does not fit into the categories of rigorous technoscience. Typically, prospects for successful treatment were significantly less, when not wholly abandoned:

> The success of medicine has created a strain: the doctor sees his role as the curer of disease and "forgets" his role as a healer of the sick, and patients wander disabled but without a culturally acceptable mantle of disease with which to clothe the nakedness of their pain. (Cassell, 1976: 27)

Notwithstanding the ongoing debate about disease and illness, there is some agreement that "disease" and "illness" are the key concepts of the opposite trends, the analytical-reductionist and the holistic-normative, in conceiving both malady and medicine. From this point of view, the mentioned demand for an integrative position concerning the status of medicine, which combines the analytic-reductionist with the normative-holistic perspective, must also be raised for the distinction between disease and illness.

1.3 Can a Human Science Be a Science? Human Sciences and Objectivity

As just noted, clarifying the relation between the two faces of malady expressed by the terms "disease" and "illness" is crucial for creating an integrative view capable of overcoming the opposition between the reductionistic-analytical and the holistic-normative perspective of medicine. However, the demands for such an integrative view must not only be connected with ethical reasons or reasons of practical desirability (which will not be taken into consideration here), but first of all with epistemological and methodological reasons, intrinsically linked to the status of medicine as a science: human, yet science. Medicine, in every aspect of its activity (including its disciplinary and institutional organisation), must reconcile the scientific-objective and impersonal dimension (mainly expressed by the term "disease") with its clinical and personal dimension (mainly expressed by the term "illness"). This, as we shall try to show, is required by the nature of medicine as a human science, a nature that must also be taken into account in order to increase its own degree of scientificity, objectivity, or intersubjective controllability. Obviously, medicine is not to be considered a human science in the restrictive sense that it must be "humanitarian" (a welcome property, of course), but as having a methodologically specific object of investigation, that is, beings which are biological organisms, and, at the same time, operate as more than mere biological organisms, that is—to remain as neutral as possible with respect to a difficult philosophical debate—as cultural-biological entities.

We propose to show that there is an aspect of "illness" that is central in both dimensions of medicine. This aspect is at once objective and intersubjectively controllable, yet also characteristic of the human sciences and not completely reducible to the natural sciences. Specifically, illness depends not only on the physical and biochemical reality of the patient's body, but it includes the cultural side, which is always involved in the particular way in which patients live, and respond to their own experience of illness. The way in which patients live their illness, both on a personal and social level, is not only influenced by natural constraints (such as the fact of possessing a certain body structure or, to put it as Boorse does, a certain "species design"), but also by law-like cultural constraints. Even the domain of the experience of illness, like those dealt with by the other human sciences, is subjected to regularities that add to, and interfere with, those that physical and biochemical investigations highlight. It is only by taking them into account, therefore, that it is possible to increase the ability of medicine to prevent, diagnose and treat illnesses in an increasingly effective way. In other words, the interpersonal and social side of illness can and must be investigated from the point of view of medicine as a science. Both the bio-physician and the clinician cannot ignore, each from their own point of view, the patient's attitude towards their illness and the development of the patient's ways of coping with the illness, which are not only of paramount importance for any therapy, but also for any diagnosis. An approach to medicine that would try to leverage exclusively the biological aspect of the disease (as was for example the explicit intent of Boorse's famous essay of 1977, in accordance with a naturalistic-statistical view of biological normality inaugurated by Claude Bernard in 1865), would not only be a medicine that would break that physician–patient alliance, which obviously no medicine can disregard, but it would be a medicine unaware of its nature as a human science.

1.4 The Treatment Plan

The present paper is organized as follows. Section 2 describes two important but opposite conceptions: health and malady (understood as a lack of, or contrary to, health: see, e.g., Sadegh-Zadeh, 2012: 153–154). It explores certain difficulties with each, which together lead to a kind of antinomy, which, as shown in Sect. 3, also affects the few attempts to offer an integrative view of health and malady (the most important of which is perhaps that of Wakefield: see e.g., 1992, 2007, 2014, and 2015). Section 4 shows that the strengths of both the naturalistic, analytic-reductionist and the normative, holistic-humanistic conceptions can be preserved, on different levels, if one understands in what sense medicine is a human science. Section 5 considers the case of statistics (seen as a process of generalising from individual observations) when applied to human sciences. Section 6 builds a proposal for epistemic reconciliation and integration of naturalistic and normative standpoints.

2 Health, Disease, Illness: Analytic-Naturalistic Versus Holistic-Normative Perspectives

2.1 Two Accounts of Health and Malady

Two main accounts of health and malady are distinguishable in the literature, even if this distinction runs the risk of a certain oversimplification (cf. Simon, 2007 and Kingma, 2014). The two key concepts that correspond to these two different conceptions of health and malady are "illness" and "disease."

On the one hand, the first view—sometimes defined as naturalism, or an analytic-naturalistic perspective—puts particular emphasis on the term "disease", something which can and ought to be objectively, scientifically ascertained and localized within the patient's body. This is very clear in the thought of Christian Boorse, the best-known and most discussed exponent of such a naturalistic view. According to Boorse, we have to distinguish between "illness", which is a concept dependent on a cultural context, and "disease" (or "pathology", as he prefers since 2014: see Boorse, 2014), a descriptive, non-normative concept, whose main elements are biological function and statistical normality. "Disease" could be defined as

> a type of internal state which is either an impairment of normal functional ability, i.e., a reduction of one or more functional abilities below typical efficiency, or a limitation on functional ability caused by environmental agents. (Boorse, 1977: 567; cf. Boorse, 2014: 683–684)

Typical or normal performance is defined by the concept of "species design":

> Our species and others are in fact highly uniform in structure and function; otherwise there would be no point to the extreme detail in textbooks of human physiology. This uniformity of functional organization I call the species design. (Boorse 1997: 557; see also Boorse, 2014: 39)

From this point of view, Boorse opts to understand the concept of health as the absence of disease:

> Health as freedom from disease is then statistical normality of function, i.e., the ability to perform all typical physiological functions with at least typical efficiency. This conception of health is as value-free as statements of biological function.[1]

More generally, it might be said that, as far as "disease" is concerned, it is possible to distinguish two main components: from the point of view of its contents, essential ingredients of "disease" are biochemical, genetic, and functional-physiological (in short: biological) elements, so that 'disease' may be observed, examined, and measured in an intersubjectively testable way; however, from the point of view of its social-cultural-linguistic classification, what a 'disease' is, is determined by the theoretical lenses and the particular practices of health professionals: in this sense, because they are the undesirable conditions that health professionals de facto happen to treat, diseases vary over time with evolving historical and social conditions (see e.g., Boorse, 1977, Kleinman, 1988: 4; Aho & Aho, 2008; Sadegh-Zadeh, 2012: 151–153; on this point see also Grmek's important notion of "pathocoenosis" (Grmek, 1983, Engl. Transl., 2–3).

On the other hand, according to a second view—which finds its pivot in the term "illness" and is sometimes defined using a normativist or holistic-humanistic perspective—both health and malady must be defined by an explicit or implicit choice or convention concerning the goals we have for our own lives. In this connection it is generally held that the term "illness" encompasses feelings, beliefs, and attitudes regarding bodily and mental conditions that overtake and impede us in performing the tasks and in the interpersonal interactions of our life (e.g., we may be unable to walk or drive because of back pain) (see e.g., Kleinman, 1988: 3; Aho & Aho, 2008: 3; Hofmann, 2017: 16). This is the reason why disability (more precisely, any limitation in whole-person activities, as per the World Health Organisation definition: World Health Organisation, 2001) has to be considered here as a form of illness.

[1] Boorse (1977): 542. Though interpreted variously, Boorse's conception has been taken up by several authors: see, e.g. Taljedahl (1997), Williams (2007) (who treats disease in terms of distortions of standard cellular network processes), and Ananth (2008).

Here too we may distinguish two main components. On the one side (from the point of view of its contents) important ingredients of "illness" are psychological states such as pain, suffering, anxiety, fear, and behavioural disturbances like disability: all phenomena being undesired and medically treated, not less than bone fractures and myocardial infarctions. In this respect, the relationship between illness and health surfaces clearly. On the other hand (from the point of view of its social-cultural-linguistic classification), what an illness is, is determined by a lived experience, both at the personal and social levels.[2]

An important point in this context is that illness is usually connected to the wholeness of the individual person's feelings (which in turn largely reflect the interpersonal and social relationships in which the person is involved). This point was finally transposed by the World Health Organization (WHO) into its famous definition of health ("a state of complete physical, mental and social well-being and not merely the absence of disease or infirmity"), whose defects do not exclude that it had the merit of connecting the 'normal functioning' of the body to the more general well-being of human life, considered both in its psychological and social dimension.[3]

The distinction we have been drawing has been concisely expressed by Cassell in writing that "[d]isease [...] is something an organ has; illness is something a man has."[4] With a more explicit reference to the subject of our paper, Nordenfelt spoke of two perspectives from which 'health', 'disease', and other similar concepts may be interpreted: the "analytic" (or "atomistic-biological")—sometimes defined as naturalism—and the "holistic" (or "holistic-humanistic") perspective. From the former

[2] In this last case, in accordance with the work of Parsons about the "sick role" (Parsons 1951, 1975), "sickness" is the more often used term, but it will be included here under the umbrella of "illness", which emphasizes the interpersonal and social aspect (consistent with this, in common parlance, is calling work absences due to health care problems "sick" leave). On this point see also Twaddle (1968, 1994a, 1994b), who was one of the first authors to distinguish between disease, illness, and sickness.

[3] Callahan (1973: 86). For the second view, see e.g., Veatch (1973: 524–5), Engelhardt (1975, 1984, and 1996), Margolis (1976); Schaefer (1976: 20–21); Engel (1978); Whitbeck (1981a; 1981b; 1981c), Nordenfelt (1984, 1986, 1995, 1997a, 1997b), Reznek (1987), Pörn (1993), Wieland (1995, 2002), Fulford (2001). As far as the notion of "illness" is concerned, phenomenological literature is also very important, starting from Merleau-Ponty (1945) up to the most recent works, among which see at least Zaner (1981), Toombs (1992), Aho and Aho (2008), and Svenaeus (2000, 2014). But already in antiquity Seneca had given voice to "illness" by trying to define maladies in subjective terms: "Triae haec in omni morbo gravia sunt, metus mortis, dolor corporis, intermissio voluptatum." (Epistulae morales ad Lucilium, 78, 6.).

[4] See Cassell (1976: 27). The relation between a focal localization of malady and its pervasive affection of the self, reflects itself in the English language, which distinguishes between "pain" and "ache": one speaks of knee pain and skin pain, but of headache, toothache and stomachache, and depending on his/her experience, the patient may speak of back pain or backache. Neo-Latin languages have difficulties in recognizing this distinction: for instance, there were problems translating into Italian a famous English questionnaire asking whether your pain is of "aching" quality. The proposed solution was translating "aching" as "dà sofferenza" ("it makes you suffer") (Maiani and Sanavio 1985). However, both in English and in Italian, "suffering" defines a domain much wider than "suffering because of pain" does. In building a back pain questionnaire including the "aching pain" item, the Authors found that only some of the Italian participants perceived "dà sofferenza" as a specific qualifier, rather than a synonym, for pain (Tesio et al., 1997).

perspective, a person is mainly regarded "as a complicated biological organism with a vast number of interacting parts", and the central concepts are biological, chemical, and statistical; from the latter, "man is taken to be fundamentally a social agent, a complete human being acting in society. On such a platform the theory construction will primarily use humanistic or social concepts. The concept of a person is central, so are the concepts of action and goal." (Nordenfelt, 1986: 281) From the former perspective, again, "one directs one's attention to particular parts of the human organism, and considers their structure and function"; from the latter, "one focuses on the state of the human being as a whole, and judges whether he or she is healthy." (Nordenfelt, 1995: xiii) In this latter sense, health

> has its locus on the level of the whole person. A human being as a whole can be healthy. It is not molecules or tissues which are healthy. We may say metaphorically that a heart or a lung is healthy, but what we then mean is that this heart or lung contributes to the health of the whole person. (Nordenfelt, 1997a: 244)

2.2 *Irreconcilable Perspectives?*

Despite attempts to heal the disease/illness dichotomy, it remains deeply rooted in contemporary medicine. Take the cases of disability and mental health. The World Health Organisation periodically updates the International Classification of Diseases (ICD), used worldwide in epidemiology and clinical practice (for the latest, 11th version see: https://www.who.int/classifications/classification-of-diseases; accessed Dec28th 2021). Biological "diseases" are loosely related to functional consequences at the whole-person level, i.e., the level of behaviours and psychological states. Therefore, in 1980, the WHO strove to separate the "functioning" of body parts (related to "impairments") from the functioning of the person as a whole with respect to the outer world (related to "disabilities" and "handicaps"). In this sense, a brain junction gap caused by an infarction would be an impairment, while the reduced capacity to walk or to communicate (i.e., needs shared by all human beings) would represent disabilities, and finally, if an individual person needed to abandon a job, or be absent from school, this would be a handicap. In 2001, for the WHO, "disabilities" became "activity limitations", and "handicaps" became "participation restrictions", but the distinction with respect to impairments (a whole-parts distinction) remained sharp. And again, both the WHO and the American Psychiatric Association (2013) do not speak of mental "diseases" but of mental "disorders" instead.

It is therefore no accident that the ensuing discussion has brought to the fore a kind of antinomy, which arises from considering one side of health or malady in abstraction from the other. Against the internal coherence of a merely descriptive-naturalistic account of disease, it has been argued that having a particular digestive or breathing system, and/or particular reproductive organs, represents already a potential set of behavioural patterns or norms to be followed by an organism, and these norms are indirectly brought to light whenever the 'normal' functional ability of an organ is reduced or limited to some extent.[5]

In other words, simply claiming that some body functions are fundamental already assumes that they were embedded in a teleological system of aims or values, and the assertion that an organ functions well already presupposes that this is something which we *should* preserve in its present state. No appeal to a particular species design can avoid a rationally unjustifiable jump from 'is' to 'ought'.

2.3 Splitting the Normal and the Pathological

As we have just mentioned, the inherently normative nature of any kind of "normality" has already been asserted by some authors. However, to be fair, Boorse is much more aware of the difficulties facing his account than his critics have usually assumed. His proposal is that neither the mere recourse to the species design (ultimately appealing to a Darwinian fitness-selection model, see Garson, 2016) nor to statistical normality provides a satisfactory definition of health or disease. Both are necessary to do this.

But here it can be said that two mistakes fail to make a truth. It is not so much a question of solving the problems raised by some counterexamples, some of which Boorse admits that he has to solve with an ad hoc exclusion (specifically, the case of "universal diseases", such as dental caries and some geriatric or epidemic diseases, which are usually seen as pathological in spite of their being statistically very frequent phenomena: cf. Boorse, 1977: 566–567). Instead, it is a question of recognising that, in order to establish what health and disease are, the construction of a statistic is never value-free.

You cannot statistically determine what the nature of 'health' is in itself, since the production of statistics necessarily presupposes certain choices concerning the variables to be normalized and the "weight" to be assigned to these variables: these procedures, in turn, depend on evaluative elements (to say nothing of the choice of more sophisticated statistical models relying on complex assumptions). Any modelling is in itself a theoretic-normative process: for example, see the statistical 'war' between

[5] Cf. above all Canguilhem (1972) (which, in a certain sense, can be regarded as a worthy *ante litteram* critique of Boorse), Toulmin (1976), and Fedoryka (1997). Important pages have, however, been written on several occasions by Engel on this point (see for example 1978 and 1979). Engel put into question a biomedical model which requires both that disease be dealt with as an entity independent of social behaviour (which implicitly assumes mind–body dualism), and that behavioural deviations from the population's mean are to be explained only on the basis of biochemical or neurophysiological processes (which implicitly assumes reductionism). O'Leary's thesis that Engel's "bizarre insistence that dualism and reductionism are one and the same view" (O'Leary 2020), is a misunderstanding of Engel's basic idea, which is very simple: if one accepts a dualistic paradigm such as the Cartesian one, one also accepts the idea that only the body (mechanistically conceived), as opposed to the soul (which is immaterial), can fall ill, but then, the tendency naturally follows according to which the doctor must only deal with the body and neglect the mind: which is precisely, in medicine, the reductionism combined with the dualism of which Engel spoke.

frequentists and Bayesians (cf. Mayo, 2018), and the recent surge against researchers' faith in arbitrary levels of "statistical significance", which are still the benchmark for decisions in most biomedical research (Amrhein et al., 2019).

2.3.1 Statistics Is Value-Laden

The natural environment is constantly changing (though, usually, very slowly compared to the duration of a human life), so that statistical 'norms', in a certain sense, must adapt: consider, as an example, how climate change and food availability impact the biological features and "diseases" of all living beings. However, the normative nature of statistics has in medicine further and distinct reasons, acting along the time-frame of human (and even individual) history (on this point see above all Canguilhem, 1972 and Wieland, 1995). First, humans modify—in accordance with their interests and values—not only the natural environment, but also one another. It follows that the range of 'normal' values changes depending on new discoveries, social attitudes and political contexts, and the recognition itself of a given phenomenon as 'normal' (or, as a 'disease') is historically variable. In a society where body weight was assigned a great biological value there would be more tolerance of higher values than of lower ones. You would accept as 'normal' a positive statistical deviation from the average body weight much more in a society where thinness is an aesthetic ideal. To take another example, only in the late 1970's did people begin to seriously consider "essential" hypertension as a "disease" in itself (the cause is still unknown, for which reason it is defined "essential"), but 'normal' values were very lenient, and very high pressure levels were accepted in elderly people. Nowadays, "essential" hypertension is considered as one of the most important causes of mortality and disability (Saklayen & Deshpande, 2016), and has generated huge statistical studies, which have set a much more restrictive range of 'normality'. But perhaps the best example to show that the recognition of a given phenomenon as 'normal' (or, as a 'disease') is historically variable, is the "disease" drapetomania—which was supposed to induce slaves to run away and abscond (Cartwright, 1851/2004). Nowadays, no Ethical Committee would authorize studies on drapetomania (although forms of slavery, unfortunately, still exist; see https://news.un.org/en/tags/modern-slavery, accessed Dec 28 2021). In the same vein, homosexuality has long been classified as a psychiatric disorder. Nowadays, no Ethical Committee would authorize studies on the effectiveness of drugs claimed to correct "sexual orientations". The latter are no longer classified as "diseases" or "mental disorders", for either the World Health Organisation or the American Psychiatric Association.

In other words, not only does each statistical parameter depend on interests and values, both of which vary from individual to individual and from society to society, but—more crucially—interests and values can never be captured entirely by statistics, because these interests and values are key driving forces in changing the social environment as an essential ingredient of the criterion of normality. "Normality" is a value judgment as far as it describes ranges of societal tolerance not less than frequency ranges. An important lesson to be drawn from the examples provided above is not only that statistics cannot produce data that are completely neutral because they are based on the "nature" of things, but also that every statistic and every medical classification presupposes moral choices that, if they are to be made responsibly, require a rational debate not only on what the natural data or regularities are, but also on the values that we want to place at the basis of social (co)existence.

2.3.2 Statistics and Individuals Depend on One Another

However, there is at least one more reason—and perhaps this is even more important in our epistemological and methodological context—that in one way unites, but in another distinguishes (again as normative) the use of statistics in the human sciences. There are no statistics concerning human behaviour (including those concerning illness or disease) that are not built from and controlled by resorting to, individual cases, that is, to the *interpretation* of individual behaviours or actions. To know whether a statistically significant number of patients have reacted positively to a therapy, one has to interpret not only signs, but also the patient's own answers to the clinician's questions. And this is not only the case when checking the effectiveness of a particular psychotherapy, but also the effectiveness of a vaccine or of a surgical operation.[6]

2.3.3 Is Paralysis of Lower Limbs a Pathology?

A simple thought experiment may serve to illustrate both points. Let us assume a highly developed technological society where people move, for short as well as long distances, by transport beams that send an individual's molecules from one place to another and reassemble the molecules upon arrival (for this example, cf. Buzzoni, 2003). As for the first point—that is, that statistical "normality" is value-laden, it can be easily deduced from what has been said: in such a society, an individual with both lower limbs paralyzed—or somebody suffering from any "pathology" involving a walking impairment—may be considered healthy or "normal" to the extent that he or she achieves the main objectives assigned to its individual members by that society; stated otherwise, he/she is healthy from the standpoint of "activity limitations". More specifically, the person might not be disabled, according to the WHO glossary (Prodinger et al., 2016). As regards the second point—that statistics depend on considerations of individuals about themselves and other people-suppose you need to measure that person's level of independence in daily life (Tesio et al., 2002, 168–176) or satisfaction with life (Franchignoni et al., 1999). Cumulative questionnaires are needed (either self-administered or not) like in any measurement of behaviours, attitudes or perceptions (Tesio, 2003). These measures run the risk

[6] The relevance of patients' perspective is indeed more and more acknowledged in biomedical research: see the development of "patient reported outcome measures-PROMS" (Crossnohere et al., 2020).

of being biased by the author's perspective (e.g., which items are selected for the questionnaire, and who determines their scores?) yet, they cannot be replaced by, say, biomechanical or neurophysiological measurements at the lower limbs (see Sect. 5 below for development of this idea). In using this perspective to ascertain whether a given person with paralysed or otherwise injured lower limbs has to be regarded as really healthy, one would already have overstepped the bounds of a merely naturalistic dimension of health and illness, for one would need to interpret reflections given by human beings about themselves and other people.

2.4 Illness and Disease Are Not Fully Independent

One advantage of a biological definition of disease is that it can explain why a patient can be confident in estimations of their health, despite the fact that pathological changes are taking place in the body. A tumour can remain asymptomatic or regress spontaneously without necessarily having any effect on the subjective well-being of a person. The function of the immune organs is another illustration of the fact that the concept of health ought to refer to more than the level of subjective well-being. As Taljedahl (1997) noted against Nordenfelt (1993), when immune organs exert their capacity to combat infection,

> they may give rise to symptoms that are transiently incapacitating, i.e., the expressions of bad health. Yet, these symptoms of bad health are in a sense also the expressions of good health. (Taljedahl, 1997: 68)

However, it can be easily shown that the relative 'objectivity', or even independence, of certain natural processes that occur in us, cannot justify any purely biological concept of disease. One can only conceive of the possibility of objective pathologies existing in the nature of things by indirectly connecting them with a subjective illness, which—in accordance with our past experience—indicates an objectively detectable disease. In a word, a disease is defined as a disease because it is acknowledged that sooner or later, in at least some of the affected people, it will lead to an illness. A skin nevus is not called a disease: although it is when it is recognised that it can become a skin tumour. A laboratory finding enables a physician to diagnose a disease not yet noticed by the patient, because the physician—in contrast to the patient—knows how to rank that finding in the context of a typical medical history. The physician can come to an assessment without having to rely on symptoms which occur later and are then felt subjectively, only because they know that, in the past, other individuals with similar, at first equally inconspicuous findings developed (inter)subjectively detectable symptoms after a certain period of time: this remains the necessary methodological starting point for an objective concept of disease. Without a direct or an indirect reference to these symptoms, that is, by entirely excluding the subjective feelings of illness, no laboratory result would ever be a meaningful (bio)marker of anything that might be called a "disease".[7] Before one can methodologically reconstruct a disease as the biological counterpart of illness, illness must be felt as such by some person, who, moreover, may have reached the awareness of being ill only through other people. If these "others" are scientists, the person's illness can be framed in a taxonomic system of diseases (let us encase here also syndromes, i.e., sets of signs and symptoms, into the disease concept).

In a pathoanatomical dissection one does not see a disease or a diseased organ. One can just see an anatomical peculiarity, a difference from reference norms, to which one can attach the sense of a disease exclusively because, consciously or unconsciously, one relates them to ideas of unwanted suffering and death, which epistemologically precede the corresponding biological, metrical and behavioural reality (in this sense, pathology, as a normative concept, precedes physiology, as a naturalistic concept). But these ideas of suffering and death cannot be formed apart from the interpretation of the reports and the conduct of other people with whom one is in personal and social interaction. In this sense, the scientific-reductionist or atomistic-biological notion of diseased cells or organs depends on the holistic-clinical or holistic-humanistic conception of the illnesses of human persons.

Finally, as far as the functioning of the immune system is recalled, a high fever is anything but a state of subjective well-being, but the fact remains that even in this case symptoms may be considered as expressions of good health. With Canguilhem, one might say that "anomaly" (coming from a Greek word meaning "irregular", "uneven", "rugged") is not "abnormality". Physicians are worried about "anergic" patients, unable to increase their temperature when needed. Fever may be welcome whenever we know that it may be designed to preserve the patient's health as subjective well-being, and the ability to pursue the fundamental goals of their life, including survival.

2.5 *Disease Is Not Enough for a Human Science; Nor Is Illness*

So far, we have seen the difficulties inherent to the objective biological conception of malady as disease. At first glance, these difficulties might seem to point in the direction of a more normative and socially conditioned concept of health. As we have already mentioned, many attempts in recent decades to move in the direction of a holistic-humanistic perspective of malady are based on the concept of an ability to act in such a way as to enable the attainment of one's goals. For example, health can be defined as an equilibrium between personal capacities, social environment, and a person's "high-ranking projects in the life-plan" (Pörn, 1984, 1993), or a person's "vital goals" (Nordenfelt, 1984; see also Nordenfelt, 1986, 1995, 1997a, 1997b; Engelhardt, 1975 and 1984; Whitbeck 1981a, 1981b). Given that projects

[7] On this, see Kumbhare and Tesio (2020); see also Buzzoni (2003). In particular, as far as "biomarkers" are concerned, see Buchner et al. (1999), and Taylor and Elston (2009).

and goals are predicates of the person as a whole—and not of isolated molecules, cells or organs—, these theories of health, as already mentioned, are regarded as "holistic-humanistic", rivals to "atomistic-biological" ones.

However, on closer inspection, even a holistic-humanistic and culturally oriented conception of health and malady lends itself to serious objections. Paradoxically, one could even say that certain aspects of a cultural conception of health and malady, which privilege it over naturalistic accounts, also constitute a weakness. If one admits that health and malady only depend on historically changing cultural values, they seem to lose all scientific relevance. In other words, if one admits that a patient's attitude towards their own disease frequently influences the success of the therapy to such an extent that a biological therapy itself becomes ancillary if not useless, the fundamental pillar of any objective control of the procedures that medicine uses to reliably treat patients falls. Moreover, this variability is increased by other problems that have their origin in what distinguishes the human sciences (or even the 'humanities') from the natural sciences, the "soft" from the "hard" sciences, that is, in the mediation by human consciousness.

Patients can report symptoms or present with behaviours unsupported by biological alterations, e.g., reflecting malingering or unconscious somatisations. These represent a severe challenge, mostly because the wealth of modern instrumental diagnostics inflates the risk of false positives for a known "disease". In other cases, the therapy could be prejudiced by the patient's knowledge of it. The widespread knowledge of psychoanalytic theory, for example, could be a great obstacle to this form of therapy, for patients can use it to reinforce their resistance to the unveiling of their actual unconscious motives.[8] Surely, one seeks to obviate this problem by control groups and double, or even triple blind experiments, where neither physician nor patient (nor the researcher who evaluates the results) know who gets what. Yet, no matter what degree of sophistication one's methods of experimental control attain, the placebo effect will always interfere to some degree with the effectiveness of a treatment (on the difficulties of a definition of placebo, see e.g. Howick, 2016; for technicalities, see Benedetti, 2021). Faith-healing through pilgrimages has long been recognized by medicine as effecting unexplained recovery from somatic manifestations of psychological illnesses (Charcot, 1892). On the other side of the same coin, cases have long been recognized in which illness or even death may follow curses, an extreme variant of the so-called nocebo effect (Cannon, 1942).

All this is a serious weakness in the holistic-humanistic conception of malady. No doubt, we need an objective, intersubjectively verifiable, socially (and perhaps legally) recognized concept of health. But given the important and unavoidable point made by the holistic-cultural perspective about health and malady, on what basis

[8] See Legrand (1975). The psychiatric disorders leading to various clinical conditions, sometimes very severe, sometimes mimicking familiar diseases or impairments, and sometimes rather weird, have been variously defined since the first distinction between malingering and "hysteria" made by Jean-Martin Charcot in 1890 (Harris 2005). From the Freudian concept of symptoms representing a "conversion" of unconscious conflicts to the contemporary concept of semi-organic, "neurofunctional" disorders (Demartini et al., 2016), the pendulum never ceased to swing between a psychological and a neurologic interpretation.

can we assert that a person is healthy or unhealthy in some intersubjectively testable sense?

The opposition of the two views about health and malady seems to end in an antinomy without any way out. The holistic-humanistic perspective apparently could, on one hand, show the naturalistic notion of disease to be untenable because it is unavoidably value-laden; but on the other hand, the demand for culturally changing decisions, values and norms seems to deprive the notions of health and illness of any genuinely scientific value.

2.6 A Recent Attempt to Heal the Old Gap

A recent debate has refreshed the illness-disease antinomy. The debate originated from a concern about the symptom of central fatigue, i.e., fatigue unrelated to exercise. This symptom is a common component of many illnesses, among which is Chronic Fatigue Syndrome (which, embracing the "somatic" perspective, is also called myalgic encephalomyelitis). On the one hand, discussing this concept, Sharpe and Greco (2019) insisted on the possibility of an "illness without disease". This is a concept whose validity is repeatedly admitted in the literature. For example, Hofmann (2017) notes, "there are no necessary connections" among disease, illness, and sickness, any of which, though *de facto* often occurring jointly, may perfectly well "occur without the others" (Hofmann, 2017: 18).

On the other hand, Wilshire and Ward (2020) claimed that the notion of 'illness without disease' is methodologically problematic and accused Sharpe and Greco of using the distinction between illness and disease "to identify a problem space that is not amenable to medical interventions at all, but rather must be addressed through social and/or psychologically based interventions". The notion of "illness without disease" "can lead to unwarranted causal assumptions" as it seems to assume that "any experience not directly predicted by a disease model is necessarily of psychosocial origin" (Wilshire & Ward, 2020: 532; for more details on this debate, see Tesio & Buzzoni, 2020).

It might be objected that Wilshire and Ward's criticism of Sharpe and Greco glosses over several important distinctions. In particular, instead of thinking of illness and disease "in terms of the hierarchical difference between subjective (or mental) and objective (or physical) realities," Sharpe and Greco "propose that we could think of them in terms of different degrees and forms of abstraction from the totality of what is real." (Sharpe & Greco, 2019: 185) However, though one may accept that both the experience of illness and what is usually called disease are "abstractions", it is still necessary to raise the question about the nature of the relationship between such abstractions as well as between them and the rest of reality: how are such abstractions connected with one another? And how can their connection be investigated in an intersubjectively reproducible and testable way?

Precise answers to these questions will be sought in vain in Sharpe and Greco's paper. In order to fill this gap they ought to provide a clear view about the causal relationship that exists between the different results of our abstractions, and especially between, on the one hand, the experience of "illness", and on the other, that aspect of reality which is the organic correlate of what is usually referred to as "disease" (and it goes without saying that if we do not wish to forsake science for magic and fiction, then this causal relationship should be such that, at least in principle, it can be made the object of an intersubjectively testable investigation).

On the other hand, however, Wilshire and Ward cannot accommodate one of the most important claims made by Sharpe and Greco, namely, that what is usually designated as the subjective, or better, normative-humanistic, dimension of health is relatively autonomous. Experiences of illness are real as long as they are 'lived through': in this sense, illness cannot be understood adequately only in terms, for example, of biological mechanisms, even if it is always possible to find some biological mechanism (at least some neural electric or metabolic activity) that is related to it and can be the subject of scientific investigation.

2.7 *Healing the Gap by Rethinking Causality*

Now, a first necessary step to answer this question in a satisfactory way is a more flexible, context- or perspective-oriented conception of causality. It is an important step, since it allows a psychological factor to be the cause of another psychological or even biological factor, and vice versa, rather than only speaking in terms of organic causes. To claim that this (bi-directional) interaction is not possible would be contrary to available evidence. It is a well-known fact that 'subjective' states can influence biological features, and this influence can be more or less direct. In the more direct version, it could take the form of changed biological parameters. For instance, 'stress' (admittedly a form of 'illness') can cause detectable changes in blood steroid concentration and immune markers (for a review, see Yaribeygi et al., 2017). In other cases, the influence of 'illnesses without disease' on the purely biological aspect of malady can be mediated by the personal, behavioural, and social context; for instance, depression appears to be more a cause than an effect of adolescent obesity, a condition associated with a broad series of related diseases.[9]

It is important to stress that no direction of the causal vector should be privileged a priori. Instead, it can only be determined following experimental evidence. The

[9] See Byrne et al. (2015). From this point of view, O'Leary's idea that a nonreductive dualism requires a clear distinction between "psychiatric care" and "medical care" is untenable. It can be dangerous to provide only psychiatric care to patients in need of medical care (O'Leary 2020), but also to provide *only* medical care to patients in need of additional psychiatric care. By "psychiatric", here, it is intended any form of "words only", relational treatment, from psychological counselling to formal psychoanalytic approaches. O'Leary's nonreductive dualism, as in the case of Wilshire and Ward (2020), also seems to lack a contextual and pragmatic conception of causality, i.e., one which does not limit a priori the directions of the causal vector. If we change the point of view and the interests of the investigation, it is in principle possible to trace a clinical picture back to causal links that refer, in different proportions to both biological and cultural reality.

experience of illness is a real thing that manifests within, and interacts with, the context of human existence. On the one hand, we see that there are decisive causal chains that go from the molecular to the cellular level to processes at the tissue level, which in turn are causally related to processes at the organ level, which again influence processes at the perceptual, behavioural, psychological and social level. On the other hand, however, in other circumstances and thanks to other pragmatic interests, we can see that processes at the biochemical level of cells are causally conditioned by processes at the tissue or organ level (see e.g., the "Tissue Organization Field Theory" of carcinogenesis developed by Soto & Sonnenschein, 2004, 2006, 2011), which in turn may be influenced by those at the interpersonal and social level.

Identifying the chemical or electrophysiological correlates of a psychiatric 'disorder' can be of fundamental therapeutic importance, but the possibility that words (which modify biological correlates) may be an essential ingredient of a therapy can never be excluded. The importance of observation and experiment should not make us forget that to cast aside the use of the spoken word in medicine was a unilateral, although perhaps historically necessary choice, which Virgil famously expressed by his definition of medicine as "the silent art" (*muta ars*).[10]

In this sense, the concept of illness encompasses the "pathology" not only of "psychologically-intangible" entities like feelings, emotions, and perceptions, but also of any observable behaviour that can be ascribed to a person as a whole. (By the way, there are no psychological states we can know, if not manifested through motor behaviour: consider, at a minimum, the blinking eye of dramatic "locked-in" syndromes). Any form of disability is, actually, a form of illness: it is "lived through" by the person. A unidirectional bottom-up view of causality may distort the interpretation of behavioural results associated with biological treatments. For example, electrostimulation of the spinal cord has been claimed to allow the recovery of autonomous walking (with the help of rollators or parallel bars) in some chronic spinal cord injured patients (Angeli et al., 2014). It was objected, however, that learning a skilful use of the upper limbs might foster the transmission of force to the lower limbs, thus eliciting proper muscular reflexes (Tesio & Scarano, 2021): learning is a property of the person, not of the spinal cord.

2.8 An Open Challenge: Intersubjective Control

This viewpoint solves the problem of the possibility of causal influences between different levels of biological organization and in an important sense returns to the point of view of common sense. However, as we have just mentioned, this still

[10] On this aspect of the history of Western Medicine, see Laín Entralgo (1970: xxi). A word of caution is in order, however. Care must be taken in avoiding extreme reliance on "word-based" medical approaches, at the expenses of biological approaches. For instance, the interesting proposal of "narrative Medicine" (Charon, 2006) should not be embraced uncritically. In any case, it must be left to experimental science deciding to which ring of the causal chain the available treatments can be optimally applied.

does not solve the problem of intersubjective controllability which afflicts cultural conceptions of health and malady. There is no doubt that, if diagnosis and assessment of potential treatments are to be based only on the symptoms explicitly felt by the patients, or on their overall behaviour, or on the aims that they set themselves, or say they set themselves, they would be very variable, to the point of escaping generalisation and intersubjective controllability. To this should be added that malady is not something we can establish completely on our own, as isolated individuals, but always involves a social element. As noted by Talcott Parsons, illness may legitimately be regarded as a type of deviant behaviour (see the above comment on disability) which involves a particular social role: it is a claim upon others to be "taken care of" and relieves one of blame, shame and of certain social obligations, but it imposes duties to seek therapy from experts (Parsons, 1951: 283–297). Now, this dependence on the social recognition of malady by the community is also a factor of relativity, which must be somehow reconciled with the intersubjectively controllable nature of formal medical assessment.

Some authors have attempted to resolve this problem with an eclectic position, conceding something to the opposite perspective. In the next section, we shall briefly examine one of the most important attempts to develop a hybrid model of disease, that of Jerome C. Wakefield's "harmful dysfunction model".

3 Wakefield's "Harmful Dysfunction Model" of Health and Disease. A Way Out?

3.1 Against the Nurture-Nature Dichotomy

According to Wakefield's model, originally developed for mental disorders but then generalised to all types of medically treated conditions, two requirements have to be met in order for something to count as a disorder: (1) an (evolutionarily determined) objective biological failure or dysfunction of an organ or body part has occurred; and (2) the dysfunction must cause a particular social harm (Wakefield, 1992: 3, 2007: 149–156; Wakefield, 2014; Wakefield, 2015).[11]

Even in the case of Wakefield's model it can be said that, paradoxically, its greatest value is also its greatest weakness. It tries to integrate into a single model both the objective-organic and the interpersonal and social dimensions of malady, but because of its eclecticism, it is ultimately exposed to the objections that both parties raise one against the other.

On the one hand, the strength of this model lies in the fact that Wakefield makes every effort not to hypostatize the distinction between the natural and cultural spheres:

[11] It should be noted that to some extent the main idea of Wakefield's model was anticipated by Robert M. Veatch, who also argued that health should be defined as "an organic condition of the body judged by the social system of meaning and value to be good." (Veatch 1973: 78).

> the contemporary understanding of brain plasticity means there is plenty of room for socially sculpted neurobiological changes not only since antiquity, but locally. The understanding of normality and pathology is [...] necessarily an understanding of the dance between evolved human nature and cultural influences. Moreover, evolutionary accounts are not limited to genetic and brain-matter levels. Thoughts and emotions are as biologically real as genes and neurons and have naturally selected features that operate at the representational level. (Wakefield, 2015: 351)

In the endeavour to overcome the nurture-nature (or the genes-environment) dichotomy, Wakefield is more and more supported by the growing evidence that acquired phenotypes (including behaviours and competences) can indeed be transmitted through "epigenetic" mechanisms working much faster than foreseen by the classic Darwinian paradigm (see on this, e.g., Crews et al., 2014, and Jablonka & Lamb, 2014).

On the other hand, however, he fails to answer satisfactorily the question of when a social change leads to a disease in the properly medical sense. His answer is that,

> [w]hen culture exploits human variability and malleability—mental or physical—to sculpt human beings in socially desired ways, [...] the socially desired result is not a disorder if there is no socially defined harm. (Wakefield, 2015: 352)

It is clear that this only shifts the problem without solving it: we do not yet have a criterion to distinguish between health and disease, because we do not yet know how to distinguish between the changes of human beings that involve and those that do not involve social harm.

Wakefield therefore tries to identify such a criterion, which he believes he has found in overly rapid social changes, which do not allow for the evolutionary adaptation of human cognitive structures:

> the construction process can be pursued so relentlessly that damaging side effects occur that constitute true disorders. For example, the chronic stress of contemporary competitive educational and occupational environments that wring as much productivity as possible from the naturally talented can cause anxiety disorders in the vulnerable. Castel's characterization of at least some standard OCD [*sc.*: obsessive-compulsive disorder] cases would presumably fit here as genuinely disordered casualties of autonomy training. (Wakefield, 2015: 352)

As we can see, what causes the disorder is the fact that cultural evolution affecting biological structures does so much faster than the subsequent biological adaptation process, which, so to speak, cannot keep pace. But the criterion is clearly unsatisfactory. How can we distinguish between changes that are too rapid from those that are not? The only criterion here seems to be the emergence of a social harm, without which no health problems arise. A vicious circle seems evident.

Moreover, Wakefield rightly wants us to distinguish between correct and incorrect assessments of social harm: psychiatry becomes an oppressive social control if one incorrectly labels socially valued outcomes as natural and then classifies variations that fail to manifest the socially desirable features as disorders (see the above example of 'drapetomania') (Wakefield, 2015: 353). But this, on reflection, shows that it is only a value judgement that ultimately determines what we consider or do not consider a disease and, secondly, that biological modifications induced by a different cultural environment can and sometimes must be practically encouraged or, on the contrary, rejected and fought against.[12]

From this point of view, all the difficulties mentioned above regarding a normative and cultural notion of what health and malady are, remain. Wakefield's model, in fact, does not tell us how it is possible to limit the problematic factors that undermine any attempt to establish in an intersubjectively controllable way whether something produces an authentic or only apparent "social harm". The difficulty is not gotten rid of, but merely shelved.

3.2 *Opening a Crack in the Vicious Circle: Margolis's Model*

An interesting variant of hybrid models of disease, which has not been discussed in the literature as much as its theoretical depth would warrant, was developed by Joseph Margolis. He rightly insisted that distinctions are smaller than they might seem at first glance: since the human body has changed over millennia relatively little as compared with social institutions, "the functional norms of somatic medicine are relatively conservative (unlike the norms of law)" because they are intimately connected with fundamental human abilities (Margolis, 1976: 575).

This model admits that one cannot speak of health or illness/disease unless one assumes the existence of fundamental objectives of human life with respect to which the functioning or non-functioning of our organism is a necessary condition. But in this model the achievement of these fundamental objectives requires, in a fairly uniform way, in the various societies, the possession of certain skills related to the use of our body in the environment in which we live, which would explain the relatively transcultural and ahistorical value of certain pathologies: the use of the hands to grasp objects and intervene for a thousand different reasons on the environment that surrounds us is so important that hardly an injury, deterioration or decrease in the functionality of our hands will not be considered as pathologic, and supported as a disease in all societies, across all ages.

Now, one can object to this position that the uniformity we are talking about is, in fact, only prevalently transcultural and ahistorical. Strictly speaking, there is no natural mechanism (be it physical, biological, chemical, etc.) that is not immersed in a cultural dimension that changes it in various ways. And it is always with an implicit or explicit reference to this cultural mediation that we can speak of health and malady in humans. Take the example of the difficulty or incapacity to read (dyslexia), clearly a neural biological problem, which could be detected and named a "disorder" only

[12] Deformities resulting from binding of women's feet (Chan, 1970) and children emasculation to obtain castrati singers (Hatzinger et al., 2012) were not considered social harms for many centuries, before these practices were formally banned. The same holds for female genital mutilation, which is still widely practiced in various socio-cultural contexts (see: https://www.unwomen.org/en/digital-library/publications/2020/07/a-75-279-sg-report-female-genital-mutilation, accessed Dec 28 2021).

in a social context where some literacy is expected from the population: in fact, it was "discovered" in Germany, only in 1887.[13]

To sum up, despite some biological constraints, health remains largely relative to persons, to their interpersonal and societal environment, to a meaningful relation with it, to custom, and convention. And it is precisely this measure of wide variability that raises the problem of the intersubjective controllability of our judgements about health and malady in different societies and in different historical periods. Moreover, as Engel never tired of pointing out, social values do not merely infuse the appraisal of biological dysfunctions, but they also infuse the various and individual ways that one responds to those dysfunctions (Engel, 1960, e.g. pp. 466–467). We seem to have fallen back again into the antinomy highlighted already in Sect. 2.

As we shall see in the next section, in order to solve this problem satisfactorily—in addition to a pragmatic and contextual theory of causality, which we cannot dwell on here (see in particular Buzzoni, 2014 and Tesio & Buzzoni, 2021)—it is necessary not only to allow that illness, in some cases, can powerfully influence its organic basis (and therefore the "disease"), but also to understand how this interaction is not arbitrary, but responds to regularities which, although different from those of the natural sciences and proper instead to human sciences, can be ascertained in an intersubjectively controllable way.

4 Medicine as a Human Science

4.1 Human Sciences: Habits as Law-Like Regularities

Everything we have said so far about the cultural aspect of health and malady seems to suggest that it is impossible to satisfy one of the basic prerequisites of any empirical and scientific notion, namely that of being intersubjectively controllable. The problem now is: How far, and in what form, is it possible to reconcile the unpredictable variability of the way in which the individual patient or particular groups of patients experience their health and illness with the equally undeniable organic-objective basis of their lives (and whose importance lies at the basis of the very notion of "disease" as opposed to illness). It has already been hinted that the answer to this question lies, at least to some extent, in the answer to another question: In what sense, and to what extent, is medicine a "human science"? As already mentioned, the term human should not be taken as a synonym for "humanitarian", but in the more classical sense instead, meaning a science that studies the ways in which human beings (classified as *Homo sapiens sapiens*), throughout their history, manifest themselves in various ways of coping with the natural and social environment.

Now, the answer to the question just raised (and the justification of our claim about the unity and distinction between analytical and holistic perspectives) can only

[13] For a history of dyslexia, "both an ongoing psychological diagnosis and a social construct", see https://dyslexiahistory.web.ox.ac.uk/brief-history-dyslexia, accessed Dec 28th 2021.

be satisfactorily given if a fundamental characteristic of medicine, which it shares with other human sciences (such as psychology, sociology and, as a borderline case, historiography itself) is brought to light. This is not the place to give a comprehensive outline of the epistemological and methodological status of the human sciences. But we do want to insist on a point without which the link defended here between unity and distinction, and between the analytical and holistic perspectives on health and disease would remain suspended in the void, and specifically, without philosophical justification (for a more adequate justification of the brief hints that follow, see Buzzoni, 1989, 2010).

The subject matter of the human sciences is the result of a peculiar process of sedimentation of that which was wanted or done in the past by human beings. Typical focii include habits concerning action and thinking established by frequent repetition, more or less consciously transmitted to descendants. Human relationships are based on a more or less unconscious tissue of habits, regarding acting or thinking conditions both at the individual and the collective level. These habits are quasi-mechanisms on which institutions, traditions and customs are based, or, to put it another way, usually we follow these habits quasi-mechanically and unconsciously, for which reason they are very difficult to avoid. The regularity and predictability of human habit grounds talk of the "bureaucratic (mega)machine", the "machinery of justice", the "market machinery", and the like. More precisely, human actions depart mostly to a negligible extent from these habits, so that one is able to subsume such actions under general—psychological, sociological, etc.—law-like regularities in order to explain them.

While, in this respect, the rules concerning human actions are analogous to the scientific laws of nature, they can at any time be revoked by becoming aware of them. This makes it possible for human beings to modify, improve or sometimes even entirely (or better, almost entirely) suspend or change psychological, sociological, ethnological, medical, etc., regularities. Psychoanalysis is paradigmatic of this, but everyday life also repeatedly confirms the possibility of suspending routines as fixed ways of thinking or doing things. By contrast, you can change a Ptolemaic model of astronomy in favour of a Copernican one, but no change of attitude can suspend the relative Sun-Earth motion itself.

In other words, human actions always have two sides. One side consists largely of unconscious routines and quasi-automatisms governed by law-like regularities or rules, and this allows a scientific explanation of human actions; the other side persistently dissolves the routines or quasi-automatisms in new courses of action. The two sides are not separate, but dialectically connected. Far from being inconsistent with one another, each of the sides presupposes the other, and each is necessary for the appropriate interpretation of the other: the possibility in principle of suspending (relatively) unconscious rules or habits is indeed the other side of a de facto dependence of our will on these rules or habits, without which, on reflection, no free action would be possible.

Now, if we apply these considerations to the above theses regarding the relationship between the analytical-reductionist and the holistic-humanistic perspective on health and malady, as well as the relationship between "illness" and "disease", we are in a position to identify an important, but usually neglected characteristic of medicine as a human science.

4.2 Why Medicine Can Be a Science

As we have sketched above, the cultural dimension of human actions consists to a very great extent in unconscious, law-like habits, and this makes it possible to recognise intersubjectively testable connections not only between some symptoms and some objectively ascertainable processes in the ill body, but also in the various, individual as well as collective ways to respond to illness (as the lived experience of bodily and mentally painful and/or disabling processes) and to the often, but not always, diagnosed disease (and thus to the associated statistics and/or organic dysfunctions). The powerful influences exerted by the cultural, subjective-interpersonal and social, dimension on the organic dimension of malady and health are not arbitrary; on the contrary, they obey laws which, even though different from natural or empirical laws (because they are made, accepted, modified, or rejected by human beings), they are sufficiently stable to make possible predictions and explanations similar to those of the experimental sciences. For example, on the one side, dyslexia would not be a medical problem in an imaginary future society where information might be transmitted electronically from brain to brain with no mediation of written words. But in spite of the fact that perhaps reading will not be a human task in a science fictional future, dyslexia will likely be considered a medical problem deserving a rigorous scientific approach for many generations to come.

From this point of view, if we define illness as the lived experience of bodily and mentally unwanted painful and/or disabling processes, medicine must be concerned with more than "diseases" and "illness" in the most common (for example in Boorse's) sense. Medicine must also be explicitly concerned with *the more or less law-like ways in which patients as persons (as well as the members of their families or wider social environment) respond to illness and disease.* In other words, not only illness as the lived experience of bodily and mental processes, such as respiratory wheezes, abdominal cramps, painful joints, and stuffed sinuses (see e.g., Kleinman, 1988: 3–4), but also the law-like attitudes of patients about how best to deal with illness (in its wider sense, including disabilities) and with the associated practical problems in daily living. To the extent that the illness and disease, as well as their classifications, depends upon such attitudes and reciprocal expectations (and upon conventions, policies, social norms and roles constituted by those attitudes and expectations), they should be studied as part of the genuine subject-matter of medicine as a human science, with the technical specificities that this nature requires.

4.3 The Specificity of a Human Science

The cultural influences on illnesses and diseases (as well as on their classifications) we have stressed in the preceding section do not exclude the possibility of formulating law-like regularities concerning the object of medicine. However, they impose some methodological precautions and constraints. As already mentioned, while in one sense the rules concerning human actions are analogous to the scientific laws of nature, in another sense they differ in principle from them, because they can at any time—in principle entirely, but de facto only to some extent—be suspended by any individual becoming aware of them (this possibility is confirmed not only by psychoanalytic and neuroscientific inquires into tacit knowledge, but also by our everyday experience).

The methodological counterbalance of this possibility that is typically used in the humanities is, so to speak, a second-level use of the statistical tool, which is of interest here only insofar as it depends on reasons other than those for which statistics has been adopted in the natural sciences. In fact, the conventional bio-statistics aimed at summarizing data and making inferences on the observed measures may be regarded as a 'first-level' strategy: necessary, but not sufficient, in the human sciences (including medicine). The further second-level, here, includes two peculiarities:

(a) the need for a circular statistical approach moving from individuals to populations and vice versa, and
(b) the need for statistical inference on the nature of the variable under study.

The biological effect of an intervention (e.g., a given drug) interacts with the socially-influenced, yet highly individual, subject's psychology. Consistently enough, drug research necessarily requires experimentation on humans. Of course, tests on humans are needed because biological specificities must be taken into account. Another reason, however, and of more interest here, is that the person's variables must be taken into account: they converge in determining effect modifiers such as treatment compliance, life habits, proneness to the placebo effect, and the like. Trial designs, therefore, should include such variables, which are of little to no use in studies on animals. One may object that individual peculiarities, both biological and behavioural, can be conditioned out through proper use of population statistics (e.g., the placebo effect can be neutralized by randomisation to the treatment and placebo arms of the study), thus making research practice on humans similar to practice in the natural sciences. However, statistics on data (e.g., means, standard deviations, etc.) wipe out fundamental information on the reasons why individual responses did vary: a critical flaw given that as a rule clinicians treat single cases. For this reason, medicine needs statistical methods to estimate what is the degree of uncertainty (modelled as error) surrounding single measurements and their changes in individuals: an uncertainty-error typically larger than that affecting population means (see Tesio, 2012a). Whereas statistics on forces and temperatures can give us a glimpse into the nature of the variable, the same does not hold, for instance, for statistics on "depression", "quality of life", "pain", "balance" and the like: for an investigation into the existence and nature of these requires a "second-level" of inference.

The necessity for all human sciences (including medicine) to resort to specific statistics is not the result of overly complex subject matters. In other words, it is not due to some unavoidable ignorance of all the relevant factors and the relations between them (as is usually the case in the natural sciences). Rather, the necessity results from something intrinsically connected to the nature of the human sciences' subject matter: one must resort to statistics because one cannot abstract entirely from personal consciousness, which is a relevant causal factor unknown *per definitionem*. Since the rules we find in the human sciences can change or dissolve, these sciences, including medicine, are under an obligation to inspect constantly the validity of the rules on which their predictions and explanations are based, both by indirect, statistical, and direct, clinical-empirical methods. The statistical approach is, as it were, the best available counterweight to the risk, which always hangs over the human sciences, of investigating the 'wrong' subject-matter, a subject-matter that could at any time change because it is partially self-made, and therefore escapes the generalisations so far successfully applied to it. The particular character of the "laws" of the human sciences in general is thus closely connected to the duty of these disciplines to test and support their assertions not only directly, with reference to particular cases, but also indirectly, by means of specific statistics. It should be clear that the issue of statistics is of critical relevance to our discourse, so that a digression is needed.

5 Replicating the Illness/Disease Dichotomy: Statistics from Biology to Behaviour

It may be useful to recall that nowadays, by "statistics", two wide domains of human knowledge are intended. Boldly stated, the first includes the algebraic techniques used to describe, summarize, or predict some measures and their uncertainty (descriptive and inferential statistics, respectively). The second domain includes the logic of the experimental set-up, fostering reliable inferences about the causal rather than the merely associative nature of the relationships across variables ("trial design", in medical jargon). For instance, computing means and tracing a regression line are algebraic; adopting a double-blind protocol is an essential component of trial design.

Looking at the algebra, it must be noticed that Claude Bernard, the founder of contemporary "experimental" medicine (see e.g., Bernard, 1865), did not like inferential statistics (already highly developed at his time). He saw biological processes as deterministic (after all, they were seen as results of chemical/physical phenomena), so that the predictions allowed by authentic laws should be free from uncertainty. He conceived the deviance of observed results from expected ones (paradoxically) as the effect of imperfections in the experiments and/or the assumed law of nature.

To many contemporary clinicians, algebraic statistics still appear as a sort of complicated cosmetic surgery done to embellish imperfect data, or to over-simplify empirical reality.[14] The alarm has been raised that "in medicine today uncertainty is generally suppressed and ignored, consciously and subconsciously" (Simpkin & Schwartzstein, 2016). Uncertainty in medicine has been the topic of philosophical discourse (Djulbegovic et al., 2011), and of proposals for new training regimens for medical students (Tonelli & Ross Upshur, 2019).

The problem cannot be solved as long as two key points are overlooked.

First, biological statistics, the one (scarcely, indeed) taught to clinicians, is based on population summaries (e.g., means or medians) whereas clinical practice faces single and entire, much less predictable, individuals.[15]

Second, whole-person variables are much less regular (i.e., there are more peculiarities/idiosyncrasies), compared to biological variables, for the reasons expressed above. To overcome this difficulty, in the early twentieth century, psychology married statistics and generated a branch now called "psychometrics", although the word "personmetrics" might be more appropriate (Tesio, 2003). Psychometrics, in turn, coined the term "latent variables" (or latent traits) to indicate variables (such as perceptions, abilities, attitudes) than can be ascribed to a whole person, only. These variables have an inherent variability-instability, within and between subjects, going beyond the variability caused by both biological instability and measurement errors, and related to the "interaction between person and situation" (Steyer et al., 1999). Boldly stated, "noises" of both biological and relational origin interact in the manifestations of such variables. Pain, depression, memory, language, continence, balance, voluntary force, fatigue, all may manifest themselves in potentially infinite circumstances, and with variable intensity. These properties are much more indirectly observable than body weight, nerve conduction velocity, or glucose concentration, and inferences must be done on a very limited set of observations, usually lumped together in cumulative questionnaires. These provide scores that simply report counts of observations (e.g., how many yes or no answers one gives to questions listed in a questionnaire). How much of the latent variable is represented by "yes = 1" to different questions is unknown, so that models are required to infer true linear measures from the so-called raw scores. Uncertainty is increased by the fact that the very existence of the

[14] Statistical verbiage does not help. The word "regression" has a disagreeable flavour, although it defines perhaps the most popular procedure. By "regressing" the data you lose dimensions: points dispersed in a volume can be regressed to a plane; point dispersed in a plane can be regressed to a line. Means and medians themselves are a form of regression: you lose a line and obtain a point. The statistical language of uncertainty is no more reassuring: "standard error", "confidence limits", "hypothesis testing", all apparently point towards unavoidable approximation in attaining scientific "truth".

[15] It may look paradoxical that both chemistry, physics and biology on the one side, and epidemiology, on the other side, deal with abstract "mean" entities, be they molecules, cells or citizens. Individual differences are managed as sources of "variance" with respect to the primary object of study, using more or less the same statistical principles (see Tesio2019). By contrast, single persons are the object of clinical practice. Applying to persons the same experimental paradigms valid for electrons and elections may be highly misleading. Renouncing any statistical control, however, confines observations in the limbo of anecdotes.

"latent" variables is debatable: there is always the risk that we are cramming into the questionnaire items that just reflect the author's opinion (if not prejudice), so that the variable is imagined rather than discovered (items "form" rather than "reflect" the latent variable, according to the psychometric jargon). This ontological problem thus adds to the problem of quantitative estimation (Borsboom et al., 2003).

The trial design also requires a particular approach. The armamentarium typical for biological research (e.g., randomization to "true" vs "control/placebo" treatment; double blind treatment/assessment) does not fit the individual, customized, multifactorial and relational (in short: clinical) approach to illness.

All of the above peculiarities, however, represent technicalities and not ontologically irreconcilable differences between a truly quantitative/experimental and a purely qualitative/descriptive approach. These problems are well known to the world of the so-called "soft" or "human" sciences (from psychology to education and marketing), and elegant formal solutions have been proposed that allow investigators to apply scientific rigour to both the algebra (e.g., Tesio, 2003) and the trial design of "statistics" (Shadish, Cook, & Campbell, 2002) in these fields of human knowledge. Simply put, medicine is eager to snub the know-how of the "soft sciences", thus unnecessarily slowing down the possibility of scientific discovery in conditions where illness is just as relevant as (and often more accessible than) disease (Tesio, 2019). Measuring the effects of treatments in terms of both biological and behavioural changes, and in terms of number of patients changed rather than "mean" changes, may lead to more rational decisions (for an example, see Zamboni et al., 2018).

The illness standpoint on human suffering is prioritized by "alternative"/"complementary" forms of medicine. "Alternative" is the more adequate adjective for those approaches that do not accept the link to contemporary experimental method (Tesio, 2012b). In so doing, they free the treatment of illness from any formal obligations with respect to the treatment of disease. Not surprisingly, "alternative" medicine is rising in popularity, whereas "alternative" biology, physics, chemistry, if they exist at all, are much less popular.

6 Reconciling Individual Observations and Statistics

6.1 The Circle of Extraclinical and Clinical Knowledge

We can now better formulate the scope and limits of Boorse's definition of disease. On the one hand, there is an element of truth in his insistence on the use of statistics in the definition of the concepts of disease and health. As we have said, the dependence in principle of any general 'rules' on the consciousness of the individual patient represents a factor of uncertainty that cannot be completely eliminated in medicine, insofar as it is a science of humans. In some cases, cultural and social variability will be minimal (when highly effective methods and instruments are found to solve problems related to the functions of our bodies that are fundamental in most cultural circumstances). In other cases, however, for example in the psychiatric context, it will usually be very difficult to minimize uncertainty (especially in cases of psychological features that are easily exposed to the influence of culture and political power). But in all cases this uncertainty can be (to varying degrees) limited by applying the appropriate statistical tools as a methodical counterweight, in order to ascertain that the conjectured rules are true of at least a significant number of individual cases ('significant' with respect to our purposes).

But, on the other hand, the converse is also true, and it brings to light the limits of Boorse's naturalistic concept of disease. A definition of disease or health cannot be based on statistics and/or biology alone because it cannot be entirely separated from considerations of the overall behaviour, whether analytic or holistic, of the individuals to whom we wish to ascribe a state of health or disease. The main reason is simple. There is no human statistic that is not based on the interpretation of individual cases. The statistical test of the efficacy of a biological therapy cannot entirely take place without data obtained in the clinical setting by means of the clinical method, a vast domain in itself (Piantadosi, 2017), because one can ascertain only in the clinical encounter whether the patient's change occurred or not. For example, *pace* Grünbaum (1984), to ascertain whether a single case of mental illness can be classed as a case of paranoia, one must presuppose a definition and an operationalization of 'paranoia' that is at least implicitly clinical (Buzzoni, 1989). Defining not only a "disease", but also a "syndrome" or a "disorder" implies extracting regular associations of signs and symptoms from many single patients, in which different sets of signs and symptoms occur. Extraclinical tests can never entirely free themselves from clinical ones. There are no statistics on human subjects that can leave aside an interpretative understanding of the single cases on which they must ultimately be grounded. A pathologist or a physiologist cannot even begin their research without presupposing the existence and at least the partial reliability of clinical results (a reliability, as we have pointed out, that is also based on the law-like connections that medicine shares with the other human sciences).

We come here to a general conclusion: even though there are many ways to reduce the subjective-cultural variability of a patient's behaviour and to measure appropriately the quantity of their "latent" traits, one must bear in mind that all of this must be in the end tested by resorting to single cases again, all of which are mediated and partially obfuscated by the presence of consciousness. The search for objective relations clashes with difficulties that one can always try to minimize, but never wholly eliminate. If, in a sense, the reliability of clinical tests presupposes the reliability of extraclinical tests, in another sense, extraclinical tests presuppose the reliability of clinical ones. Or, to put it another way, the growing success of population- and evidence-based medicine (Greenhalgh et al., 2014) is epistemologically and methodologically well-grounded, but only if one does not neglect that this approach, in an important sense, in inherently dependent on clinically-based evidence.

6.2 Making Virtuous a Vicious Circle

Does the swinging from individual to statistical population means imply a vicious circle? The answer is no: not only from a hermeneutic, but also from an operational point of view, it is easy to acknowledge that there are not only vicious, but also virtuous circles, in which self-correcting or spiraliform procedures take place, with qualitatively new results, which each procedure alone could not produce. There is an obvious but important sense in which a reciprocal presupposing is not vicious. A circle is not operationally vicious if each of its elements (or actions)—even though similar in almost all respects—is different from at least one viewpoint. Such a difference makes it possible for each element to support the other (or others) in obtaining a novel effect, in a spiraliform progress. Everyday life offers plenty of examples. One cannot make a playing-card stand on a table in a slanted position, but this happens when one playing-card leans against another, and vice versa. For playing-cards (or the relative actions that are to be performed on them) are similar from many viewpoints, but are different at least with respect to their inclination at a particular time and place. In this sense, building a house of cards is a good example of procedures in which an action presupposes another action (and vice versa), and yet each action could not attain the intended result without the reciprocal one. The same organs (e.g., brain, muscles, sensory organs, etc.) can be studied from the perspective of their homeostatic biology within the body or, from the perspective of their capacity to provide the individual with active interactions with the external world (Tesio, 2020). In a similar way, even though similar in many respects (both approaches aim at the maintenance or restoration of health of concrete people, both recognise experimental evidence as the ultimate criterion for the reliability of their statements, etc.), the analytic-reductionist perspective (based on biochemical or physical research and statistical analysis) and the holistic-humanistic perspective (based on clinical methods) are different from various viewpoints: they operate in different contexts, on different variables, give different importance to generalized truths and particular events, etc. More precisely, if on the one hand the analytical-naturalistic perspective on health and malady must be subordinated (so to speak ethically and 'teleologically') to the holistic-humanistic one, on the other hand the latter must be filled (so to speak 'mechanistically') with intersubjectively controllable empirical contents.

There is therefore no vicious circle in using clinical results to formulate statistics or in using laboratory values to correct the subjective-cultural elements that are involved in the maintenance or restoration of health. As one does not need a hammer to forge a hammer, so considerations about the illness of the patient and the ways they respond to it need not be definitively established in order to reinforce extraclinical (biochemical and statistical) considerations, and vice versa. They must only claim a provisional degree of certainty, tentatively assumed to acquire a further, additional one.[16]

[16] It is of interest that biostatistics is revitalizing the study of single cases as an important source of knowledge (Gabler et al., 2011), taking up a long tradition in psychometrics (Tesio, 2012a). Both in biostatistics and psychometrics, knowledge acquired from population studies may be fed back,

6.3 Reconciling Singularities and Regularities

We have tried to show that the variability of individual clinical assessments is not unlimited. Even the clinical response of the individual patient can in reality never completely escape (and indeed is usually influenced by) factors exhibiting the same kind of regularity that is typical of the laws of the human sciences. The interpersonal-cultural dimension does not exert an arbitrary or unpredictable influence on the organic dimension of health and malady; on the contrary, its influence obeys regularities which (although different from the natural ones, because they are constantly modified by human beings) are sufficiently stable to make possible intersubjective statements about human health and all the concepts that are closely connected to it: prevention, diagnosis, prognosis, therapy, etc. Only by adding this piece does Wakefield's position (see above, Sect. 3) become sustainable. It is only by adding this element that we can understand in what sense the "social harm" in Wakefield's model of "harmful dysfunction" also possesses scientific and intersubjective value. Without this element, his position remains close to common sense, but it also remains hopelessly eclectic and insufficiently well-supported. The same holds for Canguilhem's position: if any statement about "the normal and the pathological" is irremediably value-laden, medicine will never be a science. We contend that it can be, if essentially the same scientific method is applied to the person, though with the methodical corrections (and the proper modesty) made necessary by the different variables observed (i.e., behavioural-holistic vs biological-analytic).

From this point of view, the problems connected with the possibility of an intersubjectively controllable definition of health and malady are, at least in principle, solved: they do not necessarily make such a definition impossible, since it is possible to exploit regularities in contexts that are more strongly conditioned by the material-organic base, as well as in those that are more sensitive to cultural influence. Exploiting these regularities can guarantee a certain intersubjective controllability. How far this intersubjective controllability extends cannot be decided a priori. It is decided by the researcher (including the clinician themself), i.e., the scientist at work, when they give us reproducible and therefore intersubjectively controllable results in fields that until then have not yet been included in scientific knowledge, precisely because they lacked this fundamental property.

This makes it possible to limit to a great extent the uncertainty that is natural to the human sciences, although we must be aware that it can never be completely eliminated. No matter how sophisticated control methods become—to limit ourselves to what is perhaps the sharpest example—the placebo effect will always interfere to some extent with the therapeutic efficacy of a treatment. In the same vein, we cannot exclude a priori that certain antibiotics may have different effects with respect to certain social groups (and a fortiori with respect to particular patients) than those

through proper algorithms, into single-case designs. This creates a virtuous circle bridging the gap between the 'scientific' status of population studies and the merely anecdotal evidence of individual observations, which are so relevant in clinical practice. The issue has been expanded upon for the specific case of physical medicine and rehabilitation medicine in (Tesio, 2019).

that have been shown in the populations where they have been used until now: which is because behavioural and social features are no less important than genetics. But just the awareness that medicine is a human science in the sense defended here may lead to concrete improvements for real medical practice, both clinical-individual and biomedical-statistical: what is called for is some awareness that we must establish from time to time if and how much the status of human science has or has not influenced our categorisations, diagnoses and therapies, and therefore in which direction we have to look for improvements both of biomedical knowledge and clinical practice.

7 Conclusion

In the literature of the last decades about the status of medicine, a new awareness has grown that an adequate notion of medical praxis requires an integrative position, which combines the analytic-reductionist with the normative-holistic perspectives on health and illness/disease. We have tried to show that it is possible to take an important step towards such an integrative view, if the relationship of unity and distinction between what we usually designate as 'illness' and 'disease' is correctly set up.

The demand for such an integrative view is a result not only of ethical considerations (which were not discussed here), but perhaps even more fundamentally for epistemological and methodological reasons, intimately connected with the status of medicine as a science of humans. In this connection, the key idea has been that the variability in the ways in which patients experience and respond to their illnesses—which is emphasized by the holistic-normative views about health and malady—is subject to regularities that may be investigated in an intersubjectively testable way. On the one hand, this variability—which originates in what distinguishes all human from natural sciences, that is, in the personal human consciousness on which all law-like relations of human and cultural reality depend in principle—undermines the scientificity, i.e., the intersubjective controllability, of medicine. On the other hand, however, this variability can be methodologically counterbalanced by the fact that the cultural domain of illness, like the objects of other human sciences, is subject to regularities that may be investigated in an intersubjectively testable way. Because these regularities add to, and may interfere with, those that biological research investigates, medicine cannot ignore its nature as a human science. It must be concerned not only with "diseases" and "illness" in the most common (for example in Boorse's) sense, but also with the ways in which patients as persons (as well as the members of their families or wider social environment) respond to malady and associated practical problems in everyday life. Insofar as such attitudes and expectations (as well as conventions, policies, social norms and roles constituted by those attitudes and expectations) influence the criteria of illness (and disease), they must be studied as part of the genuine subject-matter of medicine as a science.

Although there is a factor of uncertainty that cannot be completely eliminated in medicine, insofar as it is a science of humans, this uncertainty can be (to varying degrees) limited by applying appropriate statistical tools as a methodical counterweight, in order to ascertain that the conjectured rules are true of at least a significant number of individual cases ('significant' with respect to our purposes). In this connection, it is important to note that a definition of disease or health cannot be based on statistics and biology alone because it cannot be entirely separated from considerations of the overall behaviour, whether analytic or holistic, of the individuals to whom we wish to ascribe a state of health or disease. The main reason is that there is no human statistic that is not based on the interpretation of individual-clinical cases. The statistical test of the efficacy of a biological therapy cannot entirely take place without data obtained in the clinical setting by means of the clinical method, because one can ascertain only in the clinical encounter whether the patient's change occurred or not. Extraclinical tests can never entirely free themselves from clinical ones. Even though there are many ways to reduce the subjective-cultural variability of a patient's behaviours (which are always also interpersonal and social answers to their illness or disability), and to measure appropriately the quantity of their "latent" traits, one must bear in mind that all this must be tested, in the end, by resorting to single cases, where the blurring mediation by consciousness reappears. The search for objective relations clashes with limits that one can always shift away from, but never wholly eliminate. It follows that the reliability of clinical tests presupposes the reliability of extraclinical tests, and extraclinical tests presuppose the reliability of clinical ones (and vice-versa); or, said otherwise, population- and evidence-based medicine is epistemologically and methodologically well-grounded only in its synergy with clinically based evidence.

Funding This work was supported by the Italian Ministry of Education, University and Research through the PRIN 2017 program "The Manifest Image and the Scientific Image" prot. 2017ZNWW7F_004 and by the Italian Ministry of Health, "ricerca corrente", Istituto Auxologico Italiano-IRCCS, RESET project).

References

Aho, J., & Aho, K. (2008). *Body matters. A phenomenology of sickness, disease, and illness*. Rowman & Littlefield, Lanham.

American Psychiatric Association. (2013). *DSM-5. Diagnostic and statistical manual of mental disorders* (5th ed.).

Amrhein, V., Greenland, S., & Mcshane, B. (2019). Retire statistical significance. *Nature, 567*, 305–307. https://www.nature.com/articles/d41586-019-00857-9.

Ananth, M. (2008). *In defense of an evolutionary concept of health: Nature, norms, and human biology*. Ashgate.

Angeli, C. A., Reggie, E. V., Gerasimenko, Y. P., & Harkema, S. J. (2014). Altering spinal cord excitability enables voluntary movements after chronic complete paralysis in humans. *Brain, 137*, 1394–1409. https://doi.org/10.1093/brain/awu038

Benedetti, F. (2021). *Placebo effects* (3rd ed.). Oxford University Press. https://doi.org/10.1093/OSO/9780198843177.001.0001.
Bernard, C. (1865). Introduction à l'étude de La Médecine Expérimentale. Baillière, Paris. Engl. Transl. by H. C. Greene, *An introduction to the study of experimental medicine*. Schuman, 1949.
Boorse, C. (1977). Health as a theoretical concept. *Philosophy of Science, 44*, 542–573.
Boorse, C. (2014). A second rebuttal on health. *Journal of Medicine and Philosophy, 39*, 683–724.
Borsboom, D., Mellenbergh, G. J., & van Heerden, J. (2003). The theoretical status of latent variables. *Psychological Review, 110*, 203–219. https://doi.org/10.1037/0033-295X.110.2.203
Bucher, H. C., Guyatt, G. G., Cook, D. J., Holbrook, A., & McAlister, F. A. (1999). Users' guides to the medical literature: XIX. Applying clinical trial results A. How to use an article measuring the effect of an intervention on surrogate end points. *Journal of the American Medical Association, 282*, 771–778.
Buzzoni, M. (2003). On medicine as a human science. *Theoretical Medicine, 24*, 79–94.
Buzzoni, M. (2014). The agency theory of causality, anthropomorphism, and simultaneity. *International Studies in the Philosophy of Science, 28*, 375–395.
Buzzoni, M. (1989). *Operazionismo ed ermeneutica. Saggio sullo statuto epistemologico della psicoanalisi.* Angeli, Milano
Buzzoni, M. (2010). The unconscious as the object of the human sciences. In E. Agazzi & G. Di Bernardo (Eds.), *Relations between natural sciences and human sciences, special issue of Epistemologia. An Italian Journal for the Philosophy of Sciences* (pp. 227–246).
Byrne, M. L., O'Brien-Simpson, N. M., Mitchell, S. A., et al. (2015). Adolescent-onset depression: Are obesity and inflammation developmental mechanisms or outcomes? *Child Psychiatry and Human Development, 46*, 839–850. https://doi.org/10.1007/s10578-014-0524-9
Callahan, D. (1973). The WHO definition of 'Health'. *The Hastings Center Studies, 1*(3), 77–87. (The Concept of Health)
Canguilhem, G. (1972). *Le Normale et le pathologique*. PUF.
Cannon, W. B. (1942). Voodoo death. *American Anthropologist, 44*(NS), 169–181
Cartwright, S. A. (1851/2004). Report on the diseases and physical peculiarities of the Negro Race. In A. Caplan, J. Mccartney, & D. Sisti (Eds.), *Health, disease, and illness: Concepts in medicine* (pp. 28–39). Georgetown University Press.
Cassell, E. J. (1976). Illness and disease. *The Hastings Center Report, 6*(2), 27–37.
Chan, L. M. (1970). Foot binding in Chinese women and its psycho-social implications. *Canadian Pscyhiatric Association, 15*, 229–231. https://doi.org/10.1177/070674377001500218.
Charcot, J.-M. (1892). La Foi Qui Guérit. *Révue Hebdomadaire, 7*, 112–132.
Charon, R. (2006). *Narrative medicine. Honoring the stories of illness*. Oxford University Press.
Christian, P. (1989). *Anthropologische Medizin*. Springer.
Cole, T. R., & Carlin, N. (2009). The suffering of physicians. *Lancet, 374*, 1414–1415. https://doi.org/10.1016/S0140-6736(09)61851-1
Crews, D., Gillette, R., Miller-Crews, I., Gore, A. C., & Skinner, M. K. (2014). Nature, nurture and epigenetics. *Molecular and Cellular Endocrinology, 398*, 42–52. https://doi.org/10.1016/j.mce.2014.07.013
Crossnohere, N. L., Brundage, M., Calvert, M. J., King, M., Reeve, B. B., Thorner, E., Wu, A. W., & Snyder, C. (2020). International guidance on the selection of patient-reported outcome measures in clinical trials: A review. *Quality of Life Research*. https://doi.org/10.1007/s11136-020-02625-z
Demartini, B., D'Agostino, A., & Gambini, O. (2016). From conversion disorder (DSM-IV-TR) to functional neurological symptom disorder (DSM-5): When a label changes the perspective for the neurologist, the psychiatrist and the patient. *Journal of the Neurological Sciences, 360*, 55–56. https://doi.org/10.1016/j.jns.2015.11.026
Djulbegovic, B., Hozo, I., & Greenland, S. (2011). Uncertainty in clinical medicine. *Philosophy of Medicine, 16*, 259–356. https://doi.org/10.1016/B978-0-444-51787-6.50011-8
Engel, G. L. (1978). The biopsychosocial model and the education of health professionals. *Annals of the New York Academy of Sciences, 310*, 169–181.

Engel G. L. (1960). A unified concept of health and disease. *Perspectives in Biology and Medicine, 3*, 459–485

Engelhardt, H. T., Jr. (1975). The concepts of health and disease. In H. T. Engelhardt Jr. & S. F. Spicker (Eds.), *Evaluation and explanation in the biomedical sciences* (pp. 125–141). Reidel.

Engelhardt, H. T., Jr. (1996). *The foundations of bioethics* (2nd ed.). Oxford University Press.

Engelhardt, H. T. Jr. (1984). Clinical problems and the concept of disease. In L Nordenfelt & B. L. B. Ingemar (Eds.), *Health, disease, and causal explanation in medicine* (pp. 27–41). Reidel.

Fedoryka, K. (1997). Health as a normative concept: Towards a new conceptual framework. *The Journal of Medicine and Philosophy, 22*, 143–160.

Franchignoni, F., Tesio, L., Ottonello, M., & Benevolo, E. (1999). Life satisfaction index. *American Journal of Physical Medicine and Rehabilitation, 78*, 509–515.

Fulford, K. W. M. (2001). What is (mental) disease? An open letter to Christopher Boorse. *Journal of Medical Ethics, 27*, 80–85.

Gabler, N. B., Duan, N., Vohra, S., & Kravitz, R. L. (2011). N-of-1 Trials in the medical literature: A systematic review. *Medical Care, 49*, 761–768. https://doi.org/10.1097/MLR.0b013e318215d90d

Garson, J. (2016). *A critical overview of biological functions*. Springer.

Glick, S. M. (1981). Humanistic medicine in a modern age. *New England Journal of Medicine, 304*, 1036–1038.

Greenhalgh, T., Howick, J., Maskrey, N., Brassey, J., Burch, D., Burton, M., Chang, H., et al. (2014). Evidence based medicine: A movement in crisis? *BMJ, 348*, 1–7. https://doi.org/10.1136/bmj.g3725

Grmek, M. D. (1983). Les Maladies à l'aube de La Civilisation Occidentale, Payot, Paris, 1983. Engl. Transl. by M. and L. Mueller, *Diseases in the Ancient Greek World*. John Hopkins, 1989.

Grünbaum, A. (1984). *The foundations of psychoanalysis*. University of California Press.

Hahn, P. (2000). Wissenschaft und Wissenschaftlichkeit in der Medizin. In W. Pieringer & F. Ebner (Eds.), *Zur Philosophie der Medizin* (pp. 35–53). Springer.

Harris, J. C. (2005). A clinical lesson at the Salpêtrière. *Archives of General Psychiatry, 62*, 470–472. https://doi.org/10.1001/archpsyc.62.5.470

Hatzinger, M., Vöge, D., Stastny, M., Moll, F., & Sohn, M. (2012). Castrati singers—All for fame. *The Journal of Sexual Medicine, 9*, 2233–2237. https://doi.org/10.1111/j.1743-6109.2012.02844.x

Hofmann, B. (2017). Disease, illness, and sickness. In M. Solomon, J. R. Simon, & H. Kincaid (Eds.), *The Routledge companion to philosophy of medicine* (pp. 15–25).

Howick, J. (2016). The relativity of 'Placebos': Defending a modified version of Grünbaum's definition. *Synthese, 194*, 1363–1396.

Jablonka, E., & Lamb, M. J. (2014). Evolution in four dimensions. Revised Edition. *Genetic, epigenetic, behavioral, and symbolic variation in the history of life*. MIT Press.

Jaspers, K. (1919). *Psychologie der Weltanschauungen*. Springer.

Jaspers K. 1958. Der Arzt im technischen Zeitalter. Klinische Wochenschrift. 36: 1037–1043. Engl. Transl. by Arthur A. Grugan, The physician in the technological age. *Theoretical Medicine 10*, 251–267.

Kingma, E. (2014). Naturalism about health and disease: Adding nuance for progress. *Journal of Medicine and Philosophy, 39*, 590–608.

Kleinman, A. (1988). *Suffering, healing, and the human condition*. Basic Books.

Kumbhare, D., & Tesio, L. (2020). A theoretical framework to improve the construct for chronic pain disorders using fibromyalgia as an example. *Therapeutic Advances in Muskuloskeletal Diseases, 12*, 1–9. https://doi.org/10.1177/1759720X20966490

Laín Entralgo, P. (1970). The Therapy of the word in classical antiquity. Ed. and transl. by L. J. Rather & J. M. Sharp. Yale University Press.

Larkin, M., Eatough, V., & Osborn, M. (2011). Interpretative phenomenological analysis and embodied, active, situated cognition. *Theory & Psychology, 21*, 318–337.

Legrand, M. (1975). Hypothèses pour une histoire de la psychanalyse. *Dialectica, 29*, 189–207.

Maiani, G., & Sanavio, E. (1985). Semantics of pain in Italy: The Italian version of the McGill Pain Questionnaire. *Pain, 22*, 399–405. https://doi.org/10.1016/0304-3959(85)90045-4
Marcum, J. A. (2008). *An introductory philosophy of medicine*. Humanizing Modern Medicine.
Margolis, J. (1976). The concept of disease. *Journal of Medicine and Philosophy 1*, 238–255. Repr. In A. L. Caplan, H. Tristram Engelhardt, Jr. & J. J. McCartney (Eds.), *Concepts of health and disease. Interdisciplinary perspectives* (pp. 561–577). Addison-Wesley, 1981 (quotations are from this edition).
Mayo, D. G. (2018). *Statistical inference as severe testing*. Cambridge University Press.
Nordenfelt, L. (1997a). Holism reconsidered: a reply to Täljedahl]. *Scandinavian Journal of Social Medicine, 25*, 243–245.
Nordenfelt, L. (1997b). On holism and conceptual structures. *Scandinavian Journal of Social Medicine, 25*, 247–248.
Nordenfelt, L. (2013). Standard circumstances and vital goals: Comments on Venkata-puram's critique. *Bioethics, 27*, 280–284.
Nordenfelt, L. (1984). Comments on Pörn's 'an equilibrium model of health'. In L. Nordenfelt & B. I. B Lindhahl (Eds.), *Health, disease, and causal explanations in medicine* (pp. 11–13). Reidel.
Nordenfelt, L. (1986). Health and disease: Two philosophical perspectives. *Journal of Epidemiology and Community Health (1979–), 40*, 281–284
Nordenfelt, L. (1995). *On the nature of health. An action-theoretic approach* (Second revised edition) Kluwer.
Parsons, T. (1975). The sick role and the role of the physician reconsidered. *Milbank Memorial Fund Quarterly, 53*, 257–278.
Parsons, T. (1951). *The social system*. Collier-Macmillan, London (quotations are from the 1964 edition).
Piantadosi, S. (2017). *Clinical trials. A methodologic perspective* (3rd ed.) Hoboken, Wiley & Sons
Pieringer, W., & Fazekas, Ch. (2000). Grundzüge einer theoretischen Pathologie. In W. Pieringer & F. Ebner (Eds.), *Zur Philosophie der Medizin* (pp. 89–111). Springer.
Pörn, I. (1984). An Equilibrium Model of Health. In L. Nordenfelt & B. I. B. Lindahl (Eds.), *Health, disease, and causal explanations in medicine* (pp. 3–9). Springer.
Pörn, I. (1993). Health and adaptedness. *Theoretical Medicine, 14*, 295–303.
Prodinger, B., Cieza, A., Oberhauser, C., Bickenbach, J., Üstün, T. B., Chatterji, S., & Stucki, G. (2016). Toward the international classification of functioning, disability and health (ICF) rehabilitation set: A minimal generic set of domains for rehabilitation as a health strategy. *Archives of Physical Medicine and Rehabilitation, 97*, 875–884. https://doi.org/10.1016/j.apmr.2015.12.030
Reznek, L. (1987). *The nature of disease*. Routledge & Kegan Paul.
Sadegh-Zadeh, K. (2012). *Handbook of analytic philosophy of medicine*. Springer.
Saklayen, M. G., & Deshpande, N. W. (2016). Timeline of history of hypertension treatment. *Frontiers in Cardiovascular Medicine, 3*, 3. https://doi.org/10.3389/fcvm.2016.00003
Schaefer, H. (1976). Der Krankheitsbegriff. In M. Blohmke, C. von Ferber, K. P. Kisker, & H. Schaefer (Eds.), *Sozialmedizin in der Praxis* (pp. 15–31). Enke.
Shadish, W. R., Cook, T. D., & Campbell, D. T. (2002). *Experimental and Quasi-experimental designs for generalized casual inference*. Ed. by Houghton Mifflin Company. https://doi.org/10.1198/jasa.2005.s22.
Sharpe, M., & Greco, G. (2019). Chronic fatigue syndrome and an illness-focused approach to care: Controversy, morality and paradox. *Medical Humanities, 45*, 183–187.
Simon, J. (2007). Beyond naturalism and normativism: Reconceiving the 'disease' debate. *Philosophical Papers, 36*, 343–370.
Simpkin, A. L., & Schwartzstein, R. M. (2016). Tolerating uncertainty—The next medical revolution? *New England Journal of Medicine, 375*, 1713–1715. https://doi.org/10.1056/NEJMp1606402
Soto, A. M., & Sonnenschein, C. (2004). The somatic mutation theory of cancer: Growing problems with the paradigm? *BioEssays, 26*, 1097–1110.

Soto, A. M., & Sonnenschein, C. (2006). Emergentism by default: A view from the bench. *Synthese, 151*, 361–376.

Soto, A. M., & Sonnenschein, C. (2011). A testable replacement for the somatic mutation theory. *BioEssays, 33*, 332–340.

Steyer, R., Schmitt, M., & Michael, E. (1999). Latent state-trait theory and re-search in personality and individual differences. *European Journal of Personality, 13*, 389–408. https://doi.org/10.1002/(sici)1099-0984(199909/10)13:5%3c389::aid-per361%3e3.0.co;2-a

Svenaeus, F. (2000). *The Hermeneutics of medicine and the phenomenology of health: Steps towards a philosophy of medical practice.* Kluwer.

Svenaeus, F. (2014). The phenomenology of suffering in medicine and bioethics. *Theoretical Medicine and Bioethics, 35*, 407–420.

Taljedahl, I. B. (1997). Weak and strong holism. *Scandinavian Journal of Social Medicine, 25*, 67–69.

Taylor, R. S., & Elston, J. (2009). The use of surrogate outcomes in model-based cost-effectiveness analyses: A survey of UK health technology assessment reports. *Health Technology Assessment, 13*(8).

Tesio, L. (2003). Measuring behaviours and perceptions: Rasch analysis as a tool for rehabilitation research. *Journal of Rehabilitation Medicine, 35*, 105–115.

Tesio, L. (2012a). Outcome measurement in behavioural sciences: A view on how to shift attention from means to individuals and why. *International Journal of Rehabilitation Research, 35*, 1–12. https://doi.org/10.1097/MRR.0b013e32834fbe89.

Tesio, L. (2012b). Alternative medicines: Yes; alternatives to medicines. *American Journal of Physical Medicine and Rehabilitation, , 92*, 542–545. https://doi.org/10.1097/PHM.0b013e318282c937.

Tesio, L. (2019). 6.3B scientific background of physical and rehabilitation medicine: Specificity of a clinical science. *The Journal of the International Society of Physical and Rehabilitation Medicine, 2*, S113–S120. https://doi.org/10.4103/jisprm.jisprm_27_19

Tesio, L., & Buzzoni, M. (2021). *Medical Humanities, 47*(4), 507–512 https://doi.org/10.1136/medhum-2020-011873

Tesio, L., Granger, C. V., & Fiedler, R. C. (1997). A unidimensional pain/disability measure for low-back pain syndrome. *Pain, 69*, 269–278. https://doi.org/10.1016/S0304-3959(96)03268-X

Tesio, L., Granger, C. V., Perucca, L., Franchignoni, F. P., Battaglia, M. A., & Russell, C. F. (2002). The FIM™ instrument in the United States and Italy: A comparative study. *American Journal of Physical Medicine and Rehabilitation, 81*, 168–176.

Tesio, L., & Scarano, S. (2021). Ground Walking in Chronic Complete Spinal Cord Injury: Does Epidural Stimulation Allow "Awakening" of Corticospinal Circuits? A Wide-Ranging Epistemic Criticism. *American Journal of Physical Medicine & Rehabilitation. 100*(4), e43–e47. https://doi.org/10.1097/PHM.0000000000001520.

Tesio, L. (2020). Physical and rehabilitation medicine targets relational organs. *International Journal of Rehabilitation Research, 43*(3),193–194 https://doi.org/10.1097/mrr.0000000000000404.

Tonelli, M. R., & Ross Upshur, R. E. (2019). A philosophical approach to addressing uncertainty in medical education. *Academic Medicine, 94*, 507–511. https://doi.org/10.1097/ACM.0000000000002512

Toombs, S. K. (1992). *The meaning of illness a phenomenological account of the different perspectives of physician and patient*. Kluwer.

Toulmin, S. (1976). On the nature of physician's understanding. *The Journal of Medicine and Philosophy, 1*, 32–50.

Twaddle, A. (1994b). Disease, illness, sickness and health: A response to Nordenfelt. In: A. Twaddle & L. Nordenfelt (Eds.), *Disease, illness and sickness: Three central concepts in the theory of health* (pp. 37–53). Linkoping University Press

Twaddle, A. (1994a). Disease, illness and sickness revisited. In: Twaddle, A. & Nordenfelt, L. (Eds.), *Disease, illness and sickness: Three central concepts in the theory of health* (pp. 1–18). Linkoping University Press.

Twaddle, A. (1968). Influence and illness: Definitions and definers of illness behavior among older males in Providence, Rhode Island. Ph.D. dissertation, Brown University, Providence, RI.

Veatch R.M. 1973. The Medical Model: Its Nature & Problems. The Hastings Center Studies, 1(3) (The concept of health): 59–76. Repr. In A. L. Caplan, H. T. Engelhardt, Jr. J. J. McCartney (Eds.), Concepts of health and disease (pp. 523–544). Addison-Wesley.

Wakefield, J. C. (1992). The concept of mental disorder: On the boundary between Biological Facts and Social Values. *American Psychologist, 47*, 373–388.

Wakefield, J. C. (2007). The concept of mental disorder: Diagnostic implications of the harmful dysfunction analysis. *World Psychiatry, 6*, 149–156.

Wakefield, J. C. (2014). The biostatistical theory versus the harmful dysfunction analysis, part 1: Is part-dysfunction a sufficient condition for medical disorder? *Journal of Medicine and Philosophy, 39*, 648–682.

Wakefield, J. C. (2015). Social construction, biological design, and mental disorder. *Philosophy, Psychiatry & Psychology, 21*, 349–355.

Whitbeck, C. (1981c). What is diagnosis? Some critical reflections. *Metamedicine, 2*, 319–329

Whitbeck, C. (1981b). On the aims of medicine: Comments on 'philosophy of medicine as the source for medical ethics. *Metamedicine, 2*, 35–41

Whitbeck, C. (1981a). A theory of health. In A. L. Caplan, H. T. Engelhardt, Jr., & J. J. McCartney (eds.), *Concepts of health and disease* (pp. 611–626). Addison-Wesley

Wieland, W. (1995). Philosophische Aspekte des Krankheitsbegriffs. In V. Becker & H. Schipperges (Eds.), *Krankheitsbegriff Krankheitsforschung, Krankheitswesen* (pp. 59–76). Springer.

Wieland, W. (2002). The character and mission of the practical sciences, as exemplified by medicine. *Poiesis & Praxis, 1*, 123–134.

Williams, N. E. (2007). The factory model of disease. *The Monist, 90*, 555–584.

Willis, E. (1989). Complementary healers. In G. M. Lupton & J. M. Najman (Eds.), *Sociology of health and illness: Australian Readings* (pp. 259–279). MacMillan.

Wilshire, C., & Ward, T. (2020). Conceptualising illness and disease: Reflections on Sharpe and Greco. *Medical Humanities, 46*, 532–536.

World Health Organisation. (2001). *Overview of ICF components*. International Classification of Functioning, Disability and Health (ICF). https://doi.org/10.1055/s-0033-1338283.

Wyss, D. (1986). *Erkranktes Leben - Kranker Leib*. Vandenhoeck & Ruprecht.

Yaribeygi, H., Panahi, Y., Sahraei, H., Johnston, T. P., & Sahebkar, A. (2017). The impact of stress on body function: A review. *EXCLI Journal, 16*, 1057–1072.

Zamboni, P., Tesio, L., Galimberti, S., Massacesi, L., Salvi, F., D'Alessandro, R., Cenni, P. P., et al. (2018). Efficacy and safety of extracranial vein angioplasty in multiple sclerosis. *JAMA Neurology, 75*(1), 35–43. https://doi.org/10.1001/jamaneurol.2017.3825

Zaner, R. M. (1981). *The context of the self*. Ohio University Press.

About Context, Fiction, and Schizophrenia

Manuel Rebuschi

1 Introduction

The present paper explores the hypothesis that conversation context management for persons with schizophrenia not only represents a general challenge but could potentially serve as a basis for deepening our understanding of the pathology itself.[1] It builds on a joint interdisciplinary study of pathological conversations between schizophrenic patients and psychologists (Amblard et al., 2011; Musiol & Rebuschi, 2007, 2011; Rebuschi et al., 2013, 2014). Research for this study relied on empirical data and several stages of analyses. In a first stage, written transcriptions of recorded conversations were analyzed to identify relevant discontinuities or inconsistencies. These selected excerpts were then formally categorized using the semantic and pragmatic framework of the Segmented Discourse Representation Theory (SDRT) (Asher & Lascarides, 2003).

Linguistic analysis based on SDRT relies on a narrow notion of context which is a set of accessible possibilities to allow for the continued interpretation of a conversation. Other notions of contexts include general background, presuppositions and attitudes of the interlocutors, or even larger sets of features of the material and social environments (Kleiber, 2009). Individuals with schizophrenia, however, sometimes exhibit a tendency to use the narrowest level of context in a deviant manner. The challenge, therefore, is exploring whether this dysfunction can be generalized to other levels or aspects of conversation contexts to account for general features of the schizophrenic mind and reasoning. In this work, I propose approaching this specific issue by transposing certain insights from a philosophy of language approach to fiction.

[1] The paper is an updated and extended version of Rebuschi (2017), except for Sect. 2 which has been reduced.

M. Rebuschi (✉)
AHP-PReST (UMR 7117), University of Lorraine, Nancy, France
e-mail: manuel.rebuschi@univ-lorraine.fr

S. Wuppuluri and I. Stewart (eds.), *From Electrons to Elephants and Elections*,
The Frontiers Collection, https://doi.org/10.1007/978-3-030-92192-7_40

The paper is structured in three parts. In Sect. 2, I present the main results of the interdisciplinary study mentioned above on pathological conversations. In Sect. 3, I propose an informal hierarchy comprised of three levels of contexts that can be used for general conversation analysis. In Sect. 4, I describe the fourth level, *pragmatic context*. Applied in parallel with semantics of fictional discourse and its use of contexts, this new and final level may shed new light on our current understanding of schizophrenia.

2 Breaks in Pathological Conversations

Conversations with people diagnosed as schizophrenic can sometimes seem contradictory.[2] In our linguistic analyses we propose to take into account two viewpoints by building two representations of the same conversation, one for each interlocutor (Musiol & Rebuschi, 2007). The next issue is then to locate the apparent inconsistencies of the schizophrenic speaker. What consistently emerges from conversations with schizophrenic people is the *apparent* occurrence of contradictions, which manifest in frequent conversational *breaks* or discontinuities. In some cases, these breaks occur at times when the schizophrenic person gives the *appearance* of accepting (and generating) contradictory judgments. How can we account for this?

2.1 Locating Failures

Locating conversational breaks depends on perspective. From the ordinary speaker's point of view, failures are spontaneously placed in semantics and seen as mere contradictions in the semantic content of their utterances. However, postulating logicality for schizophrenics leads us to take into account their own viewpoints on conversation, where failures must be grounded elsewhere.

In line with other theorists, Sass (1994) denies that the *reality-testing deficit*, usually included among the symptoms of schizophrenia, adequately characterizes the thinking of schizophrenic subjects. The reality-testing deficit can act as an obstacle to reality that would result in the production of false and contradictory beliefs. Sass disputes this notion since it brings the deficit to the *content* of mental states, whereas we should consider that the defect involves the *states* themselves. To put it in other words, what is at stake is the mode of presentation of the content rather than the content itself. According to Sass, where we see *beliefs*, the schizophrenic entertains *states* of a type far less committed vis-à-vis reality. For Campbell (2001), these are *framework propositions*, a concept which can be approached through Searle's *background capacities* (Henriksen, 2013; Searle, 1992).

[2] This section is taken from a joint paper with Rebuschi et al. (2014). Those interested by formal or empirical details should refer to this article.

According to Sass, the mental attitude of schizophrenics is closed to that underlying philosophical solipsism as per Wittgenstein. Let us call *schizo-beliefs* such belief-like attitudes of schizophrenics. The idea is that, far from objectifying the contents of his or her schizo-beliefs, the subject would tend to subjectivize them, that is to say, deny them any genuine status. This is consistent with widespread questioning of perceptions implied by the radical skepticism of solipsism. The delusional thoughts and states resulting from perceptions are treated in the same fashion, as schizo-beliefs rather than beliefs.

How does playing on the container (the type of mental state) allow us to remove contradictions from the content? This is difficult to describe given that schizo-beliefs are characteristic of schizophrenic thinking. They belong to a type of mental state that non-schizophrenics do not have, which explains the difficulty in understanding (e.g. through empathy) schizophrenic subjects.

Pragmatic inconsistencies. The strategy we develop for the analysis of conversations is not based on a new classification of mental states.[3] However, we agree with Sass that the problem of schizophrenic thinking, as expressed in conversation, is not a problem of inconsistency of content. We postulate that schizophrenic speakers conform perfectly to classical logic. But we place the deviance of rationality in the *rules of language use*, i.e. in language conventions of rhetorical and pragmatic types. The way contents are structured (for a particular type of mental state in Sass's approach to delusion, by such pragmatic relations in the analysis of pathological conversations we develop) is an essential component of rationality. In short, rationality is not reducible to logicality.

Our empirical analyses focus on transcripts of one-on-one conversations between schizophrenic individuals (extraordinary subjects) and a psychologist (ordinary subject). These conversations lead to breaks which are perceived by the ordinary subjects, but not necessarily by the schizophrenic interlocutor causing them. The analysis involves constructing representations of conversations based on the formalism of SDRT. These representations include two levels: semantic representation (i.e. the content of the conversation), and pragmatic representation (i.e. a tree modeling the hierarchical structure of the speech acts that constitute the conversation).

To analyze pathological conversations, we propose the systematic construction of two simultaneous conversational representations, one for each interlocutor. On the schizophrenic's side, according to the principle of charity, there are no semantic contradictions. If there are failures, they occur at the pragmatic level, via violation of SDRT tree construction rules. The situation is different relative to the other side. In the conversations studied, the ordinary speaker is a psychologist asked to continue the interview. She does so in such a way as to repair the conversational structure after a break that would normally cause the interruption of a conversation. We then assume a corresponding postulate according to which the construction of a representation must respect pragmatic constraints. This option causes the appearance of inconsistencies at the semantic level.

[3] However we will come back to Sass's idea in Sect. 4.

The duality of conversational representations reflects the duality of views of the conversation: the schizophrenic subject seems to contradict ordinary subjects, so the conversation works, but the representation of the co-constructed world is inconsistent (in third-person terms). Conversely, because the schizophrenic person's conversational dysfunction is pragmatic in nature, their representation of the world built through the conversation does not suffer from this defect (first-person point of view).

2.2 Formalizing Empirical Data

SDRT combines two levels of analysis in order to account for the interpretive process at work in conversations: semantic content and conversational pragmatics. The first is analyzed via Segmented Discourse Representation Structures (SDRSs) inspired by the DRSs of Discourse Representation Theory (DRT), which is a syntactic construction updated by conversational flow (Kamp & Reyle, 1993). Conversation also implies pragmatic relations between speech acts, the complexity of which gives rise to a hierarchical structure first described in linguistics in the 1980s (Roulet et al., 1985). We formalize this structure with the rhetorical relations in SDRT.

The rhetorical structures of SDRT link the actions of speakers and are represented as hierarchical trees with vertical, horizontal and diagonal relations depending on the type under consideration. The tree structure (hierarchical ordering) encodes properties of the discourse and can be used to resolve semantic effects (e.g. prediction of attachment sites or resolution of anaphora). A discourse relation is viewed as a binary relation between speech acts. A narration is thus typically a horizontal relationship (same hierarchical level), as well as the answer to a question, while an elaboration is a vertical relationship (subordinated to what it elaborates on) and a question an oblique relationship (vertical, and thus subordinated, but also horizontal because requiring an answer).

The tree is updated throughout the discourse. Each subsequent intervention by one of the interlocutors is supposed to be related to the conversational representation already built. The structure offers general constraints affecting the attachment sites. The main constraint is the so-called right-frontier constraint, forcing the connection to the nodes located on the right side of the tree.

In order to formalize pathological conversations, we made two conjectures:

1. Schizophrenics are logically consistent; therefore, conversational breaks occur in the construction process of the pragmatic structure of conversations (i.e., on the rhetorical relationships between SDRSs); and
2. Underspecification (ambiguity) plays a central role in these failures, which could be summarized by the slogan: *A choice is never definitive!*

The first conjecture is nothing but the implementation of the principle of charity. The second conjecture, which is primarily based on empirical observation, is a heuristic for the location of remedial strategies in action by the ordinary speaker. When there is the appearance of discontinuity, the speaker uses the underspecified relations in order to maintain the pragmatic consistency of the dialogue. In other words, the flexibility of underspecified relations enables one to build a conversational representation under any circumstance.

The formalization of conversations is reduced to the elements relevant to our analysis, which means that we abandon anything that does not seem to play a role in explaining the breaks. The representation of semantic content is thus stripped to a minimum, namely to the conversational topic. Each conversational sequence is indeed built around a *theme* or *Question Under Discussion* (QUD), which is the main contextual element relevant to disambiguating the underspecified terms.[4] In ordinary conversations, the conversational theme usually changes after a conventional signal (e.g., "Well, but..." or "Moreover..."), or another form of closure of the current conversational sequence. Maintaining the ongoing QUD enables the continuation of a tree, while a QUD shift implies a rise through the tree to relate to a dominant node which corresponds to a sequence preceding the exchange.

In order to analyze pathological conversations, our formal framework provides the simultaneous construction of two representations, one for each speaker. For the schizophrenic person, the postulate of logicality means that the representation is devoid of contradictions at the semantic level. If there are breakdowns, they operate at the pragmatic level, with a departure from the rules for constructing the SDRT tree. For the ordinary speaker, we assume that the construction of the SDRT tree complies with the usual pragmatic rules. This option causes the appearance of inconsistencies on the semantic side. According to the "ordinary" subject, the schizophrenic speaker apparently contradicts the dialogical behavior so that the conversation works, but the representation of the co-constructed world is inconsistent. Conversely, when we assume dysfunction in a schizophrenic's management of pragmatic relations, the representation of the world built by the conversation does not suffer from this defect.

Analyses of excerpts led us to highlight two transgressions of the standard SDRT rules: *breaks of the right frontier* and *rises through the structure without any acceptable closing* (inconsistency of representation). For the second phenomenon, it is indeed common in corpora to identify items that are used both to close a part of the exchange and to open a new one. But the schizophrenic sometimes does not respect this dual effect and creates an incomplete representation that is not interpretable in a usual way.

Both kinds of transgressions correspond to a problem of management of the discursive context, which in the SDRT framework is constituted by the sites of attachment. Schizophrenic individuals patently shift the context where ordinary people would not, or where the latter would not do so without warning their interlocutor. This is basically a pragmatic inconsistency for it plays at the level of the whole structure of speech acts. Of course, for an interlocutor who sticks to the standard pragmatic rules, the schizophrenic way of conversing appears as semantically inconsistent.

[4] The fact that many ruptures take place around underspecified expressions reinforces our choice to represent the thematic element in the formalization.

3 Three Levels of Context

As mentioned in the previous section, empirical data collected from pathological conversations with schizophrenic people most often involve a play on ambiguities. Let us give a few examples[5]:

- the polysemy of "dead", alternating between literal and symbolic meanings;
- the polysemy of "lost", alternating between "x is lost" ("I'm lost") and "x lost y" ("I lost my friends");
- indexicals like "here" (underspecification between "here in the room" and "here in hospital");
- over-ambiguisation, with a play on syntax and neologism ("provocation" that became "pro-by-vocation").

This puts underspecification at the core of such pathological conversations. It appears that according to schizophrenic people, an interpretative choice is never definitive. There is always the possibility that it will occur again, without warning, whereas for an ordinary speaker it would usually be considered conclusive. Since underspecification resolution depends on contexts at a general level, we can hypothesize that pathological conversations expose a general problem of context management relative to interaction. Of course, this idea is not new and can be related to many analyses according to which schizophrenia is denoted by deficits in context processing (see e.g. Beaune, 2005; Cohen et al., 1999; Green et al., 2005; Phillips & Silverstein, 2003).

As is recalled by García-Carpintero (2015), there are two major notions of context in contemporary semantics. The first notion is due to Kaplan (1989) who defines contexts as sets of parameters such as agent, time, location, possible world, etc. This is a relatively narrow notion, and well-suited to provide semantic values for indexicals like "I", "now", "here", "actually", etc. Predelli (2005) is reluctant to qualify such sets as context and prefers talking about *index*. The second notion comes from Stalnaker (1978). Here context is viewed as "common ground" for conversation. Such a context corresponds to the common knowledge and presuppositions of the participants in communication, and it is construed as a set of propositions.

My proposal will roughly draw on the second notion of context. However, I will not consider context as an indefinite whole and I will not exclusively stay confined to Stalnaker's propositional account. I will introduce three kinds or levels of context, each of which are likely to play a role in resolving underspecification:

1. Discursive context (narrow)
2. Presuppositional context (intermediate)
3. Material and social context (large).

[5] For a detailed analysis of these examples, see Amblard et al. (2015). An overview of unusual language impairments displayed by patients with schizophrenia is presented in Convington et al. (2005).

This hierarchy of contexts[6] should be considered as a rough guide to make useful distinctions about what arises in discourse and conversation rather than a clear-cut theoretical proposal. As I will argue, a conversational break or an inconsistency may occur as soon as the context is not fully shared between interlocutors. Schizophrenia would then represent a specific case of this general scheme. In Sect. 4, I will introduce a fourth and final level which will play a specific role in my proposal.

3.1 Discursive Context (Narrow)

The first level under consideration is the most narrow type of context and is referred to as *discursive context*. This is the basic level considered in dynamic conversation analyses, and is that which we most often referred to in SDRT-based formalizations. It consists of the previous steps of a discourse available for further interpretation. Some of its characteristics include:

- discursive context is internal to discourse, i.e. it is a purely linguistic kind of context (sometimes called *cotext*);
- it involves all the dimensions of discourse, i.e. syntactic, semantic and pragmatic;
- it includes an interactional aspect: questions require answers, speech-turns, etc.;
- it contains judgments: representations have a semantic content, or at least a thematic one.

Within the SDRT theoretical framework, this level of context corresponds exactly to the active Segmented Discourse Representation Structure (SDRS), to which subsequent language acts can be attached. As such, it offers a number of attachment sites along the right frontier, which are made available to the speakers. This level of context is used to solve anaphora and, in some cases, polysemy and homonymy.

For polysemy and homonymy, underspecified expressions can be disambiguated by the discursive context in a straightforward manner when the context provides the QUD like in the following example:

(1) a "Max wanted to see the river. He went to the bank."
b "Max needed money. He went to the bank."

For anaphora resolution, a well-known example by Asher and Lascarides (2003) is the short (artificial) discourse:

(2) "Max had a lovely evening. He had a great meal. He ate salmon. He ate a lot of cheese. He found *it* really wonderful."

Here the anaphoric pronoun "it" cannot refer to the salmon, but could refer to the cheese, the meal, or the evening. Due to the right-frontier constraint, the sentence "He ate salmon" is indeed no longer available in the discursive context.

[6] Below, I will indifferently write about different *contexts* or different *levels of context*.

As was mentioned in Sect. 2, schizophrenic persons can have trouble handling this level of context. Basically, it's not uncommon for them to occasionally produce utterances that defy "normal" discursive constraints, which leads to an altered view of the overarching discursive context.

3.2 Presuppositional Context (Intermediate)

A broader level or kind of context more or less corresponds to what Lewis (1979) labeled as *conversational score* or, as mentioned earlier, to Stalnaker's *common ground*. This is the context that makes a language-game possible. Following Gárcia-Carpintero (2015) we consider this common ground as composed of several "set[s] of propositions to which speakers are committed", including not only knowledge or beliefs and propositions accepted for the purpose of conversation, but also shared questions that guide their inquiry as well as other possible directives:

- presuppositional context consists in a set of presuppositions shared by the interlocutors, with the beliefs or assumptions of the speakers about the world and about their interlocutors, their common knowledge, shared questions, etc.;
- it is made up of implicit propositional contents, either common or distributed, with respective shared commitments (either beliefs, or questions, etc.);
- it presupposes a common social context, or a common form of life in its background.

Presuppositional context is pivotal to mutual understanding between speakers. One of its functions is to constrain the thematic possibilities or QUDs. It enables the interlocutors to solve some cases of homonymy and polysemy.

Thematic constraints obviously appear in different social contexts; an academic will not share the same presuppositional context speaking with her colleagues as she would, for example, conversing with an interior decorator. As such, an utterance like the following:

(3) "I found a nice paper this morning"

would take on different meanings depending on the context.

To consider another example, speaking about Max's last holidays, and knowing that Max's vacation home is near a river or a lake, interlocutors would accommodate: "He went to the bank", as in example (1.a) above, even though no explicit link was introduced in previous discourse, thanks to the presuppositional context.

3.3 Material and Social Context (Large)

This is the largest kind or level of context. It corresponds to the general background of the conversation, and it is potentially unlimited (Kleiber, 2009). Unlike the two previous kinds of context, this one is neither linguistic nor propositional. It involves all the mundane features of the environment likely to play a role in determining the meaning of the conversation:

- material and social context can involve locations, immediate physical environment, interlocutors' mutual positions and postures, etc.;
- it includes the social context of the interaction (medical, professional...), the relationships between interlocutors (parents, patients and nursing staff...), cultural environment, etc.;
- it finally includes interlocutors relevant properties: children/adults, pathologies or not, cognitive skills, memory capacities, etc.

This general background context plays a fundamental role in discourse interpretation, both in indirect and direct manners. It partly determines the presuppositional context (VanDijk, 2006) and thus plays an indirect role in interpretation. For instance, conversations with children will involve a set of presuppositions distinctly different from those characteristic of conversations between adults. However, the material context obviously plays a direct role in that it offers the referents of deictics ("this", "that"...) and pure indexicals ("I", "you", "here"...). And some cases of homonymy or of polysemy can be solved using the same level. Ultimately, this is where we find the parameters constitutive of Kaplan's contexts or of Predelli's indices, even if material and social context cannot be reduced to this narrow and specific notion.[7]

4 Schizophrenia and Context Management

It could seem as though the three levels just presented suffice to account for conversations in general and for pathological conversations in particular. Analysis of empirical data shows us that schizophrenic speakers adhere to their own set of rules regarding the discursive context, effectively enabling them to shift the interpretation by discounting the usual rules. But there seems to be another factor at play at the presuppositional context level. Do the interlocutors share all of their presuppositions? Perhaps not. It would appear that this relies instead on a more basic level, a pragmatic one, that determines the presuppositional context. This pragmatic level may explain some breaks and gaps occurring in conversations with schizophrenics.

[7] However, as is noted by Recanati (2001, p. 86), the semantic value of "here" and "now" might not exclusively depend on the narrow context but also on the speaker's intended referent, thus on the wide context, where "a pragmatic process take[s] place … to determine *which* narrow context, among indefinitely many candidates compatible with the facts of the utterance, serves as argument to the character function".

4.1 A Fourth Level: Pragmatic Context (Intermediate)

Our fourth level determines which *kind of language-game* is played. It is neither linguistic, nor propositional, nor mundane, but *conventional*. As we will see, two interlocutors need not share the same pragmatic context and this gap can lead to subtle effects. Let us briefly present its characteristics:

- the pragmatic context can depend on the (material and) social context: a restaurant, a court, a theater... all these social contexts standardly trigger specific language-games;
- it determines the kind of language-game[8] that is at stake, hence the kind of speech acts produced in a conversation: assertions and questions, pretend assertions and pretend questions, avowals, notifications, etc.;
- it determines the kind of mental states that are expressed by the speakers' speech acts: beliefs, pretend beliefs or imaginings, commitments, etc.

The pragmatic context plays two crucial roles: it can shift the whole presuppositional context, and as a consequence, it can change the resolution of deictics, indexicals, and some cases of homonymy and polysemy.

What I call pragmatic contexts are used by philosophers of language to account for fiction (Predelli, 1998). Since pure indexicals like "I" or "here" are rigid designators, i.e. they denote the same object in every possible world, one cannot explain their use with a mere fictional modality à la Lewis (1978). The idea is then to conceive fictional discourse as being a *context-shifter*. This shifter makes the interlocutors (or the writer and their readers) switch from a serious pragmatic context to a fictional pragmatic context.

As expected, several features of interpretation change as a result of such a switch:

1. the value of indexicals: the fictional speaker referred to by the first-person pronoun "I" is no longer the actual speaker but is instead a fictional narrator;
2. the presuppositional context: the presuppositions, QUDs, etc. are now those shared by the fictional characters, and no longer those of the speaker and listeners;
3. the value of deictics, i.e. the interpretation of elements belonging to the material and social context: in the case of fictional context (like theater), one can stage surrounding objects to assign them with a new function.

Voltolini (2006) considers fiction as a paradigmatic case that can be translated to non-standard uses of indexicals: messages of the form "I am not here now" either written or recorded on answering machines; historical present as in "Now Hitler begins his invasion of USSR"; the use of a first-person pronoun "I" by the English translator of an Italian politician... all these examples can be construed as resembling fictional discourse: the context relevant for semantic interpretation is not the (Kaplanian) context of utterance "but a *fictional* context, i.e., a context which has at least one fictional parameter: a *pretended* agent, or a *pretended* space, or a *pretended* time, or a *pretended* world..." (Voltolini, 2006, p. 27)

[8] Or the kind of *interaction* in general, which may not be reduced to linguistic interaction.

Rather than committing to a uniform fictionalist account of extraordinary contexts, I will consider the possibility of distinct pragmatic contexts which might include both ordinary and fictional contexts, and maybe others. Pragmatic contexts roughly correspond to Goffman's *frames*, which can be primary or transformed (Goffman, 1974). Pragmatic contexts need not be shared by the interlocutors for the conversation to continue. Simple examples would be speakers lying to trusting listeners,[9] or some cases of fiction telling, e.g. telling stories about Santa Claus to trusting children. The conversation can continue naturally but the gap at the pragmatic level entails a gap at the presuppositional level. In both examples, the speaker can continue without expressing personal beliefs, but he or she is interpreted as doing so. Hence, according to the listeners the speaker's assertions broaden the set of common beliefs, which might not necessarily be shared by the speaker.

4.2 From Fiction to Schizophrenia

Switching from a serious to a fictional pragmatic context implies several fundamental shifts. Basically, there is a change of *language-game* which entails a shift in both speech acts and mental states:

- *speech acts*: assertions are replaced by pretend assertions, questions by pretend questions, etc.;
- *mental states* expressed by fictional speech acts: beliefs are replaced by pretend beliefs, i.e. *imaginings* or even mere *suppositions* (Whitt, 1985).

This roughly conforms to Walton's (1990) conception of fiction as *make-believe*. Within some fictional pragmatic context, i.e., following the rules of some conventional fictional language-game, apparent assertions are not real assertions. Instead, these are prescriptions to imagine fictional situations. Consequently, they will not fill the usual presuppositional context, at least not the main subsets of common knowledge and shared beliefs. Pretend assertions, rather, fill a subset of imagined propositions specific to the fictional work under consideration. Entering a fictional language-game thus entails a shifting of the presuppositional context.

At this point, a general hypothesis can be proposed using the notion of pragmatic context. Let us assume that there is a *schizophrenic language-game*. This is a specific language-game, which is neither serious, nor fictional, nor poetic, nor humorous – even if it might share several features with all of them. As with any language-game, the schizophrenic language-game involves specific speech acts and mental states:

- *speech acts*: assertions, or at least a part of them, are replaced by *schizo-assertions*, questions by *schizo-questions*, etc.;

[9] For a systematic comparison of pragmatic features of lying and fiction in a framework similar to that of the present paper, see Maier (2018).

- *mental states* expressed by schizophrenic speech acts: beliefs (expressed by assertions) are replaced by *schizo-beliefs*[10] (expressed by schizo-assertions).

We thus hypothesize that schizophrenic speakers introduce a specific *schizophrenic pragmatic context*, implying a schizophrenic language-game, and in particular schizo-assertions expressing schizo-beliefs. This new pragmatic context potentially allows for a switch to occur relative not only to the material and social context, but to the presuppositional context. In conversation processing, it potentially leads to a reinterpretation of indexicals, deictics, homonyms and polysemic terms.

Pathological conversations between ordinary and extraordinary schizophrenic speakers are thus expected to introduce a context gap. This gap is similar to that which might occur between the pragmatic context of an actor and the corresponding context of a passer-by with a hidden camera. This is what Goffman labeled *misframing* (Goffman, 1974). The two interlocutors in this case do not play the same language-game, and conversation breaks can occur as general conversational expectations will be mismatched.

4.3 More About Schizophrenic Pragmatic Contexts

The idea of a pragmatic context specific to schizophrenia could account for several phenomena linked to the pathology, like the patients' deviant phenomenology and possible hallucinations.[11] Before concluding this paper, we can briefly examine how fruitful this hypothesis might be.

Schizophrenic language-games can fairly easily be compared to fictional language-games in that imagination plays a certain role in both. In fiction, imaginings or pretend beliefs emerge from pretend assertions, the content of which updates a dedicated subset of the presuppositional context (common ground). On the other hand, according to some authors like Currie, delusion and hallucination can be compared to "the misidentification of the subject's imaginings as perceptions and beliefs" (Currie, 2000). If we were to follow this track, the description of the informational content of a particular delusion or hallucination experienced by an individual with schizophrenia would be mistakenly added to the main subset of common beliefs that make up the common ground – at least from his or her own first-person perspective. This impairment in context management need not be considered as a rational failure. It could instead stem from more basic mechanisms and consist of a metacognitive error at a subpersonal level, that might enhance a feeling of experiencing real informational states that are in fact internally generated (Dokic, 2016).

Nevertheless, it seems that delusional states associated with schizophrenia do not systematically result in *beliefs* about what is "perceived" by the subject. In many

[10] See Sect. 2.

[11] It is of course not claimed here that schizophrenia could in some way be *reduced* to an impairment in context processing. However, pragmatic contexts offer an insight into conversation analysis that might be translated to other phenomena.

cases, schizophrenic patients behave as though this part of their internal life were private and relatively independent from their public behavior. Several illustrative cases are reported by Sass:

> One of Eugen Bleuler's patients was well aware that the voices he heard originated in his own ears, for he compared these voices to the sound of the sea that can be heard by placing a shell to one's ear. It seems that something about the hallucinations and delusions of such patients sets their delusional worlds apart from the realm of normal, consensual reality. (Sass, 1994, p. 21)
>
> Schreber insists, however, that such beliefs – he calls them "my so-called delusions" – refer to a separate realm, one that does not motivate him to act and wherein the usual criteria of worldly proof or belief do not apply. (Sass, 1994, p. 31)

It would consequently be inconsistent to postulate that every so-called assertion made by someone with schizophrenia is genuine, nor that these represent true expressions of the speaker's true belief. Schizophrenic assertions, however, cannot be equated to fictional or pretend assertions either, because their content is not felt or experienced as freely created and imagined by the subject. Their status lies somewhere between serious and fictional assertions, and the contents correlatively update a specific subset of the given presuppositional context.

Another interesting point is the possible co-occurrence of two or more pragmatic contexts in the same conversation or interaction. In a conversation about a work of fiction, interlocutors can alternate and even mix fictional and serious pragmatic contexts, e.g. "On page 72, James Bond killed a Russian spy." Processing such conversations is not particularly difficult but does require differentiating fictional (James Bond killed a Russian spy) from non-fictional (that fictional fact is reported on page 72 of the book) narratives. A similar juxtaposition of contexts has been noted in schizophrenia:

> [A] feature of schizophrenic patients is what has been called their "double bookkeeping." ... Rather than mistaking the imaginary for the real, they often seem to live in two parallel but separate worlds: consensual reality and the realms of their hallucinations and delusions. A patient who claims that the doctors and nurses are trying to torture and poison her may nevertheless happily consume the food they give her; a patient who asserts that the people around him are phantoms of automatons still interacts with them as if they were reals. (Sass, 1994, p. 21)

An exceptional pictorial illustration of the phenomenon is offered by the painting "Le Pays des météores" (The Land of meteors) by Le Voyageur Français, a schizophrenic patient of Dr Auguste Marie at the beginning of 20th Century.[12] As can be clearly seen on the reproduction below, the painting is divided into two distinct parts. In the background there is a natural landscape, painted in a sensible and realistic manner; in the foreground, we can see a meteor depicted in a radically different style, with imaginary and almost dreamlike colors. This painting need not be reduced to the simple representation of an imaginary land. The co-occurrence of two styles

[12] This photograph was taken by the author during the exhibition "La Folie en tête" at the Maison de Victor Hugo in 2017. The painting belongs to the Collection de l'Art Brut (Lausanne, Switzerland), and I am grateful for their permission to reproduce it for this work.

in the same work strongly suggests that it should be interpreted as a representation of two parallel universes: the actual world with its natural river and trees, and the imaginary sphere of the painter with a fantastic meteor.

Looking at such a painting, one could say that "There are meteors and there are no meteors" so as to reflect the painter's thought. With our framework we can construe this utterance without assuming any semantic contradiction. The two parts of the sentence are not involved in the same speech-act or language game. Whereas "There are meteors" is relevant for the schizophrenic pragmatic context, "there are no meteors" pertains to the ordinary (serious) pragmatic context. The apparent contradiction can thus be resolved by the play on contexts.

"Le Pays des météores" (Le Voyageur français, 1902)
(Collection de l'Art Brut, Lausanne)

The juxtaposition or superposition of views means that schizophrenics can manipulate several pragmatic contexts in a rather subtle way. Their impairment in context processing would not come from an inability to play with differently valued contexts. It is more certainly related to the involuntary production of the specific pragmatic context I introduced, involving uncontrolled imaginings not recognized as such.

To sum up: by postulating a *schizophrenic pragmatic context*, it seems that not only conversational breaks but other phenomena such as delusion and hallucination can be accounted for without denying the subject's consistency.[13]

[13] Other strange cases like depersonalization (Chauvier, 2009) or Cotard's syndrome (Billon, 2015) could be accommodated with the same kind of analysis. These cases are paradoxical from a semantic

5 Conclusion

Psycholinguistic and formal approaches to understanding pathological conversations with schizophrenic people can account for conversational discontinuities. These are perceived as genuine contradictions expressed by ordinary interlocutors, which is the third-person viewpoint. Assuming the principle of charity, i.e. if we presuppose that the extraordinary, schizophrenic interlocutors are semantically consistent, effectively brings us closer to being able to reconstruct a first-person viewpoint.

Linguistic modeling sheds light on the major role underspecified phrases play in the context of such breaks. Schizophrenic speakers have the capacity to change the meaning of the same expression multiple times (at least twice, often more) in conversational situations where ordinary speakers would consider the situational context stabilized. This means that for schizophrenic speakers, the context that enables conversational interpretation is not as firmly grounded as it is for other speakers.

In this work, I introduced a distinction between four levels of context with the objective of gaining a more accurate view of the way schizophrenic interlocutors proceed in conversations. The idea of a schizophrenic pragmatic context provides an explanation of some specificities in terms of a specific language-game, which presents several similarities with fictional discourse. Moreover, it explains the occurrence of shifts between interlocutors in terms of context gaps. Finally, in terms of contexts and context gaps, applying this new pragmatic context may enable us to draw relevant connections between psycholinguistic analyses and phenomenological accounts of schizophrenia.

References

Amblard, M., Musiol, M., & Rebuschi. M. (2011). Une analyse basée sur la S-DRT pour la modélisation de dialogues pathologiques. In *Actes de la 18e conférence sur le Traitement Automatique des Langues Naturelles – TALN*.

Amblard, M., Musiol, M., & Rebuschi, M. (2015). L'interaction conversationnelle à l'épreuve du handicap schizophrénique. *Recherches sur la philosophie et le langage, 31*, 67–89.

Asher, N., & Lascarides, A. (2003). *Logics of conversation*. Cambridge University Press.

Billon, A. (2015). Why are we certain that we exist? *Philosophy and Phenomenological Research, 91*(3), 723–759.

Campbell, J. (2001). Rationality, meaning, and the analysis of delusion. *Philosophy, Psychiatry, & Psychology, 8*(2/3), 89–100.

Chauvier, S. (2009). Auto-cognition défaillante ou subjectivation déviante. *L'Évolution psychiatrique, 74*, 353–362.

Cohen, J. D., Barch, D. M., Carter, C., & Servan-Schreiber, D. (1999). Context-processing deficits in schizophrenia: Converging evidence from three theoretically motivated cognitive tasks. *Journal of Abnormal Psychology, 108*(1), 120–133.

point of view since the first-person pronoun is still expected to refer even though the speaker believes he or she does not exist. However, this semantic puzzle is not far from those generated by anti-substantialist accounts of the subject, and can be solved using context shifters (Rebuschi, 2011).

Covington, M. A., He, C., Brown, C., Naçi, L., McClain, J. T., Fjordbak, B. S., Semple, J., & Brown, J. (2005). Schizophrenia and the structure of language: The linguist's view. *Schizophrenia Research, 77*(1), 85–98.
Currie, G. (2000). Imagination, delusion and hallucination. *Mind & Language, 15*(1), 168–183.
Dokic, J. (2016). Toward a unified account of hallucinations. *Journal of Consciousness Studies, 23*(7–8), 82–99.
García-Carpintero, M. (2015). Contexts as shared commitments. *Frontiers in Psychology, 6*, 1–13.
Goffman, E. (1974). *Frame analysis: An essay on the organization of experience*. Harper and Row.
Green, M. J., Uhlhaas, P. J., & Coltheart, M. (2005). Context processing and social cognition in schizophrenia. *Current Psychiatry Reviews, 1*(1), 11–22.
Henriksen, M. G. (2013). On incomprehensibility in schizophrenia. *Phenomenology and the Cognitive Sciences, 12*(1), 105–129.
Kamp, H., & Reyle, U. (1993). *From discourse to logic: Introduction to model theoretic semantics of natural language, formal logic and discourse representation theory. Studies in linguistics and philosophy*. Kluwer Academic.
Kaplan, D. (1989). Demonstratives. In J. Almog, J. Perry, & H. Wettstein (Eds.), *Themes from Kaplan* (pp. 481–563). Oxford University Press.
Kleiber, G. (2009). D'un contexte à l'autre: aspects et dimensions du contexte. *L'Information grammaticale, 123*, 17–32.
Leroy, F., & Beaune, D. (2005). Langage et schizophrénie: intention, contexte et pseudo-concepts. *Bulletin de psychologie, 479*, 567–577.
Lewis, D. (1978). Truth in fiction. *American Philosophical Quarterly, 15*(1), 37–46.
Lewis, D. (1979). Scorekeeping in a language game. *Journal of Philosophical Logic, 8*, 338–359.
Maier, E. (2018). Lying and fiction. In J. Meibauer (Ed.). Oxford Handbook of Lying, OUP.
Musiol, M., & Rebuschi, M. (2007). La rationalité de l'incohérence en conversation schizophrène (analyse pragmatique conversationnelle et sémantique formelle). *Psychologie française, 52*(2), 137–169.
Musiol, M., & Rebuschi, M. (2011). Toward a two-step formalization of verbal interaction in schizophrenia: A case study. In A. Trognon, M. Batt, J. Caelen, & D. Vernant (Eds.), *Logical properties of dialogue* (pp. 187–225). PUN.
Phillips, W. A., & Silverstein, S. M. (2003). Convergence of biological and psychological perspectives on cognitive coordination in schizophrenia. *Behavioral and Brain Sciences, 26*(1), 65–138.
Predelli, S. (1998). I am not here now. *Analysis, 58*, 107–115.
Predelli, S. (2005). *Contexts. meaning, truth, and the use of language*. Clarendon Press.
Rebuschi, M. (2011). Le cogito sans engagement. *Igitur, 3*(2), 1–25.
Rebuschi, M. (2017). Schizophrenic conversations and context shifting. In P. Brézillon et al. (Eds.), *CONTEXT 2017, Lecture Notes in Computer Science* (Vol. 10257, pp. 708–721).
Rebuschi, M., Amblard, M., & Musiol, M. (2013). Schizophrénie, logicité et compréhension en première personne. *L'Evolution psychiatrique, 78*(1), 127–141.
Rebuschi, M., Amblard, M., & Musiol, M. (2014). Using SDRT to analyze pathological conversations. Logicality, rationality and pragmatic deviances. In M. Rebuschi et al. (Eds.), *Interdisciplinary works in logic, epistemology, psychology and linguistics. Dialogue, rationality, formalism.* (pp. 343–368). Springer.
Recanati, F. (2001). What is said. *Synthese, 128*, 75–91.
Roulet, E., Auchlin, A., Schelling, M., Moeschler, J., & Rubattel, C. (1985). *L'articulation du discours en français contemporain*. Peter Lang.
Sass, L. A. (1994). *The paradoxes of delusion: Wittgenstein, Schreber, and the schizophrenic mind.* Cornell.
Searle, J. R. (1992). *The rediscovery of the mind*. The MIT Press.
Stalnaker, R. (1978). Assertion. In P. Cole (Ed.), *Syntax and semantics* (Vol. 9, pp. 315–332). Academic Press.
Van Dijk, T. A. (2006). Discourse, context and cognition. *Discourse Studies, 8*(1), 159–177.
Voltolini, A. (2006). Fiction as a base of interpretation contexts. *Synthese, 153*, 23–47.

Walton, K. L. (1990). *Mimesis as make-believe. On the foundations of the representational arts*. Harvard University Press.

Whitt, L. A. (1985). Fictional contexts and referential opacity. *Canadian Journal of Philosophy, 15–2*, 327–338.

Humanities and Social Sciences

On the Explanation of Social and Societal Facts

Friedel Weinert

1 Introduction

A long-standing dispute has been raging in the Social Sciences about the level at which social science explanations should be sought. Traditionally two positions have opposed each other. **Methodological Individualism** holds that the explanation of social phenomena must ultimately be reducible to the level of the actions and motives of individual social actors in a given social setting. According to this view only individual social agents exist. Its aim is to explain *social* facts by reference to individual facts. **Methodological Holism** argues that at least some macro-phenomena can only be explained by reference to *societal* facts. Such societal facts (groups, institutions, states, traditions) or societal wholes have a presence over and above a summation of the activities of individual social actors. Societal wholes do not exist like tables and chairs but they constrain individual agents in social life. Over the years many different versions of these positions have emerged but the above are the standard characterizations. These opposite positions are qualified as 'methodological' because they characterize explanatory rather than ontological claims. Everyone is in agreement that individuals exist but not whether societies, states, institutions 'exist' over and above their individual members. Individuals have motivations, reasons and wills but it is questionable whether societal wholes do. The emphasis on *methodological* individualism/holism sidesteps this metaphysical question. The issue is about explanation: can collective facts be sufficiently explained by reducing them to individual facts about agents?

Such concerns are not the prerogative of arcane philosophical preoccupations. They are not confined to the ivory tower of philosophical contemplation. For explanations of social and societal facts must be appropriate in order to be successful. An explanation is inappropriate if it fails to explain the explanandum. It makes an

F. Weinert (✉)
University of Bradford, Bradford BD7 1DP, West Yorkshire, UK
e-mail: f.weinert@bradford.ac.uk

S. Wuppuluri and I. Stewart (eds.), *From Electrons to Elephants and Elections*, The Frontiers Collection, https://doi.org/10.1007/978-3-030-92192-7_41

explanatory difference if explanations refer to reducible or irreducible facts. On the other hand, entrenched positions may be unhelpful. In order to overcome the exclusive dichotomy between methodological individualism and holism a pragmatic approach is more appropriate. It is the social science problems at hand which should determine the level at which solutions are pitched. A consideration of the opposing positions will show that in many cases successful explanations must remain on the macro-level. Explanations in the social sciences employ generalizing models. Therein lies a certain affinity between the natural and the social sciences.

2 Traditional Methodological Individualism

Methodological individualism insists that social facts must be reduced to the micro-level; to the level of individual social agents. But a closer analysis shows that it is neither invariably possible nor desirable to explain social facts in terms of individual facts. This impossibility is already built into the assumptions of traditional methodological individualism. That is, the proponents of methodological individualism themselves make reference to elements, which do not strictly fall within the remit of their methodology. Consider, for instance, the contributions of Hayek and Watkins, both of whom were close to Popper, who also defended methodological individualism. Watkins (1952b: 186) characterizes methodological individualism as the precept that 'social processes should be explained by being deduced from principles governing the behaviour of participating individuals and from analyses of their situations and *not* from superindividual "holistic" sociological laws.' As the last part of his sentence indicates he associates methodological holism with the belief that (a) 'macroscopic laws govern societal wholes' and (b) that its 'components can be deduced from the functions of the components within the whole.' Let us leave aside the assertion that methodological holism requires societal laws. It is questionable whether societal laws even exist. Patterns of behavior or trends do exist but are not to be confused with natural laws. Social trends can be modified or reversed, but natural laws cannot. Human beings are incapable of influencing the orbits of the planets but can change their customs, norms and traditions. Watkins (1955: 58) holds that social phenomena are generated by individuals and must be explained individualistically. Furthermore, understanding social structures can only be derived from more empirical beliefs about concrete individuals. Unfortunately, Watkins commits a *non-sequitur*. Even if it is admitted that 'social events are brought about by people' it does not follow that 'they must be explained in terms of people.' We must not confuse methodological individualism with ontological individualism or methodological holism with ontological holism. There is no need to dispute the obvious fact that 'social events are brought about by people' and that 'social structures are created by people.' That is, we can concede that 'societal wholes' do not lead an independent Platonic existence, irrespective of the activities of social agents. But as Durkheim (1982: Ch. V.2) pointed out, societal systems exercise a constraining function on social agents. Social agents are necessary but not sufficient for this function. The question is one of methodology, not ontology. Are the actions and interactions of social agents alone sufficient to explain social events, processes and societal structures? The methodological individualist affirms that rock-bottom explanations are sufficient, a claim which the methodological holist denies.

In a further publication Watkins (1952a) discusses Weber's methodology of ideal types. Max Weber was also committed to methodological individualism. According to Watkins, Weber held two conceptions of ideal types in succession. In his early work he employed holistic ideal types which depict salient features of historically complex states or situations. But in his monumental work *Wirtschaft und Gesellschaft* (1921; *Economy and Society* 1978), a key text of modern sociology, Weber introduced individualistic ideal types. They model social agents as hypothetical rational actors in idealized social settings. Both ideal types are models of social situations, which make abstraction from the messiness of real life and emphasize the typicality in social life. Holistic ideal types may depict a model of feudalism, capitalism or democracy, highlighting typical features of such societal institutions. An individualistic ideal type may depict a model of a typical consumer, a typical conservative voter or a typical middle-class male. But what is typical in one context may not be typical in another. That is, the consideration of types may require the inclusion of national or local characteristics. A typical conservative voter is not the same in every country, as religious affiliations differ. However, there will be similarities, like an appreciation of tradition and respect for existing institutions. Watkins adds (1952a: 29 [fn2]) that an explanation may be in terms of 'typical' dispositions or in terms of specific individuals.[1] Assertions, like the latter, are reminiscent of J. S. Mill's psychologistic individualism. For Mill the 'laws of the phenomena of society are, and can be, nothing but the laws of the actions and passions of human beings united in the social state.' (Quoted in Hollis, 1994: 10) Hence he derives the phenomena of society from the phenomena of human nature. But the latter suggestion would reduce the social sciences to psychology, which was not Weber's intention. For Weber ideal types are conceptual models, which describe salient features of the target of the description (Weinert, 1996). As models they necessarily idealize and abstract, as all models do in the natural and social sciences. Weber constructed holistic ideal types (feudalism, capitalism), which ignore individual particularities. Even his construction of the Protestant Ethic displays general features. But Watkins (1952a: 42–3) is right that individual ideal types are reconstructed from typically significant dispositions and typical situations. Thus despite Watkins' best efforts he has not shown that explanations of social phenomena are reducible to psychological terms, since ideal typical explanations involve models which focus on typical features. In his analysis of Weber's causal model Ringer comes to a similar conclusion. Weber's ideal types can causally explain holistic developments in history.

> Weber's line of analysis allows him to move from methodological individualism to the study of complex social interactions and organizations. He can stipulate that a state "exists" or

---

[1] Watkins (1952a: 36–40) accepts that psychology makes generalizations but adds that general knowledge of human behavior must be supplemented by 'knowledge of peculiar personalities', which, however, refer to 'public and institutional dispositions.'.

> "has ceased to exist," and that is surely to make a statement about a structured collectivity. (Ringer, 2002: 177)

We encounter a similar vacillation between an avowal of methodological individualism and the inability to carry it through in other proponents of this methodology. Both Hayek's and Popper's espousal of methodological individualism occurred against the backdrop of their experience of totalitarianism in the 1930s. Like Watkins, they associate holistic thinking with an overpowering state, which controls individual lives. Hayek characterizes 'collectivism' as the conscious direction of all forces of society and warns that methodological collectivism leads to political collectivism (Hayek, 1964: 91–2). For Hayek methodological individualism also starts from individuals and their actions (1964: Pt. I, Ch. IV) but he adds that the social sciences must infer complex phenomena from their constitutive parts. Hayek does not fall into the trap of psychologism, which always threatens pure forms of methodological individualism. Hayek affirms that 'the aim of the social sciences is not to explain conscious action'. 'Individual beliefs and attributes' are not the object of social science explanations. They constitute the elements from which we 'build up the *structure* of possible relationships between individuals.' (Hayek, 1964: 39, emphasis added) Hayek follows Weber in holding that the social sciences construct models of social phenomena. Hayek returns repeatedly to this emphasis on societal structures, at which the social sciences aim. Societal wholes are not observable, he states, they are constructions of the mind.

> The social sciences do not deal with 'given' wholes but their task is to *constitute* these wholes by constructing models from the familiar elements - models which reproduce the structure of relationships between some of the many phenomena which we always simultaneously observe in real life. (Hayek, 1964: 56, 68–73; italics in original)

Hayek thus rejects ontological holism, since societal wholes do not 'exist'. Like Popper, he recommends the use of methodological individualism as a methodology of the social sciences. But in order to avoid psychologism and affirm the autonomy of the social from the natural sciences, he recommends the use of abstracting and idealizing models in the social sciences: '*all* thought must be to some degree abstract.' (Hayek, 1964: 68; italics in original)

3 Ideal Types

We have seen that traditional methodological individualism is motivated by a fear of totalitarianism, which it associates with holism. Hayek does not clearly distinguish political collectivism from methodological holism. He does not mention Weber but he endorses the use of models, which capture the 'structure' of social relationships. Methodological individualism, despite its protestations, cannot fulfill its promise to explain social phenomena solely in terms of individuals—'their properties, goals and beliefs'. (Elster, 1982: 453) Even such an attempt will involve abstractions and idealizations, leading to 'individualistic ideal types'. But if individualism arose out a concern for the individual, now that totalitarianism has receded in the Western World, what does methodological individualism mean today? The term was introduced by J. Schumpeter (1908) but elaborated by Max Weber. (Heath, 2020: §1) Weber's methodology of ideal types is particularly important in this connection because ideal types combine understanding and explanation. Weber sought a synthesis between the French School and the German School. According to the French School, as defended in the writings of Saint Simon and A. Comte, sociology had to become 'social physics'. It was modeled on the natural sciences and its aim was a causal explanation of social phenomena. The German School, as defended by W. Dilthey and H. Rickert, opposed this scientistic view of the social sciences. According to Dilthey the methodology of the social sciences was analogous to the discipline of history, whose main business was the understanding of the past. Hayek also opposed 'scientism' in the social sciences: the task of the social sciences was the understanding (*Verstehen*) of social life. Weber was influenced by the German school. He was committed to an interpretative sociology whose methodology was based on the notion of understanding. He stated that 'social collectivities result from particular acts of individuals, who alone are meaning-using agents. Only action is subjectively understandable.' (Quoted in Heath, 2020: §1) Yet Weber did not reject the call for causal explanations. He defines sociology as a

> ...science concerning itself with the interpretive understanding of social action and thereby with a causal explanation of its course and consequences. (Weber, 1978: 4)

For Weber, not every type of action is social in character. Social character is 'confined to cases where the actor's behaviour is meaningfully oriented to that of others.' For instance, the collision of two cyclists or people's simultaneous opening of umbrellas does not constitute social action. (Weber *Economy & Society* 1978: 23) But the reaction of an individual to other individuals in a social setting constitutes social action. For Weber social action must both be understandable and explainable. He dubs this combination of understanding and explanation *explanatory understanding*. In addition, Weber proposes a notion of *adequate causation*, which he considers to be an appropriate view of causation in the social sciences. It is also based on his ideal-type methodology.

In order to adhere to explanatory understanding Weber relies on 'models' of social action. Ideal types provide such models. To illustrate how explanatory understanding works, consider the case of a social scientist who wishes to explain why members of particular social-economic groups tend to vote for conservative parties in a particular country. The first task of the social scientist is to *understand* the political system of the country under investigation, including the make-up of the electorate, political constituencies, political programmes, socio-economic conditions, and prevailing traditions in this country. As the social scientist is not concerned with a particular voter, *A*, but with typical conservative voters in that country (or constituency) a model of such voters, *A'*, must be constructed. Such a model will depend on such criteria as age, education, gender, income but at the exclusion of idiosyncratic characteristics which are irrelevant for understanding the model voter's tendency to cast her/his vote for the conservative party. Although it is a model, it must capture important features of the typical conservative voter. In this respect the model will differ from country to country. In some countries the model may have to include religious affiliations, republican traditions; in other countries support for the monarchy may need to be added. Once such a model is in place, it will adequately explain why *A'*, a typical voter, will typically vote conservative. It is not a statement about individual *A*, who may have further reasons to vote for a conservative candidate, such as personal sympathy or loyalty to the party. But the model explains why any individual *A*, who shares characteristics with *A'*, is likely to vote conservative. It is a statement about the most likely behaviour from the point of view of the social scientist, and not a prediction. To construct the model *A'* the social scientist needs understanding of typical behaviour which in turn then provides an adequate causal explanation of the model voter's voting preferences. The model constructs expected, not predicted behaviour. Built into Weber's ideal type methodology is a notion of testing. Like any other model an ideal type must 'fit' reality as closely as possible. The social scientist may misconstrue typical *A'* as a result of misunderstanding the political situation in the chosen research area. If this occurs and typical *A'* fails to correspond to real *A*, the model must be adjusted. Weber's explanatory understanding therefore avoids psychologism, because it requires understanding of social situations rather than individual psychology.

But Weber did not only believe that the social sciences could achieve explanatory understanding of social action. He also believed that the social sciences could provide causal understanding.

A causal model seeks to relate some occurrence in the social world, which is regarded as an effect, to prior causal conditions. Looking at Weber's own work, his attempt to explain the emergence of capitalism in the West as a result of the adoption of puritan lifestyles is a striking example of causal analysis in the social world. But social scientists also seek to explain, say, the origin of the Slave Trade, the outbreak of World War Two, juvenile delinquency and differential educational performances. The aim in each case is to isolate as far as possible the actual determinant factors, which are likely to have caused some event in history or the social world. Social systems, however, are open-ended. The social scientist is faced with a cluster of potentially determining factors, which could be possible causal conditions. Out of the complex of potential determining factors, the social scientist must distil a complex of causal relations, which 'should culminate in a synthesis of the "real" causal complex'. (Weber 1905/1949: 173) Weber speaks of 'adequate causation' when the social science model meets several conditions: (a) the social scientist has isolated a number of conditions which are regarded as statistically relevant to the effect in question; (b) the reconstruction of the social or historical event, on the part of the social scientist, probably isolates the 'likely cause of an actual historical event or events of that type.' The ideal type model of the causal sequence of social events therefore depicts an objective possibility. That is, it is objectively possible, even likely, that the isolated conditions are causally responsible for the occurrence of the event. The model of the social scientist, which has some claim to probability, provides the most adequate causal conditions which are likely to have brought about the social event in question. How can a social scientist be relatively certain that a proposed causal model of, say, the outbreak of World War Two, captures the most adequate conditions, which are most likely to have brought about this event? Weber insists that 'it is possible to determine, even in the field of history, with a certain degree of certainty which conditions are more likely to bring about an effect than others.' (Weber 1905/1949: 183) The way to achieve this aim is to submit the ideal typical model of some causal sequence to factual knowledge of a historical or social event. Thus Weber tests the model against reality. In this manner Weber hopes to throw light on the 'historical significance' of the actual determinant factors in the emergence of some historical event. It is well known that historians and sociologists disagree about the relevant factors, which can be held responsible for some event in history or society. But certain factors will be so improbable that they can be omitted from the causal account. It is implausible, for instance, that the eruption of a distant volcano will have had an effect on the outbreak of World War II. On the other hand, new empirical data regarding the outbreak of World War II may well credit some factors as the relevant ones and discredit others as irrelevant. Thus both explanatory understanding and adequate causation avoid psychologism.

4 Situational Analysis

Amongst methodological individualists Karl Popper stands out for persistently warning social scientists of the dangers of psychologism. Popper praises Marx for having shown the autonomy of the social sciences, and sociology, from psychology. For Popper (1960: §29, p. 142)

> …the social sciences are comparatively independent of psychological assumptions, and (…) psychology can be treated, not as the basis of all social sciences, but as one social science among others.

Social relations are not reducible to psychological relations, since social or societal facts refer to 'large-scale features of society.' (Papineau, 1978: 6–9) Such facts concern the 'forms of organization in society.' (Mandelbaum, in Gardiner, 1959: 478; Gellner, in Gardiner, 1959: 501) For if societal facts were reducible, the social sciences would be in the business of describing the psychological make-up of individuals. But such psychological make-ups are either not generalizable or, if generalizable, they again make use of models. This is also Hayek's line of argument. Reliance on psychology would force the social sciences to revert to vague notions like 'human nature' about which little agreement exists. Popper characterizes the 'fundamental task' of the social sciences as the description of 'man's social environment'. A social science oriented towards objective understanding can be developed independently of all subjective or psychological ideas. (Popper, 1976: 101–2) The main task of the social sciences, according to a modern view, lies in the explanation of the behaviour of social systems and phenomena. (Coleman, 1990: 2).

In order to avert psychologism Popper relies on his method of *situational analysis* or *situational logic*. It is a reconstruction of the social situation in terms of the rationality principle. Popper implicitly uses one of Weber's ideal types: purposive-rational action, to promote his version of methodological individualism. Popper writes that

> …in most situations, if not in all, there is an element of *rationality*. Admittedly, human beings hardly ever act quite rationally (i.e. as they would if they could make the optimal use of all available information for the attainment of whatever end they many have), but they act, none the less, more or less rationally; and this makes it possible to construct comparatively simple models of their actions and inter-actions and use these models as approximations. (1960: §29, p. 140–1, italics in original)

The social sciences should adopt, according to Popper, 'what may be called the method of logical or rational reconstruction or perhaps the zero method'.

> By this I mean the method of constructing a model on the assumption of the possession of complete rationality (and perhaps also on the assumption of the possession of complete information) on the part of all the individuals concerned, and of estimating the deviation of the actual behaviour of people from the model behaviour, using the latter as a kind of zero-co-ordinate' (1960: 141)

The contrast between model and actual behaviour not only allows gauging the deviation of the latter from the former but also constitutes a test of the adequacy of the model in comparison to observed social behaviour.

We can conclude that methodological individualism does not reduce to psychology or mental phenomena. It starts from individuals, their actions and interrelationships. But these provide only the data from which social science explanations are inferred (see Hayek, 1964: Part I, Ch. IV). Today's emphasis in the social sciences is either on models (ideal types) or social mechanisms as modes of explanation. Both methodologies, according to methodological individualism, are grounded in micro-explanations. The bone of contention in the methodological debates is whether such micro-explanations are sufficient or necessary to explain social phenomena.

5 Methodological Individualism at Work

We have now left ontological considerations firmly behind and operate purely on an explanatory level. If institutional aspects are required at this level, the question of some form of macro-explanation arises. Even Watkins (1952b: 187) admitted that individual behaviour need not entail that there is no overall system. But he insisted that methodological individualism could provide adequate explanations. We have seen, however, that even methodological individualism requires models to abstract from individual idiosyncracies. So the question returns whether methodological individualism, broadly understood, has the resources to account for societal relations and their effects on individuals.

There are certainly social situations, which can be handled by methodological individualism even when it omits reference to psychological dispositions. Weber's analysis of the rise of modern capitalism is often cited as an example of the need for micro-explanations regarding social phenomena. The basic idea is that the Protestant Ethic aided the rise of modern capitalism. It was made possible through the central notion of *calling*, the fulfillment of one's duty in worldly affairs.

> The only way of living acceptably to God was not to surpass worldly morality in monastic asceticism, but solely through the fulfillment of the obligations imposed upon the individual by his position in the world. That was his calling. (Weber, 1974: 80; cf. Coleman, 1990: 6–9)

As Weber adheres to methodological individualism, he cites individual attitudes and adherence to strict moral codes as causal factors, which paved the way for the rise of modern capitalist attitudes. Individuals, following the Protestant Ethic, adopted an austere lifestyle, spurning all luxury. He characterizes modern capitalism, in contrast to 'adventure capitalism', as the systematic pursuit of profit. Weber matches characteristic features of capitalism (profit maximization and investment) with personal traits (austerity and saving) and argues that many Protestant individuals displaying these traits—austere lifestyles—contributed to the rise of modern capitalism. The money, which individuals did not spend on luxuries, was invested in commercial activities. Weber uses a bottom-up approach, like rational choice theory, but his explanation still displays holistic elements. They reside in the attitude of asceticism and the economic institution of capitalism. At this point it could be asked why Protestant denominations encourage ascetic attitudes and why their collective behaviour led to capitalism. People were born into a pre-existing religious faith, which shaped their ascetic outlook on life. Modern capitalism itself was a pre-existing institution, which determined the commercial behaviour of individuals. Weber's analysis, then, makes use of links between micro- and macro-levels.

We need a clearer example of methodological individualism, which is free from holistic assumptions. Rational choice theory, in its basic form, fits the bill. Individuals are modeled as rational agents who possess complete information, clear preferences, make their decisions in terms of a cost–benefit analysis and opt for the most beneficial choice of action. Their uncoordinated individual actions lead to a social situation. Consider, for instance, the poor teaching standards in inner city schools. This social phenomenon is explained by a rock-bottom account, which is the hallmark of methodological individualism. When public transport to the outskirts of large cities becomes available, individual middle-class families decide to move to the suburbs, commuting to their inner-city jobs and leaving poorer communities behind in the inner cities. Public resources and amenities are then withdrawn from inner-city schools and invested in educational facilities in the outlying areas. Apart from the mention of public transport, this bottom-up explanation makes no reference to macro-phenomena. Nor does it refer to the psychological motivations of the people involved. There is no need to read their minds. The explanation appeals to the uncoordinated action of individual families who rationally consider their options and possess the means to move to the suburbs. It achieves what Weber calls 'adequate explanation' of this social fact. That is, it provides a mixture of necessary and sufficient conditions (=public transport, financial means) of the state of inner-city education. The drop in educational levels is the unintended consequence of the uncoordinated decision of middle-class families to leave the inner cities (Little, 1991: 15–6).

6 Institutional Individualism

Methodological individualism is a bottom-up approach. In order to be successful, it needs to explain, not just individual actions and relations, but also institutional facts. For culture, institutions, traditions, social values and social norms exist and their functioning needs to be explained. In 1975, Joseph Agassi, a former student of Karl Popper's at the *London School of Economics*, introduced the idea of *institutional individualism,* which constituted real progress in the evolution of methodological individualism. Agassi's contribution achieved several things: (1) it marked a definite departure of methodological individualism from psychologism; (2) it affirmed the autonomy of the social sciences; (3) it embraced the 'existence' or better the 'reality' of distinct social entities. Agassi also highlights the shift, in this discussion, from ontology to methodology and the need for methodological individualism to include institutional aspects. The position accepts that only intentional individuals exist in a primary sense; societal entities exist in a secondary sense; they are constraining realities. Intentional agents are necessary for the existence of societal entities but not sufficient for their functioning. As Popper (1976: 104) puts it: 'Institutions do not act; rather, only individuals act, in or for or through institutions.' Agassi developed this approach into *institutional individualism.* According to this view only individuals have aims and responsibilities but institutions (culture, states, and traditions) are of primary importance to the social sciences. They exercise constraints on individuals but individuals also have the capacity to change social conditions. It is for this reason that trends exits in the social world, but social 'laws' do not. Institutional individualism therefore presents a middle-way between psychological individualism and old-fashioned holism. In addition, Agassi seems to have abandoned the commitment to the rationality principle, which is still present in Popper and Weber. Popper's situational analysis requires the attention to both institutional situations and the rationality principle. The emphasis now shifts to the institutional aspects of human interactions. Not all human interactions can be explained by reference to a rationality principle, as crowd behaviour shows.

The Spanish economist Fernando Toboso extended Agassi's institutional individualism into a 'middle mode of explanation', which makes the abandonment of the rationality principle explicit. According to Toboso (2001: 9–11; italics in original), three rules characterize institutional individualism.

- *Only persons can pursue aims and promote interests.*
- *Formal and informal sets of institutional rules affecting interactions among persons must be part of the explanatory variables.*

• *Marginal institutional changes always result from the independent or collective actions of some persons and always take place within the wider institutional frameworks.*

Institutional individualism is a middle way because it

> ...yields non-systematic and non-reductionist explanations at the same time as it allows for the incorporation into economic theories and models many formal and informal institutional aspects surrounding all human interactions, whether these interactions take place within stable structures of legal rules and social norms or whether they attempt to change the said rules and norms.' (Toboso, 2001: 11)

Toboso's second rule relaxes institutional individualism still further because it requires that institutional aspects of society (both formal and informal) be incorporated as explanatory variables. For instance, in the famous Milgram experiment (1961) on obedience to authority such institutional aspects—like the 'authority of scientists'—come into play.

Gustave LeBon and Sigmund Freud implicitly appealed to informal rules in their analysis of crowd behaviour. They established the causal role which crowds have on the behaviour of individuals. Individuals will often behave differently under the influence of a crowd than they would otherwise. Freud explained mass behaviour through the effect of the mass on the individual's psychic economy. (Freud, 1921: 117) He observed that crowd behaviour is qualitatively different from the behaviour of the sum of many individuals. The interaction and interrelation of many individuals ('the crowd') are used as explanatory variables to account for individual behaviour. The interdependence of social actions produces 'macro-level outcomes', which range widely: from the effects of individual actions on others, bilateral exchanges, collective decisions to formal organizations and social control. (Coleman, 1990: 20–3).

By contrast, methodological individualism insists, according to Elster (1982: 453; italics in original), 'that *all* social phenomena (their structure and their change) are in principle explicable in terms of individuals—their properties, goals and beliefs.' He adds that it is 'compatible with beliefs about supra-individual entities or reference to other individuals.' Elster's claim raises the question (a) whether *all* social phenomena can thus be explained and (b) whether it is even possible in principle to find such explanations. We have seen that institutional individualism clearly denies the first question. Methodological holism in addition denies the second question. Institutional individualism enhances methodological individualism by adding institutional frameworks but does not treat institutions as reducible to individual beliefs and beliefs. Methodological holism, where applicable, focuses on institutions and societal structures without reference to individual agents, whose primary existence it does not deny.

7 Explanatory Holism

Both Hayek and Popper gave holism a bad name by associating it with political totalitarianism. This opposition has rubbed off on 'explanatory holism'. Holism can either be understood in the traditional way as the affirmation that societal 'wholes' exist and have distinct aims and interests of their own. It is this strong, implausible sense, which Hayek, Popper, Watkins and Elster reject. But a weaker sense of holism holds that the societal set-up shapes individual behaviour. Durkheim defended this weaker sense of holism when he held that individual social agents are necessary but not sufficient to create social life. Societal entities exercise a 'force' on individuals. This force comes in the guise of traditions, values and norms. Agassi follows Durkheim when he characterizes institutionalism as the view that 'social entities are of primary importance for the social sciences' and that they amount to more than the sum of individuals. Institutionalism shares with methodological individualism the view that only individuals have aims and responsibilities. (Agassi, 1975: 155) But it also holds that societal entities possess reality. The interdependence of actions amounts to more than the mere aggregation of individual actions. It is felt in the force which they exercise. Marx, for instance, regarded the economic structure of society as fundamental and thought that it determined individual actions. The sociologist Nobert Elias shifted the emphasis to societal aggregates in his famous study of the 'civilizing process'. Elias analyzes the emergence of civility (civil behavior, self-constraint and concern for others) in Western societies. His thesis is that with the establishment of the modern state and the rational organization of political and social affairs, there developed parallel internal constraints in individual members of these societies. According to Elias neither the sociogenesis of the state nor the control of emotions are the results of deliberate plans. One of the striking aspects of his study is that the civilizing process—the emergence of civil behavior—is not driven by rational motives, for instance hygienic concerns. His thesis is that the greater the interdependence of people the greater the control of psychological drives. He sees a close link between the emergence of modern societies and changes in personality structures towards civility.

> Each "increase" in restraints and interdependencies is an expression of the fact that the ties between people, the way they depend on one another, are changing, and changing qualitatively. This is what is meant by differences in social structure. And with the dynamic network of dependencies into which a human life is woven, the drives and behavior of people take on a *different* form. (Elias, 1982: Ch. 3, VIII., §37, p. 87; italics in original)

Such macro-analyses show that explanations are sometimes needed in the social sciences, which operate purely on a collective or institutional level. This view could therefore be called 'emergent holism' because societal features can no longer be reduced, in explanatory terms, to individual behaviour.[2] Social aggregates exercise an effect on typical individuals.

---

[2] In an earlier publication I described Durkheim's stance as 'emergent holism' (Weinert, 2009: 135, 252) because the societal wholes emerge from the relations of individuals and become autonomous in the sense that they constitute a reality, which has social effects and is more than the sum of

The existence of human languages illustrates this weaker sense of holism. A human language needs speakers, human beings, for its practice and survival. Language speakers are necessary for its continuation. Without them, the language will become extinct. Note that a language need not be actively spoken in a human population to survive. Both Latin and ancient Greek survive because they exist in written form, which experts can read. But a language exists prior to and beyond the existence of a particular group of speakers. Any individual is born into a linguistic culture and adopts the language spoken around them. No particular individual is necessary for the survival of a language. The existence of language rules is the result of linguistic evolution. They lay down how a language should be spoken at any particular moment in time. These rules change over time but they permit to distinguish correct from indirect usage of the language at any particular stage. Non-native speakers are corrected by native speakers. Even native speakers make mistakes, as judged by grammar books. A human language is a social institution, which exists in the secondary sense. It enjoys an independent reality and exercises constraints on individual behaviour. But languages evolve, due to the interaction and participation of the community of language users.

When societal aggregates do all the explanatory work, we arrive at methodological or explanatory holism. It is not a political doctrine. For, as St. Lukes (1968: 127; cf. Mandelbaum, in Gardiner, 1959) points out social situations and interrelations between individuals 'can be described in non-individual terms without reference to political holism.' The issue is methodological holism: whether social relations and societal institutions can be explained without reference to individual terms. I have already argued that if methodological individualism is taken to its logical conclusion, it can explain only some social facts, like the drop in teaching standards in inner city schools, following the exodus of the middle classes. There are societal facts which cannot be adequately explained without reference to macro-phenomena. Such supra-individual explanations may involve ideal types or reference to social mechanisms. Consider two typical examples of 'holistic' explanations.

8 Explanatory Holism at Work

(A) In 2001, two American economists, following a Swedish study, provided an explanation of why there was a significant drop in juvenile crime rates in certain parts of the United States in the late 1980s and early 1990s (Levitt & Donohue, 2001; Fig. 1). They related this drop in crime rates to the legalization of abortion in many parts of the US in 1973. The Donohue/Levitt hypothesis argues that the legalization of abortion led to the birth of fewer unwanted children or children whose parents could not support them. They further argued

individuals at any one time. I was not aware at the time that Coleman (1990: 28) also employs the language of emergence when he argues that 'the "system level" exists solely of emergent properties characterizing the system of action as a whole.'

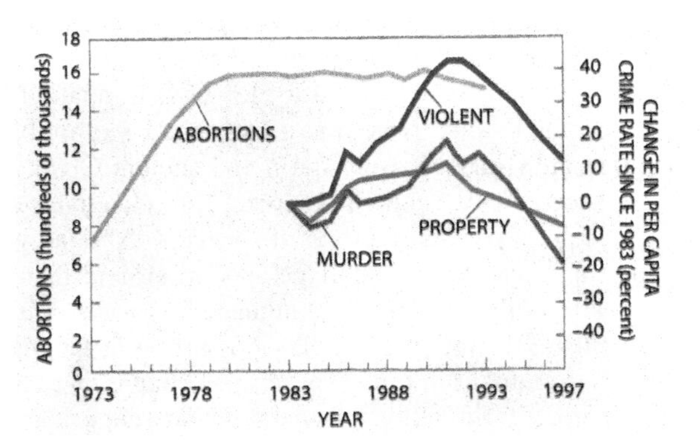

Fig. 1 The Donohue/Levitt hypothesis. *Source* Scientific American (December 1999: 14)

that unwanted children are more likely to commit crimes as young adults. And thus the drop in delinquency rates, starting in 1992, can be linked to the availability of abortion. This hypothesis was and remains controversial. But whether true or not, it is an example of an explanation of a societal fact—the reduction in crime rates—due to another societal fact—the legalization of abortion. In both cases individuals are involved—legislators and women—but the explanation concerns macro-phenomena, i.e. abortion and crime rates. The Donohue/Levitt hypothesis refers to a general social mechanism. If the authorities in a given society enable certain social or legal situations, certain predictable effects are likely to happen. In 1999 the Home Office in Britain predicted that the number of burglaries and thefts would increase by almost a third in a short time span of two to three years. The expected rise in crime was a consequence of the rise in the young male population in Britain. In this instance a causal analysis combines with a prediction. The predicted increase in the number of burglaries and thefts is blamed on an increase in the number of young adults and a growth in the amount of stealable goods. The latter are the causal conditions, which are said to lead to the effect, if no other conditions interfere.

These examples show that the social sciences are able to make statistical predictions. Such predictions become possible when social mechanisms are at hand, which can account for the production of social facts. The connection between societal facts and statistical regularities is revealed whenever population statistics are cited.

The argument in favour of "social facts" is historically connected with the presence of statistical regularities where none can be found at the molecular, individual level. The statistical regularity can be explained in terms of features of the social situation as a whole, but in practice it is seldom possible to trace the nexus in individual cases. (Gellner, in Gardiner, 1959: 493)

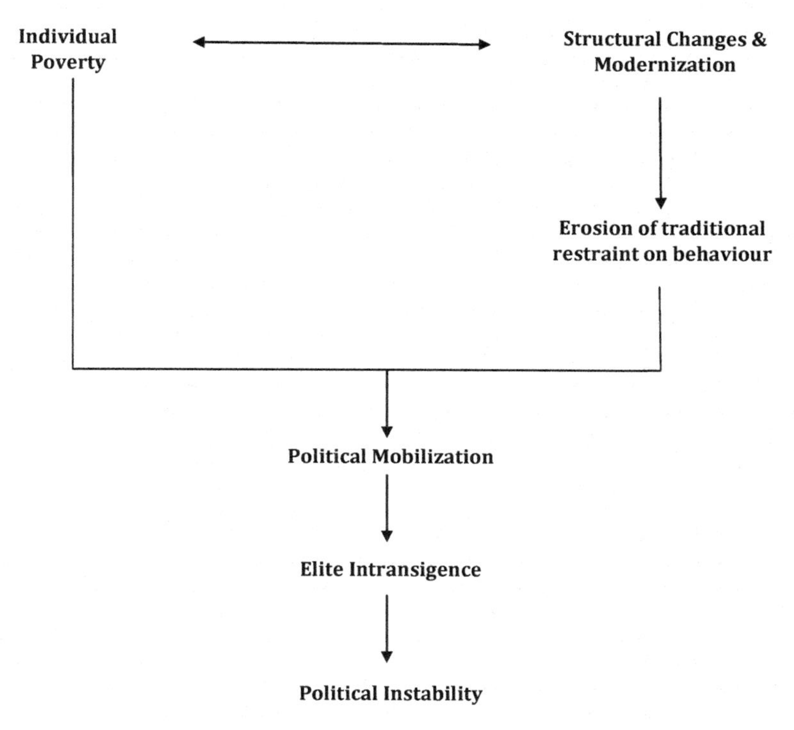

Fig. 2 Poverty as a cause of political instability

(B) Another example of a social mechanism explanation refers to social unrest. It may be due to a number of factors, such as social, political, economic and environmental causes. My particular example deals with social unrest in Albania in 1997, which occurred as a result of a combination of various factors: poverty in the population is not sufficient to cause social unrest. Individual poverty must be combined with structural changes in society, which lead to the erosion of traditional restraint upon behavior. These two factors lead to the development of political awareness. But it is only when this political mobilization is met by intransigence on the part of the political class that it results in high-level instability (Fig. 2; cf. Little, 1991: 28).

9 Social Mechanisms

What arguments do methodological holists offer to support their claim that macro-explanations of social phenomena are inevitable? Recall Popper's notion of a problem situation. A *problem situation* consists of a perceived problem, background knowledge, available problem-solving techniques and tentative solutions. I say 'perceived' problems because the perception of a problem changes over time. What was once regarded as a problem may have disappeared and be replaced by new problems. The background knowledge consists of the state of knowledge of a discipline at a given moment in time. The available techniques consist of statistical analyses, observations, interviews and commitment to a methodology. Due to his commitment to methodological individualism and Enlightenment rationality Popper includes in his model of situational analysis a rationality principle, adopted from Rational Choice theory. But, as Toboso (2001: 8) points out, institutional individualism need not make any reference to a rationality principle. The rules of institutional individualism spell out which institutional aspects are to be considered and how they affect human interactions. But they do not specify types of rationality, preferences on the part of individuals or whether consequences of human interactions are intended or unintended. If the emphasis shifts from the individual to the institutional level, we turn to explanatory holism. Methodological holism affirms that situational analysis requires societal facts to be explained by macro-explanations.

In the literature various arguments have been put forward in favour of methodological holism. They concern either the indispensability of holistic explanations, the impossibility of translating macro-level theories into individualist theories or arguments which refer to social-level mechanisms (Zahle, 2016). As I have considered social science explanations either in terms of Weber's ideal types or social mechanisms I will focus here on these types of explanation. I will then ask how they are related.

The appeal to social mechanisms is useful because they can be adapted to different levels and situations. Mechanisms do not possess the unrestricted universality of laws in the hard sciences. But they exist at a deeper level than mere descriptions or statistical correlations. There is no agreement in the literature on what constitutes a social mechanism. But the general idea motivating the search for mechanisms is that they provide causal explanations. Mechanisms constitute causal links between surface phenomena and their underlying causes. The term 'mechanism' refers to recurrent processes linking specified initial conditions to specific outcomes. Statements about mechanisms are generalizing causal propositions (Mayntz, 2004: §2). It is useful in this context to recall Weber's notion of adequate causal explanations in the social sciences. The identification of certain factors in a given situation is regarded as the most likely explanation of a social phenomenon. It is a feature of the social sciences that such explanations are often contested or subject to modifications. The asceticism of the Protestant Ethic could be an adequate explanation of the rise of modern capitalism in mostly Protestant countries. The baby boom of the 1960s could be an adequate explanation of the rise in juvenile crime, as predicted by the British Home Office. Elias's link between structural changes in society and the emergence of civility could be an adequate explanation of the civilizing process. The Donohue/Levitt hypothesis could be an adequate explanation of the drop in delinquency twenty years after the legalization of abortion. Poverty (etc.) could be the cause of political unrest. Such explanations may be adequate, but they do not possess the certainty of causal explanations in physics. They are nevertheless important because the social sciences need to explain social and societal facts.

Causal explanations with generative mechanisms constitute important types of social science explanations. According to Bunge (2004), there are no universal mechanisms in the social world, only system-specific ones. Bunge proposes his *CESM* model $\mu(\sigma)$ of a concrete system, σ:

$$\mu(\sigma) = \langle \mathrm{C}(\sigma), \mathrm{E}(\sigma), \mathrm{S}(\sigma), \mathrm{M}(\sigma) \rangle$$

(where $\mu(\sigma)$ is the model of the system σ, $\mathrm{C}(\sigma)$ is the set of its parts, $\mathrm{E}(\sigma)$ are environmental conditions, $\mathrm{S}(\sigma)$ is the structure of the system and $\mathrm{M}(\sigma)$ is the mechanism, which spells out a sequence of states). As an example Bunge (2004: 189) cites the nuclear family:

> The most familiar example of a social system is the traditional nuclear family. Its components are the parents and children; the relevant environment is the immediate physical environment, the neighborhood, and the workplace; the structure is made up of such biological and psychological bonds as love, sharing, and relations with others; and the mechanism consists essentially of domestic chores, marital encounters of various kinds, and child rearing. If any of the mechanisms breaks down, so does the system.

Mechanisms can operate at various levels. Bunge refers to mechanisms at the individual level; the avoidance of cognitive dissonance also operates at the micro-level. When faced with equally attractive choices (say, a holiday in Spain or Italy) individuals must choose one of the available options. Once the choice has been made individuals typically find reasons why their choice is better than the rejected alternative. (The sour grapes phenomenon belongs to this category.) Note that the mechanism of cognitive dissonance explains what typical or average individuals are expected to do, not what particular individuals will do. The question here is whether such mechanisms exist at the macro-level. One hint is given by methodological individualism itself, which accepts that interpersonal relations must be included even if this methodology is adopted. Popper (1976: 103ff) explains that

> …this programme consists of building a model of the social situation, particularly incorporating the institutional situation the agent confronts when acting, so that the rationality of his action may be accounted for.

But institutional individualism no longer insists on the rationality principle. As it also holds that the 'institutional situation' presents a 'reality' over and above individual actions, it would be a 'fatal mistake' to infer macro-phenomena from 'motivated individual behaviour' (Mayntz, 2004: §4). The reason is that interpersonal relations involve institutional and structural factors, which, according to both institutional individualism and methodological holism, put constraints on individual behaviour. This phenomenon is clearly at work when political parties decide on their political programmes. It transforms individual views into collective decisions. Consider, for instance, the Conservative Party in the United Kingdom and its decision to support Brexit, that is the departure of the UK from the European Union. At first individuals in the Party represented a variety of opinions, between the two extremes 'leave' and 'remain'. But once the Party had agreed on the 'leave' option, it became its official policy. Although some individuals in the party continued to oppose the collective decision, officials began to represent and defend it publicly. The collective decision constituted constraints on individual members and those who continued to oppose the collective decision were sanctioned. (For instance, remainer MPs lost their seats in parliament.) There never was a unanimous consensus even amongst Conservative MPs regarding Brexit. But once the decision had been adopted, individual opinions became irrelevant: the individual views of some had morphed into the collective decision. Such processes are widespread.

Social mechanisms play an important part in understanding the social world. They cover economic mechanisms:

> The mechanism generating the macro effect "market equilibrium" also depends on structural features, such as the existence of a plurality of competing producers and the absence of political price fixing; the rational decisions of individuals to offer or buy are the "material stuff" of the process, but its shape is determined by these structural elements. (Mayntz, 2004: 251)

And they cover political mechanisms:

> Democracy is a social mechanism for resolving the problem of societal decision-making among conflicting interest groups with minimum force and maximum consensus. (Lipset, 1959: 92; quoted in Bunge, 2004: 185)

In all cases, as Mayntz (2004: 251) points out, 'specific structural (or institutional) features are decisive for the generation of aggregate macro-effects by the motivated action of individuals.' Economic and political mechanisms represent societal mechanisms.

These examples show that 'explanatory holism' does not commit its proponents to 'social fact holism' (List & Spiekermann, 2013: 630). In fact, ontological holism belongs to the field of metaphysics, explanatory holism belongs to methodology. In the early debates, as we have seen, there often lurked the danger of confusing these two issues. But Durkheim's claim that societal facts act as constraints on individual members of society is compatible with the view that society is made up of individuals who alone have intentionality. Individuals are, in Durkheim's phrase, necessary but not sufficient for the existence of societal facts. For interpersonal relations give rise to institutional structures and group-level phenomena. The exercise of these constraints on individuals requires explanation, so does the reality of institutions.

According to a detailed defense of explanatory holism a 'social system requires explanatory holism if and only if three jointly necessary and sufficient conditions are satisfied' (List & Spiekermann, 2013: 639–40):

- Multiple levels of description exist: The system admits lower and higher levels of description. 'The lower level of description typically refers to individuals and their properties, whilst the higher level refers to the properties of social aggregates.' As we have seen there are cases where either of these approaches or a mixture of them is required. If a pragmatic approach is adopted, it becomes the problem-situation, which determines which level of description is appropriate.

- Higher-level properties can be realized in multiple ways: 'The system's higher-level properties are determined by the lower-level properties', which means that individuals are necessary for social aggregates. But these higher-level properties 'can be realized by numerous different configurations of them and hence cannot feasibly be re-described in terms of lower-level properties.' If, for instance, 'we take a long-term view of social facts or processes over time, which could have been brought about by different individual actions, then an explanation in terms of individual-level properties becomes implausible.' Statistical regularities spring to mind. In this connection it is convenient to recall that the principle 'same cause, same effect', which works well in the natural sciences, does not work in the social sciences. For instance, political parties typically aim at or promise the same effect—namely, the improvement of the lives of the citizens. But the way they intend to achieve this aim differs widely according to their political agenda, be it socialist, liberal or conservative. This is also illustrated in the last condition.
- Higher-level causal relations remain invariable through micro-permutations: 'The causal relations in which some of the system's higher-level properties stand are robust to changes in their lower-level realization.' For instance, the commitment of political parties to citizens' welfare is consistent throughout history and different political systems. It continues to hold throughout the rotation of individual political actors. The survival of a natural language is independent of the employment of the language by individual speakers. The community of speakers undergoes constant fluctuations, but the higher-level properties of the language are robust to changes in individual speakers.

The authors cite the study of ethnic conflict and civil war as one example where explanatory holism is required. It involves the weakness of the state and insurgency of the population, as was the case in Albania in 1999. A famous example of explanatory holism is Durkheim's suicide study (1897). Durkheim found that in certain parts of Europe the suicide rate rose by 100% between 1856 and 1878. But the recorded reasons for suicide, i.e. illness, poverty, jealousy remained the same during this period. So the same 'causes' obtained but the effects differed. Durkheim concluded that the aforementioned psychological factors could not be the true causes of suicide. The explanation had to be sought in societal factors. The members of a society are governed by social forces which exist independently of each of them. Durkheim identified three different sorts of suicides in terms of three distinct social causes: Egoistic suicide results from too little social integration; altruistic suicide results from too much social integration and anomic suicide results from great and rapid changes in the degree of social integration which leave social agents normless and disoriented. Durkheim explained each of the social facts about differences in suicide rates by a societal fact about differences in the degree of social integration (see Rosenberg, 1995: 132). According to a more recent study, published by the Office for National Statistics in the UK (2015), there is a clear link between austerity and suicide, between poor employment or underemployment and mental health problems.

Thus there are societal facts, which require macro-explanations. Such higher-level explanations are needed, for instance, when social scientists consider the evolution of modern society. Weber led the way by analyzing structural aspects of traditional and modern societies. But modern industrial societies are now themselves transitioning to 'risk societies', according to Beck (1992) and Giddens (1990). Whilst danger has always been a feature of life, it mainly struck as natural disasters. A new situation has emerged; industrialization and modernization have led to self-induced risks: the threat of global warming, the fear of nuclear war and the risk of overpopulation are striking examples. Risk becomes transnational. These self-inflicted risks result in reflexive modernization, which forces societies to reflect on the very effects of their activities on nature and society. It leads to a questioning of traditional authorities and an underlying trust in established institutions. This skepticism produces new forms of risk. If modern societies do indeed evolve towards 'risk societies', this evolution requires macro-level explanations.

10 Social Mechanisms and Ideal Types

Although there are different types of explanations in the social sciences—functional, statistical, descriptive, causal, partly reflecting the different social science disciplines—I have focused on ideal–typical and mechanistic explanations. These types of explanation best illustrate the methodological debates around methodological individualism and holism. The question arises how Weber's ideal type methodology and the method of social mechanisms are related. Both are models of social systems. It seems to me that ideal–typical models possess more generality and flexibility than mechanism models. Explanations in terms of social mechanisms can without much effort be included in ideal–typical models. But this implies that ideal types can also be used to explain micro- and macro-phenomena. The latter are exactly the ideal types, which Watkins dismissed as 'holistic'. Weber uses such models to describe and explain macro-phenomena, like feudalism and forms of capitalism. (Cf. Ringer, 2002) But his use of 'holistic' models was much more extensive than Watkins considered. It covered types of bureaucracy and authority, power, world religions and social structures. Elias's macro-analysis of the emergence of Western civilization also employs ideal–typical constructions. They focus on salient features of a historical situation or social phenomena. It is easy to see that social mechanisms may be included in the salient features, which the model describes. According to Bunge innovation is the central mechanism which drives a capitalist economy. Bunge's *CESM* model also includes the components of the system, which are related by the mechanism. Bunge (2004: 186–8) defines a system as 'a complex object whose parts or components are held together by a bond of some kind.' Such a general characterization of systems applies both in the natural and social sciences. A nice illustration of a natural system is the solar system: the planets are the components of the system; the system is held together by the bond of gravity and Kepler's laws. Darwin also has systems in mind when he speaks of interdependent systems in the natural world. A plant supports an insect and the insect needs the plant for its survival. Social and societal systems are held together by social bonds: institutions, laws, norms, values, traditions. The constituents of the systems are either macro-components, like institutions, or micro-components, like social agents. Their interactions create the reality of societal aggregates, both of which require macro-level explanations. Hence we can conclude that Bunge's model is an ideal–typical model, which includes the mechanisms of various social systems (family, capitalist economy). Social mechanisms can thus become part of ideal–typical explanations.

References

Agassi, J. (1975). Institutional individualism. *British Journal of Sociology, 26*(2), 144–155.

Beck, U. (1992). *Risk society*. Sage Publications. Translation of the German original: *Risikogesellschaft*. Frankfurt a. M.: Suhrkamp (1986).

Bunge, M. (2004). How does it work: The search for explanatory mechanisms. *Philosophy of the Social Sciences, 34*(2), 169–181.

Coleman, J. (1990). *Foundations of social theory*. Cambridge University Press.

Durkheim, E. (1982/1895). *The rules of sociological method*. London: Macmillan.

Durkheim, E. (2006/1897). *On suicide*. Penguin Classics.

Elias, N. (1982). *State formation and civilization—The civilizing process* (Vol. II). Blackwell Publishers.

Elster, J. (1982). The case for methodological individualism. *Theory and Society, 11*(4), 453–482.

Freud, S. (1921). Group psychology and the analysis of the ego. *Complete Works, XVIII* (1955), 69–143.

Gardiner, P. (Ed.). (1959). *Theories of history*. The Free Press.

Gellner, E. (1959). Holism versus individualism in history and sociology. In P. Gardiner (Ed.) (pp, 488–503).

Giddens, A. (1990). *Consequences of modernity*. Polity Press.

Hayek, F. A. (1964/1955). *The counter-revolution of science*. The Free Press of Glencoe.

Heath, J. (2020). Methodological individualism. *Stanford Encyclopedia of Philosophy*.

Hollis, M. (1994). *The philosophy of social science*. Cambridge University Press.

Levitt, St., & Donohue, J. (2001). The impact of legalized abortion on crime. *The Quarterly Journal of Economics, 116*(2), 379–420.

Lipset, S. M. (1959). Political sociology. In R. K. Merton, L. Broom, & L. S. Cottrell (Eds.), *Sociology today: Problems and prospects* (pp. 81–114). Basic Books.

List, C., & Spiekermann, K. (2013). Methodological individualism and holism in political science: A reconciliation. *American Political Science Review, 107*(4), 629–643

Little, D. (1991). *Varieties of social explanations*. Westview Press.

Lukes, St. (1968). Methodological individualism reconsidered. *British Journal of Sociology, 19*(2), 453–482.

Mandelbaum, M. (1959). History and the social sciences. P. Gardiner (Ed.) (pp. 476–488).

Mayntz, R. (2004). Mechanisms in the analysis of social macro-phenomena. *Philosophy of the Social Sciences, 34*(2), 237–259.

Papineau, D. (1978). *For science in the social sciences*. Palgrave Macmillan.

Popper, K., et al. (1976). The logic of the social sciences. In Th. W. Adorno (Ed.), *The positivist dispute in German sociology* (pp. 87–104). Heinemann.

Popper, K. (^{2}1960/1957). *The poverty of historicism*. Routledge.

Ringer, F. (2002). Max Weber on causal analysis, interpretation and comparison. *History and Theory, 41*(2), 163–178.

Rosenberg, A. (^{2}1995). *Philosophy of social science*. Westview Press.

Toboso, F. (2001). Institutional individualism and institutional change: A middle way. *Cambridge Journal of Economics, 25*(6), 765–783.
Watkins, J. W. N. (1952a). Ideal types and historical explanation. *The British Journal for the Philosophy of Science, 3*(9), 22–53.
Watkins, J. W. N. (1952b). The principle of methodological individualism. *The British Journal for the Philosophy of Science, 3*(10), 186–189.
Watkins, J. W. N. (1955). Methodological individualism: A reply. *Philosophy of Science, 22*(1), 58–62.
Weber, M. (1978). *Wirtschaft und Gesellschaft* (1921). English Translation: *Economy and society*. University of California Press.
Weber, M. (1905/1949). Objective possibility and adequate causation in historical explanation. Reprinted in E. A. Shils, & H. A. Finch (Eds.), *The Methodology of the Social Sciences: Max Weber*. Glencoe (Ill.) (pp. 164–188). The Free Press (1949).
Weber, M. (1974). Die Protestantische Ethik und der Geist des Kapitalismus (1904–5). English Translation: *The protestant ethic and the spirit of capitalism*. Unwin University Books.
Weinert, F. (2009). *Copernicus, Darwin and Freud*. Wiley.
Weinert, F. (1996). Weber's ideal types as models in the social sciences. In A. O'Hear, (Ed.), *Verstehen and Humane Understanding* (pp. 73–93). Royal Institute of Philosophy Supplement: 41. Cambridge UP (1996),
Zahle, J. (2016). Methodological holism. *Stanford Encyclopedia of Philosophy*.

On the Irreversible Journey of Matter, Life and Human Culture

Diederik Aerts and Massimiliano Sassoli de Bianchi

1 Introduction

Year 2021 could be the one in the course of which some of the mysteries enveloping antimatter will start to be addressed. Not only about its electromagnetic behavior (for instance, does antihydrogen have the same spectral lines as hydrogen?) but also about its gravitational behavior. For the latter, three different experiments are already planned at CERN: AEgIS (Kellerbauer et al., 2008), ALPHA-g (Bertsche, 2017) and Gbar (Indelicato et al., 2014), to determine if antimatter, likewise matter, has a positive gravitational charge, or a negative one.

Most physicists believe antimatter normally falls downwards and not upwards, but the latter hypothesis is not so unlikely, considering it is the main ingredient of the Dirac-Milne cosmological model (Benoit-Lévy & Chardin, 2012), providing possible explanations for dark matter and dark energy (Chardin & Manfredi, 2018). Also, antigravity could explain the "victory" of matter over antimatter. Indeed, being antimatter forced to rapidly move away from matter, the latter would have remained protected from the former, by escaping its deadly embrace.

A different possible mechanism, also explaining the existence of our matter-universe, where structures were able to form despite the destructive menace of antimatter, was proposed many years ago by Sakharov (1967), with the hypothesis of a tiny difference in the production of matter and antimatter, i.e., of a violation of the baryon number conservation law. According to it, a colossal—out of thermal equilibrium (Shaposhnikov & Farrar, 1993)—battle between matter and antimatter

D. Aerts (✉) · M. Sassoli de Bianchi
Center Leo Apostel for Interdisciplinary Studies, Brussels Free University, Brussels, Belgium
e-mail: diraerts@vub.ac.be

M. Sassoli de Bianchi
e-mail: msassoli@vub.ac.be

M. Sassoli de Bianchi
Laboratorio di Autoricerca di Base, 6917 Barbengo, Switzerland

S. Wuppuluri and I. Stewart (eds.), *From Electrons to Elephants and Elections*,
The Frontiers Collection, https://doi.org/10.1007/978-3-030-92192-7_42

saw the almost complete destruction of both "armies," with the survival of only a residual fraction of matter, which then formed the universe in which we live today, where antimatter only exists in traces and no longer constitutes a danger for present and future structures.

We have no idea of course if one of the above two scenarios correctly addresses what is considered to be one of the big problems of modern cosmology. Seeing that the survival of matter over antimatter has allowed for the construction of complex structures, thus of life itself, we could say that this question is part of another big question: that of the very emergence of life and human culture, and its struggle not to succumb to the omnipresence of disruptive forces, which in classical physics can be associated with the second law of thermodynamics, and in quantum physics with the mechanism of decoherence. One of the ideas expressed in this chapter is that in the context of human culture, both can be connected with the ancient notion of 'evil'.

In Aerts et al. (2019b), one of us tried to express the above in somehow poetic terms, with the following suggestive words:

> Nobody ever promised that things would be easy, and they are not.
>
> Nobody ever promised that suffering would not be intrinsically part of life, including part of your life, and it is.
>
> Nobody ever promised differently, namely that each time again problems would arise, some hard ones, maybe some terribly hard ones, and some less hard ones, and that is how it is.
>
> Nobody ever promised that it would be possible to solve even the minor ones of these problems at a first try, let stand the harder ones, and so it is.
>
> Nobody ever promised that evil would not be lurking around, mostly even in those places where it can best hide its nature, and so it is.
>
> The above short description is a good characterization of life in depth and we know why it is, because we, as human beings, have awaken to it in consciousness.
>
> This is its nature because life is the moment-to-moment irreversible choice and constant fight against the spontaneous regression to 'just being', which on the local level of a human body we call death.
>
> Every breath, every piece of food, every step, every smile, every embrace, every sprinkle of love, compassion, collaboration, construction, creation, are little parts of this constant fight against 'just being', or local death, the content of the second law of thermodynamics, and on a deeper physics level the separation of matter from anti-matter in time.
>
> As little individual humans we are humble participants in this great endeavor of life, this great enterprise of struggle and fight, this deep endurance of suffering, this never-ending attempting of solving problems, failing and trying again, and sometimes succeeding in a local and modest victory.
>
> The deep secret of life is that if you pursue a noble goal all the characteristics of life described above become the substance of meaning, which is the food of the human mind.

It is the purpose of this chapter to explore the view synthetized in the above lines, which at first sight might be considered to be mainly a "psychological metaphor."

We believe instead that they have to be taken quite literally, not only psychologically, but also physically, and biologically. They indeed adequately capture the very special condition in which all living organisms are, like plants, animals, but also microorganisms, like bacteria, although only we humans have become fully aware of it and, as we will emphasize, this puts us in a special condition with respect to it.

From the point of view of physics, the potential presence of antimatter, menacing to annihilate all material structures into light, is an aspect we will consider, as well as the manifestation of the second law of thermodynamics, which also plays a fundamental role. Indeed, following this law, isolated systems spontaneously evolve towards a state of maximum entropy, which is a state of (thermal) equilibrium. Hence, life, as initially pointed out by Schrödinger (1944) in his famous book "What is Life?", is about struggling to stay away from isolation and from the menace of all sorts of equilibria.

2 Stable an Unstable Equilibrium

When we bring together an electron and a proton, we can easily form a hydrogen atom. What we tend to forget is that this creation of structure can only happen because we live today, cosmologically speaking, in a relatively peaceful realm, which emerged from the primordial separation of matter from antimatter. As we mentioned already, different mechanisms could be at the origin of this separation, which in turn could be the consequence of an even deeper separation.

One can speculate that the matter–antimatter split was the result, at a much more fundamental level, of the coming into existence of the very arrow of time, i.e., of the distinction between past and future, forcing matter and antimatter to move in opposite temporal directions, as suggested for instance by Feynman's diagrams (Feynman, 1949). Hence, spatial separation would have stemmed from an even deeper separation, taking place at the temporal level. And we can associate such primordial level of separation also with the moment when the Higgs field conferred a rest mass to some of the physical entities evolving in those ancient times (Kibble, 2009), hence allowing for the matter–antimatter duality split.

So, if a hydrogen atom can easily form, this is because the "mother" aspect of matter (the meaning of the word "matter" comes from the Latin "mater," i.e., "mother") can today easily play her role in promoting the birth of all sorts of structures, similarly to a fertile ground. The cosmos, freed from the presence of antimatter, has become a relatively peaceful island. And as far as we can say, the process is irreversible, hence, at the global level of evolution of our universe, the process of 'construction of new structures' is winning over (is stronger than) the process that tends to destroy them. But this is not always the case when the 'construction' and 'destruction' aspects face each other within a same reality layer. In that situation, the opposite appears to be more likely: we observe a 'fragility of construction' and a remarkable 'robustness of destruction'.

This we might not realize when, say, in our garden, we plant a healthy seed in a fertile ground and see it grow. Again, this is because our garden is a relatively peaceful island of stability, where everything threatening the blossoming and growth of the new seed has been previously evacuated (assuming here that we are good gardeners). In the wild, that same fertile ground could face moments of extreme dryness, or become too wet, and generally would have been colonized by numerous other plants, which being older and stronger can easily take the sun light away from the newborn, impeding its growth (it is of course just an example).

In other words, with apparent no effort, the universe can bring about new structures, new forms of complexity, but one should not be fooled by this apparent effortlessness, as this can happen only when the right circumstances have already been set in place, i.e., when local islands of stability have already been created, playing the role of 'steppingstones for future progress'.

At the local level, when we are outside of these peaceful islands, what can possibly come into existence will in general never come into existence, because each new step is the result of a battle against the pervasive destructuring forces, always at work, hence, every new structure, every new life form, is to be considered as a true achievement.

There is here a subtle point to understand. Life requires peaceful islands to unfold and advance, but at the same time, these very peaceful islands are akin to an aspect which is precisely what life must constantly fight. As indicated in the above quote, there is an aspect in reality which can be described as its level of 'pure existence', or 'pure being', or 'just being'. Then, there is a completely different level, that of 'change', of 'evolution', of 'becoming', which is the level inhabited by life itself, to be here understood in the largest possible meaning of the word.

'Being' is always in balance, but 'life' is intrinsically 'not in balance' and, more precisely, it is because 'being is in balance' that 'life is not in balance'. The level of 'being' manifests as the tendency of reality to always seek a condition of balance, of sameness. This is really the main characteristic of what we call 'being', whose great ally is therefore the second law of thermodynamics, with its inexorable drive towards obtaining stable states of equilibrium. On the other hand, 'life' manifests as a constant struggle in always seeking conditions of deep unbalance.

Of course, life is also creation and construction, hence, while trying to promote conditions of unbalance, without which it could not create anything new, at the same time life also tries to stabilize its own creations, and in doing so it somehow imitates the 'balance seeking behavior' of the level of 'being', although only provisionally.

In other words, life, while seeking unbalance, also creates temporary domains of stability, which are immediately used as new steppingstones to explore new states of unbalance, and the creation of all these subsequent steppingstones corresponds to the ancient 'instruction mechanism' of life, and before 'life', of 'matter', as we consider here 'matter', in the absence of 'antimatter', the starting phase of life itself in this universe.

3 Irreversible Steppingstones

So, reality has a spontaneous tendency to reach states of balance, and this is a characteristic of the 'level of being', which as we said has the second law of thermodynamics as its main ally, and therefore the latter is a major obstacle to life, which constantly seeks to keep itself from reaching equilibrium. Since it does so by also creating provisional domains of stability, the illusion of 'life as a manifestation of balance' could be easily fostered at the local level, but is just an illusion.

Different from matter, what we generically call 'light' (we use here the term in an extended sense, i.e., to refer to the entire electromagnetic radiation and not just the tiny portion of it that can be perceived by our human eyes) describes the level of 'pure being' (or at least, it is very close to it). This because a photon, being massless, has no "antimaterial counterpart." It therefore moves in a domain of reality that is not touched (or very weakly touched) by the matter–antimatter dichotomy. From that standpoint, it is not a material entity and does not participate in the life struggle.

Another domain where, possibly, such struggle can be avoided, is the mental domain. Many ancient traditions, for instance Buddhism, aim in their teachings at reducing, apparently also with some success, any forms of conflict that might emerge at the mental/consciential level. So, although it is clear that the human mind can reflect conflicts that are manifested at the level of the physical body, it is also possible to assume that minds (but maybe we should better say 'consciousnesses') can free themselves from the struggle that characterize biological life. In other words, it is not impossible that human minds (a notion that needs here not to be conflated with that of human brains), and possibly also non-human minds, like animal minds, could belong to a different realm of our reality, one where 'being and perfect balance' could be the default state.

Now, it is because life is an asymmetric 'out of balance' state of affairs that it always seeks new constructions and creations and is intrinsically irreversible. When the choice of matter over antimatter is made, i.e., of unbalance over balance, of instability over stability, there is no going back. Of course, the choice always remains of either using the past steppingstones, i.e., the past unstable local regions of peace and balance, to move forward and create more advanced ones, or to simply destroy them.

The biological processes that keep a human body alive and healthy constitutes a perfect example of such a region of illusionary peace and balance. Changes in the outer environment can produce a sudden departure from it, implying the abandonment of the state of health, which can mean the onset of disease, or even the death of the organism in question, or the promotion instead of a higher state of health (for example, via transitional processes of hormesis).

4 War and Peace

An interesting example of a local steppingstone is the situation where a group of nations are at peace with one another. Again, the equilibrium is not stable. It is indeed sufficient that one among these nations starts a war that many of these nations, if not all, will be automatically pushed away from their previous state of peace. In other words, there is clearly an asymmetry between the 'state of war' and the 'state of peace'. The former is much more stable than the latter.

Consider the opposite situation where a group of nations are at war with one another and that one of them decides to cease fighting and to relate to the other nations in a peaceful way. Will this affect the latter? Almost certainly not. Indeed, in a situation of global war, until a sufficiently large level of destruction is reached (the entropic equilibrium), the different parties cannot individually escape its destructive logic (unless they decide to do so all together), penalty of seeing their own territory invaded and conquered by the other entities participating in the conflict.

In other words, 'peace' is like a pencil standing on its tip (unstable equilibrium) whereas 'war' is like a pendulum (stable equilibrium). If a small impulse is transferred to the pendulum, it will quickly return to its original position, while even the smallest impulse transferred to a pencil balanced on its tip will make it irremediably lose its original state.

The question that naturally arises is then: If war is a stable attractor state, and peace is an unstable state, why don't we always find ourselves at war? The reason is simple: although 'war' describes a stable state, it is in the nature of such state to destroy the previous islands of stability, i.e., the social structures of the countries at war and, more importantly and more radically, the bodies of the inhabitants of such countries. It is in this process of 'search for balance through destruction' that the irreversibility of the trajectory of life reveals itself.

Indeed, the only survival possible, or chance of progress, once the material structures have engaged in that path of struggle that is life, is in the constant fight for the conquering of new local regions of temporary stability, i.e., of new steppingstones, otherwise all that has been conquered can be lost. And this is why nations know that peace is never to be taken for granted, that it requires continuous collective effort to maintain it and perfect it over time.

5 Instability and Freedom

The instability of the life process, while requiring living beings to make continuous efforts for their survival, or to be able to meet evolutionary challenges, also offers a considerable advantage. To illustrate this, consider the example of humans walking on two feet, rather than a quadruped moving on its four paws. Standing on two feet is an unstable state: it is not like a pendulum, but more like a pencil standing on its tip. It is an unstable state locally stabilized by the balance system provided by the inner ear and its neuronal negative feedback mechanism (allowing the brain to know the position of the head relative to gravity and its surroundings).

The advantage of unstable states lies precisely in the fact that small disturbances can make them collapse into very different states (think of the pencil on its tip, which can fall along all possible directions). This means that unstable systems have considerable 'freedom' in exploring different possibilities. On the contrary, stable states have no freedom, and this precisely because of their steady balance, which reduces to almost zero their 'potentiality level'. And this by the way is also the reason why unstable states are intrinsically unpredictable, whereas stable states are deterministic in their evolution.

Of course, when a system opens to a vast set of potentialities, i.e., of 'collapses towards different possible outcome states', many of those will also pose a threat. But this is the price to pay for freedom. When we walk on two feet, the risk of falling is much greater than if we walk on four. But having released two out of four limbs means being able to use them to explore new regions of our reality, in ways that were previously impossible even to imagine.

Consider again a pendulum as the archetype of balance and stability. Every little force applied to it will make it spontaneously move back to its original state of equilibrium. There is almost no freedom involved in this archetype and one could even state, for the sake of clarity, 'no freedom at all' because of 'pure balance'.

So, if many 'collapses', i.e., 'actualizations of potential outcomes', are heading towards disaster, in fact the majority of them, a few will be instead towards the construction of new steppingstones, i.e., of new islands of local stability. This applies to all forms of freedom, not only freedom of movement in physical systems, but also, say, freedom of speech in sociocultural systems, so important for the development of our modern societies, increasingly focused on verbal and symbolic communication.

But the intrinsic instability of a state of freedom requires those who inhabit it to never get distracted, because in every instant of their advancement on the path of life, choices are required that could lead to the construction of new local balances, but also to self-destruction. Expressed in mythical language, it is about staying aware of the constant battle between 'good' and 'evil'. The latter will always prevail if the former takes for granted what was achieved up to that moment. To put it in a catchphrase: *The forces of good, not to succumb to evil, must make of instability their only stability.*

6 The Human Condition

Depending on the context, we can of course replace the above 'instability-stability' binomial with that of 'uncertainty-certainty', 'insecurity-security', 'vulnerability-invulnerability', etc. What is important to observe is that there are many more ways to be uncertain than to be certain, to be insecure than to be secure, to be vulnerable than to be invulnerable, and so on. In other words, uncertainty, insecurity, vulnerability, and alike, are just different variations of the instability notion, relating to different contexts of our human experience.

The notion of instability comes of course from physics, but it extends beyond the strict domain of physics, because its definition only relies on the very general notions of 'entity' (or 'system') and 'state', and it doesn't matter if the entity in question is physical, biological, cultural, economic, etc. For instance, it equally well applies if the entity being studied is, say, the stock market.

Let us recall here what the typical definition is. An entity is said to be in a 'state of stable equilibrium' if, following an interaction producing a change of its state, when the interaction stops the entity will evolve back, spontaneously, to its initial (equilibrium) state. Reciprocally, an entity is in a 'state of unstable equilibrium' if, following an interaction producing a change of its state, when the interaction stops the entity will move away, spontaneously, from its initial (unstable) state.

Our human condition is certainly different from that of other biological species evolving on this planet. This is well captured, as we mentioned already, by our leap towards the 'upright posture', which corresponds to a new steppingstone in our evolutionary path, allowing us to explore new forms of 'freedom in instability' and new 'provisional and local stabilities'. Most of the other animals on the surface of the planet have waited too long in their previous region of temporary and local stability, that of the 'crawling posture', or even that of the 'immobile posture', if we think of plants.

We humans, through the upright posture, have freed our hands, and by doing so could also keep our gaze fixed on the horizon, rather than on the ground. This produced an additional change of perspective, a new form of instability we could say: that of being able to see not only the present, what is close to us, but also the future, which is distant from us, initially in spatial terms, but then, more generally, and more abstractly, in pure temporal terms. And the future brings with it many new elements of uncertainty, thus many new instabilities.

All this fostered an evolutionary shift on a purely cognitive level, which involved brand new critical choices along the way, between 'construction' and 'destruction', between 'good' and 'evil', the former always representing only a small subset of the totality of all possible choices. Think of the uniqueness of today human condition, of its newly acquired instabilities and local stabilities. How better to do so than by considering nuclear energy. Due to progress in physics' research, humanity has found a way to harness and unleash that form of energy that was defused in the early times, when matter was separated from antimatter, in the very beginning of the matter-life journey.

The drama of our collective choice of unleashing such primordial energy, back in 1945, first in the Trinity test then on Hiroshima and Nagasaki, can be appreciated by observing that it was made in order to fight the advancement of highly destructive forces, those embodied by Nazism, which at that historical moment certainly were a faithful representative, within our human culture, of the so-called 'forces of evil', whose goal (almost reached at the time) was to destroy the former islands of stability, rather than to use them to leap forward.

The danger inherent in this new instability, which was created following the construction of the bomb, is well expressed by Born (1971), when a few years later he wrote to his friend Albert Einstein the following: "We've really put our foot in it this time, poor fools that we are, and I am truly sad for our beautiful physics! There we have been trying to puzzle things out, only to help the human race to expedite its departure from this beautiful earth!".

Today, numerous nations possess an enormous destructive power that can easily destroy all of humanity multiple times, which can be released by simply pressing a button. So, the constant fight of life against the threat of spontaneous regression to 'just being', which does not forgive any distraction, is here expressed by the constant effort required not to fall into the temptation of 'pushing that button'. This is certainly a new dramatic phase of unbalance in the journey of human life. All of humanity, from the forties of the last century to today, and until a new steppingstone will be conquered, is in a situation similar to when an alpinist walks on a small ridge between two cliffs. And of course, we could reason in similar ways when considering other global menaces that have been promoted by our human activity, such as pandemics (and we have a tragic example of this with the recent Sars-Cov-2 crisis, still ongoing), loss of biodiversity, global warming and artificial intelligence (Harari, 2018).

This intrinsic nature of matter-life, of being based on instabilities instead of stabilities, on out of equilibrium states instead of stable equilibria, is also clearly reflected in those systems where the human component is present, like in economics, where for instance the inherent instability of the stock market is well known and explains why very rigid forms of top-down economic models have not worked very well. More generally, how the different countries and federations are structured plays a fundamental role in our advancing, as a human society, along the path of matter-life. In that respect, think of the role played by the different political ideologies, always bringing with them specific aspects of stability (via the laws and regulations in force) and instability (via the freedom that these same laws and regulations allow).

7 An Ontology for Morality

Following the logic of our discussion, we can observe that the codes of conduct inscribed in human morality (or moralities), distinguishing between right and wrong behaviors, good and evil, etc., are also mechanisms aimed at creating local islands of stability, that is, steppingstones on the path of matter-life. There is of course an enormous range of possibilities to create such islands of stability, hence different morals can co-exist, playing a similar role in the creation of temporary platforms of stability, which being temporary are of course also in that sense illusionary (the illusion being that of considering them permanent when they are not).

Consider the example of robbery. To protect the population from its nefarious effects, human societies have created conditions such that stealing becomes strongly discouraged, because thieves are arrested and possibly punished for their acts through imprisonment. In other words, through the implementation of specific laws, a realm is created where robbing is not an advice anybody would give to a good friend, because of the heavy consequences that in all likelihood this would entail.

Of course, discouraging through punishment, by incarceration, is a rather crude way of creating an island of stability in relation to the destructive mechanisms of theft. We can easily imagine that as our human societies evolve, sufficient living conditions will be guaranteed for everyone, so much so that no one will have to steal anymore to live decently, hence robbery will also be discouraged in this way, through the creation of a completely different context of life.

Furthermore, it will become possible to discourage theft not just downstream, as usually done, through the punitive mechanism, but also upstream, by improving the educational mechanism. And for those who nevertheless fall into the temptation of stealing, the "punishment" would be that of having to receive a complement of education, in an environment capable of conveying principles such as for example a renewed trust in the human potential, to be understood as a force of construction (good) instead of destruction (evil). Exactly the opposite of what unfortunately happens in many penitentiaries of so-called advanced countries, which while effective in isolating the offenders from the rest of society, to protect it, they rarely succeed in re-educating them to a more positive view of human evolution.

As Nelson Mandela used to say, "It is said that no one truly knows a nation until one has been inside its jails. A nation should not be judged by how it treats its highest citizens, but its lowest ones."

8 Challenging the Common View

We can ask: how widespread are the unstable states of equilibrium of the "pencil standing on its tip" kind? And considering the importance of these states in the process of life and its evolution, we can also ask: how widespread is the potential for life in the fabric of our physical reality? The answer is that it is present at its most intimate level. This means that in the same way the threat of thermal death is omnipresent, and living systems have to constantly struggle not to succumb to it, the resources for escaping from its yoke are also always available to be exploited, at least within those islands of stability that we have built over time.

To use a quantum language, this is so also because superposition states are ubiquitous (at the micro level, they constantly emerge as a consequence of the linearity of the Hilbert space). Indeed, a superposition state expresses a situation of potentiality, where each of the states in the superposition can be actualized. This can happen each time the entity in question interacts with a suitable context, which in quantum mechanics is called a 'measurement context' and more generally can be called an 'indeterministic context' (Aerts, 2002).

In fact, a superposition state is a superposition only in relation to a context which is able to trigger a collapse into one of its components, and in that respect a quantum measurement process can be considered as a (weighted) symmetry breaking process, very similar to that of the pencil standing on its tip and collapsing to the ground along a specific (prior to the collapse only potential) direction. So, given a context, a superposition state is the genuine expression of an unstable equilibrium, where the indeterministic transition towards one of the states in the superposition is initiated by the presence of fluctuations in the context, bringing the entity in question into a more stable state (Aerts & Sassoli de Bianchi, 2014).

Similar situations exist at the macro level, for self-organizing complex systems, where the collapse phenomenon is usually described by using the notion of 'bifurcation'. As the values of some order parameters increase, the number of stable solutions may also increase, forcing the system to select only one among a number of a priori equivalent solutions, which are more stable attractor states (Heylighen, 2021). So, there is a deep correspondence between the way macroscopic complex system and microscopic quantum entities evolve, as regards the role played by indeterministic contexts producing symmetry breaking processes.

These bifurcations, or collapses, are at the core of the evolution of living systems, always struggling to find new islands of stability, which however are also always temporary, as the contexts with which an entity interacts are also constantly changing. In other words, if the path of matter-life is formed by a succession of dynamical instabilities, interspersed with momentary local stabilities, this is so because it is also a path of constantly changing contexts, the local stabilities being only defined with respect to a given context, hence they are always relative and never absolute.

Take again the paradigmatic example of the pencil on its tip. This is a very unstable state, but only because the pencil is in the context of the gravitational field generated by our planet and is initially aligned along its lines of force. If the context changes, for instance the pencil is placed in a spaceship, stabilities and instabilities will be defined by the spaceship directions of acceleration, and a spaceship constantly changing direction will clearly impose on the pencil a sequence of different indeterministic contexts. In other words, a stable state relative to a given context might suddenly become unstable relative to a newly emerging context.

We mentioned the second law of thermodynamics, pushing isolated systems towards states of equilibrium, as a main obstacle for the way matter-life evolves along its trajectory of instabilities, strewn with steppingstones of provisional local stabilities (Schrödinger, 1944). We also said that a quantum superposition state "is" precisely an unstable equilibrium state, when we consider the presence of uncontrollable fluctuations in its environment (Aerts, 1986; Aerts & Sassoli de Bianchi, 2014). This means that the mechanism of the second law of thermodynamics is already at work at the quantum mechanical scale, although research is still ongoing to fine tune its formulation (Bera et al., 2017; Binder et al., 2018; Popescu et al., 2006).

What we are challenging here is the common view that life would only be about growth, reaction to stimuli, and reproduction, and that living organisms would always seek a condition of homeostasis, i.e., of equilibrium. Again, we are not saying that homeostasis is not key for the survival of an organism. Our thesis is that this does not capture the essence of what life tries to achieve, by always seeking for new out of equilibrium states. Our emphasis is on the 'bifurcation aspect' of complex living systems, that is, the way they can access new potentialities and subsequently new islands of stability through a succession of transitions, this being governed also by the fact that their contexts are constantly changing, hence imposing incessant processes of adaptation, not only through continuous changes, but also and mostly through discontinuous ones (jumps, collapses, bifurcations).

That being said, note that even homeostasis is to be understood in a purely dynamical sense, as is clear that complex self-organizing systems, as living systems are, are open systems continuously exchanging matter and energy with their environment, supposedly also working at the so-called 'edge of chaos' (Packard, 1988). Even maintaining the status quo, i.e., being able to survive rather than die, is per se a great challenge. Preserving over time one's structural integrity is already something not to be taken for granted, and we all know how difficult this is when we struggle already in simply maintaining our house in order.

It is not by chance that Jordan Peterson has precisely phrased one of his antidotes to chaos (i.e., to disorder) by precisely referring to such task, as a way of "fighting against one's personal entropy" (Peterson, 2018): "Set your house in perfect order before you criticize the world," he writes. Because if we do not pay constant attention, everything will spontaneously fall apart. We have to constantly clean up our house in the same way that we have to constantly clean up our life, and our organisms. And the only way to do so, is to move the dirt that accumulates out of it.

The good news is that 'moving the dirt outside' is always possible, despite the diktat of the second law of thermodynamics. One of the reasons is that our physical reality has plenty of space. Here we need to emphasize again that a house (an organism) cannot be kept clean if it remains closed, i.e., if it remains isolated. Even if the entropy of the entire universe, when considered as a closed system, is doomed to increase, local regions in it are allowed to keep their entropy low, or even decrease it, by exchanging matter or energy with their outer space, so that the global increase of entropy, as per the second law, can be respected.

On our planet, life has achieved the goal of preserving and increasing the structural order that is necessary for its development by "absorbing order" from the electromagnetic waves coming from the sun, and by "eliminating disorder" via sending these same waves back to the cosmos, but with a lower frequency, hence, with a higher entropy. This is exactly what happens in the process of photosynthesis, initiated billions of years ago by the cyanobacteria (Brittin & Gamov, 1961).

Note also that, as we said already, in quantum mechanics collapses are possible because of the formation, prior to the collapse, of superposition states. If on one hand the quantum superposition of a system with its environment, i.e., its entanglement with the environment, is what will cause its thermalization (Popescu et al., 2006), the latter can also be avoided thanks to the collapse mechanisms. In quantum mechanics, thermalization is an expression of what is called 'decoherence', whereas stability islands are here to be understood as the creation of local 'domains of coherence'.

9 The Hidden Coherence

It is sometimes claimed that the final state of the universe needs to be one of equilibrium, the so called 'thermal death', and that in this sense life is doomed to fail in its battle with the second law of thermodynamics, and its quantum analog. This claim, however, cannot any longer be upheld within the cosmological scheme we presented. Indeed, it is not at all straightforward which mechanism is the more fundamental one: that of matter-life going through an irreversible journey over instabilities, using local islands of stability as steppingstones, or that of symmetry breaking processes starting from an abundance of superposition states and gradually building up a global stability island.

An indication that even inside our universe this question is not settled can be found in the overwhelming quantum coherence that is revealed when, in man-made laboratories, boson gases are cooled down to temperatures that are so close to the absolute zero that nowhere in the universe similar conditions are known to exist, allowing for the creation of the mysterious Bose–Einstein condensates (Anderson et al., 1995, Bradley et al., 1995, Ketterle & Druten, 1996). The realization of the latter, and the pure quantum coherence of the atoms participating in the condensate state, is not only a masterpiece of experimental physics, but also the proof that humans, guided by the knowledge provided by science, can penetrate into realms of reality that (as far as we know) were never realized before through natural (non-human) processes, following the Big Bang.

We will return to this later in the chapter and just want to remark here that this is a concrete example of how the human cultural evolution can reveal a perfect quantum coherence hidden in the depths of reality, in this case by shielding the atoms used in the above-mentioned experiments from the disturbance of the constant random bombardment from heat photons, which would make the construction of the Bose–Einstein condensate impossible. The question whether this is or not an essentially different process from that of human culture creating new material artifacts that had never existed before, and could never had come to light in a natural way after the Big Bang, is certainly one opening up to further issues about the nature of reality, worth to ponder and explore.

Something else we will return to in the chapter, worth mentioning here already, is that 'quantum entanglement' and 'meaning entanglement' can give rise to more powerful ways of creating order than those allowed by the second law of thermodynamics at the classical level, i.e., mass or energy exchanges with the outside world, which might shed a very different light on the question of the long distant future of the whole of reality, including human culture.

It is certainly plausible, within our view, to consider human culture as the continuation of the evolution of life, in turn to be considered as the descendant of the elementary quantum entities: hadrons and leptons, atoms and molecules. In other words, we can view the latter as the ancestors of a descent that produced life in matter. But not all lines of this descent have been able to bring about fruitful forms. In that respect, gases, liquids and crystals should be considered as evolutionary dead-ends, too stable in our actual macroscopic universe to promote the necessary instabilities for the path of matter-life to be reinitiated (Aerts & Sassoli de Bianchi, 2018; Aerts & Sozzo, 2015).

Coming back to our actual state of affairs on planet Earth, we can say that life and human culture have a good chance to win their relentless battle against disorder, thanks to the existence of a virtually unlimited amount of space. In a famous quote, Carl Sagan said that "The universe is a pretty big place. If it's just us, seems like an awful waste of space." Well, even if it would be "just for us," it would still not be a waste of space, as space is what we need, in order to dispose our present and future wastes, particularly so if humanity will end up conquering a large portion of the cosmos (Deutsch, 1997). But life can win its relentless battle against disorder also thanks to its ability of making the necessary 'discrete route changes', using the available collapse/bifurcation mechanisms. And the full potentiality for these further impulses along the road of life, dotted with steppingstones, is today to be found in human culture, in its ability to acquire additional knowledge via the protective guide of the methods of science (to be understood in a non-reductive sense).

10 The Global Scenario

So, the descendants of the microscopic constituents of matter are to be found both in living matter and in human culture (downsizing the importance of the macroscopic material universe as regards the essence of the life process and its evolutionary trajectory). But also, quantum theory indicates that the microscopic entities, in most of their states, are non-spatial. This means that matter-energy, in its macroscopic form, and our whole material universe, have only apparently reached a huge local stability. Indeed, it follows from our interpretation of quantum superposition that our spatiotemporal material universe is only a portion of a much vaster non-spatial and non-temporal reality (Aerts, 1999; Sassoli de Bianchi, 2020).

Also, as we mentioned in the beginning of this chapter, we do not have yet an established physical theory able to explain the absence of antimatter in our material universe. Such theory would probably also provide the missing explanation of the very existence of our universe. But whatever the details of such theory will be, we can generally think to the global situation as resulting from a cosmic symmetry breaking process, where a superposition between matter and antimatter has been collapsed toward the actual 'only matter state' of our universe. In that sense, our material universe would just be a "local" steppingstone of a global (cosmic) evolutionary process, in which the mechanisms of 'actualization of potential properties' plays a central role, at all levels of manifestation, no less important than that played by the classical (in the sense of non-quantum-like) Darwinian mechanisms (Aerts & Sassoli de Bianchi, 2018; Gabora & Aerts, 2005).

As we mentioned already, we probably touch here at aspects related to the nature of time itself, if we understand matter as 'moving forward in time' and antimatter as 'moving backwards in time'. A possibility is that time, space and irreversibility would have jointly come into existence with matter and life. In other words, irreversibility would start when matter irreversibly separates from antimatter, introducing in our reality a temporal permanent arrow.

Only light (i.e., electromagnetic phenomena) would be immune to that arrow, being an entity existing independently of the coming into existence of time, space, matter and energy. This explains why light behaves so atypically when we observe it from the particularity of our matter-life trajectory and of the spatial theater in which it unfolds. This atypicality is for instance manifested in the fact that the light speed is always the same in every reference frame. This is so because 'light' is just 'being', whereas time and space are constructed around those entities that participate in the matter-life trajectory. In that respect, time and space are to be considered a very peculiar emergent aspect of our reality (Aerts, 1999, 2018).

In our human journey, we tend to only focus on stabilities and balances (and we tend to forget they are only local and temporary), but we should never forget about the existence of a bigger historical perspective to our evolution as matter-energy entities. Also, we should keep in mind the crucial difference between the substance of 'being' and the substance of 'matter-life and human culture', with its very peculiar trajectory. To some extent, some of the features of 'being' can also be applied to the 'matter-life and human culture' trajectory, but this is so only because of the existence of the local regions of stability, in which for some time 'life' and also 'human culture' can resemble 'being', i.e., 'balance'. But as we have already repeated several times, it is just a local resemblance, and therefore only a partial one.

This is similar to how a curved spacetime resembles, if we only look at it locally, to a flat spacetime. Its curvature can only be felt when we look at things from a sufficiently global perspective. With the life trajectory it is the same. We need to observe it using a global perspective, to understand that it describes a very particular state, very different from the default state of thermal equilibrium.

11 Concentration of Meaning

If our considerations about the essence of the 'matter-life and human culture' phenomenon might be considered relatively credible when applied to our human bodies and to the material vehicles of manifestation of other living creatures, as we mentioned already less evident is the situation as regards human minds and human consciousnesses. Should they also be included in the same "fight or perish" evolutionary scenario of the matter-life processes? Maybe not fully, but it is likely that a similar scenario is to be considered for at least a part of our cognitive functions, if for instance we consider, and take seriously enough, the emergent field called 'quantum cognition', where cognitive phenomena are shown to be quite effectively modeled using the formalism of quantum theory (Aerts et al., 2013, 2016; Busemeyer & Bruza, 2012; Haven & Khrennikov, 2013; Wendt, 2015).

Since there are reasons to believe that thermodynamics has its roots in the quantum probabilities, i.e., that the probabilities of statistical mechanics, and therefore the second law of thermodynamics, can be obtained directly from the quantum probabilities, so that entropy would be a meaningful concept even for individual quantum systems (Beretta et al., 1984; Hatsopoulos & Gyftopoulos, 1976), this means that the menace of the second law also applies to cognitive entities, if it is true that the quantum formalism (or a generalization of it) describes them equally well than the microphysical entities (i.e., they should also be understood as bona fide quantum entities).

The more so if we also consider that there are reasons to believe that our physical world could in fact also be conceptual in nature, although distinct from the realm of human concepts. This is at least the hypothesis at the foundation of the 'conceptuality interpretation of quantum mechanics' (and of relativity theory), which is currently under investigation (Aerts & Beltran, 2020; Aerts et al., 2020).

Note that this hypothesis is not the result of an 'ad hoc and top-down' assumption, child of the metaphysical tastes of those who have formulated it. It really came about because of purely technical reasons, related to the similarity in behavior between quantum micro-entities and conceptual entities. In particular, concepts are also characterized by an ontological Heisenberg's uncertainty principle, since a concept cannot be maximally concrete and maximally abstract at the same time, and similarly to quantum entities, they can also be truly indiscernible, they can establish meaning connections, reminiscent of the entanglement phenomenon, etc. (Aerts et al., 2020).

So, in the light of the conceptuality interpretation, the existential threat that the second law represents, i.e., the increase of entropy of a system when it isolates itself and ceases to struggle for its survival and evolution, translates in the cognitive domain in the threat of knowledge destruction, of a growth of ignorance, a loss of meaning information, as in ultimate analysis 'meaning' would be the constitutive substance of 'matter, life and culture'. In that respect, the evolution of life can be truly conceived as a process of 'concentration of meaning', where 'meaning' is also to be understood as a process of 'creation of coherence', under the constant threat of the environmental 'decoherence processes'.

To better explain what we mean, let us give an example. A city, with its cultural realizations and happenings, is to be considered an environment with a high concentration of meaning, i.e., where coherence is densely present, mostly at its center. But moving from the center outwards, one can also find garbage belts, at the outskirts of the city, which are instead a good example of how things can decohere. If we walk in the town library, we can find books on the shelves and tables that can be read by visitors, all well-kept and classified. Everything in the library, both in terms of how the books are organized and their semantic content, is an expression of perfect coherence, and visitors have to follow a well-defined 'meaning trajectory' when coming to the library to find and consult a book.

Compare now the above with a walk where waste dumps are located, in the outskirts of the city. There as well one might find books, but mostly in the form of pieces of volumes that have become unreadable because of their deterioration, or in the form of single pages torn from books. These fragments have lost all their meaning, their coherence, with the exception maybe of some meaning still remaining at a more "microscopic" level, in those pages where paragraphs can still be read and the original meaning content grasped (note however that reading a paragraph, or even an entire page, does not mean being always able to grasp its full meaning, as this might require the full context provided by the missing paragraphs, or pages).

Our macroscopic, material and spatiotemporal universe, is rather similar to this waste dump. Indeed, it is only in the micro-realm of particles, atoms and molecules, that a full quantum coherence has been preserved, while in the macro-realm the heat photons are constantly randomly flying around, bouncing off the surfaces of planets, with the consequence that little quantum coherence is left there, with the exception of the surface of our planet Earth (and of the surface of planets where life and culture might have emerged in a similar way), where first biological life and then human culture came into existence, restoring in this way the presence of a high concentration of meaning-coherence. This up to the point of accessing a level of quantum coherence that was never realized following the Big Bang, when physicists created the first Bose–Einstein condensates, by protecting the boson gas from the bombardment of the heat photons with an efficiency never achieved before (Aerts & Sassoli de Bianchi, 2018; Aerts & Sozzo, 2015; Anderson et al., 1995; Bradley et al., 1995; Ketterle & Druten, 1996).

Returning to the conceptuality interpretation, note that it contemplates the existence of a fundamental duality, that of 'mind and language', which at the physical level translates into the duality of 'matter and force fields', i.e., of 'fermions and bosons'. Following the hypothesis of supersymmetry (Martin, 1988), such duality would not be fundamental, whereas for the conceptuality interpretation it is plausible (although not strictly necessary) that it could be older than our spatiotemporal universe. Note that we know from Hamiltonian mechanics that a remarkable symmetry exists between 'space and time' and 'momentum and energy'. Such symmetry is however not any more effective in the spatiotemporal universe we inhabit, where energy and momentum are not part of the "canvass," but are the properties attributed to the material entities evolving on that canvass. Hence, a fundamental and very ancient symmetry breaking must have taken place, possibly even before the beginning of our universe, and because of that our standard view of reality, based on having minds strictly connected with material macroscopic bodies, should be considered to be a parochial one.

The above discussion brings with it, as is inevitable, further big questions, like: what is the nature of the symmetry breaking that has caused us to end up into such a niche, where time and space are 'extensions', whereas momentum and energy are properties of entities living inside such extensions? Although it is difficult to directly answer this question, if we assume that the nature of our physical reality is conceptual, that is, governed by meaning, we can observe that there is a deep correspondence between the idea of symmetry breaking centered around the 'coming into existence of macroscopic matter' and the corresponding 'coming into existence of time and space', and the symmetry breaking manifesting at the level of the logical connectives 'and' and 'or' (Aerts, 2013).

Indeed, the connective 'and' tends towards extension, whereas the connective 'or' penetrates inwards. If, for example, we say 'chair and another chair', and we consider them to be objects, then the combination 'chair and another chair' needs extended space to exist compared to the situation of a single chair. On the other hand, if we say 'chair or another chair', we are immediately outside of time and space. An object A 'or' another object B is in fact no longer an object, but a concept. So, the notion of 'object' breaks the symmetry between the connective 'and' and 'or', whereas such symmetry remains intact in the more abstract realm of concepts. Indeed, connecting two concepts by means of the 'or' connective simply leads to a more abstract concept, while connecting two concepts via the 'and' connective gives rise to a more concrete concept (Aerts et al., 2020).

Our hypothesis is that the above symmetry breaking (which took place in the sedimentation process going from the more abstract level of concepts to that of objects, with the latter to be considered as the limit case of maximally concrete concepts) mirrors what took place in the creation of our spatiotemporal material universe. In other words, time and space would result from the working of the 'and' connective, bringing concepts together in a spatiotemporal way, whereas the connective 'or' allowed for the creation of entities, systems and organisms, with an inside in which the 'or' quantum-like dynamics dominate.

12 Concluding Remarks

Summing up, we have put forward the view that the split-up between matter and antimatter, following the Big Bang, marked the coming into existence of past and future and the beginning of life in our universe as we know it, with its constant struggle to find new states of imbalance, while conquering provisional domains of stability to be used as evolutionary steppingstones.

We have argued that the struggle in question is the endless fight against the second law of thermodynamics, at the scales of both classical physics and quantum physics, and the presence of antimatter, and decoherence, at the even deeper quantum scale.

We remarked already that some fine tuning is still needed at the quantum level for a complete formulation of the second law of thermodynamics (Bera et al., 2017; Binder et al., 2018), but the way we put forward its role here remains essentially valid. Additionally, in the quantum case the 'exchange flows of matter or energy between a system and its environment', necessary to keep the entropy of the system low, can take the form of the 'entanglement correlations', giving rise to different structural possibilities, as a consequence of the von Neumann entropy being applicable instead of the Boltzmann entropy.

More concretely, for quantum systems, if a system is in a product (non-entangled) state, the entropy of the system is just the sum of the entropies of the subsystems, as in the case for classical systems. But when the subsystems become entangled, the von Neumann entropy of the composite system will be generally lower than the sum of the entropies of the subsystems, as quantum entropy is subadditive (Araki & Lieb, 1970). One can easily check this by considering a bipartite system formed by two spin one-half entities in a singlet state, which has minimal quantum entropy while its two subsystems have maximum quantum entropy. This means that at the quantum level a more general and powerful mechanism of 'order creation' is available as a consequence of the possibility of 'entanglement connections' than what 'matter and energy streams' can give rise to on the classical level, i.e., a composite system can acquire more order than all of its subsystems, due to the entanglement of these subsystems.

Taking into account 'quantum cognition', this more powerful way of order creation also exists at the cultural level, and in our Brussels research group we have put forward explicit examples of 'cognitive entanglement situations' where this more general and powerful way of order construction can be identified (Aerts & Sozzo, 2011, Aerts et al., 2019a). Hence, at the cultural level, similarly to the quantum level, 'concentration of meaning' can be an expression of 'meaning entanglement', i.e., of a meaning-connection between different parts of a composite system that creates a situation of relative order and stability.

In that respect, different from the non-living entities, the living ones (the cultural layer being interpreted as a continuation of the biological one) are those that are able to protect themselves from death, not in the sense of becoming immortal (at least, not at the individual level), but in the sense of being able to shield themselves for long enough from all sorts of decoherence processes, like the incessant random bombardment of heat photons on the surface of the planet. This in order to preserve and enhance their inner organization and participate in the construction of always more advanced islands of stability.

In the case of humanity, this protective capacity has evolved to the level of exploiting the amplifying effect of our nervous systems, with the advent of human cognition, language and cultural evolution, allowing to transfer to the macro level the 'quantum coherence' that is inherent to the micro level. So, in ultimate analysis, what we have called in this chapter 'islands of stability' are 'island of concentration of meaning', 'islands of knowledge', as knowledge and meaning are what provide the necessary support to fight against the present and future adverse environmental conditions, thus allowing for an unlimited evolutionary reach.

Said this, it is worth also emphasizing that 'islands of stability' are useful only if they are also 'islands of morality', as is clear that knowledge is of little use if we do not know in which direction to apply it, that is, if it lacks the historical perspective offered to us when we contemplate and fully appreciate the cosmic battle of life against the menace of a regression to the dimension of 'pure being'; a battle that probably started with the matter–antimatter primordial separation and still continues today through our struggle against the second law of thermodynamics (and so, it would be incorrect to think that biological complexity only started from single-cell organisms).

Only with such cosmic-historical perspective it becomes possible to have access to an ontology for morality and its principles, so as to equip ourselves with a reliable moral compass, which can guide us in the choices that await us in the future.

We invite the reader, at this point, to read the quote in the Introduction once again, as a short poetic description, and suggestion, of a moral founded on the knowledge we presented in this chapter. Indeed, knowing that in our human condition we are profoundly and inevitably bound to traveling along an irreversible trajectory, dashed with steppingstones of only local and temporary stability, it undoubtedly leads us to contemplate a very different ontology than that of not possessing the perspective offered by such knowledge. The last sentence of the quote, "The deep secret of life is that if you pursue a noble goal all the characteristics of life described above become the substance of meaning, which is the food of the human mind," can now be better understood and appreciated, we hope, in the light of what we have tried to explain.

In that respect, we also hope that the present essay may represent a small contribution in the direction of a more objective demarcation between what the ancients indicated with the terms of 'good' and 'evil', i.e., between the forces of 'construction' and 'evolution', and those that instead promote the 'destruction' and 'involution' of every possible structure.

References

Aerts, D. (1986). A possible explanation for the probabilities of quantum mechanics. *Journal of Mathematical Physics, 27*, 202–210.

Aerts, D. (1999). The stuff the world is made of: Physics and reality. In D. Aerts, J. Broekaert, & E. Mathijs (Eds.), *Einstein meets Magritte: An Interdisciplinary Reflection* (pp. 129–183). Springer.

Aerts, D. (2002). Being and change: Foundations of a realistic operational formalism. In D. Aerts, M. Czachor, & T. Durt (Eds.), *Probing the structure of quantum mechanics: Nonlinearity, nonlocality, probability and axiomatics* (pp. 71–110). World Scientific.

Aerts, D. (2018). Relativity theory refounded. *Foundations of Science, 23*, 511–547. https://doi.org/10.1007/s10699-017-9538-7

Aerts, D., & Beltran, L. (2020). Quantum structure in cognition: Human language as a boson gas of entangled words. *Foundations of Science, 25*, 755–802.

Aerts, D., & Sassoli de Bianchi, M. (2014). The extended Bloch representation of quantum mechanics and the hidden-measurement solution to the measurement problem. *Annals of Physics, 351*, 975–1025.

Aerts, D., & Sassoli de Bianchi, M. (2018). Quantum perspectives on evolution. In S. Wuppuluri & F. A. Doria (Eds.), *The map and the territory: Exploring the foundations of science, thought and reality* (pp. 571–595). Springer, The Frontiers collection.

Aerts, D., Gabora, L., & Sozzo, S. (2013). Concepts and their dynamics: A quantum-theoretic modeling of human thought. *Topics in Cognitive Science, 5*, 737–772.

Aerts, D., Sassoli de Bianchi, M., & Sozzo, S. (2016). On the foundations of the Brussels operational-realistic approach to cognition. *Frontiers of Physics, 4*, 17. https://doi.org/10.3389/fphy.2016.00017

Aerts, D., Aerts Arguëlles, J., Beltran, L., Geriente, S., Sassoli de Bianchi, M., Sozzo, S., & Veloz, T. (2019a). Quantum entanglement in physical and cognitive systems: A conceptual analysis and a general representation. *The European Physical Journal Plus, 134*, 493. https://doi.org/10.1140/epjp/i2019-12987-0

Aerts, D., Ekeson, K. W., Schneider, V., & Sassoli de Bianchi, M. (2019b). The secret of life. *AutoRicerca, 18*, 21–107.

Aerts, D., & Sozzo, S. (2011). Quantum structure in cognition: Why and how concepts are entangled. In D. Song, M. Melucci, I. Frommholz, P. Zhang, I. Wang, & S. Arafat (Eds), *Quantum interaction. QI 2011. Lecture Notes in Computer Science 7052* (pp. 116–127). Springer. https://doi.org/10.1007/978-3-642-24971-6_12

Aerts, D., & Sozzo, S. (2015). What is quantum? Unifying its micro-physical and structural appearance. In H. Atmanspacher, C. Bergomi, T. Filk, & K. Kitto (Eds), *Quantum interaction. QI 2014.*

Lecture Notes in Computer Science 8951 (pp. 12–23). Springer. https://doi.org/10.1007/978-3-319-15931-7_2

Aerts, D., Sassoli de Bianchi, M., Sozzo, S., & Veloz, T. (2020). On the conceptuality interpretation of quantum and relativity theories. *Foundations of Science, 25*, 5–54.

Aerts, D. (2013). La mecànica cuántica y la conceptualidad: Sobre materia, historias, semántica y espacio-tiempo. *Scientiae Studia, 11*(2013), 75–100. https://doi.org/10.1590/S1678-316620130 00100004. Translated from: Quantum theory and conceptuality: Matter, stories, semantics and space-time. arXiv:1110.4766 [quant-ph].

Anderson, M. H., Ensher, J. R., Matthews, M. R., Wieman, C. E., & Cornell, E. A. (1995). Observation of Bose-Einstein condensation in a dilute atomic vapor. *Science, New Series, 269*, 198–201.

Araki, H., & Lieb, E. H. (1970). Entropy inequalities. *Communications in Mathematics Physics, 18*(2), 160–170.

Benoit-Lévy, A., & Chardin, G. (2012). Introducing the Dirac-Milne universe. *Astronomy & Astrophysics, 537*, A78. https://doi.org/10.1051/0004-6361/201016103

Bera, M. N., Riera, A., Lewenstein, M., & Winter, A. (2017). Generalized laws of thermodynamics in the presence of correlations. *Nature Communications, 8*, 2180. https://doi.org/10.1038/s41467-017-02370-x

Beretta, G. P., Gyftopoulos, E. P., Park, J. L., & Hatsopoulos, G. N. (1984). Quantum thermodynamics. A new equation of motion for a single constituent of matter. *Nuovo Cimento B, 82*, 169–191.

Bertsche, W. A. (2017). Prospects for comparison of matter and antimatter gravitation with ALPHA-g. *Philosophical Transactions of the Royal Society A, 376*, 20170265.

Binder, F., Correa, L. A., Gogolin, C., Anders, J., & Adesso, G. (Eds.). (2018). *Thermodynamics in the quantum regime*. Springer.

Born, M. (1971). *The Born-Einstein letters. Correspondence between Albert Einstein and Max & Hedwig Born from 1916 to 1955 with commentaries by Max Born*. The Macmillian Press Ltd.

Bradley, C. C., Sackett, C. A., Tollett, J. J., & Hulet, R. G. (1995). Evidence of Bose-Einstein condensation in an atomic gas with attractive interactions. *Physical Review Letters, 75*, 1687–1690. https://doi.org/10.1103/PhysRevLett.75.1687

Brittin, W., & Gamov, G. (1961). Negative entropy and photosynthesis. *Proceedings of the National Academy of Sciences, 47*, 724–727.

Busemeyer, J. R., & Bruza, P. D. (2012). *Quantum models of cognition and decision*. Cambridge University Press.

Chardin, G., & Manfredi, G. (2018). Gravity, antimatter and the Dirac-Milne universe. *Hyperfine Interactions, 239*, 45. https://doi.org/10.1007/s10751-018-1521-3

Deutsch, D. (1997). *The fabric of reality*. The Penguin Press.

Feynman, R. P. (1949). The theory of positrons. *Physical Review, 76*, 749.

Gabora, L., & Aerts, D. (2005). Evolution as context-driven actualization of potential. *Interdisciplinary Science Reviews, 30*, 69–88. https://doi.org/10.1179/030801805X25873

Harari, Y. N. (2018). *21 lessons for the 21st century*. Spiegel & Grau.

Hatsopoulos, G. N., & Gyftopoulos, E. P. (1976). A unified quantum theory of mechanics and thermodynamics Part i. Postulates. *Foundations of Physics, 6*, 15–31.

Haven, E., & Khrennikov, A. Y. (2013). *Quantum social science*. Cambridge University Press.

Heylighen, F. (2021). Entanglement, symmetry breaking and collapse: correspondences between quantum and self-organizing dynamics. *Foundations of Science*. https://doi.org/10.1007/s10699-021-09780-7

Indelicato, P., et al. (2014). The Gbar project, or how does antimatter fall? *Hyperfine Interactions, 228*, 141.

Kellerbauer, A., et al. (2008). Proposed antimatter gravity measurement with an antihydrogen beam. *Nuclear Instruments and Methods in Physics Research Section b: Beam Interactions with Materials and Atoms, 266*, 351.

Ketterle, W., & van Druten, N. J. (1996). Bose-Einstein condensation of a finite number of particles trapped in one or three dimensions. *Physical Review A, 54*, 656–660. https://doi.org/10.1103/PhysRevA.54.656

Kibble, T. (2009). Englert-Brout-Higgs-Guralnik-Hagen-Kibble mechanism. *Scholarpedia, 4*, 6441. https://doi.org/10.4249/scholarpedia.6441

Martin, S. P. (1988). A supersymmetry primer. In G. L. Kane (Ed.), *Advanced series on directions in high energy physics. Perspectives on supersymmetry* (pp. 1–98). World Scientific.

Packard, N. H. (1988). *Adaptation toward the edge of Chaos*. University of Illinois at Urbana-Champaign.

Peterson, J. B. (2018). *12 rules for life. An Antidote to Chaos*. Random House.

Popescu, S., Short, A. J., & Winter, A. (2006). Entanglement and the foundations of statistical mechanics. *Nature Physics, 2*, 754–758.

Sakharov, A. D. (1967). Violation of CP invariance, C asymmetry, and baryon asymmetry of the universe. *Journal of Experimental and Theoretical Physics Letters, 5*, 24–27.

Sassoli de Bianchi, M. (2020). A non-spatial reality. *Foundations of Science*. https://doi.org/10.1007/s10699-020-09719-4

Schrödinger, E. (1944). *What is life? The physical aspect of the living cell*. Based on lectures delivered under the auspices of the Dublin Institute for Advanced Studies at Trinity College, Dublin, in February 1943. Cambridge University Press 1967.

Shaposhnikov, M. E., & Farrar, G. R. (1993). Baryon asymmetry of the universe in the minimal Standard Model. *Physical Review Letters, 70*, 2833–2836. https://doi.org/10.1103/PhysRevLett.70.2833

Wendt, A. (2015). *Quantum mind and social science*. Cambridge University Press.

Architecture and Big Data: From Scale to Capacity

Nana Last

Without fanfare, the late 1990s saw the construct of scale slip quietly from its cardinal role in orchestrating architecture production to be tacitly replaced by capacity. Ushered in via computing technologies, whose spotlight masks the breadth of this transformation, this changing of the guard saw the central ordering system of architectural production effectively move from its traditional linear, physical, and visual underpinnings to processes based in nonvisible orderings. In this process, scale's inculcated architectural logic of discrete units came to be supplanted by a performance-driven capacity—the greatest amount of information processed in the least amount of time. Part of a broader scientific and cultural transformation of complex ordering principles that structure human-information interactions, within a decade, the impact of this development has become apparent: the emergence of capacity as architecture's fundamental orchestrating determinant of production has unwittingly set the stage for big data to enter in the hollow core opened up by capacity's drive for content.

The technological overtones of this transformation, however, do not tell the full story. Rather, big data proves to be a broad, supra-disciplinary concept that extends beyond computation to establish itself as a wide-ranging cultural technology. Born of the union of advanced computing technologies and the amassing of diverse and near-endless quantities of information, big data is a rapidly emerging cultural technology that goes beyond introducing new procedures into existing modes of cultural and scientific production—it embeds those into the core of operations, and in so doing, transforms them. Not confined to a set of computational methods, it orchestrates a mode of engagement with the subject matter. For architecture, this contributes something other than the next chapter in its history; the advance of big data's approach across disciplines disrupts settled domains of activity to raise the question: history of *what*? Examining the shift in architecture's guiding principles from scale to capacity

N. Last (✉)
School of Architecture, University of Virginia, Campbell Hall 421, Charlottesville, VA 22904, US
e-mail: ndl5g@virginia.edu

S. Wuppuluri and I. Stewart (eds.), *From Electrons to Elephants and Elections*,
The Frontiers Collection, https://doi.org/10.1007/978-3-030-92192-7_43

suggests an answer: big data's association with the discipline merges two of architecture's evolving histories—the ordering principles of narrative and technology[1] as they are intertwined in the relaying, reception, and formation of information.

Architecture's shift from scale to capacity both forms and is the sign of the central operative mechanism in the discipline's adoption of computational and algorithmic logics. Scale has a twofold meaning that is key to its operations in architectural production. It designates both a proportional relation between two sets of dimensions and a graduated measure of relative size, extent, or degree. This constitutive duality allows scale to connect *between* two spheres of measure or modes of being as well as to order activities or things *within* a given domain. Architectural practice has long conjoined these meanings to one another in a way that has come to be seen as natural. Acting across realms, scale's first sense determines the proportional relationship of the architectural representation to what it represents. As such, architectural plans and sections, in being drawn to specific scales, institute a form of specificity between drawing and what is represented, wherein, say, one-eighth inch or one-sixteenth inch equals one foot. The larger the scale (as in, one-quarter inch equals one foot, for example, rather than one-sixteenth inch), the larger the resulting representation. Traditionally, this practice tacitly led to scale's second role of organizing architecture's production process according to a certain ranked order in the development or incorporation of information.

With information as the hinge between the two parts, scale's traditional enabling logic is thus threefold: material, informational, and procedural. The first part of scale—the proportional representation—relays information between the physical world represented and the representation. The second acts within the realm of representation by incrementally incorporating amounts of information according to a given procedural order. Taken together, the resulting scalar logic all but dictates an order of design development. To make details visible, or to represent larger projects, requires larger drawings; the larger the representation, the larger its brute physical correlate of sheet of paper. This manner of connecting a representation to what it represents thereby yielded an order based in a progressive classification of size and importance in relation to the physical world and information. Through this process, the specific scale of the drawing effectively dictated the amount of detail or information included. In practice, then, the two components of scale merged. This routinely meant that, for example, early design stages with limited information, or site plans covering large areas, were drawn at smaller scales such as one-thirty-second inch. Conversely, details or partial views, such as building sections, necessitated representations at a larger scale to develop and make more detailed information visible. Consolidated in what can be thought of as scalar logic, the twofold workings of scale effectively created a series of correspondences that tied type of representation to

---

[1] I am thinking here of these terms, narrative and technology, with their standard dictionary definitions: *narrative* is a spoken or written account of connected events and *technology* is the application of scientific knowledge for practical purposes, or a manner of accomplishing a task especially using technical processes, methods, or knowledge. In this I am particularly concerned with how narrative produces a specified order. I address this in my forthcoming book, tentatively entitled *From Text to Algorithm: Architecture and Big Data.*

information relayed to stage of design development. In this process, scale supplies the functional lynchpin in this web of relations, correlating physical size of representation to quantity and type of information relayed. Scalar logic thus instituted a bond between architectural drawing and the world according to a fixed, readily and visually comprehensible set of principles that suture the describing and inscribing of reality to a specified set of correspondences.

Scalar logic remained dominant in architectural production until the ascendancy of computer-based drawing loosened scale's overarching grip on project development. Architecture's frequently recounted history with computing technologies[2] is one of first explicitly embracing computer-aided drawing and, later, computational logics.[3] With the adoption of the widespread use of computers, digital drawings began to replace hand drawings. Digitally produced architectural drawings, while remaining to scale, come into being in an open-ended digital space that obliterates any notion of a fixed correspondence between representation size and represented—either on screen or paper. This disrupts the material component of scalar logic, as the physical size of a drawing no longer matters. While digitally produced drawings remained conceptually proportional—to scale—the actual size of the visible image on the screen or when printed was no longer fixed. As a result, the impact of scale's first meaning, its proportional relation, was diminished. In a Deleuzean act of reterritorialization, while digital space dissociates information from any fixed physical size, it associates it to a capacity to process large amounts of information in a compressed period of time that coupled with the development of computationally driven architecture strategies to lead to capacity's emergence as a dominant ordering principle overseeing design production.

If scale acts relationally, both across and within realms, capacity, in designating maximums of containment or production, is both more self-contained and more

[2] Architecture's frequently recounted history with computing technologies dates to the early 1990s' popular software Form Z and Rhino, and its often-cited first generation of digital designers, Gregg Lynn and Bernard Cache.

[3] An early entry into this break with scalar logic's centrality appeared as a fascination with the geometry of fractals. Popularized in architecture by Peter Eisenman among others, fractal geometries are infinitely complex patterns that are self-similar across all scales so as to impart the same character to each part as to the whole. Unique in a number of ways, fractals straddle the line between being a naturally occurring state and an abstract construct. They appear in the geometry of coastlines, snowflakes, and crystals, on the one hand, and are computer generated through feedback loops (as with the Mandelbrot Set), on the other. But it is their signature self-similarity that counters long-standing scalar logic. Rather than fix a certain amount of information to a given size, self-similarity allots equal amounts of information at each scale, up or down. For architectural practice, fractals offered a tantalizing image of a nonscalar ordering principle; their distinctive geometric narrative constitutes an intermediary stage in the breakdown of scalar logic and a simultaneous shift to an emphasis on patterns, providing a mediating stage between scale and capacity. Although architecture's foray into fractals was limited, the fascination with them suggests a desire to disrupt long-held notions of scale, representation, and materiality. Any substantial impact on the discipline, however, was quickly eclipsed by the soon-to-be-had, thoroughgoing technological developments in architecture, and its turn to computers and computational logics in the following two decades. The early 2000s saw a fascination with patterns that similarly exhibited a desire to move away from scalar logic.

absolute. As digital drawings became not merely produced on computers, but also computationally driven, computational capacity became more and more of a factor in production. Greater capacity allowed for both more complex drawings and for more to be achieved in less time. Less tied to linearly ordered, sequential increments than to a liminal determinant, capacity does not predominantly hold architecture to considerations of its known physical or informational terrain. Rather, it introduces criteria that favor relations between information to speed and time, rather than to physical size and spatial distance. Although this transformation directly alters scale's proportional component, it had an even larger and more consequential impact on scale's second component: internal ordering that structures how, when, and in what form information may become incorporated into the architectural design processes. Questioning when information enters the design process couples with new technologies to force the rethinking of what information can enter, as well as at what stage it enters. Challenging or changing what information can—or should—be incorporated into architecture, and in what forms this might occur, ultimately forces the discipline to rethink the relations it bears (or defines) to the broader world.

And this is where big data enters.

"Big data" is a shorthand label for processes in which advanced computation, software algorithms, and statistical methods of analysis are applied to contemporary society's immense volumes and varieties of digital data. What gives the term its piquancy is that the label, which began gaining currency around 2008, denotes not just quantity, but also method or logic. As a method, it is not fixed to a specific discipline. In many ways, then, its adoption is unbounded; its very logic defies disciplinary or other boundaries, with its processes only limited by availability of data, speed of computation, and underlying algorithm. Distinguishing big data is its twofold potential to conglomerate diverse forms of traditional and nontraditional data sources while simultaneously instituting previously unseen approaches to digital analysis within existing practices.[4] In doing so, big data's procedures are at least as revolutionary as its information-processing capacity. Driving big data's absorption across disciplines is its inherent promise to unearth otherwise undetectable patterns within amassed data. This potential to mine data for answers to fundamental questions has led some to deem big data science's fourth paradigm.[5] Unlike modern empirical science's reliance upon compressing data from limited samples into generalized explanatory models of natural phenomena,[6] big data's processes need neither compression nor

---

[4] These sources include crowdsourced, social, and other digitally available data from websites, blogs, tweets, data trails, social network communications, sensor data, and other data, and are collected largely without quality control. They span social, economic, political, and environmental spheres to envelop traditional informational sources along with surveillance and communications technologies. The resulting agglomeration spans from scientific principles, at one end, to the production of biopolitical subjects through surveillance and tracking, at the other.

[5] See, for example, Tony Hey (2009).

[6] Big data spans two branches of predictive modeling. Both branches share the goal of using data to learn about some phenomena, but differ in the volume, velocity, and variety of data processed, and the degree and type of human involvement. At one pole of this spectrum is machine learning, which employs algorithms that learn from data without relying on rules-based programming, and, as its

generalization. In lieu of these procedures, data-driven sciences employ algorithms to either mine data for unseen patterns beyond human apprehension[7] or to directly search data to serve as precedents for the behavior of current phenomena.[8] Big data's ready popularization across disciplines and media, however, shows it to designate something beyond discrete methods of data mining. It offers disciplines the ability to do more than navigate vast quantities of disorganized data, and to utilize them to advantage.

While capacity is central to big data, determining the speed, extent, and breadth of data that can be searched and utilized, big data's logic does more than build on capacity. In a reciprocal process, the importing of big data's methods into a discipline or practice—including architecture—upsets pre-existing procedures (and their contained histories); in so doing, it transforms the disciplines embracing them. The injection of data-driven logics into architecture begins by turning scale's structured order inside out to frontload information and emphasize production. If scalar order acted to visualize fixed correspondences, big data's order is multifarious, haphazard, unseen, unvisualized, and potentially unvisualizable. Big data has no need for scale's traditional enabling logic, that connected the material, informational, and procedural to one another. Instead, by reimagining the operations of what had been the unstated glue of scalar logic (information), it disrupts the preordained correspondence of scale's constructed ordering of physical proportional relationship to information to procedural order.

For architecture, the advent of big data discloses a history of architectural responses to increasing modes and quantities of available information, one that has already yielded an array of approaches for introducing or incorporating data into its practices. In the first decades of the twenty-first century, prior to the widespread introduction of data-driven methods designed for the purpose, nascent attempts to grapple with contemporary society's overwhelming flood of information appear in the widespread efforts (both in architecture and other disciplines/practices) to corral data into maps, charts, graphs, and timelines. In reaction to the eruption and availability of digital data's vast and unruly field outside of disciplinary or other order, a host of common visualization methods appeared everywhere in attempts to process disparate information into readily legible images. In architecture, their upsurge signals a profession grappling with ways to incorporate and compress increasing

---

name suggests, it requires few assumptions and minimal human intervention. Machine learning uses computers to parse trillions of collected observations to find patterns hidden in data. Its goal is to directly transpose data into predictions so as to simultaneously learn from past events and predict future ones by discovering complex patterns beyond human comprehension. The second approach is held by statistical modeling, which employs mathematical equations to look for relationships between variables. Based on a set of assumptions, the method requires the modeler to understand the relation between variables. But there are no hard lines between the approaches, as each learns from the other.

[7] Harvey J. Miller and Michael F. Goodchild (2015).

[8] Big data thus fundamentally distinguishes itself from forms of science that relied on generalized models formed by compressing a limited number of observations into a universal model. Data-driven methods need neither limit the number of observations nor compress the accumulated data into a generalized model.

amounts of newly malleable data into its practices. By transforming the potentially meaningless—and certainly unmanageable—accumulation of data into something meaningful, constructing readily comprehensible images of data serves a number of purposes, including alleviating anxiety over the discipline's standing in face of technological, intellectual, and epistemological advances in society.

In a manner comparable to scale's proportional functioning, these methods sutured realms together. By contrast, however, they highlight the correspondence of information across realms, rather than spatial or material relations. Omnipresent timelines, for example, encapsulate and envision data as linear pairings of dates to events. In amassing relevant material and molding it into a definitive history, timelines, charts, diagrams, graphs, etc., act as mediators that situate architectural projects into a broader narrative that acknowledges the need to connect to information beyond its bounds. Acting as a frame to architectural projects, this range of procedures constitutes a state between competing modes of ordering information, in which information mediates, migrates between, or is shared by disciplines and the broader society. Such approaches, however, typically harbor an asymmetric logic: they respond to the digital amassing and availability of information with an analog mentality of turning that information into readily recognizable images of history or other contexts, images in which a discipline sits neatly within a pre-existing and surrounding field of events. If timelines and charts aim to make information visible and legible, the logic and methods of big data disrupt the correspondence of a discipline to its surrounds, visualized or otherwise. It acts on a discipline's relations from both ends—altering how we conceive of information and how we conceive a discipline, negating both scalar logic's proportional correspondence that connects across realms and its concomitant predetermined order of practice within a realm. As a result, the relation between linearly ordered timelines (which necessarily rein in information) to the unruliness and vastness of emerging data constitutes as much a disconnection as a connection.

In constituting associations across realms, scalar logic situated human beings between the realm of the represented world and the act of representing it, making the construction of human subjectivity inseparable from the design process. Against this, the reliance on big data's computer-driven, algorithm-based methods are valuable, exactly because they operate in ways humans cannot. The question, then, is: what is the effect of this on human subjectivity? Implicitly and explicitly, by employing advanced technologies to detect patterns outside of what humans can visualize, big data methods realign traditional relations between meaning and information. This dissociation severs the relationship of information to the visual and comprehensible realm. The institution of new procedures unsettles established modes of processing, ordering, relaying, and structuring information and, with them, their entrenched hierarchies. Rather than aim to curb the plenitude of information, data-driven methods readily exploit it. Unlike other methods for processing information to make it legible, big data's use of algorithms need not remake data over in the image of a more humanly comprehensible visual order associated with texts, graphs, or charts. The resulting misfit of old and new orders highlights the interface between forms of information and modes of practice.

Propelled by computer-driven algorithms and ushered in with a lexicon of data, algorithms, and performance, the shift to capacity questions what the bounds of a given discipline are, and how that boundary functions. Scale and capacity each establish distinct approaches to the subject matter that serve to mediate the ways in which the discipline responds to various modes and types of information. Whereas scalar's linear logic places the architect as director of the process and producer of meaning, capacity's data-driven logics frontload information, emphasize production, and de-emphasize subjectivity. The entry of big data, with its inherent promise to advance knowledge by detecting patterns and relations (beyond those humans can comprehend), challenges the functioning and identity of disciplines following its methods. The use of algorithms dominates these data-driven enterprises. Focusing on algorithmic operations eschews the ambiguous and thorny acts of interpretation associated with projects that focus on producing meaning or narrative, replacing the goal of engaging with a human-produced or -centered narrative with that of achieving an unbiased, scientific, data-driven approach. This is even the case when the approach is in service of a more human-centrically defined end. Data-driven approaches, in which problem solving is key, necessarily define specific goals or outcomes. The move to algorithm-based, data-driven methods thus proceeds from the emphasis on meaning and representation that dominated the 1980s' embrace of architecture-as-text to a performance-focused, algorithm-driven process. This shift in process affects the producer as well. Adhering to its logics undergirds a move away from understanding the intellectual human as an interpreter or elucidator of meaning to one who directly heeds data; from one who engages in interpretation, and controls the influx of information, to one who follows algorithms into an otherwise unnavigable territory. Pronouncing this change, architecture projects engaging these methods frequently proclaim not just a different set of procedures, but a different set of goals.

In fundamentally changing our relationship to information, big data instigates an ontological shift in who we are.[9] When architecture meets big data, it couples big data's implicit resituating of the human being in relation to knowledge with the broader spatio-social situating of humans that forms the traditional province of architecture. That is, it couples an epistemological and methodological resituating of the subject with the notably material and ontological one. Architecture projects that engage big data necessarily enter into this confluence, bringing this epistemological, methodological realm into fruition or actualization. The results so far have effectively split architecture's subjectivity into two forms (manifest as two loci of subjectivity): the subjectivity of the architect/designer/producer, on the one hand, and the subjectivity of the architecture user, on the other. This division sends the components of subjectivity (which scale's inherent duality had sutured together) on divergent paths. Architecture's responses to big data, explicit or not, have so far extended this split, generating projects and practices divided between emphasizing methods and

[9] This repositioning has, itself, been partially buried under the onslaught of discussions that grapple with architectural ontology in its material rather than epistemological form over the past two decades.

the changed subjectivity of the designer or those highlighting the roles of information and technology in society that spotlight the subjectivity of the user. Big data's broader challenge to traditional forms of subjectivity animates the rift in subjectivity that scalar logic—in tying the producing subject to the experiencing one (when presupposing a universal subjectivity)—covered over. While scalar's proportional aspect secured the relation of an internal order to a larger world—with the architect as producer at its helm—architecture's adoption of big data's logics raises the question of how the discipline's emerging methods—how its modes of describing and inscribing reality—relate to its forms of production. To ask this question another way: what information can architecture now incorporate in its practices, and how might this be done? How can the discipline define, describe, and incorporate these new, variant forms and myriad quantities of information? And with that, what information is seen to lie within architecture's domain of action? This last concern, historically discussed around issues of autonomy, is here rethought in relation not to the possible autonomy of the discipline, but rather to autonomy's theoretical impossibility in the age of big data.

Neither big data nor architecture's intersections with it, however, are monolithic. Interfaces between architecture and big data have so far manifested two main categories. Each strain presents a characteristic approach to understanding where and how information enters the project and how it relates to humans, technology, or the environment. This can be thought of as the models of association each engenders, with the first seeking to adopt data-driven methods and the second actively imagining the social situating of data in contemporary society. In seeking to adopt, incorporate, or emulate big data's methods as a central component of architecture's practice, the first group transforms the role, and with it the subjectivity, of the designer. Against this, the second approach develops architecture projects that consider and manifest complex interactions between humans, data, technology, and society, all of which highlight the subjectivity of the architecture user.

Architecture practices in this first category adopt big data's aims and methods for much the same reasons other disciplines do: to increase the amount of information they can utilize toward responding to complex situations and to optimize results. The hallmark of this category is method and the frontloading of information in the architecture process. Projects in this category combine various computational strategies and digital analytics to approach architecture through a lens of performativity, problem solving, and optimization of structural, material, and operational faculties. As with big data's methodology, these practices look to data-driven methods to optimize results, and they achieve those results while simultaneously removing or minimizing human-associated bias.

A series of consequences follow from this. A primary effect of introducing big data operations into architecture is to disrupt the very idea of how, when, and in what form information enters the design process. While scalar logic focused on linear accumulation of information, algorithm-driven big data frontloads information, throwing architecture's more established order of production into disarray. This effectively transposes information from its scalar positioning as something akin to later-stage architectural details, to an initiating premise in the form of algorithms, data sets, and parameters that span materials, structures, environmental conditions, social uses, and more. In this transposition, big data's methodology eschews scale's carefully constructed correspondences, replacing the linear accrual of information with the ability to search varied and unordered accumulations of data. Big data's injecting quantities of specific information at the start of the process, without generalizations, visualizations, or other forms of human direction, haphazardly rearranges previously ordered procedures and relations. What had been visualizable and sequential becomes amalgamated within unseen, algorithmic, and computational processes. While incorporating these processes requires practices in this group to confront the changing role and subjectivity of the designer in the wake of science's fourth paradigm, they do not, however, automatically consider the changing role of the architecture user.

Rather than subscribe to big data's methods, the second category takes varied approaches that mix processes and sources of collecting and aggregating data with data-driven technologies to resituate, enter into, and reveal relations between humans and the amassing of data. Less focused on new design methods, the projects employ data, as both a component of the project and producer of experience, to present a new locus of subjectivity—the user defined in relation to information and data. Projects in this group actively situate and make visible (or legible) data in contemporary society. Toward this, information repeatedly appears in process rather than fixed, as though it were in the middle of being absorbed, released, found, or formed over time. The resulting architectural interventions utilize information and data as design elements, frequently as constitutive of social functions. In doing so, they implicitly situate the project and its users amidst a data-saturated world.

Within each category, two stages of development are already apparent, with the latter one incorporating more technologically advanced uses of data. In some ways, the clearest impact of both the use of data-driven methods and the focus on data within a project arises most sharply in the earlier phase. Additionally, the newness of the methods and their lack of fit with existing logics (which implicitly reveal their status) couple with explicit statements that proclaim their stance. The effect of this is to highlight the mechanisms and processes at work before they become completely absorbed in later projects. This intermediate stage most sharply manifests the shift from scale to capacity that, itself, began decades prior to architecture's broad adoption of digital modes of production and well in advance of big data's methods. Capacity's centrality follows on the heels of the emergence of the hegemony of immaterial and instantaneous measures that the French cultural and urban theorist Paul Virilio addressed in his 1991 essay, "The Overexposed City."[10] The essay discusses the ways in which technical developments in computation and communication had already transformed the city, historically defined by the physical determinants that are the traditional province of architecture, into an organism organized by the immaterial measures that are the hallmark of advanced technology.It contends that advanced technology's emergence inserted a wedge between architecture's historical material and cultural functioning and the operations of the social realm that supports it.

[10] Paul Virilio (1997), 381–90.

Some of the earliest examples from the first category come from the architecture firm MVRDV, whose aspirations and reasoning provide a clear link between Virilio's observations and big data's methods. Echoing Virilio, the firm claims: "Due to ever-expanding communications networks and the immeasurable web of interrelationships they generate, the world has shed the anachronism 'global village' and is transforming into the more advanced state of the 'metacity.'"[11] Rather than strive for meaning, the rhetoric common to MVRDV and other design firms in this group looks to shed any remaining representational or narrative skeleton that formed a cornerstone of much of architecture's previous undertakings. Enacted prior to the coining of the term, MVRDV's work shares big data's focus on employing data, embracing what they lovingly dub "pure data." Behind this approach is their finding that the economic and spatial possibilities of the current world situation are so complex that "statistical techniques seem the only way to grasp its processes."[12] Turning to data allowed MVRDV to more readily integrate socio-cultural, economic, and environmental issues into the early stages of the design process, becoming the means for producing what they call "datascapes." MVRDV consolidated this approach in their 1999 book of theoretical projects *Metacity/Datatown*. The premise of this work is to utilize data and information as active creators in the design process. The book presents its projects as the inevitable and scientific result of data, predictive techniques, procedures, and research methodologies through which data becomes the consolidator of a diverse range of concerns and types of information.

While short of big data's full computational potential, MVRDV's projects' mix of data-compiling, statistically based methods couples with their goal of a project of pure data, to transform the architecture process and, with it, the role of the designer. Their method is distinctly hybrid. They first frontload information in the form of aggregated statistical data and then utilize that data to inform architectural diagramming exercises that modify and literally form the data into spatial configurations. Their process begins by investigating relevant facts, including laws, regulations, conditions, experiences, available financing, client needs, and any number of other things.[13] Used to generate the data set, these facts are, themselves, transformed into a set of diagrams, which in turn become aggregated by superimposing one on the other. The superimposed diagrams constitute the spatial and theoretical framework for the project. In transforming various types of data into spatial configurations, diagramming acts as an intermediary between the data and the resulting architectural form. MVRDV describes this process as an iterative, rational one employing statistical spatial data and diagrammatic exercises. The iterative aspect develops through feedback loops in which the results are evaluated and used to form new iterations. In developing this method, they move architecture decisively away from the realm of meaning and representation characteristic of the text-driven architecture of the 1980s and early'90 s and toward a "pure" state, scientific and data-driven—moved, that is,

[11] MVRDV (1999), 16.

[12] Winy Maas (MVRDV), (2010), 249. They state that "by selecting and connecting data according to hypothetical prescriptions, a world of numbers turns into diagrams.".

[13] Michele Costanzo (2006), 60.

from scale's linear accrual of information to capacity's (and big data's) increased computational and algorithmic strategies. They make this explicit, rejecting outright the terms underlying textual as well as more traditional approaches to architecture, proclaiming: "Datatown" is a "city that knows no given topography, no prescribed ideology, no representation, no context. Only huge, pure data."[14] While unhesitatingly introducing this thinking, they also acknowledge its uncertain outcomes, issuing the still-unanswered question: "What agenda emerges from this numerical approach?".[15]

More recently, some architecture practices have moved closer to fully adopting big data methods by employing computational methods to incorporate increased amounts of data, frontloading information and looking to machine learning as the primary driver of the design process. As with machine learning in other fields, this use of algorithms to guide the design process transforms it. Frontloading complex and varied information combined with a reliance on algorithm-driven computation further repositions the role of the designer in the design process. Such an approach, at least in principle, completely breaks with scalar logic. Rather than take existing data and statistics and turn them into architecture via familiar exercises (if data-driven/diagrammatic, as with MVRDV), such practices look to complex computation to orchestrate increasing amounts and varieties of data and optimize the design's resulting performativity. Examples include practitioners, such as Michael Hensel or Achim Menges, who each employ advanced computational methods to strive for a comprehensive, computational approach to design capable of achieving an otherwise unattainable level of complexity.

Menges, for example, has designed a series of pavilions, modeled on the functioning of an array of biological models such as lobster skeletons or the hardened forewings of flying beetles. His 2016 design for a robot-built pavilion for the V & A Museum in London models its web-like, carbon-fiber canopy on elytra, a material found in the hardened forewings of flying beetles. Far removed from the idea of adding material and other information specifics as late-stage design development, the specificity of a material modeled on elytra's performative potential becomes the starting point for conceiving a pavilion design whose canopy combines traditionally discrete functions to serve as architectural envelope, load-bearing structure, and environmental interface. From the onset, Menges's approach integrates performative goals with material and structural functioning in relation to the project's environment. This process necessarily frontloads what traditional methods developed incrementally. That information is used toward realizing specific goals in the final design. Far removed from the idea of incrementally accruing detailed and discrete information, these practices are premised on an aggregative understanding of information in the design process.[16]

Beyond the shaping of architectural outcomes, the more the discipline incorporates big data's methods, the more those methods challenge the traditional role and practice

[14] Garcia, 249.

[15] *Ibid.*

[16] See, for example, Achim Menges (2011), 198–210.

of designers. Menges both acknowledges and embraces this. Echoing the rhetoric around big data, he describes his design for a pavilion as offering "a glimpse of the transformative power of the fourth industrial revolution currently underway, and the way it again challenges established modes of design, engineering and making."[17] As Menges stresses, this places "the designer in an alternative role"—one needed in the face of computation—"one that is central to enabling, moderating, and influencing truly integral design processes and requires novel abilities and sensitivities."[18]

Likely a matter only of time, architecture's full adoption of big data's methods still remains limited, relying upon the existence of a wealth of relevant data to mine that the discipline lacks. These current limitations, however, raise interesting questions of the relation a discipline bears to types of information. While the absorption of big data's methods may largely point toward performativity and optimization, it need not be so constrained. The existence of big data leads us to the question—asked on a much wider scale than was previously possible—of what data is relevant for inclusion. The relation of the category of practices employing data-driven methods to that of ones situating data in society forms a microcosm of the concern with what information is relevant to the discipline. A clue to the relation between these two strains of architectural approach surfaced in French anthropologist Claude Lévi-Strauss's mid-twentieth century structural study of myth. At the core of Lévi-Strauss's analysis of myth are a set of diagrams that transform myth into a spatio-temporal, grid-based matrix. The diagrams' fundamental functioning hinges on Lévi-Strauss's method for converting linear narrative into a spatial array that maintains narrative order while rendering it visually secondary to its structure/meaning. The diagrams function by being readable in two ways: the temporal and linear order associated with narrative and the spatial structure tied to the myth's (underlying) meaning. To do this, the diagrams maintain narrative order, but subsume it to the myth's underlying pattern of relations.

Lévi-Strauss's method, here, is key. It begins by including all versions of the myth in the analysis, thus nullifying the search for the correct or original version. Lévi-Strauss then proceeds (à la Freud's analysis of dreams) by breaking each myth into small, discrete, narrative elements. Following his dictum that repetition makes the structure apparent, he analyzes the narrative elements in search of repeated themes and arrays them to show these newly discovered relations: the myth's pairs of repressed binaries.

Notably, Lévi-Strauss's process for re-envisioning of myth bears a twofold, and prescient, relation to big data. Both begin by multiplying the amount of available data (by including all versions of a myth), then search the expanded data pool for patterns (structure). This method allows the diagrams to unearth hidden patterns across a myth's many variants. Over half a century later, the core of big data's methods employs algorithm-driven functions to search aggregated data for hidden patterns. While this comparison may initially seem too great a leap, Lévi-Strauss explicitly recognized the association to enhanced computational ability, pronouncing

[17] Quoted in Kim Megson (2016).

[18] Menges, "Integral Formation," 199.

that his process of analysis—the detecting of underlying structure or patterns within myth—would be greatly advanced by the assistance of IBM equipment.[19]

There is a second way that Lévi-Strauss's analysis becomes relevant to architecture's relations to big data. Along with methodology, it points to how the advance of knowledge brings about potential social upheaval. By connecting method to the social standing of new information, the analysis of myth also suggests ties between the two categories of architectural responses. Getting to the core of this, Lévi-Strauss's structural analysis shows that myth fulfills a particular purpose: it acts to maintain traditional identity in the face of social disruptions wrought by scientific, technological, and other changes.[20] This happens as myth suppresses the incompatible, contradictory ideas at its core—beneath a coherent narrative surface. Lévi-Strauss's study reveals not just that new knowledge and technology question established relations and hierarchies across society, but that they do so by introducing ordering principles outside of, or in contrast to, existing ones. Architecture's status as an amalgam of spatial, temporal, and social performance presents a uniquely sensitive territory to changes and potentials across these registers. Projects in the second group addressing this provide ways of framing the issues raised by projects in the first. They show how the absorption of these methods is part and parcel of its social positioning—including upheaval—with the one necessarily invoking the other. From them it becomes clear that while architecture's relations of information, computation, and architecture, in readily appearing as a "purely" technological and practical undertaking, may mask big data's social and cultural functioning, they in no way eradicate it.

The second main strain of architectural responses to big data emphasizes data less as *method* and more as *content*, pointing to the cultural impact and possibilities stemming from the introduction of vast quantities and increased access to information into architecture, its sites and roles in society. One of the hallmarks of these is data's emergence as an explicit component of the architecture and orchestrator of the subject's experience. Manifestations of data in these projects mediate the subject's complex of social relations. While data remains integral to the design, in these instances it is less about frontloading information in the design process than moving data to the forefront of the architecture. From within the project, data emerges as formative of both project and user experience. Data moves from input to output, transformed from its operative role in the design process to an active component within the architecture project. As architectural component it appears in various guises, visible, sensible, audible, tangible, or directive of human action. In becoming explicitly utilized as a material or architectural component, data is transformed into an active, visualized, spatialized, and operative element. With this shift, the projects move away from the first group's concern with the subjectivity of the architect/producer, to a concern with the subjectivity produced by the project and experienced by the user—a subject generated in, around, or by data.

---

[19] Claude Lévi-Strauss (1986), 821.

[20] Wittgenstein raises this important issue in Sect. 125 of his *Philosophical Investigations*, which concludes by stating: "The civil status of a contradiction, or its status in civil life: there is the philosophical problem." (Ludwig Wittgenstein, 1953).

Projects in this group simultaneously question and recognize human input and interactions with data and data-driven and responsive technologies. By consciously positioning and manifesting data's, information's, and technology's evolving roles in society, these projects introduce complex intersections between machine learning, data, computation, and human-directed experience. In raising issues surrounding the social impacts of big data, more, say, than using data for a variety of ends (including complex problem solving or material optimization), they emphasize architecture's relations to big data as a *cultural* technology, rather than the use of it as a specific method. These projects develop territories concerned with data and technology's status in the world, in which architecture is not just responding to and incorporating methods developed elsewhere (or even exploiting them for specific ends), but also conditioning, visualizing, and postulating (big) data's social acting.

Diller and Scofidio's 2002 Blur Building provides an early moment of this thinking. Its mix of data, experience, and elusive form bracket social experience between computation and data while envisioning the result of that interaction. Located at the base of Switzerland's Lake Neuchatel, the Blur Building was a temporary pavilion for Swiss EXPO 2002. Measuring in extent some three hundred feet wide, two hundred feet deep, and sixty-five feet high, the pavilion was composed of an open-air platform enveloped in an artificial atmospheric cloud formed by a fine mist of filtered lake water. As if animating Duchamp's Large Glass and its two-part construction of desire and identity, the Blur pavilion is a machine for transforming data into experience. While many aspects were never instituted as designed, data underlies both the pavilion's intended material and social operations. In doing this, it implicitly acknowledges the larger discussion of data's role in society as well as in architecture. Both the pavilion's ephemeral—if overwhelming—materiality and the visitor's experience are data-directed. The precarity of its material existence testifies to its merging of layers of information and material to yield experience. This happens in distinct components that are aggregated in the experience of the pavilion. To sustain the cloud, the pavilion's fog output is governed by data gathered from its built-in weather station. The weather system monitors the lake's shifting climatic conditions—temperature, humidity, and wind direction and speed—turning environmental data into substance by using it to determine the amount of mist the system must generate to sustain the cloud. In this way, the pavilion acts as an active interface between its information-gathering ability, its self-monitoring, and its geographic condition in the lake.

Visitor experience was equally designed to be data driven. The architects created a prescribed sequence for the pavilion's visitors. It was to commence on land, with visitors logging into a computer and filling out a personality questionnaire. Their responses were then to be uploaded to the cloud's computer network and deposited in a smart raincoat, dubbed a "braincoat." Donned by visitors as protection from the wet environment, the idea of the braincoats was their dual material and data function, storing personality profiles while protecting wearers from the mist. Braincoat on, the visitors walk down the four-hundred-foot-long ramp and enter the pavilion. At that point, the cloud diminishes vision and the braincoats take over where human vision fails. By searching data for compatibility between nearby profiles, the braincoats direct anonymous interactions between visitors.[21] In paired correlative moves, they utilize technology to mine data to see what humans cannot, and then transform the data into human sensorial cultural codes, which the braincoats present as the visual array of warm to cool colors, or the audible one of accelerated or decelerated pings. The braincoats thereby first assess and then visually and audibly signal degrees of compatibility between visitors by changing from cold blue-green to warm red and accelerating its emission of audible pings as visitors near profile matches. Visitors are then left to follow the colors and sounds to meet their digitally determined match. Upon reaching the deck above the cloud, the coats are deactivated, and human vision again assumes control. Visitors then complete the sequence by returning to land, logging out, and receiving a password for future access to their experience on the pavilion's website. Blur thus stages associations between the logic of data and a visualized human experience that build on the interdependence of personal narrative and its translation into digital technology and socio-spatial relations. Through these processes, Blur actively produces itself. It becomes a machine for correlating phenomena and data, mixing information with experience, technology with narrative, initially treating each of these as discrete and mappable entities with distinct origins, and then merging them to produce a socio-spatial/materio-environmental blur.

If Blur's proposed data-framed experience was, unfortunately, incompletely fulfilled, over the past decade a growing number of practices have enlisted newly available technology to simultaneously enact and emphasize relations between humans and data. In these, inherently interactive processes such as feedback loops, sensors, social media streams, and other responsive technologies figure prominently. Still early in this phase, narrative, while frequently invoked, is done so in a limited manner, yet it is developing a set of characteristics. Notably, a number of projects lay the groundwork for open-ended narratives as extensions of modes of data collection and distribution. Based in social media, they relocate information and communication from Blur's proposed realm of personal narrative and discrete interactions between individuals to that of digitally reified social relations.

A more recent wave of projects use increasingly advanced data-driven methods to create various interactions amongst humans, data, and technological, social, material, and environmental components. Projects designed by Future Cities Lab, for example, utilize big data's aggregative basis, as in their installations Datagrove 2012 and Murmur Wall 2015.[22] These employ responsive technologies to create socio-technological experiences and feedback loops, mixing data from the immediate physical environment with more far-flung forms of social media. Surpassing in complexity the lights and sounds emitted from Blur's braincoats, the installations function by transforming data into architectural components. The projects convert sensor-based

[21] Discussing Blur, Wolfe writes that it leads to the question of "[w]ho is doing the experiencing," that is, "Who, in phenomenological and political terms, are 'we,' exactly?" (Cary Wolfe, 2010), 219.

[22] Datagrove was an installation located in the courtyard of an historic California Theater, as part of the ZERO1 "Seeking Silicon Valley" 2012 Biennial.

information and data streams into animated LED light and digital text displays, interwoven with the installations to create continuously evolving interfaces. The projects become material repositories of immaterial (and potentially nebulous) data. The force of these projects is their transforming of data collection and activation from method into architectural and atmospheric components. While Blur had fixed and individualized data input, which was then mapped through a computational database, these projects aim to inseparably intertwine individuals with one another and the data they emit—more akin to big data's aggregative, and unseen, logic. By sharing and redistributing information, they emphasize the technologies and systems themselves. As if in direct manifestation of Virilio's urban observations, the projects enact the spatial collapse of the local architectural manifestation into the broader urban and social environment. Using information and communication's immaterial entities—now visually and audibly manifest—they connect the project beyond its physical site in ways that trouble any remaining distinction between project and context.[23] Despite all the emphasis upon information and communication in these projects, the information is all but emptied of specificity, transformed largely into reference to communication, on the one hand, and potential commentary on contemporary society's relations of information, on the other.

If the first category extols big data's rhetoric of embracing machine learning and eschewing narrative, the second one, in showing data-human relations, necessarily invokes narratives of some kind. This leads to the question: *what* narratives? In building on the social amassing of information, such projects include the social as a critical component from the outset. Blur, for example, invokes overt narratives of association, emanating from a reductive data/digital version of "what is the personal," and orchestrated by lights and pings that direct the visitor through the vision-obscuring mist of the "cloud." By contrast, Future Cities Lab's use of social media less directly orchestrates social relations than it creates a place for them to be noted—a place beyond the grasp of those within the physical limits of the project, emanating from what in the past would have been seen as context. The resulting state points to the existence of a subject suspended between that of spectator and that of active contributor.

By finding patterns in accumulated data that humans cannot see, the methods central to big data signal a cultural shift in the history between human relations and what we produce. While big data's computational-algorithmic logic has no need to visualize its workings, humans frequently do for a variety of reasons and purposes. Against this status arises the interesting problem of how we might comprehend and "visualize" big data, not just use its results. This is particularly the case, as big data is founded on the idea that we no longer need to visualize such processes. That is, we no longer need to produce the universal, generalized laws that empirical science built itself upon. Those laws were, of course, visualizations of various sorts[24] that allowed

[23] This thinking extends the logic of Jacques Derrida's 1972 essay "Signature Event Context." See Jacques Derrida (1993), 1–23.

[24] For example, C. S. Peirce grouped algebraic equations with other visual images and called them "icons".

for the prediction of the behavior of natural phenomena, but also allowed for a way of seeing and understanding the phenomena themselves.[25] Big data's methods, its very usefulness, belies the intermediary processes that served as human entry points into natural and artificial systems alike.

While the focus on performance largely and vehemently rejects the recourse to representation, analogy, and meaning, those aspects nevertheless arise. An interesting issue arises here in context with the possibilities and problems for envisioning big data. One response suggests architecture invent at least temporary images or spatial constructs to describe the recently developing logics and social structures wrought by big data that challenge traditional orders or inscriptions and descriptions of reality. This need, of course, is not unique to the discipline. Consider how—in order to mediate between old and new modes of operating—a set of organizational analogies have accompanied advances in computation into new territories, as keeping pace with increases in computational capacity creates problems for description. As capacities advanced—as megabytes gave way to terabytes and petabytes—organizational analogies also worked to keep pace, developing from the neatly ordered folders, file cabinets, and library analogies[26] to today's "cloud," which describes both storage location and lack of readily visualizable formal organization of information.

In this light, Blur's cloud is both suggestive of the digital realm—materialized in ephemeral form—and supplies an architectural/environmental image to do just that. Blur's very elusiveness proves informative. It raises questions of precisely *what* is elusive and what is not—agency, vision, defining the personal, to name a few possibilities. Blur makes the question of elusiveness palpable in relation to information, data, society, individuals, and social interactions. Its prescribed sequence (akin to narration or scale) for entering and leaving the pavilion brackets the inchoate experience of the cloud between the acts of logging on and off, activating and deactivating the braincoats, revealing and obscuring personal information. An array of other projects actively associate experience with data and technology, often wielding cycles that mix varieties of input, methods of aggregation, and layers of complexity. In place of Blur's prescribed sequence that bracketed disorienting experience with precise technology, these offer no specific starting or ending point, leaving it unclear as to whether humans or technology are the driver.

More broadly, architecture's range of responses to the advent of big data questions the discipline's functioning in the world. Is it problem solving? Is it about making propositions? How does it situate the input of data and information in relation to

[25] One might here begin to think of Freud's development of a system to see the unconscious, a system that in many ways Lévi-Strauss appropriated for his analysis of myth. However, big data itself, in being a form of analysis based on computational ability beyond the human—rather than a repressed, unconscious, but still very human way of thinking—offers a whole new level of challenge.

[26] At the "petabyte scale, information is not a matter of simple three and four dimensional taxonomy and order but of dimensionally agnostic statistics." This calls for us to lose the "tether of data as something that can be visualized in its totality," to which *Wired* magazine notes that organizational analogies seem to have been exhausted. See Chris Anderson, "The End of Theory: The Data Deluge Makes the Scientific Method Obsolete," *Wired*, June 23, 2008, https://www.wired.com/2008/06/pb-theory/.

the design process or to the experience of the design? Rather than move entirely beyond questions of meaning, big data's absorption into architecture cannot help but raise new ones. The introduction of a new ordering procedure, as with big data, dislodges architecture's accustomed procedures and, hence, its potential social roles, revealing some and masking others. On a procedural level, what the frame of big data allows us to see is that *scale* provided a method to control the input of information into the design process, as a fixed scalar logic controlled not just the quantity of information and rate of its inclusion, but, effectively, the type of information as well. Big data thus queries architecture's status, the question of what stuff it is grappling with—information and materiality, order and environment, procedures and patterns. It opens up a history related to the procedures it disrupts. Through its advent, we see how scalar logic made a series of crucial correspondences seem natural, against which big data upsets that naturalness and, in so doing, provides an entry point for history.[27]

References

Hey, T. (2009). The fourth paradigm: Data-intensive scientific discovery, in S. Tansley, & K. Tolle (Eds.). Microsoft Research.

Miller, H. J., & Goodchild, M. F. (2015). Data-driven geography. *GeoJournal, 80*, 449–461.

Virilio, P. (1997). The Overexposed City. In N. Leach (Ed.), *Rethinking architecture: A reader in cultural theory* (pp. 381–390). Routledge.

MVRDV, Metacity/Datatown (1999). MVRDV/010 Publishers, p. 16.

Maas, W. (20100). (MVRDV), "Metacity/Datatown," in The diagrams of architecture, in M. Garcia (Ed.), (p. 249), Wiley.

Costanzo, M. (2006). MVRDV works and projects 1991–2006, Skira Editore S.p.A., p. 60.

Menges, A. (2011). Integral formation and materialisation: Computational form and material gestalt. In A. Menges & S. Ahlquist (Eds.), *Computational design thinking* (pp. 198–210). Wiley.

Megson, K. (20 May 2016). " 'The fourth industrial revolution is underway': Achim Menges launches robot-built pavilion at V&A," CLADnews.

Lévi-Strauss, C. (1986). "The structural study of Myth," in Critical Theory Since 1965, in H. Adams, & L. Searle (Tallahassee: Florida State University Press,), 821.

Wittgenstein, L. (1953). Philosophical Investigations, trans. G. E. M. Anscombe, Macmillan.

Wolfe, C. (2010). "Lose the building," in What is posthumanism?. University of Minnesota Press, p. 219

Derrida, J. (1993). "Signature event context," in Limited Inc., (pp. 1–23). Northwestern University Press.

[27] My forthcoming book on this topic will address the issue of the relations between ideology, history, subjectivity, and science that stem from architecture's engagement with big data as the fourth scientific paradigm. To this end, I discuss how the two components of subjectivity become reassociated in this process.

Being or Tea?

Annika Döring and José Ordóñez García

Kakuzo Okakura, in his essay *The Book of Tea*, maps several aspects of Asian lifestyle by referring to its tea-culture. Later, the German philosopher Martin Heidegger was accused of having plagiarized them in his most famous book *Being and Time*:

> Before moving back to Japan at the end of his studies, Professor Itō handed Heidegger a copy of Das Buch vom Tee, the German translation of Okakura Kakuzō's The Book of Tea, as a token of his appreciation. That was in 1919. Sein und Zeit (Being and Time) was published in 1927, and made Heidegger famous. Mr. Itō was surprised and indignant that Heidegger used Zhuangzi's concept without giving him credit. Years later in 1945, Professor Itō reminisced with me and, speaking in his Shonai dialect, said, 'Heidegger did a lot for me, but I should've laid into him for stealing'. There are other indications that Heidegger was inspired by Eastern writings, but let's leave this topic here. I have heard many stories of this kind from Professor Itō and checked their veracity. I recounted this story at a reception held after a series of lectures I gave in 1968 at the University of Heidelberg at the invitation of Hans-Georg Gadamer. Japanese exchange students attended these lectures, and I explained that there were many other elements of classical Eastern thought in Heidegger's philosophy and gave some examples. I must have said too much and may even have said that Heidegger was a plagiarist (Plagiator). Gadamer was Heidegger's favorite student, and we ended up not speaking to each other for 4 or 5 years because he was so angry with me.[1]

Indeed, regarding the content, there are topics both thinkers refer to, like death and truth.[2] But Okakura and Heidegger have a totally different understanding of what they mean when they refer to these topics. Whereas, for example, death to Okakura

[1] Imamichi, Tomonobu (2004). pp. 123–124.

[2] See Okakura, Kakuzo (1956), Heidegger, Martin (1962)[1]. [21]2001.

We thank Francesca Brencio, Jon Rea, and Marcus Novaes for helpful comments and translation.

A. Döring (✉) · J. O. García
Universidad de Sevilla, Seville, Spain

J. O. García
e-mail: ordogar@us.es

S. Wuppuluri and I. Stewart (eds.), *From Electrons to Elephants and Elections*, The Frontiers Collection, https://doi.org/10.1007/978-3-030-92192-7_44

means rebirth,[3] death to Heidegger means "[T]he 'end' of Being-in-the-world",[4] so they mention the same terms in profoundly—and irreconcilably—different contexts.

We will first provide an overview of the content of Okakura's *The Book of Tea*, with a focus on the topics that are relevant for Heidegger as well, then provide an overview of Heidegger's *Being and Time* with a focus on the same topics. Based on this, we will compare the writings with one another afterwards. We will point out that, albeit the wording is similar, the (cultural) context in which both refer to these shared topics, is totally different. The contextual differences will clarify that, despite the similarities regarding the topics, the contents of *The Book of Tea* and *Being and Time* are different as well.

1 Heidegger: An East for west[5]; Being and Time: An Orientation for Existence

The abuse of tea as a sign of weakness in the face of emotions puts us directly in relation to the ways that we find ourselves in the world: the more tea consumption, the more restlessness, the more surrendered to the affection that arises from otherness. This is what Heidegger defines as *state-of-mind,* an existential (Existentiell) characteristic of the Dasein *(being there),* fundamental and determinant of his daily life. This *state-of-mind* belongs to the ontological order, whose factual phenomenon is *mood and the attunement*. So, we can establish the following terminological arc: temple, affectivity, emotion and feeling. "Temple" is usually understood as the energetic fortress and the serene courage to face difficulties and risks[6]; "affectivity" is the set of feelings, passions and emotions of a person[7]; "emotion" is understood as the intense and temporary change of the mood, intense and fleeting, pleasant or painful, which is accompanied by a certain somatic incidence[8]; and "feeling" is the fact or effect of feeling something or feeling himself (e.g.: "he lets himself be carried away

[3] Okakura, Kakuzo (1956). p. 128.

[4] Heidegger, Martin (1962)[1]. [21]2001. p. 276 f.

[5] Among many others, there are texts in Spanish such as *El oriente de Heidegger*, of Carlo Saviani (Herder, Barcelona, 2004), *La palabra inicial*, of Hugo Mújica (Trotta, Madrid, 1995) other *Filósofos de la nada. Un ensayo sobre la escuela de Kioto*, of James W. Heisig (Herder, Barcelona, 2002), on Heidegger's influence on Eastern thought. It is well known how well received the conference *Was ist Metaphysik* (Vittorio Klostermann, Frankfurt a. M., 1981) in figures of Japanese thought such as Nishida Kitarô, one of the founders of the Kyoto School, together with Tanabe Hajime and Nishitani Keiji. However, we have no evidence that Heidegger paid much interest to Eastern thought, beyond the well-known anecdote of his attempt to approach Lao Zi's Tao te King, an approach of which we can see a certain resonance in his work *Aus der Erfahrung des Denkens* (Günter Neske, Pfullingen, 1954).

[6] Diccionario de la Real Academía Española (RAE).

[7] Ibid.

[8] Ibid.

by his feelings").[9] All are phenomena derived from the *state-of-mind*, as a generic structure of existence, which takes place in each case in a certain way. But in all of them we find that they are governed by the spontaneity, the involuntary, that is, it is about genuine phenomena, determinant and irreducible to rationality (excessive fixation on tea).

The existing is always affected by something or someone, although there are those who may or may not feel affected by a certain phenomenon: there are those who are affected by the result of a soccer match and those who not, because other things affect them. Prejudices, as well as experience, will influence in affectivity, they are its condition of possibility, hence the importance of *Das Vor-Verstehen* and *Die Auslegung*.

It is, therefore, a faculty inherent to existence. For this reason, it is possible that there are those who are deprived of it. The deprivation is only possible in those who "should have", but do not. Affective coldness is not deprivation, but precisely an affective state with certain characteristics. A different case is that of some extreme pathologies such as, for example, autism, in which there is no affectivity in the strict sense, but rather a nervous, automatic reaction. With Heidegger, we can say that the phenomenon of State-of-mind is totally radical. It is the engine of existence, the fuel of daily life.

The affective phenomena to which Dasein does not usually take much account, due to its fleeting irruption, frequently go unnoticed, despite the fact that they are decisive in daily occupation. So, the affectivity takes part of, and determines, our existence. Therefore, Heidegger can affirm: "Dasein always has some mood (gestimmt ist)".[10] However, and due to this characteristic, the existing one can experience a kind of affective indeterminacy that is commonly manifested in the form of boredom (Überdruss). This way of finding itself has the peculiarity of lacking a cause, not because there is none, but because the existing one is incapable of fixing it on some motive. Thus, boredom, experienced as a burden (Last),[11] happens, succeeds, occurs, in the same way as time and being. And the cause is ignored, because boredom (Überdruss) is not a conquest of knowledge, but a way of finding oneself in the world prior to any occupation; it lacks will and intention, as it is an intimate, irruptive experience. That is why boredom is not "resolved" with knowledge, but with description, that is to say, with the discourse that puts the tedious in front of the existing, expressing the tediousness, opening it, letting it be. Boredom takes us just like that, and selfhood ceases to be an active value to become a burden. In boredom one does not find attraction or interest in what surrounds him. It is a way of being in the world, but without mooring, without occupation, without desire. Instead of will there is laziness, which should not be confused with detachment.

One talks about moods from a certain state of mind. Hence, the mood cannot be treated as an object (*Gegenstand*) for a subject, since the own subject finds himself subjected to his existential condition, in which the mood is originally constitutive

[9] Ibid.

[10] Being and Time Blackwell Publishers Ltd. (2001. p. 172)

[11] Ibid.

and constituent. Here the experience of consciousness as self-consciousness is not valid, in the manner of Hegel, because it is not a question of the necessary recognition of otherness, but precisely of the irreducible to knowledge, since it is not a question of knowledge but of a "finding himself", and here the otherness does not allude to another existing one, but, at most, to an estrangement linked to the selfhood. This, taken from affectivity, has had itself as something that is a difficulty, a bore. In other words, this affective indeterminacy is not the absence of affect, as we have already pointed out, but the presence of its indeterminacy in the strict sense, that is to say: a lack of determination to take charge of selfhood. However, here it is important not to lose sight of that "one does not *know*"[12] in relation to *Dasein* as a burden (*Last*). Why? The meaning of this statement refers to the fact that the existing one has taken the being of his there (Da) as a burden (Last), that is to say, he ignores why the being of the "there" manifests himself as tedious (*überdrüssig*), as something indifferent to me. The answer to why Heidegger founds it in the primacy originally from moods over knowledge. This, the knowledge, is a secondary phenomenon, of second order, compared to the state of mind:

> And Dasein cannot know anything of the sort because the possibilities of disclosure which belong to cognition reach far too short a way compared with the primordial disclosure belonging to moods, in which Dasein is brought before its Being as "there".[13]

Thus, it is clear the involuntary character of the moods in the face of the decision for knowledge, that is to say: that the desire of knowing is going to be legitimized, or it will simply be possible, as a result of a certain mood, in an affective way of finding oneself. It is, then, as checked in *Being and Time* to thematize the a-thematic, to account for what "happens to me", in "how" I feel. That is why Heidegger limits himself to describing the "how" the existing one finds itself in boredom (Überdruss) and thematizes it phenomenologically: the being of "There" (Da) as "burden" (Last). Therefore, the ontological fundamentum of boredom is burden. In this regard, it must be added that the exposition on this subject is presented, perhaps, in the first person – or so it should be, not to be seen as a theorization, but as an "existential" description–, in other words, that it must be Heidegger himself who is describing "how" his state of mind is, or has been, while writing *Being and Time* in Todtnauberg's hut. From his existential experiences, he abstracts the common existentials (existentiell), from the ontic-ontological to the pre-ontological, or in other words: from the factual ontic-existential concretion derived from the generic ontological-existential temple. And this movement where the world arises is filtered by that determination; therefore, the world arises through a way of finding oneself, and, by doing so, it is interpreted: "In this «how one is», having a mood brings Being to its «there»".[14]

The avoidance before the state of mind conforms the state of openness of the existing one before him. In this way, it is as Heidegger tells us, that we find ourselves sustained in a certain temple just like that, de facto, and, therefore, in an open disposition that is not consequence of a decision, but that is an already accomplished fact

[12] Ibid.

[13] Ibid.

[14] Ibid.

to which we let it "speaks" to us or for what we silence ourselves by not taking account of him. However, not paying attention is part of our deal with the affective disposition:

> In an ontico-existentiell sense, Dasein for the most part evades the Being which is disclosed in the mood. In an ontologico-existential sense, this means that even in that to which such a mood pays no attention, Dasein is unveiled in its Being-delivered-over to the "there". In the evasion itself the "there" is something disclosed.[15]

Somehow, die state-of-mind and *die Geworfenheit* (*disposedness and* thrownness) go hand in hand, since that "one does not *know*" opens to this "Da" in its *Geworfenheit*, because it is perceives "what it is", however, it does not know its "from where" and its "where. " This *Geworfenheit* is a way of feeling derived from boredom (Überdruss). In it, experiencing existence as a burden (Last) leads to that feeling of *Geworfenheit*, which can be expressed like this: I am here, I know it, but I don't know for what or why, neither what I have to do. In der *Überdruss* (disgust) one knows what it is, but ignores what it has to be, that is, der *Entwurf* (thrown) appears there under the cloak of perplexity, astonishment, fear or impotence. There is only possibility, nothing more, that is to say: the knowing that one has that he needs to be something, or rather: doing something to be someone. It is another way of insisting on the original being thrown the existing one into its affective condition:

> Factically, Dasein can, should, and must, through knowledge and will, become master of its moods; in certain possible ways of existing, this may signify a priority of volition and cognition. Only we must not be misled by this into denying that ontologically mood is a primordial kind of Being for Dasein, in which Dasein is disclosed to itself *prior to* all cognition and volition, and *beyond* their range of disclosure.[16]

It is an interpretation of Heidegger –guided by *der Überdruss*– that, because it is mediated by *der Entwurf*, (thrown) is considered factual, but dramatic, due to the *ausweichenden Abkehr* (evasive turning away) with which Dasein behaves in relation to die state-of-mind. Therefore, and to a certain extent, the development of knowledge in Heidegger has its origin in *die Be-sorgnis* (care/concern), of which he makes his *Besorgen.* A *Be-sorgnis* that is existential compared to the common and ordinary, that is merely banal.

2 *State-Of-Mind* an Originary West

The reflection, then, can only aspire to account for experiences, of what has been lived, which is of the order of facticity, hence Heidegger's turn towards language and, specifically, of his interest in poetry, whose expressionist character (especially that of Hölderlin, but also that of G. Benn, R. M. Rilke, S. George, or G. Trakl) could seem to him a kind of experiential phenomenology, insofar as it is not limited to being a mere

[15] Ibid., 173–174

[16] Ibid., 175

description, but is a pure event: letting be what happens transforming language into a commemorative action. And here, poetry can be considered as the western version of the eastern fixation on tea. In this sense, it seems correct to affirm that, as a result of the pre-eminence of affectivity in relation to occupation, it is possible that it is precisely die state-of-mind what guides our decision towards a certain occupation. Heidegger reminds us in the lecture *What is philosophy?*, quoting Aristotle, that the temple of mood that characterized the philosophizing of the Greeks (Heraclitus and Parmenides, above all) was das *Erstaunen*[17] *(astonishment)*. Does this mean that already in *Being and Time* is determined the affective east of everything to be occupied in the world with entities? So it is. And not only that, Heidegger also understands that the thinking of our time must also be determined by a certain temple of mind. It is not clear that anguish must be that temple, despite the fact that Heidegger gives it a pre-eminence in that work and, above all, it is the affect that most accurately comes to represent our relationship with the knowledge of death. So the anguish –in the period that culminates in *Being and Time*– is at that moment the temple that sustains Heidegger's thinking, insofar as it is shown as the gateway to *Dasein* as entity:

> How far is anxiety a state-of-mind which is distinctive? How is it that in anxiety Dasein gets brought before itself through its own Being, so that we can define phenomenologically the character of the entity disclosed in anxiety, and define it as such in its Being, or make adequate preparations for doing so?[18]

And why is *die Angst* the gateway to Dasein and not others? In principle, and according to the letter of the text, because it is about Dasein as such, not about a certain Dasein in a certain occupation and with a certain sex or gender. No, here we are dealing with the privileged entity, since its privilege is based on its temporal dimension, on its destiny as such, the root of the knowledge of *(das) Vorbei*, of the knowledge of death. Dasein knows this, but chooses what to do with the knowledge that it puts before itself. This is the difference that Heidegger seems to point out to us: the anguish that is experienced occurs in everything that exists, but not all existing one takes care of it in an authentic sense, but, in most cases, the answer consists in forgetting it and even live turning your back on it: "It turns away from itself in accordance with its ownmost inertia [Zug] of falling".[19]

However, as a gateway to Dasein from factuality, that is, as it is given and shown without further (to Heidegger), anguish puts the *Leib* in the Körper, since, otherwise, it would not be a question of an affective phenomenon "lived", "felt". Thus, Heidegger establishes that *die Angst* is what puts us before the ontic dimension of Dasein. Consequently, he has to justify and give a reason why *die Angst* is a fundamental mood. It is because it places Dasein before its selfhood, its most proper being. That is, in *die Angst* it is constituted as the original encounter/return to the selfhood (of which we do not want to know in the daily occupation), a selfhood that alludes to the totality of Dasein (its mortality), since it is its own being the one who interpellates

[17] Was ist das die Philosophie? Neske, Pfullingen, 1984, s. 26.

[18] Op., cit, 228

[19] Ibid., 229

itself, a being who is now faced with the pure "possibility" (*Seinkönnen*). In *die Angst* the Dasein falls apart (*die Dasein verfällt*) to its selfhood. Thus, perhaps, two modes of the fall can be distinguished in Heidegger: in the impropriety of the they *das Man* and in the appropriation of the *Selbst*. In the first case, *Dasein* finds itself alienated, while in the second it is shown the "since" it was alienated. Because die *Selbstheit* is an original and constitutive phenomenon of Dasein –an ontological-existential phenomenon–, Dasein can be, in the falling apart of the they, deprived of it –not lacking in it–and thus flee from itself, turn its back to its selfhood, which is the common and current ontic-existential phenomenon.

All these considerations, which are relative to the unfolding of content in turning one's back on selfhood, already indicate how in that flight there is an opening to what is fleeing from and which constitutes the possibility (*Seinkönnen*) of flight. Well, here lies the justification for taking *die Angst* as a privileged *State-of-mind*. However, despite this privilege, Heidegger warns that this privilege is, in principle, only a mere statement: "It might be contended that anxiety performs some such function".[20] One difficulty in this sense is that *Angst* and *Furcht* are generally confused and identified. Due to this ambiguity, it is necessary to differentiate and clarify the singularity, the meaning, of each one, establishing the *das Wovor* of *the Furcht* and the *Wovor* of the *Angst* (fear). The following paragraph is clear about it:

> But one is not necessarily fleeing whenever one shrinks back in the face of something or turns away from it. Shrinking back in the face of what fear discloses-in the face of something threatening-is founded upon fear; and this shrinking back has the character of fleeing. Our Interpretation of fear as a state-of-mind has shown that in each case that in the face of which we fear is a detrimental entity within-the-world which comes from some definite region but is close by and is bringing itself close, and yet might stay away. In falling, Dasein turns away from itself. That in the face of which it thus shrinks back must, in any case, be an entity with the character of threatening; yet this entity has the same kind of Being as the one that shrinks back: it is Dasein itself. That in the face of which it thus shrinks back cannot be taken as something 'fearsome', for anything 'fearsome' is always encountered as an entity within-the-world.[21]

That is to say, and in short, *die Furcht* is to the ontic-existential just as *die Angst* is to the ontological-existential. So, the original thing is not fear, but *Die Angst* of which fear is a possibility. All ontic fear is possible thanks to ontological anguish. If at this point we want to bring into play *die ontologische Differenz* (ontological difference), perhaps it is possible to interpret it in this other mode of difference: anguish (in relation to being: indeterminate) and fear (in relation to entity: determined). We believe that the conviction of the *die Furcht* is a phenomenon made possible by *die Angst* is basically due to the knowledge of death, whose affect is the temporality.[22] Why? Because, for Heidegger, *die Angst* is the affective experience that is experimented facing the "encounter" (*more* event: *Ereignis*) with the being without the entity, that is, with that nothingness of entity that we represent ourselves as a result of our *Vorbei*: our no longer being in the world.

---

[20] Ibid., 230

[21] Ibid.. 230 (185).

[22] Cfr., Ordóñez-García, J (2014), pp. 155–169.

2.1 *Okakura's Introduction to Eastern (Tea-)Culture*

Okakura describes "Teaism"[23] as "a religion of aestheticism".[24] He regards tea as an artwork that requires a skilled person to make the best out of it.[25] This is "[t]he truly beautiful"[26] and it needs to be seen in a certain way,[27] so Okakura 's understanding of art strongly adresses the recipient:

> Our mind is the canvas on which the artists lay their colour; their pigments are our emotions; their chiaroscuro the light of joy, the shadow of sadness. The masterpiece is of ourselves, as we are of the masterpiece.[28]

Okakura refers to the teaplant, regarded by the Taoists as crucial for "the elixir of immortality".[29] Drinking tea is regarded by him as a catharsis of the body from "all the wrong of life".[30] Still, referring to the Tao, Okakura quotes Laotse: "[...] I do not know its name and so call it the Path. With reluctance I call it the Infinite. Infinity is the Fleeting [...].",[31] because to define, to a Taoist, means to refrain from development, and therefore ethics would not be absolute.[32] Okakura refers to historians from China, whom he quotes without giving names, saying that they regard "Taoism as the 'art of being in the world,' ".[33]

Okakura ends his essay by referring to death: The cherry blossoms gain their grace within their death, when they let themselves fall from the tree,[34] and Okakura chooses a wording that refers to the Japanese blossoms not only as sentient beings but also with a free will by stating that they would "freely surrender themselves to the winds."[35]

Tea, in the aesthetics of Eastern life, stands for the enjoyment of beauty,[36] and the ethical or religious humbleness that values the small things and living entities such as flowers.[37] Its aesthetics are to be seen in the way a human being forms a unity with the culture and society he is a member of:

> Great as has been the influence of the tea-masters in the field of art, it is as nothing compared to that which they have exerted on the conduct of life. Not only in the usages of polite society,

[23] Okakura, Kakuzo (1956). P. 3.
[24] Ibid.
[25] Ibid. p. 25.
[26] Ibid.
[27] Ibid. p. 107
[28] Ibid.
[29] Ibid. p. 28.
[30] Ibid. p. 34.
[31] Ibid. p. 50.
[32] Ibid. p. 53.
[33] Ibid. p. 58.
[34] Ibid. p. 146.
[35] Ibid.
[36] Ibid. p. 3.
[37] Ibid. p. 123 ff.

> but also in the arrangement of all our domestic details, do we feel the presence of the tea-masters. Many of our delicate dishes, as well as our way of serving food, are their inventions. They have taught us to dress only in garments of sober colors. They have instructed us in the proper spirit in which to approach flowers. They have given emphasis to our natural love of simplicity, and shown us the beauty of humility. In fact, through their teachings tea has entered the life of the people.[38]

2.2 A Comparison

Getting to the context first, it is to say that Okakura's *The Book of Tea* is an essay in which he refers to Asian culture. He mentions several aspects of this culture and explains them, although not in depths, by means of a series of narratives. Okakura aims at evoking an understanding for Asian culture: "When will the West understand, or try to understand, the East?"[39] In his *The Book of Tea* he explains the East by its tea-culture, that he links to aesthetics, Zennism and Taoism—to art, religion and a certain understanding of ethics. Heidegger, on the contrary, in *Being and Time*, avoids ethical and religious aspects completely. Okakura identifies himself with Eastern culture: "**Our** writers [...]",[40] "**We** Asiatics [...]",[41] "You may laugh at **us [the Asians]** for having 'too much tea', but may **we** not suspect that you of the West have 'no tea' in your constitution?"[42] – just to give a few examples. Heidegger does the contrary: He makes clear that the others are not only the people from whom we distinguish ourselves, but also the society with which we surround ourselves and identify with: "Everyone is the other, and no-one is himself."[43] Heidegger calls this "the real dictatorship of the 'they' "[44] Okakura appeals for self-sacrifice, by referring to Laotse´s saying about the vacuum:

> Vacuum is all potent because all containing. In vacuum alone motion becomes possible. One who could make of himself a vacuum into which others might freely enter would become master of all situations. The whole can always dominate the part.[45]

Okakura refers "to the Code of the Samurai—the Art of Death which makes our soldiers exult in self-sacrifice".[46] Death, to Okakura, means to be reborn.[47] Death, according to Heidegger, limits the lifetime of a person: "[...] Being, to be made

---

[38] Ibid. p. 154 f.
[39] Ibid. p. 8.
[40] Ibid. p. 9.
[41] Ibid. 8
[42] Ibid. p. 12.
[43] Heidegger, Martin (1962)[1]. [21]2001. p. 165.
[44] Ibid. p. 164.
[45] Okakura, Kakuzo (1956). p. 60.
[46] Ibid. p. 7.
[47] Ibid. p. 128.

definite in an existential way by Being-towards-death.".[48] The finite limitation of one 's own personal life calls for self-realisation:

> In the anticipatory revealing of this potentiality-for-Being, Dasein discloses itself to itself as regards its uttermost possibility. But to project itself on its ownmost potentiality-for-Being means to be able to understand itself in the Being of the entity so revealed-namely, to exist. Anticipation turns out to be the possibility of understanding one's *ownmost* [emphasis in original] and uttermost potentiality for-Being-that is to say, the possibility of *authentic existence* [emphasis in original]. The ontological constitution of such existence must be made visible by setting forth the concrete structure of anticipation of death.[49]

Okakura, on the contrary, writes about "polite amusements".[50] To Okakura, the commitment to traditions is aesthetical: "Teaism is a cult founded on the adoration of the beautiful among the sordid facts of everyday existence. It inculcates purity and harmony, the mystery of mutual charity, the romanticism of the social order."[51]

Okakura refers to a dialogue between a Taoist named Soshi and Soshi´s friend, in which Soshi speaks about the happiness of swimming fish.[52] When his friend replies that Soshi could not know whether the fish were happy or not, because he is a human, Soshi answers that his friend was not Soshi, so " 'how do you know that I do not know that the fishes are enjoying themselves?' ".[53] Soshi´s answer contains three logical fallacies; the *petitio principii* –the *burden of proof* and the *tu quoque*. Heidegger´s understanding of truth is different:

> Thus truth has by no means the structure of an agreement between knowing and the object in the sense of a likening of one entity (the subject) to another (the Object).[54]

To Okakura, art must be a crucial aspect of Asian tea-culture, since he dedicates a whole chapter of *The Book of Tea* to art. The beholder, to Okakura, is the important reference when he talks about artworks.[55] Okakura links aesthetics to a life that respects and is organized around traditions and social hierarchies.[56] Heidegger, on the contrary, points out the importance of minding not to lose oneself within social structures that refrain us from our own individual self-being:

> What is decisive is just that inconspicuous domination by Others which has already been taken over unawares from Dasein as Being-with. One belongs to the Others oneself and enhances their power. 'The Others' whom one thus designates in order to cover up the fact of one's belonging to them essentially oneself, are those who proximally and for the most part 'are there' in everyday Being-with-one-another.[57]

---

[48] Heidegger, Martin (1962)[1]. [21]2001. p. 277.

[49] Ibid. p. 307.

[50] Okakura, Kakuzo (1956). p. 3.

[51] Ibid.

[52] Ibid. p. 66.

[53] Ibid.

[54] Heidegger, Martin (1962)[1]. [21]2001. p. 261.

[55] Okakura, Kakuzo (1956). p. 103 ff.

[56] Ibid. 1906. p. 3.

[57] Heidegger, Martin (1962)[1]. [21]2001. p. 164.

Okakura refers to the exported tea:

> The afternoon tea is now an important function in Western society. In the delicate clatter of trays and saucers, in the soft rustle of feminine hospitality, in the common catechism about cream and sugar, we know that the Worship of Tea is established beyond question. The philosophic resignation of the guest to the fate awaiting him in the dubious decoction proclaims that in this single instance the Oriental spirit reigns supreme.[58]

In the West, tea is consumed and enjoyed, but certainly not worshipped. Certainly, Okakura appreciates the export of tea.[59] Heidegger is convinced that artworks—he never said he would consider tea as one —should not be transported like other goods[60]:

> Everyone is familiar with artworks. One finds works of architecture and sculpture erected in public places, in churches, and in private homes. Art-works from the most diverse ages and peoples are housed in collections and exhibitions. If we regard works in their pristine reality and do not deceive ourselves, the following becomes evident: works are as naturally present as things. The picture hangs on the wall like a hunting weapon or a hat. A painting- for example van Gogh's portrayal of a pair of peasant shoes- travels from one exhibition to another. Works are shipped like coal from the Ruhr or logs from the Black Forest. During the war Hölderlin's hymns were packed in the soldier's knapsack along with cleaning equipment. Beethoven's quartets lie in the publisher's storeroom like potatoes in a cellar.[61]

For Okakaura, "the art of living"[62] means to fit into the social surrounding,[63] while Heidegger writes about the "real dictatorship of the 'they'.".[64]

As shown, Okakura in *The Book of Tea* and Heidegger in *Being and Time* both refer to topics like truth and art, but they do this in completely different contexts and therefore write about different contents.

3 Conclusion

Sein und Zeit revolutionized the philosophical panorama of Europe, although its influence also reached countries in the American continent and some intellectuals in Asia. Until then, no one had been interested in the human being in such original manner, as rigorous and profound, at least within the university world. Thus, with this work, Heidegger picks up the challenge of the Delphic oracle proposed to Socrates and takes charge of "know yourself", thus establishing the singular characteristics—which he will call formal or existentiell indicators—of all "existing one" where it can be found and wherever it is. But the most significant thing, in our view, is that, with

---

[58] Okakura, Kakuzo (1956). p. 13.

[59] Ibid.

[60] Regarding Heidegger´s view on art and home see also: Döring, Annika; Horden, Peregrine (2018).

[61] Heidegger, Martin (2002). : p. 2f.

[62] Okakura, Kakuzo (1956). p. 61.

[63] Ibid. p. 3.

[64] Heidegger, Martin (1962)[1]. [21]2001. p. 164.

this work, the disciple of Husserl puts before our eyes what constitutes us as such and which, at the same time, is constituent. To put it in some way, this ouvre tells us "how" we are and that, for this very reason, we cannot go "beyond" our singularities, from our ways of being. Hence we cannot consider *Sein und Zeit* as a metaphysics, an anthropology, or a kind of basic psychology. That existence (*ec-sistere*) is what defines and determines *Dasein* in its factual existence, already supposes an authentic declaration of intent regarding the path that Heidegger is going to travel alone and on the fringes of the gloomy and sparse academic world, in that period, dominated by neo-Kantianism. The thematization of ordinary and basic life as it is shown, uncovering the unthinkable of what "is" considered to be self-evident, constitutes Heidegger's undeniable merit: he thematized the historically a-thematized, in the same way that Freud dealt with something so unedifying morally, in his time, like sexuality.

With the establishment of existentiells we find the guiding referents, which have to indicate and determine the meaning and the response in relation to our choices and decisions. It is, therefore, the guide to development along life. Because, it seems, our entire vital adventure must not lose sight of the *Wie* of our way of being, that which represents, from an ethical perspective, the horizon of our possibilities. What does this reveal to us? In my view, something as simple as it is difficult: that we cannot—and in most cases we should not—go against what constitutes us or, in other words, try to live, act and update in each case, those indicative indicators that are formal as common, but unique as their own and appropriate to our character. The fact that we do not know "how" (*Wie*) we are does not mean that we are not determined by what we are. Not knowing that we are mortal does not prevent us from dying from any circumstance, so we do not die from a disease, but because we are mortal, although it is true that, without the disease, without a fortuitous accident, it is not possible for that fundamental characteristic to be revealed. The *Sein zum Tode* is our east, while *die Zeitlichkeit* is precisely the *Wie* of existing while our end does not occur, our *Vorbei*. Thus, living is going through temporality knowing our destiny, which is none other than that indicated by the possibility of the impossible. The orientation of existentiells is not aimed at a happy life, but a serene life. Faced with the merely irruptive states of happy moments, which are basically Dionysian and, as such, alien to the principle of individuation, *die Gelassenheit*, on the contrary, is the ethical conviction capable of shaping existence by establishing it in temporality. It, *die Gelassenheit*, is the affective response, from the choice and the decision (in front of the *das Man*), revealing itself, in this way, as the true east of the existing one, that is, of the one who knows what *wie* is and seeks not to forget it by making your life the testimony of its possibility.

References

Döring, A., Horden, P. (2018). *Heidegger as Mediterraneanist*. In Y. Elhariry, & E. Tamalet Talbayev (Eds.), *Critically Mediterranean. Temporalities, Aesthetics, and Deployments of a Sea in Crisis* (pp. 25–43). Cham: Palgrave Macmillan.
Heidegger, M. (1954). *Aus der Erfahrung des Denkens*. Pfullingen: Günter Neske.
Heidegger, M. (1962). *Being and Time*. Translated by J. Macquarrie & E. Robinson. Oxford: Blackwell. [21]2001.
Heidegger, M. (1981). *Was ist Metaphysik?* Frankfurt a. M.: Vittorio Klostermann.
Heidegger, M. (1984). *Was ist das die Philosophie?* Neske.
Heidegger, M. (2002). The Origin of the Work of Art. In Heidegger, M (Ed.), Off the Beaten Track. Translated by J. Young & K. Haynes. Cambridge: Cambridge University Press.
Heisig, J. W. (2002). *Filósofos de la nada*. Un ensayo sobre la escuela de Kioto. Barcelona: Herder.
Imamichi, T. (2004). In search of wisdom: One philosopher's journey. Translated by Mary E. Foster. Tokyo: International House of Japan.
Lao Zi: Tao te King.
Mújica, H. (1995). *La palabra inicial*. Madrid: Trotta.
Okakura, K. (1956). *The Book of Tea*. Rutland: C. E. Tuttle Co.
Ordóñez-García, J. (2014). Heidegger and the notion of prescience (Vorwissenschaft) as existential propaedeutic. (Part 1: The affection of the time), *Estudios de Filosofía*, nº 50, Universidad de Antioquia (Colombia), pp. 155–169.
REAL ACADEMIA ESPAÑOLA. (2021). *Diccionario de la lengua española, 23.5*. https://dle.rae.es.
Saviani, C. (2004). *El oriente de Heidegger*. Barcelona: Herder.

Art is Critical

John D. Barrow

Human beings are good at finding all the ways in which to be creative within prescribed limits—painting inside a rectangular frame, writing only in iambic pentameters or in sonnet form. Scientists sometimes like to study how that creativity occurs, what it achieves, and where to look for inspiration. The scientific method has two facets. Coming up with ideas and hypotheses are processes that have no rules at all. But the process by which those ideas are subsequently tested again the evidence has a fairly rigid set of rules. Whereas artistic creation strives to be individualistic and unique, science is not like that. Most scientific discoveries would have been made soon by someone else if the first discoverers had not done their work. Sometimes that discovery by others would have happened very soon afterwards but in other cases a very long interval of time might need to pass before others passed that way again. This is why 'science is We but art is I'.

Many artists are nervous about scientific analysis. They fear its success, worried that art might lose its power, or they might be diminished, if the psychological roots of their work and its impact on us were exposed. They might even be right to be worried. Unbridled reductionism—'music is nothing but the trace of an air-pressure curve'—is a surprisingly common world view that should be given no encouragement. However, one also finds the equally mistaken contrary view that science has nothing to offer the arts: that they transcend all attempts to capture them objectively. Indeed, many scientists see the creative arts as entirely subjective activities but enjoy them no less for all that.

As science has started to come to grips with the study of complexity, it is natural that it will encounter artistic creations, like music or abstract art, because they have interesting things to teach us about the development of complexity in the forms that we find so appealing. E. O. Wilson has suggested that the link between science and art can be made closest when both are viewed from the vantage point of the study and

J. D. Barrow (✉)
DAMTP, Centre for Mathematical Sciences, Cambridge University, Cambridge, UK

S. Wuppuluri and I. Stewart (eds.), *From Electrons to Elephants and Elections*,
The Frontiers Collection, https://doi.org/10.1007/978-3-030-92192-7_45

appreciation of complexity: 'The love of complexity without reductionism makes art; the love of complexity with reductionism makes science.'[1]

There is a further interesting feature of complex phenomena that sheds light upon what it is we like about many forms of art that we value most. If we allow a stream of grains to fall vertically downwards on to a tabletop then a pile of sand (or rice grains) will steadily grow. The falling grains will have haphazard trajectories as they tumble over the others. Yet, steadily, the chaotically unpredictable falls of the individual grains will build up a large orderly pile. Its sides steepen gradually until they reach a particular slope. After that, the slope gets no steeper. The special 'critical' slope will continue to be maintained by regular avalanches of all sizes, some involving just one or two grains, but other rarer events producing a collapse of the whole side of the pile. The overall result is striking. The haphazard falls of the individual grains have organised themselves to produce a stable, orderly pile. In this 'critical' state the overall order is maintained by the chaotic sensitivity of the individual grain trajectories. If the pile was on an open tabletop then eventually the grains would start falling off the sides of the table at the same rate that they drop from above. The pile would be always composed of different grains: it is a transient steady state.

The robustness of the overall shape of the pile that exists despite the sensitivity of the individual grain trajectories is very suggestive of what it is we like about many artistic creations: a sensitivity to small changes in the face of overall structural robustness. A 'good' book, film, play or piece of music is one that we want to experience again. The 'bad' work is one that we don't. Why do we want to see a great theatrical work like *The Tempest*, or hear a Beethoven symphony, more than once? It is because small changes in production—different actors, modern dress, new styles of direction, or different orchestras and conductors—create a completely new experience for the audience. Great works are sensitive to small changes in ways that enable them to give you a new and pleasurable aesthetic experience. There is novelty et the overall order is maintained. They seem to exhibit a type of criticality. And this combination of unpredictability in the presence of predictability—stability maintained by instability—is something that we seem to find artistically very appealing.[2]

Notes

1. Wilson (1998).
2. Further a more extensive discussion of the interactions between mathematics and many aspects of the arts see also Barrow (2014).

References

Wilson, E. O. (1998). *Consilience*. Knopf.
Barrow, J. D. (2014). *100 Essential things you didn't know about maths and the arts*. Bodley Head.